ESSENTIALS OF PRECALCULUS

Essentials of Precalculus

Richard N. Aufmann
Richard D. Nation
Palomar College

Houghton Mifflin Company
Boston New York

Publisher: Jack Shira
Senior Sponsoring Editor: Lynn Cox
Associate Editor: Jennifer King
Assistant Editor: Melissa Parkin
Senior Project Editor: Tamela Ambush
Editorial Assistant: Sage Anderson
Manufacturing Manager: Karen Banks
Senior Marketing Manager: Danielle Potvin
Marketing Coordinator: Nicole Mollica

Cover photograph: PunchStock

PHOTO CREDITS:

Chapter 1: *p. 1* Charles O'Rear / CORBIS; *p. 8* The Granger Collection; *p. 77* CORBIS. **Chapter 2:** *p. 115* Sonda Dawes / The Image Works, Inc.; *p. 116* David Young-Wolff / PhotoEdit, Inc.; *p. 136* Syndicated Features Limited / The Image Works, Inc.; *p. 157* The Granger Collection; *p. 167* Bettmann/CORBIS; *p. 169* Bettman / CORBIS; *p. 190* Richard T. Nowitz / CORBIS; **Chapter 3:** *p. 201* Chris McLaughlin / CORBIS; *p. 221* Bettman / CORBIS; *p. 227* Charles O'Rear / CORBIS; *p. 227* David James / Getty Images; *p. 231* Bettman / CORBIS. **Chapter 4:** *p. 304* Courtesy of NASA and STScI; *p. 318* Reuters / CORBIS. **Chapter 5:** *p. 357* Massimo Listri / CORBIS; *p. 373* Art; *p. 373* Tony Craddock / Getty Images; *p. 402* Courtesy of Richard Nation. **Chapter 6:** *p. 461* Stephen Johnson / Getty Images; *p. 461* Stephen Johnson / Getty Images.

Printed in the U.S.A.

Library of Congress Control Number: 2003110130

ISBNs:
Student's Edition: 0-618-44702-4
Instructor's Annotated Edition: 0-618-44703-2

1 2 3 4 5 6 7 8 9-VH-09 08 07 06 05

CONTENTS

3 Exponential and Logarithmic Functions 201

4 Trigonometric Functions 289

5 Applications of Trigonometry and Trigonometric Identities 357

PREFACE

Essentials of Precalculus provides students with material that focuses on selected key concepts of precalculus and how those concepts can be applied to a variety of problems. To help students master these concepts, we have tried to maintain a balance among theory, application, modeling, and drill. Carefully developed mathematics is complemented by applications that are both contemporary and representative of a wide range of disciplines. Many application exercises are accompanied by a diagram that helps the student visualize the mathematics of the application.

Technology is introduced naturally to support and advance better understanding of a concept. The optional *Integrating Technology* boxes and graphing calculator exercises are designed to promote an appreciation of both the power and the limitations of technology.

Features

Interactive Presentation *Essentials of Precalculus* is written in a style that encourages the student to interact with the textbook. At various places throughout the text, we pose a question to the student about the material being presented. This question encourages the reader to pause and think about the current discussion and to answer the question. To ensure that the student does not miss important information, the answer to the question is provided as a footnote on the same page.

Each section contains a variety of worked examples. Each example is given a title so that the student can see at a glance the type of problem being illustrated. Most examples are accompanied by annotations that assist the student in moving from step to step. Following the worked example is a suggested exercise for the student to work. The *complete solution* to that exercise can be found in an appendix in the text. This feature allows students to self-assess their progress and to get immediate feedback by means of not just an answer, but a complete solution.

Focus on Problem Solving Each chapter begins with a *Focus on Problem Solving* that demonstrates various strategies that are used by successful problem solvers. At the completion of the *Focus on Problem Solving*, the student is directed to an exercise in the text that can be solved using the problem-solving strategy that was just discussed.

Mathematics and Technology Technology is introduced in the text to illustrate or enhance a concept. We attempt to foster the idea that technology, combined with analytical thinking, can lead to deeper understanding of a concept. The optional *Integrating Technology* boxes and graphing calculator exercises are designed to develop an awareness of technology's capabilities and limitations.

Topics for Discussion are found at the end of each section of the text. These topics can form the basis for a group discussion or serve as writing assignments.

Extensive Exercise Sets The exercise sets of *Essentials of Precalculus* are carefully developed to provide the student with a variety of exercises. The exercises range from drill and practice to interesting challenges and were chosen to illustrate the many facets of the topics discussed in the text. Each exercise set emphasizes concept building, skill building and maintenance, and, as appropriate, applications.

Projects are included at the end of each exercise set and are designed to encourage students to research and write about mathematics and its applications. These projects encourage critical thinking beyond the scope of the regular problem sets. Responses to the projects are given in the *Instructor's Solutions Manual.*

Prepare for Next Section Exercises are found at the end of each exercise set, except for the last section of a chapter. These exercises concentrate on topics from previous sections of the text that are particularly relevant to the next section of the text. By completing these exercises, the student reviews some of the concepts and skills necessary for success in the next section.

Chapter Review Exercises allow the student to review concepts and skills presented in the chapter. Answers to all chapter review exercises are included in the student answer section. If a student incorrectly answers an exercise, there is a section reference next to each answer that directs the student to the section from which that exercise was taken. Using this reference, the student can review the concepts that are required to correctly solve the exercise.

Chapter Tests provide students with an opportunity to self-assess their understanding of the concepts presented in the chapter. The answers to all exercises in the chapter tests are included in the student answer section. As with the answers to the chapter review exercises, there is a section reference next to each answer that directs the student to the section from which that exercise was taken.

Cumulative Review Exercises end every chapter after Chapter 1. These exercises allow students to refresh their knowledge of previously studied skills and concepts and help them maintain skills that promote success in precalculus. Answers to all cumulative review exercises are included in the student answer section. As with the answers to the chapter review exercises, there is a section reference next to each answer that directs the student to the section from which that exercise was taken.

CHAPTER OPENER FEATURES

CHAPTER OPENER

Each chapter begins with a **Chapter Opener** that illustrates a specific application of a concept from the chapter. There is a reference to a particular exercise within the chapter that asks the student to solve a problem related to the chapter opener topic.

The icons   
at the bottom of the page let students know of additional resources available on CD, video/DVD, in the *Student Study Guide*, and online at math.college.hmco.com/students.

Page 1

CHAPTER 1

FUNCTIONS AND GRAPHS

Functions as Models

The Golden Gate Bridge spans the Golden Gate Strait, which is the entrance to the San Francisco Bay from the Pacific Ocean. Designed by Joseph Strass, the Golden Gate Bridge is a *suspension* bridge. A *quadratic function*, one of the topics of this chapter, can be used to model a cable of this bridge. See **Exercise 74 on page 79.**

Strass had many skeptics who did not believe the bridge could be built. Nonetheless, the bridge opened on May 27, 1937, a little over 4 years after construction began. When it was completed, Strass composed the following poem.

The Mighty Task Is Done

At last the mighty task is done;
Resplendent in the western sun;
The Bridge looms mountain high.

On its broad decks in rightful pride,
The world in swift parade shall ride
Throughout all time to be.

Launched midst a thousand hopes and fears,
Damned by a thousand hostile sneers,
Yet ne'er its course was stayed.
But ask of those who met the foe,
Who stood alone when faith was low,
Ask them the price they paid.
High over
Far, far be
Unceasing

Page 79

74. Golden Gate Bridge The suspension cables of the main span of the Golden Gate Bridge are in the shape of a parabola. If a coordinate system is drawn as shown, find the quadratic function that models a suspension cable for the main span of the bridge.

(−2100, 500) (2100, 500) (10, 6)

Page 116

FOCUS ON PROBLEM SOLVING

Find and Use Clues to Narrow the Search

The game of Clue is a classic whodunit game. At the beginning of the game you are informed that Mr. Boddy has been murdered! It is your job to find and use clues to determine the murderer, the weapon, and the room in which the murder was committed. Was it Miss Scarlet in the billiard room with the revolver? Or did Professor Plum commit the murder in the conservatory with the rope? There are six suspects, six possible murder weapons, and nine rooms in Mr. Boddy's mansion. There are a total of $6 \times 6 \times 9 = 324$ possible solutions to each game.

In this chapter you will often need to find the zeros of a polynomial function. Finding the zeros of a polynomial function can be more complicated than solving a game of Clue. After all, any real or complex number is a possible zero. Quite often the zeros of a polynomial function are found by using several theorems to narrow the search. In most cases no single theorem can be used to find the zeros of a given polynomial function, but by combining the results of several theorems, we are often able to gather enough clues to find the zeros. In **Exercise 52, page 165,** you will apply several theorems from this chapter to find the zeros of a polynomial function.

FOCUS ON PROBLEM SOLVING

A **Focus on Problem Solving** follows the Chapter Opener. This feature highlights and demonstrates a problem-solving strategy that may be used to successfully solve some of the problems presented in the chapter.

AUFMANN INTERACTIVE METHOD (AIM)

INTERACTIVE PRESENTATION

Essentials of Precalculus is written in a style that encourages the student to interact with the textbook.

EXAMPLES

Each section contains a variety of worked examples. Examples are **titled** so that the student can see at a glance the type of problem being illustrated, often accompanied by **annotations** that assist the student in moving from step to step; and offers the **final answer in color** so that it is readily identifiable.

Page 208

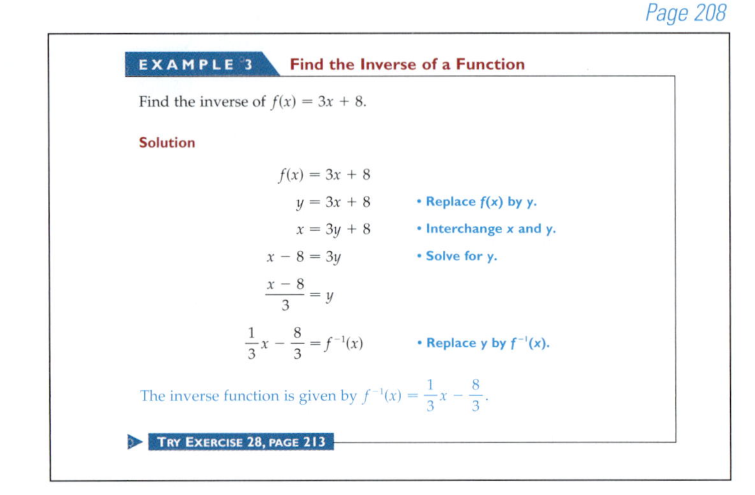

EXAMPLE 3 Find the Inverse of a Function

Find the inverse of $f(x) = 3x + 8$.

Solution

$f(x) = 3x + 8$

$y = 3x + 8$ • Replace $f(x)$ by y.

$x = 3y + 8$ • Interchange x and y.

$x - 8 = 3y$ • Solve for y.

$\frac{x - 8}{3} = y$

$\frac{1}{3}x - \frac{8}{3} = f^{-1}(x)$ • Replace y by $f^{-1}(x)$.

The inverse function is given by $f^{-1}(x) = \frac{1}{3}x - \frac{8}{3}$.

▶ TRY EXERCISE 28, PAGE 213

TRY EXERCISES

Following every example is a suggested **Try Exercise** from that section's exercise set for the student to work. The exercises are color coded by number in the exercise set and the *complete solution* to that exercise can be found in an appendix to the text.

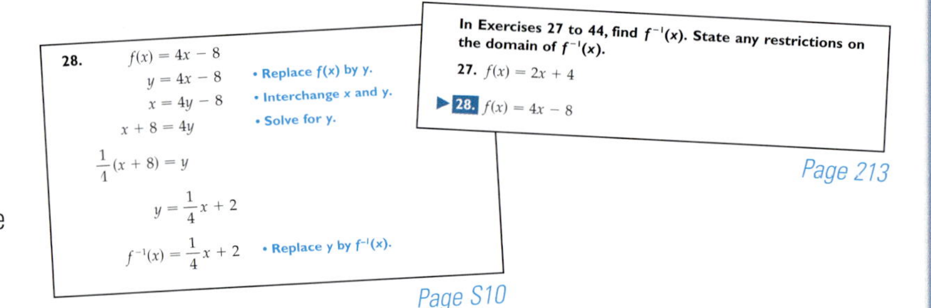

28. $f(x) = 4x - 8$

$y = 4x - 8$ • Replace $f(x)$ by y.

$x = 4y - 8$ • Interchange x and y.

$x + 8 = 4y$ • Solve for y.

$\frac{1}{4}(x + 8) = y$

$y = \frac{1}{4}x + 2$

$f^{-1}(x) = \frac{1}{4}x + 2$ • Replace y by $f^{-1}(x)$.

Page S10

In Exercises 27 to 44, find $f^{-1}(x)$. State any restrictions on the domain of $f^{-1}(x)$.

27. $f(x) = 2x + 4$

▶ **28.** $f(x) = 4x - 8$

Page 213

We can use the properties of exponents to establish the following additional logarithmic properties.

Properties of Logarithms

In the following properties, b, M, and N are positive real numbers $(b \neq 1)$.

Product property	$\log_b(MN) = \log_b M + \log_b N$
Quotient property	$\log_b \frac{M}{N} = \log_b M - \log_b N$
Power property	$\log_b(M^p) = p \log_b M$
Logarithm-of-each-side property	$M = N$ implies $\log_b M = \log_b N$
One-to-one property	$\log_b M = \log_b N$ implies $M = N$

take note

Pay close attention to these properties. Note that

$\log_b(MN) \neq \log_b M \cdot \log_b N$

and

$\log_b \frac{M}{N} \neq \frac{\log_b M}{\log_b N}$

Also,

$\log_b(M + N) \neq \log_b M + \log_b N$

In fact, the expression $\log_b(M + N)$ cannot be expanded at all.

QUESTION Is it true that $\ln 5 + \ln 10 = \ln 50$?

The above properties of logarithms are often used to rewrite logarithmic expressions in an equivalent form.

ANSWER Yes. By the product property, $\ln 5 + \ln 10 = \ln(5 \cdot 10)$.

Page 243

QUESTION/ANSWER

In every section, we pose at least one **Question** to the student about the material being presented. This question encourages the reader to pause and think about the current discussion and to answer the question. To make sure that the student does not miss important information, the **Answer** to the question is provided as a footnote on the same page.

REAL DATA AND APPLICATIONS

APPLICATIONS

One way to motivate an interest in mathematics is through applications. Applications require the student to use problem-solving strategies, along with the skills covered in a section, to solve practical problems. This careful integration of applications generates student awareness of the value of algebra as a real-life tool.

Applications are taken from many disciplines including agriculture, business, chemistry, construction, Earth science, education, economics, manufacturing, nutrition, real estate, and sociology.

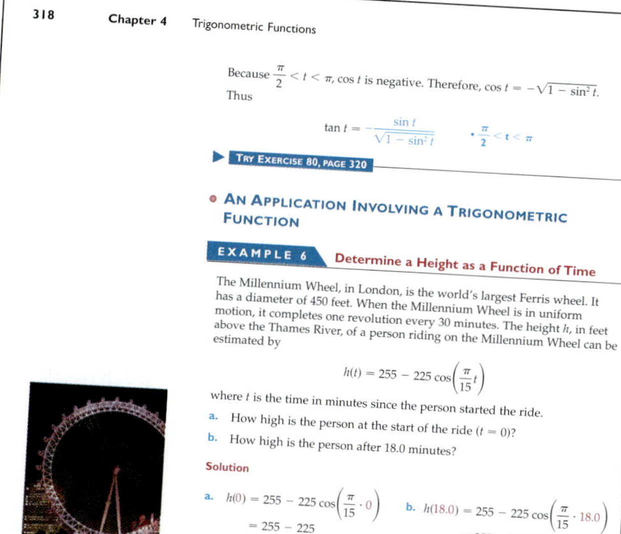

318 Chapter 4 Trigonometric Functions

Because $\frac{\pi}{2} < t < \pi$, $\cos t$ is negative. Therefore, $\cos t = -\sqrt{1 - \sin^2 t}$.

Thus

$$\tan t = -\frac{\sin t}{\sqrt{1 - \sin^2 t}} \qquad \cdot \frac{\pi}{2} < t < \pi$$

▶ TRY EXERCISE 80, PAGE 320

AN APPLICATION INVOLVING A TRIGONOMETRIC FUNCTION

EXAMPLE 6 Determine a Height as a Function of Time

The Millennium Wheel, in London, is the world's largest Ferris wheel. It has a diameter of 450 feet. When the Millennium Wheel is in uniform motion, it completes one revolution every 30 minutes. The height h, in feet above the Thames River, of a person riding on the Millennium Wheel can be estimated by

$$h(t) = 255 - 225\cos\left(\frac{\pi}{15}t\right)$$

where t is the time in minutes since the person started the ride.

a. How high is the person at the start of the ride ($t = 0$)?

b. How high is the person after 18.0 minutes?

The Millennium Wheel, on the banks of the Thames River, London.

Solution

a. $h(0) = 255 - 225\cos\left(\frac{\pi}{15} \cdot 0\right)$
$= 255 - 225$
$= 30$

At the start of the ride, the person is 30 feet above the Thames.

b. $h(18.0) = 255 - 225\cos\left(\frac{\pi}{15} \cdot 18.0\right)$
$\approx 255 - (-182)$
$= 437$

After 18.0 minutes, the person is about 437 feet above the Thames.

▶ TRY EXERCISE 84, PAGE 320

TOPICS FOR DISCUSSION

1. Is $W(t)$ a number? Explain.
2. Explain how to find the exact value of $\cos\left(\frac{13\pi}{6}\right)$.
3. Is $f(x) = \cos^3 x$ an even function or an odd function? Explain how you made your decision.
4. Explain how to make use of a unit circle to show that $\sin(-t) = -\sin t$.

Page 318

Page 62

62 Chapter 1 Functions and Graphs

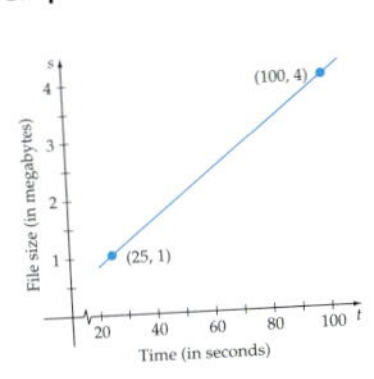

55. AUTOMOTIVE TECHNOLOGY The table below shows the EPA fuel economy values for selected two-seater cars for the 2003 model year. (*Source:* www.fueleconomy.gov.)

EPA Fuel Economy Values for Selected Two-Seater Cars

Car	City mpg	Highway mpg
Audi, TT Roadster	20	29
BMW, Z8	13	21
Ferrari, 360 Spider	11	16
Lamborghini, L-174	9	13
Lotus, Esprit V8	15	22
Maserati, Spider GT	11	17

a. Using the data for the Lamborghini and the Audi, find a linear function that predicts highway miles per gallon in terms of city miles per gallon. Round the slope to the nearest hundredth.

b. Using your model, predict the highway miles per gallon for a Porsche Boxer, whose city fuel efficiency is 18 miles per gallon. Round to the nearest whole number.

▶ 56. CONSUMER CREDIT The amount of revolving consumer credit (such as credit cards and department store cards) for the years 1997 to 2003 is given in the table below. (*Source:* www.nber.org, Board of Governor's of the Federal Reserve System.)

Year	Consumer Credit (in billions of $)
1997	531.0
1998	562.5
1999	598.0
2000	667.4
2001	701.3
2002	712.0
2003	725.0

a. Using the data for 1997 and 2003, find a linear model that predicts the amount of revolving consumer credit (in billions) for year t. Round the slope to the nearest tenth.

b. Using this model, in what year will consumer credit first exceed $850 billion?

57. LABOR MARKET According to the Bureau of Labor Statistics (BLS), there were 38,000 desktop publishing jobs in the United States in the year 2000. The BLS projects that there will be 63,000 desktop publishing jobs in 2010.

a. Using the BLS data, find the number of desktop publishing jobs as a linear function of the year.

b. Using your model, in what year will the number of desktop publishing jobs first exceed 60,000?

58. POTTERY A piece of pottery is removed from a kiln and allowed to cool in a controlled environment. The temperature (in degrees Fahrenheit) of the pottery after it is removed from the kiln for various times (in minutes) is shown in the table below.

Time, min	Temperature, °F
15	2200
20	2150
30	2050
60	1750

a. Find a linear model for the temperature of the pottery after t minutes.

b. Explain the meaning of the slope of this line in the context of the problem.

c. Assuming temperature continues to decrease at the same rate, what will be the temperature of the pottery in 3 hours?

59. LUMBER INDUSTRY For a log, the number of board-feet (bf) that can be obtained from the log depends on the diameter, in inches, of the log and its length. The table below shows the number of board-feet of lumber that can be obtained from a log that is 32 feet long.

Diameter, inches	bf
16	180
18	240
20	300
22	360

REAL DATA

Real data examples and exercises, identified by [pie chart icon], ask students to analyze and create mathematical models from actual situations. Students are often required to work with tables, graphs, and charts drawn from a variety of disciplines.

TECHNOLOGY

INTEGRATING TECHNOLOGY

The **Integrating Technology** feature contains optional discussions that can be used to further explore a concept using technology. Some introduce technology as an alternative way to solve certain problems and others provide suggestions for using a calculator to solve certain problems and applications. Additionally, optional graphing calculator examples and exercises (identified by) are presented throughout the text.

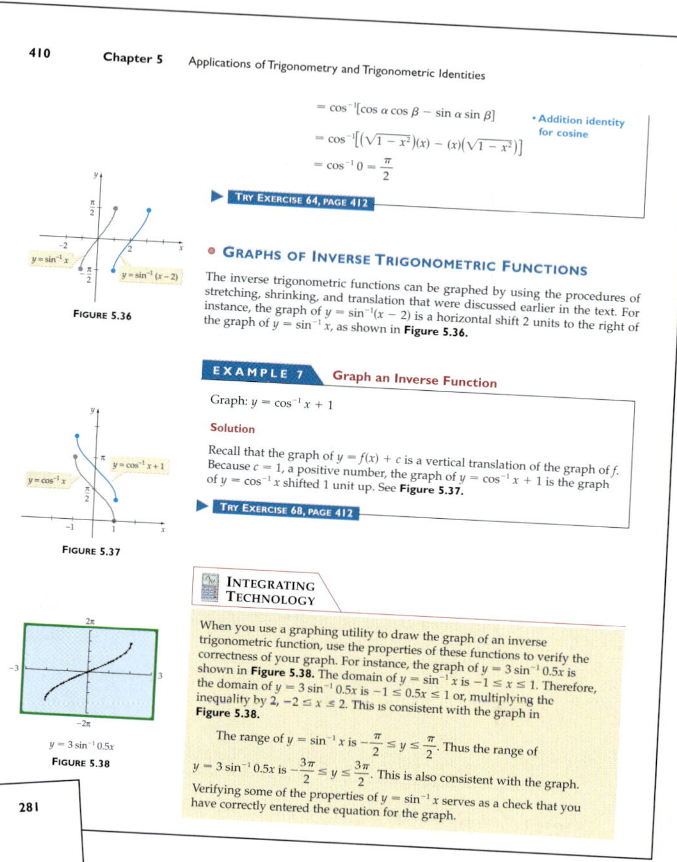

410 Chapter 5 Applications of Trigonometry and Trigonometric Identities

$= \cos^{-1}[\cos\alpha\cos\beta - \sin\alpha\sin\beta]$ • Addition identity for cosine

$= \cos^{-1}[(\sqrt{1-x^2})(x) - (x)(\sqrt{1-x^2})]$

$= \cos^{-1}0 = \frac{\pi}{2}$

▶ Try Exercise 64, page 412

Figure 5.36

• Graphs of Inverse Trigonometric Functions

The inverse trigonometric functions can be graphed by using the procedures of stretching, shrinking, and translation that were discussed earlier in the text. For instance, the graph of $y = \sin^{-1}(x-2)$ is a horizontal shift 2 units to the right of the graph of $y = \sin^{-1}x$, as shown in **Figure 5.36.**

Example 7 Graph an Inverse Function

Graph: $y = \cos^{-1}x + 1$

Solution

Recall that the graph of $y = f(x) + c$ is a vertical translation of the graph of f. Because $c = 1$, a positive number, the graph of $y = \cos^{-1}x + 1$ is the graph of $y = \cos^{-1}x$ shifted 1 unit up. See **Figure 5.37.**

▶ Try Exercise 68, page 412

Figure 5.37

Integrating Technology

When you use a graphing utility to draw the graph of an inverse trigonometric function, use the properties of these functions to verify the correctness of your graph. For instance, the graph of $y = 3\sin^{-1}0.5x$ is shown in **Figure 5.38.** The domain of $y = \sin^{-1}x$ is $-1 \le x \le 1$. Therefore, the domain of $y = 3\sin^{-1}0.5x$ is $-1 \le 0.5x \le 1$ or, multiplying the inequality by 2, $-2 \le x \le 2$. This is consistent with the graph in **Figure 5.38.**

The range of $y = \sin^{-1}x$ is $-\frac{\pi}{2} \le y \le \frac{\pi}{2}$. Thus the range of $y = 3\sin^{-1}0.5x$ is $-\frac{3\pi}{2} \le y \le \frac{3\pi}{2}$. This is also consistent with the graph.

Verifying some of the properties of $y = \sin^{-1}x$ serves as a check that you have correctly entered the equation for the graph.

$y = 3\sin^{-1}0.5x$

Figure 5.38

Page 281

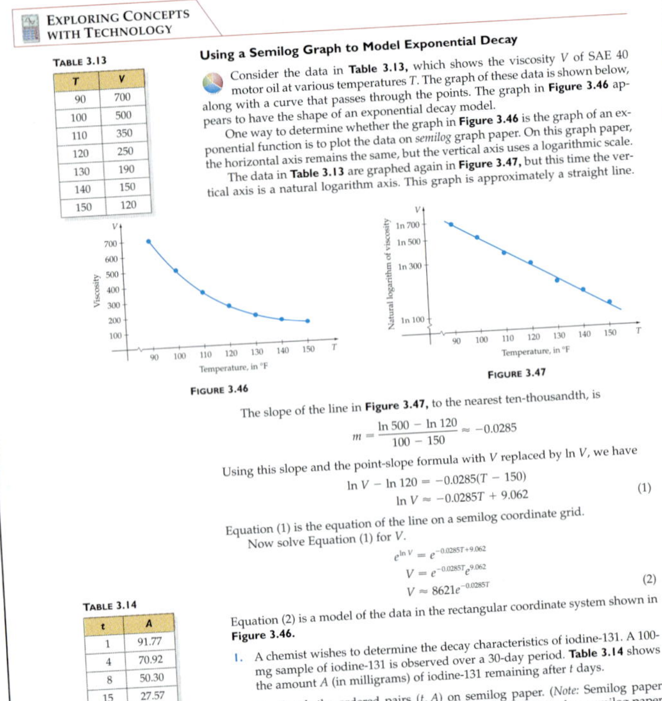

Exploring Concepts with Technology 281

Exploring Concepts with Technology

Using a Semilog Graph to Model Exponential Decay

Consider the data in **Table 3.13,** which shows the viscosity V of SAE 40 motor oil at various temperatures T. The graph of these data is shown below, along with a curve that passes through the points. The graph in **Figure 3.46** appears to have the shape of an exponential decay model.

One way to determine whether the graph in **Figure 3.46** is the graph of an exponential function is to plot the data on *semilog* graph paper. On this graph paper, the horizontal axis remains the same, but the vertical axis uses a logarithmic scale.

The data in **Table 3.13** are graphed again in **Figure 3.47,** but this time the vertical axis is a natural logarithm axis. This graph is approximately a straight line.

Table 3.13

T	V
90	700
100	500
110	350
120	250
130	190
140	150
150	120

Figure 3.46

Figure 3.47

The slope of the line in **Figure 3.47,** to the nearest ten-thousandth, is

$$m = \frac{\ln 500 - \ln 120}{100 - 150} \approx -0.0285$$

Using this slope and the point-slope formula with V replaced by $\ln V$, we have

$$\ln V - \ln 120 = -0.0285(T - 150)$$

$$\ln V \approx -0.0285T + 9.062 \quad (1)$$

Equation (1) is the equation of the line on a semilog coordinate grid.

Now solve Equation (1) for V.

$$e^{\ln V} = e^{-0.0285T + 9.062}$$

$$V = e^{-0.0285T}e^{9.062}$$

$$V \approx 8621e^{-0.0285T} \quad (2)$$

Equation (2) is a model of the data in the rectangular coordinate system shown in **Figure 3.46.**

1. A chemist wishes to determine the decay characteristics of iodine-131. A 100-mg sample of iodine-131 is observed over a 30-day period. **Table 3.14** shows the amount A (in milligrams) of iodine-131 remaining after t days.
   a. Graph the ordered pairs (t, A) on semilog paper. (*Note:* Semilog paper comes in different varieties. Our calculations are based on semilog paper that has a natural logarithm scale on the vertical axis.)

Table 3.14

t	A
1	91.77
4	70.92
8	50.30
15	27.57
20	17.95
30	7.60

Page 410

EXPLORING CONCEPTS WITH TECHNOLOGY

A special end-of-chapter feature, **Exploring Concepts with Technology**, extends ideas introduced in the text by using technology (graphing calculator, CAS, etc.) to investigate extended applications or mathematical topics. These explorations can serve as group projects, class discussions, or extra-credit assignments.

STUDENT PEDAGOGY

TOPIC LIST

At the beginning of each section is a list of the major topics covered in the section.

KEY TERMS AND CONCEPTS

Key terms, in bold, emphasize important terms. Key concepts are presented in blue boxes in order to highlight these important concepts and to provide for easy reference.

MATH MATTERS

These margin notes contain interesting sidelights about mathematics, its history, or its application.

TAKE NOTE

These margin notes alert students to a point requiring special attention or are used to amplify the concept under discussion.

REVIEW NOTES

A  directs the student to the place in the text where the student can review a concept that was previously discussed.

Page 463

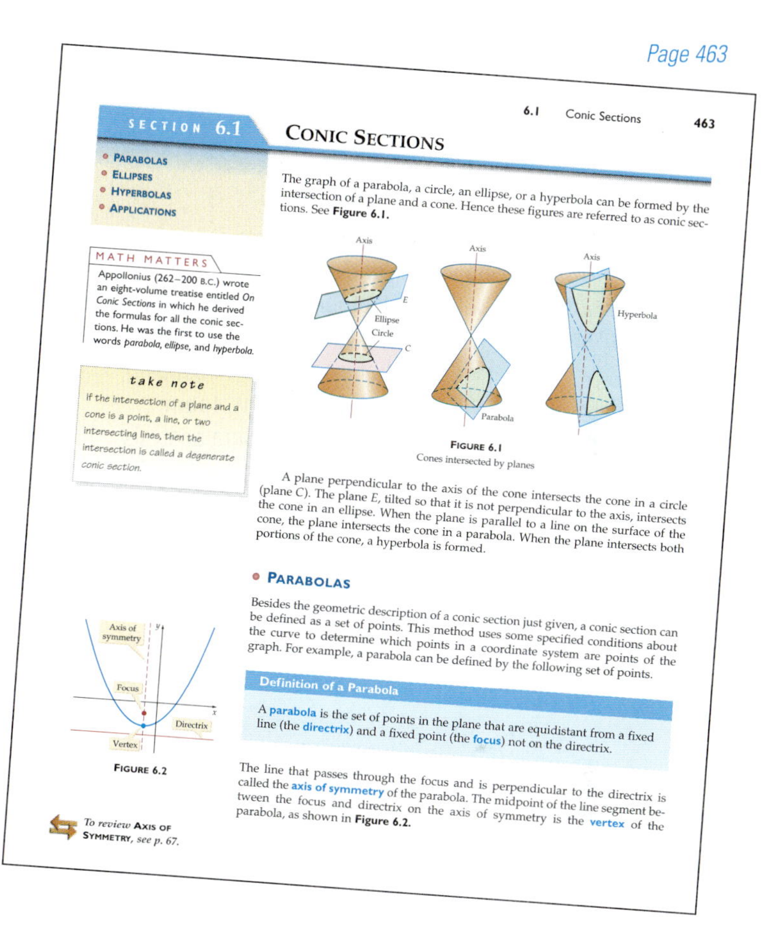

6.1 Conic Sections 463

SECTION 6.1 CONIC SECTIONS

- Parabolas
- Ellipses
- Hyperbolas
- Applications

The graph of a parabola, a circle, an ellipse, or a hyperbola can be formed by the intersection of a plane and a cone. Hence these figures are referred to as conic sections. See **Figure 6.1**.

MATH MATTERS

Apollonius (262–200 B.C.) wrote an eight-volume treatise entitled *On Conic Sections* in which he derived the formulas for all the conic sections. He was the first to use the words *parabola*, *ellipse*, and *hyperbola*.

take note

If the intersection of a plane and a cone is a point, a line, or two intersecting lines, then the intersection is called a degenerate conic section.

Figure 6.1
Cones intersected by planes

A plane perpendicular to the axis of the cone intersects the cone in a circle (plane *C*). The plane *E*, tilted so that it is not perpendicular to the axis, intersects the cone in an ellipse. When the plane is parallel to a line on the surface of the cone, the plane intersects the cone in a parabola. When the plane intersects both portions of the cone, a hyperbola is formed.

PARABOLAS

Besides the geometric description of a conic section just given, a conic section can be defined as a set of points. This method uses some specified conditions about the curve to determine which points in a coordinate system are points of the graph. For example, a parabola can be defined by the following set of points.

Definition of a Parabola

A **parabola** is the set of points in the plane that are equidistant from a fixed line (the **directrix**) and a fixed point (the **focus**) not on the directrix.

Figure 6.2

The line that passes through the focus and is perpendicular to the directrix is called the **axis of symmetry** of the parabola. The midpoint of the line segment between the focus and directrix on the axis of symmetry is the **vertex** of the parabola, as shown in **Figure 6.2**.

To review Axis of Symmetry, *see p. 67.*

Page 55

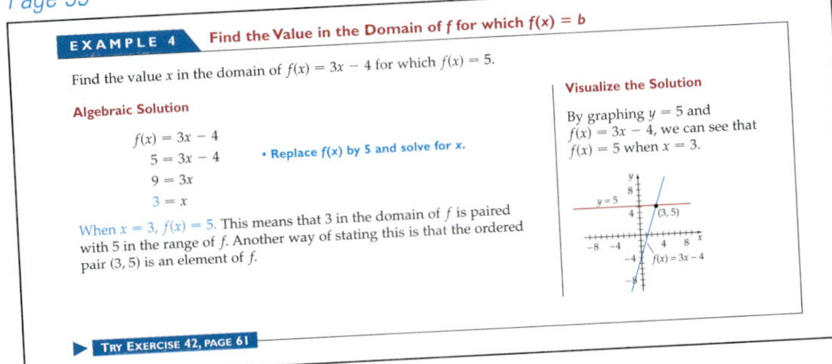

EXAMPLE 4 Find the Value in the Domain of f for which $f(x) = b$

Find the value x in the domain of $f(x) = 3x - 4$ for which $f(x) = 5$.

Algebraic Solution

$$f(x) = 3x - 4$$
$$5 = 3x - 4 \quad \text{• Replace } f(x) \text{ by 5 and solve for } x.$$
$$9 = 3x$$
$$3 = x$$

When $x = 3$, $f(x) = 5$. This means that 3 in the domain of f is paired with 5 in the range of f. Another way of stating this is that the ordered pair (3, 5) is an element of f.

Visualize the Solution

By graphing $y = 5$ and $f(x) = 3x - 4$, we can see that $f(x) = 5$ when $x = 3$.

Try Exercise 42, page 61

VISUALIZE THE SOLUTION

For appropriate examples within the text, we have provided both an algebraic solution and a graphical representation of the solution. This approach creates a link between the algebraic and visual components of a solution.

EXERCISES

TOPICS FOR DISCUSSION

These special exercises provide questions related to key concepts in the section. Instructors can use these to initiate class discussions or to ask students to write about concepts presented.

EXERCISES

The exercise sets in *Essentials of Precalculus* were carefully developed to provide a wide variety of exercises. The exercises range from drill and practice to interesting challenges. They were chosen to illustrate the many facets of topics discussed in the text. Each exercise set emphasizes skill building, skill maintenance, and, as appropriate, applications. **Icons** identify appropriate writing , group , data analysis

, web ,and graphing calculator 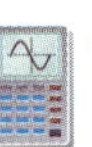 exercises.

Page 124

Topics for Discussion

1. What is an imaginary number? What is a complex number?
2. How are the real numbers related to the complex numbers?
3. Is zero a complex number?
4. What is the conjugate of a complex number?
5. Explain how you know that the equation $x^2 + bx - 2 = 0$ always has real number solutions regardless of the value of the real number b.

Year (00 represents 1900)

Time (in hours)

Use $M(t)$ and a graphing utility to estimate

a. during what year the U.S. marriage rate reached its maximum for the period from 1900 to 1999.

b. the relative minimum marriage rate, rounded to the nearest 0.1, during the period from 1950 to 1970.

54. **Gazelle Population** A herd of 204 African gazelles is introduced into a wild animal park. The population of the gazelles, $P(t)$, after t years is given by $P(t) = -0.7t^3 + 18.7t^2 - 69.5t + 204$, where $0 < t \le 18$.

a. Use a graph of P to determine the absolute minimum gazelle population (rounded to the nearest single gazelle) that is attained during this time period.

b. Use a graph of P to determine the absolute maximum gazelle population (rounded to the nearest single gazelle) that is attained during this time period.

55. **Medication Level** Pseudoephedrine hydrochloride is an allergy medication. The function

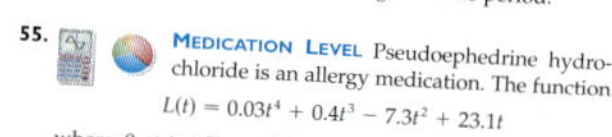

$$L(t) = 0.03t^4 + 0.4t^3 - 7.3t^2 + 23.1t$$

where $0 \le t \le 5$, models the level of pseudoephedrine hydrochloride, in milligrams, in the bloodstream of a patient t hours after 30 milligrams of the medication have been taken.

a. Use a graphing utility and the function $L(t)$ to determine the maximum level of pseudoephedrine hydrochloride in the patient's bloodstream. Round your result to the nearest 0.01 milligram.

b. At what time t, to the nearest minute, is this maximum level of pseudoephedrine hydrochloride reached?

56. **Squirrel Population** The population P of squirrels in a wilderness area is given by

$$P(t) = 0.6t^4 - 13.7t^3 + 104.5t^2 - 243.8t + 360,$$

where $0 \le t \le 12$ years.

a. What is the absolute minimum number of squirrels (rounded to the nearest single squirrel) attained on the interval $0 \le t \le 12$?

b. The absolute maximum of P is attained at the endpoint, where $t = 12$. What is this absolute maximum (rounded to the nearest single squirrel)?

57. **Beam Deflection** The deflection D, in feet, of an 8-foot beam that is center loaded is given by

$$D(x) = (-0.0025)(4x^3 - 3 \cdot 8x^2), \quad 0 < x \le 4$$

where x is the distance, in feet, from one end of the beam.

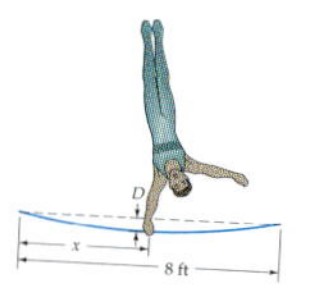

a. Determine the deflection of the beam when $x = 3$ feet. Round to the nearest hundredth of an inch.

b. At what point does the beam achieve its maximum deflection? What is the maximum deflection? Round to the nearest hundredth of an inch.

c. What is the deflection at $x = 5$ feet?

Page 387

5.2 Verification of Trigonometric Identities 387

CONNECTING CONCEPTS

In Exercises 91 to 95, verify the identity.

91. $\dfrac{1 - \sin x + \cos x}{1 + \sin x + \cos x} = \dfrac{\cos x}{\sin x + 1}$

92. $\dfrac{1 - \tan x + \sec x}{1 + \tan x - \sec x} = \dfrac{1 + \sec x}{\tan x}$

93. $\cos(x + y + z) = \cos x \cos y \cos z - \sin x \sin y \cos z - \sin x \cos y \sin z - \cos x \sin y \sin z$

94. $\dfrac{\sin(x + h) - \sin x}{h} = \cos x \dfrac{\sin h}{h} + \sin x \dfrac{(\cos h - 1)}{h}$

95. $\dfrac{\cos(x + h) - \cos x}{h} = \cos x \dfrac{(\cos h - 1)}{h} - \sin x \dfrac{\sin h}{h}$

96. **Model Resistance** The drag (resistance) on a fish when it is swimming is two to three times the drag when it is gliding. To compensate for this, some fish swim in a sawtooth pattern, as shown in the accompanying figure. The ratio of the amount of energy the fish expends when swimming upward at angle β and then gliding down at angle α to the energy it expends swimming horizontally is given by

$$E_R = \frac{k \sin \alpha + \sin \beta}{k \sin(\alpha + \beta)}$$

where k is a value such that $2 \le k \le 3$, and k depends on the assumptions we make about the amount of drag experienced by the fish. Find E_R for $k = 2$, $\alpha = 10°$, and $\beta = 20°$.

PREPARE FOR SECTION 5.3

97. Use the identity for $\sin(\alpha + \beta)$ to rewrite $\sin 2\alpha$. [5.2]

98. Use the identity for $\cos(\alpha + \beta)$ to rewrite $\cos 2\alpha$. [5.2]

99. Use the identity for $\tan(\alpha + \beta)$ to rewrite $\tan 2\alpha$. [5.2]

100. Compare $\tan \dfrac{\alpha}{2}$ and $\dfrac{\sin \alpha}{1 + \cos \alpha}$ for $\alpha = 60°$, $\alpha = 90°$, and $\alpha = 120°$. [5.1]

101. Verify that $\sin 2\alpha = 2 \sin \alpha$ is *not* an identity. *Hint:* Find a value of α for which the left side of the equation does not equal the right side. [5.1]

102. Verify that $\cos \dfrac{\alpha}{2} = \dfrac{1}{2} \cos \alpha$ is *not* an identity. [5.1]

PROJECTS

1. **Grading a Quiz** Suppose that you are a teacher's assistant. You are to assist the teacher of a trigonometry class by grading a four-question quiz. Each question asks the student to find a trigonometric expression that models a given application. The teacher has prepared an answer key. These answers are shown in the next column. A student gives as answers the expressions shown in the far right column. Determine for which problems the student has given a correct response.

Answer Key	Student's Response
1. $\csc x \sec x$	1. $\cot x + \tan x$
2. $\cos^2 x$	2. $(1 + \sin x)(1 - \sin x)$
3. $\cos x \cot x$	3. $\csc x - \sec x$
4. $\csc x \cot x$	4. $\sin x(\cot x + \cot^3 x)$

Page 152

Included in each exercise set are **Connecting Concepts** exercises. These exercises extend some of the concepts discussed in the section and require students to connect ideas studied earlier with new concepts.

EXERCISES TO PREPARE FOR THE NEXT SECTION

Every section's exercise set (except for the last section of a chapter) contains exercises that allow students to practice the previously-learned skills they will need to be successful in the next section. Next to each question, in brackets, is a reference to the section of the text that contains the concepts related to the question for students to easily review. All answers are provided in the Answer Appendix.

PROJECTS

Projects are provided at the end of each exercise set. They are designed to encourage students to do research and write about what they have learned. These Projects generally emphasize critical thinking skills and can be used as collaborative learning exercises or as extra-credit assignments.

END OF CHAPTER

Page 454

454 Chapter 5 Applications of Trigonometry and Trigonometric Identities

CHAPTER 5 SUMMARY

5.1 Trigonometric Functions of Angles

- Let θ be an acute angle of a right triangle. The six trigonometric functions of θ are given by

$$\sin\theta = \frac{\text{opp}}{\text{hyp}} \quad \csc\theta = \frac{\text{hyp}}{\text{opp}}$$
$$\cos\theta = \frac{\text{adj}}{\text{hyp}} \quad \sec\theta = \frac{\text{hyp}}{\text{adj}}$$
$$\tan\theta = \frac{\text{opp}}{\text{adj}} \quad \cot\theta = \frac{\text{adj}}{\text{opp}}$$

- Let $P(x, y)$ be a point, except the origin, on the terminal side of an angle θ in standard position. The six trigonometric functions of θ are

$$\sin\theta = \frac{y}{r} \quad \csc\theta = \frac{r}{y},\ y \neq 0$$
$$\cos\theta = \frac{x}{r} \quad \sec\theta = \frac{r}{x},\ x \neq 0$$
$$\cot\theta = \frac{x}{y},\ y \neq 0$$

- The cofunction identities are

$$\sin(90^\circ - \theta) = \cos\theta \quad \cos(90^\circ - \theta) = \sin\theta$$
$$\tan(90^\circ - \theta) = \cot\theta \quad \cot(90^\circ - \theta) = \tan\theta$$
$$\sec(90^\circ - \theta) = \csc\theta \quad \csc(90^\circ - \theta) = \sec\theta$$

where θ is in degrees. If θ is in radian measure, replace 90° with $\frac{\pi}{2}$.

5.3 More on Trigonometric Identities

- The double-angle identities are

$$\sin 2\alpha = 2\sin\alpha\cos\alpha$$
$$\cos 2\alpha = \cos^2\alpha - \sin^2\alpha = 1 - 2\sin^2\alpha = 2\cos^2\alpha - 1$$
$$\tan 2\alpha = \frac{2\tan\alpha}{1 - \tan^2\alpha}$$

- The half-angle identities are

CHAPTER SUMMARY

At the end of each chapter there is a Chapter Summary that provides a concise section-by-section review of the chapter topics.

TRUE/FALSE EXERCISES

Following each chapter summary are true/false exercises. These exercises are intended to help students understand concepts and can be used to initiate class discussions.

CHAPTER 5 TRUE/FALSE EXERCISES

In Exercises 1 to 12, answer true or false. If the statement is false, give a reason or an example to show that the statement is false.

1. The angle θ measured in degrees is in standard position with the terminal side in the second quadrant. The reference angle of θ is $180^\circ - \theta$.
2. $\frac{\sin x}{\cos y} = \tan\frac{x}{y}$
3. $\sin^{-1} x = \csc x^{-1}$
4. $\sin 2\alpha = 2\sin\alpha$ for all α
5. $\sin(\alpha + \beta) = \sin\alpha + \sin\beta$
6. If $\tan\alpha = \tan\beta$, then $\alpha = \beta$.
7. $\cos^{-1}(\cos x) = x$
8. If $-1 \le x \le 1$, then $\cos(\cos^{-1} x) = x$.
9. The Law of Cosines can be used to solve any triangle given two sides and an angle.

Page 455

Page 456

CHAPTER 5 REVIEW EXERCISES

1. Find the values of the six trigonometric functions of an angle in standard position with the point $P(1, -3)$ on the terminal side of the angle.
2. Find the exact value of
   a. $\sec 150^\circ$ b. $\tan\left(-\frac{3\pi}{4}\right)$
   c. $\cot(-225^\circ)$ d. $\cos\left(\frac{2\pi}{3}\right)$
3. Find the value of each of the following to the nearest ten-thousandth.
   a. $\cos 123^\circ$ b. $\cot 4.22$
   c. $\sec 612^\circ$ d. $\tan\frac{2\pi}{5}$
4. A car climbs a hill that has a constant angle of 4.5° for a distance of 1.14 miles. What is the car's increase in altitude?

13. $\tan\left(67\frac{1}{2}\right)^\circ$
14. $\sin 112.5^\circ$

In Exercises 15 to 18, find the exact values of the given functions.

15. Given $\sin\alpha = \frac{1}{2}$, α in Quadrant I, and $\cos\beta = \frac{1}{2}$, β in Quadrant IV, find $\cos(\alpha - \beta)$.
16. Given $\sin\alpha = \frac{\sqrt{3}}{2}$, α in Quadrant II, and $\cos\beta = -\frac{1}{2}$, β in Quadrant III, find $\sin(\alpha + \beta)$.
17. Given $\sin\alpha = -\frac{1}{2}$, α in Quadrant IV, and $\cos\beta = -\frac{\sqrt{3}}{2}$, β in Quadrant III, find $\tan 2\alpha$.
18. Given $\sin\alpha = \frac{\sqrt{2}}{2}$, α in Quadrant I, and $\cos\beta = \frac{\sqrt{3}}{2}$, β in Quadrant IV, find $\sin 2\alpha$.

CHAPTER 5 TEST

1. Find the exact value of $\tan\frac{\pi}{6}\cos\frac{\pi}{3} - \sin\frac{\pi}{2}$.
2. The angle of elevation from point A to the top of a tree is 42.2°. At point B, 5.24 meters from A and on a line through the base of the tree and A, the angle of elevation is 37.4°. Find the height of the tree.
3. Verify the identity $1 + \sin^2 x \sec^2 x = \sec^2 x$.
4. Verify the identity
$$\frac{1}{\sec x - \tan x} - \frac{1}{\sec x + \tan x} = 2\tan x$$
5. Verify the identity $\csc x - \cot x = \frac{1 - \cos x}{\sin x}$.
6. Find the exact value of $\sin 195^\circ$.
7. Given $\sin\alpha = -\frac{3}{5}$, α in Quadrant III, and $\cos\beta = -\frac{\sqrt{2}}{2}$, β in Quadrant II, find $\sin(\alpha + \beta)$.
8. Verify the identity $\tan\frac{\theta}{2} + \frac{\cos\theta}{\sin\theta} = \csc\theta$.
9. Find the exact value of $\sin 15^\circ \cos 75^\circ$.
10. Write $y = -\frac{\sqrt{3}}{2}\sin x + \frac{1}{2}\cos x$ in the form $y = k\sin(x + \alpha)$, where α is measured in radians.
11. Find the exact value of $\sin\left(\cos^{-1}\frac{12}{13}\right)$.

Page 458

CHAPTER REVIEW EXERCISES

Review exercises are found at the end of each chapter. These exercises are selected to help the student integrate all of the topics presented in the chapter.

CHAPTER TEST

The Chapter Test exercises are designed to simulate a possible test of the material in the chapter.

CUMULATIVE REVIEW EXERCISES

1. Explain how to use the graph of $y = f(x)$ to produce the graph of $y = f(x + 1) + 2$.
2. Explain how to use the graph of $y = f(x)$ to produce the graph of $y = -f(x)$.
3. Find the vertical asymptote for the graph of $f(x) = \frac{x + 3}{x - 2}$.
4. Determine whether $f(x) = x - \sin x$ is an even function or an odd function.
5. Find the inverse of $f(x) = \frac{5x}{x - 1}$.
12. Evaluate: $\sin^{-1}\frac{1}{2}$
13. Use interval notation to state the domain of $f(x) = \cos^{-1} x$.
14. Use interval notation to state the range of $f(x) = \tan^{-1} x$.
15. Evaluate $\tan\left(\sin^{-1}\left(\frac{12}{13}\right)\right)$.
16. Solve: $2\cos^2 x + \sin x - 1 = 0$, for $0 \le x < 2\pi$
17. Find the magnitude and direction angle for the vector $\langle -3, 4\rangle$. Round the angle to the nearest tenth of a degree.

Page 459

CUMULATIVE REVIEW EXERCISES

Cumulative Review Exercises, which appear at the end of each chapter (except Chapter 1), help students maintain skills learned in previous chapters.

The answers to all **Chapter Review Exercises,** all **Chapter Test Exercises,** and all **Cumulative Review Exercises** are given in the Answer Section. Along with the answer, there is a reference to the section that pertains to each exercise.

Instructor Resources

Essentials of Precalculus has a complete set of support materials for the instructor.

Instructor's Annotated Edition This edition contains a replica of the student text with additional resources for the instructor. These include: *Instructor Notes, Alternative Example* notes, *PowerPoint* icons, *Suggested Assignments,* and answers to all exercises.

Instructor's Solutions Manual The *Instructor's Solutions Manual* contains worked-out solutions for all exercises in the text.

Instructor's Resource Manual with Testing This resource includes six ready-to-use printed *Chapter Tests* per chapter, and a *Printed Test Bank* providing a printout of one example of each of the algorithmic items on the *HM Testing* CD-ROM program.

HM ClassPrep w/ HM Testing CD-ROM *HM ClassPrep* contains a multitude of text-specific resources for instructors to use to enhance the classroom experience. These resources can be easily accessed by chapter or resource type and also can link you to the text's website. *HM Testing* is our computerized test generator and contains a database of algorithmic test items, as well as providing **online testing** and **gradebook** functions.

Instructor Text-specific Website The resources available on the *ClassPrep CD* are also available on the instructor website at math.college.hmco.com/instructors. Appropriate items are password protected. Instructors also have access to the student part of the text's website.

Student Resources

Student Study Guide The *Student Study Guide* contains complete solutions to all odd-numbered exercises in the text, as well as study tips and a practice test for each chapter.

Math Study Skills Workbook *by Paul D. Nolting* This workbook is designed to reinforce skills and minimize frustration for students in any math class, lab, or study skills course. It offers a wealth of study tips and sound advice on note taking, time management, and reducing math anxiety. In addition, numerous opportunities for self assessment enable students to track their own progress.

HM Eduspace® Online Learning Environment *Eduspace* is a text-specific, web-based learning environment that combines an algorithmic tutorial program with homework capabilities. Specific content is available 24 hours a day to help you further understand your textbook.

HM mathSpace® Tutorial CD-ROM This tutorial CD-ROM allows students to practice skills and review concepts as many times as necessary by providing algorithmically-generated exercises and step-by-step solutions for practice.

SMARTHINKING™ Live, Online Tutoring Houghton Mifflin has partnered with SMARTHINKING to provide an easy-to-use and effective online tutorial service. **Whiteboard Simulations** and **Practice Area** promote real-time visual interaction.

Three levels of service are offered.

- **Text-specific Tutoring** provides real-time, one-on-one instruction with a specially qualified 'e-structor.'
- **Questions Any Time** allows students to submit questions to the tutor outside the scheduled hours and receive a reply within 24 hours.
- **Independent Study Resources** connect students with around-the-clock access to additional educational services, including interactive websites, diagnostic tests, and Frequently Asked Questions posed to SMARTHINKING e-structors.

Houghton Mifflin Instructional Videos and DVDs Text-specific videos and DVDs, hosted by Dana Mosely, cover all sections of the text and provide a valuable resource for further instruction and review.

Student Text-specific Website Online student resources can be found at this text's website at math.college.hmco.com/students.

Acknowledgments

The authors would like to thank the people who have provided many valuable suggestions during the development of this text.

Ioannis K. Argyros, *Cameron University, OK*
Timothy D. Beaver, *Isothermal Community College, NC*
Norma Bisulca, *The University of Maine–Augusta, ME*
C. Allen Brown, *Wabash Valley College, IL*
Larry Buess, *Trevecca Nazarene University, TN*
Dr. Warren J. Burch, *Brevard Community College, FL*
Alice Burstein, *Middlesex Community College, CT*
Sharon Rose Butler, *Pikes Peak Community College, CO*
Michael Button, *The Master's College, CA*
Harold Carda, *South Dakota School of Mines and Technology, SD*
Charles Cheney, *Spring Hill College, AL*
Oiyin Pauline Chow, *Harrisburg Area Community College, PA*
Dr. Greg Clements, *Midland Lutheran College, NE*
Jacqueline Coomes, *Eastern Washington University, WA*
Anna Cox, *Kellogg Community College, MI*
Ellen Cunningham, *Saint Mary-of-the-Woods College, IN*
Rohan Dalpatadu, *University of Nevada, NV*
Jerry Davis, *Johnson State College, VT*
Lucy S. DeComo, *Walsh University, OH*
Blair T. Dietrich, *Georgia College and State University, GA*
Sharon Dunn, *Southside Virginia Community College, VA*
Heather Van Dyke, *Manhattan Christian College, NY*
Noelle Eckley, *Lassen Community College, CA*
David Ellis, *San Francisco State University, CA*
Theodore S. Erickson, *Wheeling Jesuit University, WV*
Dr. Hamidullah Farhat, *Hampton University, VA*
Cathy Ferrer, *Valencia Community College, FL*
Dr. William P. Fox, *Francis Marion University, SC*
Tarsh Freeman, *Bevill State Community College–Brewer Campus, AL*

Allen G. Fuller, *Gordon College, GA*
John D. Gieringer, *Alvernia College, PA*
Earl Gladue, *Roger Williams University, RI*
John T. Gordon, *Southern Polytechnic State University, GA*
Cornell Grant, *DeKalb Technical College, GA*
Allen Hamlin, *Palm Beach Community College, FL*
Ronald E. Harrell, *Allegheny College, PA*
Gregory P. Henderson, *Hillsborough Community College–Plant City Campus, FL*
Dr. Shahryar Heydari, *Piedmont College, GA*
Kaat Higham, *Bergen Community College, NJ*
Dr. Philip Holladay, *Geneva College, PA*
Matthew Hudock, *St. Philip's College, TX*
John Jacobs, *MassBay Community College, MA*
Dr. Jay M. Jahangiri, *Kent State University–Geauga Campus, OH*
David W. Jessee, Jr., *Manatee Community College, FL*
John M. Johnson, *George Fox University, OR*
Glen A. Just, *The Franciscan University, IA*
Regina Keller, *Suffolk County Community College, NY*
Dr. Gary Kimball, *Southeastern College, FL*
Ellen Knapp, *Philadelphia University, PA*
Dr. F. Kostanyan, *DeVry University, NY*
Carlos de la Lama, *San Diego City College, CA*
Mike Lavinder, *Wenatchee Valley College, WA*
Dr. Shinemin Lin, *Savannah State University, GA*
Domingo Litong, *Houston Community College, TX*
Nicholas Loudin, *Alderson-Broaddus College, WV*
Rich Maresh, *Viterbo University, WI*
Dr. Chris Masters, *Doane College, NE*
Frank Mattero, *Quinsigamond Community College, MA*
Richard McCall, *St. Louis College of Pharmacy, MO*
Robbie McKelvy, *Cossatot Community College, AR*
Dr. Helen Medley
Robert Messer, *Albion College, MI*
Dr. Beverly K. Michael, *University of Pittsburgh, PA*
Dr. Cheryl Chute Miller, *SUNY Potsdam, NY*
Deborah Mirdamadi, *Pennsylvania State University–Mont Alto, PA*
Gregory A. Mitchell, *Penn Valley Community College, MO*
Charles D. Mooney, *Reinhardt College, GA*
Dr. John C. Nardo, *Oglethorpe University, GA*
Al Niemier, *Holy Cross College, IN*
Steve Nimmo, *Morningside College, IA*
Neal Ninteman, *George Fox University, OR*
Earl Packard, *Kutztown University of Pennsylvania, PA*
Ron Palcic, *Johnson County Community College, KS*
Leslie A. Palmer, *Mercyhurst College, PA*
Vadim Ponomarenko, *Trinity University, TX*
Stephen J. Ramirez, *Crafton Hills College, CA*
Jane Roads, *Moberly Area Community College, MO*
Pascal Roubides, *Georgia Institute of Technology, GA*
Carol Roush, *Community College of Baltimore County–Catonsville Campus, MD*

Fred Safier, *City College of San Francisco, CA*
Dr. Davis A. Santos, *Community College of Philadelphia, PA*
Kurt Scholz, *University of St. Thomas, MN*
Dr. Bill Schwendner, *University of Connecticut–Stamford, CT*
Lauri Semarne
Melody Shipley, *North Central Missouri College, MO*
Caroline Shook, *Bellevue Community College, WA*
Jelinda Spotorno, *Clayton College and State University, GA*
Daryl Stephens, *East Tennessee State University, TN*
David Tanenbaum, *Harford Community College, MD*
Dr. Stephen J. Tillman, *Wilkes University, PA*
William K. Tomhave, *Concordia College, MN*
Dr. Rajah P. Varatharajah, *North Carolina A&T State University, NC*
Dr. Paul Vaz, *Arizona State University, AZ*
Dr. August Waltmann, *Wartburg College, IA*
Connie Wappes, *Fond du Lac Tribal and Community College, MN*
Dr. Peter R. Weidner, *Edinboro University of Pennsylvania, PA*
LeAnn Werner, *South Dakota State University, SD*
Denise Widup, *University of Wisconsin–Parkside, WI*
Robert R. Young, *Brevard Community College, FL*

CHAPTER 1

FUNCTIONS AND GRAPHS

Functions as Models

The Golden Gate Bridge spans the Golden Gate Strait, which is the entrance to the San Francisco Bay from the Pacific Ocean. Designed by Joseph Strass, the Golden Gate Bridge is a *suspension* bridge. A *quadratic function,* one of the topics of this chapter, can be used to model a cable of this bridge. See **Exercise 74 on page 79.**

Strass had many skeptics who did not believe the bridge could be built. Nonetheless, the bridge opened on May 27, 1937, a little over 4 years after construction began. When it was completed, Strass composed the following poem.

The Mighty Task Is Done

At last the mighty task is done;
Resplendent in the western sun;
The Bridge looms mountain high.

On its broad decks in rightful pride,
The world in swift parade shall ride
Throughout all time to be.

Launched midst a thousand hopes and fears,
Damned by a thousand hostile sneers,
Yet ne'er its course was stayed.
But ask of those who met the foe,
Who stood alone when faith was low,
Ask them the price they paid.
High overhead its lights shall gleam,
Far, far below life's restless stream,
Unceasingly shall flow....

Difference Tables

When devising a plan to solve a problem, it may be helpful to organize information in a table. One particular type of table that can be used to discern some patterns is called a *difference table*. For instance, suppose that we want to determine the number of square tiles in the tenth figure of a pattern whose first four figures are

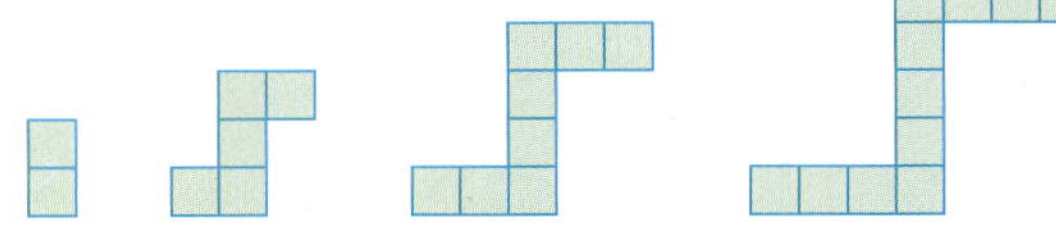

We begin by creating a difference table by listing the number of tiles in each figure and the differences between the numbers of tiles on the next line.

Tiles	2		5		8		11
Differences		3		3		3	

From the difference table, note that each succeeding figure has three more tiles. Therefore, we can find the number of tiles in the tenth figure by extending the difference table.

Tiles	2		5		8		11		14		17		20		23		26		29
Differences		3		3		3		3		3		3		3		3		3	

There are 29 tiles in the tenth figure.

Sometimes the first differences are not constant as they were in the preceding example. In this case we find the second differences. For instance, consider the pattern at the right.

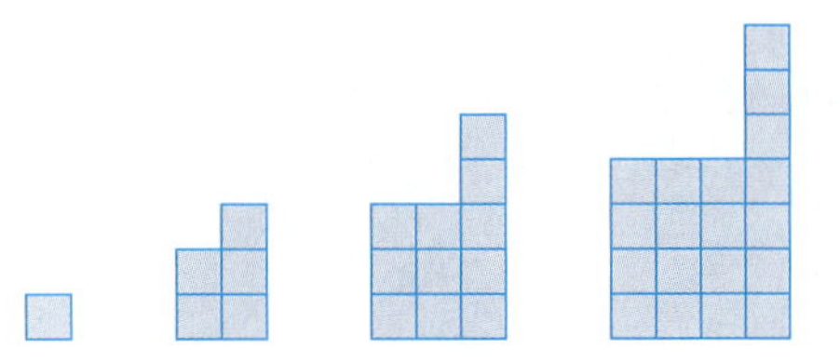

The difference table is shown below.

Tiles	1		5		11		19
First differences		4		6		8	
Second differences			2		2		

In this case the first differences are not constant, but the second differences are constant. With this information we can determine the number of tiles in succeeding figures.

Tiles	1		5		11		19		29		41		55		71		89		109
First differences		4		6		8		10		12		14		16		18		20	
Second differences			2		2		2		2		2		2		2		2		

In this case there are 109 tiles in the tenth figure.

If second differences are not constant, try third differences.[1] If third differences are not constant, try fourth differences, and so on.

[1] Not all lists of numbers will end with a difference row of constants. For instance, consider 1, 1, 2, 3, 5, 8, 13, 21, 34, ...

SECTION 1.1

EQUATIONS AND INEQUALITIES

- THE REAL NUMBERS
- ABSOLUTE VALUE AND DISTANCE
- LINEAR AND QUADRATIC EQUATIONS
- LINEAR AND QUADRATIC INEQUALITIES
- ABSOLUTE VALUE INEQUALITIES

MATH MATTERS

Archimedes (c. 287–212 B.C.) was the first to calculate π with any degree of precision. He was able to show that

$$3\frac{10}{71} < \pi < 3\frac{1}{7}$$

from which we get the approximation $3\frac{1}{7} \approx \pi$. The use of the symbol π for this quantity was introduced by Leonhard Euler (1707–1783) in 1739, approximately 2000 years after Archimedes.

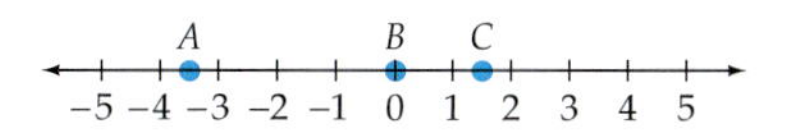

FIGURE 1.1

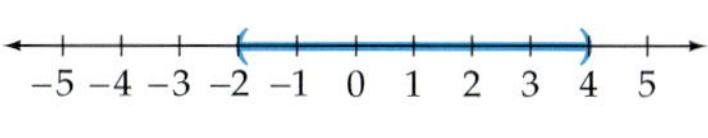

The open interval (−2, 4)

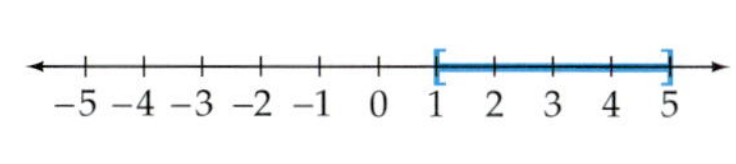

The closed interval [1, 5]

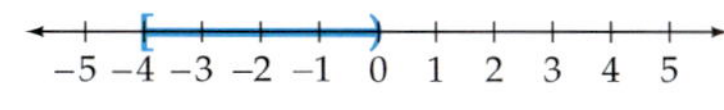

The half-open interval [−4, 0)

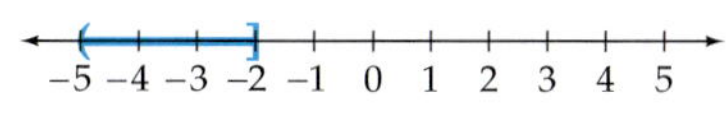

The half-open interval (−5, −2]

FIGURE 1.2

THE REAL NUMBERS

The real numbers are used extensively in mathematics. The set of real numbers is quite comprehensive and contains several unique sets of numbers.

The **integers** are the set of numbers

$$\{\ldots, -4, -3, -2, -1, 0, 1, 2, 3, 4, \ldots\}$$

Recall that the brace symbols, { }, are used to identify a set. The positive integers are called **natural numbers.**

The **rational numbers** are the set of numbers of the form $\frac{a}{b}$, where a and b are integers and $b \neq 0$. Thus the rational numbers include $-\frac{3}{4}$ and $\frac{5}{2}$. Because each integer can be expressed in the form $\frac{a}{b}$ with denominator $b = 1$, the integers are included in the set of rational numbers. Every rational number can be written as either a terminating or a repeating decimal.

A number written in decimal form that does not repeat or terminate is called an **irrational number.** Some examples of irrational numbers are $0.141141114\ldots$, $\sqrt{2}$, and π. These numbers cannot be expressed as quotients of integers. The set of **real numbers** is the union of the sets of rational and irrational numbers.

A real number can be represented geometrically on a **coordinate axis** called a **real number line.** Each point on this line is associated with a real number called the **coordinate** of the point. Conversely, each real number can be associated with a point on a real number line. In **Figure 1.1,** the coordinate of A is $-\frac{7}{2}$, the coordinate of B is 0, and the coordinate of C is $\sqrt{2}$.

Given any two real numbers a and b, we say that a is **less than** b, denoted by $a < b$, if $a - b$ is a negative number. Similarly, we say that a is **greater than** b, denoted by $a > b$, if $a - b$ is a positive number. When a **equals** b, $a - b$ is zero. The symbols $<$ and $>$ are called **inequality symbols.** Two other inequality symbols, $\leq$ (less than or equal to) and $\geq$ (greater than or equal to), are also used.

The inequality symbols can be used to designate sets of real numbers. If $a < b$, the **interval notation** (a, b) is used to indicate the set of real numbers between a and b. This set of numbers also can be described using **set-builder notation:**

$$(a, b) = \{x \mid a < x < b\}$$

When reading a set written in set-builder notation, we read $\{x \mid$ as "the set of x such that." The expression that follows the vertical bar designates the elements in the set.

The set (a, b) is called an **open interval.** The graph of the open interval consists of all the points on the real number line between a and b, not including a and b. A **closed interval,** denoted by $[a, b]$, consists of all points between a and b, including a and b. We can also discuss **half-open intervals.** An example of each type of interval is shown in **Figure 1.2.**

> ***take note***
>
> The interval notation $[1, \infty)$ represents all real numbers greater than or equal to 1. The interval notation $(-\infty, 4)$ represents all real numbers less than 4.

$(-2, 4) = \{x \mid -2 < x < 4\}$	An open interval
$[1, 5] = \{x \mid 1 \le x \le 5\}$	A closed interval
$[-4, 0) = \{x \mid -4 \le x < 0\}$	A half-open interval
$(-5, -2] = \{x \mid -5 < x \le -2\}$	A half-open interval

Absolute Value and Distance

The *absolute value* of a real number is a measure of the distance from zero to the point associated with the number on a real number line. Therefore, the absolute value of a real number is always positive or zero. We now give a more formal definition of absolute value.

Absolute Value

For a real number a, the **absolute value** of a, denoted by $|a|$, is

$$|a| = \begin{cases} a & \text{if } a \ge 0 \\ -a & \text{if } a < 0 \end{cases}$$

The distance d between the points A and B with coordinates -3 and 2, respectively, on a real number line is the absolute value of the difference between the coordinates. See **Figure 1.3.**

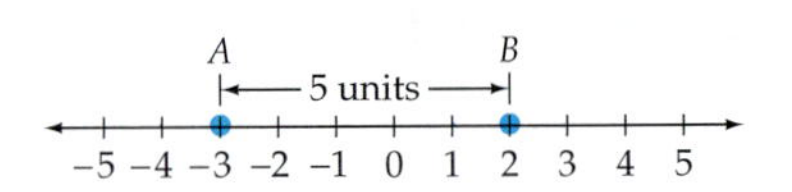

Figure 1.3

$$d = |2 - (-3)| = 5$$

Because the absolute value is used, we could also write

$$d = |(-3) - 2| = 5$$

In general, we define the *distance* between any two points A and B on a real number line as the absolute value of the difference between the coordinates of the points.

Distance Between Two Points on a Coordinate Line

Let a and b be the coordinates of the points A and B, respectively, on a real number line. Then the **distance** between A and B, denoted $d(A, B)$, is

$$d(A, B) = |a - b|$$

This formula applies to any real number line. It can be used to find the distance between two points on a vertical real number line, as shown in **Figure 1.4.**

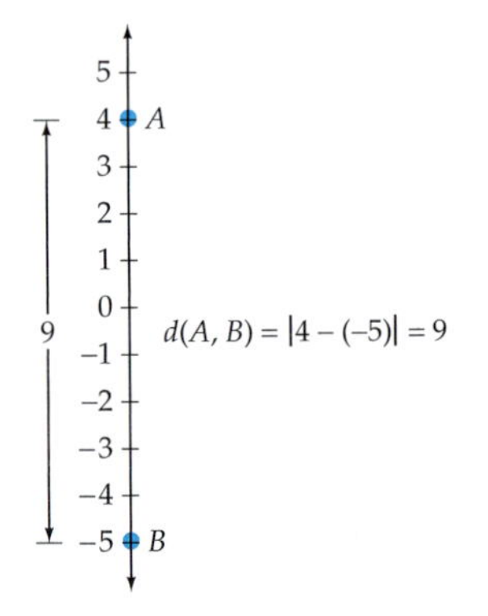

Figure 1.4

Linear and Quadratic Equations

An **equation** is a statement about the equality of two expressions. Examples of equations follow.

$$7 = 2 + 5 \qquad x^2 = 4x + 5 \qquad 3x - 2 = 2(x + 1) + 3$$

The values of the variable that make an equation a true statement are the **roots** or **solutions** of the equation. To **solve** an equation means to find the solutions of the equation. The number 2 is said to satisfy the equation $2x + 1 = 5$, because substituting 2 for x produces $2(2) + 1 = 5$, which is a true statement.

Definition of a Linear Equation

A **linear equation** in the single variable x is an equation of the form $ax + b = 0$, where $a \neq 0$.

To solve a linear equation in one variable, isolate the variable on one side of the equals sign.

EXAMPLE 1 Solve a Linear Equation

Solve: $3x - 5 = 2$

Solution

$$3x - 5 = 2$$

$$3x - 5 + 5 = 2 + 5 \qquad \text{• Add 5 to each side of the equation.}$$

$$3x = 7$$

$$\frac{3x}{3} = \frac{7}{3} \qquad \text{• Divide each side of the equation by 3.}$$

$$x = \frac{7}{3}$$

The solution is $\frac{7}{3}$.

▶ TRY EXERCISE 6, PAGE 13

Definition of a Quadratic Equation

A **quadratic equation** in x is an equation that can be written in the **standard quadratic form**

$$ax^2 + bx + c = 0$$

where a, b, and c are real numbers and $a \neq 0$.

MATH MATTERS

The term *quadratic* is derived from the Latin word *quadrare*, which means "to make square." Because the area of a square that measures x units on each side is x^2, we refer to equations that can be written in the form $ax^2 + bx + c = 0$ as equations that are quadratic in x.

Several methods can be used to solve a quadratic equation. For instance, if you can factor $ax^2 + bx + c$ into linear factors, then $ax^2 + bx + c = 0$ can be solved by applying the following property.

The Zero Product Principle

If A and B are algebraic expressions such that $AB = 0$, then $A = 0$ or $B = 0$.

The zero product principle states that if the product of two factors is zero, then at least one of the factors must be zero. In Example 2, the zero product principle is used to solve a quadratic equation.

EXAMPLE 2 Solve by Factoring

Solve by factoring: $2x^2 - 5x = 12$

Solution

$$\begin{aligned} 2x^2 - 5x &= 12 \\ 2x^2 - 5x - 12 &= 0 && \bullet \text{ Write in standard quadratic form.} \\ (x-4)(2x+3) &= 0 && \bullet \text{ Factor.} \\ x - 4 = 0 \qquad 2x + 3 &= 0 && \bullet \text{ Set each factor equal to zero.} \\ x = 4 \qquad 2x &= -3 && \bullet \text{ Solve each linear equation.} \\ x &= -\frac{3}{2} \end{aligned}$$

To review **FACTORING**, *see the Review Appendix, p. 556.*

A check shows that 4 and $-\dfrac{3}{2}$ are both solutions of $2x^2 - 5x = 12$.

TRY EXERCISE 22, PAGE 13

To review **RADICAL EXPRESSIONS**, *see the Review Appendix, p. 543.*

The solutions of $x^2 = 25$ can be found by taking the square root of each side of the equation.

$$\begin{aligned} x^2 &= 25 \\ \sqrt{x^2} &= \sqrt{25} \\ |x| &= 5 && \bullet \text{ Recall that } \sqrt{x^2} = |x|. \\ x &= \pm 5 \\ x = 5 \quad &\text{or} \quad x = -5 \end{aligned}$$

We will refer to the preceding method of solving a quadratic equation as the *square root procedure.*

The Square Root Procedure

If $x^2 = c$, then $x = \sqrt{c}$ or $x = -\sqrt{c}$, which can also be written as $x = \pm\sqrt{c}$.

This procedure can be used to solve $(x + 1)^2 = 49$.

$$\begin{aligned} (x+1)^2 &= 49 \\ x + 1 &= \pm\sqrt{49} && \bullet \text{ Apply the Square Root Procedure.} \\ x &= -1 \pm 7 && \bullet \text{ Simplify.} \\ x &= -8 \quad \text{or} \quad 6 \end{aligned}$$

The solutions are -8 and 6.

Note that in each of the following perfect-square trinomials, the coefficient of x^2 is 1, and the constant term is the square of half the coefficient of the x term.

$$(x + 5)^2 = x^2 + 10x + 25, \qquad \left(\frac{1}{2} \cdot 10\right)^2 = 25$$

$$(x - 3)^2 = x^2 - 6x + 9, \qquad \left(\frac{1}{2} \cdot (-6)\right)^2 = 9$$

Adding to a binomial of the form $x^2 + bx$ the constant term that makes the binomial a perfect-square trinomial is called **completing the square.** For example, to complete the square of $x^2 + 8x$, add

$$\left(\frac{1}{2} \cdot 8\right)^2 = 16$$

to produce the perfect-square trinomial $x^2 + 8x + 16$.

Completing the square is a powerful procedure because it can be used to solve *any* quadratic equation.

Completing the square by adding the square of half the coefficient of the x term requires that the coefficient of the x^2 term be 1. If the coefficient of the x^2 term is not 1, then first multiply each term on each side of the equation by the reciprocal of the coefficient of x^2 to produce a coefficient of 1 for the x^2 term.

EXAMPLE 3 Solve by Completing the Square

Solve $2x^2 + 8x - 1 = 0$ by completing the square.

Solution

$$2x^2 + 8x - 1 = 0$$

$$2x^2 + 8x = 1 \qquad \text{• Isolate the constant term.}$$

$$\frac{1}{2}(2x^2 + 8x) = \frac{1}{2}(1) \qquad \text{• Multiply both sides of the equation by the reciprocal of the coefficient of } x^2.$$

$$x^2 + 4x = \frac{1}{2}$$

$$x^2 + 4x + 4 = \frac{1}{2} + 4 \qquad \text{• Complete the square.}$$

$$(x + 2)^2 = \frac{9}{2} \qquad \text{• Factor and simplify.}$$

$$x + 2 = \pm\sqrt{\frac{9}{2}} \qquad \text{• Apply the square root procedure.}$$

$$x = -2 \pm \frac{3\sqrt{2}}{2} \qquad \text{• Solve for } x.$$

$$x = \frac{-4 \pm 3\sqrt{2}}{2} \qquad \text{• Simplify.}$$

The solutions are $x = \dfrac{-4 + 3\sqrt{2}}{2}$ and $x = \dfrac{-4 - 3\sqrt{2}}{2}$.

▶ TRY EXERCISE 28, PAGE 13

To review **SIMPLIFYING RADICAL EXPRESSIONS**, *see the Review Appendix, p. 543.*

Completing the square on $ax^2 + bx + c = 0$ $(a \neq 0)$ produces a formula for x in terms of the coefficients a, b, and c. The formula is known as the *quadratic formula*, and it can be used to solve *any* quadratic equation.

MATH MATTERS

Evariste Galois (1811–1832)

The quadratic formula provides the solutions to the general quadratic equation

$$ax^2 + bx + c = 0$$

and formulas have been developed to solve the general cubic

$$ax^3 + bx^2 + cx + d = 0$$

and the general quartic

$$ax^4 + bx^3 + cx^2 + dx + e = 0$$

However, the French mathematician Evariste Galois, shown above, was able to prove that there are no formulas that can be used to solve "by radicals" general equations of degree 5 or larger.

Shortly after completion of his remarkable proof, Galois was shot in a duel. It has been reported that as Galois lay dying, he asked his brother, Alfred, to "Take care of my work. Make it known. Important." When Alfred broke into tears, Evariste said, "Don't cry Alfred. I need all my courage to die at twenty." (*Source: Whom the Gods Love*, by Leopold Infeld, The National Council of Teachers of Mathematics, 1978, p. 299.)

The Quadratic Formula

If $ax^2 + bx + c = 0$, $a \neq 0$, then

$$x = \frac{-b \pm \sqrt{b^2 - 4ac}}{2a}$$

Proof We assume a is a positive real number. If a were a negative real number, then we could multiply each side of the equation by -1 to make it positive.

$$ax^2 + bx + c = 0 \quad (a \neq 0)$$ • Given.

$$ax^2 + bx = -c$$ • Isolate the constant term.

$$x^2 + \frac{b}{a}x = -\frac{c}{a}$$ • Multiply each term on each side of the equation by $\frac{1}{a}$.

$$x^2 + \frac{b}{a}x + \left(\frac{b}{2a}\right)^2 = \left(\frac{b}{2a}\right)^2 - \frac{c}{a}$$ • Complete the square.

$$\left(x + \frac{b}{2a}\right)^2 = \frac{b^2}{4a^2} - \frac{c}{a}$$ • Factor the left side. Simplify the powers on the right side.

$$\left(x + \frac{b}{2a}\right)^2 = \frac{b^2}{4a^2} - \frac{4a}{4a} \cdot \frac{c}{a}$$ • Use a common denominator to simplify the right side.

$$x + \frac{b}{2a} = \pm\sqrt{\frac{b^2 - 4ac}{4a^2}}$$ • Apply the square root procedure.

$$x + \frac{b}{2a} = \pm\frac{\sqrt{b^2 - 4ac}}{2a}$$ • Because $a > 0$, $\sqrt{4a^2} = 2a$.

$$x = -\frac{b}{2a} \pm \frac{\sqrt{b^2 - 4ac}}{2a}$$ • Add $-\frac{b}{2a}$ to each side.

$$x = \frac{-b \pm \sqrt{b^2 - 4ac}}{2a}$$

As a general rule, you should first try to solve quadratic equations by factoring. If the factoring process proves difficult, then solve by using the quadratic formula.

EXAMPLE 4 Solve by Using the Quadratic Formula

Use the quadratic formula to solve $x^2 = 3x + 5$.

Solution

The standard form of $x^2 = 3x + 5$ is $x^2 - 3x - 5 = 0$. Substituting $a = 1$, $b = -3$, and $c = -5$ in the quadratic formula produces

$$x = \frac{-(-3) \pm \sqrt{(-3)^2 - 4(1)(-5)}}{2(1)}$$

$$= \frac{3 \pm \sqrt{29}}{2}$$

The solutions are $x = \frac{3 + \sqrt{29}}{2}$ and $x = \frac{3 - \sqrt{29}}{2}$.

TRY EXERCISE 30, PAGE 13

QUESTION Can the quadratic formula be used to solve any quadratic equation $ax^2 + bx + c = 0$ with real coefficients and $a \neq 0$?

LINEAR AND QUADRATIC INEQUALITIES

A statement that contains the symbol $<$, $>$, $\leq$, or $\geq$ is called an **inequality.** An inequality expresses the relative order of two mathematical expressions. The **solution set of an inequality** is the set of real numbers each of which, when substituted for the variable, results in a true inequality. The inequality $x > 4$ is true for any value of x greater than 4. For instance, 5, $\sqrt{21}$, and $\frac{17}{3}$ are all solutions of $x > 4$. The solution set of the inequality can be written in set-builder notation as $\{x \mid x > 4\}$ or in interval notation as $(4, \infty)$.

Equivalent inequalities have the same solution set. We solve an inequality by producing *simpler* but equivalent inequalities until the solutions are found. To produce these simpler but equivalent inequalities, we apply the following properties.

Properties of Inequalities

Let a, b, and c be real numbers.

1. *Addition Property* Adding the same real number to each side of an inequality preserves the direction of the inequality symbol.

$a < b$ and $a + c < b + c$ are equivalent inequalities.

2. *Multiplication Properties*

a. Multiplying each side of an inequality by the same *positive* real number *preserves* the direction of the inequality symbol.

If $c > 0$, then $a < b$ and $ac < bc$ are equivalent inequalities.

b. Multiplying each side of an inequality by the same *negative* real number *changes* the direction of the inequality symbol.

If $c < 0$, then $a < b$ and $ac > bc$ are equivalent inequalities.

ANSWER Yes. However, it is sometimes easier to find the solutions by factoring, by the square root procedure, or by completing the square.

Note the difference between Property 2a and Property 2b. Property 2a states that an equivalent inequality is produced when each side of a given inequality is multiplied by the same *positive* real number and that the direction of the inequality symbol is *not* changed. By contrast, Property 2b states that when each side of a given inequality is multiplied by a *negative* real number, we must *reverse* the direction of the inequality symbol to produce an equivalent inequality.

A **linear inequality in one variable** is one that can be written in the form $ax + b < 0$ or $ax + b > 0$, where $a \neq 0$. The inequality symbols $\leq$ and $\geq$ can also be used.

EXAMPLE 5 Solve a Linear Inequality

Solve: $2(x + 3) < 4x + 10$. Write the solution set in interval notation.

Solution

$2(x + 3) < 4x + 10$

$2x + 6 < 4x + 10$ • **Use the distributive property.**

$-2x < 4$ • **Subtract 4x and 6 from each side of the inequality.**

$x > -2$ • **Divide each side by −2 and reverse the inequality symbol.**

Thus the original inequality is true for all real numbers greater than −2. The solution set is $(-2, \infty)$.

take note

Solutions of inequalities are often stated using set-builder notation or interval notation. For instance, the real numbers that are solutions of the inequality in Example 5 can be written in set-builder notation as $\{x \mid x > -2\}$ or in interval notation as $(-2, \infty)$.

▶ TRY EXERCISE 50, PAGE 13

A **quadratic inequality in one variable** is one that can be written in the form $ax^2 + bx + c < 0$ or $ax^2 + bx + c > 0$, where $a \neq 0$. The symbols $\leq$ and $\geq$ can also be used.

Quadratic inequalities can be solved by algebraic means. However, it is often easier to use a graphical method to solve these inequalities. The graphical method is used in the example that follows.

To solve $x^2 - x - 6 < 0$, factor the trinomial.

$$x^2 - x - 6 < 0$$
$$(x - 3)(x + 2) < 0$$

On a number line, draw vertical lines at the numbers that make each factor equal to zero. See **Figure 1.5.**

$$x - 3 = 0 \qquad x + 2 = 0$$
$$x = 3 \qquad x = -2$$

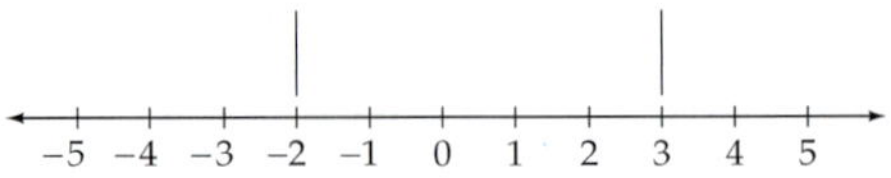

−5 −4 −3 −2 −1 0 1 2 3 4 5

FIGURE 1.5

For each factor, place plus signs above the number line for those regions where the factor is positive and negative signs where the factor is negative, as in **Figure 1.6.**

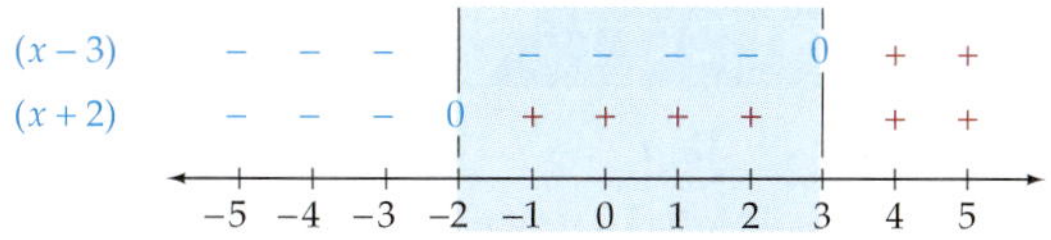

Figure 1.6

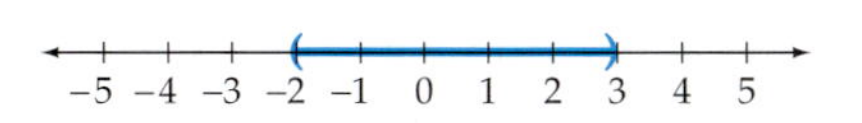

Figure 1.7

Because $x^2 - x - 6 < 0$, the solution set will be the regions where one factor is positive and the other factor is negative. This occurs when $-2 < x < 3$ or $(-2, 3)$. The graph of the solution set of the inequality $x^2 - x - 6 < 0$ is shown in **Figure 1.7.**

EXAMPLE 6 Solve a Quadratic Inequality

Solve and graph the solution set of $2x^2 - x - 3 \geq 0$.

Solution

$$2x^2 - x - 3 \geq 0$$
$$(2x - 3)(x + 1) \geq 0$$
$$2x - 3 = 0 \qquad x + 1 = 0$$

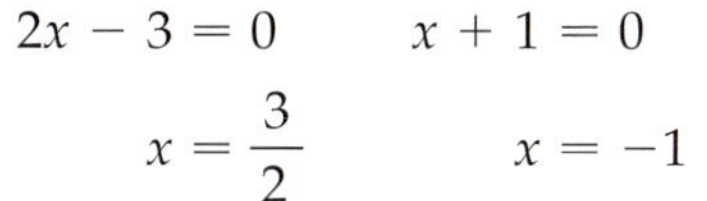

$$x = \frac{3}{2} \qquad x = -1$$

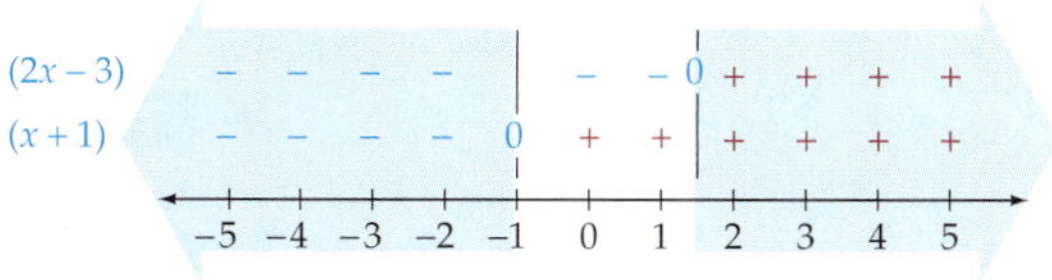

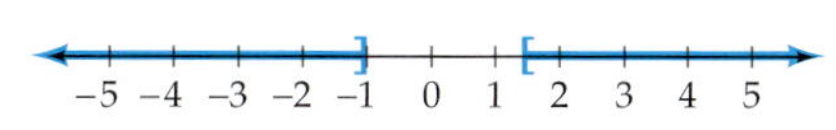

The solution set is $(-\infty, -1] \cup \left[\frac{3}{2}, \infty\right)$.

▶ **Try Exercise 58, page 13**

> **take note**
>
> We use the set union symbol ∪ to join intervals that contain solutions to inequalities, as in Example 6.

Absolute Value Inequalities

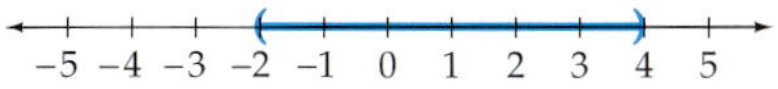

The solution set of the absolute value inequality $|x - 1| < 3$ is the set of all real numbers whose distance from 1 is *less than* 3. Therefore, the solution set consists of all numbers between -2 and 4. See **Figure 1.8.** In interval notation, the solution set is $(-2, 4)$.

Figure 1.8

The solution set of the absolute value inequality $|x - 1| > 3$ is the set of all real numbers whose distance from 1 is *greater than* 3. Therefore, the solution set consists of all real numbers less than -2 *or* greater than 4. See **Figure 1.9.** In interval notation, the solution set is $(-\infty, -2) \cup (4, \infty)$.

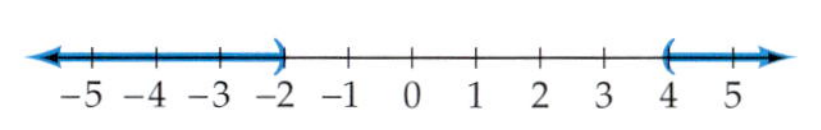

Figure 1.9

The following properties are used to solve absolute value inequalities.

Properties of Absolute Value Inequalities

For any variable expression E and any nonnegative real number k,

$$|E| \leq k \quad \text{if and only if} \quad -k \leq E \leq k$$
$$|E| \geq k \quad \text{if and only if} \quad E \leq -k \quad \text{or} \quad E \geq k$$

EXAMPLE 7 Solve an Absolute Value Inequality

Solve: $|2 - 3x| < 7$

Solution

$|2 - 3x| < 7$ implies $-7 < 2 - 3x < 7$. Solve this compound inequality.

$$-7 < 2 - 3x < 7$$

$$-9 < \quad -3x \quad < 5$$ • Subtract 2 from each of the three parts of the inequality.

$$3 > \quad x \quad > -\frac{5}{3}$$ • Multiply each part of the inequality by $-\frac{1}{3}$ and reverse the inequality symbols.

In interval notation, the solution set is given by $\left(-\frac{5}{3}, 3\right)$. See **Figure 1.10**.

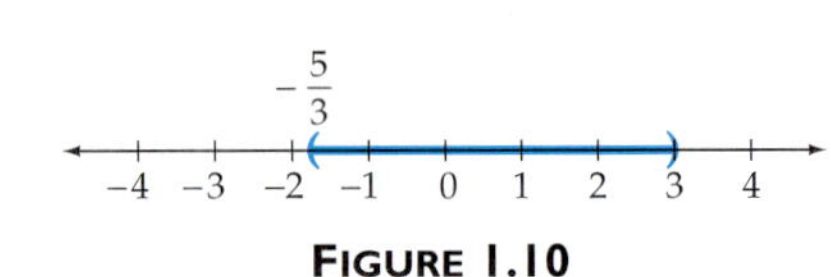

FIGURE 1.10

▶ TRY EXERCISE 70, PAGE 13

EXAMPLE 8 Solve an Absolute Value Inequality

Solve: $|4x - 3| \geq 5$

Solution

$|4x - 3| \geq 5$ implies $4x - 3 \leq -5$ or $4x - 3 \geq 5$. Solving each of these inequalities produces

$$4x - 3 \leq -5 \qquad \text{or} \qquad 4x - 3 \geq 5$$

$$4x \leq -2 \qquad\qquad 4x \geq 8$$

$$x \leq -\frac{1}{2} \qquad\qquad x \geq 2$$

Therefore, the solution set is $\left(-\infty, -\frac{1}{2}\right] \cup [2, \infty)$. See **Figure 1.11**.

take note

Some inequalities have a solution set that consists of all real numbers. For example, $|x + 9| \geq 0$ is true for all values of x. Because an absolute value is always nonnegative, the equation is always true.

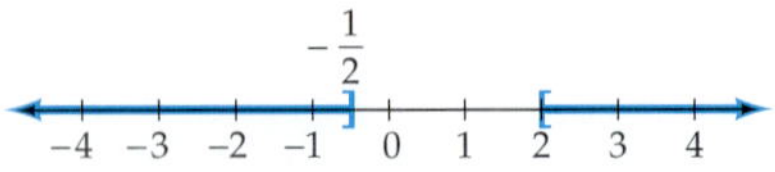

FIGURE 1.11

▶ TRY EXERCISE 72, PAGE 13

TOPICS FOR DISCUSSION

1. Discuss the similarities and differences among natural numbers, integers, rational numbers, and real numbers.
2. Discuss the differences among an equation, an inequality, and an expression.
3. Is it possible for an equation to have no solution? If not, explain why. If so, give an example of an equation with no solution.
4. Is the statement $|x| = -x$ ever true? Explain why or why not.
5. How do quadratic equations in one variable differ from linear equations in one variable? Explain how the method used to solve an equation depends on whether it is a linear or a quadratic equation.

EXERCISE SET 1.1

In Exercises 1 to 18, solve and check each equation.

1. $2x + 10 = 40$
2. $-3y + 20 = 2$
3. $5x + 2 = 2x - 10$
4. $4x - 11 = 7x + 20$
5. $2(x - 3) - 5 = 4(x - 5)$
6. $6(5s - 11) - 12(2s + 5) = 0$
7. $\frac{3}{4}x + \frac{1}{2} = \frac{2}{3}$
8. $\frac{x}{4} - 5 = \frac{1}{2}$
9. $\frac{2}{3}x - 5 = \frac{1}{2}x - 3$
10. $\frac{1}{2}x + 7 - \frac{1}{4}x = \frac{19}{2}$
11. $0.2x + 0.4 = 3.6$
12. $0.04x - 0.2 = 0.07$
13. $\frac{3}{5}(n + 5) - \frac{3}{4}(n - 11) = 0$
14. $-\frac{5}{7}(p + 11) + \frac{2}{5}(2p - 5) = 0$
15. $3(x + 5)(x - 1) = (3x + 4)(x - 2)$
16. $5(x + 4)(x - 4) = (x - 3)(5x + 4)$
17. $0.08x + 0.12(4000 - x) = 432$
18. $0.075y + 0.06(10{,}000 - y) = 727.50$

In Exercises 19 to 26, solve each quadratic equation by factoring and applying the zero product property.

19. $x^2 - 2x - 15 = 0$
20. $y^2 + 3y - 10 = 0$
21. $8y^2 + 189y - 72 = 0$
22. $12w^2 - 41w + 24 = 0$
23. $3x^2 - 7x = 0$
24. $5x^2 = -8x$
25. $(x - 5)^2 - 9 = 0$
26. $(3x + 4)^2 - 16 = 0$

In Exercises 27 to 40, solve by completing the square or by using the quadratic formula.

27. $x^2 - 2x - 15 = 0$
28. $x^2 - 5x - 24 = 0$
29. $x^2 + x - 1 = 0$
30. $x^2 + x - 2 = 0$
31. $2x^2 + 4x + 1 = 0$
32. $2x^2 + 4x - 1 = 0$
33. $3x^2 - 5x - 3 = 0$
34. $3x^2 - 5x - 4 = 0$
35. $\frac{1}{2}x^2 + \frac{3}{4}x - 1 = 0$
36. $\frac{2}{3}x^2 - 5x + \frac{1}{2} = 0$
37. $\sqrt{2}x^2 + 3x + \sqrt{2} = 0$
38. $2x^2 + \sqrt{5}x - 3 = 0$
39. $x^2 = 3x + 5$
40. $-x^2 = 7x - 1$

In Exercises 41 to 50, use the properties of inequalities to solve each inequality. Write answers using interval notation.

41. $2x + 3 < 11$
42. $3x - 5 > 16$
43. $x + 4 > 3x + 16$
44. $5x + 6 < 2x + 1$
45. $-6x + 1 \ge 19$
46. $-5x + 2 \le 37$
47. $-3(x + 2) \le 5x + 7$
48. $-4(x - 5) \ge 2x + 15$
49. $-4(3x - 5) > 2(x - 4)$
50. $3(x + 7) \le 5(2x - 8)$

In Exercises 51 to 58, solve each quadratic inequality. Use interval notation to write each solution set.

51. $x^2 + 7x > 0$
52. $x^2 - 5x \le 0$
53. $x^2 + 7x + 10 < 0$
54. $x^2 + 5x + 6 < 0$
55. $x^2 - 3x \ge 28$
56. $x^2 < -x + 30$
57. $6x^2 - 4 \le 5x$
58. $12x^2 + 8x \ge 15$

In Exercises 59 to 76, use interval notation to express the solution set of each inequality.

59. $|x| < 4$
60. $|x| > 2$
61. $|x - 1| < 9$
62. $|x - 3| < 10$
63. $|x + 3| > 30$
64. $|x + 4| < 2$
65. $|2x - 1| > 4$
66. $|2x - 9| < 7$
67. $|x + 3| \ge 5$
68. $|x - 10| \ge 2$
69. $|3x - 10| \le 14$
70. $|2x - 5| \ge 1$
71. $|4 - 5x| \ge 24$
72. $|3 - 2x| \le 5$
73. $|x - 5| \ge 0$
74. $|x - 7| \ge 0$
75. $|x - 4| \le 0$
76. $|2x + 7| \le 0$

77. **GEOMETRY** The perimeter of a rectangle is 27 centimeters, and its area is 35 square centimeters. Find the length and the width of the rectangle.

78. GEOMETRY The perimeter of a rectangle is 34 feet and its area is 60 square feet. Find the length and the width of the rectangle.

79. RECTANGULAR ENCLOSURE A gardener wishes to use 600 feet of fencing to enclose a rectangular region and subdivide the region into two smaller rectangles. The total enclosed area is 15,000 square feet. Find the dimensions of the enclosed region.

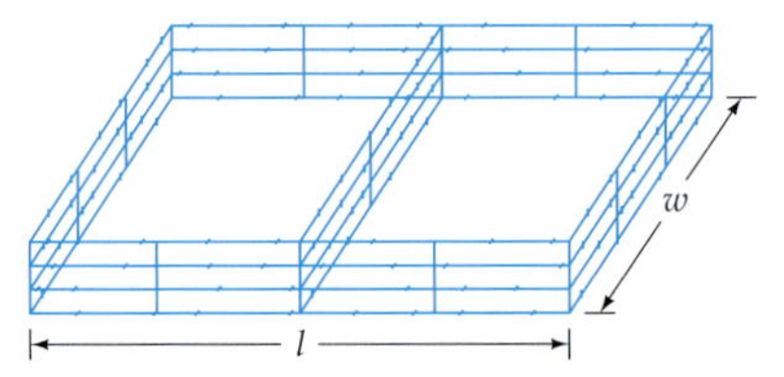

80. RECTANGULAR ENCLOSURE A farmer wishes to use 400 yards of fencing to enclose a rectangular region and subdivide the region into three smaller rectangles. If the total enclosed area is 4800 square yards, find the dimensions of the enclosed region.

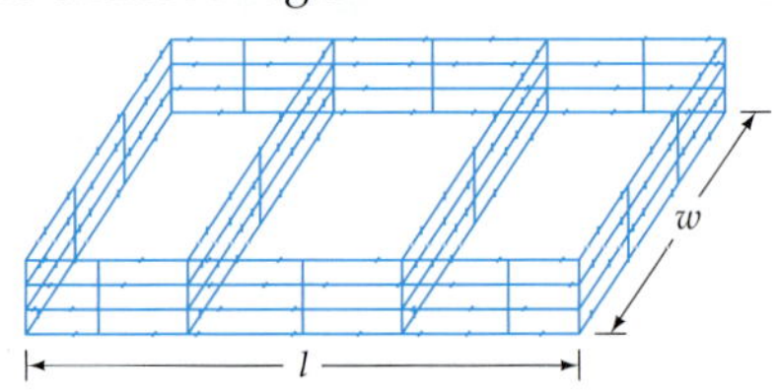

81. PERSONAL FINANCE A bank offers two checking account plans. The monthly fee and charge per check for each plan are shown below. Under what conditions is it less expensive to use the LowCharge plan?

Account Plan	*Monthly Fee*	*Charge per Check*
LowCharge	\$5.00	\$.01
FeeSaver	\$1.00	\$.08

82. PERSONAL FINANCE You can rent a car for the day from company A for \$29.00 plus \$0.12 a mile. Company B charges \$22.00 plus \$0.21 a mile. Find the number of miles m (to the nearest mile) per day for which it is cheaper to rent from company A.

83. PERSONAL FINANCE A sales clerk has a choice between two payment plans. Plan A pays \$100.00 a week plus \$8.00 a sale. Plan B pays \$250.00 a week plus \$3.50 a sale. How many sales per week must be made for plan A to yield the greater paycheck?

84. PERSONAL FINANCE A video store offers two rental plans. The yearly membership fee and the daily charge per video are shown below. How many videos can be rented per year if the No-fee plan is to be the less expensive of the plans?

THE VIDEO STORE

Rental Plan	Yearly Fee	Daily Charge per Video
Low-rate	\$15.00	\$1.49
No-fee	None	\$1.99

85. AVERAGE TEMPERATURES The average daily minimum-to-maximum temperatures for the city of Palm Springs during the month of September are 68°F to 104°F. What is the corresponding temperature range measured on the Celsius temperature scale?

CONNECTING CONCEPTS

86. A GOLDEN RECTANGLE The ancient Greeks defined a rectangle as a "golden rectangle" if its length l and its width w satisfied the equation

$$\frac{l}{w} = \frac{w}{l - w}$$

a. Solve this formula for w.

b. If the length of a golden rectangle is 101 feet, determine its width. Round to the nearest hundredth.

87. SUM OF NATURAL NUMBERS The sum S of the first n natural numbers $1, 2, 3, \ldots, n$ is given by the formula

$$S = \frac{n}{2}(n + 1)$$

How many consecutive natural numbers starting with 1 produce a sum of 253?

88. Number of Diagonals The number of diagonals D of a polygon with n sides is given by the formula

$$D = \frac{n}{2}(n - 3)$$

a. Determine the number of sides of a polygon with 464 diagonals.

b. Can a polygon have 12 diagonals? Explain.

89. Revenue The monthly revenue R for a product is given by $R = 420x - 2x^2$, where x is the price in dollars of each unit produced. Find the interval in terms of x for which the monthly revenue is greater than zero.

90. Write an absolute value inequality to represent all real numbers within

a. 8 units of 3

b. k units of j (assume $k > 0$)

91. Height of a Projectile The equation

$$s = -16t^2 + v_0 t + s_0$$

gives the height s, in feet above ground level, of an object t seconds after the object is thrown directly upward from a height s_0 feet above the ground with an initial velocity of v_0 feet per second. A ball is thrown directly upward from ground level with an initial velocity of 64 feet per second. Find the time interval during which the ball has a height of more than 48 feet.

92. Height of a Projectile A ball is thrown directly upward from a height of 32 feet above the ground with an initial velocity of 80 feet per second. Find the time interval during which the ball will be more than 96 feet above the ground. (*Hint:* See Exercise 91.)

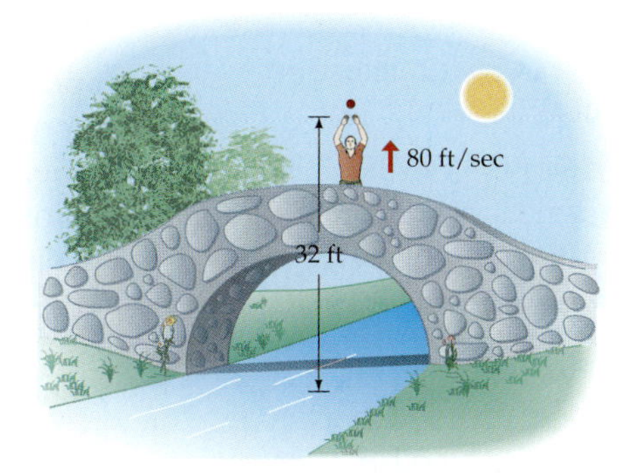

93. Geometry The length of the side of a square has been measured accurately to within 0.01 foot. This measured length is 4.25 feet.

a. Write an absolute value inequality that describes the relationship between the actual length of each side of the square s and its measured length.

b. Solve the absolute value inequality you found in part **a.** for s.

PREPARE FOR SECTION 1.2

94. Evaluate $\frac{x_1 + x_2}{2}$ when $x_1 = 4$ and $x_2 = -7$.

95. Simplify $\sqrt{50}$. [A.1]

96. Is $y = 3x - 2$ a true equation when $y = 5$ and $x = -1$? [1.1]

97. If $y = x^2 - 3x + 2$, find x when $y = 0$. [1.1]

98. Evaluate $|-x - y|$ when $x = 3$ and $y = -1$. [1.1]

99. Evaluate $\sqrt{a^2 + b^2}$ when $a = -3$ and $b = 4$. [A.1]

PROJECTS

1. Teaching Mathematics Prepare a lesson that you could use to explain to someone how to solve linear and quadratic equations. Be sure to include an explanation of the differences between these two types of equations and the different methods that are used to solve them.

2. Cubic Equations Write an essay on the development of the solution of the cubic equation. An excellent source of information is the chapter "Cardano and the Solution of the Cubic" in *Journey Through Genius* by William Dunham (New York: Wiley, 1990). Another excellent source is *A History of Mathematics: An Introduction* by Victor J. Katz (New York: Harper Collins, 1993).

SECTION 1.2

A Two-Dimensional Coordinate System and Graphs

- Cartesian Coordinate Systems
- The Distance and Midpoint Formulas
- Graph of an Equation
- Intercepts
- Circles, Their Equations, and Their Graphs

take note

Abscissa comes from the same root word as scissors. An open pair of scissors looks like an x.

MATH MATTERS

The concepts of *analytic geometry* developed over an extended period of time, culminating in 1637 with the publication of two works: *Discourse on the Method for Rightly Directing One's Reason and Searching for Truth in the Sciences* by René Descartes (1596–1650) and *Introduction to Plane and Solid Loci* by Pierre de Fermat. Each of these works was an attempt to integrate the study of geometry with the study of algebra. Of the two mathematicians, Descartes is usually given most of the credit for developing analytic geometry. In fact, Descartes became so famous in La Haye, the city in which he was born, that it was renamed La Haye-Descartes.

Cartesian Coordinate Systems

Each point on a coordinate axis is associated with a number called its **coordinate.** Each point on a flat, two-dimensional surface, called a **coordinate plane** or xy-plane, is associated with an **ordered pair** of numbers called **coordinates** of the point. Ordered pairs are denoted by (a, b), where the real number a is the **x-coordinate** or **abscissa** and the real number b is the **y-coordinate** or **ordinate.**

The coordinates of a point are determined by the point's position relative to a horizontal coordinate axis called the **x-axis** and a vertical coordinate axis called the **y-axis.** The axes intersect at the point $(0, 0)$, called the **origin.** In **Figure 1.12,** the axes are labeled such that positive numbers appear to the right of the origin on the x-axis and above the origin on the y-axis. The four regions formed by the axes are called **quadrants** and are numbered counterclockwise. This two-dimensional coordinate system is referred to as a **Cartesian coordinate system** in honor of René Descartes.

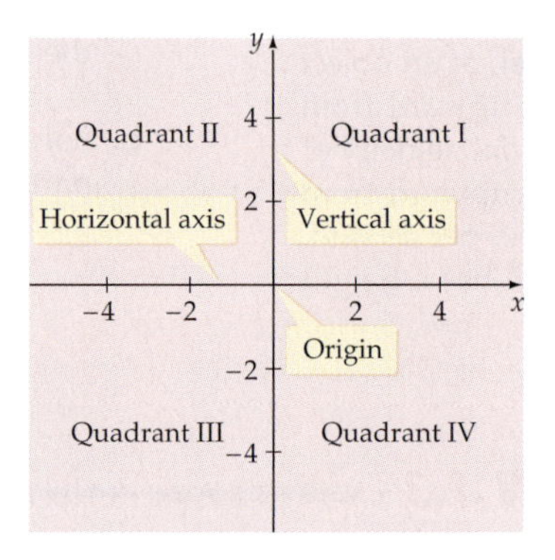

Figure 1.12

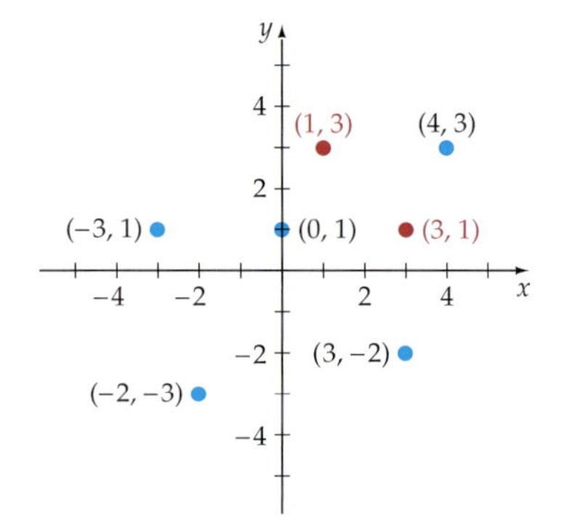

Figure 1.13

To **plot a point** $P(a, b)$ means to draw a dot at its location in the coordinate plane. In **Figure 1.13** we have plotted the points $(4, 3)$, $(-3, 1)$, $(-2, -3)$, $(3, -2)$, $(0, 1)$, $(1, 3)$, and $(3, 1)$. The order in which the coordinates of an ordered pair are listed is important. **Figure 1.13** shows that $(1, 3)$ and $(3, 1)$ do not denote the same point.

Data often are displayed in visual form as a set of points called a *scatter diagram* or *scatter plot*. For instance, the scatter diagram in **Figure 1.14** shows the number of Internet virus incidents from 1993 to 2003. The point whose coordinates are approximately $(7, 21{,}000)$ means that in the year 2000 there were approximately 21,000 Internet virus incidents. The line segments that connect the points in **Figure 1.14** help illustrate trends.

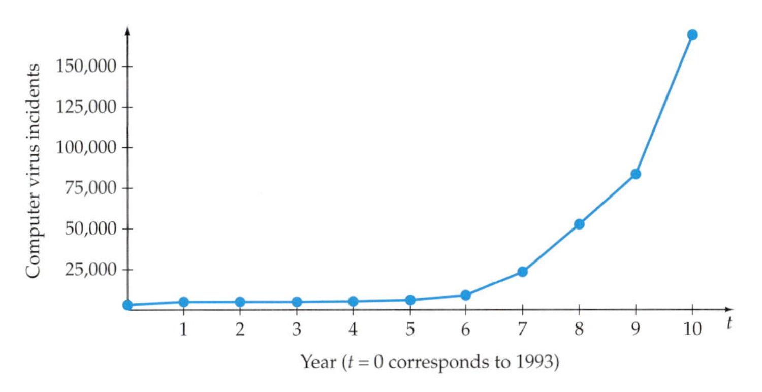

FIGURE 1.14

Source: www.cert.org

> **take note**
>
> The notation (a, b) was used earlier to denote an interval on a one-dimensional number line. In this section, (a, b) denotes an ordered pair in a two-dimensional plane. This should not cause confusion in future sections because as each mathematical topic is introduced, it will be clear whether a one-dimensional or a two-dimensional coordinate system is involved.

? QUESTION If the trend in **Figure 1.14** continues, will the number of virus incidents in 2004 be more or less than 200,000?

In some instances, it is important to know when two ordered pairs are equal.

Equality of Ordered Pairs

The ordered pairs (a, b) and (c, d) are equal if and only if $a = c$ and $b = d$.

For instance, if $(3, y) = (x, -2)$, then $x = 3$ and $y = -2$.

The Distance and Midpoint Formulas

The Cartesian coordinate system makes it possible to combine the concepts of algebra and geometry into a branch of mathematics called *analytic geometry*.

The distance between two points on a horizontal line is the absolute value of the difference between the x-coordinates of the two points. The distance between two points on a vertical line is the absolute value of the difference between the y-coordinates of the two points. For example, as shown in **Figure 1.15,** the distance d between the points with coordinates $(1, 2)$ and $(1, -3)$ is $d = |2 - (-3)| = 5$.

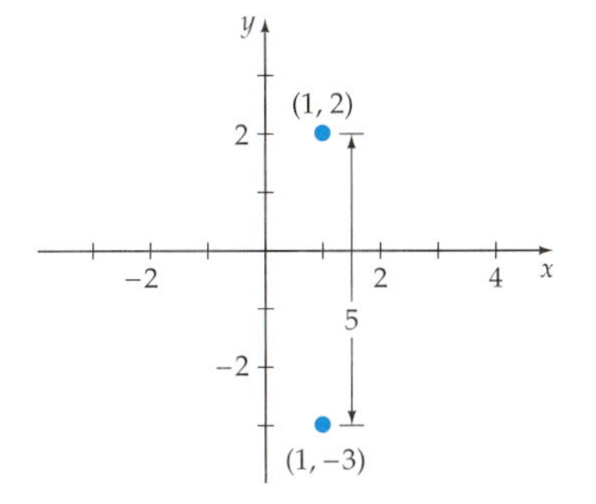

FIGURE 1.15

If two points are not on a horizontal or vertical line, then a *distance formula* for the distance between the two points can be developed as follows.

The distance between the points $P_1(x_1, y_1)$ and $P_2(x_2, y_2)$ in **Figure 1.16** is the length of the hypotenuse of a right triangle whose sides are horizontal and verti-

---

? ANSWER More. The increase between 2002 and 2003 was more than 70,000. If this trend continues, the increase between 2003 and 2004 will be at least 70,000 more than 150,000. That is, the number of virus incidents in 2004 will be at least 220,000.

cal line segments that measure $|x_2 - x_1|$ and $|y_2 - y_1|$, respectively. Applying the Pythagorean Theorem to this triangle produces

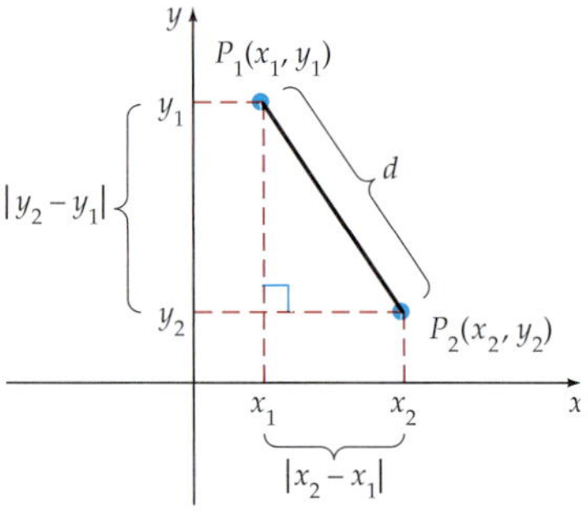

$$d^2 = |x_2 - x_1|^2 + |y_2 - y_1|^2$$

$$d = \sqrt{|x_2 - x_1|^2 + |y_2 - y_1|^2}$$

- **The square root theorem. Because d is nonnegative, the negative root is not listed.**

$$= \sqrt{(x_2 - x_1)^2 + (y_2 - y_1)^2}$$

- **Because $|x_2 - x_1|^2 = (x_2 - x_1)^2$ and $|y_2 - y_1|^2 = (y_2 - y_1)^2$**

FIGURE 1.16

Thus we have established the following theorem.

The Distance Formula

The distance d between the points $P_1(x_1, y_1)$ and $P_2(x_2, y_2)$ is

$$d = \sqrt{(x_2 - x_1)^2 + (y_2 - y_1)^2}$$

The distance d between the points whose coordinates are $P_1(x_1, y_1)$ and $P_2(x_2, y_2)$ is denoted by $d(P_1, P_2)$. To find the distance $d(P_1, P_2)$ between the points $P_1(-3, 4)$ and $P_2(7, 2)$, we apply the distance formula with $x_1 = -3$, $y_1 = 4$, $x_2 = 7$, and $y_2 = 2$.

$$\begin{aligned} d(P_1, P_2) &= \sqrt{(x_2 - x_1)^2 + (y_2 - y_1)^2} \\ &= \sqrt{[7 - (-3)]^2 + (2 - 4)^2} \\ &= \sqrt{104} = 2\sqrt{26} \approx 10.2 \end{aligned}$$

The **midpoint** M of a line segment is the point on the line segment that is equidistant from the endpoints $P_1(x_1, y_1)$ and $P_2(x_2, y_2)$ of the segment. See **Figure 1.17.**

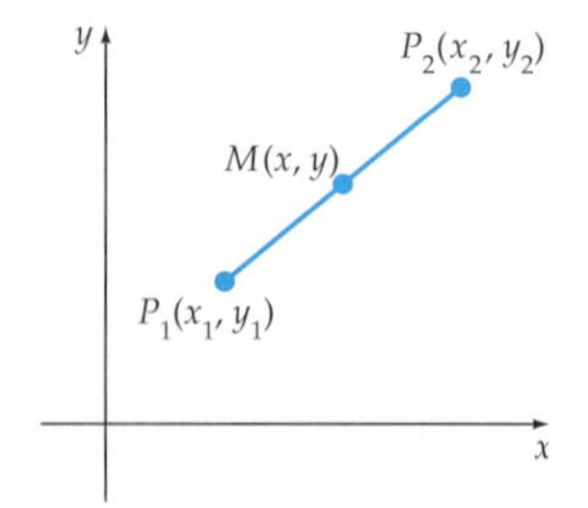

FIGURE 1.17

The Midpoint Formula

The midpoint M of the line segment from $P_1(x_1, y_1)$ to $P_2(x_2, y_2)$ is given by

$$\left(\frac{x_1 + x_2}{2}, \frac{y_1 + y_2}{2}\right)$$

The midpoint formula states that the x-coordinate of the midpoint of a line segment is the *average* of the x-coordinates of the endpoints of the line segment and that the y-coordinate of the midpoint of a line segment is the *average* of the y-coordinates of the endpoints of the line segment.

The midpoint M of the line segment connecting $P_1(-2, 6)$ and $P_2(3, 4)$ is

$$M = \left(\frac{x_1 + x_2}{2}, \frac{y_1 + y_2}{2}\right) = \left(\frac{(-2) + 3}{2}, \frac{6 + 4}{2}\right) = \left(\frac{1}{2}, 5\right)$$

EXAMPLE 1 Find the Midpoint and Length of a Line Segment

Find the midpoint and the length of the line segment connecting the points whose coordinates are $P_1(-4, 3)$ and $P_2(4, -2)$.

Solution

$$\text{Midpoint} = \left(\frac{x_1 + x_2}{2}, \frac{y_1 + y_2}{2}\right) = \left(\frac{-4 + 4}{2}, \frac{3 + (-2)}{2}\right) = \left(0, \frac{1}{2}\right)$$

$$d(P_1, P_2) = \sqrt{(x_2 - x_1)^2 + (y_2 - y_1)^2}$$
$$= \sqrt{(4 - (-4))^2 + (-2 - 3)^2} = \sqrt{(8)^2 + (-5)^2}$$
$$= \sqrt{64 + 25} = \sqrt{89}$$

TRY EXERCISE 6, PAGE 27

Graph of an Equation

The equations below are equations in two variables.

$$y = 3x^3 - 4x + 2 \qquad x^2 + y^2 = 25 \qquad y = \frac{x}{x + 1}$$

The solution of an equation in two variables is an ordered pair (x, y) whose coordinates satisfy the equation. For instance, the ordered pairs $(3, 4)$, $(4, -3)$, and $(0, 5)$ are some of the solutions of $x^2 + y^2 = 25$. Generally, there are an infinite number of solutions of an equation in two variables. These solutions can be displayed in a *graph*.

Graph of an Equation

The **graph of an equation** in the two variables x and y is the set of all points whose coordinates satisfy the equation.

Consider $y = 2x - 1$. Substituting various values of x into the equation and solving for y produces some of the ordered pairs of the equation. It is convenient to record the results in a table similar to the one shown on the following page. The graph of the ordered pairs is shown in **Figure 1.18.**

x	y = 2x − 1	y	(x, y)
−2	2(−2) − 1	−5	(−2, −5)
−1	2(−1) − 1	−3	(−1, −3)
0	2(0) − 1	−1	(0, −1)
1	2(1) − 1	1	(1, 1)
2	2(2) − 1	3	(2, 3)

Choosing some noninteger values of x produces more ordered pairs to graph, such as $\left(-\frac{3}{2}, -4\right)$ and $\left(\frac{5}{2}, 4\right)$, as shown in **Figure 1.19.** Using still other values of x would result in more and more ordered pairs being graphed. The result would be so many dots that the graph would appear as the straight line shown in **Figure 1.20,** which is the graph of $y = 2x - 1$.

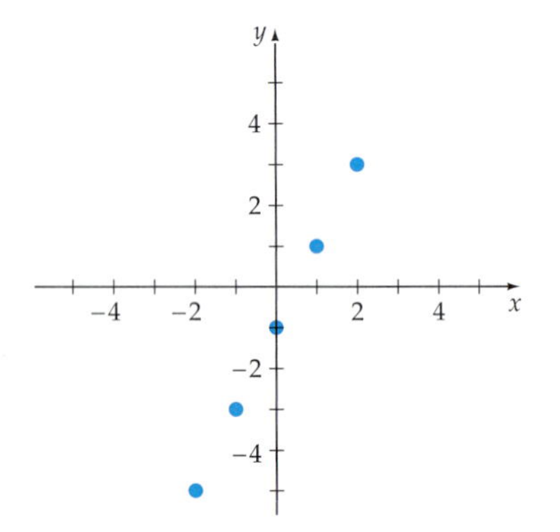

FIGURE 1.18

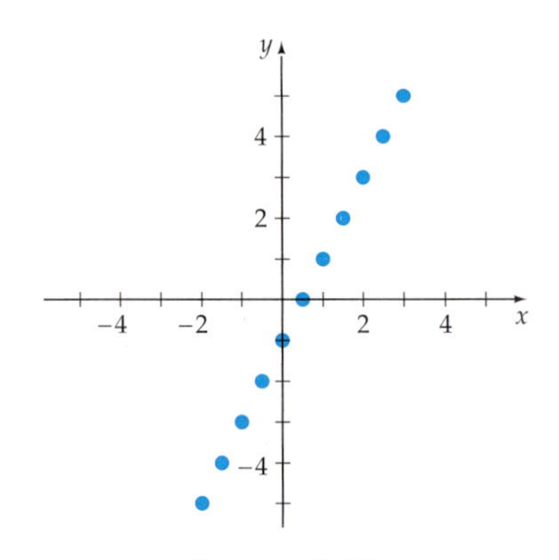

FIGURE 1.19

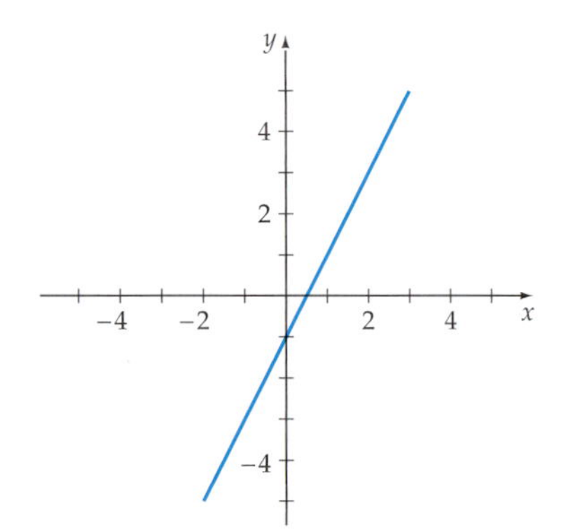

FIGURE 1.20

EXAMPLE 2 Draw a Graph by Plotting Points

Graph: $-x^2 + y = 1$

Solution

Solve the equation for y.

$$y = x^2 + 1$$

Select values of x and use the equation to calculate y. Choose enough values of x so that an accurate graph can be drawn. Plot the points and draw a curve through them. See **Figure 1.21.**

x	$y = x^2 + 1$	y	(x, y)
−2	$(-2)^2 + 1$	5	(−2, 5)
−1	$(-1)^2 + 1$	2	(−1, 2)
0	$(0)^2 + 1$	1	(0, 1)
1	$(1)^2 + 1$	2	(1, 2)
2	$(2)^2 + 1$	5	(2, 5)

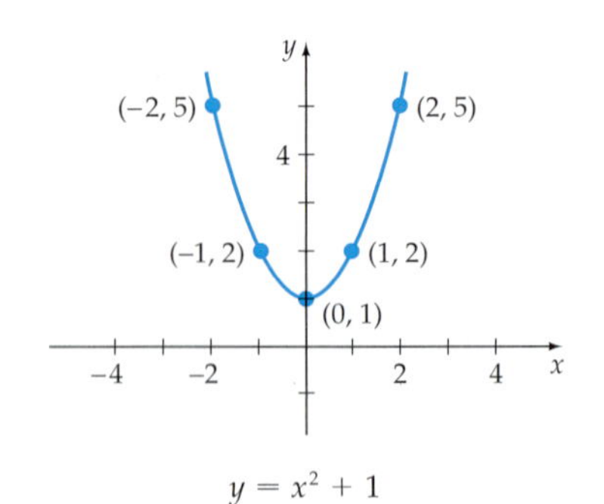

$y = x^2 + 1$

FIGURE 1.21

TRY EXERCISE 26, PAGE 28

MATH MATTERS

Maria Agnesi (1718–1799) wrote *Foundations of Analysis for the Use of Italian Youth*, one of the most successful textbooks of the 18th century. The French Academy authorized a translation into French in 1749, noting that "there is no other book, in any language, which would enable a reader to penetrate as deeply, or as rapidly, into the fundamental concepts of analysis." A curve that she discusses in her text is given by the equation $y = \dfrac{a^3}{x^2 + a^2}$. Unfortunately, due to a translation error from Italian to English, the curve became known as the "witch of Agnesi."

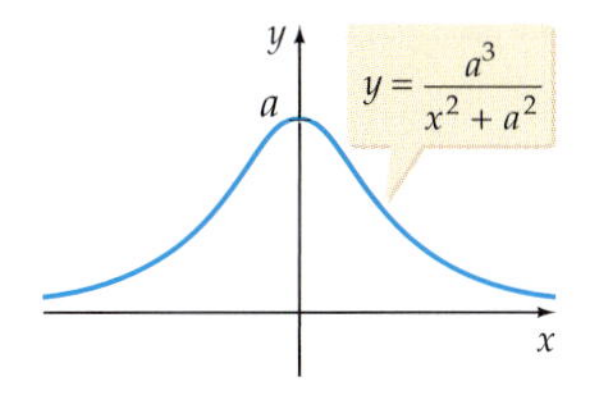

INTEGRATING TECHNOLOGY

Some graphing calculators, such as the *TI-83*, have a TABLE feature that allows you to create a table similar to the one shown in Example 2. Enter the equation to be graphed, the first value for x, and the increment (the difference between successive values of x). For instance, entering $y_1 = x^2 + 1$, an initial value of -2 for x, and an increment of 1 yields a display similar to the one in **Figure 1.22.** Changing the initial value to -6 and the increment to 2 gives the table in **Figure 1.23.**

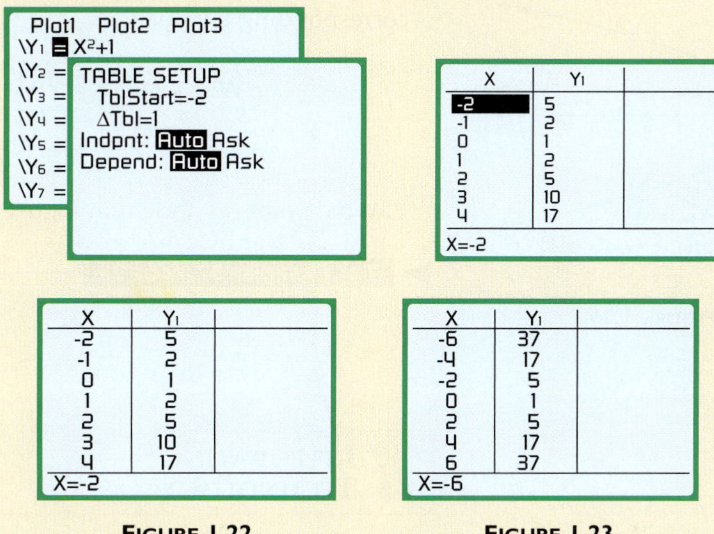

FIGURE 1.22 **FIGURE 1.23**

With some calculators, you may scroll through the table by using the up- or down-arrow keys. In this way, you can determine many more ordered pairs of the graph.

EXAMPLE 3 Graph by Plotting Points

Graph: $y = |x - 2|$

Solution

This equation is already solved for y, so start by choosing an x value and using the equation to determine the corresponding y value. For example, if $x = -3$, then $y = |(-3) - 2| = |-5| = 5$. Continuing in this manner produces the following table:

When x is	−3	−2	−1	0	1	2	3	4	5
y is	5	4	3	2	1	0	1	2	3

Now plot the points listed in the table. Connecting the points forms a V shape, as shown in **Figure 1.24.**

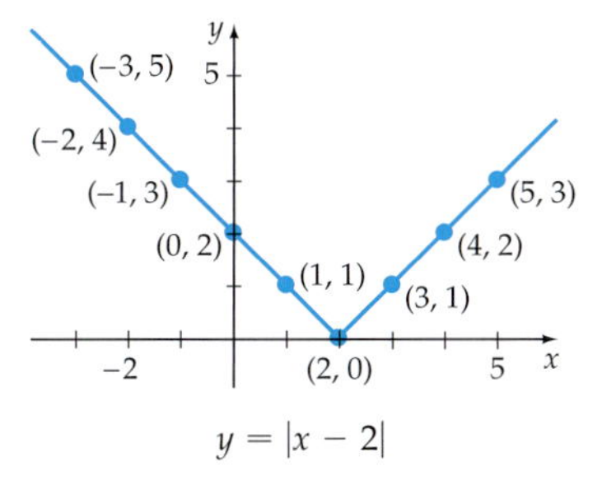

$y = |x - 2|$

FIGURE 1.24

TRY EXERCISE 30, PAGE 28

EXAMPLE 4 Graph by Plotting Points

Graph: $y^2 = x$

Solution

Solve the equation for y.

$$y^2 = x$$
$$y = \pm\sqrt{x}$$

Choose several x values, and use the equation to determine the corresponding y values.

When x is	0	1	4	9	16
y is	0	±1	±2	±3	±4

Plot the points as shown in **Figure 1.25.** The graph is a *parabola*.

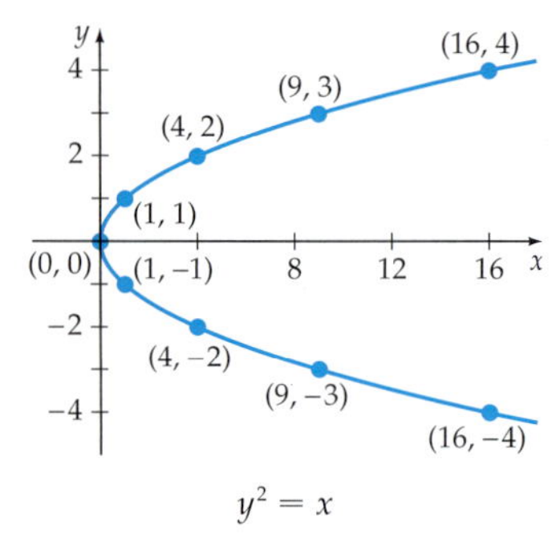

FIGURE 1.25

TRY EXERCISE 32, PAGE 28

INTEGRATING TECHNOLOGY

A graphing calculator or computer graphing software can be used to draw the graphs in Examples 3 and 4. These graphing utilities graph a curve in much the same way as you would, by selecting values of x and calculating the corresponding values of y. A curve is then drawn through the points.

If you use a graphing utility to graph $y = |x - 2|$, you will need to use the *absolute value* function that is built into the utility. The equation you enter will look similar to Y1=abs(X−2).

To graph the equation in Example 4, you will enter two equations. The equations you enter will be similar to

Y1=√(X)

Y2=−√(X)

The graph of the first equation will be the top half of the parabola; the graph of the second equation will be the bottom half.

INTERCEPTS

Any point that has an x- or a y-coordinate of zero is called an **intercept** of the graph of an equation because it is at these points that the graph intersects the x- or the y-axis.

Definition of x-Intercepts and y-Intercepts

If $(x_1, 0)$ satisfies an equation, then the point $(x_1, 0)$ is called an **x-intercept** of the graph of the equation.

If $(0, y_1)$ satisfies an equation, then the point $(0, y_1)$ is called a **y-intercept** of the graph of the equation.

To find the x-intercepts of the graph of an equation, let $y = 0$ and solve the equation for x. To find the y-intercepts of the graph of an equation, let $x = 0$ and solve the equation for y.

EXAMPLE 5 Find x- and y-Intercepts

Find the x- and y-intercepts of the graph of $y = x^2 - 2x - 3$.

Algebraic Solution

To find the y-intercept, let $x = 0$ and solve for y.

$$y = 0^2 - 2(0) - 3 = -3$$

To find the x-intercepts, let $y = 0$ and solve for x.

$$0 = x^2 - 2x - 3$$
$$0 = (x - 3)(x + 1)$$
$$(x - 3) = 0 \quad \text{or} \quad (x + 1) = 0$$
$$x = 3 \quad \text{or} \quad x = -1$$

Because $y = -3$ when $x = 0$, $(0, -3)$ is a y-intercept. Because $x = 3$ or -1 when $y = 0$, $(3, 0)$ and $(-1, 0)$ are x-intercepts. **Figure 1.26** confirms that these three points are intercepts.

Visualize the Solution

The graph of $y = x^2 - 2x - 3$ is shown below. Observe that the graph intersects the x-axis at $(-1, 0)$ and $(3, 0)$, the x-intercepts. The graph also intersects the y-axis at $(0, -3)$, the y-intercept.

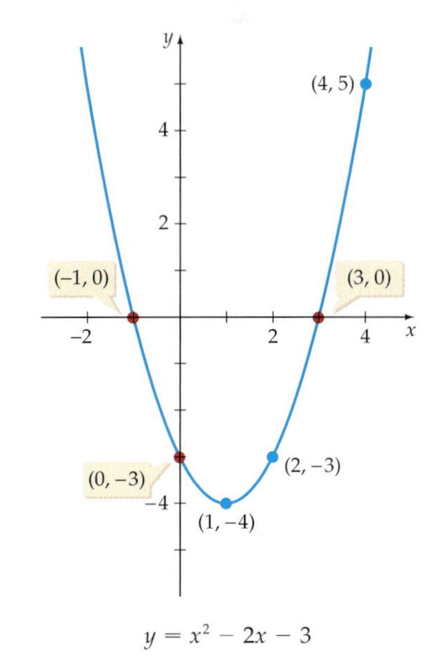

FIGURE 1.26

▶ TRY EXERCISE 40, PAGE 28

In Example 5 it was possible to find the x-intercepts by solving a quadratic equation. In some instances, however, solving an equation to find the intercepts may be very difficult. In these cases, a graphing calculator can be used to estimate the x-intercepts.

The x-intercepts of the graph of $y = x^3 + x + 4$ can be estimated using the INTERCEPT feature of a TI-83 calculator. The keystrokes and some sample screens for this procedure are shown below.

Press [Y=]. Now enter X^3+X+4. Press [ZOOM] and select the standard viewing window.

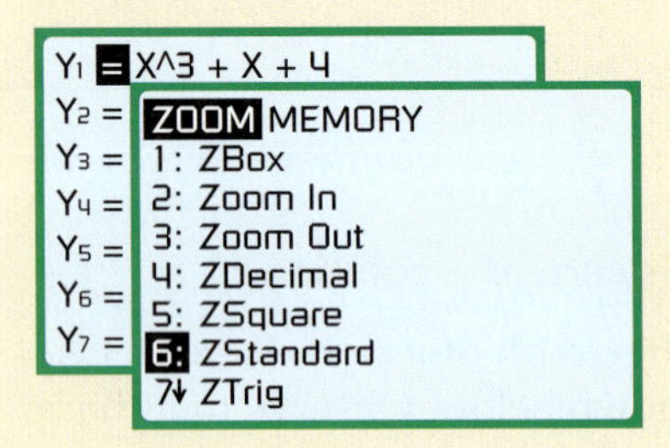

Press [2nd] CALC to access the CALCULATE menu. The y-coordinate of an x-intercept is zero. Therefore, select 2:zero. Press [ENTER].

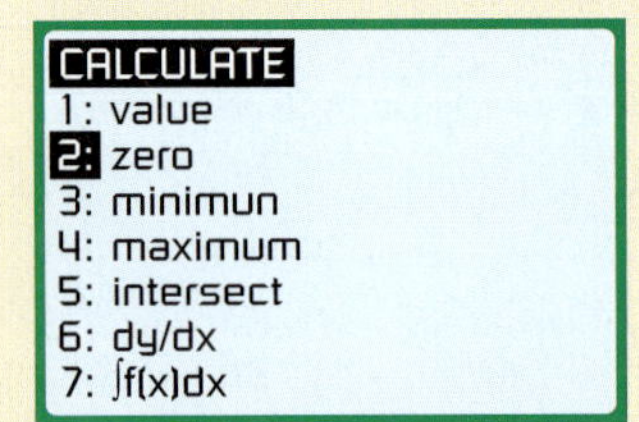

The "Left Bound?" shown on the bottom of the screen means to move the cursor until it is to the left of an x-intercept. Press [ENTER].

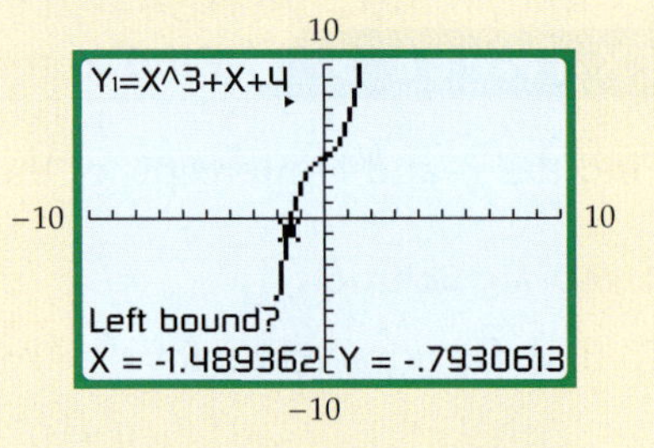

The "Right Bound?" shown on the bottom of the screen means to move the cursor until it is to the right of the desired x-intercept. Press [ENTER].

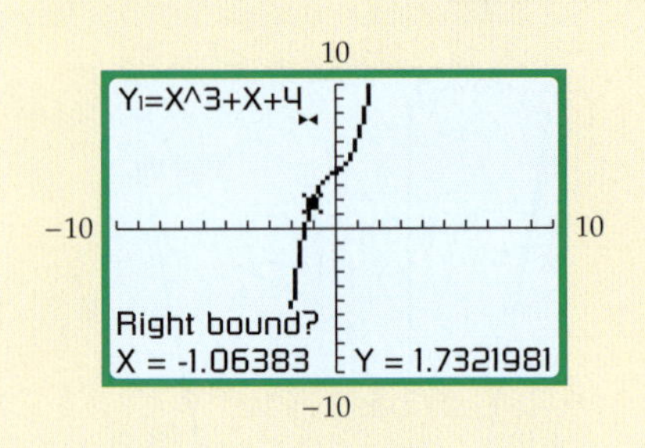

"Guess?" is shown on the bottom of the screen. Move the cursor until it is approximately on the x-intercept. Press [ENTER].

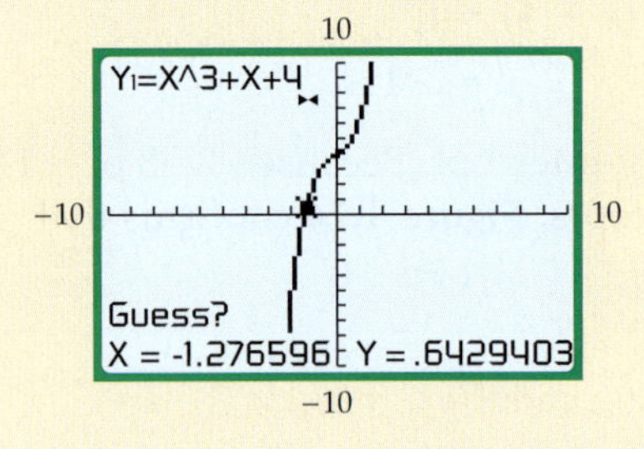

The "Zero" shown on the bottom of the screen means that the value of y is 0 when $x = -1.378797$. The x-intercept is about $(-1.378797, 0)$.

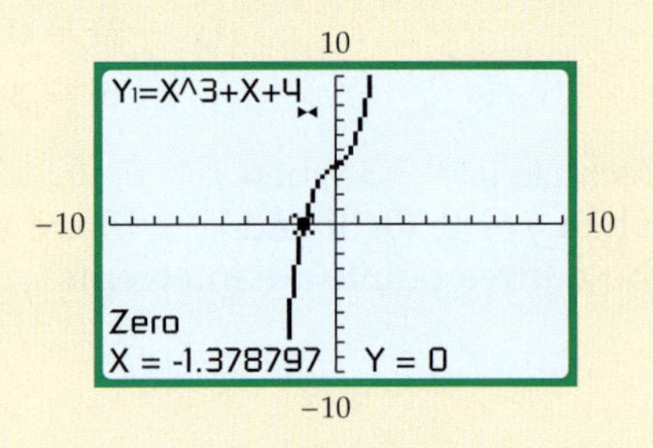

Circles, Their Equations, and Their Graphs

Frequently you will sketch graphs by plotting points. However, some graphs can be sketched merely by recognizing the form of the equation. A *circle* is an example of a curve whose graph you can sketch after you have inspected its equation.

Definition of a Circle

A **circle** is the set of points in a plane that are a fixed distance from a specified point. The distance is the **radius** of the circle, and the specified point is the **center** of the circle.

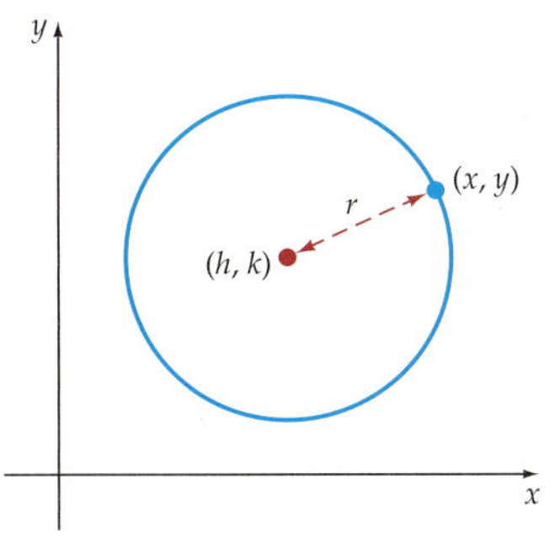

FIGURE 1.27

The standard form of the equation of a circle is derived by using this definition. To derive the standard form, we use the distance formula. **Figure 1.27** is a circle with center (h, k) and radius r. The point (x, y) is on the circle if and only if it is a distance of r units from the center (h, k). Thus (x, y) is on the circle if and only if

$$\sqrt{(x-h)^2+(y-k)^2}=r$$
$$(x-h)^2+(y-k)^2=r^2 \quad \text{• Square each side.}$$

Standard Form of the Equation of a Circle

The **standard form of the equation of a circle** with center at (h, k) and radius r is

$$(x-h)^2+(y-k)^2=r^2$$

For example, the equation $(x-3)^2+(y+1)^2=4$ is the equation of a circle. The standard form of the equation is

$$(x-3)^2+(y-(-1))^2=2^2$$

from which it can be determined that $h=3$, $k=-1$, and $r=2$. Thus the graph is a circle centered at $(3, -1)$ with a radius of 2.

If a circle is centered at the origin $(0, 0)$ (that is, if $h=0$ and $k=0$), then the standard form of the equation of the circle simplifies to

$$x^2+y^2=r^2$$

For example, the graph of $x^2+y^2=9$ is a circle with center at the origin and radius of 3.

QUESTION What are the radius and the coordinates of the center of the circle with equation $x^2+(y-2)^2=10$?

EXAMPLE 6 Find the Standard Form of the Equation of a Circle

Find the standard form of the equation of the circle that has center $C(-4, -2)$ and contains the point $P(-1, 2)$.

Solution

See the graph of the circle in **Figure 1.28.** Because the point P is on the circle, the radius r of the circle must equal the distance from C to P. Thus

$$r=\sqrt{(-1-(-4))^2+(2-(-2))^2}$$
$$=\sqrt{9+16}=\sqrt{25}=5$$

Using the standard form with $h=-4$, $k=-2$, and $r=5$, we obtain

$$(x+4)^2+(y+2)^2=5^2$$

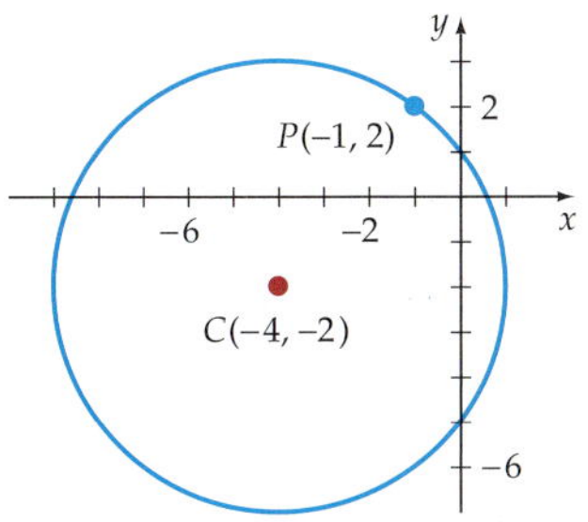

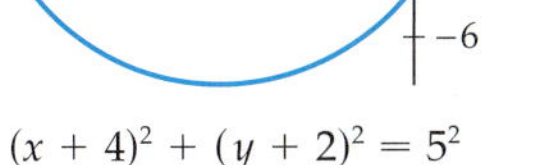
$(x+4)^2+(y+2)^2=5^2$

FIGURE 1.28

TRY EXERCISE 64, PAGE 28

ANSWER The radius is $\sqrt{10}$ and the coordinates of the center are $(0, 2)$.

If we rewrite $(x + 4)^2 + (y + 2)^2 = 5^2$ by squaring and combining like terms, we produce

$$x^2 + 8x + 16 + y^2 + 4y + 4 = 25$$
$$x^2 + y^2 + 8x + 4y - 5 = 0$$

This form of the equation is known as the **general form of the equation of a circle.** By completing the square, it is always possible to write the general form $x^2 + y^2 + Ax + By + C = 0$ in the standard form

$$(x - h)^2 + (y - k)^2 = s$$

for some number s. If $s > 0$, the graph is a circle with radius $r = \sqrt{s}$. If $s = 0$, the graph is the point (h, k), and if $s < 0$, the equation has no real solutions and there is no graph.

EXAMPLE 7 Find the Center and Radius of a Circle by Completing the Square

Find the center and the radius of the circle given by

$$x^2 + y^2 - 6x + 4y - 3 = 0$$

Solution

To review **COMPLETING THE SQUARE**, *see p. 7.*

First rearrange and group the terms as shown.

$$(x^2 - 6x) + (y^2 + 4y) = 3$$

Now complete the squares of $(x^2 - 6x)$ and $(y^2 + 4y)$.

$$(x^2 - 6x + 9) + (y^2 + 4y + 4) = 3 + 9 + 4$$

• Add 9 and 4 to each side of the equation.

$$(x - 3)^2 + (y + 2)^2 = 16$$
$$(x - 3)^2 + (y - (-2))^2 = 4^2$$

This equation is the standard form of the equation of a circle and indicates that the graph of the original equation is a circle centered at $(3, -2)$ with radius 4. See **Figure 1.29.**

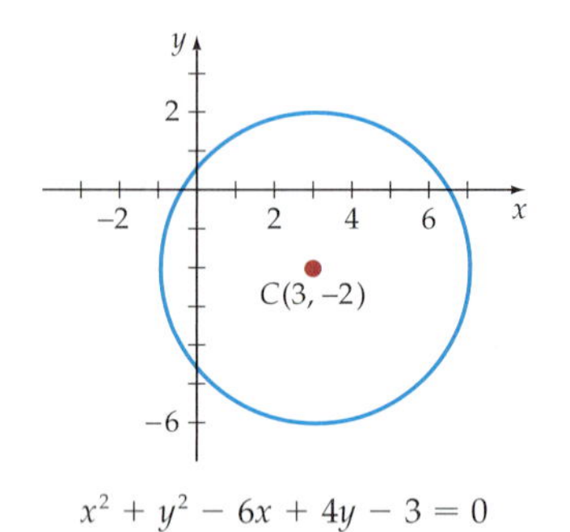

$x^2 + y^2 - 6x + 4y - 3 = 0$

FIGURE 1.29

TRY EXERCISE 66, PAGE 28

TOPICS FOR DISCUSSION

1. The distance formula states that the distance d between the points $P_1(x_1, y_1)$ and $P_2(x_2, y_2)$ is $d = \sqrt{(x_2 - x_1)^2 + (y_2 - y_1)^2}$. Can the distance formula also be written as follows? Explain.

$$d = \sqrt{(x_1 - x_2)^2 + (y_1 - y_2)^2}$$

2. Does the equation $(x - 3)^2 + (y + 4)^2 = -6$ have a graph that is a circle? Explain.

3. Explain why the graph of $|x| + |y| = 1$ does not contain any points that have
   a. a y-coordinate that is greater than 1 or less than -1
   b. an x-coordinate that is greater than 1 or less than -1
4. Discuss the graph of $xy = 0$.
5. Explain how to determine the x- and y-intercepts of a graph defined by an equation (without using the graph).

EXERCISE SET 1.2

In Exercises 1 and 2, plot the points whose coordinates are given on a Cartesian coordinate system.

1. $(2, 4), (0, -3), (-2, 1), (-5, -3)$

2. $(-3, -5), (-4, 3), (0, 2), (-2, 0)$

3. **HEALTH** A study at the Ohio State University measured the changes in heart rates of students doing stepping exercises. Students stepped onto a platform that was approximately 11 inches high at a rate of 14 steps per minute. The heart rates, in beats per minute, before and after the exercise are given in the table below.

Before	After	Before	After
63	84	96	141
72	99	69	93
87	111	81	96
90	129	75	90
90	108	84	90

   a. Draw a scatter diagram for these data.

   b. For these students, what is the average increase in heart rate?

4. **AVERAGE INCOME** The following graph, based on data from the Bureau of Economic Analysis, shows per capita personal income in the United States.

   a. From the data, does it appear that per capita personal income is increasing, decreasing, or remaining the same?

   b. If per capita personal income continues to increase by the same percent as the percent increase between 2001 and 2002, what will be the per capita income in 2004?

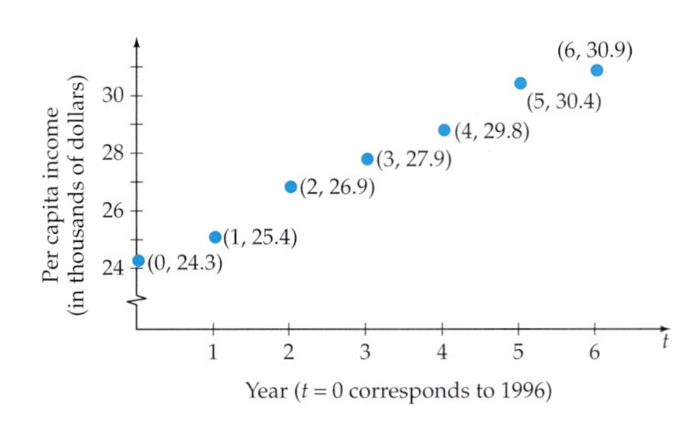

In Exercises 5 to 16, find the distance between the points whose coordinates are given.

5. $(6, 4), (-8, 11)$
6. $(-5, 8), (-10, 14)$
7. $(-4, -20), (-10, 15)$
8. $(40, 32), (36, 20)$
9. $(5, -8), (0, 0)$
10. $(0, 0), (5, 13)$
11. $(\sqrt{3}, \sqrt{8}), (\sqrt{12}, \sqrt{27})$
12. $(\sqrt{125}, \sqrt{20}), (6, 2\sqrt{5})$
13. $(a, b), (-a, -b)$
14. $(a - b, b), (a, a + b)$
15. $(x, 4x), (-2x, 3x)$ given that $x < 0$
16. $(x, 4x), (-2x, 3x)$ given that $x > 0$

17. Find all points on the x-axis that are 10 units from $(4, 6)$. (*Hint:* First write the distance formula with $(4, 6)$ as one of the points and $(x, 0)$ as the other point.)

18. Find all points on the y-axis that are 12 units from $(5, -3)$.

In Exercises 19 to 24, find the midpoint of the line segment with the following endpoints.

19. $(1, -1), (5, 5)$
20. $(-5, -2), (6, 10)$

21. $(6, -3), (6, 11)$

22. $(4, 7), (-10, 7)$

23. $(1.75, 2.25), (-3.5, 5.57)$

24. $(-8.2, 10.1), (-2.4, -5.7)$

In Exercises 25 to 38, graph each equation by plotting points that satisfy the equation.

25. $x - y = 4$

26. $2x + y = -1$

27. $y = 0.25x^2$

28. $3x^2 + 2y = -4$

29. $y = -2|x - 3|$

30. $y = |x + 3| - 2$

31. $y = x^2 - 3$

32. $y = x^2 + 1$

33. $y = \frac{1}{2}(x - 1)^2$

34. $y = 2(x + 2)^2$

35. $y = x^2 + 2x - 8$

36. $y = x^2 - 2x - 8$

37. $y = -x^2 + 2$

38. $y = -x^2 - 1$

In Exercises 39 to 48, find the x- and y-intercepts of the graph of each equation. Use the intercepts and additional points as needed to draw the graph of the equation.

39. $2x + 5y = 12$

40. $3x - 4y = 15$

41. $x = -y^2 + 5$

42. $x = y^2 - 6$

43. $x = |y| - 4$

44. $x = y^3 - 2$

45. $x^2 + y^2 = 4$

46. $x^2 = y^2$

47. $|x| + |y| = 4$

48. $|x - 4y| = 8$

In Exercises 49 to 56, determine the center and radius of the circle with the given equation.

49. $x^2 + y^2 = 36$

50. $x^2 + y^2 = 49$

51. $(x - 1)^2 + (y - 3)^2 = 49$

52. $(x - 2)^2 + (y - 4)^2 = 25$

53. $(x + 2)^2 + (y + 5)^2 = 25$

54. $(x + 3)^2 + (y + 5)^2 = 121$

55. $(x - 8)^2 + y^2 = \frac{1}{4}$

56. $x^2 + (y - 12)^2 = 1$

In Exercises 57 to 64, find an equation of a circle that satisfies the given conditions. Write your answer in standard form.

57. Center $(4, 1)$, radius $r = 2$

58. Center $(5, -3)$, radius $r = 4$

59. Center $\left(\frac{1}{2}, \frac{1}{4}\right)$, radius $r = \sqrt{5}$

60. Center $\left(0, \frac{2}{3}\right)$, radius $r = \sqrt{11}$

61. Center $(0, 0)$, passing through $(-3, 4)$

62. Center $(0, 0)$, passing through $(5, 12)$

63. Center $(1, 3)$, passing through $(4, -1)$

64. Center $(-2, 5)$, passing through $(1, 7)$

In Exercises 65 to 72, find the center and the radius of the graph of the circle. The equations of the circles are written in the general form.

65. $x^2 + y^2 - 6x + 5 = 0$

66. $x^2 + y^2 - 6x - 4y + 12 = 0$

67. $x^2 + y^2 - 14x + 8y + 56 = 0$

68. $x^2 + y^2 - 10x + 2y + 25 = 0$

69. $4x^2 + 4y^2 + 4x - 63 = 0$

70. $9x^2 + 9y^2 - 6y - 17 = 0$

71. $x^2 + y^2 - x + 3y - \frac{15}{4} = 0$

72. $x^2 + y^2 + 3x - 5y + \frac{25}{4} = 0$

73. Find an equation of a circle that has a diameter with endpoints $(2, 3)$ and $(-4, 11)$. Write your answer in standard form.

74. Find an equation of a circle that has a diameter with endpoints $(7, -2)$ and $(-3, 5)$. Write your answer in standard form.

75. Find an equation of a circle that has its center at $(7, 11)$ and is tangent to the x-axis. Write your answer in standard form.

76. Find an equation of a circle that has its center at $(-2, 3)$ and is tangent to the y-axis. Write your answer in standard form.

CONNECTING CONCEPTS

In Exercises 77 to 86, graph the set of all points whose *x*- and *y*-coordinates satisfy the given conditions.

77. $x = 1, y \geq 1$

78. $y = -3, x \geq -2$

79. $y \leq 3$

80. $x \geq 2$

81. $xy \geq 0$

82. $|y| \geq 1, \dfrac{x}{y} \leq 0$

83. $|x| = 2, |y| = 3$

84. $|x| = 4, |y| = 1$

85. $|x| \leq 2, y \geq 2$

86. $x \geq 1, |y| \leq 3$

In Exercises 87 to 90, find the other endpoint of the line segment that has the given endpoint and midpoint.

87. Endpoint $(5, 1)$, midpoint $(9, 3)$

88. Endpoint $(4, -6)$, midpoint $(-2, 11)$

89. Endpoint $(-3, -8)$, midpoint $(2, -7)$

90. Endpoint $(5, -4)$, midpoint $(0, 0)$

91. Find a formula for the set of all points (x, y) for which the distance from (x, y) to $(3, 4)$ is 5.

92. Find a formula for the set of all points (x, y) for which the distance from (x, y) to $(-5, 12)$ is 13.

93. Find a formula for the set of all points (x, y) for which the sum of the distances from (x, y) to $(4, 0)$ and from (x, y) to $(-4, 0)$ is 10.

94. Find a formula for the set of all points for which the absolute value of the difference of the distances from (x, y) to $(0, 4)$ and from (x, y) to $(0, -4)$ is 6.

95. Find an equation of a circle that is tangent to both axes, has its center in the second quadrant, and has a radius of 3.

96. Find an equation of a circle that is tangent to both axes, has its center in the third quadrant, and has a diameter of $\sqrt{5}$.

PREPARE FOR SECTION 1.3

97. Evaluate $x^2 + 3x - 4$ when $x = -3$. [A.2]

98. From the set of ordered pairs $A = \{(-3, 2), (-2, 4), (-1, 1), (0, 4), (2, 5)\}$, create two new sets, D and R, where D is the set of the first coordinates of the ordered pairs of A and R is the set of the second coordinates of the ordered pairs of A.

99. Find the length of the line segment connecting $P_1(-4, 1)$ and $P_2(3, -2)$. [1.2]

100. For what values of x is $\sqrt{2x - 6}$ a real number?

101. For what values of x is $\dfrac{x + 3}{x^2 - x - 6}$ not a real number?

102. If $a = 3x + 4$ and $a = 6x - 5$, find the value of a.

PROJECTS

1. Verify a Geometric Theorem Use the midpoint formula and the distance formula to prove that the midpoint M of the hypotenuse of a right triangle is equidistant from each of the vertices of the triangle. (*Hint:* Label the vertices of the triangle as shown in the figure at the right.)

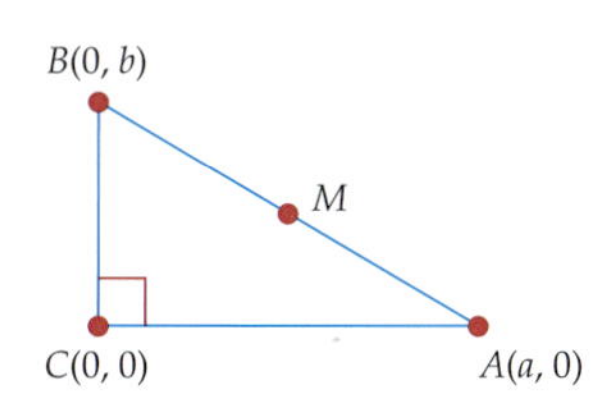

2. **SOLVE A QUADRATIC EQUATION GEOMETRICALLY** In the 17th century, Descartes (and others) solved equations by using both algebra and geometry. This project outlines the method Descartes used to solve certain quadratic equations.

   a. Consider the equation $x^2 = 2ax + b^2$. Construct a right triangle ABC with $d(A, C) = a$ and $d(C, B) = b$. Now draw a circle with center at A and radius a. Let P be the point at which the circle intersects the hypotenuse of the right triangle and Q the point at which an extension of the hypotenuse intersects the circle. Your drawing should be similar to the one at the right.

   b. Show that a solution of the equation $x^2 = 2ax + b^2$ is $d(Q, B)$.

   c. Show that $d(P, B)$ is a solution of the equation $x^2 = -2ax + b^2$.

   d. Construct a line parallel to AC and passing through B. Let S and T be the points at which the line intersects the circle. Show that $d(S, B)$ and $d(T, B)$ are solutions of the equation $x^2 = 2ax - b^2$.

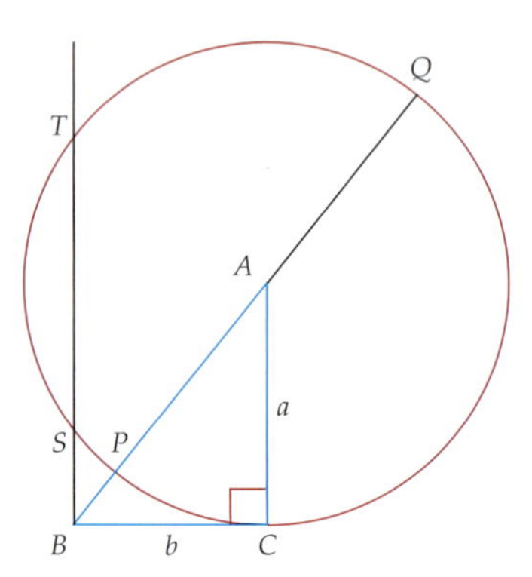

SECTION 1.3 INTRODUCTION TO FUNCTIONS

- RELATIONS
- FUNCTIONS
- FUNCTIONAL NOTATION
- IDENTIFYING FUNCTIONS
- GRAPHS OF FUNCTIONS
- THE GREATEST INTEGER FUNCTION (FLOOR FUNCTION)
- APPLICATIONS

RELATIONS

In many situations in science, business, and mathematics, a correspondence exists between two sets. The correspondence is often defined by a *table*, an *equation*, or a *graph*, each of which can be viewed from a mathematical perspective as a set of ordered pairs. In mathematics, any set of ordered pairs is called a **relation.**

Table 1.1 defines a correspondence between a set of percent scores and a set of letter grades. For each score from 0 to 100, there corresponds only one letter grade. The score 94% corresponds to the letter grade of A. Using ordered-pair notation, we record this correspondence as (94, A).

TABLE 1.1

Score	Grade
[90, 100]	A
[80, 90)	B
[70, 80)	C
[60, 70)	D
[0, 60)	F

The *equation* $d = 16t^2$ indicates that the distance d that a rock falls (neglecting air resistance) corresponds to the time t that it has been falling. For each nonnegative value t, the equation assigns only one value for the distance d. According to this equation, in 3 seconds a rock will fall 144 feet, which we record as (3, 144). Some of the other ordered pairs determined by $d = 16t^2$ are (0, 0), (1, 16), (2, 64), and (2.5, 100).

Equation: $d = 16t^2$

If $t = 3$, then $d = 16(3)^2 = 144$

The *graph* in **Figure 1.30** defines a correspondence between the length of a pendulum and the time it takes the pendulum to complete one oscillation. For each nonnegative pendulum length, the graph yields only one time. According to the graph, a pendulum length of 2 feet yields an oscillation time of 1.6 seconds, and a length of 4 feet yields an oscillation time of 2.2 seconds, where the time is measured to the nearest tenth of a second. These results can be recorded as the ordered pairs (2, 1.6) and (4, 2.2).

Graph: A pendulum's oscillation time

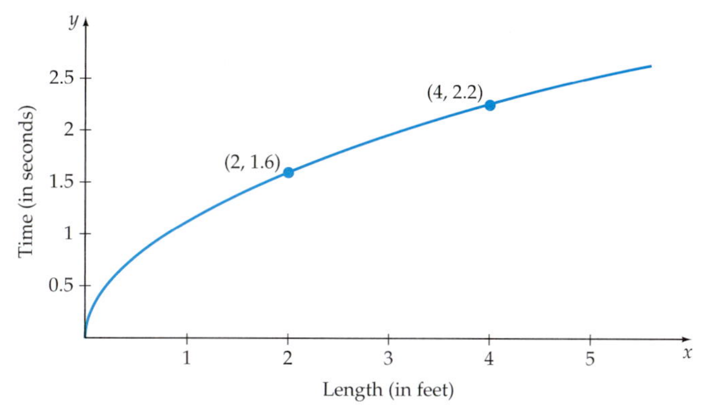

FIGURE 1.30

FUNCTIONS

The preceding table, equation, and graph each determines a special type of relation called a *function.*

> **MATH MATTERS**
>
> It is generally agreed among historians that Leonhard Euler (1707–1783) was the first person to use the word *function*. His definition of function occurs in his book *Introduction to Analysis of the Infinite*, published in 1748. Euler contributed to many areas of mathematics and was one of the most prolific expositors of mathematics.

Definition of a Function

A **function** is a set of ordered pairs in which no two ordered pairs have the same first coordinate and different second coordinates.

Although every function is a relation, not every relation is a function. For instance, consider (94, A) from the grading correspondence. The first coordinate, 94, is paired with a second coordinate of A. It would not make sense to have 94 paired with A, (94, A), and 94 paired with B, (94, B). The same first coordinate would be paired with two different second coordinates. This would mean that two students with the same score received different grades, one student an A and the other a B!

Functions may have ordered pairs with the same second coordinate. For instance, (94, A) and (95, A) are both ordered pairs that belong to the function defined by **Table 1.1.** A function may have different first coordinates and the same second coordinate.

The equation $d = 16t^2$ represents a function because for each value of t there is only one value of d. Not every equation, however, represents a function. For instance, $y^2 = 25 - x^2$ does not represent a function. The ordered pairs $(-3, 4)$ and $(-3, -4)$ are both solutions of the equation. But these ordered pairs do not satisfy the definition of a function; there are two ordered pairs with the same first coordinate but *different* second coordinates.

QUESTION Does the set $\{(0, 0), (1, 0), (2, 0), (3, 0), (4, 0)\}$ define a function?

The **domain** of a function is the set of all the first coordinates of the ordered pairs. The **range** of a function is the set of all the second coordinates. In the func-

---

ANSWER Yes. There are no two ordered pairs with the same first coordinate that have different second coordinates.

tion determined by the grading correspondence in **Table 1.1,** the domain is the interval [0, 100]. The range is {A, B, C, D, F}. In a function, each domain element is paired with one and only one range element.

If a function is defined by an equation, the variable that represents elements of the domain is the **independent variable.** The variable that represents elements of the range is the **dependent variable.** In the free-fall experiment, we used the equation $d = 16t^2$. The elements of the domain represented the time the rock fell, and the elements of the range represented the distance the rock fell. Thus, in $d = 16t^2$, the independent variable is t and the dependent variable is d.

The specific letters used for the independent and the dependent variable are not important. For example, $y = 16x^2$ represents the same function as $d = 16t^2$. Traditionally, x is used for the independent variable and y for the dependent variable. Any time we use the phrase "y is a function of x" or a similar phrase with different letters, the variable that follows "function of" is the independent variable.

Functional Notation

Functions can be named by using a letter or a combination of letters, such as f, g, A, log, or tan. If x is an element of the domain of f, then $f(x)$, which is read "f of x" or "the value of f at x," is the element in the range of f that corresponds to the domain element x. The notation "f" and the notation "$f(x)$" mean different things. "f" is the name of the function, whereas "$f(x)$" is the value of the function at x. Finding the value of $f(x)$ is referred to as *evaluating* f at x. To evaluate $f(x)$ at $x = a$, substitute a for x, and simplify.

EXAMPLE 1 Evaluate Functions

Let $f(x) = x^2 - 1$, and evaluate.

a. $f(-5)$ **b.** $f(3b)$ **c.** $3f(b)$ **d.** $f(a + 3)$ **e.** $f(a) + f(3)$

Solution

a. $f(-5) = (-5)^2 - 1 = 25 - 1 = 24$ • **Substitute −5 for x, and simplify.**

b. $f(3b) = (3b)^2 - 1 = 9b^2 - 1$ • **Substitute 3b for x, and simplify.**

c. $3f(b) = 3(b^2 - 1) = 3b^2 - 3$ • **Substitute b for x, and simplify.**

d. $f(a + 3) = (a + 3)^2 - 1$ • **Substitute a + 3 for x.**

$= a^2 + 6a + 8$ • **Simplify.**

e. $f(a) + f(3) = (a^2 - 1) + (3^2 - 1)$ • **Substitute a for x; substitute 3 for x.**

$= a^2 + 7$ • **Simplify.**

Try Exercise 2, page 43

take note

In Example 1, observe that

$f(3b) \neq 3f(b)$

and that

$f(a + 3) \neq f(a) + f(3)$

Piecewise-defined functions are functions represented by more than one expression. The function shown below is an example of a piecewise-defined function.

$$f(x) = \begin{cases} 2x, & x < -2 \\ x^2, & -2 \le x < 1 \\ 4 - x, & x \ge 1 \end{cases}$$

• **This function is made up of different pieces, $2x$, x^2, and $4 - x$, depending on the value of x.**

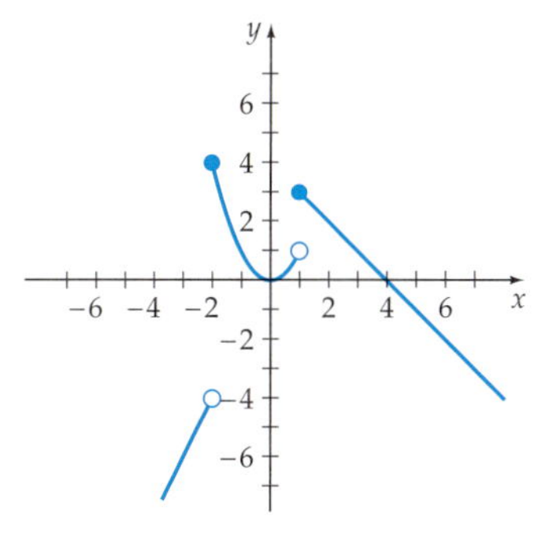

FIGURE 1.31

The expression that is used to evaluate this function depends on the value of x. For instance, to find $f(-3)$, we note that $-3 < -2$ and therefore use the expression $2x$ to evaluate the function.

$$f(-3) = 2(-3) = -6 \qquad \bullet \text{ When } x < -2, \text{ use the expression } 2x.$$

Here are some additional instances of evaluating this function:

$$f(-1) = (-1)^2 = 1 \qquad \bullet \text{ When } x \text{ satisfies } -2 \le x < 1, \text{ use the expression } x^2.$$

$$f(4) = 4 - 4 = 0 \qquad \bullet \text{ When } x \ge 1, \text{ use the expression } 4 - x.$$

The graph of this function is shown in **Figure 1.31.** Note the use of the open and closed circles at the endpoints of the intervals. These circles are used to show the evaluation of the function at the endpoints of each interval. For instance, because -2 is in the interval $-2 \le x < 1$, the value of the function at -2 is 4 $[f(-2) = (-2)^2 = 4]$. Therefore a closed dot is placed at $(-2, 4)$. Similarly, when $x = 1$, because 1 is in the interval $x \ge 1$, the value of the function at 1 is 3 $(f(1) = 4 - 1 = 3)$.

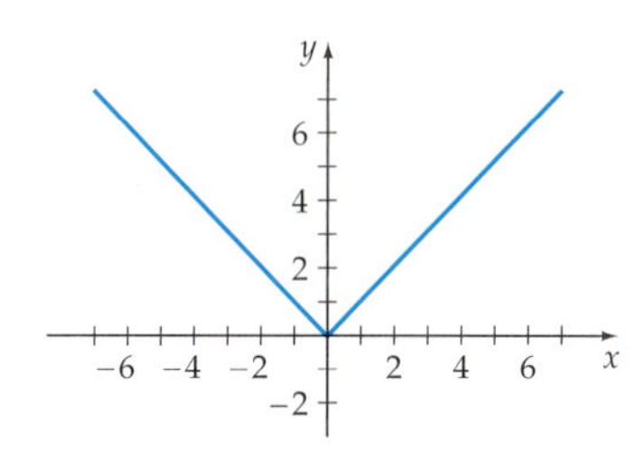

FIGURE 1.32

 **QUESTION** Evaluate the function f defined at the bottom of page 32 when $x = 0.5$.

The absolute value function is another example of a piecewise-defined function. Below is the definition of this function, which is sometimes abbreviated abs(x). Its graph **(Figure 1.32)** is shown at the left.

$$\text{abs}(x) = \begin{cases} -x, & x < 0 \\ x, & x \ge 0 \end{cases}$$

EXAMPLE 2 Evaluate a Piecewise-Defined Function

The number of monthly spam email attacks is shown in **Figure 1.33.**

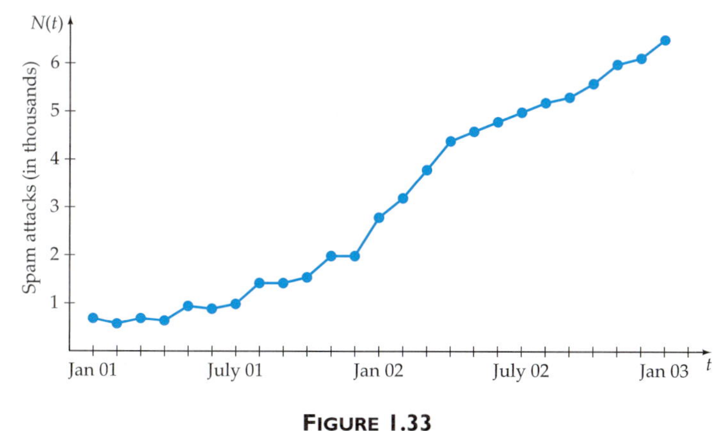

FIGURE 1.33

Source: www.brightmail.com

Continued ▶

ANSWER 0.5 is in the interval $-2 \le x < 1$. Therefore, $f(0.5) = 0.5^2 = 0.25$.

The data in the graph can be approximated by

$$N(t) = \begin{cases} 24.68t^2 - 170.47t + 957.73, & 0 \le t < 17 \\ 196.9t + 1164.6, & 17 \le t \le 26 \end{cases}$$

where $N(t)$ is the number of spam attacks in thousands for month t, where $t = 0$ corresponds to January 2001. Use this function to estimate, to the nearest hundred thousand, the number of monthly spam attacks for the following months.

a. October 2001 **b.** December 2002

Solution

a. The month October 2001 corresponds to $t = 9$. Because $t = 9$ is in the interval $0 \le t < 17$, evaluate $24.68t^2 - 170.47t + 957.73$ at $t = 9$.

$$24.68t^2 - 170.47t + 957.73$$
$$24.68(9)^2 - 170.47(9) + 957.73 = 1422.58$$

There were approximately 1,423,000 spam attacks in October 2001.

b. The month December 2002 corresponds to $t = 23$. Because $t = 23$ is in the interval $17 \le t \le 26$, evaluate $196.9t + 1164.6$ at 23.

$$196.9t + 1164.6$$
$$196.9(23) + 1164.6 = 5693.3$$

There were approximately 5,693,000 spam attacks in December 2002.

TRY EXERCISE 10, PAGE 44

IDENTIFYING FUNCTIONS

Recall that although every function is a relation, not every relation is a function. In the next example we examine four relations to determine which are functions.

EXAMPLE 3 Identify Functions

Which relations define y as a function of x?

a. $\{(2, 3), (4, 1), (4, 5)\}$ **b.** $3x + y = 1$ **c.** $-4x^2 + y^2 = 9$

d. The correspondence between the x values and the y values in **Figure 1.34.**

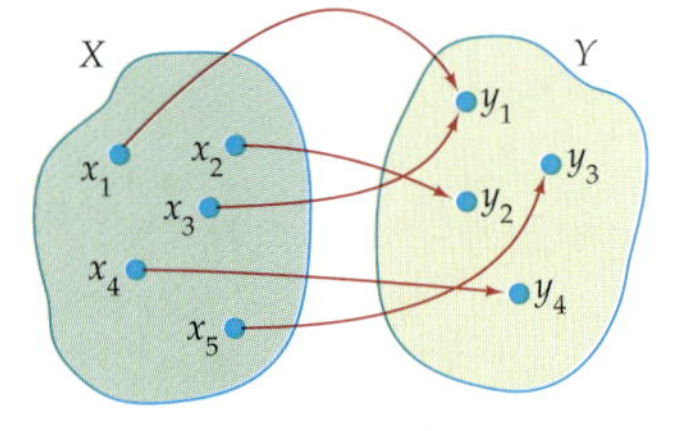

FIGURE 1.34

Solution

a. There are two ordered pairs, (4, 1) and (4, 5), with the same first coordinate and different second coordinates. This set does not define y as a function of x.

b. Solving $3x + y = 1$ for y yields $y = -3x + 1$. Because $-3x + 1$ is a unique real number for each x, this equation defines y as a function of x.

c. Solving $-4x^2 + y^2 = 9$ for y yields $y = \pm\sqrt{4x^2 + 9}$. The right side $\pm\sqrt{4x^2 + 9}$ produces two values of y for each value of x. For example, when $x = 0$, $y = 3$ or $y = -3$. Thus $-4x^2 + y^2 = 9$ does not define y as a function of x.

d. Each x is paired with one and only one y. The correspondence in **Figure 1.34** defines y as a function of x.

TRY EXERCISE 14, PAGE 44

take note

You may indicate the domain of a function using set notation or interval notation. For instance, the domain of $f(x) = \sqrt{x - 3}$ may be given in each of the following ways:

Set notation: $\{x \mid x \geq 3\}$

Interval notation: $[3, \infty)$

Sometimes the domain of a function is stated explicitly. For example, each of f, g, and h below is given by an equation, followed by a statement that indicates the domain of the function.

$$f(x) = x^2, x > 0 \qquad g(t) = \frac{1}{t^2 + 4}, 0 \leq t \leq 5 \qquad h(x) = x^2, x = 1, 2, 3$$

Although f and h have the same equation, they are different functions because they have different domains. If the domain of a function is not explicitly stated, then its domain is determined by the following convention.

Domain of a Function

Unless otherwise stated, the domain of a function is the set of all real numbers for which the function makes sense and yields real numbers.

EXAMPLE 4 Determine the Domain of a Function

Determine the domain of each function.

a. $G(t) = \dfrac{1}{t - 4}$ b. $f(x) = \sqrt{x + 1}$

c. $A(s) = s^2$, where $A(s)$ is the area of a square whose sides are s units.

Solution

a. The number 4 is not an element of the domain because G is undefined when the denominator $t - 4$ equals 0. The domain of G is all real numbers except 4. In interval notation the domain is $(-\infty, 4) \cup (4, \infty)$.

b. The radical $\sqrt{x + 1}$ is a real number only when $x + 1 \geq 0$ or when $x \geq -1$. Thus, in set-builder notation, the domain of f is $\{x \mid x \geq -1\}$.

c. Because s represents the length of the side of a square, s must be positive. In interval notation the domain of A is $(0, \infty)$.

TRY EXERCISE 28, PAGE 44

Graphs of Functions

If a is an element of the domain of a function, then $(a, f(a))$ is an ordered pair that belongs to the function.

Graph of a Function

The **graph of a function** is the graph of all the ordered pairs that belong to the function.

EXAMPLE 5 Graph a Function by Plotting Points

Graph each function. State the domain and the range of each function.

a. $f(x) = |x - 1|$ **b.** $n(x) = \begin{cases} 2, & \text{if } x \le 1 \\ x, & \text{if } x > 1 \end{cases}$

Solution

a. The domain of f is the set of all real numbers. Write the function as $y = |x - 1|$. Evaluate the function for several domain values. We have used $x = -3, -2, -1, 0, 1, 2, 3$, and 4.

x	−3	−2	−1	0	1	2	3	4
y = |x − 1|	4	3	2	1	0	1	2	3

Plot the points determined by the ordered pairs. Connect the points to form the graph in **Figure 1.35.**

Because $|x - 1| \ge 0$, we can conclude that the graph of f extends from a height of 0 upward, so the range is $\{y \mid y \ge 0\}$.

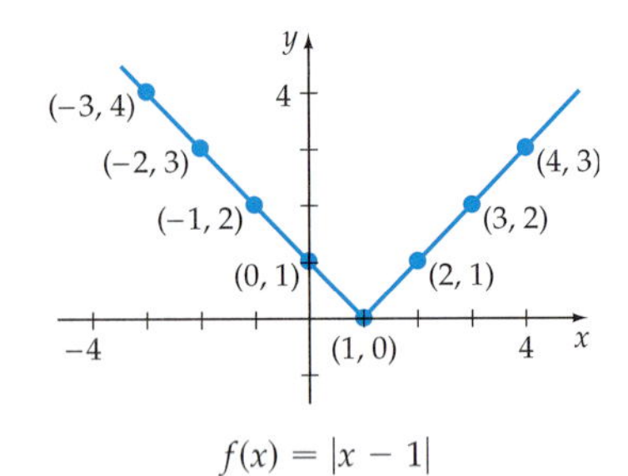

Figure 1.35

b. The domain is the union of the inequalities $x \le 1$ and $x > 1$. Thus the domain of n is the set of all real numbers. For $x \le 1$, graph $n(x) = 2$. This results in the horizontal ray in **Figure 1.36.** The solid circle indicates that the point (1, 2) *is* part of the graph. For $x > 1$, graph $n(x) = x$. This produces the second ray in **Figure 1.36.** The open circle indicates that the point (1, 1) *is not* part of the graph.

Examination of the graph shows that it includes only points whose y values are greater than 1. Thus the range of n is $\{y \mid y > 1\}$.

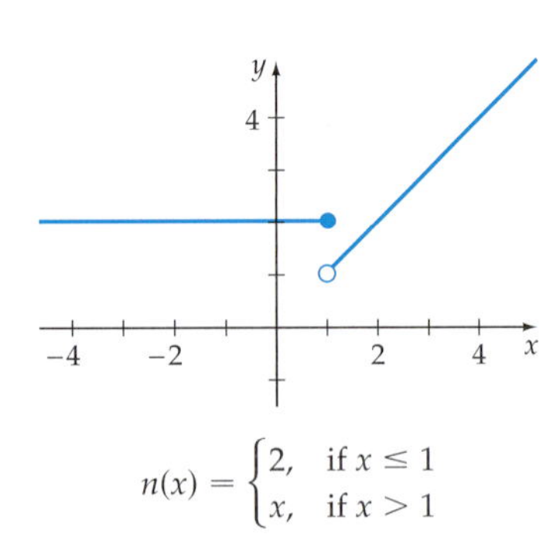

Figure 1.36

▶ Try Exercise 40, page 44

INTEGRATING TECHNOLOGY

A graphing utility also can be used to draw the graph of a function. For instance, to graph $f(x) = x^2 - 1$, you will enter an equation similar to Y1=x²–1. The graph is shown in **Figure 1.37.**

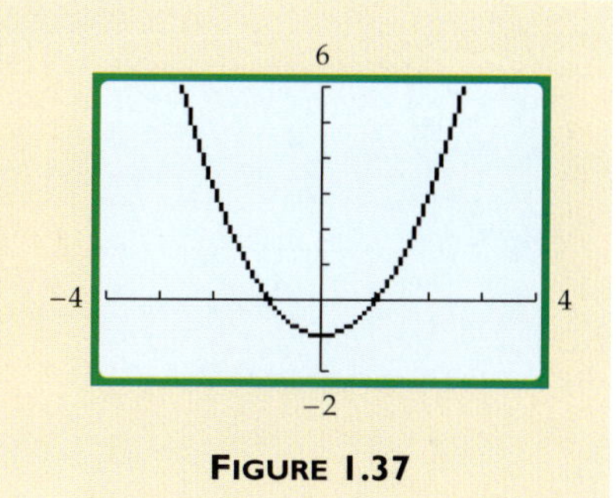

FIGURE 1.37

The definition that a function is a set of ordered pairs in which no two ordered pairs that have the same first coordinate have different second coordinates implies that any vertical line intersects the graph of a function at no more than one point. This is known as the *vertical line test.*

The Vertical Line Test for Functions

A graph is the graph of a function if and only if no vertical line intersects the graph at more than one point.

EXAMPLE 6 Apply the Vertical Line Test

Which of the following graphs are graphs of functions?

a.

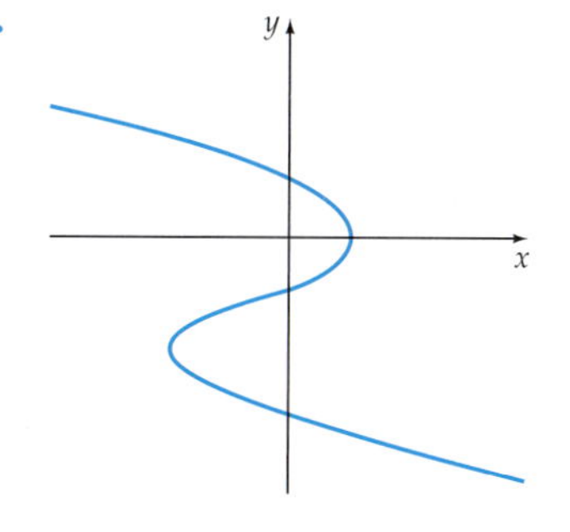

b.

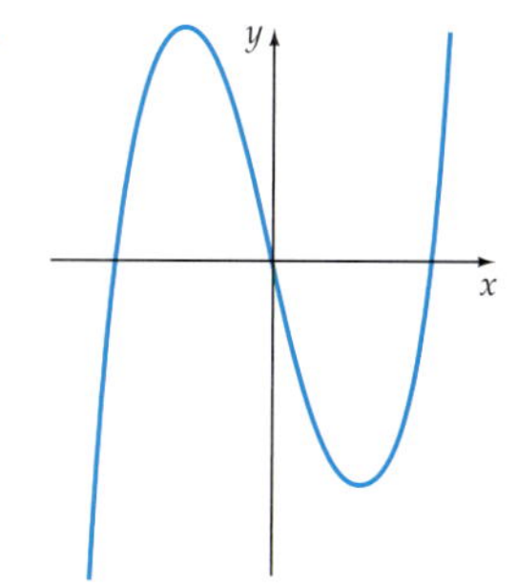

Solution

a. This graph *is not* the graph of a function because some vertical lines intersect the graph in more than one point.

b. This graph *is* the graph of a function because every vertical line intersects the graph in at most one point.

▶ TRY EXERCISE 50, PAGE 45

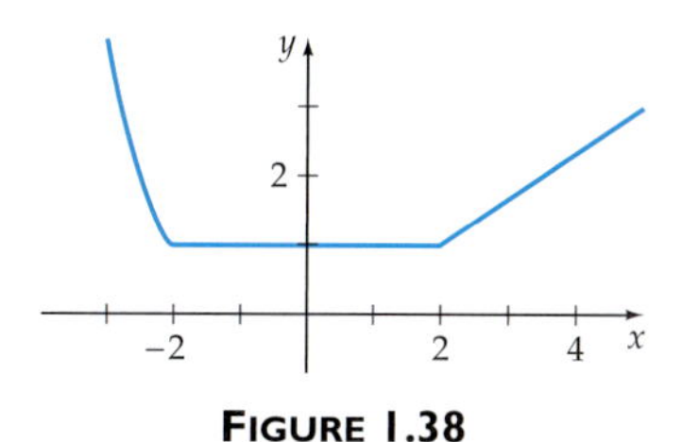

FIGURE 1.38

Consider the graph in **Figure 1.38.** As a point on the graph moves from left to right, this graph falls for values of $x \leq -2$, it remains the same height from

$x = -2$ to $x = 2$, and it rises for $x \geq 2$. The function represented by the graph is said to be *decreasing* on the interval $(-\infty, -2]$, *constant* on the interval $[-2, 2]$, and *increasing* on the interval $[2, \infty)$.

Definition of Increasing, Decreasing, and Constant Functions

If a and b are elements of an interval I that is a subset of the domain of a function f, then

- f is **increasing** on I if $f(a) < f(b)$ whenever $a < b$.
- f is **decreasing** on I if $f(a) > f(b)$ whenever $a < b$.
- f is **constant** on I if $f(a) = f(b)$ for all a and b.

Recall that a function is a relation in which no two ordered pairs that have the same first coordinate have different second coordinates. This means that given any x, there is only one y that can be paired with that x. A **one-to-one function** satisfies the additional condition that given any y, there is only one x that can be paired with that y. In a manner similar to applying the vertical line test, we can apply a *horizontal line test* to identify one-to-one functions.

Horizontal Line Test for a One-To-One Function

If every horizontal line intersects the graph of a function at most once, then the graph is the graph of a one-to-one function.

For example, some horizontal lines intersect the graph in **Figure 1.39** at more than one point. It is *not* the graph of a one-to-one function. Every horizontal line intersects the graph in **Figure 1.40** at most once. This is the graph of a one-to-one function.

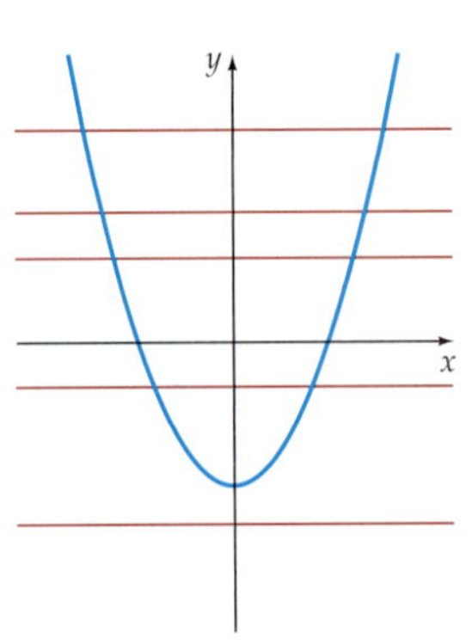

FIGURE 1.39
Some horizontal lines intersect this graph at more than one point. It is *not* the graph of a one-to-one function.

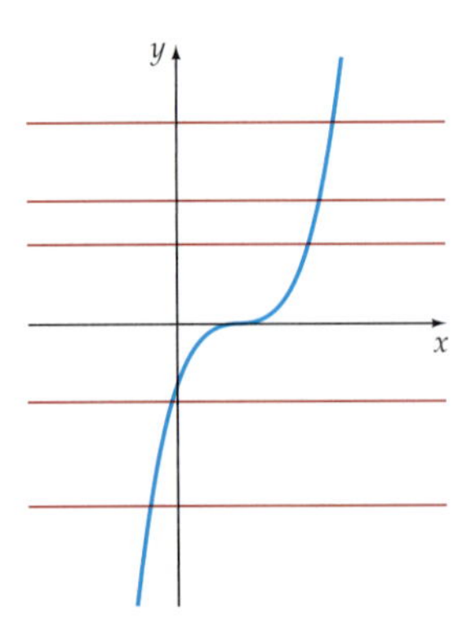

FIGURE 1.40
Every horizontal line intersects this graph at most once. It is the graph of a one-to-one function.

THE GREATEST INTEGER FUNCTION (FLOOR FUNCTION)

take note

The greatest integer function is an important function that is often used in advanced mathematics and also in computer science.

To this point, the graphs of the functions have not had any breaks or gaps. These functions whose graphs can be drawn without lifting the pencil off the paper are called *continuous functions*. The graphs of some functions do have breaks or *discontinuities*. One such function is the **greatest integer function** or **floor function.** This function is denoted by various symbols such as $[\![x]\!]$, $\lfloor x \rfloor$, and $\text{int}(x)$.

The value of the greatest integer function at x is the greatest integer that is less than or equal to x. For instance,

$$\lfloor -1.1 \rfloor = -2 \qquad [\![-3]\!] = -3 \qquad \text{int}\left(\frac{5}{2}\right) = 2 \qquad \lfloor 5 \rfloor = 5 \qquad [\![\pi]\!] = 3$$

INTEGRATING TECHNOLOGY

Many graphing calculators use the notation $\text{int}(x)$ for the greatest integer function. Here is a screen from a TI-83 Plus.

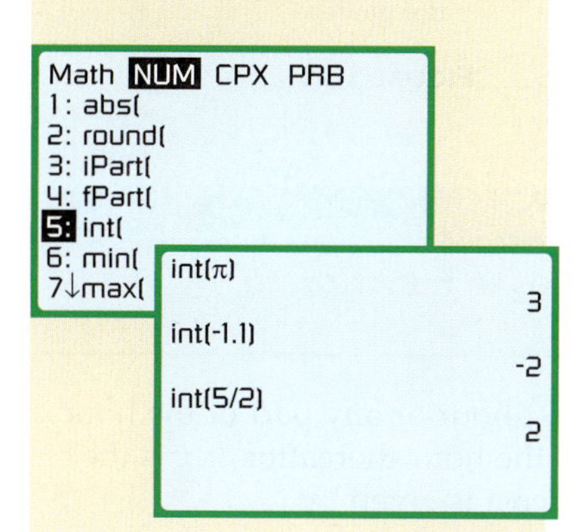

QUESTION Evaluate. **a.** $\text{int}\left(-\frac{3}{2}\right)$ **b.** $\lfloor 2 \rfloor$

To graph the floor function, first observe that the value of the floor function is constant between any two consecutive integers. For instance, between the integers 1 and 2, we have

$$\text{int}(1.1) = 1 \qquad \text{int}(1.35) = 1 \qquad \text{int}(1.872) = 1 \qquad \text{int}(1.999) = 1$$

Between -3 and -2, we have

$$\text{int}(-2.98) = -3 \qquad \text{int}(-2.4) = -3 \qquad \text{int}(-2.35) = -3 \qquad \text{int}(-2.01) = -3$$

Using this property of the floor function, we can create a table of values and then graph the floor function (**Figure 1.41**).

x	y = int(x)
$-5 \le x < -4$	-5
$-4 \le x < -3$	-4
$-3 \le x < -2$	-3
$-2 \le x < -1$	-2
$-1 \le x < 0$	-1
$0 \le x < 1$	0
$1 \le x < 2$	1
$2 \le x < 3$	2
$3 \le x < 4$	3
$4 \le x < 5$	4

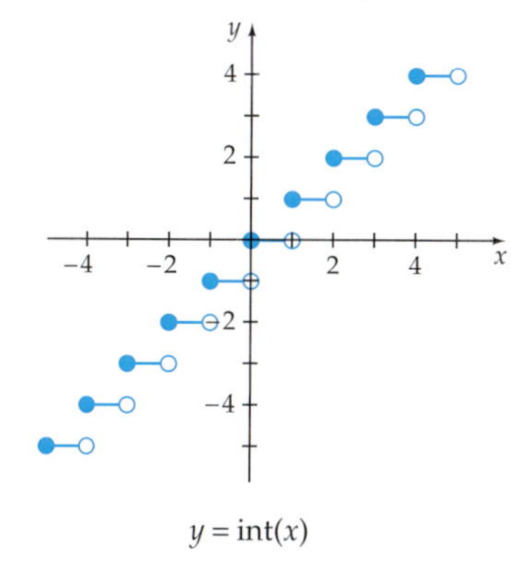

$y = \text{int}(x)$

FIGURE 1.41

The graph of the floor function has discontinuities (breaks) whenever x is an integer. The domain of the floor function is the set of real numbers; the range is the set of integers. Because the graph appears to be a series of steps, sometimes the floor function is referred to as a **step function.**

ANSWER **a.** Because -2 is the greatest integer that is less than $-\frac{3}{2}$, $\text{int}\left(-\frac{3}{2}\right) = -2$.

b. Because 2 is the greatest integer less than or *equal* to 2, $\lfloor 2 \rfloor = 2$.

INTEGRATING TECHNOLOGY

A graphing calculator also can be used to graph the floor function. The graph in **Figure 1.42** was drawn in "connected" mode. This graph does not show the discontinuities that occur whenever x is an integer.

The graph in **Figure 1.43** was constructed by graphing the floor function in "dot" mode. In this case the discontinuities at the integers are apparent.

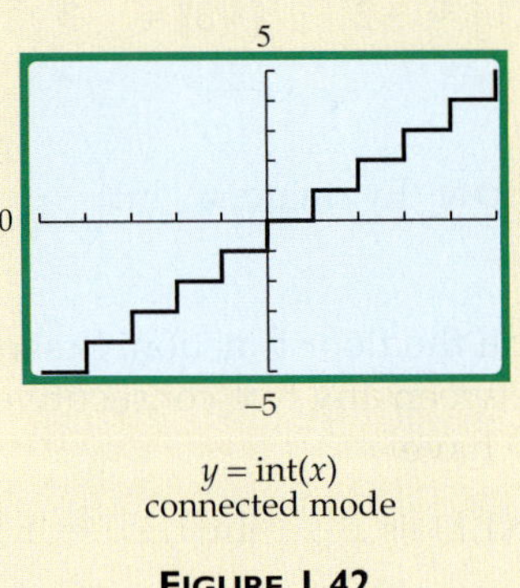

$y = \text{int}(x)$
connected mode

FIGURE 1.42

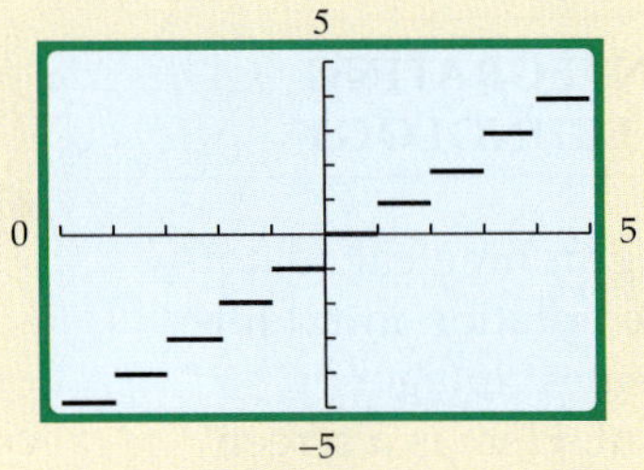

5
0
5
−5

$y = \text{int}(x)$
dot mode

FIGURE 1.43

EXAMPLE 7 Use the Greatest Integer Function to Model Expenses

The cost of parking in a garage is \$3 for the first hour or any part of the hour and \$2 for each additional hour or any part of the hour thereafter. If x is the time in hours that you park your car, then the cost is given by

$$C(x) = 3 - 2\,\text{int}(1 - x), \quad x > 0$$

a. Evaluate $C(2)$ and $C(2.5)$. **b.** Graph $y = C(x)$ for $0 < x \le 5$.

Solution

a.
$$\begin{aligned} C(2) &= 3 - 2\,\text{int}(1 - 2) \\ &= 3 - 2\,\text{int}(-1) \\ &= 3 - 2(-1) \\ &= \$5 \end{aligned} \qquad \begin{aligned} C(2.5) &= 3 - 2\,\text{int}(1 - 2.5) \\ &= 3 - 2\,\text{int}(-1.5) \\ &= 3 - 2(-2) \\ &= \$7 \end{aligned}$$

b. To graph $C(x)$ for $0 < x \le 5$, consider the value of $\text{int}(1 - x)$ for each of the intervals $0 < x \le 1$, $1 < x \le 2$, $2 < x \le 3$, $3 < x \le 4$, and $4 < x \le 5$. For instance, when $0 < x \le 1$, $0 \le 1 - x < 1$. Thus $\text{int}(1 - x) = 0$ when $0 < x \le 1$. Now consider $1 < x \le 2$. When $1 < x \le 2$, $1 \le 1 - x < 2$. Thus $\text{int}(1 - x) = 1$ when $1 < x \le 2$. Applying the same reasoning to each of the other intervals gives the following table of values and the corresponding graph of C (**Figure 1.44**).

x	C(x) = 3 − 2 int(1 − x)
$0 < x \le 1$	3
$1 < x \le 2$	5
$2 < x \le 3$	7
$3 < x \le 4$	9
$4 < x \le 5$	11

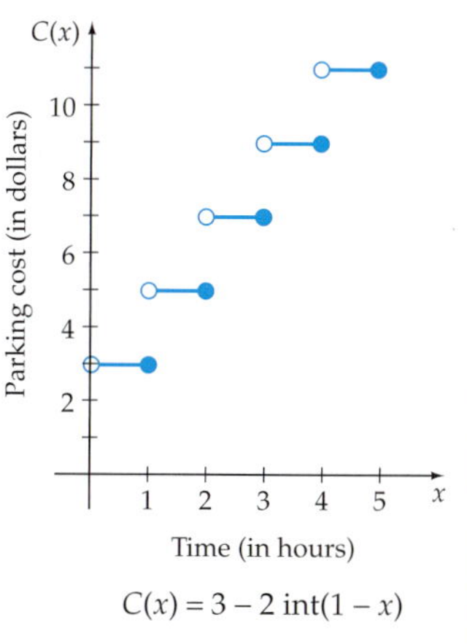

FIGURE 1.44

Because $C(1) = 3$, $C(2) = 5$, $C(3) = 7$, $C(4) = 9$, and $C(5) = 11$, we can use a solid circle at the right endpoint of each "step" and an open circle at each left endpoint.

TRY EXERCISE 48, PAGE 45

INTEGRATING TECHNOLOGY

Example 7 illustrates that a graphing calculator may not produce a graph that is a good representation of a function. You may be required to *make adjustments* in the MODE, SET UP, or WINDOW of the graphing calculator so that it will produce a better representation of the function. Some graphs may also require some *fine tuning*, such as open or solid circles at particular points, to accurately represent the function.

APPLICATIONS

EXAMPLE 8 Solve an Application

A car was purchased for \$16,500. Assuming the car depreciates at a constant rate of \$2200 per year (*straight-line depreciation*) for the first 7 years, write the value v of the car as a function of time, and calculate the value of the car 3 years after purchase.

Solution

Let t represent the number of years that have passed since the car was purchased. Then $2200t$ is the amount that the car has depreciated after t years. The value of the car at time t is given by

$$v(t) = 16{,}500 - 2200t, \quad 0 \le t \le 7$$

Continued ►

When $t = 3$, the value of the car is

$$v(3) = 16{,}500 - 2200(3) = 16{,}500 - 6600 = \$9900$$

▶ TRY EXERCISE 66, PAGE 46

Often in applied mathematics, formulas are used to determine the functional relationship that exists between two variables.

EXAMPLE 9 Solve an Application

A lighthouse is 2 miles south of a port. A ship leaves port and sails east at a rate of 7 mph. Express the distance d between the ship and the lighthouse as a function of time, given that the ship has been sailing for t hours.

Solution

Draw a diagram and label it as shown in **Figure 1.45.** Note that because distance = (rate)(time) and the rate is 7, in t hours the ship has sailed a distance of $7t$.

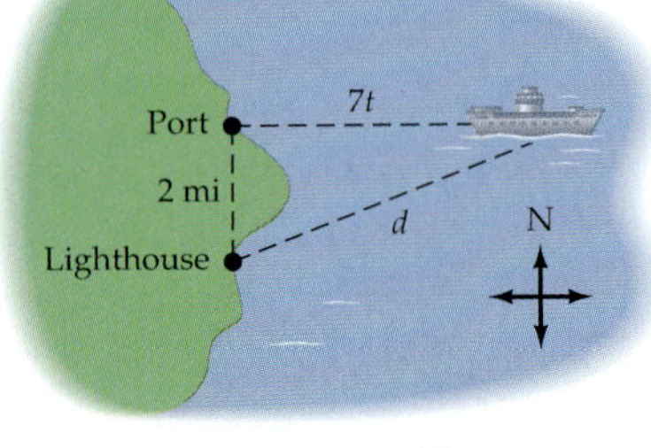

FIGURE 1.45

$$[d(t)]^2 = (7t)^2 + 2^2 \qquad \text{• The Pythagorean Theorem}$$

$$[d(t)]^2 = 49t^2 + 4$$

$$d(t) = \sqrt{49t^2 + 4} \qquad \text{• The } \pm \text{ sign is not used because } d \text{ must be nonnegative.}$$

▶ TRY EXERCISE 72, PAGE 47

EXAMPLE 10 Solve an Application

An open box is to be made from a square piece of cardboard that measures 40 inches on each side. To construct the box, squares that measure x inches on each side are cut from each corner of the cardboard as shown in **Figure 1.46.**

a. Express the volume V of the box as a function of x.

b. Determine the domain of V.

Solution

a. The length l of the box is $40 - 2x$. The width w is also $40 - 2x$. The height of the box is x. The volume V of a box is the product of its length, its width, and its height. Thus

$$V = (40 - 2x)^2 x$$

b. The squares that are cut from each corner require x to be larger than 0 inches but less than 20 inches. Thus the domain is $\{x \mid 0 < x < 20\}$.

FIGURE 1.46

▶ TRY EXERCISE 68, PAGE 46

Topics for Discussion

1. Discuss the definition of *function*. Give examples of some relationships that are functions and some that are not functions.
2. What is the difference between the domain and range of a function?
3. How many y-intercepts can a function have? How many x-intercepts can a function have?
4. Discuss how the vertical line test is used to determine whether or not a graph is the graph of a function. Explain why the vertical line test works.
5. What is the domain of $f(x) = \dfrac{\sqrt{1-x}}{x^2-9}$? Explain.
6. Is 2 in the range of $g(x) = \dfrac{6x-5}{3x+1}$? Explain the process you used to make your decision.
7. Suppose that f is a function and that $f(a) = f(b)$. Does this imply that $a = b$? Explain your answer.

Exercise Set 1.3

In Exercises 1 to 8, evaluate each function.

1. Given $f(x) = 3x - 1$, find
   a. $f(2)$ b. $f(-1)$ c. $f(0)$
   d. $f\left(\dfrac{2}{3}\right)$ e. $f(k)$ f. $f(k + 2)$

▶ 2. Given $g(x) = 2x^2 + 3$, find
   a. $g(3)$ b. $g(-1)$ c. $g(0)$
   d. $g\left(\dfrac{1}{2}\right)$ e. $g(c)$ f. $g(c + 5)$

3. Given $A(w) = \sqrt{w^2 + 5}$, find
   a. $A(0)$ b. $A(2)$ c. $A(-2)$
   d. $A(4)$ e. $A(r + 1)$ f. $A(-c)$

4. Given $J(t) = 3t^2 - t$, find
   a. $J(-4)$ b. $J(0)$ c. $J\left(\dfrac{1}{3}\right)$
   d. $J(-c)$ e. $J(x + 1)$ f. $J(x + h)$

5. Given $f(x) = \dfrac{1}{|x|}$, find
   a. $f(2)$ b. $f(-2)$ c. $f\left(\dfrac{-3}{5}\right)$
   d. $f(2) + f(-2)$ e. $f(c^2 + 4)$ f. $f(2 + h)$

6. Given $T(x) = 5$, find
   a. $T(-3)$ b. $T(0)$ c. $T\left(\dfrac{2}{7}\right)$
   d. $T(3) + T(1)$ e. $T(x + h)$ f. $T(3k + 5)$

7. Given $s(x) = \dfrac{x}{|x|}$, find
   a. $s(4)$ b. $s(5)$ c. $s(-2)$
   d. $s(-3)$ e. $s(t), t > 0$ f. $s(t), t < 0$

8. Given $r(x) = \dfrac{x}{x+4}$, find
   a. $r(0)$ b. $r(-1)$ c. $r(-3)$

d. $r\left(\frac{1}{2}\right)$ **e.** $r(0.1)$ **f.** $r(10{,}000)$

In Exercises 9 and 10, evaluate each piecewise-defined function for the indicated values.

9. $P(x) = \begin{cases} 3x + 1, & \text{if } x < 2 \\ -x^2 + 11, & \text{if } x \geq 2 \end{cases}$

a. $P(-4)$ **b.** $P(\sqrt{5})$

c. $P(c), \quad c < 2$ **d.** $P(k + 1), \quad k \geq 1$

▶ **10.** $Q(t) = \begin{cases} 4, & \text{if } 0 \leq t \leq 5 \\ -t + 9, & \text{if } 5 < t \leq 8 \\ \sqrt{t - 7}, & \text{if } 8 < t \leq 11 \end{cases}$

a. $Q(0)$ **b.** $Q(e), \quad 6 < e < 7$

c. $Q(n), \quad 1 < n < 2$ **d.** $Q(m^2 + 7), \quad 1 < m \leq 2$

In Exercises 11 to 20, identify the equations that define y as a function of x.

11. $2x + 3y = 7$ **12.** $5x + y = 8$

13. $-x + y^2 = 2$ ▶ **14.** $x^2 - 2y = 2$

15. $y = 4 \pm \sqrt{x}$ **16.** $x^2 + y^2 = 9$

17. $y = \sqrt[3]{x}$ **18.** $y = |x| + 5$

19. $y^2 = x^2$ **20.** $y^3 = x^3$

In Exercises 21 to 26, identify the sets of ordered pairs (x, y) that define y as a function of x.

21. $\{(2, 3), (5, 1), (-4, 3), (7, 11)\}$

22. $\{(5, 10), (3, -2), (4, 7), (5, 8)\}$

23. $\{(4, 4), (6, 1), (5, -3)\}$

24. $\{(2, 2), (3, 3), (7, 7)\}$

25. $\{(1, 0), (2, 0), (3, 0)\}$

26. $\left\{\left(-\frac{1}{3}, \frac{1}{4}\right), \left(-\frac{1}{4}, \frac{1}{3}\right), \left(\frac{1}{4}, \frac{2}{3}\right)\right\}$

In Exercises 27 to 38, determine the domain of the function represented by the given equation.

27. $f(x) = 3x - 4$ ▶ **28.** $f(x) = -2x + 1$

29. $f(x) = x^2 + 2$ **30.** $f(x) = 3x^2 + 1$

31. $f(x) = \dfrac{4}{x + 2}$ **32.** $f(x) = \dfrac{6}{x - 5}$

33. $f(x) = \sqrt{7 + x}$ **34.** $f(x) = \sqrt{4 - x}$

35. $f(x) = \sqrt{4 - x^2}$ **36.** $f(x) = \sqrt{12 - x^2}$

37. $f(x) = \dfrac{1}{\sqrt{x + 4}}$ **38.** $f(x) = \dfrac{1}{\sqrt{5 - x}}$

In Exercises 39 to 46, graph each function. Insert solid circles or hollow circles where necessary to indicate the true nature of the function.

39. $f(x) = \begin{cases} |x|, & \text{if } x \leq 1 \\ 2, & \text{if } x > 1 \end{cases}$

▶ **40.** $g(x) = \begin{cases} -4, & \text{if } x \leq 0 \\ x^2 - 4, & \text{if } 0 < x \leq 1 \\ -x, & \text{if } x > 1 \end{cases}$

41. $J(x) = \begin{cases} 4, & \text{if } x \leq -1 \\ x^2, & \text{if } -1 < x < 1 \\ -x + 5, & \text{if } x \geq 1 \end{cases}$

42. $K(x) = \begin{cases} 1, & \text{if } x \leq -2 \\ x^2 - 4, & \text{if } -2 < x < 2 \\ \frac{1}{2}x, & \text{if } x \geq 2 \end{cases}$

43. $L(x) = \left[\!\left[\frac{1}{3}x\right]\!\right] \quad \text{for } -6 \leq x \leq 6$

44. $M(x) = [\![x]\!] + 2 \quad \text{for } 0 \leq x \leq 4$

45. $N(x) = \text{int}(-x) \quad \text{for } -3 \leq x \leq 3$

46. $P(x) = \text{int}(x) + x \quad \text{for } 0 \leq x \leq 4$

47. **FIRST-CLASS MAIL** In 2003, the cost to mail a first-class letter is given by

$$C(w) = 0.37 - 0.34\ \text{int}(1 - w), \quad w > 0$$

where C is in dollars and w is the weight of the letter in ounces.

a. What is the cost to mail a letter that weighs 2.8 ounces?

b. Graph C for $0 < w \leq 5$.

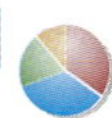

48. **INCOME TAX** The amount of federal income tax $T(x)$ a person owed in 2003 is given by

$$T(x) = \begin{cases} 0.10x, & 0 \le x < 6000 \\ 0.15(x - 6000) + 600, & 6000 \le x < 27{,}950 \\ 0.27(x - 27{,}950) + 3892.50, & 27{,}950 \le x < 67{,}700 \\ 0.30(x - 67{,}700) + 14{,}625, & 67{,}700 \le x < 141{,}250 \\ 0.35(x - 141{,}250) + 36{,}690, & 141{,}250 \le x < 307{,}050 \\ 0.386(x - 307{,}050) + 94{,}720, & x \ge 307{,}050 \end{cases}$$

where x is the adjusted gross income of the taxpayer.

a. What is the domain of this function?

b. Find the income tax owed by a taxpayer whose adjusted gross income was \$31,250.

c. Find the income tax owed by a taxpayer whose adjusted gross income was \$72,000.

49. Use the vertical line test to determine which of the following graphs are graphs of functions.

a.

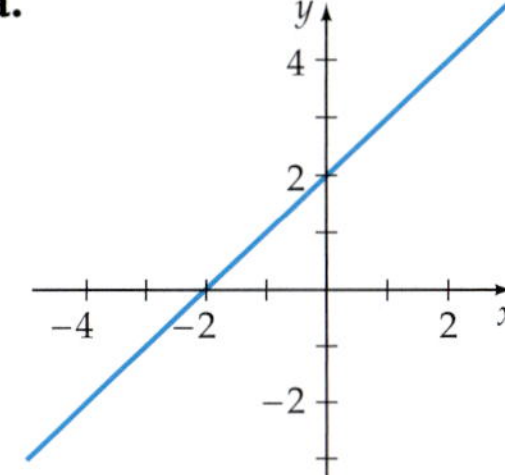

b.

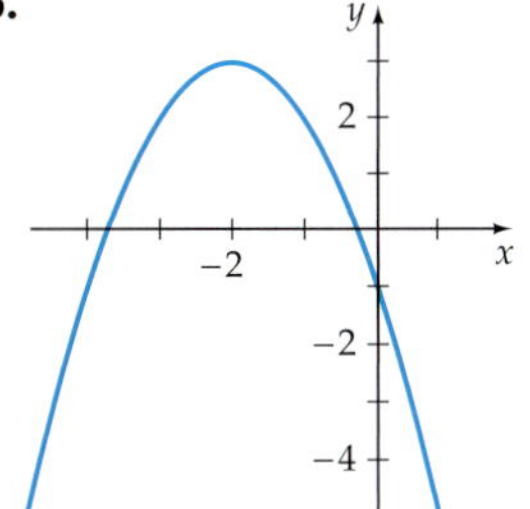

c.

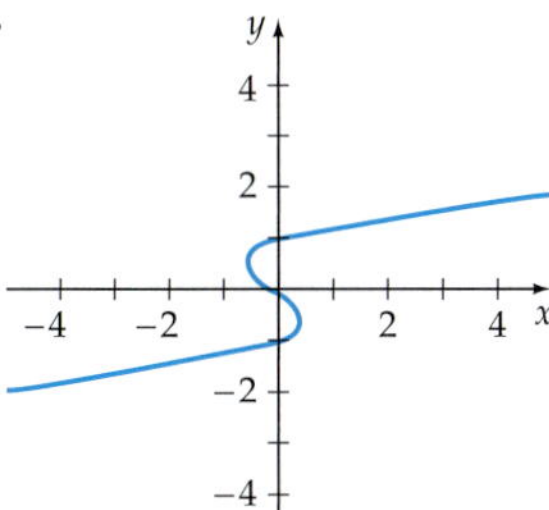

d.

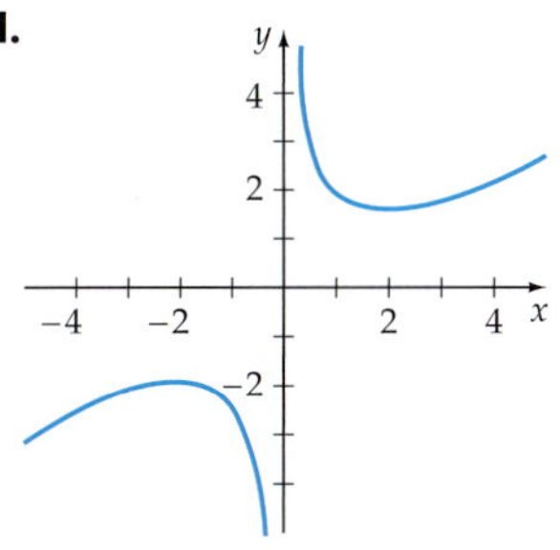

50. Use the vertical line test to determine which of the following graphs are graphs of functions.

a.

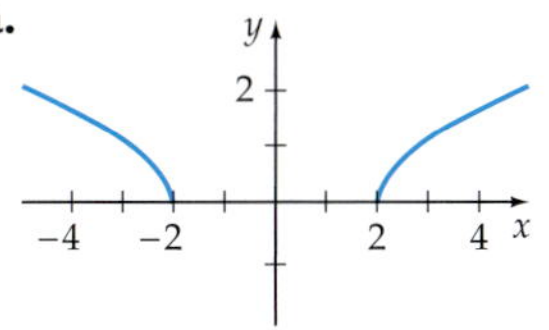

b.

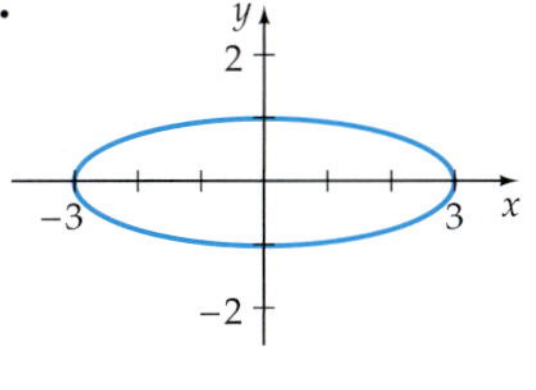

c.

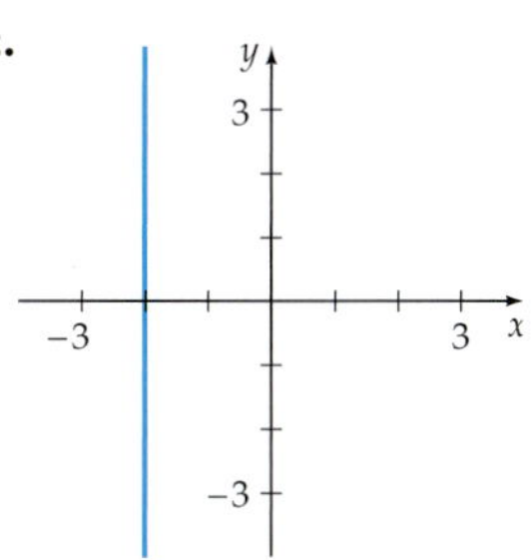

d.

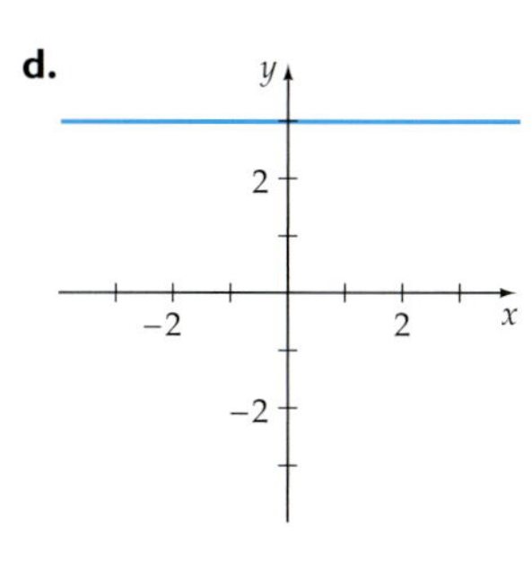

In Exercises 51 to 60, use the indicated graph to identify the intervals over which the function is increasing, constant, or decreasing.

51.

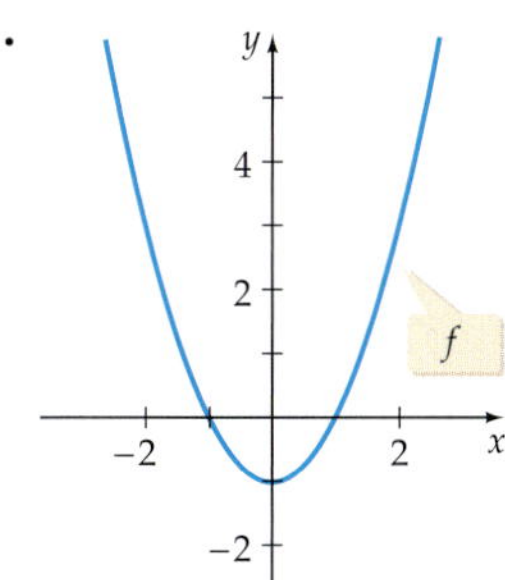

52.

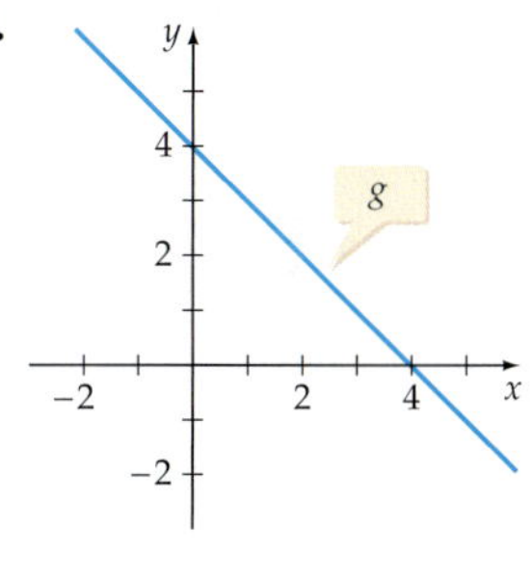

53.

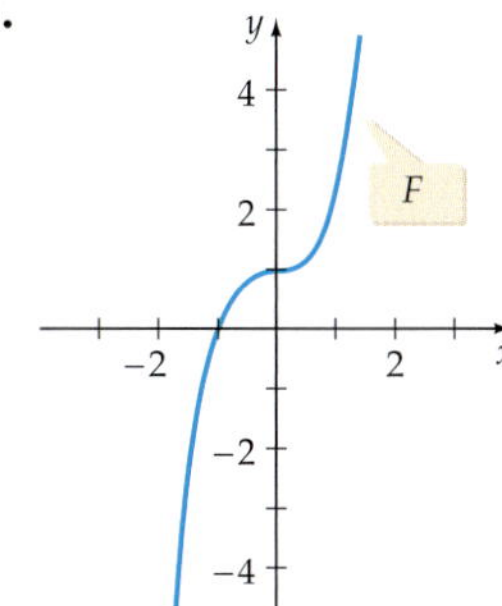

54.

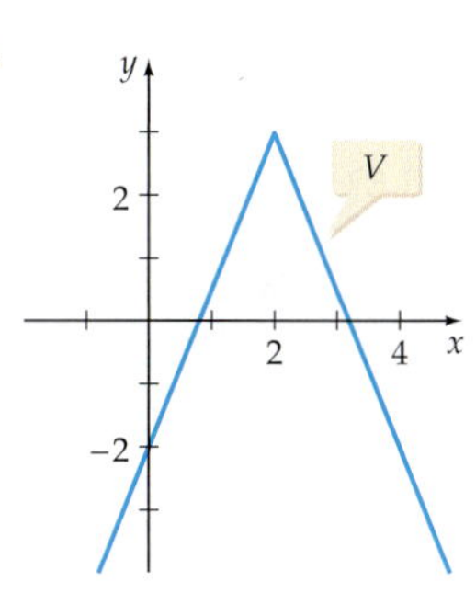

55.

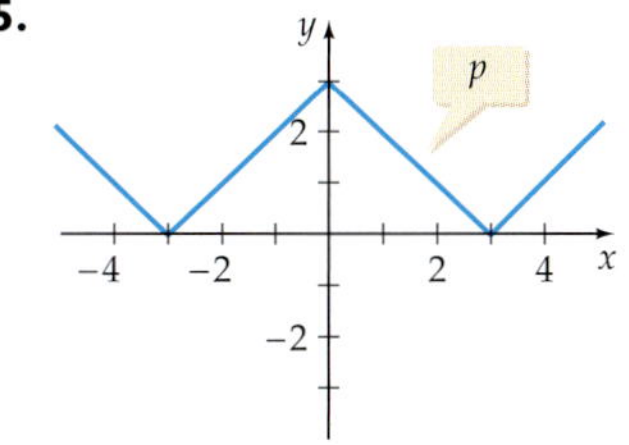

56.

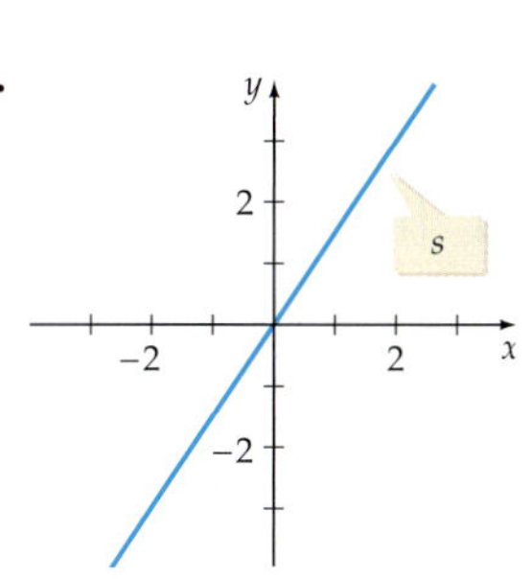

57.

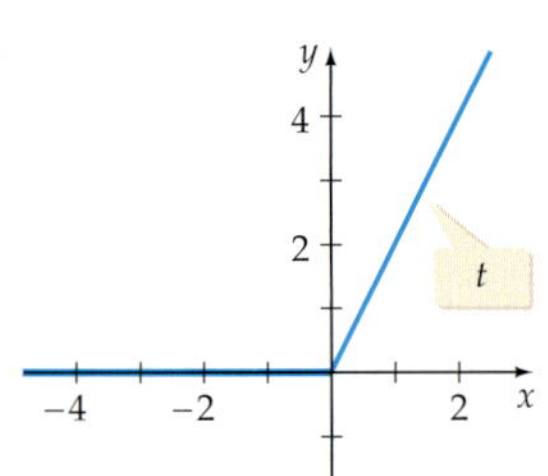

58.

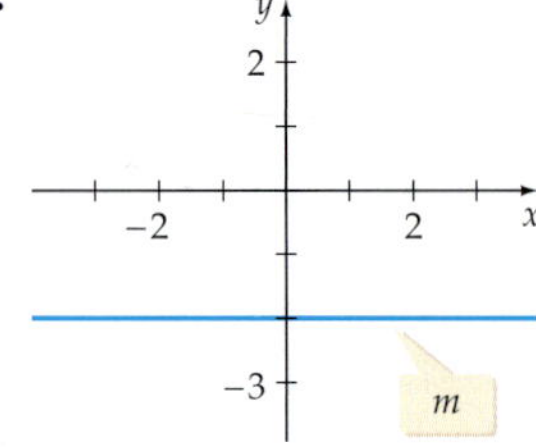

59.

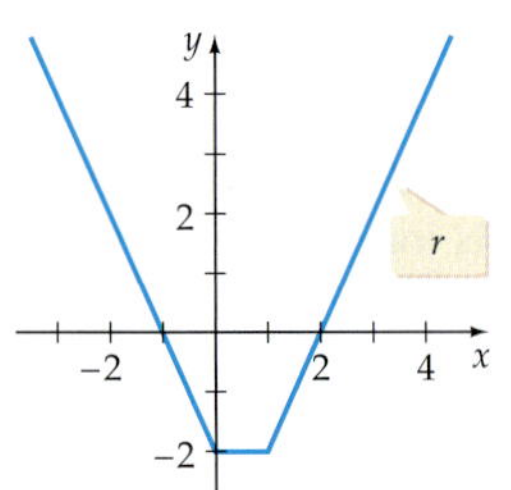

60.

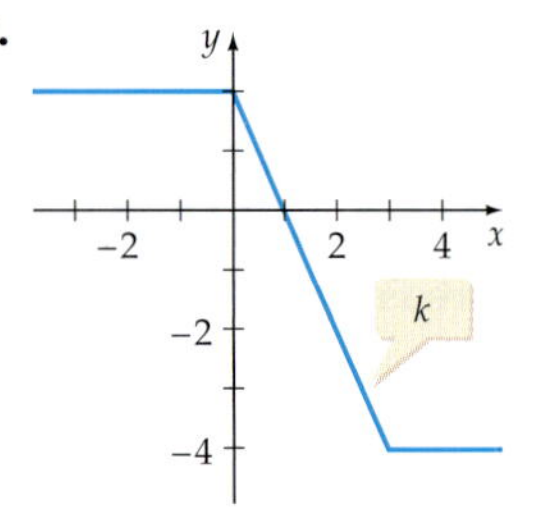

61. Use the horizontal line test to determine which of the following functions are one-to-one.

f as shown in Exercise 51
g as shown in Exercise 52
F as shown in Exercise 53
V as shown in Exercise 54
p as shown in Exercise 55

62. Use the horizontal line test to determine which of the following functions are one-to-one.

s as shown in Exercise 56
t as shown in Exercise 57
m as shown in Exercise 58
r as shown in Exercise 59
k as shown in Exercise 60

63. A rectangle has a length of l feet and a perimeter of 50 feet.

a. Write the width w of the rectangle as a function of its length.

b. Write the area A of the rectangle as a function of its length.

64. The sum of two numbers is 20. Let x represent one of the numbers.

a. Write the second number y as a function of x.

b. Write the product P of the two numbers as a function of x.

65. **Depreciation** A bus was purchased for $80,000. Assuming the bus depreciates at a rate of $6500 per year (*straight-line depreciation*) for the first 10 years, write the value v of the bus as a function of the time t (measured in years) for $0 \le t \le 10$.

66. **Depreciation** A boat was purchased for $44,000. Assuming the boat depreciates at a rate of $4200 per year (*straight-line depreciation*) for the first 8 years, write the value v of the boat as a function of the time t (measured in years) for $0 \le t \le 8$.

67. **Cost, Revenue, and Profit** A manufacturer produces a product at a cost of $22.80 per unit. The manufacturer has a fixed cost of $400.00 per day. Each unit retails for $37.00. Let x represent the number of units produced in a 5-day period.

a. Write the total cost C as a function of x.

b. Write the revenue R as a function of x.

c. Write the profit P as a function of x. [*Hint:* The profit function is given by $P(x) = R(x) - C(x)$.]

68. **Volume of a Box** An open box is to be made from a square piece of cardboard having dimensions 30 inches by 30 inches by cutting out squares of area x^2 from each corner, as shown in the figure.

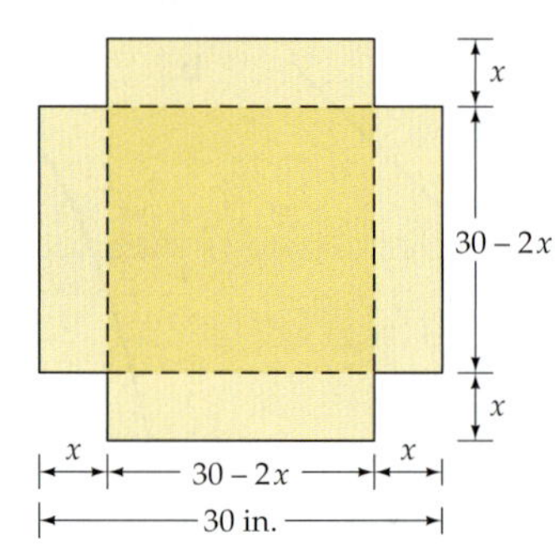

a. Express the volume V of the box as a function of x.

b. State the domain of V.

69. **Height of an Inscribed Cylinder** A cone has an altitude of 15 centimeters and a radius of 3 centimeters. A right circular cylinder of radius r and height h is inscribed in the cone as shown in the figure. Use similar triangles to write h as a function of r.

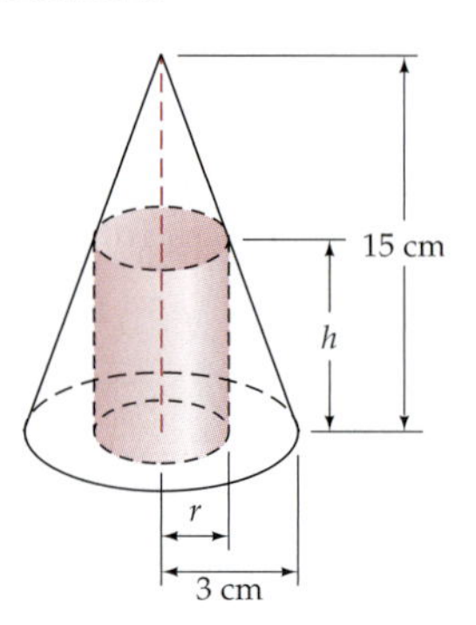

70. Volume of Water Water is flowing into a conical drinking cup that has an altitude of 4 inches and a radius of 2 inches, as shown in the figure.

a. Write the radius r of the water as a function of its depth h.

b. Write the volume V of the water as a function of its depth h.

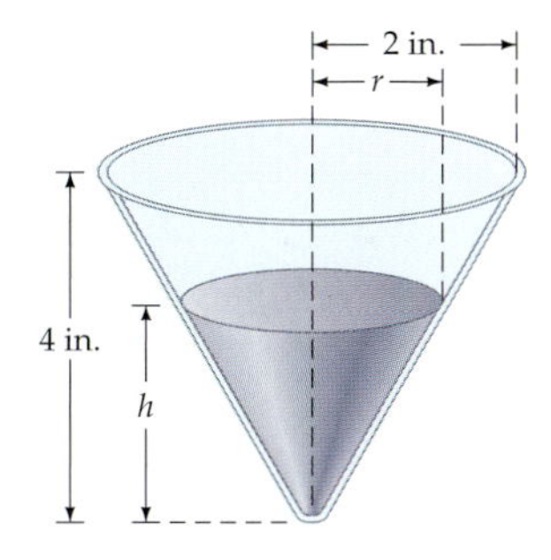

71. Distance from a Balloon For the first minute of flight, a hot air balloon rises vertically at a rate of 3 meters per second. If t is the time in seconds that the balloon has been airborne, write the distance d between the balloon and a point on the ground 50 meters from the point of lift-off as a function of t.

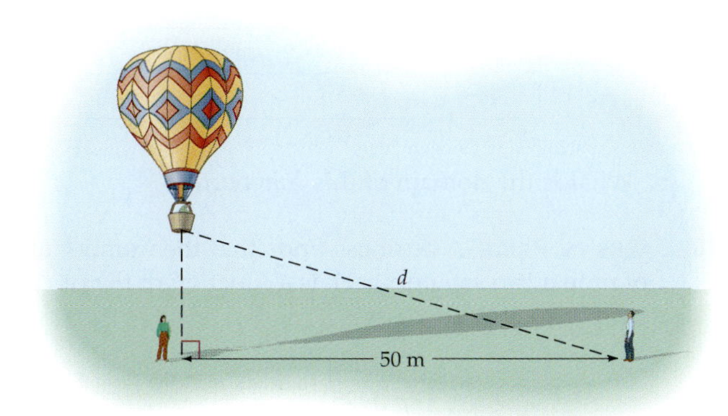

72. Time for a Swimmer An athlete swims from point A to point B at the rate of 2 mph and runs from point B to point C at a rate of 8 mph. Use the dimensions in the figure to write the time t required to reach point C as a function of x.

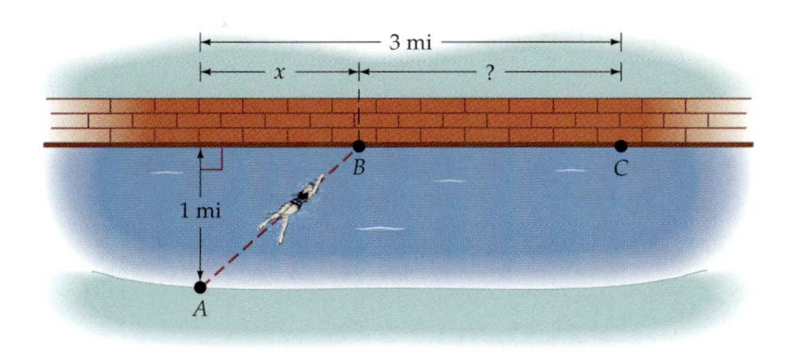

73. Distance Between Ships At 12:00 noon Ship A is 45 miles due south of ship B and is sailing north at a rate of 8 mph. Ship B is sailing east at a rate of 6 mph. Write the distance d between the ships as a function of the time t, where $t = 0$ represents 12:00 noon.

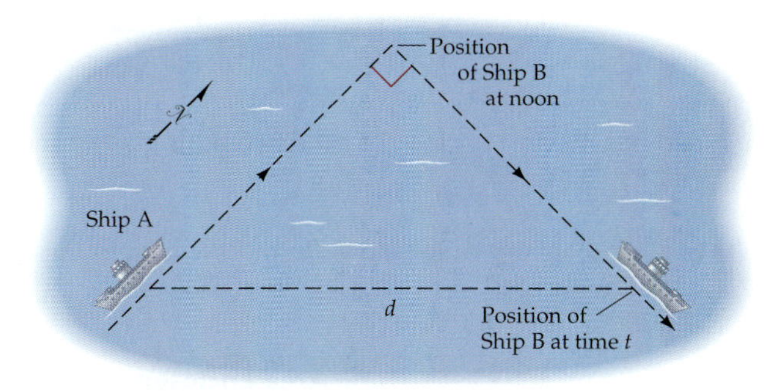

74. Area A rectangle is bounded by the x- and y-axes and the graph of $y = -\frac{1}{2}x + 4$.

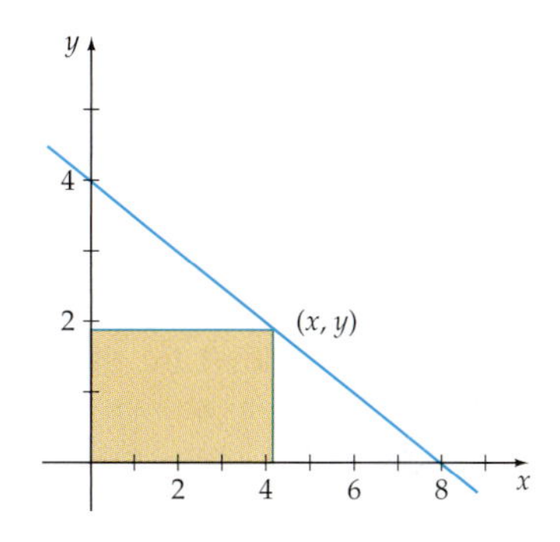

a. Find the area of the rectangle as a function of x.

b. Complete the table below.

x	Area
1	
2	
4	
6	
7	

c. What is the domain of this function?

75. Area A piece of wire 20 centimeters long is cut at a point x centimeters from the left end. The left-hand piece is formed into the shape of a circle and the right-hand piece is formed into a square.

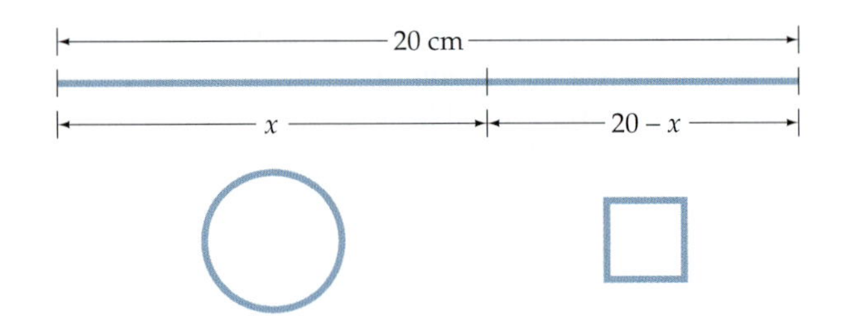

a. Find the area enclosed by the two figures as a function of x.

b. Complete the table below. Round the area to the nearest hundredth.

x	Total Area Enclosed
0	
4	
8	
12	
16	
20	

c. What is the domain of this function?

76. AREA A triangle is bounded by the x- and y-axes and must pass through $P(2, 2)$, as shown below.

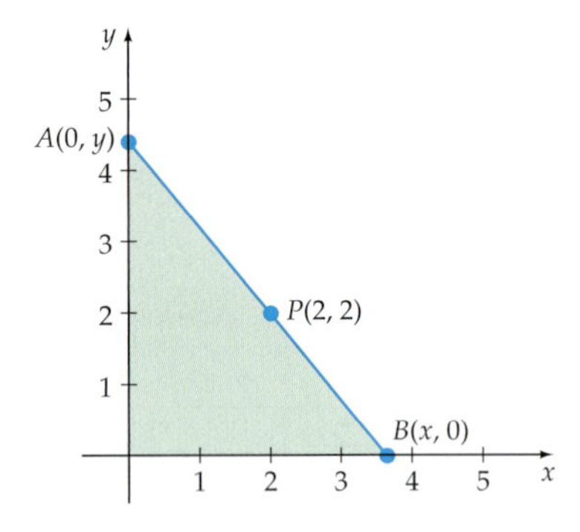

a. Find the area of the triangle as a function of x. (*Suggestion:* The slope of the line between points A and P equals the slope of the line between P and B.)

b. What is the domain of the function you found in part **a.**?

77. LENGTH Two guy wires are attached to utility poles that are 40 feet apart, as shown in the following diagram.

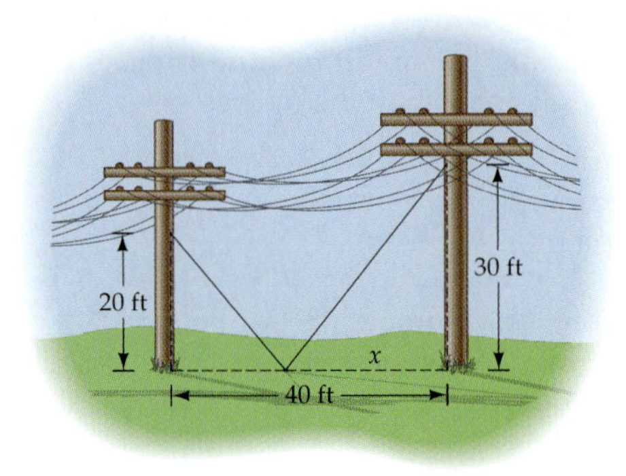

a. Find the total length of the two guy wires as a function of x.

b. Complete the table below. Round the length to the nearest hundredth.

x	Total Length of Wires
0	
10	
20	
30	
40	

c. What is the domain of this function?

78. SALES VS. PRICE A business finds that the number of feet f of pipe it can sell per week is a function of the price p in cents per foot as given by

$$f(p) = \frac{320{,}000}{p + 25}, \quad 40 \le p \le 90$$

Complete the following table by evaluating f (to the nearest hundred feet) for the indicated values of p.

p	40	50	60	75	90
f(p)					

79. MODEL YIELD The yield Y of apples per tree is related to the amount x of a particular type of fertilizer applied (in pounds per year) by the function

$$Y(x) = 400[1 - 5(x - 1)^{-2}], \quad 5 \le x \le 20$$

Complete the following table by evaluating Y (to the nearest apple) for the indicated applications.

x	5	10	12.5	15	20
Y(x)					

80. **Model Cost** A manufacturer finds that the cost C in dollars of producing x items of a product is given by

$$C(x) = \left(225 + 1.4\sqrt{x}\right)^2, \quad 100 \le x \le 1000$$

Complete the following table by evaluating C (to the nearest dollar) for the indicated numbers of items.

x	100	200	500	750	1000
$C(x)$					

81. If $f(x) = x^2 - x - 5$ and $f(c) = 1$, find c.

82. If $g(x) = -2x^2 + 4x - 1$ and $g(c) = -4$, find c.

83. Determine whether 1 is in the range of $f(x) = \dfrac{x-1}{x+1}$.

84. Determine whether 0 is in the range of $g(x) = \dfrac{1}{x-3}$.

In Exercises 85 to 90, use a graphing utility.

85. Graph $f(x) = \dfrac{[\![x]\!]}{|x|}$ for $-4.7 \le x \le 4.7$ and $x \ne 0$.

86. Graph $f(x) = \dfrac{[\![2x]\!]}{|x|}$ for $-4 \le x \le 4$ and $x \ne 0$.

87. Graph: $f(x) = x^2 - 2|x| - 3$

88. Graph: $f(x) = x^2 - |2x - 3|$

89. Graph: $f(x) = |x^2 - 1| - |x - 2|$

90. Graph: $f(x) = |x^2 - 2x| - 3$

CONNECTING CONCEPTS

The notation $f(x)|_a^b$ is used to denote the difference $f(b) - f(a)$. That is,

$$f(x)|_a^b = f(b) - f(a)$$

In Exercises 91 to 94, evaluate $f(x)|_a^b$ for the given function f and the indicated values of a and b.

91. $f(x) = x^2 - x;\ f(x)|_2^3$

92. $f(x) = -3x + 2;\ f(x)|_4^7$

93. $f(x) = 2x^3 - 3x^2 - x;\ f(x)|_0^2$

94. $f(x) = \sqrt{8 - x};\ f(x)|_0^8$

In Exercises 95 to 98, each function has two or more independent variables.

95. Given $f(x, y) = 3x + 5y - 2$, find

a. $f(1, 7)$ **b.** $f(0, 3)$ **c.** $f(-2, 4)$

d. $f(4, 4)$ **e.** $f(k, 2k)$ **f.** $f(k + 2, k - 3)$

96. Given $g(x, y) = 2x^2 - |y| + 3$, find

a. $g(3, -4)$ **b.** $g(-1, 2)$

c. $g(0, -5)$ **d.** $g\left(\dfrac{1}{2}, -\dfrac{1}{4}\right)$

e. $g(c, 3c), c > 0$ **f.** $g(c + 5, c - 2), c < 0$

97. **Area of a Triangle** The area of a triangle with sides a, b, and c is given by the function

$$A(a, b, c) = \sqrt{s(s-a)(s-b)(s-c)}$$

where s is the semiperimeter

$$s = \frac{a + b + c}{2}$$

Find $A(5, 8, 11)$.

98. **Cost of a Painter** The cost in dollars to hire a house painter is given by the function

$$C(h, g) = 15h + 14g$$

where h is the number of hours it takes to paint the house and g is the number of gallons of paint required to paint the house. Find $C(18, 11)$.

A *fixed point* of a function is a number a such that $f(a) = a$. In Exercises 99 and 100, find all fixed points for the given function.

99. $f(x) = x^2 + 3x - 3$

100. $g(x) = \dfrac{x}{x+5}$

In Exercises 101 and 102, sketch the graph of the piecewise-defined function.

101. $s(x) = \begin{cases} 1 & \text{if } x \text{ is an integer} \\ 2 & \text{if } x \text{ is not an integer} \end{cases}$

102. $v(x) = \begin{cases} 2x - 2 & \text{if } x \neq 3 \\ 1 & \text{if } x = 3 \end{cases}$

PREPARE FOR SECTION 1.4

103. Find the distance on a real number line between the points whose coordinates are -2 and 5. [1.1]

104. Find the product of a nonzero number and its negative reciprocal.

105. Given the points $P_1(-3, 4)$ and $P_2(2, -4)$, evaluate $\dfrac{y_2 - y_1}{x_2 - x_1}$.

106. Solve $y - 3 = -2(x - 3)$ for y.

107. Solve $3x - 5y = 15$ for y.

108. Given $y = 3x - 2(5 - x)$, find the value of x for which $y = 0$. [1.1]

PROJECTS

1. **Day of the Week** A formula known as Zeller's Congruence makes use of the greatest integer function $[\![x]\!]$ to determine the day of the week on which a given day fell or will fall. To use Zeller's Congruence, we first compute the integer z given by

$$z = \left[\!\!\left[\frac{13m - 1}{5}\right]\!\!\right] + \left[\!\!\left[\frac{y}{4}\right]\!\!\right] + \left[\!\!\left[\frac{c}{4}\right]\!\!\right] + d + y - 2c$$

The variables c, y, d, and m are defined as follows:

c = the century
y = the year of the century
d = the day of the month
m = the month, using 1 for March, 2 for April, ..., 10 for December. January and February are assigned the values 11 and 12 of the previous year.

For example, for the date September 12, 2001, we use $c = 20$, $y = 1$, $d = 12$, and $m = 7$. The remainder of z divided by 7 gives the day of the week. A remainder of 0 represents a Sunday, a remainder of 1 a Monday, ..., and a remainder of 6 a Saturday.

a. Verify that December 7, 1941 was a Sunday.

b. Verify that January 1, 2010 will fall on a Friday.

c. Determine on what day of the week Independence Day (July 4, 1776) fell.

d. Determine on what day of the week you were born.

SECTION 1.4

LINEAR FUNCTIONS

- SLOPES OF LINES
- FIND THE EQUATION OF A LINE
- APPLICATIONS
- PARALLEL AND PERPENDICULAR LINES

The following function has many applications.

Definition of a Linear Function

A function of the form

$$f(x) = mx + b, \quad m \neq 0$$

where m and b are real numbers, is a **linear function** of x.

SLOPES OF LINES

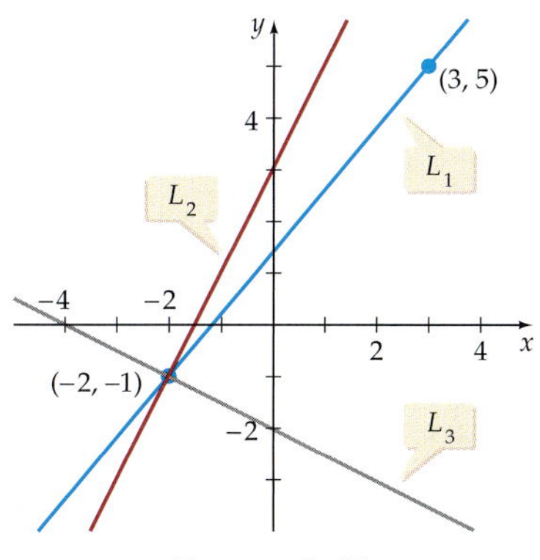

FIGURE 1.47

The graph of $f(x) = mx + b$, or $y = mx + b$, is a nonvertical straight line.

The graphs shown in **Figure 1.47** are the graphs of $f(x) = mx + b$ for various values of m. The graphs intersect at the point $(-2, -1)$, but they differ in *steepness*. The steepness of a line is called the *slope* of the line and is denoted by the symbol m. The slope of a line is the ratio of the change in the y values of any two points on the line to the change in the x values of the same two points. For example, the graph of the line L_1 in **Figure 1.47** passes through the points $(-2, -1)$ and $(3, 5)$. The change in the y values is determined by subtracting the two y-coordinates.

$$\text{Change in } y = 5 - (-1) = 6$$

The change in the x values is determined by subtracting the two x-coordinates.

$$\text{Change in } x = 3 - (-2) = 5$$

The slope m of L_1 is the ratio of the change in the y values of the two points to the change in the x values of the two points. That is,

$$m = \frac{\text{change in } y}{\text{change in } x} = \frac{6}{5}$$

Because the slope of a nonvertical line can be calculated by using any two arbitrary points on the line, we have the following formula.

Slope of a Nonvertical Line

The **slope** m of the line passing through the points $P_1(x_1, y_1)$ and $P_2(x_2, y_2)$ with $x_1 \neq x_2$ is given by

$$m = \frac{y_2 - y_1}{x_2 - x_1}$$

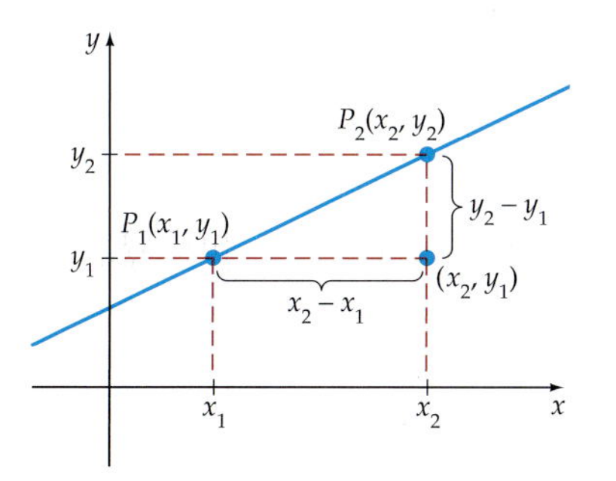

FIGURE 1.48

Because the numerator $y_2 - y_1$ is the vertical **rise** and the denominator $x_2 - x_1$ is the horizontal **run** from P_1 to P_2, slope is often referred to as the *rise over the run* or the *change in y divided by the change in x*. See **Figure 1.48.** Lines that have a positive slope slant upward from left to right. Lines that have a negative slope slant downward from left to right.

EXAMPLE 1 Find the Slope of a Line

Find the slope of the line passing through the points whose coordinates are given.

a. (1, 2) and (3, 6) **b.** (−3, 4) and (1, −2)

Solution

a. The slope of the line passing through (1, 2) and (3, 6) is

$$m = \frac{y_2 - y_1}{x_2 - x_1} = \frac{6 - 2}{3 - 1} = \frac{4}{2} = 2$$

Because $m > 0$, the line slants upward from left to right. See **Figure 1.49.**

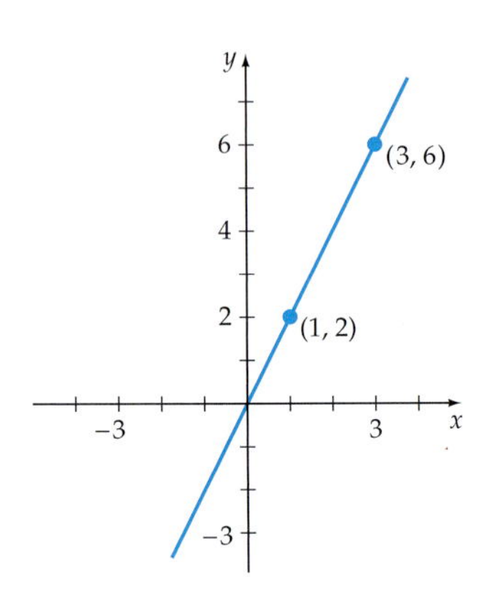

FIGURE 1.49

b. The slope of the line passing through (−3, 4) and (1, −2) is

$$m = \frac{y_2 - y_1}{x_2 - x_1} = \frac{-2 - 4}{1 - (-3)} = \frac{-6}{4} = -\frac{3}{2}$$

Because $m < 0$, the line slants downward from left to right. See **Figure 1.50.**

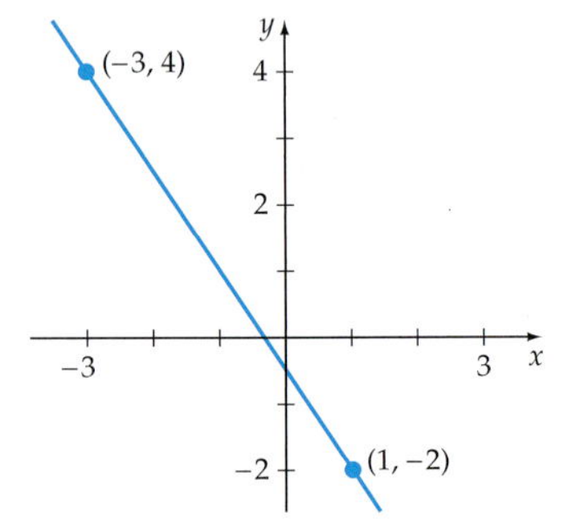

FIGURE 1.50

TRY EXERCISE 2, PAGE 60

The definition of slope does not apply to vertical lines. Consider, for example, the points (2, 1) and (2, 3) on the vertical line l_1 in **Figure 1.51.** Applying the definition of slope to this line produces

$$m = \frac{3 - 1}{2 - 2}$$

which is undefined because it requires division by zero. Because division by zero is undefined, we say that the slope of any vertical line is undefined.

Every point on the vertical line through $(a, 0)$ has an x-coordinate of a. The equation of the vertical line through $(a, 0)$ is $x = a$. See **Figure 1.52.**

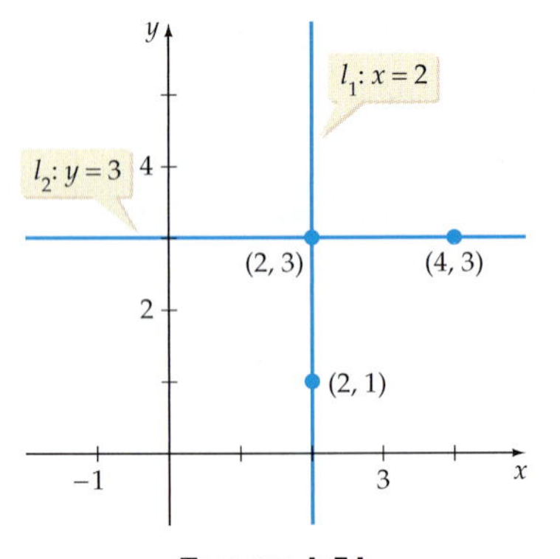

FIGURE 1.51

? QUESTION Is the graph of a vertical line the graph of a function?

All horizontal lines have 0 slope. For example, the line l_2 through (2, 3) and (4, 3) in **Figure 1.51** is a horizontal line. Its slope is given by

$$m = \frac{3 - 3}{4 - 2} = \frac{0}{2} = 0$$

When computing the slope of a line, it does not matter which point we label P_1 and which P_2 because

$$\frac{y_2 - y_1}{x_2 - x_1} = \frac{y_1 - y_2}{x_1 - x_2}$$

? ANSWER No. For example, the vertical line passing through $x = 2$ contains the ordered pairs (2, 3) and (2, −5). Thus there are two ordered pairs with the same first coordinate but different second coordinates.

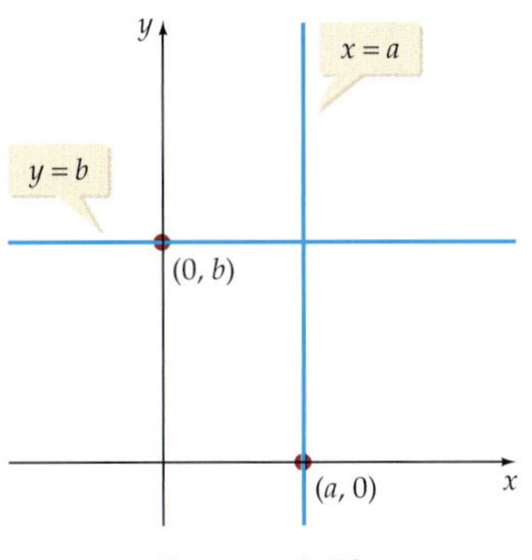

FIGURE 1.52

In functional notation, the points P_1 and P_2 can be represented by

$$(x_1, f(x_1)) \qquad \text{and} \qquad (x_2, f(x_2))$$

In this notation, the slope formula

$$m = \frac{y_2 - y_1}{x_2 - x_1} \quad \text{is expressed as} \quad m = \frac{f(x_2) - f(x_1)}{x_2 - x_1} \tag{1}$$

If $m = 0$, then $f(x) = mx + b$ can be written as $f(x) = b$, or $y = b$. The graph of $y = b$ is the horizontal line through $(0, b)$. See **Figure 1.52.** Because every point on the graph of $y = b$ has a y-coordinate of b, the function $f(x) = b$ is called a **constant function.**

Horizontal Lines and Vertical Lines

The graph of $x = a$ is a vertical line through $(a, 0)$.

The graph of $y = b$ is a horizontal line through $(0, b)$.

The equation $f(x) = mx + b$ is called the **slope-intercept form** of the equation of a line because of the following theorem.

Slope-Intercept Form

The graph of $f(x) = mx + b$ is a line with slope m and y-intercept $(0, b)$.

Proof The slope of the graph of $f(x) = mx + b$ is given by Equation (1).

$$\frac{f(x_2) - f(x_1)}{x_2 - x_1} = \frac{(mx_2 + b) - (mx_1 + b)}{x_2 - x_1} = \frac{m(x_2 - x_1)}{x_2 - x_1} = m, \quad x_1 \neq x_2$$

The y-intercept of the graph of $f(x) = mx + b$ is found by letting $x = 0$.

$$f(0) = m(0) + b = b$$

Thus $(0, b)$ is the y-intercept, and m is the slope, of the graph of $f(x) = mx + b$. ◆

If a function is written in the form $f(x) = mx + b$, then its graph can be drawn by first plotting the y-intercept $(0, b)$ and then using the slope m to determine another point on the line.

EXAMPLE 2 Graph a Linear Function

Graph: $f(x) = 2x - 1$

Solution

The equation $y = 2x - 1$ is in slope-intercept form, with $b = -1$ and $m = 2$. Thus the y-intercept is $(0, -1)$, and the slope is 2. Write the slope as

$$m = \frac{2}{1} = \frac{\text{change in } y}{\text{change in } x}$$

Continued ▶

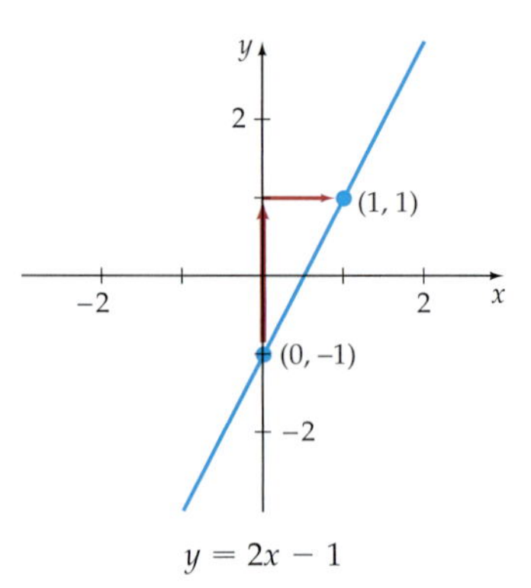

$y = 2x - 1$

FIGURE 1.53

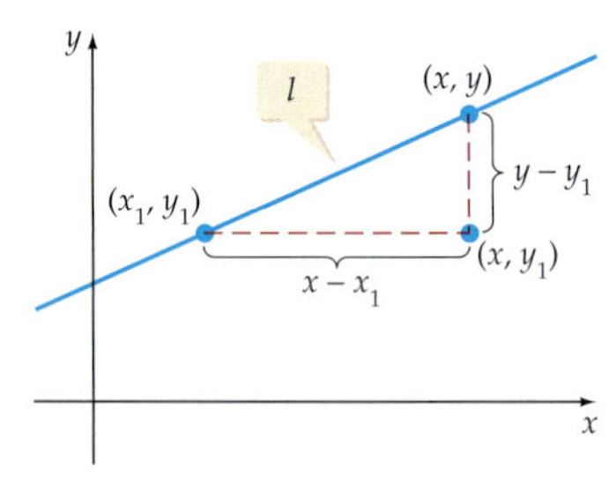

The slope of line l is $m = \dfrac{y - y_1}{x - x_1}$.

FIGURE 1.54

To graph the equation, first plot the y-intercept, and then use the slope to plot a second point. This second point is 2 units up (change in y) and 1 unit to the right (change in x) of the y-intercept. See **Figure 1.53.**

▶ **TRY EXERCISE 16, PAGE 61**

FIND THE EQUATION OF A LINE

We can find an equation of a line provided we know its slope and at least one point on the line. **Figure 1.54** suggests that if (x_1, y_1) is a point on a line l of slope m, and (x, y) is *any other* point on the line, then

$$\frac{y - y_1}{x - x_1} = m, \quad x \neq x_1$$

Multiplying each side by $x - x_1$ produces $y - y_1 = m(x - x_1)$. This equation is called the **point-slope form** of the equation of line l.

Point-Slope Form

The graph of

$$y - y_1 = m(x - x_1)$$

is a line that has slope m and passes through (x_1, y_1).

EXAMPLE 3 Use the Point-Slope Form

Find an equation of the line with slope -3 that passes through $(-1, 4)$.

Solution

Use the point-slope form with $m = -3$, $x_1 = -1$, and $y_1 = 4$.

$$y - y_1 = m(x - x_1)$$

$$y - 4 = -3[x - (-1)] \quad \text{• Substitute.}$$

$$y - 4 = -3x - 3 \quad \text{• Solve for } y.$$

$$y = -3x + 1 \quad \text{• Slope-intercept form}$$

▶ **TRY EXERCISE 28, PAGE 61**

take note

To determine an equation of a nonvertical line that passes through two points, first determine the slope of the line and then use the coordinates of either one of the points in the point-slope form.

An equation of the form $Ax + By = C$, where A, B, and C are real numbers and both A and B are not zero, is called the **general form of the equation of a line.** For example, the equation $y = -3x + 1$ in Example 3 can be written in general form as $3x + y = 1$.

> **take note**
>
> It is not always possible to solve a linear equation in general form for y in terms of x. For instance, the linear equation $x = 4$ is in general form with $A = 1$, $B = 0$, and $C = 4$.

One way to graph a linear equation that is written in general form is to first solve the equation for y in terms of x. For instance, to graph $3x - 2y = 4$, solve for y.

$$3x - 2y = 4$$
$$-2y = -3x + 4$$
$$y = \frac{3}{2}x - 2$$

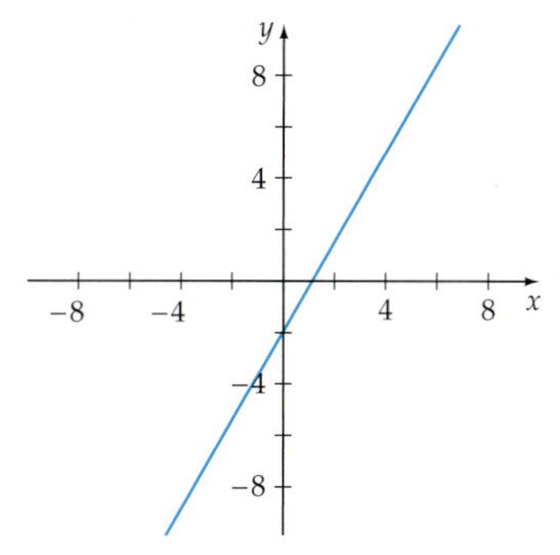

FIGURE 1.55

The y-intercept is $(0, -2)$ and the slope is $\frac{3}{2}$. The graph is shown in **Figure 1.55.**

By solving a first-degree equation, the specific relationship between an element of the domain and an element of the range of a linear function can be determined.

EXAMPLE 4 Find the Value in the Domain of f for which $f(x) = b$

Find the value x in the domain of $f(x) = 3x - 4$ for which $f(x) = 5$.

Algebraic Solution

$$f(x) = 3x - 4$$
$$5 = 3x - 4 \quad \text{• Replace } f(x) \text{ by 5 and solve for } x.$$
$$9 = 3x$$
$$3 = x$$

When $x = 3$, $f(x) = 5$. This means that 3 in the domain of f is paired with 5 in the range of f. Another way of stating this is that the ordered pair $(3, 5)$ is an element of f.

Visualize the Solution

By graphing $y = 5$ and $f(x) = 3x - 4$, we can see that $f(x) = 5$ when $x = 3$.

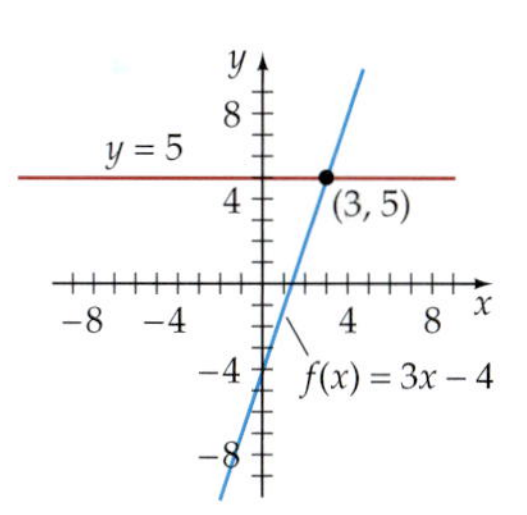

▶ TRY EXERCISE 42, PAGE 61

Although we are mainly concerned with linear functions in this section, the following theorem applies to all functions. It illustrates a powerful relationship between the real solutions of $f(x) = 0$ and the x-intercepts of the graph of $y = f(x)$.

Real Solutions and x-Intercepts Theorem

For every function f, the real number c is a solution of $f(x) = 0$ if and only if $(c, 0)$ is an x-intercept of the graph of $y = f(x)$.

? QUESTION Is $(-2, 0)$ an x-intercept of $f(x) = x^3 - x + 6$?

The real solutions and x-intercepts theorem tells us that we can find real solutions of $f(x) = 0$ by graphing. The following example illustrates the theorem for a linear function of x.

? ANSWER Yes. $f(-2) = (-2)^3 - (-2) + 6 = 0$

EXAMPLE 5 Verify the Real Solutions and x-Intercepts Theorem

Let $f(x) = -2x + 6$. Find the real solution of $f(x) = 0$ and then graph $f(x)$. Compare the solution of $f(x) = 0$ with the x-intercept of the graph of f.

Algebraic Solution

To find the real solution of $f(x) = 0$, replace $f(x)$ by $-2x + 6$ and solve for x.

$$
\begin{aligned}
f(x) &= 0 \\
-2x + 6 &= 0 \\
-2x &= -6 \\
x &= 3
\end{aligned}
$$

The x-coordinate of the x-intercept is 3. The real solution of $f(x) = 0$ is 3.

Visualize the Solution

Graph $f(x) = -2x + 6$ (see **Figure 1.56**).

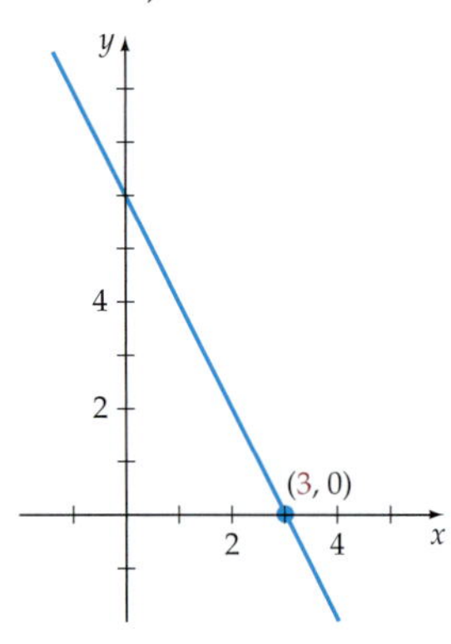

FIGURE 1.56

The x-intercept is (3, 0).

▶ TRY EXERCISE 46, PAGE 61

EXAMPLE 6 Solve $f_1(x) = f_2(x)$

Let $f_1(x) = 2x - 1$ and $f_2(x) = -x + 11$. Find the values x for which $f_1(x) = f_2(x)$.

Algebraic Solution

$$
\begin{aligned}
f_1(x) &= f_2(x) \\
2x - 1 &= -x + 11 \\
3x &= 12 \\
x &= 4
\end{aligned}
$$

When $x = 4$, $f_1(x) = f_2(x)$.

Visualize the Solution

The graphs of $y = f_1(x)$ and $y = f_2(x)$ are shown on the same coordinate axes (see **Figure 1.57**). Note that the point of intersection is (4, 7).

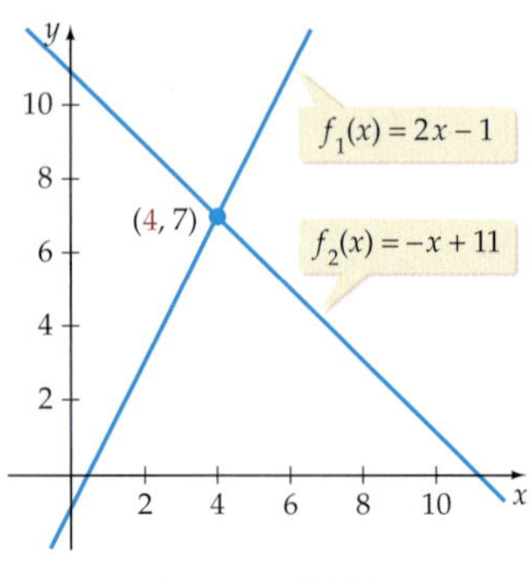

FIGURE 1.57

▶ TRY EXERCISE 50, PAGE 61

APPLICATIONS

EXAMPLE 7 Find a Linear Model of Data

The bar graph in **Figure 1.58** is based on data from the Nevada Department of Motor Vehicles. The graph illustrates the distance (in feet) a car travels between the time (in seconds) a driver recognizes an emergency and the time the brakes are applied.

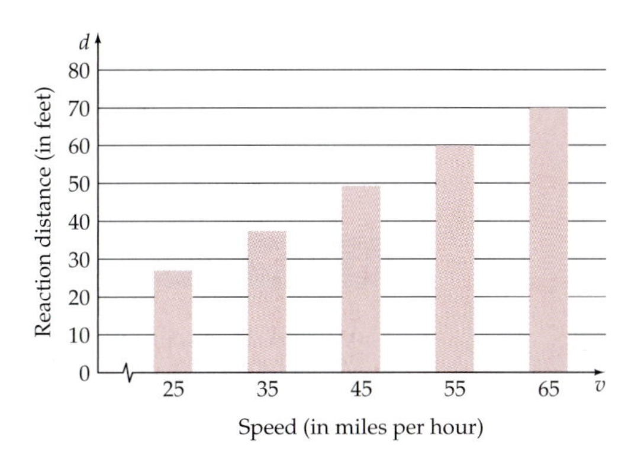

FIGURE 1.58

a. Find a linear function that models the reaction distance in terms of the speed of the car by using the ordered pairs (25, 27) and (55, 60).

b. What reaction distance does the model predict for a car traveling at 50 miles per hour?

Solution

a. First, calculate the slope of the line. Then use the point-slope formula to find the equation of the line.

$$m = \frac{d_2 - d_1}{v_2 - v_1} = \frac{60 - 27}{55 - 25} = \frac{33}{30} = 1.1$$ • **Find the slope.**

$$d - d_1 = m(v - v_1)$$ • **Use the point-slope formula.**

$$d - 27 = 1.1(v - 25)$$ • $d_1 = 27$, $v_1 = 25$, $m = 1.1$

$$d = 1.1v - 0.5$$

In functional notation, the linear model is $d(v) = 1.1v - 0.5$.

b. To find the reaction distance for a car traveling at 50 miles per hour, evaluate $d(v)$ when $v = 50$.

$$d(v) = 1.1v - 0.5$$
$$d(50) = 1.1(50) - 0.5$$
$$= 54.5$$

The reaction distance is 54.5 feet.

▶ TRY EXERCISE 56, PAGE 62

If a manufacturer produces x units of a product that sells for p dollars per unit, then the **cost function** C, the **revenue function** R, and the **profit function** P are defined as follows:

$$C(x) = \text{cost of producing and selling } x \text{ units}$$

$$R(x) = xp = \text{revenue from the sale of } x \text{ units at } p \text{ dollars each}$$

$$P(x) = \text{profit from selling } x \text{ units}$$

Because profit equals the revenue less the cost, we have

$$P(x) = R(x) - C(x)$$

The value of x for which $R(x) = C(x)$ is called the **break-even point.** At the break-even point, $P(x) = 0$.

EXAMPLE 8 Find the Profit Function and the Break-even Point

A manufacturer finds that the costs incurred in the manufacture and sale of a particular type of calculator are \$180,000 plus \$27 per calculator.

a. Determine the profit function, P, given that x calculators are manufactured and sold at \$59 each.

b. Determine the break-even point.

Solution

a. The cost function is $C(x) = 27x + 180{,}000$. The revenue function is $R(x) = 59x$. Thus the profit function is

$$\begin{aligned} P(x) &= R(x) - C(x) \\ &= 59x - (27x + 180{,}000) \\ &= 32x - 180{,}000, \quad x \geq 0 \text{ and } x \text{ is an integer} \end{aligned}$$

b. At the break-even point, $R(x) = C(x)$.

$$\begin{aligned} 59x &= 27x + 180{,}000 \\ 32x &= 180{,}000 \\ x &= 5625 \end{aligned}$$

The manufacturer will break even when 5625 calculators are sold.

▶ TRY EXERCISE 66, PAGE 64

take note

The graphs of C, R, and P are shown below. Observe that the graphs of C and R intersect at the break-even point, where $x = 5625$ and $P(5625) = 0$.

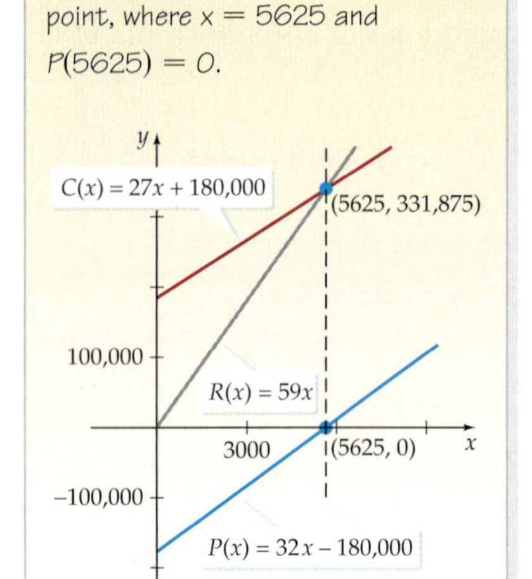

• PARALLEL AND PERPENDICULAR LINES

Two nonintersecting lines in a plane are **parallel.** All vertical lines are parallel to each other. All horizontal lines are parallel to each other.

Two lines are **perpendicular** if and only if they intersect and form adjacent angles, each of which measures 90°. In a plane, vertical and horizontal lines are perpendicular to one another.

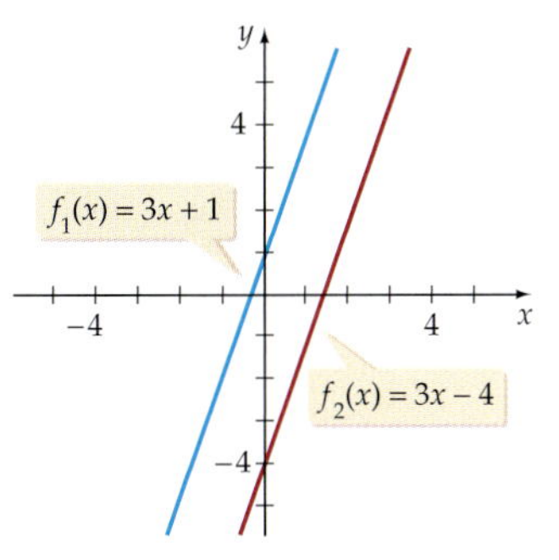

FIGURE 1.59

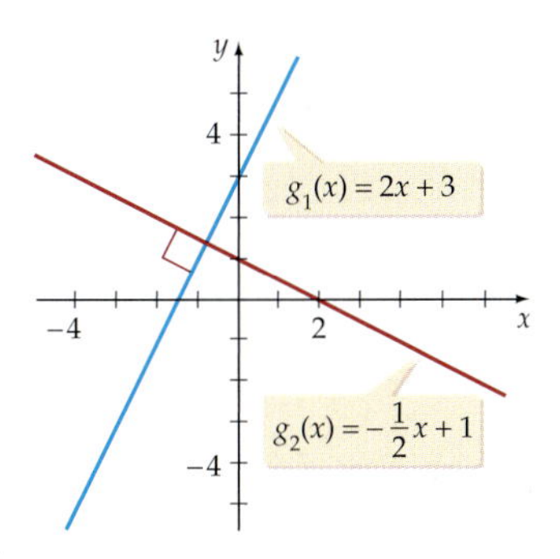

FIGURE 1.60

Parallel and Perpendicular Lines

Let l_1 be the graph of $f_1(x) = m_1x + b$ and l_2 be the graph of $f_2(x) = m_2x + b$. Then

- l_1 and l_2 are parallel if and only if $m_1 = m_2$.
- l_1 and l_2 are perpendicular if and only if $m_1 = -\dfrac{1}{m_2}$.

The graphs of $f_1(x) = 3x + 1$ and $f_2(x) = 3x - 4$ are shown in **Figure 1.59.** Because $m_1 = m_2 = 3$, the lines are parallel.

If $m_1 = -\dfrac{1}{m_2}$, then m_1 and m_2 are negative reciprocals of each other. The graphs of $g_1(x) = 2x + 3$ and $g_2(x) = -\dfrac{1}{2}x + 1$ are shown in **Figure 1.60.** Because 2 and $-\dfrac{1}{2}$ are negative reciprocals of each other, the lines are perpendicular. The symbol ⌉ indicates an angle of 90°. In **Figure 1.60** it is used to indicate that the lines are perpendicular.

EXAMPLE 9 Determine a Point of Impact

A rock attached to a string is whirled horizontally in a circular counter-clockwise path about the origin. When the string breaks, the rock travels on a linear path perpendicular to the radius $\overline{OP}$ and hits a wall located at

$$y = x + 12 \tag{2}$$

If the string breaks when the rock is at $P(4, 3)$, determine the point at which the rock hits the wall. See **Figure 1.61.**

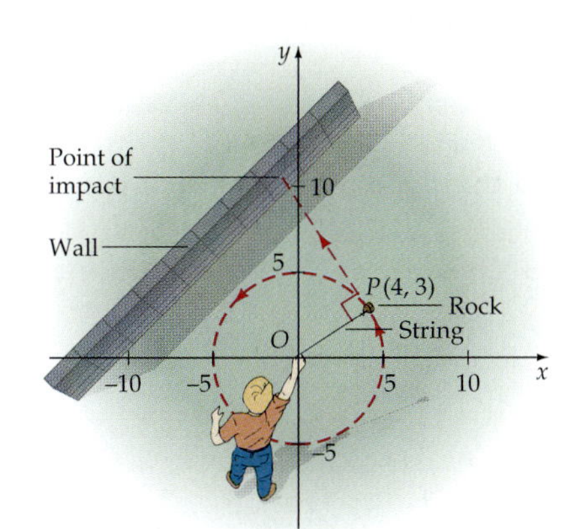

FIGURE 1.61

Solution

The slope of the radius from (0, 0) to (4, 3) is $\dfrac{3}{4}$. The negative reciprocal of $\dfrac{3}{4}$ is $-\dfrac{4}{3}$. Therefore, the linear path of the rock is given by

$$y - 3 = -\frac{4}{3}(x - 4)$$

$$y = -\frac{4}{3}x + \frac{25}{3} \tag{3}$$

To find the point at which the rock hits the wall, set the right side of Equation (2) equal to the right side of Equation (3) and solve for x. This is the procedure explained in Example 6.

Continued ▶

$$-\frac{4}{3}x + \frac{25}{3} = x + 12$$

$$-4x + 25 = 3x + 36 \qquad \text{• Multiply all terms by 3.}$$

$$-7x = 11$$

$$x = -\frac{11}{7}$$

For every point on the wall, x and y are related by $y = x + 12$. Therefore, substituting $-\frac{11}{7}$ for x in $y = x + 12$ yields $y = -\frac{11}{7} + 12 = \frac{73}{7}$, and the rock hits the wall at $\left(-\frac{11}{7}, \frac{73}{7}\right)$.

▶ TRY EXERCISE 78, PAGE 65

Topics for Discussion

1. Can the graph of a linear function contain points in only one quadrant? only two quadrants? only four quadrants?
2. Is a "break-even point" a point or a number? Explain.
3. Some perpendicular lines do not have the property that their slopes are negative reciprocals of each other. Characterize these lines.
4. Does the real solutions and x-intercepts theorem apply only to linear functions?
5. Explain why the function $f(x) = x$ is referred to as the identity function.

Exercise Set 1.4

In Exercises 1 to 10, find the slope of the line that passes through the given points.

1. $(3, 4)$ and $(1, 7)$

▶ **2.** $(-2, 4)$ and $(5, 1)$

3. $(4, 0)$ and $(0, 2)$

4. $(-3, 4)$ and $(2, 4)$

5. $(0, 0)$ and $(0, 4)$

6. $(0, 0)$ and $(3, 0)$

7. $(-3, 4)$ and $(-4, -2)$

8. $(-5, -1)$ and $(-3, 4)$

9. $\left(-4, \frac{1}{2}\right)$ and $\left(\frac{7}{3}, \frac{7}{2}\right)$

10. $\left(\frac{1}{2}, 4\right)$ and $\left(\frac{7}{4}, 2\right)$

In Exercises 11 to 14, find the slope of the line that passes through the given points.

11. $(3, f(3))$ and $(3 + h, f(3 + h))$

12. $(-2, f(-2 + h))$ and $(-2 + h, f(-2 + h))$

13. $(0, f(0))$ and $(h, f(h))$

14. $(a, f(a))$ and $(a + h, f(a + h))$

In Exercises 15 to 26, graph y as a function of x by finding the slope and y-intercept of each line.

15. $y = 2x - 4$

16. $y = -x + 1$

17. $y = -\frac{1}{3}x + 4$

18. $y = \frac{2}{3}x - 2$

19. $y = 3$

20. $y = x$

21. $y = 2x$

22. $y = -3x$

23. $2x + y = 5$

24. $x - y = 4$

25. $4x + 3y - 12 = 0$

26. $2x + 3y + 6 = 0$

In Exercises 27 to 38, find the equation of the indicated line. Write the equation in the form $y = mx + b$.

27. y-intercept $(0, 3)$, slope 1

28. y-intercept $(0, 5)$, slope -2

29. y-intercept $\left(0, \frac{1}{2}\right)$, slope $\frac{3}{4}$

30. y-intercept $\left(0, \frac{3}{4}\right)$, slope $-\frac{2}{3}$

31. y-intercept $(0, 4)$, slope 0

32. y-intercept $(0, -1)$, slope $\frac{1}{2}$

33. Through $(-3, 2)$, slope -4

34. Through $(-5, -1)$, slope -3

35. Through $(3, 1)$ and $(-1, 4)$

36. Through $(5, -6)$ and $(2, -8)$

37. Through $(7, 11)$ and $(2, -1)$

38. Through $(-5, 6)$ and $(-3, -4)$

39. Find the value of x in the domain of $f(x) = 2x + 3$ for which $f(x) = -1$.

40. Find the value of x in the domain of $f(x) = 4 - 3x$ for which $f(x) = 7$.

41. Find the value of x in the domain of $f(x) = 1 - 4x$ for which $f(x) = 3$.

42. Find the value of x in the domain of $f(x) = \frac{2x}{3} + 2$ for which $f(x) = 4$.

43. Find the value of x in the domain of $f(x) = 3 - \frac{x}{2}$ for which $f(x) = 5$.

44. Find the value of x in the domain of $f(x) = 4x - 3$ for which $f(x) = -2$.

In Exercises 45 to 48, find the solution of $f(x) = 0$. Verify that the solution of $f(x) = 0$ is the same as the x-coordinate of the x-intercept of the graph of $y = f(x)$.

45. $f(x) = 3x - 12$

46. $f(x) = -2x - 4$

47. $f(x) = \frac{1}{4}x + 5$

48. $f(x) = -\frac{1}{3}x + 2$

In Exercises 49 to 52, solve $f_1(x) = f_2(x)$ by an algebraic method and by graphing.

49. $f_1(x) = 4x + 5$ $\quad f_2(x) = x + 6$

50. $f_1(x) = -2x - 11$ $\quad f_2(x) = 3x + 7$

51. $f_1(x) = 2x - 4$ $\quad f_2(x) = -x + 12$

52. $f_1(x) = \frac{1}{2}x + 5$ $\quad f_2(x) = \frac{2}{3}x - 7$

53. **Oceanography** The graph below shows the relationship between the speed of sound in water and the temperature of the water. Find the slope of this line, and write a sentence that explains the meaning of the slope in the context of this problem.

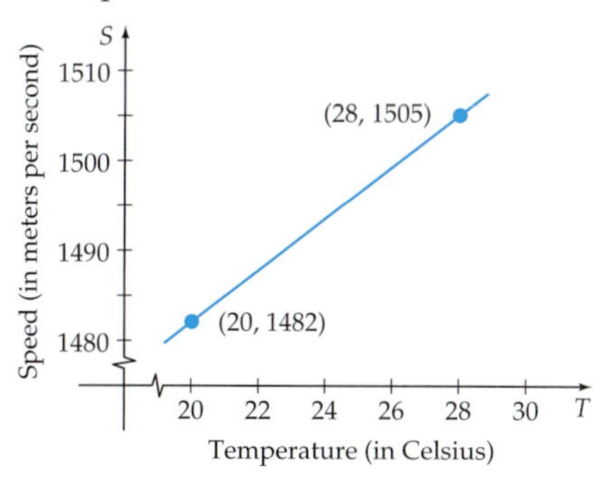

54. **Computer Science** The graph on the following page shows the relationship between the time, in seconds, it takes to download a file and the size of the file in megabytes. Find the slope of the line between the two points shown on the graph. Write a sentence that states the meaning of the slope in the context of this problem.

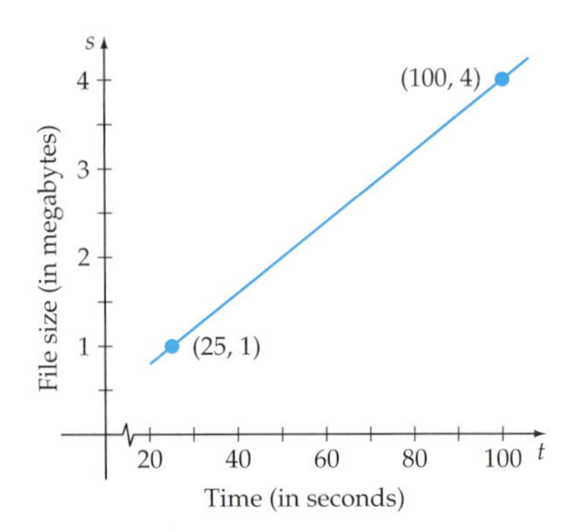

55. **AUTOMOTIVE TECHNOLOGY** The table below shows the EPA fuel economy values for selected two-seater cars for the 2003 model year. (*Source:* www.fueleconomy.gov.)

EPA Fuel Economy Values for Selected Two-Seater Cars

Car	City mpg	Highway mpg
Audi, TT Roadster	20	29
BMW, Z8	13	21
Ferrari, 360 Spider	11	16
Lamborghini, L-174	9	13
Lotus, Esprit V8	15	22
Maserati, Spider GT	11	17

a. Using the data for the Lamborghini and the Audi, find a linear function that predicts highway miles per gallon in terms of city miles per gallon. Round the slope to the nearest hundredth.

b. Using your model, predict the highway miles per gallon for a Porsche Boxer, whose city fuel efficiency is 18 miles per gallon. Round to the nearest whole number.

56. **CONSUMER CREDIT** The amount of revolving consumer credit (such as credit cards and department store cards) for the years 1997 to 2003 is given in the table below. (*Source:* www.nber.org, Board of Governor's of the Federal Reserve System.)

Year	Consumer Credit (in billions of $)
1997	531.0
1998	562.5
1999	598.0
2000	667.4
2001	701.3
2002	712.0
2003	725.0

a. Using the data for 1997 and 2003, find a linear model that predicts the amount of revolving consumer credit (in billions) for year t. Round the slope to the nearest tenth.

b. Using this model, in what year will consumer credit first exceed $850 billion?

57. **LABOR MARKET** According to the Bureau of Labor Statistics (BLS), there were 38,000 desktop publishing jobs in the United States in the year 2000. The BLS projects that there will be 63,000 desktop publishing jobs in 2010.

a. Using the BLS data, find the number of desktop publishing jobs as a linear function of the year.

b. Using your model, in what year will the number of desktop publishing jobs first exceed 60,000?

58. **POTTERY** A piece of pottery is removed from a kiln and allowed to cool in a controlled environment. The temperature (in degrees Fahrenheit) of the pottery after it is removed from the kiln for various times (in minutes) is shown in the table below.

Time, min	Temperature, °F
15	2200
20	2150
30	2050
60	1750

a. Find a linear model for the temperature of the pottery after t minutes.

b. Explain the meaning of the slope of this line in the context of the problem.

c. Assuming temperature continues to decrease at the same rate, what will be the temperature of the pottery in 3 hours?

59. **LUMBER INDUSTRY** For a log, the number of board-feet (bf) that can be obtained from the log depends on the diameter, in inches, of the log and its length. The table below shows the number of board-feet of lumber that can be obtained from a log that is 32 feet long.

Diameter, inches	bf
16	180
18	240
20	300
22	360

a. Find a linear model for the number of board-feet as a function of tree diameter.

b. Write a sentence explaining the meaning of the slope of this line in the context of the problem.

c. Using this model, how many board-feet of lumber can be obtained from a log 32 feet long with a diameter of 19 inches?

60. ECOLOGY The rate at which water evaporates from a certain reservoir depends on air temperature. The table below shows the number of acre-feet (af) of water per day that evaporate from the reservoir for various temperatures in degrees Fahrenheit.

Temperature, °F	af
40	800
60	1640
70	2060
85	2690

a. Find a linear model for the number of acre-feet of water that evaporate as a function of temperature.

b. Write a sentence that explains the meaning of the slope of this line in the context of this problem.

c. Assuming that water continues to evaporate at the same rate, how many acre-feet of water will evaporate per day when the temperature is 75°F?

61. CYCLING SPEEDS Michelle and Amanda start from the same place on a cycling course. Michelle is riding at 15 miles per hour and Amanda is cycling at 12 miles per hour. The graphs below show the total distance traveled by each cyclist and the total distance between Michelle and Amanda after t hours. Which graphs represent which distances?

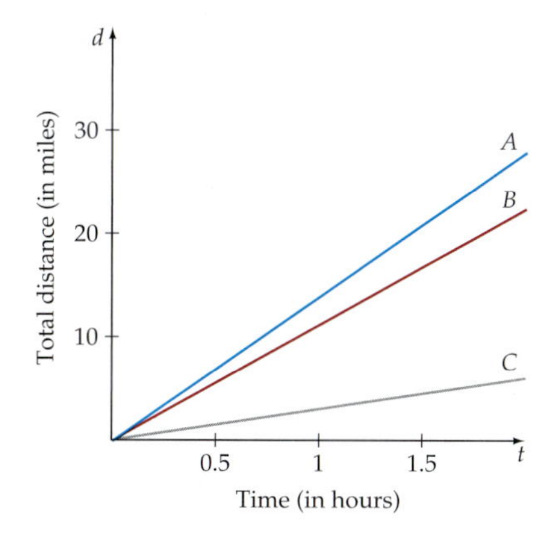

62. TEMPERATURE The graph below shows the temperature changes, in degrees Fahrenheit, over a 12-hour period at a weather station.

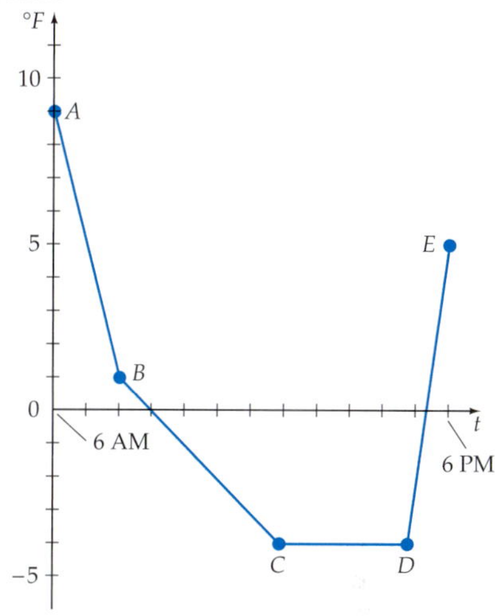

a. How many degrees per hour did the temperature change between A and B?

b. Between which two points did the temperature change most rapidly?

c. Between which two points was the temperature constant?

63. HEALTH Framingham Heart Study is an ongoing research project that is attempting to identify risk factors associated with heart disease. Selected blood pressure data from that study are shown in the table at the right.

Selected Framingham Blood Pressure Statistics

Diastolic	Systolic
100	135
88	154
80	110
70	110
80	114
108	180
85	135
75	115

a. Find the equation of a linear model of these data, given that the graph of the line passes through $P_1(70, 110)$ and $P_2(108, 180)$.

b. What systolic blood pressure does the model you found in part **a.** predict for a diastolic pressure of 90?

64. HEALTH The table on the following page shows average remaining lifetime, by age, of all people in the United States in 1997. (*Source:* National Institutes of Health)

Average Remaining Lifetime by Age in the United States

Current Age	Remaining Years
0	76.5
15	62.3
35	43.4
65	17.7
75	11.2

a. Find the equation of a linear model of these data, given that the graph of the line passes through (0, 76.5) and (75, 11.2).

b. Based on your model, what is the average remaining lifetime of a person whose current age is 25?

In Exercises 65 to 68, determine the profit function for the given revenue function and cost function. Also determine the break-even point.

65. $R(x) = 92.50x$; $C(x) = 52x + 1782$

66. $R(x) = 124x$; $C(x) = 78.5x + 5005$

67. $R(x) = 259x$; $C(x) = 180x + 10{,}270$

68. $R(x) = 14{,}220x$; $C(x) = 8010x + 1{,}602{,}180$

69. Marginal Cost In business, *marginal cost* is a phrase used to represent the rate of change or slope of a cost function that relates the cost C to the number of units x produced. If a cost function is given by $C(x) = 8x + 275$, find

a. $C(0)$ **b.** $C(1)$ **c.** $C(10)$ **d.** marginal cost

70. Marginal Revenue In business, *marginal revenue* is a phrase used to represent the rate of change or slope of a revenue function that relates the revenue R to the number of units x sold. If a revenue function is given by the function $R(x) = 210x$, find

a. $R(0)$ **b.** $R(1)$ **c.** $R(10)$ **d.** marginal revenue

71. Break-Even Point for a Rental Truck A rental company purchases a truck for \$19,500. The truck requires an average of \$6.75 per day for maintenance.

a. Find the linear function that expresses the total cost C of owning the truck after t days.

b. The truck rents for \$55.00 a day. Find the linear function that expresses the revenue R when the truck has been rented for t days.

c. The profit after t days, $P(t)$, is given by the function $P(t) = R(t) - C(t)$. Find the linear function $P(t)$.

d. Use the function $P(t)$ that you obtained in **c.** to determine how many days it will take the company to break even on the purchase of the truck. Assume that the truck is always in use.

72. Break-Even Point for a Publisher A magazine company had a profit of \$98,000 per year when it had 32,000 subscribers. When it obtained 35,000 subscribers, it had a profit of \$117,500. Assume that the profit P is a linear function of the number of subscribers s.

a. Find the function P.

b. What will the profit be if the company obtains 50,000 subscribers?

c. What is the number of subscribers needed to break even?

In Exercises 73 to 76, find the equation of the indicated line. Write the equation in the form $y = mx + b$.

73. Through (1, 3) and parallel to $3x + 4y = -24$

74. Through $(2, -1)$ and parallel to $x + y = 10$

75. Through (1, 2) and perpendicular to $x + y = 4$

76. Through $(-3, 4)$ and perpendicular to $2x - y = 7$

77. Point of Impact A rock attached to a string is whirled horizontally, in a counterclockwise circular path with radius 5 feet, about the origin. When the string breaks, the rock travels on a linear path perpendicular to the radius $\overline{OP}$ and hits a wall located at $y = 10$ feet.

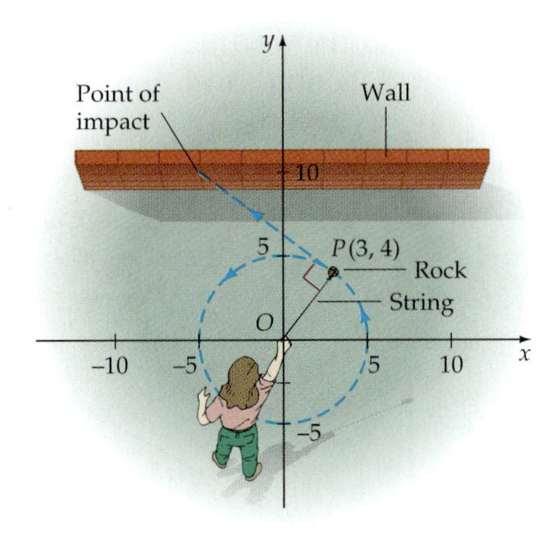

If the string breaks when the rock is at P(3 feet, 4 feet), find the x-coordinate of the point at which the rock hits the wall.

78. Point of Impact A rock attached to a string is whirled horizontally, in a counterclockwise circular path with radius 4 feet, about the origin. When the string breaks, the rock travels on a linear path perpendicular to the radius $\overline{OP}$ and hits a wall located at $y = 14$ feet. If the string breaks when the rock is at $P(\sqrt{15}$ feet, 1 foot$)$, find the x-coordinate of the point at which the rock hits the wall.

79. Slope of a Secant Line The graph of $y = x^2 + 1$ is shown below with $P(2, 5)$ and $Q(2 + h, [2 + h]^2 + 1)$ points on the graph.

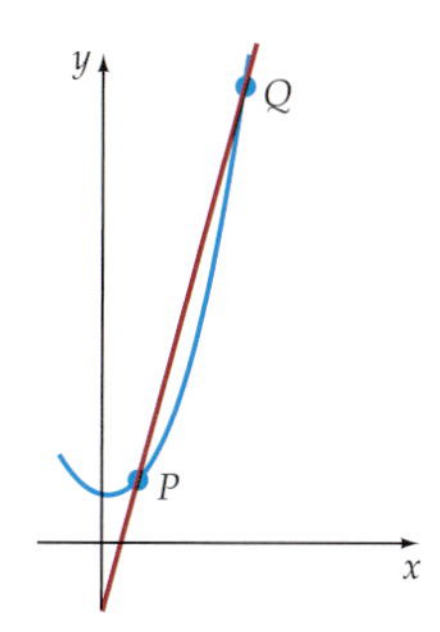

a. If $h = 1$, determine the coordinates of Q and the slope of the line PQ.

b. If $h = 0.1$, determine the coordinates of Q and the slope of the line PQ.

c. If $h = 0.01$, determine the coordinates of Q and the slope of the line PQ.

d. As h approaches 0, what value does the slope of the line PQ seem to be approaching?

e. Verify that the slope of the line passing through $(2, 5)$ and $(2 + h, [2 + h]^2 + 1)$ is $4 + h$.

80. Slope of a Secant Line The graph of $y = 3x^2$ is shown below with $P(-1, 3)$ and $Q(-1 + h, 3[-1 + h]^2)$ points on the graph.

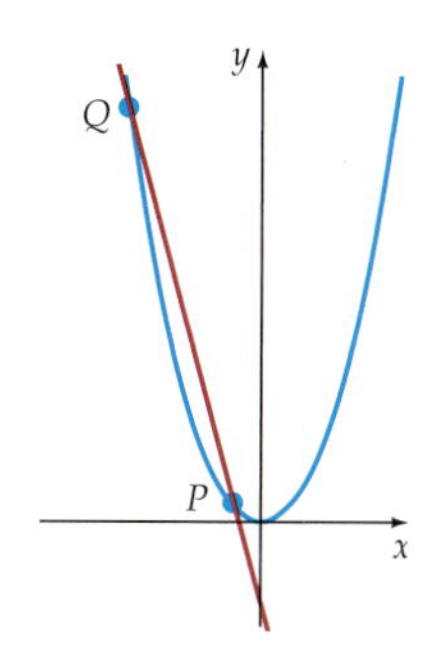

a. If $h = 1$, determine the coordinates of Q and the slope of the line PQ.

b. If $h = 0.1$, determine the coordinates of Q and the slope of the line PQ.

c. If $h = 0.01$, determine the coordinates of Q and the slope of the line PQ.

d. As h approaches 0, what value does the slope of the line PQ seem to be approaching?

e. Verify that the slope of the line passing through $(-1, 3)$ and $(-1 + h, 3[-1 + h]^2)$ is $-6 + 3h$.

81. Verify that the slope of the line passing through (x, x^2) and $(x + h, [x + h]^2)$ is $2x + h$.

82. Verify that the slope of the line passing through $(x, 4x^2)$ and $(x + h, 4[x + h]^2)$ is $8x + 4h$.

CONNECTING CONCEPTS

83. The Two-Point Form Use the point-slope form to derive the following equation, which is called the two-point form.

$$y - y_1 = \left(\frac{y_2 - y_1}{x_2 - x_1}\right)(x - x_1)$$

84. The Intercept Form Use the two-point form from Exercise 83 to show that the line with intercepts $(a, 0)$ and $(0, b)$, $a \neq 0$ and $b \neq 0$, has the equation

$$\frac{x}{a} + \frac{y}{b} = 1$$

In Exercises 85 and 86, use the two-point form to find an equation of the line that passes through the indicated points. Write your answers in slope-intercept form.

85. $(5, 1), (4, 3)$

86. $(2, 7), (-1, 6)$

In Exercises 87 to 90, use the equation from Exercise 84 (called the intercept form) to write an equation of the line with the indicated intercepts.

87. x-intercept $(3, 0)$, y-intercept $(0, 5)$

88. x-intercept $(-2, 0)$, y-intercept $(0, 7)$

89. x-intercept $(a, 0)$, y-intercept $(0, 3a)$, point on the line $(5, 2)$, $a \neq 0$

90. x-intercept $(-b, 0)$, y-intercept $(0, 2b)$, point on the line $(-3, 10)$, $b \neq 0$

91. Verify that the slope of the line passing through $(1, 3)$ and $(1 + h, 3[1 + h]^3)$ is $9 + 9h + 3h^2$.

92. Find the two points on the circle given by $x^2 + y^2 = 25$ such that the slope of the radius from $(0, 0)$ to each point is 0.5.

93. Find a point $P(x, y)$ on the graph of the equation $y = x^2$ such that the slope of the line through the point $(3, 9)$ and P is $\frac{15}{2}$.

94. Determine whether there is a point $P(x, y)$ on the graph of the equation $y = \sqrt{x + 1}$ such that the slope of the line through the point $(3, 2)$ and P is $\frac{3}{8}$.

PREPARE FOR SECTION 1.5

95. Factor: $3x^2 + 10x - 8$ [A.3]

96. Complete the square of $x^2 - 8x$. Write the resulting trinomial as the square of a binomial. [1.1]

97. Find $f(-3)$ for $f(x) = 2x^2 - 5x - 7$. [1.3]

In Exercises 98 and 99, solve for x.

98. $2x^2 - x = 1$ [1.1]

99. $x^2 + 3x - 2 = 0$ [1.1]

100. Suppose that $h = -16t^2 + 64t + 5$. Find two values of t for which $h = 53$. [1.1]

PROJECTS

1. Visual Insight

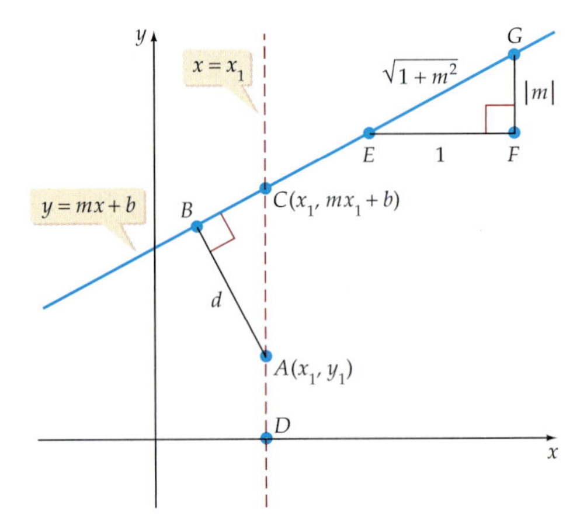

The distance d between the point $A(x_1, y_1)$ and the line given by $y = mx + b$ is $d = \frac{|mx_1 + b - y_1|}{\sqrt{1 + m^2}}$

Write a paragraph that explains how to make use of the figure above to verify the formula for the distance d.

2. Verify Geometric Theorems

a. Prove that in any triangle, the line segment that joins the midpoints of two sides of the triangle is parallel to the third side. (*Hint:* Assign coordinates to the vertices of the triangle as shown in the figure at the left below.)

b. Prove that in any square, the diagonals are perpendicular bisectors of each other. (*Hint:* Assign coordinates to the vertices of the square as shown in the figure at the right below.)

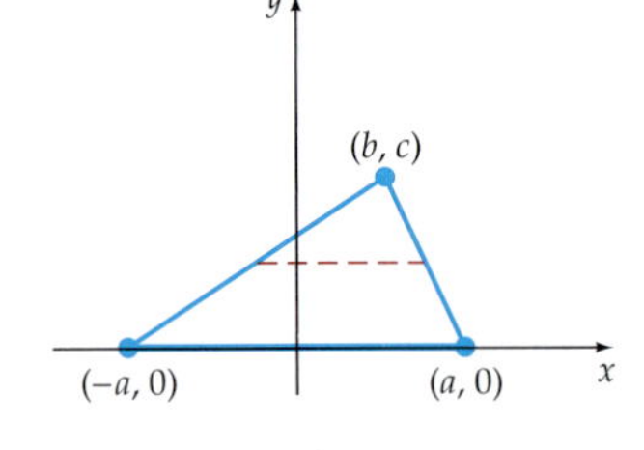

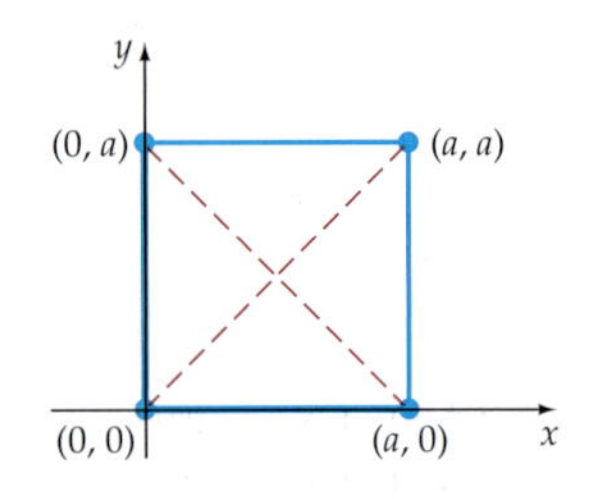

SECTION 1.5

QUADRATIC FUNCTIONS

- VERTEX OF A PARABOLA
- MAXIMUM AND MINIMUM OF A QUADRATIC FUNCTION
- APPLICATIONS

Some applications can be modeled by a *quadratic function.*

Definition of a Quadratic Function

A **quadratic function** of x is a function that can be represented by an equation of the form

$$f(x) = ax^2 + bx + c$$

where a, b, and c are real numbers and $a \neq 0$.

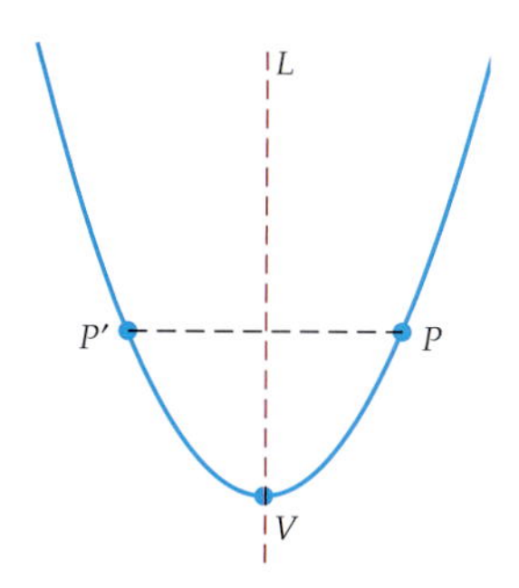

FIGURE 1.62

The graph of $f(x) = ax^2 + bx + c$ is a *parabola*. The graph opens up if $a > 0$, and it opens down if $a < 0$. The **vertex of a parabola** is the lowest point on a parabola that opens up or the highest point on a parabola that opens down. Point V is the vertex of the parabola in **Figure 1.62.**

The graph of $f(x) = ax^2 + bx + c$ is *symmetric* with respect to a vertical line through its vertex.

Definition of Symmetry with Respect to a Line

A graph is **symmetric with respect to a line** L if for each point P on the graph there is a point P' on the graph such that the line L is the perpendicular bisector of the line segment PP'.

take note

The axis of symmetry is a line. When asked to determine the axis of symmetry, the answer is an equation, not just a number.

In **Figure 1.62,** the parabola is symmetric with respect to the line L. The line L is called the **axis of symmetry.** The points P and P' are reflections or images of each other with respect to the axis of symmetry.

If $b = 0$ and $c = 0$, then $f(x) = ax^2 + bx + c$ simplifies to $f(x) = ax^2$. The graph of $f(x) = ax^2$ $(a \neq 0)$ is a parabola with its vertex at the origin, and the y-axis is its axis of symmetry. The graph of $f(x) = ax^2$ can be constructed by plotting a few points and drawing a smooth curve that passes through these points, with the origin as the vertex and the y-axis as its axis of symmetry. The graphs of $f(x) = x^2$, $g(x) = 2x^2$, and $h(x) = -\frac{1}{2}x^2$ are shown in **Figure 1.63.**

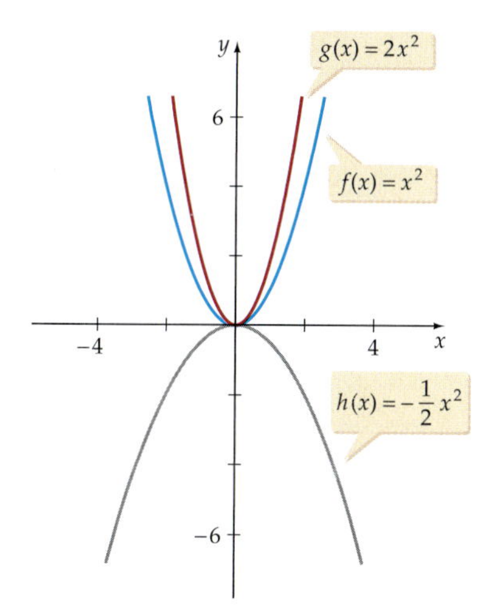

FIGURE 1.63

take note

The equation $z = x^2 - y^2$ defines z as a quadratic function of x and y. You might think that the graph of every quadratic function is a parabola. However, the graph of $z = x^2 - y^2$ is the *saddle* shown in the figure below. You will study quadratic functions involving two or more independent variables in calculus.

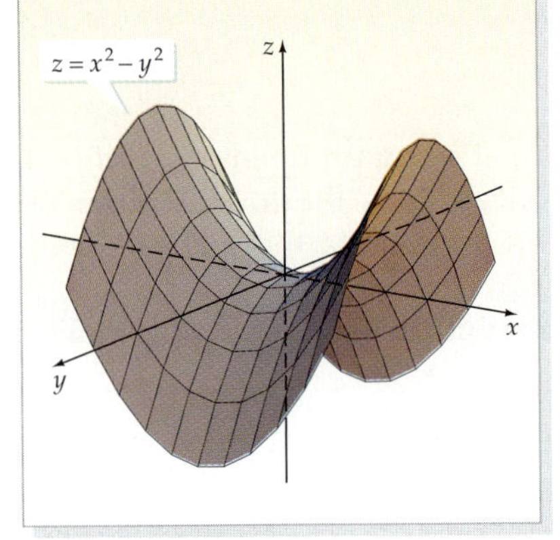

Standard Form of Quadratic Functions

Every quadratic function f given by $f(x) = ax^2 + bx + c$ can be written in the **standard form of a quadratic function:**

$$f(x) = a(x - h)^2 + k, \quad a \neq 0$$

The graph of f is a parabola with vertex (h, k). The parabola opens up if $a > 0$, and it opens down if $a < 0$. The vertical line $x = h$ is the axis of symmetry of the parabola.

The standard form is useful because it readily gives information about the vertex of the parabola and its axis of symmetry. For example, note that the graph of $f(x) = 2(x - 4)^2 - 3$ is a parabola. The coordinates of the vertex are $(4, -3)$, and the line $x = 4$ is its axis of symmetry. Because a is the positive number 2, the parabola opens upward.

EXAMPLE 1 Find the Standard Form of a Quadratic Function

Use the technique of completing the square to find the standard form of $g(x) = 2x^2 - 12x + 19$. Sketch the graph.

Solution

$$\begin{aligned}
g(x) &= 2x^2 - 12x + 19 \\
&= 2(x^2 - 6x) + 19 && \text{• Factor 2 from the variable terms.} \\
&= 2(x^2 - 6x + 9 - 9) + 19 && \text{• Complete the square.} \\
&= 2(x^2 - 6x + 9) - 2(9) + 19 && \text{• Regroup.} \\
&= 2(x - 3)^2 - 18 + 19 && \text{• Factor and simplify.} \\
&= 2(x - 3)^2 + 1 && \text{• Standard form}
\end{aligned}$$

The vertex is $(3, 1)$. The axis of symmetry is $x = 3$. Because $a > 0$, the parabola opens up. See **Figure 1.64.**

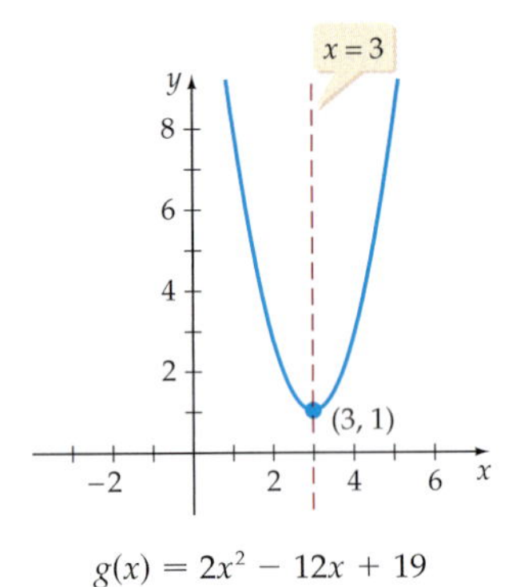

$g(x) = 2x^2 - 12x + 19$

FIGURE 1.64

▶ **TRY EXERCISE 10, PAGE 75**

• VERTEX OF A PARABOLA

We can write $f(x) = ax^2 + bx + c$ in standard form by completing the square of $ax^2 + bx + c$. This will allow us to derive a general expression for the x- and y-coordinates of the graph of $f(x) = ax^2 + bx + c$.

$$\begin{aligned}
f(x) &= ax^2 + bx + c \\
&= a\left(x^2 + \frac{b}{a}x\right) + c && \text{• Factor } a \text{ from } ax^2 + bx. \\
&= a\left(x^2 + \frac{b}{a}x + \frac{b^2}{4a^2}\right) + c - \frac{b^2}{4a} && \text{• Complete the square by adding and subtracting } \left(\frac{1}{2}\cdot\frac{b}{a}\right)^2 = \frac{b^2}{4a^2}. \\
&= a\left(x + \frac{b}{2a}\right)^2 + \frac{4ac - b^2}{4a} && \text{• Factor and simplify.}
\end{aligned}$$

Thus $f(x) = ax^2 + bx + c$ in standard form is $f(x) = a\left(x + \frac{b}{2a}\right)^2 + \frac{4ac - b^2}{4a}$. Comparing this last expression with $f(x) = a(x - h)^2 + k$, we see that the coordinates of the vertex are $\left(-\frac{b}{2a}, \frac{4ac - b^2}{4a}\right)$.

Note that by evaluating $f(x) = a\left(x + \frac{b}{2a}\right)^2 + \frac{4ac - b^2}{4a}$ at $x = -\frac{b}{2a}$, we have

$$\begin{aligned} f(x) &= a\left(x + \frac{b}{2a}\right)^2 + \frac{4ac - b^2}{4a} \\ f\left(-\frac{b}{2a}\right) &= a\left(-\frac{b}{2a} + \frac{b}{2a}\right)^2 + \frac{4ac - b^2}{4a} = a(0) + \frac{4ac - b^2}{4a} \\ &= \frac{4ac - b^2}{4a} \end{aligned}$$

That is, the y-coordinate of the vertex is $f\left(-\frac{b}{2a}\right)$. This is summarized by the following formula.

Vertex Formula

The coordinates of the vertex of $f(x) = ax^2 + bx + c$ are $\left(-\frac{b}{2a}, f\left(-\frac{b}{2a}\right)\right)$.

The vertex formula can be used to write the standard form of the equation of a parabola. We have

$$h = -\frac{b}{2a} \quad \text{and} \quad k = f\left(-\frac{b}{2a}\right)$$

EXAMPLE 2 **Find the Vertex and Standard Form of a Quadratic Function**

Use the vertex formula to find the vertex and standard form of $f(x) = 2x^2 - 8x + 3$. See **Figure 1.65.**

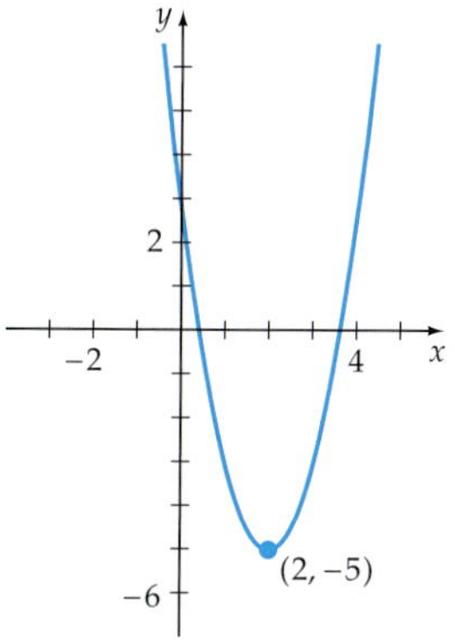

Solution

$f(x) = 2x^2 - 8x + 3$ • $a = 2, b = -8, c = 3$

$h = -\frac{b}{2a} = -\frac{-8}{2(2)} = 2$ • x-coordinate of the vertex

$k = f\left(-\frac{b}{2a}\right) = 2(2)^2 - 8(2) + 3 = -5$ • y-coordinate of the vertex

The vertex is $(2, -5)$. Substituting into the standard form equation $f(x) = a(x - h)^2 + k$ yields the standard form $f(x) = 2(x - 2)^2 - 5$.

TRY EXERCISE 20, PAGE 76

$f(x) = 2x^2 - 8x + 3$

FIGURE 1.65

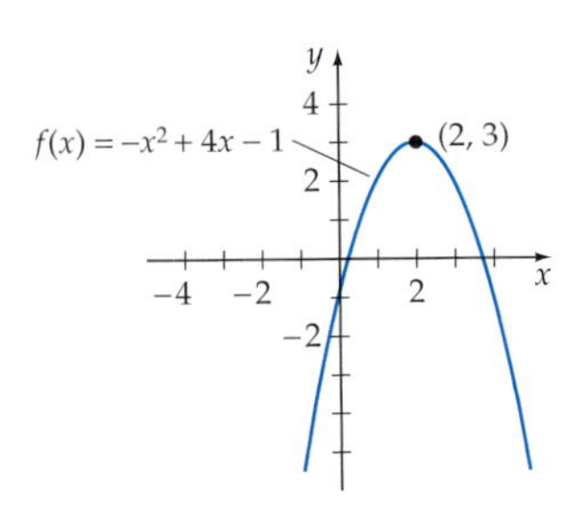

Maximum and Minimum of a Quadratic Function

Note from Example 2 that the graph of the parabola opens up, and the vertex is the *lowest* point on the graph of the parabola. Therefore, the y-coordinate of the vertex is the *minimum* value of that function. This information can be used to determine the range of $f(x) = 2x^2 - 8x + 3$. The range is $\{y \mid y \geq -5\}$. Similarly, if the graph of a parabola opened down, the vertex would be the *highest* point on the graph, and the y-coordinate of the vertex would be the *maximum* value of the function. For instance, the maximum value of $f(x) = -x^2 + 4x - 1$, graphed at the left, is 3, the y-coordinate of the vertex. The range of the function is $\{y \mid y \leq 3\}$. For the function in Example 2 and the function whose graph is shown at the left, the domain is the set of real numbers.

EXAMPLE 3 Find the Range of $f(x) = ax^2 + bx + c$

Find the range of $f(x) = -2x^2 - 6x - 1$. Determine the values of x for which $f(x) = 3$.

Algebraic Solution

To find the range of f, determine the y-coordinate of the vertex of the graph of f.

$f(x) = -2x^2 - 6x - 1$ • **$a = -2$, $b = -6$, $c = -1$**

$$h = -\frac{b}{2a} = -\frac{-6}{2(-2)} = -\frac{3}{2}$$ • **Find the x-coordinate of the vertex.**

$$k = f\left(-\frac{3}{2}\right) = -2\left(-\frac{3}{2}\right)^2 - 6\left(-\frac{3}{2}\right) - 1 = \frac{7}{2}$$ • **Find the y-coordinate of the vertex.**

The vertex is $\left(-\frac{3}{2}, \frac{7}{2}\right)$. Because the parabola opens down, $\frac{7}{2}$ is the maximum value of f. Therefore, the range of f is $\left\{y \mid y \leq \frac{7}{2}\right\}$.

To determine the values of x for which $f(x) = 3$, replace $f(x)$ by $-2x^2 - 6x - 1$ and solve for x.

$f(x) = 3$

$-2x^2 - 6x - 1 = 3$ • **Replace $f(x)$ by $-2x^2 - 6x - 1$.**

$-2x^2 - 6x - 4 = 0$ • **Solve for x.**

$-2(x + 1)(x + 2) = 0$ • **Factor.**

$x + 1 = 0$ or $x + 2 = 0$ • **Use the Principle of Zero Products to solve for x.**

$x = -1 \qquad x = -2$

The values of x for which $f(x) = 3$ are -1 and -2.

Visualize the Solution

The graph of f is shown below. The vertex of the graph is $\left(-\frac{3}{2}, \frac{7}{2}\right)$. Note that the line $y = 3$ intersects the graph of f when $x = -2$ and when $x = -1$.

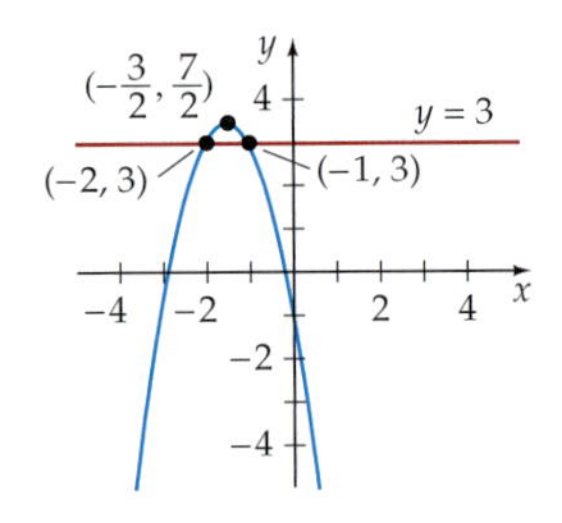

▶ Try Exercise 32, page 76

The following theorem can be used to determine the maximum value or the minimum value of a quadratic function.

Maximum or Minimum Value of a Quadratic Function

If $a > 0$, then the vertex (h, k) is the lowest point on the graph of $f(x) = a(x - h)^2 + k$, and the y-coordinate k of the vertex is the **minimum value** of the function f. See **Figure 1.66a.**

If $a < 0$, then the vertex (h, k) is the highest point on the graph of $f(x) = a(x - h)^2 + k$, and the y-coordinate k is the **maximum value** of the function f. See **Figure 1.66b.**

In either case, the maximum or minimum is achieved when $x = h$.

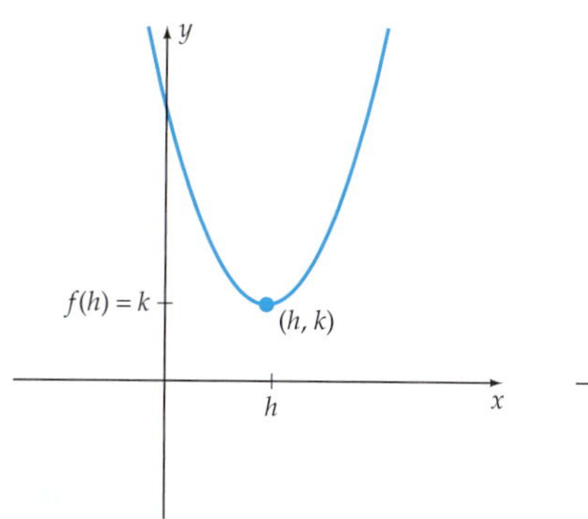

a. k is the minimum value of f.

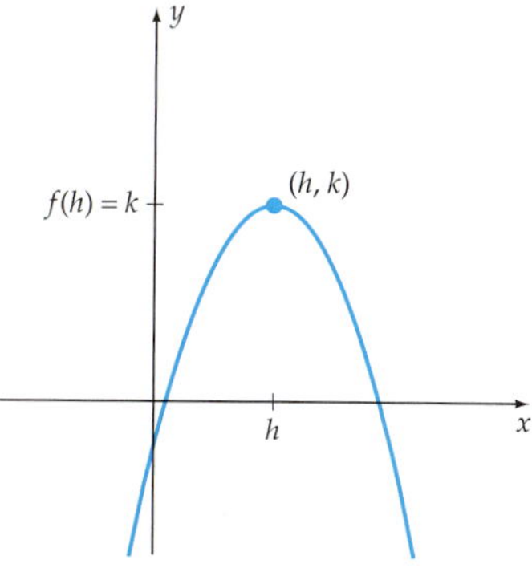

b. k is the maximum value of f.

FIGURE 1.66

EXAMPLE 4 Find the Maximum or Minimum of a Quadratic Function

Find the maximum or minimum value of each quadratic function. State whether the value is a maximum or a minimum.

a. $F(x) = -2x^2 + 8x - 1$ **b.** $G(x) = x^2 - 3x + 1$

Solution

The maximum or minimum value of a quadratic function is the y-coordinate of the vertex of the graph of the function.

a. $h = -\dfrac{b}{2a} = -\dfrac{8}{2(-2)} = 2$ • **x-coordinate of the vertex**

$k = F\left(-\dfrac{b}{2a}\right) = -2(2)^2 + 8(2) - 1 = 7$ • **y-coordinate of the vertex**

Because $a < 0$, the function has a maximum value but no minimum value. The maximum value is 7. See **Figure 1.67.**

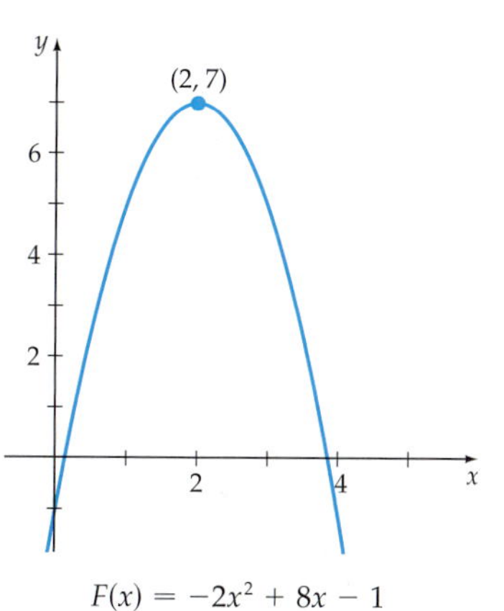

$F(x) = -2x^2 + 8x - 1$

FIGURE 1.67

Continued ▶

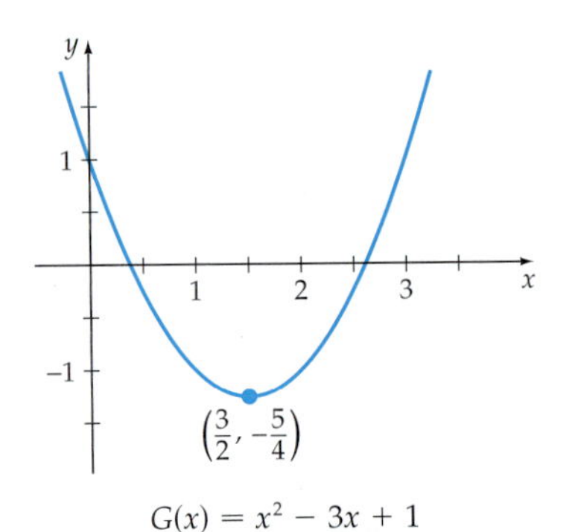

$G(x) = x^2 - 3x + 1$

FIGURE 1.68

b. $h = -\dfrac{b}{2a} = -\dfrac{-3}{2(1)} = \dfrac{3}{2}$ • **x-coordinate of the vertex**

$k = G\left(-\dfrac{b}{2a}\right) = \left(\dfrac{3}{2}\right)^2 - 3\left(\dfrac{3}{2}\right) + 1$

$= -\dfrac{5}{4}$ • **y-coordinate of the vertex**

Because $a > 0$, the function has a minimum value but no maximum value. The minimum value is $-\dfrac{5}{4}$. See **Figure 1.68.**

▶ **TRY EXERCISE 36, PAGE 76**

APPLICATIONS

EXAMPLE 5 Find the Maximum of a Quadratic Function

A long sheet of tin 20 inches wide is to be made into a trough by bending up two sides until they are perpendicular to the bottom. How many inches should be turned up so that the trough will achieve its maximum carrying capacity?

Solution

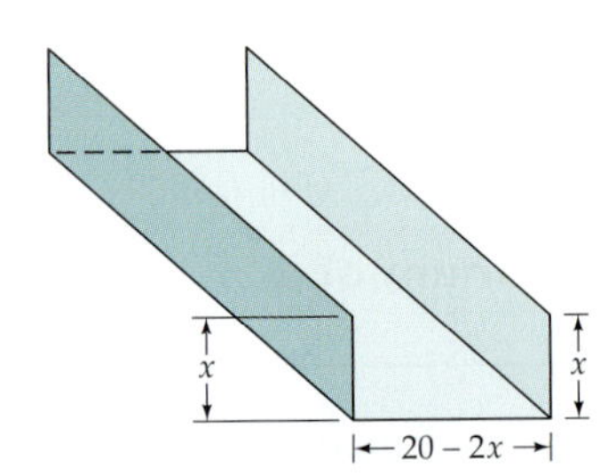

FIGURE 1.69

The trough is shown in **Figure 1.69.** If x is the number of inches to be turned up on each side, then the width of the base is $20 - 2x$ inches. The maximum carrying capacity of the trough will occur when the cross-sectional area is a maximum. The cross-sectional area $A(x)$ is given by

$$A(x) = x(20 - 2x) \quad \text{• Area = (length)(width)}$$
$$= -2x^2 + 20x$$

To find the point at which A obtains its maximum value, find the x-coordinate of the vertex of the graph of A. Using the vertex formula with $a = -2$ and $b = 20$, we have

$$x = -\frac{b}{2a} = -\frac{20}{2(-2)} = 5$$

Therefore, the maximum carrying capacity will be achieved when 5 inches are turned up. See **Figure 1.70.**

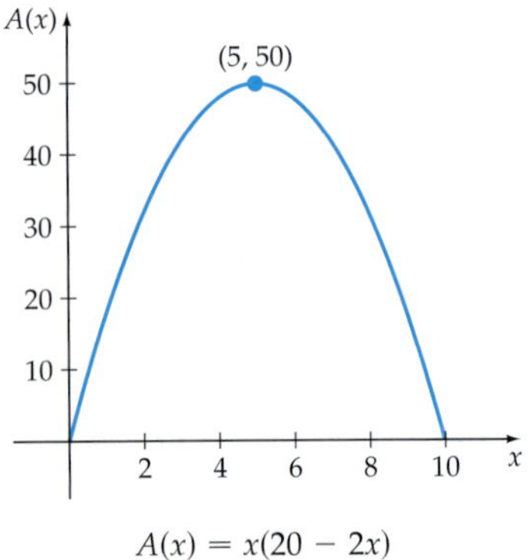

$A(x) = x(20 - 2x)$

FIGURE 1.70

▶ **TRY EXERCISE 46, PAGE 76**

EXAMPLE 6 Solve a Business Application

The owners of a travel agency have determined that they can sell all 160 tickets for a tour if they charge \$8 (their cost) for each ticket. For each \$0.25 increase in the price of a ticket, they estimate they will sell 1 ticket less. A business manager determines that their cost function is $C(x) = 8x$ and that the customer's price per ticket is

$$p(x) = 8 + 0.25(160 - x) = 48 - 0.25x$$

where x represents the number of tickets sold. Determine the maximum profit and the cost per ticket that yields the maximum profit.

Solution

The profit from selling x tickets is $P(x) = R(x) - C(x)$, where P, R, and C are the profit function, the revenue function, and the cost function as defined in Section 1.4. Thus

$$\begin{aligned} P(x) &= R(x) - C(x) \\ &= x[p(x)] - C(x) \\ &= x(48 - 0.25x) - 8x \\ &= 40x - 0.25x^2 \end{aligned}$$

The graph of the profit function is a parabola that opens down. Thus the maximum profit occurs when

$$x = -\frac{b}{2a} = -\frac{40}{2(-0.25)} = 80$$

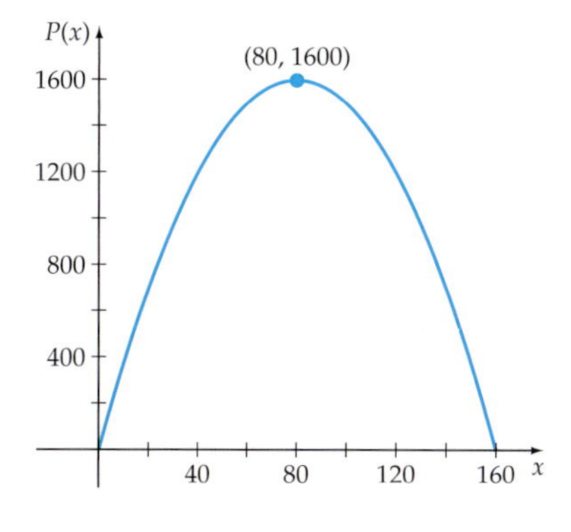

$P(x) = 40x - 0.25x^2$

FIGURE 1.71

The maximum profit is determined by evaluating $P(x)$ at $x = 80$.

$$P(80) = 40(80) - 0.25(80)^2 = 1600$$

The maximum profit is \$1600.

To find the price per ticket that yields the maximum profit, we evaluate $p(x)$ at $x = 80$.

$$p(80) = 48 - 0.25(80) = 28$$

Thus the travel agency can expect a maximum profit of \$1600 when 80 people take the tour at a ticket price of \$28 per person. The graph of the profit function is shown in **Figure 1.71.**

QUESTION In **Figure 1.71,** why have we shown only the portion of the graph that lies in quadrant I?

TRY EXERCISE 68, PAGE 78

ANSWER Since x represents the number of tickets sold, x must be greater than or equal to zero but less than or equal to 160. $P(x)$ is nonnegative for $0 \leq x \leq 160$.

EXAMPLE 7 Solve a Projectile Application

In **Figure 1.72,** a ball is thrown vertically upward with an initial velocity of 48 feet per second. If the ball started its flight at a height of 8 feet, then its height s at time t can be determined by $s(t) = -16t^2 + 48t + 8$, where $s(t)$ is measured in feet above ground level and t is the number of seconds of flight.

a. Determine the time it takes the ball to attain its maximum height.

b. Determine the maximum height the ball attains.

c. Determine the time it takes the ball to hit the ground.

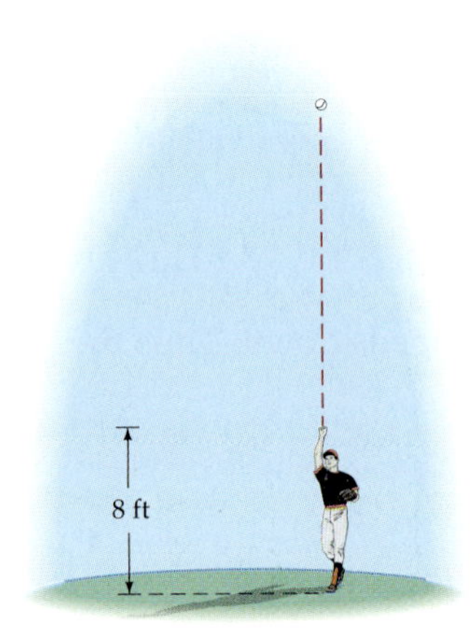

FIGURE 1.72

Solution

a. The graph of $s(t) = -16t^2 + 48t + 8$ is a parabola that opens downward. See **Figure 1.73.** Therefore, s will attain its maximum value at the vertex of its graph. Using the vertex formula with $a = -16$ and $b = 48$, we get

$$t = -\frac{b}{2a} = -\frac{48}{2(-16)} = \frac{3}{2}$$

Therefore, the ball attains its maximum height $1\frac{1}{2}$ seconds into its flight.

b. When $t = \frac{3}{2}$, the height of the ball is

$$s\left(\frac{3}{2}\right) = -16\left(\frac{3}{2}\right)^2 + 48\left(\frac{3}{2}\right) + 8 = 44 \text{ feet}$$

c. The ball will hit the ground when its height $s(t) = 0$. Therefore, solve $-16t^2 + 48t + 8 = 0$ for t.

$$-16t^2 + 48t + 8 = 0$$

$$-2t^2 + 6t + 1 = 0 \qquad \text{• Divide each side by 8.}$$

$$t = \frac{-(6) \pm \sqrt{6^2 - 4(-2)(1)}}{2(-2)} \qquad \text{• Use the quadratic formula.}$$

$$= \frac{-6 \pm \sqrt{44}}{-4} = \frac{-3 \pm \sqrt{11}}{-2}$$

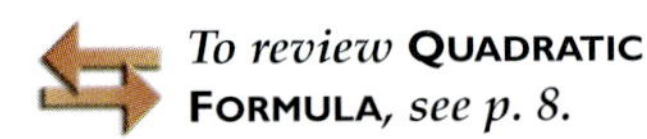
To review QUADRATIC FORMULA, *see p. 8.*

Using a calculator to approximate the positive root, we find that the ball will hit the ground in $t \approx 3.16$ seconds. This is also the value of the t-coordinate of the t-intercept in **Figure 1.73.**

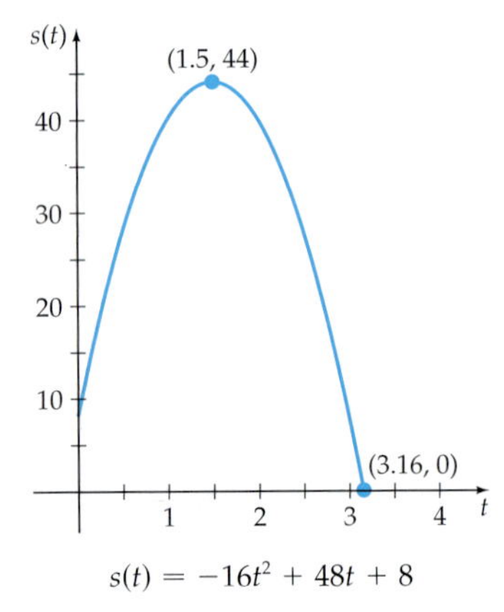

$s(t) = -16t^2 + 48t + 8$

FIGURE 1.73

TRY EXERCISE 70, PAGE 78

TOPICS FOR DISCUSSION

1. Does the graph of every quadratic function of the form

$$f(x) = ax^2 + bx + c$$

have a y-intercept? If so, what are the coordinates of the y-intercept?

2. The graph of $f(x) = -x^2 + 6x + 11$ has a vertex of $(3, 20)$. Is this vertex point the highest point or the lowest point on the graph of f?

3. A tutor states that the graph of $f(x) = ax^2 + bx + c$ $(a \neq 0)$ is a parabola and that its axis of symmetry is $y = -\dfrac{b}{2a}$. Do you agree?

4. Every quadratic function of the form $f(x) = ax^2 + bx + c$ has a domain of all real numbers. Do you agree?

5. A classmate states that the graph of every quadratic function of the form

$$f(x) = ax^2 + bx + c$$

must contain points from at least two quadrants. Do you agree?

Exercise Set 1.5

In Exercises 1 to 8, match each graph in *a.* through *h.* with the proper quadratic function.

1. $f(x) = x^2 - 3$
2. $f(x) = x^2 + 2$
3. $f(x) = (x - 4)^2$
4. $f(x) = (x + 3)^2$
5. $f(x) = -2x^2 + 2$
6. $f(x) = -\dfrac{1}{2}x^2 + 3$
7. $f(x) = (x + 1)^2 + 3$
8. $f(x) = -2(x - 2)^2 + 2$

a.
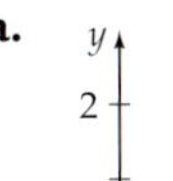
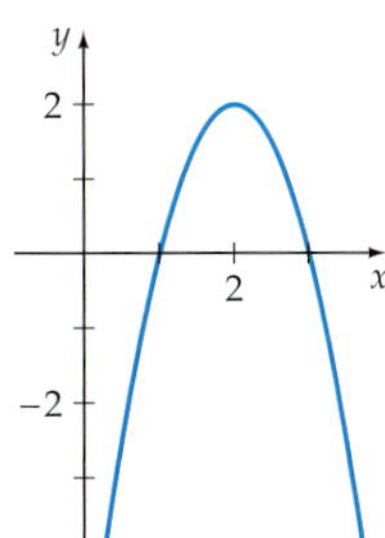

b.

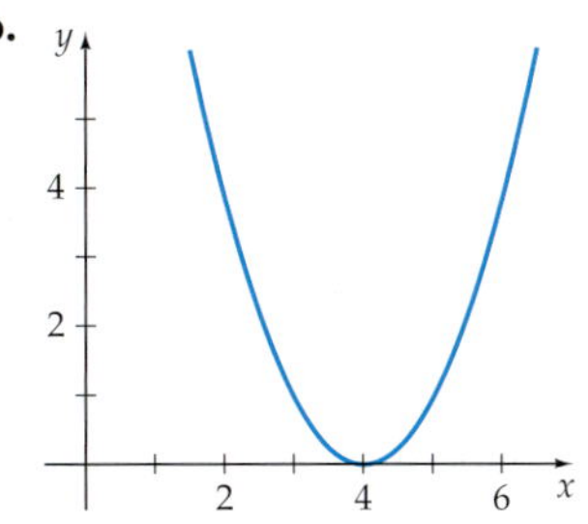

c.

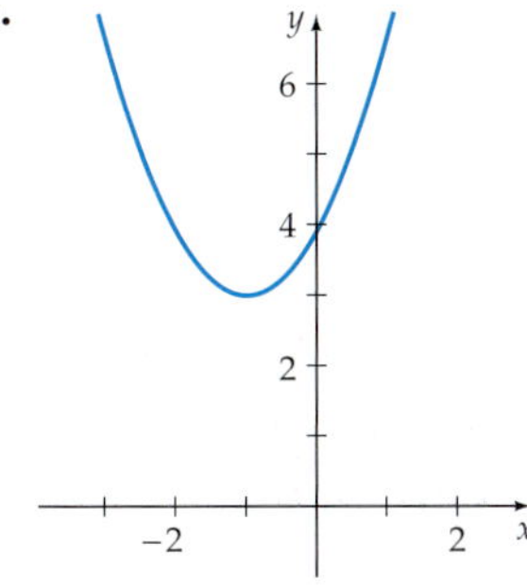

d.
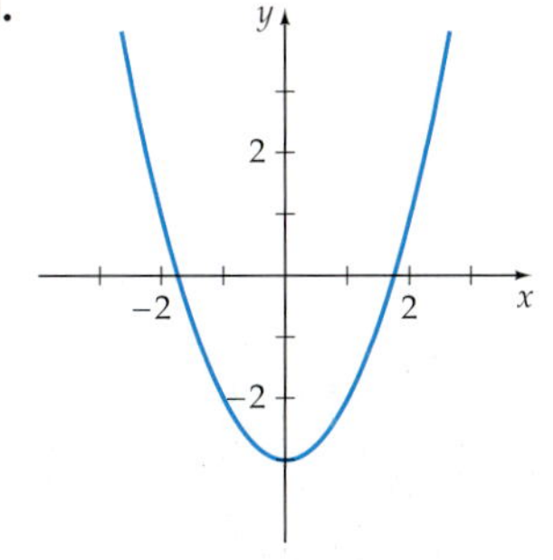

e.
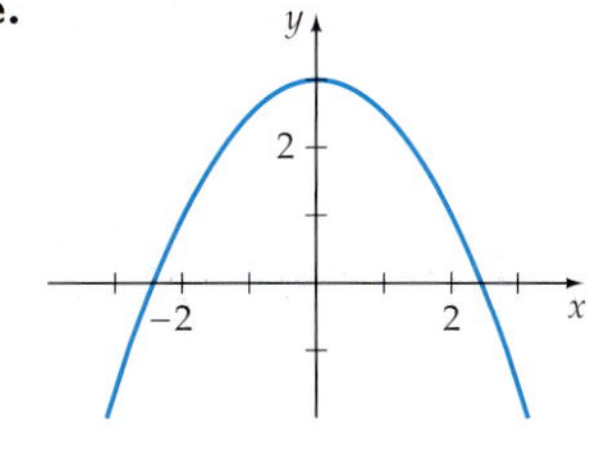

f.
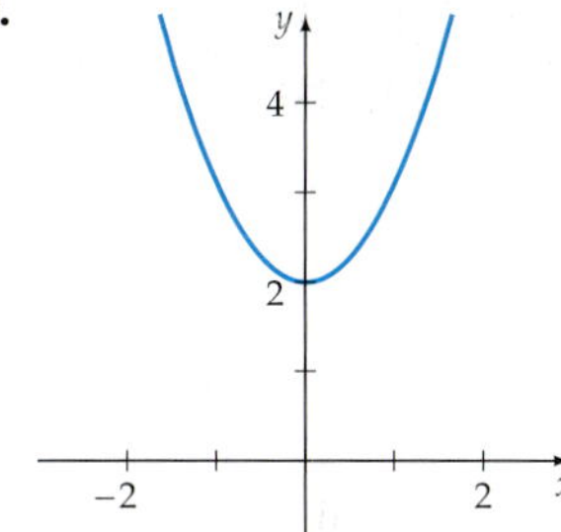

g.
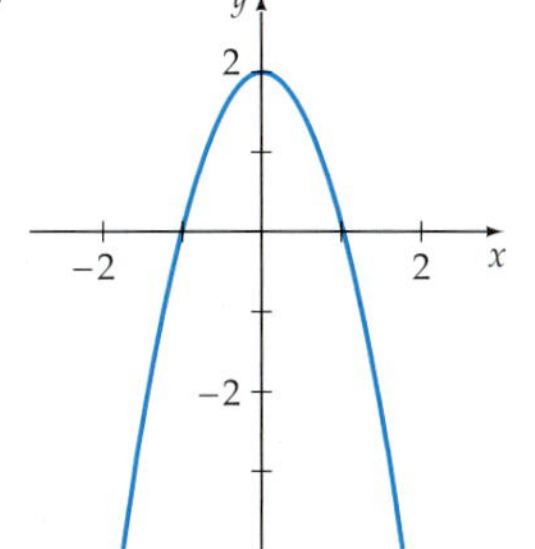

h.
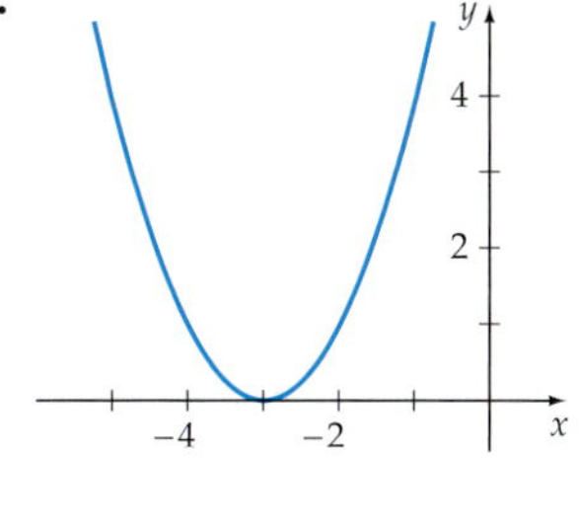

In Exercises 9 to 18, use the method of completing the square to find the standard form of the quadratic function, and then sketch its graph. Label its vertex and axis of symmetry.

9. $f(x) = x^2 + 4x + 1$
▶ **10.** $f(x) = x^2 + 6x - 1$
11. $f(x) = x^2 - 8x + 5$
12. $f(x) = x^2 - 10x + 3$
13. $f(x) = x^2 + 3x + 1$
14. $f(x) = x^2 + 7x + 2$

15. $f(x) = -x^2 + 4x + 2$

16. $f(x) = -x^2 - 2x + 5$

17. $f(x) = -3x^2 + 3x + 7$

18. $f(x) = -2x^2 - 4x + 5$

In Exercises 19 to 28, use the vertex formula to determine the vertex of the graph of the function and write the function in standard form.

19. $f(x) = x^2 - 10x$

▶ **20.** $f(x) = x^2 - 6x$

21. $f(x) = x^2 - 10$

22. $f(x) = x^2 - 4$

23. $f(x) = -x^2 + 6x + 1$

24. $f(x) = -x^2 + 4x + 1$

25. $f(x) = 2x^2 - 3x + 7$

26. $f(x) = 3x^2 - 10x + 2$

27. $f(x) = -4x^2 + x + 1$

28. $f(x) = -5x^2 - 6x + 3$

29. Find the range of $f(x) = x^2 - 2x - 1$. Determine the values of x in the domain of f for which $f(x) = 2$.

30. Find the range of $f(x) = -x^2 - 6x - 2$. Determine the values of x in the domain of f for which $f(x) = 3$.

31. Find the range of $f(x) = -2x^2 + 5x - 1$. Determine the values of x in the domain of f for which $f(x) = 2$.

▶ **32.** Find the range of $f(x) = 2x^2 + 6x - 5$. Determine the values of x in the domain of f for which $f(x) = 15$.

33. Is 3 in the range of $f(x) = x^2 + 3x + 6$? Explain your answer.

34. Is -2 in the range of $f(x) = -2x^2 - x + 1$? Explain your answer.

In Exercises 35 to 44, find the maximum or minimum value of the function. State whether this value is a maximum or a minimum.

35. $f(x) = x^2 + 8x$

▶ **36.** $f(x) = -x^2 - 6x$

37. $f(x) = -x^2 + 6x + 2$

38. $f(x) = -x^2 + 10x - 3$

39. $f(x) = 2x^2 + 3x + 1$

40. $f(x) = 3x^2 + x - 1$

41. $f(x) = 5x^2 - 11$

42. $f(x) = 3x^2 - 41$

43. $f(x) = -\frac{1}{2}x^2 + 6x + 17$

44. $f(x) = -\frac{3}{4}x^2 - \frac{2}{5}x + 7$

45. HEIGHT OF AN ARCH The height of an arch is given by the equation

$$h(x) = -\frac{3}{64}x^2 + 27, \quad -24 \le x \le 24$$

where $|x|$ is the horizontal distance in feet from the center of the arch.

a. What is the maximum height of the arch?

b. What is the height of the arch 10 feet to the right of center?

c. How far from the center is the arch 8 feet tall?

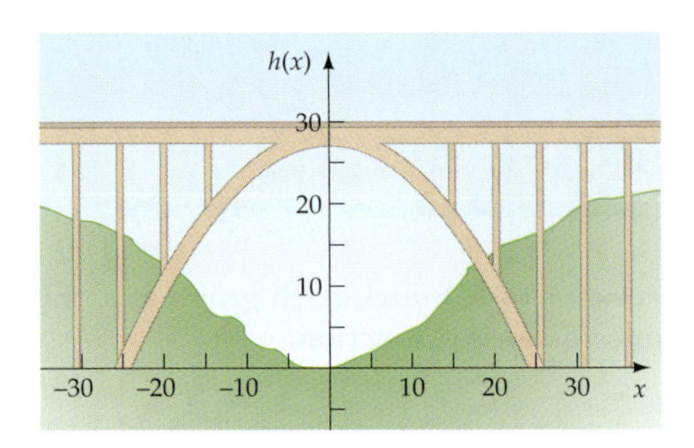

▶ **46.** The sum of the length l and the width w of a rectangular area is 240 meters.

a. Write w as a function of l.

b. Write the area A as a function of l.

c. Find the dimensions that produce the greatest area.

47. RECTANGULAR ENCLOSURE A veterinarian uses 600 feet of chain-link fencing to enclose a rectangular region and also to subdivide the region into two smaller rectangular regions by placing a fence parallel to one of the sides, as shown in the figure.

a. Write the width w as a function of the length l.

b. Write the total area A as a function of l.

c. Find the dimensions that produce the greatest enclosed area.

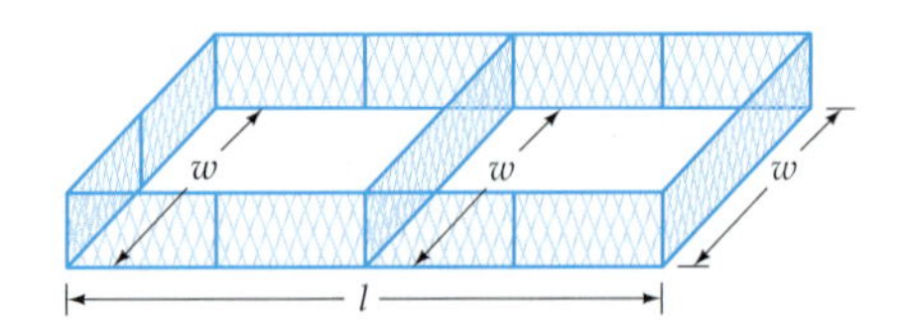

48. RECTANGULAR ENCLOSURE A farmer uses 1200 feet of fence to enclose a rectangular region and also to subdivide the region into three smaller rectangular regions by placing the fences parallel to one of the sides. Find the dimensions that produce the greatest enclosed area.

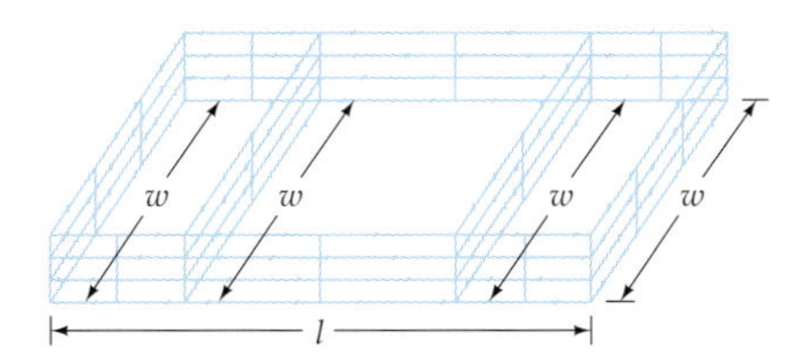

49. TEMPERATURE FLUCTUATIONS The temperature $T(t)$, in degrees Fahrenheit, during the day can be modeled by the equation $T(t) = -0.7t^2 + 9.4t + 59.3$, where t is the number of hours after 6:00 A.M.

a. At what time is the temperature a maximum? Round to the nearest minute.

b. What is the maximum temperature? Round to the nearest degree.

50. LARVAE SURVIVAL Soon after insect larvae are hatched, they must begin to search for food. The survival rate of the larvae depends on many factors, but the temperature of the environment is one of the most important. For a certain species of insect, a model of the number of larvae, $N(T)$, that survive this searching period is given by

$$N(T) = -0.6T^2 + 32.1T - 350$$

where T is the temperature in degrees Celsius.

a. At what temperature will the maximum number of larvae survive? Round to the nearest degree.

b. What is the maximum number of surviving larvae? Round to the nearest whole number.

c. Find the x-intercepts, to the nearest whole number, for the graph of this function.

d. Write a sentence that describes the meaning of the x-intercepts in the context of this problem.

51. REAL ESTATE The number of California homes that have sold for over $1,000,000 between 1989 and 2002 can be modeled by

$$N(t) = 1.43t^2 - 11.44t + 47.68$$

where $N(t)$ is the number (in hundreds) of homes that were sold in year t, with $t = 0$ corresponding to 1989. According to this model, in what year were the least number of million-dollar homes sold? How many million-dollar homes, to the nearest hundred, were sold that year?

52. GEOLOGY In June 2001, Mt. Etna in Sicily, Italy erupted, sending volcanic bombs (masses of molten lava ejected from the volcano) into the air. A model of the height h, in meters, of a volcanic bomb above the crater of the volcano t seconds after the eruption is given by $h(t) = -9.8t^2 + 100t$. Find the maximum height of a volcanic bomb above the crater for this eruption. Round to the nearest meter.

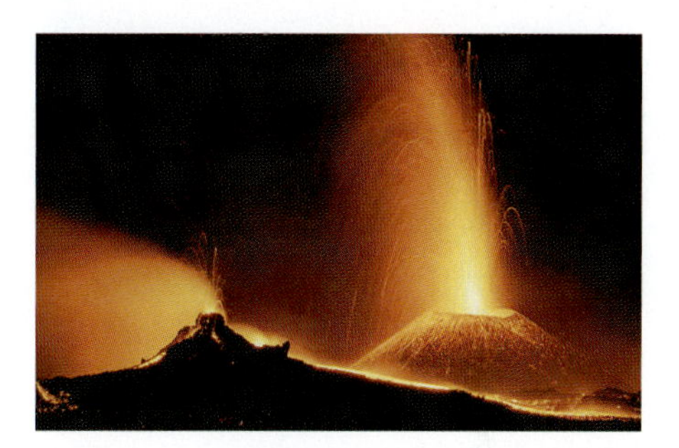

53. SPORTS For a serve to be legal in tennis, the ball must be at least 3 feet high when it is 39 feet from the server, and it must land in a spot that is less than 60 feet from the server. Does the path of a ball given by $h(x) = -0.002x^2 - 0.03x + 8$, where $h(x)$ is the height of the ball (in feet) x feet from the server, satisfy the conditions of a legal serve?

54. SPORTS A pitcher releases a baseball 6 feet above the ground at a speed of 132 feet per second (90 miles per hour) toward home plate, which is 60.5 feet away. The height $h(x)$, in feet, of the ball x feet from home plate can be approximated by $h(x) = -0.0009x^2 + 6$. To be considered a strike, the ball must cross home plate and be at least 2.5 feet high and less than 5.4 feet high. Assuming the ball crosses home plate, is this particular pitch a strike? Explain.

55. AUTOMOTIVE ENGINEERING The fuel efficiency for a certain midsize car is given by

$$E(v) = -0.018v^2 + 1.476v + 3.4$$

where $E(v)$ is the fuel efficiency in miles per gallon for a car traveling v miles per hour.

a. What speed will yield the maximum fuel efficiency? Round to the nearest mile per hour.

b. What is the maximum fuel efficiency for this car? Round to the nearest mile per gallon.

56. SPORTS Some football fields are built in a parabolic mound shape so that water will drain off the field. A model for the shape of a certain field is given by

$$h(x) = -0.0002348x^2 + 0.0375x$$

where $h(x)$ is the height, in feet, of the field at a distance of x feet from one sideline. Find the maximum height of the field. Round to the nearest tenth of a foot.

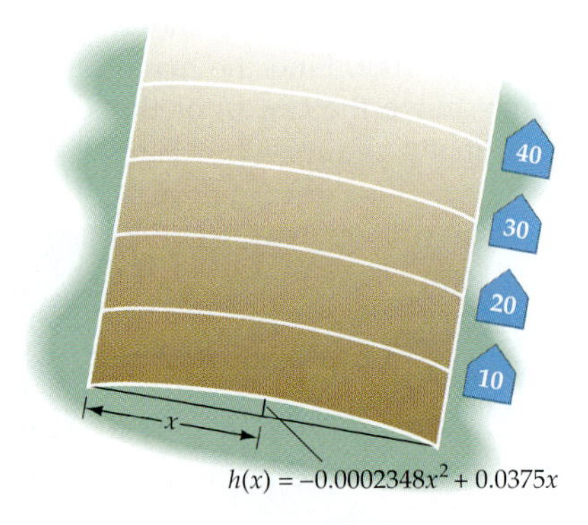

In Exercises 57 to 60, determine the y- and x-intercepts (if any) of the quadratic function.

57. $f(x) = x^2 + 6x$

58. $f(x) = -x^2 + 4x$

59. $f(x) = -3x^2 + 5x - 6$

60. $f(x) = 2x^2 + 3x + 4$

In Exercises 61 and 62, determine the number of units x that produce a maximum revenue for the given revenue function. Also determine the maximum revenue.

61. $R(x) = 296x - 0.2x^2$

62. $R(x) = 810x - 0.6x^2$

In Exercises 63 and 64, determine the number of units x that produce a maximum profit for the given profit function. Also determine the maximum profit.

63. $P(x) = -0.01x^2 + 1.7x - 48$

64. $P(x) = -\dfrac{x^2}{14{,}000} + 1.68x - 4000$

In Exercises 65 and 66, determine the profit function for the given revenue function and cost function. Also determine the break-even point(s).

65. $R(x) = x(102.50 - 0.1x)$; $C(x) = 52.50x + 1840$

66. $R(x) = x(210 - 0.25x)$; $C(x) = 78x + 6399$

67. **Tour Cost** A charter bus company has determined that the cost of providing x people a tour is

$$C(x) = 180 + 2.50x$$

A full tour consists of 60 people. The ticket price per person is \$15 plus \$0.25 for each unsold ticket. Determine

a. the revenue function **b.** the profit function

c. the company's maximum profit

d. the number of ticket sales that yields the maximum profit

▶ **68.** **Delivery Cost** An air freight company has determined that the cost, in dollars, of delivering x parcels per flight is

$$C(x) = 2025 + 7x$$

The price per parcel, in dollars, the company charges to send x parcels is

$$p(x) = 22 - 0.01x$$

Determine

a. the revenue function **b.** the profit function

c. the company's maximum profit

d. the price per parcel that yields the maximum profit

e. the minimum number of parcels the air freight company must ship to break even

69. **Projectile** If the initial velocity of a projectile is 128 feet per second, then its height h in feet is a function of time t in seconds given by the equation $h(t) = -16t^2 + 128t$.

a. Find the time t when the projectile achieves its maximum height.

b. Find the maximum height of the projectile.

c. Find the time t when the projectile hits the ground.

▶ **70.** **Projectile** The height in feet of a projectile with an initial velocity of 64 feet per second and an initial height of 80 feet is a function of time t in seconds given by

$$h(t) = -16t^2 + 64t + 80$$

a. Find the maximum height of the projectile.

b. Find the time t when the projectile achieves its maximum height.

c. Find the time t when the projectile has a height of 0 feet.

71. **Fire Management** The height of a stream of water from the nozzle of a fire hose can be modeled by

$$y(x) = -0.014x^2 + 1.19x + 5$$

where $y(x)$ is the height, in feet, of the stream x feet from the firefighter. What is the maximum height that the stream of water from this nozzle can reach? Round to the nearest foot.

72. **OLYMPIC SPORTS** In 1988, Louise Ritter of the United States set the women's Olympic record for the high jump. A mathematical model that approximates her jump is given by

$$h(t) = -204.8t^2 + 256t$$

where $h(t)$ is her height in inches t seconds after beginning her jump. Find the maximum height of her jump.

73. **NORMAN WINDOW** A Norman window has the shape of a rectangle surmounted by a semicircle. The exterior perimeter of the window shown in the figure is 48 feet. Find the height h and the radius r that will allow the maximum amount of light to enter the window. (*Hint:* Write the area of the window as a quadratic function of the radius r.)

74. **GOLDEN GATE BRIDGE** The suspension cables of the main span of the Golden Gate Bridge are in the shape of a parabola. If a coordinate system is drawn as shown, find the quadratic function that models a suspension cable for the main span of the bridge.

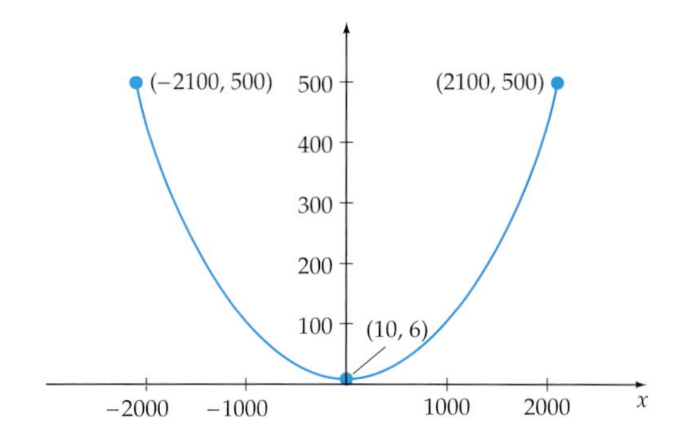

CONNECTING CONCEPTS

75. Let $f(x) = x^2 - (a + b)x + ab$, where a and b are real numbers.

a. Show that the x-intercepts are $(a, 0)$ and $(b, 0)$.

b. Show that the minimum value of the function occurs at the x-value of the midpoint of the line segment defined by the x-intercepts.

76. Let $f(x) = ax^2 + bx + c$, where a, b, and c are real numbers.

a. What conditions must be imposed on the coefficients so that f has a maximum?

b. What conditions must be imposed on the coefficients so that f has a minimum?

c. What conditions must be imposed on the coefficients so that the graph of f intersects the x-axis?

77. Find the quadratic function of x whose graph has a minimum at $(2, 1)$ and passes through $(0, 4)$.

78. Find the quadratic function of x whose graph has a maximum at $(-3, 2)$ and passes through $(0, -5)$.

79. **AREA OF A RECTANGLE** A wire 32 inches long is bent so that it has the shape of a rectangle. The length of the rectangle is x and the width is w.

a. Write w as a function of x.

b. Write the area A of the rectangle as a function of x.

80. **MAXIMIZE AREA** Use the function A from part **b.** in Exercise 79 to prove that the area A is greatest if the rectangle is a square.

81. Show that the function $f(x) = x^2 + bx - 1$ has a real zero for any value b.

82. Show that the function $g(x) = -x^2 + bx + 1$ has a real zero for any value b.

83. What effect does increasing the constant c have on the graph of $f(x) = ax^2 + bx + c$?

84. If $a > 0$, what effect does decreasing the coefficient a have on the graph of $f(x) = ax^2 + bx + c$?

85. Find two numbers whose sum is 8 and whose product is a maximum.

86. Find two numbers whose difference is 12 and whose product is a minimum.

87. Verify that the slope of the line passing through (x, x^3) and $(x + h, [x + h]^3)$ is $3x^2 + 3xh + h^2$.

88. Verify that the slope of the line passing through $(x, 4x^3 + x)$ and $(x + h, 4[x + h]^3 + [x + h])$ is given by $12x^2 + 12xh + 4h^2 + 1$.

PREPARE FOR SECTION 1.6

89. For the graph of the parabola whose equation is $f(x) = x^2 + 4x - 6$, what is the equation of the axis of symmetry? [1.5]

90. For $f(x) = \dfrac{3x^4}{x^2 + 1}$, show that $f(-3) = f(3)$. [1.3]

91. For $f(x) = 2x^3 - 5x$, show that $f(-2) = -f(2)$. [1.3]

92. Let $f(x) = x^2$ and $g(x) = x + 3$. Find $f(a) - g(a)$ for $a = -2, -1, 0, 1, 2$. [1.3]

93. What is the midpoint of the line segment between $P(-a, b)$ and $Q(a, b)$? [1.2]

94. What is the midpoint of the line segment between $P(-a, -b)$ and $Q(a, b)$? [1.2]

PROJECTS

1. **THE CUBIC FORMULA** Write an essay on the development of the cubic formula. An excellent source of information is the chapter "Cardano and the Solution of the Cubic" in *Journey Through Genius*, by William Dunham (New York: Wiley, 1990).

2. **SIMPSON'S RULE** In calculus a procedure known as *Simpson's Rule* is often used to approximate the area under a curve. The figure at the right shows the graph of a parabola that passes through $P_0(-h, y_0)$, $P_1(0, y_1)$, and $P_2(h, y_2)$. The equation of the parabola is of the form $y = Ax^2 + Bx + C$. Using calculus procedures, we can show that the area bounded by the parabola, the x-axis, and the vertical lines $x = -h$ and $x = h$ is

$$\frac{h}{3}(2Ah^2 + 6C)$$

Use algebra to show that $y_0 + 4y_1 + y_2 = 2Ah^2 + 6C$, from which we can deduce that the area of the bounded region can also be written as

$$\frac{h}{3}(y_0 + 4y_1 + y_2)$$

(*Hint:* Evaluate $Ax^2 + Bx + C$ at $x = -h$, $x = 0$, and $x = h$ to determine values of y_0, y_1, and y_2, respectively. Then compute $y_0 + 4y_1 + y_2$.)

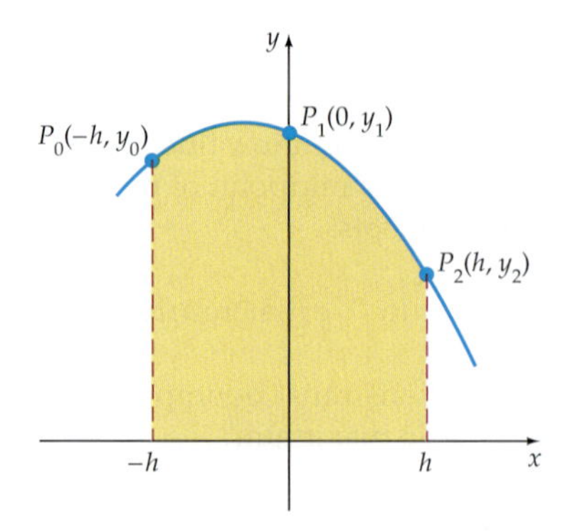

SECTION 1.6 PROPERTIES OF GRAPHS

- SYMMETRY
- EVEN AND ODD FUNCTIONS
- TRANSLATIONS OF GRAPHS
- REFLECTIONS OF GRAPHS
- COMPRESSING AND STRETCHING OF GRAPHS

SYMMETRY

The graph in **Figure 1.74** is symmetric with respect to the line l. Note that the graph has the property that if the paper is folded along the dotted line l, the point A' will coincide with the point A, the point B' will coincide with the point B, and the point C' will coincide with the point C. One part of the graph is a *mirror image* of the rest of the graph across the line l.

A graph is **symmetric with respect to the y-axis** if, whenever the point given by (x, y) is on the graph, then $(-x, y)$ is also on the graph. The graph in **Figure 1.75** is symmetric with respect to the y-axis. A graph is **symmetric with respect to the x-axis** if, whenever the point given by (x, y) is on the graph, then $(x, -y)$ is also on the graph. The graph in **Figure 1.76** is symmetric with respect to the x-axis.

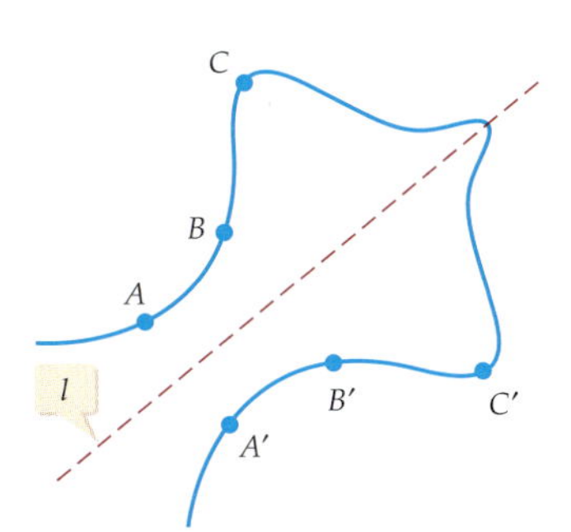

FIGURE 1.74

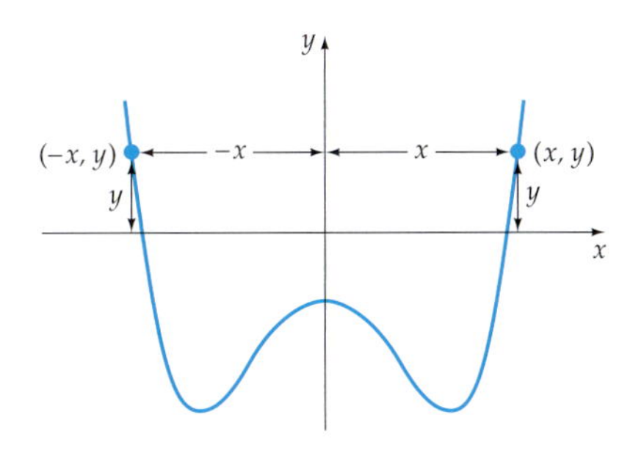

FIGURE 1.75
Symmetry with respect to the y-axis

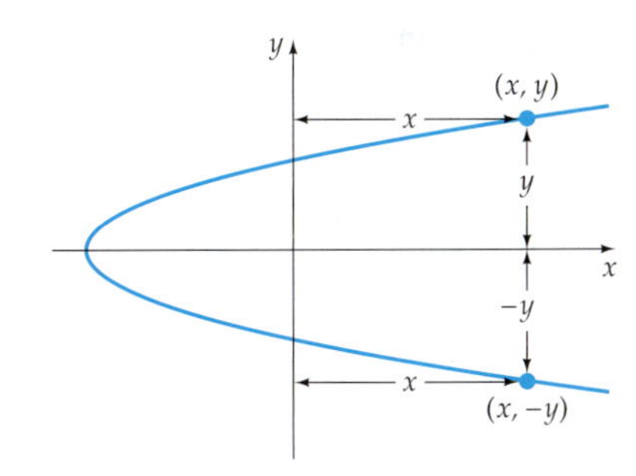

FIGURE 1.76
Symmetry with respect to the x-axis

Tests for Symmetry with Respect to a Coordinate Axis

The graph of an equation is symmetric with respect to

- the y-axis if the replacement of x with $-x$ leaves the equation unaltered.
- the x-axis if the replacement of y with $-y$ leaves the equation unaltered.

? QUESTION Which of the graphs below, I, II, or III, is **a.** symmetric with respect to the x-axis? **b.** symmetric with respect to the y-axis?

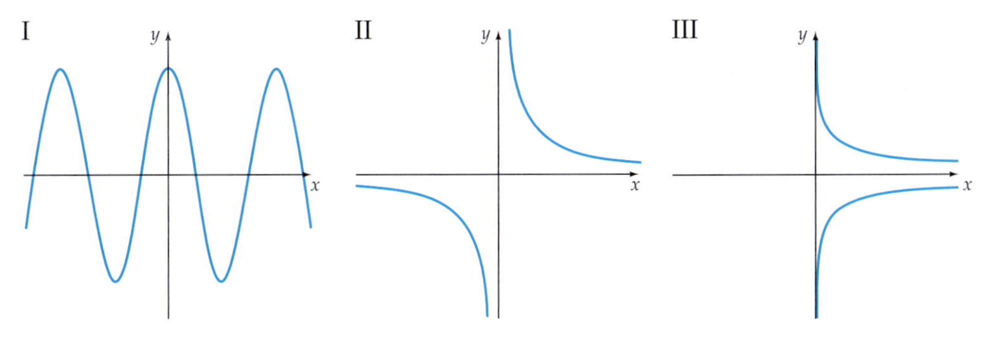

? ANSWER **a.** I is symmetric with respect to the y-axis.
b. III is symmetric with respect to the x-axis.

EXAMPLE 1 Determine Symmetries of a Graph

Determine whether the graph of the given equation has symmetry with respect to either the x- or the y-axis.

a. $y = x^2 + 2$ **b.** $x = |y| - 2$

Solution

a. The equation $y = x^2 + 2$ *is unaltered* by the replacement of x with $-x$. That is, the simplification of $y = (-x)^2 + 2$ yields the original equation $y = x^2 + 2$. Thus the graph of $y = x^2 + 2$ is symmetric with respect to the y-axis. However, the equation $y = x^2 + 2$ *is altered* by the replacement of y with $-y$. That is, the simplification of $-y = x^2 + 2$, which is $y = -x^2 - 2$, *does not* yield the original equation $y = x^2 + 2$. The graph of $y = x^2 + 2$ is not symmetric with respect to the x-axis. See **Figure 1.77.**

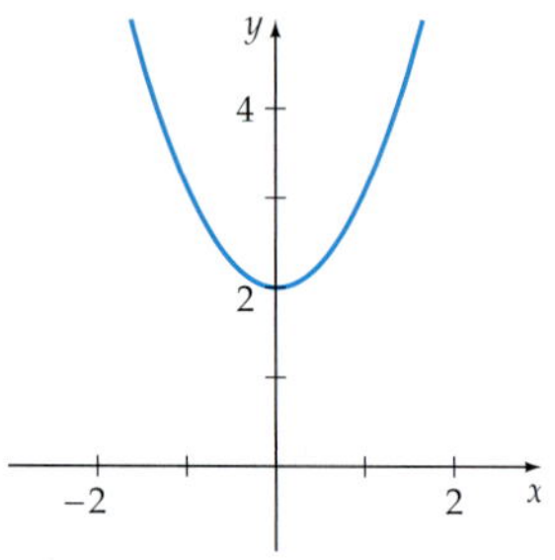

$y = x^2 + 2$

FIGURE 1.77

b. The equation $x = |y| - 2$ *is altered* by the replacement of x with $-x$. That is, the simplification of $-x = |y| - 2$, which is $x = -|y| + 2$, *does not* yield the original equation $x = |y| - 2$. This implies that the graph of $x = |y| - 2$ is not symmetric with respect to the y-axis. However, the equation $x = |y| - 2$ *is unaltered* by the replacement of y with $-y$. That is, the simplification of $x = |-y| - 2$ yields the original equation $x = |y| - 2$. The graph of $x = |y| - 2$ is symmetric with respect to the x-axis. See **Figure 1.78.**

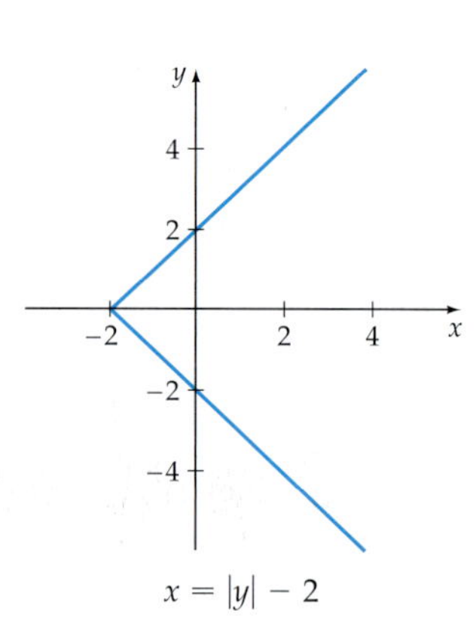

$x = |y| - 2$

FIGURE 1.78

▶ **TRY EXERCISE 14, PAGE 91**

Symmetry with Respect to a Point

A graph is **symmetric with respect to a point** Q if for each point P on the graph there is a point P' on the graph such that Q is the midpoint of the line segment PP'.

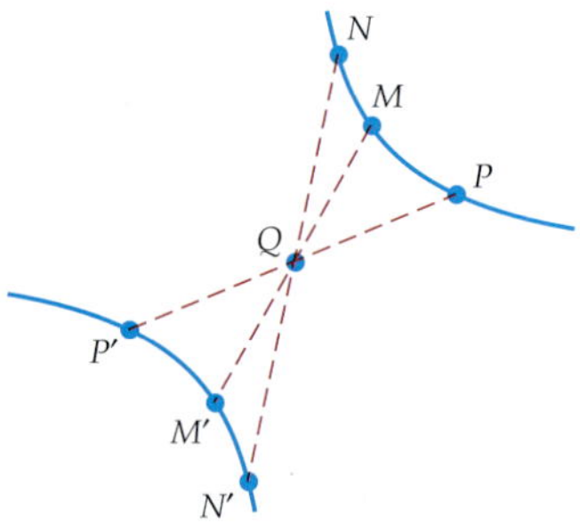

FIGURE 1.79

The graph in **Figure 1.79** is symmetric with respect to the point Q. For any point P on the graph, there exists a point P' on the graph such that Q is the midpoint of $P'P$.

When we discuss symmetry with respect to a point, we frequently use the origin. A graph is symmetric with respect to the origin if, whenever the point given by (x, y) is on the graph, then $(-x, -y)$ is also on the graph. The graph in **Figure 1.80** is symmetric with respect to the origin.

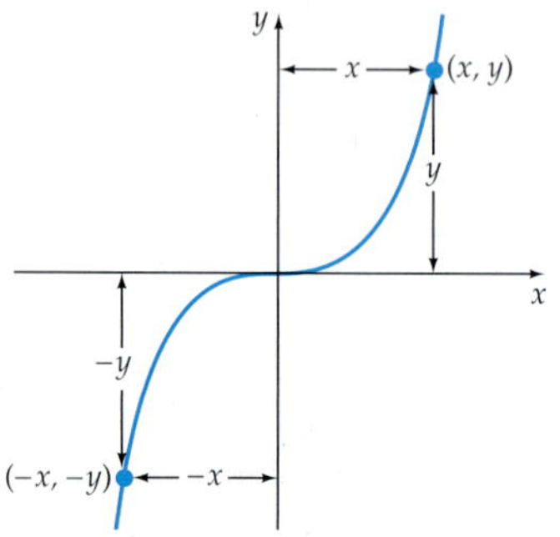

Symmetry with respect to the origin

FIGURE 1.80

Test for Symmetry with Respect to the Origin

The graph of an equation is symmetric with respect to the origin if the replacement of x with $-x$ and of y with $-y$ leaves the equation unaltered.

EXAMPLE 2 Determine Symmetry with Respect to the Origin

Determine whether the graph of each equation has symmetry with respect to the origin.

a. $xy = 4$ **b.** $y = x^3 + 1$

Solution

a. The equation $xy = 4$ is unaltered by the replacement of x with $-x$ and of y with $-y$. That is, the simplification of $(-x)(-y) = 4$ yields the original equation $xy = 4$. Thus the graph of $xy = 4$ is symmetric with respect to the origin. See **Figure 1.81.**

b. The equation $y = x^3 + 1$ *is altered* by the replacement of x with $-x$ and of y with $-y$. That is, the simplification of $-y = (-x)^3 + 1$, which is $y = x^3 - 1$, *does not* yield the original equation $y = x^3 + 1$. Thus the graph of $y = x^3 + 1$ is not symmetric with respect to the origin. See **Figure 1.82.**

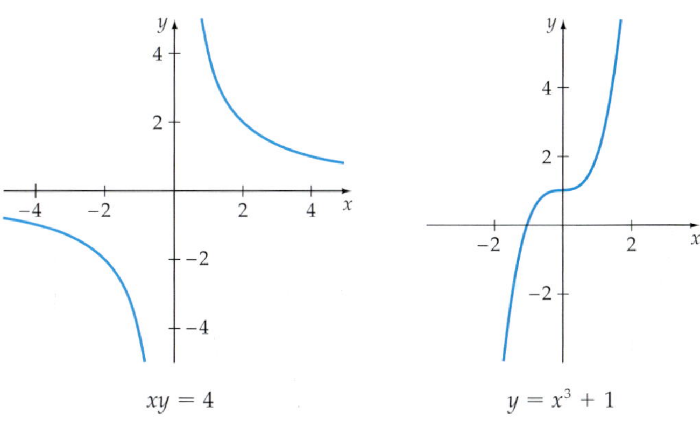

$xy = 4$

FIGURE 1.81

$y = x^3 + 1$

FIGURE 1.82

TRY EXERCISE 24, PAGE 91

Some graphs have more than one symmetry. For example, the graph of $|x| + |y| = 2$ has symmetry with respect to the x-axis, the y-axis, and the origin. **Figure 1.83** is the graph of $|x| + |y| = 2$.

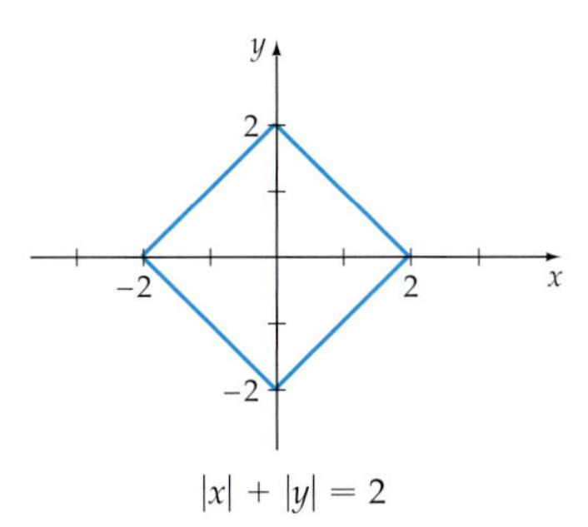

$|x| + |y| = 2$

FIGURE 1.83

EVEN AND ODD FUNCTIONS

Some functions are classified as either *even* or *odd*.

Definition of Even and Odd Functions

The function f is an **even function** if

$$f(-x) = f(x) \quad \text{for all } x \text{ in the domain of } f$$

The function f is an **odd function** if

$$f(-x) = -f(x) \quad \text{for all } x \text{ in the domain of } f$$

EXAMPLE 3 Identify Even or Odd Functions

Determine whether each function is even, odd, or neither.

a. $f(x) = x^3$ **b.** $F(x) = |x|$ **c.** $h(x) = x^4 + 2x$

Solution

Replace x with $-x$ and simplify.

a. $f(-x) = (-x)^3 = -x^3 = -(x^3) = -f(x)$
Because $f(-x) = -f(x)$, this function is an odd function.

b. $F(-x) = |-x| = |x| = F(x)$
Because $F(-x) = F(x)$, this function is an even function.

c. $h(-x) = (-x)^4 + 2(-x) = x^4 - 2x$
This function is neither an even nor an odd function because

$$h(-x) = x^4 - 2x,$$

which is not equal to either $h(x)$ or $-h(x)$.

▶ TRY EXERCISE 44, PAGE 91

The following properties are a result of the tests for symmetry:

- The graph of an even function is symmetric with respect to the y-axis.
- The graph of an odd function is symmetric with respect to the origin.

The graph of f in **Figure 1.84** is symmetric with respect to the y-axis. It is the graph of an even function. The graph of g in **Figure 1.85** is symmetric with respect to the origin. It is the graph of an odd function. The graph of h in **Figure 1.86** is not symmetric with respect to the y-axis and is not symmetric with respect to the origin. It is neither an even nor an odd function.

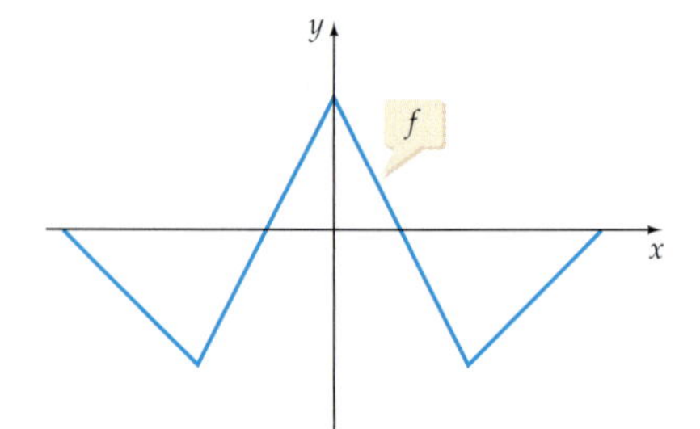

FIGURE 1.84
The graph of an even function is symmetric with respect to the y-axis.

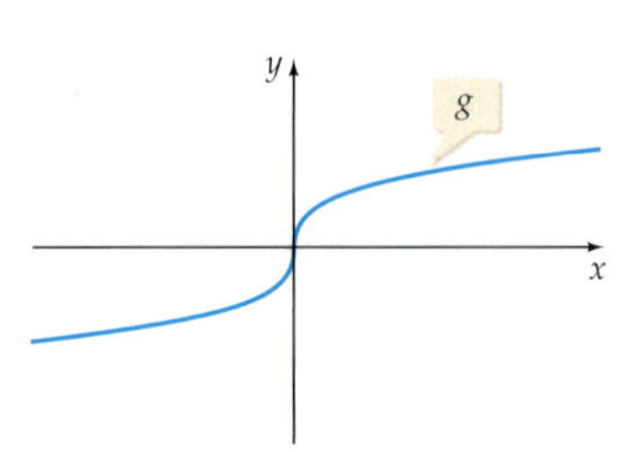

FIGURE 1.85
The graph of an odd function is symmetric with respect to the origin.

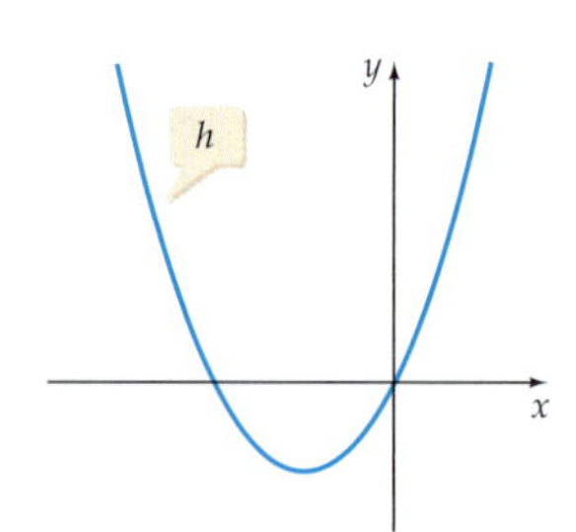

FIGURE 1.86
If the graph of a function is not symmetric to the y-axis or to the origin, then the function is neither even nor odd.

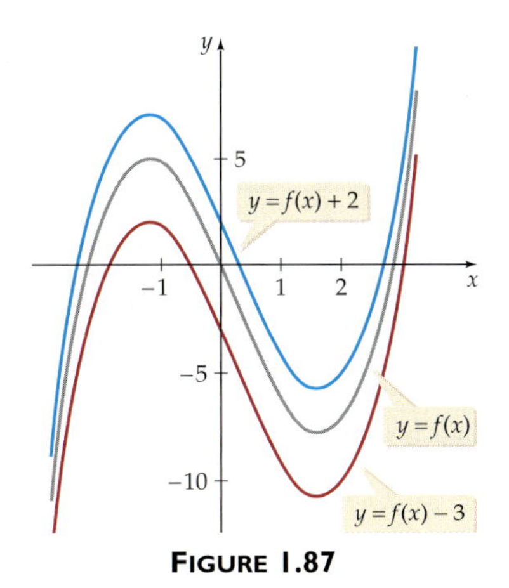

FIGURE 1.87

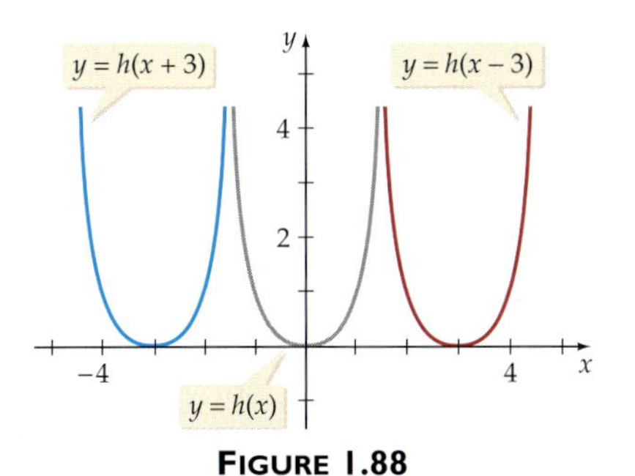

FIGURE 1.88

TRANSLATIONS OF GRAPHS

The shape of a graph may be exactly the same as the shape of another graph; only their positions in the xy-plane may differ. For example, the graph of $y = f(x) + 2$ is the graph of $y = f(x)$ with each point moved up vertically 2 units. The graph of $y = f(x) - 3$ is the graph of $y = f(x)$ with each point moved down vertically 3 units. See **Figure 1.87.**

The graphs of $y = f(x) + 2$ and $y = f(x) - 3$ in **Figure 1.87** are called *vertical translations* of the graph of $y = f(x)$.

Vertical Translations

If f is a function and c is a positive constant, then the graph of

- $y = f(x) + c$ is the graph of $y = f(x)$ shifted up *vertically* c units.
- $y = f(x) - c$ is the graph of $y = f(x)$ shifted down *vertically* c units.

In **Figure 1.88,** the graph of $y = h(x + 3)$ is the graph of $y = h(x)$ with each point shifted to the left horizontally 3 units. Similarly, the graph of $y = h(x - 3)$ is the graph of $y = h(x)$ with each point shifted to the right horizontally 3 units.

The graphs of $y = h(x + 3)$ and $y = h(x - 3)$ in **Figure 1.88** are called *horizontal translations* of the graph of $y = h(x)$.

Horizontal Translations

If f is a function and c is a positive constant, then the graph of

- $y = f(x + c)$ is the graph of $y = f(x)$ shifted left *horizontally* c units.
- $y = f(x - c)$ is the graph of $y = f(x)$ shifted right *horizontally* c units.

INTEGRATING TECHNOLOGY

A graphing calculator can be used to draw the graphs of a *family* of functions. For instance, $f(x) = x^2 + c$ constitutes a family of functions with **parameter** c. The only feature of the graph that changes is the value of c.

A graphing calculator can be used to produce the graphs of a family of curves for specific values of the parameter. The LIST feature of the calculator can be used. For instance, to graph $f(x) = x^2 + c$ for $c = -2, 0$, and 1, we will create a list and use that list to produce the family of curves. The keystrokes for a TI-83 calculator are given below.

2nd { -2 , 0 , 1 2nd } STO 2nd L1

Now use the Y= key to enter

Y= X x² + 2nd L1 ZOOM 6

Continued ▶

Sample screens for the keystrokes and graphs are shown here. You can use similar keystrokes for Exercises 75–82 of this section.

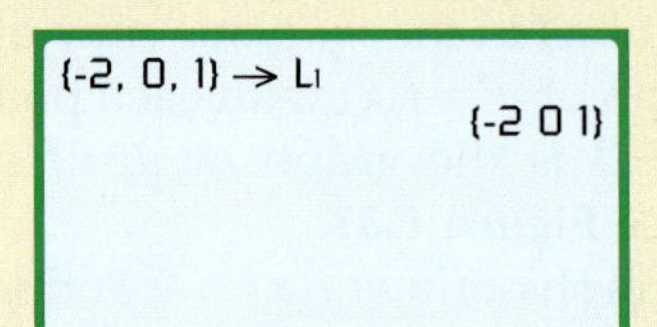

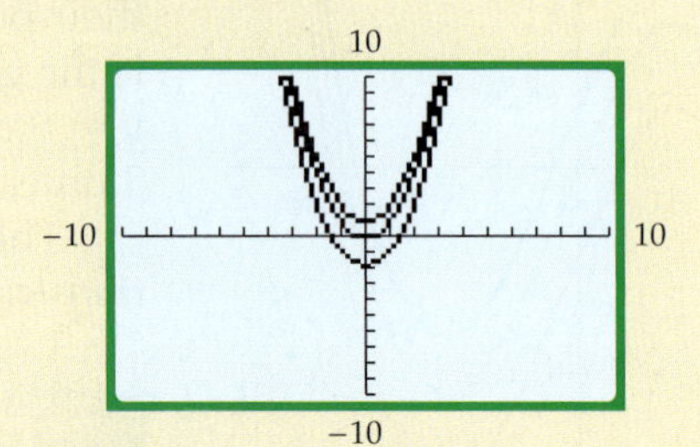

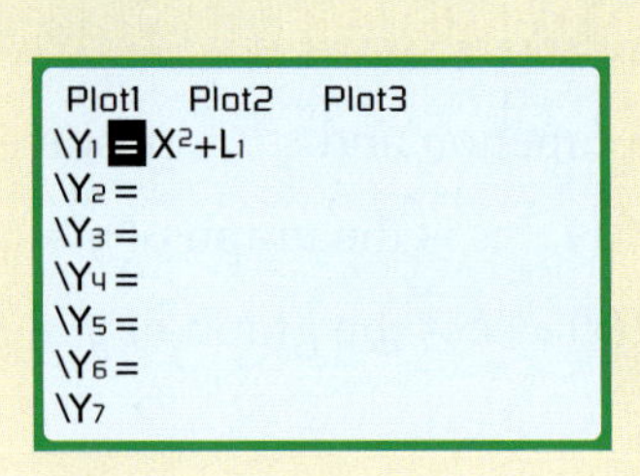

EXAMPLE 4 Graph by Using Translations

Use vertical and horizontal translations of the graph of $f(x) = x^3$, shown in **Figure 1.89,** to graph

a. $g(x) = x^3 - 2$ **b.** $h(x) = (x + 1)^3$

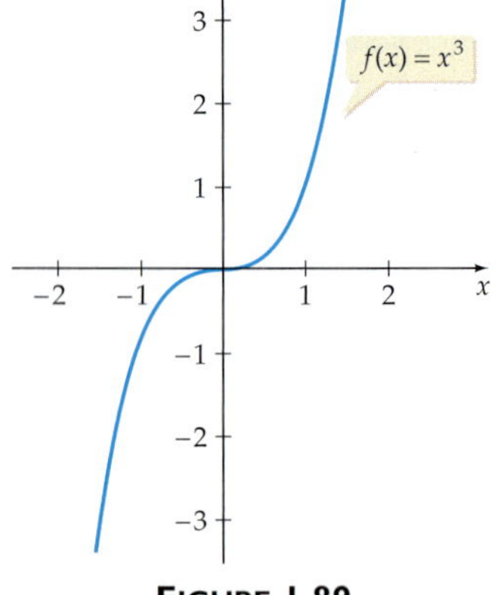

FIGURE 1.89

Solution

a. The graph of $g(x) = x^3 - 2$ is the graph of $f(x) = x^3$ shifted down vertically 2 units. See **Figure 1.90.**

b. The graph of $h(x) = (x + 1)^3$ is the graph of $f(x) = x^3$ shifted to the left horizontally 1 unit. See **Figure 1.91.**

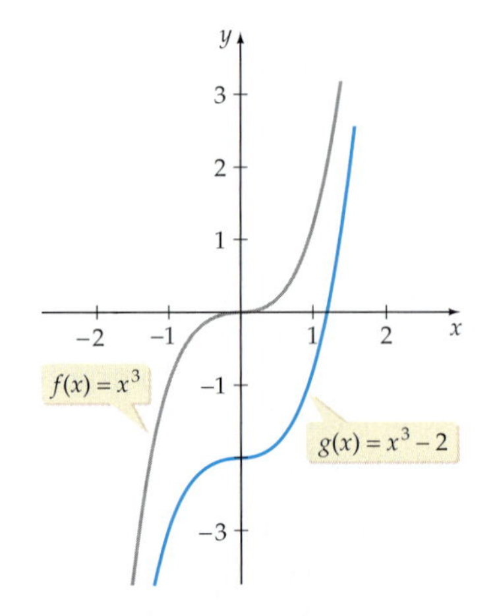

FIGURE 1.90

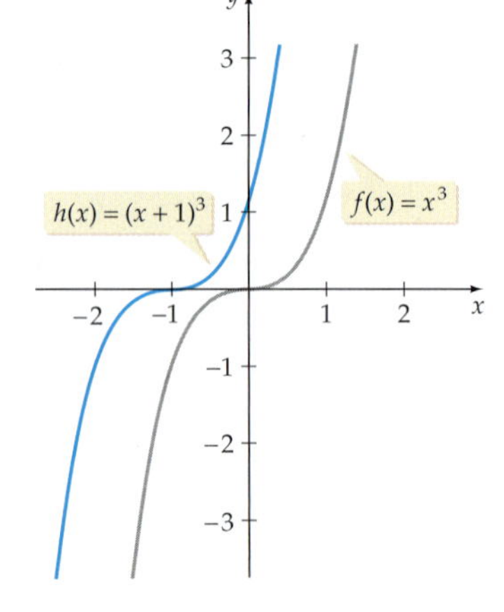

FIGURE 1.91

▶ TRY EXERCISE 58, PAGE 92

Reflections of Graphs

The graph of $y = -f(x)$ cannot be obtained from the graph of $y = f(x)$ by a combination of vertical and/or horizontal shifts. **Figure 1.92** illustrates that the graph of $y = -f(x)$ is the reflection of the graph of $y = f(x)$ across the x-axis.

The graph of $y = f(-x)$ is the reflection of the graph of $y = f(x)$ across the y-axis, as shown in **Figure 1.93.**

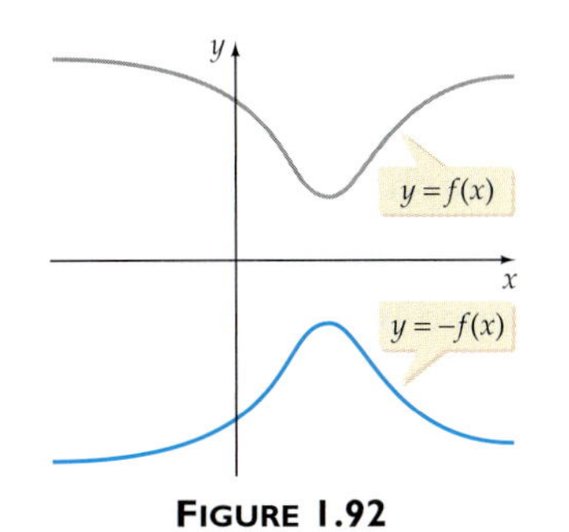

Figure 1.92

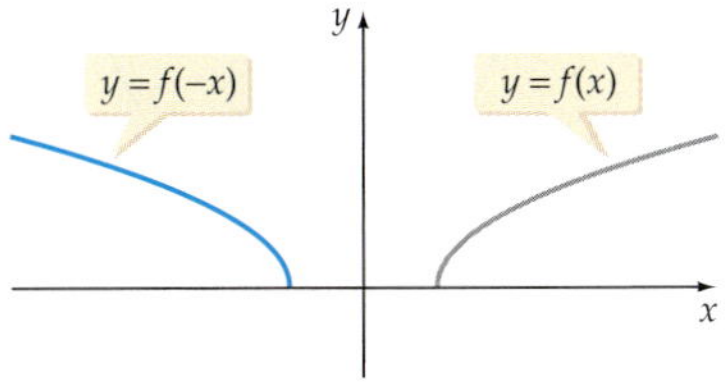

Figure 1.93

Reflections

The graph of

- $y = -f(x)$ is the graph of $y = f(x)$ reflected across the x-axis.
- $y = f(-x)$ is the graph of $y = f(x)$ reflected across the y-axis.

EXAMPLE 5 Graph by Using Reflections

Use reflections of the graph of $f(x) = \sqrt{x-1} + 1$, shown in **Figure 1.94,** to graph

a. $g(x) = -\left(\sqrt{x-1} + 1\right)$ **b.** $h(x) = \sqrt{-x-1} + 1$

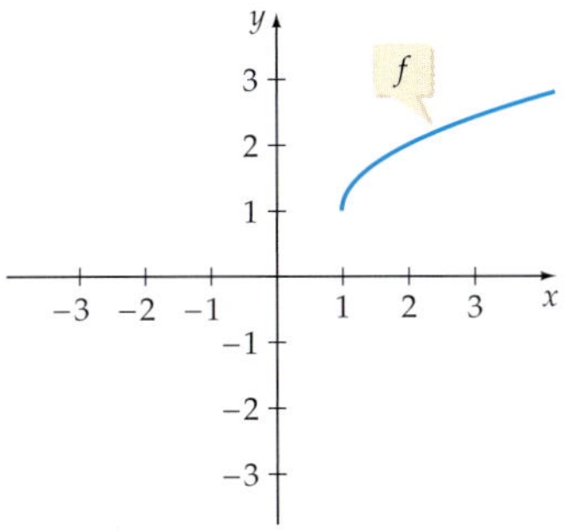

Figure 1.94

Solution

a. Because $g(x) = -f(x)$, the graph of g is the graph of f reflected across the x-axis. See **Figure 1.95.**

b. Because $h(x) = f(-x)$, the graph of h is the graph of f reflected across the y-axis. See **Figure 1.96.**

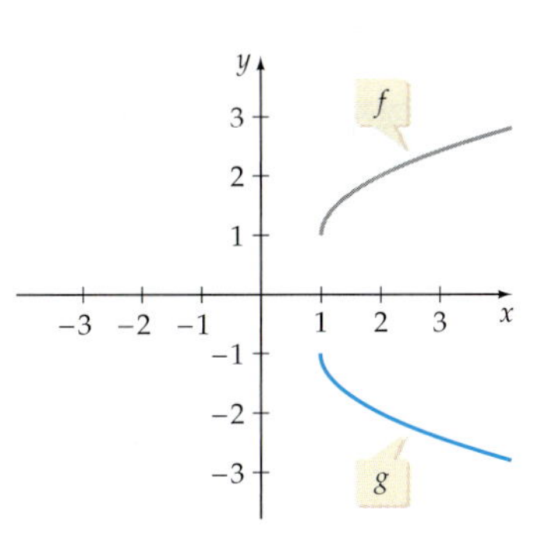

Figure 1.95

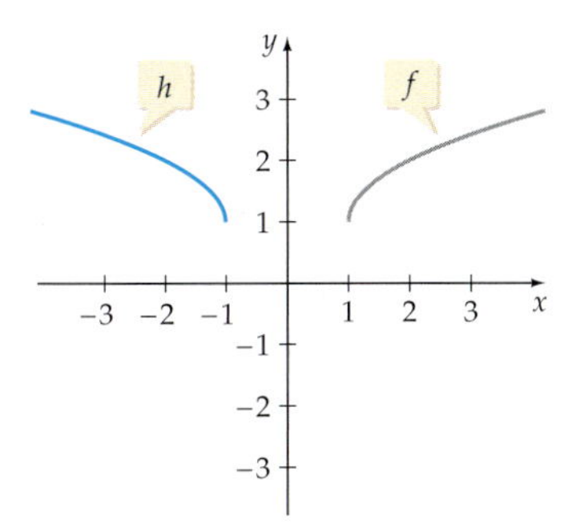

Figure 1.96

Try Exercise 68, page 93

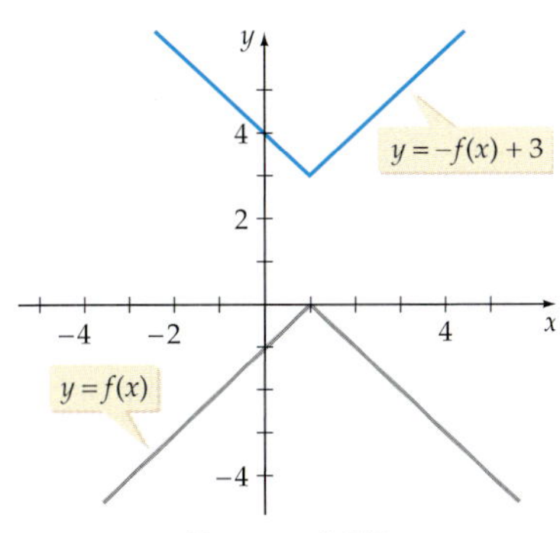

Figure 1.97

Some graphs of functions can be constructed by using a combination of translations and reflections. For instance, the graph of $y = -f(x) + 3$ in **Figure 1.97** was obtained by reflecting the graph of $y = f(x)$ in **Figure 1.97** across the x-axis and then shifting that graph up vertically 3 units.

Compressing and Stretching of Graphs

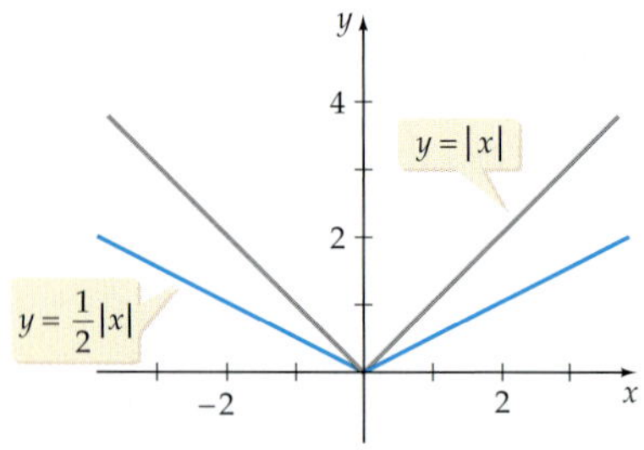

The graph of the equation $y = c \cdot f(x)$ for $c \neq 1$ vertically compresses or stretches the graph of $y = f(x)$. To determine the points on the graph of $y = c \cdot f(x)$, multiply each y-coordinate of the points on the graph of $y = f(x)$ by c. For example, **Figure 1.98** shows that the graph of $y = \frac{1}{2}|x|$ can be obtained by plotting points that have a y-coordinate that is one-half of the y-coordinate of those found on the graph of $y = |x|$.

Figure 1.98

If $0 < c < 1$, then the graph of $y = c \cdot f(x)$ is obtained by *compressing* the graph of $y = f(x)$. **Figure 1.98** illustrates the vertical compressing of the graph of $y = |x|$ toward the x-axis to form the graph of $y = \frac{1}{2}|x|$.

If $c > 1$, then the graph of $y = c \cdot f(x)$ is obtained by *stretching* the graph of $y = f(x)$. For example, if $f(x) = |x|$, then we obtain the graph of

$$y = 2f(x) = 2|x|$$

by stretching the graph of f away from the x-axis. See **Figure 1.99.**

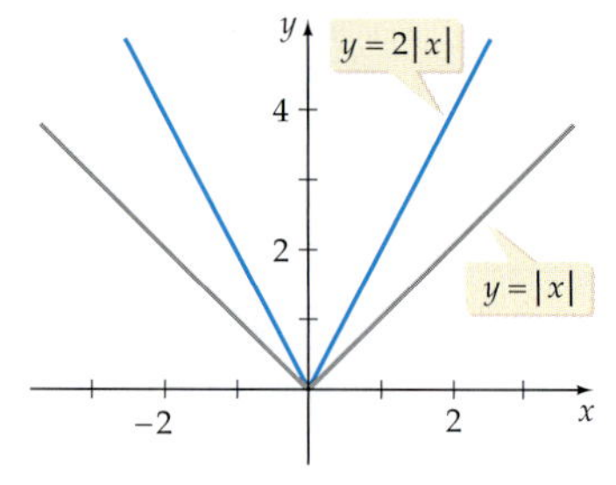

Figure 1.99

Vertical Stretching and Compressing of Graphs

If f is a function and c is a positive constant, then

- if $c > 1$, the graph of $y = c \cdot f(x)$ is the graph of $y = f(x)$ *stretched* vertically by a factor of c away from the x-axis.
- if $0 < c < 1$, the graph of $y = c \cdot f(x)$ is the graph of $y = f(x)$ *compressed* vertically by a factor of c toward the x-axis.

Example 6 Graph by Using Vertical Compressing and Shifting

Graph: $H(x) = \frac{1}{4}|x| - 3$

Solution

The graph of $y = |x|$ has a V shape that has its lowest point at (0, 0) and passes through (4, 4) and (−4, 4). The graph of $y = \frac{1}{4}|x|$ is a compression of the graph of $y = |x|$. The y-coordinates of the ordered pairs (0, 0), (4, 1), and (−4, 1) are obtained by multiplying the y-coordinates of the ordered pairs (0, 0), (4, 4), and (−4, 4) by $\frac{1}{4}$. To find the points on the graph of H, we still need to subtract 3 from each y-coordinate. Thus the graph of H is a V shape that has its lowest point at (0, −3) and passes through (4, −2) and (−4, −2). See **Figure 1.100.**

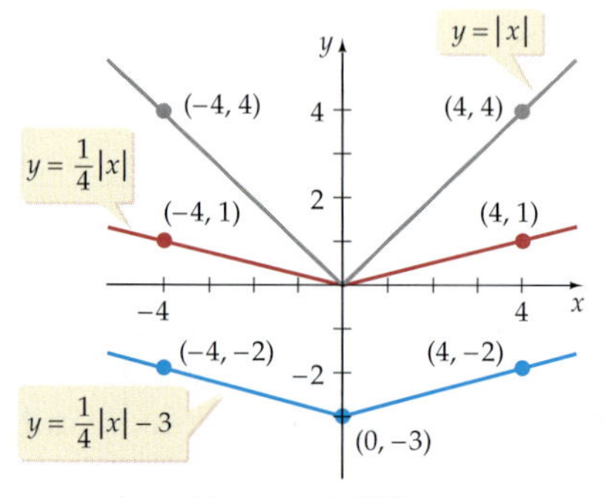

Figure 1.100

▶ Try Exercise 70, page 93

Some functions can be graphed by using a horizontal compressing or stretching of a given graph. The procedure makes use of the following concept.

Horizontal Compressing and Stretching of Graphs

If f is a function and c is a positive constant, then

- if $c > 1$, the graph of $y = f(c \cdot x)$ is the graph of $y = f(x)$ *compressed* horizontally by a factor of $\frac{1}{c}$ toward the y-axis.
- if $0 < c < 1$, the graph of $y = f(c \cdot x)$ is the graph of $y = f(x)$ *stretched* horizontally by a factor of $\frac{1}{c}$ away from the y-axis.

If the point (x, y) is on the graph of $y = f(x)$, then the graph of $y = f(cx)$ will contain the point $\left(\frac{1}{c}x, y\right)$.

EXAMPLE 7 Graph by Using Horizontal Compressing and Stretching

Use the graph of $y = f(x)$ shown in **Figure 1.101** to graph

a. $y = f(2x)$ **b.** $y = f\left(\frac{1}{3}x\right)$

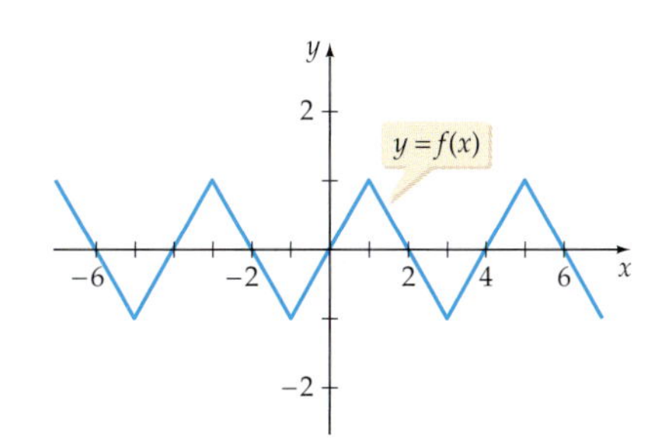

FIGURE 1.101

Solution

a. Because $2 > 1$, the graph of $y = f(2x)$ is a horizontal compression of the graph of $y = f(x)$ by a factor of $\frac{1}{2}$. For example, the point $(2, 0)$ on the graph of $y = f(x)$ becomes the point $(1, 0)$ on the graph of $y = f(2x)$. See **Figure 1.102.**

b. Since $0 < \frac{1}{3} < 1$, the graph of $y = f\left(\frac{1}{3}x\right)$ is a horizontal stretching of the graph of $y = f(x)$ by a factor of 3. For example, the point $(1, 1)$ on the

Continued ►

graph of $y = f(x)$ becomes the point (3, 1) on the graph of $y = f\left(\frac{1}{3}x\right)$. See **Figure 1.103.**

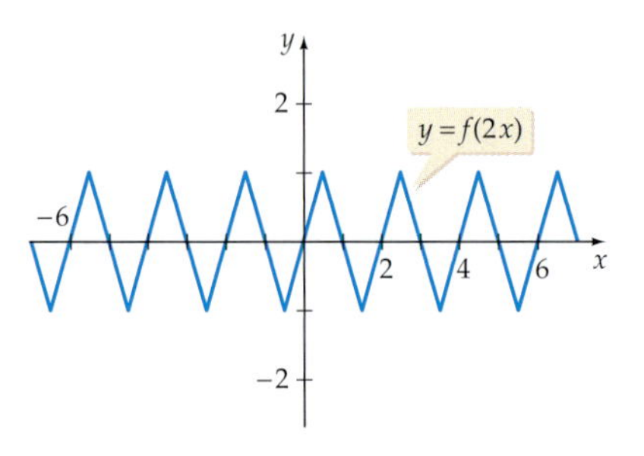

FIGURE 1.102

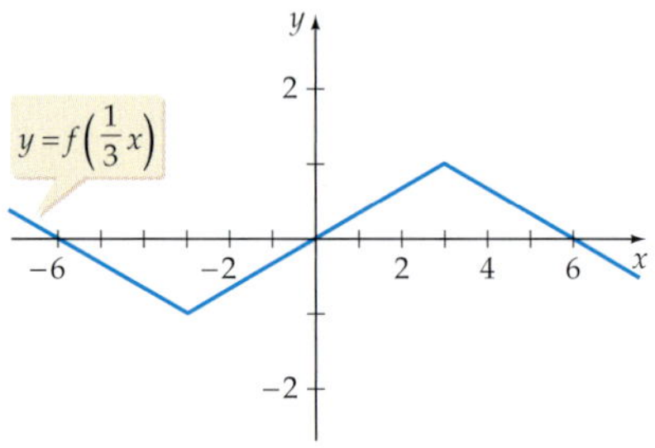

FIGURE 1.103

TRY EXERCISE 72, PAGE 93

TOPICS FOR DISCUSSION

1. Discuss the meaning of symmetry of a graph with respect to a line. How do you determine whether a graph has symmetry with respect to the x-axis? with respect to the y-axis?
2. Discuss the meaning of symmetry of a graph with respect to a point. How do you determine whether a graph has symmetry with respect to the origin?
3. What does it mean to reflect a graph across the x-axis or across the y-axis?
4. Explain how the graphs of $y_1 = 2x^3 - x^2$ and $y_2 = 2(-x)^3 - (-x)^2$ are related.
5. Given the graph of $y_3 = f(x)$, explain how to obtain the graph of $y_4 = f(x - 3) + 1$.
6. The graph of the *step function* $y_5 = [\![x]\!]$ has steps that are 1 unit wide. Determine how wide the steps are in the graph of $y_6 = \left[\!\left[\frac{1}{3}x\right]\!\right]$.

EXERCISE SET 1.6

In Exercises 1 to 6, plot the image of the given point with respect to
a. **the *y*-axis. Label this point *A*.**
b. **the *x*-axis. Label this point *B*.**
c. **the origin. Label this point *C*.**

1. $P(5, -3)$ **2.** $Q(-4, 1)$ **3.** $R(-2, 3)$

4. $S(-5, 3)$ **5.** $T(-4, -5)$ **6.** $U(5, 1)$

In Exercises 7 and 8, sketch a graph that is symmetric to the given graph with respect to the *x*-axis.

7.

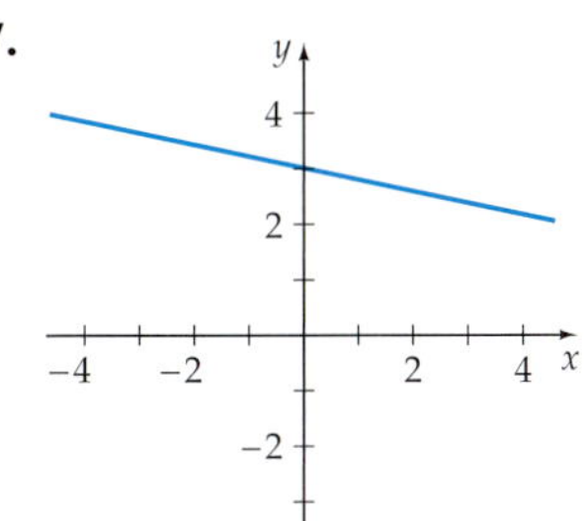

8.

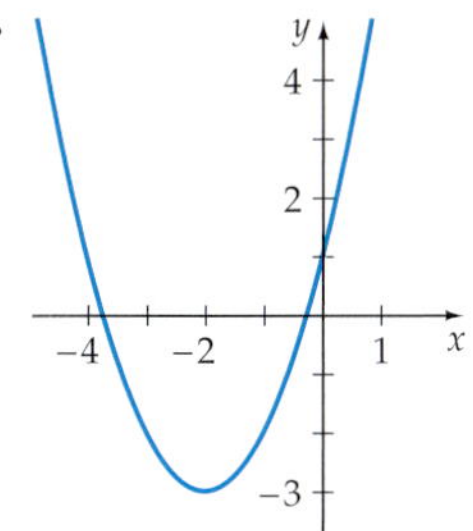

In Exercises 9 and 10, sketch a graph that is symmetric to the given graph with respect to the *y*-axis.

9.

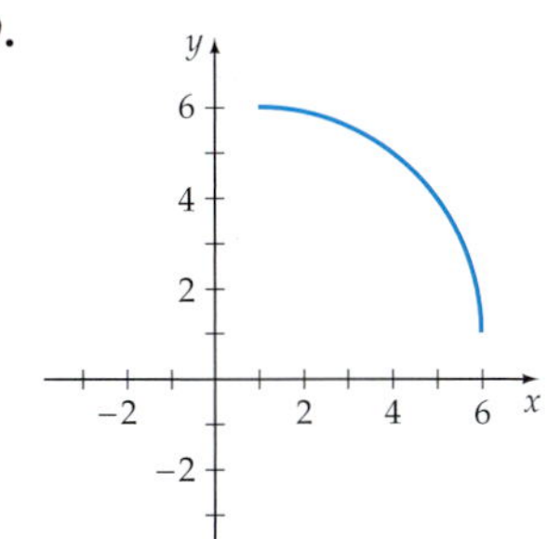

10.

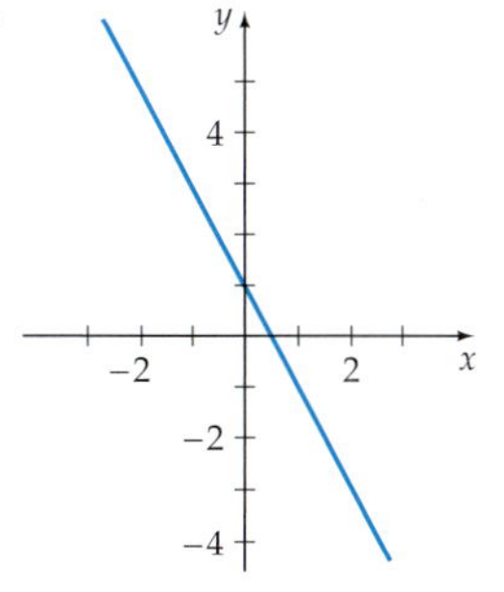

In Exercises 11 and 12, sketch a graph that is symmetric to the given graph with respect to the origin.

11.

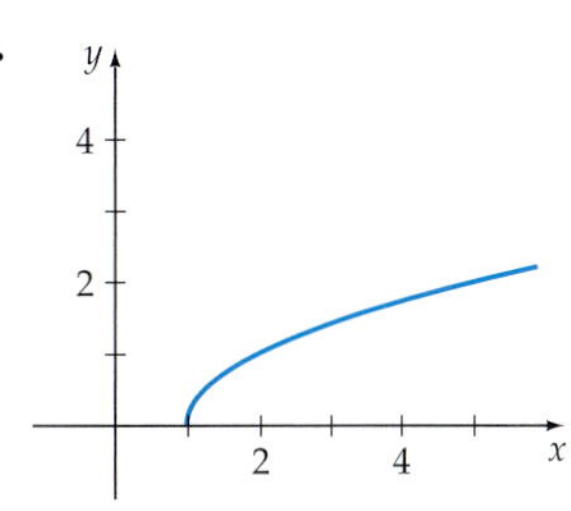

12.

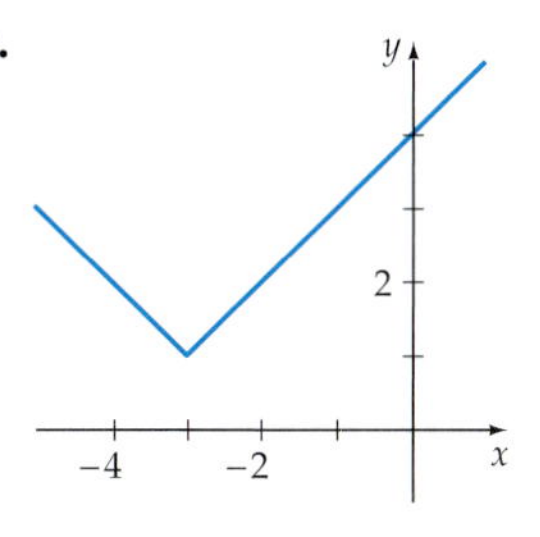

In Exercises 13 to 21, determine whether the graph of each equation is symmetric with respect to the *a.* *x*-axis, *b.* *y*-axis.

13. $y = 2x^2 - 5$ ▶**14.** $x = 3y^2 - 7$ **15.** $y = x^3 + 2$

16. $y = x^5 - 3x$ **17.** $x^2 + y^2 = 9$ **18.** $x^2 - y^2 = 10$

19. $x^2 = y^4$ **20.** $xy = 8$ **21.** $|x| - |y| = 6$

In Exercises 22 to 30, determine whether the graph of each equation is symmetric with respect to the origin.

22. $y = x + 1$ **23.** $y = 3x - 2$ ▶**24.** $y = x^3 - x$

25. $y = -x^3$ **26.** $y = \dfrac{9}{x}$ **27.** $x^2 + y^2 = 10$

28. $x^2 - y^2 = 4$ **29.** $y = \dfrac{x}{|x|}$ **30.** $|y| = |x|$

In Exercises 31 to 42, graph the given equation. Label each intercept. Use the concept of symmetry to confirm that the graph is correct.

31. $y = x^2 - 1$ **32.** $x = y^2 - 1$

33. $y = x^3 - x$ **34.** $y = -x^3$

35. $xy = 4$ **36.** $xy = -8$

37. $y = 2|x - 4|$ **38.** $y = |x - 2| - 1$

39. $y = (x - 2)^2 - 4$ **40.** $y = (x - 1)^2 - 4$

41. $y = x - |x|$ **42.** $|y| = |x|$

In Exercises 43 to 56, identify whether the given function is an even function, an odd function, or neither.

43. $g(x) = x^2 - 7$ ▶**44.** $h(x) = x^2 + 1$

45. $F(x) = x^5 + x^3$ **46.** $G(x) = 2x^5 - 10$

47. $H(x) = 3|x|$ **48.** $T(x) = |x| + 2$

49. $f(x) = 1$ **50.** $k(x) = 2 + x + x^2$

51. $r(x) = \sqrt{x^2 + 4}$ **52.** $u(x) = \sqrt{3 - x^2}$

53. $s(x) = 16x^2$ **54.** $v(x) = 16x^2 + x$

55. $w(x) = 4 + \sqrt[3]{x}$

56. $z(x) = \dfrac{x^3}{x^2 + 1}$

57. Use the graph of f to sketch the graph of

a. $y = f(x) + 3$ **b.** $y = f(x - 3)$

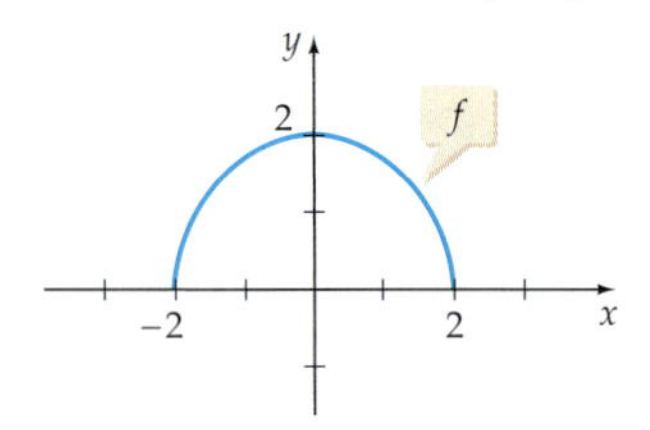

58. Use the graph of g to sketch the graph of

a. $y = g(x) - 2$ **b.** $y = g(x - 3)$

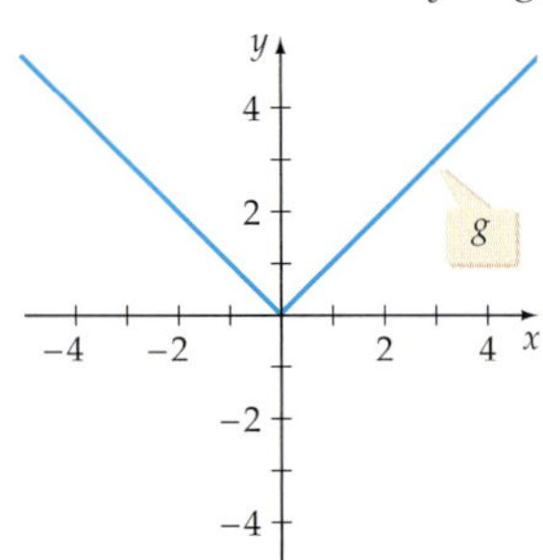

59. Use the graph of f to sketch the graph of

a. $y = f(x + 2)$ **b.** $y = f(x) + 2$

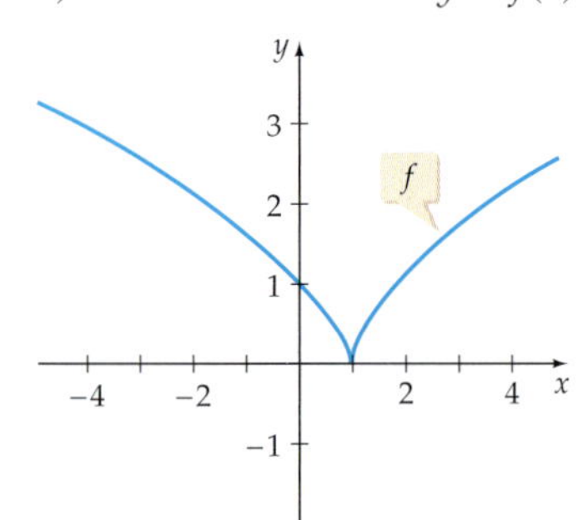

60. Use the graph of g to sketch the graph of

a. $y = g(x - 1)$ **b.** $y = g(x) - 1$

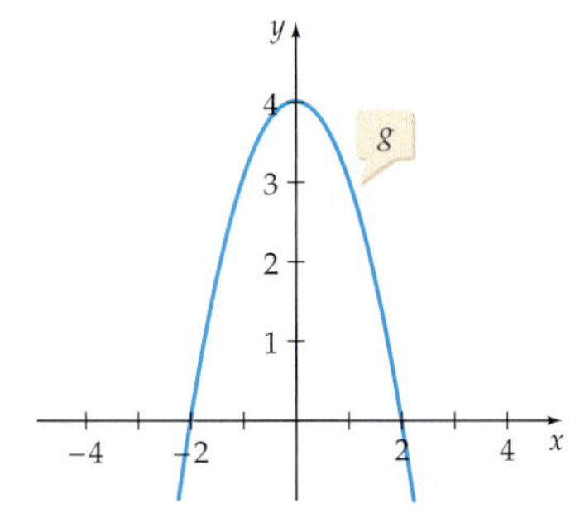

61. Let f be a function such that $f(-2) = 5$, $f(0) = -2$, and $f(1) = 0$. Give the coordinates of three points on the graph of

a. $y = f(x + 3)$ **b.** $y = f(x) + 1$

62. Let g be a function such that $g(-3) = -1$, $g(1) = -3$, and $g(4) = 2$. Give the coordinates of three points on the graph of

a. $y = g(x - 2)$ **b.** $y = g(x) - 2$

63. Use the graph of f to sketch the graph of

a. $y = f(-x)$ **b.** $y = -f(x)$

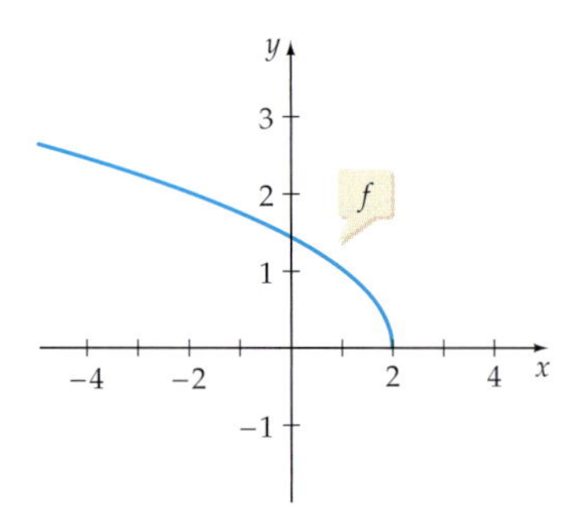

64. Use the graph of g to sketch the graph of

a. $y = -g(x)$ **b.** $y = g(-x)$

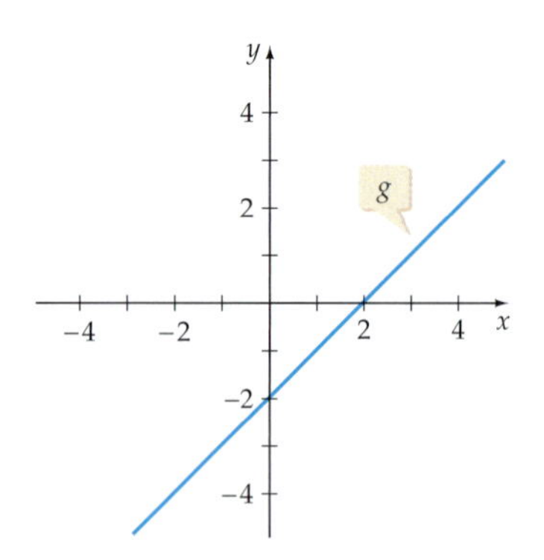

65. Let f be a function such that $f(-1) = 3$ and $f(2) = -4$. Give the coordinates of two points on the graph of

a. $y = f(-x)$ **b.** $y = -f(x)$

66. Let g be a function such that $g(4) = -5$ and $g(-3) = 2$. Give the coordinates of two points on the graph of

a. $y = -g(x)$ **b.** $y = g(-x)$.

67. Use the graph of F to sketch the graph of

a. $y = -F(x)$ **b.** $y = F(-x)$

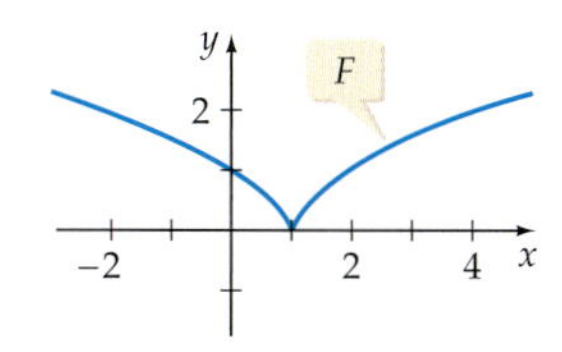

68. Use the graph of E to sketch the graph of

a. $y = -E(x)$ **b.** $y = E(-x)$

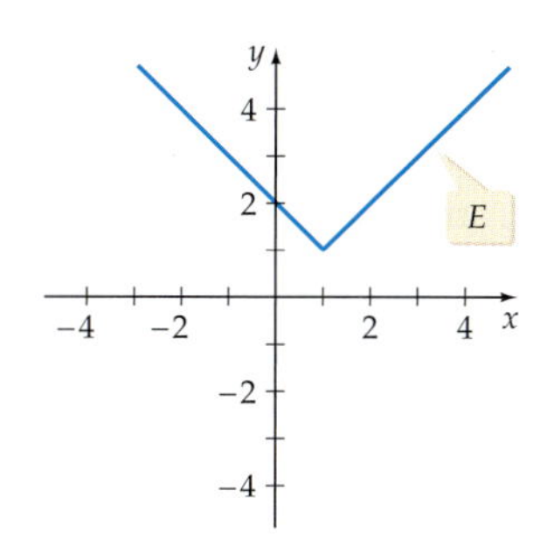

69. Use the graph of $m(x) = x^2 - 2x - 3$ to sketch the graph of $y = -\frac{1}{2}m(x) + 3$.

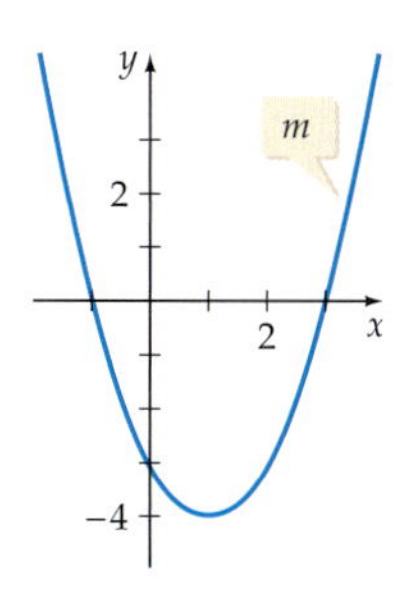

70. Use the graph of $n(x) = -x^2 - 2x + 8$ to sketch the graph of $y = \frac{1}{2}n(x) + 1$.

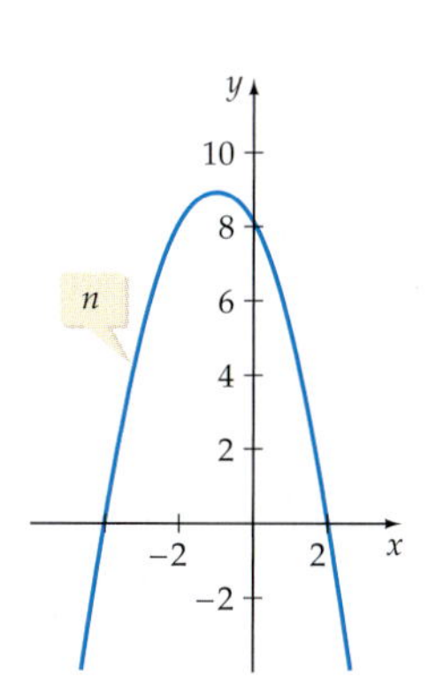

71. Use the graph of $y = f(x)$ to sketch the graph of

a. $y = f(2x)$ **b.** $y = f\left(\frac{1}{3}x\right)$

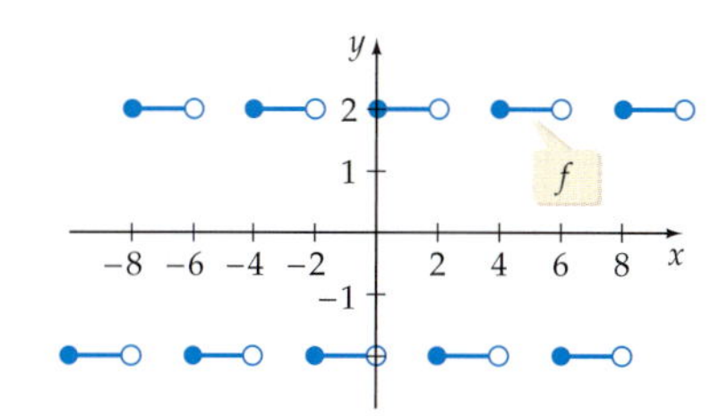

72. Use the graph of $y = g(x)$ to sketch the graph of

a. $y = g(2x)$ **b.** $y = g\left(\frac{1}{2}x\right)$

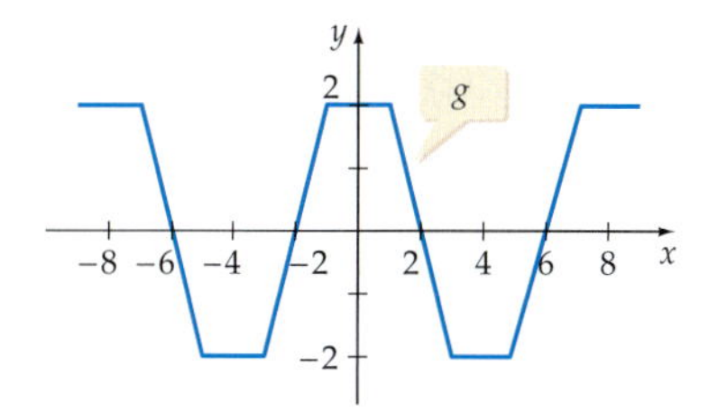

73. Use the graph of $y = h(x)$ to sketch the graph of

a. $y = h(2x)$ **b.** $y = h\left(\frac{1}{2}x\right)$

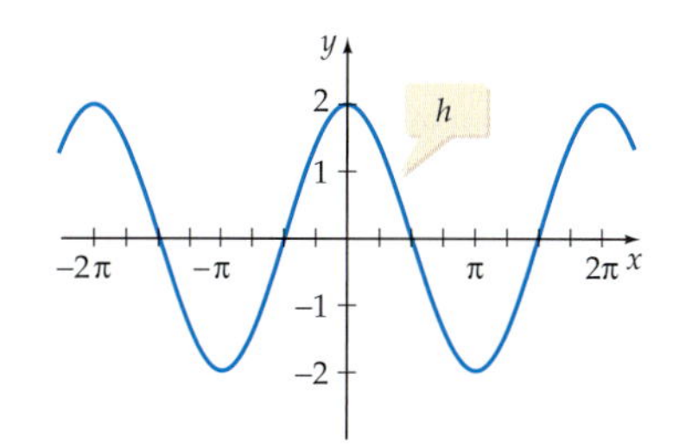

74. Use the graph of $y = j(x)$ to sketch the graph of

a. $y = j(2x)$ **b.** $y = j\left(\frac{1}{3}x\right)$

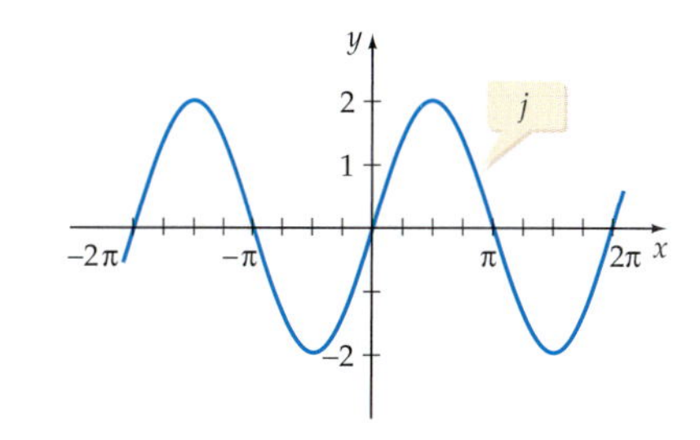

In Exercises 75 to 82, use a graphing utility.

75. On the same coordinate axes, graph

$$G(x) = \sqrt[3]{x} + c$$

for $c = 0, -1$, and 3.

76. On the same coordinate axes, graph

$$H(x) = \sqrt[3]{x + c}$$

for $c = 0, -1$, and 3.

77. On the same coordinate axes, graph

$$J(x) = |2(x + c) - 3| - |x + c|$$

for $c = 0, -1$, and 2.

78. On the same coordinate axes, graph

$$K(x) = |x - 1| - |x| + c$$

for $c = 0, -1$, and 2.

79. On the same coordinate axes, graph

$$L(x) = cx^2$$

for $c = 1, \dfrac{1}{2}$, and 2.

80. On the same coordinate axes, graph

$$M(x) = c\sqrt{x^2 - 4}$$

for $c = 1, \dfrac{1}{3}$, and 3.

81. On the same coordinate axes, graph

$$S(x) = c(|x - 1| - |x|)$$

for $c = 1, \dfrac{1}{4}$, and 4.

82. On the same coordinate axes, graph

$$T(x) = c\left(\frac{x}{|x|}\right)$$

for $c = 1, \dfrac{2}{3}$, and $\dfrac{3}{2}$.

83. Graph $V(x) = [\![cx]\!]$, $0 \le x \le 6$, for each value of c.

a. $c = 1$ **b.** $c = \dfrac{1}{2}$ **c.** $c = 2$

84. Graph $W(x) = [\![cx]\!] - cx$, $0 \le x \le 6$, for each value of c.

a. $c = 1$ **b.** $c = \dfrac{1}{3}$ **c.** $c = 3$

CONNECTING CONCEPTS

85. Use the graph of $f(x) = 2/(x^2 + 1)$ to determine an equation for the graphs shown in **a.** and **b.**

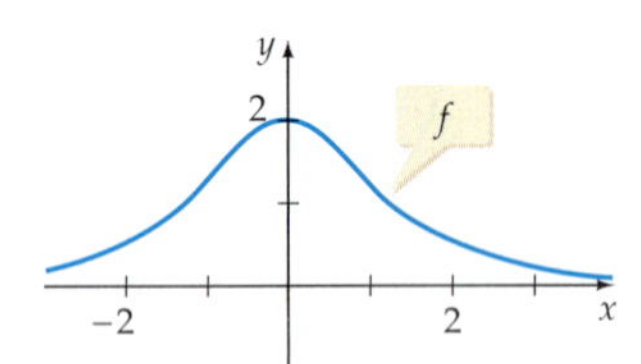

a.

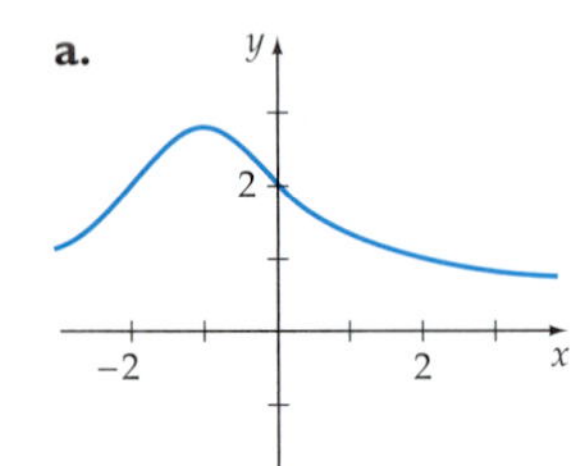

b.

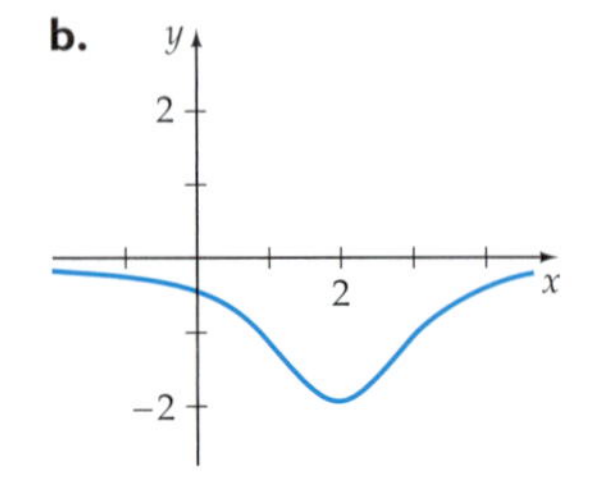

86. Use the graph of $f(x) = x\sqrt{2 + x}$ to determine an equation for the graphs shown in **a.** and **b.**

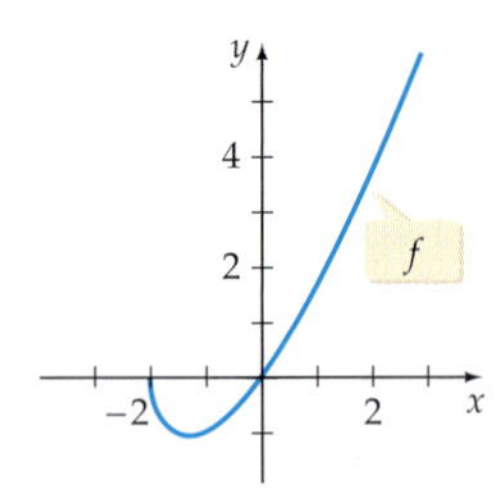

a.

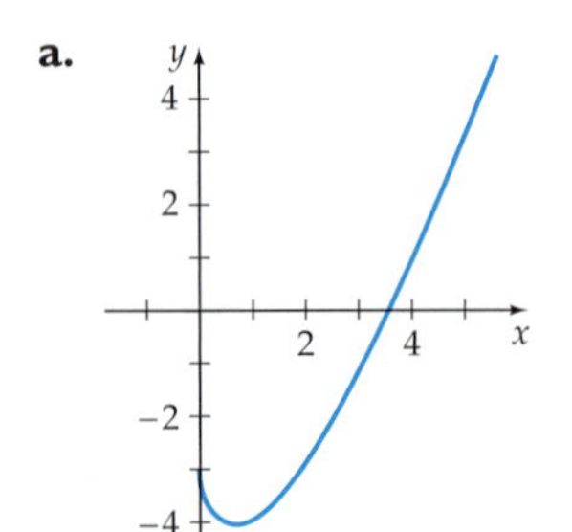

b.

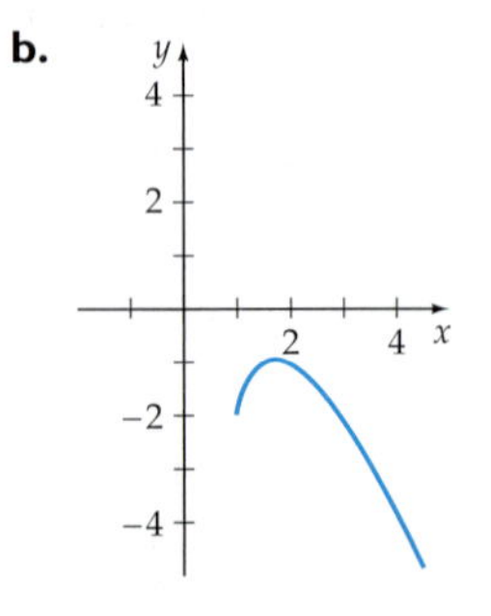

PREPARE FOR SECTION 1.7

87. Subtract: $(2x^2 + 3x - 4) - (x^2 + 3x - 5)$ [A.2]

88. Multiply: $(3x^2 - x + 2)(2x - 3)$ [A.2]

In Exercises 89 and 90, find each of the following for $f(x) = 2x^2 - 5x + 2$.

89. $f(3a)$ [1.3]

90. $f(2 + h)$ [1.3]

In Exercises 91 and 92, find the domain of each function.

91. $F(x) = \dfrac{x}{x - 1}$ [1.3]

92. $r(x) = \sqrt{2x - 8}$ [1.3]

PROJECTS

1. **DIRICHLET FUNCTION** We owe our present-day definition of a function to the German mathematician Peter Gustav Dirichlet (1805–1859). He created the following unusual function, which is now known as the *Dirichlet function*.

$$f(x) = \begin{cases} 0, & \text{if } x \text{ is a rational number} \\ 1, & \text{if } x \text{ is an irrational number} \end{cases}$$

Answer the following questions about the Dirichlet function.

a. What is its domain? **b.** What is its range?

c. What are its x-intercepts?

d. What is its y-intercept?

e. Is it an even or an odd function?

f.  Explain why a graphing calculator cannot be used to produce an accurate graph of the function.

g.  Write a sentence or two that describes its graph.

2. 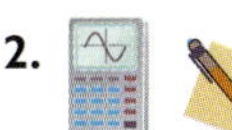  **ISOLATED POINT** Consider the function given by

$$y = \sqrt{(x - 1)^2(x - 2)} + 1$$

Verify that the point $(1, 1)$ is a solution of the equation. Now use a graphing utility to graph the function. Does your graph include the isolated point at $(1, 1)$, as shown at the right? If the graphing utility you used failed to include the point $(1, 1)$, explain at least one reason for the omission of this isolated point.

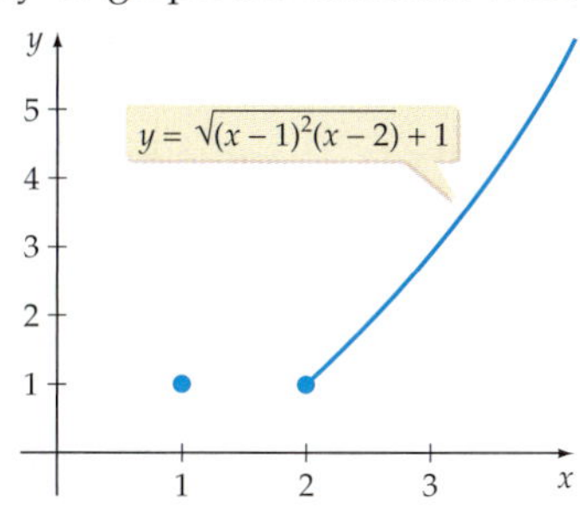

3. 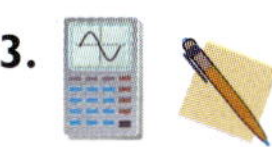 **A LINE WITH A HOLE** The function

$$f(x) = \frac{(x - 2)(x + 1)}{(x - 2)}$$

graphs as a line with a y-intercept of 1, a slope of 1, and a hole at $(2, 3)$. Use a graphing utility to graph f. Explain why a graphing utility might not show the hole at $(2, 3)$.

4. **FINDING A COMPLETE GRAPH** Use a graphing utility to graph the function $f(x) = 3x^{5/3} - 6x^{4/3} + 2$ for $-2 \le x \le 10$. Compare your graph with the graph below. Does your graph include the part to the left of the y-axis? If not, how might you enter the function so that the graphing utility you used would include this part?

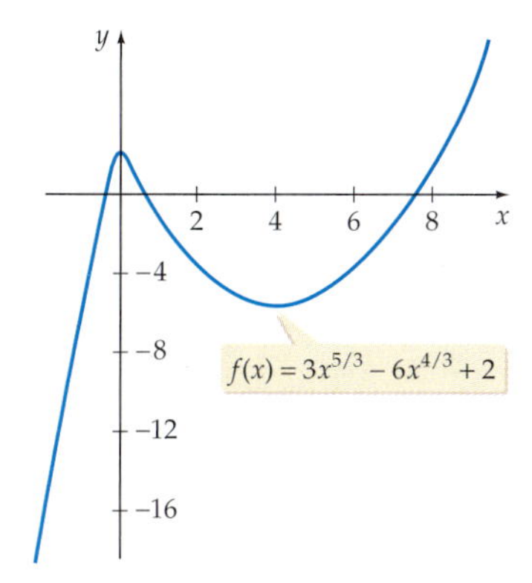

SECTION 1.7 THE ALGEBRA OF FUNCTIONS

OPERATIONS ON FUNCTIONS

Functions can be defined in terms of other functions. For example, the function defined by $h(x) = x^2 + 8x$ is the sum of

$$f(x) = x^2 \qquad \text{and} \qquad g(x) = 8x$$

Thus, if we are given any two functions f and g, we can define the four new functions $f + g$, $f - g$, fg, and $\dfrac{f}{g}$ as follows.

Operations on Functions

For all values of x for which both $f(x)$ and $g(x)$ are defined, we define the following functions.

Sum	$(f + g)(x) = f(x) + g(x)$
Difference	$(f - g)(x) = f(x) - g(x)$
Product	$(fg)(x) = f(x) \cdot g(x)$
Quotient	$\left(\dfrac{f}{g}\right)(x) = \dfrac{f(x)}{g(x)}, \quad g(x) \neq 0$

Domain of $f + g, f - g, fg, f/g$

For the given functions f and g, the domains of $f + g$, $f - g$, and $f \cdot g$ consist of all real numbers formed by the intersection of the domains of f and g. The domain of $\dfrac{f}{g}$ is the set of all real numbers formed by the intersection of the domains of f and g, except for those real numbers x such that $g(x) = 0$.

EXAMPLE 1 Determine the Domain of a Function

If $f(x) = \sqrt{x - 1}$ and $g(x) = x^2 - 4$, find the domains of $f + g$, $f - g$, fg, and $\dfrac{f}{g}$.

Solution

Note that f has the domain $\{x \mid x \geq 1\}$ and g has the domain of all real numbers. Therefore, the domain of $f + g$, $f - g$, and fg is $\{x \mid x \geq 1\}$. Because $g(x) = 0$ when $x = -2$ or $x = 2$, neither -2 nor 2 is in the domain of $\dfrac{f}{g}$. The domain of $\dfrac{f}{g}$ is $\{x \mid x \geq 1 \text{ and } x \neq 2\}$.

▶ TRY EXERCISE 10, PAGE 104

EXAMPLE 2 Evaluate Functions

Let $f(x) = x^2 - 9$ and $g(x) = 2x + 6$. Find

a. $(f + g)(5)$ **b.** $(fg)(-1)$ **c.** $\left(\frac{f}{g}\right)(4)$

Solution

a. $(f + g)(x) = f(x) + g(x) = (x^2 - 9) + (2x + 6) = x^2 + 2x - 3$
Therefore, $(f + g)(5) = (5)^2 + 2(5) - 3 = 25 + 10 - 3 = 32.$

b. $(fg)(x) = f(x) \cdot g(x) = (x^2 - 9)(2x + 6) = 2x^3 + 6x^2 - 18x - 54$
Therefore, $(fg)(-1) = 2(-1)^3 + 6(-1)^2 - 18(-1) - 54$
$= -2 + 6 + 18 - 54 = -32.$

c. $\left(\frac{f}{g}\right)(x) = \frac{f(x)}{g(x)} = \frac{x^2 - 9}{2x + 6} = \frac{\cancel{(x + 3)}(x - 3)}{2\cancel{(x + 3)}} = \frac{x - 3}{2}, \quad x \neq -3$
Therefore, $\left(\frac{f}{g}\right)(4) = \frac{4 - 3}{2} = \frac{1}{2}.$

▶ TRY EXERCISE 14, PAGE 104

THE DIFFERENCE QUOTIENT

> **take note**
> The difference quotient is an important concept that plays a fundamental role in calculus.

The expression

$$\frac{f(x + h) - f(x)}{h}, \quad h \neq 0$$

is called the **difference quotient** of f. It enables us to study the manner in which a function changes in value as the independent variable changes.

EXAMPLE 3 Determine a Difference Quotient

Determine the difference quotient of $f(x) = x^2 + 7$.

Solution

$$\frac{f(x + h) - f(x)}{h} = \frac{[(x + h)^2 + 7] - [x^2 + 7]}{h}$$

• Apply the difference quotient.

$$= \frac{[x^2 + 2xh + h^2 + 7] - [x^2 + 7]}{h}$$

$$= \frac{x^2 + 2xh + h^2 + 7 - x^2 - 7}{h}$$

$$= \frac{2xh + h^2}{h} = \frac{\cancel{h}(2x + h)}{\cancel{h}} = 2x + h$$

▶ TRY EXERCISE 30, PAGE 104

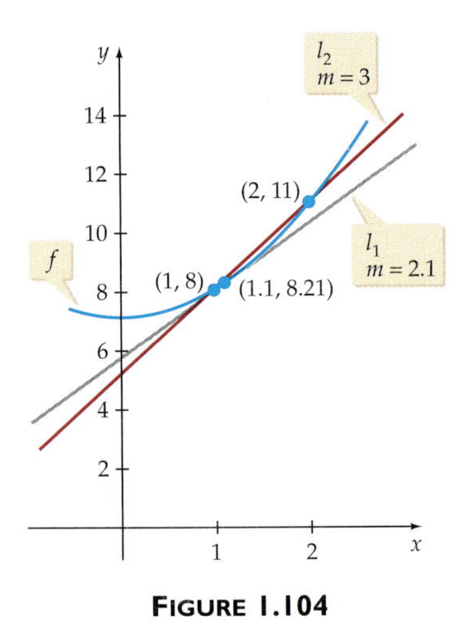

FIGURE 1.104

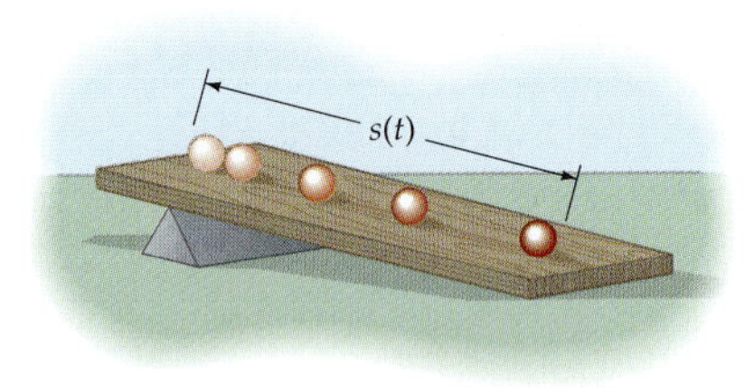

FIGURE 1.105

The difference quotient $2x + h$ of $f(x) = x^2 + 7$ from Example 3 is the slope of the secant line through the points

$$(x, f(x)) \qquad \text{and} \qquad (x + h, f(x + h))$$

For instance, let $x = 1$ and $h = 1$. Then the difference quotient is

$$2x + h = 2(1) + 1 = 3$$

This is the slope of the secant line l_2 through (1, 8) and (2, 11), as shown in **Figure 1.104.** If we let $x = 1$ and $h = 0.1$, then the difference quotient is

$$2x + h = 2(1) + 0.1 = 2.1$$

This is the slope of the secant line l_1 through (1, 8) and (1.1, 8.21).

The difference quotient

$$\frac{f(x + h) - f(x)}{h}$$

can be used to compute *average velocities.* In such cases it is traditional to replace f with s (for distance), the variable x with the variable a (for the time at the start of an observed interval of time), and the variable h with Δt (read as "delta t"), where Δt is the difference between the time at the end of an interval and the time at the start of the interval. For example, if an experiment is observed over the time interval from $t = 3$ seconds to $t = 5$ seconds, then the time interval is denoted as [3, 5] with $a = 3$ and $\Delta t = 5 - 3 = 2$. Thus if the distance traveled by a ball that rolls down a ramp is given by $s(t)$, where t is the time in seconds after the ball is released (see **Figure 1.105**), then the **average velocity** of the ball over the interval $t = a$ to $t = a + \Delta t$ is the difference quotient

$$\frac{s(a + \Delta t) - s(a)}{\Delta t}$$

EXAMPLE 4 Evaluate Average Velocities

The distance traveled by a ball rolling down a ramp is given by $s(t) = 4t^2$, where t is the time in seconds after the ball is released, and $s(t)$ is measured in feet. Evaluate the average velocity of the ball for each time interval.

a. [3, 5] **b.** [3, 4] **c.** [3, 3.5] **d.** [3, 3.01]

Solution

a. In this case, $a = 3$ and $\Delta t = 2$. Thus the average velocity over this interval is

$$\frac{s(a + \Delta t) - s(a)}{\Delta t} = \frac{s(3 + 2) - s(3)}{2} = \frac{s(5) - s(3)}{2} = \frac{100 - 36}{2}$$
$$= 32 \text{ feet per second}$$

b. Let $a = 3$ and $\Delta t = 4 - 3 = 1$.

$$\frac{s(a + \Delta t) - s(a)}{\Delta t} = \frac{s(3 + 1) - s(3)}{1} = \frac{s(4) - s(3)}{1} = \frac{64 - 36}{1}$$
$$= 28 \text{ feet per second}$$

c. Let $a = 3$ and $\Delta t = 3.5 - 3 = 0.5$.

$$\frac{s(a + \Delta t) - s(a)}{\Delta t} = \frac{s(3 + 0.5) - s(3)}{0.5} = \frac{49 - 36}{0.5} = 26 \text{ feet per second}$$

d. Let $a = 3$ and $\Delta t = 3.01 - 3 = 0.01$.

$$\frac{s(a + \Delta t) - s(a)}{\Delta t} = \frac{s(3 + 0.01) - s(3)}{0.01} = \frac{36.2404 - 36}{0.01}$$

$$= 24.04 \text{ feet per second}$$

TRY EXERCISE 72, PAGE 106

Composition of Functions

Composition of functions is another way in which functions can be combined. This method of combining functions uses the output of one function as the input for a second function.

Suppose that the spread of oil from a leak in a tanker can be approximated by a circle with the tanker at its center. The radius r (in feet) of the spill t hours after the leak begins is given by $r(t) = 150\sqrt{t}$. The area of the spill is the area of a circle and is given by the formula $A(r) = \pi r^2$. To find the area of the spill 4 hours after the leak begins, we first find the radius of the spill and then use that number to find the area of the spill.

$$\begin{aligned} r(t) &= 150\sqrt{t} \\ r(4) &= 150\sqrt{4} && \bullet\ t = 4 \text{ hours} \\ &= 150(2) \\ &= 300 \end{aligned} \qquad \begin{aligned} A(r) &= \pi r^2 \\ A(300) &= \pi(300^2) && \bullet\ r = 300 \text{ feet} \\ &= 90{,}000\pi \\ &\approx 283{,}000 \end{aligned}$$

The area of the spill after 4 hours is approximately 283,000 square feet.

There is an alternative way to solve this problem. Because the area of the spill depends on the radius and the radius depends on the time, there is a relationship between area and time. We can determine this relationship by evaluating the formula for the area of a circle using $r(t) = 150\sqrt{t}$. This will give the area of the spill as a function of time.

$$\begin{aligned} A(r) &= \pi r^2 \\ A[r(t)] &= \pi[r(t)]^2 && \bullet \text{ Replace } r \text{ by } r(t). \\ &= \pi\left[150\sqrt{t}\right]^2 && \bullet\ r(t) = 150\sqrt{t} \\ A(t) &= 22{,}500\pi t && \bullet \text{ Simplify.} \end{aligned}$$

The area of the spill as a function of time is $A(t) = 22{,}500\pi t$. To find the area of the oil spill after 4 hours, evaluate this function at $t = 4$.

$$\begin{aligned} A(t) &= 22{,}500\pi t \\ A(4) &= 22{,}500\pi(4) && \bullet\ t = 4 \text{ hours} \\ &= 90{,}000\pi \\ &\approx 283{,}000 \end{aligned}$$

This is the same result we calculated earlier.

The function $A(t) = 22{,}500\pi t$ is referred to as the *composition* of A with r. The notation $A \circ r$ is used to denote this composition of functions. That is,

$$(A \circ r)(t) = 22{,}500\pi t$$

Definition of the Composition of Two Functions

Let f and g be two functions such that $g(x)$ is in the domain of f for all x in the domain of g. Then the composition of the two functions, denoted by $f \circ g$, is the function whose value at x is given by $(f \circ g)(x) = f[g(x)]$.

The function defined by $(f \circ g)(x)$ is also called the *composite* of f and g. We read $(f \circ g)(x)$ as "f circle g of x" and $f[g(x)]$ as "f of g of x."

Consider the functions $f(x) = 2x - 1$ and $g(x) = x^2 - 3$. The expression $(f \circ g)(-1)$ (or, equivalently, $f[g(-1)]$) means to evaluate the function f at $g(-1)$.

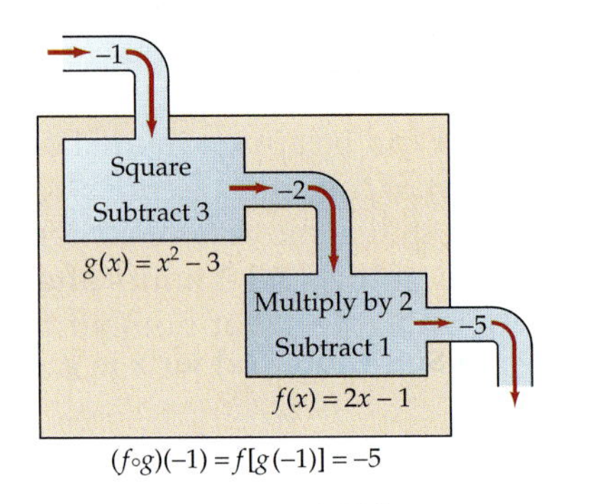

$(f \circ g)(-1) = f[g(-1)] = -5$

FIGURE 1.106

$$g(x) = x^2 - 3$$
$$g(-1) = (-1)^2 - 3 \qquad \text{• Evaluate } g \text{ at } -1.$$
$$= -2$$

$$f(x) = 2x - 1$$
$$f(-2) = 2(-2) - 1 = -5 \qquad \text{• Evaluate } f \text{ at } g(-1) = -2.$$

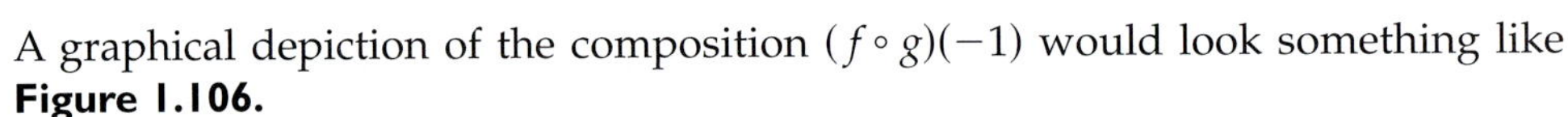
A graphical depiction of the composition $(f \circ g)(-1)$ would look something like **Figure 1.106.**

The requirement in the definition of the composition of two functions that $g(x)$ be in the domain of f for all x in the domain of g is important. For instance, let

$$f(x) = \frac{1}{x - 1} \quad \text{and} \quad g(x) = 3x - 5$$

When $x = 2$,

$$g(2) = 3(2) - 5 = 1$$
$$f[g(2)] = f(1) = \frac{1}{1 - 1} = \frac{1}{0} \qquad \text{• Undefined}$$

In this case, $g(2)$ is not in the domain of f. Thus the composition $(f \circ g)(x)$ is not defined at 2.

We can find a general expression for $f[g(x)]$ by evaluating f at $g(x)$. For instance, using $f(x) = 2x - 1$ and $g(x) = x^2 - 3$ as in **Figure 1.106,** we have

$$f(x) = 2x - 1$$
$$f[g(x)] = 2[g(x)] - 1 \qquad \text{• Replace } x \text{ by } g(x).$$
$$= 2[x^2 - 3] - 1 \qquad \text{• Replace } g(x) \text{ by } x^2 - 3.$$
$$= 2x^2 - 7 \qquad \text{• Simplify.}$$

In general, the composition of functions is not a commutative operation. That is, $(f \circ g)(x) \neq (g \circ f)(x)$. To verify this, we will compute the composition

$(g \circ f)(x) = g[f(x)]$, again using the functions $f(x) = 2x - 1$ and $g(x) = x^2 - 3$.

$$g(x) = x^2 - 3$$
$$g[f(x)] = [f(x)]^2 - 3 \qquad \text{• Replace } x \text{ by } f(x).$$
$$= [2x - 1]^2 - 3 \qquad \text{• Replace } f(x) \text{ by } 2x - 1.$$
$$= 4x^2 - 4x - 2 \qquad \text{• Simplify.}$$

Thus $f[g(x)] = 2x^2 - 7$, which is not equal to $g[f(x)] = 4x^2 - 4x - 2$. Therefore, $(f \circ g)(x) \neq (g \circ f)(x)$ and composition is not a commutative operation.

? QUESTION Let $f(x) = x - 1$ and $g(x) = x + 1$. Then $f[g(x)] = g[f(x)]$. (You should verify this statement.) Does this contradict the statement we made that composition is not a commutative operation?

EXAMPLE 5 Form Composite Functions

If $f(x) = x^2 - 3x$ and $g(x) = 2x + 1$, find

a. $(g \circ f)$ **b.** $(f \circ g)$

Solution

a.

$$(g \circ f) = g[f(x)] = 2(f(x)) + 1 \qquad \text{• Substitute } f(x) \text{ for } x \text{ in } g.$$
$$= 2(x^2 - 3x) + 1 \qquad \text{• } f(x) = x^2 - 3x$$
$$= 2x^2 - 6x + 1$$

b.

$$(f \circ g) = f[g(x)] = (g(x))^2 - 3(g(x)) \qquad \text{• Substitute } g(x) \text{ for } x \text{ in } f.$$
$$= (2x + 1)^2 - 3(2x + 1) \qquad \text{• } g(x) = 2x + 1$$
$$= 4x^2 - 2x - 2$$

▶ TRY EXERCISE 38, PAGE 105

Note that in this example $(f \circ g) \neq (g \circ f)$. As stated earlier, in general, the composition of functions is not a commutative operation.

Caution Some care must be used when forming the composition of functions. For instance, if $f(x) = x + 1$ and $g(x) = \sqrt{x - 4}$, then

$$(g \circ f)(2) = g[f(2)] = g(3) = \sqrt{3 - 4} = \sqrt{-1}$$

which is not a real number. We can avoid this problem by imposing suitable restrictions on the domain of f so that the range of f is part of the domain of g. If the

---

? ANSWER No. When we say that composition is not a commutative operation, we mean that generally, given any two functions, $(f \circ g)(x) \neq (g \circ f)(x)$. However, there may be particular instances in which $(f \circ g)(x) = (g \circ f)(x)$. It turns out that these particular instances are quite important, as we shall see later.

domain of f is restricted to $[3, \infty)$, then the range of f is $[4, \infty)$. But this is precisely the domain of g. Note that $2 \notin [3, \infty)$, and thus we avoid the problem of $(g \circ f)(2)$ not being a real number.

To evaluate $(f \circ g)(c)$ for some constant c, you can use either of the following methods.

Method 1 First evaluate $g(c)$. Then substitute this result for x in $f(x)$.

Method 2 First determine $f[g(x)]$ and then substitute c for x.

> **take note**
>
> In Example 6, both Method 1 and Method 2 produce the same result. Although Method 2 is longer, it is the better method if you must evaluate $(f \circ g)(x)$ for several values of x.

EXAMPLE 6 Evaluate a Composite Function

Evaluate $(f \circ g)(3)$, where $f(x) = 2x - 7$ and $g(x) = x^2 + 4$.

Solution

Method 1

$$(f \circ g)(3) = f[g(3)]$$
$$= f[(3)^2 + 4] \qquad \text{• Evaluate } g(3).$$
$$= f(13)$$
$$= 2(13) - 7 = 19 \qquad \text{• Substitute 13 for } x \text{ in } f.$$

Method 2

$$(f \circ g)(x) = 2[g(x)] - 7 \qquad \text{• Form } f[g(x)].$$
$$= 2[x^2 + 4] - 7$$
$$= 2x^2 + 1$$
$$(f \circ g)(3) = 2(3)^2 + 1 = 19 \qquad \text{• Substitute 3 for } x.$$

TRY EXERCISE 50, PAGE 105

Figures 1.107 and **1.108** graphically illustrate the difference between Method 1 and Method 2.

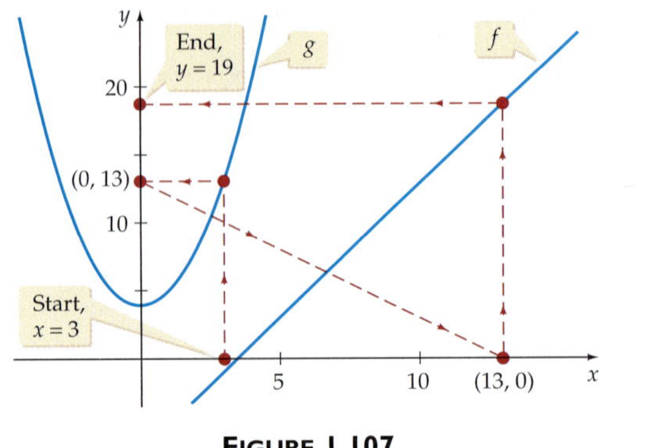

FIGURE 1.107
Method 1

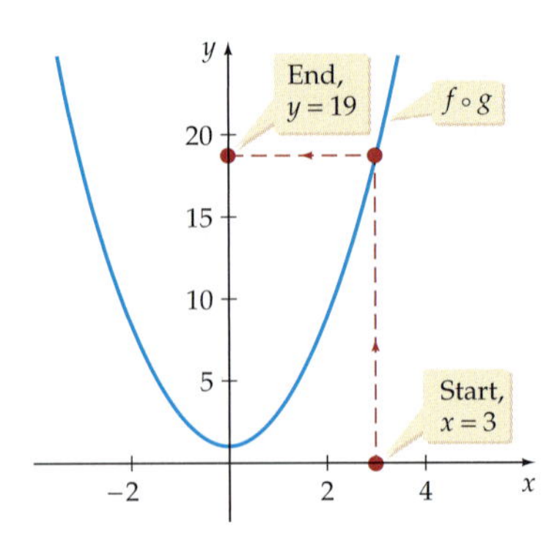

FIGURE 1.108
Method 2

EXAMPLE 7 Use a Composite Function to Solve an Application

A graphic artist has drawn a 3-inch by 2-inch rectangle on a computer screen. The artist has been scaling the size of the rectangle for t seconds in such a way that the upper right corner of the original rectangle is moving to the right at the rate of 0.5 inch per second and downward at the rate of 0.2 inch per second. See **Figure 1.109.**

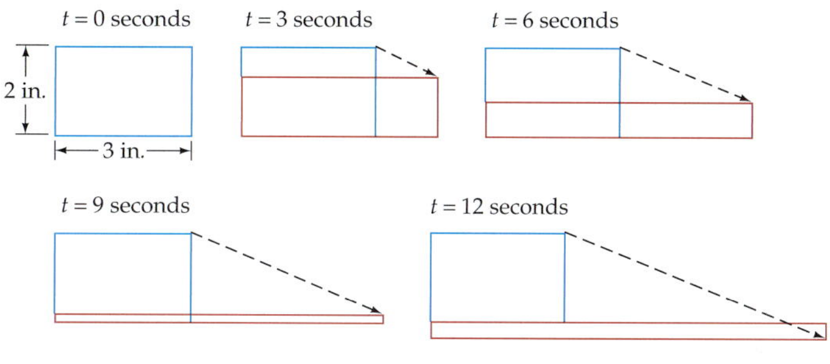

FIGURE 1.109

a. Write the lengths l and the widths w of the scaled rectangles as functions of t.

b. Write the area A of the scaled rectangle as a function of t.

c. Find the intervals on which A is an increasing function for $0 \leq t \leq 14$. Also find the intervals on which A is a decreasing function.

d. Find the value of t (where $0 \leq t \leq 14$) that maximizes $A(t)$.

Solution

a. Because *distance* = *rate* · *time*, we see that the change in l is given by $0.5t$. Therefore, the length at any time t is $l = 3 + 0.5t$. For $0 \leq t \leq 10$, the width is given by $w = 2 - 0.2t$. For $10 < t \leq 14$, the width is $w = -2 + 0.2t$. In either case the width can be determined by finding $w = |2 - 0.2t|$. (The absolute value symbol is needed to keep the width positive for $10 < t \leq 14$.)

b. $A = lw = (3 + 0.5t)|2 - 0.2t|$

c. Use a graphing utility to determine that A is increasing on $[0, 2]$ and on $[10, 14]$ and that A is decreasing on $[2, 10]$. See **Figure 1.110.**

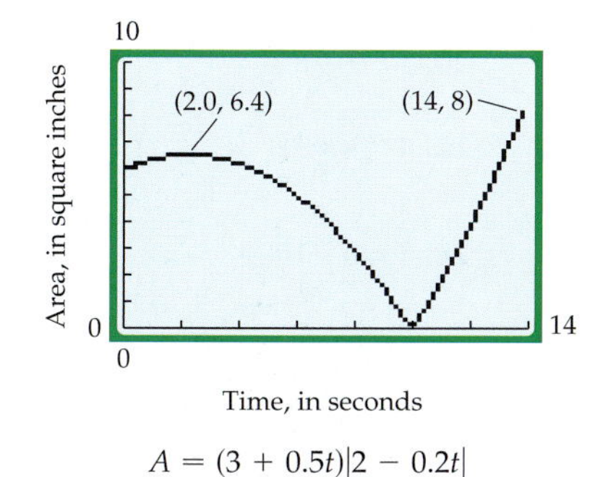

$A = (3 + 0.5t)|2 - 0.2t|$

FIGURE 1.110

d. The highest point on the graph of A occurs when $t = 14$ seconds. See **Figure 1.110.**

TRY EXERCISE 66, PAGE 105

You may be inclined to think that if the area of a rectangle is decreasing, then its perimeter is also decreasing, but this is not always the case. For example, the area of the scaled rectangle in Example 7 was shown to decrease on [2, 10] even though its perimeter is always increasing. See Exercise 68 in Exercise Set 1.7.

Topics for Discussion

1. The domain of $f + g$ consists of all real numbers formed by the *union* of the domain of f and the domain of g. Do you agree?
2. Given $f(x) = 3x - 2$ and $g(x) = \frac{1}{3}x + \frac{2}{3}$, determine $f \circ g$ and $g \circ f$. Does this show that composition of functions is a commutative operation?
3. A tutor states that the difference quotient of $f(x) = x^2$ and the difference quotient of $g(x) = x^2 + 4$ are the same. Do you agree?
4. A classmate states that the difference quotient of any linear function $f(x) = mx + b$ is always m. Do you agree?
5. When we use a difference quotient to determine an average velocity, we generally replace the variable h with the variable Δt. What does Δt represent?

Exercise Set 1.7

In Exercises 1 to 12, use the given functions f and g to find $f + g$, $f - g$, fg, and $\frac{f}{g}$. State the domain of each.

1. $f(x) = x^2 - 2x - 15, \quad g(x) = x + 3$
2. $f(x) = x^2 - 25, \quad g(x) = x - 5$
3. $f(x) = 2x + 8, \quad g(x) = x + 4$
4. $f(x) = 5x - 15, \quad g(x) = x - 3$
5. $f(x) = x^3 - 2x^2 + 7x, \quad g(x) = x$
6. $f(x) = x^2 - 5x - 8, \quad g(x) = -x$
7. $f(x) = 2x^2 + 4x - 7, \quad g(x) = 2x^2 + 3x - 5$
8. $f(x) = 6x^2 + 10, \quad g(x) = 3x^2 + x - 10$
9. $f(x) = \sqrt{x - 3}, \quad g(x) = x$
10. ▶ $f(x) = \sqrt{x - 4}, \quad g(x) = -x$
11. $f(x) = \sqrt{4 - x^2}, \quad g(x) = 2 + x$
12. $f(x) = \sqrt{x^2 - 9}, \quad g(x) = x - 3$

In Exercises 13 to 28, evaluate the indicated function, where $f(x) = x^2 - 3x + 2$ and $g(x) = 2x - 4$.

13. $(f + g)(5)$
14. ▶ $(f + g)(-7)$
15. $(f + g)\left(\frac{1}{2}\right)$
16. $(f + g)\left(\frac{2}{3}\right)$
17. $(f - g)(-3)$
18. $(f - g)(24)$
19. $(f - g)(-1)$
20. $(f - g)(0)$
21. $(fg)(7)$
22. $(fg)(-3)$
23. $(fg)\left(\frac{2}{5}\right)$
24. $(fg)(-100)$
25. $\left(\frac{f}{g}\right)(-4)$
26. $\left(\frac{f}{g}\right)(11)$
27. $\left(\frac{f}{g}\right)\left(\frac{1}{2}\right)$
28. $\left(\frac{f}{g}\right)\left(\frac{1}{4}\right)$

In Exercises 29 to 36, find the difference quotient of the given function.

29. $f(x) = 2x + 4$
30. ▶ $f(x) = 4x - 5$
31. $f(x) = x^2 - 6$
32. $f(x) = x^2 + 11$

33. $f(x) = 2x^2 + 4x - 3$

34. $f(x) = 2x^2 - 5x + 7$

35. $f(x) = -4x^2 + 6$

36. $f(x) = -5x^2 - 4x$

In Exercises 37 to 48, find $g \circ f$ and $f \circ g$ for the given functions f and g.

37. $f(x) = 3x + 5, \quad g(x) = 2x - 7$

38. $f(x) = 2x - 7, \quad g(x) = 3x + 2$

39. $f(x) = x^2 + 4x - 1, \quad g(x) = x + 2$

40. $f(x) = x^2 - 11x, \quad g(x) = 2x + 3$

41. $f(x) = x^3 + 2x, \quad g(x) = -5x$

42. $f(x) = -x^3 - 7, \quad g(x) = x + 1$

43. $f(x) = \dfrac{2}{x+1}, \quad g(x) = 3x - 5$

44. $f(x) = \sqrt{x+4}, \quad g(x) = \dfrac{1}{x}$

45. $f(x) = \dfrac{1}{x^2}, \quad g(x) = \sqrt{x-1}$

46. $f(x) = \dfrac{6}{x-2}, \quad g(x) = \dfrac{3}{5x}$

47. $f(x) = \dfrac{3}{|5-x|}, \quad g(x) = -\dfrac{2}{x}$

48. $f(x) = |2x + 1|, \quad g(x) = 3x^2 - 1$

In Exercises 49 to 64, evaluate each composite function, where $f(x) = 2x + 3$, $g(x) = x^2 - 5x$, and $h(x) = 4 - 3x^2$.

49. $(g \circ f)(4)$

50. $(f \circ g)(4)$

51. $(f \circ g)(-3)$

52. $(g \circ f)(-1)$

53. $(g \circ h)(0)$

54. $(h \circ g)(0)$

55. $(f \circ f)(8)$

56. $(f \circ f)(-8)$

57. $(h \circ g)\left(\dfrac{2}{5}\right)$

58. $(g \circ h)\left(-\dfrac{1}{3}\right)$

59. $(g \circ f)(\sqrt{3})$

60. $(f \circ g)(\sqrt{2})$

61. $(g \circ f)(2c)$

62. $(f \circ g)(3k)$

63. $(g \circ h)(k + 1)$

64. $(h \circ g)(k - 1)$

65. **Water Tank** A water tank has the shape of a right circular cone, with height 16 feet and radius 8 feet. Water is running into the tank so that the radius r (in feet) of the surface of the water is given by $r = 1.5t$, where t is the time (in minutes) that the water has been running.

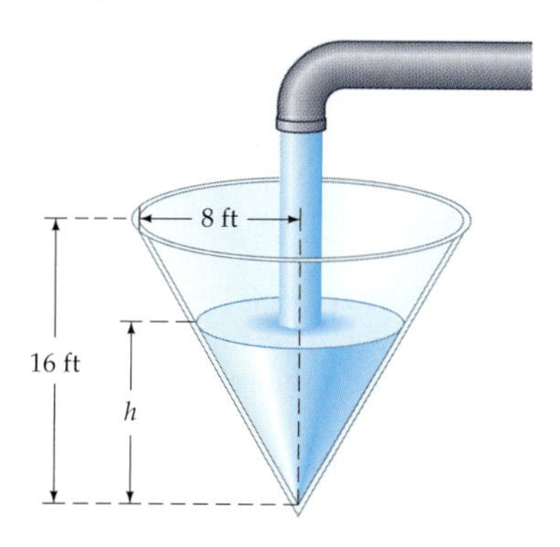

a. The area A of the surface of the water is $A = \pi r^2$. Find $A(t)$ and use it to determine the area of the surface of the water when $t = 2$ minutes.

b. The volume V of the water is given by $V = \dfrac{1}{3}\pi r^2 h$. Find $V(t)$ and use it to determine the volume of the water when $t = 3$ minutes. (*Hint:* The height of the water in the cone is always twice the radius of the water.)

66. **Scaling a Rectangle** Work Example 7 of this section with the scaling as follows. The upper right corner of the original rectangle is pulled to the *left* at 0.5 inch per second and downward at 0.2 inch per second.

67. **Towing a Boat** A boat is towed by a rope that runs through a pulley that is 4 feet above the point where the rope is tied to the boat. The length (in feet) of the rope from the boat to the pulley is given by $s = 48 - t$, where t is the time in seconds that the boat has been in tow. The horizontal distance from the pulley to the boat is d.

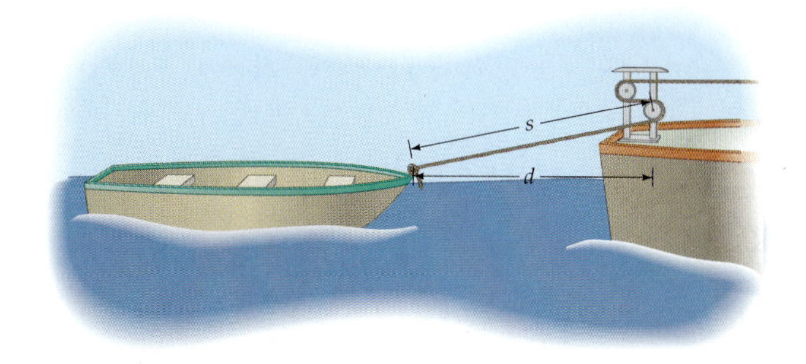

a. Find $d(t)$.

b. Evaluate $s(35)$ and $d(35)$.

68. **PERIMETER OF A SCALED RECTANGLE** Show by a graph that the perimeter

$$P = 2(3 + 0.5t) + 2|2 - 0.2t|$$

of the scaled rectangle in Example 7 of this section is an increasing function over $0 \le t \le 14$.

69. **CONVERSION FUNCTIONS** The function $F(x) = \dfrac{x}{12}$ converts x inches to feet. The function $Y(x) = \dfrac{x}{3}$ converts x feet to yards. Explain the meaning of $(Y \circ F)(x)$.

70. **CONVERSION FUNCTIONS** The function $F(x) = 3x$ converts x yards to feet. The function $I(x) = 12x$ converts x feet to inches. Explain the meaning of $(I \circ F)(x)$.

71. **CONCENTRATION OF A MEDICATION** The concentration $C(t)$ (in milligrams per liter) of a medication in a patient's blood is given by the data in the following table.

Concentration of Medication in Patient's Blood

t hours	$C(t)$ mg/l
0	0
0.25	47.3
0.50	78.1
0.75	94.9
1.00	99.8
1.25	95.7
1.50	84.4
1.75	68.4
2.00	50.1
2.25	31.6
2.50	15.6
2.75	4.3

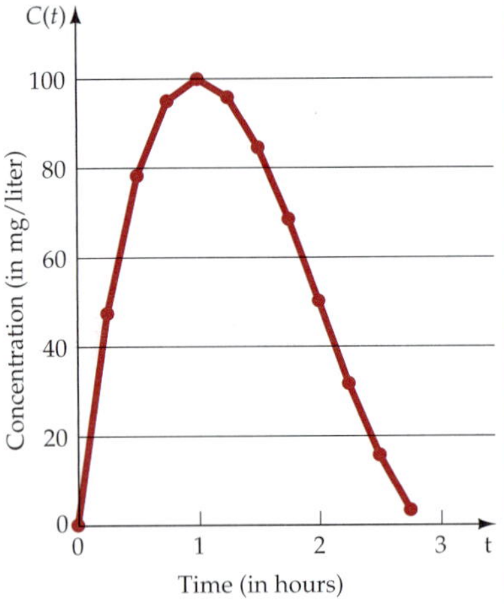

C(t)
100
80
60
40
20
0
Concentration (in mg/liter)
0 1 2 3 t
Time (in hours)

The **average rate of change** of the concentration over the time interval from $t = a$ to $t = a + \Delta t$ is

$$\frac{C(a + \Delta t) - C(a)}{\Delta t}$$

Use the data in the table to evaluate the average rate of change for each of the following time intervals.

a. $[0, 1]$ (*Hint:* In this case, $a = 0$ and $\Delta t = 1$.) Compare this result to the slope of the line through $(0, C(0))$ and $(1, C(1))$.

b. $[0, 0.5]$ **c.** $[1, 2]$ **d.** $[1, 1.5]$ **e.** $[1, 1.25]$

f. The data in the table can be modeled by the function $Con(t) = 25t^3 - 150t^2 + 225t$. Use $Con(t)$ to verify that the average rate of change over $[1, 1 + \Delta t]$ is $-75(\Delta t) + 25(\Delta t)^2$. What does the average rate of change over $[1, 1 + \Delta t]$ seem to approach as Δt approaches 0?

72. **BALL ROLLING ON A RAMP** The distance traveled by a ball rolling down a ramp is given by $s(t) = 6t^2$, where t is the time in seconds after the ball is released, and $s(t)$ is measured in feet. The ball travels 6 feet in 1 second and it travels 24 feet in 2 seconds. Use the difference quotient for average velocity given on page 98 to evaluate the average velocity for each of the following time intervals.

a. $[2, 3]$ (*Hint:* In this case, $a = 2$ and $\Delta t = 1$.) Compare this result to the slope of the line through $(2, s(2))$ and $(3, s(3))$.

b. $[2, 2.5]$ **c.** $[2, 2.1]$ **d.** $[2, 2.01]$ **e.** $[2, 2.001]$

f. Verify that the average velocity over $[2, 2 + \Delta t]$ is $24 + 6(\Delta t)$. What does the average velocity seem to approach as Δt approaches 0?

CONNECTING CONCEPTS

In Exercises 73 to 76, show that $(f \circ g)(x) = (g \circ f)(x)$.

73. $f(x) = 2x + 3$; $g(x) = 5x + 12$

74. $f(x) = 4x - 2$; $g(x) = 7x - 4$

75. $f(x) = \dfrac{6x}{x - 1}$; $g(x) = \dfrac{5x}{x - 2}$

76. $f(x) = \dfrac{5x}{x + 3}$; $g(x) = -\dfrac{2x}{x - 4}$

In Exercises 77 to 82, show that

$$(g \circ f)(x) = x \quad \text{and} \quad (f \circ g)(x) = x$$

77. $f(x) = 2x + 3, \quad g(x) = \dfrac{x - 3}{2}$

78. $f(x) = 4x - 5, \quad g(x) = \dfrac{x + 5}{4}$

79. $f(x) = \dfrac{4}{x + 1}, \quad g(x) = \dfrac{4 - x}{x}$

80. $f(x) = \dfrac{2}{1 - x}, \quad g(x) = \dfrac{x - 2}{x}$

81. $f(x) = x^3 - 1, \quad g(x) = \sqrt[3]{x + 1}$

82. $f(x) = -x^3 + 2, \quad g(x) = \sqrt[3]{2 - x}$

PROJECTS

1. **A Graphing Utility Project** For any two different real numbers x and y, the larger of the two numbers is given by

$$\text{Maximum}(x, y) = \frac{x + y}{2} + \frac{|x - y|}{2} \tag{1}$$

a. Verify Equation (1) for $x = 5$ and $y = 9$.

b. Verify Equation (1) for $x = 201$ and $y = 80$.

For any two different functional values $f(x)$ and $g(x)$, the larger of the two is given by

$$\text{Maximum}(f(x), g(x)) = \frac{f(x) + g(x)}{2} + \frac{|f(x) - g(x)|}{2} \tag{2}$$

To illustrate how we might make use of Equation (2), consider the functions $y_1 = x^2$ and $y_2 = \sqrt{x}$ on the interval from Xmin $= -1$ to Xmax $= 6$. The graphs of y_1 and y_2 are shown below.

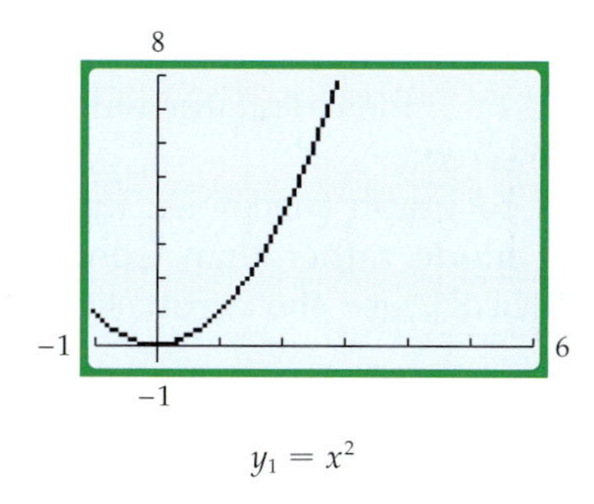

$y_1 = x^2$

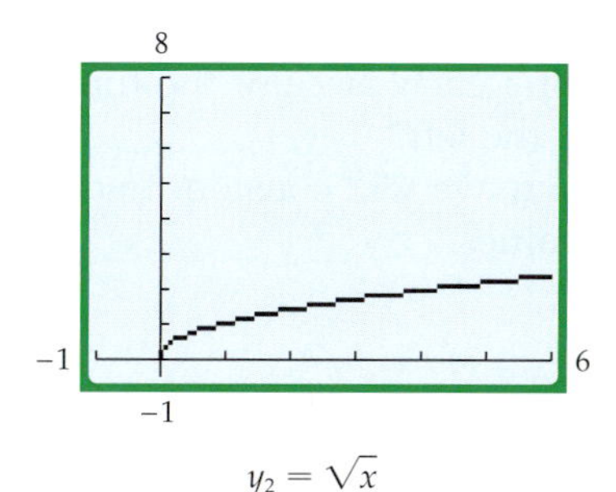

$y_2 = \sqrt{x}$

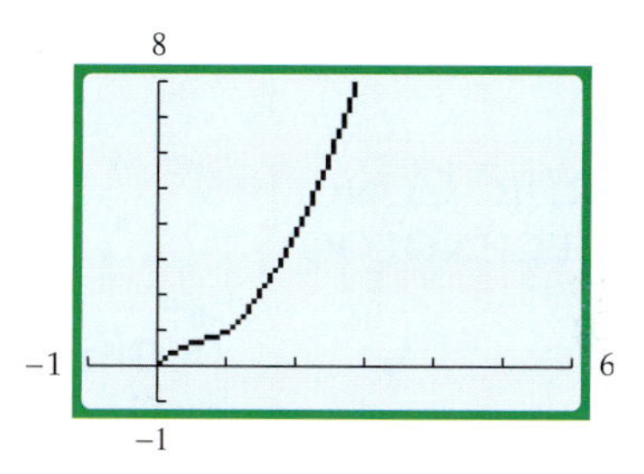

$y_3 = (y_1 + y_2)/2 + (\text{abs}\,(y_1 - y_2))/2$

Now consider the function

$$y_3 = (y_1 + y_2)/2 + (\text{abs}(y_1 - y_2))/2$$

where "abs" represents the absolute value function. The graph of y_3 is shown above.

c. Write a sentence or two that explains why the graph of y_3 is as shown.

d. What is the domain of y_1? of y_2? of y_3? Write a sentence that explains how to determine the domain of y_3, given the domain of y_1 and the domain of y_2.

e. Determine a formula for the function Minimum$(f(x), g(x))$.

2. **The Never-Negative Function** The author J. D. Murray describes a function f_+ that is defined in the following manner.[2]

$$f_+ = \begin{cases} f & \text{if } f \geq 0 \\ 0 & \text{if } f < 0 \end{cases}$$

We will refer to this function as a **never-negative** function. Never-negative functions can be graphed by using Equation (2) in Project 1. For example, if we let $g(x) = 0$, then Equation (2) simplifies to

$$\text{Maximum}(f(x), 0) = \frac{f(x)}{2} + \frac{|f(x)|}{2} \tag{3}$$

The graph of $y = \text{Maximum}(f(x), 0)$ is the graph of $y = f(x)$ provided that $f(x) \geq 0$, and it is the graph of $y = 0$ provided that $f(x) < 0$.

[2] *Mathematical Biology* (New York: Springer-Verlag, 1989), p. 101.

An Application The mosquito population per acre of a large resort is controlled by spraying on a monthly basis. A biologist has determined that the mosquito population can be approximated by the never-negative function M_+ with

$$M(t) = -35{,}400(t - \text{int}(t))^2 + 35{,}400(t - \text{int}(t)) - 4000$$

Here t represents the month, and $t = 0$ corresponds to June 1, 2004.

a. Use a graphing utility to graph M for $0 \le t \le 3$.

b. Use a graphing utility to graph M_+ for $0 \le t \le 3$.

c. Write a sentence or two that explains how the graph of M_+ differs from the graph of M.

d. What is the maximum mosquito population per acre for $0 \le t \le 3$? When does this maximum mosquito population occur?

e. Explain when would be the best time to visit the resort, provided that you wished to minimize your exposure to mosquitos.

Exploring Concepts with Technology

Graphing Piecewise Functions with a Graphing Calculator

A graphing calculator can be used to graph piecewise functions by including as part of the function the interval on which each piece of the function is defined. The method is based on the fact that a graphing calculator "evaluates" inequalities. For purposes of this Exploration, we will use keystrokes for a TI-83 calculator.

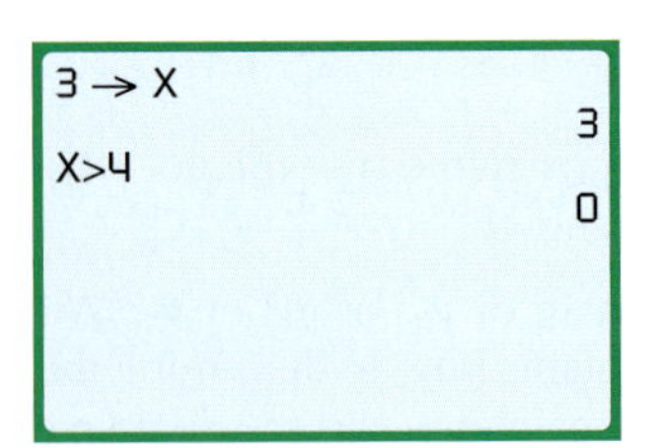

For instance, store 3 in X by pressing 3 [STO▶] [X,T,Θ,*n*] [ENTER]. Now enter the inequality $x > 4$ by pressing [X,T,Θ,*n*] [2ND] TEST 3 4 [ENTER]. Your screen should look like the one at the left. Note that the value of the inequality is 0. This occurs because the calculator replaced X by 3 and then determined whether the inequality $3 > 4$ was true or false. The calculator expresses the fact that the inequality is false by placing a zero on the screen. If we repeat the sequence of steps above, except that we store 5 in X instead of 3, the calculator will determine that the inequality is true and place a 1 on the screen.

This property of calculators is used to graph piecewise functions. Graphs of these functions work best when Dot mode rather than Connected mode is used. To switch to Dot mode, select [MODE], use the arrow keys to highlight [DOT], and then press [ENTER].

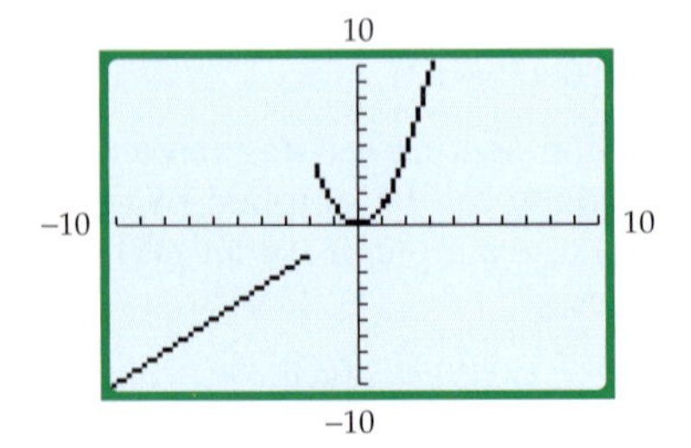

Now we will graph the piecewise function defined by $f(x) = \begin{cases} x, & x \le -2 \\ x^2, & x > -2 \end{cases}$.

Enter the function[3] as Y1=X*(X≤-2)+X²*(X>-2) and graph this in the standard viewing window. Note that you are multiplying each piece of the function by its domain. The graph will appear as shown at the left.

To understand how the graph is drawn, we will consider two values of x, -8 and 2, and evaluate Y1 for each of these values.

[3]Note that pressing [2ND] TEST will display the inequality menu.

Y1=X*(X≤-2)+X²*(X>-2)

$= -8(-8 \le -2) + (-8)^2(-8 > -2)$

$= -8(1) + 64(0) = -8$

• **When $x = -8$, the value assigned to $-8 \le -2$ is 1; the value assigned to $-8 > -2$ is 0.**

Y1=X*(X≤-2)+X²*(X>-2)

$=2(2 \le -2) + 2^2(2 > -2)$

$=2(0) + 4(1) = 4$

• **When $x = 2$, the value assigned to $2 \le -2$ is 0; the value assigned to $2 > -2$ is 1.**

In a similar manner, for any value of x for which $x \le -2$, the value assigned to (X≤−2) is 1 and the value assigned to (X>−2) is 0. Thus Y1=X*1+X²*0=X on that interval. This means that only the $f(x) = x$ piece of the function is graphed. When $x > -2$, the value assigned to (X≤−2) is 0 and the value assigned to (X>−2) is 1. Thus Y1=X*0+X²*1=X² on that interval. This means that only the $f(x) = x^2$ piece of the function is graphed on that interval.

1. Graph: $f(x) = \begin{cases} x^2, & x < 2 \\ -x, & x \ge 2 \end{cases}$

2. Graph: $f(x) = \begin{cases} x^2 - x, & x < 2 \\ -x + 4, & x \ge 2 \end{cases}$

3. Graph: $f(x) = \begin{cases} -x^2 + 1, & x < 0 \\ x^2 - 1, & x \ge 0 \end{cases}$

4. Graph: $f(x) = \begin{cases} x^3 - 4x, & x < 1 \\ x^2 - x + 2, & x \ge 1 \end{cases}$

CHAPTER 1 SUMMARY

1.1 Equations and Inequalities

- The *integers* are the set $\{\ldots, -3, -2, -1, 0, 1, 2, 3, \ldots\}$. The positive integers are called *natural numbers*. The *rational numbers* are $\left\{ \dfrac{a}{b} \,\middle|\, a, b \text{ integers and } b \ne 0 \right\}$. *Irrational numbers* are nonterminating and nonrepeating decimals. The *real numbers* are the set of rational and irrational numbers.
- The interval of real numbers $(a, b) = \{x \mid a < x < b\}$.
- A *linear equation* is one that can be written in the form $ax + b = 0$, $a \ne 0$. A *quadratic equation* is one that can be written in the form $ax^2 + bx + c = 0$, $a \ne 0$.
- An *inequality* is a statement that contains the symbol $<$, $\le$, $>$, or $\ge$.

1.2 A Two-Dimensional Coordinate System and Graphs

- *The Distance Formula* The distance d between the points represented by (x_1, y_1) and (x_2, y_2) is

$$d = \sqrt{(x_2 - x_1)^2 + (y_2 - y_1)^2}$$

- The midpoint of the line segment from $P_1(x_1, y_1)$ to $P_2(x_2, y_2)$ is

$$\left(\frac{x_1 + x_2}{2}, \frac{y_1 + y_2}{2} \right)$$

- The standard form of the equation of a circle with center at (h, k) and radius r is $(x - h)^2 + (y - k)^2 = r^2$.

1.3 Introduction to Functions

- *Definition of a Function* A function is a set of ordered pairs in which no two ordered pairs that have the same first coordinate have different second coordinates.
- A graph is the graph of a function if and only if no vertical line intersects the graph at more than one point. If every horizontal line intersects the graph of a function at most once, then the graph is the graph of a one-to-one function.

1.4 Linear Functions

- A function is a linear function of x if it can be written in the form $f(x) = mx + b$, where m and b are real numbers and $m \neq 0$.
- The slope m of the line passing through the points $P_1(x_1, y_1)$ and $P_2(x_2, y_2)$ with $x_1 \neq x_2$ is given by

$$m = \frac{y_2 - y_1}{x_2 - x_1}$$

- The graph of the equation $f(x) = mx + b$ has slope m and y-intercept $(0, b)$.
- Two nonvertical lines are parallel if and only if their slopes are equal. Two lines with slopes m_1 and m_2 are perpendicular if and only if $m_1 = -\dfrac{1}{m_2}$.

1.5 Quadratic Functions

- A quadratic function of x is a function that can be represented by an equation of the form $f(x) = ax^2 + bx + c$, where a, b, and c are real numbers and $a \neq 0$.
- The vertex of the graph of $f(x) = ax^2 + bx + c$ is

$$\left(-\frac{b}{2a}, f\left(-\frac{b}{2a}\right)\right)$$

- Every quadratic function $f(x) = ax^2 + bx + c$ can be written in the standard form $f(x) = a(x - h)^2 + k$, $a \neq 0$. The graph of f is a parabola with vertex (h, k). The parabola is symmetric with respect to the vertical line $x = h$, which is called the axis of symmetry of the parabola. The parabola opens up if $a > 0$; it opens down if $a < 0$.

1.6 Properties of Graphs

- The graph of an equation is symmetric with respect to

  the y-axis if the replacement of x with $-x$ leaves the equation unaltered.

  the x-axis if the replacement of y with $-y$ leaves the equation unaltered.

  the origin if the replacement of x with $-x$ and y with $-y$ leaves the equation unaltered.

- If f is a function and c is a positive constant, then

  $y = f(x) + c$ is the graph of $y = f(x)$ shifted up *vertically* c units

  $y = f(x) - c$ is the graph of $y = f(x)$ shifted down *vertically* c units

  $y = f(x + c)$ is the graph of $y = f(x)$ shifted left *horizontally* c units

  $y = f(x - c)$ is the graph of $y = f(x)$ shifted right *horizontally* c units

- The graph of

  $y = -f(x)$ is the graph of $y = f(x)$ reflected across the x-axis.

  $y = f(-x)$ is the graph of $y = f(x)$ reflected across the y-axis.

- If $a > 1$, then the graph of $y = f(ax)$ is a horizontal compressing of $y = f(x)$.
- If $0 < a < 1$, then the graph of $y = f(ax)$ is a horizontal stretching of the graph of $y = f(x)$.

1.7 The Algebra of Functions

- For all values of x for which both $f(x)$ and $g(x)$ are defined, we define the following functions.

Sum	$(f + g)(x) = f(x) + g(x)$
Difference	$(f - g)(x) = f(x) - g(x)$
Product	$(fg)(x) = f(x) \cdot g(x)$
Quotient	$\left(\dfrac{f}{g}\right)(x) = \dfrac{f(x)}{g(x)}, \quad g(x) \neq 0$

- The expression

$$\frac{f(x + h) - f(x)}{h}, \quad h \neq 0$$

  is called the difference quotient of f. The difference quotient is an important function because it can be used to compute the *average rate of change* of f over the time interval $[x, x + h]$.

- For the functions f and g, the composite function, or composition, of f by g is given by $(g \circ f)(x) = g[f(x)]$ for all x in the domain of f such that $f(x)$ is in the domain of g.

CHAPTER 1 TRUE/FALSE EXERCISES

In Exercises 1 to 13, answer true or false. If the statement is false, give an example or a reason to show that the statement is false.

1. Let f be any function. Then $f(a) = f(b)$ implies that $a = b$.
2. If f and g are two functions, then $(f \circ g)(x) = (g \circ f)(x)$.
3. If f is not a one-to-one function, then there are at least two numbers u and v in the domain of f for which $f(u) = f(v)$.
4. Let f be a function such that $f(x) = f(x + 4)$ for all real numbers x. If $f(2) = 3$, then $f(18) = 3$.
5. For all functions f, $[f(x)]^2 = f[f(x)]$.
6. Let f be any function. Then for all a and b in the domain of f such that $f(b) \neq 0$ and $b \neq 0$,
$$\frac{f(a)}{f(b)} = \frac{a}{b}$$
7. The **identity function** $f(x) = x$ is its own inverse.
8. If f is a function, then $f(a + b) = f(a) + f(b)$ for all real numbers a and b in the domain of f.
9. If f is defined by $f(x) = |x|$, then $f(ab) = f(a)f(b)$ for all real numbers a and b.
10. If f is a one-to-one function and a and b are real numbers in the domain of f with $a < b$, then $f(a) \neq f(b)$.
11. The coordinates of a point on the graph of $y = f(x)$ are (a, b). If k is a positive constant, then (a, kb) are the coordinates of a point on the graph of $y = kf(x)$.
12. For every function f, the real number c is a solution of $f(x) = 0$ if and only if $(c, 0)$ is an x-intercept of the graph of $y = f(x)$.
13. The domain of every polynomial function is the set of real numbers.

CHAPTER 1 REVIEW EXERCISES

In Exercises 1 to 14, solve each equation or inequality.

1. $3 - 4z = 12$
2. $4y - 3 = 6y + 5$
3. $2x - 3(2 - 3x) = 14x$
4. $5 - 2(3m + 2) = 3(1 - m)$
5. $y^2 - 3y - 18 = 0$
6. $2z^2 - 9z + 4 = 0$
7. $3v^2 + v = 1$
8. $3s = 4 - 2s^2$
9. $3c - 5 \leq 5c + 7$
10. $7a > 5 - 2(3a - 4)$
11. $x^2 - x - 12 \geq 0$
12. $2x^2 - x < 1$
13. $|2x - 5| > 3$
14. $|1 - 3x| \leq 4$

In Exercises 15 and 16, find the distance between the points whose coordinates are given.

15. $(-3, 2)$ $(7, 11)$
16. $(5, -4)$ $(-3, -8)$

In Exercises 17 and 18, find the midpoint of the line segment with the given endpoints.

17. $(2, 8)$ $(-3, 12)$
18. $(-4, 7)$ $(8, -11)$

In Exercises 19 and 20, determine the center and radius of the circle with the given equation.

19. $(x - 3)^2 + (y + 4)^2 = 81$
20. $x^2 + y^2 + 10x + 4y + 20 = 0$

In Exercises 21 and 22, find the equation in standard form of the circle that satisfies the given conditions.

21. Center $C = (2, -3)$, radius $r = 5$

22. Center $C = (-5, 1)$, passing through $(3, 1)$

23. If $f(x) = 3x^2 + 4x - 5$, find

a. $f(1)$ **b.** $f(-3)$ **c.** $f(t)$

d. $f(x + h)$ **e.** $3f(t)$ **f.** $f(3t)$

24. If $g(x) = \sqrt{64 - x^2}$, find

a. $g(3)$ **b.** $g(-5)$ **c.** $g(8)$

d. $g(-x)$ **e.** $2g(t)$ **f.** $g(2t)$

25. If $f(x) = x^2 + 4x$ and $g(x) = x - 8$, find

a. $(f \circ g)(3)$ **b.** $(g \circ f)(-3)$

c. $(f \circ g)(x)$ **d.** $(g \circ f)(x)$

26. If $f(x) = 2x^2 + 7$ and $g(x) = |x - 1|$, find

a. $(f \circ g)(-5)$ **b.** $(g \circ f)(-5)$

c. $(f \circ g)(x)$ **d.** $(g \circ f)(x)$

27. If $f(x) = 4x^2 - 3x - 1$, find the difference quotient

$$\frac{f(x + h) - f(x)}{h}$$

28. If $g(x) = x^3 - x$, find the difference quotient

$$\frac{g(x + h) - g(x)}{h}$$

In Exercises 29 to 34, sketch the graph of *f*. Find the interval(s) on which *f* is *a.* increasing, *b.* constant, *c.* decreasing.

29. $f(x) = |x - 3| - 2$ **30.** $f(x) = x^2 - 5$

31. $f(x) = |x + 2| - |x - 2|$ **32.** $f(x) = [\![x + 3]\!]$

33. $f(x) = \frac{1}{2}x - 3$ **34.** $f(x) = \sqrt[3]{x}$

In Exercises 35 to 38, determine the domain of the function represented by the given equation.

35. $f(x) = -2x^2 + 3$ **36.** $f(x) = \sqrt{6 - x}$

37. $f(x) = \sqrt{25 - x^2}$ **38.** $f(x) = \dfrac{3}{x^2 - 2x - 15}$

In Exercises 39 and 40, find the slope-intercept form of the equation of the line through the two points.

39. $(-1, 3)$ $(4, -7)$ **40.** $(0, 0)$ $(7, 11)$

41. Find the slope-intercept form of the equation of the line that is parallel to the graph of $3x - 4y = 8$ and passes through $(2, 11)$.

42. Find the slope-intercept form of the equation of the line that is perpendicular to the graph of $2x = -5y + 10$ and passes through $(-3, -7)$.

In Exercises 43 to 48, use the method of completing the square to write each quadratic equation in its standard form.

43. $f(x) = x^2 + 6x + 10$ **44.** $f(x) = 2x^2 + 4x + 5$

45. $f(x) = -x^2 - 8x + 3$ **46.** $f(x) = 4x^2 - 6x + 1$

47. $f(x) = -3x^2 + 4x - 5$ **48.** $f(x) = x^2 - 6x + 9$

In Exercises 49 to 52, find the vertex of the graph of the quadratic function.

49. $f(x) = 3x^2 - 6x + 11$ **50.** $h(x) = 4x^2 - 10$

51. $k(x) = -6x^2 + 60x + 11$ **52.** $m(x) = 14 - 8x - x^2$

53. Use the formula

$$d = \frac{|mx_1 + b - y_1|}{\sqrt{1 + m^2}}$$

to find the distance from the point $(1, 3)$ to the line given by $y = 2x - 3$.

54. A freight company has determined that its cost per delivery of delivering x parcels is

$$C(x) = 1050 + 0.5x$$

The price it charges to send a parcel is \$13.00 per parcel. Determine

a. the revenue function

b. the profit function

c. the minimum number of parcels the company must ship to break even

In Exercises 55 and 56, sketch a graph that is symmetric to the given graph with respect to the *a.* x-axis, *b.* y-axis, *c.* origin.

55.

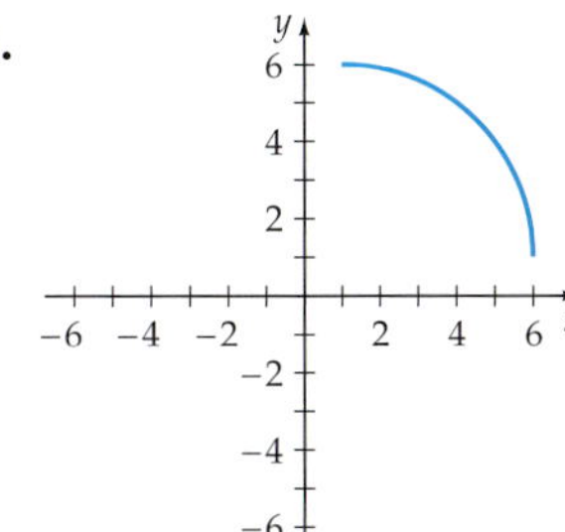

56.

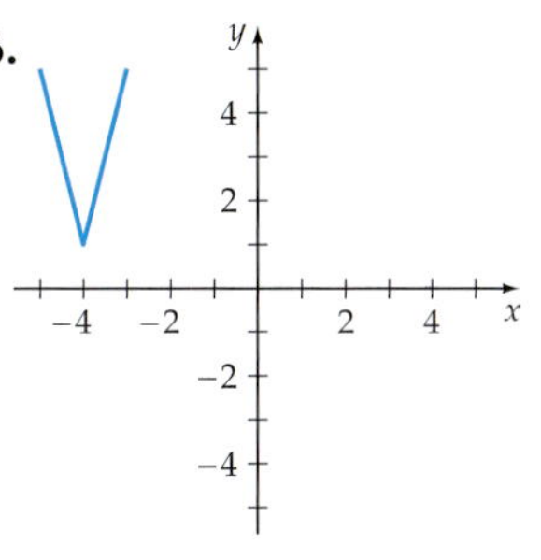

In Exercises 57 to 64, determine whether the graph of each equation is symmetric with respect to the *a.* x-axis, *b.* y-axis, *c.* origin.

57. $y = x^2 - 7$

58. $x = y^2 + 3$

59. $y = x^3 - 4x$

60. $y^2 = x^2 + 4$

61. $\frac{x^2}{3^2} + \frac{y^2}{4^2} = 1$

62. $xy = 8$

63. $|y| = |x|$

64. $|x + y| = 4$

In Exercises 65 to 70, sketch the graph of g. *a.* Find the domain and the range of g. *b.* State whether g is even, odd, or neither even nor odd.

65. $g(x) = -x^2 + 4$

66. $g(x) = -2x - 4$

67. $g(x) = |x - 2| + |x + 2|$

68. $g(x) = \sqrt{16 - x^2}$

69. $g(x) = x^3 - x$

70. $g(x) = 2[\![x]\!]$

In Exercises 71 to 76, first write the quadratic function in standard form, and then make use of translations to graph the function.

71. $F(x) = x^2 + 4x - 7$

72. $A(x) = x^2 - 6x - 5$

73. $P(x) = 3x^2 - 4$

74. $G(x) = 2x^2 - 8x + 3$

75. $W(x) = -4x^2 - 6x + 6$

76. $T(x) = -2x^2 - 10x$

77. On the same set of coordinate axes, sketch the graph of $p(x) = \sqrt{x} + c$ for $c = 0, -1$, and 2.

78. On the same set of coordinate axes, sketch the graph of $q(x) = \sqrt{x + c}$ for $c = 0, -1$, and 2.

79. On the same set of coordinate axes, sketch the graph of $r(x) = c\sqrt{9 - x^2}$ for $c = 1, \frac{1}{2}$, and -2.

80. On the same set of coordinate axes, sketch the graph of $s(t) = [\![cx]\!]$ for $c = 1, \frac{1}{4}$, and 4.

In Exercises 81 and 82, graph each piecewise-defined function.

81. $f(x) = \begin{cases} x, & \text{if } x \le 0 \\ \frac{1}{2}x, & \text{if } x > 0 \end{cases}$

82. $g(x) = \begin{cases} -2, & \text{if } x < -3 \\ \frac{2}{3}x, & \text{if } -3 \le x \le 3 \\ 2, & \text{if } x > 3 \end{cases}$

In Exercises 83 and 84, use the given functions *f* and *g* to find *f* + *g*, *f* − *g*, *fg*, and $\frac{f}{g}$. State the domain of each.

83. $f(x) = x^2 - 9, \quad g(x) = x + 3$

84. $f(x) = x^3 + 8, \quad g(x) = x^2 - 2x + 4$

85. Find two numbers whose sum is 50 and whose product is a maximum.

86. Find two numbers whose difference is 10 and the sum of whose squares is a minimum.

87. The distance traveled by a ball rolling down a ramp is given by $s(t) = 3t^2$, where t is the time in seconds after the ball is released and $s(t)$ is measured in feet. Evaluate the average velocity of the ball for each of the following time intervals.

a. [2, 4] **b.** [2, 3] **c.** [2, 2.5] **d.** [2, 2.01]

e. What appears to be the average velocity of the ball for the time interval $[2, 2 + \Delta t]$ as Δt approaches 0?

88. The distance traveled by a ball that is pushed down a ramp is given by $s(t) = 2t^2 + t$, where t is the time in seconds after the ball is released and $s(t)$ is measured in feet. Evaluate the average velocity of the ball for each of the following time intervals.

a. [3, 5] **b.** [3, 4] **c.** [3, 3.5] **d.** [3, 3.01]

e. What appears to be the average velocity of the ball for the time interval $[3, 3 + \Delta t]$ as Δt approaches 0?

CHAPTER 1 TEST

1. Solve: $2x - 3(1 - x) = 3x + 5$

2. Solve: $x^2 - 3x = 5$

3. Solve: $4x - 5 \geq 6x + 7$

4. Solve: $x^2 + 4x - 12 \leq 0$

5. Find the midpoint and the length of the line segment with endpoints $(-2, 3)$ and $(4, -1)$.

6. Determine the x- and y-intercepts, and then graph the equation $x = 2y^2 - 4$.

7. Graph the equation $y = |x + 2| + 1$.

8. Find the center and radius of the circle that has the general form $x^2 - 4x + y^2 + 2y - 4 = 0$.

9. Determine the domain of the function

$$f(x) = -\sqrt{x^2 - 16}$$

10. Graph $f(x) = -2|x - 2| + 1$. Identify the intervals over which the function is

    a. increasing

    b. constant

    c. decreasing

11. An air freight company has determined that its cost per flight of delivering x parcels is

$$C(x) = 875 + 0.75x$$

The price it charges to send a parcel is $12.00 per parcel. Determine

a. the revenue function

b. the profit function

c. the minimum number of parcels the company must ship to break even

12. Use the graph of $f(x) = |x|$ to graph $y = -f(x + 2) - 1$.

13. Classify each of the following as an even function, an odd function, or neither an even nor an odd function.

    a. $f(x) = x^4 - x^2$ **b.** $f(x) = x^3 - x$

    c. $f(x) = x - 1$

14. Find the slope-intercept form of the equation of the line that passes through $(4, -2)$ and is perpendicular to the graph of $3x - 2y = 4$.

15. Find the maximum or minimum value of the function $f(x) = x^2 - 4x - 8$. State whether this value is a maximum or a minimum value.

16. Let $f(x) = x^2 - 1$ and $g(x) = x - 2$. Find $(f + g)$ and (f/g).

17. Find the difference quotient of the function

$$f(x) = x^2 + 1$$

18. Evaluate $(f \circ g)(x)$, where

$$f(x) = x^2 - 2x \quad \text{and} \quad g(x) = 2x + 5$$

CHAPTER 2

POLYNOMIAL AND RATIONAL FUNCTIONS

DVD players have become very popular in the last few years. More than 31 million DVD players have been sold as of January 2002. In the second week of June 2003, the number of DVDs rented in the United States in a one-week span exceeded the number of VHS rentals.

Source: http://shumans.com/archives/000024.php

Production Cost and Average Cost

In this chapter you will study polynomial and rational functions. These types of functions have many practical applications. For instance, they can be used to model production costs and average costs associated with the manufacture of DVD players.

The cost, in dollars, of producing x DVD players is given by the polynomial function

$$C(x) = 0.001x^2 + 101x + 245{,}000$$

The average cost per DVD player is given by the rational function

$$\overline{C}(x) = \frac{C(x)}{x} = \frac{0.001x^2 + 101x + 245{,}000}{x}$$

The following graph of $\overline{C}$ shows that the minimum average cost per DVD player is obtained by producing 15,652 DVD players.

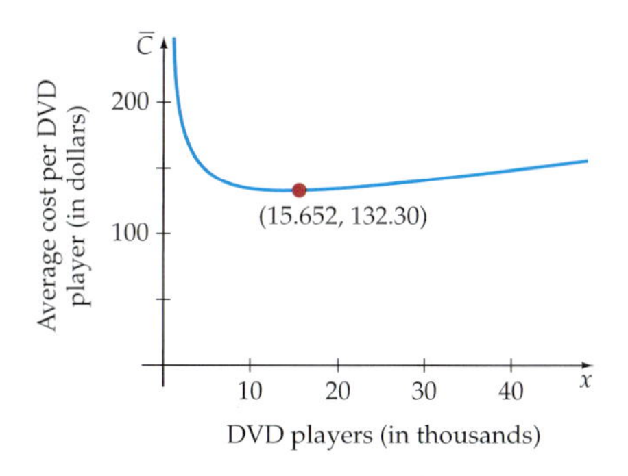

Additional average cost applications are given in **Exercises 53 and 54 on page 190.**

Find and Use Clues to Narrow the Search

The game of Clue is a classic whodunit game. At the beginning of the game you are informed that Mr. Boddy has been murdered! It is your job to find and use clues to determine the murderer, the weapon, and the room in which the murder was committed. Was it Miss Scarlet in the billiard room with the revolver? Or did Professor Plum commit the murder in the conservatory with the rope? There are six suspects, six possible murder weapons, and nine rooms in Mr. Boddy's mansion. There are a total of $6 \times 6 \times 9 = 324$ possible solutions to each game.

In this chapter you will often need to find the zeros of a polynomial function. Finding the zeros of a polynomial function can be more complicated than solving a game of Clue. After all, any real or complex number is a possible zero. Quite often the zeros of a polynomial function are found by using several theorems to narrow the search. In most cases no single theorem can be used to find the zeros of a given polynomial function, but by combining the results of several theorems, we are often able to gather enough clues to find the zeros. In **Exercise 52, page 165,** you will apply several theorems from this chapter to find the zeros of a polynomial function.

SECTION 2.1 COMPLEX NUMBERS

- Introduction to Complex Numbers
- Addition and Subtraction of Complex Numbers
- Multiplication of Complex Numbers
- Division of Complex Numbers
- Powers of *i*
- Solve Quadratic Equations with Complex Solutions

MATH MATTERS

It may seem strange to just invent new numbers, but that is how mathematics evolves. For instance, negative numbers were not an accepted part of mathematics until well into the 13th century. In fact, these numbers often were referred to as "fictitious numbers."

In the 17th century, Rene Descartes called square roots of negative numbers "imaginary numbers," an unfortunate choice of words, and started using the letter *i* to denote these numbers. These numbers were subjected to the same skepticism as negative numbers.

It is important to understand that these numbers are not *imaginary* in the dictionary sense of the word. It is similar to the situation of negative numbers being called fictitious.

If you think of a number line, then the numbers to the right of zero are positive numbers and the numbers to the left of zero are negative numbers. One way to think of an imaginary number is to visualize it as *up* or *down* from zero. See the Project on page 126 for more information on this topic.

Introduction to Complex Numbers

Recall that $\sqrt{9} = 3$ because $3^2 = 9$. Now consider the expression $\sqrt{-9}$. To find $\sqrt{-9}$, we need to find a number c such that $c^2 = -9$. However, the square of any real number c (except zero) is a *positive* number. Consequently, we must expand our concept of number to include numbers whose squares are negative numbers.

Around the 17th century, a new number, called an *imaginary number*, was defined so that a negative number would have a square root. The letter i was chosen to represent the number whose square is -1.

Definition of *i*

The number i, called the **imaginary unit,** is the number such that $i^2 = -1$.

The principal square root of a negative number is defined in terms of i.

Principal Square Root of a Negative Number

If a is a positive real number, then $\sqrt{-a} = i\sqrt{a}$. The number $i\sqrt{a}$ is called an **imaginary number.**

Here are some examples of imaginary numbers.

$$\sqrt{-36} = i\sqrt{36} = 6i \qquad \sqrt{-18} = i\sqrt{18} = 3i\sqrt{2}$$
$$\sqrt{-23} = i\sqrt{23} \qquad \sqrt{-1} = i\sqrt{1} = i$$

It is customary to write i in front of a radical sign, as we did for $i\sqrt{23}$, to avoid confusing $\sqrt{a}\,i$ with $\sqrt{ai}$.

Complex Numbers

A **complex number** is a number of the form $a + bi$, where a and b are real numbers and $i = \sqrt{-1}$. The number a is the **real part** of $a + bi$, and b is the **imaginary part.**

Here are some examples of complex numbers.

$-3 + 5i$ • Real part: -3; imaginary part: 5

$2 - 6i$ • Real part: 2; imaginary part: -6

5 • Real part: 5; imaginary part: 0

$7i$ • Real part: 0; imaginary part: 7

Note from these examples that a real number is a complex number whose imaginary part is zero and that an imaginary number is a complex number whose real part is zero.

MATH MATTERS

The imaginary unit i is important in the field of electrical engineering. However, because the letter i is used by engineers as the symbol for electric current, these engineers use j for the complex unit.

QUESTION What are the real part and the imaginary part of $3 - 5i$?

Note from the following diagram that the real numbers are a subset of the complex numbers, and imaginary numbers are a subset of the complex numbers. The real numbers and imaginary numbers are disjoint sets.

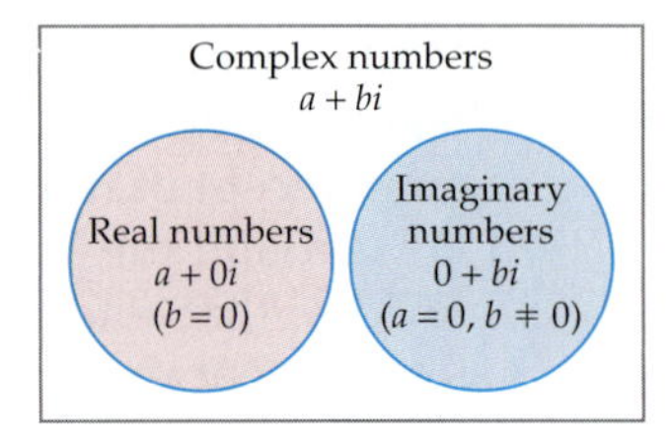

Example 1 illustrates writing a complex number in the standard form $a + bi$.

EXAMPLE 1 Write a Complex Number in Standard Form

Write $7 + \sqrt{-45}$ in the form $a + bi$.

Solution

$$\begin{aligned} 7 + \sqrt{-45} &= 7 + i\sqrt{45} \\ &= 7 + i\sqrt{9} \cdot \sqrt{5} \\ &= 7 + 3i\sqrt{5} \end{aligned}$$

TRY EXERCISE 8, PAGE 124

ADDITION AND SUBTRACTION OF COMPLEX NUMBERS

All the standard arithmetic operations that are applied to real numbers can be applied to complex numbers.

Definition of Addition and Subtraction of Complex Numbers

If $a + bi$ and $c + di$ are complex numbers, then

Addition $(a + bi) + (c + di) = (a + c) + (b + d)i$

Subtraction $(a + bi) - (c + di) = (a - c) + (b - d)i$

Basically, these rules say that to add two complex numbers, add the real parts and add the imaginary parts. To subtract two complex numbers, subtract the real parts and subtract the imaginary parts.

ANSWER Real part: 3; imaginary part: -5

EXAMPLE 2 **Add or Subtract Complex Numbers**

Simplify.

a. $(7 - 2i) + (-2 + 4i)$ **b.** $(-9 + 4i) - (2 - 6i)$

Solution

a. $(7 - 2i) + (-2 + 4i) = (7 + (-2)) + (-2 + 4)i = 5 + 2i$

b. $(-9 + 4i) - (2 - 6i) = (-9 - 2) + (4 - (-6))i = -11 + 10i$

TRY EXERCISE 18, PAGE 125

MULTIPLICATION OF COMPLEX NUMBERS

When multiplying complex numbers, the term i^2 is frequently a part of the product. Recall that $i^2 = -1$. Therefore,

$$3i(5i) = 15i^2 = 15(-1) = -15$$
$$-2i(6i) = -12i^2 = -12(-1) = 12$$
$$4i(3 - 2i) = 12i - 8i^2 = 12i - 8(-1) = 8 + 12i$$

> **take note**
>
> Recall that the definition of the product of radical expressions requires that the radicand be a *positive* number. Therefore, when multiplying expressions containing *negative* radicands, we must first rewrite the expression using *i* and a positive radicand.

When multiplying square roots of negative numbers, first rewrite the radical expressions using i. For instance,

$$\begin{aligned}\sqrt{-6} \cdot \sqrt{-24} &= i\sqrt{6} \cdot i\sqrt{24} \\ &= i^2\sqrt{144} = -1 \cdot 12 \\ &= -12\end{aligned}$$

• $\sqrt{-6} = i\sqrt{6}$, $\sqrt{-24} = i\sqrt{24}$

Note from this example that it would have been incorrect to multiply the radicands of the two radical expressions. To illustrate:

$$\sqrt{-6} \cdot \sqrt{-24} \neq \sqrt{(-6)(-24)}$$

QUESTION What is the product of $\sqrt{-2}$ and $\sqrt{-8}$?

To multiply two complex numbers, we use the following definition.

Definition of Multiplication of Complex Numbers

If $a + bi$ and $c + di$ are complex numbers, then

$$(a + bi)(c + di) = (ac - bd) + (ad + bc)i$$

Because every complex number can be written as a sum of two terms, it is natural to perform multiplication on complex numbers in a manner consistent with

ANSWER $\sqrt{-2} \cdot \sqrt{-8} = i\sqrt{2} \cdot i\sqrt{8} = i^2\sqrt{16} = -1 \cdot 4 = -4$

the operation defined on binomials and the definition $i^2 = -1$. By using this analogy, you can multiply complex numbers without memorizing the definition.

EXAMPLE 3 Multiply Complex Numbers

Simplify. **a.** $(3 - 4i)(2 + 5i)$ **b.** $(2 + \sqrt{-3})(4 - 5\sqrt{-3})$

Solution

a.
$$\begin{aligned}(3 - 4i)(2 + 5i) &= 6 + 15i - 8i - 20i^2 \\ &= 6 + 15i - 8i - 20(-1) && \text{• Replace } i^2 \text{ by } -1. \\ &= 6 + 15i - 8i + 20 && \text{• Simplify.} \\ &= 26 + 7i\end{aligned}$$

b.
$$\begin{aligned}(2 + \sqrt{-3})(4 - 5\sqrt{-3}) &= (2 + i\sqrt{3})(4 - 5i\sqrt{3}) \\ &= 8 - 10i\sqrt{3} + 4i\sqrt{3} - 5i^2\sqrt{9} \\ &= 8 - 10i\sqrt{3} + 4i\sqrt{3} - 5(-1)(3) \\ &= 8 - 10i\sqrt{3} + 4i\sqrt{3} + 15 = 23 - 6i\sqrt{3}\end{aligned}$$

TRY EXERCISE 28, PAGE 125

Division of Complex Numbers

Recall that the number $\dfrac{3}{\sqrt{2}}$ is not in simplest form because there is a radical expression in the denominator. Similarly, $\dfrac{3}{i}$ is not in simplest form because $i = \sqrt{-1}$. To write this expression in simplest form, multiply the numerator and denominator by i.

$$\frac{3}{i} \cdot \frac{i}{i} = \frac{3i}{i^2} = \frac{3i}{-1} = -3i$$

Here is another example.

$$\begin{aligned}\frac{3 - 6i}{2i} &= \frac{3 - 6i}{2i} \cdot \frac{i}{i} = \frac{3i - 6i^2}{2i^2} = \frac{3i - 6(-1)}{2(-1)} \\ &= \frac{3i + 6}{-2} = -3 - \frac{3}{2}i\end{aligned}$$

Recall that to simplify the quotient $\dfrac{2 + \sqrt{3}}{5 + 2\sqrt{3}}$, we multiply the numerator and denominator by the conjugate of $5 + 2\sqrt{3}$, which is $5 - 2\sqrt{3}$. In a similar manner, to find the quotient of two complex numbers, we multiply the numerator and denominator by the conjugate of the denominator.

The complex numbers $a + bi$ and $a - bi$ are called **complex conjugates** or **conjugates** of each other. The conjugate of the complex number z is denoted by $\overline{z}$. For instance,

$$\overline{2 + 5i} = 2 - 5i \qquad \text{and} \qquad \overline{3 - 4i} = 3 + 4i$$

Some graphing calculators can be used to perform operations on complex numbers. Here are some typical screens for a TI-83 Plus.

Press MODE. Use the down arrow key to highlight $a + bi$.

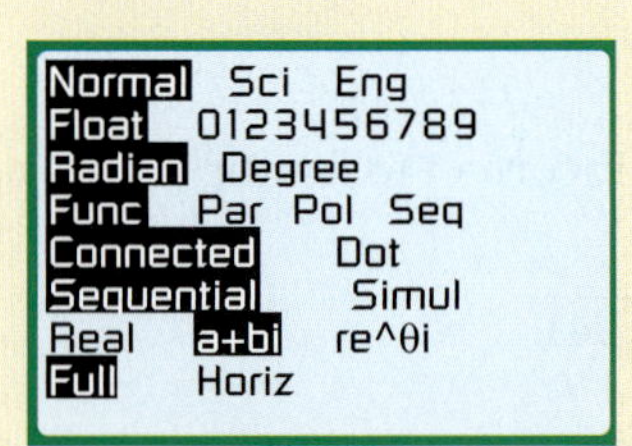

Press ENTER 2nd [QUIT].

Here are two examples of computations on complex numbers. To enter an i, use 2nd [i], which is above the decimal point key.

```
(3-4i)(2+5i)
                26+7i
(16-11i)/(5+2i)
                 2-3i
```

Consider the product of a complex number and its conjugate. For instance,

$$\begin{aligned}(2 + 5i)(2 - 5i) &= 4 - 10i + 10i - 25i^2 \\ &= 4 - 25(-1) = 4 + 25 \\ &= 29\end{aligned}$$

Note that the product is a *real* number. This is always true.

Product of Complex Conjugates

The product of a complex number and its conjugate is a real number. That is, $(a + bi)(a - bi) = a^2 + b^2$.

For instance, $(5 + 3i)(5 - 3i) = 5^2 + 3^2 = 25 + 9 = 34$.

The next example shows how the quotient of two complex numbers is determined by using conjugates.

EXAMPLE 4 Divide Complex Numbers

Simplify: $\dfrac{16 - 11i}{5 + 2i}$

Solution

$$\begin{aligned}\frac{16 - 11i}{5 + 2i} &= \frac{16 - 11i}{5 + 2i} \cdot \frac{5 - 2i}{5 - 2i} \\ &= \frac{80 - 32i - 55i + 22i^2}{5^2 + 2^2} \\ &= \frac{80 - 32i - 55i + 22(-1)}{25 + 4} \\ &= \frac{80 - 87i - 22}{29} \\ &= \frac{58 - 87i}{29} \\ &= \frac{29(2 - 3i)}{29} = 2 - 3i\end{aligned}$$

- Multiply numerator and denominator by the conjugate of the denominator.

▶ TRY EXERCISE 42, PAGE 125

Powers of *i*

The following powers of i illustrate a pattern:

$$\begin{aligned}&i^1 = i && i^5 = i^4 \cdot i = 1 \cdot i = i \\ &i^2 = -1 && i^6 = i^4 \cdot i^2 = 1(-1) = -1 \\ &i^3 = i^2 \cdot i = (-1)i = -i && i^7 = i^4 \cdot i^3 = 1(-i) = -i \\ &i^4 = i^2 \cdot i^2 = (-1)(-1) = 1 && i^8 = (i^4)^2 = 1^2 = 1\end{aligned}$$

Because $i^4 = 1$, $(i^4)^n = 1^n = 1$ for any integer n. Thus it is possible to evaluate powers of i by factoring out powers of i^4, as shown in the following:

$$i^{27} = (i^4)^6 \cdot i^3 = 1^6 \cdot i^3 = 1 \cdot (-i) = -i$$

The following theorem can be used to evaluate powers of i.

Powers of *i*

If n is a positive integer, then $i^n = i^r$, where r is the remainder of the division of n by 4.

EXAMPLE 5 **Evaluate a Power of *i***

Evaluate: i^{153}

Solution

Use the powers of i theorem.

$$i^{153} = i^1 = i$$ • **Remainder of 153 ÷ 4 is 1.**

▶ **TRY EXERCISE 48, PAGE 125**

Solve Quadratic Equations with Complex Solutions

Some quadratic equations have complex number solutions. For instance, the quadratic equation $x^2 = -25$ has $5i$ and $-5i$ as solutions. These solutions were found by taking the square root of each side of the equation and prefixing the square root of the constant with the plus-or-minus symbol.

$$x^2 = -25$$
$$\sqrt{x^2} = \sqrt{-25}$$
$$x = \pm\sqrt{-25}$$
$$x = \pm 5i$$

The notation $x = \pm 5i$ means $x = 5i$ or $x = -5i$.

In the following example we use the process of taking square roots to solve the equation $(x - 5)^2 = -9$.

$$(x-5)^2 = -9$$
$$\sqrt{(x-5)^2} = \sqrt{-9}$$ • **Take the square root of each side of the equation.**
$$x - 5 = \pm\sqrt{-9}$$ • **Prefix the square root of the constant with the ± symbol.**
$$x - 5 = \pm 3i$$ • **Solve for x.**
$$x = 5 \pm 3i$$

If you find that you are not able to solve a quadratic equation by taking square roots, then you should solve the equation by factoring, by completing the square, or by using the quadratic formula. In the follwing example we use the quadratic formula to solve two quadratic equations.

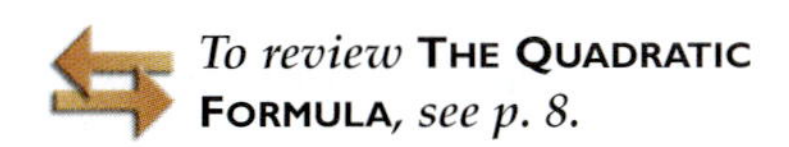
To review **The Quadratic Formula**, *see p. 8.*

EXAMPLE 6 **Solve by Using the Quadratic Formula**

Use the quadratic formula to solve each equation.

a. $2x^2 - 2x + 5 = 0$ **b.** $4x^2 - 2x + 1 = 0$

Solution

a. For the equation $2x^2 - 2x + 5 = 0$, we have $a = 2$, $b = -2$, and $c = 5$. Substitute these values into the quadratic formula and simplify.

$$x = \frac{-b \pm \sqrt{b^2 - 4ac}}{2a}$$

$$= \frac{-(-2) \pm \sqrt{(-2)^2 - 4(2)(5)}}{2(2)}$$

$$= \frac{2 \pm \sqrt{-36}}{4} = \frac{2 \pm 6i}{4} = \frac{1}{2} \pm \frac{3}{2}i$$

The solutions are $\frac{1}{2} - \frac{3}{2}i$ and $\frac{1}{2} + \frac{3}{2}i$.

b. For the equation $4x^2 - 2x + 1 = 0$, we have $a = 4$, $b = -2$, and $c = 1$. Substitute into the quadratic formula and simplify.

$$x = \frac{-b \pm \sqrt{b^2 - 4ac}}{2a}$$

$$= \frac{-(-2) \pm \sqrt{(-2)^2 - 4(4)(1)}}{2(4)}$$

$$= \frac{2 \pm \sqrt{-12}}{8} = \frac{2 \pm 2\sqrt{3}i}{8} = \frac{2}{8} \pm \frac{2\sqrt{3}}{8}i = \frac{1}{4} \pm \frac{\sqrt{3}}{4}i$$

The solutions are $\frac{1}{4} - \frac{\sqrt{3}}{4}i$ and $\frac{1}{4} + \frac{\sqrt{3}}{4}i$.

Try Exercise 66, page 125

The solutions of $ax^2 + bx + c = 0$, $a \neq 0$, are given by

$$x = \frac{-b \pm \sqrt{b^2 - 4ac}}{2a}$$

The expression under the radical, $b^2 - 4ac$, is called the **discriminant** of the equation $ax^2 + bx + c = 0$. If $b^2 - 4ac \geq 0$, then $\sqrt{b^2 - 4ac}$ is a real number. If $b^2 - 4ac < 0$, then $\sqrt{b^2 - 4ac}$ is not a real number. Thus the sign of the discriminant can be used to determine whether the solutions of a quadratic equation are real numbers.

The Discriminant and the Solutions of a Quadratic Equation

The equation $ax^2 + bx + c = 0$, with real coefficients and $a \neq 0$, has as its discriminant $b^2 - 4ac$.

- If $b^2 - 4ac > 0$, then $ax^2 + bx + c = 0$ has *two distinct real solutions.*
- If $b^2 - 4ac = 0$, then $ax^2 + bx + c = 0$ has *one real solution*. The solution is a double solution.
- If $b^2 - 4ac < 0$, then $ax^2 + bx + c = 0$ has *two distinct nonreal complex solutions.* The solutions are conjugates of each other.

take note

The equation $x^2 + 6x + 9 = 0$ can be solved by factoring as shown below.

$$x^2 + 6x + 9 = 0$$
$$(x + 3)(x + 3) = 0$$
$$x + 3 = 0 \quad \text{or} \quad x + 3 = 0$$
$$x = -3 \qquad\qquad x = -3$$

The two solutions are both the same real number, -3. When both solutions of a quadratic equation are the same real number, the solution is called a **double solution** or a **double root.**

In the following examples we find the discriminant of each quadratic equation to determine whether its solutions are real numbers or nonreal complex numbers.

- The discriminant of $2x^2 - 5x + 1 = 0$ is $b^2 - 4ac = (-5)^2 - 4(2)(1) = 17$. Because the discriminant is positive, $2x^2 - 5x + 1 = 0$ has two distinct real number solutions.
- The discriminant of $3x^2 + 6x + 7 = 0$ is $b^2 - 4ac = 6^2 - 4(3)(7) = -48$. Because the discriminant is negative, the equation $3x^2 + 6x + 7 = 0$ has two distinct nonreal complex solutions.
- The discriminant of $x^2 + 6x + 9 = 0$ is $b^2 - 4ac = 6^2 - 4(1)(9) = 0$. Because the discriminant is zero, the equation $x^2 + 6x + 9 = 0$ has one real solution. The solution is a double root. See the Take Note at the left.

Topics for Discussion

1. What is an imaginary number? What is a complex number?
2. How are the real numbers related to the complex numbers?
3. Is zero a complex number?
4. What is the conjugate of a complex number?
5. Explain how you know that the equation $x^2 + bx - 2 = 0$ always has real number solutions regardless of the value of the real number b.

Exercise Set 2.1

In Exercises 1 to 10, write the complex number in standard form.

1. $\sqrt{-81}$
2. $\sqrt{-64}$
3. $\sqrt{-98}$
4. $\sqrt{-27}$
5. $\sqrt{16} + \sqrt{-81}$
6. $\sqrt{25} + \sqrt{-9}$
7. $5 + \sqrt{-49}$
8. $6 - \sqrt{-1}$
9. $8 - \sqrt{-18}$
10. $11 + \sqrt{-48}$

In Exercises 11 to 30, simplify and write the complex number in standard form.

11. $(5 + 2i) + (6 - 7i)$
12. $(4 - 8i) + (5 + 3i)$

13. $(-2 - 4i) - (5 - 8i)$

14. $(3 - 5i) - (8 - 2i)$

15. $(1 - 3i) + (7 - 2i)$

16. $(2 - 6i) + (4 - 7i)$

17. $(-3 - 5i) - (7 - 5i)$

▶ **18.** $(5 - 3i) - (2 + 9i)$

19. $3(2 + 5i) - 2(3 - 2i)$

20. $3i(2 + 5i) + 2i(3 - 4i)$

21. $(4 + 2i)(3 - 4i)$

22. $(6 + 5i)(2 - 5i)$

23. $(-3 - 4i)(2 + 7i)$

24. $(-5 - i)(2 + 3i)$

25. $(4 - 5i)(4 + 5i)$

26. $(3 + 7i)(3 - 7i)$

27. $(3 + \sqrt{-4})(2 - \sqrt{-9})$

▶ **28.** $(5 + 2\sqrt{-16})(1 - \sqrt{-25})$

29. $(3 + 2\sqrt{-18})(2 + 2\sqrt{-50})$

30. $(5 - 3\sqrt{-48})(2 - 4\sqrt{-27})$

In Exercises 31 to 42, write each expression as a complex number in standard form.

31. $\dfrac{6}{i}$

32. $\dfrac{-8}{2i}$

33. $\dfrac{6 + 3i}{i}$

34. $\dfrac{4 - 8i}{4i}$

35. $\dfrac{1}{7 + 2i}$

36. $\dfrac{5}{3 + 4i}$

37. $\dfrac{2i}{1 + i}$

38. $\dfrac{5i}{2 - 3i}$

39. $\dfrac{5 - i}{4 + 5i}$

40. $\dfrac{4 + i}{3 + 5i}$

41. $\dfrac{3 + 2i}{3 - 2i}$

▶ **42.** $\dfrac{8 - i}{2 + 3i}$

In Exercises 43 to 50, evaluate the power of *i*.

43. i^{15}

44. i^{66}

45. $-i^{40}$

46. $-i^{51}$

47. $\dfrac{1}{i^{25}}$

▶ **48.** $\dfrac{1}{i^{83}}$

49. i^{-34}

50. i^{-52}

In Exercises 51 to 60, take square roots to solve each quadratic equation.

51. $x^2 + 49 = 0$

52. $x^2 + 4 = 0$

53. $2x^2 + 18 = 0$

54. $3x^2 + 48 = 0$

55. $(x - 3)^2 = -36$

56. $(x - 5)^2 = -64$

57. $(x + 2)^2 + 5 = 0$

58. $(x + 7)^2 + 3 = 0$

59. $(2x + 3)^2 + 25 = 0$

60. $(3x + 2)^2 + 100 = 0$

In Exercises 61 to 70, use the quadratic formula to solve each quadratic equation.

61. $x^2 - 4x = -29$

62. $x^2 + 6x = -25$

63. $2x^2 + 2x + 1 = 0$

64. $8x^2 + 12x = -17$

65. $x^2 - 4x = 1$

▶ **66.** $8x^2 - 4x + 5 = 0$

67. $4x^2 - 8x + 13 = 0$

68. $2x^2 + 2x + 13 = 0$

69. $4x^2 - 4x = -9$

70. $4x^2 + 4x + 5 = 0$

CONNECTING CONCEPTS

The property that the product of conjugates of the form $(a + bi)(a - bi)$ is equal to $a^2 + b^2$ can be used to factor the sum of two perfect squares over the set of complex numbers. For example, $x^2 + y^2 = (x + yi)(x - yi)$. In Exercises 71 to 74, factor the binomial over the set of complex numbers.

71. $x^2 + 16$

72. $x^2 + 9$

73. $4x^2 + 81$

74. $9x^2 + 1$

75. Show that if $x = 1 + 2i$, then $x^2 - 2x + 5 = 0$.

76. Show that if $x = 1 - 2i$, then $x^2 - 2x + 5 = 0$.

77. When we think of the cube root of 8, $\sqrt[3]{8}$, we normally mean the *real* cube root of 8 and write $\sqrt[3]{8} = 2$. However, there are two other cube roots of 8 that are complex numbers. Verify that $-1 + i\sqrt{3}$ and $-1 - i\sqrt{3}$ are cube roots of 8 by showing that $(-1 + i\sqrt{3})^3 = 8$ and $(-1 - i\sqrt{3})^3 = 8$.

78. It is possible to find the square root of a complex number.

Verify that $\sqrt{i} = \frac{\sqrt{2}}{2}(1 + i)$ by showing that

$$\left[\frac{\sqrt{2}}{2}(1+i)\right]^2 = i.$$

79. Simplify $i + i^2 + i^3 + i^4 + \cdots + i^{28}$.

80. Simplify $i + i^2 + i^3 + i^4 + \cdots + i^{100}$.

PREPARE FOR SECTION 2.2

81. Simplify: $(3x^2 + 2x) - (3x^2 - 6x)$ [1.7]

82. Simplify: $(6x^3 - 16x^2) - (6x^3 - 4x^2)$ [1.7]

83. Given $f(x) = 4x^3 - 10x^2 - 8x + 6$, find $f(3)$. [1.3]

84. Given $g(x) = 4x^4 - 6x^2 + 5$, find $g(-2)$. [1.3]

85. Given $f(x) = x^3 + 4x^2 - x - 4$ and $g(x) = x + 1$, find $(fg)(x)$. [1.7]

86. Given $f(x) = x - 3$ and $g(x) = x^2 + 3x + 9$, find $(fg)(x)$. [1.7]

PROJECTS

ARGAND DIAGRAM Just as we can graph a real number on a real number line, we can graph a complex number. This is accomplished by using one number line for the real part of the complex number and one number line for the imaginary part of the complex number. These two number lines are drawn perpendicular to each other and pass through their respective origins, as shown below.

The result is called the *complex plane* or an *Argand diagram* after Jean-Robert Argand (1768–1822), an accountant and amateur mathematician. Although he is given credit for this representation of complex numbers, Caspar Wessel (1745–1818) actually conceived the idea before Argand.

To graph the complex number $3 + 4i$, start at 3 on the real axis. Now move 4 units up (for positive numbers move up, for negative numbers move down) and place a dot at that point, as shown in the diagram. Graphs of several other complex numbers are also shown.

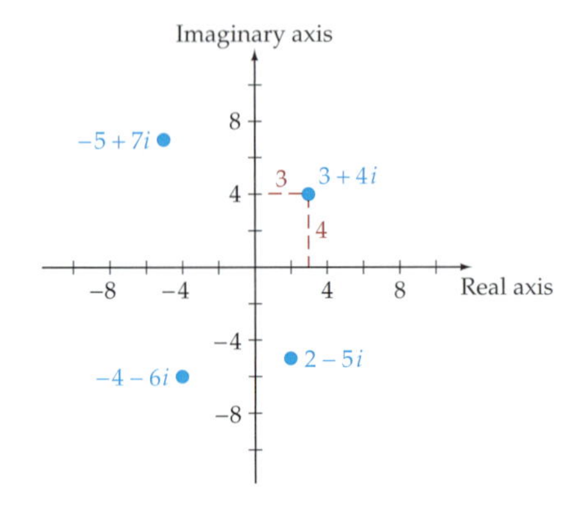

In Exercises 1 to 8, graph the complex number.

1. $2 + 5i$ **2.** $4 - 3i$ **3.** $-2 + 6i$ **4.** $-3 - 5i$

5. 4 **6.** $-2i$ **7.** $3i$ **8.** -5

The absolute value of a complex number is given by $|a + bi| = \sqrt{a^2 + b^2}$. In Exercises 9 to 12, find the absolute value of the complex number.

9. $2 + 5i$ **10.** $4 - 3i$ **11.** $-2 + 6i$ **12.** $-3 - 5i$

13. The additive inverse of $a + bi$ is $-a - bi$. Show that the absolute value of a complex number and the absolute value of its additive inverse are equal.

14. A *real* number and its additive inverse are the same distance from zero but on opposite sides of zero on a real number line. Describe the relationship between the graphs of a complex number and its additive inverse.

SECTION 2.2

The Remainder Theorem and the Factor Theorem

- Division of Polynomials
- The Remainder Theorem
- The Factor Theorem
- Reduced Polynomials

If $P(x)$ is a polynomial, then the values of x for which $P(x)$ is equal to 0 are called the **zeros** of $P(x)$. For instance, -1 is a zero of $P(x) = 2x^3 - x + 1$ because

$$\begin{aligned} P(-1) &= 2(-1)^3 - (-1) + 1 \\ &= -2 + 1 + 1 \\ &= 0 \end{aligned}$$

QUESTION Is 0 a zero of $P(x) = 2x^3 - x + 1$?

Much of the work in this chapter concerns finding the zeros of a polynomial. Sometimes the zeros of a polynomial $P(x)$ are determined by dividing $P(x)$ by another polynomial.

Division of Polynomials

take note

Recall that a fraction bar acts as a grouping symbol. Division of a polynomial by a monomial is an application of the distributive property.

To divide a polynomial by a monomial, divide each term of the polynomial by the monomial. For instance,

$$\frac{16x^3 - 8x^2 + 12x}{4x} = \frac{16x^3}{4x} - \frac{8x^2}{4x} + \frac{12x}{4x}$$

• Divide each term in the numerator by the denominator.

$$= 4x^2 - 2x + 3$$

• Simplify.

To divide a polynomial by a binomial, we use a method similar to that used to divide whole numbers. For instance, consider $(6x^3 - 16x^2 + 23x - 5) \div (3x - 2)$.

$$3x - 2\overline{)6x^3 - 16x^2 + 23x - 5}$$

$$\begin{array}{r} 2x^2 \phantom{{}- 16x^2 + 23x - 5} \\ 3x - 2\overline{)6x^3 - 16x^2 + 23x - 5} \\ \underline{6x^3 - 4x^2} \phantom{{}+ 23x - 5} \\ -12x^2 + 23x \phantom{{}- 5} \end{array}$$

• Think $\frac{6x^3}{3x} = 2x^2$.

• Multiply: $2x^2(3x - 2) = 6x^3 - 4x^2$

• Subtract and bring down the next term, $23x$.

$$\begin{array}{r} 2x^2 - 4x \phantom{{}+ 23x - 5} \\ 3x - 2\overline{)6x^3 - 16x^2 + 23x - 5} \\ \underline{6x^3 - 4x^2} \phantom{{}+ 23x - 5} \\ -12x^2 + 23x \phantom{{}- 5} \\ \underline{-12x^2 + 8x} \phantom{{}- 5} \\ 15x - 5 \end{array}$$

• Think $\frac{-12x^2}{3x} = -4x$.

• Multiply: $-4x(3x - 2) = -12x^2 + 8x$

• Subtract and bring down the next term, -5.

ANSWER No. $P(0) = 2(0)^3 - 0 + 1 = 1$. Because $P(0) \neq 0$, we know that 0 is not a zero of $P(x)$.

$$\begin{array}{r} 2x^2 - 4x + 5 \\ 3x - 2\overline{)6x^3 - 16x^2 + 23x - 5} \\ \underline{6x^3 - 4x^2} \qquad\qquad \\ -12x^2 + 23x \qquad \\ \underline{-12x^2 + 8x} \qquad \\ 15x - 5 \\ \underline{15x - 10} \\ 5 \end{array}$$

- **Think $\frac{15x}{3x} = 5$.**
- **Multiply: $5(3x - 2) = 15x - 10$**
- **Subtract to produce the remainder, 5.**

Thus $(6x^3 - 16x^2 + 23x - 5) \div (3x - 2) = 2x^2 - 4x + 5$ with a remainder of 5.

Although there is nothing wrong with writing the answer as we did above, it is more common to write the answer as the quotient plus the remainder divided by the divisor. (See the Take Note at the left.) Using this method, we write

> **take note**
>
> $\frac{20}{3}$ written as a mixed number is $6\frac{2}{3}$. Recall, however, that $6\frac{2}{3}$ means $6 + \frac{2}{3}$, which is in the form quotient $+ \frac{\text{remainder}}{\text{divisor}}$.

$$\underbrace{\frac{6x^3 - 16x^2 + 23x - 5}{\underbrace{3x - 2}_{\text{Divisor}}}}_{\text{Dividend}} = \underbrace{2x^2 - 4x + 5}_{\text{Quotient}} + \frac{5}{3x - 2} \begin{array}{l} \leftarrow \text{Remainder} \\ \leftarrow \text{Divisor} \end{array}$$

In this example, $6x^3 - 16x^2 + 23x - 5$ is called the **dividend,** $3x - 2$ is the **divisor,** $2x^2 - 4x + 5$ is the **quotient,** and 5 is the **remainder.** In every division, the dividend is equal to the product of the quotient and divisor, plus the remainder. That is,

$$\underbrace{6x^3 - 16x^2 + 23x - 5}_{\text{Dividend}} = \underbrace{(2x^2 - 4x + 5)}_{\text{quotient}} \cdot \underbrace{(3x - 2)}_{\text{divisor}} + \underbrace{5}_{\text{remainder}}$$

Before dividing polynomials, make sure that each polynomial is written in descending order. In some cases, it is helpful to insert a 0 in the dividend for a missing term (one whose coefficient is 0) so that like terms align in the same column. This is demonstrated in Example 1.

QUESTION What is the first step you should perform to find the quotient of $(2x + 1 + x^2) \div (x - 1)$?

EXAMPLE 1 Divide Polynomials

Divide: $\frac{-5x^2 - 8x + x^4 + 3}{x - 3}$

Solution

Write the numerator in descending order. Then divide.

$$\frac{-5x^2 - 8x + x^4 + 3}{x - 3} = \frac{x^4 - 5x^2 - 8x + 3}{x - 3}$$

ANSWER Write the dividend in descending order as $x^2 + 2x + 1$.

$$\begin{array}{r}
x^3 + 3x^2 + 4x + 4 \\
x - 3\overline{)x^4 + 0x^3 - 5x^2 - 8x + 3} \\
\underline{x^4 - 3x^3} \qquad\qquad\qquad \\
3x^3 - 5x^2 \qquad\qquad \\
\underline{3x^3 - 9x^2} \qquad\qquad \\
4x^2 - 8x \qquad \\
\underline{4x^2 - 12x} \qquad \\
4x + 3 \\
\underline{4x - 12} \\
15
\end{array}$$

• **Inserting $0x^3$ for the missing term helps align like terms in the same column.**

Thus $\dfrac{-5x^2 - 8x + x^4 + 3}{x - 3} = x^3 + 3x^2 + 4x + 4 + \dfrac{15}{x - 3}$.

▶ TRY EXERCISE 2, PAGE 135

A procedure called **synthetic division** can expedite the division process. To apply the synthetic division procedure, the divisor must be a polynomial of the form $x - c$, where c is a constant. In the synthetic division procedure, the variables that occur in the polynomials are not listed. To understand how synthetic division is performed, examine the following **long division** on the left and the related synthetic division on the right.

Long Division

Coefficients of the quotient

$$\begin{array}{r}
4x^2 + 3x + 8 \\
x - 2\overline{)4x^3 - 5x^2 + 2x - 10} \\
\underline{4x^3 - 8x^2} \qquad\qquad \\
3x^2 + 2x \qquad \\
\underline{3x^2 - 6x} \qquad \\
8x - 10 \\
\underline{8x - 16} \\
6
\end{array}$$

Remainder

Synthetic Division

2	4	−5	2	−10	← First row
		8	6	16	← Second row
	4	3	8	6	← Third row

4 3 8: Coefficients of the quotient; 6: Remainder

In the long division above, the dividend is $4x^3 - 5x^2 + 2x - 10$, and the divisor is $x - 2$. Because the divisor is of the form $x - c$, with $c = 2$, the division can be performed by the synthetic division procedure. Observe that in the accompanying synthetic division

1. The constant c is listed as the first number in the first row, followed by the coefficients of the dividend.
2. The first number in the third row is the leading coefficient of the dividend.
3. Each number in the second row is determined by computing the product of c and the number in the third row of the preceding column.
4. Each of the numbers in the third row, other than the first number, is determined by adding the numbers directly above it.

The following explanation illustrates the steps used to find the quotient and remainder of $(2x^3 - 8x + 7) \div (x + 3)$ by using synthetic division. The divisor $x + 3$ is written in $x - c$ form as $x - (-3)$, which indicates that $c = -3$. The dividend $2x^3 - 8x + 7$ is missing an x^2 term. If we insert $0x^2$ for the missing term, the dividend becomes $2x^3 + 0x^2 - 8x + 7$.

Coefficients of the dividend

$$\begin{array}{r|rrrr} -3 & 2 & 0 & -8 & 7 \\ & & & & \\ \hline & 2 & & & \end{array}$$

- Write the constant c, -3, followed by the coefficients of the dividend. Bring down the first coefficient in the first row, 2, as the first number of the third row.

$$\begin{array}{r|rrrr} -3 & 2 & 0 & -8 & 7 \\ & & -6 & & \\ \hline & 2 & -6 & & \end{array}$$

- Multiply c times the first number in the third row, 2, to produce the first number of the second row, -6. Add the 0 and the -6 to produce the next number of the third row, -6.

$$\begin{array}{r|rrrr} -3 & 2 & 0 & -8 & 7 \\ & & -6 & 18 & \\ \hline & 2 & -6 & 10 & \end{array}$$

- Multiply c times the second number in the third row, -6, to produce the next number of the second row, 18. Add the -8 and the 18 to produce the next number of the third row, 10.

$$\begin{array}{r|rrrr} -3 & 2 & 0 & -8 & 7 \\ & & -6 & 18 & -30 \\ \hline & 2 & -6 & 10 & -23 \end{array}$$

- Multiply c times the third number in the third row, 10, to produce the next number of the second row, -30. Add the 7 and the -30 to produce the last number of the third row, -23.

Coefficients of the quotient — Remainder

> **take note**
>
> $2x^2 - 6x + 10 + \dfrac{-23}{x + 3}$
>
> can also be written as
>
> $2x^2 - 6x + 10 - \dfrac{23}{x + 3}$

The last number in the bottom row, -23, is the remainder. The other numbers in the bottom row are the coefficients of the quotient. The quotient of a synthetic division always has a degree that is *one less* than the degree of the dividend. Thus the quotient in this example is $2x^2 - 6x + 10$. The results of the above synthetic division can be expressed in **fractional form** as

$$\frac{2x^3 - 8x + 7}{x + 3} = 2x^2 - 6x + 10 + \frac{-23}{x + 3}$$

or as

$$2x^3 - 8x + 7 = (x + 3)(2x^2 - 6x + 10) - 23$$

In Example 2 we illustrate the compact form of synthetic division, obtained by condensing the process explained above.

EXAMPLE 2 Use Synthetic Division to Divide Polynomials

Use synthetic division to divide $x^4 - 4x^2 + 7x + 15$ by $x + 4$.

Solution

Because the divisor is $x + 4$, we perform synthetic division with $c = -4$.

$$\begin{array}{r|rrrrr} -4 & 1 & 0 & -4 & 7 & 15 \\ & & -4 & 16 & -48 & 164 \\ \hline & 1 & -4 & 12 & -41 & 179 \end{array}$$

The quotient is $x^3 - 4x^2 + 12x - 41$, and the remainder is 179.

$$\frac{x^4 - 4x^2 + 7x + 15}{x + 4} = x^3 - 4x^2 + 12x - 41 + \frac{179}{x + 4}$$

TRY EXERCISE 12, PAGE 135

INTEGRATING TECHNOLOGY

A TI-82/83 synthetic-division program called SYDIV is available on the Internet at math.college.hmco.com. The program prompts you to enter the degree of the dividend, the coefficients of the dividend, and the constant c from the divisor $x - c$. For instance, to perform the synthetic division in Example 2, enter 4 for the degree of the dividend, followed by the coefficients 1, 0, −4, 7, and 15. See **Figure 2.1.** Press ENTER followed by −4 to produce the display in **Figure 2.2.** Press ENTER to produce the display in **Figure 2.3.** Press ENTER again to produce the display in **Figure 2.4.**

```
prgmSYDIV
DEGREE? 4
DIVIDEND COEF
?1
?0
?-4
?7
?15
```

FIGURE 2.1

```
C? -4
```

FIGURE 2.2

```
COEF OF QUOTIENT
                   1
                  -4
                  12
                 -41
```

FIGURE 2.3

```
REMAINDER
                 179
QUIT? PRESS 1
NEW C? PRESS 2
```

FIGURE 2.4

THE REMAINDER THEOREM

The following theorem shows that synthetic division can be used to determine the value $P(c)$ for a given polynomial P and constant c.

The Remainder Theorem

If a polynomial $P(x)$ is divided by $x - c$, then the remainder equals $P(c)$.

The following example illustrates the Remainder Theorem by showing that the remainder of $(x^2 + 9x - 16) \div (x - 3)$ is the same as $P(x) = x^2 + 9x - 16$ evaluated at $x = 3$.

Let $x = 3$ and $P(x) = x^2 + 9x - 16$.

$$\begin{aligned} \text{Then } P(3) &= (3)^2 + 9(3) - 16 \\ &= 9 + 27 - 16 \\ &= 20 \end{aligned}$$

$$\begin{array}{r} x + 12 \\ x - 3\overline{)x^2 + 9x - 16} \\ \underline{x^2 - 3x} \\ 12x - 16 \\ \underline{12x - 36} \\ 20 \end{array}$$

$P(3)$ is equal to the remainder of $P(x)$ divided by $(x - 3)$.

In Example 3 we use synthetic division and the Remainder Theorem to evaluate a polynomial function.

EXAMPLE 3 Use the Remainder Theorem to Evaluate a Polynomial Function

Let $P(x) = 2x^3 + 3x^2 + 2x - 2$. Use the Remainder Theorem to find $P(c)$ for $c = -2$ and $c = \frac{1}{2}$.

Algebraic Solution

Perform synthetic division with $c = -2$ and $c = \frac{1}{2}$ and examine the remainders.

$$\begin{array}{r|rrrr} -2 & 2 & 3 & 2 & -2 \\ & & -4 & 2 & -8 \\ \hline & 2 & -1 & 4 & -10 \end{array}$$

The remainder is -10. Therefore, $P(-2) = -10$.

$$\begin{array}{r|rrrr} \frac{1}{2} & 2 & 3 & 2 & -2 \\ & & 1 & 2 & 2 \\ \hline & 2 & 4 & 4 & 0 \end{array}$$

The remainder is 0. Therefore, $P\left(\frac{1}{2}\right) = 0$.

Visualize the Solution

A graph of P shows that the points $(-2, -10)$ and $\left(\frac{1}{2}, 0\right)$ are on the graph.

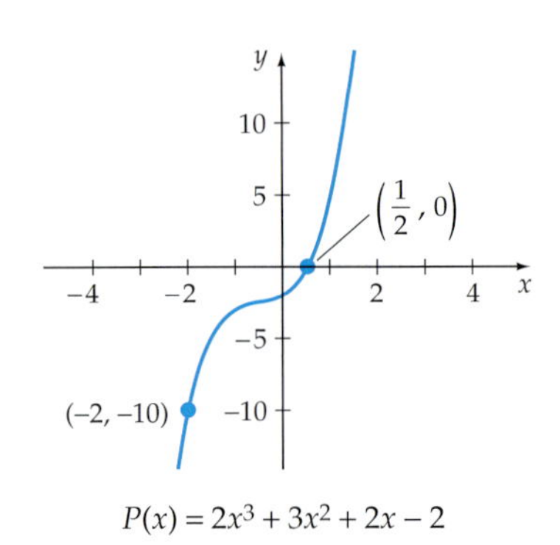

$P(x) = 2x^3 + 3x^2 + 2x - 2$

▶ TRY EXERCISE 26, PAGE 135

Using the Remainder Theorem to evaluate a polynomial function is often faster than evaluating the polynomial function by direct substitution. For instance, to evaluate $P(x) = x^5 - 10x^4 + 35x^3 - 50x^2 + 24x$ by substituting 7 for x, we must do the following work.

$$\begin{aligned} P(7) &= (7)^5 - 10(7)^4 + 35(7)^3 - 50(7)^2 + 24(7) \\ &= 16{,}807 - 10(2401) + 35(343) - 50(49) + 24(7) \\ &= 16{,}807 - 24{,}010 + 12{,}005 - 2450 + 168 \\ &= 2520 \end{aligned}$$

> **take note**
>
> Because $P(x)$ has a constant term of 0, we must include 0 as the last number in the first row of the synthetic division at the right.

Using the Remainder Theorem to perform the above evaluation requires only the following work.

$$\begin{array}{r|rrrrrr} 7 & 1 & -10 & 35 & -50 & 24 & 0 \\ & & 7 & -21 & 98 & 336 & 2520 \\ \hline & 1 & -3 & 14 & 48 & 360 & 2520 \end{array} \longleftarrow P(7)$$

THE FACTOR THEOREM

Note from Example 3 that $P\left(\frac{1}{2}\right) = 0$. Recall that $\frac{1}{2}$ is a zero of P because $P(x) = 0$ when $x = \frac{1}{2}$.

The following theorem is a direct result of the Remainder Theorem. It points out the important relationship between a zero of a given polynomial function and a factor of the polynomial function.

The Factor Theorem

A polynomial function $P(x)$ has a factor $(x - c)$ if and only if $P(c) = 0$. That is, $(x - c)$ is a factor of $P(x)$ if and only if c is a zero of P.

EXAMPLE 4 Apply the Factor Theorem

Use synthetic division and the Factor Theorem to determine whether $(x + 5)$ or $(x - 2)$ is a factor of $P(x) = x^4 + x^3 - 21x^2 - x + 20$.

Solution

$$\begin{array}{r|rrrrr} -5 & 1 & 1 & -21 & -1 & 20 \\ & & -5 & 20 & 5 & -20 \\ \hline & 1 & -4 & -1 & 4 & 0 \end{array}$$

The remainder of 0 indicates that $(x + 5)$ is a factor of $P(x)$.

$$\begin{array}{r|rrrrr} 2 & 1 & 1 & -21 & -1 & 20 \\ & & 2 & 6 & -30 & -62 \\ \hline & 1 & 3 & -15 & -31 & -42 \end{array}$$

The remainder of -42 indicates that $(x - 2)$ is not a factor of $P(x)$.

TRY EXERCISE 36, PAGE 135

QUESTION Is -5 a zero of $P(x)$ given in Example 4?

Here is a summary of the important role played by the remainder in the division of a polynomial by $(x - c)$.

The Remainder of a Polynomial Division

In the division of the polynomial function $P(x)$ by $(x - c)$, the remainder is

- equal to $P(c)$.
- 0 if and only if $(x - c)$ is a factor of P.
- 0 if and only if c is a zero of P.

Also, if c is a real number, then the remainder of $P(x) \div (x - c)$ is 0 if and only if $(c, 0)$ is an x-intercept of the graph of P.

ANSWER Yes. Because $(x + 5)$ is a factor of $P(x)$, the Factor Theorem states that $P(-5) = 0$, and thus -5 is a zero of $P(x)$.

Reduced Polynomials

In Example 4 we determined that $(x + 5)$ is a factor of the polynomial function $P(x) = x^4 + x^3 - 21x^2 - x + 20$ and that the quotient of $x^4 + x^3 - 21x^2 - x + 20$ divided by $(x + 5)$ is $Q(x) = x^3 - 4x^2 - x + 4$. Thus

$$P(x) = (x + 5)(x^3 - 4x^2 - x + 4)$$

The quotient $Q(x) = x^3 - 4x^2 - x + 4$ is called a **reduced polynomial** or a **depressed polynomial** of $P(x)$ because it is a factor of $P(x)$ and its degree is 1 less than the degree of $P(x)$. Reduced polynomials will play an important role in Sections 2.4 and 2.5.

EXAMPLE 5 Find a Reduced Polynomial

Verify that $(x - 3)$ is a factor of $P(x) = 2x^3 - 3x^2 - 4x - 15$, and write $P(x)$ as the product of $(x - 3)$ and the reduced polynomial $Q(x)$.

Solution

$$\begin{array}{r|rrrr} 3 & 2 & -3 & -4 & -15 \\ & & 6 & 9 & 15 \\ \hline & 2 & 3 & 5 & 0 \end{array}$$

Coefficients of the reduced polynomial $Q(x)$

Thus $(x - 3)$ and the reduced polynomial $2x^2 + 3x + 5$ are both factors of $P(x)$. That is,

$$P(x) = 2x^3 - 3x^2 - 4x - 15 = (x - 3)(2x^2 + 3x + 5)$$

▶ Try Exercise 56, page 136

Topics for Discussion

1. Explain the meaning of the phrase *zero of a polynomial.*
2. If $P(x)$ is a polynomial of degree 3, what is the degree of the quotient of $\dfrac{P(x)}{x - c}$?
3. Discuss how the Remainder Theorem can be used to determine whether a number is a zero of a polynomial.
4. A zero of $P(x) = x^3 - x^2 - 14x + 24$ is -4. Discuss how this information and the Factor Theorem can be used to solve $x^3 - x^2 - 14x + 24 = 0$.
5. Discuss the advantages and disadvantages of using synthetic division rather than substitution to evaluate a polynomial function at $x = c$.

Exercise Set 2.2

In Exercises 1 to 10, use long division to divide the first polynomial by the second.

1. $5x^3 + 6x^2 - 17x + 20, \quad x + 3$

▶ **2.** $6x^3 + 15x^2 - 8x + 2, \quad x + 4$

3. $x^4 - 5x^2 + 3x - 1, \quad x - 2$

4. $x^4 - 5x^3 + x - 4, \quad x - 1$

5. $x^2 + x^3 - 2x - 5, \quad x - 3$

6. $4x + 3x^2 + x^3 - 5, \quad x - 2$

7. $x^4 + 3x^3 - 5x + 3x^2 - 1, \quad x - 4$

8. $2x^4 + x^3 - 5x^2 + 2x - 8, \quad x + 4$

9. $x^5 + x^4 - 2x^3 + 2x^2 - 3x - 7, \quad x - 1$

10. $x^5 - 2x^4 - x^3 + 3x^2 - 5x + 8, \quad x + 4$

In Exercises 11 to 24, use synthetic division to divide the first polynomial by the second.

11. $4x^3 - 5x^2 + 6x - 7, \quad x - 2$

▶ **12.** $5x^3 + 6x^2 - 8x + 1, \quad x - 5$

13. $4x^3 - 2x + 3, \quad x + 1$

14. $6x^3 - 4x^2 + 17, \quad x + 3$

15. $x^5 - 10x^3 + 5x - 1, \quad x - 4$

16. $6x^4 - 2x^3 - 3x^2 - x, \quad x - 5$

17. $x^5 - 1, \quad x - 1$

18. $x^4 + 1, \quad x + 1$

19. $8x^3 - 4x^2 + 6x - 3, \quad x - \frac{1}{2}$

20. $12x^3 + 5x^2 + 5x + 6, \quad x + \frac{3}{4}$

21. $x^8 + x^6 + x^4 + x^2 + 4, \quad x - 2$

22. $-x^7 - x^5 - x^3 - x - 5, \quad x + 1$

23. $x^6 + x - 10, \quad x + 3$

24. $2x^5 - 3x^4 - 5x^2 - 10, \quad x - 4$

In Exercises 25 to 34, use the Remainder Theorem to find *P*(*c*).

25. $P(x) = 3x^3 + x^2 + x - 5, c = 2$

▶ **26.** $P(x) = 2x^3 - x^2 + 3x - 1, c = 3$

27. $P(x) = 4x^4 - 6x^2 + 5, c = -2$

28. $P(x) = 6x^3 - x^2 + 4x, c = -3$

29. $P(x) = -2x^3 - 2x^2 - x - 20, c = 10$

30. $P(x) = -x^3 + 3x^2 + 5x + 30, c = 8$

31. $P(x) = -x^4 + 1, c = 3$

32. $P(x) = x^5 - 1, c = 1$

33. $P(x) = x^4 - 10x^3 + 2, c = 3$

34. $P(x) = x^5 + 20x^2 - 1, c = -5$

In Exercises 35 to 44, use synthetic division and the Factor Theorem to determine whether the given binomial is a factor of *P*(*x*).

35. $P(x) = x^3 + 2x^2 - 5x - 6, x - 2$

▶ **36.** $P(x) = x^3 + 4x^2 - 27x - 90, x + 6$

37. $P(x) = 2x^3 + x^2 - 3x - 1, x + 1$

38. $P(x) = 3x^3 + 4x^2 - 27x - 36, x - 4$

39. $P(x) = x^4 - 25x^2 + 144, x + 3$

40. $P(x) = x^4 - 25x^2 + 144, x - 3$

41. $P(x) = x^5 + 2x^4 - 22x^3 - 50x^2 - 75x, x - 5$

42. $P(x) = 9x^4 - 6x^3 - 23x^2 - 4x + 4, x + 1$

43. $P(x) = 16x^4 - 8x^3 + 9x^2 + 14x + 4, x - \frac{1}{4}$

44. $P(x) = 10x^4 + 9x^3 - 4x^2 + 9x + 6, x + \frac{1}{2}$

In Exercises 45 to 54, use synthetic division to show that c is a zero of $P(x)$.

45. $P(x) = 3x^3 - 8x^2 - 10x + 28, c = 2$

46. $P(x) = 4x^3 - 10x^2 - 8x + 6, c = 3$

47. $P(x) = x^4 - 1, c = 1$

48. $P(x) = x^3 + 8, c = -2$

49. $P(x) = 3x^4 + 8x^3 + 10x^2 + 2x - 20, c = -2$

50. $P(x) = x^4 - 2x^2 - 100x - 75, c = 5$

51. $P(x) = 2x^3 - 18x^2 - 50x + 66, c = 11$

52. $P(x) = 2x^4 - 34x^3 + 70x^2 - 153x + 45, c = 15$

53. $P(x) = 3x^2 - 8x + 4, c = \frac{2}{3}$

54. $P(x) = 5x^2 + 12x + 4, c = -\frac{2}{5}$

In Exercises 55 to 58, verify that the given binomial is a factor of $P(x)$, and write $P(x)$ as the product of the binomial and its reduced polynomial $Q(x)$.

55. $P(x) = x^3 + x^2 + x - 14, x - 2$

▶ **56.** $P(x) = x^4 + 5x^3 + 3x^2 - 5x - 4, x + 1$

57. $P(x) = x^4 - x^3 - 9x^2 - 11x - 4, x - 4$

58. $P(x) = 2x^5 - x^4 - 7x^3 + x^2 + 7x - 10, x - 2$

59. Cost of a Wedding The average cost of a wedding, in dollars, is modeled by

$$C(t) = 38t^2 + 291t + 15{,}208$$

where $t = 0$ represents the year 1990 and $0 \le t \le 12$. Use the Remainder Theorem to estimate the average cost of a wedding in

a. 1998.

b. 2001.

60. Selection of Bridesmaids A bride-to-be has several girlfriends, but she has decided to have only five bridesmaids, including the maid of honor. The number of different ways n girlfriends can be chosen and assigned a position, such as maid of honor, first matron, second matron, and so on, is given by the polynomial function

$$P(n) = n^5 - 10n^4 + 35n^3 - 50n^2 + 24n, \quad n \ge 5$$

a. Use the Remainder Theorem to determine the number of ways the bride can select her bridesmaids if she chooses from $n = 7$ girlfriends.

b. Evaluate $P(n)$ for $n = 7$ by substituting 7 for n. How does this result compare with the result obtained in part **a.**?

61. Selection of Cards The number of ways you can select three cards from a stack of n cards, in which the order of selection is important, is given by

$$P(n) = n^3 - 3n^2 + 2n, \quad n \ge 3$$

a. Use the Remainder Theorem to determine the number of ways you can select three cards from a stack of $n = 8$ cards.

b. Evaluate $P(n)$ for $n = 8$ by substituting 8 for n. How does this result compare with the result obtained in part **a.**?

62. Rocket Launch A model rocket is projected upward from an initial height of 4 feet with an initial velocity of 158 feet per second. The height of the rocket, in feet, is given by $s = -16t^2 + 158t + 4$, where $0 \le t \le 9.9$ seconds. Use the Remainder Theorem to determine the height of the rocket at

a. $t = 5$ seconds.

b. $t = 8$ seconds.

63. House of Cards The number of cards C needed to build a house of cards with r rows (levels) is given by the function $C(r) = 1.5r^2 + 0.5r$.

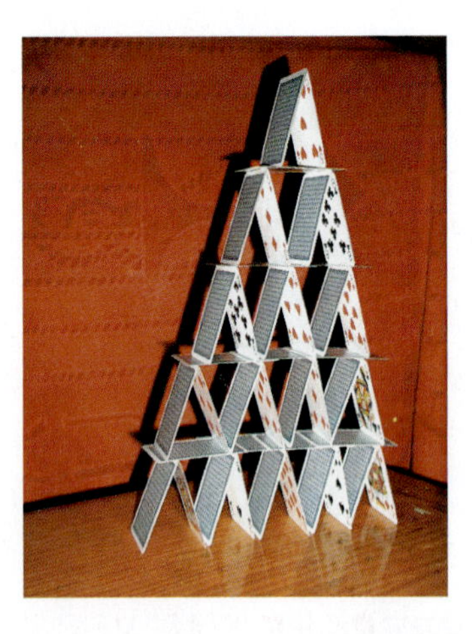

Use the Remainder Theorem to determine the number of cards needed to build a house of cards with

a. $r = 8$ rows.

b. $r = 20$ rows.

64. Election of Class Officers The number of ways a class of n students can elect a president, a vice president, a secretary, and a treasurer is given by the function

$P(n) = n^4 - 6n^3 + 11n^2 - 6n$, where $n \geq 4$. Use the Remainder Theorem to determine the number of ways the class can elect officers if the class consists of

a. $n = 12$ students.

b. $n = 24$ students.

65. **Population Density of a City** The population density D, in people per square mile, of a city is related to the distance x, in miles, from the center of the city by $D = -45x^2 + 190x + 200$, $0 < x < 5$. Use the Remainder Theorem to determine the population density of the city at a distance of

a. $x = 2$ miles.

b. $x = 4$ miles.

66. **Volume of a Solid** The volume of the solid at the right is given by $V(x) = x^3 + 3x^2$.

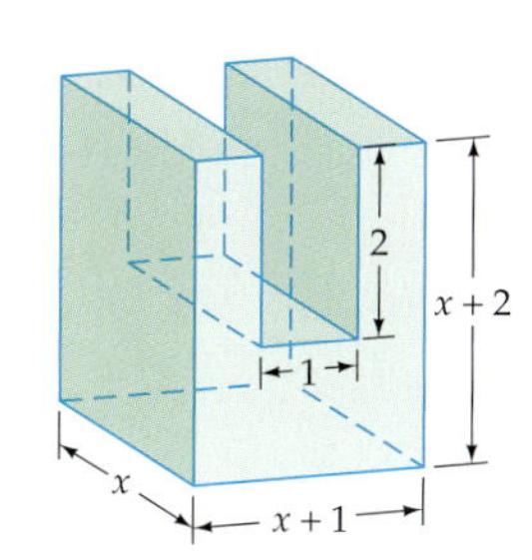

Use the Remainder Theorem to determine the volume of the solid if

a. $x = 7$ inches.

b. $x = 11$ inches.

67. **Volume of a Solid** The volume of the following solid is given by $V(x) = x^3 + x^2 + 10x - 8$.

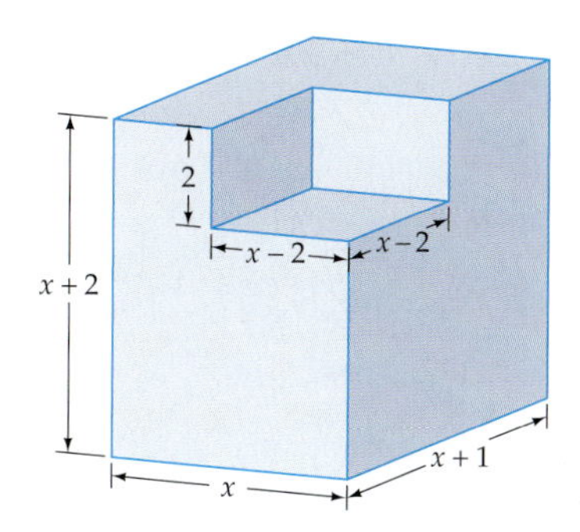

Use the Remainder Theorem to determine the volume of the solid if

a. $x = 6$ inches.

b. $x = 9$ inches.

CONNECTING CONCEPTS

68. Use the Factor Theorem to show that for any positive integer n, $P(x) = x^n - 1$ has $x - 1$ as a factor.

69. Find the remainder of $5x^{48} + 6x^{10} - 5x + 7$ divided by $x - 1$.

70. Find the remainder of $18x^{80} - 6x^{50} + 4x^{20} - 2$ divided by $x + 1$.

71. Determine whether i is a zero of $P(x) = x^3 - 3x^2 + x - 3$.

72. Determine whether $-2i$ is a zero of $P(x) = x^4 - 2x^3 + x^2 - 8x - 12$.

PREPARE FOR SECTION 2.3

73. Find the minimum value of $P(x) = x^2 - 4x + 6$. [1.5]

74. Find the maximum value of $P(x) = -2x^2 - x + 1$. [1.5]

75. Find the interval on which $P(x) = x^2 + 2x + 7$ is increasing. [1.3]

76. Find the interval on which $P(x) = -2x^2 + 4x + 5$ is decreasing. [1.3]

77. Factor: $x^4 - 5x^2 + 4$ [A.3]

78. Find the x-intercepts of the graph of $P(x) = 6x^2 - x - 2$. [1.5]

PROJECTS

1. **HORNER'S POLYNOMIAL FORM** William Horner (1786–1837) devised a method of writing a polynomial in a form that does not involve any exponents other than 1. For instance, $4x^4 + 2x^3 - 5x^2 + 7x - 11$ can be written in each of the following forms.

$4x^4 + 2x^3 - 5x^2 + 7x - 11$

$= (4x^3 + 2x^2 - 5x + 7)x - 11$ • Factor an x from the first four terms.

$= [(4x^2 + 2x - 5)x + 7]x - 11$ • Factor an x from the first three terms inside the innermost parentheses.

$= \{[(4x + 2)x - 5]x + 7\}x - 11$ • Factor an x from the first two terms inside the innermost parentheses.

Horner's form, $\{[(4x + 2)x - 5]x + 7\}x - 11$, is easier to evaluate than the descending exponent form, $4x^4 + 2x^3 - 5x^2 + 7x - 11$. Horner's form is sometimes used by computer programmers to make their programs run faster.

a. Let $P(x) = 3x^5 - 4x^4 + 5x^3 - 2x^2 + 3x - 8$. Find $P(6)$ by direct substitution.

b. Use Horner's method to write $P(x)$ in a form that does not involve any exponents other than 1. Now use this form to evaluate $P(6)$. Which was easier to perform, the evaluation in part **a.** or in part **b.**?

SECTION 2.3

POLYNOMIAL FUNCTIONS OF HIGHER DEGREE

- FAR-LEFT AND FAR-RIGHT BEHAVIOR
- MAXIMUM AND MINIMUM VALUES
- REAL ZEROS OF A POLYNOMIAL FUNCTION
- EVEN AND ODD POWERS OF $(x - c)$ THEOREM
- A PROCEDURE FOR GRAPHING POLYNOMIAL FUNCTIONS

Table 2.1 summarizes information developed in Chapter 1 about graphs of polynomial functions of degree 0, 1, or 2.

TABLE 2.1

Polynomial Function $P(x)$	Graph
$P(x) = a$ (degree 0)	Horizontal line through $(0, a)$
$P(x) = ax + b$ (degree 1), $a \neq 0$	Line with y-intercept $(0, b)$ and slope a.
$P(x) = ax^2 + bx + c$ (degree 2), $a \neq 0$	Parabola with vertex $\left(-\frac{b}{2a}, P\left(-\frac{b}{2a}\right)\right)$

Polynomial functions of degree 3 or higher can be graphed by the technique of plotting points; however, some additional knowledge about polynomial functions will make graphing easier.

All polynomial functions have graphs that are **smooth continuous curves.** The terms *smooth* and *continuous* are defined rigorously in calculus, but for the

> **take note**
>
> The **general form of a polynomial** is given by $a_nx^n + a_{n-1}x^{n-1} + \cdots + a_0$. In this text the coefficients $a_n, a_{n-1}, \ldots, a_0$ are all real numbers unless specifically stated otherwise.

present, a smooth curve is a curve that does not have sharp corners such as that shown in **Figure 2.5a.** A continuous curve does not have a break or hole such as those shown in **Figure 2.5b.**

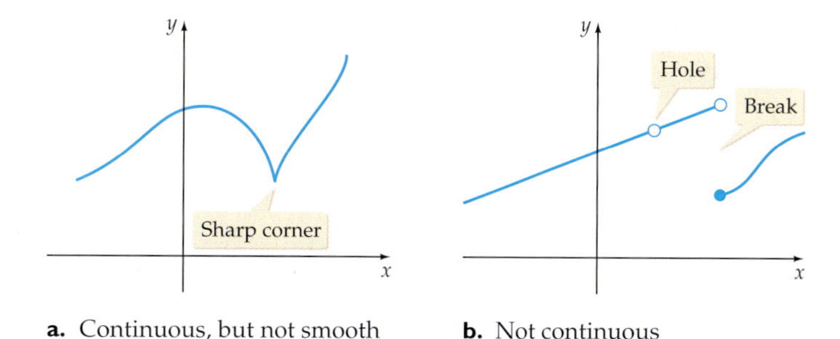

a. Continuous, but not smooth **b.** Not continuous

FIGURE 2.5

Far-Left and Far-Right Behavior

> **take note**
>
> The leading term of a polynomial function in x is the nonzero term that contains the largest power of x. The leading coefficient of a polynomial function is the coefficient of the leading term.

The graph of a polynomial function may have several up and down fluctuations; however, the graph of every polynomial function eventually will increase or decrease without bound as $|x|$ becomes large. The **leading term** a_nx^n is said to be the **dominate term** of the polynomial function $P(x) = a_nx^n + a_{n-1}x^{n-1} + \cdots + a_1x + a_0$ because as $|x|$ becomes large, the absolute value of a_nx^n will be much larger than the absolute value of any of the other terms. Because of this condition, you can determine the **far-left and far-right behavior** of the polynomial by examining the **leading coefficient** a_n and the degree n of the polynomial.

Table 2.2 indicates the far-left and far-right behavior of a polynomial function $P(x)$ with leading term a_nx^n.

TABLE 2.2 Far-Right and Far-Left Behavior of the Graph of a Polynomial Function with Leading Term a_nx^n

	n is even	*n* is odd
$a_n > 0$	Up to far left and up to far right	Down to far left and up to far right
$a_n < 0$	Down to far left and down to far right	Up to far left and down to far right

EXAMPLE 1 Determine the Far-Left and Far-Right Behavior of a Polynomial Function

Examine the leading term to determine the far-left and far-right behavior of the graph of each polynomial function.

a. $P(x) = x^3 - x$ **b.** $S(x) = \dfrac{1}{2}x^4 - \dfrac{5}{2}x^2 + 2$

c. $T(x) = -2x^3 + x^2 + 7x - 6$ **d.** $U(x) = 9 + 8x^2 - x^4$

Continued ▶

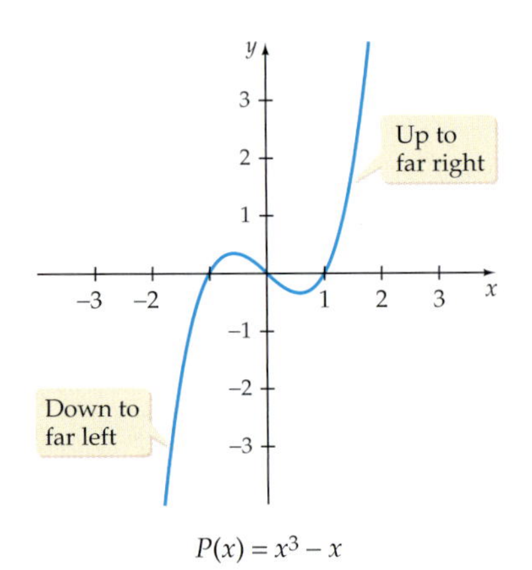

$P(x) = x^3 - x$

FIGURE 2.6

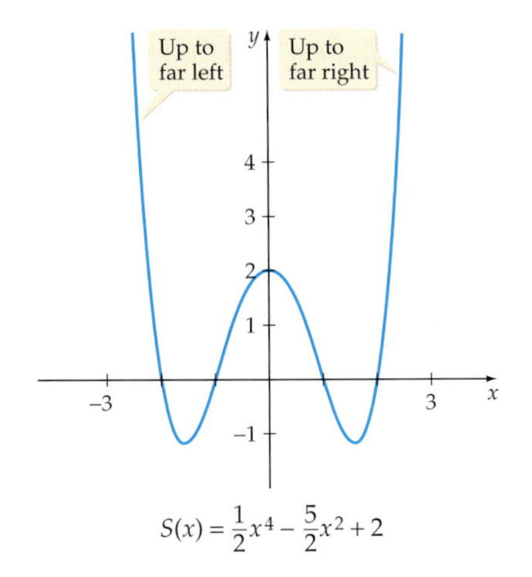

$S(x) = \frac{1}{2}x^4 - \frac{5}{2}x^2 + 2$

FIGURE 2.7

Solution

a. Because $a_n = 1$ is *positive* and $n = 3$ is *odd*, the graph of P goes down to its far left and up to its far right. See **Figure 2.6.**

b. Because $a_n = \dfrac{1}{2}$ is *positive* and $n = 4$ is *even*, the graph of S goes up to its far left and up to its far right. See **Figure 2.7.**

c. Because $a_n = -2$ is *negative* and $n = 3$ is *odd*, the graph of T goes up to its far left and down to its far right. See **Figure 2.8.**

d. The leading term of $U(x)$ is $-x^4$ and the leading coefficient is -1. Because $a_n = -1$ is *negative* and $n = 4$ is *even*, the graph of U goes down to its far left and down to its far right. See **Figure 2.9.**

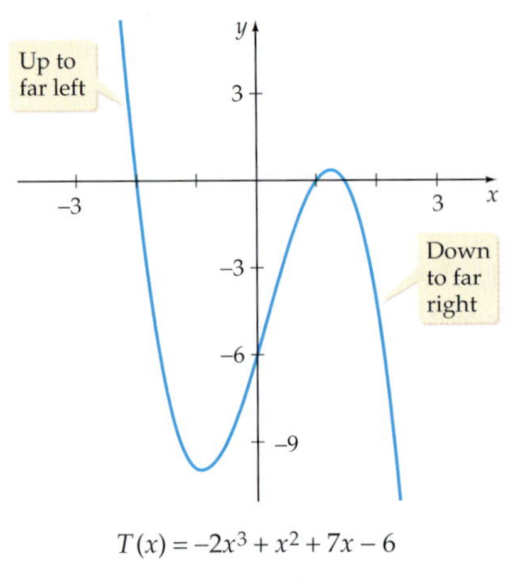

$T(x) = -2x^3 + x^2 + 7x - 6$

FIGURE 2.8

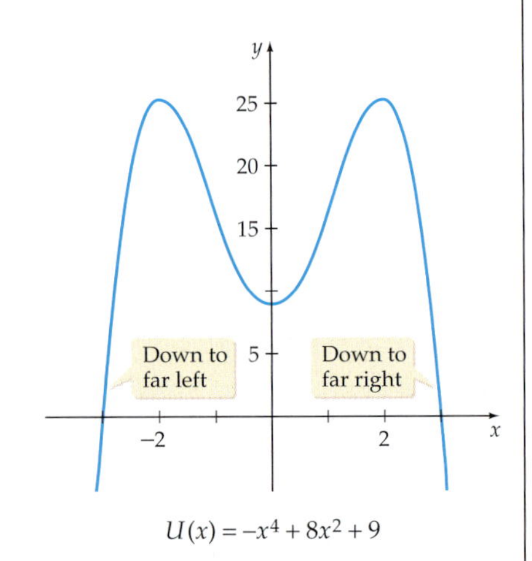

$U(x) = -x^4 + 8x^2 + 9$

FIGURE 2.9

▶ TRY EXERCISE 2, PAGE 149

MAXIMUM AND MINIMUM VALUES

Figure 2.10 illustrates the graph of a polynomial function of degree 3 with two **turning points,** points at which the function changes from an increasing function to a decreasing function, or vice versa. In general, the graph of a polynomial function of degree n has at most $n - 1$ turning points.

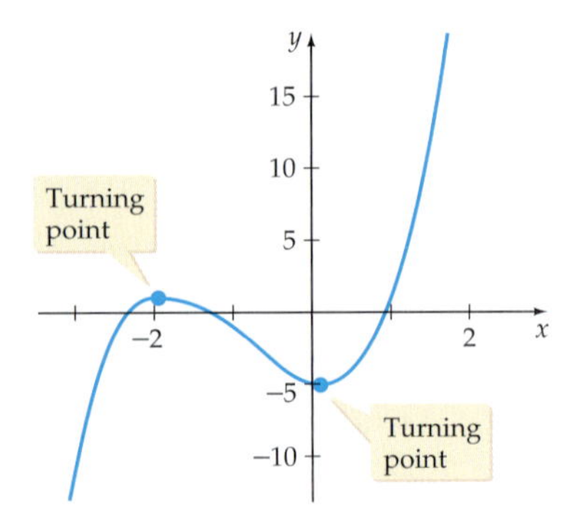

$P(x) = 2x^3 + 5x^2 - x - 5$

FIGURE 2.10

Turning points can be related to the concepts of maximum and minimum values of a function. These concepts were introduced in the discussion of graphs of second-degree equations in two variables earlier in the text. Recall that the minimum value of a function f is the smallest range value of f. It is often called the **absolute minimum.** The maximum value of a function f is the largest range value of f. The maximum value of a function is also called the **absolute maximum.** For the function whose graph is shown in **Figure 2.11,** the y value of point E is the absolute minimum. There are no y values less than y_5.

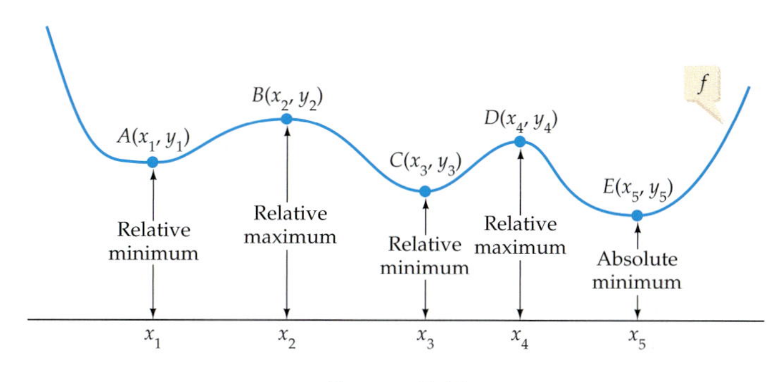

FIGURE 2.11

Now consider y_1, the y value of turning point A in **Figure 2.11.** It is not the smallest y value of every point on the graph of f; however, it is the smallest y value if we *localize* our field of view to a small open interval containing x_1. It is for this reason that we refer to y_1 as a **local minimum,** or **relative minimum,** of f. The y value of point C is also a relative minimum of f.

The function does not have an absolute maximum because it goes up both to its far left and to its far right. The y value of point B is a relative maximum, as is the y value of point D. The formal definitions of **relative maximum** and **relative minimum** are presented below.

Relative Minimum and Relative Maximum

If there is an open interval I containing c on which

- $f(c) \leq f(x)$ for all x in I, then $f(c)$ is a **relative minimum** of f.
- $f(c) \geq f(x)$ for all x in I, then $f(c)$ is a **relative maximum** of f.

QUESTION Is the absolute minimum y_5 shown in **Figure 2.11** also a relative minimum of f?

ANSWER Yes, the absolute minimum y_5 also satisfies the requirements of a relative minimum.

A graphing utility can estimate the minimum and maximum values of a function. To use a TI-83 calculator to estimate the relative maximum of

$$P(x) = 0.3x^3 - 2.8x^2 + 6.4x + 2$$

use the following steps:

1. Enter the function in the Y = menu. Choose your window settings.
2. Select 4:maximum from the CALC menu, which is located above the TRACE key. The graph of Y1 is displayed.
3. PRESS ◀ or ▶ repeatedly to select an x-value that is to the left of the relative maximum point. Press ENTER. A left bound is displayed in the bottom left corner.
4. Press ▶ repeatedly to select an x-value that is to the right of the relative maximum point. Press ENTER. A right bound is displayed in the bottom left corner.
5. The word **Guess?** is now displayed in the bottom left corner. Press ◀ repeatedly to move to a point near the maximum point. Press ENTER.
6. The cursor appears on the relative maximum point and the coordinates of the relative maximum point are displayed. In this example, the y value 6.312608 is the approximate relative maximum of the function P. *Note:* If your window settings, bounds, or your guess are different from those shown below, then your final results may differ slightly from the final results shown below in step 6.

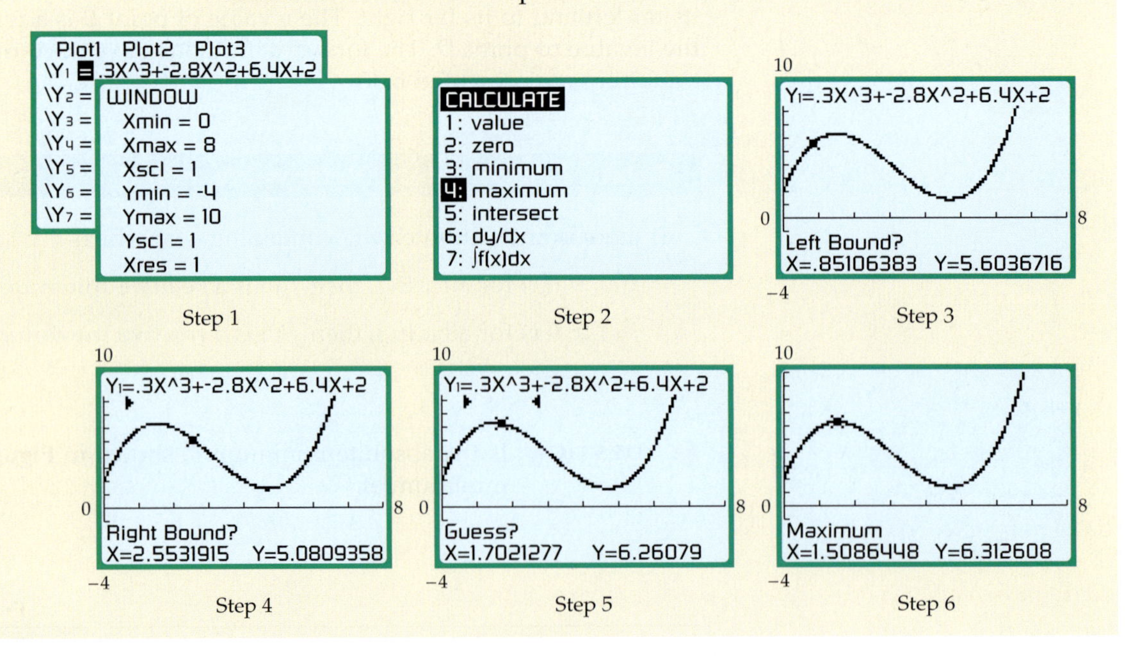

Step 1 Step 2 Step 3

Step 4 Step 5 Step 6

The following example illustrates the role a maximum may play in an application.

EXAMPLE 2 Solve an Application

A rectangular piece of cardboard measures 12 inches by 16 inches. An open box is formed by cutting congruent squares that measure x inches by x inches from each of the corners of the cardboard and folding up the sides as shown below.

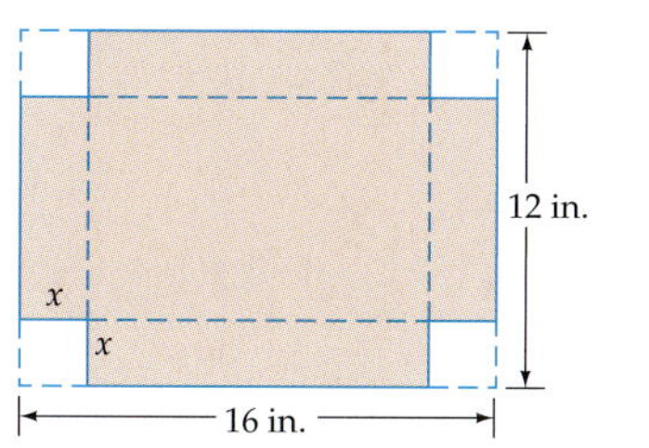

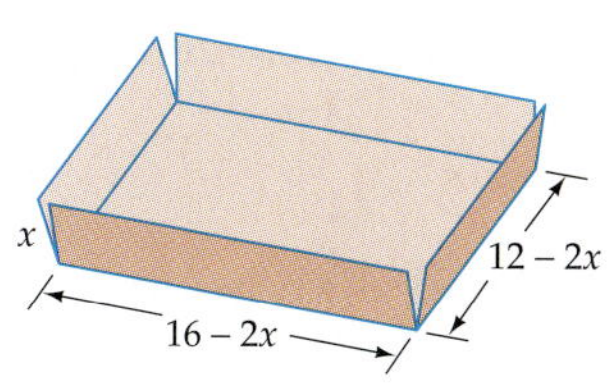

a. Express the volume V of the box as a function of x.

b. Determine (to the nearest tenth of an inch) the x value that maximizes the volume.

250

−2 10

−100

$y = 4x^3 - 56x^2 + 192x$

FIGURE 2.12

Solution

a. The height, width, and length of the open box are x, $12 - 2x$, and $16 - 2x$. The volume is given by

$$V(x) = x(12 - 2x)(16 - 2x)$$
$$= 4x^3 - 56x^2 + 192x$$

b. Use a graphing utility to graph $y = V(x)$. The graph is shown in **Figure 2.12.**

Note that we are interested only in the part of the graph for which $0 < x < 6$. This is so because the length of each side of the box must be positive. In other words,

$$x > 0, \quad 12 - 2x > 0 \quad \text{and} \quad 16 - 2x > 0$$
$$x < 6 \qquad\qquad x < 8$$

The domain of V is the intersection of the solution sets of the three inequalities. Thus the domain is $\{x \mid 0 < x < 6\}$.

Now use a graphing utility to find that V attains its maximum of about 194.06736 when $x \approx 2.3$. See **Figure 2.13.**

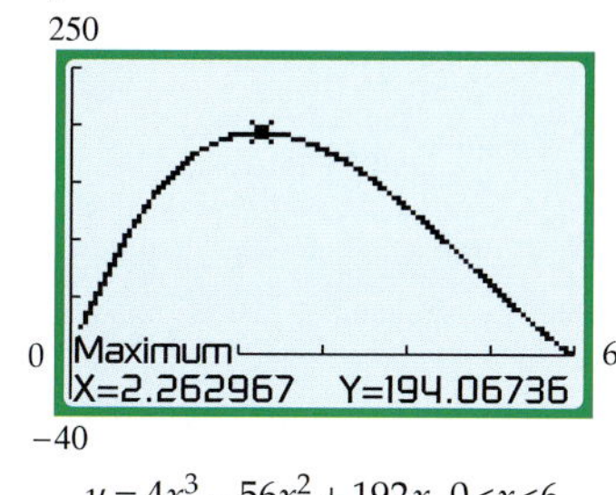

$y = 4x^3 - 56x^2 + 192x, 0<x<6$

FIGURE 2.13

▶ TRY EXERCISE 48, PAGE 150

A TI graphing calculator program is available that simulates the construction of a box by cutting out squares from each corner of a rectangular piece of cardboard. This program, CUTOUT, can be found on our website at
math.college.hmco.com

Real Zeros of a Polynomial Function

Sometimes the real zeros of a polynomial function can be determined by using factoring procedures. We illustrate this concept in the next example.

EXAMPLE 3 Factor to Find the Real Zeros of a Polynomial Function

Factor to find the three real zeros of $P(x) = x^3 + 3x^2 - 4x$.

Algebraic Solution

$P(x)$ can be factored as shown below.

$$\begin{aligned} P(x) &= x^3 + 3x^2 - 4x \\ &= x(x^2 + 3x - 4) && \text{• Factor out the common factor } x. \\ &= x(x - 1)(x + 4) && \text{• Factor the trinomial } x^2 + 3x - 4. \end{aligned}$$

The real zeros of $P(x)$ are $x = 0$, $x = 1$, and $x = -4$.

Visualize the Solution

The graph of $P(x)$ has x-intercepts at (0, 0), (1, 0), and (−4, 0).

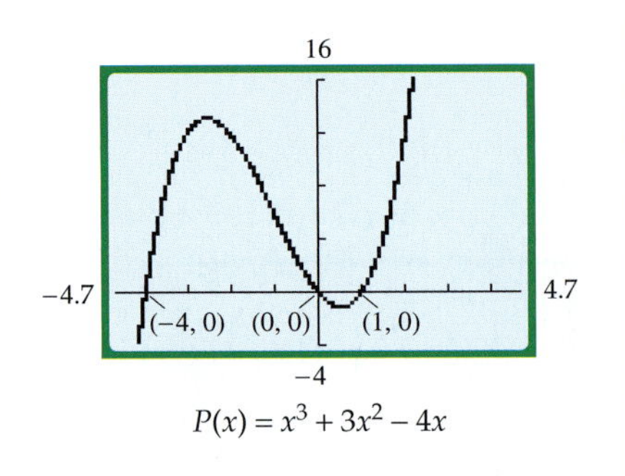

$P(x) = x^3 + 3x^2 - 4x$

▶ TRY EXERCISE 22, PAGE 150

The graph of every polynomial function P is a smooth continuous curve, and if the value of P changes sign on an interval, then $P(c)$ must equal zero for at least one real number c in the interval. This result is known as the *Zero Location Theorem.*

The Zero Location Theorem

Let $P(x)$ be a polynomial function and let a and b be two distinct real numbers. If $P(a)$ and $P(b)$ have opposite signs, then there is at least one real number c between a and b such that $P(c) = 0$.

For instance, if the value of P is negative at $x = a$ and positive at $x = b$, then there is at least one real number c between a and b such that $P(c) = 0$. See **Figure 2.14.**

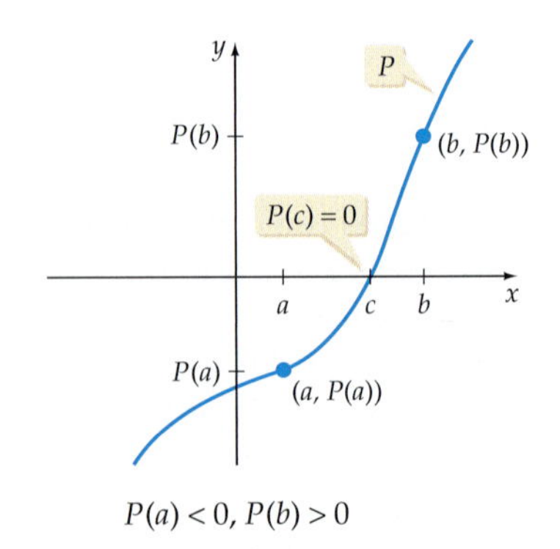

$P(a) < 0$, $P(b) > 0$

Figure 2.14

EXAMPLE 4 Apply the Zero Location Theorem

Use the Zero Location Theorem to verify that $S(x) = x^3 - x - 2$ has a real zero between 1 and 2.

Algebraic Solution

Use synthetic division to evaluate S for $x = 1$ and $x = 2$. If S changes sign between these two values, then S has a real zero between 1 and 2.

$$\begin{array}{r|rrrr} 1 & 1 & 0 & -1 & -2 \\ & & 1 & 1 & 0 \\ \hline & 1 & 1 & 0 & -2 \end{array}$$ • $S(1)$ is negative.

$$\begin{array}{r|rrrr} 2 & 1 & 0 & -1 & -2 \\ & & 2 & 4 & 6 \\ \hline & 1 & 2 & 3 & 4 \end{array}$$ • $S(2)$ is positive.

The graph of S is continuous because S is a polynomial function. Also, $S(1)$ is negative and $S(2)$ is positive. Thus the Zero Location Theorem indicates that there is a real zero between 1 and 2.

Visualize the Solution

The graph of S crosses the x-axis between $x = 1$ and $x = 2$. Thus S has a real zero between 1 and 2.

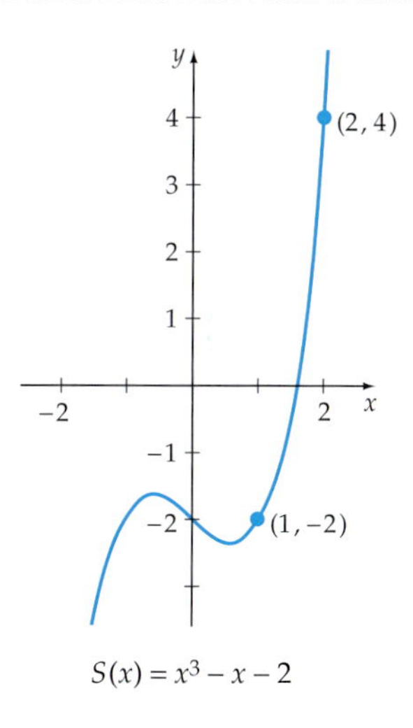

$S(x) = x^3 - x - 2$

TRY EXERCISE 28, PAGE 150

The following theorem summarizes important relationships among the real zeros of a polynomial function, the x-intercepts of its graph, and its factors that can be written in the form $(x - c)$, where c is a real number.

Polynomial Functions, Real Zeros, Graphs, and Factors $(x - c)$

If P is a polynomial function and c is a real number, then all the following statements are equivalent in the sense that if any one statement is true, then they are all true, and if any one statement is false, then they are all false.

- $(x - c)$ is a factor of P.
- $x = c$ is a real solution of $P(x) = 0$.
- $x = c$ is a real zero of P.
- $(c, 0)$ is an x-intercept of the graph of $y = P(x)$.

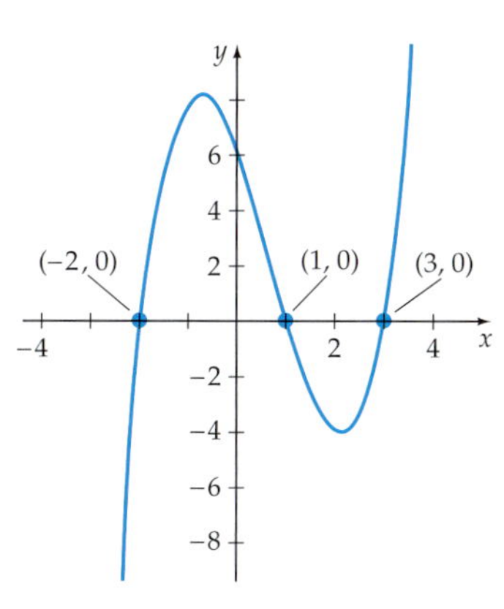

$S(x) = x^3 - 2x^2 - 5x + 6$

FIGURE 2.15

Sometimes it is possible to make use of the preceding theorem and a graph of a polynomial function to find factors of a function. For example, the graph of

$$S(x) = x^3 - 2x^2 - 5x + 6$$

is shown in **Figure 2.15.** The x-intercepts are $(-2, 0)$, $(1, 0)$, and $(3, 0)$. Hence -2, 1, and 3 are zeros of S, and $[x - (-2)]$, $(x - 1)$, and $(x - 3)$ are all factors of S.

Even and Odd Powers of $(x - c)$ Theorem

Use a graphing utility to graph $P(x) = (x + 3)(x - 4)^2$. Compare your graph with **Figure 2.16.** Examine the graph near the x-intercepts $(-3, 0)$ and $(4, 0)$. Observe that the graph of P

- crosses the x-axis at $(-3, 0)$.
- intersects the x-axis but does not cross the x-axis at $(4, 0)$.

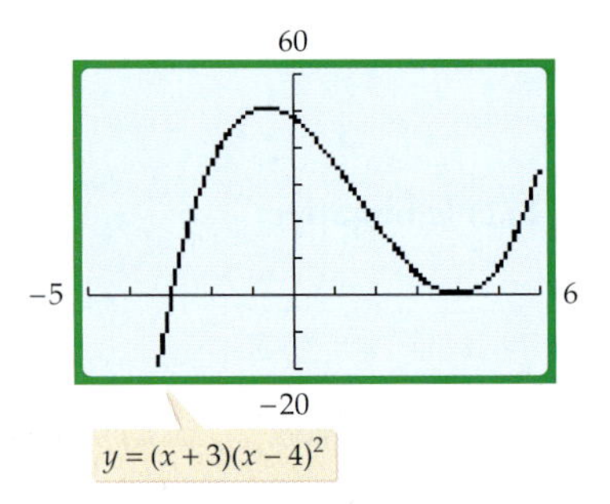

Figure 2.16

The following theorem can be used to determine at which x-intercepts the graph of a polynomial function will cross the x-axis and at which x-intercepts the graph will intersect but not cross the x-axis.

Even and Odd Powers of $(x - c)$ Theorem

If c is a real number and the polynomial function $P(x)$ has $(x - c)$ as a factor exactly k times, then the graph of P will

- intersect but not cross the x-axis at $(c, 0)$, provided k is an even positive integer.
- cross the x-axis at $(c, 0)$, provided k is an odd positive integer.

EXAMPLE 5 Apply the Even and Odd Powers of $(x - c)$ Theorem

Determine where the graph of $P(x) = (x + 3)(x - 2)^2(x - 4)^3$ crosses the x-axis and where the graph intersects but does not cross the x-axis.

Solution

The exponents of the factors $(x + 3)$ and $(x - 4)$ are odd integers. Therefore, the graph of P will cross the x-axis at the x-intercepts $(-3, 0)$ and $(4, 0)$.

The exponent of the factor $(x - 2)$ is an even integer. Therefore, the graph of P will intersect but not cross the x-axis at $(2, 0)$.

Use a graphing utility to check these results.

▶ Try Exercise 34, page 150

A Procedure for Graphing Polynomial Functions

You may find that you can sketch the graph of a polynomial function just by plotting several points; however, the following procedure will help you sketch the graphs of many polynomial functions in an efficient manner.

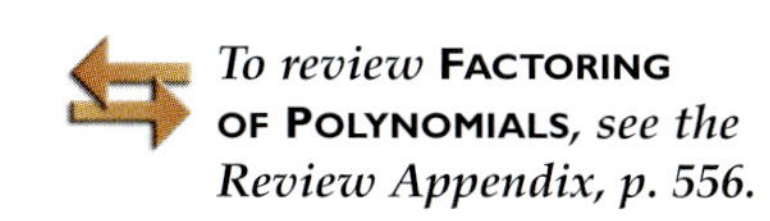
To review **FACTORING OF POLYNOMIALS**, *see the Review Appendix, p. 556.*

A Procedure for Graphing Polynomial Functions

$$P(x) = a_n x^n + a_{n-1}x^{n-1} + \cdots + a_1 x + a_0, \quad a_n \neq 0$$

To graph P:

1. ***Determine the far-left and the far-right behavior.*** Examine the leading coefficient $a_n x^n$ to determine the far-left and the far-right behavior of the graph.
2. ***Find the y-intercept.*** Determine the y-intercept by evaluating $P(0)$.
3. ***Find the x-intercept(s) and determine the behavior of the graph near the x-intercept(s).*** If possible, find the x-intercepts by factoring. If $(x - c)$, where c is a real number, is a factor of P, then $(c, 0)$ is an x-intercept of the graph. Use the Even and Odd Powers of $(x - c)$ Theorem to determine where the graph crosses the x-axis and where the graph intersects but does not cross the x-axis.
4. ***Find additional points on the graph.*** Find a few additional points (in addition to the intercepts).
5. ***Check for symmetry.***
   a. The graph of an even function is symmetric with respect to the y-axis.
   b. The graph of an odd function is symmetric with respect to the origin.
6. ***Sketch the graph.*** Use all the information obtained above to sketch the graph of the polynomial function. The graph should be a smooth continuous curve that passes through the points determined in steps 2 to 4. The graph should have a maximum of $n - 1$ turning points.

EXAMPLE 6 Graph a Polynomial Function

Sketch the graph of $P(x) = x^3 - 4x^2 + 4x$.

Solution

Step 1 ***Determine the far-left and the far-right behavior.*** The leading term is $1x^3$. Because the leading coefficient 1 is positive and the degree of the polynomial 3 is odd, the graph of P goes down to its far left and up to its far right.

Step 2 ***Find the y-intercept.*** $P(0) = 0^3 - 4(0)^2 + 4(0) = 0$. The y-intercept is $(0, 0)$.

Continued ►

Step 3 ***Find the x-intercept(s) and determine the behavior of the graph near the x-intercept(s).*** Try to factor $x^3 - 4x^2 + 4x$.

$$\begin{aligned} x^3 - 4x^2 + 4x &= x(x^2 - 4x + 4) \\ &= x(x-2)(x-2) \\ &= x(x-2)^2 \end{aligned}$$

Because $(x - 2)$ is a factor of P, the point $(2, 0)$ is an x-intercept of the graph of P. Because x is a factor of P (think of x as $x - 0$), the point $(0, 0)$ is an x-intercept of the graph of P. Applying the Even and Odd Powers of $(x - c)$ Theorem allows us to determine that the graph of P crosses the x-axis at $(0, 0)$ and intersects but does not cross the x-axis at $(2, 0)$.

Step 4 ***Find additional points on the graph.***

x	P(x)
−1	−9
0.5	1.125
1	1
3	3

Step 5 ***Check for symmetry.*** The function P is not an even or an odd function, so the graph of P is *not* symmetric to either the y-axis or the origin.

Step 6 ***Sketch the graph.***

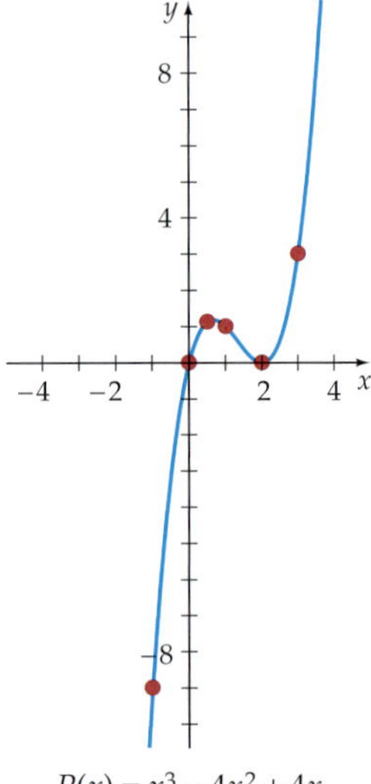

$P(x) = x^3 - 4x^2 + 4x$

▶ TRY EXERCISE 42, PAGE 150

TOPICS FOR DISCUSSION

1. Give an example of a polynomial function and a function that is not a polynomial function.

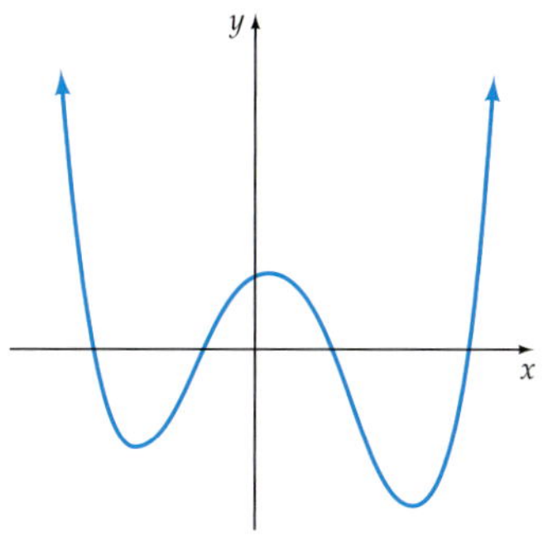

FIGURE 2.17

2. Is it possible for the graph of the polynomial function shown in **Figure 2.17** to be the graph of a polynomial function of odd degree? If so, explain how. If not, explain why not.

3. Explain the difference between a relative minimum and an absolute minimum.

4. Discuss how the Zero Location Theorem can be used to find a real zero of a polynomial function.

5. Let $P(x)$ be a polynomial function with real coefficients. Explain the relationships among the real zeros of the polynomial function, the x-coordinates of the x-intercepts of the graph of the polynomial function, and the solutions of the equation $P(x) = 0$.

EXERCISE SET 2.3

In Exercises 1 to 8, examine the leading term and determine the far-left and far-right behavior of the graph of the polynomial function.

1. $P(x) = 3x^4 - 2x^2 - 7x + 1$

 2. $P(x) = -2x^3 - 6x^2 + 5x - 1$

3. $P(x) = 5x^5 - 4x^3 - 17x^2 + 2$

4. $P(x) = -6x^4 - 3x^3 + 5x^2 - 2x + 5$

5. $P(x) = 2 - 3x - 4x^2$

6. $P(x) = -16 + x^4$

7. $P(x) = \frac{1}{2}(x^3 + 5x^2 - 2)$

8. $P(x) = -\frac{1}{4}(x^4 + 3x^2 - 2x + 6)$

9. The following graph is the graph of a third-degree (cubic) polynomial function. What does the far-left and far-right behavior of the graph say about the leading coefficient a?

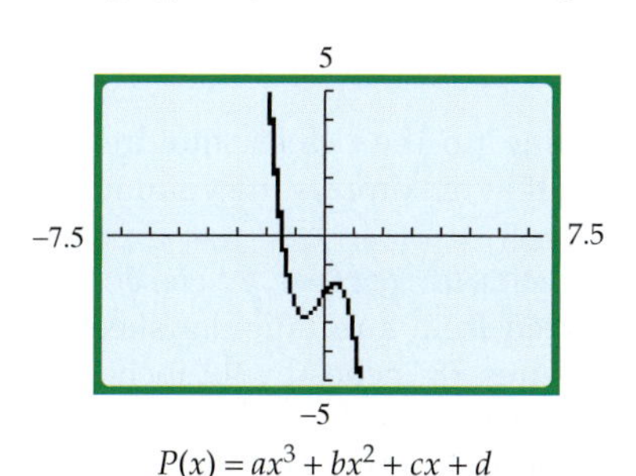

$P(x) = ax^3 + bx^2 + cx + d$

10. The following graph is the graph of a fourth-degree (quartic) polynomial function. What does the far-left and far-right behavior of the graph say about the leading coefficient a?

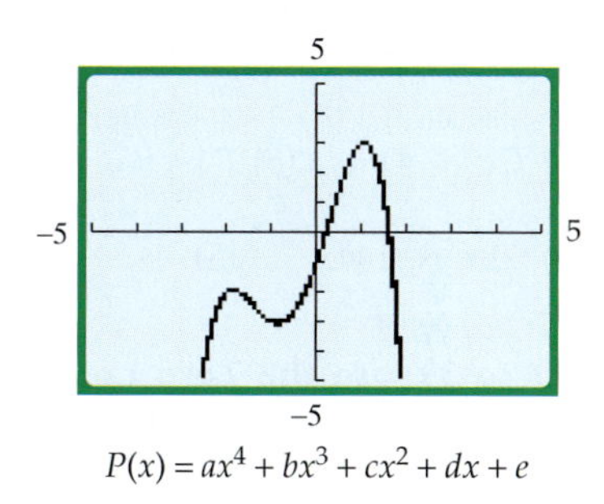

$P(x) = ax^4 + bx^3 + cx^2 + dx + e$

In Exercises 11 to 14, state the vertex of the graph of the function and use your knowledge of the vertex of a parabola to find the maximum or minimum of each function.

11. $P(x) = x^2 + 4x - 1$

12. $P(x) = x^2 + 6x + 1$

13. $P(x) = -x^2 - 8x + 1$

14. $P(x) = -2x^2 + 8x - 1$

In Exercises 15 to 20, use a graphing utility to graph each polynomial. Use the maximum and minimum features of the graphing utility to estimate, to the nearest tenth, the coordinates of the points where $P(x)$ has a relative maximum or a relative minimum. For each point, indicate whether the y value is a relative maximum or a relative minimum. The number in parentheses to the

right of the polynomial is the total number of relative maxima and minima.

15. $P(x) = x^3 + x^2 - 9x - 9$ (2)

16. $P(x) = x^3 + 4x^2 - 4x - 16$ (2)

17. $P(x) = x^3 - 3x^2 - 24x + 3$ (2)

18. $P(x) = -2x^3 - 3x^2 + 12x + 1$ (2)

19. $P(x) = x^4 - 4x^3 - 2x^2 + 12x - 5$ (3)

20. $P(x) = x^4 - 10x^2 + 9$ (3)

In Exercises 21 to 26, find the real zeros of each polynomial function by factoring. The number in parentheses to the right of each polynomial indicates the number of real zeros of the given polynomial function.

21. $P(x) = x^3 - 2x^2 - 15x$ (3)

▶ **22.** $P(x) = x^3 - 6x^2 + 8x$ (3)

23. $P(x) = x^4 - 13x^2 + 36$ (4)

24. $P(x) = 4x^4 - 37x^2 + 9$ (4)

25. $P(x) = x^5 - 5x^3 + 4x$ (5)

26. $P(x) = x^5 - 25x^3 + 144x$ (5)

In Exercises 27 to 32, use the Zero Location Theorem to verify that *P* has a zero between *a* and *b*.

27. $P(x) = 2x^3 + 3x^2 - 23x - 42$; $a = 3, b = 4$

▶ **28.** $P(x) = 4x^3 - x^2 - 6x + 1$; $a = 0, b = 1$

29. $P(x) = 3x^3 + 7x^2 + 3x + 7$; $a = -3, b = -2$

30. $P(x) = 2x^3 - 21x^2 - 2x + 25$; $a = 1, b = 2$

31. $P(x) = 4x^4 + 7x^3 - 11x^2 + 7x - 15$; $a = 1, b = 1\frac{1}{2}$

32. $P(x) = 5x^3 - 16x^2 - 20x + 64$; $a = 3, b = 3\frac{1}{2}$

In Exercises 33 to 40, determine the *x*-intercepts of the graph of *P*. For each *x*-intercept, use the Even and Odd Powers of $(x - c)$ Theorem to determine whether the graph of *P* crosses the *x*-axis or intersects but does not cross the *x*-axis.

33. $P(x) = (x - 1)(x + 1)(x - 3)$

▶ **34.** $P(x) = (x + 2)(x - 6)^2$

35. $P(x) = -(x - 3)^2(x - 7)^5$

36. $P(x) = (x + 2)^3(x - 6)^{10}$

37. $P(x) = (2x - 3)^4(x - 1)^{15}$

38. $P(x) = (5x + 10)^6(x - 2.7)^5$

39. $P(x) = x^3 - 6x^2 + 9x$

40. $P(x) = x^4 + 3x^3 + 4x^2$

In Exercises 41 to 46, sketch the graph of the polynomial function.

41. $P(x) = x^3 - x^2 - 2x$

▶ **42.** $P(x) = x^3 + 2x^2 - 3x$

43. $P(x) = -x^3 - 2x^2 + 5x + 6$ [*Hint:* In factored form $P(x) = (x + 3)(x + 1)(x - 2)$.]

44. $P(x) = -x^3 - 3x^2 + x + 3$ [*Hint:* In factored form $P(x) = (x + 3)(x + 1)(x - 1)$.]

45. $P(x) = x^4 - 4x^3 + 2x^2 + 4x - 3$ [*Hint:* In factored form $P(x) = (x + 1)(x - 1)^2(x - 3)$.]

46. $P(x) = x^4 - 6x^3 + 8x^2$

47. **Construction of a Box** A company constructs boxes from rectangular pieces of cardboard that measure 10 inches by 15 inches. An open box is formed by cutting squares that measure x inches by x inches from each corner of the cardboard and folding up the sides, as shown in the following figure.

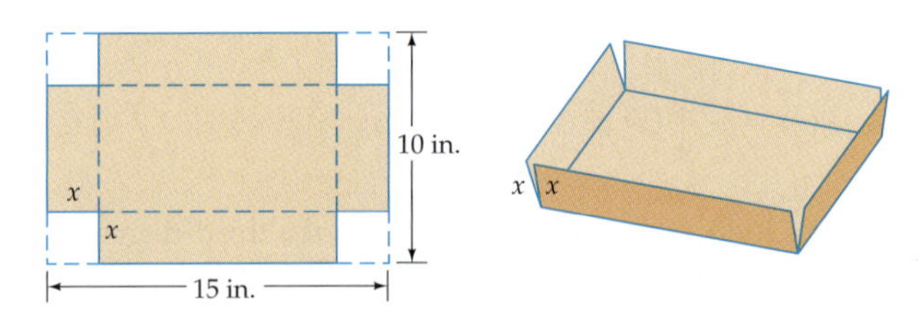

a. Express the volume V of the box as a function of x.

b. Determine (to the nearest hundredth of an inch) the x value that maximizes the volume of the box.

▶ **48.** **Maximizing Volume** A closed box is to be constructed from a rectangular sheet of cardboard that measures 18 inches by 42 inches. The box is made by cutting rectangles that measure x inches by $2x$ inches from two of the corners and by cutting two squares that measure x inches by x inches from the top and from the bottom of the rectangle, as shown in the following figure. What value of x (to the nearest thousandth of an inch) will produce a box with maximum volume?

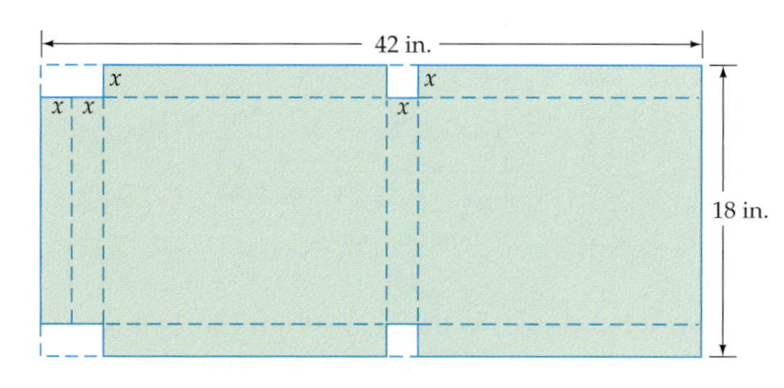

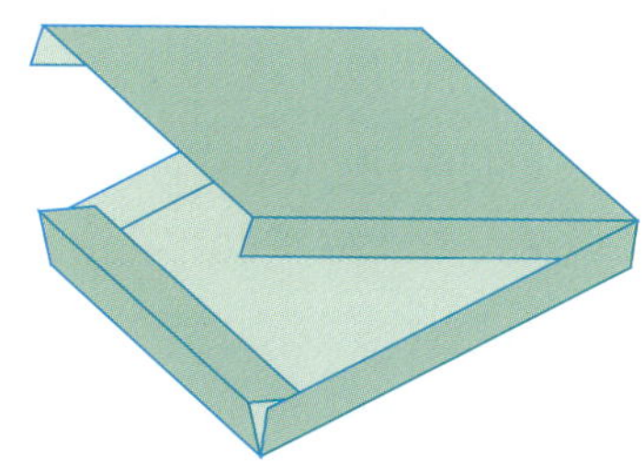

49. **MAXIMIZING VOLUME** An open box is to be constructed from a rectangular sheet of cardboard that measures 16 inches by 22 inches. To assemble the box, make the four cuts shown in the figure below and then fold on the dashed lines. What value of x (to the nearest thousandth of an inch) will produce a box with maximum volume?

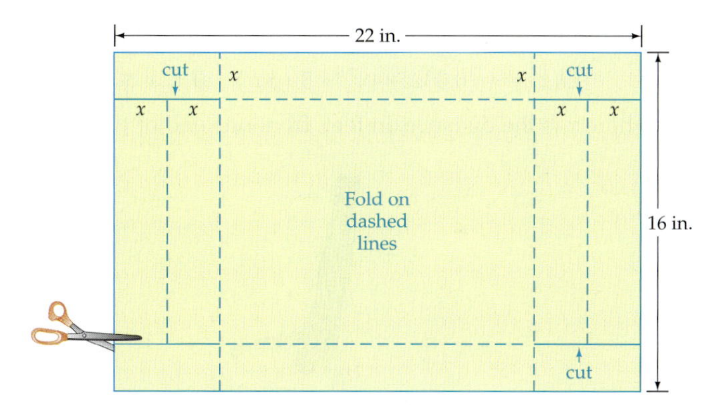

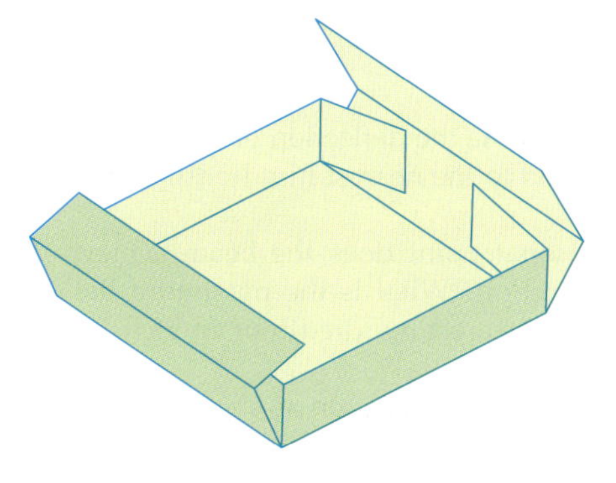

50. **PROFIT** A software company produces a computer game. The company has determined that its profit P, in dollars, from the manufacture and sale of x games is given by

$$P(x) = -0.000001x^3 + 96x - 98{,}000$$

where $0 < x \leq 9000$.

a. What is the maximum profit, to the nearest thousand dollars, the company can expect from the sale of its games?

b. How many games, to the nearest unit, does the company need to produce and sell to obtain the maximum profit?

51. **ADVERTISING EXPENSES** A company manufactures digital cameras. The company estimates that the profit from camera sales is

$$P(x) = -0.02x^3 + 0.01x^2 + 1.2x - 1.1$$

where P is the profit in millions of dollars and x is the amount, in hundred-thousands of dollars, spent on advertising.

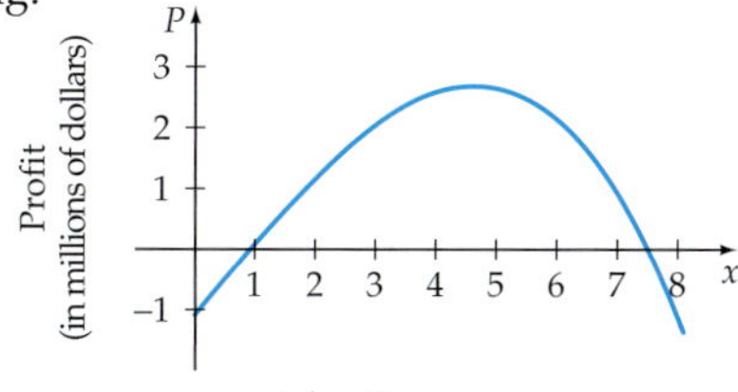

Determine the amount, rounded to the nearest thousand dollars, the company needs to spend on advertising if it is to generate the maximum profit.

52. **DIVORCE RATE** The divorce rate for a given year is defined as the number of divorces per thousand population. The function

$$D(t) = 0.00001807t^4 - 0.001406t^3 + 0.02884t^2 - 0.003466t + 2.1148$$

approximates the U.S. divorce rate for the years 1960 ($t = 0$) to 1999 ($t = 39$). Use $D(t)$ and a graphing utility to estimate

a. the year during which the U.S. divorce rate reached its absolute maximum for the period from 1960 to 1999.

b. the absolute minimum divorce rate, rounded to the nearest 0.1, during the period from 1960 to 1999.

53. **MARRIAGE RATE** The marriage rate for a given year is defined as the number of marriages per thousand population. The function

$$M(t) = -0.00000115t^4 + 0.000252t^3 - 0.01827t^2 + 0.4438t + 9.1829$$

approximates the U.S. marriage rate for the years 1900 ($t = 0$) to 1999 ($t = 99$).

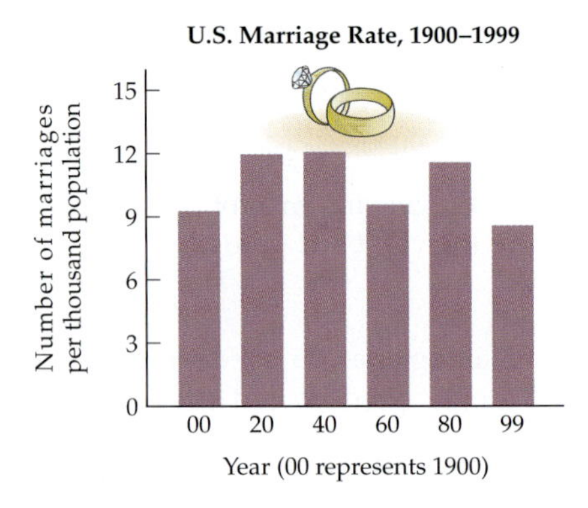

Use $M(t)$ and a graphing utility to estimate

a. during what year the U.S. marriage rate reached its maximum for the period from 1900 to 1999.

b. the relative minimum marriage rate, rounded to the nearest 0.1, during the period from 1950 to 1970.

54. **GAZELLE POPULATION** A herd of 204 African gazelles is introduced into a wild animal park. The population of the gazelles, $P(t)$, after t years is given by $P(t) = -0.7t^3 + 18.7t^2 - 69.5t + 204$, where $0 < t \le 18$.

a. Use a graph of P to determine the absolute minimum gazelle population (rounded to the nearest single gazelle) that is attained during this time period.

b. Use a graph of P to determine the absolute maximum gazelle population (rounded to the nearest single gazelle) that is attained during this time period.

55. 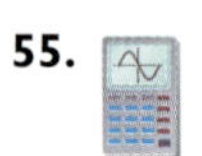  **MEDICATION LEVEL** Pseudoephedrine hydrochloride is an allergy medication. The function

$$L(t) = 0.03t^4 + 0.4t^3 - 7.3t^2 + 23.1t$$

where $0 \le t \le 5$, models the level of pseudoephedrine hydrochloride, in milligrams, in the bloodstream of a patient t hours after 30 milligrams of the medication have been taken.

a. Use a graphing utility and the function $L(t)$ to determine the maximum level of pseudoephedrine hydrochloride in the patient's bloodstream. Round your result to the nearest 0.01 milligram.

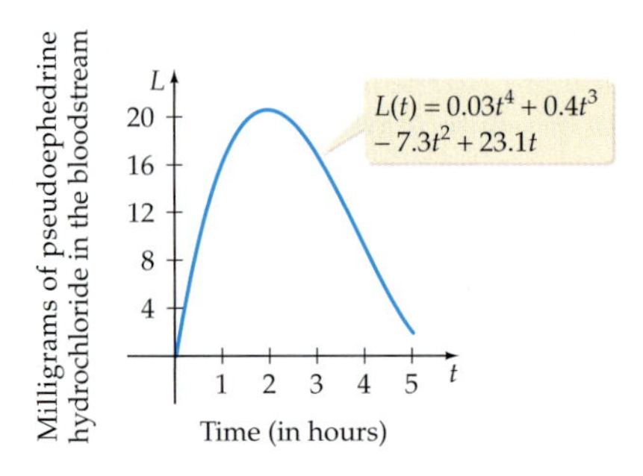

b. At what time t, to the nearest minute, is this maximum level of pseudoephedrine hydrochloride reached?

56. **SQUIRREL POPULATION** The population P of squirrels in a wilderness area is given by

$$P(t) = 0.6t^4 - 13.7t^3 + 104.5t^2 - 243.8t + 360,$$

where $0 \le t \le 12$ years.

a. What is the absolute minimum number of squirrels (rounded to the nearest single squirrel) attained on the interval $0 \le t \le 12$?

b. The absolute maximum of P is attained at the endpoint, where $t = 12$. What is this absolute maximum (rounded to the nearest single squirrel)?

57. **BEAM DEFLECTION** The deflection D, in feet, of an 8-foot beam that is center loaded is given by

$$D(x) = (-0.0025)(4x^3 - 3 \cdot 8x^2), \quad 0 < x \le 4$$

where x is the distance, in feet, from one end of the beam.

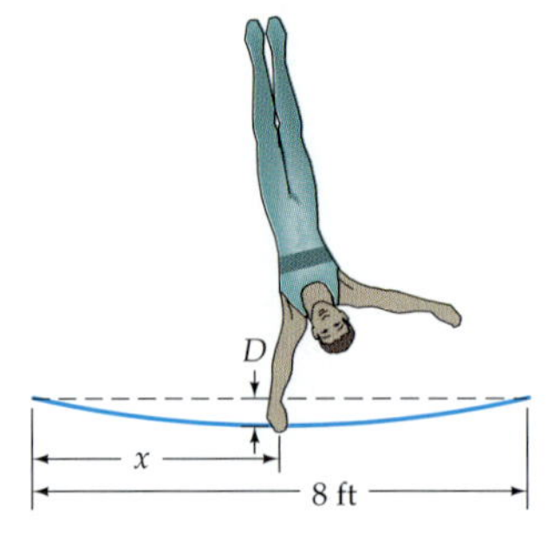

a. Determine the deflection of the beam when $x = 3$ feet. Round to the nearest hundredth of an inch.

b. At what point does the beam achieve its maximum deflection? What is the maximum deflection? Round to the nearest hundredth of an inch.

c. What is the deflection at $x = 5$ feet?

58. **ENGINEERING** A cylindrical log with a diameter of 22 inches is to be cut so that it will yield a beam that has a rectangular cross section of depth d and width w.

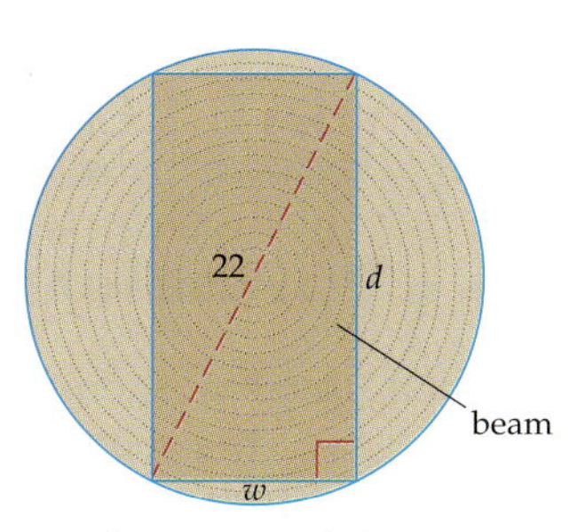

Cross section of a log

An engineer has determined that the stiffness S of the resulting beam is given by $S = 1.15wd^2$, where $0 < w < 22$ inches. Find the width and the depth that will maximize the stiffness of the beam. Round each result to the nearest hundredth of an inch. (*Hint:* Use the Pythagorean Theorem to solve for d^2 in terms of w^2.)

CONNECTING CONCEPTS

59. Use a graph of $P(x) = x^3 - x - 25$ to determine between which two consecutive integers P has a real zero.

60. Use a graph of the polynomial function $P(x) = 4x^4 - 12x^3 + 13x^2 - 12x + 9$ to determine between which two consecutive integers P has a real zero.

61. The point $(2, 0)$ is on the graph of $P(x)$. What point must be on the graph of $P(x - 3)$?

62. The point $(3, 5)$ is on the graph of $P(x)$. What point must be on the graph of $P(x + 1) - 2$?

63. Explain how to use the graph of $y = x^3$ to produce the graph of $P(x) = (x - 2)^3 + 1$.

64. Consider the following conjecture. Let $P(x)$ be a polynomial function. If a and b are real numbers such that $a < b$, $P(a) > 0$, and $P(b) > 0$, then $P(x)$ does not have a real zero between a and b. Is this conjecture true or false? Support your answer.

PREPARE FOR SECTION 2.4

65. Find the zeros of $P(x) = 6x^2 - 25x + 14$. [1.5]

66. Use synthetic division to divide $2x^3 + 3x^2 + 4x - 7$ by $x + 2$. [2.2]

67. Use synthetic division to divide $3x^4 - 21x^2 - 3x - 5$ by $x - 3$. [2.2]

68. List all natural numbers that are factors of 12.

69. List all integers that are factors of 27.

70. Given $P(x) = 4x^3 - 3x^2 - 2x + 5$, find $P(-x)$. [1.6]

PROJECTS

1. A student thinks that $P(n) = n^3 - n$ is always a multiple of 6 for all natural numbers n. What do you think? Provide a mathematical argument to show that the student is correct or a counterexample to show that the student is wrong.

SECTION 2.4

ZEROS OF POLYNOMIAL FUNCTIONS

MULTIPLE ZEROS OF A POLYNOMIAL FUNCTION

Recall that if $P(x)$ is a polynomial function, then the values of x for which $P(x)$ is equal to 0 are called the *zeros* of $P(x)$ or the **roots** of the equation $P(x) = 0$. A zero of a polynomial function may be a **multiple zero.** For example, $P(x) = x^2 + 6x + 9$ can be expressed in factored form as $(x + 3)(x + 3)$. Setting each factor equal to zero yields $x = -3$ in both cases. Thus $P(x) = x^2 + 6x + 9$ has a zero of -3 that occurs twice. The following definition will be most useful when we are discussing multiple zeros.

Definition of Multiple Zeros of a Polynomial Function

If a polynomial function $P(x)$ has $(x - r)$ as a factor exactly k times, then r is a **zero of multiplicity k** of the polynomial function $P(x)$.

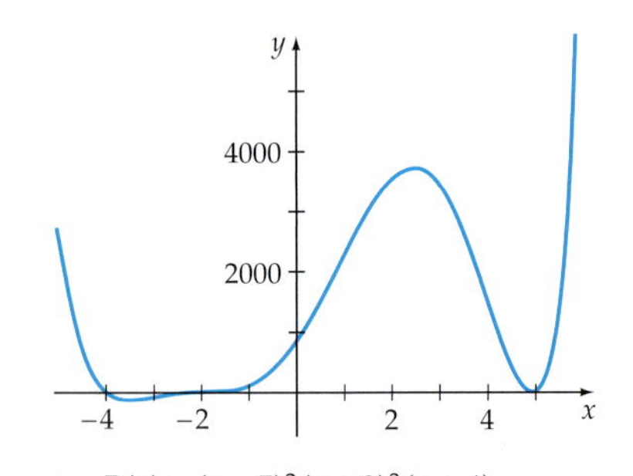

$P(x) = (x - 5)^2(x + 2)^3(x + 4)$

FIGURE 2.18

The graph of the polynomial function

$$P(x) = (x - 5)^2(x + 2)^3(x + 4)$$

is shown in **Figure 2.18.** This polynomial function has

- 5 as a zero of multiplicity 2.
- -2 as a zero of multiplicity 3.
- -4 as a zero of multiplicity 1.

A zero of multiplicity 1 is generally referred to as a **simple zero.**

When searching for the zeros of a polynomial function, it is important that we know how many zeros to expect. This question is answered completely in Section 2.5. For the work in this section, the following result is valuable.

Number of Zeros of a Polynomial Function

A polynomial function P of degree n has at most n zeros, where each zero of multiplicity k is counted k times.

THE RATIONAL ZERO THEOREM

The rational zeros of polynomial functions with integer coefficients can be found with the aid of the following theorem.

take note

The Rational Zero Theorem is one of the most important theorems of this chapter. It enables us to narrow the search for rational zeros to a finite list.

The Rational Zero Theorem

If $P(x) = a_nx^n + a_{n-1}x^{n-1} + \cdots + a_1x + a_0$ has *integer* coefficients ($a_n \neq 0$) and $\frac{p}{q}$ is a rational zero (in lowest terms) of P, then

- p is a factor of the constant term a_0 and
- q is a factor of the leading coefficient a_n.

The Rational Zero Theorem often is used to make a list of all possible rational zeros of a polynomial function. The list consists of all rational numbers of the form $\frac{p}{q}$, where p is an integer factor of the constant term a_0 and q is an integer factor of the leading coefficient a_n.

EXAMPLE 1 Apply the Rational Zero Theorem

Use the Rational Zero Theorem to list all possible rational zeros of

$$P(x) = 4x^4 + x^3 - 40x^2 + 38x + 12$$

Solution

List all integers p that are factors of 12 and all integers q that are factors of 4.

$$p: \quad \pm1, \pm2, \pm3, \pm4, \pm6, \pm12$$
$$q: \quad \pm1, \pm2, \pm4$$

Form all possible rational numbers using ±1, ±2, ±3, ±4, ±6, and ±12 for the numerator and ±1, ±2, and ±4 for the denominator. By the Rational Zero Theorem, the possible rational zeros are

$$\pm1, \pm\frac{1}{2}, \pm\frac{1}{4}, \pm2, \pm3, \pm\frac{3}{2}, \pm\frac{3}{4}, \pm4, \pm6, \pm12$$

It is not necessary to list a factor that is already listed in reduced form. For example, $\pm\frac{6}{4}$ is not listed because it is equal to $\pm\frac{3}{2}$.

▶ TRY EXERCISE 10, PAGE 164

take note

The Rational Zero Theorem gives the *possible* rational zeros of a polynomial function. That is, if P has a rational zero, then it must be one indicated by the theorem. However, P may not have any rational zeros. In the case of the polynomial function in Example 1, the only rational zeros are $-\frac{1}{4}$ and 2. The remaining rational numbers in the list are not zeros of P.

QUESTION If $P(x) = a_nx^n + a_{n-1}x^{n-1} + \cdots + a_1x + a_0$ has integer coefficients and a leading coefficient of $a_n = 1$, must all the rational zeros of P be integers?

ANSWER Yes. By the Rational Zero Theorem, the rational zeros of P are of the form $\frac{p}{q}$, where p is an integer factor of a_0 and q is an integer factor of a_n. Thus $q = \pm1$ and $\frac{p}{q} = \frac{p}{\pm1} = \pm p$.

Upper and Lower Bounds for Real Zeros

A real number b is called an **upper bound** of the zeros of the polynomial function P if no zero is greater than b. A real number b is called a **lower bound** of the zeros of P if no zero is less than b. The following theorem is often used to find positive upper bounds and negative lower bounds for the real zeros of a polynomial function.

Upper- and Lower-Bound Theorem

Let $P(x)$ be a polynomial function with real coefficients. Use synthetic division to divide $P(x)$ by $x - b$, where b is a nonzero real number.

Upper bound

a. If $b > 0$ and the leading coefficient of P is positive, then b is an upper bound for the real zeros of P provided none of the numbers in the bottom row of the synthetic division are negative.

b. If $b > 0$ and the leading coefficient of P is negative, then b is an upper bound for the real zeros of P provided none of the numbers in the bottom row of the synthetic division are positive.

Lower bound If $b < 0$ and the numbers in the bottom row of the synthetic division alternate in sign (the number zero can be considered positive or negative as needed to produce an alternating sign pattern), then b is a lower bound for the real zeros of P.

Upper and lower bounds are not unique. For example, if b is an upper bound for the real zeros of P, then any number greater than b is also an upper bound. Likewise, if a is a lower bound for the real zeros of P, then any number less than a is also a lower bound.

EXAMPLE 2 Find Upper and Lower Bounds

According to the Upper- and Lower-Bound Theorem, what is the smallest positive integer that is an upper bound and the largest negative integer that is a lower bound of the real zeros of $P(x) = 2x^3 + 7x^2 - 4x - 14$?

Solution

To find the smallest positive-integer upper bound, use synthetic division with 1, 2, . . . , as test values.

$$\begin{array}{r|rrrr} 1 & 2 & 7 & -4 & -14 \\ & & 2 & 9 & 5 \\ \hline & 2 & 9 & 5 & -9 \end{array} \qquad \begin{array}{r|rrrr} 2 & 2 & 7 & -4 & -14 \\ & & 4 & 22 & 36 \\ \hline & 2 & 11 & 18 & 22 \end{array}$$

• No negative numbers

Thus 2 is the smallest positive-integer upper bound.

> **take note**
>
> When you check for bounds, you do not need to limit your choices to the possible zeros given by the Rational Zero Theorem. For instance, in Example 2 the integer -4 is a lower bound; however, -4 is not one of the possible zeros of P as given by the Rational Zero Theorem.

Now find the largest negative-integer lower bound.

-1	2	7	-4	-14
		-2	-5	9
	2	5	-9	-5

-2	2	7	-4	-14
		-4	-6	20
	2	3	-10	6

-3	2	7	-4	-14
		-6	-3	21
	2	1	-7	7

-4	2	7	-4	-14
		-8	4	0
	2	-1	0	-14

• **Alternating signs**

Thus -4 is the largest negative-integer lower bound.

▶ **TRY EXERCISE 18, PAGE 164**

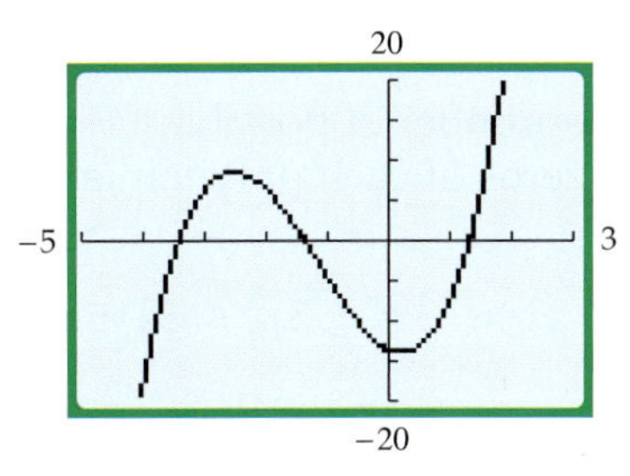

$P(x) = 2x^3 + 7x^2 - 4x - 14$

FIGURE 2.19

INTEGRATING TECHNOLOGY

You can use the Upper- and Lower-Bound Theorem to determine Xmin (the lower bound) and Xmax (the upper bound) for the viewing window of a graphing utility. This will ensure that all the real zeros, which are the x-coordinates of the x-intercepts of the polynomial function, will be shown. Note in **Figure 2.19** that the zeros of $P(x) = 2x^3 + 7x^2 - 4x - 14$ are between -4 (a lower bound) and 2 (an upper bound).

DESCARTES' RULE OF SIGNS

Descartes' Rule of Signs is another theorem that is often used to obtain information about the zeros of a polynomial function. In Descartes' Rule of Signs, the number of **variations in sign** of the coefficients of $P(x)$ or $P(-x)$ refers to sign changes of the coefficients from positive to negative or from negative to positive that we find when we examine successive terms of the function. The terms are assumed to appear in order of descending powers of x. For example, the polynomial function

$$P(x) = \underbrace{+3x^4 - 5x^3}_{1} \underbrace{- 7x^2 + x}_{2} \underbrace{\,+ x - 7}_{3}$$

has three variations in sign. The polynomial function

$$\begin{aligned} P(-x) &= +3(-x)^4 - 5(-x)^3 - 7(-x)^2 + (-x) - 7 \\ &= +\ 3x^4 \ \underbrace{+\ 5x^3 \ -}_{1} \ 7x^2 \ - \ x \ - 7 \end{aligned}$$

has one variation in sign.

Terms that have a coefficient of 0 are not counted as variations in sign and may be ignored. For example,

$$P(x) = \underbrace{-x^5 + 4x^2}_{1} + 1$$

has one variation in sign.

> **MATH MATTERS**
>
> 
>
> Descartes' Rule of Signs first appeared in his *La Géométrie* (1673). Although a proof of Descartes' Rule of Signs is beyond the scope of this course, we can see that a polynomial function with no variations in sign cannot have a positive zero. For instance, consider $P(x) = x^3 + x^2 + x + 1$. Each term of P is positive for any positive value of x. Thus P is never zero for $x > 0$.

Descartes' Rule of Signs

Let $P(x)$ be a polynomial function with real coefficients and with the terms arranged in order of decreasing powers of x.

1. The number of positive real zeros of $P(x)$ is equal to the number of variations in sign of $P(x)$, or to that number decreased by an even integer.
2. The number of negative real zeros of $P(x)$ is equal to the number of variations in sign of $P(-x)$, or to that number decreased by an even integer.

EXAMPLE 3 Apply Descartes' Rule of Signs

Use Descartes' Rule of Signs to determine both the number of possible positive and the number of possible negative real zeros of each polynomial function.

a. $P(x) = x^4 - 5x^3 + 5x^2 + 5x - 6$ **b.** $P(x) = 2x^5 + 3x^3 + 5x^2 + 8x + 7$

Solution

a.

$$P(x) = +x^4 - 5x^3 + 5x^2 + 5x - 6$$

1 2 3

There are three variations in sign. By Descartes' Rule of Signs, there are either three or one positive real zeros. Now examine the variations in sign of $P(-x)$.

$$P(-x) = x^4 + 5x^3 + 5x^2 - 5x - 6$$

1

There is one variation in sign of $P(-x)$. By Descartes' Rule of Signs, there is one negative real zero.

b. $P(x) = 2x^5 + 3x^3 + 5x^2 + 8x + 7$ has no variation in sign, so there are no positive real zeros.

$$P(-x) = -2x^5 - 3x^3 + 5x^2 - 8x + 7$$

1 2 3

$P(-x)$ has three variations in sign, so there are either three or one negative real zeros.

▶ TRY EXERCISE 28, PAGE 164

QUESTION If $P(x) = ax^2 + bx + c$ has two variations in sign, must $P(x)$ have two positive real zeros?

ANSWER No. According to Descartes' Rule of Signs, $P(x)$ will have either two positive real zeros or no positive real zeros.

In applying Descartes' Rule of Signs, we count each zero of multiplicity k as k zeros. For instance,

$$P(x) = x^2 - 10x + 25$$

has two variations in sign. Thus, by Descartes' Rule of Signs, $P(x)$ must have either two or no positive real zeros. Factoring $P(x)$ produces $(x - 5)^2$, from which it can be observed that 5 is a positive zero of multiplicity 2.

Zeros of a Polynomial Function

Guidelines for Finding the Zeros of a Polynomial Function with Integer Coefficients

1. ***Gather general information.*** Determine the degree n of the polynomial function. The number of distinct zeros of the polynomial function is at most n. Apply Descartes' Rule of Signs to find the possible number of positive zeros and also the possible number of negative zeros.
2. ***Check suspects.*** Apply the Rational Zero Theorem to list rational numbers that are possible zeros. Use synthetic division to test numbers in your list. If you find an upper or a lower bound, then eliminate from your list any number that is greater than the upper bound or less than the lower bound.
3. ***Work with the reduced polynomials.*** Each time a zero is found, you obtain a reduced polynomial.
   - If a reduced polynomial is of degree 2, find its zeros either by factoring or by applying the quadratic formula.
   - If the degree of a reduced polynomial is 3 or greater, repeat the above steps for this polynomial.

Example 4 illustrates the procedure discussed in the above guidelines.

EXAMPLE 4 Find the Zeros of a Polynomial Function

Find the zeros of $P(x) = 3x^4 + 23x^3 + 56x^2 + 52x + 16$.

Solution

1. *Gather general information.* The degree of P is 4. Thus the number of zeros of P is at most 4. By Descartes' Rule of Signs, there are no positive zeros, and there are either four, two, or no negative zeros.
2. *Check suspects.* By the Rational Zero Theorem, the possible negative rational zeros of P are

$$\frac{p}{q}: \quad -1, -2, -4, -8, -16, -\frac{1}{3}, -\frac{2}{3}, -\frac{4}{3}, -\frac{8}{3}, -\frac{16}{3}$$

Continued ▶

INTEGRATING TECHNOLOGY

If you have a graphing utility, you can produce a graph similar to the one below. By looking at the x-intercepts of the graph, you can reject as possible zeros some of the values suggested by the Rational Zero Theorem. This will reduce the amount of work that is necessary to find the zeros of the polynomial function.

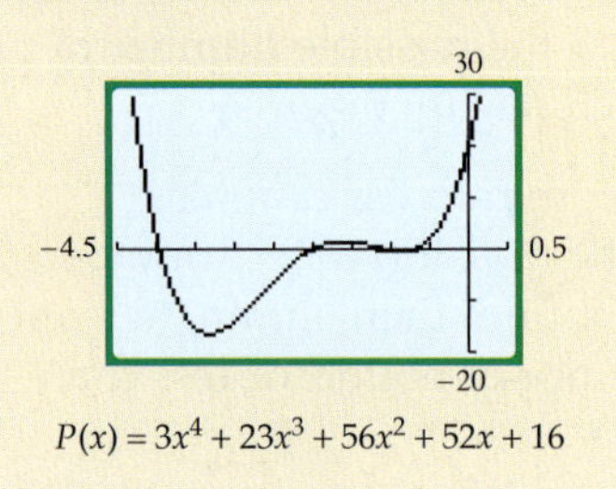

$P(x) = 3x^4 + 23x^3 + 56x^2 + 52x + 16$

Use synthetic division to test the possible rational zeros. The following work shows that -4 is a zero of P.

$$\begin{array}{r|rrrrr} -4 & 3 & 23 & 56 & 52 & 16 \\ & & -12 & -44 & -48 & -16 \\ \hline & 3 & 11 & 12 & 4 & 0 \end{array}$$

Coefficients of the first reduced polynomial

3. *Work with the reduced polynomials.* Because -4 is a zero, $(x + 4)$ and the first reduced polynomial $(3x^3 + 11x^2 + 12x + 4)$ are both factors of P. Thus

$$P(x) = (x + 4)(3x^3 + 11x^2 + 12x + 4)$$

All remaining zeros of P must be zeros of $3x^3 + 11x^2 + 12x + 4$. The Rational Zero Theorem indicates that the only possible negative rational zeros of $3x^3 + 11x^2 + 12x + 4$ are

$$\frac{p}{q}: \quad -1, -2, -4, -\frac{1}{3}, -\frac{2}{3}, -\frac{4}{3}$$

Synthetic division is again used to test possible zeros.

$$\begin{array}{r|rrrr} -2 & 3 & 11 & 12 & 4 \\ & & -6 & -10 & -4 \\ \hline & 3 & 5 & 2 & 0 \end{array}$$

Coefficients of the second reduced polynomial

Because -2 is a zero, $(x + 2)$ is also a factor of P. Thus

$$P(x) = (x + 4)(x + 2)(3x^2 + 5x + 2)$$

The remaining zeros of P must be zeros of $3x^2 + 5x + 2$.

$$3x^2 + 5x + 2 = 0$$
$$(3x + 2)(x + 1) = 0$$
$$x = -\frac{2}{3} \quad \text{and} \quad x = -1$$

The zeros of $P(x) = 3x^4 + 23x^3 + 56x^2 + 52x + 16$ are -4, -2, $-\frac{2}{3}$, and -1.

TRY EXERCISE 38, PAGE 164

APPLICATIONS OF POLYNOMIAL FUNCTIONS

In the following example we make use of an upper bound to eliminate several of the possible zeros that are given by the Rational Zero Theorem.

EXAMPLE 5 Solve an Application

Glasses can be stacked to form a triangular pyramid.

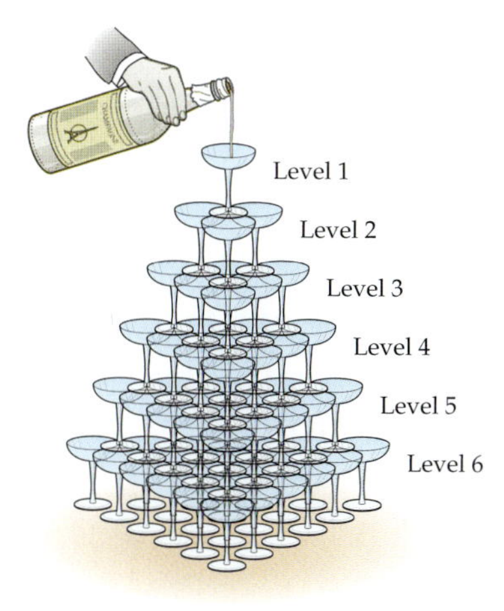

The total number of glasses in one of these pyramids is given by

$$T = \frac{1}{6}(k^3 + 3k^2 + 2k)$$

where k is the number of levels in the pyramid. If 220 glasses are used to form a triangular pyramid, how many levels are in the pyramid?

Solution

We need to solve $220 = \frac{1}{6}(k^3 + 3k^2 + 2k)$ for k. Multiplying each side of the equation by 6 produces $1320 = k^3 + 3k^2 + 2k$, which can be written as $k^3 + 3k^2 + 2k - 1320 = 0$. The number 1320 has many natural number divisors, but we can eliminate many of these by showing that 12 is an upper bound.

$$\begin{array}{r|rrrr} 12 & 1 & 3 & 2 & -1320 \\ & & 12 & 180 & 2184 \\ \hline & 1 & 15 & 182 & 864 \end{array}$$

No number in the bottom row is negative. Thus 12 is an upper bound.

The only natural number divisors of 1320 that are less than 12 are 1, 2, 3, 4, 5, 6, 8, 10, and 11. The following synthetic division shows that 10 is a zero of $k^3 + 3k^2 + 2k - 1320$.

$$\begin{array}{r|rrrr} 10 & 1 & 3 & 2 & -1320 \\ & & 10 & 130 & 1320 \\ \hline & 1 & 13 & 132 & 0 \end{array}$$

The pyramid has 10 levels. There is no need to seek additional solutions, because the number of levels is uniquely determined by the number of glasses.

take note

The reduced polynomial $k^2 + 13k + 132$ has zeros of $k = \frac{-13 \pm i\sqrt{359}}{2}$. These zeros are not solutions of this application because the number of levels must be a natural number.

▶ TRY EXERCISE 72, PAGE 167

The procedures developed in this section will not find all solutions of every polynomial equation. However, a graphing utility can be used to estimate the real solutions of any polynomial equation. In Example 6 we utilize a graphing utility to solve an application.

EXAMPLE 6 Use a Graphing Utility to Solve an Application

A CO_2 (carbon dioxide) cartridge for a paintball rifle has the shape of a right circular cylinder with a hemisphere at each end. The cylinder is 4 inches long, and the volume of the cartridge is 2π cubic inches (approximately 6.3 cubic inches). In the figure at the right, the common interior radius of the cylinder and the hemispheres is denoted by x. Use a graphing utility to estimate, to the nearest hundredth of an inch, the length of the radius x.

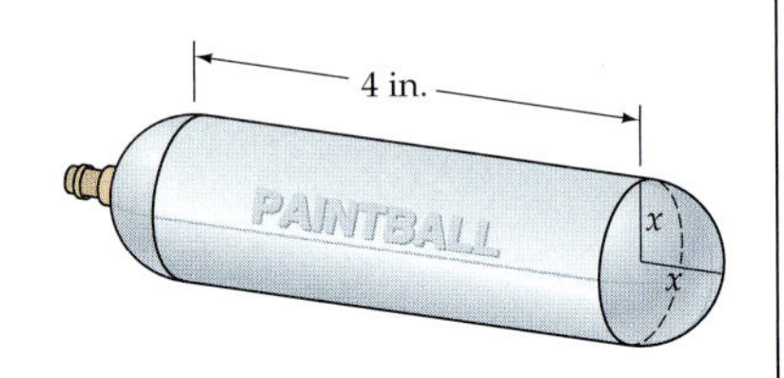

Solution

The volume of the cartridge is equal to the volume of the two hemispheres plus the volume of the cylinder. Recall that the volume of a sphere of radius x is given by $\frac{4}{3}\pi x^3$. Therefore, the volume of a hemisphere is $\frac{1}{2}\left(\frac{4}{3}\pi x^3\right)$. The volume of a right circular cylinder is $\pi x^2 h$, where x is the radius of the base and h is the height of the cylinder. Thus the volume V of the cartridge is given by

$$V = \frac{1}{2}\left(\frac{4}{3}\pi x^3\right) + \frac{1}{2}\left(\frac{4}{3}\pi x^3\right) + \pi x^2 h$$

$$= \frac{4}{3}\pi x^3 + \pi x^2 h$$

Replacing V with 2π and h with 4 yields

$$2\pi = \frac{4}{3}\pi x^3 + 4\pi x^2$$

$$2 = \frac{4}{3}x^3 + 4x^2 \quad \text{• Divide by } \pi.$$

$$3 = 2x^3 + 6x^2 \quad \text{• Multiply by } \frac{3}{2}.$$

Here are two methods that can be used to solve

$$3 = 2x^3 + 6x^2 \tag{1}$$

for x with the aid of a graphing utility.

1. **Intersection Method** Use a graphing utility to graph $y = 2x^3 + 6x^2$ and $y = 3$ on the same screen, with $x > 0$. The x-coordinate of the point of intersection of the two graphs is the desired solution. The graphs intersect at $x \approx 0.64$ inch. See the following figures.

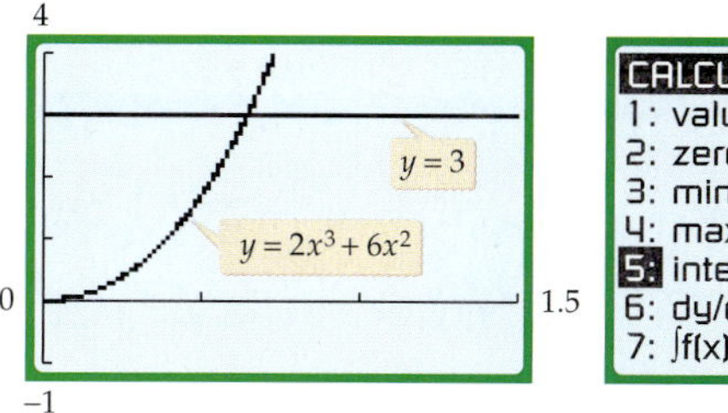

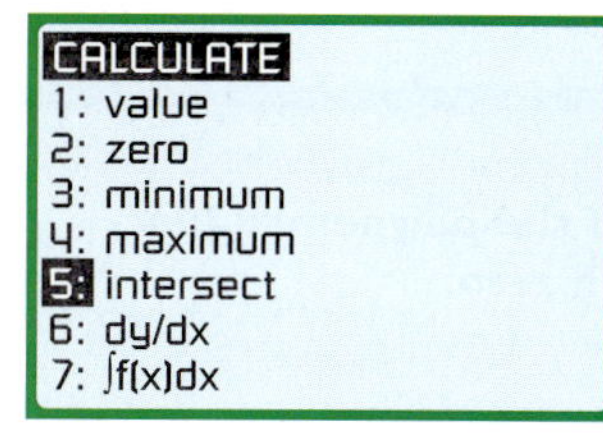

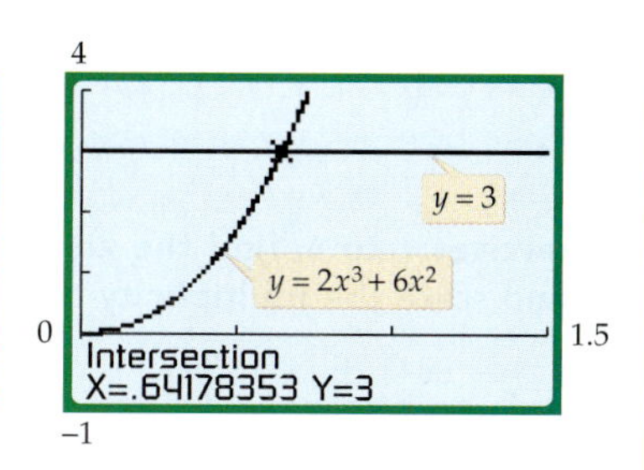

The length of the radius is approximately 0.64 inch.

2. **Intercept Method** Rewrite Equation (1) as $2x^3 + 6x^2 - 3 = 0$. Graph $y = 2x^3 + 6x^2 - 3$ with $x > 0$. Use a graphing utility to find the x-intercept of the graph. This method also shows that $x \approx 0.64$ inch.

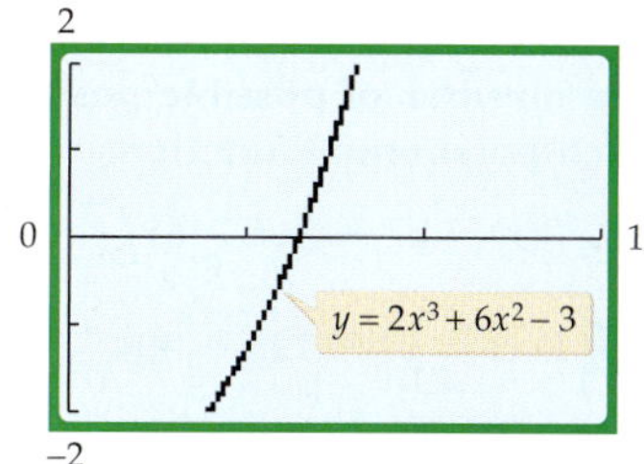

CALCULATE
1: value
2: zero
3: minimum
4: maximum
5: intersect
6: dy/dx
7: ∫f(x)dx

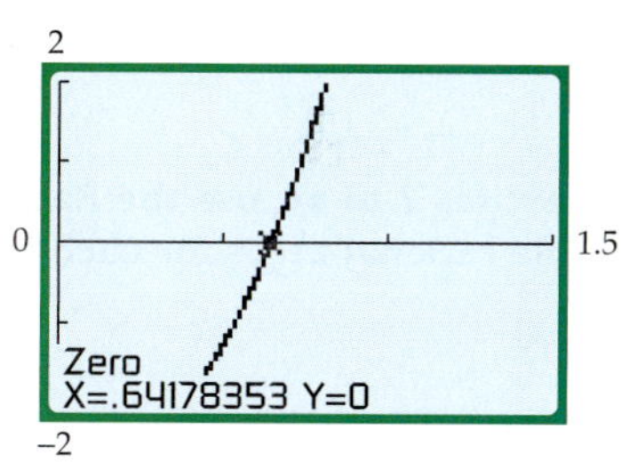

The length of the radius is approximately 0.64 inch.

TRY EXERCISE 68, PAGE 166

TOPICS FOR DISCUSSION

1. What is a multiple zero of a polynomial function? Give an example of a polynomial function that has -2 as a multiple zero.
2. Discuss how the Rational Zero Theorem is used.
3. Let $P(x)$ be a polynomial function with real coefficients. Explain why $(a, 0)$ is an x-intercept of the graph of $P(x)$ if a is a real zero of $P(x)$.
4. Let $P(x)$ be a polynomial function with integer coefficients. Suppose that the Rational Zero Theorem is applied to $P(x)$ and that after testing each possible rational zero, it is determined that $P(x)$ has no rational zeros. Does this mean that all of the zeros of $P(x)$ are irrational numbers?

EXERCISE SET 2.4

In Exercises 1 to 6, find the zeros of the polynomial function and state the multiplicity of each zero.

1. $P(x) = (x - 3)^2(x + 5)$

2. $P(x) = (x + 4)^3(x - 1)^2$

3. $P(x) = x^2(3x + 5)^2$

4. $P(x) = x^3(2x + 1)(3x - 12)^2$

5. $P(x) = (x^2 - 4)(x + 3)^2$

6. $P(x) = (x + 4)^3(x^2 - 9)^2$

In Exercises 7 to 16, use the Rational Zero Theorem to list possible rational zeros for each polynomial function.

7. $P(x) = x^3 + 3x^2 - 6x - 8$

8. $P(x) = x^3 - 19x - 30$

9. $P(x) = 2x^3 + x^2 - 25x + 12$

▶ **10.** $P(x) = 3x^3 + 11x^2 - 6x - 8$

11. $P(x) = 6x^4 + 23x^3 + 19x^2 - 8x - 4$

12. $P(x) = 2x^3 + 9x^2 - 2x - 9$

13. $P(x) = 4x^4 - 12x^3 - 3x^2 + 12x - 7$

14. $P(x) = x^5 - x^4 - 7x^3 + 7x^2 - 12x - 12$

15. $P(x) = x^5 - 32$

16. $P(x) = x^4 - 1$

In Exercises 17 to 26, find the smallest positive integer and the largest negative integer that, by the Upper- and Lower-Bound Theorem, are upper and lower bounds for the real zeros of each polynomial function.

17. $P(x) = x^3 + 3x^2 - 6x - 6$

▶ **18.** $P(x) = x^3 - 19x - 28$

19. $P(x) = 2x^3 + x^2 - 25x + 10$

20. $P(x) = 3x^3 + 11x^2 - 6x - 9$

21. $P(x) = 6x^4 + 23x^3 + 19x^2 - 8x - 4$

22. $P(x) = -2x^3 - 9x^2 + 2x + 9$

23. $P(x) = -4x^4 + 12x^3 + 3x^2 - 12x + 7$

24. $P(x) = x^5 - x^4 - 7x^3 + 7x^2 - 12x - 12$

25. $P(x) = x^5 - 32$

26. $P(x) = x^4 - 1$

In Exercises 27 to 36, use Descartes' Rule of Signs to state the number of possible positive and negative real zeros of each polynomial function.

27. $P(x) = x^3 + 3x^2 - 6x - 8$

▶ **28.** $P(x) = x^3 - 19x - 30$

29. $P(x) = 2x^3 + x^2 - 25x + 12$

30. $P(x) = 3x^3 + 11x^2 - 6x - 8$

31. $P(x) = 6x^4 + 23x^3 + 19x^2 - 8x - 4$

32. $P(x) = 2x^3 + 9x^2 - 2x - 9$

33. $P(x) = 4x^4 - 12x^3 - 3x^2 + 12x - 7$

34. $P(x) = x^5 - x^4 - 7x^3 + 7x^2 - 12x - 12$

35. $P(x) = x^5 - 32$

36. $P(x) = x^4 - 1$

In Exercises 37 to 58, find the zeros of each polynomial function. If a zero is a multiple zero, state its multiplicity.

37. $P(x) = x^3 + 3x^2 - 6x - 8$

▶ **38.** $P(x) = x^3 - 19x - 30$

39. $P(x) = 2x^3 + x^2 - 25x + 12$

40. $P(x) = 3x^3 + 11x^2 - 6x - 8$

41. $P(x) = 6x^4 + 23x^3 + 19x^2 - 8x - 4$

42. $P(x) = 2x^3 + 9x^2 - 2x - 9$

43. $P(x) = 2x^4 - 9x^3 - 2x^2 + 27x - 12$

44. $P(x) = 3x^3 - x^2 - 6x + 2$

45. $P(x) = x^3 - 8x^2 + 8x + 24$

46. $P(x) = x^3 - 7x^2 - 7x + 69$

47. $P(x) = 2x^4 - 19x^3 + 51x^2 - 31x + 5$

48. $P(x) = 4x^4 - 35x^3 + 71x^2 - 4x - 6$

49. $P(x) = 3x^6 - 10x^5 - 29x^4 + 34x^3 + 50x^2 - 24x - 24$

50. $P(x) = 2x^4 + 3x^3 - 4x^2 - 3x + 2$

51. $P(x) = x^3 - 3x - 2$

52. $P(x) = 3x^4 - 4x^3 - 11x^2 + 16x - 4$

53. $P(x) = x^4 - 5x^2 - 2x$

54. $P(x) = x^3 - 2x + 1$

55. $P(x) = x^4 + x^3 - 3x^2 - 5x - 2$

56. $P(x) = 6x^4 - 17x^3 - 11x^2 + 42x$

57. $P(x) = 2x^4 - 17x^3 + 4x^2 + 35x - 24$

58. $P(x) = x^5 + 5x^4 + 10x^3 + 10x^2 + 5x + 1$

59. **Find the Dimensions** A cube measures n inches on each edge. If a slice 2 inches thick is cut from one face of the cube, the resulting solid has a volume of 567 cubic inches. Find n.

60. **Find the Dimensions** A cube measures n units on each edge. If a slice 1 inch thick is cut from one face of the cube, and then a slice 3 inches thick is cut from another face of the cube as shown, the resulting solid has a volume of 1560 cubic inches. Find the dimensions of the original cube.

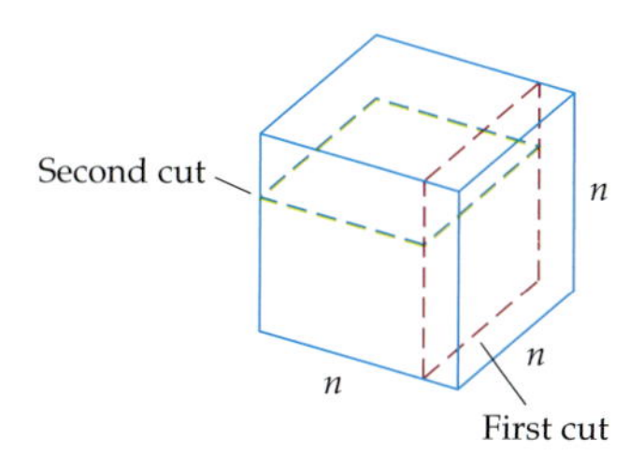

61. **Dimensions of a Solid** For what value of x will the volume of the following solid be 112 cubic inches?

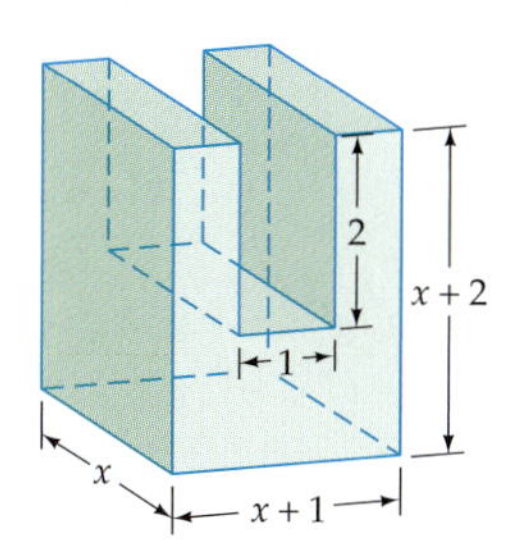

62. **Dimensions of a Box** The length of a rectangular box is 1 inch more than twice the height of the box, and the width is 3 inches more than the height. If the volume of the box is 126 cubic inches, find the dimensions of the box.

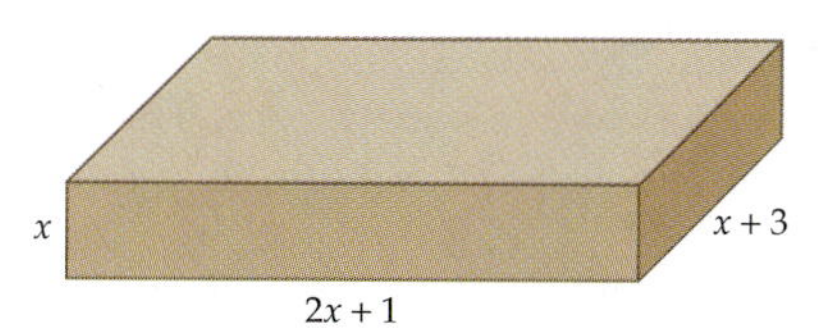

63. **Pieces and Cuts** One straight cut through a thick piece of cheese produces two pieces. Two straight cuts can produce a maximum of four pieces. Three straight cuts can produce a maximum of eight pieces.

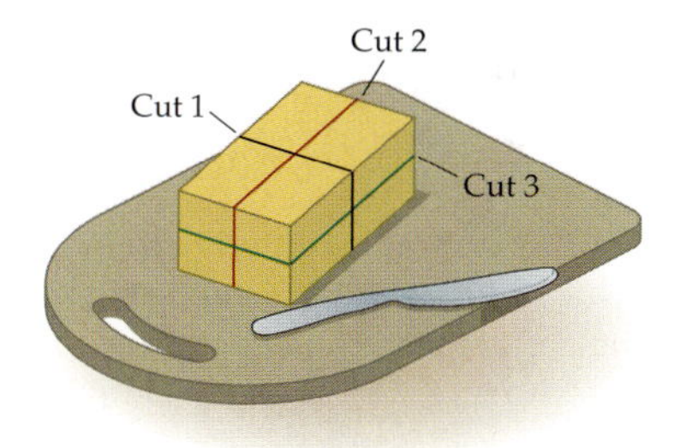

You might be inclined to think that every additional cut doubles the previous number of pieces. However, for four straight cuts, you get a maximum of 15 pieces. The maximum number of pieces P that can be produced by n straight cuts is given by

$$P(n) = \frac{n^3 + 5n + 6}{6}$$

a. Use the above function to determine the maximum number of pieces that can be produced by five straight cuts.

b. What is the fewest number of straight cuts that are needed to produce 64 pieces?

64. **INSCRIBED QUADRILATERAL** Isaac Newton discovered that if a quadrilateral with sides of lengths a, b, c, and x is inscribed in a semicircle with diameter x, then the lengths of the sides are related by the following equation.

$$x^3 - (a^2 + b^2 + c^2)x - 2abc = 0$$

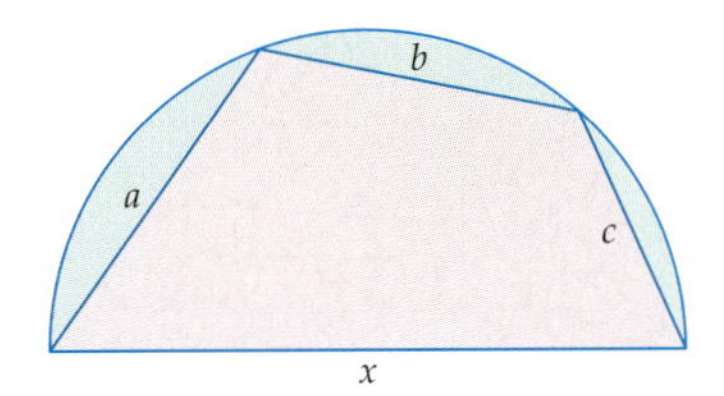

Given $a = 6$, $b = 5$, and $c = 4$, find x. Round to the nearest hundredth.

65. **CANNONBALL STACKS** Cannonballs can be stacked to form a pyramid with a square base. The total number of cannonballs T in one of these square pyramids is

$$T = \frac{1}{6}(2n^3 + 3n^2 + n)$$

where n is the number of rows (levels). If 140 cannonballs are used to form a square pyramid, how many rows are in the pyramid?

66. **ADVERTISING EXPENSES** A company manufactures digital cameras. The company estimates that the profit from camera sales is

$$P(x) = -0.02x^3 + 0.01x^2 + 1.2x - 1.1$$

where P is the profit in millions of dollars and x is the amount, in hundred-thousands of dollars, spent on advertising.

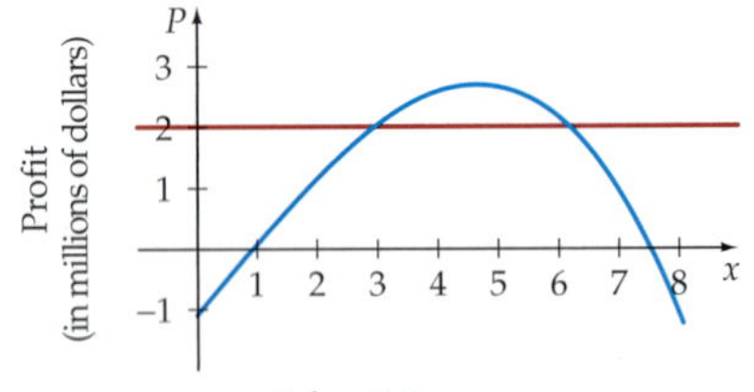

Determine the minimum amount, rounded to the nearest thousand dollars, the company needs to spend on advertising if it is to receive a profit of $2,000,000.

67. **COST CUTTING** At the present time, a nutrition bar in the shape of a rectangular solid measures 0.75 inch by 1 inch by 5 inches.

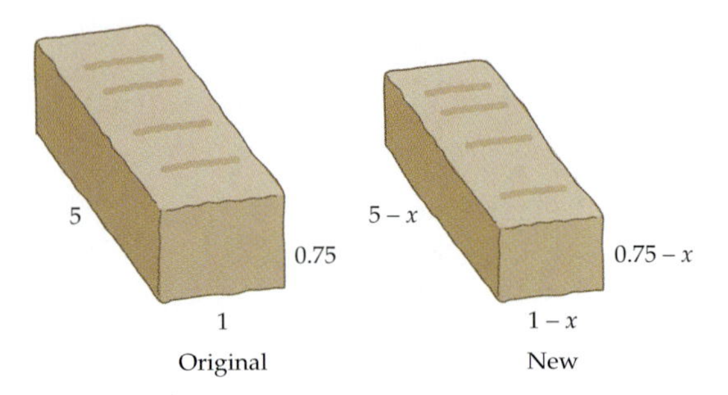

To reduce costs the manufacturer has decided to decrease each of the dimensions of the nutrition bar by x inches. What value of x, rounded to the nearest thousandth of an inch, will produce a new nutrition bar with a volume that is 0.75 cubic inch less than the present bar's volume?

68. **PROPANE TANK DIMENSIONS** A propane tank has the shape of a circular cylinder with a hemisphere at each end. The cylinder is 6 feet long and the volume of the tank is 9π cubic feet. Find, to the nearest thousandth of a foot, the length of the radius x.

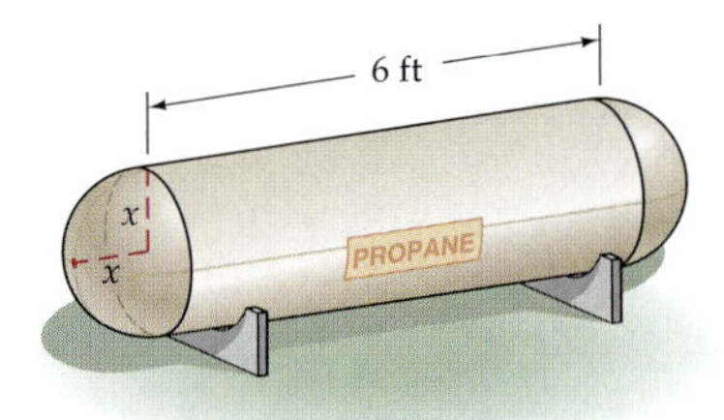

69. **DIVORCE RATE** The divorce rate for a given year is defined as the number of divorces per thousand population. The polynomial function

$$D(t) = 0.00001807t^4 - 0.001406t^3 + 0.02884t^2 - 0.003466t + 2.1148$$

approximates the U.S. divorce rate for the years 1960 ($t = 0$) to 1999 ($t = 39$). Use $D(t)$ and a graphing utility to determine during what years the U.S. divorce rate attained a level of 5.0.

70.  **MEDICATION LEVEL** Pseudoephedrine hydrochloride is an allergy medication. The polynomial function

$$L(t) = 0.03t^4 + 0.4t^3 - 7.3t^2 + 23.1t$$

where $0 \le t \le 5$, models the level of pseudoephedrine hydrochloride, in milligrams, in the bloodstream of a patient t hours after 30 milligrams of the medication have been taken.

At what times, to the nearest minute, does the level of pseudoephedrine hydrochloride in the bloodstream reach 12 milligrams?

71. **Weight and Height of Giraffes** A veterinarian at a wild animal park has determined that the average weight w, in pounds, of an adult male giraffe is closely approximated by the function

$$w = 8.3h^3 - 307.5h^2 + 3914h - 15{,}230$$

where h is the giraffe's height in feet, and $15 \le h \le 18$. Use the above function to estimate the height of a giraffe that weighs 3150 pounds. Round to the nearest tenth of a foot.

72. **Selection of Cards** The number of ways one can select three cards from a group of n cards (the order of the selection matters), where $n \ge 3$, is given by $P(n) = n^3 - 3n^2 + 2n$. For a certain card trick a magician has determined that there are exactly 504 ways to choose three cards from a given group. How many cards are in the group?

73. **Digits of Pi** In 1999, Professor Yasumasa Kanada of the University of Tokyo used a supercomputer to compute 206,158,430,000 digits of pi (π). (*Source: Guinness World Records 2001,* Bantam Books, p. 252.) Computer scientists often try to find mathematical models that approximate the time a computer program takes to complete a calculation or mathematical procedure. Procedures for which the completion time can be closely modeled by a polynomial are called *polynomial time procedures.* Here is an example. A student finds that the time, in seconds, required to compute $n \times 10{,}000$ digits of pi on a personal computer using the mathematical program MAPLE is closely approximated by

$$T(n) = 0.23245n^3 + 0.53797n^2 + 7.88932n - 8.53299$$

a. Evaluate $T(n)$ to estimate how long, to the nearest second, the computer takes to compute 50,000 digits of pi.

b. About how many digits of pi can the computer compute in 5 minutes? Round to the nearest thousand digits.

Connecting Concepts

74. If p is a prime number, prove that $\sqrt{p}$ is an irrational number. (*Hint:* Start with the equation $x = \sqrt{p}$, and square each side to produce the equivalent equation $x^2 = p$, which can be written as $x^2 - p = 0$. Then apply the Rational Zero Theorem to show that $P(x) = x^2 - p$ has no rational zeros.)

The mathematician Augustin Louis Cauchy (1789–1857) proved the following theorem, which can be used to quickly establish a bound *B* for *all* the zeros (both real and complex) of a given polynomial function.

Cauchy's Bound Theorem

Let $P(x) = a_n x^n + a_{n-1}x^{n-1} + \cdots + a_1 + a_0$ be a polynomial function with complex coefficients. The absolute value of each zero of P is less than

$$B = \left(\frac{\text{maximum of } (|a_{n-1}|, |a_{n-2}|, \cdots, |a_1|, |a_0|)}{|a_n|} + 1\right)$$

In Exercises 75 to 78, a polynomial function and its zeros are given. For each polynomial function, apply Cauchy's Bound Theorem to determine the bound *B* for the polynomial and determine whether the absolute value of each of the given zeros is less than *B*. (*Hint:* $|a + bi| = \sqrt{a^2 + b^2}$)

75. $P(x) = 2x^3 - 5x^2 - 28x + 15$, zeros: $-3, \frac{1}{2}, 5$

76. $P(x) = x^3 - 5x^2 + 2x + 8$, zeros: $-1, 2, 4$

77. $P(x) = x^4 - 2x^3 + 9x^2 + 2x - 10$, zeros: $1 + 3i, 1 - 3i, 1, -1$

78. $P(x) = x^4 - 4x^3 + 14x^2 - 4x + 13$, zeros: $2 + 3i, 2 - 3i, i, -i$

PREPARE FOR SECTION 2.5

79. What is the conjugate of $3 - 2i$? [2.1]

80. What is the conjugate of $2 + i\sqrt{5}$? [2.1]

81. Find $(x - 1)(x - 3)(x - 4)$. [A.2]

82. Find $[x - (2 + i)][x - (2 - i)]$. [2.1]

83. Solve: $x^2 + 9 = 0$ [2.1]

84. Solve: $x^2 - x + 5 = 0$ [2.1]

PROJECTS

1.  **RELATIONSHIPS BETWEEN ZEROS AND COEFFICIENTS** Consider the polynomial function

$$P(x) = x^n + C_1x^{n-1} + C_2x^{n-2} + \cdots + C_n$$

with zeros $r_1, r_2, r_3, \ldots, r_n$. The following equations illustrate important relationships between the zeros of the polynomial function and the coefficients of the polynomial.

- The sum of the zeros.

$$r_1 + r_2 + r_3 + \cdots + r_{n-1} + r_n = -C_1$$

- The sum of the products of the zeros taken two at a time.

$$r_1r_2 + r_1r_3 + \cdots + r_{n-2}r_n + r_{n-1}r_n = C_2$$

- The sum of the products of the zeros taken three at a time.

$$r_1r_2r_3 + r_1r_2r_4 + \cdots + r_{n-2}r_{n-1}r_n = -C_3$$

$$\vdots$$

- The product of the zeros.

$$r_1r_2r_3r_4 \cdots r_{n-1}r_n = (-1)^nC_n$$

a. Show that each of the previous equations holds true for the polynomial function

$$P(x) = x^3 - 6x^2 + 11x - 6$$

which has zeros of 1, 2, and 3.

b. Create a polynomial function of degree 4 with four real zeros. Illustrate that each of the above equations holds true for your polynomial function. (*Hint:* The polynomial function

$$P(x) = (x - a)(x - b)(x - c)(x - d)$$

has a, b, c, and d as zeros.)

SECTION 2.5 THE FUNDAMENTAL THEOREM OF ALGEBRA

- The Fundamental Theorem of Algebra
- The Number of Zeros of a Polynomial Function
- The Conjugate Pair Theorem
- Find a Polynomial Function with Given Zeros

THE FUNDAMENTAL THEOREM OF ALGEBRA

The German mathematician Carl Friedrich Gauss (1777–1855) was the first to prove that every polynomial function has at least one complex zero. This concept is so basic to the study of algebra that it is called the **Fundamental Theorem of Algebra.** The proof of the Fundamental Theorem is beyond the scope of this text; however, it is important to understand the theorem and its consequences. As you consider each of the following theorems, keep in mind that the terms *complex coefficients* and *complex zeros* include real coefficients and real zeros because the set of real numbers is a subset of the set of complex numbers.

MATH MATTERS

Carl Friedrich Gauss (1777–1855) has often been referred to as the Prince of Mathematics. His work covered topics in algebra, calculus, analysis, probability, number theory, non-Euclidean geometry, astronomy, and physics, to name but a few. The following quote by Eric Temple Bell gives credence to the fact that Gauss was one of the greatest mathematicians of all time. "Archimedes, Newton, and Gauss, these three, are in a class by themselves among the great mathematicians, and it is not for ordinary mortals to attempt to range them in order of merit."*

Men of Mathematics, by E. T. Bell, New York, Simon and Schuster, 1937.

The Fundamental Theorem of Algebra

If $P(x)$ is a polynomial function of degree $n \geq 1$ with complex coefficients, then $P(x)$ has at least one complex zero.

The Number of Zeros of a Polynomial Function

Let $P(x)$ be a polynomial function of degree $n \geq 1$ with complex coefficients. The Fundamental Theorem implies that $P(x)$ has a complex zero—say, c_1. The Factor Theorem implies that

$$P(x) = (x - c_1)Q(x)$$

where $Q(x)$ is a polynomial of degree one less than the degree of $P(x)$. Recall that the polynomial $Q(x)$ is called a *reduced polynomial.* Assuming that the degree of $Q(x)$ is 1 or more, the Fundamental Theorem implies that it also must have a zero. A continuation of this reasoning process leads to the following theorem.

The Linear Factor Theorem

If $P(x)$ is a polynomial function of degree $n \geq 1$ with leading coefficient $a_n \neq 0$,

$$P(x) = a_nx^n + a_{n-1}x^{n-1} + \cdots + a_1x^1 + a_0$$

then $P(x)$ has exactly n linear factors

$$P(x) = a_n(x - c_1)(x - c_2)\cdots(x - c_n)$$

where $c_1, c_2, \ldots, c_n$ are complex numbers.

The following theorem follows directly from the Linear Factor Theorem.

The Number of Zeros of a Polynomial Function Theorem

If $P(x)$ is a polynomial function of degree $n \geq 1$, then $P(x)$ has exactly n complex zeros, provided each zero is counted according to its multiplicity.

The Linear Factor Theorem and the Number of Zeros of a Polynomial Function Theorem are referred to as **existence theorems.** They state that an nth degree polynomial will have n linear factors and n complex zeros, but they do not provide any information on how to determine the linear factors or the zeros. In Example 1 we make use of previously developed methods to actually find the linear factors and zeros of some polynomial functions.

EXAMPLE 1 Find the Zeros and Linear Factors of a Polynomial Function

Find all the zeros of each of the following polynomial functions, and write each polynomial as a product of linear factors.

a. $P(x) = x^4 - 4x^3 + 8x^2 - 16x + 16$

b. $S(x) = x^4 - 6x^3 + 10x^2 + 2x - 15$

Solution

a. We know that $P(x)$ will have four zeros and four linear factors. The possible rational zeros are $\pm 1, \pm 2, \pm 4, \pm 8, \pm 16$. Synthetic division can be used to show that 2 is a zero of multiplicity 2.

$$\begin{array}{r|rrrrr} 2 & 1 & -4 & 8 & -16 & 16 \\ & & 2 & -4 & 8 & -16 \\ \hline & 1 & -2 & 4 & -8 & 0 \end{array}$$

$$\begin{array}{r|rrrr} 2 & 1 & -2 & 4 & -8 \\ & & 2 & 0 & 8 \\ \hline & 1 & 0 & 4 & 0 \end{array}$$

The final reduced polynomial is $x^2 + 4$. The zeros of $x^2 + 4$ can be found by solving $x^2 + 4 = 0$, as shown below.

$$\begin{aligned} x^2 + 4 &= 0 \\ x^2 &= -4 \\ x &= \pm\sqrt{-4} \\ x &= \pm 2i \end{aligned}$$

Thus the four zeros of $P(x)$ are $2, 2, -2i$, and $2i$. The linear factored form of $P(x)$ is

$$P(x) = (x - 2)(x - 2)[x - (-2i)][x - 2i]$$

or

$$P(x) = (x - 2)^2(x + 2i)(x - 2i)$$

b. We know that $S(x)$ will have four zeros and four linear factors. The possible rational zeros are $\pm 1, \pm 3, \pm 5, \pm 15$. Synthetic division can be used to show that 3 and -1 are zeros of $S(x)$.

$$\begin{array}{r|rrrrr} 3 & 1 & -6 & 10 & 2 & -15 \\ & & 3 & -9 & 3 & 15 \\ \hline & 1 & -3 & 1 & 5 & 0 \end{array}$$

$$\begin{array}{r|rrrr} -1 & 1 & -3 & 1 & 5 \\ & & -1 & 4 & -5 \\ \hline & 1 & -4 & 5 & 0 \end{array}$$

The final reduced polynomial is $x^2 - 4x + 5$. We can find the remaining zeros by using the quadratic formula to solve $x^2 - 4x + 5 = 0$.

$$x = \frac{-(-4) \pm \sqrt{(-4)^2 - 4(1)(5)}}{2(1)}$$
$$= \frac{4 \pm \sqrt{-4}}{2}$$
$$= 2 \pm i$$

Thus the four zeros of $S(x)$ are $3, -1, 2 + i,$ and $2 - i$. The linear factored form of $S(x)$ is

$$S(x) = (x - 3)[x - (-1)][x - (2 + i)][x - (2 - i)]$$

or $$S(x) = (x - 3)(x + 1)(x - 2 - i)(x - 2 + i)$$

▶ TRY EXERCISE 2, PAGE 175

THE CONJUGATE PAIR THEOREM

You may have noticed that the complex zeros of the polynomial function in Example 1 were complex conjugates. The following theorem shows that this is not a coincidence.

The Conjugate Pair Theorem

If $a + bi$ $(b \neq 0)$ is a complex zero of a polynomial function *with real coefficients,* then the conjugate $a - bi$ is also a complex zero of the polynomial function.

EXAMPLE 2 Use the Conjugate Pair Theorem to Find Zeros

Find all the zeros of $P(x) = x^4 - 4x^3 + 14x^2 - 36x + 45$ given that $2 + i$ is a zero.

Solution

Because the coefficients are real numbers and $2 + i$ is a zero, the Conjugate Pair Theorem implies that $2 - i$ also must be a zero. Using synthetic division with $2 + i$ and then $2 - i$, we have

$$\begin{array}{r|rrrrr} 2+i & 1 & -4 & 14 & -36 & 45 \\ & & 2+i & -5 & 18+9i & -45 \\ \hline & 1 & -2+i & 9 & -18+9i & 0 \end{array}$$

• The coefficients of the reduced polynomial

$$\begin{array}{r|rrrr} 2-i & 1 & -2+i & 9 & -18+9i \\ & & 2-i & 0 & 18-9i \\ \hline & 1 & 0 & 9 & 0 \end{array}$$

• The coefficients of the next reduced polynomial

The resulting reduced polynomial is $x^2 + 9$, which has $3i$ and $-3i$ as zeros. Therefore, the four zeros of $x^4 - 4x^3 + 14x^2 - 36x + 45$ are $2 + i, 2 - i, 3i,$ and $-3i$.

▶ TRY EXERCISE 12, PAGE 175

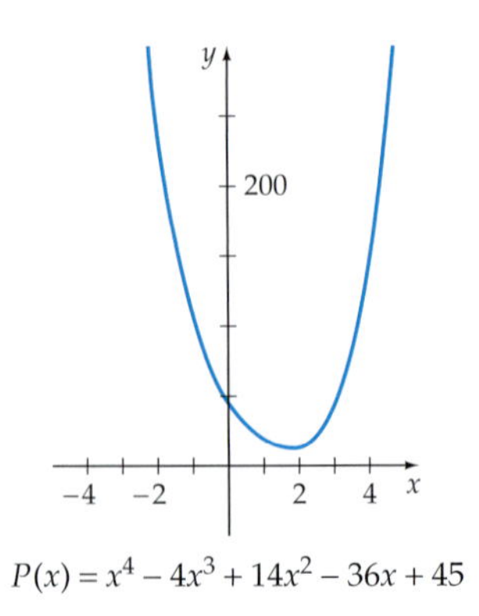

$P(x) = x^4 - 4x^3 + 14x^2 - 36x + 45$

FIGURE 2.20

A graph of $P(x) = x^4 - 4x^3 + 14x^2 - 36x + 45$ is shown in **Figure 2.20.** Because the polynomial in Example 2 is a fourth-degree polynomial and because we have verified that $P(x)$ has four imaginary solutions, it comes as no surprise that the graph does not intersect the x-axis.

When performing synthetic division with complex numbers, it is helpful to write the coefficients of the given polynomial as complex coefficients. For instance, -10 can be written as $-10 + 0i$. This technique is illustrated in the next example.

INTEGRATING TECHNOLOGY

Many graphing calculators can be used to do computations with complex numbers. The following TI-83 screen display shows that the product of $3 - 5i$ and $-7 - 5i$ is $-46 + 20i$. The i symbol is located above the decimal point key.

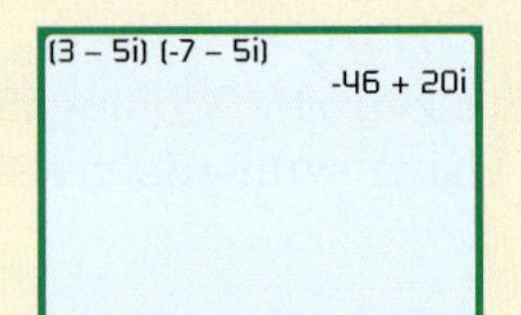

EXAMPLE 3 Apply the Conjugate Pair Theorem

Find all the zeros of $P(x) = x^5 - 10x^4 + 65x^3 - 184x^2 + 274x - 204$ given that $3 - 5i$ is a zero.

Solution

Because the coefficients are real numbers and $3 - 5i$ is a zero, $3 + 5i$ also must be a zero. Use synthetic division to produce

$$\begin{array}{r|rrrrrr} 3-5i & 1 & -10+0i & 65+0i & -184+0i & 274+0i & -204 \\ & & 3-5i & -46+20i & 157-35i & -256+30i & 204 \\ \hline 3+5i & 1 & -7-5i & 19+20i & -27-35i & 18+30i & 0 \\ & & 3+5i & -12-20i & 21+35i & -18-30i & \\ \hline & 1 & -4 & 7 & -6 & 0 & \end{array}$$

Descartes' Rule of Signs can be used to show that the reduced polynomial $x^3 - 4x^2 + 7x - 6$ has three or one positive zeros and no negative zeros. Using the Rational Zero Theorem, we have

$$\frac{p}{q} = 1, 2, 3, 6$$

Use synthetic division to determine that 2 is a zero.

$$\begin{array}{r|rrrr} 2 & 1 & -4 & 7 & -6 \\ & & 2 & -4 & 6 \\ \hline & 1 & -2 & 3 & 0 \end{array}$$

Use the quadratic formula to solve $x^2 - 2x + 3 = 0$.

$$x = \frac{-(-2) \pm \sqrt{(-2)^2 - 4(1)(3)}}{2(1)} = \frac{2 \pm \sqrt{-8}}{2} = \frac{2 \pm 2\sqrt{2}i}{2} = 1 \pm \sqrt{2}i$$

The zeros of $P(x) = x^5 - 10x^4 + 65x^3 - 184x^2 + 274x - 204$ are $3 - 5i$, $3 + 5i$, 2, $1 + \sqrt{2}i$, and $1 - \sqrt{2}i$.

▶ **TRY EXERCISE 16, PAGE 175**

? QUESTION Is it possible for a third-degree polynomial function with real coefficients to have two real zeros and one complex zero?

? ANSWER No. Because the coefficients of the polynomial are real numbers, the complex zeros of the polynomial function must occur as conjugate pairs.

Recall that the real zeros of a polynomial function P are the x-coordinates of the x-intercepts of the graph of P. This important connection between the real zeros of a polynomial function and the x-intercepts of the graph of the polynomial function is the basis for using a graphing utility to solve equations. Careful analysis of the graph of a polynomial function and your knowledge of the properties of polynomial functions can be used to solve many polynomial equations.

EXAMPLE 4 **Solve a Polynomial Equation**

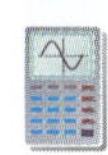

Solve: $x^4 - 5x^3 + 4x^2 + 3x + 9 = 0$

Solution

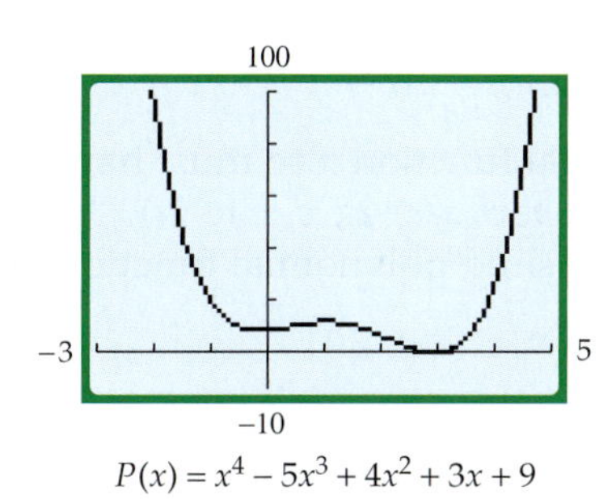

$P(x) = x^4 - 5x^3 + 4x^2 + 3x + 9$

FIGURE 2.21

Let $P(x) = x^4 - 5x^3 + 4x^2 + 3x + 9$. The x-intercepts of the graph of P are the real solutions of the equation. Use a graphing utility to graph P. See **Figure 2.21.**

From the graph, it appears that $(3, 0)$ is an x-intercept and the only x-intercept. Because the graph of P intersects but does not cross the x-axis at $(3, 0)$, we know that 3 is a multiple zero of P with an even multiplicity.

$$\begin{array}{r|rrrrr} 3 & 1 & -5 & 4 & 3 & 9 \\ & & 3 & -6 & -6 & -9 \\ \hline & 1 & -2 & -2 & -3 & 0 \end{array}$$

• Coefficients of P

• The remainder is zero. Thus 3 is a zero.

By the Number of Zeros Theorem, there are three more zeros of P. Use synthetic division to show that 3 is also a zero of the reduced polynomial $x^3 - 2x^2 - 2x - 3$.

$$\begin{array}{r|rrrr} 3 & 1 & -2 & -2 & -3 \\ & & 3 & 3 & 3 \\ \hline & 1 & 1 & 1 & 0 \end{array}$$

• Coefficients of reduced polynomial

• The remainder is zero. Thus 3 is a zero of multiplicity 2.

We now have 3 as a double root of the original equation, and from the last line of the preceding synthetic division, the remaining solutions must be solutions of $x^2 + x + 1 = 0$. Use the quadratic formula to solve this equation.

$$x = \frac{-1 \pm \sqrt{1^2 - 4(1)(1)}}{2(1)} = \frac{-1 \pm \sqrt{-3}}{2} = \frac{-1 \pm i\sqrt{3}}{2}$$

The solutions of $x^4 - 5x^3 + 4x^2 + 3x + 9 = 0$ are 3, 3, $-\frac{1}{2} + \frac{\sqrt{3}}{2}i$, and $-\frac{1}{2} - \frac{\sqrt{3}}{2}i$.

▶ TRY EXERCISE 24, PAGE 175

FIND A POLYNOMIAL FUNCTION WITH GIVEN ZEROS

Many of the problems in this section and in Section 2.4 dealt with the process of finding the zeros of a given polynomial function. Example 5 considers the reverse process, finding a polynomial function when the zeros are given.

EXAMPLE 5 Determine a Polynomial Function Given Its Zeros

Find each polynomial function.

a. A polynomial function of degree 3 that has 1, 2, and -3 as zeros

b. A polynomial function of degree 4 that has real coefficients and zeros $2i$ and $3 - 7i$

Solution

a. Because 1, 2, and -3 are zeros, $(x - 1)$, $(x - 2)$, and $(x + 3)$ are factors. The product of these factors produces a polynomial function that has the indicated zeros.

$$P(x) = (x - 1)(x - 2)(x + 3) = (x^2 - 3x + 2)(x + 3) = x^3 - 7x + 6$$

b. By the Conjugate Pair Theorem, the polynomial function also must have $-2i$ and $3 + 7i$ as zeros. The product of the factors $x - 2i$, $x - (-2i)$, $x - (3 - 7i)$, and $x - (3 + 7i)$ produces the desired polynomial function.

$$\begin{aligned} P(x) &= (x - 2i)(x + 2i)[x - (3 - 7i)][x - (3 + 7i)] \\ &= (x^2 + 4)(x^2 - 6x + 58) \\ &= x^4 - 6x^3 + 62x^2 - 24x + 232 \end{aligned}$$

▶ TRY EXERCISE 42, PAGE 175

A polynomial function that has a given set of zeros is not unique. For example, $P(x) = x^3 - 7x + 6$ has zeros 1, 2, and -3, but so does any nonzero multiple of $P(x)$, such as $S(x) = 2x^3 - 14x + 12$. This concept is illustrated in **Figure 2.22.** The graphs of the two polynomial functions are different, but they have the same x-intercepts.

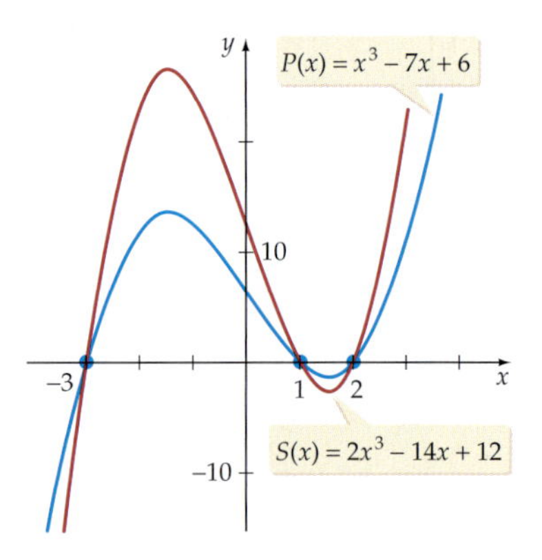

FIGURE 2.22

TOPICS FOR DISCUSSION

1. What is the Fundamental Theorem of Algebra, and why is this theorem so important?

2. Let $P(x)$ be a polynomial function of degree n with real coefficients. Discuss the number of *possible* real zeros of this polynomial function. Include in your discussion the cases n is even and n is odd.

3. Consider the graph of a polynomial function in **Figure 2.23.** Is it possible that the degree of the polynomial is 3? Explain.

4. If two polynomial functions have exactly the same zeros, do the graphs of the polynomial functions look exactly the same?

5. Does the graph of every polynomial function have at least one x-intercept?

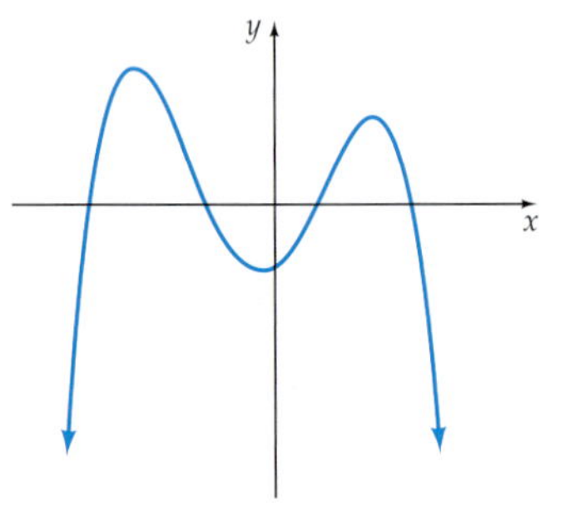

FIGURE 2.23

EXERCISE SET 2.5

In Exercises 1 to 10, find all the zeros of the polynomial function and write the polynomial as a product of linear factors. (*Hint:* First determine the rational zeros.)

1. $P(x) = x^4 + x^3 - 2x^2 + 4x - 24$
2. $P(x) = x^3 - 3x^2 + 7x - 5$
3. $P(x) = 2x^4 + x^3 + 39x^2 + 136x - 78$
4. $P(x) = x^3 - 13x^2 + 65x - 125$
5. $P(x) = x^5 - 9x^4 + 34x^3 - 58x^2 + 45x - 13$
6. $P(x) = x^4 - 4x^3 + 53x^2 - 196x + 196$
7. $P(x) = 2x^4 - x^3 - 15x^2 + 23x + 15$
8. $P(x) = 3x^4 - 17x^3 - 39x^2 + 337x + 116$
9. $P(x) = 2x^4 - 14x^3 + 33x^2 - 46x + 40$
10. $P(x) = 3x^4 - 10x^3 + 15x^2 + 20x - 8$

In Exercises 11 to 22, use the given zero to find the remaining zeros of each polynomial function.

11. $P(x) = 2x^3 - 5x^2 + 6x - 2;\quad 1 + i$
12. $P(x) = 3x^3 - 29x^2 + 92x + 34;\quad 5 + 3i$
13. $P(x) = x^3 + 3x^2 + x + 3;\quad -i$
14. $P(x) = x^4 - 6x^3 + 71x^2 - 146x + 530;\quad 2 + 7i$
15. $P(x) = x^4 - 4x^3 + 14x^2 - 4x + 13;\quad 2 - 3i$
16. $P(x) = x^5 - 6x^4 + 22x^3 - 64x^2 + 117x - 90;\quad 3i$
17. $P(x) = x^4 - 4x^3 + 19x^2 - 30x + 50;\quad 1 + 3i$
18. $P(x) = x^5 - x^4 - 4x^3 - 4x^2 - 5x - 3;\quad i$
19. $P(x) = x^5 - 3x^4 + 7x^3 - 13x^2 + 12x - 4;\quad -2i$
20. $P(x) = x^4 - 8x^3 + 18x^2 - 8x + 17;\quad i$
21. $P(x) = x^4 - 17x^3 + 112x^2 - 333x + 377;\quad 5 + 2i$
22. $P(x) = 2x^5 - 8x^4 + 61x^3 - 99x^2 + 12x + 182;\quad 1 - 5i$

In Exercises 23 to 30, use a graph and your knowledge of the zeros of polynomial functions to determine the *exact* values of all the solutions of each equation.

23. $2x^3 - x^2 + x - 6 = 0$
24. $4x^3 + 3x^2 + 16x + 12 = 0$
25. $24x^3 - 62x^2 - 7x + 30 = 0$
26. $12x^3 - 52x^2 + 27x + 28 = 0$
27. $x^4 - 4x^3 + 5x^2 - 4x + 4 = 0$
28. $x^4 + 4x^3 + 8x^2 + 16x + 16 = 0$
29. $x^4 + 4x^3 - 2x^2 - 12x + 9 = 0$
30. $x^4 + 3x^3 - 6x^2 - 28x - 24 = 0$

In Exercises 31 to 40, find a polynomial function of lowest degree with integer coefficients that has the given zeros.

31. $4, -3, 2$
32. $-1, 1, -5$
33. $3, 2i, -2i$
34. $0, i, -i$
35. $3 + i, 3 - i, 2 + 5i, 2 - 5i$
36. $2 + 3i, 2 - 3i, -5, 2$
37. $6 + 5i, 6 - 5i, 2, 3, 5$
38. $\frac{1}{2}, 4 - i, 4 + i$
39. $\frac{3}{4}, 2 + 7i, 2 - 7i$
40. $\frac{1}{4}, -\frac{1}{5}, i, -i$

In Exercises 41 to 46, find a polynomial function $P(x)$ that has the indicated zeros.

41. Zeros: $2 - 5i, -4$; degree 3
42. Zeros: $3 + 2i, 7$; degree 3
43. Zeros: $4 + 3i, 5 - i$; degree 4
44. Zeros: $i, 3 - 5i$; degree 4
45. Zeros: $-2, 1, 3, 1 + 4i, 1 - 4i$; degree 5
46. Zeros: $-5, 3$ (multiplicity 2), $2 + i, 2 - i$; degree 5

CONNECTING CONCEPTS

In Exercises 47 to 50, find a polynomial function $P(x)$ with real coefficients that has the indicated zeros and satisfies the given conditions.

47. Zeros: $-1, 2, 3$; degree 3; $P(1) = 12$

48. Zeros: $3i, 2$; degree 3; $P(3) = 27$

49. Zeros: $3, -5, 2 + i$; degree 4; $P(1) = 48$

50. Zeros: $\frac{1}{2}, 1 - i$; degree 3; $P(4) = 140$

51. Verify that $P(x) = x^3 - x^2 - ix^2 - 9x + 9 + 9i$ has $1 + i$ as a zero and that its conjugate $1 - i$ is not a zero. Explain why this does not contradict the Conjugate Pair Theorem.

52. Verify that $P(x) = x^3 - x^2 - ix^2 - 20x + ix + 20i$ has a zero of i and that its conjugate $-i$ is not a zero. Explain why this does not contradict the Conjugate Pair Theorem.

PREPARE FOR SECTION 2.6

53. Simplify: $\dfrac{x^2 - 9}{x^2 - 2x - 15}$

54. Evaluate $\dfrac{x + 4}{x^2 - 2x - 5}$ for $x = -1$.

55. Evaluate $\dfrac{2x^2 + 4x - 5}{x + 6}$ for $x = -3$.

56. For what values of x does the denominator of $\dfrac{x^2 - x - 5}{2x^3 + x^2 - 15x}$ equal zero? [2.4]

57. Determine the degree of the numerator and the degree of the denominator of $\dfrac{x^3 + 3x^2 - 5}{x^2 - 4}$. [A.2]

58. Write $\dfrac{x^3 + 2x^2 - x - 11}{x^2 - 2x}$ in $Q(x) + \dfrac{R(x)}{x^2 - 2x}$ form. [2.2]

PROJECTS

1. **INVESTIGATE THE ROOTS OF A CUBIC EQUATION** Hieronimo Cardano, using a technique he learned from Nicolo Tartaglia, was able to solve some cubic equations.

a. Show that the cubic equation $x^3 + bx^2 + cx + d = 0$ can be transformed into the "reduced" cubic $y^3 + my = n$, where m and n are constants, depending on b, c, and d, by using the substitution $x = y - \frac{b}{3}$.

b. Cardano then showed that a solution of the reduced cubic is given by

$$\sqrt[3]{\frac{n}{2} + \sqrt{\frac{n^2}{4} + \frac{m^3}{27}}} - \sqrt[3]{-\frac{n}{2} + \sqrt{\frac{n^2}{4} + \frac{m^3}{27}}}$$

Use Cardano's procedure to solve the equation $x^3 - 6x^2 + 20x - 33 = 0$.

SECTION 2.6

Graphs of Rational Functions and Their Applications

- Vertical and Horizontal Asymptotes
- A Sign Property of Rational Functions
- A General Graphing Procedure
- Slant Asymptotes
- Graph Rational Functions That Have a Common Factor
- Applications of Rational Functions

Vertical and Horizontal Asymptotes

If $P(x)$ and $Q(x)$ are polynomials, then the function F given by

$$F(x) = \frac{P(x)}{Q(x)}$$

is called a **rational function.** The domain of F is the set of all real numbers except those for which $Q(x) = 0$. For example, let

$$F(x) = \frac{x^2 - x - 5}{2x^3 + x^2 - 15x}$$

Setting the denominator equal to zero, we have

$$2x^3 + x^2 - 15x = 0$$
$$x(2x - 5)(x + 3) = 0$$

The denominator is 0 for $x = 0$, $x = \frac{5}{2}$, and $x = -3$. Thus the domain of F is the set of all real numbers except 0, $\frac{5}{2}$, and -3.

The graph of $G(x) = \frac{x+1}{x-2}$ is given in **Figure 2.24.** The graph shows that G has the following properties:

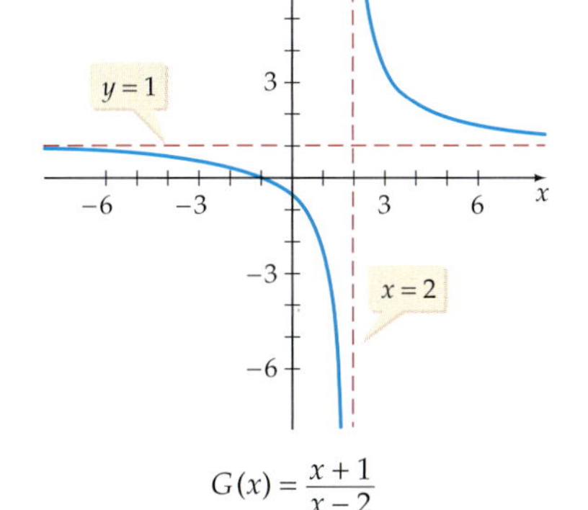

$G(x) = \frac{x+1}{x-2}$

Figure 2.24

- The graph has an x-intercept at $(-1, 0)$ and a y-intercept at $\left(0, -\frac{1}{2}\right)$.
- The graph does not exist when $x = 2$.

Note the behavior of the graph as x takes on values that are close to 2 but *less* than 2. Mathematically, we say that "x approaches 2 from the left."

x	1.9	1.95	1.99	1.995	1.999
G(x)	−29	−59	−299	−599	−2999

From this table and the graph, it appears that as x approaches 2 from the left, the functional values $G(x)$ decrease without bound.

- In this case, we say that "$G(x)$ approaches negative infinity."

Now observe the behavior of the graph as x takes on values that are close to 2 but *greater* than 2. Mathematically, we say that "x approaches 2 from the right."

x	2.1	2.05	2.01	2.005	2.001
G(x)	31	61	301	601	3001

From this table and the graph, it appears that as x approaches 2 from the right, the functional values $G(x)$ increase without bound.

- In this case, we say that "$G(x)$ approaches positive infinity."

Now consider the values of $G(x)$ as x *increases* without bound. The following table gives values of $G(x)$ for selected values of x.

x	1000	5000	10,000	50,000	100,000
G(x)	1.00301	1.00060	1.00030	1.00006	1.00003

- As x increases without bound, the values of $G(x)$ become closer to 1.

Now let the values of x *decrease* without bound. The table below gives the values of $G(x)$ for selected values of x.

x	−1000	−5000	−10,000	−50,000	−100,000
G(x)	0.997006	0.999400	0.999700	0.999940	0.999970

- As x decreases without bound, the values of $G(x)$ become closer to 1.

When we are discussing graphs that increase or decrease without bound, it is convenient to use mathematical notation. The notation

$$f(x) \to \infty \quad \text{as} \quad x \to a^+$$

means that the functional values $f(x)$ increase without bound as x approaches a from the right. Recall that the symbol ∞ does not represent a real number but is used merely to describe the concept of a variable taking on larger and larger values without bound. See **Figure 2.25a.**

The notation

$$f(x) \to \infty \quad \text{as} \quad x \to a^-$$

means that the function values $f(x)$ increase without bound as x approaches a from the left. See **Figure 2.25b.**

The notation

$$f(x) \to -\infty \quad \text{as} \quad x \to a^+$$

means that the functional values $f(x)$ decrease without bound as x approaches a from the right. See **Figure 2.25c.**

The notation

$$f(x) \to -\infty \quad \text{as} \quad x \to a^-$$

means that the functional values $f(x)$ decrease without bound as x approaches a from the left. See **Figure 2.25d.**

Each graph in **Figure 2.25** approaches a vertical line through $(a, 0)$ as $x \to a^+$ or a^-. The line is said to be a *vertical asymptote* of the graph.

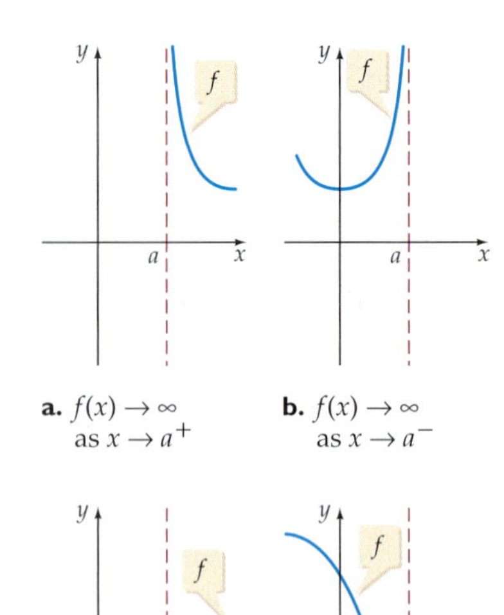

a. $f(x) \to \infty$ as $x \to a^+$ **b.** $f(x) \to \infty$ as $x \to a^-$ **c.** $f(x) \to -\infty$ as $x \to a^+$ **d.** $f(x) \to -\infty$ as $x \to a^-$

FIGURE 2.25

Definition of a Vertical Asymptote

The line $x = a$ is a **vertical asymptote** of the graph of a function F provided

$$F(x) \to \infty \quad \text{or} \quad F(x) \to -\infty$$

as x approaches a from either the left or right.

In **Figure 2.24,** the line $x = 2$ is a vertical asymptote of the graph of G. Note that the graph of G in **Figure 2.24** also approaches the horizontal line $y = 1$ as $x \to \infty$ and as $x \to -\infty$. The line $y = 1$ is a *horizontal asymptote* of the graph of G.

Definition of a Horizontal Asymptote

The line $y = b$ is a **horizontal asymptote** of the graph of a function F provided

$$F(x) \to b \quad \text{as} \quad x \to \infty \quad \text{or} \quad x \to -\infty$$

Figure 2.26 illustrates some of the ways in which the graph of a rational function may approach its horizontal asymptote. It is common practice to display the asymptotes of the graph of a rational function by using dashed lines. Although a rational function may have several vertical asymptotes, it can have at most one horizontal asymptote. The graph may intersect its horizontal asymptote.

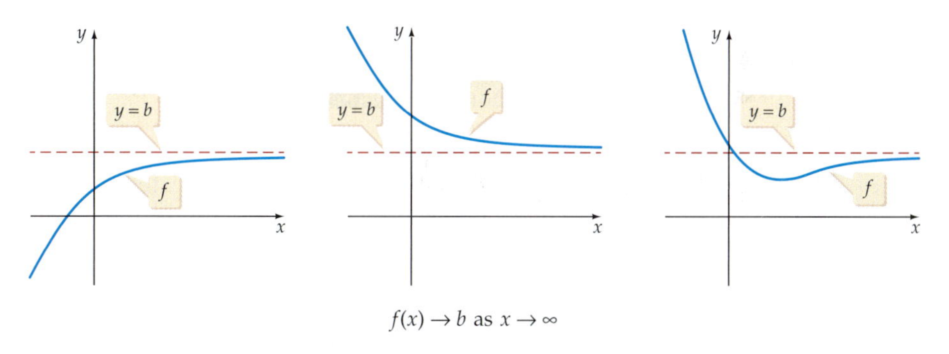

$f(x) \to b$ as $x \to \infty$

FIGURE 2.26

? QUESTION Can a graph of a rational function cross its vertical asymptote? Why or why not?

Geometrically, a line is an asymptote of a curve if the distance between the line and a point $P(x, y)$ on the curve approaches zero as the distance between the origin and the point P increases without bound.

? ANSWER No. If $x = a$ is a vertical asymptote of a rational function R, then $R(a)$ is undefined.

Vertical asymptotes of the graph of a rational function can be found by using the following theorem.

Theorem on Vertical Asymptotes

If the real number a is a zero of the denominator $Q(x)$, then the graph of $F(x) = P(x)/Q(x)$, where $P(x)$ and $Q(x)$ have no common factors, has the vertical asymptote $x = a$.

EXAMPLE 1 Find the Vertical Asymptotes of a Rational Function

Find the vertical asymptotes of each rational function.

a. $f(x) = \dfrac{x^3}{x^2 + 1}$ **b.** $g(x) = \dfrac{x}{x^2 - x - 6}$

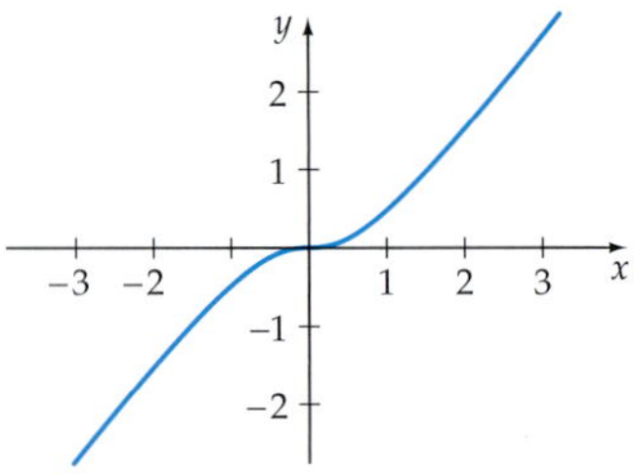

$f(x) = \dfrac{x^3}{x^2+1}$

Figure 2.27

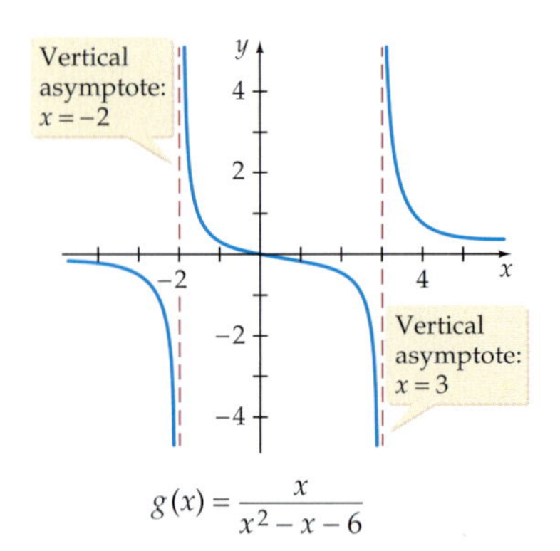

$g(x) = \dfrac{x}{x^2 - x - 6}$

Figure 2.28

Solution

a. To find the vertical asymptotes, determine the real zeros of the denominator. The denominator $x^2 + 1$ has no real zeros, so the graph of f has no vertical asymptotes. See **Figure 2.27.**

b. The denominator $x^2 - x - 6 = (x - 3)(x + 2)$ has zeros of 3 and -2. The numerator has no common factors with the denominator, so $x = 3$ and $x = -2$ are both vertical asymptotes of the graph of g, as shown in **Figure 2.28.**

▶ **Try Exercise 2, page 189**

The following theorem indicates that a horizontal asymptote can be determined by examining the leading terms of the numerator and the denominator of a rational function.

Theorem on Horizontal Asymptotes

Let
$$F(x) = \frac{a_n x^n + a_{n-1}x^{n-1} + \cdots + a_1 x + a_0}{b_m x^m + b_{m-1}x^{m-1} + \cdots + b_1 x + b_0}$$

be a rational function with numerator of degree n and denominator of degree m.

1. If $n < m$, then the x-axis, which is the line given by $y = 0$, is the horizontal asymptote of the graph of F.
2. If $n = m$, then the line given by $y = a_n/b_m$ is the horizontal asymptote of the graph of F.
3. If $n > m$, the graph of F has no horizontal asymptote.

EXAMPLE 2 Find the Horizontal Asymptote of a Rational Function

Find the horizontal asymptote of each rational function.

a. $f(x) = \dfrac{2x + 3}{x^2 + 1}$ **b.** $g(x) = \dfrac{4x^2 + 1}{3x^2}$ **c.** $h(x) = \dfrac{x^3 + 1}{x - 2}$

Solution

a. The degree of the numerator $2x + 3$ is less than the degree of the denominator $x^2 + 1$. By the Theorem on Horizontal Asymptotes, the x-axis is the horizontal asymptote of f. See the graph of f in **Figure 2.29.**

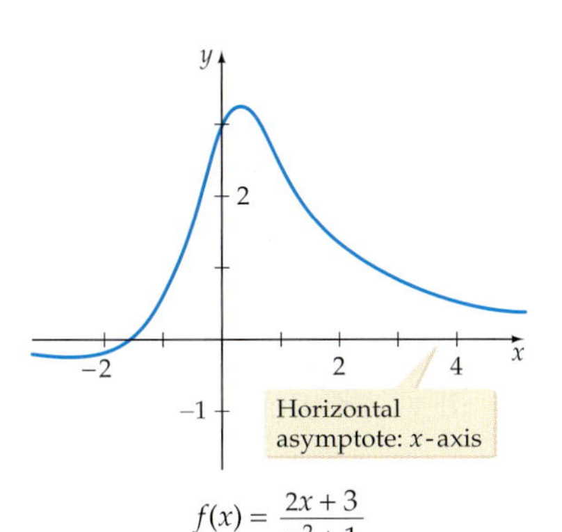

$f(x) = \dfrac{2x+3}{x^2+1}$

FIGURE 2.29

b. The numerator $4x^2 + 1$ and the denominator $3x^2$ of g are both of degree 2. By the Theorem on Horizontal Asymptotes, the line $y = \dfrac{4}{3}$ is the horizontal asymptote of g. See the graph of g in **Figure 2.30.**

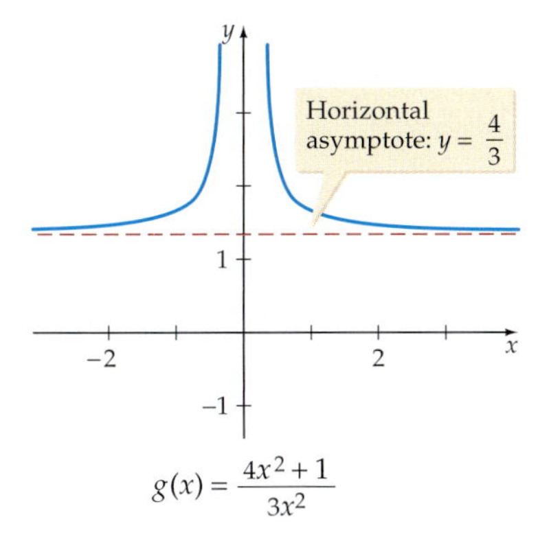

$g(x) = \dfrac{4x^2+1}{3x^2}$

FIGURE 2.30

c. The degree of the numerator $x^3 + 1$ is larger than the degree of the denominator $x - 2$, so by the Theorem on Horizontal Asymptotes, the graph of h has no horizontal asymptotes.

▶ **TRY EXERCISE 6, PAGE 189**

The proof of the Theorem on Horizontal Asymptotes makes use of the technique employed in the following verification. To verify that

$$y = \frac{5x^2 + 4}{3x^2 + 8x + 7}$$

has a horizontal asymptote of $y = \dfrac{5}{3}$, divide the numerator and the denominator by the largest power of the variable x (x^2 in this case).

$$y = \frac{\dfrac{5x^2 + 4}{x^2}}{\dfrac{3x^2 + 8x + 7}{x^2}} = \frac{5 + \dfrac{4}{x^2}}{3 + \dfrac{8}{x} + \dfrac{7}{x^2}}, \quad x \neq 0$$

As x increases without bound or decreases without bound, the fractions $\dfrac{4}{x^2}$, $\dfrac{8}{x}$, and $\dfrac{7}{x^2}$ approach zero. Thus

$$y \to \frac{5 + 0}{3 + 0 + 0} = \frac{5}{3} \quad \text{as} \quad x \to \pm\infty$$

and hence the line $y = \dfrac{5}{3}$ is a horizontal asymptote of the graph.

A Sign Property of Rational Functions

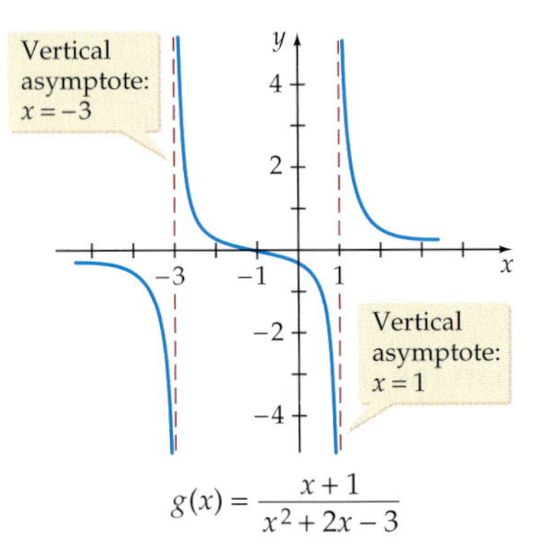

Figure 2.31

The zeros and vertical asymptotes of a rational function F divide the x-axis into intervals. In each interval, $F(x)$ is positive for all x in the interval or $F(x)$ is negative for all x in the interval. For example, consider the rational function

$$g(x) = \frac{x+1}{x^2 + 2x - 3}$$

which has vertical asymptotes of $x = -3$ and $x = 1$ and a zero of -1. These three numbers divide the x-axis into the four intervals $(-\infty, -3)$, $(-3, -1)$, $(-1, 1)$, and $(1, \infty)$. Note in **Figure 2.31** that the graph of g is negative for all x such that $x < -3$, positive for all x such that $-3 < x < -1$, negative for all x such that $-1 < x < 1$, and positive for all x such that $x > 1$.

A General Graphing Procedure

If $F(x) = P(x)/Q(x)$, where $P(x)$ and $Q(x)$ are polynomials that have no common factors, then the following general procedure offers useful guidelines for graphing F.

General Procedure for Graphing Rational Functions That Have No Common Factors

1. *Asymptotes* Find the real zeros of the denominator $Q(x)$. For each zero a, draw the dashed line $x = a$. Each line is a vertical asymptote of the graph of F. Also graph any horizontal asymptotes.
2. *Intercepts* Find the real zeros of the numerator $P(x)$. For each real zero c, plot the point $(c, 0)$. Each such point is an x-intercept of the graph of F. For each x-intercept use the even and odd powers of $(x - c)$ to determine if the graph crosses the x-axis at the intercept or intersects but does not cross the x-axis. Also evaluate $F(0)$. Plot $(0, F(0))$, the y-intercept of the graph of F.
3. *Symmetry* Use the tests for symmetry to determine whether the graph of the function has symmetry with respect to the y-axis or symmetry with respect to the origin.
4. *Additional points* Plot some points that lie in the intervals between and beyond the vertical asymptotes and the x-intercepts.
5. *Behavior near asymptotes* If $x = a$ is a vertical asymptote, determine whether $F(x) \to \infty$ or $F(x) \to -\infty$ as $x \to a^-$ and also as $x \to a^+$.
6. *Complete the sketch* Use all the information obtained above to sketch the graph of F.

EXAMPLE 3 Graph a Rational Function

Sketch a graph of $f(x) = \dfrac{2x^2 - 18}{x^2 + 3}$.

Solution

Asymptotes The denominator $x^2 + 3$ has no real zeros, so the graph of f has no vertical asymptotes. The numerator and denominator both are of degree 2. The leading coefficients are 2 and 1, respectively. By the Theorem on Horizontal Asymptotes, the graph of f has a horizontal asymptote of $y = \dfrac{2}{1} = 2$.

Intercepts The zeros of the numerator occur when $2x^2 - 18 = 0$ or, solving for x, when $x = -3$ and $x = 3$. Therefore, the x-intercepts are $(-3, 0)$ and $(3, 0)$. The factored numerator is $2(x + 3)(x - 3)$. Each linear factor has an exponent of 1, an odd number. Thus the graph crosses the x-axis at its x-intercepts. To find the y-intercept, evaluate f when $x = 0$. This gives $y = -6$. Therefore, the y-intercept is $(0, -6)$.

Symmetry Below we show that $f(-x) = f(x)$, which means that f is an even function and therefore its graph is symmetric with respect to the y-axis.

$$f(-x) = \frac{2(-x)^2 - 18}{(-x)^2 + 3} = \frac{2x^2 - 18}{x^2 + 3} = f(x)$$

Additional Points The intervals determined by the x-intercepts are $x < -3$, $-3 < x < 3$, and $x > 3$. Generally, it is necessary to determine points in all intervals. However, because f is an even function, its graph is symmetric with respect to the y-axis. The following table lists a few points for $x > 0$. Symmetry can be used to locate corresponding points for $x < 0$.

x	1	2	6
f(x)	-4	$-\frac{10}{7} \approx -1.43$	$\frac{18}{13} \approx 1.38$

Behavior Near Asymptotes As x increases or decreases without bound, $f(x)$ approaches the horizontal asymptote $y = 2$.

To determine whether the graph of f intersects the horizontal asymptote at any point, solve the equation $f(x) = 2$.

There are no solutions of $f(x) = 2$ because

$$\frac{2x^2 - 18}{x^2 + 3} = 2 \quad \text{implies} \quad 2x^2 - 18 = 2x^2 + 6 \quad \text{implies} \quad -18 = 6$$

This is not possible. Thus the graph of f does not intersect the horizontal asymptote but approaches it from below as x increases or decreases without bound.

Continued ▶

Complete the Sketch Use the summary in **Table 2.3,** to the left, to finish the sketch. The completed graph is shown in **Figure 2.32.**

TABLE 2.3

Vertical Asymptote	None
Horizontal Asymptote	$y = 2$
***x*-Intercepts**	crosses at $(-3, 0)$, crosses at $(3, 0)$
***y*-Intercept**	$(0, -6)$
Additional Points	$(1, -4)$, $(2, -1.43)$, $(6, 1.38)$

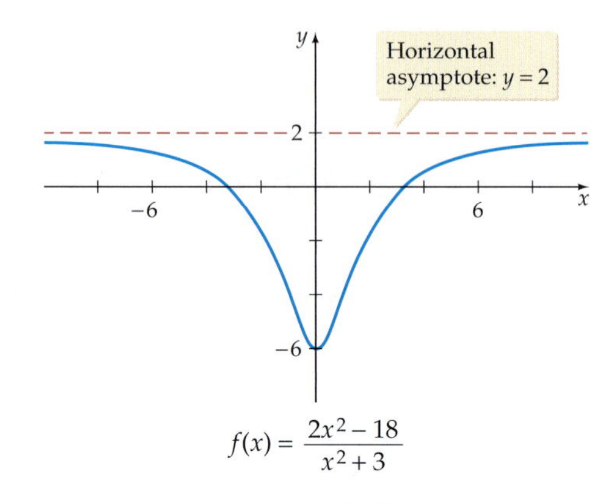

FIGURE 2.32

▶ TRY EXERCISE 10, PAGE 189

EXAMPLE 4 Graph a Rational Function

Sketch a graph of $h(x) = \dfrac{x^2 + 1}{x^2 + x - 2}$.

Solution

Asymptotes The denominator $x^2 + x - 2 = (x + 2)(x - 1)$ has zeros -2 and 1; because there are no common factors of the numerator and the denominator, the lines $x = -2$ and $x = 1$ are vertical asymptotes.

The numerator and denominator both are of degree 2. The leading coefficients of the numerator and denominator are both 1. Thus h has the horizontal asymptote $y = \dfrac{1}{1} = 1$.

Intercepts The numerator $x^2 + 1$ has no real zeros, so the graph of h has no x-intercepts. Because $h(0) = -0.5$, h has the y-intercept $(0, -0.5)$.

Symmetry By applying the tests for symmetry, we can determine that the graph of h is not symmetric with respect to the origin or to the y-axis.

Additional Points The intervals determined by the vertical asymptotes are $(-\infty, -2)$, $(-2, 1)$, and $(1, \infty)$. Plot a few points from each interval.

x	-5	-3	-1	0.5	2	3	4
h(x)	$\frac{13}{9}$	$\frac{5}{2}$	-1	-1	$\frac{5}{4}$	1	$\frac{17}{18}$

The graph of h will intersect the horizontal asymptote $y = 1$ exactly once. This can be determined by solving the equation $h(x) = 1$.

$$\frac{x^2 + 1}{x^2 + x - 2} = 1$$

$$x^2 + 1 = x^2 + x - 2 \qquad \text{• Multiply both sides by } x^2 + x - 2.$$

$$1 = x - 2$$

$$3 = x$$

The only solution is $x = 3$. Therefore, the graph of h intersects the horizontal asymptote at (3, 1).

Behavior Near Asymptotes As x approaches -2 from the left, the denominator $(x + 2)(x - 1)$ approaches 0 but remains positive. The numerator $x^2 + 1$ approaches 5, which is positive, so the quotient $h(x)$ increases without bound. Stated in mathematical notation,

$$h(x) \to \infty \quad \text{as} \quad x \to -2^-$$

Similarly, it can be determined that

$$h(x) \to -\infty \quad \text{as} \quad x \to -2^+$$
$$h(x) \to -\infty \quad \text{as} \quad x \to 1^-$$
$$h(x) \to \infty \quad \text{as} \quad x \to 1^+$$

Complete the Sketch Use the summary in **Table 2.4** to obtain the graph sketched in **Figure 2.33**.

TABLE 2.4

Vertical Asymptote	$x = -2, x = 1$
Horizontal Asymptote	$y = 1$
x-Intercepts	None
y-Intercept	$(0, -0.5)$
Additional Points	$(-5, 1.\overline{4})$, $(-3, 2.5)$, $(-1, -1)$, $(0.5, -1)$, $(2, 1.2\overline{5})$, $(3, 1)$, $(4, 0.9\overline{4})$

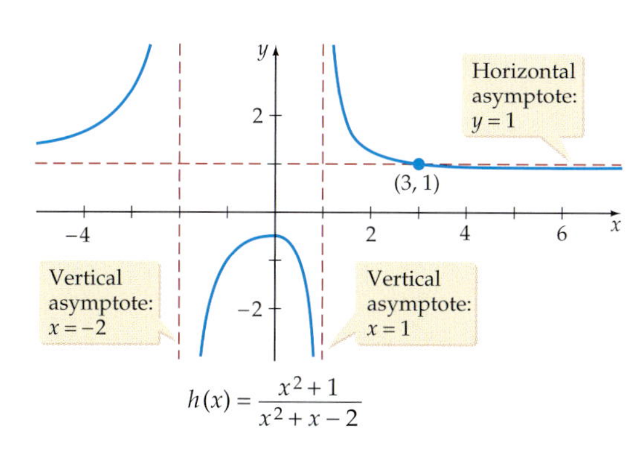

FIGURE 2.33

TRY EXERCISE 26, PAGE 189

SLANT ASYMPTOTES

Some rational functions have an asymptote that is neither vertical nor horizontal, but slanted.

Theorem on Slant Asymptotes

The rational function given by $F(x) = P(x)/Q(x)$, where $P(x)$ and $Q(x)$ have no common factors, has a **slant asymptote** if the degree of the polynomial $P(x)$ in the numerator is one greater than the degree of the polynomial $Q(x)$ in the denominator.

To find the slant asymptote, divide $P(x)$ by $Q(x)$ and write $F(x)$ in the form

$$F(x) = \frac{P(x)}{Q(x)} = (mx + b) + \frac{r(x)}{Q(x)}$$

where the degree of $r(x)$ is less than the degree of $Q(x)$. Because

$$\frac{r(x)}{Q(x)} \longrightarrow 0 \quad \text{as} \quad x \longrightarrow \pm\infty$$

we know that $F(x) \longrightarrow mx + b$ as $x \longrightarrow \pm\infty$.

The line represented by $y = mx + b$ is the slant asymptote of the graph of F.

EXAMPLE 5 Find the Slant Asymptote of a Rational Function

Find the slant asymptote of $f(x) = \dfrac{2x^3 + 5x^2 + 1}{x^2 + x + 3}$.

Solution

Because the degree of the numerator $2x^3 + 5x^2 + 1$ is exactly one larger than the degree of the denominator $x^2 + x + 3$ and f is in simplest form, f has a slant asymptote. To find the asymptote, divide $2x^3 + 5x^2 + 1$ by $x^2 + x + 3$.

$$\begin{array}{r} 2x + 3 \\ x^2 + x + 3 \overline{)2x^3 + 5x^2 + 0x + 1} \\ \underline{2x^3 + 2x^2 + 6x } \\ 3x^2 - 6x + 1 \\ \underline{3x^2 + 3x + 9} \\ -9x - 8 \end{array}$$

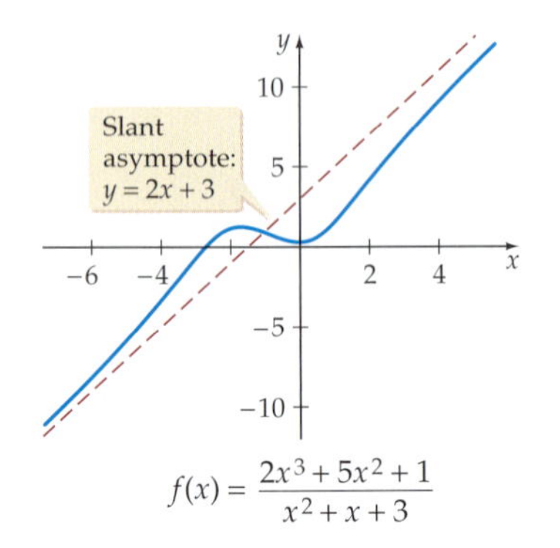

$f(x) = \dfrac{2x^3 + 5x^2 + 1}{x^2 + x + 3}$

FIGURE 2.34

Therefore,

$$f(x) = \frac{2x^3 + 5x^2 + 1}{x^2 + x + 3} = 2x + 3 + \frac{-9x - 8}{x^2 + x + 3}$$

and the line given by $y = 2x + 3$ is the slant asymptote for the graph of f. **Figure 2.34** shows the graph of f and its slant asymptote.

▶ TRY EXERCISE 32, PAGE 189

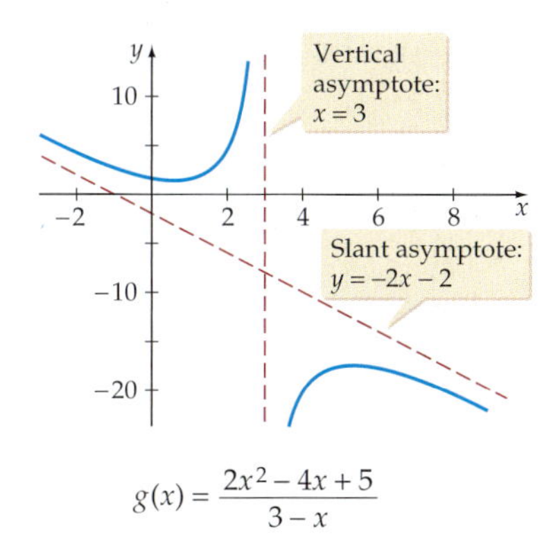

FIGURE 2.35

The function f in Example 5 does not have a vertical asymptote because the denominator $x^2 + x + 3$ does not have any real zeros. However, the function

$$g(x) = \frac{2x^2 - 4x + 5}{3 - x}$$

has both a slant asymptote and a vertical asymptote. The vertical asymptote is $x = 3$, and the slant asymptote is $y = -2x - 2$. **Figure 2.35** shows the graph of g and its asymptotes.

GRAPH RATIONAL FUNCTIONS THAT HAVE A COMMON FACTOR

If a rational function has a numerator and denominator that have a common factor, then you should reduce the rational function to lowest terms before you apply the general procedure for sketching the graph of a rational function.

EXAMPLE 6 Graph a Rational Function That Has a Common Factor

Sketch the graph of $f(x) = \frac{x^2 - 3x - 4}{x^2 - 6x + 8}$.

Solution

Factor the numerator and denominator to obtain

$$f(x) = \frac{x^2 - 3x - 4}{x^2 - 6x + 8} = \frac{(x + 1)(x - 4)}{(x - 2)(x - 4)}, \quad x \neq 2, x \neq 4$$

Thus for all x values other than $x = 4$, the graph of f is the same as the graph of

$$G(x) = \frac{x + 1}{x - 2}$$

Figure 2.24 on page 177 shows a graph of G. The graph of f will be the same as this graph, except that it will have an open circle at (4, 2.5) to indicate that it is undefined at $x = 4$. See the graph of f in **Figure 2.36.** The height of the open circle was found by evaluating the resulting reduced rational function $G(x) = \frac{x + 1}{x - 2}$ at $x = 4$.

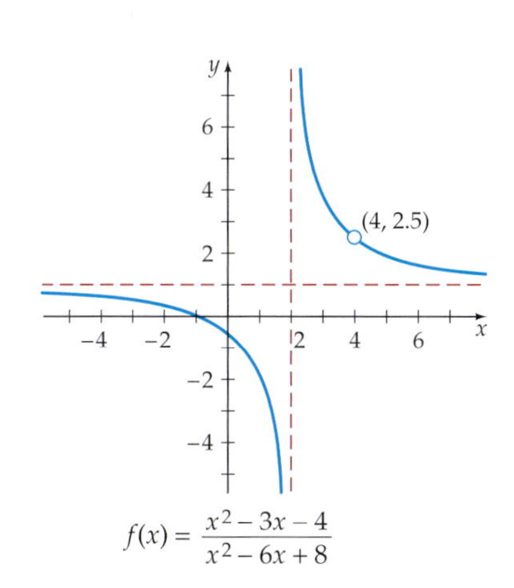

FIGURE 2.36

▶ TRY EXERCISE 48, PAGE 190

QUESTION Does $F(x) = \frac{x^2 - x - 6}{x^2 - 9}$ have a vertical asymptote at $x = 3$?

ANSWER No. $F(x) = \frac{x^2 - x - 6}{x^2 - 9} = \frac{(x - 3)(x + 2)}{(x - 3)(x + 3)} = \frac{x + 2}{x + 3}, x \neq 3$. As $x \to 3$, $F(x) \to \frac{5}{6}$.

Applications of Rational Functions

EXAMPLE 7 Solve an Application

A cylindrical soft drink can is to be constructed so that it will have a volume of 21.6 cubic inches. See **Figure 2.37.**

Figure 2.37

a. Write the total surface area A of the can as a function of r, where r is the radius of the can in inches.

b. Use a graphing utility to estimate the value of r (to the nearest tenth of an inch) that produces the minimum surface area.

A web applet is available to explore the relationship between the radius of a cylinder with a given volume and the surface area of the cylinder. This applet, CYLINDER, can be found on our website at **math.college.hmco.com**

Solution

a. The formula for the volume of a cylinder is $V = \pi r^2 h$, where r is the radius and h is the height. Because we are given that the volume is 21.6 cubic inches, we have

$$21.6 = \pi r^2 h$$

$$\frac{21.6}{\pi r^2} = h \qquad \text{• Solve for } h.$$

The surface area of the cylinder is given by

$$A = 2\pi r^2 + 2\pi r h$$

$$A = 2\pi r^2 + 2\pi r\left(\frac{21.6}{\pi r^2}\right) \qquad \text{• Substitute for } h.$$

$$A = 2\pi r^2 + \frac{2(21.6)}{r} \qquad \text{• Simplify.}$$

$$A = \frac{2\pi r^3 + 43.2}{r} \qquad (1)$$

b. Use Equation (1) with $y = A$ and $x = r$ and a graphing utility to determine that A is a minimum when $r \approx 1.5$ inches. See **Figure 2.38.**

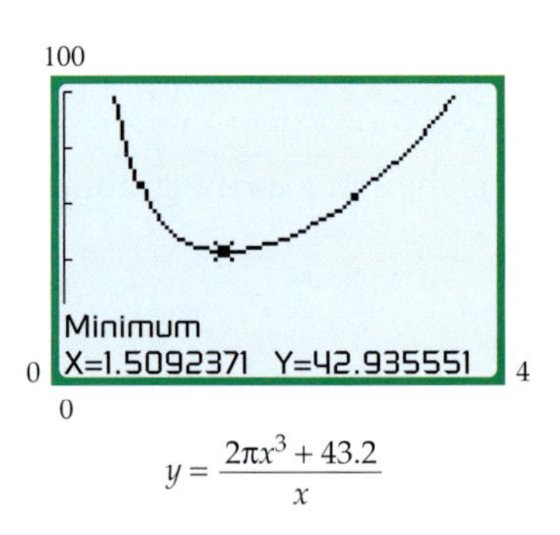

$y = \dfrac{2\pi x^3 + 43.2}{x}$

Figure 2.38

▶ TRY EXERCISE 56, PAGE 190

Topics for Discussion

1. What is a rational function? Give examples of functions that are rational functions and of functions that are not rational functions.

2. Does the graph of every rational function have at least one vertical asymptote? If so, explain why. If not, give an example of a rational function without a vertical asymptote.

3. Does the graph of every rational function have a horizontal asymptote? If so, explain why. If not, give an example of a rational function without a horizontal asymptote.

4. Can the graph of a polynomial function have a vertical asymptote? a horizontal asymptote?

Exercise Set 2.6

In Exercises 1 to 4, find all vertical asymptotes of each rational function.

1. $F(x) = \dfrac{2x - 1}{x^2 + 3x}$

▶ 2. $F(x) = \dfrac{3x^2 + 5}{x^2 - 4}$

3. $F(x) = \dfrac{x^2 + 11}{6x^2 - 5x - 4}$

4. $F(x) = \dfrac{3x - 5}{x^3 - 8}$

In Exercises 5 to 8, find the horizontal asymptote of each rational function.

5. $F(x) = \dfrac{4x^2 + 1}{x^2 + x + 1}$

▶ 6. $F(x) = \dfrac{3x^3 - 27x^2 + 5x - 11}{x^5 - 2x^3 + 7}$

7. $F(x) = \dfrac{15{,}000x^3 + 500x - 2000}{700 + 500x^3}$

8. $F(x) = 6000\left(1 - \dfrac{25}{(x + 5)^2}\right)$

In Exercises 9 to 30, determine the vertical and horizontal asymptotes and sketch the graph of the rational function *F*. Label all intercepts and asymptotes.

9. $F(x) = \dfrac{1}{x + 4}$

▶ 10. $F(x) = \dfrac{1}{x - 2}$

11. $F(x) = \dfrac{-4}{x - 3}$

12. $F(x) = \dfrac{-3}{x + 2}$

13. $F(x) = \dfrac{4}{x}$

14. $F(x) = \dfrac{-4}{x}$

15. $F(x) = \dfrac{x}{x + 4}$

16. $F(x) = \dfrac{x}{x - 2}$

17. $F(x) = \dfrac{x + 4}{2 - x}$

18. $F(x) = \dfrac{x + 3}{1 - x}$

19. $F(x) = \dfrac{1}{x^2 - 9}$

20. $F(x) = \dfrac{-2}{x^2 - 4}$

21. $F(x) = \dfrac{1}{x^2 + 2x - 3}$

22. $F(x) = \dfrac{1}{x^2 - 2x - 8}$

23. $F(x) = \dfrac{x^2}{x^2 + 4x + 4}$

24. $F(x) = \dfrac{2x^2}{x^2 - 1}$

25. $F(x) = \dfrac{10}{x^2 + 2}$

▶ 26. $F(x) = \dfrac{x^2}{x^2 - 6x + 9}$

27. $F(x) = \dfrac{2x^2 - 2}{x^2 - 9}$

28. $F(x) = \dfrac{6x^2 - 5}{2x^2 + 6}$

29. $F(x) = \dfrac{x^2 + x + 4}{x^2 + 2x - 1}$

30. $F(x) = \dfrac{2x^2 - 14}{x^2 - 6x + 5}$

In Exercises 31 to 34, find the slant asymptote of each rational function.

31. $F(x) = \dfrac{3x^2 + 5x - 1}{x + 4}$

▶ 32. $F(x) = \dfrac{x^3 - 2x^2 + 3x + 4}{x^2 - 3x + 5}$

33. $F(x) = \dfrac{x^3 - 1}{x^2}$

34. $F(x) = \dfrac{4000 + 20x + 0.0001x^2}{x}$

In Exercises 35 to 44, determine the vertical and slant asymptotes and sketch the graph of the rational function *F*.

35. $F(x) = \dfrac{x^2 - 4}{x}$

36. $F(x) = \dfrac{x^2 + 10}{2x}$

37. $F(x) = \dfrac{x^2 - 3x - 4}{x + 3}$

38. $F(x) = \dfrac{x^2 - 4x - 5}{2x + 5}$

39. $F(x) = \dfrac{2x^2 + 5x + 3}{x - 4}$

40. $F(x) = \dfrac{4x^2 - 9}{x + 3}$

41. $F(x) = \dfrac{x^2 - x}{x + 2}$

42. $F(x) = \dfrac{x^2 + x}{x - 1}$

43. $F(x) = \dfrac{x^3 + 1}{x^2 - 4}$

44. $F(x) = \dfrac{x^3 - 1}{3x^2}$

In Exercises 45 to 52, sketch the graph of the rational function *F*. (*Hint:* First examine the numerator and denominator to determine whether there are any common factors.)

45. $F(x) = \dfrac{x^2 + x}{x + 1}$

46. $F(x) = \dfrac{x^2 - 3x}{x - 3}$

47. $F(x) = \dfrac{2x^3 + 4x^2}{2x + 4}$

▶ **48.** $F(x) = \dfrac{x^2 - x - 12}{x^2 - 2x - 8}$

49. $F(x) = \dfrac{-2x^3 + 6x}{2x^2 - 6x}$

50. $F(x) = \dfrac{x^3 + 3x^2}{x(x + 3)(x - 1)}$

51. $F(x) = \dfrac{x^2 - 3x - 10}{x^2 + 4x + 4}$

52. $F(x) = \dfrac{2x^2 + x - 3}{x^2 - 2x + 1}$

53. 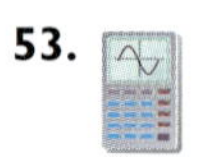  **Average Cost of Golf Balls** The cost, in dollars, of producing x golf balls is given by

$$C(x) = 0.43x + 76{,}000$$

The average cost per golf ball is given by

$$\overline{C}(x) = \frac{C(x)}{x} = \frac{0.43x + 76{,}000}{x}$$

a. Find the average cost of producing 1000, 10,000, and 100,000 golf balls.

b. What is the equation of the horizontal asymptote of the graph of $\overline{C}$? Explain the significance of the horizontal asymptote as it relates to this application.

54. **Average Cost of CD Players** The cost, in dollars, of producing x CD players is given by

$$C(x) = 0.001x^2 + 54x + 175{,}000$$

The average cost per CD player is given by

$$\overline{C}(x) = \frac{C(x)}{x} = \frac{0.001x^2 + 54x + 175{,}000}{x}$$

a. Find the average cost of producing 1000, 10,000, and 100,000 CD players.

b. What is the minimum average cost per CD player? How many CD players should be produced to minimize the average cost per CD player?

55. **Desalinization** The cost C, in dollars, to remove $p\%$ of the salt in a tank of seawater is given by

$$C(p) = \frac{2000p}{100 - p}, \quad 0 \le p < 100$$

a. Find the cost of removing 40% of the salt.

b. Find the cost of removing 80% of the salt.

c. Sketch the graph of C.

▶ **56.** **Production Costs** The cost, in dollars, of producing x cellular telephones is given by

$$C(x) = 0.0006x^2 + 9x + 401{,}000$$

The average cost per telephone is

$$\overline{C}(x) = \frac{C(x)}{x} = \frac{0.0006x^2 + 9x + 401{,}000}{x}$$

a. Find the average cost per telephone when 1000, 10,000, and 100,000 telephones are produced.

b. What is the minimum average cost per telephone? How many cellular telephones should be produced to minimize the average cost per telephone?

57.   **Wedding Expenses** The function $C(t) = 17t^2 + 128t + 5900$ models the average cost of a wedding reception, and the function $W(t) = 38t^2 + 291t + 15{,}208$ models the average cost of a wedding, where $t = 0$ represents the year 1990 and $0 \le t \le 12$. The rational function

$$R(t) = \frac{C(t)}{W(t)} = \frac{17t^2 + 128t + 5900}{38t^2 + 291t + 15{,}208}$$

gives the relative cost of the reception compared to the cost of a wedding.

a. Use $R(t)$ to estimate the relative cost of the reception compared to the cost of a wedding for the years $t = 0$, $t = 7$, and $t = 12$. Round your results to the nearest tenth of a percent.

b. According to the function $R(t)$, what percent of the total cost of a wedding, to the nearest tenth of a percent, will the cost of the reception approach as the years go by?

58. **INCOME TAX THEORY** The economist Arthur Laffer conjectured that if taxes were increased starting from very low levels, then the tax revenue received by the government would increase. But as tax rates continued to increase, there would be a point at which the tax revenue would start to decrease. The underlying concept was that if taxes were increased too much, people would not work as hard because much of their additional income would be taken from them by the increase in taxes. Laffer illustrated his concept by drawing a curve similar to the following.

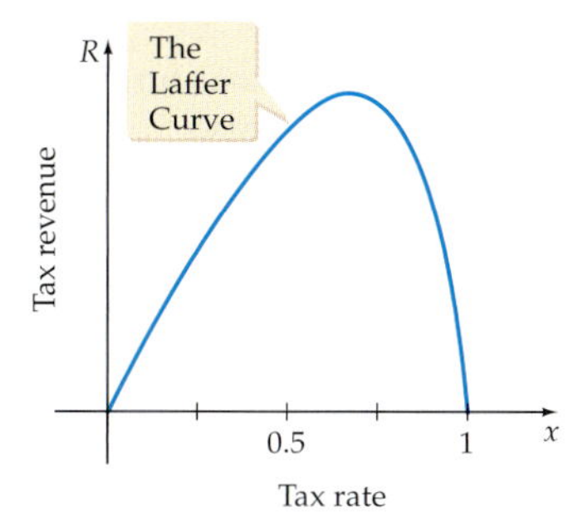

Laffer's curve shows that if the tax rate is 0%, the tax revenue will be \$0, and if the tax rate is 100%, the tax revenue also will be \$0. Laffer assumed that most people would not work if all their income went for taxes.

Most economists agree with Laffer's basic concept, but there is much disagreement about the equation of the actual tax revenue curve R and the tax rate x that will maximize the government's tax revenues.

a. Assume that Laffer's curve is given by

$$R(x) = \frac{-6.5(x^3 + 2x^2 - 3x)}{x^2 + x + 1}$$

where R is measured in trillions of dollars. Use a graphing utility to determine the tax rate x, to the nearest tenth of a percent, that would produce the maximum tax revenue. (*Hint:* Use a domain of [0, 1] and a range of [0, 4].)

b. Assume that Laffer's curve is given by

$$R(x) = \frac{-1500(x^3 + 2x^2 - 3x)}{x^2 + x + 400}$$

where R is measured in trillions of dollars. Use a graphing utility to determine the tax rate x, to the nearest tenth of a percent, that would produce the maximum tax revenue.

59. **A POPULATION MODEL** The population of a suburb, in thousands, is given by

$$P(t) = \frac{420t}{0.6t^2 + 15}$$

where t is the time in years after June 1, 1996.

a. Find the population of the suburb for $t = 1, 4$, and 10 years.

b. In what year will the population of the suburb reach its maximum?

c. What will happen to the population as $t \to \infty$?

60. **A MEDICATION MODEL** The rational function

$$M(t) = \frac{0.5t + 400}{0.04t^2 + 1}$$

models the number of milligrams of medication in the bloodstream of a patient t hours after 400 milligrams of the medication have been injected into the patient's bloodstream.

a. Find $M(5)$ and $M(10)$. Round to the nearest milligram.

b. What will M approach as $t \to \infty$?

61. **MINIMIZING SURFACE AREA** A cylindrical soft drink can is to be made so that it will have a volume of 354 milliliters. If r is the radius of the can in centimeters, then the total surface area A of the can is given by the rational function

$$A(r) = \frac{2\pi r^3 + 708}{r}$$

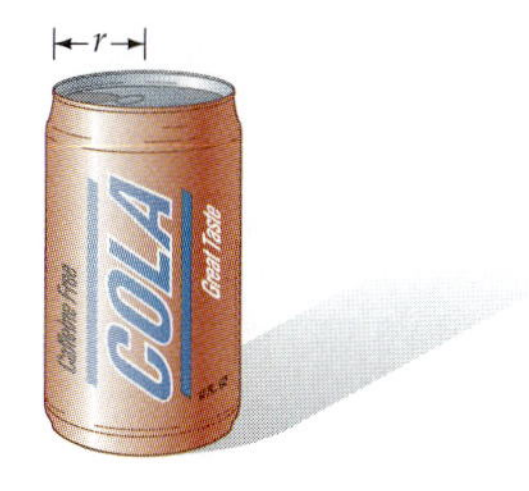

a. Graph A and use the graph to estimate (to the nearest tenth of a centimeter) the value of r that produces the minimum value of A.

b. Does the graph of A have a slant asymptote?

c. 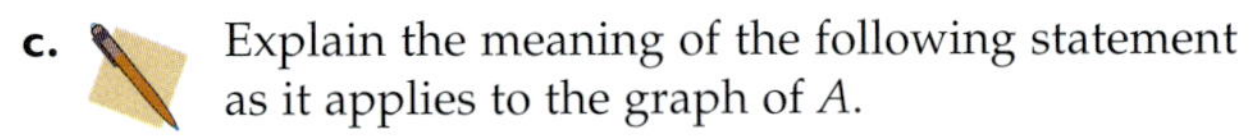 Explain the meaning of the following statement as it applies to the graph of A.

$$\text{As } r \to \infty, A \to 2\pi r^2.$$

62. **Resistors in Parallel** The electronic circuit at the right shows two resistors connected in parallel.

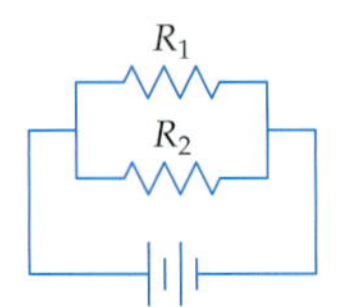

One resistor has a resistance of R_1 ohms and the other has a resistance of R_2 ohms. The total resistance for the circuit, measured in ohms, is given by the formula

$$R_T = \frac{R_1 R_2}{R_1 + R_2}$$

Assume R_1 has a fixed resistance of 10 ohms.

a. Compute R_T for $R_2 = 2$ ohms and for $R_2 = 20$ ohms.

b. What happens to R_T as $R_2 \to \infty$?

CONNECTING CONCEPTS

63. Determine the point at which the graph of

$$F(x) = \frac{2x^2 + 3x + 4}{x^2 + 4x + 7}$$

intersects its horizontal asymptote.

64. Determine the point at which the graph of

$$F(x) = \frac{3x^3 + 2x^2 - 8x - 12}{x^2 + 4}$$

intersects its slant asymptote.

65. Determine the two points at which the graph of

$$F(x) = \frac{x^3 + x^2 + 4x + 1}{x^3 + 1}$$

intersects its horizontal asymptote.

66. Give an example of a rational function that intersects its slant asymptote at two points.

67. Write a rational function that has vertical asymptotes at $x = -2$ and $x = 3$ and a horizontal asymptote at $y = 1$.

68. Write a rational function that has vertical asymptotes at $x = -3$ and $x = 1$ and a horizontal asymptote at $y = 0$.

PROJECTS

1.   **Parabolic Asymptotes** It can be shown that the rational function $F(x) = R(x)/S(x)$, where $R(x)$ and $S(x)$ have no common factors, has a parabolic asymptote provided the degree of $R(x)$ is *two* greater than the degree of $S(x)$. For instance, the rational function

$$F(x) = \frac{x^3 + 2}{x + 1}$$

has a parabolic asymptote given by $y = x^2 - x + 1$.

a. Use a graphing utility to graph $F(x)$ and the parabola given by $y = x^2 - x + 1$ in the same viewing window. Does the parabola appear to be an asymptote for the graph of F? Explain.

b. Write a paragraph that explains how to determine the equation of the parabolic asymptote for a rational function $F(x) = R(x)/S(x)$, where $R(x)$ and $S(x)$ have no common factors and the degree of $R(x)$ is two greater than the degree of $S(x)$.

c. What is the equation of the parabolic asymptote for the rational function $G(x) = \dfrac{x^4 + x^2 + 2}{x^2 - 1}$? Use a graphing utility to graph $G(x)$ and the parabolic asymptote in the same viewing window. Does the parabola appear to be an asymptote for the graph of G?

d. Create a rational function that has $y = x^2 + x + 2$ as its parabolic asymptote. Explain the procedure you used to create your rational function.

Finding Zeros of a Polynomial Using *Mathematica*

Computer algebra systems (CAS) are computer programs that are used to solve equations, graph functions, simplify algebraic expressions, and help us perform many other mathematical tasks. In this exploration, we will demonstrate how to use one of these programs, *Mathematica,* to find the zeros of a polynomial function.

Recall that a zero of a function P is a number x for which $P(x) = 0$. The idea behind finding a zero of a polynomial function by using a CAS is to solve the polynomial equation $P(x) = 0$ for x.

Two commands in *Mathematica* that can be used to solve an equation are **Solve** and **NSolve**. [*Mathematica* is sensitive about syntax (the way in which an expression is typed). You *must* use upper-case and lower-case letters as we indicate.] **Solve** will attempt to find an *exact* solution of the equation; **NSolve** attempts to find *approximate* solutions. Here are some examples.

To find the exact values of the zeros of $P(x) = x^3 + 5x^2 + 11x + 15$, input the following. *Note:* The two equals signs are necessary.

```
Solve[x^3+5x^2+11x+15==0]
```

Press Enter. The result should be

```
{{x->-3}, {x->-1-2 I}, {x->-1+2 I}}
```

Thus the three zeros of P are -3, $-1 - 2i$, and $-1 + 2i$.

To find the approximate values of the zeros of $P(x) = x^4 - 3x^3 + 4x^2 + x - 4$, input the following.

```
NSolve[x^4-3x^3+4x^2+x-4==0]
```

Press Enter. The result should be

```
{{x->-0.821746}, {x->1.2326}, {x->1.29457-1.50771 I},
{x->1.29457+1.50771 I}}
```

The four zeros are (approximately) -0.821746, 1.2326, $1.29457 - 1.50771i$, and $1.29457 + 1.50771i$.

Not all polynomial equations can be solved exactly. This means that **Solve** will not always give solutions with *Mathematica*. Consider the two examples below.

Input `NSolve[x^5-3x^3+2x^2-5==0]`

Output `{{x->-1.80492}, {x->-1.12491}, {x->0.620319-1.03589 I}, {x->0.620319+1.03589 I}, {x->1.68919}}`

These are the approximate zeros of the polynomial.

Input `Solve[x^5-3x^3+2x^2-5==0]`

Output `{ToRules[Roots[2x²-3x³+x⁵==5]]}`

In this case, no exact solution could be found. In general, there are no formulas like the quadratic formula, for instance, that yield exact solutions for fifth- or higher-degree polynomial equations.

Use *Mathematica* (or another CAS) to find the zeros of each of the following polynomial functions.

1. $P(x) = x^4 - 3x^3 + x - 5$

2. $P(x) = 3x^3 - 4x^2 + x - 3$

3. $P(x) = 4x^5 - 3x^3 + 2x^2 - x + 2$

4. $P(x) = -3x^4 - 6x^3 + 2x - 8$

CHAPTER 2 SUMMARY

2.1 Complex Numbers

- The number i, called the *imaginary unit,* is the number such that $i^2 = -1$.
- If a is a positive real number, then $\sqrt{-a} = i\sqrt{a}$. The number $i\sqrt{a}$ is called an *imaginary number.*
- A *complex number* is a number of the form $a + bi$, where a and b are real numbers and $i = \sqrt{-1}$. The number a is the *real part* of $a + bi$, and b is the *imaginary part*.
- The complex numbers $a + bi$ and $a - bi$ are called *complex conjugates* or *conjugates* of each other.
- **Operations on Complex Numbers**

$$(a + bi) + (c + di) = (a + c) + (b + d)i$$

$$(a + bi) - (c + di) = (a - c) + (b - d)i$$

$$(a + bi)(c + di) = (ac - bd) + (ad + bc)i$$

$$\frac{a + bi}{c + di} = \frac{a + bi}{c + di} \cdot \frac{c - di}{c - di}$$

- **Multiply numerator and denominator by the conjugate of the denominator.**

2.2 The Remainder Theorem and the Factor Theorem

- *The Remainder Theorem* If a polynomial function $P(x)$ is divided by $(x - c)$, then the remainder equals $P(c)$.
- *The Factor Theorem* A polynomial function $P(x)$ has a factor $(x - c)$ if and only if $P(c) = 0$.

2.3 Polynomial Functions of Higher Degree

- Characteristics and properties used in graphing polynomial functions include:
  1. Continuity—Polynomial functions are smooth continuous curves.
  2. Leading term test—Determines the behavior of the graph of a polynomial function at the far right and at the far left.
  3. The real zeros of the function determine the x-intercepts.
- *Relative Minimum and Relative Maximum* If there is an open interval I containing c on which

  $f(c) \le f(x)$ for all x in I, then $f(c)$ is a relative minimum of f.

  $f(c) \ge f(x)$ for all x in I, then $f(c)$ is a relative maximum of f.
- *The Zero Location Theorem* Let $P(x)$ be a polynomial function. If $a < b$, and if $P(a)$ and $P(b)$ have opposite signs, then there is at least one real number c between a and b such that $P(c) = 0$.

2.4 Zeros of Polynomial Functions

- Values of x that satisfy $P(x) = 0$ are called zeros of P.
- *Definition of Multiple Zeros of a Polynomial* If a polynomial function $P(x)$ has $(x - r)$ as a factor exactly k times, then r is said to be a zero of multiplicity k of the polynomial function $P(x)$.
- *The Rational Zero Theorem* If

$$P(x) = a_n x^n + a_{n-1}x^{n-1} + \cdots + a_1 x + a_0,\ a_n \ne 0$$

has integer coefficients, and $\frac{p}{q}$ (where p and q have no common factors) is a rational zero of P, then p is a factor of a_0 and q is a factor of a_n.

- *Upper- and Lower-Bound Theorem*
  Let $P(x)$ be a polynomial function with real coefficients. Use synthetic division to divide $P(x)$ by $x - b$, where b is a nonzero real number.

  Upper Bound

  a. If $b > 0$ and the leading coefficient of P is positive, then b is an upper bound for the real zeros of P provided none of the numbers in the bottom row of the synthetic division are negative.

  b. If $b > 0$ and the leading coefficient of P is negative, then b is an upper bound for the real zeros of P provided none of the numbers in the bottom row of the synthetic division are positive.

Lower Bound If $b < 0$ and the numbers in the bottom row of the synthetic division of P by $x - b$ alternate in sign, then b is a lower bound for the real zeros of P.

- *Descartes' Rule of Signs* Let $P(x)$ be a polynomial function with real coefficients and with terms arranged in order of decreasing powers of x.

  1. The number of positive real zeros of $P(x)$ is equal to the number of variations in sign of $P(x)$, or is equal to that number decreased by an even integer.
  2. The number of negative real zeros of $P(x)$ is equal to the number of variations in sign of $P(-x)$, or is equal to that number decreased by an even integer.

- The zeros of some polynomial functions with integer coefficients can be found by using the guidelines stated on page 159.

2.5 The Fundamental Theorem of Algebra

- *The Fundamental Theorem of Algebra* If $P(x)$ is a polynomial function of degree $n \geq 1$ with complex coefficients, then $P(x)$ has at least one complex zero.

- *The Conjugate Pair Theorem* If $a + bi$ $(b \neq 0)$ is a complex zero of the polynomial function $P(x)$, with real coefficients, then the conjugate $a - bi$ is also a complex zero of the polynomial function.

2.6 Graphs of Rational Functions and Their Applications

- If $P(x)$ and $Q(x)$ are polynomials, then the function F given by

$$F(x) = \frac{P(x)}{Q(x)}$$

is called a rational function.

- *General Procedure for Graphing Rational Functions That Have No Common Factors*

  1. Find the real zeros of the denominator. For each zero a, the vertical line $x = a$ will be a vertical asymptote. Use the Theorem on Horizontal Asymptotes to determine if the function has a horizontal asymptote. Graph the horizontal asymptote.
  2. Find the real zeros of the numerator. For each real zero a, plot $(a, 0)$. These points are the x-intercepts. The y-intercept of the graph of $F(x)$ is the point $(0, F(0))$.
  3. Use the tests for symmetry to determine whether the graph has symmetry with respect to the y-axis or to the origin.
  4. Find additional points that lie in the intervals between the x-intercepts and the vertical asymptotes.
  5. Determine the behavior of the graph near the asymptotes.
  6. Use the information obtained in the above steps to sketch the graph.

- *Theorem on Slant Asymptotes* The rational function given by $F(x) = P(x)/Q(x)$, where $P(x)$ and $Q(x)$ have no common factors, has a slant asymptote if the degree of the polynomial $P(x)$ in the numerator is one greater than the degree of the polynomial $Q(x)$ in the denominator.

CHAPTER 2 TRUE/FALSE EXERCISES

In Exercises 1 to 12, answer true or false. If the statement is false, explain why the statement is false or give an example to show that the statement is false.

1. The complex zeros of a polynomial function with complex coefficients always occur in conjugate pairs.
2. Descartes' Rule of Signs indicates that the polynomial function $P(x) = x^3 - x^2 + x - 1$ must have three positive zeros.
3. The polynomial $2x^5 + x^4 - 7x^3 - 5x^2 + 4x + 10$ has two variations in sign.
4. If 4 is an upper bound of the zeros of the polynomial function P, then 5 is also an upper bound of the zeros of P.
5. The graph of every rational function has a vertical asymptote.
6. The graph of the rational function $F(x) = \dfrac{x^2 - 4x + 4}{x^2 - 5x + 6}$ has a vertical asymptote of $x = 2$.
7. If 7 is a zero of the polynomial function P, then $x - 7$ is a factor of P.

8. According to the Zero Location Theorem, the polynomial function $P(x) = x^3 + 6x - 2$ has a real zero between 0 and 1.

9. Every fourth-degree polynomial function with complex coefficients has exactly four complex zeros, provided each zero is counted according to its multiplicity.

10. The graph of a rational function can have at most one horizontal asymptote.

11. Descartes' Rule of Signs indicates that the polynomial function $P(x) = x^3 + 2x^2 + 4x - 7$ does have a positive zero.

12. Every polynomial function has at least one real zero.

CHAPTER 2 REVIEW EXERCISES

In Exercises 1 and 2, write the complex number in standard form.

1. $5 + \sqrt{-64}$
2. $2 - \sqrt{-18}$

In Exercises 3 to 10, perform the indicated operation and write the answer in simplest form.

3. $(2 - 3i) + 4 + 2i$
4. $(4 + 7i) - (6 - 3i)$
5. $2i(3 - 4i)$
6. $(4 - 3i)(2 + 7i)$
7. $(3 + i)^2$
8. i^{345}
9. $\dfrac{4 - 6i}{2i}$
10. $\dfrac{2 - 5i}{3 + 4i}$

In Exercises 11 to 14, use the quadratic formula to solve each quadratic equation.

11. $9x^2 - 12x = -5$
12. $16x^2 - 24x + 13 = 0$
13. $x^2 - x = -1$
14. $4x^2 - 8x = -5$

In Exercises 15 to 20, use synthetic division to divide the first polynomial by the second.

15. $4x^3 - 11x^2 + 5x - 2, x - 3$
16. $5x^3 - 18x + 2, x - 1$
17. $3x^3 - 5x + 1, x + 2$
18. $2x^3 + 7x^2 + 16x - 10, x - \dfrac{1}{2}$
19. $3x^3 - 10x^2 - 36x + 55, x - 5$
20. $x^4 + 9x^3 + 6x^2 - 65x - 63, x + 7$

In Exercises 21 to 24, use the Remainder Theorem to find $P(c)$.

21. $P(x) = x^3 + 2x^2 - 5x + 1, c = 4$
22. $P(x) = -4x^3 - 10x + 8, c = -1$
23. $P(x) = 6x^4 - 12x^2 + 8x + 1, c = -2$
24. $P(x) = 5x^5 - 8x^4 + 2x^3 - 6x^2 - 9, c = 3$

In Exercises 25 to 28, use synthetic division to show that c is a zero of the given polynomial function.

25. $P(x) = x^3 + 2x^2 - 26x + 33, c = 3$
26. $P(x) = 2x^4 + 8x^3 - 8x^2 - 31x + 4, c = -4$
27. $P(x) = x^5 - x^4 - 2x^2 + x + 1, c = 1$
28. $P(x) = 2x^3 + 3x^2 - 8x + 3, c = \dfrac{1}{2}$

In Exercises 29 to 34, graph the polynomial function.

29. $P(x) = x^3 - x$
30. $P(x) = -x^3 - x^2 + 8x + 12$
31. $P(x) = x^4 - 6$
32. $P(x) = x^5 - x$
33. $P(x) = x^4 - 10x^2 + 9$
34. $P(x) = x^5 - 5x^3$

In Exercises 35 to 40, use the Rational Zero Theorem to list all possible rational zeros for each polynomial function.

35. $P(x) = x^3 - 7x - 6$
36. $P(x) = 2x^3 + 3x^2 - 29x - 30$

37. $P(x) = 15x^3 - 91x^2 + 4x + 12$

38. $P(x) = x^4 - 12x^3 + 52x^2 - 96x + 64$

39. $P(x) = x^3 + x^2 - x - 1$

40. $P(x) = 6x^5 + 3x - 2$

In Exercises 41 to 44, use Descartes' Rule of Signs to state the number of possible positive and negative real zeros of each polynomial function.

41. $P(x) = x^3 + 3x^2 + x + 3$

42. $P(x) = x^4 - 6x^3 - 5x^2 + 74x - 120$

43. $P(x) = x^4 - x - 1$

44. $P(x) = x^5 - 4x^4 + 2x^3 - x^2 + x - 8$

In Exercises 45 to 50, find the zeros of the polynomial function.

45. $P(x) = x^3 + 6x^2 + 3x - 10$

46. $P(x) = x^3 - 10x^2 + 31x - 30$

47. $P(x) = 6x^4 + 35x^3 + 72x^2 + 60x + 16$

48. $P(x) = 2x^4 + 7x^3 + 5x^2 + 7x + 3$

49. $P(x) = x^4 - 4x^3 + 6x^2 - 4x + 1$

50. $P(x) = 2x^3 - 7x^2 + 22x + 13$

In Exercises 51 and 52, use the given zero to find the remaining zeros of each polynomial function.

51. $P(x) = x^4 - 4x^3 + 6x^2 - 4x - 15;\ 1 - 2i$

52. $P(x) = x^4 - x^3 - 17x^2 + 55x - 50;\ 2 + i$

53. Find a third-degree polynomial function with integer coefficients and zeros of 4, -3, and $\frac{1}{2}$.

54. Find a fourth-degree polynomial function with zeros of 2, -3, i, and $-i$.

55. Find a fourth-degree polynomial function with real coefficients that has zeros of 1, 2, and $5i$.

56. Find a fourth-degree polynomial function with real coefficients that has -2 as a zero of multiplicity 2 and also has $1 + 3i$ as a zero.

In Exercises 57 to 60, find the vertical, horizontal, and slant asymptotes for each rational function.

57. $f(x) = \dfrac{3x + 5}{x + 2}$

58. $f(x) = \dfrac{2x^2 + 12x + 2}{x^2 + 2x - 3}$

59. $f(x) = \dfrac{2x^2 + 5x + 11}{x + 1}$

60. $f(x) = \dfrac{6x^2 - 1}{2x^2 + x + 7}$

In Exercises 61 to 68, graph each rational function.

61. $f(x) = \dfrac{3x - 2}{x}$

62. $f(x) = \dfrac{x + 4}{x - 2}$

63. $f(x) = \dfrac{6}{x^2 + 2}$

64. $f(x) = \dfrac{4x^2}{x^2 + 1}$

65. $f(x) = \dfrac{2x^3 - 4x + 6}{x^2 - 4}$

66. $f(x) = \dfrac{x}{x^3 - 1}$

67. $f(x) = \dfrac{3x^2 - 6}{x^2 - 9}$

68. $f(x) = \dfrac{-x^3 + 6}{x^2}$

69. **Average Cost of Skateboards** The cost, in dollars, of producing x skateboards is given by

$$C(x) = 5.75x + 34{,}200$$

The average cost per skateboard is given by

$$\overline{C}(x) = \frac{C(x)}{x} = \frac{5.75x + 34{,}200}{x}$$

a. Find the average cost per skateboard, to the nearest cent, of producing 5000 and 50,000 skateboards.

b. What is the equation of the horizontal asymptote of the graph of $\overline{C}$? Explain the significance of the horizontal asymptote as it relates to this application.

70. **Food Temperature** The temperature F, in degrees Fahrenheit, of a dessert placed in a freezer for t hours is given by the rational function

$$F(t) = \frac{60}{t^2 + 2t + 1}, \quad t \geq 0$$

a. Find the temperature of the dessert after it has been in the freezer for 1 hour.

b. Find the temperature of the dessert after 4 hours.

c. What temperature will the dessert approach as $t \to \infty$?

71. **PHYSIOLOGY** One of Poiseuille's Laws states that the resistance R encountered by blood flowing through a blood vessel is given by

$$R(r) = C\frac{L}{r^4}$$

where C is a positive constant determined by the viscosity of the blood, L is the length of the blood vessel, and r is its radius.

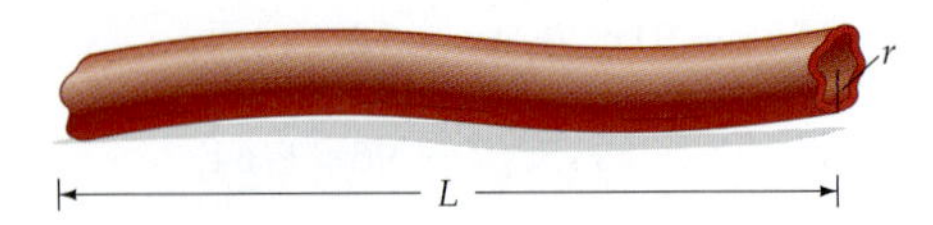

a. Explain the meaning of $R(r) \to \infty$ as $r \to 0$.

b. Explain the meaning of $R(r) \to 0$ as $r \to \infty$.

CHAPTER 2 TEST

In Exercises 1 and 2, simplify and write each complex number in standard form.

1. a. $2 - \sqrt{-54}$ **b.** $3(2 - 5i) - i(3 + 2i)$

2. a. $(2 - 5i)(3 - 7i)$ **b.** $\dfrac{4 + 2i}{1 - i}$

3. Use synthetic division to divide:

$$(3x^3 + 5x^2 + 4x - 1) \div (x + 2)$$

4. Use the Remainder Theorem to find $P(-2)$ if

$$P(x) = -3x^3 + 7x^2 + 2x - 5$$

5. Show that $x - 1$ is a factor of

$$x^4 - 4x^3 + 7x^2 - 6x + 2$$

6. Find the real solutions of $3x^3 + 7x^2 - 6x = 0$.

7. Use the Zero Location Theorem to verify that

$$P(x) = 2x^3 - 3x^2 - x + 1$$

has a zero between 1 and 2.

8. Find the zeros of

$$P(x) = (x^2 - 4)^2(2x - 3)(x + 1)^3$$

and state the multiplicity of each.

9. Use the Rational Zero Theorem to list the possible rational zeros of

$$P(x) = 6x^3 - 3x^2 + 2x - 3$$

10. Find, by using the Upper- and Lower-Bound Theorem, the smallest positive integer and the largest negative integer that are upper and lower bounds for the real zeros of the polynomial function

$$P(x) = 2x^4 + 5x^3 - 23x^2 - 38x + 24$$

11. Use Descartes' Rule of Signs to state the number of possible positive and negative real zeros of

$$P(x) = x^4 - 3x^3 + 2x^2 - 5x + 1$$

12. Find the zeros of $P(x) = 2x^3 - 3x^2 - 11x + 6$

13. Given that $2 + 3i$ is a zero of

$$P(x) = 6x^4 - 5x^3 + 12x^2 + 207x + 130$$

find the remaining zeros.

14. Find all the zeros of

$$P(x) = x^5 - 6x^4 + 14x^3 - 14x^2 + 5x$$

15. Find a polynomial of smallest degree that has real coefficients and zeros $1 + i$, 3, and 0.

16. a. Find all vertical asymptotes of the graph of

$$f(x) = \frac{3x^2 - 2x + 1}{x^2 - 5x + 6}$$

b. Find the horizontal asymptote of the graph of

$$f(x) = \frac{3x^2 - 2x + 1}{2x^2 - 1}$$

17. Graph $f(x) = \dfrac{x^2 - 1}{x^2 - 2x - 3}$. Use an open circle to show the hole in the graph of f.

18. Graph $f(x) = \dfrac{2x^2 + 2x + 1}{x + 1}$ and label the slant asymptote with its equation.

19. The rational function

$$w(t) = \frac{70t + 120}{t + 40}, \quad t \geq 0$$

models Rene's typing speed, in words per minute, after t hours of typing lessons.

a. Find $w(1)$, $w(10)$, and $w(20)$. Round to the nearest word per minute.

b. What will Rene's typing speed approach as $t \to \infty$?

20. **MAXIMIZING VOLUME** You are to construct an open box from a rectangular sheet of cardboard that measures 18 inches by 25 inches. To assemble the box, make the four cuts shown in the figure at the right and then fold on the dashed lines. What value of x (to the nearest hundredth of an inch) will produce a box with maximum volume? What is the maximum volume (to the nearest tenth of a cubic inch)?

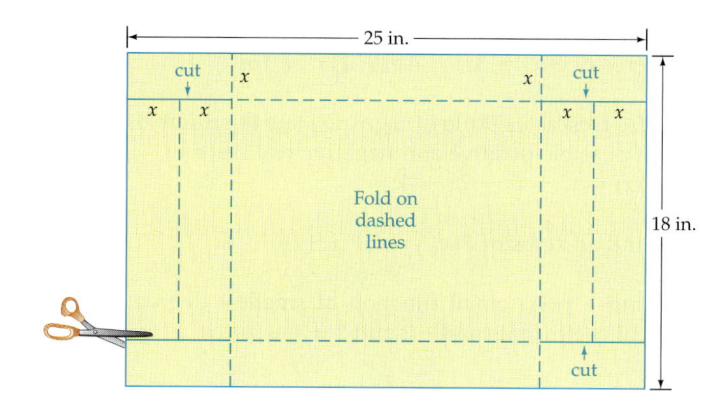

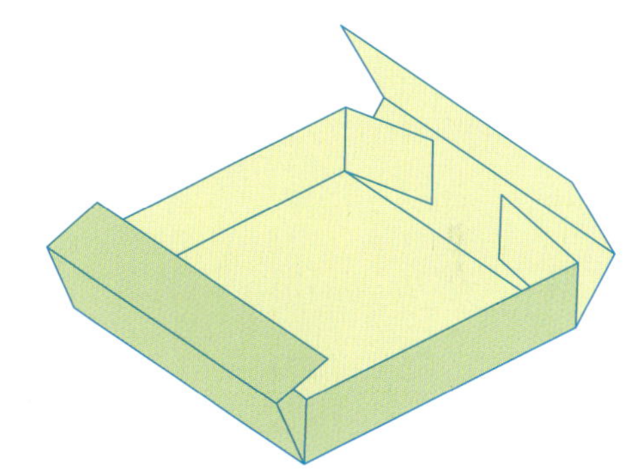

CUMULATIVE REVIEW EXERCISES

1. Write $\dfrac{3 + 4i}{1 - 2i}$ in $a + bi$ form.

2. Use the quadratic formula to solve $x^2 - x - 1 = 0$.

3. Use the quadratic formula to solve $2x^2 - 3x + 2 = 0$.

4. Find the midpoint of the line segment with endpoints $(2, -5)$ and $(6, 8)$.

5. Find the distance between the points $(2, 5)$ and $(7, -11)$.

6. Explain how to use the graph of $y = x^2$ to produce the graph of $y = (x - 2)^2 + 4$.

7. Find the difference quotient for the function $P(x) = x^2 - 2x - 3$.

8. Given $f(x) = 2x^2 + 5x - 3$ and $g(x) = 4x - 7$, find $(f \circ g)(x)$.

9. Given $f(x) = x^3 - 2x + 7$ and $g(x) = x^2 - 3x - 4$, find $(f - g)(x)$.

10. Use synthetic division to divide $(4x^4 - 2x^2 - 4x - 5)$ by $(x + 2)$.

11. Use the Remainder Theorem to find $P(3)$ for $P(x) = 2x^4 - 3x^2 + 4x - 6$.

12. Determine the far-right behavior of the graph of $P(x) = -3x^4 - x^2 + 7x - 6$.

13. Determine the relative maximum of the polynomial function $P(x) = -3x^3 - x^2 + 4x - 1$. Round to the nearest ten thousandth.

14. Use the Rational Zero Theorem to list all possible rational zeros of $P(x) = 3x^4 - 4x^3 - 11x^2 + 16x - 4$.

15. Use Descartes' Rule of Signs to state the number of possible positive and negative real zeros of $P(x) = x^3 + x^2 + 2x + 4$.

16. Find all zeros of $P(x) = x^3 + x + 10$.

17. Find a polynomial function of smallest degree that has real coefficients and -2 and $3 + i$ as zeros.

18. Write $P(x) = x^3 - 2x^2 + 9x - 18$ as a product of linear factors.

19. Determine the vertical and horizontal asymptotes of the graph of $F(x) = \dfrac{4x^2}{x^2 + x - 6}$.

20. Find the equation of the slant asymptote for the graph of $F(x) = \dfrac{x^3 + 4x^2 + 1}{x^2 + 4}$.

CHAPTER 3

EXPONENTIAL AND LOGARITHMIC FUNCTIONS

Modeling Data with an Exponential Function

The following table shows the time, in hours, before the body of a scuba diver, wearing a 5-millimeter-thick wet suit, reaches hypothermia (95°F) for various water temperatures.

Water Temperature, °F	Time, hours
36	1.5
41	1.8
46	2.6
50	3.1
55	4.9

Source: Data extracted from the *American Journal of Physics*, vol. 71, no. 4 (April 2003), Fig. 3, p. 336.

The following function, which is an example of an exponential function, closely models the data in the table:

$$T(F) = 0.1509(1.0639)^F$$

In this function F represents the Fahrenheit temperature of the water, and T represents the time in hours. A diver can use the function to determine the time it takes to reach hypothermia for water temperatures that are not included in the table.

Exponential functions can be used to model many other situations. **Exercise 43 on page 226** uses an exponential function to estimate the growth of broadband Internet connections.

Use Two Methods to Solve and Compare Results

Sometimes it is possible to solve a problem in two or more ways. In such situations it is recommended that you use at least two methods to solve the problem, and compare your results. Here is an example of an application that can be solved in more than one way.

Example

In a league of eight basketball teams, each team plays every other team in the league exactly once. How many league games will take place?

Solution

Method 1: *Use an analytic approach.* Each of the eight teams must play the other seven teams. Using this information, you might be tempted to conclude that there will be $8 \cdot 7 = 56$ games, but this result is too large because it counts each game between two individual teams as two different games. Thus the number of league games will be

$$\frac{8 \cdot 7}{2} = \frac{56}{2} = 28$$

Method 2: *Make an organized list.* Use the letters A, B, C, D, E, F, G, and H to represent the eight teams. Use the notation AB to represent the game between team A and team B. Do not include BA in your list because it represents the same game between team A and team B.

AB	AC	AD	AE	AF	AG	AH
	BC	BD	BE	BF	BG	BH
		CD	CE	CF	CG	CH
			DE	DF	DG	DH
				EF	EG	EH
					FG	FH
						GH

The list shows that there will be 28 league games.

The procedure of using two different solution methods and comparing results is employed often in this chapter. For instance, see **Example 2, page 257.** In this example, a solution is found by applying algebraic procedures and also by graphing. Notice that both methods produce the same result.

SECTION 3.1

INVERSE FUNCTIONS

- Introduction to Inverse Functions
- Graphs of Inverse Functions
- Composition of a Function and Its Inverse
- Find an Inverse Function
- Application

Introduction to Inverse Functions

Consider the "doubling" function $f(x) = 2x$ that doubles every input. Some of the ordered pairs of this function are

$$\left\{(-4, -8), (-1.5, -3), (1, 2), \left(\frac{5}{3}, \frac{10}{3}\right), (7, 14)\right\}$$

Now consider the "halving" function $g(x) = \frac{1}{2}x$ that takes one-half of every input. Some of the ordered pairs of this function are

$$\left\{(-8, -4), (-3, -1.5), (2, 1), \left(\frac{10}{3}, \frac{5}{3}\right), (14, 7)\right\}$$

Observe that the coordinates of the ordered pairs of g are the reverse of the coordinates of the ordered pairs of f. This is always the case for f and g. Here are two more examples.

$$f(5) = 2(5) = 10 \qquad g(10) = \frac{1}{2}(10) = 5$$

Ordered pair: (5, 10) **Ordered pair: (10, 5)**

$$f(a) = 2(a) = 2a \qquad g(2a) = \frac{1}{2}(2a) = a$$

Ordered pair: (a, $2a$) **Ordered pair: ($2a$, a)**

For these functions, f and g are called *inverse functions* of one another.

Inverse Function

If the coordinates of the ordered pairs of a function g are the reverse of the coordinates of the ordered pairs of a function f, then g is said to be the **inverse function** of f.

take note

It is important to remember the information in the paragraph at the right. If f is a function and g is the inverse of f, then

Domain of g = range of f

and

Range of g = domain of f

Because the coordinates of the ordered pairs of the inverse function g are the reverse of the coordinates of the ordered pairs of the function f, the domain of g is the range of f, and the range of g is the domain of f.

Not all functions have an inverse that is a function. Consider, for instance, the "square" function $S(x) = x^2$. Some of the ordered pairs of S are

$$\{(-3, 9), (-1, 1), (0, 0), (1, 1), (3, 9), (5, 25)\}$$

If we reverse the coordinates of the ordered pairs, we have

$$\{(9, -3), (1, -1), (0, 0), (1, 1), (9, 3), (25, 5)\}$$

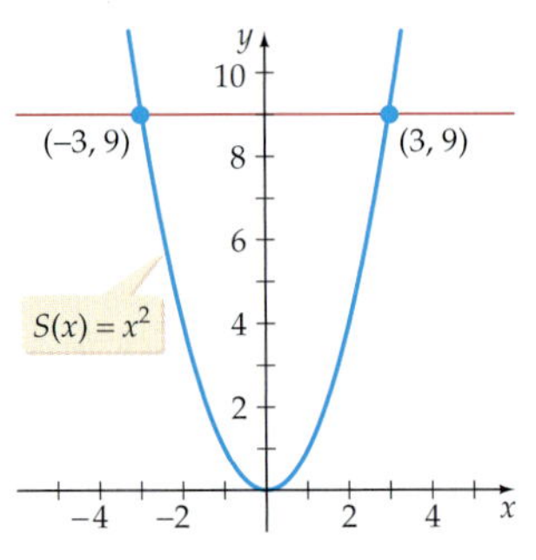

FIGURE 3.1

This set of ordered pairs is not a function because there are ordered pairs, for instance (9, −3) and (9, 3), with the same first coordinate and different second coordinates. In this case, S has an inverse *relation* but not an inverse *function.*

A graph of S is shown in **Figure 3.1.** Note that $x = -3$ and $x = 3$ produce the same value of y. Thus the graph of S fails the horizontal line test, and therefore S is not a one-to-one function. This observation is used in the following theorem.

Condition for an Inverse Function

A function f has an inverse function if and only if f is a one-to-one function.

Recall that increasing functions or decreasing functions are one-to-one functions. Thus we can state the following theorem.

Alternative Condition for an Inverse Function

If f is an increasing function or a decreasing function, then f has an inverse function.

QUESTION Which of the functions graphed below has an inverse function?

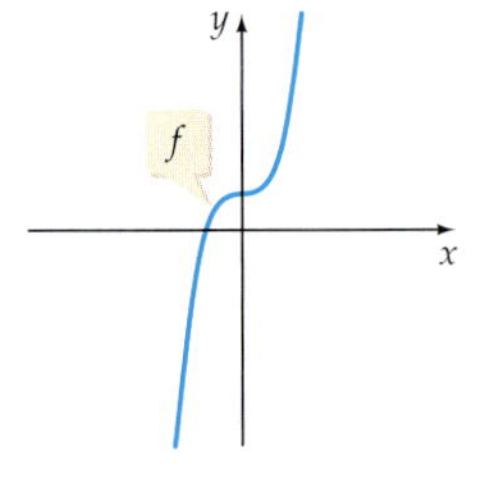

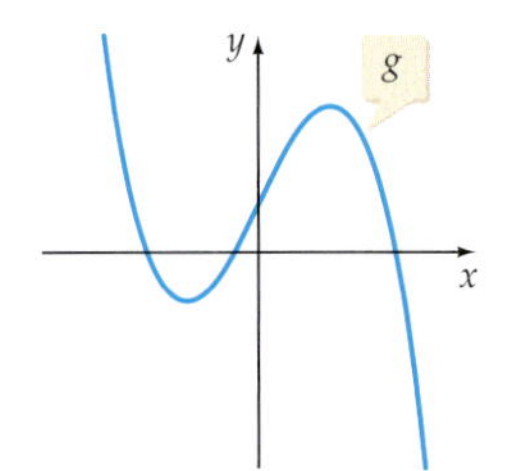

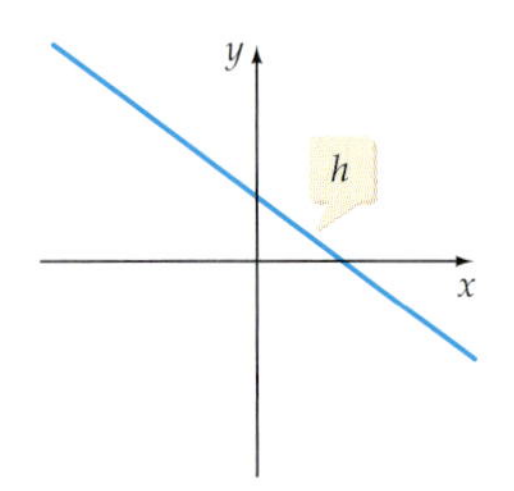

take note

$f^{-1}(x)$ does not mean $\frac{1}{f(x)}$. For $f(x) = 2x$, $f^{-1}(x) = \frac{1}{2}x$ but $\frac{1}{f(x)} = \frac{1}{2x}$.

If a function g is the inverse of a function f, we usually denote the inverse function by f^{-1} rather than g. For the doubling and halving functions f and g discussed on page 203, we write

$$f(x) = 2x \qquad f^{-1}(x) = \frac{1}{2}x$$

Graphs of Inverse Functions

Because the coordinates of the ordered pairs of the inverse of a function f are the reverse of the coordinates of f, we can use them to create a graph of f^{-1}.

ANSWER The graph of f is the graph of an increasing function. Therefore, f is a one-to-one function and has an inverse function. The graph of h is the graph of a decreasing function. Therefore, h is a one-to-one function and has an inverse function. The graph of g is not the graph of a one-to-one function. g does not have an inverse function.

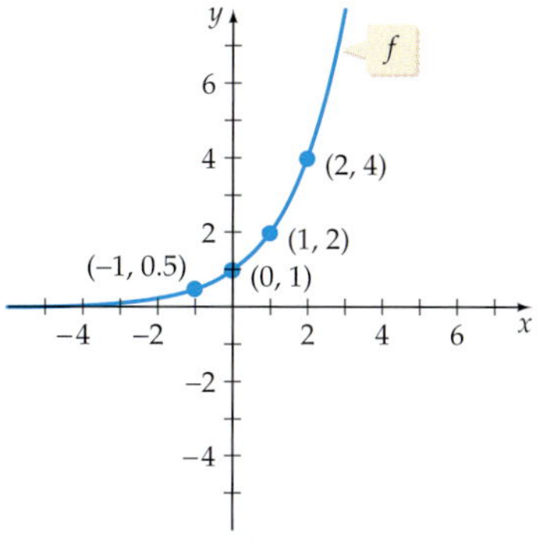

FIGURE 3.2

EXAMPLE 1 Sketch the Graph of the Inverse of a Function

Sketch the graph of f^{-1} given that f is the function shown in **Figure 3.2.**

Solution

Because the graph of f passes through $(-1, 0.5)$, $(0, 1)$, $(1, 2)$, and $(2, 4)$, the graph of f^{-1} must pass through $(0.5, -1)$, $(1, 0)$, $(2, 1)$, and $(4, 2)$. Plot the points and then draw a smooth graph through the points, as shown in **Figure 3.3.**

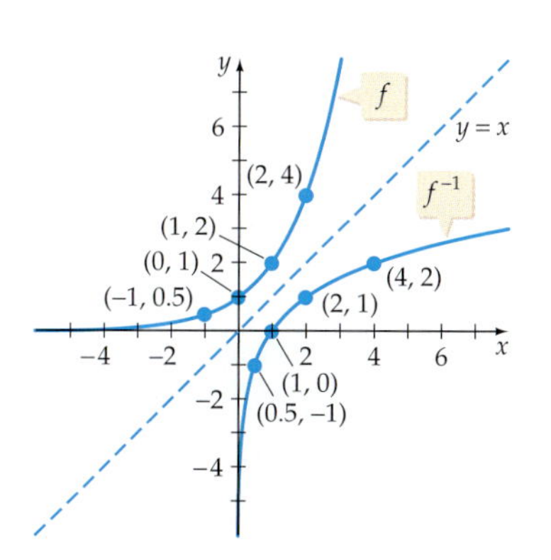

FIGURE 3.3

TRY EXERCISE 10, PAGE 212

The graph from the solution to Example 1 is shown again in **Figure 3.4.** Note that the graph of f^{-1} is symmetric to the graph of f with respect to the graph of $y = x$. If the graph were folded along the dashed line, the graph of f would lie on top of the graph of f^{-1}. This is a characteristic of all graphs of functions and their inverses. In **Figure 3.5,** although S does not have an inverse that is a function, the graph of the inverse relation S^{-1} is symmetric to S with respect to the graph of $y = x$.

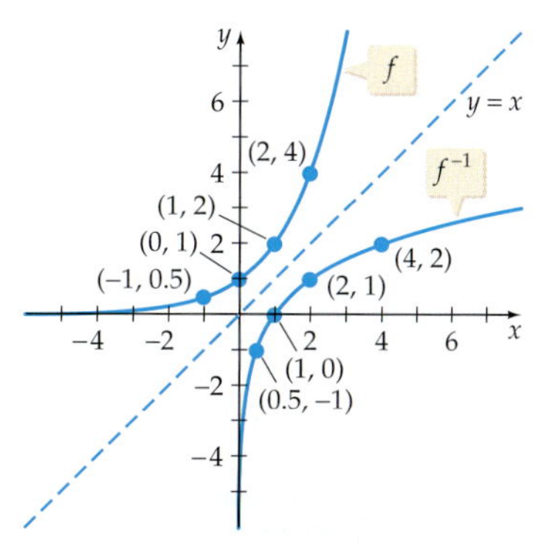

FIGURE 3.4

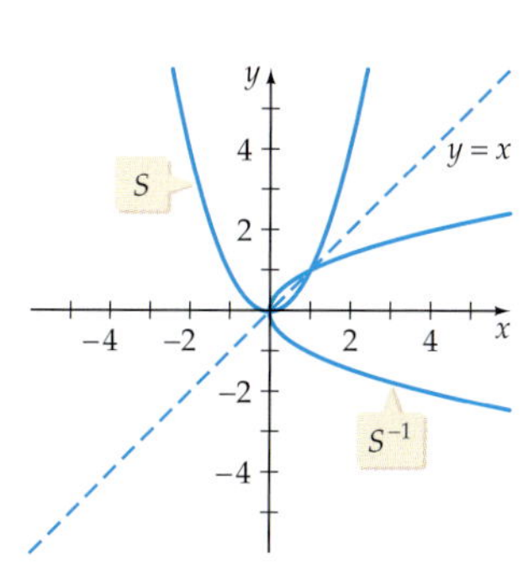

FIGURE 3.5

Composition of a Function and Its Inverse

Observe the effect, as shown below, of taking the composition of functions that are inverses of one another.

$$f(x) = 2x \qquad\qquad g(x) = \frac{1}{2}x$$

$$f[g(x)] = 2\left[\frac{1}{2}x\right] \quad \text{• Replace } x \text{ by } g(x). \qquad g[f(x)] = \frac{1}{2}[2x] \quad \text{• Replace } x \text{ by } f(x).$$

$$f[g(x)] = x \qquad\qquad g[f(x)] = x$$

This property of the composition of inverse functions always holds true. When taking the composition of inverse functions, the inverse function reverses the effect of the original function. For the two functions above, f doubles a number, and g halves a number. If you double a number and then take one-half of the result, you are back to the original number.

take note

If we think of a function as a machine, then the Composition of Inverse Functions Property can be represented as shown below. Take any input x for f. Use the output of f as the input for f^{-1}. The result is the original input, x.

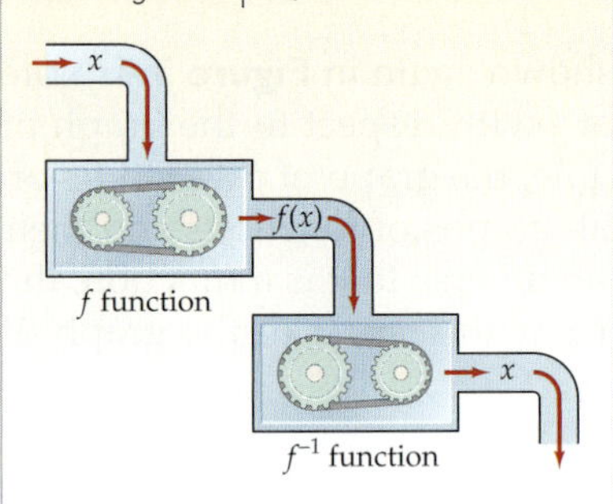

Composition of Inverse Functions Property

If f is a one-to-one function, then f^{-1} is the inverse function of f if and only if

$$(f \circ f^{-1})(x) = f[f^{-1}(x)] = x \qquad \text{for all } x \text{ in the domain of } f^{-1}$$

and

$$(f^{-1} \circ f)(x) = f^{-1}[f(x)] = x \qquad \text{for all } x \text{ in the domain of } f.$$

EXAMPLE 2 Use the Composition of Inverse Functions Property

Use composition of functions to show that $f^{-1}(x) = 3x - 6$ is the inverse function of $f(x) = \frac{1}{3}x + 2$.

Solution

We must show that $f[f^{-1}(x)] = x$ and $f^{-1}[f(x)] = x$.

$$f(x) = \frac{1}{3}x + 2 \qquad\qquad f^{-1}(x) = 3x - 6$$

$$f[f^{-1}(x)] = \frac{1}{3}[3x - 6] + 2 \qquad f^{-1}[f(x)] = 3\left[\frac{1}{3}x + 2\right] - 6$$

$$f[f^{-1}(x)] = x \qquad\qquad f^{-1}[f(x)] = x$$

▶ TRY EXERCISE 20, PAGE 213

INTEGRATING TECHNOLOGY

In the standard viewing window of a calculator, the distance between two tic marks on the x-axis is not equal to the distance between two tic marks on the y-axis. As a result, the graph of $y = x$ does not appear to bisect the first and third quadrants. See **Figure 3.6.** This anomaly is important if a graphing calculator is being used to check whether two functions are inverses of one another. Because the graph of $y = x$ does not appear to bisect the first and third quadrants, the graphs of f and f^{-1} will not appear to be symmetric about the graph of $y = x$. The graphs of $f(x) = \dfrac{1}{3}x + 2$ and $f^{-1}(x) = 3x - 6$ from Example 2 are shown in **Figure 3.7.** Notice that the graphs do not appear to be quite symmetric about the graph of $y = x$.

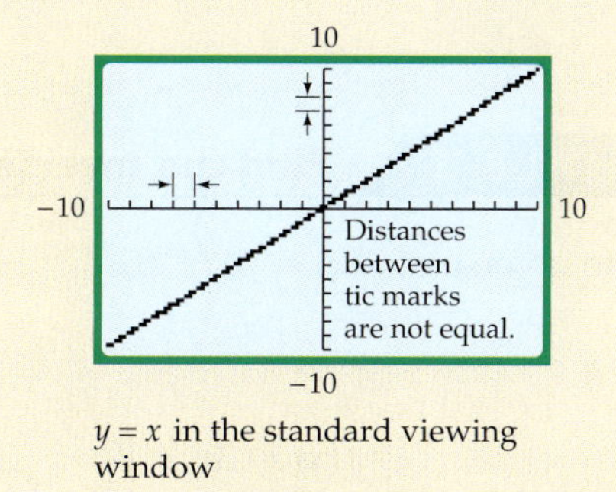

$y = x$ in the standard viewing window

FIGURE 3.6

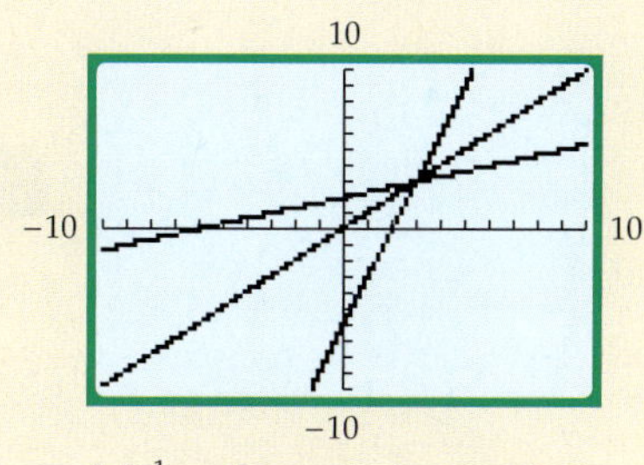

f, f^{-1}, and $y = x$ in the standard viewing window

FIGURE 3.7

To get a better view of a function and its inverse, it is necessary to use the SQUARE viewing window, as in **Figure 3.8.** In this window, the distance between two tic marks on the x-axis is equal to the distance between two tic marks on the y-axis.

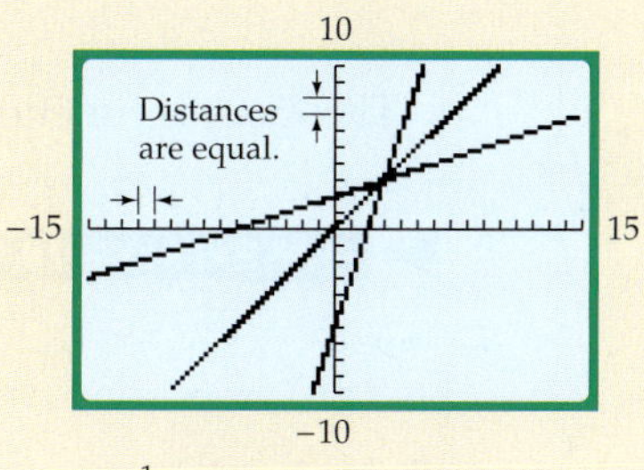

f, f^{-1}, and $y = x$ in a square viewing window

FIGURE 3.8

Find an Inverse Function

If a one-to-one function f is defined by an equation, then we can use the following method to find the equation for f^{-1}.

take note

If the ordered pairs of f are given by (x, y), then the ordered pairs of f^{-1} are given by (y, x). That is, x and y are interchanged. This is the reason for Step 2 at the right.

Steps for Finding the Inverse of a Function

To find the equation of the inverse f^{-1} of the one-to-one function f:

1. Substitute y for $f(x)$.
2. Interchange x and y.
3. Solve, if possible, for y in terms of x.
4. Substitute $f^{-1}(x)$ for y.

EXAMPLE 3 Find the Inverse of a Function

Find the inverse of $f(x) = 3x + 8$.

Solution

$$f(x) = 3x + 8$$

$$y = 3x + 8 \qquad \text{• Replace } f(x) \text{ by } y.$$

$$x = 3y + 8 \qquad \text{• Interchange } x \text{ and } y.$$

$$x - 8 = 3y \qquad \text{• Solve for } y.$$

$$\frac{x - 8}{3} = y$$

$$\frac{1}{3}x - \frac{8}{3} = f^{-1}(x) \qquad \text{• Replace } y \text{ by } f^{-1}(x).$$

The inverse function is given by $f^{-1}(x) = \dfrac{1}{3}x - \dfrac{8}{3}$.

▶ TRY EXERCISE 28, PAGE 213

EXAMPLE 4 Find the Inverse of a Function

Find the inverse of $f(x) = \dfrac{2x + 1}{x}, x \neq 0$.

Solution

$$f(x) = \frac{2x + 1}{x}$$

$$y = \frac{2x + 1}{x}$$ • Replace $f(x)$ by y.

$$x = \frac{2y + 1}{y}$$ • Interchange x and y.

$$xy = 2y + 1$$ • Solve for y.

$$xy - 2y = 1$$

$$y(x - 2) = 1$$ • Factor the left side.

$$y = \frac{1}{x - 2}$$

$$f^{-1}(x) = \frac{1}{x - 2},\ x \neq 2$$ • Replace y by $f^{-1}(x)$.

▶ TRY EXERCISE 34, PAGE 213

? QUESTION If f is a one-to-one function and $f(4) = 5$, what is $f^{-1}(5)$?

The graph of $f(x) = x^2 + 4x + 3$ is shown in **Figure 3.9a.** The function f is not a one-to-one function and therefore does not have an inverse function. However, the function given by $G(x) = x^2 + 4x + 3$, shown in **Figure 3.9b,** for which the domain is restricted to $\{x \mid x \geq -2\}$, is a one-to-one function and has an inverse function G^{-1}. This is shown in Example 5.

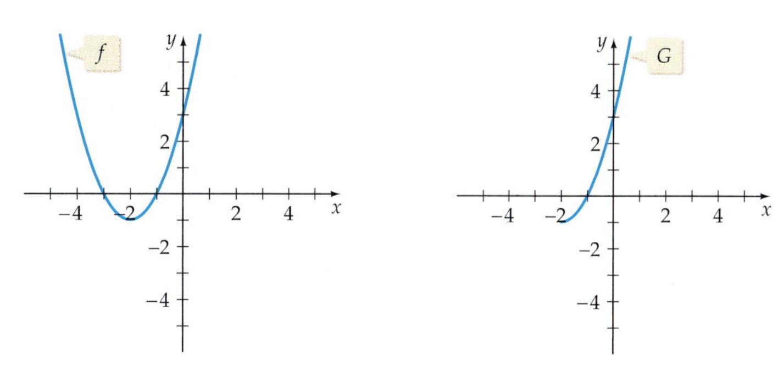

FIGURE 3.9a **FIGURE 3.9b**

? ANSWER Because f^{-1} is the inverse function of f, the coordinates of the ordered pairs of f^{-1} are the reverse of the coordinates of the ordered pairs of f. Therefore, $f^{-1}(5) = 4$.

take note

Recall that the range of a function f is the domain of f^{-1}, and the domain of f is the range of f^{-1}.

EXAMPLE 5 Find the Inverse of a Function with a Restricted Domain

Find the inverse of $G(x) = x^2 + 4x + 3$, where the domain of G is $\{x \mid x \geq -2\}$.

Solution

$$G(x) = x^2 + 4x + 3$$

$$y = x^2 + 4x + 3$$ • Replace $G(x)$ by y.

$$x = y^2 + 4y + 3$$ • Interchange x and y.

$$x = (y^2 + 4y + 4) - 4 + 3$$ • Solve for y by completing the square of $y^2 + 4y$.

$$x = (y + 2)^2 - 1$$ • Factor.

$$x + 1 = (y + 2)^2$$ • Add 1 to each side of the equation.

$$\sqrt{x + 1} = \sqrt{(y + 2)^2}$$ • Take the square root of each side of the equation.

$$\pm\sqrt{x + 1} = y + 2$$ • Recall that if $a^2 = b$, then $a = \pm\sqrt{b}$.

$$\pm\sqrt{x + 1} - 2 = y$$

Because the domain of G is $\{x \mid x \geq -2\}$, the range of G^{-1} is $\{y \mid y \geq -2\}$. This means that we must choose the positive value of $\pm\sqrt{x + 1}$. Thus $G^{-1}(x) = \sqrt{x + 1} - 2$. See **Figure 3.10.**

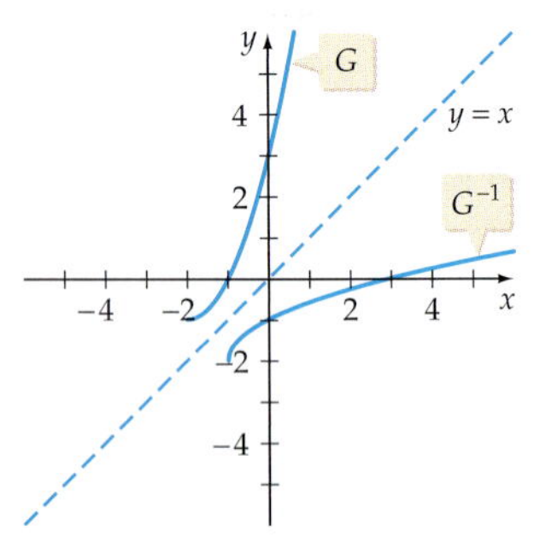

FIGURE 3.10

TRY EXERCISE 40, PAGE 213

APPLICATION

There are practical applications of finding the inverse of a function. Here is one in which a shirt size in the United States is converted to a shirt size in Italy. Finding the inverse function gives the function that converts a shirt size in Italy to a shirt size in the United States.

EXAMPLE 6 Solve an Application

The function $IT(x) = 2x + 8$ converts a men's shirt size x in the United States to the equivalent shirt size in Italy.

a. Use IT to determine the equivalent Italian shirt size for a size 16.5 U.S. shirt.

b. Find IT^{-1} and use IT^{-1} to determine the U.S. men's shirt size that is equivalent to an Italian shirt size of 36.

Solution

a. $IT(16.5) = 2(16.5) + 8 = 33 + 8 = 41$

A size 16.5 U.S. shirt is equivalent to a size 41 Italian shirt.

b. To find the inverse function, begin by substituting y for $IT(x)$.

$$IT(x) = 2x + 8$$
$$y = 2x + 8$$
$$x = 2y + 8 \qquad \text{• Interchange } x \text{ and } y.$$
$$x - 8 = 2y \qquad \text{• Solve for } y.$$
$$\frac{x - 8}{2} = y$$

In inverse notation, the above equation can be written as

$$IT^{-1}(x) = \frac{x - 8}{2} \qquad \text{or} \qquad IT^{-1}(x) = \frac{1}{2}x - 4$$

Substitute 36 for x to find the equivalent U.S. shirt size.

$$IT^{-1}(36) = \frac{1}{2}(36) - 4 = 18 - 4 = 14$$

A size 36 Italian shirt is equivalent to a size 14 U.S. shirt.

TRY EXERCISE 50, PAGE 214

INTEGRATING TECHNOLOGY

Some graphing utilities can be used to draw the graph of the inverse of a function without the user having to find the inverse function. For instance, **Figure 3.11** shows the graph of $f(x) = 0.1x^3 - 4$. The graphs of f and f^{-1} are both shown in **Figure 3.12,** along with the graph of $y = x$. Note that the graph of f^{-1} is the reflection of the graph of f with respect to the graph of $y = x$. The display shown in **Figure 3.12** was produced on a TI-83 graphing calculator by using the DrawInv command, which is in the DRAW menu.

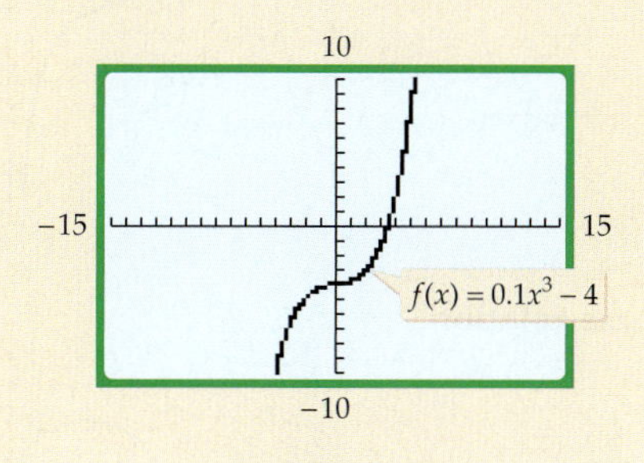

FIGURE 3.11

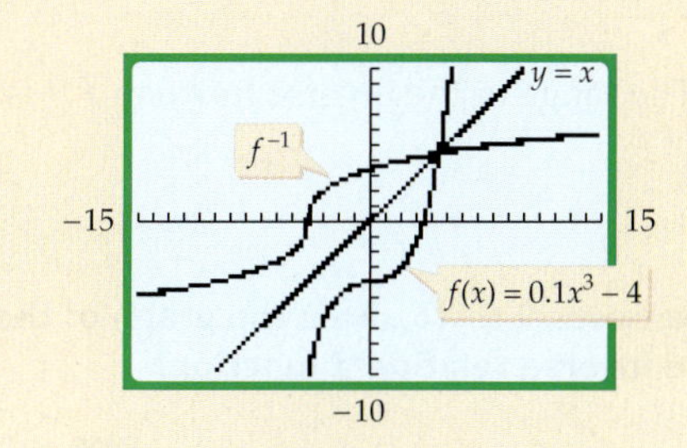

FIGURE 3.12

TOPICS FOR DISCUSSION

1. If $f(x) = 3x + 1$, what are the values of $f^{-1}(2)$ and $[f(2)]^{-1}$?
2. How are the domain and range of a one-to-one function f related to the domain and range of the inverse function of f?

3. How is the graph of the inverse of a function f related to the graph of f?
4. The function $f(x) = -x$ is its own inverse. Find at least two other functions that are their own inverses.
5. What are the steps in finding the inverse of a one-to-one function?

EXERCISE SET 3.1

In Exercises 1 to 4, assume that the given function has an inverse function.

1. Given $f(3) = 7$, find $f^{-1}(7)$.
2. Given $g(-3) = 5$, find $g^{-1}(5)$.
3. Given $h^{-1}(-3) = -4$, find $h(-4)$.
4. Given $f^{-1}(7) = 0$, find $f(0)$.
5. If 3 is in the domain of f^{-1}, find $f[f^{-1}(3)]$.
6. If f is a one-to-one function and $f(0) = 5$, $f(1) = 2$, and $f(2) = 7$, find:

   a. $f^{-1}(5)$ **b.** $f^{-1}(2)$
7. The domain of the inverse function f^{-1} is the ________ of f.
8. The range of the inverse function f^{-1} is the ________ of f.

In Exercises 9 to 16, draw the graph of the inverse relation. Is the inverse relation a function?

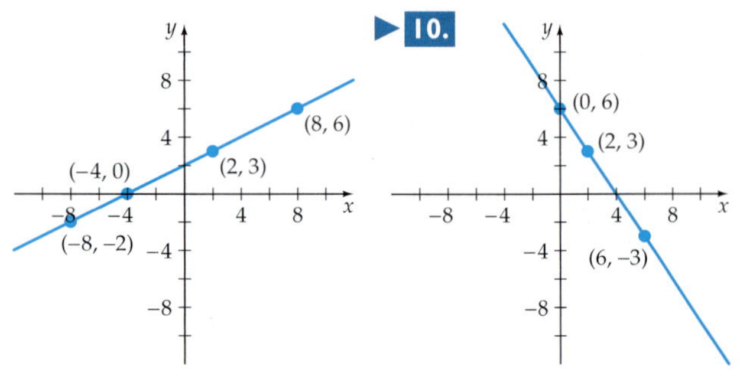

11.

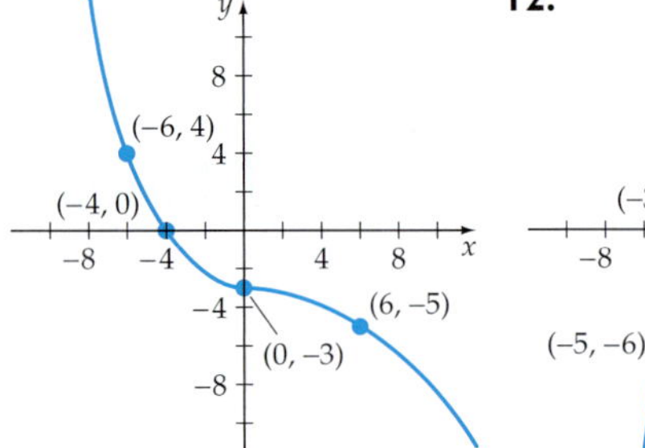

12.

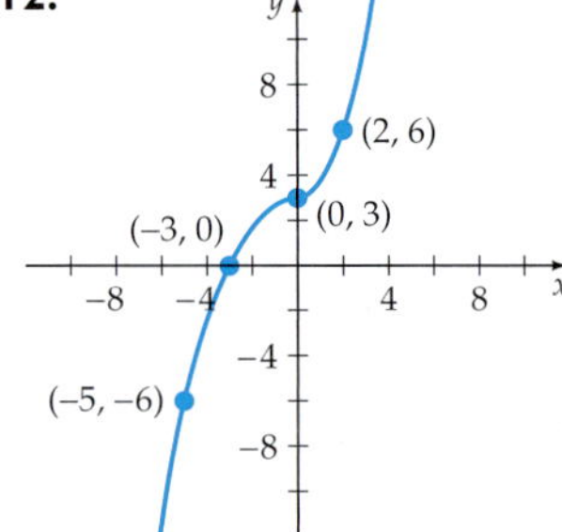

13.

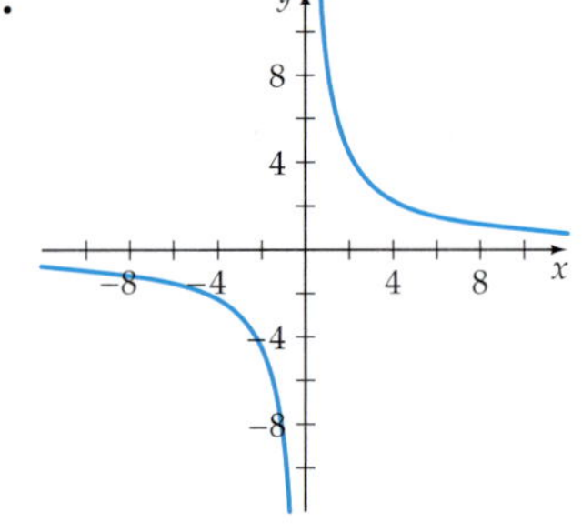

14.

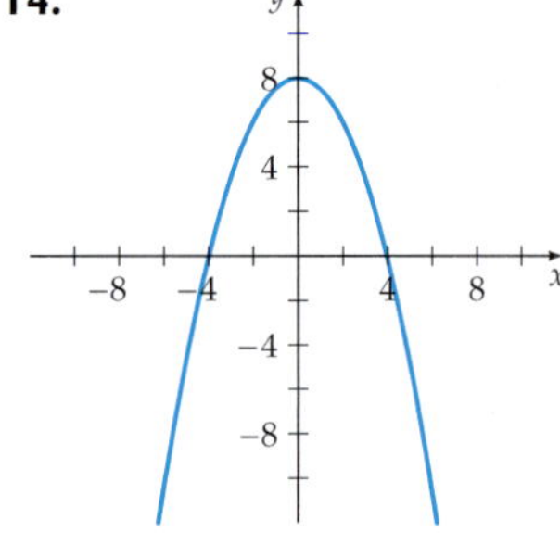

15.

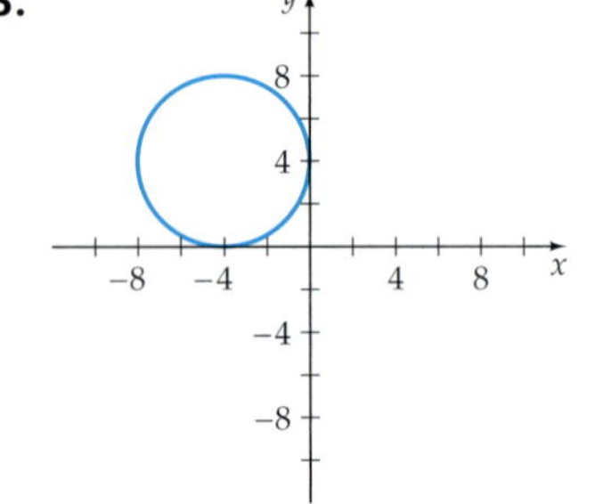

16.

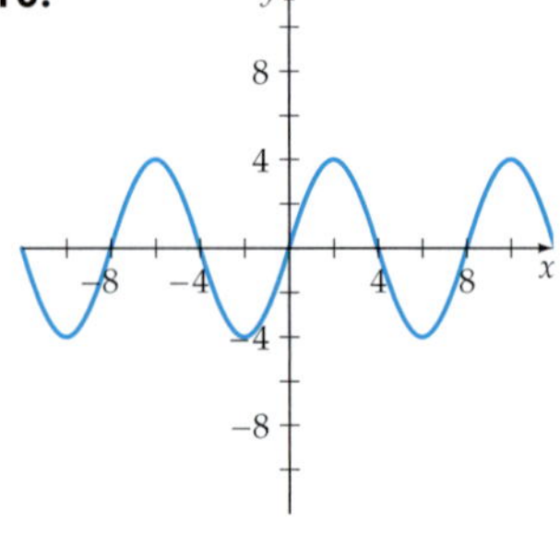

In Exercises 17 to 22, use composition of functions to determine whether f and g are inverses of one another.

17. $f(x) = 4x; g(x) = \dfrac{x}{4}$

18. $f(x) = 3x; g(x) = \dfrac{1}{3x}$

19. $f(x) = 4x - 1; g(x) = \dfrac{1}{4}x + \dfrac{1}{4}$

20. $f(x) = \dfrac{1}{2}x - \dfrac{3}{2}; g(x) = 2x + 3$

21. $f(x) = -\dfrac{1}{2}x - \dfrac{1}{2}; g(x) = -2x + 1$

22. $f(x) = 3x + 2; g(x) = \dfrac{1}{3}x - \dfrac{2}{3}$

In Exercises 23 to 26, find the inverse of the function. If the function does not have an inverse function, write "no inverse function."

23. $\{(-3, 1), (-2, 2), (1, 5), (4, -7)\}$

24. $\{(-5, 4), (-2, 3), (0, 1), (3, 2), (7, 11)\}$

25. $\{(0, 1), (1, 2), (2, 4), (3, 8), (4, 16)\}$

26. $\{(1, 0), (10, 1), (100, 2), (1000, 3), (10{,}000, 4)\}$

In Exercises 27 to 44, find $f^{-1}(x)$. State any restrictions on the domain of $f^{-1}(x)$.

27. $f(x) = 2x + 4$

28. $f(x) = 4x - 8$

29. $f(x) = 3x - 7$

30. $f(x) = -3x - 8$

31. $f(x) = -2x + 5$

32. $f(x) = -x + 3$

33. $f(x) = \dfrac{2x}{x - 1}, \quad x \neq 1$

34. $f(x) = \dfrac{x}{x - 2}, \quad x \neq 2$

35. $f(x) = \dfrac{x - 1}{x + 1}, \quad x \neq -1$

36. $f(x) = \dfrac{2x - 1}{x + 3}, \quad x \neq -3$

37. $f(x) = x^2 + 1, \quad x \geq 0$

38. $f(x) = x^2 - 4, \quad x \geq 0$

39. $f(x) = \sqrt{x - 2}, \quad x \geq 2$

40. $f(x) = \sqrt{4 - x}, \quad x \leq 4$

41. $f(x) = x^2 + 4x, \quad x \geq -2$

42. $f(x) = x^2 - 6x, \quad x \leq 3$

43. $f(x) = x^2 + 4x - 1, \quad x \leq -2$

44. $f(x) = x^2 - 6x + 1, \quad x \geq 3$

45. **Geometry** The volume of a cube is given by $V(x) = x^3$, where x is the measure of the length of a side of the cube. Find $V^{-1}(x)$ and explain what it represents.

46. **Unit Conversions** The function $f(x) = 12x$ converts feet, x, into inches, $f(x)$. Find $f^{-1}(x)$ and explain what it determines.

47. **Unit Conversions** A conversion function such as the one in Exercise 46 converts a measurement in one unit into another unit. Is a conversion function always a one-to-one function? Does a conversion function always have an inverse function? Explain your answer.

48. **Grading Scale** Does the grading scale function given below have an inverse function? Explain your answer.

Score	Grade
90–100	A
80–89	B
70–79	C
60–69	D
0–59	F

49. **Fashion** The function $s(x) = 2x + 24$ can be used to convert a U.S. women's shoe size into an Italian women's shoe size. Determine the function $s^{-1}(x)$ that can be used to convert an Italian women's shoe size to its equivalent U.S. shoe size.

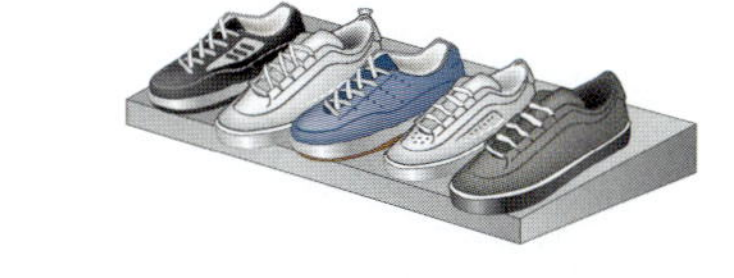

50. **FASHION** The function $K(x) = 1.3x - 4.7$ converts a men's shoe size in the United States to the equivalent shoe size in the United Kingdom. Determine the function $K^{-1}(x)$ that can be used to convert a United Kingdom men's shoe size to its equivalent U.S. shoe size.

51. **COMPENSATION** The monthly earnings $E(s)$, in dollars, of a software sales executive is given by $E(s) = 0.05s + 2500$, where s is the value, in dollars, of the software sold by the executive during the month. Find $E^{-1}(s)$, and explain how the executive could use this function.

52. **POSTAGE** Does the first-class postage rate function given below have an inverse function? Explain your answer.

Weight (in ounces)	Cost
$0 < w \le 1$	\$.37
$1 < w \le 2$	\$.60
$2 < w \le 3$	\$.83
$3 < w \le 4$	\$1.06

53. **INTERNET COMMERCE** Functions and their inverses can be used to create secret codes that are used to secure business transactions made over the Internet. Let A = 10, B = 11, . . . , and Z = 35. Let $f(x) = 2x - 1$ define a coding function. Code the word MATH (M—22, A—10, T—29, H—17), which is 22102917, by finding $f(22102917)$. Now find the inverse of f and show that applying f^{-1} to the output of f returns the original word.

54. **CRYPTOGRAPHY** A friend is using the letter-number correspondence in Exercise 53 and the coding function $f(x) = 2x + 3$. Suppose this friend sends you the coded message 5658602671. Decode this message.

In Exercises 55 to 60, answer the question without finding the equation of the linear function.

55. Suppose that f is a linear function, $f(2) = 7$, and $f(5) = 12$. If $f(4) = c$, then is c less than 7, between 7 and 12, or greater than 12? Explain your answer.

56. Suppose that f is a linear function, $f(1) = 13$, and $f(4) = 9$. If $f(3) = c$, then is c less than 9, between 9 and 13, or greater than 13? Explain your answer.

57. Suppose that f is a linear function, $f(2) = 3$, and $f(5) = 9$. Between which two numbers is $f^{-1}(6)$?

58. Suppose that f is a linear function, $f(5) = -1$, and $f(9) = -3$. Between which two numbers is $f^{-1}(-2)$?

59. Suppose that g is a linear function, $g^{-1}(3) = 4$, and $g^{-1}(7) = 8$. Between which two numbers is $g(5)$?

60. Suppose that g is a linear function, $g^{-1}(-2) = 5$, and $g^{-1}(0) = -3$. Between which two numbers is $g(0)$?

CONNECTING CONCEPTS

In Exercises 61 and 62, find the inverse of the given function.

61. $f(x) = ax + b, \quad a \neq 0$

62. $f(x) = ax^2 + bx + c, \quad a \neq 0, \quad x \ge -\dfrac{b}{2a}$

63. Use a graph of $f(x) = -x + 3$ to explain why f is its own inverse.

64. Use a graph of $f(x) = \sqrt{16 - x^2}$, with $0 \le x \le 4$, to explain why f is its own inverse.

Only one-to-one functions have inverses that are functions. In Exercises 65 to 68, determine if the given function is a one-to-one function.

65. $p(t) = \sqrt{9 - t}$

66. $v(t) = \sqrt{16 + t}$

67. $F(x) = |x| + x$

68. $T(x) = |x^2 - 6|, \quad x \ge 0$

PREPARE FOR SECTION 3.2

69. Evaluate: 2^3 [A.1]

70. Evaluate: 3^{-4} [A.1]

71. Evaluate: $\dfrac{2^2 + 2^{-2}}{2}$ [A.1]

72. Evaluate: $\dfrac{3^2 - 3^{-2}}{2}$ [A.1]

73. Evaluate $f(x) = 10^x$ for $x = -1, 0, 1,$ and 2. [1.3]

74. Evaluate $f(x) = \left(\dfrac{1}{2}\right)^x$ for $x = -1, 0, 1,$ and 2. [1.3]

PROJECTS

1. **INTERSECTION POINTS FOR THE GRAPHS OF f AND f^{-1}** For each of the following, graph f and its inverse.

i. $f(x) = 2x - 4$

ii. $f(x) = -x + 2$

iii. $f(x) = x^3 + 1$

iv. $f(x) = x - 3$

v. $f(x) = -3x + 2$

vi. $f(x) = \dfrac{1}{x}$

a. Do the graphs of a function and its inverse always intersect?

b. If the graphs of a function and its inverse intersect at one point, what is true about the coordinates of the point of intersection?

c. Can the graphs of a function and its inverse intersect at more than one point?

SECTION 3.2 EXPONENTIAL FUNCTIONS AND THEIR APPLICATIONS

- EXPONENTIAL FUNCTIONS
- GRAPHS OF EXPONENTIAL FUNCTIONS
- THE NATURAL EXPONENTIAL FUNCTION
- APPLICATIONS OF EXPONENTIAL FUNCTIONS

EXPONENTIAL FUNCTIONS

In 1965, Gordon Moore, one of the cofounders of Intel Corporation, observed that the maximum number of transistors that could be placed on a microprocessor seemed to be doubling every 18 to 24 months. **Table 3.1** below shows how the maximum number of transistors on various Intel processors has changed over time. (*Source:* Intel Museum home page.)

TABLE 3.1

Year	1971	1979	1983	1985	1990	1993	1995	1998	2000
Number of transistors per microprocessor (in thousands)	2.3	31	110	280	1200	3100	5500	14,000	42,000

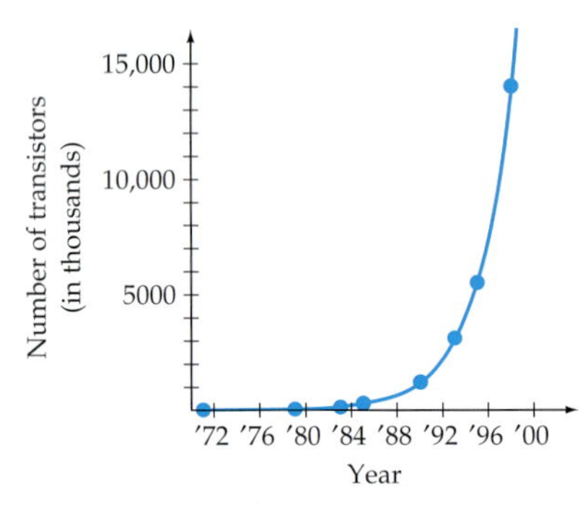

FIGURE 3.13
Moore's Law

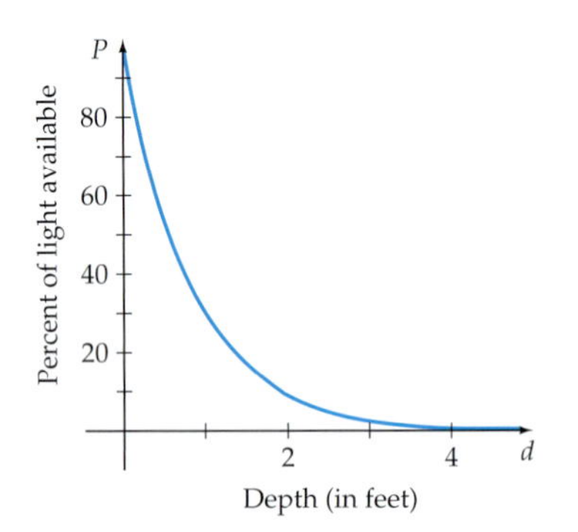

FIGURE 3.14

The curve that approximately passes through the points is a mathematical model of the data. See **Figure 3.13.** The model is based on an *exponential* function.

When light enters water, the intensity of the light decreases with the depth of the water. The graph in **Figure 3.14** shows a model, for Lake Michigan, of the decrease in the percentage of available light as the depth of the water increases. This model is also based on an exponential function.

Definition of an Exponential Function

The **exponential function with base *b*** is defined by

$$f(x) = b^x$$

where $b > 0$, $b \neq 1$, and x is a real number.

The base b of $f(x) = b^x$ is required to be positive. If the base were a negative number, the value of the function would be a complex number for some values of x. For instance, if $b = -4$ and $x = \frac{1}{2}$, then $f\left(\frac{1}{2}\right) = (-4)^{1/2} = 2i$. To avoid complex number values of a function, the base of any exponential function must be a nonnegative number. Also, b is defined such that $b \neq 1$ because $f(x) = 1^x = 1$ is a constant function.

In the following examples we evaluate $f(x) = 2^x$ at $x = 3$ and $x = -2$.

$$f(3) = 2^3 = 8 \qquad f(-2) = 2^{-2} = \frac{1}{2^2} = \frac{1}{4}$$

To evaluate the exponential function $f(x) = 2^x$ at an irrational number such as $x = \sqrt{2}$, we use a rational approximation of $\sqrt{2}$, such as 1.4142, and a calculator to obtain an approximation of the function. For instance, if $f(x) = 2^x$, then $f(\sqrt{2}) = 2^{\sqrt{2}} \approx 2^{1.4142} \approx 2.6651$.

EXAMPLE 1 Evaluate an Exponential Function

Evaluate $f(x) = 3^x$ at $x = 2$, $x = -4$, and $x = \pi$.

Solution

$f(2) = 3^2 = 9$

$f(-4) = 3^{-4} = \frac{1}{3^4} = \frac{1}{81}$

$f(\pi) = 3^\pi \approx 3^{3.1415927} \approx 31.54428$ • **Evaluate with the aid of a calculator.**

▶ **TRY EXERCISE 2, PAGE 224**

GRAPHS OF EXPONENTIAL FUNCTIONS

The graph of $f(x) = 2^x$ is shown in **Figure 3.15.** The coordinates of some of the points on the curve are given in **Table 3.2.**

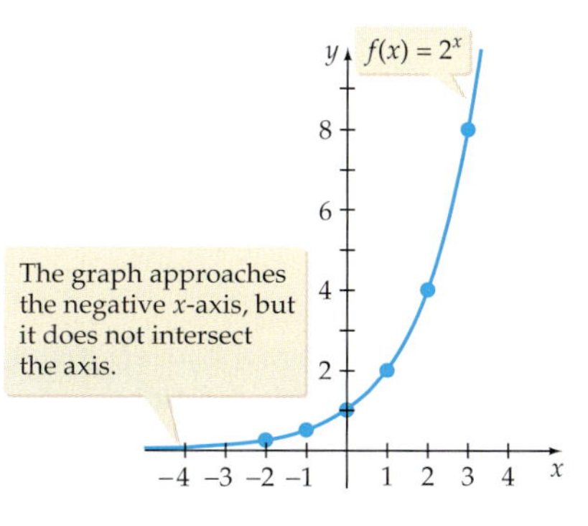

FIGURE 3.15

TABLE 3.2

x	$y = f(x) = 2^x$	(x, y)
−2	$f(-2) = 2^{-2} = \frac{1}{4}$	$\left(-2, \frac{1}{4}\right)$
−1	$f(-1) = 2^{-1} = \frac{1}{2}$	$\left(-1, \frac{1}{2}\right)$
0	$f(0) = 2^0 = 1$	$(0, 1)$
1	$f(1) = 2^1 = 2$	$(1, 2)$
2	$f(2) = 2^2 = 4$	$(2, 4)$
3	$f(3) = 2^3 = 8$	$(3, 8)$

Note the following properties of the graph of the exponential function $f(x) = 2^x$.

- The y-intercept is $(0, 1)$.
- The graph passes through $(1, 2)$.
- As x decreases without bound (that is, as $x \to -\infty$), $f(x) \to 0$.
- The graph is a smooth continuous increasing curve.

Now consider the graph of an exponential function for which the base is between 0 and 1. The graph of $f(x) = \left(\frac{1}{2}\right)^x$ is shown in **Figure 3.16.** The coordinates of some of the points on the curve are given in **Table 3.3.**

TABLE 3.3

x	$y = f(x) = \left(\frac{1}{2}\right)^x$	(x, y)
−3	$f(-3) = \left(\frac{1}{2}\right)^{-3} = 8$	$(-3, 8)$
−2	$f(-2) = \left(\frac{1}{2}\right)^{-2} = 4$	$(-2, 4)$
−1	$f(-1) = \left(\frac{1}{2}\right)^{-1} = 2$	$(-1, 2)$
0	$f(0) = \left(\frac{1}{2}\right)^{0} = 1$	$(0, 1)$
1	$f(1) = \left(\frac{1}{2}\right)^{1} = \frac{1}{2}$	$\left(1, \frac{1}{2}\right)$
2	$f(2) = \left(\frac{1}{2}\right)^{2} = \frac{1}{4}$	$\left(2, \frac{1}{4}\right)$

The graph approaches the positive x-axis, but it does not intersect the axis.

$f(x) = \left(\frac{1}{2}\right)^x$

FIGURE 3.16

Note the following properties of the graph of $f(x) = \left(\frac{1}{2}\right)^x$ in **Figure 3.16.**

- The y-intercept is $(0, 1)$.
- The graph passes through $\left(1, \frac{1}{2}\right)$.

- As x increases without bound, the y-values decrease toward 0. That is, as $x \to \infty$, $f(x) \to 0$.
- The graph is a smooth continuous decreasing curve.

The basic properties of exponential functions are provided in the following summary.

Properties of $f(x) = b^x$

For positive real numbers b, $b \neq 1$, the exponential function defined by $f(x) = b^x$ has the following properties:

1. The function f is a one-to-one function. It has the set of real numbers as its domain and the set of positive real numbers as its range.
2. The graph of f is a smooth continuous curve with a y-intercept of $(0, 1)$, and the graph passes through $(1, b)$.
3. If $b > 1$, f is an increasing function and the graph of f is asymptotic to the negative x-axis. [As $x \to \infty$, $f(x) \to \infty$, and as $x \to -\infty$, $f(x) \to 0$.] See **Figure 3.17a.**
4. If $0 < b < 1$, f is a decreasing function and the graph of f is asymptotic to the positive x-axis. [As $x \to -\infty$, $f(x) \to \infty$, and as $x \to \infty$, $f(x) \to 0$.] See **Figure 3.17b.**

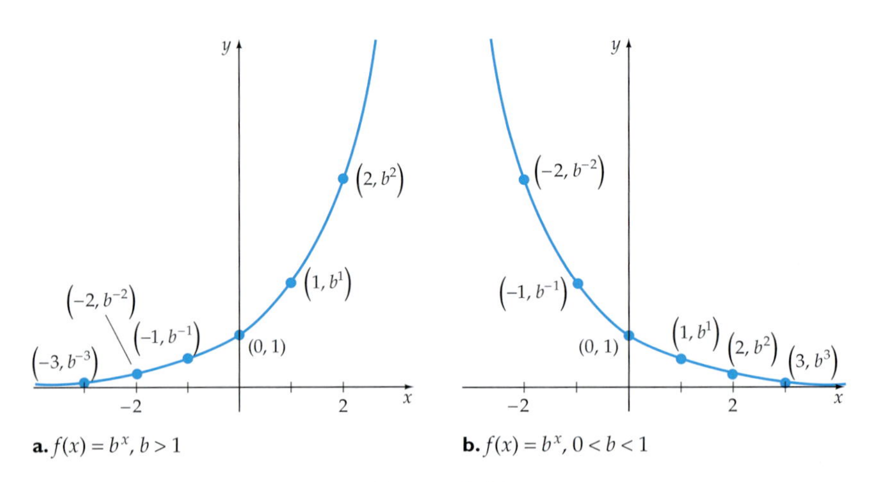

a. $f(x) = b^x$, $b > 1$ **b.** $f(x) = b^x$, $0 < b < 1$

FIGURE 3.17

QUESTION What is the x-intercept of the graph of $f(x) = \left(\frac{1}{3}\right)^x$?

ANSWER The graph does not have an x-intercept. As x increases, the graph approaches the x-axis, but it does not intersect the x-axis.

EXAMPLE 2 Graph an Exponential Function

Graph $g(x) = \left(\frac{3}{4}\right)^x$.

Solution

Because the base $\frac{3}{4}$ is less than 1, we know that the graph of g is a decreasing function that is asymptotic to the positive x-axis. The y-intercept of the graph is the point $(0, 1)$, and the graph also passes through $\left(1, \frac{3}{4}\right)$. Plot a few additional points (see **Table 3.4**), and then draw a smooth curve through the points as in **Figure 3.18.**

TABLE 3.4

x	$y = g(x) = \left(\frac{3}{4}\right)^x$	(x, y)
−3	$\left(\frac{3}{4}\right)^{-3} = \frac{64}{27}$	$\left(-3, \frac{64}{27}\right)$
−2	$\left(\frac{3}{4}\right)^{-2} = \frac{16}{9}$	$\left(-2, \frac{16}{9}\right)$
−1	$\left(\frac{3}{4}\right)^{-1} = \frac{4}{3}$	$\left(-1, \frac{4}{3}\right)$
2	$\left(\frac{3}{4}\right)^{2} = \frac{9}{16}$	$\left(2, \frac{9}{16}\right)$
3	$\left(\frac{3}{4}\right)^{3} = \frac{27}{64}$	$\left(3, \frac{27}{64}\right)$

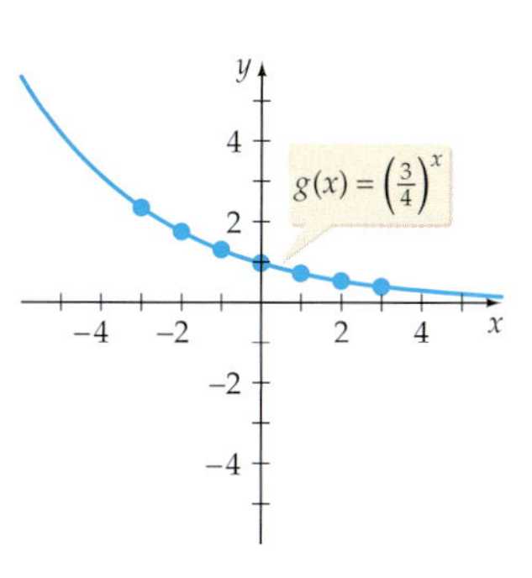

FIGURE 3.18

▶ TRY EXERCISE 22, PAGE 225

Consider the functions $F(x) = 2^x - 3$ and $G(x) = 2^{x-3}$. You can construct the graphs of these functions by plotting points; however, it is easier to construct their graphs by using translations of the graph of $f(x) = 2^x$, as shown in Example 3.

EXAMPLE 3 Use a Translation to Produce a Graph

a. Explain how to use the graph of $f(x) = 2^x$ to produce the graph of $F(x) = 2^x - 3$.

b. Explain how to use the graph of $f(x) = 2^x$ to produce the graph of $G(x) = 2^{x-3}$.

Solution

a. $F(x) = 2^x - 3 = f(x) - 3$. The graph of F is a vertical translation of f down 3 units, as shown in **Figure 3.19.**

Continued ▶

b. $G(x) = 2^{x-3} = f(x - 3)$. The graph of G is a horizontal translation of f to the right 3 units, as shown in **Figure 3.20.**

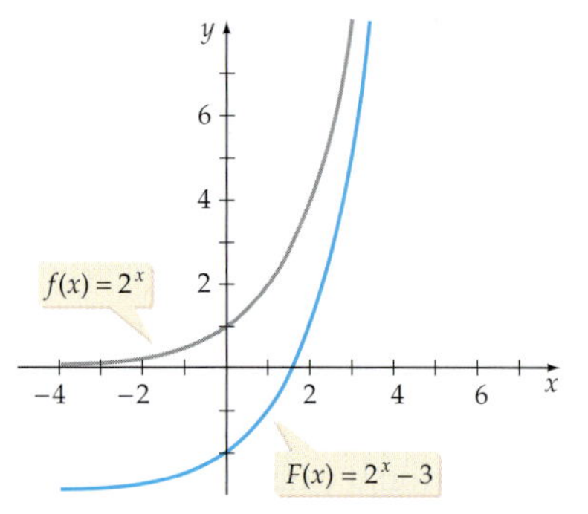

FIGURE 3.19

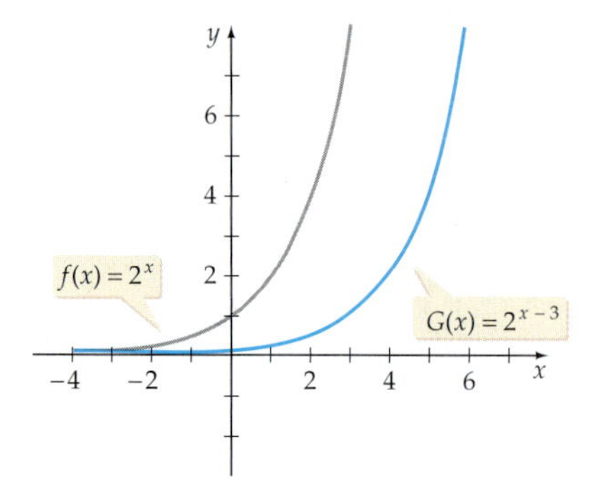

FIGURE 3.20

TRY EXERCISE 28, PAGE 225

The graphs of some functions can be constructed by stretching, compressing, or reflecting the graph of an exponential function.

EXAMPLE 4 Use Stretching or Reflecting Procedures to Produce a Graph

a. Explain how to use the graph of $f(x) = 2^x$ to produce the graph of $M(x) = 2(2^x)$.

b. Explain how to use the graph of $f(x) = 2^x$ to produce the graph of $N(x) = 2^{-x}$.

Solution

a. $M(x) = 2(2^x) = 2f(x)$. The graph of M is a vertical stretching of f, as shown in **Figure 3.21.** If (x, y) is a point on the graph of $f(x) = 2^x$, then $(x, 2y)$ is a point on the graph of M.

b. $N(x) = 2^{-x} = f(-x)$. The graph of N is the graph of f reflected across the y-axis, as shown in **Figure 3.22.** If (x, y) is a point on the graph of $f(x) = 2^x$, then $(-x, y)$ is a point on the graph of N.

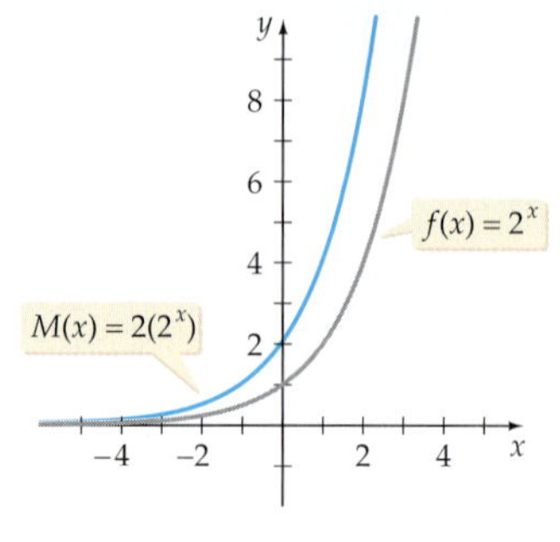

FIGURE 3.21

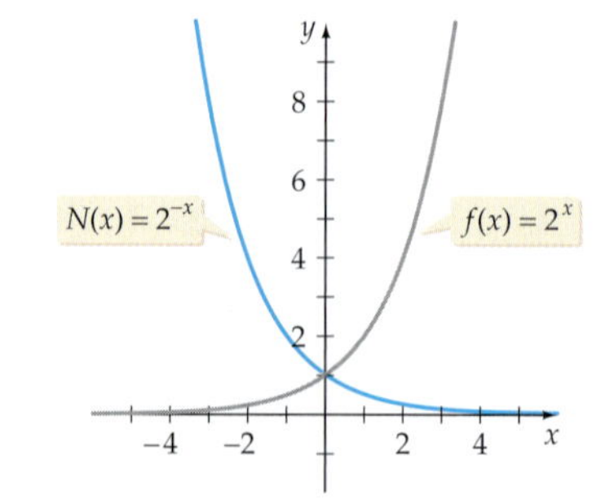

FIGURE 3.22

TRY EXERCISE 30, PAGE 225

The Natural Exponential Function

The irrational number π is often used in applications that involve circles. Another irrational number, denoted by the letter e, is useful in applications that involve growth or decay.

Definition of e

The **number e** is defined as the number that

$$\left(1 + \frac{1}{n}\right)^n$$

approaches as n increases without bound.

MATH MATTERS

Leonhard Euler (1707–1783)

Some mathematicians consider Euler to be the greatest mathematician of all time. He certainly was the most prolific writer of mathematics of all time. He made substantial contributions in the areas of number theory, geometry, calculus, differential equations, differential geometry, topology, complex variables, and analysis, to name but a few. Euler was the first to introduce many of the mathematical notations that we use today. For instance, he introduced the symbol i for the square root of -1, the symbol π for pi, the functional notation $f(x)$, and the letter e for the base of the natural exponential function. Euler's computational skills were truly amazing. The mathematician François Arago remarked, "Euler calculated without apparent effort, as men breathe, or as eagles sustain themselves in the wind."

The letter e was chosen in honor of the Swiss mathematician Leonhard Euler. He was able to compute the value of e to several decimal places by evaluating $\left(1 + \frac{1}{n}\right)^n$ for large values of n, as shown in **Table 3.5.**

Table 3.5

Value of n	Value of $\left(1 + \frac{1}{n}\right)^n$
1	2
10	2.59374246
100	2.704813829
1000	2.716923932
10,000	2.718145927
100,000	2.718268237
1,000,000	2.718280469
10,000,000	2.718281693

The value of e accurate to eight decimal places is 2.71828183.

The Natural Exponential Function

For all real numbers x, the function defined by

$$f(x) = e^x$$

is called the **natural exponential function.**

A calculator can be used to evaluate e^x for specific values of x. For instance,

$$e^2 \approx 7.389056, \quad e^{3.5} \approx 33.115452, \quad \text{and} \quad e^{-1.4} \approx 0.246597$$

On a TI-83 calculator the e^x function is located above the LN key.

The graph below was produced on a TI-83 graphing calculator by entering e^x in the Y= menu.

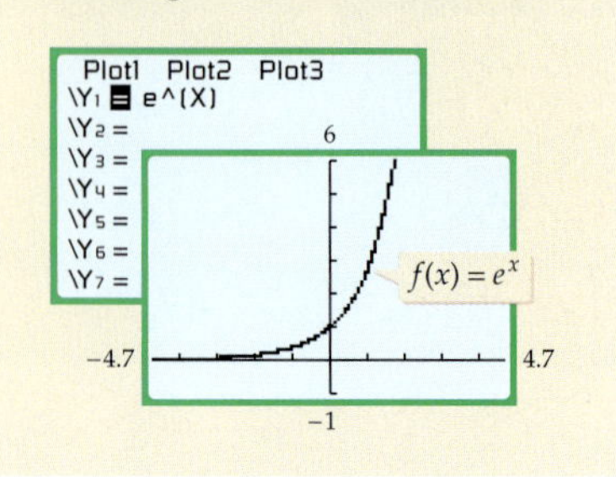

To graph $f(x) = e^x$, use a calculator to find the range values for a few domain values. The range values in **Table 3.6** have been rounded to the nearest tenth.

Table 3.6

x	−2	−1	0	1	2
$f(x) = e^x$	0.1	0.4	1.0	2.7	7.4

Plot the points given in **Table 3.6,** and then connect the points with a smooth curve. Because $e > 1$, we know that the graph is an increasing function. To the far left, the graph will approach the x-axis. The y-intercept is $(0, 1)$. See **Figure 3.23.** Note in **Figure 3.24** how the graph of $f(x) = e^x$ compares with the graphs of $g(x) = 2^x$ and $h(x) = 3^x$. You may have anticipated that the graph of $f(x) = e^x$ would lie between the two other graphs because e is between 2 and 3.

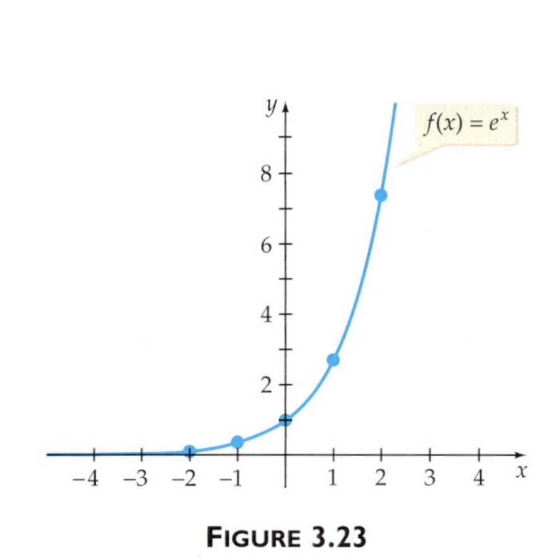

Figure 3.23

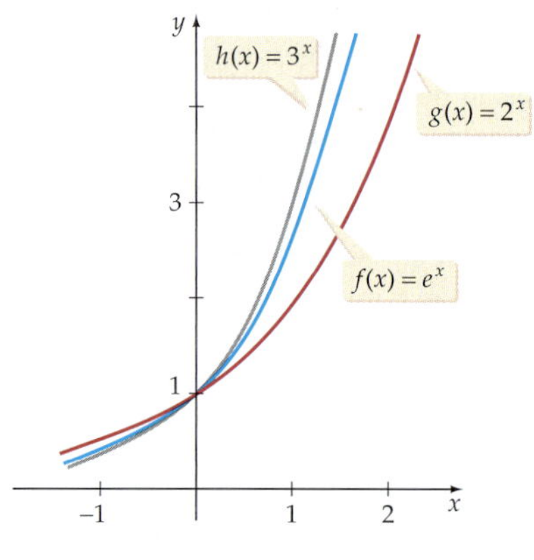

Figure 3.24

Applications of Exponential Functions

Many applications can be effectively modeled by functions that involve an exponential function. For instance, in Example 5 we make use of a function that involves an exponential function to model the temperature of a cup of coffee.

EXAMPLE 5 Use a Mathematical Model

A cup of coffee is heated to 160°F and placed in a room that maintains a temperature of 70°F. The temperature T of the coffee, in degrees Fahrenheit, after t minutes is given by

$$T = 70 + 90e^{-0.0485t}$$

a. Find the temperature of the coffee, to the nearest degree, 20 minutes after it is placed in the room.

b. Use a graphing utility to determine when the temperature of the coffee will reach 90°F.

take note

In Example 5**b.**, we use a graphing utility to solve the equation $90 = 70 + 90e^{-0.0485t}$. Analytic methods of solving this type of equation without the use of a graphing utility will be developed in Section 3.5.

Solution

a. $T = 70 + 90e^{-0.0485t}$

$= 70 + 90e^{-0.0485\cdot(20)}$ • **Substitute 20 for *t*.**

$\approx 70 + 34.1$

≈ 104.1

After 20 minutes the temperature of the coffee is about 104°F.

b. Graph $T = 70 + 90e^{-0.0485t}$ and $T = 90$. See the following figure.

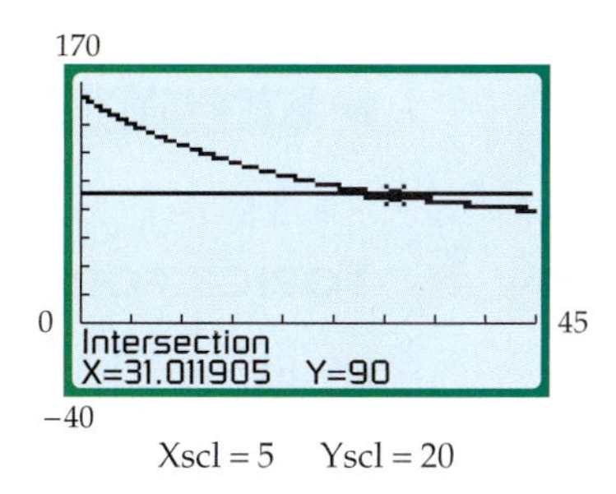

Xscl = 5 Yscl = 20

The graphs intersect at about (31.01, 90). It takes the coffee about 31 minutes to cool to 90°F.

▶ **TRY EXERCISE 44, PAGE 226**

EXAMPLE 6 Use a Mathematical Model

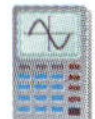

The weekly revenue R, in dollars, from the sale of a product varies with time according to the function

$$R(x) = \frac{1760}{8 + 14e^{-0.03x}}$$

where x is the number of weeks that have passed since the product was put on the market. What will the weekly revenue approach as time goes by?

Solution

Method I Use a graphing utility to graph $R(x)$, and use the TRACE feature to see what happens to the revenue as the time increases. The following graph shows that as the weeks go by, the weekly revenue will increase and approach \$220.00 per week.

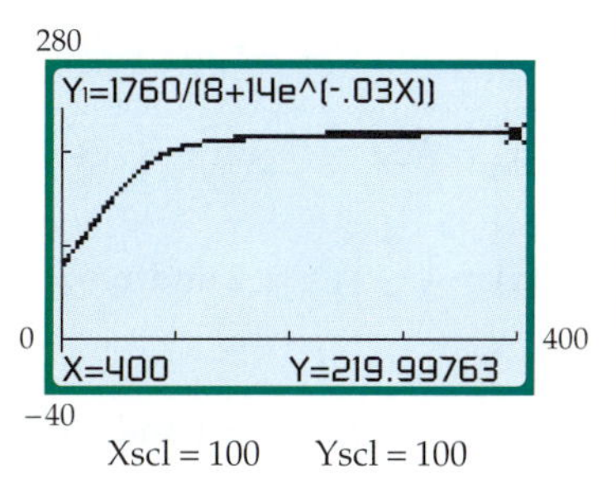

Xscl = 100 Yscl = 100

Continued ▶

Method 2 Write the revenue function in the following form.

$$R(x) = \frac{1760}{8 + \frac{14}{e^{0.03x}}}$$ • $14e^{-0.03x} = \frac{14}{e^{0.03x}}$

As x increases without bound, $e^{0.03x}$ increases without bound, and the fraction $\frac{14}{e^{0.03x}}$ approaches 0. Therefore, as $x \to \infty$, $R(x) \to \frac{1760}{8 + 0} = 220$. Both methods indicate that as the number of weeks increases, the revenue approaches \$220 per week.

▶ TRY EXERCISE 54, PAGE 227

TOPICS FOR DISCUSSION

1. Explain how to use the graph of $f(x) = 2^x$ to produce the graph of $g(x) = 2^{(x-3)} + 4$.
2. At what point does the function $g(x) = e^{-x^2/2}$ take on its maximum value?
3. Without using a graphing utility, determine whether the revenue function $R(t) = 10 + e^{-0.05t}$ is an increasing function or a decreasing function.
4. Discuss the properties of the graph of $f(x) = b^x$ when $b > 1$.
5. What is the base of the natural exponential function? How is it calculated? What is its approximate value?

EXERCISE SET 3.2

In Exercises 1 to 8, evaluate the exponential function for the given x-values.

1. $f(x) = 3^x$; $x = 0$ and $x = 4$
2. $f(x) = 5^x$; $x = 3$ and $x = -2$
3. $g(x) = 10^x$; $x = -2$ and $x = 3$
4. $g(x) = 4^x$; $x = 0$ and $x = -1$
5. $h(x) = \left(\frac{3}{2}\right)^x$; $x = 2$ and $x = -3$
6. $h(x) = \left(\frac{2}{5}\right)^x$; $x = -1$ and $x = 3$
7. $j(x) = \left(\frac{1}{2}\right)^x$; $x = -2$ and $x = 4$
8. $j(x) = \left(\frac{1}{4}\right)^x$; $x = -1$ and $x = 5$

In Exercises 9 to 14, use a calculator to evaluate the exponential function for the given x-value. Round to the nearest hundredth.

9. $f(x) = 2^x$, $x = 3.2$
10. $f(x) = 3^x$, $x = -1.5$
11. $g(x) = e^x$, $x = 2.2$
12. $g(x) = e^x$, $x = -1.3$
13. $h(x) = 5^x$, $x = \sqrt{2}$
14. $h(x) = 0.5^x$, $x = \pi$

15. Examine the following four functions and the graphs labeled **a, b, c,** and **d.** For each graph, determine which function has been graphed.

$$f(x) = 5^x \qquad g(x) = 1 + 5^{-x}$$
$$h(x) = 5^{x+3} \qquad k(x) = 5^x + 3$$

a.

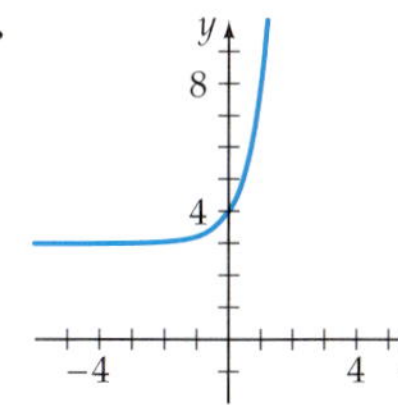

b.

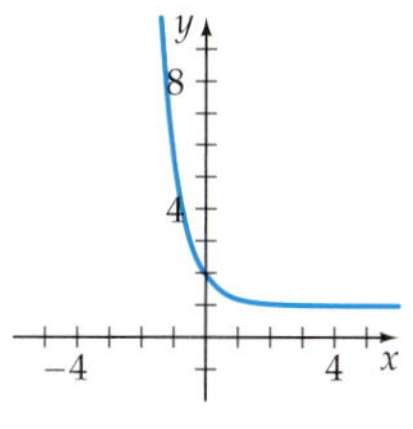

c.

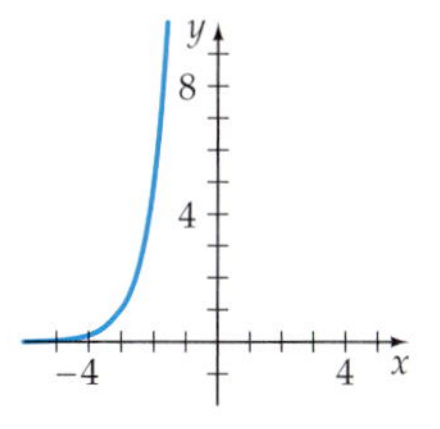

d.

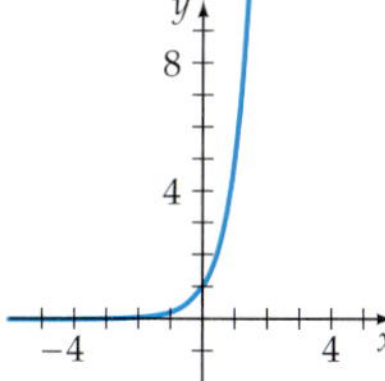

16. Examine the following four functions and the graphs labeled **a, b, c,** and **d.** For each graph, determine which function has been graphed.

$$f(x) = \left(\frac{1}{4}\right)^x \qquad g(x) = \left(\frac{1}{4}\right)^{-x}$$
$$h(x) = \left(\frac{1}{4}\right)^{x-2} \qquad k(x) = 3\left(\frac{1}{4}\right)^x$$

a.

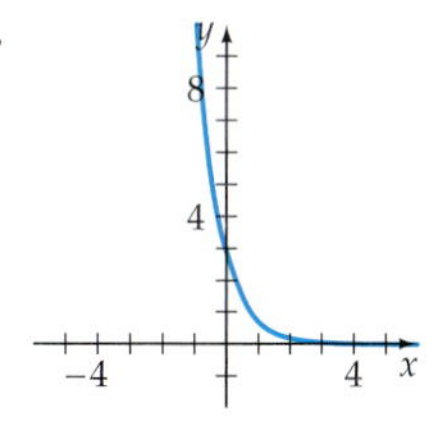

b.

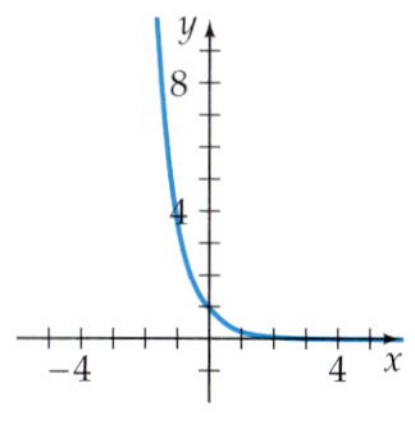

c.

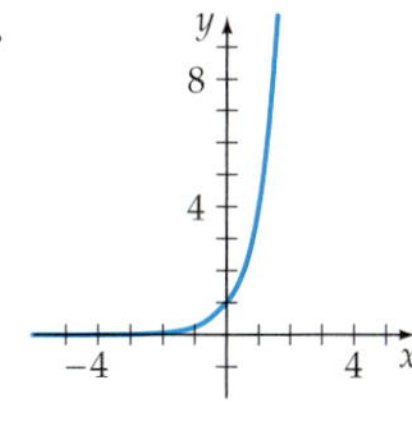

d.

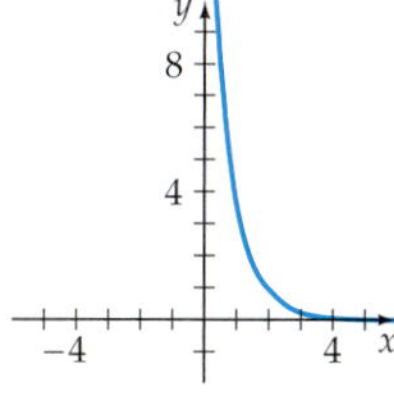

In Exercises 17 to 24, sketch the graph of each function.

17. $f(x) = 3^x$

18. $f(x) = 4^x$

19. $f(x) = 10^x$

20. $f(x) = 6^x$

21. $f(x) = \left(\frac{3}{2}\right)^x$

▶ **22.** $f(x) = \left(\frac{5}{2}\right)^x$

23. $f(x) = \left(\frac{1}{3}\right)^x$

24. $f(x) = \left(\frac{2}{3}\right)^x$

In Exercises 25 to 34, explain how to use the graph of the first function *f* to produce the graph of the second function *F*.

25. $f(x) = 3^x, F(x) = 3^x + 2$

26. $f(x) = 4^x, F(x) = 4^x - 3$

27. $f(x) = 10^x, F(x) = 10^{x-2}$

▶ **28.** $f(x) = 6^x, F(x) = 6^{x+5}$

29. $f(x) = \left(\frac{3}{2}\right)^x, F(x) = \left(\frac{3}{2}\right)^{-x}$

▶ **30.** $f(x) = \left(\frac{5}{2}\right)^x, F(x) = -\left[\left(\frac{5}{2}\right)^x\right]$

31. $f(x) = \left(\frac{1}{3}\right)^x, F(x) = 2\left[\left(\frac{1}{3}\right)^x\right]$

32. $f(x) = \left(\frac{2}{3}\right)^x, F(x) = \frac{1}{2}\left[\left(\frac{2}{3}\right)^x\right]$

33. $f(x) = e^x, F(x) = e^{-x} + 2$

34. $f(x) = e^x, F(x) = e^{x-3} + 1$

In Exercises 35 to 42, use a graphing utility to graph each function. If the function has a horizontal asymptote, state the equation of the horizontal asymptote.

35. $f(x) = \dfrac{3^x + 3^{-x}}{2}$

36. $f(x) = 4 \cdot 3^{-x^2}$

37. $f(x) = \dfrac{e^x - e^{-x}}{2}$

38. $f(x) = \dfrac{e^x + e^{-x}}{2}$

39. $f(x) = -e^{(x-4)}$

40. $f(x) = 0.5e^{-x}$

41. $f(x) = \dfrac{10}{1 + 0.4e^{-0.5x}}, \; x \geq 0$

42. $f(x) = \dfrac{10}{1 + 1.5e^{-0.5x}}, \; x \geq 0$

43. **INTERNET CONNECTIONS** Data from Forrester Research suggest that the number of broadband [cable and digital subscriber line (DSL)] connections to the Internet can be modeled by $f(x) = 1.353(1.9025)^x$, where x is the number of years after January 1, 1998, and $f(x)$ is the number of connections in millions.

a. How many broadband Internet connections, to the nearest million, does this model predict will exist on January 1, 2005?

b. According to the model, in what year will the number of broadband connections first reach 300 million? [*Hint:* Use the intersect feature of a graphing utility to determine the x-coordinate of the point of intersection of the graphs of $f(x)$ and $y = 300$.]

44. **MEDICATION IN BLOODSTREAM** The function $A(t) = 200e^{-0.014t}$ gives the amount of medication, in milligrams, in a patient's bloodstream t minutes after the medication has been injected into the patient's bloodstream.

a. Find the amount of medication, to the nearest milligram, in the patient's bloodstream after 45 minutes.

b. Use a graphing utility to determine how long it will take, to the nearest minute, for the amount of medication in the patient's bloodstream to reach 50 milligrams.

45. **DEMAND FOR A PRODUCT** The demand d for a specific product, in items per month, is given by

$$d(p) = 25 + 880e^{-0.18p}$$

where p is the price, in dollars, of the product.

a. What will be the monthly demand, to the nearest unit, when the price of the product is \$8 and when the price is \$18?

b. What will happen to the demand as the price increases without bound?

46. **SALES** The monthly income I, in dollars, from a new product is given by

$$I(t) = 24{,}000 - 22{,}000e^{-0.005t}$$

where t is the time, in months, since the product was first put on the market.

a. What was the monthly income after the 10th month and after the 100th month?

b. What will the monthly income from the product approach as the time increases without bound?

47. **A PROBABILITY FUNCTION** The manager of a home improvement store finds that between 10 A.M. and 11 A.M., customers enter the store at the average rate of 45 customers per hour. The following function gives the probability that a customer will arrive within t minutes of 10 A.M. (*Note:* A probability of 0.6 means there is a 60% chance that a customer will arrive during a given time period.)

$$P(t) = 1 - e^{-0.75t}$$

a. Find the probability, to the nearest hundredth, that a customer will arrive within 1 minute of 10 A.M.

b. Find the probability, to the nearest hundredth, that a customer will arrive within 3 minutes of 10 A.M.

c. Use a graph of $P(t)$ to determine how many minutes, to the nearest tenth of a minute, it takes for $P(t)$ to equal 98%.

d. Write a sentence that explains the meaning of the answer in part **c.**

48. **A PROBABILITY FUNCTION** The owner of a sporting goods store finds that between 9 A.M. and 10 A.M., customers enter the store at the average rate of 12 customers per hour. The following function gives the probability that a customer will arrive within t minutes of 9 A.M.

$$P(t) = 1 - e^{-0.2t}$$

a. Find the probability, to the nearest hundredth, that a customer will arrive within 5 minutes of 9 A.M.

b. Find the probability, to the nearest hundredth, that a customer will arrive within 15 minutes of 9 A.M.

c. Use a graph of $P(t)$ to determine how many minutes, to the nearest 0.1 minute, it takes for $P(t)$ to equal 90%.

d. Write a sentence that explains the meaning of the answer in part **c.**

Exercises 49 and 50 involve the factorial function $x!$, which is defined for whole numbers x as

$$x! = \begin{cases} 1, & \text{if } x = 0 \\ x \cdot (x-1) \cdot (x-2) \cdot \cdots \cdot 3 \cdot 2 \cdot 1, & \text{if } x \geq 1 \end{cases}$$

For example, $3! = 3 \cdot 2 \cdot 1 = 6$ and $5! = 5 \cdot 4 \cdot 3 \cdot 2 \cdot 1 = 120$.

49. **QUEUING THEORY** During the 30-minute period before a Broadway play begins, the members of the audience arrive at the theater at the average rate of 12 people per minute. The probability that x people will arrive during a particular minute is given by $P(x) = \frac{12^x e^{-12}}{x!}$. Find the probability, to the nearest 0.1%, that

a. 9 people will arrive during a given minute.

b. 18 people will arrive during a given minute.

50. **QUEUING THEORY** During the period from 2:00 P.M. to 3:00 P.M., a bank finds that an average of seven people enter the bank every minute. The probability that x people will enter the bank during a particular minute is given by $P(x) = \frac{7^x e^{-7}}{x!}$. Find the probability, to the nearest 0.1%, that

a. only two people will enter the bank during a given minute.

b. 11 people will enter the bank during a given minute.

51. **E. COLI INFECTION** *Escherichia coli (E. coli)* is a bacterium that can reproduce at an exponential rate. The *E. coli* reproduce by dividing. A small number of *E. coli* bacteria in the large intestine of a human can trigger a serious infection within a few hours. Consider a particular *E. coli* infection that starts with 100 *E. coli* bacteria. Each bacterium splits into two parts every half hour. Assuming none of the bacteria die, the size of the *E. coli* population after t hours is given by $P(t) = 100 \cdot 2^{2t}$, where $0 \le t \le 16$.

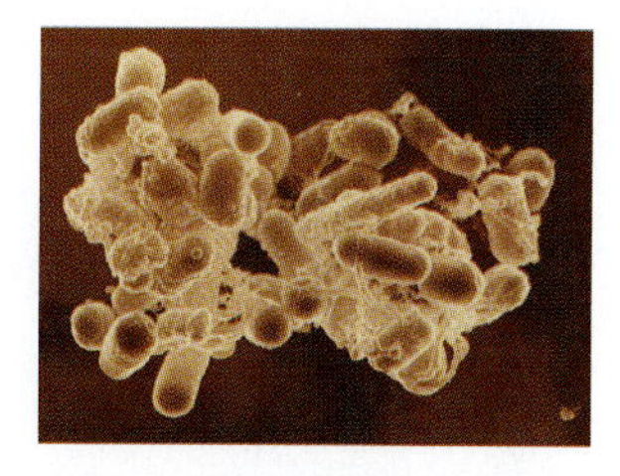

a. Find $P(3)$ and $P(6)$.

b. Use a graphing utility to find the time, to the nearest tenth of an hour, it takes for the *E. coli* population to number 1 billion.

52. **RADIATION** Lead shielding is used to contain radiation. The percentage of a certain radiation that can penetrate x millimeters of lead shielding is given by $I(x) = 100e^{-1.5x}$.

a. What percentage of radiation, to the nearest tenth of a percent, will penetrate a lead shield that is 1 millimeter thick?

b. How many millimeters of lead shielding are required so that less than 0.05% of the radiation penetrates the shielding? Round to the nearest millimeter.

53.  **AIDS** An exponential function that approximates the number of people in the United States who have been infected with AIDS is given by $N(t) = 138{,}000(1.39)^t$, where t is the number of years after January 1, 1990.

a. According to this function, how many people had been infected with AIDS as of January 1, 1994? Round to the nearest thousand.

b. Use a graph to estimate during what year the number of people in the United States who had been infected with AIDS first reached 1.5 million.

54. **FISH POPULATION** The number of bass in a lake is given by

$$P(t) = \frac{3600}{1 + 7e^{-0.05t}},$$

where t is the number of months that have passed since the lake was stocked with bass.

a. How many bass were in the lake immediately after it was stocked?

b. How many bass were in the lake 1 year after the lake was stocked?

c. What will happen to the bass population as t increases without bound?

55. **THE PAY IT FORWARD MODEL** In the movie *Pay It Forward*, Trevor McKinney, played by Haley Joel Osment, is given a school assignment to "think of an idea to change the world—and then put it into action." In response to this assignment, Trevor develops a *pay it forward* project. In this project, anyone who benefits from another person's good deed must do a good deed for

three additional people. Each of these three people is then obligated to do a good deed for another three people, and so on.

The following diagram shows the number of people who have been a beneficiary of a good deed after 1 round and after 2 rounds of this project.

Three beneficiaries after one round.

A total of 12 beneficiaries after two rounds ($3 + 9 = 12$).

A mathematical model for the number of pay it forward beneficiaries after n rounds is given by $B(n) = \dfrac{3^{n+1} - 3}{2}$. Use this model to determine

a. the number of beneficiaries after 5 rounds and after 10 rounds. Assume that no person is a beneficiary of more than one good deed.

b. how many rounds are required to produce at least 2 million beneficiaries.

56. **INTENSITY OF LIGHT** The percent $I(x)$ of the original intensity of light striking the surface of a lake that is available x feet below the surface of the lake is given by $I(x) = 100e^{-0.95x}$.

a. What percentage of the light, to the nearest tenth of a percent, is available 2 feet below the surface of the lake?

b. At what depth, to the nearest hundredth of a foot, is the intensity of the light one-half the intensity at the surface?

57. **A TEMPERATURE MODEL** A cup of coffee is heated to 180°F and placed in a room that maintains a temperature of 65°F. The temperature of the coffee after t minutes is given by $T(t) = 65 + 115e^{-0.042t}$.

a. Find the temperature, to the nearest degree, of the coffee 10 minutes after it is placed in the room.

b. Use a graphing utility to determine when, to the nearest tenth of a minute, the temperature of the coffee will reach 100°F.

58. **A TEMPERATURE MODEL** Soup that is at a temperature of 170°F is poured into a bowl in a room that maintains a constant temperature. The temperature of the soup decreases according to the model given by $T(t) = 75 + 95e^{-0.12t}$, where t is time in minutes after the soup is poured.

a. What is the temperature, to the nearest tenth of a degree, of the soup after 2 minutes?

b. A certain customer prefers soup at a temperature of 110°F. How many minutes, to the nearest 0.1 minute, after the soup is poured does the soup reach that temperature?

c. What is the temperature of the room?

59. **MUSICAL SCALES** Starting on the left side of a standard 88-key piano, the frequency, in vibrations per second, of the nth note is given by $f(n) = (27.5)2^{(n-1)/12}$.

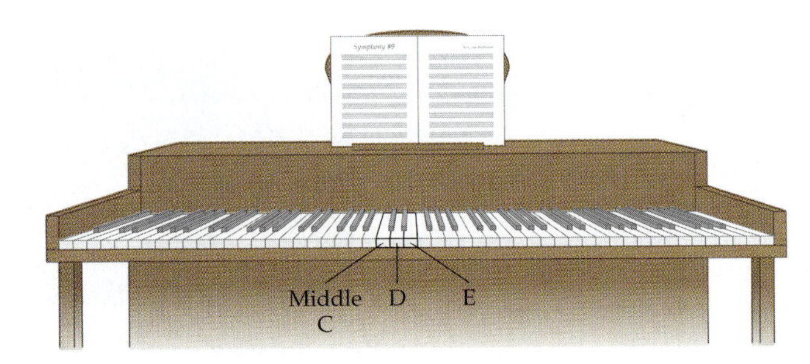

a. Using this formula, determine the frequency, to the nearest hundredth of a vibration per second, of middle C, key number 40 on an 88-key piano.

b. Is the difference in frequency between middle C (key number 40) and D (key number 42) the same as the difference in frequency between D (key number 42) and E (key number 44)? Explain.

CONNECTING CONCEPTS

60. Verify that the hyperbolic cosine function $\cosh(x) = \dfrac{e^x + e^{-x}}{2}$ is an even function.

61. Verify that the hyperbolic sine function $\sinh(x) = \dfrac{e^x - e^{-x}}{2}$ is an odd function.

62. Graph $g(x) = 10^x$, and then sketch the graph of g reflected across the line given by $y = x$.

63. Graph $f(x) = e^x$, and then sketch the graph of f reflected across the line given by $y = x$.

In Exercises 64 to 67, determine the domain of the given function. Write the domain using interval notation.

64. $f(x) = \dfrac{e^x - e^{-x}}{e^x + e^{-x}}$

65. $f(x) = \dfrac{e^{|x|}}{1 + e^x}$

66. $f(x) = \sqrt{1 - e^x}$

67. $f(x) = \sqrt{e^x - e^{-x}}$

PREPARE FOR SECTION 3.3

68. If $2^x = 16$, determine the value of x. [3.2]

69. If $3^{-x} = \dfrac{1}{27}$, determine the value of x. [3.2]

70. If $x^4 = 625$, determine the value of x. [3.2]

71. Find the inverse of $f(x) = \dfrac{2x}{x + 3}$. [3.1]

72. State the domain of $g(x) = \sqrt{x - 2}$. [1.3]

73. If the range of $h(x)$ is the set of all positive real numbers, then what is the domain of $h^{-1}(x)$? [3.1]

PROJECTS

1. 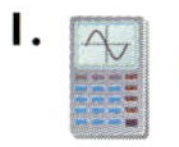 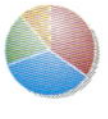 **THE SAINT LOUIS GATEWAY ARCH** The Gateway Arch in Saint Louis was designed in the shape of an inverted **catenary,** as shown by the red curve in the drawing at the right. The Gateway Arch is one of the largest optical illusions ever created. As you look at the arch (and its basic shape defined by the catenary curve), it appears to be much taller than it is wide. However, this is not the case. The height of the catenary is given by

$$h(x) = 693.8597 - 68.7672\left(\frac{e^{0.0100333x} + e^{-0.0100333x}}{2}\right)$$

where x and $h(x)$ are measured in feet and $x = 0$ represents the position at ground level that is directly below the highest point of the catenary.

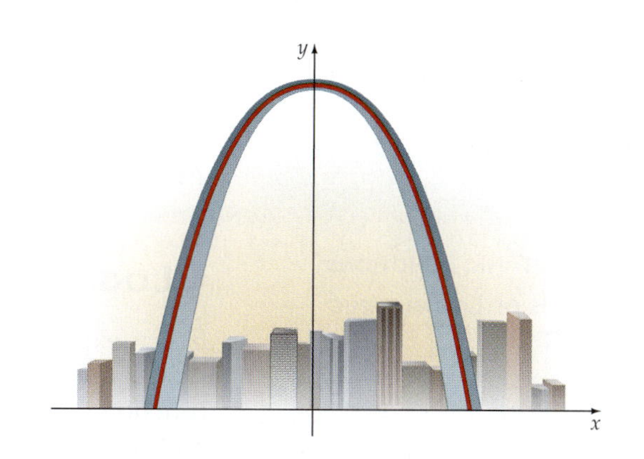

a. Use a graphing utility to graph $h(x)$.

b. Use your graph to find the height of the catenary for $x = 0$, 100, 200, and 299 feet. Round each result to the nearest tenth of a foot.

c. What is the width of the catenary at ground level and what is the maximum height of the catenary? Round each result to the nearest tenth of a foot.

d. By how much does the maximum height of the catenary exceed its width at ground level? Round to the nearest tenth of a foot.

2. **AN EXPONENTIAL REWARD** According to legend, when Sissa Ben Dahir of India invented the game of chess, King Shirham was so impressed with the game that he summoned the game's inventor and offered him the reward of his choosing. The inventor pointed to the chessboard and requested, for his reward, one grain of wheat on the first square, two grains of wheat on the second square, four grains on the third square, eight grains on the fourth square, and so on for all 64 squares on the chessboard. The King considered this a very modest reward and said he would grant the inventor's wish. The following table shows how many grains of wheat are on each of the first six squares and the total number of grains of wheat needed to cover squares 1 to n for $n \leq 6$.

Square number, n	Number of grains of wheat on square n	Total number of grains of wheat on squares 1 through n
1	1	1
2	2	3
3	4	7
4	8	15
5	16	31
6	32	63

a. If all 64 squares of the chessboard are piled with wheat as requested by Sissa Ben Dahir, how many grains of wheat are on the board?

b. A grain of wheat weighs approximately 0.000008 kilogram. Find the total weight of the wheat requested by Sissa Ben Dahir.

c. In a recent year, a total of 6.5×10^8 metric tons of wheat were produced in the world. At this level, how many years, to the nearest year, of wheat production would be required to fill the request of Sissa Ben Dahir? One metric ton equals 1000 kilograms.

SECTION 3.3

LOGARITHMIC FUNCTIONS AND THEIR APPLICATIONS

- Logarithmic Functions
- Graphs of Logarithmic Functions
- Domains of Logarithmic Functions
- Common and Natural Logarithms
- Applications of Logarithmic Functions

Logarithmic Functions

Every exponential function of the form $g(x) = b^x$ is a one-to-one function and therefore has an inverse function. Sometimes we can determine the inverse of a function represented by an equation by interchanging the variables of its equation and then solving for the dependent variable. If we attempt to use this procedure for $g(x) = b^x$, we obtain

$$g(x) = b^x$$
$$y = b^x$$
$$x = b^y \quad \text{• Interchange the variables.}$$

None of our previous methods can be used to solve the equation $x = b^y$ for the exponent y. Thus we need to develop a new procedure. One method would be to merely write

$$y = \text{the power of } b \text{ that produces } x$$

Although this would work, it is not very concise. We need a compact notation to represent "y is the power of b that produces x." This more compact notation is given in the following definition.

MATH MATTERS

Logarithms were developed by John Napier (1550–1617) as a means of simplifying the calculations of astronomers. One of his ideas was to devise a method by which the product of two numbers could be determined by performing an addition.

Definition of a Logarithm and a Logarithmic Function

If $x > 0$ and b is a positive constant $(b \neq 1)$, then

$$y = \log_b x \quad \text{if and only if} \quad b^y = x$$

The notation $\log_b x$ is read "the **logarithm** (or log) base b of x." The function defined by $f(x) = \log_b x$ is a **logarithmic function** with base b. This function is the inverse of the exponential function $g(x) = b^x$.

It is essential to remember that $f(x) = \log_b x$ is the inverse function of $g(x) = b^x$. Because these functions are inverses and because functions that are inverses have the property that $f(g(x)) = x$ and $g(f(x)) = x$, we have the following important relationships.

Composition of Logarithmic and Exponential Functions

Let $g(x) = b^x$ and $f(x) = \log_b x$ $(x > 0, b > 0, b \neq 1)$. Then

$$g(f(x)) = b^{\log_b x} = x \quad \text{and} \quad f(g(x)) = \log_b b^x = x$$

take note

The notation $\log_b x$ replaces the phrase "the power of b that produces x." For instance, "3 is the power of 2 that produces 8" is abbreviated $3 = \log_2 8$. In your work with logarithms, remember that a logarithm is an *exponent*.

As an example of these relationships, let $g(x) = 2^x$ and $f(x) = \log_2 x$. Then

$$2^{\log_2 x} = x \quad \text{and} \quad \log_2 2^x = x$$

The equations

$$y = \log_b x \quad \text{and} \quad b^y = x$$

are different ways of expressing the same concept.

Exponential Form and Logarithmic Form

The **exponential form** of $y = \log_b x$ is $b^y = x$.

The **logarithmic form** of $b^y = x$ is $y = \log_b x$.

These concepts are illustrated in the next two examples.

EXAMPLE 1 **Change from Logarithmic to Exponential Form**

Write each equation in its exponential form.

a. $3 = \log_2 8$ **b.** $2 = \log_{10}(x + 5)$ **c.** $\log_e x = 4$ **d.** $\log_b b^3 = 3$

Solution

Use the definition $y = \log_b x$ if and only if $b^y = x$.

Logarithms are exponents.

a. $3 = \log_2 8$ if and only if $2^3 = 8$

Base

b. $2 = \log_{10}(x + 5)$ if and only if $10^2 = x + 5$.

c. $\log_e x = 4$ if and only if $e^4 = x$.

d. $\log_b b^3 = 3$ if and only if $b^3 = b^3$.

TRY EXERCISE 4, PAGE 239

EXAMPLE 2 **Change from Exponential to Logarithmic Form**

Write each equation in its logarithmic form.

a. $3^2 = 9$ **b.** $5^3 = x$ **c.** $a^b = c$ **d.** $b^{\log_b 5} = 5$

Solution

The logarithmic form of $b^y = x$ is $y = \log_b x$.

Exponent

a. $3^2 = 9$ if and only if $2 = \log_3 9$

Base

b. $5^3 = x$ if and only if $3 = \log_5 x$.

c. $a^b = c$ if and only if $b = \log_a c$.

d. $b^{\log_b 5} = 5$ if and only if $\log_b 5 = \log_b 5$.

TRY EXERCISE 12, PAGE 239

The definition of a logarithm and the definition of inverse functions can be used to establish many properties of logarithms. For instance:

- $\log_b b = 1$ because $b = b^1$.
- $\log_b 1 = 0$ because $1 = b^0$.
- $\log_b(b^x) = x$ because $b^x = b^x$.
- $b^{\log_b x} = x$ because $f(x) = \log_b x$ and $g(x) = b^x$ are inverse functions. Thus $g[f(x)] = x$.

We will refer to the preceding properties as the *basic logarithmic properties.*

Basic Logarithmic Properties

1. $\log_b b = 1$ **2.** $\log_b 1 = 0$ **3.** $\log_b(b^x) = x$ **4.** $b^{\log_b x} = x$

EXAMPLE 3 Apply the Basic Logarithmic Properties

Evaluate each of the following logarithms.

a. $\log_8 1$ **b.** $\log_5 5$ **c.** $\log_2(2^4)$ **d.** $3^{\log_3 7}$

Solution

a. By Property 2, $\log_8 1 = 0$.

b. By Property 1, $\log_5 5 = 1$.

c. By Property 3, $\log_2(2^4) = 4$.

d. By Property 4, $3^{\log_3 7} = 7$.

TRY EXERCISE 28, PAGE 239

Some logarithms can be evaluated just by remembering that a logarithm is an exponent. For instance, $\log_5 25$ equals 2 because the base 5 raised to the second power equals 25.

- $\log_{10} 100 = 2$ because $10^2 = 100$.
- $\log_4 64 = 3$ because $4^3 = 64$.
- $\log_7 \frac{1}{49} = -2$ because $7^{-2} = \frac{1}{7^2} = \frac{1}{49}$.

QUESTION What is the value of $\log_5 625$?

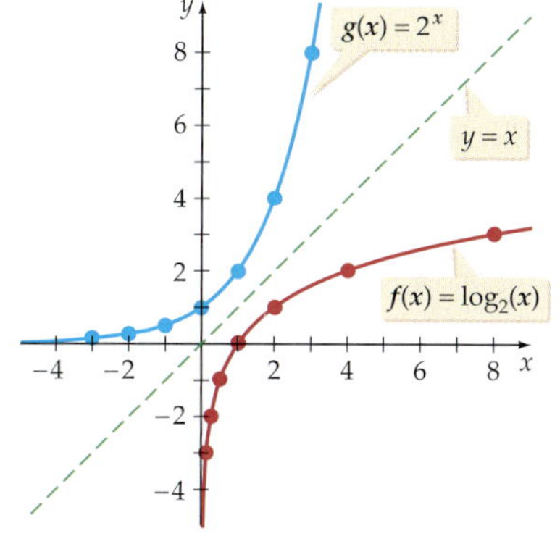

FIGURE 3.25

GRAPHS OF LOGARITHMIC FUNCTIONS

Because $f(x) = \log_b x$ is the inverse function of $g(x) = b^x$, the graph of f is a reflection of the graph of g across the line given by $y = x$. The graph of $g(x) = 2^x$ is shown in **Figure 3.25. Table 3.7** below shows some of the ordered pairs on the graph of g.

TABLE 3.7

x	−3	−2	−1	0	1	2	3
$g(x) = 2^x$	$\frac{1}{8}$	$\frac{1}{4}$	$\frac{1}{2}$	1	2	4	8

ANSWER $\log_5 625 = 4$ because $5^4 = 625$.

The graph of the inverse of g, which is $f(x) = \log_2 x$, is also shown in **Figure 3.25.** Some of the ordered pairs of f are shown in **Table 3.8.** Note that if (x, y) is a point on the graph of g, then (y, x) is a point on the graph of f. Also notice that the graph of f is a reflection of the graph of g across the line given by $y = x$.

TABLE 3.8

x	$\frac{1}{8}$	$\frac{1}{4}$	$\frac{1}{2}$	1	2	4	8
$f(x) = \log_2 x$	−3	−2	−1	0	1	2	3

The graph of a logarithmic function can be drawn by first rewriting the function in its exponential form. This procedure is illustrated in Example 4.

EXAMPLE 4 Graph a Logarithmic Function

Graph $f(x) = \log_3 x$.

Solution

To graph $f(x) = \log_3 x$, consider the equivalent exponential equation $x = 3^y$. Because this equation is solved for x, choose values of y and calculate the corresponding values of x, as shown in **Table 3.9.**

TABLE 3.9

$x = 3^y$	$\frac{1}{9}$	$\frac{1}{3}$	1	3	9
y	−2	−1	0	1	2

Now plot the ordered pairs and connect the points with a smooth curve, as shown in **Figure 3.26.**

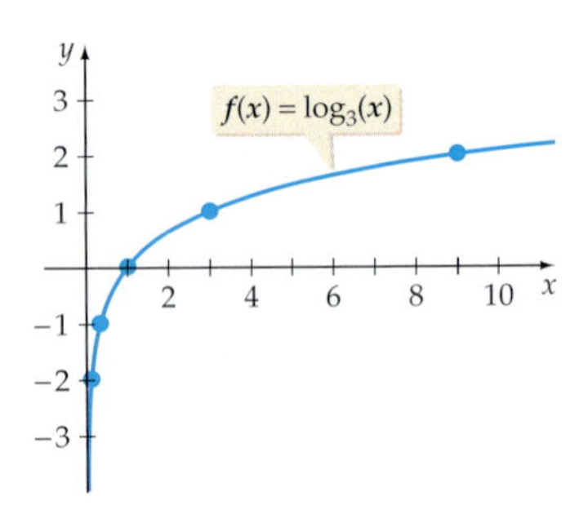

FIGURE 3.26

TRY EXERCISE 32, PAGE 239

We can use a similar procedure to draw the graph of a logarithmic function with a fractional base. For instance, consider $y = \log_{2/3} x$. Rewriting this in expo-

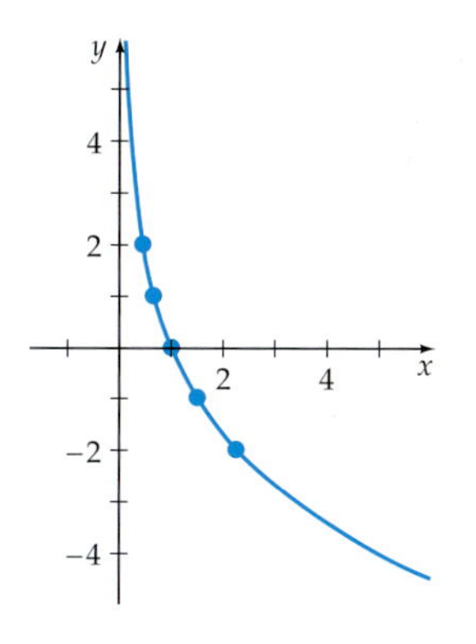

$y = \log_{2/3} x$

FIGURE 3.27

nential form gives us $\left(\frac{2}{3}\right)^y = x$. Choose values of y and calculate the corresponding x-values. See **Table 3.10.** Plot the points corresponding to the ordered pairs (x, y), and then draw a smooth curve through the points, as shown in **Figure 3.27.**

TABLE 3.10

$x = \left(\frac{2}{3}\right)^y$	$\left(\frac{2}{3}\right)^{-2} = \frac{9}{4}$	$\left(\frac{2}{3}\right)^{-1} = \frac{3}{2}$	$\left(\frac{2}{3}\right)^{0} = 1$	$\left(\frac{2}{3}\right)^{1} = \frac{2}{3}$	$\left(\frac{2}{3}\right)^{2} = \frac{4}{9}$
y	-2	-1	0	1	2

Properties of $f(x) = \log_b x$

For all positive real numbers b, $b \neq 1$, the function $f(x) = \log_b x$ has the following properties:

1. The domain of f consists of the set of positive real numbers and its range consists of the set of all real numbers.
2. The graph of f has an x-intercept of $(1, 0)$ and passes through $(b, 1)$.
3. If $b > 1$, f is an increasing function and its graph is asymptotic to the negative y-axis. [As $x \to \infty, f(x) \to \infty$, and as $x \to 0$ from the right, $f(x) \to -\infty$.] See **Figure 3.28a.**
4. If $0 < b < 1$, f is a decreasing function and its graph is asymptotic to the positive y-axis. [As $x \to \infty, f(x) \to -\infty$, and as $x \to 0$ from the right, $f(x) \to \infty$.] See **Figure 3.28b.**

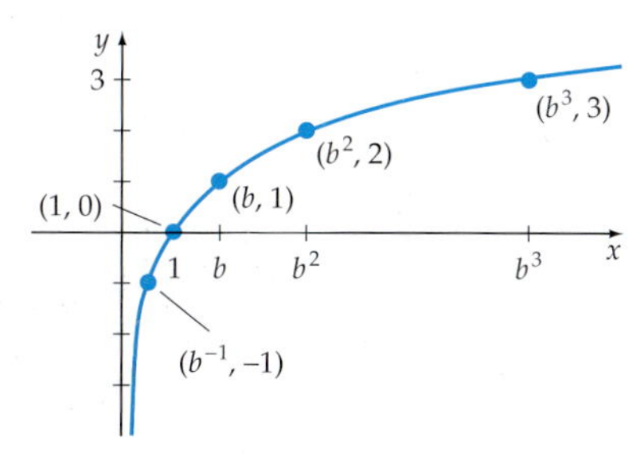

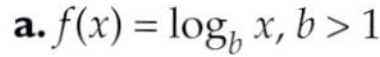

a. $f(x) = \log_b x, b > 1$

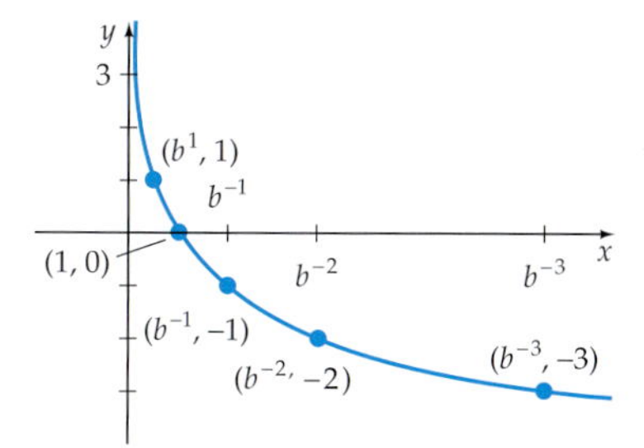

b. $f(x) = \log_b x, 0 < b < 1$

FIGURE 3.28

DOMAINS OF LOGARITHMIC FUNCTIONS

The function $f(x) = \log_b x$ has as its domain the set of positive real numbers. The function $f(x) = \log_b(g(x))$ has as its domain the set of all x for which $g(x) > 0$. To determine the domain of a function such as $f(x) = \log_b(g(x))$, we must determine the values of x that make $g(x)$ positive. This process is illustrated in Example 5.

EXAMPLE 5 Find the Domain of a Logarithmic Function

Find the domain of each of the following logarithmic functions.

a. $f(x) = \log_6(x - 3)$ **b.** $F(x) = \log_2|x + 2|$ **c.** $R(x) = \log_5\left(\dfrac{x}{8 - x}\right)$

Solution

a. Solving $(x - 3) > 0$ for x gives us $x > 3$. The domain of f consists of all real numbers greater than 3. In interval notation the domain is $(3, \infty)$.

b. The solution set of $|x + 2| > 0$ consists of all real numbers x except $x = -2$. The domain of F consists of all real numbers $x \neq -2$. In interval notation the domain is $(-\infty, -2) \cup (-2, \infty)$.

c. Solving $\left(\dfrac{x}{8 - x}\right) > 0$ yields the set of all real numbers x between 0 and 8. The domain of R is all real numbers x such that $0 < x < 8$. In interval notation the domain is $(0, 8)$.

▶ **TRY EXERCISE 40, PAGE 239**

Some logarithmic functions can be graphed by using horizontal and/or vertical translations of a previously drawn graph.

EXAMPLE 6 Use Translations to Graph Logarithmic Functions

Graph: **a.** $f(x) = \log_4(x + 3)$ **b.** $f(x) = \log_4 x + 3$

Solution

a. The graph of $f(x) = \log_4(x + 3)$ can be obtained by shifting the graph of $g(x) = \log_4 x$ to the left 3 units. See **Figure 3.29.** Note that the domain of f consists of all real numbers x greater than -3 because $x + 3 > 0$ for $x > -3$. The graph of f is asymptotic to the vertical line $x = -3$.

b. The graph of $f(x) = \log_4 x + 3$ can be obtained by shifting the graph of $g(x) = \log_4 x$ upward 3 units. See **Figure 3.30.**

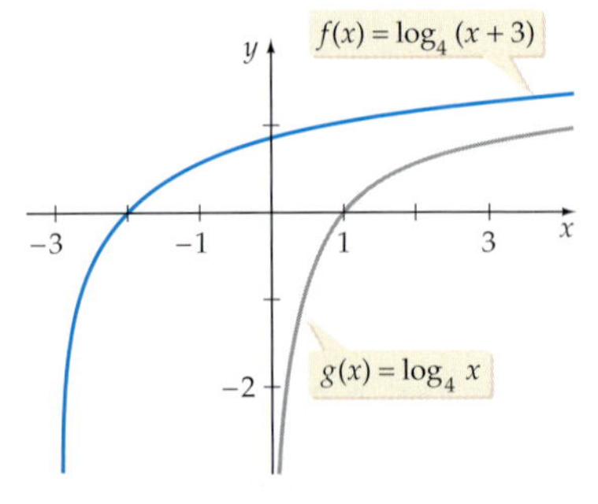

FIGURE 3.29

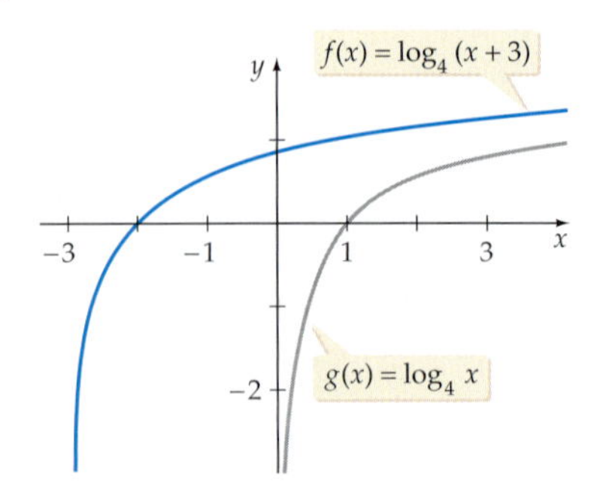

FIGURE 3.30

▶ **TRY EXERCISE 50, PAGE 239**

Common and Natural Logarithms

Two of the most frequently used logarithmic functions are *common logarithms,* which have base 10, and *natural logarithms,* which have base e (the base of the natural exponential function).

Definition of Common and Natural Logarithms

The function defined by $f(x) = \log_{10} x$ is called the **common logarithmic function.** It is customarily written without stating the base as $f(x) = \log x$.

The function defined by $f(x) = \log_e x$ is called the **natural logarithmic function.** It is customarily written as $f(x) = \ln x$.

Most scientific or graphing calculators have a LOG key for evaluating common logarithms and an LN key to evaluate natural logarithms. For instance, using a graphing calculator,

$$\log 24 \approx 1.3802112 \qquad \text{and} \qquad \ln 81 \approx 4.3944492$$

The graphs of $f(x) = \log x$ and $f(x) = \ln x$ can be drawn using the same techniques we used to draw the graphs in the preceding examples. However, these graphs also can be produced with a graphing calculator by entering $\log x$ and $\ln x$ into the Y= menu. See **Figure 3.31** and **Figure 3.32.**

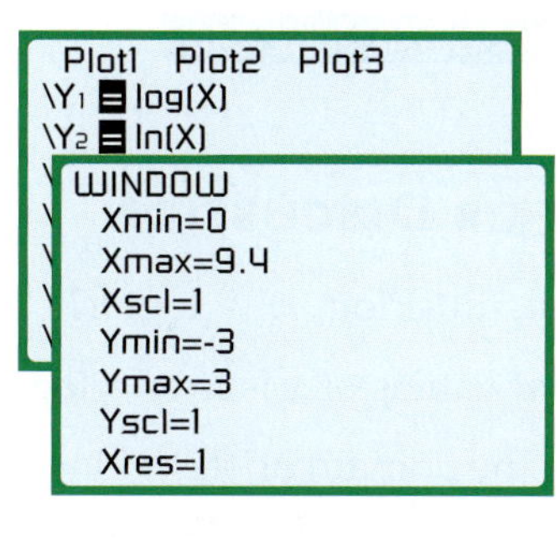

Figure 3.31

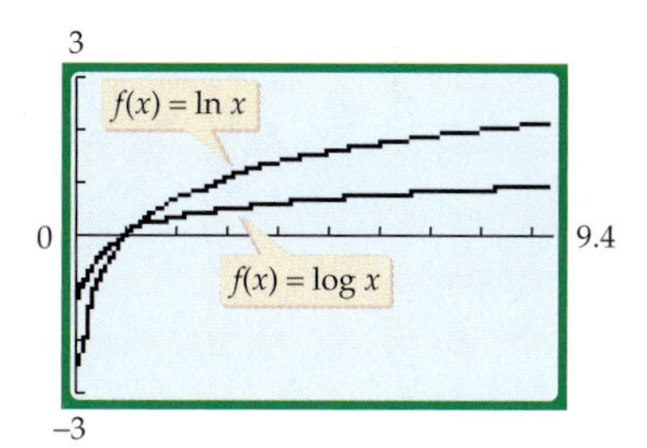

Figure 3.32

Observe that each graph passes through (1, 0). Also note that as $x \to 0$ from the right, the functional values $f(x) \to -\infty$. Thus the y-axis is a vertical asymptote for each of the graphs. The domain of both $f(x) = \log x$ and $f(x) = \ln x$ is the set of positive real numbers. Each of these functions has a range consisting of the set of real numbers.

Applications of Logarithmic Functions

Many applications can be modeled by logarithmic functions.

EXAMPLE 7 Average Time of a Major League Baseball Game

From 1981 to 1999, the average time of a major league baseball game tended to increase each year. If the year 1981 is represented by $x = 1$, then the function

$$T(x) = 149.57 + 7.63 \ln x$$

approximates the average time T, in minutes, of a major league baseball game for the years 1981 to 1999—that is, for $x = 1$ to $x = 19$.

a. Use the function T to determine the average time of a major league baseball game during the 1981 season and during the 1999 season.

b. By how much did the average time of a major league baseball game increase during the years 1981 to 1999?

Solution

a. The year 1981 is represented by $x = 1$ and the year 1999 by $x = 19$.

$$T(1) = 149.57 + 7.63 \ln(1) = 149.57$$

In 1981 the average time of a baseball game was about 149.57 minutes.

$$T(19) = 149.57 + 7.63 \ln(19) \approx 172.04$$

In 1999 the average time of a baseball game was about 172.04 minutes.

b. $T(19) - T(1) \approx 172.04 - 149.57 = 22.47$. During the years 1981 to 1999, the average time of a baseball game increased by about 22.47 minutes.

TRY EXERCISE 70, PAGE 240

MATH MATTERS

Although logarithms were originally developed to assist with computations, logarithmic functions have a much broader use today. They are often used in such disciplines as geology, acoustics, chemistry, physics, and economics, to name a few.

TOPICS FOR DISCUSSION

1. If $m > n$, must $\log_b m > \log_b n$?

2. For what values of x is $\ln x > \log x$?

3. What is the domain of $f(x) = \log(x^2 + 1)$? Explain why the graph of f does not have a vertical asymptote.

4. The subtraction $3 - 5$ does not have an answer if we require that the answer be positive. Keep this idea in mind as you work the rest of this exercise.

   Press the MODE key of a TI-83 graphing calculator, and choose "Real" from the menu. Now use the calculator to evaluate $\log(-2)$. What output is given by the calculator? Press the MODE key, and choose "a + bi" from the menu. Now use the calculator to evaluate $\log(-2)$. What output is given by the calculator? Write a sentence or two that explain why the output is different for these two evaluations.

Exercise Set 3.3

In Exercises 1 to 10, change each equation to its exponential form.

1. $\log 10 = 1$

2. $\log 10{,}000 = 4$

3. $\log_8 64 = 2$

▶ **4.** $\log_4 64 = 3$

5. $\log_7 x = 0$

6. $\log_3 \frac{1}{81} = -4$

7. $\ln x = 4$

8. $\ln e^2 = 2$

9. $\ln 1 = 0$

10. $\ln x = -3$

In Exercises 11 to 20, change each equation to its logarithmic form. Assume $y > 0$ and $b > 0$.

11. $3^2 = 9$

▶ **12.** $5^3 = 125$

13. $4^{-2} = \frac{1}{16}$

14. $10^0 = 1$

15. $b^x = y$

16. $2^x = y$

17. $y = e^x$

18. $5^1 = 5$

19. $100 = 10^2$

20. $2^{-4} = \frac{1}{16}$

In Exercises 21 to 30, evaluate each logarithm. Do not use a calculator.

21. $\log_4 16$

22. $\log_{3/2} \frac{8}{27}$

23. $\log_3 \frac{1}{243}$

24. $\log_b 1$

25. $\ln e^3$

26. $\log_b b$

27. $\log \frac{1}{100}$

▶ **28.** $\log 1{,}000{,}000$

29. $\log_{0.5} 16$

30. $\log_{0.3} \frac{100}{9}$

In Exercises 31 to 38, graph each function by using its exponential form.

31. $f(x) = \log_4 x$

▶ **32.** $f(x) = \log_6 x$

33. $f(x) = \log_{12} x$

34. $f(x) = \log_8 x$

35. $f(x) = \log_{1/2} x$

36. $f(x) = \log_{1/4} x$

37. $f(x) = \log_{5/2} x$

38. $f(x) = \log_{7/3} x$

In Exercises 39 to 48, find the domain of the function. Write the domain using interval notation.

39. $f(x) = \log_5(x - 3)$

▶ **40.** $k(x) = \log_4(5 - x)$

41. $k(x) = \log_{2/3}(11 - x)$

42. $H(x) = \log_{1/4}(x^2 + 1)$

43. $P(x) = \ln(x^2 - 4)$

44. $J(x) = \ln\left(\frac{x - 3}{x}\right)$

45. $h(x) = \ln\left(\frac{x^2}{x - 4}\right)$

46. $R(x) = \ln(x^4 - x^2)$

47. $N(x) = \log_2(x^3 - x)$

48. $s(x) = \log_7(x^2 + 7x + 10)$

In Exercises 49 to 56, use translations of the graphs in Exercises 31 to 38 to produce the graph of the given function.

49. $f(x) = \log_4(x - 3)$

▶ **50.** $f(x) = \log_6(x + 3)$

51. $f(x) = \log_{12} x + 2$

52. $f(x) = \log_8 x - 4$

53. $f(x) = 3 + \log_{1/2} x$

54. $f(x) = 2 + \log_{1/4} x$

55. $f(x) = 1 + \log_{5/2}(x - 4)$

56. $f(x) = \log_{7/3}(x - 3) - 1$

57. Examine the following four functions and the graphs labeled **a**, **b**, **c**, and **d**. Determine which graph is the graph of each function.

$$f(x) = \log_5(x - 2) \qquad g(x) = 2 + \log_5 x$$
$$h(x) = \log_5(-x) \qquad k(x) = -\log_5(x + 3)$$

a.

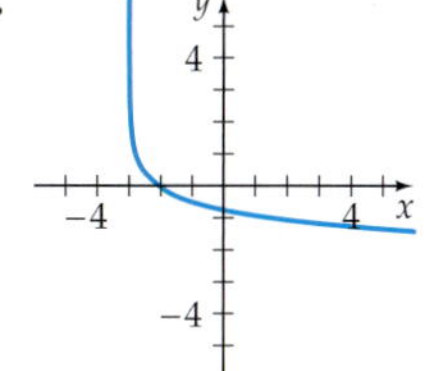

b.

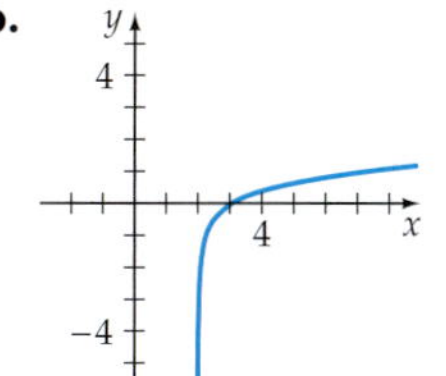

c.

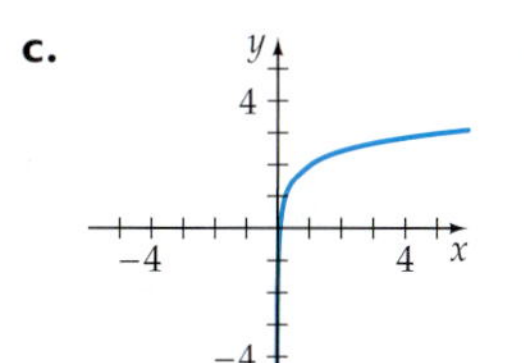

d.

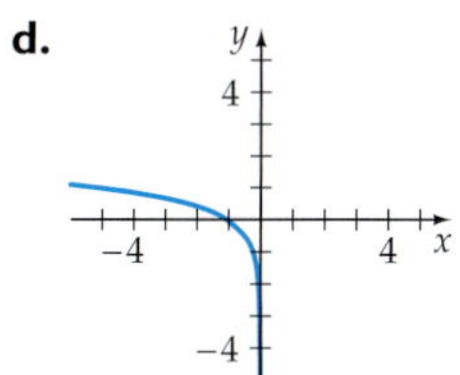

58. Examine the following four functions and the graphs labeled **a, b, c,** and **d.** Determine which graph is the graph of each function.

$f(x) = \ln x + 3 \qquad g(x) = \ln(x - 3)$

$h(x) = \ln(3 - x) \qquad k(x) = -\ln(-x)$

a.

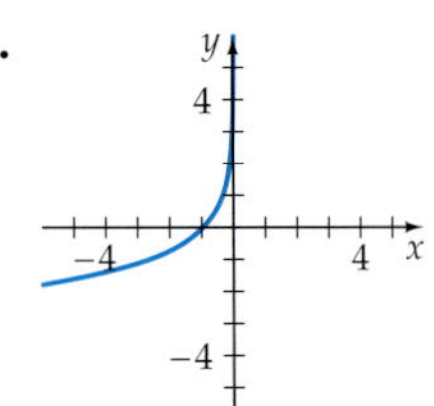

b.

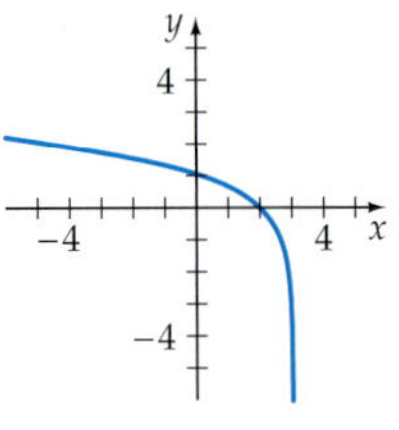

c.

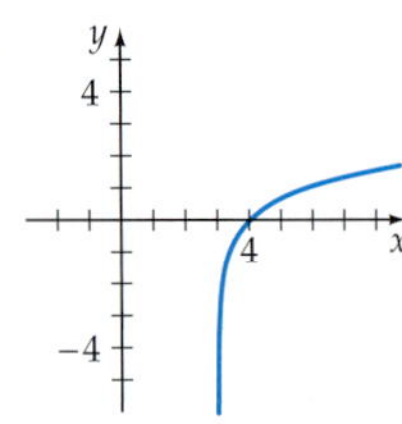

d.

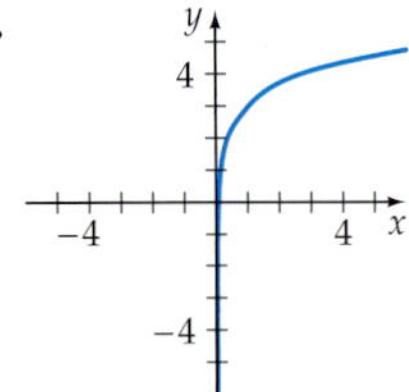

In Exercises 59 to 68, use a graphing utility to graph the function.

59. $f(x) = -2 \ln x$

60. $f(x) = -\log x$

61. $f(x) = |\ln x|$

62. $f(x) = \ln |x|$

63. $f(x) = \log \sqrt[3]{x}$

64. $f(x) = \ln \sqrt{x}$

65. $f(x) = \log(x + 10)$

66. $f(x) = \ln(x + 3)$

67. $f(x) = 3 \log |2x + 10|$

68. $f(x) = \frac{1}{2} \ln |x - 4|$

69.  **MONEY MARKET RATES** The function

$$r(t) = 0.69607 + 0.60781 \ln t$$

gives the annual interest rate r, as a percent, a bank will pay on its money market accounts, where t is the term (the time the money is invested) in months.

a. What interest rate, to the nearest tenth of a percent, will the bank pay on a money market account with a term of 9 months?

b. What is the minimum number of complete months during which a person must invest to receive an interest rate of at least 3%?

70. AVERAGE TYPING SPEED The following function models the average typing speed S, in words per minute, of a student who has been typing for t months.

$$S(t) = 5 + 29 \ln(t + 1), \quad 0 \le t \le 16$$

a. What was the student's average typing speed, to the nearest word per minute, when the student first started to type? What was the student's average typing speed, to the nearest word per minute, after 3 months?

b. Use a graph of S to determine how long, to the nearest tenth of a month, it will take the student to achieve an average typing speed of 65 words per minute.

71.  **ADVERTISING COSTS AND SALES** The function

$$N(x) = 2750 + 180 \ln\left(\frac{x}{1000} + 1\right)$$

models the relationship between the dollar amount x spent on advertising a product and the number of units N that a company can sell.

a. Find the number of units that will be sold with advertising expenditures of \$20,000, \$40,000, and \$60,000.

b. How many units will be sold if the company does not pay to advertise the product?

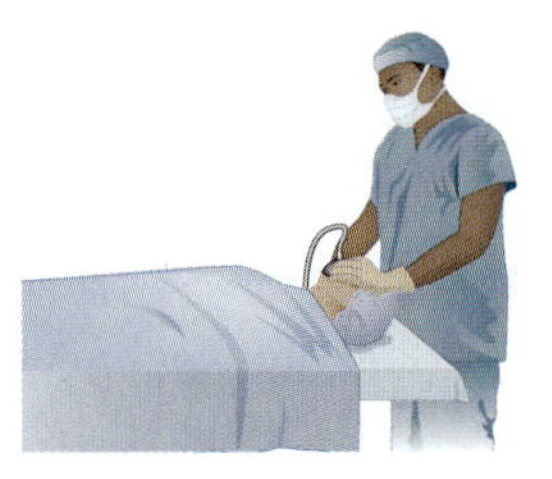

In anesthesiology it is necessary to accurately estimate the body surface area of a patient. One formula for estimating body surface area (*BSA*) was developed by Edith Boyd (University of Minnesota Press, 1935). Her formula for the *BSA* (in square meters) of a patient of height H (in centimeters) and weight W (in grams) is

$$BSA = 0.0003207 \cdot H^{0.3} \cdot W^{(0.7285 - 0.0188 \log W)}$$

MEDICINE In Exercises 72 and 73, use Boyd's formula to estimate the body surface area of a patient with the given weight and height. Round to the nearest hundredth of a square meter.

72. $W = 110$ pounds (49,895.2 grams); $H = 5$ feet 4 inches (162.56 centimeters)

73. $W = 180$ pounds (81,646.6 grams); $H = 6$ feet 1 inch (185.42 centimeters)

74.   **ASTRONOMY** Astronomers measure the apparent brightness of a star by a unit called the **apparent magnitude.** This unit was created in the second century B.C. when the Greek astronomer Hipparchus classified the relative brightness of several stars. In his list he assigned the number 1 to the stars that appeared to be the brightest (Sirius, Vega, and Deneb). They are first-magnitude stars. Hipparchus assigned the number 2 to all the stars in the Big Dipper. They are second-magnitude stars. The following table shows the relationship between a star's brightness relative to a first-magnitude star and the star's apparent magnitude. Notice from the table that a first-magnitude star appears in the sky to be about 2.51 times as bright as a second-magnitude star.

Brightness relative to a first-magnitude star x	Apparent magnitude $M(x)$
1	1
$\frac{1}{2.51}$	2
$\frac{1}{6.31} \approx \frac{1}{2.51^2}$	3
$\frac{1}{15.85} \approx \frac{1}{2.51^3}$	4
$\frac{1}{39.82} \approx \frac{1}{2.51^4}$	5
$\frac{1}{100} \approx \frac{1}{2.51^5}$	6

The following logarithmic function gives the apparent magnitude $M(x)$ of a star as a function of its brightness x.

$$M(x) = -2.51 \log x + 1, \quad 0 < x \le 1$$

a. Use $M(x)$ to find the apparent magnitude of a star that is $\frac{1}{10}$ as bright as a first-magnitude star. Round to the nearest hundredth.

b. Find the approximate apparent magnitude of a star that is $\frac{1}{400}$ as bright as a first-magnitude star. Round to the nearest hundredth.

c. Which star appears brighter: a star with an apparent magnitude of 12 or a star with an apparent magnitude of 15?

d. Is $M(x)$ an increasing function or a decreasing function?

75. **NUMBER OF DIGITS IN b^x** An engineer has determined that the number of digits N in the expansion of b^x, where both b and x are positive integers, is $N = \text{int}(x \log b) + 1$, where $\text{int}(x \log b)$ denotes the greatest integer of $x \log b$. (*Note:* The greatest integer of the real number x is x if x is an integer and is the largest integer less than x if x is not an integer. For example, the greatest integer of 5 is 5 and the greatest integer of 7.8 is 7.)

a. Because $2^{10} = 1024$, we know that 2^{10} has four digits. Use the equation $N = \text{int}(x \log b) + 1$ to verify this result.

b. Find the number of digits in 3^{200}.

c. Find the number of digits in 7^{4005}.

d. The largest known prime number as of November 17, 2003 was $2^{20996011} - 1$. Find the number of digits in this prime number. (*Hint:* Because $2^{20996011}$ is not a power of 10, both $2^{20996011}$ and $2^{20996011} - 1$ have the same number of digits.)

76. **NUMBER OF DIGITS IN $9^{(9^9)}$** A science teacher has offered 10 points extra credit to any student who will write out all the digits in the expansion of $9^{(9^9)}$.

a. Use the formula from Exercise 75 to determine the number of digits in this number.

b. Assume that you can write 1000 digits per page and that 500 pages of paper are in a ream of paper. How many reams of paper, to the nearest tenth of a ream, are required to write out the expansion of $9^{(9^9)}$? Assume that you write on only one side of each page.

CONNECTING CONCEPTS

77. Use a graphing utility to graph $f(x) = \frac{e^x - e^{-x}}{2}$ and $g(x) = \ln\left(x + \sqrt{x^2 + 1}\right)$ on the same screen. Use a square viewing window. What appears to be the relationship between f and g?

78. Use a graphing utility to graph $f(x) = \frac{e^x + e^{-x}}{2}$, for $x \ge 0$, and $g(x) = \ln\left(x + \sqrt{x^2 - 1}\right)$, for $x \ge 1$, on the same screen. Use a square viewing window. What appears to be the relationship between f and g?

79. The functions $f(x) = \frac{e^x - e^{-x}}{e^x + e^{-x}}$ and $g(x) = \frac{1}{2} \ln \frac{1 + x}{1 - x}$ are inverse functions. The domain of f is the set of all real numbers. The domain of g is $\{x \mid -1 < x < 1\}$. Use this information to determine the range of f and the range of g.

80. Use a graph of $f(x) = \frac{2}{e^x + e^{-x}}$ to determine the domain and the range of f.

PREPARE FOR SECTION 3.4

In Exercises 81 to 86, use a calculator to compare each of the given expressions.

81. $\log 3 + \log 2; \log 6$ [3.2]

82. $\ln 8 - \ln 3; \ln\left(\frac{8}{3}\right)$ [3.2]

83. $3 \log 4; \log(4^3)$ [3.2]

84. $2 \ln 5; \ln(5^2)$ [3.2]

85. $\ln 5; \frac{\log 5}{\log e}$ [3.2]

86. $\log 8; \frac{\ln 8}{\ln 10}$ [3.2]

PROJECTS

1.  **BENFORD'S LAW** The authors of this text know some interesting details about your finances. For instance, of the last 150 checks you have written, about 30% are for amounts that start with the number 1. Also, you have written about 3 times as many checks for amounts that start with the number 2 than you have for amounts that start with the number 7.

We are sure of these results because of a mathematical formula known as **Benford's Law.** This law was first discovered by the mathematician Simon Newcomb in 1881 and then rediscovered by the physicist Frank Benford in 1938. Benford's Law states that the probability P that the first digit of a number selected from a wide range of numbers is d is given by

$$P(d) = \log\left(1 + \frac{1}{d}\right)$$

a. Use Benford's Law to complete the table below and the bar graph at the top of the next column.

d	$P(d) = \log\left(1 + \frac{1}{d}\right)$
1	0.301
2	0.176
3	0.125
4	
5	
6	
7	
8	
9	

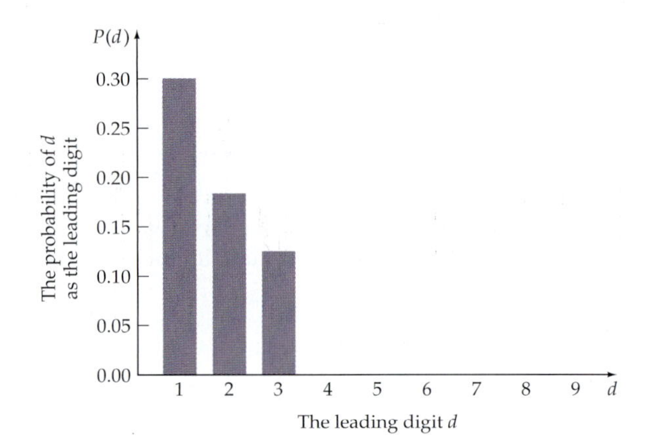

Benford's Law applies to most data with a wide range. For instance, it applies to

- the populations of the cities in the U.S.
- the numbers of dollars in the savings accounts at your local bank.
- the number of miles driven during a month by each person in a state.

b. Use the table in part **a.** to find the probability that in a U.S. city selected at random, the number of telephones in that city will be a number starting with 6.

c. Use the table in part **a.** to estimate how many times as many purchases you have made for dollar amounts that start with a 1 than for dollar amounts that start with a 9.

d. Explain why Benford's Law would not apply to the set of telephone numbers of the people living in a small city such as Le Mars, Iowa.

e. Explain why Benford's Law would not apply to the set of all the ages, in years, of students at a local high school.

An Application of Benford's Law Benford's Law has been used to identify fraudulent accountants. In most cases these accountants are unaware of Benford's Law and have replaced valid numbers with numbers selected at random. Their numbers do not conform to Benford's Law. Hence an audit is warranted.

Section 3.4 Logarithms and Logarithmic Scales

- Properties of Logarithms
- Change-of-Base Formula
- Logarithmic Scales

Properties of Logarithms

In Section 3.3 we introduced the following basic properties of logarithms.

$$\log_b b = 1 \qquad \text{and} \qquad \log_b 1 = 0$$

Also, because exponential functions and logarithmic functions are inverses of each other, we observed the relationships

$$\log_b(b^x) = x \qquad \text{and} \qquad b^{\log_b x} = x$$

We can use the properties of exponents to establish the following additional logarithmic properties.

Properties of Logarithms

In the following properties, b, M, and N are positive real numbers ($b \neq 1$).

Product property	$\log_b(MN) = \log_b M + \log_b N$
Quotient property	$\log_b \frac{M}{N} = \log_b M - \log_b N$
Power property	$\log_b(M^p) = p \log_b M$
Logarithm-of-each-side property	$M = N$ implies $\log_b M = \log_b N$
One-to-one property	$\log_b M = \log_b N$ implies $M = N$

take note

Pay close attention to these properties. Note that

$$\log_b(MN) \neq \log_b M \cdot \log_b N$$

and

$$\log_b \frac{M}{N} \neq \frac{\log_b M}{\log_b N}$$

Also,

$$\log_b(M + N) \neq \log_b M + \log_b N$$

In fact, the expression $\log_b(M + N)$ cannot be expanded at all.

? QUESTION Is it true that $\ln 5 + \ln 10 = \ln 50$?

The above properties of logarithms are often used to rewrite logarithmic expressions in an equivalent form.

? ANSWER Yes. By the product property, $\ln 5 + \ln 10 = \ln(5 \cdot 10)$.

EXAMPLE 1 **Rewrite Logarithmic Expressions**

Use the properties of logarithms to express the following logarithms in terms of logarithms of x, y, and z.

a. $\log_5(xy^2)$ **b.** $\log_b \dfrac{2\sqrt{y}}{z^5}$

Solution

a.
$$\begin{aligned}\log_5(xy^2) &= \log_5 x + \log_5 y^2 && \text{• Product property}\\ &= \log_5 x + 2\log_5 y && \text{• Power property}\end{aligned}$$

b.
$$\begin{aligned}\log_b \frac{2\sqrt{y}}{z^5} &= \log_b(2\sqrt{y}) - \log_b z^5 && \text{• Quotient property}\\ &= \log_b 2 + \log_b \sqrt{y} - \log_b z^5 && \text{• Product property}\\ &= \log_b 2 + \log_b y^{1/2} - \log_b z^5 && \text{• Replace } \sqrt{y} \text{ with } y^{1/2}.\\ &= \log_b 2 + \frac{1}{2}\log_b y - 5\log_b z && \text{• Power property}\end{aligned}$$

▶ TRY EXERCISE 2, PAGE 251

The properties of logarithms are also used to rewrite expressions that involve several logarithms as a single logarithm.

EXAMPLE 2 **Rewrite Logarithmic Expressions**

Use the properties of logarithms to rewrite each expression as a single logarithm with a coefficient of 1.

a. $2\log_b x + \dfrac{1}{2}\log_b(x+4)$ **b.** $4\log_3(x+2) - 3\log_3(x-5)$

Solution

a.
$$\begin{aligned}&2\log_b x + \frac{1}{2}\log_b(x+4)\\ &\quad= \log_b x^2 + \log_b(x+4)^{1/2} && \text{• Power property}\\ &\quad= \log_b[x^2(x+4)^{1/2}] && \text{• Product property}\\ &\quad= \log_b\left(x^2\sqrt{x+4}\right)\end{aligned}$$

b.
$$\begin{aligned}&4\log_3(x+2) - 3\log_3(x-5)\\ &\quad= \log_3(x+2)^4 - \log_3(x-5)^3 && \text{• Power property}\\ &\quad= \log_3\frac{(x+2)^4}{(x-5)^3} && \text{• Quotient property}\end{aligned}$$

▶ TRY EXERCISE 10, PAGE 251

CHANGE-OF-BASE FORMULA

Recall that to determine the value of y in $\log_3 81 = y$, we are basically asking, "What power of 3 is equal to 81?" Because $3^4 = 81$, we have $\log_3 81 = 4$. Now suppose that we need to determine the value of $\log_3 50$. In this case we need to find the power of 3 that produces 50. Because $3^3 = 27$ and $3^4 = 81$, the value we are seeking is somewhere between 3 and 4. The following procedure can be used to produce an estimate of $\log_3 50$.

The exponential form of $\log_3 50 = y$ is $3^y = 50$. Applying logarithmic properties gives us

$$3^y = 50$$

$$\ln 3^y = \ln 50 \qquad \text{• Logarithm-of-each-side property}$$

$$y \ln 3 = \ln 50 \qquad \text{• Power property}$$

$$y = \frac{\ln 50}{\ln 3} \approx 3.56088 \qquad \text{• Solve for } y.$$

Thus $\log_3 50 \approx 3.56088$. In the above procedure we could just as well have used logarithms of any base and arrived at the same value. Thus any logarithm can be expressed in terms of logarithms of any base we wish. This general result is summarized in the following formula.

Change-of-Base Formula

If x, a, and b are positive real numbers with $a \neq 1$ and $b \neq 1$, then

$$\log_b x = \frac{\log_a x}{\log_a b}$$

Because most calculators use only common logarithms ($a = 10$) or natural logarithms ($a = e$), the change-of-base formula is used most often in the following form.

If x and b are positive real numbers and $b \neq 1$, then

$$\log_b x = \frac{\log x}{\log b} = \frac{\ln x}{\ln b}$$

EXAMPLE 3 Use the Change-of-Base Formula

Evaluate each logarithm. Round to the nearest hundred thousandth.

a. $\log_3 18$ **b.** $\log_{12} 400$

Solution

To approximate these logarithms, we may use the change-of-base formula with $a = 10$ or $a = e$. For this example we choose to use the change-of-base formula with $a = e$. That is, we will evaluate these logarithms by using the LN key on a scientific or graphing calculator.

a. $\log_3 18 = \dfrac{\ln 18}{\ln 3} \approx 2.63093$ **b.** $\log_{12} 400 = \dfrac{\ln 400}{\ln 12} \approx 2.41114$

TRY EXERCISE 16, PAGE 252

take note

If common logarithms had been used for the calculation in Example 3**a.**, the final result would have been the same.

$$\log_3 18 = \frac{\log 18}{\log 3} \approx 2.63093$$

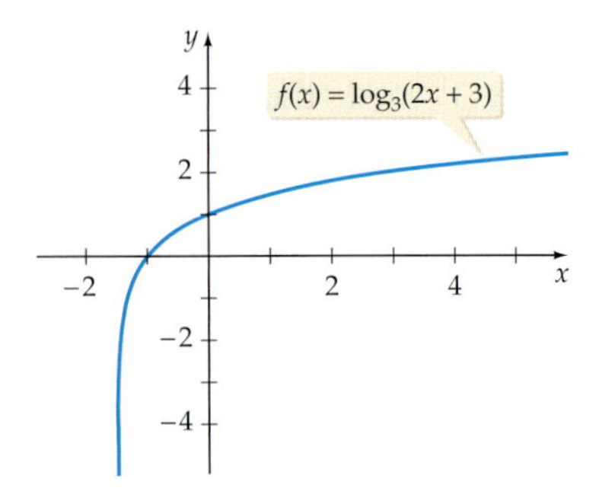

FIGURE 3.33

The change-of-base formula and a graphing calculator can be used to graph logarithmic functions that have a base other than 10 or e. For instance, to graph $f(x) = \log_3(2x + 3)$, we rewrite the function in terms of base 10 or base e. Using base 10 logarithms, we have $f(x) = \log_3(2x + 3) = \dfrac{\log(2x + 3)}{\log 3}$. The graph is shown in **Figure 3.33.**

EXAMPLE 4 Use the Change-of-Base Formula to Graph a Logarithmic Function

Graph $f(x) = \log_2|x - 3|$.

Solution

Rewrite f using the change-of-base formula. We will use the natural logarithmic function; however, the common logarithmic function could be used instead.

$$f(x) = \log_2|x - 3| = \frac{\ln|x - 3|}{\ln 2}$$

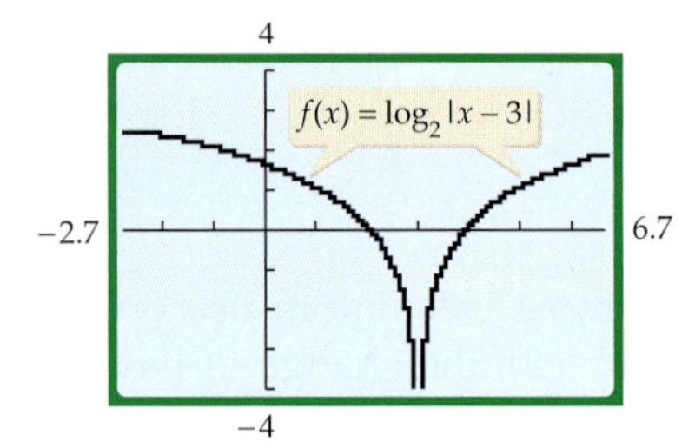

- **Enter $\dfrac{\ln|x - 3|}{\ln 2}$ into Y1. Note that the domain of $f(x) = \log_2|x - 3|$ is all real numbers except 3, because $|x - 3| = 0$ when $x = 3$ and $|x - 3|$ is positive for all other values of x.**

▶ TRY EXERCISE 24, PAGE 252

LOGARITHMIC SCALES

Logarithmic functions are often used to scale very large (or very small) numbers into numbers that are easier to comprehend. For instance, the *Richter scale* magnitude of an earthquake uses a logarithmic function to convert the intensity of the earthquake's shock waves I into a number M, which for most earthquakes is in the range of 0 to 10. The intensity I of an earthquake is often given in terms of the constant I_0, where I_0 is the intensity of the smallest earthquake (called a **zero-level earthquake**) that can be measured on a seismograph near the earthquake's epicenter. The following formula is used to compute the Richter scale magnitude of an earthquake.

MATH MATTERS

The Richter scale was created by the seismologist Charles F. Richter in 1935. Notice that a tenfold increase in the intensity level of an earthquake only increases the Richter scale magnitude of the earthquake by 1.

The Richter Scale Magnitude of an Earthquake

An earthquake with an intensity of I has a **Richter scale magnitude** of

$$M = \log\left(\frac{I}{I_0}\right)$$

where I_0 is the measure of the intensity of a zero-level earthquake.

EXAMPLE 5 Determine the Magnitude of an Earthquake

Find the Richter scale magnitude (to the nearest 0.1) of the 1999 Joshua Tree, California earthquake that had an intensity of $I = 12{,}589{,}254I_0$.

Solution

$$M = \log\left(\frac{I}{I_0}\right) = \log\left(\frac{12{,}589{,}254I_0}{I_0}\right) = \log(12{,}589{,}254) \approx 7.1$$

The 1999 Joshua Tree earthquake had a Richter scale magnitude of 7.1.

▶ TRY EXERCISE 56, PAGE 253

take note

Notice in Example 5 that we didn't need to know the value of I_0 to determine the Richter scale magnitude of the quake.

If you know the Richter scale magnitude of an earthquake, you can determine the intensity of the earthquake.

EXAMPLE 6 Determine the Intensity of an Earthquake

Find the intensity of the 1999 Taiwan earthquake, which measured 7.6 on the Richter scale.

Solution

$$\log\left(\frac{I}{I_0}\right) = 7.6$$

$$\frac{I}{I_0} = 10^{7.6}$$ • Write in exponential form.

$$I = 10^{7.6}I_0$$ • Solve for I.

$$I \approx 39{,}810{,}717I_0$$

The 1999 Taiwan earthquake had an intensity that was approximately 39,811,000 times the intensity of a zero-level earthquake.

▶ TRY EXERCISE 58, PAGE 253

In Example 7 we make use of the Richter scale magnitudes of two earthquakes to compare the intensities of the earthquakes.

EXAMPLE 7 Compare Earthquakes

The 1960 Chile earthquake had a Richter scale magnitude of 9.5. The 1989 San Francisco earthquake had a Richter scale magnitude of 7.1. Compare the intensities of the earthquakes.

Continued ▶

take note

The results of Example 7 show that if an earthquake has a Richter scale magnitude of M_1 and a smaller earthquake has a Richter scale magnitude of M_2, then the larger earthquake is $10^{M_1-M_2}$ times as intense as the smaller earthquake.

Solution

Let I_1 be the intensity of the Chilean earthquake and I_2 the intensity of the San Francisco earthquake. Then

$$\log\left(\frac{I_1}{I_0}\right) = 9.5 \qquad \text{and} \qquad \log\left(\frac{I_2}{I_0}\right) = 7.1$$

$$\frac{I_1}{I_0} = 10^{9.5} \qquad\qquad \frac{I_2}{I_0} = 10^{7.1}$$

$$I_1 = 10^{9.5}I_0 \qquad\qquad I_2 = 10^{7.1}I_0$$

To compare the intensities of the earthquakes, we compute the ratio I_1/I_2.

$$\frac{I_1}{I_2} = \frac{10^{9.5}I_0}{10^{7.1}I_0} = \frac{10^{9.5}}{10^{7.1}} = 10^{9.5-7.1} = 10^{2.4} \approx 251$$

The earthquake in Chile was approximately 251 times as intense as the San Francisco earthquake.

▶ TRY EXERCISE 60, PAGE 253

Seismologists generally determine the Richter scale magnitude of an earthquake by examining a *seismogram.* See **Figure 3.34.**

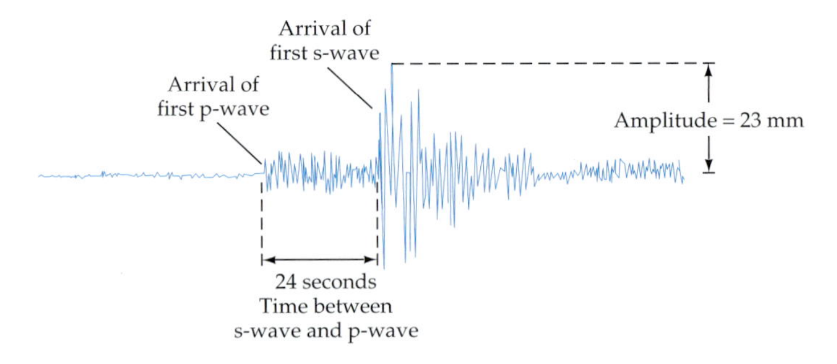

FIGURE 3.34

The magnitude of an earthquake cannot be determined just by examining the amplitude of a seismogram because this amplitude decreases as the distance between the epicenter of the earthquake and the observation station increases. To account for the distance between the epicenter and the observation station, a seismologist examines a seismogram for both small waves called **p-waves** and larger waves called **s-waves.** The Richter scale magnitude M of the earthquake is a function of both the amplitude A of the s-waves and the difference in time t between the occurrence of the s-waves and the p-waves. In the 1950s, Charles Richter developed the following formula to determine the magnitude of an earthquake from the data in a seismogram.

Amplitude-Time-Difference Formula

The Richter scale magnitude M of an earthquake is given by

$$M = \log A + 3 \log 8t - 2.92$$

where A is the amplitude, in millimeters, of the s-waves on a seismogram and t is the difference in time, in seconds, between the s-waves and the p-waves.

EXAMPLE 8 Determine the Magnitude of an Earthquake from Its Seismogram

Find the Richter scale magnitude of the earthquake that produced the seismogram in **Figure 3.34.**

Solution

$$\begin{aligned} M &= \log A + 3 \log 8t - 2.92 \\ &= \log 23 + 3 \log[8 \cdot 24] - 2.92 \\ &\approx 1.36173 + 6.84990 - 2.92 \\ &\approx 5.3 \end{aligned}$$

• Substitute 23 for A and 24 for t.

The earthquake had a magnitude of about 5.3 on the Richter scale.

▶ TRY EXERCISE 64, PAGE 253

take note

The Richter scale magnitude is usually rounded to the nearest tenth.

Logarithmic scales are also used in chemistry. One example concerns the pH of a liquid, which is a measure of the liquid's **acidity** or **alkalinity.** (You may have tested the pH of a swimming pool or an aquarium.) Pure water, which is considered neutral, has a pH of 7.0. The pH scale ranges from 0 to 14, with 0 corresponding to the most acidic solutions and 14 to the most alkaline. Lemon juice has a pH of about 2, whereas household ammonia measures about 11.

Specifically, the pH of a solution is a function of the hydronium-ion concentration of the solution. Because the hydronium-ion concentration of a solution can be very small (with values such as 0.00000001), pH uses a logarithmic scale.

take note

One mole is equivalent to 6.022×10^{23} ions.

The pH of a Solution

The **pH of a solution** with a hydronium-ion concentration of H^+ moles per liter is given by

$$\text{pH} = -\log[H^+]$$

EXAMPLE 9 Find the pH of a Solution

Find the pH of each liquid. Round to the nearest tenth.

a. Orange juice with $H^+ = 2.8 \times 10^{-4}$ mole per liter

b. Milk with $H^+ = 3.97 \times 10^{-7}$ mole per liter

c. Rainwater with $H^+ = 6.31 \times 10^{-5}$ mole per liter

d. A baking soda solution with $H^+ = 3.98 \times 10^{-9}$ mole per liter

MATH MATTERS

The pH scale was created by the Danish biochemist Søren Sørensen in 1909 to measure the acidity of water used in the brewing of beer. pH is an abbreviation for *pondus hydrogenii,* which translates as "potential hydrogen."

Solution

a. $pH = -\log[H^+] = -\log(2.8 \times 10^{-4}) \approx 3.6$
The orange juice has a pH of 3.6.

b. $pH = -\log[H^+] = -\log(3.97 \times 10^{-7}) \approx 6.4$
The milk has a pH of 6.4.

c. $pH = -\log[H^+] = -\log(6.31 \times 10^{-5}) \approx 4.2$
The rainwater has a pH of 4.2.

d. $pH = -\log[H^+] = -\log(3.98 \times 10^{-9}) \approx 8.4$
The baking soda solution has a pH of 8.4.

▶ TRY EXERCISE 48, PAGE 252

Figure 3.35 illustrates the pH scale, along with the corresponding hydronium-ion concentrations. A solution on the left half of the scale, with a pH of less than 7, is an **acid,** and a solution on the right half of the scale is an **alkaline solution** or a **base.** Because the scale is logarithmic, a solution with a pH of 5 is 10 times more acidic than a solution with a pH of 6. From Example 9 we see that the orange juice, rainwater, and milk are acids, whereas the baking soda solution is a base.

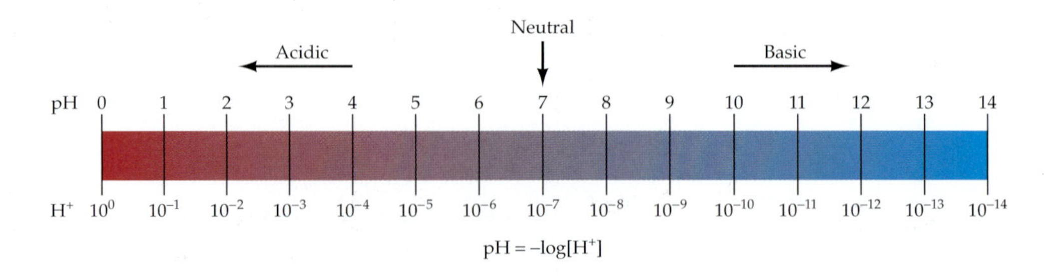

FIGURE 3.35

EXAMPLE 10 Find the Hydronium-Ion Concentration

A sample of blood has a pH of 7.3. Find the hydronium-ion concentration of the blood.

Solution

$$\begin{aligned} pH &= -\log[H^+] \\ 7.3 &= -\log[H^+] && \text{• Substitute 7.3 for pH.} \\ -7.3 &= \log[H^+] && \text{• Multiply both sides by } -1. \\ 10^{-7.3} &= H^+ && \text{• Change to exponential form.} \\ 5.0 \times 10^{-8} &\approx H^+ \end{aligned}$$

The hydronium-ion concentration of the blood is about 5.0×10^{-8} mole per liter.

▶ TRY EXERCISE 50, PAGE 253

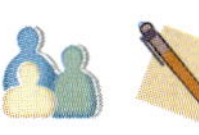

Topics for Discussion

1. The function $f(x) = \log_b x$ is defined only for $x > 0$. Explain why this condition is imposed.
2. If p and q are positive numbers, explain why $\ln(p + q)$ isn't normally equal to $\ln p + \ln q$.
3. If $f(x) = \log_b x$ and $f(c) = f(d)$, can we conclude that $c = d$?
4. Give examples of situations in which it is advantageous to use logarithmic scales.

Exercise Set 3.4

In Exercises 1 to 8, write the given logarithm in terms of logarithms of x, y, and z.

1. $\log_b(xyz)$
2. $\ln \dfrac{z^3}{\sqrt{xy}}$
3. $\ln \dfrac{x}{z^4}$
4. $\log_5 \dfrac{xy^2}{z^4}$
5. $\log_2 \dfrac{\sqrt{x}}{y^3}$
6. $\log_b(x\sqrt[3]{y})$
7. $\log_7 \dfrac{\sqrt{xz}}{y^2}$
8. $\ln \sqrt[3]{x^2\sqrt{y}}$

In Exercises 9 to 14, write each logarithmic expression as a single logarithm with a coefficient of 1. Simplify when possible.

9. $\log(x + 5) + 2\log x$
10. $3\log_2 t - \dfrac{1}{3}\log_2 u + 4\log_2 v$
11. $\ln(x^2 - y^2) - \ln(x - y)$
12. $\dfrac{1}{2}\log_8(x + 5) - 3\log_8 y$

13. $3 \log x + \frac{1}{3} \log y + \log(x + 1)$

14. $\ln(xz) - \ln(x\sqrt{y}) + 2 \ln \frac{y}{z}$

In Exercises 15 to 22, use the change-of-base formula to approximate the logarithm accurate to the nearest ten thousandth.

15. $\log_7 20$

16. $\log_5 37$

17. $\log_{11} 8$

18. $\log_{50} 22$

19. $\log_6 \frac{1}{3}$

20. $\log_3 \frac{7}{8}$

21. $\log_9 \sqrt{17}$

22. $\log_4 \sqrt{7}$

In Exercises 23 to 30, use a graphing utility and the change-of-base formula to graph the logarithmic function.

23. $f(x) = \log_4 x$

24. $g(x) = \log_8(5 - x)$

25. $g(x) = \log_8(x - 3)$

26. $t(x) = \log_9(5 - x)$

27. $h(x) = \log_3(x - 3)^2$

28. $J(x) = \log_{12}(-x)$

29. $F(x) = -\log_5|x - 2|$

30. $n(x) = \log_2\sqrt{x - 8}$

In Exercises 31 to 40, determine if the statement is true or false for all $x > 0$, $y > 0$. If it is false, write an example that disproves the statement.

31. $\log_b(x + y) = \log_b x + \log_b y$

32. $\log_b(xy) = \log_b x \cdot \log_b y$

33. $\log_b(xy) = \log_b x + \log_b y$

34. $\log_b x \cdot \log_b y = \log_b x + \log_b y$

35. $\log_b x - \log_b y = \log_b(x - y), \quad x > y$

36. $\log_b \frac{x}{y} = \frac{\log_b x}{\log_b y}$

37. $\frac{\log_b x}{\log_b y} = \log_b x - \log_b y$

38. $\log_b(x^n) = n \log_b x$

39. $(\log_b x)^n = n \log_b x$

40. $\log_b \sqrt{x} = \frac{1}{2} \log_b x$

41. Evaluate the following *without* using a calculator.

$$\log_3 5 \cdot \log_5 7 \cdot \log_7 9$$

42. Evaluate the following *without* using a calculator.

$$\log_5 20 \cdot \log_{20} 60 \cdot \log_{60} 100 \cdot \log_{100} 125$$

43. Which is larger, 500^{501} or 506^{500}? These numbers are too large for most calculators to handle. (They each have 1353 digits!) (*Hint:* Let $x = 500^{501}$ and $y = 506^{500}$ and then compare $\ln x$ with $\ln y$.)

44. Which number is smaller, $\frac{1}{50^{300}}$ or $\frac{1}{151^{233}}$?

45. ANIMATED MAPS A software company that creates interactive maps for Web sites has designed an animated zooming feature so that when a user selects the zoom-in option, the map appears to expand on a location. This is accomplished by displaying several intermediate maps to give the illusion of motion. The company has determined that zooming in on a location is more informative and pleasing to observe when the scale of each step of the animation is determined using the equation

$$S_n = S_0 \cdot 10^{\frac{n}{N}(\log S_f - \log S_0)}$$

where S_n represents the scale of the current step n ($n = 0$ corresponds to the initial scale), S_0 is the starting scale of the map, S_f is the final scale, and N is the number of steps in the animation following the initial scale. (If the initial scale of the map is 1:200, then $S_0 = 200$.) Determine the scales to be used at each intermediate step if a map is to start with a scale of 1:1,000,000 and proceed through five intermediate steps to end with a scale of 1:500,000.

46. ANIMATED MAPS Use the equation in Exercise 45 to determine the scales for each stage of an animated map zoom that goes from a scale of 1:250,000 to a scale of 1:100,000 in four steps (following the initial scale).

47. pH Milk of magnesia has a hydronium-ion concentration of about 3.97×10^{-11} mole per liter. Determine the pH of milk of magnesia and state whether it is an acid or a base.

48. pH Vinegar has a hydronium-ion concentration of 1.26×10^{-3} mole per liter. Determine the pH of vinegar and state whether it is an acid or a base.

49. HYDRONIUM-ION CONCENTRATION A morphine solution has a pH of 9.5. Determine the hydronium-ion concentration of the morphine solution.

50. **Hydronium-Ion Concentration** A rainstorm in New York City produced rainwater with a pH of 5.6. Determine the hydronium-ion concentration of the rainwater.

51. **Decibel Level** The range of sound intensities that the human ear can detect is so large that a special decibel scale (named after Alexander Graham Bell) is used to measure and compare sound intensities. The **decibel level** dB of a sound is given by

$$dB(I) = 10 \log\left(\frac{I}{I_0}\right)$$

where I_0 is the intensity of sound that is barely audible to the human ear. Find the decibel level for the following sounds. Round to the nearest tenth of a decibel.

Sound	Intensity
a. Automobile traffic	$I = 1.58 \times 10^8 \cdot I_0$
b. Quiet conversation	$I = 10{,}800 \cdot I_0$
c. Fender guitar	$I = 3.16 \times 10^{11} \cdot I_0$
d. Jet engine	$I = 1.58 \times 10^{15} \cdot I_0$

52. **Comparison of Sound Intensities** A team in Arizona installed a 48,000-watt sound system in a Ford Bronco that it claims can output 175-decibel sound. The human pain threshold for sound is 125 decibels. How many times more intense is the sound from the Bronco than the human pain threshold?

53. **Comparison of Sound Intensities** How many times more intense is a sound that measures 120 decibels than a sound that measures 110 decibels?

54. **Decibel Level** If the intensity of a sound is doubled, what is the increase in the decibel level? [*Hint:* Find $dB(2I) - dB(I)$.]

55. **Earthquake Magnitude** What is the Richter scale magnitude of an earthquake with an intensity of $I = 100{,}000I_0$?

56. **Earthquake Magnitude** The Colombia earthquake of 1906 had an intensity of $I = 398{,}107{,}000I_0$. What did it measure on the Richter scale?

57. **Earthquake Intensity** The Coalinga, California, earthquake of 1983 had a Richter scale magnitude of 6.5. Find the intensity of this earthquake.

58. **Earthquake Intensity** The earthquake that occurred just south of Concepción, Chile, in 1960 had a Richter scale magnitude of 9.5. Find the intensity of this earthquake.

59. **Comparison of Earthquakes** Compare the intensity of an earthquake that measures 5.0 on the Richter scale to the intensity of an earthquake that measures 3.0 on the Richter scale by finding the ratio of the larger intensity to the smaller intensity.

60. **Comparison of Earthquakes** How many times more intense was the 1960 earthquake in Chile, which measured 9.5 on the Richter scale, than the San Francisco earthquake of 1906, which measured 8.3 on the Richter scale?

61. **Comparison of Earthquakes** On March 2, 1933, an earthquake of magnitude 8.9 on the Richter scale struck Japan. In October 1989, an earthquake of magnitude 7.1 on the Richter scale struck San Francisco, California. Compare the intensity of the larger earthquake to the intensity of the smaller earthquake by finding the ratio of the larger intensity to the smaller intensity.

62. **Comparison of Earthquakes** An earthquake that occurred in China in 1978 measured 8.2 on the Richter scale. In 1988, an earthquake in California measured 6.9 on the Richter scale. Compare the intensity of the larger earthquake to the intensity of the smaller earthquake by finding the ratio of the larger intensity to the smaller intensity.

63. **Earthquake Magnitude** Find the Richter scale magnitude of the earthquake that produced the seismogram in the following figure.

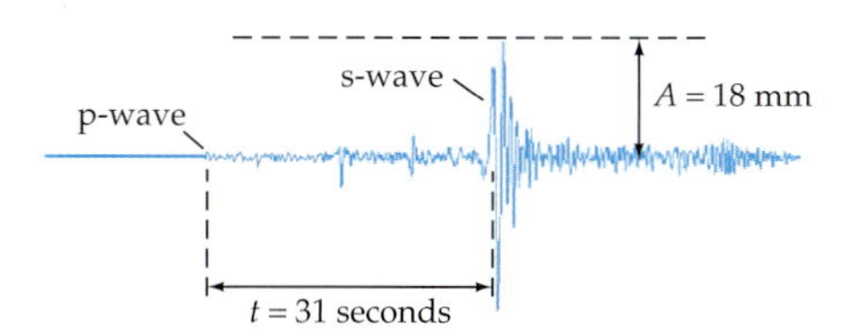

64. **Earthquake Magnitude** Find the Richter scale magnitude of the earthquake that produced the seismogram in the following figure.

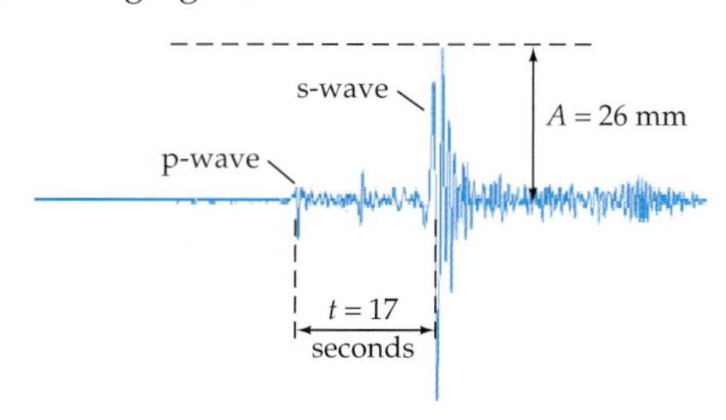

CONNECTING CONCEPTS

65. Nomograms and Logarithmic Scales A **nomogram** is a diagram used to determine a numerical result by drawing a line across numerical scales. The following nomogram, used by Richter, determines the magnitude of an earthquake from its seismogram. To use the nomogram, mark the amplitude of a seismogram on the amplitude scale and mark the time between the s-wave and the p-wave on the S-P scale. Draw a line between these marks. The Richter scale magnitude of the earthquake that produced the seismogram is shown by the intersection of the line and the center scale. The example below shows that an earthquake with a seismogram amplitude of 23 millimeters and an S-P time of 24 seconds has a Richter scale magnitude of about 5.

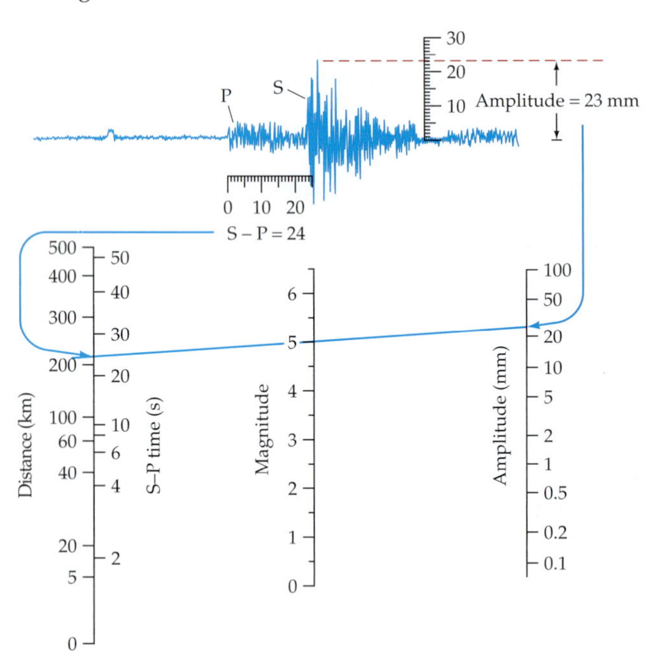

Richter's earthquake nomogram

The amplitude and the S-P time are shown on logarithmic scales. On the amplitude scale, the distance from 1 to 10 is the same as the distance from 10 to 100, because $\log 100 - \log 10 = \log 10 - \log 1$.

Use the nomogram at the left to determine the Richter scale magnitude of an earthquake with a seismogram

a. amplitude of 50 millimeters and S-P time of 40 seconds.

b. amplitude of 1 millimeter and S-P time of 30 seconds.

c. How do the results in parts **a.** and **b.** compare with the Richter scale magnitudes produced by using the amplitude-time-difference formula?

PREPARE FOR SECTION 3.5

66. Use the definition of a logarithm to write the exponential equation $3^6 = 729$ in logarithmic form. [3.2]

67. Use the definition of a logarithm to write the logarithmic equation $\log_5 625 = 4$ in exponential form. [3.2]

68. Use the definition of a logarithm to write the exponential equation $a^{x+2} = b$ in logarithmic form. [3.2]

69. Solve for x: $4a = 7bx + 2cx$.

70. Solve for x: $165 = \dfrac{300}{1 + 12x}$.

71. Solve for x: $A = \dfrac{100 + x}{100 - x}$.

PROJECTS

1. **LOGARITHMIC SCALES** Sometimes **logarithmic scales** are used to better view a collection of data that span a wide range of values. For instance, consider the table below, which lists the approximate masses of various marine creatures in grams. Next we have attempted to plot the masses on a number line.

Animal	Mass (g)
Rotifer	0.000000006
Dwarf goby	0.30
Lobster	15,900
Leatherback turtle	851,000
Giant squid	1,820,000
Whale shark	4,700,000
Blue whale	120,000,000

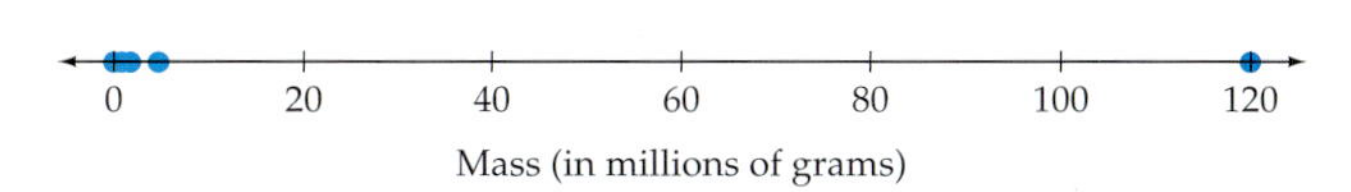

As you can see, we had to use such a large span of numbers that the data for most of the animals are bunched up at the left. Visually, this number line isn't very helpful for any comparisons.

a. Make a new number line, this time plotting the logarithm (base 10) of each of the masses.

b. Which number line is more helpful to compare the masses of the different animals?

c. If the data points for two animals on the logarithmic number line are 1 unit apart, how do the animals' masses compare? What if the points are 2 units apart?

2. **LOGARITHMIC SCALES** The distances of the planets in our solar system from the sun are given in the table at the top of the next column.

a. Draw a number line with an appropriate scale to plot the distances.

b. Draw a second number line, this time plotting the logarithm (base 10) of each distance.

c. Which number line do you find more helpful to compare the different distances?

d. If two distances are 3 units apart on the logarithmic number line, how do the distances of the corresponding planets compare?

Planet	Distance (million km)
Mercury	58
Venus	108
Earth	150
Mars	228
Jupiter	778
Saturn	1427
Uranus	2871
Neptune	4497
Pluto	5913

3. **BIOLOGIC DIVERSITY** To discuss the variety of species that live in a certain environment, a biologist needs a precise definition of *diversity*. Let $p_1, p_2, \ldots, p_n$ be the proportions of n species that live in an environment. The biologic diversity D of this system is

$$D = -(p_1 \log_2 p_1 + p_2 \log_2 p_2 + \cdots + p_n \log_2 p_n)$$

Suppose that an ecosystem has exactly five different varieties of grass: rye (R), bermuda (B), blue (L), fescue (F), and St. Augustine (A).

a. Calculate the diversity of this ecosystem if the proportions of these grasses are as shown in Table 1. Round to the nearest hundredth.

Table 1

R	B	L	F	A
$\frac{1}{5}$	$\frac{1}{5}$	$\frac{1}{5}$	$\frac{1}{5}$	$\frac{1}{5}$

b. Because bermuda and St. Augustine are virulent grasses, after a time the proportions will be as shown in Table 2. Calculate the diversity of this system. Does this system have more or less diversity than the system given in Table 1?

Table 2

R	B	L	F	A
$\frac{1}{8}$	$\frac{3}{8}$	$\frac{1}{16}$	$\frac{1}{8}$	$\frac{5}{16}$

c. After an even longer time period, the bermuda and St. Augustine grasses completely overrun the environment and the proportions are as shown in Table 3. Calculate the diversity of this system. (*Note:* Although the equation is not technically correct, for purposes of the diversity definition, we may say that $0 \log_2 0 = 0$. By using very small values of p_i, we can demonstrate that this definition makes sense.) Does this system have more or less diversity than the system given in Table 2?

d. Finally, the St. Augustine grasses overrun the bermuda grasses and the proportions are as shown in Table 4. Calculate the diversity of this system. Write a sentence that explains the meaning of the value you obtained.

Table 3

R	B	L	F	A
0	$\frac{1}{4}$	0	0	$\frac{3}{4}$

Table 4

R	B	L	F	A
0	0	0	0	1

SECTION 3.5

EXPONENTIAL AND LOGARITHMIC EQUATIONS

- SOLVE EXPONENTIAL EQUATIONS
- SOLVE LOGARITHMIC EQUATIONS
- APPLICATION

SOLVE EXPONENTIAL EQUATIONS

If a variable appears in an exponent of a term of an equation, such as $2^{x+1} = 32$, then the equation is called an **exponential equation.** Example 1 uses the following Equality of Exponents Theorem to solve $2^{x+1} = 32$.

Equality of Exponents Theorem

If $b^x = b^y$, then $x = y$, provided that $b > 0$ and $b \neq 1$.

EXAMPLE 1 Solve an Exponential Equation

Use the Equality of Exponents Theorem to solve $2^{x+1} = 32$.

Solution

$$2^{x+1} = 32$$

$$2^{x+1} = 2^5 \quad \text{• Write each side as a power of 2.}$$

$$x + 1 = 5 \quad \text{• Equate the exponents.}$$

$$x = 4$$

Check: Let $x = 4$, then $2^{x+1} = 2^{4+1}$

$$= 2^5$$

$$= 32$$

TRY EXERCISE 2, PAGE 263

A graphing utility can also be used to find solutions of an equation of the form $f(x) = g(x)$. Either of the following two methods can be employed.

Using a Graphing Utility to Find the Solutions of $f(x) = g(x)$

Intersection Method Graph $y_1 = f(x)$ and $y_2 = g(x)$ on the same screen. The solutions of $f(x) = g(x)$ are the x-coordinates of the points of intersection of the graphs.

Intercept Method The solutions of $f(x) = g(x)$ are the x-coordinates of the x-intercepts of the graph of $y = f(x) - g(x)$.

Figures 3.36 and **3.37** illustrate the graphical methods for solving $2^{x+1} = 32$.

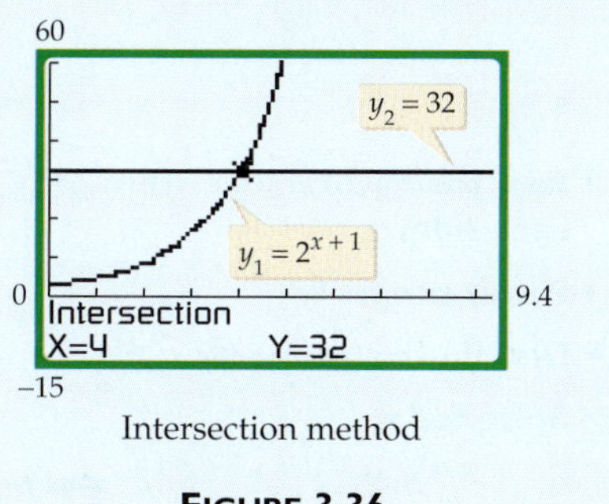

Intersection method

FIGURE 3.36

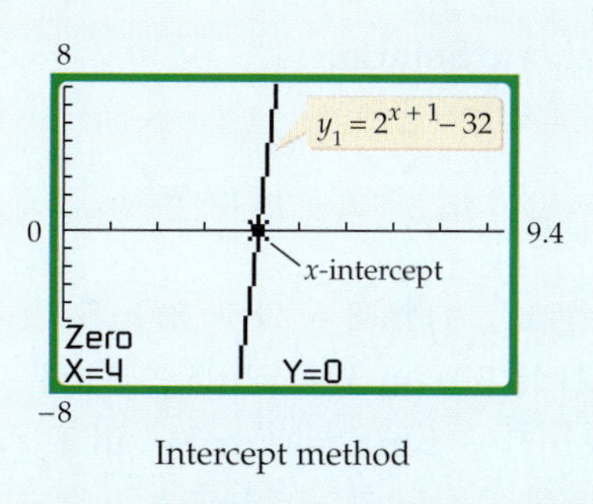

Intercept method

FIGURE 3.37

In Example 1 we were able to write both sides of the equation as a power of the same base. If you find it difficult to write both sides of an exponential equation in terms of the same base, then try the procedure of taking the logarithm of each side of the equation. This procedure is used in Example 2.

EXAMPLE 2 Solve an Exponential Equation

Solve: $5^x = 40$

Algebraic Solution

$$5^x = 40$$
$$\log(5^x) = \log 40 \quad \text{• Take the logarithm of each side.}$$
$$x \log 5 = \log 40 \quad \text{• Power property}$$
$$x = \frac{\log 40}{\log 5} \quad \text{• Exact solution}$$
$$x \approx 2.3 \quad \text{• Decimal approximation}$$

To the nearest tenth, the solution is 2.3.

Visualize the Solution

Intersection Method The solution of $5^x = 40$ is the x-coordinate of the point of intersection of $y = 5^x$ and $y = 40$ (see **Figure 3.38**).

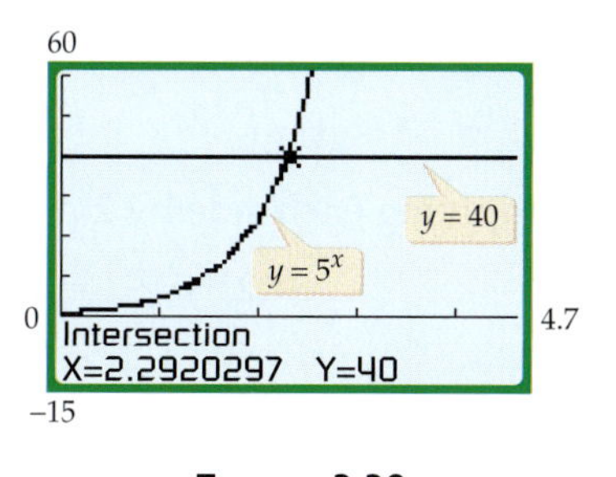

FIGURE 3.38

TRY EXERCISE 10, PAGE 263

An alternative approach to solving the equation in Example 2 is to rewrite the exponential equation in logarithmic form: $5^x = 40$ is equivalent to the logarithmic equation $\log_5 40 = x$. Using the change-of-base formula, we find that $x = \log_5 40 = \dfrac{\log 40}{\log 5}$. In the following example, however, we must take logarithms of both sides to reach a solution.

EXAMPLE 3 Solve an Exponential Equation

Solve: $3^{2x-1} = 5^{x+2}$

Algebraic Solution

$$3^{2x-1} = 5^{x+2}$$

$$\ln 3^{2x-1} = \ln 5^{x+2}$$ • Take the natural logarithm of each side.

$$(2x - 1)\ln 3 = (x + 2)\ln 5$$ • Power property

$$2x\ln 3 - \ln 3 = x\ln 5 + 2\ln 5$$ • Distributive property

$$2x\ln 3 - x\ln 5 = 2\ln 5 + \ln 3$$ • Solve for *x*.

$$x(2\ln 3 - \ln 5) = 2\ln 5 + \ln 3$$

$$x = \frac{2\ln 5 + \ln 3}{2\ln 3 - \ln 5}$$ • Exact solution

$$x \approx 7.3$$ • Decimal approximation

To the nearest tenth, the solution is 7.3.

Visualize the Solution

Intercept Method The solution of $3^{2x-1} = 5^{x+2}$ is the x-coordinate of the x-intercept of $y = 3^{2x-1} - 5^{x+2}$ (see **Figure 3.39**).

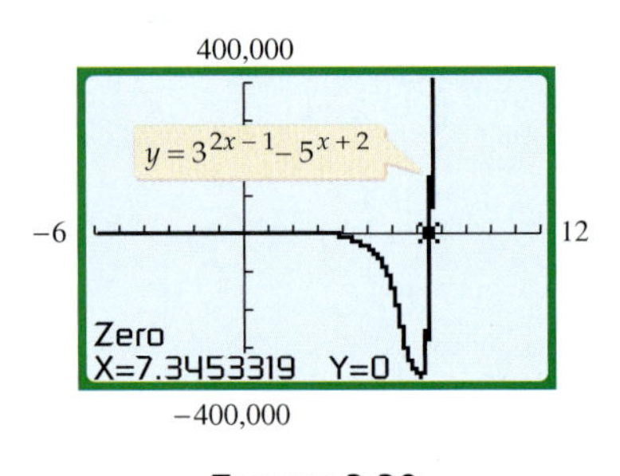

FIGURE 3.39

▶ TRY EXERCISE 18, PAGE 263

In Example 4 we solve an exponential equation that has two solutions.

EXAMPLE 4 Solve an Exponential Equation Involving $b^x + b^{-x}$

Solve: $\dfrac{2^x + 2^{-x}}{2} = 3$

Algebraic Solution

Multiplying each side by 2 produces

$$2^x + 2^{-x} = 6$$

$$2^{2x} + 2^0 = 6(2^x)$$ • Multiply each side by 2^x to clear negative exponents.

$$(2^x)^2 - 6(2^x) + 1 = 0$$ • Write in quadratic form.

$$(u)^2 - 6(u) + 1 = 0$$ • Substitute u for 2^x.

Visualize the Solution

Intersection Method The solutions of $\dfrac{2^x + 2^{-x}}{2} = 3$ are the x-coordinates of the points of intersection of

By the quadratic formula,

$$u = \frac{6 \pm \sqrt{36 - 4}}{2} = \frac{6 \pm 4\sqrt{2}}{2} = 3 \pm 2\sqrt{2}$$

$$2^x = 3 \pm 2\sqrt{2}$$ • Replace u with 2^x.

$$\log 2^x = \log(3 \pm 2\sqrt{2})$$ • Take the common logarithm of each side.

$$x \log 2 = \log(3 \pm 2\sqrt{2})$$ • Power property

$$x = \frac{\log(3 \pm 2\sqrt{2})}{\log 2} \approx \pm 2.54$$

The approximate solutions are -2.54 and 2.54.

$y = \frac{2^x + 2^{-x}}{2}$ and $y = 3$ (see **Figure 3.40**).

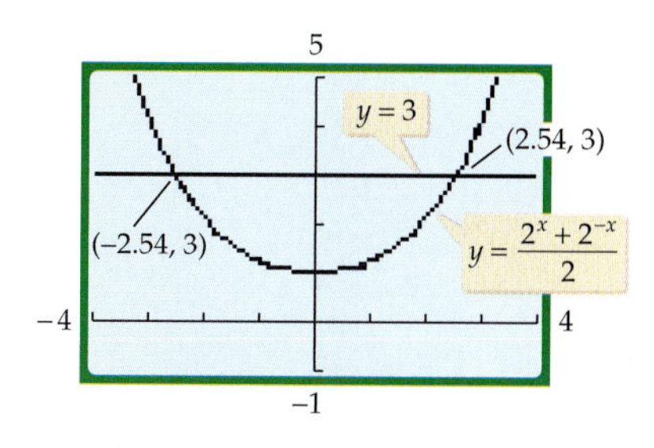

FIGURE 3.40

▶ TRY EXERCISE 40, PAGE 263

SOLVE LOGARITHMIC EQUATIONS

Equations that involve logarithms are called **logarithmic equations.** The properties of logarithms, along with the definition of a logarithm, are often used to find the solutions of a logarithmic equation.

EXAMPLE 5 Solve a Logarithmic Equation

Solve: $\log(3x - 5) = 2$

Solution

$$\log(3x - 5) = 2$$

$$3x - 5 = 10^2$$ • Definition of a logarithm

$$3x = 105$$ • Solve for x.

$$x = 35$$

Check: $\log[3(35) - 5] = \log 100 = 2$

▶ TRY EXERCISE 22, PAGE 263

QUESTION Can a negative number be a solution of a logarithmic equation?

ANSWER Yes. For instance, -10 is a solution of $\log(-x) = 1$.

EXAMPLE 6 Solve a Logarithmic Equation

Solve: $\log 2x - \log(x - 3) = 1$

Solution

$$\log 2x - \log(x - 3) = 1$$

$$\log \frac{2x}{x-3} = 1 \qquad \text{• Quotient property}$$

$$\frac{2x}{x-3} = 10^1 \qquad \text{• Definition of logarithm}$$

$$2x = 10x - 30 \qquad \text{• Solve for } x.$$

$$-8x = -30$$

$$x = \frac{15}{4}$$

Check the solution by substituting $\frac{15}{4}$ into the original equation.

▶ TRY EXERCISE 26, PAGE 263

In Example 7 we make use of the one-to-one property of logarithms to find the solution of a logarithmic equation. This example illustrates that the process of solving a logarithmic equation by using logarithmic properties may introduce an extraneous solution.

EXAMPLE 7 Solve a Logarithmic Equation

Solve: $\ln(3x + 8) = \ln(2x + 2) + \ln(x - 2)$

Algebraic Solution

$$\ln(3x + 8) = \ln(2x + 2) + \ln(x - 2)$$

$$\ln(3x + 8) = \ln[(2x + 2)(x - 2)] \qquad \text{• Product property}$$

$$\ln(3x + 8) = \ln(2x^2 - 2x - 4)$$

$$3x + 8 = 2x^2 - 2x - 4 \qquad \text{• One-to-one property of logarithms}$$

$$0 = 2x^2 - 5x - 12$$

$$0 = (2x + 3)(x - 4) \qquad \text{• Solve for } x.$$

$$x = -\frac{3}{2} \quad \text{or} \quad x = 4$$

Thus $-\frac{3}{2}$ and 4 are possible solutions. A check will show that 4 is a solution, but $-\frac{3}{2}$ is not a solution.

Visualize the Solution

The graph of $y = \ln(3x + 8) - \ln(2x + 2) - \ln(x - 2)$ has only one x-intercept (see **Figure 3.41**). Thus there is only one real solution.

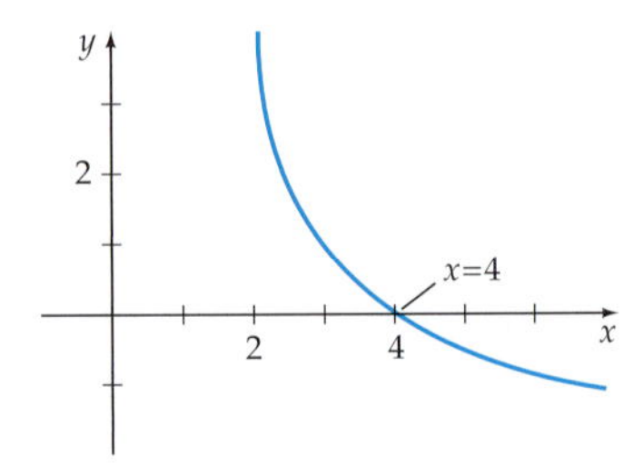

$y = \ln(3x + 8) - \ln(2x + 2) - \ln(x - 2)$

FIGURE 3.41

▶ TRY EXERCISE 36, PAGE 263

? QUESTION Why does $x = -\dfrac{3}{2}$ not check in Example 7?

Application

EXAMPLE 8 Velocity of a Sky Diver Experiencing Air Resistance

During the free-fall portion of a jump, the time t in seconds required for a sky diver to reach a velocity of v feet per second is given by

$$t = -\frac{175}{32}\ln\left(1 - \frac{v}{175}\right)$$

a. Determine the velocity of the diver after 5 seconds.

b. The graph of the above function has a vertical asymptote at $v = 175$. Explain the meaning of the vertical asymptote in the context of this example.

take note

If air resistance is not considered, then the time in seconds required for a sky diver to reach a given velocity (in feet per second) is $t = \frac{v}{32}$. The function in Example 8 is a more realistic model of the time required to reach a given velocity during the free-fall of a sky diver who is experiencing air resistance.

Solution

a. Substitute 5 for t and solve for v.

$$t = -\frac{175}{32}\ln\left(1 - \frac{v}{175}\right)$$

$$5 = -\frac{175}{32}\ln\left(1 - \frac{v}{175}\right) \qquad \text{• Replace } t \text{ with 5.}$$

$$\left(-\frac{32}{175}\right)5 = \ln\left(1 - \frac{v}{175}\right) \qquad \text{• Solve for } v.$$

$$-\frac{32}{35} = \ln\left(1 - \frac{v}{175}\right)$$

$$e^{-32/35} = 1 - \frac{v}{175} \qquad \text{• Write in exponential form.}$$

$$e^{-32/35} - 1 = -\frac{v}{175}$$

$$v = 175(1 - e^{-32/35})$$

$$v \approx 104.86$$

Continued ▶

? ANSWER If $x = -\dfrac{3}{2}$, the original equation becomes $\ln\left(\dfrac{7}{2}\right) = \ln(-1) + \ln\left(-\dfrac{7}{2}\right)$. This cannot be true, because the function $f(x) = \ln x$ is not defined for negative values of x.

After 5 seconds the velocity of the sky diver will be about 104.9 feet per second. See **Figure 3.42.**

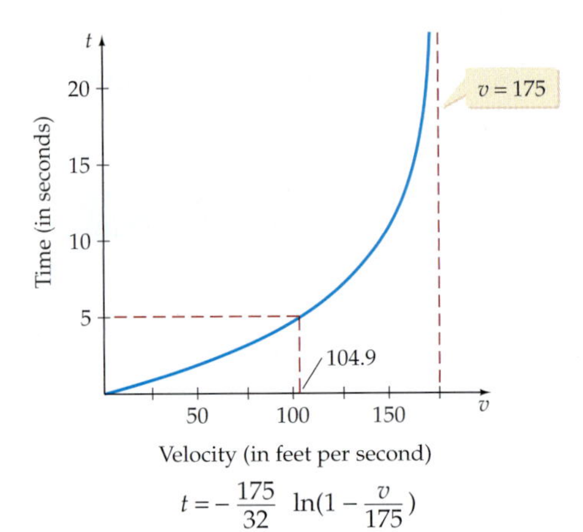

FIGURE 3.42

b. The vertical asymptote $v = 175$ indicates that the sky diver will not attain a velocity greater than 175 feet per second. In **Figure 3.42,** note that as $v \to 175$ from the left, $t \to \infty$.

TRY EXERCISE 68, PAGE 265

TOPICS FOR DISCUSSION

1. Discuss how to solve the equation $a = \log_b x$ for x.
2. What is the domain of $y = \log_4(2x - 5)$? Explain why this means that the equation $\log_4(x - 3) = \log_4(2x - 5)$ has no real number solution.
3. -8 is not a solution of the equation $\log_2 x + \log_2(x + 6) = 4$. Discuss at which step in the following solution the extraneous solution -8 was introduced.

$$\begin{aligned} \log_2 x + \log_2(x + 6) &= 4 \\ \log_2 x(x + 6) &= 4 \\ x(x + 6) &= 2^4 \\ x^2 + 6x &= 16 \\ x^2 + 6x - 16 &= 0 \\ (x + 8)(x - 2) &= 0 \\ x = -8 \quad &\text{or} \quad x = 2 \end{aligned}$$

Exercise Set 3.5

In Exercises 1 to 46, solve for x algebraically.

1. $2^x = 64$

2. $3^x = 243$

3. $49^x = \frac{1}{343}$

4. $9^x = \frac{1}{243}$

5. $2^{5x+3} = \frac{1}{8}$

6. $3^{4x-7} = \frac{1}{9}$

7. $\left(\frac{2}{5}\right)^x = \frac{8}{125}$

8. $\left(\frac{2}{5}\right)^x = \frac{25}{4}$

9. $5^x = 70$

10. $6^x = 50$

11. $3^{-x} = 120$

12. $7^{-x} = 63$

13. $10^{2x+3} = 315$

14. $10^{6-x} = 550$

15. $e^x = 10$

16. $e^{x+1} = 20$

17. $2^{1-x} = 3^{x+1}$

18. $3^{x-2} = 4^{2x+1}$

19. $2^{2x-3} = 5^{-x-1}$

20. $5^{3x} = 3^{x+4}$

21. $\log(4x - 18) = 1$

22. $\log(x^2 + 19) = 2$

23. $\ln(x^2 - 12) = \ln x$

24. $\log(2x^2 + 3x) = \log(10x + 30)$

25. $\log_2 x + \log_2(x - 4) = 2$

26. $\log_3 x + \log_3(x + 6) = 3$

27. $\log(5x - 1) = 2 + \log(x - 2)$

28. $1 + \log(3x - 1) = \log(2x + 1)$

29. $\ln(1 - x) + \ln(3 - x) = \ln 8$

30. $\log(4 - x) = \log(x + 8) + \log(2x + 13)$

31. $\log \sqrt{x^3 - 17} = \frac{1}{2}$

32. $\log(x^3) = (\log x)^2$

33. $\log(\log x) = 1$

34. $\ln(\ln x) = 2$

35. $\ln(e^{3x}) = 6$

36. $\ln x = \frac{1}{2}\ln\left(2x + \frac{5}{2}\right) + \frac{1}{2}\ln 2$

37. $e^{\ln(x-1)} = 4$

38. $10^{\log(2x+7)} = 8$

39. $\frac{10^x - 10^{-x}}{2} = 20$

40. $\frac{10^x + 10^{-x}}{2} = 8$

41. $\frac{10^x + 10^{-x}}{10^x - 10^{-x}} = 5$

42. $\frac{10^x - 10^{-x}}{10^x + 10^{-x}} = \frac{1}{2}$

43. $\frac{e^x + e^{-x}}{2} = 15$

44. $\frac{e^x - e^{-x}}{2} = 15$

45. $\frac{1}{e^x - e^{-x}} = 4$

46. $\frac{e^x + e^{-x}}{e^x - e^{-x}} = 3$

In Exercises 47 to 56, use a graphing utility to approximate the solutions of the equation to the nearest hundredth.

47. $2^{-x+3} = x + 1$

48. $3^{x-2} = -2x - 1$

49. $e^{3-2x} - 2x = 1$

50. $2e^{x+2} + 3x = 2$

51. $3 \log_2(x - 1) = -x + 3$

52. $2 \log_3(2 - 3x) = 2x - 1$

53. $\ln(2x + 4) + \frac{1}{2}x = -3$

54. $2 \ln(3 - x) + 3x = 4$

55. $2^{x+1} = x^2 - 1$

56. $\ln x = -x^2 + 4$

57. **Population Growth** The population P of a city grows exponentially according to the function

$$P(t) = 8500(1.1)^t, \quad 0 \le t \le 8$$

where t is measured in years.

a. Find the population at time $t = 0$ and also at time $t = 2$.

b. When, to the nearest year, will the population reach 15,000?

58. **Physical Fitness** After a race, a runner's pulse rate R in beats per minute decreases according to the function

$$R(t) = 145e^{-0.092t}, \quad 0 \le t \le 15$$

where t is measured in minutes.

a. Find the runner's pulse rate at the end of the race and also 1 minute after the end of the race.

b. How long, to the nearest minute, after the end of the race will the runner's pulse rate be 80 beats per minute?

59. Rate of Cooling A can of soda at 79°F is placed in a refrigerator that maintains a constant temperature of 36°F. The temperature T of the soda t minutes after it is placed in the refrigerator is given by

$$T(t) = 36 + 43e^{-0.058t}$$

a. Find the temperature, to the nearest degree, of the soda 10 minutes after it is placed in the refrigerator.

b. When, to the nearest minute, will the temperature of the soda be 45°F?

60. Medicine During surgery, a patient's circulatory system requires at least 50 milligrams of an anesthetic. The amount of anesthetic present t hours after 80 milligrams of anesthetic is administered is given by

$$T(t) = 80(0.727)^t$$

a. How much, to the nearest milligram, of the anesthetic is present in the patient's circulatory system 30 minutes after the anesthetic is administered?

b. How long, to the nearest minute, can the operation last if the patient does not receive additional anesthetic?

61. Psychology Industrial psychologists study employee training programs to assess the effectiveness of the instruction. In one study, the percent score P on a test for a person who had completed t hours of training was given by

$$P = \frac{100}{1 + 30e^{-0.088t}}$$

a. Use a graphing utility to graph the equation for $t \geq 0$.

b. Use the graph to estimate (to the nearest hour) the number of hours of training necessary to achieve a 70% score on the test.

c. From the graph, determine the horizontal asymptote.

d. Write a sentence that explains the meaning of the horizontal asymptote.

62. Psychology An industrial psychologist has determined that the average percent score for an employee on a test of the employee's knowledge of the company's product is given by

$$P = \frac{100}{1 + 40e^{-0.1t}}$$

where t is the number of weeks on the job and P is the percent score.

a. Use a graphing utility to graph the equation for $t \geq 0$.

b. Use the graph to estimate (to the nearest week) the number of weeks of employment that are necessary for the average employee to earn a 70% score on the test.

c. Determine the horizontal asymptote of the graph.

d. Write a sentence that explains the meaning of the horizontal asymptote.

63. Ecology A herd of bison was placed in a wildlife preserve that can support a maximum of 1000 bison. A population model for the bison is given by

$$B = \frac{1000}{1 + 30e^{-0.127t}}$$

where B is the number of bison in the preserve and t is time in years, with the year 1999 represented by $t = 0$.

a. Use a graphing utility to graph the equation for $t \geq 0$.

b. Use the graph to estimate (to the nearest year) the number of years before the bison population reaches 500.

c. Determine the horizontal asymptote of the graph.

d. Write a sentence that explains the meaning of the horizontal asymptote.

64. Population Growth A yeast culture grows according to the equation

$$Y = \frac{50{,}000}{1 + 250e^{-0.305t}}$$

where Y is the number of yeast and t is time in hours.

a. Use a graphing utility to graph the equation for $t \geq 0$.

b. Use the graph to estimate (to the nearest hour) the number of hours before the yeast population reaches 35,000.

c. From the graph, estimate the horizontal asymptote.

d. Write a sentence that explains the meaning of the horizontal asymptote.

65. Consumption of Natural Resources A model for how long our coal resources will last is given by

$$T = \frac{\ln(300r + 1)}{\ln(r + 1)}$$

where r is the percent increase in consumption from current levels of use and T is the time (in years) before the resource is depleted.

a. Use a graphing utility to graph this equation.

b. If our consumption of coal increases by 3% per year, in how many years will we deplete our coal resources?

c. What percent increase in consumption of coal will deplete the resource in 100 years? Round to the nearest tenth of a percent.

66. **CONSUMPTION OF NATURAL RESOURCES** A model for how long our aluminum resources will last is given by

$$T = \frac{\ln(20{,}500r + 1)}{\ln(r + 1)}$$

where r is the percent increase in consumption from current levels of use and T is the time (in years) before the resource is depleted.

a. Use a graphing utility to graph this equation.

b. If our consumption of aluminum increases by 5% per year, in how many years (to the nearest year) will we deplete our aluminum resources?

c. What percent increase in consumption of aluminum will deplete the resource in 100 years? Round to the nearest tenth of a percent.

67. **VELOCITY OF A MEDICAL CARE PACKAGE** A medical care package is air lifted and dropped to a disaster area. During the free-fall portion of the drop, the time, in seconds, required for the package to obtain a velocity of v feet per second is given by the function

$$t = 2.43 \ln \frac{150 + v}{150 - v}, \quad 0 \le v < 150$$

a. Determine the velocity of the package 5 seconds after it is dropped. Round to the nearest foot per second.

b. Determine the vertical asymptote of the function.

c. Write a sentence that explains the meaning of the vertical asymptote in the context of this application.

68. **EFFECTS OF AIR RESISTANCE ON VELOCITY** If we assume that air resistance is proportional to the square of the velocity, then the time t in seconds required for an object to reach a velocity of v feet per second is given by

$$t = \frac{9}{24} \ln \frac{24 + v}{24 - v}, 0 \le v < 24$$

a. Determine the velocity, to the nearest hundredth of a foot per second, of the object after 1.5 seconds.

b. Determine the vertical asymptote for the graph of this function.

c. Write a sentence that describes the meaning of the vertical asymptote in the context of this problem.

69. **TERMINAL VELOCITY WITH AIR RESISTANCE** The velocity v of an object t seconds after it has been dropped from a height above the surface of the earth is given by the equation $v = 32t$ feet per second, assuming no air resistance. If we assume that air resistance is proportional to the square of the velocity, then the velocity after t seconds is given by

$$v = 100\left(\frac{e^{0.64t} - 1}{e^{0.64t} + 1}\right)$$

a. In how many seconds will the velocity be 50 feet per second?

b. Determine the horizontal asymptote for the graph of this function.

c. Write a sentence that describes the meaning of the horizontal asymptote in the context of this problem.

70. **TERMINAL VELOCITY WITH AIR RESISTANCE** If we assume that air resistance is proportional to the square of the velocity, then the velocity v in feet per second of an object t seconds after it has been dropped is given by

$$v = 50\left(\frac{e^{1.6t} - 1}{e^{1.6t} + 1}\right)$$

(See Exercise 69. The reason for the difference in the equations is that the proportionality constants are different.)

a. In how many seconds will the velocity be 20 feet per second?

b. Determine the horizontal asymptote for the graph of this function.

c. Write a sentence that describes the meaning of the horizontal asymptote in the context of this problem.

71. **EFFECTS OF AIR RESISTANCE ON DISTANCE** The distance s, in feet, that the object in Exercise 69 will fall in t seconds is given by

$$s = \frac{100^2}{32} \ln\left(\frac{e^{0.32t} + e^{-0.32t}}{2}\right)$$

a. Use a graphing utility to graph this equation for $t \ge 0$.

b. How long does it take for the object to fall 100 feet? Round to the nearest tenth of a second.

72. **EFFECTS OF AIR RESISTANCE ON DISTANCE** The distance s, in feet, that the object in Exercise 70 will fall in t seconds is given by

$$s = \frac{50^2}{40}\ln\left(\frac{e^{0.8t} + e^{-0.8t}}{2}\right)$$

a. Use a graphing utility to graph this equation for $t \geq 0$.

b. How long does it take for the object to fall 100 feet? Round to the nearest tenth of a second.

73. **RETIREMENT PLANNING** The retirement account for a graphic designer contains $250,000 on January 1, 2002, and earns interest at a rate of 0.5% per month. On February 1, 2002, the designer withdraws $2000 and plans to continue these withdrawals as retirement income each month. The value V of the account after x months is

$$V = 400{,}000 - 150{,}000(1.005)^x$$

If the designer wishes to leave $100,000 to a scholarship foundation, what is the maximum number of withdrawals (to the nearest month) the designer can make from this account and still have $100,000 to donate?

74. **HANGING CABLE** The height h, in feet, of any point P on the cable shown is given by

$$h(x) = 10(e^{x/20} + e^{-x/20}), \quad -15 \leq x \leq 15$$

where $|x|$ is the horizontal distance in feet between P and the y-axis.

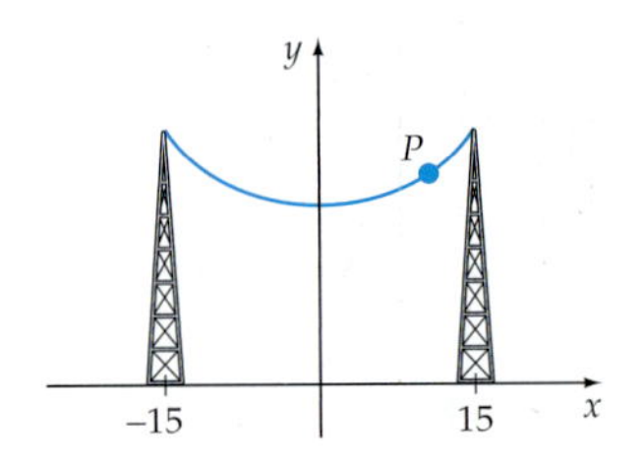

a. What is the lowest height of the cable?

b. What is the height of the cable 10 feet to the right of the y-axis? Round to the nearest tenth of a foot.

c. How far to the right of the y-axis is the cable 24 feet in height? Round to the nearest tenth of a foot.

CONNECTING CONCEPTS

75. The following argument seems to indicate that $0.125 > 0.25$. Find the first incorrect statement in the argument.

$$3 > 2$$
$$3(\log 0.5) > 2(\log 0.5)$$
$$\log 0.5^3 > \log 0.5^2$$
$$0.5^3 > 0.5^2$$
$$0.125 > 0.25$$

76. The following argument seems to indicate that $4 = 6$. Find the first incorrect statement in the argument.

$$4 = \log_2 16$$
$$4 = \log_2(8 + 8)$$
$$4 = \log_2 8 + \log_2 8$$
$$4 = 3 + 3$$
$$4 = 6$$

77. A common mistake that students make is to write $\log(x + y)$ as $\log x + \log y$. For what values of x and y does $\log(x + y) = \log x + \log y$? (*Hint:* Solve for x in terms of y.)

78. Let $f(x) = 2 \ln x$ and $g(x) = \ln x^2$. Does $f(x) = g(x)$ for all real numbers x?

79. Explain why the functions $F(x) = 1.4^x$ and $G(x) = e^{0.336x}$ represent essentially the same function.

80. Find the constant k that will make $f(t) = 2.2^t$ and $g(t) = e^{-kt}$ represent essentially the same function.

PREPARE FOR SECTION 3.6

81. Evaluate $A = 1000\left(1 + \frac{0.1}{12}\right)^{12t}$ for $t = 2$. Round to the nearest hundredth. [3.2]

82. Evaluate $A = 600\left(1 + \frac{0.04}{4}\right)^{4t}$ for $t = 8$. Round to the nearest hundredth. [3.2]

83. Solve $0.5 = e^{14k}$ for k. Round to the nearest ten-thousandth. [3.5]

84. Solve $0.85 = 0.5^{t/5730}$ for t. Round to the nearest ten. [3.5]

85. Solve $6 = \frac{70}{5 + 9e^{-k \cdot 12}}$ for k. Round to the nearest thousandth. [3.5]

86. Solve $2{,}000{,}000 = \frac{3^{n+1} - 3}{2}$ for n. Round to the nearest tenth. [3.5]

PROJECTS

1. **NAVIGATING** The pilot of a boat is trying to cross a river to a point O two miles due west of the boat's starting position by always pointing the nose of the boat toward O. Suppose the speed of the current is w miles per hour and the speed of the boat is v miles per hour. If point O is the origin and the boat's starting position is (2, 0) (see the diagram at the right), then the equation of the boat's path is given by

$$y = \left(\frac{x}{2}\right)^{1-(w/v)} - \left(\frac{x}{2}\right)^{1+(w/v)}$$

a. If the speed of the current and the speed of the boat are the same, can the pilot reach point O by always having the nose of the boat pointed toward O? If not, at what point will the pilot arrive? Explain your answer.

b. If the speed of the current is greater than the speed of the boat, can the pilot reach point O by always pointing the nose of the boat toward point O? If not, where will the pilot arrive? Explain.

c. If the speed of the current is less than the speed of the boat, can the pilot reach point O by always pointing the nose of the boat toward point O? If not, where will the pilot arrive? Explain.

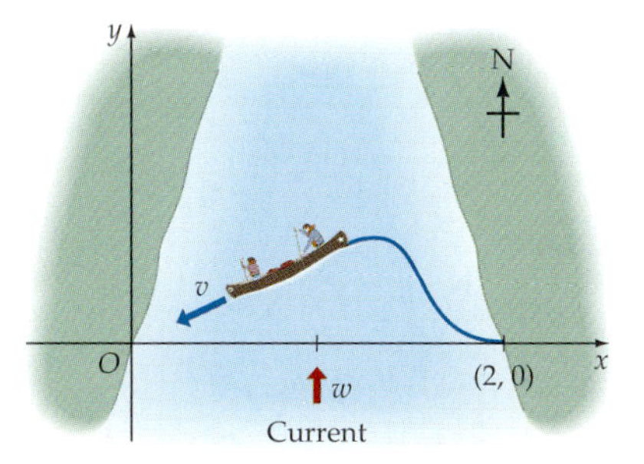

SECTION 3.6

Exponential Growth and Decay

- Compound Interest
- Exponential Growth
- Exponential Decay
- Carbon Dating

In many applications, a quantity N grows or decays according to the function $N(t) = N_0e^{kt}$. In this function, N is a function of time t, and N_0 is the value of N at time $t = 0$. If k is a *positive* constant, then $N(t) = N_0e^{kt}$ is called an **exponential growth function.** If k is a *negative* constant, then $N(t) = N_0e^{kt}$ is called an **exponential decay function.** The following examples illustrate how growth and decay functions arise naturally in the investigation of certain phenomena.

Interest is money paid for the use of money. The interest I is called **simple interest** if it is a fixed percent r, per time period t, of the amount of money invested. The amount of money invested is called the **principal** P. Simple interest is computed using the formula $I = Prt$. For example, if \$1000 is invested at 12% for 3 years, the simple interest is

$$I = Prt = \$1000(0.12)(3) = \$360$$

The balance after t years is $A = P + I = P + Prt$. In the previous example, the \$1000 invested for 3 years produced \$360 interest. Thus the balance after 3 years is \$1000 + \$360 = \$1360.

Compound Interest

In many financial transactions, interest is added to the principal at regular intervals so that interest is paid on interest as well as on the principal. Interest earned in this manner is called **compound interest.** For example, if \$1000 is invested at 12% annual interest compounded annually for 3 years, then the total interest after 3 years is

First-year interest	$1000(0.12) = $120.00	
Second-year interest	$1120(0.12) = $134.40	
Third-year interest	$1254.40(0.12) ≈ $150.53	
	$404.93	• Total interest

This method of computing the balance can be tedious and time-consuming. A *compound interest formula* that can be used to determine the balance due after t years of compounding can be developed as follows.

Note that if P dollars is invested at an interest rate of r per year, then the balance after one year is $A_1 = P + Pr = P(1 + r)$, where Pr represents the interest earned for the year. Observe that A_1 is the product of the original principal P and $(1 + r)$. If the amount A_1 is reinvested for another year, then the balance after the second year is

$$A_2 = (A_1)(1 + r) = P(1 + r)(1 + r) = P(1 + r)^2$$

Successive reinvestments lead to the results shown in **Table 3.11.** The equation $A_t = P(1 + r)^t$ is valid if r is the annual interest rate paid during each of the t years.

Table 3.11

Number of Years	Balance
3	$A_3 = P(1 + r)^3$
4	$A_4 = P(1 + r)^4$
⋮	⋮
t	$A_t = P(1 + r)^t$

If r is an annual interest rate and n is the number of compounding periods per year, then the interest rate each period is r/n and the number of compounding periods after t years is nt. Thus the compound interest formula is expressed as follows:

The Compound Interest Formula

A principal P invested at an annual interest rate r, expressed as a decimal and compounded n times per year for t years, produces the balance

$$A = P\left(1 + \frac{r}{n}\right)^{nt}$$

EXAMPLE 1 Solve a Compound Interest Application

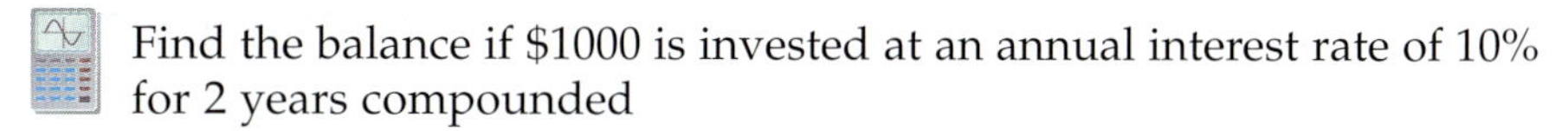

Find the balance if \$1000 is invested at an annual interest rate of 10% for 2 years compounded

a. annually **b.** monthly **c.** daily

Solution

a. Use the compound interest formula with $P = 1000$, $r = 0.1$, $t = 2$, and $n = 1$.

$$A = \$1000\left(1 + \frac{0.1}{1}\right)^{1 \cdot 2} = \$1000(1.1)^2 = \$1210.00$$

b. Because there are 12 months in a year, use $n = 12$.

$$A = \$1000\left(1 + \frac{0.1}{12}\right)^{12 \cdot 2} \approx \$1000(1.008333333)^{24} \approx \$1220.39$$

c. Because there are 365 days in a year, use $n = 365$.

$$A = \$1000\left(1 + \frac{0.1}{365}\right)^{365 \cdot 2} \approx \$1000(1.000273973)^{730} \approx \$1221.37$$

▶ TRY EXERCISE 4, PAGE 276

To **compound continuously** means to increase the number of compounding periods without bound.

To derive a continuous compounding interest formula, substitute $\frac{1}{m}$ for $\frac{r}{n}$ in the compound interest formula

$$A = P\left(1 + \frac{r}{n}\right)^{nt} \qquad (1)$$

to produce

$$A = P\left(1 + \frac{1}{m}\right)^{nt} \qquad (2)$$

This substitution is motivated by the desire to express $\left(1+\frac{r}{n}\right)^n$ as $\left[\left(1+\frac{1}{m}\right)^m\right]^r$, which approaches e^r as m gets larger without bound.

Solving the equation $\frac{1}{m}=\frac{r}{n}$ for n yields $n = mr$, so the exponent nt can be written as mrt. Therefore Equation (2) can be expressed as

$$A = P\left(1+\frac{1}{m}\right)^{mrt} = P\left[\left(1+\frac{1}{m}\right)^m\right]^{rt} \qquad (3)$$

By the definition of e, we know that as m increases without bound,

$$\left(1+\frac{1}{m}\right)^m \quad \text{approaches} \quad e$$

Thus, using continuous compounding, Equation (3) simplifies to $A = Pe^{rt}$.

Continuous Compounding Interest Formula

If an account with principal P and annual interest rate r is compounded continuously for t years, then the balance is $A = Pe^{rt}$.

EXAMPLE 2 Solve a Continuous Compound Interest Application

Find the balance after 4 years on \$800 invested at an annual rate of 6% compounded continuously.

Algebraic Solution

Use the continuous compounding formula with $P = 800$, $r = 0.06$, and $t = 4$.

$$\begin{aligned} A &= Pe^{rt} \\ &= 800e^{0.06(4)} \\ &= 800e^{0.24} \\ &\approx 800(1.27124915) \\ &\approx 1017.00 \end{aligned}$$

• To the nearest cent

The balance after 4 years will be \$1017.00.

Visualize the Solution

Figure 3.43, a graph of $A = 800e^{0.06t}$, shows that the balance is about \$1017.00 when $t = 4$.

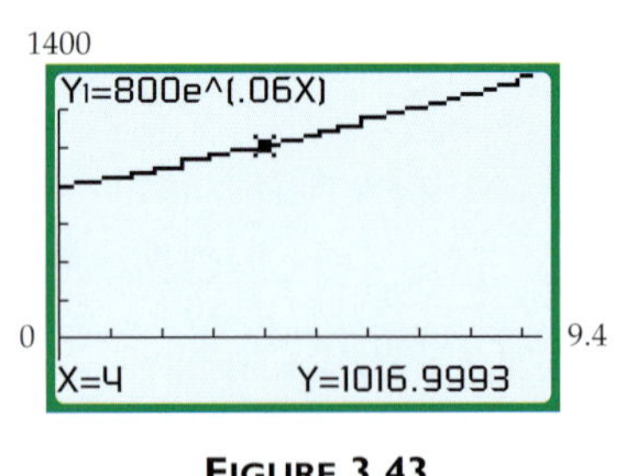

FIGURE 3.43

TRY EXERCISE 6, PAGE 276

You have probably heard it said that time is money. In fact, many investors ask the question "How long will it take to double my money?" The following example answers this question for two different investments.

EXAMPLE 3 Double Your Money

Find the time required for money invested at an annual rate of 6% to double in value if the investment is compounded

a. semiannually

b. continuously

Solution

a. Use $A = P\left(1 + \dfrac{r}{n}\right)^{nt}$ with $r = 0.06$, $n = 2$, and the balance A equal to twice the principal $(A = 2P)$.

$$2P = P\left(1 + \frac{0.06}{2}\right)^{2t}$$

$$2 = \left(1 + \frac{0.06}{2}\right)^{2t}$$ • **Divide each side by *P*.**

$$\ln 2 = \ln\left(1 + \frac{0.06}{2}\right)^{2t}$$ • **Take the natural logarithm of each side.**

$$\ln 2 = 2t \ln\left(1 + \frac{0.06}{2}\right)$$ • **Apply the power property.**

$$2t = \frac{\ln 2}{\ln\left(1 + \frac{0.06}{2}\right)}$$ • **Solve for *t*.**

$$t = \frac{1}{2} \cdot \frac{\ln 2}{\ln\left(1 + \frac{0.06}{2}\right)}$$

$$t \approx 11.72$$

If the investment is compounded semiannually, it will double in value in about 11.72 years.

b. Use $A = Pe^{rt}$ with $r = 0.06$ and $A = 2P$.

$$2P = Pe^{0.06t}$$

$$2 = e^{0.06t}$$ • **Divide each side by *P*.**

$$\ln 2 = 0.06t$$ • **Write in logarithm form.**

$$t = \frac{\ln 2}{0.06}$$ • **Solve for *t*.**

$$t \approx 11.55$$

If the investment is compounded continuously, it will double in value in about 11.55 years.

TRY EXERCISE 10, PAGE 276

Exponential Growth

Given any two points on the graph of $N(t) = N_0e^{kt}$, you can use the given data to solve for the constants N_0 and k.

EXAMPLE 4 Find the Exponential Growth Function That Models Given Data

a. Find the exponential growth function for a town whose population was 16,400 in 1990 and 20,200 in 2000.

b. Use the function from part **a.** to predict, to the nearest 100, the population of the town in 2005.

Solution

a. We need to determine N_0 and k in $N(t) = N_0e^{kt}$. If we represent the year 1990 by $t = 0$, then our given data are $N(0) = 16{,}400$ and $N(10) = 20{,}200$. Because N_0 is defined to be $N(0)$, we know that $N_0 = 16{,}400$. To determine k, substitute $t = 10$ and $N_0 = 16{,}400$ into $N(t) = N_0e^{kt}$ to produce

$$N(10) = 16{,}400e^{k\cdot 10}$$

$$20{,}200 = 16{,}400e^{10k} \qquad \text{• Substitute 20,200 for } N(10).$$

$$\frac{20{,}200}{16{,}400} = e^{10k} \qquad \text{• Solve for } e^{10k}.$$

$$\ln\frac{20{,}200}{16{,}400} = 10k \qquad \text{• Write in logarithmic form.}$$

$$\frac{1}{10}\ln\frac{20{,}200}{16{,}400} = k \qquad \text{• Solve for } k.$$

$$0.0208 \approx k$$

The exponential growth function is $N(t) \approx 16{,}400e^{0.0208t}$.

b. The year 1990 was represented by $t = 0$, so we will use $t = 15$ to represent the year 2005.

$$N(t) \approx 16{,}400e^{0.0208t}$$

$$N(15) \approx 16{,}400e^{0.0208\cdot 15}$$

$$\approx 22{,}400 \quad \text{(nearest 100)}$$

The exponential growth function yields 22,400 as the approximate population of the town in 2005.

▶ TRY EXERCISE 18, PAGE 277

take note

Because $e^{0.0208} \approx 1.021$, the growth equation can also be written as

$$N(t) \approx 16{,}400(1.021)^t$$

In this form we see that the population is growing by 2.1% $(1.021 - 1 = 0.021 = 2.1\%)$ per year.

Exponential Decay

Many radioactive materials *decrease* in mass exponentially over time. This decrease, called radioactive decay, is measured in terms of **half-life,** which is defined as the time required for the disintegration of half the atoms in a sample of a radioactive substance. **Table 3.12** shows the half-lives of selected radioactive isotopes.

Table 3.12

Isotope	Half-Life
Carbon (^{14}C)	5730 years
Radium (^{226}Ra)	1660 years
Polonium (^{210}Po)	138 days
Phosphorus (^{32}P)	14 days
Polonium (^{214}Po)	1/10,000th of a second

Example 5 Find the Exponential Decay Function That Models Given Data

Find the exponential decay function for the amount of phosphorus (^{32}P) that remains in a sample after t days.

Solution

When $t = 0$, $N(0) = N_0e^{k(0)} = N_0$. Thus $N(0) = N_0$. Also, because the phosphorus has a half-life of 14 days (from **Table 3.12**), $N(14) = 0.5N_0$. To find k, substitute $t = 14$ into $N(t) = N_0e^{kt}$ and solve for k.

$$N(14) = N_0 \cdot e^{k \cdot 14}$$

$$0.5N_0 = N_0e^{14k}$$ • Substitute $0.5N_0$ for $N(14)$.

$$0.5 = e^{14k}$$ • Divide each side by N_0.

$$\ln 0.5 = 14k$$ • Write in logarithmic form.

$$\frac{1}{14}\ln 0.5 = k$$ • Solve for k.

$$-0.0495 \approx k$$

The exponential decay function is $N(t) = N_0e^{-0.0495t}$.

Try Exercise 20, page 277

take note

Because $e^{-0.0495} \approx (0.5)^{1/14}$, the decay function $N(t) = N_0e^{-0.0495t}$ can also be written as $N(t) = N_0(0.5)^{t/14}$. In this form it is easy to see that if t is increased by 14, then N will decrease by a factor of 0.5.

EXAMPLE 6 Application to Air Resistance

Assuming that air resistance is proportional to the velocity of a falling object, the velocity (in feet per second) of the object t seconds after it has been dropped is given by $v = 82(1 - e^{-0.39t})$.

a. Determine when the velocity will be 70 feet per second.

b. Write a sentence that explains the meaning of the horizontal asymptote, which is $v = 82$, in the context of this example.

Algebraic Solution

a.

$$v = 82(1 - e^{-0.39t})$$

$$70 = 82(1 - e^{-0.39t}) \quad \text{• Replace } v \text{ by 70.}$$

$$\frac{70}{82} = 1 - e^{-0.39t} \quad \text{• Divide each side by 82.}$$

$$e^{-0.39t} = 1 - \frac{70}{82} \quad \text{• Solve for } e^{-0.39t}.$$

$$-0.39t = \ln \frac{6}{41} \quad \text{• Write in logarithmic form.}$$

$$t = \frac{\ln(6/41)}{-0.39} \approx 4.9277246 \quad \text{• Solve for } t.$$

The time is approximately 4.9 seconds.

b. The horizontal asymptote $v = 82$ means that as time increases, the velocity of the object will approach but never reach or exceed 82 feet per second.

Visualize the Solution

a. A graph of $y = 82(1 - e^{-0.39x})$ and $y = 70$ shows that the x-coordinate of the point of intersection is about 4.9.

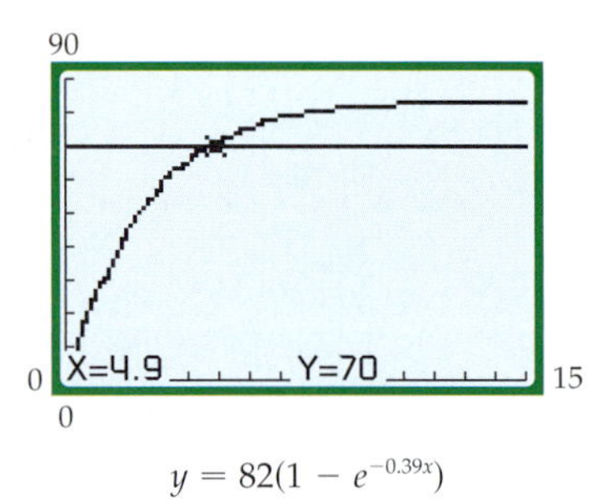

$y = 82(1 - e^{-0.39x})$

FIGURE 3.44

Note: The x value shown is rounded to the nearest tenth.

▶ TRY EXERCISE 32, PAGE 278

• CARBON DATING

The bone tissue in all living animals contains both carbon-12, which is nonradioactive, and carbon-14, which is radioactive with a half-life of approximately 5730 years. See **Figure 3.45.** As long as the animal is alive, the ratio of carbon-14 to carbon-12 remains constant. When the animal dies ($t = 0$), the carbon-14 begins to decay. Thus a bone that has a smaller ratio of carbon-14 to carbon-12 is older than a bone that has a larger ratio. The percent of carbon-14 present at time t is

$$P(t) = 0.5^{t/5730}$$

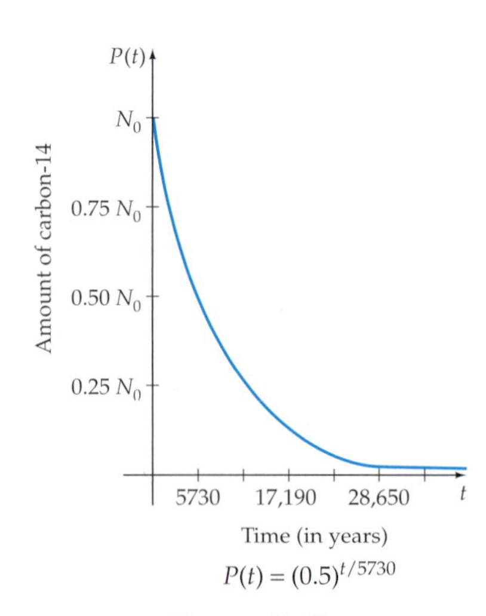

FIGURE 3.45

EXAMPLE 7 Application to Archeology

Find the age of a bone if it now has 85% of the carbon-14 it had when $t = 0$.

Solution

Let t be the time at which $P(t) = 0.85$.

$$0.85 = 0.5^{t/5730}$$

$$\ln 0.85 = \ln 0.5^{t/5730}$$ • Take the natural logarithm of each side.

$$\ln 0.85 = \frac{t}{5730} \ln 0.5$$ • Power property

$$5730\left(\frac{\ln 0.85}{\ln 0.5}\right) = t$$ • Solve for t.

$$1340 \approx t$$

The bone is about 1340 years old.

▶ TRY EXERCISE 24, PAGE 277

MATH MATTERS

The chemist Willard Frank Libby developed the carbon-14 dating technique in 1947. In 1960 he was awarded the Nobel Prize in chemistry for this achievement.

TOPICS FOR DISCUSSION

1. Explain the difference between compound interest and simple interest.
2. What is the continuous compounding interest formula and when is it used?
3. What is an exponential growth model? Give an example of an application for which the exponential growth model might be appropriate.
4. What is an exponential decay model? Give an example of an application for which the exponential decay model might be appropriate.

Exercise Set 3.6

1. **Compound Interest** If \$8000 is invested at an annual interest rate of 5% and compounded annually, find the balance after

   a. 4 years **b.** 7 years

2. **Compound Interest** If \$22,000 is invested at an annual interest rate of 4.5% and compounded annually, find the balance after

   a. 2 years **b.** 10 years

3. **Compound Interest** If \$38,000 is invested at an annual interest rate of 6.5% for 4 years, find the balance if the interest is compounded

   a. annually **b.** daily **c.** hourly

4. **Compound Interest** If \$12,500 is invested at an annual interest rate of 8% for 10 years, find the balance if the interest is compounded

   a. annually **b.** daily **c.** hourly

5. **Compound Interest** Find the balance if \$15,000 is invested at an annual rate of 10% for 5 years, compounded continuously.

6. **Compound Interest** Find the balance if \$32,000 is invested at an annual rate of 8% for 3 years, compounded continuously.

7. **Compound Interest** How long will it take \$4000 to double if it is invested in a certificate of deposit that pays 7.84% annual interest compounded continuously? Round to the nearest tenth of a year.

8. **Compound Interest** How long will it take \$25,000 to double if it is invested in a savings account that pays 5.88% annual interest compounded continuously? Round to the nearest tenth of a year.

9. **Continuous Compounding Interest** Use the Continuous Compounding Interest Formula to derive an expression for the time it will take money to triple when invested at an annual interest rate of r compounded continuously.

10. **Continuous Compounding Interest** How long will it take \$1000 to triple if it is invested at an annual interest rate of 5.5% compounded continuously? Round to the nearest year.

11. **Continuous Compounding Interest** How long will it take \$6000 to triple if it is invested in a savings account that pays 7.6% annual interest compounded continuously? Round to the nearest year.

12. **Continuous Compounding Interest** How long will it take \$10,000 to triple if it is invested in a savings account that pays 5.5% annual interest compounded continuously? Round to the nearest year.

13. **Population Growth** The number of bacteria $N(t)$ present in a culture at time t hours is given by

$$N(t) = 2200(2)^t$$

   Find the number of bacteria present when

   a. $t = 0$ hours **b.** $t = 3$ hours

14. **Population Growth** The population of a town grows exponentially according to the function

$$f(t) = 12{,}400(1.14)^t$$

   for $0 \le t \le 5$ years. Find, to the nearest hundred, the population of the town when t is

   a. 3 years **b.** 4.25 years

15. **Population Growth** A town had a population of 22,600 in 1990 and a population of 24,200 in 1995.

   a. Find the exponential growth function for the town. Use $t = 0$ to represent the year 1990.

   b. Use the growth function to predict the population of the town in 2005. Round to the nearest hundred.

16. **Population Growth** A town had a population of 53,700 in 1996 and a population of 58,100 in 2000.

   a. Find the exponential growth function for the town. Use $t = 0$ to represent the year 1996.

   b. Use the growth function to predict the population of the town in 2008. Round to the nearest hundred.

17. **Population Growth** The growth of the population of Los Angeles, California, for the years 1992 through 1996 can be approximated by the equation

$$P = 10{,}130(1.005)^t$$

   where $t = 0$ corresponds to January 1, 1992 and P is in thousands.

   a. Assuming this growth rate continues, what will be the population of Los Angeles on January 1 in the year 2004?

b. In what year will the population of Los Angeles first exceed 13,000,000?

18. **POPULATION GROWTH** The growth of the population of Mexico City, Mexico, for the years 1991 through 1998 can be approximated by the equation

$$P = 20{,}899(1.027)^t$$

where $t = 0$ corresponds to 1991 and P is in thousands.

a. Assuming this growth rate continues, what will be the population of Mexico City in the year 2003?

b. Assuming this growth rate continues, in what year will the population of Mexico City first exceed 35,000,000?

19. **MEDICINE** Sodium-24 is a radioactive isotope of sodium that is used to study circulatory dysfunction. Assuming that 4 micrograms of sodium-24 is injected into a person, the amount A in micrograms remaining in that person after t hours is given by the equation $A = 4e^{-0.046t}$.

a. Graph this equation.

b. What amount of sodium-24 remains after 5 hours?

c. What is the half-life of sodium-24?

d. In how many hours will the amount of sodium-24 be 1 microgram?

20. **RADIOACTIVE DECAY** Polonium (^{210}Po) has a half-life of 138 days. Find the decay function for the amount of polonium (^{210}Po) that remains in a sample after t days.

21. **GEOLOGY** Geologists have determined that Crater Lake in Oregon was formed by a volcanic eruption. Chemical analysis of a wood chip that is assumed to be from a tree that died during the eruption has shown that it contains approximately 45% of its original carbon-14. Determine how long ago the volcanic eruption occurred. Use 5730 years as the half-life of carbon-14.

22. **RADIOACTIVE DECAY** Use $N(t) = N_0(0.5)^{t/138}$, where t is measured in days, to estimate the percentage of polonium (^{210}Po) that remains in a sample after 2 years. Round to the nearest hundredth of a percent.

23. **ARCHEOLOGY** The Rhind papyrus, named after A. Henry Rhind, contains most of what we know today of ancient Egyptian mathematics. A chemical analysis of a sample from the papyrus has shown that it contains approximately 75% of its original carbon-14. What is the age of the Rhind papyrus? Use 5730 years as the half-life of carbon-14.

24. **ARCHEOLOGY** Determine the age of a bone if it now contains 65% of its original amount of carbon-14. Round to the nearest 100 years.

25. **PHYSICS** Newton's Law of Cooling states that if an object at temperature T_0 is placed into an environment at constant temperature A, then the temperature of the object, $T(t)$ (in degrees Fahrenheit), after t minutes is given by $T(t) = A + (T_0 - A)e^{-kt}$, where k is a constant that depends on the object.

a. Determine the constant k (to the nearest thousandth) for a canned soda drink that takes 5 minutes to cool from 75°F to 65°F after being placed in a refrigerator that maintains a constant temperature of 34°F.

b. What will be the temperature (to the nearest degree) of the soda drink after 30 minutes?

c. When (to the nearest minute) will the temperature of the soda drink be 36°F?

26. **PSYCHOLOGY** According to a software company, the users of its typing tutorial can expect to type $N(t)$ words per minute after t hours of practice with the product, according to the function $N(t) = 100(1.04 - 0.99^t)$.

a. How many words per minute can a student expect to type after 2 hours of practice?

b. How many words per minute can a student expect to type after 40 hours of practice?

c. According to the function N, how many hours (to the nearest hour) of practice will be required before a student can expect to type 60 words per minute?

27. **PSYCHOLOGY** In the city of Whispering Palms, which has a population of 80,000 people, the number of people $P(t)$ exposed to a rumor in t hours is given by the function $P(t) = 80{,}000(1 - e^{-0.0005t})$.

a. Find the number of hours until 10% of the population has heard the rumor.

b. Find the number of hours until 50% of the population has heard the rumor.

28. **LAW** A lawyer has determined that the number of people $P(t)$ in a city of 1,200,000 people who have been exposed to a news item after t days is given by the function

$$P(t) = 1{,}200{,}000(1 - e^{-0.03t})$$

a. How many days after a major crime has been reported has 40% of the population heard of the crime?

b. A defense lawyer knows it will be very difficult to pick an unbiased jury after 80% of the population has heard of the crime. After how many days will 80% of the population have heard of the crime?

29. DEPRECIATION An automobile depreciates according to the function $V(t) = V_0(1 - r)^t$, where $V(t)$ is the value in dollars after t years, V_0 is the original value, and r is the yearly depreciation rate. A car has a yearly depreciation rate of 20%. Determine, to the nearest 0.1 year, in how many years the car will depreciate to half its original value.

30. PHYSICS The current $I(t)$ (measured in amperes) of a circuit is given by the function $I(t) = 6(1 - e^{-2.5t})$, where t is the number of seconds after the switch is closed.

a. Find the current when $t = 0$.

b. Find the current when $t = 0.5$.

c. Solve the equation for t.

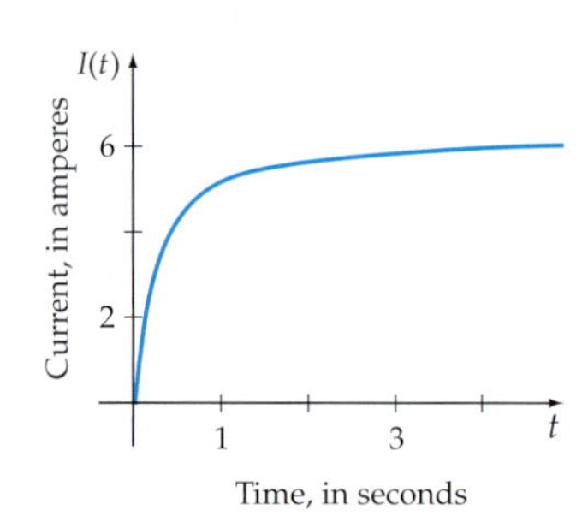

31. AIR RESISTANCE Assuming that air resistance is proportional to velocity, the velocity v, in feet per second, of a falling object after t seconds is given by $v = 32(1 - e^{-t})$.

a. Graph this equation for $t \geq 0$.

b. Determine algebraically, to the nearest 0.01 second, when the velocity is 20 feet per second.

c. Determine the horizontal asymptote of the graph of v.

d. Write a sentence that explains the meaning of the horizontal asymptote in the context of this application.

32. AIR RESISTANCE Assuming that air resistance is proportional to velocity, the velocity v, in feet per second, of a falling object after t seconds is given by $v = 64(1 - e^{-t/2})$.

a. Graph this equation for $t \geq 0$.

b. Determine algebraically, to the nearest 0.1 second, when the velocity is 50 feet per second.

c. Determine the horizontal asymptote of the graph of v.

d. Write a sentence that explains the meaning of the horizontal asymptote in the context of this application.

33. The distance s (in feet) that the object in Exercise 31 will fall in t seconds is given by $s = 32t + 32(e^{-t} - 1)$.

a. Use a graphing utility to graph this equation for $t \geq 0$.

b. Determine, to the nearest 0.1 second, the time it takes the object to fall 50 feet.

c. Calculate the slope of the secant line through $(1, s(1))$ and $(2, s(2))$.

d. Write a sentence that explains the meaning of the slope of the secant line you calculated in **c.**

34. The distance s (in feet) that the object in Exercise 32 will fall in t seconds is given by $s = 64t + 128(e^{-t/2} - 1)$.

a. Use a graphing utility to graph this equation for $t \geq 0$.

b. Determine, to the nearest 0.1 second, the time it takes the object to fall 50 feet.

c. Calculate the slope of the secant line through $(1, s(1))$ and $(2, s(2))$.

d. Write a sentence that explains the meaning of the slope of the secant line you calculated in **c.**

CONNECTING CONCEPTS

35. **MEDICATION LEVEL** A patient is given three dosages of aspirin. Each dosage contains 1 gram of aspirin. The second and third dosages are each taken 3 hours after the previous dosage is administered. The half-life of the aspirin is 2 hours. The amount of aspirin, A, in the patient's body t hours after the first dosage is administered is

$$A(t) = \begin{cases} 0.5^{t/2} & 0 \le t < 3 \\ 0.5^{t/2} + 0.5^{(t-3)/2} & 3 \le t < 6 \\ 0.5^{t/2} + 0.5^{(t-3)/2} + 0.5^{(t-6)/2} & t \ge 6 \end{cases}$$

Find, to the nearest hundredth of a gram, the amount of aspirin in the patient's body when

a. $t = 1$ **b.** $t = 4$ **c.** $t = 9$

36. **MEDICATION LEVEL** Use a graphing calculator and the dosage formula in Exercise 35 to determine when, to the nearest tenth of an hour, the amount of aspirin in the patient's body first reaches 0.25 gram.

Exercises 37 to 39 make use of the factorial function, which is defined as follows. For whole numbers n, the number $n!$ (which is read "n factorial") is given by

$$n! = \begin{cases} n(n-1)(n-2)\cdots 1, & \text{if } n \ge 1 \\ 1, & \text{if } n = 0 \end{cases}$$

Thus, $0! = 1$ and $4! = 4 \cdot 3 \cdot 2 \cdot 1 = 24$.

37. **QUEUEING THEORY** A study shows that the number of people who arrive at a bank teller's window averages 4.1 people every 10 minutes. The probability P that exactly x people will arrive at the teller's window in a given 10-minute period is

$$P(x) = \frac{4.1^x e^{-4.1}}{x!}$$

Find, to the nearest 0.1%, the probability that in a given 10-minute period, exactly

a. 0 people arrive at the window.

b. 2 people arrive at the window.

c. 3 people arrive at the window.

d. 4 people arrive at the window.

e. 9 people arrive at the window.

As $x \to \infty$, what does P approach?

38. **STIRLING'S FORMULA** *Stirling's Formula* (after James Stirling, 1692–1770),

$$n! \approx \left(\frac{n}{e}\right)^n \sqrt{2\pi n}$$

is often used to approximate very large factorials. Use Stirling's Formula to approximate 10!, and then compute the ratio of Stirling's approximation of 10! divided by the actual value of 10!, which is 3,628,800.

39. **RUBIK'S CUBE** The Rubik's cube shown here was invented by Erno Rubik in 1975. The small outer cubes are held together in such a way that they can be rotated around three axes. The total number of positions in which the Rubik's cube can be arranged is

$$\frac{3^8 2^{12} 8!\, 12!}{2 \cdot 3 \cdot 2}$$

If you can arrange a Rubik's cube into a new arrangement every second, how many centuries would it take to place the cube into each of its arrangements? Assume that there are 365 days in a year.

40. **OIL SPILLS** Crude oil leaks from a tank at a rate that depends on the amount of oil that remains in the tank. Because $\frac{1}{8}$ of the oil in the tank leaks out every 2 hours, the volume of oil $V(t)$ in the tank after t hours is given by $V(t) = V_0(0.875)^{t/2}$, where $V_0 = 350{,}000$ gallons is the number of gallons in the tank at the time the tank started to leak ($t = 0$).

a. How many gallons does the tank hold after 3 hours?

b. How many gallons does the tank hold after 5 hours?

c. How long, to the nearest hour, will it take until 90% of the oil has leaked from the tank?

PROJECTS

THE RULE OF 72 The rule of 72 states that the number of years n needed to double an investment can be approximated by $n = \frac{72}{r}$, where r is the annual interest rate, as a percent, on the investment. The actual number of years can be found by solving the compound interest equation $A = P\left(1 + \frac{r}{100}\right)^n$ for n when $A = 2P$. By completing this project, you will see why the approximation works.

1. Solve $A = P\left(1 + \frac{r}{100}\right)^n$ for n when $A = 2P$. Here is a start.

$$A = P\left(1 + \frac{r}{100}\right)^n$$ • **Compound interest formula**

$$2P = P\left(1 + \frac{r}{100}\right)^n$$ • ***A* = 2*P***

$$2 = \left(1 + \frac{r}{100}\right)^n$$ • **Divide by *P*.**

$$\ln(2) = \ln\left(1 + \frac{r}{100}\right)^n$$ • **Take the natural logarithm of each side.**

continue... (See part 3 to check your answer.)

2. Complete the following table using your solution to part 1.

Annual interest rate, *r*	Time to double money using Rule of 72	Actual time to double money
7		
8		
9		
10		
11		
12		
13		

3. The actual number of years is given by $n = \dfrac{\ln 2}{\ln\left(1 + \frac{r}{100}\right)}$.

   To approximate this number, we want to find a "easy" fraction $\frac{K}{r}$, where K is a constant and r is the annual interest rate as a percent, that is approximately equal to $\dfrac{\ln 2}{\ln\left(1 + \frac{r}{100}\right)}$. Let $\dfrac{K}{r} = \dfrac{\ln 2}{\ln\left(1 + \frac{r}{100}\right)}$. Solve this equation for K when $r = 10\%$.

4. Solve $\dfrac{K}{r} = \dfrac{\ln 2}{\ln\left(1 + \frac{r}{100}\right)}$ for K when $r = 8\%, 9\%, 11\%,$ and 12%.

5. On the basis of your answers to parts 3 and 4, explain why the Rule of 72 works.

6. Graph $Y_1 = \dfrac{72}{r}$ and $Y_2 = \dfrac{\ln 2}{\ln\left(1 + \frac{r}{100}\right)}$ for $2 \le r \le 15$.

   On the basis of your graph, explain why the Rule of 72 works.

Exploring Concepts with Technology

Using a Semilog Graph to Model Exponential Decay

Table 3.13

T	V
90	700
100	500
110	350
120	250
130	190
140	150
150	120

Consider the data in **Table 3.13,** which shows the viscosity V of SAE 40 motor oil at various temperatures T. The graph of these data is shown below, along with a curve that passes through the points. The graph in **Figure 3.46** appears to have the shape of an exponential decay model.

One way to determine whether the graph in **Figure 3.46** is the graph of an exponential function is to plot the data on *semilog* graph paper. On this graph paper, the horizontal axis remains the same, but the vertical axis uses a logarithmic scale.

The data in **Table 3.13** are graphed again in **Figure 3.47,** but this time the vertical axis is a natural logarithm axis. This graph is approximately a straight line.

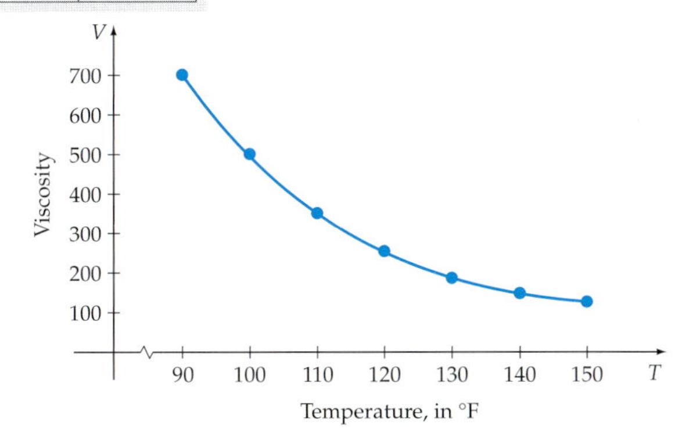

Figure 3.46

Figure 3.47

The slope of the line in **Figure 3.47,** to the nearest ten-thousandth, is

$$m = \frac{\ln 500 - \ln 120}{100 - 150} \approx -0.0285$$

Using this slope and the point-slope formula with V replaced by $\ln V$, we have

$$\ln V - \ln 120 = -0.0285(T - 150)$$

$$\ln V \approx -0.0285T + 9.062 \qquad (1)$$

Equation (1) is the equation of the line on a semilog coordinate grid.

Now solve Equation (1) for V.

$$e^{\ln V} = e^{-0.0285T+9.062}$$

$$V = e^{-0.0285T}e^{9.062}$$

$$V \approx 8621e^{-0.0285T} \qquad (2)$$

Equation (2) is a model of the data in the rectangular coordinate system shown in **Figure 3.46.**

Table 3.14

t	A
1	91.77
4	70.92
8	50.30
15	27.57
20	17.95
30	7.60

1. A chemist wishes to determine the decay characteristics of iodine-131. A 100-mg sample of iodine-131 is observed over a 30-day period. **Table 3.14** shows the amount A (in milligrams) of iodine-131 remaining after t days.

   a. Graph the ordered pairs (t, A) on semilog paper. (*Note:* Semilog paper comes in different varieties. Our calculations are based on semilog paper that has a natural logarithm scale on the vertical axis.)

b. Use the points (4, 4.3) and (15, 3.3) to approximate the slope of the line that passes through the points.

c. Using the slope calculated in part **b.** and the point (4, 4.3), determine the equation of the line.

d. Solve the equation you derived in part **c.** for A.

e. Graph the equation you derived in part **d.** in a rectangular coordinate system.

f. What is the half-life of iodine-131?

2. The live birth rates B per thousand births in the United States are given in **Table 3.15** for the years 1986 through 1990 ($t = 0$ corresponds to 1986).

TABLE 3.15

t	B
0	15.5
1	15.7
2	15.9
3	16.2
4	16.7

a. Graph the ordered pairs $(t, \ln B)$. (You will need to adjust the scale so that you can discriminate between plotted points. A suggestion is given in **Figure 3.48.**)

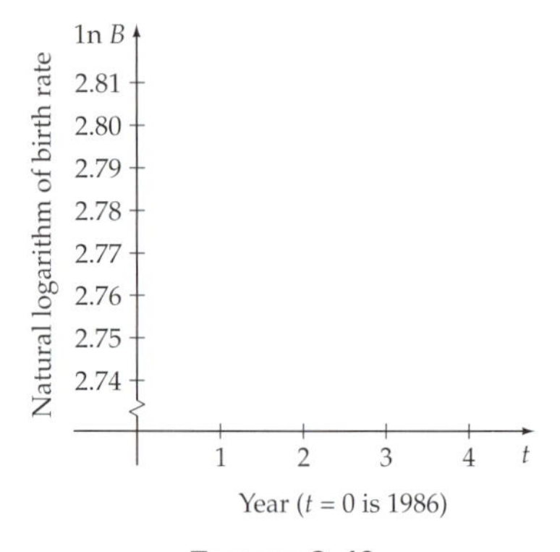

FIGURE 3.48

b. Use the points (1, 2.754) and (3, 2.785) to approximate the slope of the line that passes through the points.

c. Using the slope calculated in part **b.** and the point (1, 2.754), determine the equation of the line.

d. Solve the equation you derived in part **c.** for B.

e. Graph the equation you derived in part **d.** in a rectangular coordinate system.

f. If the live birth rate continues as predicted by your model, in what year will the live birth rate be 17.5 per thousand births?

The difference in graphing strategies between Exercise 1 and Exercise 2 is that in Exercise 1, semilog paper was used. When a point is graphed on this coordinate paper, the y-coordinate is $\ln y$. In Exercise 2, graphing a point $(x, \ln y)$ in a rectangular coordinate system has the same effect as graphing (x, y) in a semilog coordinate system.

CHAPTER 3 SUMMARY

3.1 Inverse Functions

- If f is a one-to-one function with domain X and range Y, and g is a function with domain Y and range X, then g is the inverse function of f if and only if $(f \circ g)(x) = x$ for all x in the domain of g and $(g \circ f)(x) = x$ for all x in the domain of f.
- A function f has an inverse function if and only if it is a one-to-one function. The graph of a function f and the graph of the inverse function f^{-1} are symmetric with respect to the line given by $y = x$.

3.2 Exponential Functions and Their Applications

- For all positive real numbers b, $b \neq 1$, the exponential function defined by $f(x) = b^x$ has the following properties:

  1. f has the set of real numbers as its domain.

  2. f has the set of positive real numbers as its range.

  3. f has a graph with a y-intercept of (0, 1).

  4. f has a graph asymptotic to the x-axis.

  5. f is a one-to-one function.

6. f is an increasing function if $b > 1$.
7. f is a decreasing function if $0 < b < 1$.

- As n increases without bound, $(1 + 1/n)^n$ approaches an irrational number denoted by e. The value of e accurate to eight decimal places is 2.71828183.
- The function defined by $f(x) = e^x$ is called the natural exponential function.

3.3 Logarithmic Functions and Their Applications

- *Definition of a Logarithm* If $x > 0$ and b is a positive constant ($b \neq 1$), then

$$y = \log_b x \quad \text{if and only if} \quad b^y = x$$

- For all positive real numbers b, $b \neq 1$, the function defined by $f(x) = \log_b x$ has the following properties:

1. f has the set of positive real numbers as its domain.
2. f has the set of real numbers as its range.
3. f has a graph with an x-intercept of $(1, 0)$.
4. f has a graph asymptotic to the y-axis.
5. f is a one-to-one function.
6. f is an increasing function if $b > 1$.
7. f is a decreasing function if $0 < b < 1$.

- The exponential form of $y = \log_b x$ is $b^y = x$.
- The logarithmic form of $b^y = x$ is $y = \log_b x$.
- *Basic Logarithmic Properties*

1. $\log_b b = 1$ 2. $\log_b 1 = 0$ 3. $\log_b (b^p) = p$

- The function $f(x) = \log_{10} x$ is the common logarithmic function. It is customarily written as $f(x) = \log x$.
- The function $f(x) = \log_e x$ is the natural logarithmic function. It is customarily written as $f(x) = \ln x$.

3.4 Logarithms and Logarithmic Scales

- If b, M, and N are positive real numbers ($b \neq 1$), and p is any real number, then

$$\log_b(MN) = \log_b M + \log_b N$$

$$\log_b \frac{M}{N} = \log_b M - \log_b N$$

$$\log_b(M^p) = p \log_b M$$

$$\log_b M = \log_b N \quad \text{implies} \quad M = N$$

$$M = N \quad \text{implies} \quad \log_b M = \log_b N$$

$$b^{\log_b p} = p \quad (\text{for } p > 0)$$

- *Change-of-Base Formula* If x, a, and b are positive real numbers with $a \neq 1$ and $b \neq 1$, then

$$\log_b x = \frac{\log_a x}{\log_a b}$$

- An earthquake with an intensity of I has a Richter scale magnitude of $M = \log\left(\frac{I}{I_0}\right)$, where I_0 is the measure of the intensity of a zero-level earthquake.
- The pH of a solution with a hydronium-ion concentration of H^+ mole per liter is given by $\text{pH} = -\log[H^+]$.

3.5 Exponential and Logarithmic Equations

- *Equality of Exponents Theorem* If b is a positive real number ($b \neq 1$) such that $b^x = b^y$, then $x = y$.
- Exponential equations of the form $b^x = b^y$ can be solved by using the Equality of Exponents Theorem.
- Exponential equations of the form $b^x = c$ can be solved by taking either the common logarithm or the natural logarithm of each side of the equation.
- Logarithmic equations can often be solved by using the properties of logarithms and the definition of a logarithm.

3.6 Exponential Growth and Decay

- The function defined by $N(t) = N_0 e^{kt}$ is called an exponential growth function if k is a positive constant, and it is called an exponential decay function if k is a negative constant.
- *The Compound Interest Formula* A principal P invested at an annual interest rate r, expressed as a decimal and compounded n times per year for t years, produces the balance

$$A = P\left(1 + \frac{r}{n}\right)^{nt}$$

- *Continuous Compounding Interest Formula* If an account with principal P and annual interest rate r is compounded continuously for t years, then the balance is $A = Pe^{rt}$.

CHAPTER 3 TRUE/FALSE EXERCISES

In Exercises 1 to 16, answer true or false. If the statement is false, give an example or state a reason to demonstrate that the statement is false.

1. Every function has an inverse function.
2. If $(f \circ g)(a) = a$ and $(g \circ f)(a) = a$ for some constant a, then f and g are inverse functions.
3. If $7^x = 40$, then $\log_7 40 = x$.
4. If $\log_4 x = 3.1$, then $4^{3.1} = x$.
5. If $f(x) = \log x$ and $g(x) = 10^x$, then $f[g(x)] = x$ for all real numbers x.
6. If $f(x) = \log x$ and $g(x) = 10^x$, then $g[f(x)] = x$ for all real numbers x.
7. The exponential function $h(x) = b^x$ is an increasing function.
8. The logarithmic function $j(x) = \log_b x$ is an increasing function.
9. The exponential function $h(x) = b^x$ is a one-to-one function.
10. The logarithmic function $j(x) = \log_b x$ is a one-to-one function.
11. The graph of $f(x) = \dfrac{2^x + 2^{-x}}{2}$ is symmetric with respect to the y-axis.
12. The graph of $f(x) = \dfrac{2^x - 2^{-x}}{2}$ is symmetric with respect to the origin.
13. If $x > 0$ and $y > 0$, then $\log(x + y) = \log x + \log y$.
14. If $x > 0$, then $\log x^2 = 2 \log x$.
15. If M and N are positive real numbers, then
$$\ln \frac{M}{N} = \ln M - \ln N$$
16. For all $p > 0$, $e^{\ln p} = p$.

CHAPTER 3 REVIEW EXERCISES

In Exercises 1 to 4, determine whether the given functions are inverses.

1. $F(x) = 2x - 5 \qquad G(x) = \dfrac{x + 5}{2}$
2. $h(x) = \sqrt{x} \qquad k(x) = x^2, \quad x \geq 0$
3. $l(x) = \dfrac{x + 3}{x} \qquad m(x) = \dfrac{3}{x - 1}$
4. $p(x) = \dfrac{x - 5}{2x} \qquad q(x) = \dfrac{2x}{x - 5}$

In Exercises 5 to 8, find the inverse of the function. Sketch the graph of the function and its inverse on the same set of coordinate axes.

5. $f(x) = 3x - 4$
6. $g(x) = -2x + 3$
7. $h(x) = -\dfrac{1}{2}x - 2$
8. $k(x) = \dfrac{1}{x}$

In Exercises 9 to 20, solve each equation. Do not use a calculator.

9. $\log_5 25 = x$
10. $\log_3 81 = x$
11. $\ln e^3 = x$
12. $\ln e^{\pi} = x$
13. $3^{2x+7} = 27$
14. $5^{x-4} = 625$
15. $2^x = \dfrac{1}{8}$
16. $27(3^x) = 3^{-1}$
17. $\log x^2 = 6$
18. $\dfrac{1}{2} \log |x| = 5$
19. $10^{\log 2x} = 14$
20. $e^{\ln x^2} = 64$

In Exercises 21 to 30, sketch the graph of each function.

21. $f(x) = (2.5)^x$
22. $f(x) = \left(\dfrac{1}{4}\right)^x$
23. $f(x) = 3^{|x|}$
24. $f(x) = 4^{-|x|}$
25. $f(x) = 2^x - 3$
26. $f(x) = 2^{(x-3)}$

27. $f(x) = \frac{1}{3} \log x$

28. $f(x) = 3 \log x^{1/3}$

29. $f(x) = -\frac{1}{2} \ln x$

30. $f(x) = -\ln |x|$

In Exercises 31 and 32, use a graphing utility to graph each function.

31. $f(x) = \frac{4^x + 4^{-x}}{2}$

32. $f(x) = \frac{3^x - 3^{-x}}{2}$

In Exercises 33 to 36, change each logarithmic equation to its exponential form.

33. $\log_4 64 = 3$

34. $\log_{1/2} 8 = -3$

35. $\log_{\sqrt{2}} 4 = 4$

36. $\ln 1 = 0$

In Exercises 37 to 40, change each exponential equation to its logarithmic form.

37. $5^3 = 125$

38. $2^{10} = 1024$

39. $10^0 = 1$

40. $8^{1/2} = 2\sqrt{2}$

In Exercises 41 to 44, write the given logarithm in terms of logarithms of x, y, and z.

41. $\log_b \frac{x^2 y^3}{z}$

42. $\log_b \frac{\sqrt{x}}{y^2 z}$

43. $\ln xy^3$

44. $\ln \frac{\sqrt{xy}}{z^4}$

In Exercises 45 to 48, write each logarithmic expression as a single logarithm with a coefficient of 1.

45. $2 \log x + \frac{1}{3} \log (x + 1)$

46. $5 \log x - 2 \log (x + 5)$

47. $\frac{1}{2} \ln 2xy - 3 \ln z$

48. $\ln x - (\ln y - \ln z)$

In Exercises 49 to 52, use the change-of-base formula and a calculator to approximate each logarithm accurate to six significant digits.

49. $\log_5 101$

50. $\log_3 40$

51. $\log_4 0.85$

52. $\log_8 0.3$

In Exercises 53 to 68, solve each equation for x. Give exact answers. Do not use a calculator.

53. $4^x = 30$

54. $5^{x+1} = 41$

55. $\ln 3x - \ln(x - 1) = \ln 4$

56. $\ln 3x + \ln 2 = 1$

57. $e^{\ln(x+2)} = 6$

58. $10^{\log(2x+1)} = 31$

59. $\frac{4^x + 4^{-x}}{4^x - 4^{-x}} = 2$

60. $\frac{5^x + 5^{-x}}{2} = 8$

61. $\log(\log x) = 3$

62. $\ln(\ln x) = 2$

63. $\log \sqrt{x - 5} = 3$

64. $\log x + \log(x - 15) = 1$

65. $\log_4(\log_3 x) = 1$

66. $\log_7(\log_5 x^2) = 0$

67. $\log_5 x^3 = \log_5 16x$

68. $25 = 16^{\log_4 x}$

69. **Earthquake Magnitude** Determine, to the nearest 0.1, the Richter scale magnitude of an earthquake with an intensity of $I = 51{,}782{,}000 I_0$.

70. **Earthquake Magnitude** A seismogram has an amplitude of 18 millimeters and a time delay of 21 seconds. Find, to the nearest tenth, the Richter scale magnitude of the earthquake that produced the seismogram.

71. **Comparison of Earthquakes** An earthquake had a Richter scale magnitude of 7.2. Its aftershock had a Richter scale magnitude of 3.7. Compare the intensity of the earthquake to the intensity of the aftershock by finding, to the nearest unit, the ratio of the larger intensity to the smaller intensity.

72. **Comparison of Earthquakes** An earthquake has an intensity 600 times the intensity of a second earthquake. Find, to the nearest tenth, the difference between the Richter scale magnitudes of the earthquakes.

73. **Chemistry** Find the pH of tomatoes that have a hydronium-ion concentration of 6.28×10^{-5}. Round to the nearest tenth.

74. **Chemistry** Find the hydronium-ion concentration of rainwater that has a pH of 5.4.

75. **Compound Interest** Find the balance when \$16,000 is invested at an annual rate of 8% for 3 years if the interest is compounded

a. monthly **b.** continuously

76. **Compound Interest** Find the balance when \$19,000 is invested at an annual rate of 6% for 5 years if the interest is compounded

a. daily **b.** continuously

77. **DEPRECIATION** The scrap value S of a product with an expected life span of n years is given by $S(n) = P(1 - r)^n$, where P is the original purchase price of the product and r is the annual rate of depreciation. A taxicab is purchased for \$12,400 and is expected to last 3 years. What is its scrap value if it depreciates at a rate of 29% per year?

78. **MEDICINE** A skin wound heals according to the function given by $N(t) = N_0 e^{-0.12t}$, where N is the number of square centimeters of unhealed skin t days after the injury, and N_0 is the number of square centimeters covered by the original wound.

a. What percentage of the wound will be healed after 10 days?

b. How many days, to the nearest day, will it take for 50% of the wound to heal?

c. How long, to the nearest day, will it take for 90% of the wound to heal?

In Exercises 79 to 82, find the exponential growth/decay function $N(t) = N_0 e^{kt}$ that satisfies the given conditions.

79. $N(0) = 1, N(2) = 5$

80. $N(0) = 2, N(3) = 11$

81. $N(1) = 4, N(5) = 5$

82. $N(-1) = 2, N(0) = 1$

83. **POPULATION GROWTH**

a. Find the exponential growth function for a city whose population was 25,200 in 2002 and 26,800 in 2003. Use $t = 0$ to represent the year 2002.

b. Use the growth function to predict, to the nearest hundred, the population of the city in 2009.

84. **CARBON DATING** Determine, to the nearest ten years, the age of a bone if it now contains 96% of its original amount of carbon-14. The half-life of carbon-14 is 5730 years.

CHAPTER 3 TEST

1. Find the inverse of $f(x) = 2x - 3$. Graph f and f^{-1} on the same coordinate axes.

2. Find the inverse of $f(x) = \dfrac{x}{4x - 8}$. State the domain and the range of f^{-1}.

3. a. Write $\log_b(5x - 3) = c$ in exponential form.

b. Write $3^{x/2} = y$ in logarithmic form.

4. Write $\log_b \dfrac{z^2}{y^3\sqrt{x}}$ in terms of logarithms of x, y, and z.

5. Write $\log(2x + 3) - 3\log(x - 2)$ as a single logarithm with a coefficient of 1.

6. Use the change-of-base formula and a calculator to approximate $\log_4 12$. Round your result to the nearest ten thousandth.

7. Graph: $f(x) = 3^{-x/2}$

8. Graph: $f(x) = -\ln(x + 1)$

9. Solve: $5^x = 22$. Round your solution to the nearest ten thousandth.

10. Find the *exact* solution of $4^{5-x} = 7^x$.

11. Solve: $\log(x + 99) - \log(3x - 2) = 2$

12. Solve: $\ln(2 - x) + \ln(5 - x) = \ln(37 - x)$

13. Find the balance on \$20,000 invested at an annual interest rate of 7.8% for 5 years

a. compounded monthly.

b. compounded continuously.

14. Find the time required for money invested at an annual rate of 4% to double in value if the investment is compounded monthly. Round to the nearest hundredth of a year.

15. a. What, to the nearest tenth, will an earthquake measure on the Richter scale if it has an intensity of $I = 42{,}304{,}000I_0$?

b. Compare the intensity of an earthquake that measures 6.3 on the Richter scale to the intensity of an earthquake that measures 4.5 on the Richter scale by finding the ratio of the larger intensity to the smaller intensity. Round to the nearest whole number.

16. a. Find the exponential growth function for a city whose population was 34,600 in 1996 and 39,800 in 1999. Use $t = 0$ to represent the year 1996.

b. Use the growth function to predict the population of the city in 2006. Round to the nearest thousand.

17. Determine, to the nearest ten years, the age of a bone if it now contains 92% of its original amount of carbon-14. The half-life of carbon-14 is 5730 years.

18. An investor places $3500 into an account that earns 6% annual interest compounded annually. To the nearest tenth of a year, how long will it take for the investment to double in value?

19. A glass of milk at 71°F is placed in a refrigerator that maintains a constant temperature of 35°F. The temperature T of the milk t minutes after it is placed in the refrigerator is given by $T(t) = 35 + 36e^{-0.047t}$. To the nearest minute, how long will it take before the temperature of the milk is 50°F?

20. The velocity v, in feet per second, of an object t seconds after it has been dropped from a plane is given by $v = 125\left(\dfrac{e^{0.32t} - 1}{e^{0.32t} + 1}\right)$. Determine the horizontal asymptote of the graph of this function. Write a sentence that describes the meaning of the horizontal asymptote in the context of this problem.

CUMULATIVE REVIEW EXERCISES

1. Solve $|x - 4| \leq 2$. Write the solution set using interval notation.

2. Solve $x^2 - x \geq 20$. Write the solution set using interval notation.

3. Find, to the nearest tenth, the distance between the points $(5, 2)$ and $(11, 7)$.

4. The height, in feet, of a ball released with an initial upward velocity of 44 feet per second and at an initial height of 8 feet is given by $h(t) = -16t^2 + 44t + 8$, where t is the time in seconds after the ball is released. Find the maximum height the ball will reach.

5. Given $f(x) = 2x + 1$ and $g(x) = x^2 - 5$, find $(g \circ f)$.

6. Find the inverse of $f(x) = 3x - 5$.

7. What is the number of zeros, including multiplicities, of the polynomial function $p(x) = 20x^{10} - 3x^8 + 5x - 7$?

8. Use Descartes' Rule of Signs to determine the number of possible real zeros of $P(x) = x^4 - 3x^3 + x^2 - x - 6$.

9. Find the zeros of $P(x) = x^4 - 5x^3 + x^2 + 15x - 12$.

10. Find a polynomial function of lowest degree that has 2, $1 - i$, and $1 + i$ as zeros.

11. Find the equations of the vertical and horizontal asymptotes of the graph of $r(x) = \dfrac{3x - 5}{x - 4}$.

12. Determine the domain and the range of the rational function $R(x) = \dfrac{4}{x^2 + 1}$.

13. State whether $f(x) = 0.4^x$ is an increasing function or a decreasing function.

14. Write $\log_4 x = y$ in exponential form.

15. Write $5^3 = 125$ in logarithmic form.

16. Find, to the nearest tenth, the Richter scale magnitude of an earthquake with an intensity of $I = 11{,}650{,}600I_0$.

17. Solve $2e^x = 15$. Round to the nearest ten thousandth.

18. Find the age of a bone if it now has 94% of the carbon-14 it had at time $t = 0$. Round to the nearest ten years.

19. The internal temperature of a bread dough is 85°F before it is placed in an oven that maintains a constant temperature of 375°F. The temperature T of the dough t minutes after it is placed in the oven is given by $T(t) = 375 - 290e^{-0.008t}$. To the nearest minute, how long will it take before the temperature of the bread is 125°F?

20. A study by a psychologist determined that the percent of a number of random words a person could remember t hours after hearing the words can be approximated by $P = 90 - 25 \ln t$. After how many hours, to the nearest hour, will a person first remember less than 50% of the words?

CHAPTER 4

TRIGONOMETRIC FUNCTIONS

Applications of Trigonometric Functions

In this chapter we introduce trigonometric functions. The following illustrations are from three of the applications of trigonometric functions that are considered in this chapter. The path of a satellite is modeled by a trigonometric function in **Exercise 83, page 320.**

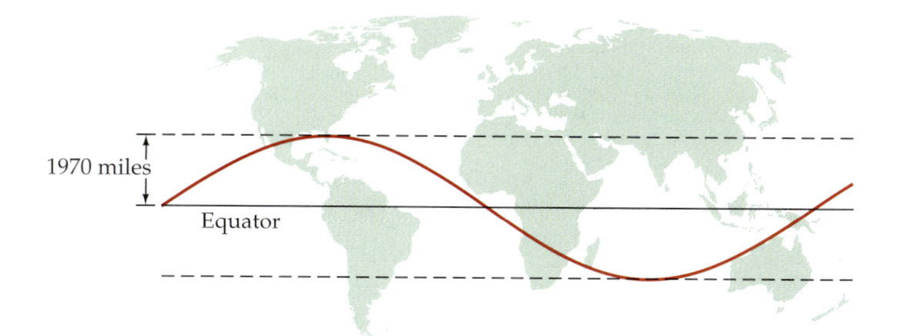

In **Exercise 70, page 348,** trigonometric functions are used to model the duration of daylight, in hours, for a period of one year at various latitudes. See the figure at the left.

In **Exercise 72, page 349,** a trigonometric function is used to approximate the temperature, in degrees Fahrenheit, at a desert location during a summer day.

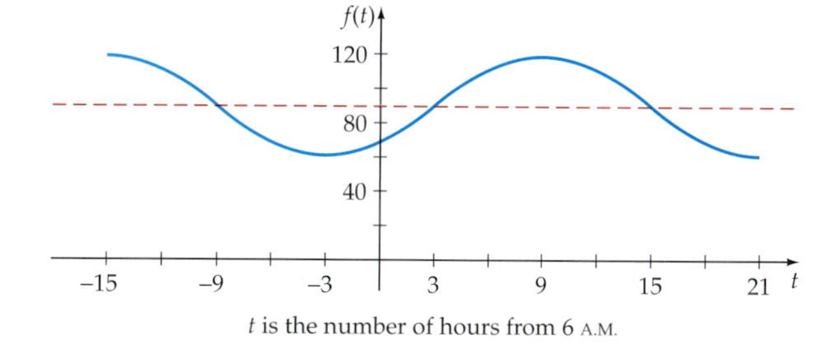

Generalizing Problem-Solving Strategies

Many times in mathematics a strategy for solving one type of problem can be applied in other circumstances. For instance, earlier in this text we discussed graphing techniques that involved compressing, stretching, and reflecting the graphs of various types of functions. For instance, suppose that f is defined by the graph at the right.

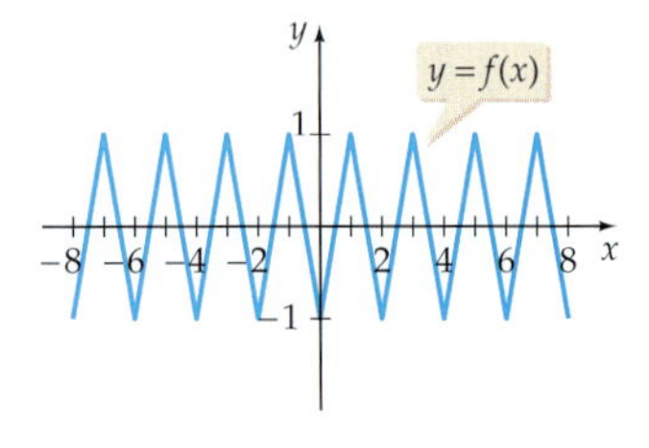

Then, using the techniques described earlier, the graphs of $y = f(2x)$, $y = f\left(\frac{1}{2}x\right)$, and $y = -f(x)$ are

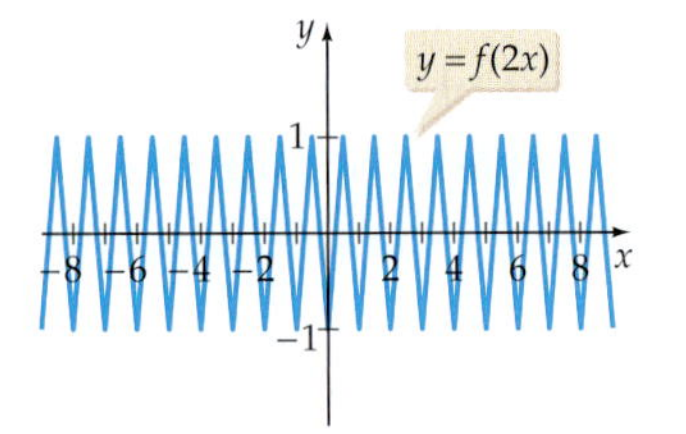

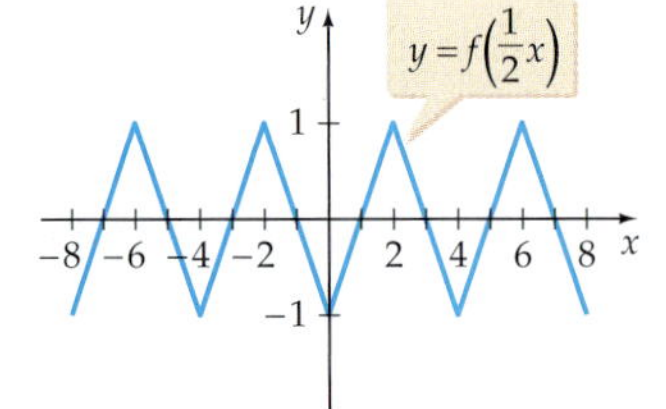

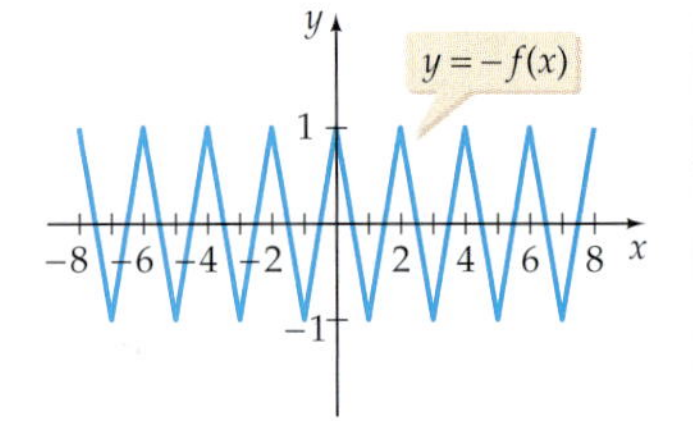

In this chapter we begin the study of trigonometric functions. One such trigonometric function is the sine function. As you will learn in this chapter, the graph of the sine function is as shown at the right.

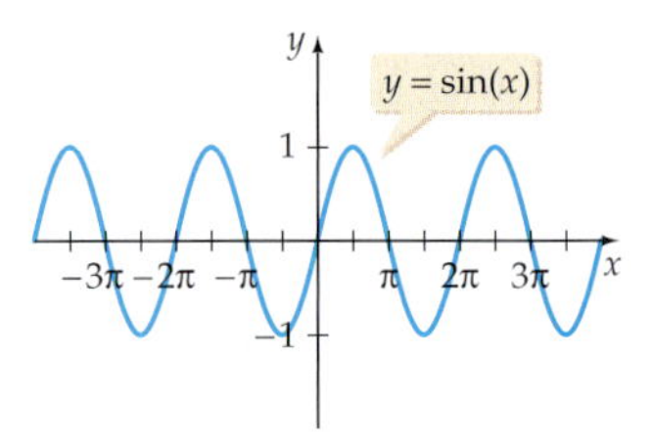

Applying the graphing techniques presented earlier, we have

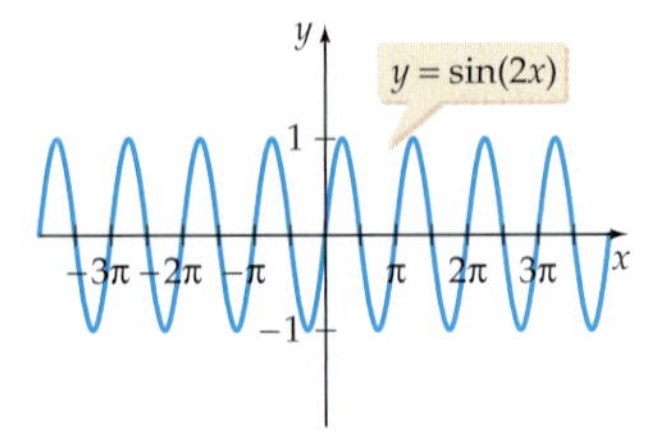

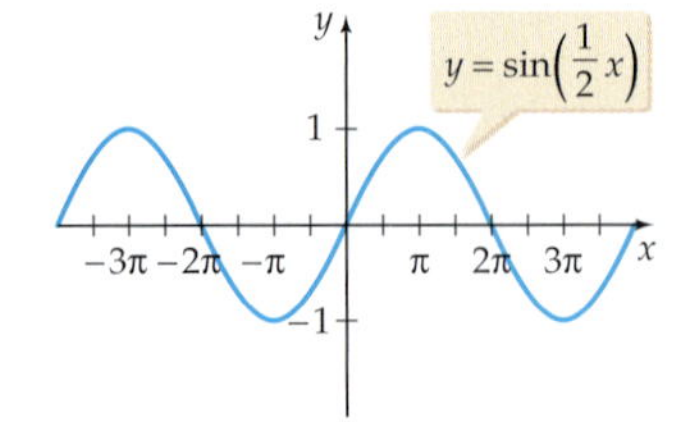

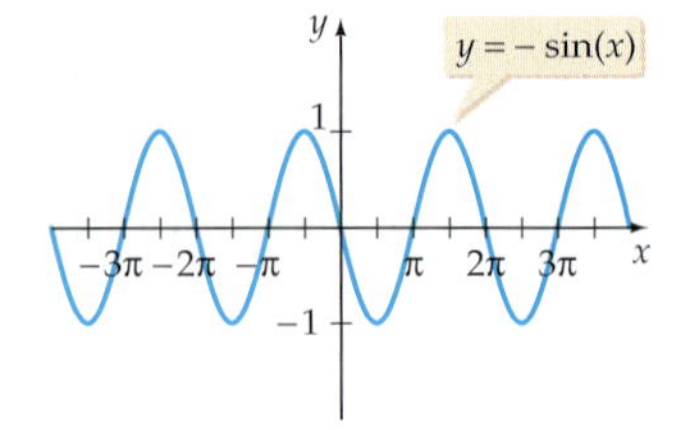

For exercises similar to these, see **Exercises 33 to 38 on page 346.**

SECTION 4.1 ANGLES AND ARCS

- Degree Measure
- Classification of Angles
- Conversion Between Units
- Radian Measure
- Arcs and Arc Length
- Linear and Angular Speed

A point P on a line separates the line into two parts, each of which is called a **half-line.** The union of point P and the half-line formed by P that includes point A is called a **ray,** and it is represented as $\overrightarrow{PA}$. The point P is the **endpoint** of ray $\overrightarrow{PA}$. **Figure 4.1** shows the ray $\overrightarrow{PA}$ and a second ray $\overrightarrow{QR}$.

In geometry, an *angle* is defined simply as the union of two rays that have a common endpoint. In trigonometry and many advanced mathematics courses, it is beneficial to define an angle in terms of a rotation.

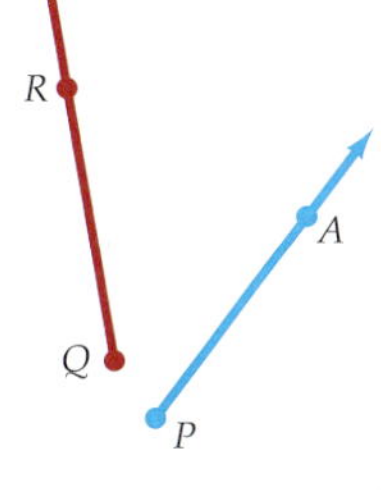
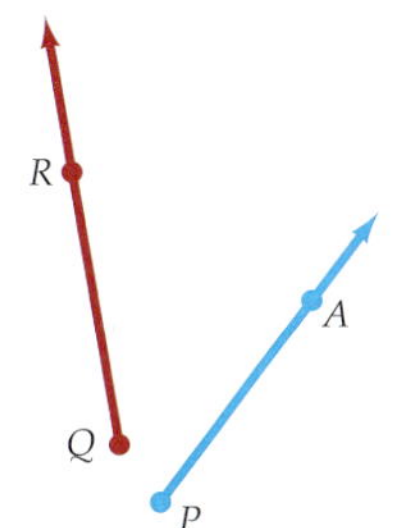

Figure 4.1

Definition of an Angle

An **angle** is formed by rotating a given ray about its endpoint to some terminal position. The original ray is the **initial side** of the angle, and the second ray is the **terminal side** of the angle. The common endpoint is the **vertex** of the angle.

Figure 4.2

There are several methods used to name an angle. One way is to employ Greek letters. For example, the angle shown in **Figure 4.2** can be designated as α or as $\angle\alpha$. It also can be named $\angle O$, $\angle AOB$, or $\angle BOA$. If you name an angle by using three points, such as $\angle AOB$, it is traditional to list the vertex point between the other two points.

Angles formed by a counterclockwise rotation are considered **positive angles,** and angles formed by a clockwise rotation are considered **negative angles.** See **Figure 4.3.**

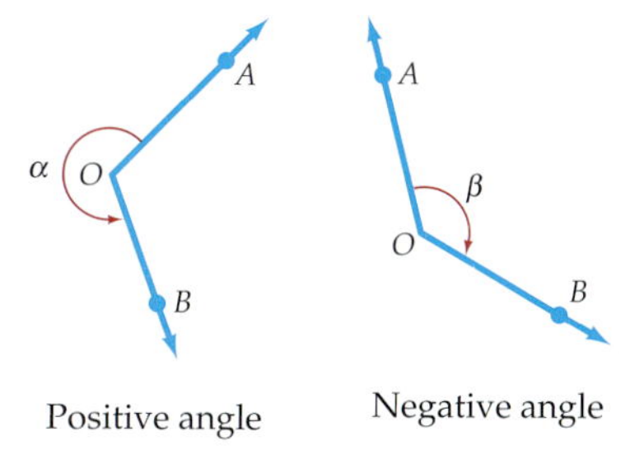

Figure 4.3

Degree Measure

The **measure** of an angle is determined by the amount of rotation of the initial ray. The concept of measuring angles in *degrees* grew out of the belief of the early Sumerians and Babylonians that the seasons repeated every 360 days.

Definition of Degree

One **degree** is the measure of an angle formed by rotating a ray $\frac{1}{360}$ of a complete revolution. The symbol for degree is °.

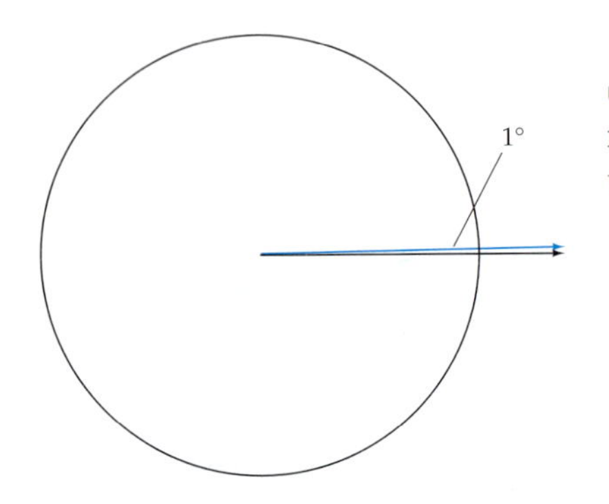

FIGURE 4.4
$1° = \frac{1}{360}$ of a revolution

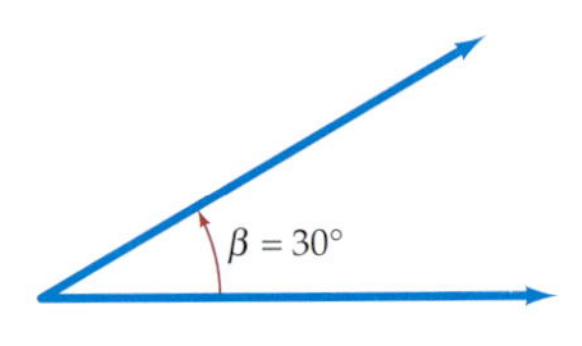

FIGURE 4.5

The angle shown in **Figure 4.4** has a measure of 1°. The angle β shown in **Figure 4.5** has a measure of 30°. We will use the notation $\beta = 30°$ to denote that the measure of angle β is 30°. The protractor shown in **Figure 4.6** can be used to measure an angle in degrees or to draw an angle with a given degree measure.

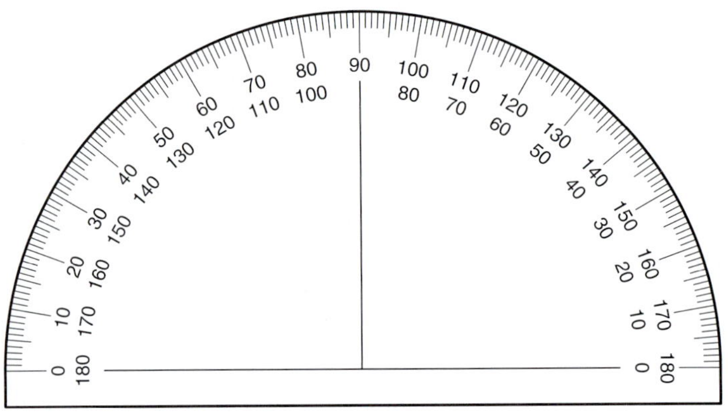

FIGURE 4.6
Protractor for measuring angles in degrees

Classification of Angles

Angles are often classified according to their measure.

- 180° angles are **straight angles.** See **Figure 4.7a.**
- 90° angles are **right angles.** See **Figure 4.7b.**
- Angles that have a measure greater than 0° but less than 90° are **acute angles.** See **Figure 4.7c.**
- Angles that have a measure greater than 90° but less than 180° are **obtuse angles.** See **Figure 4.7d.**

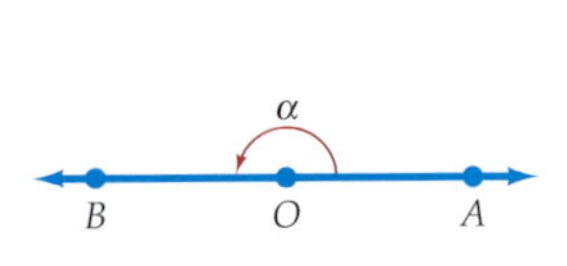

a. Straight angle ($\alpha = 180°$)

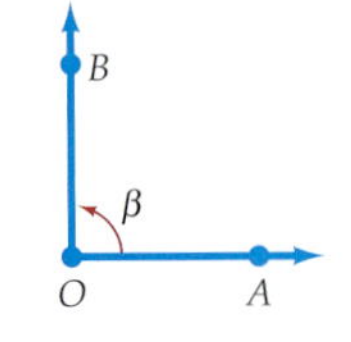

b. Right angle ($\beta = 90°$)

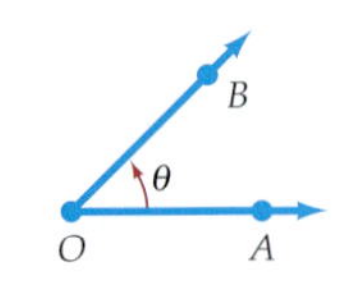

c. Acute angle ($0° < \theta < 90°$)

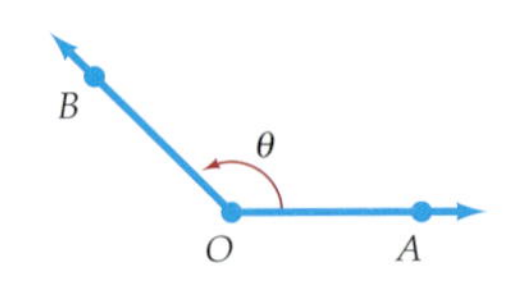

d. Obtuse angle ($90° < \theta < 180°$)

FIGURE 4.7

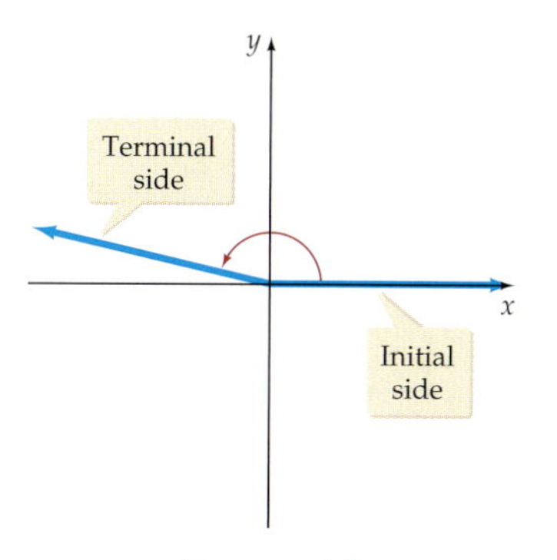

FIGURE 4.8
An angle in standard position

An angle superimposed in a Cartesian coordinate system is in **standard position** if its vertex is at the origin and its initial side is on the positive x-axis. See **Figure 4.8.**

Two positive angles are **complementary angles** **(Figure 4.9a)** if the sum of the measures of the angles is 90°. Each angle is the *complement* of the other angle. Two positive angles are **supplementary angles** **(Figure 4.9b)** if the sum of the measures of the angles is 180°. Each angle is the *supplement* of the other angle.

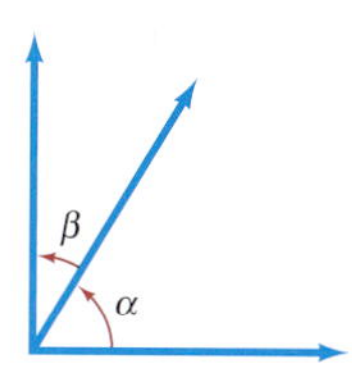

a. Complementary angles
$\alpha + \beta = 90°$

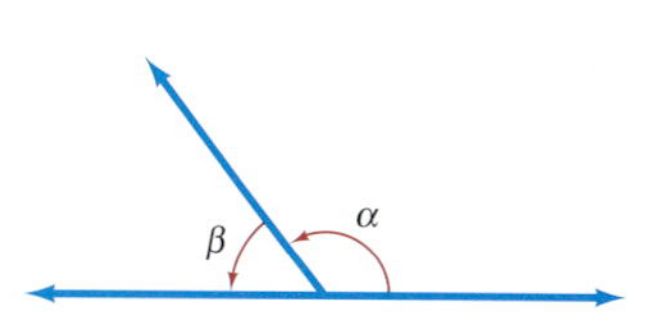

b. Supplementary angles
$\alpha + \beta = 180°$

FIGURE 4.9

FIGURE 4.10

FIGURE 4.11

EXAMPLE 1 Find the Measure of the Complement and the Supplement of an Angle

For each angle, find the measure (if possible) of its complement and of its supplement.

a. $\theta = 40°$ **b.** $\theta = 125°$

Solution

a. **Figure 4.10** shows $\angle\theta = 40°$ in standard position. The measure of its complement is $90° - 40° = 50°$. The measure of its supplement is $180° - 40° = 140°$.

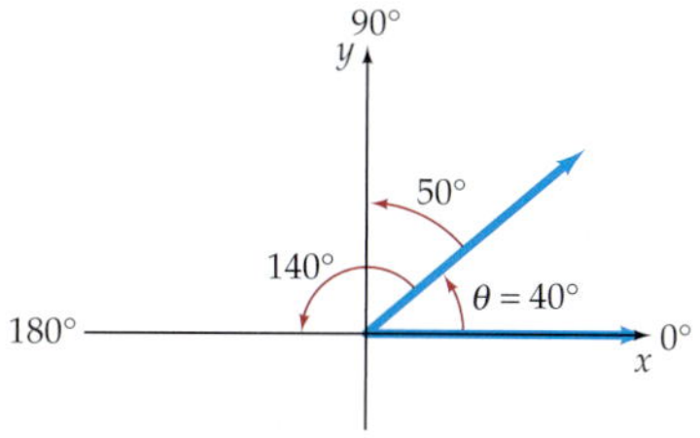

b. **Figure 4.11** shows $\angle\theta = 125°$ in standard position. Angle θ does not have a complement because there is no positive number x such that

$$x° + 125° = 90°$$

The measure of its supplement is $180° - 125° = 55°$.

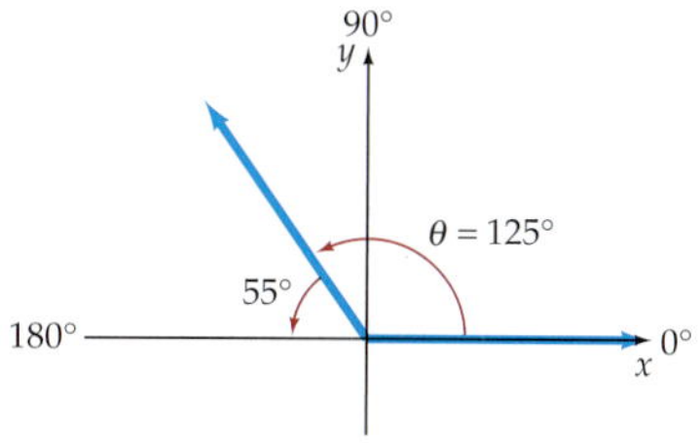

▶ TRY EXERCISE 2, PAGE 302

? QUESTION Are the two acute angles of any right triangle complementary angles? Explain.

Some angles have a measure greater than 360°. See **Figure 4.12a** and **Figure 4.12b.** The angle shown in **Figure 4.12c** has a measure less than −360°,

---

? ANSWER Yes. The sum of the measures of the angles of any triangle is 180°. The right angle has a measure of 90°. Thus the measure of the sum of the two acute angles must be $180° - 90° = 90°$.

because it is formed by a clockwise rotation of more than one revolution of the initial ray.

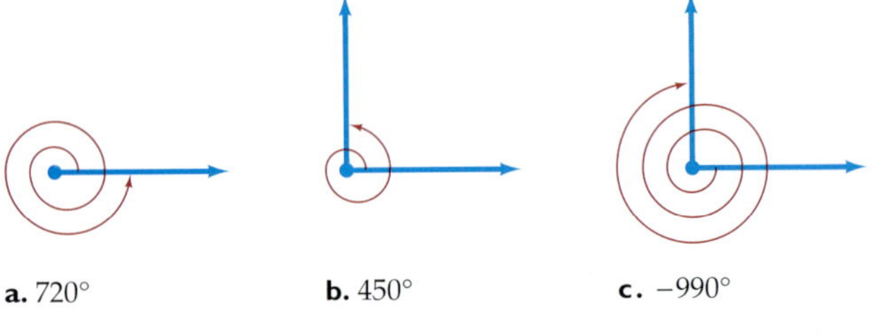

FIGURE 4.12

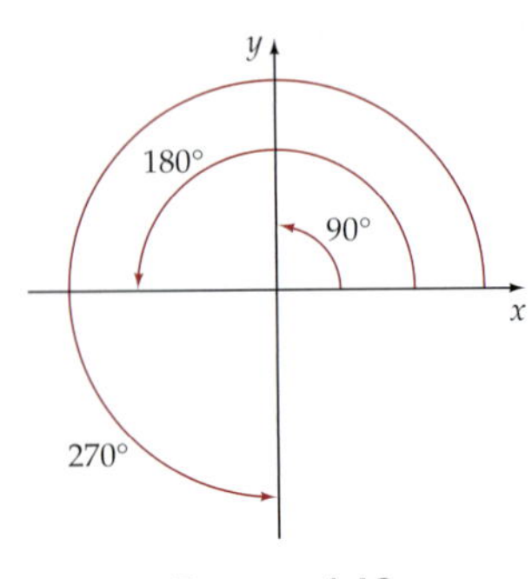

FIGURE 4.13

If the terminal side of an angle in standard position lies on a coordinate axis, then the angle is classified as a **quadrantal angle.** For example, the 90° angle, the 180° angle, and the 270° angle shown in **Figure 4.13** are all quadrantal angles.

If the terminal side of an angle in standard position does not lie on a coordinate axis, then the angle is classified according to the quadrant that contains the terminal side. For example, $\angle\beta$ in **Figure 4.14** is a Quadrant III angle.

Angles in standard position that have the same terminal sides are **coterminal angles.** Every angle has an unlimited number of coterminal angles. **Figure 4.15** shows $\angle\theta$ and two of its coterminal angles, labeled $\angle 1$ and $\angle 2$.

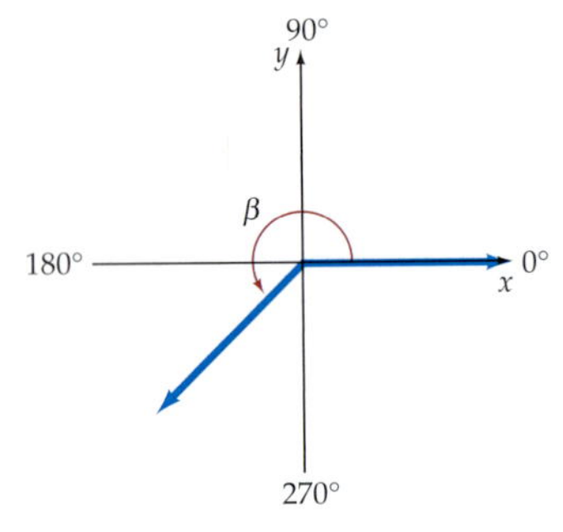

FIGURE 4.14

Measures of Coterminal Angles

Given $\angle\theta$ in standard position with measure $x°$, then the measures of the angles that are coterminal with $\angle\theta$ are given by

$$x° + k \cdot 360°$$

where k is an integer.

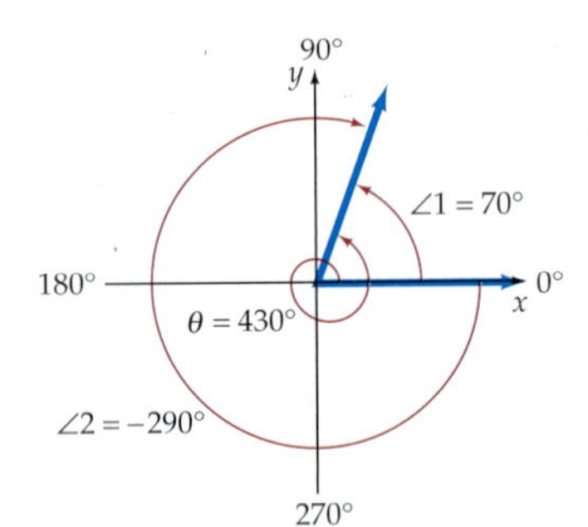

FIGURE 4.15

This theorem states that the measures of any two coterminal angles differ by an integer multiple of 360°. For instance, in **Figure 4.15,** $\theta = 430°$,

$$\angle 1 = 430° + (-1) \cdot 360° = 70°, \quad \text{and}$$
$$\angle 2 = 430° + (-2) \cdot 360° = -290°$$

If we add positive multiples of 360° to 430°, we find that the angles with measures 790°, 1150°, 1510°, ... are also coterminal with $\angle\theta$.

EXAMPLE 2 Classify by Quadrant and Find a Coterminal Angle

Assume the following angles are in standard position. Classify each angle by quadrant, and then determine the measure of the positive angle with measure less than 360° that is coterminal with the given angle.

a. $\alpha = 550°$ **b.** $\beta = -225°$ **c.** $\gamma = 1105°$

a.

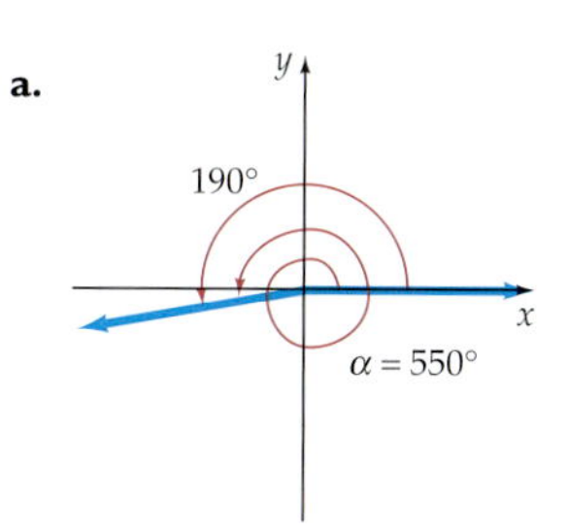

b.

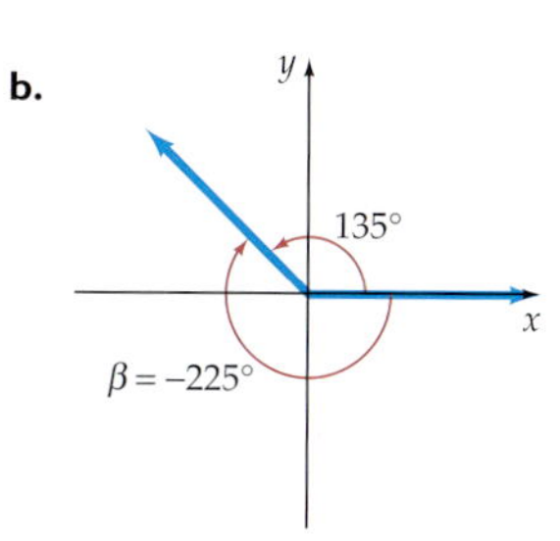

c.

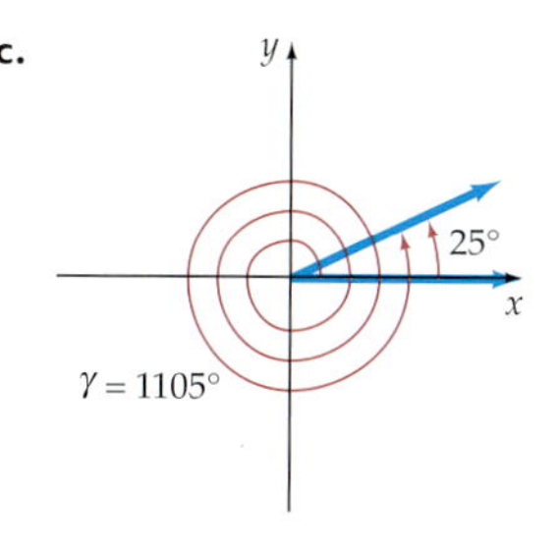

FIGURE 4.16

Solution

a. Because $550° = 190° + 360°$, $\angle\alpha$ is coterminal with an angle that has a measure of $190°$. $\angle\alpha$ is a Quadrant III angle. See **Figure 4.16a.**

b. Because $-225° = 135° + (-1) \cdot 360°$, $\angle\beta$ is coterminal with an angle that has a measure of $135°$. $\angle\beta$ is a Quadrant II angle. See **Figure 4.16b.**

c. $1105° \div 360° = 3\frac{5}{72}$. Thus $\angle\gamma$ is an angle formed by three complete counterclockwise rotations, plus $\frac{5}{72}$ of a rotation. To convert $\frac{5}{72}$ of a rotation to degrees, multiply $\frac{5}{72}$ times $360°$.

$$\frac{5}{72} \cdot 360° = 25°$$

Thus $1105° = 25° + 3 \cdot 360°$. Hence $\angle\gamma$ is coterminal with an angle that has a measure of $25°$. $\angle\gamma$ is a Quadrant I angle. See **Figure 4.16c.**

TRY EXERCISE 14, PAGE 302

CONVERSION BETWEEN UNITS

There are two popular methods for representing a fractional part of a degree. One is the decimal degree method. For example, the measure 29.76° is a decimal degree. It means

29° plus 76 hundredths of 1°

A second method of measurement is known as the DMS (**D**egree, **M**inute, **S**econd) method. In the DMS method, a degree is subdivided into 60 equal parts, each of which is called a *minute*, denoted by ′. Thus $1° = 60'$. Furthermore, a minute is subdivided into 60 equal parts, each of which is called a *second*, denoted by ″. Thus $1' = 60''$ and $1° = 3600''$. The fractions

$$\frac{1°}{60'} = 1, \qquad \frac{1'}{60''} = 1, \quad \text{and} \quad \frac{1°}{3600''} = 1$$

are another way of expressing the relationships among degrees, minutes, and seconds. Each of the fractions is known as a **unit fraction** or a **conversion factor.** Because all conversion factors are equal to 1, you can multiply a magnitude by a conversion factor and not change the magnitude, even though you change the units used to express the magnitude. The following illustrates the process of multiplying by conversion factors to write $126°12'27''$ as a decimal degree.

$$\begin{aligned} 126°12'27'' &= 126° + 12' + 27'' \\ &= 126° + 12'\left(\frac{1°}{60'}\right) + 27''\left(\frac{1°}{3600''}\right) \\ &= 126° + 0.2° + 0.0075° = 126.2075° \end{aligned}$$

Many graphing calculators can be used to convert a decimal degree measure to its equivalent DMS measure, and vice versa. For instance, **Figure 4.17** shows that 31.57° is equivalent to 31°34′12″. On a TI-83 graphing calculator, the degree symbol, °, and the DMS function are in the ANGLE menu.

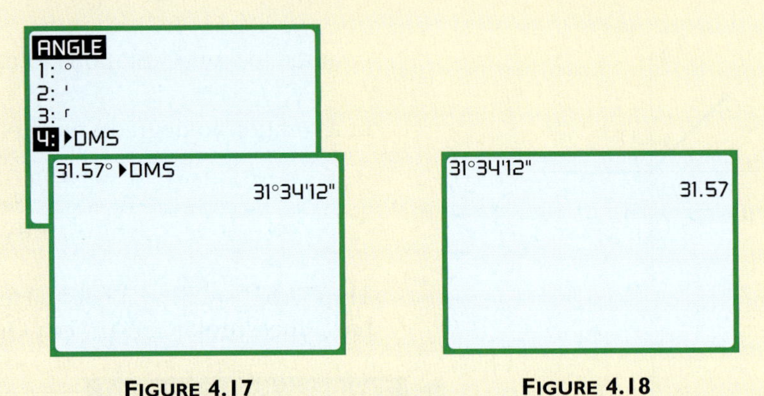

FIGURE 4.17 **FIGURE 4.18**

To convert a DMS measure to its equivalent decimal degree measure, enter the DMS measure and press ENTER. The calculator screen in **Figure 4.18** shows that 31°34′12″ is equivalent to 31.57°. A TI-83 needs to be in degree mode to produce the results displayed in **Figures 4.17** and **4.18.** On a TI-83, the degree symbol, °, and the minute symbol, ′, are both in the ANGLE menu; however, the second symbol, ″, is entered by pressing ALPHA +.

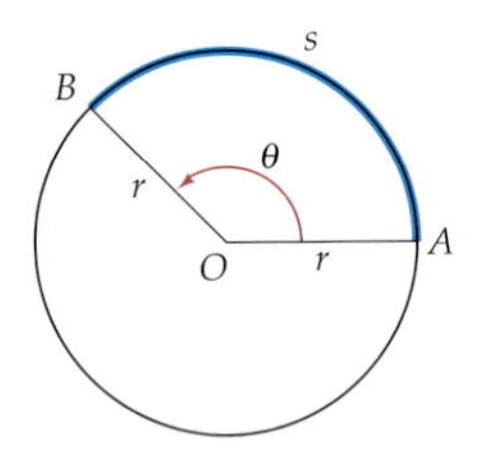

FIGURE 4.19

RADIAN MEASURE

Another commonly used angle measurement is the *radian*. To define a radian, first consider a circle of radius r and two radii $\overline{OA}$ and $\overline{OB}$. The angle θ formed by the two radii is a **central angle.** The portion of the circle between A and B is an **arc** of the circle and is written $\overset{\frown}{AB}$. We say that $\overset{\frown}{AB}$ *subtends* the angle θ. The length of $\overset{\frown}{AB}$ is s (see **Figure 4.19**).

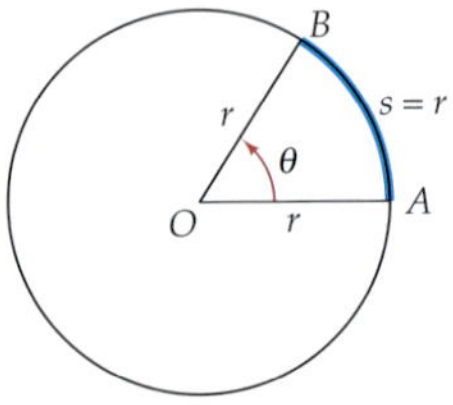

Definition of Radian

One **radian** is the measure of the central angle subtended by an arc of length r on a circle of radius r. See **Figure 4.20.**

FIGURE 4.20
Central angle θ has a measure of 1 radian.

Figure 4.21 shows a protractor that can be used to measure angles in radians or to construct angles given in radian measure.

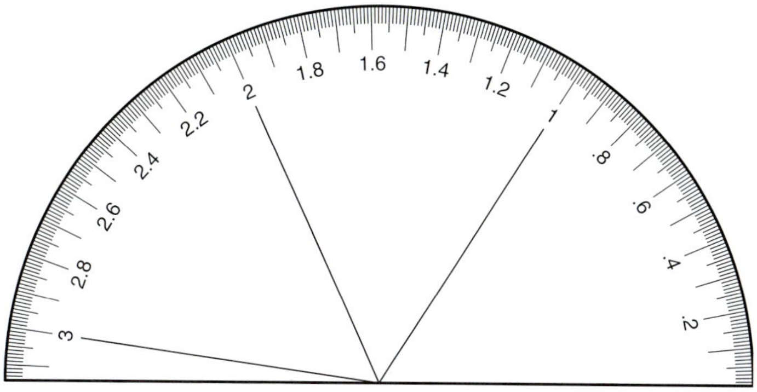

FIGURE 4.21
Protractor for measuring angles in radians

Radian Measure

Given an arc of length s on a circle of radius r, the measure of the central angle subtended by the arc is $\theta = \dfrac{s}{r}$ radians.

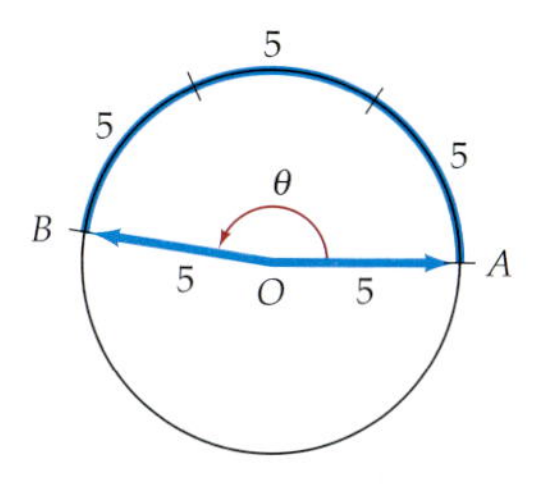

FIGURE 4.22
Central angle θ has a measure of 3 radians.

As an example, consider that an arc of length 15 centimeters on a circle with a radius of 5 centimeters subtends an angle of 3 radians, as shown in **Figure 4.22.** The same result can be found by dividing 15 centimeters by 5 centimeters.

To find the measure in radians of any central angle θ, divide the length s of the arc that subtends θ by the length of the radius of the circle. Using the formula for radian measure, we find that an arc of length 12 centimeters on a circle of radius 8 centimeters subtends a central angle θ whose measure is

$$\theta = \frac{s}{r} \text{ radians} = \frac{12 \text{ centimeters}}{8 \text{ centimeters}} \text{ radians} = \frac{3}{2} \text{ radians}$$

Note that the centimeter units are *not* part of the final result. The radian measure of a central angle formed by an arc of length 12 miles on a circle of radius 8 miles would be the same, $\dfrac{3}{2}$ radians. We say that radian is a *dimensionless* quantity because there are no units of measurement associated with a radian.

Recall that the circumference of a circle is given by the equation $C = 2\pi r$. The radian measure of the central angle θ subtended by the circumference is $\theta = \dfrac{2\pi r}{r} = 2\pi$. In degree measure, the central angle θ subtended by the circumference is $360°$. Thus we have the relationship $360° = 2\pi$ radians. Dividing each side of the equation by 2 gives $180° = \pi$ radians. From this last equation, we can establish the following conversion factors.

INTEGRATING TECHNOLOGY

A calculator shows that

$1 \text{ radian} \approx 57.29577951°$

and

$1° \approx 0.017453293 \text{ radian}$

Radian-Degree Conversion

- To convert from radians to degrees, multiply by $\left(\dfrac{180°}{\pi \text{ radians}}\right)$.
- To convert from degrees to radians, multiply by $\left(\dfrac{\pi \text{ radians}}{180°}\right)$.

EXAMPLE 3 **Convert Degree Measure to Radian Measure**

Convert 300° to radians.

Solution

$$300° = 300°\left(\frac{\pi \text{ radians}}{180°}\right)$$
$$= \frac{5}{3}\pi \text{ radians} \qquad \bullet \text{ Exact answer}$$
$$\approx 5.23598776 \text{ radians} \qquad \bullet \text{ Approximate answer}$$

▶ TRY EXERCISE 32, PAGE 302

In Example 3, note that $\frac{5}{3}\pi$ radians is an exact result. Many times it will be convenient to leave the measure of an angle in terms of π and not change it to a decimal approximation.

EXAMPLE 4 **Convert Radian Measure to Degree Measure**

Convert $-\frac{3}{4}\pi$ radians to degrees.

Solution

$$-\frac{3}{4}\pi \text{ radians} = -\frac{3}{4}\pi \text{ radians}\left(\frac{180°}{\pi \text{ radians}}\right) = -135°$$

▶ TRY EXERCISE 40, PAGE 302

Table 4.1 lists the degree and radian measures of selected angles. **Figure 4.23** illustrates each angle as measured from the positive x-axis.

TABLE 4.1

Degrees	Radians
0	0
30	$\pi/6$
45	$\pi/4$
60	$\pi/3$
90	$\pi/2$
120	$2\pi/3$
135	$3\pi/4$
150	$5\pi/6$
180	π
210	$7\pi/6$
225	$5\pi/4$
240	$4\pi/3$
270	$3\pi/2$
300	$5\pi/3$
315	$7\pi/4$
330	$11\pi/6$
360	2π

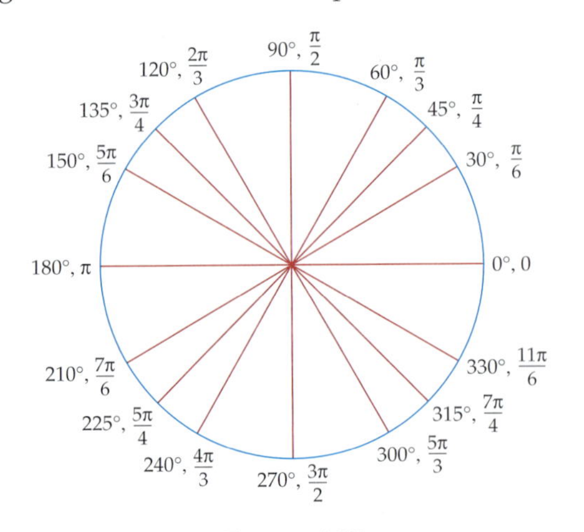

FIGURE 4.23
Degree and radian measures of selected angles

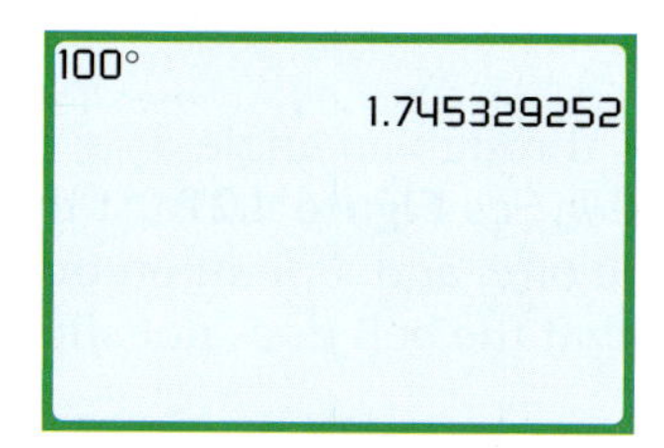

Figure 4.24

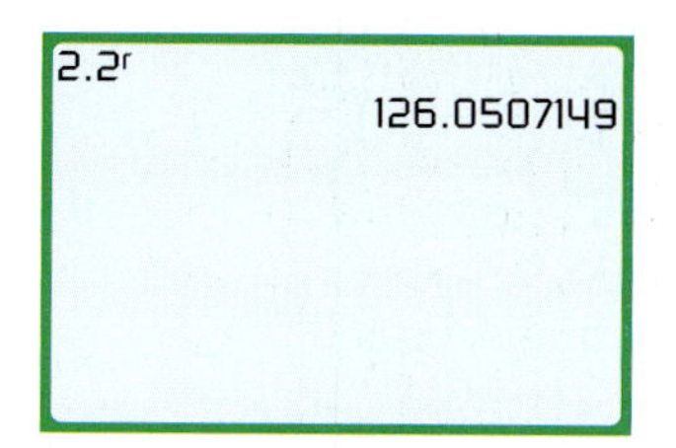

Figure 4.25

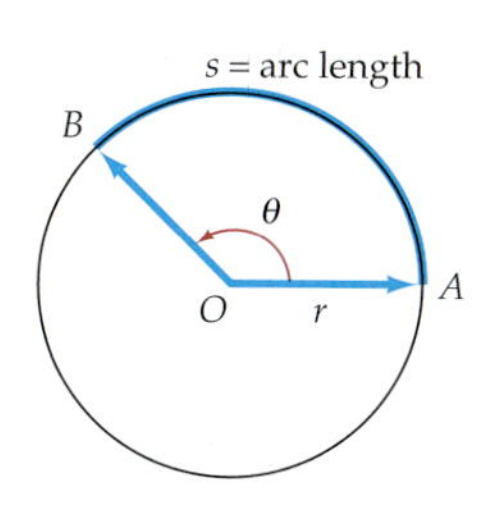

Figure 4.26
$s = r\theta$

Integrating Technology

A graphing calculator can convert degree measure to radian measure, and vice versa. For example, the calculator display in **Figure 4.24** shows that 100° is approximately 1.74533 radians. The calculator must be in radian mode to convert from degrees to radians. The display in **Figure 4.25** shows that 2.2 radians is approximately 126.051°. The calculator must be in degree mode to convert from radians to degrees.

On a TI-83, the symbol for radian measure is r, and it is in the ANGLE menu.

Arcs and Arc Length

Consider a circle of radius r. By solving the formula $\theta = \frac{s}{r}$ for s, we have an equation for arc length.

Arc Length Formula

Let r be the length of the radius of a circle and θ the nonnegative radian measure of a central angle of the circle. Then the length of the arc s that subtends the central angle is $s = r\theta$. See **Figure 4.26.**

Example 5 Find the Length of an Arc

Find the length of an arc that subtends a central angle of 120° in a circle of radius 10 centimeters.

Solution

The formula $s = r\theta$ requires that θ be expressed in radians. We first convert 120° to radian measure and then use the formula $s = r\theta$.

$$\theta = 120° = 120°\left(\frac{\pi \text{ radians}}{180°}\right) = \frac{2\pi}{3} \text{ radians} = \frac{2\pi}{3}$$

$$s = r\theta = (10 \text{ centimeters})\left(\frac{2\pi}{3}\right) = \frac{20\pi}{3} \text{ centimeters}$$

▶ Try Exercise 62, page 303

> **take note**
> The formula $s = r\theta$ is valid only when θ is expressed in radians.

Example 6 Solve an Application

A pulley with a radius of 10 inches uses a belt to drive a pulley with a radius of 6 inches. Find the angle through which the smaller pulley turns as the 10-inch pulley makes one revolution. State your answer in radians and also in degrees.

Continued ▶

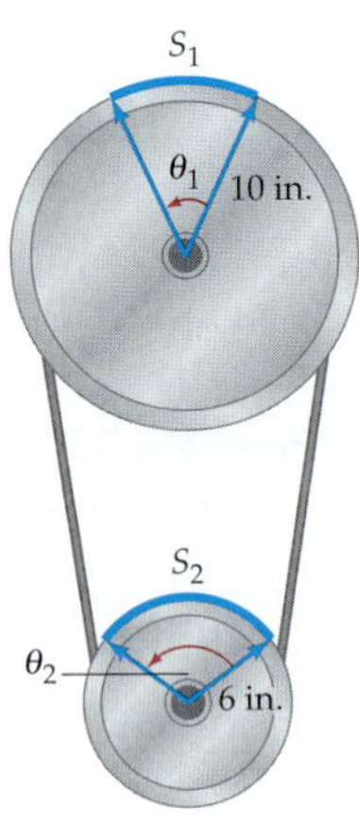

FIGURE 4.27

Solution

Use the formula $s = r\theta$. As the 10-inch pulley turns through an angle θ_1, a point on that pulley moves s_1 inches, where $s_1 = 10\theta_1$. See **Figure 4.27.** At the same time, the 6-inch pulley turns through an angle of θ_2 and a point on that pulley moves s_2 inches, where $s_2 = 6\theta_2$. Assuming that the belt does not slip on the pulleys, we have $s_1 = s_2$. Thus

$$10\theta_1 = 6\theta_2$$

$$10(2\pi) = 6\theta_2 \quad \text{• Solve for } \theta_2 \text{, when } \theta_1 = 2\pi \text{ radians.}$$

$$\frac{10}{3}\pi = \theta_2$$

The 6-inch pulley turns through an angle of $\frac{10}{3}\pi$ radians, or 600°.

TRY EXERCISE 66, PAGE 303

LINEAR AND ANGULAR SPEED

A car traveling at a speed of 55 miles per hour covers a distance of 55 miles in 1 hour. **Linear speed** v is *distance* traveled per unit time. In equation form,

$$v = \frac{s}{t}$$

where v is the linear speed, s is the distance traveled, and t is the time.

The floppy disk in a computer disk drive revolving at 300 revolutions per minute makes 300 complete revolutions in 1 minute. **Angular speed** ω is the *angle* through which a point on a circle moves per unit time. In equation form,

$$\omega = \frac{\theta}{t}$$

where ω is the angular speed, θ is the measure (in radians) of the angle through which a point has moved, and t is the time. Some common units of angular speed are revolutions per second, revolutions per minute, radians per second, and radians per minute.

EXAMPLE 7 Convert an Angular Speed

A hard disk in a computer rotates at 3600 revolutions per minute. Find the angular speed of the disk in radians per second.

Solution

As a point on the disk rotates 1 revolution (rev), the angle through which the point moves is 2π radians. Thus $\frac{2\pi \text{ radians}}{1 \text{ rev}}$ will be the unit fraction we will use to convert from revolutions to radians. To convert from minutes to seconds, use the unit fraction $\frac{1 \text{ minute}}{60 \text{ seconds}}$.

$$3600 \text{ rev/minute} = \frac{3600 \text{ rev}}{1 \text{ minute}}\left(\frac{2\pi \text{ radians}}{1 \text{ rev}}\right)\left(\frac{1 \text{ minute}}{60 \text{ seconds}}\right)$$

$$= 120\pi \text{ radians/second}$$ • **Exact answer**

$$\approx 377 \text{ radians/second}$$ • **Approximate answer**

▶ TRY EXERCISE 68, PAGE 303

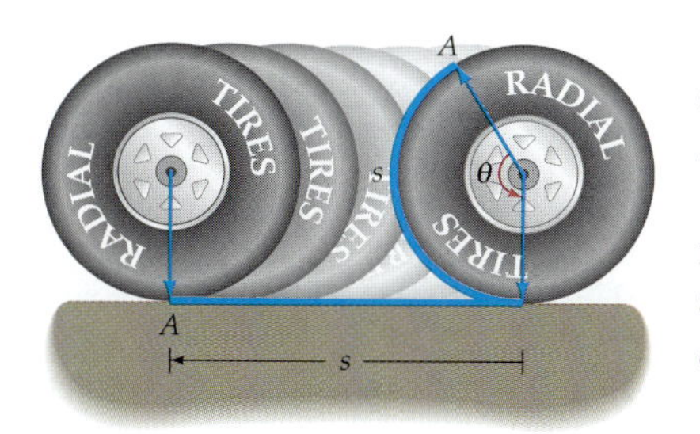

FIGURE 4.28

The tire on a car traveling along a road has both linear speed and angular speed. The relationship between linear and angular speed can be expressed by an equation.

Assume that the wheel in **Figure 4.28** is rolling without slipping. As the wheel moves a distance s, point A moves through an angle θ. The arc length subtending angle θ is also s, the distance traveled by the wheel. From the equations for linear and angular speed, we have

$$v = \frac{s}{t} = \frac{r\theta}{t} = r\frac{\theta}{t}$$ • $s = r\theta$

$$v = r\omega$$ • $\omega = \frac{\theta}{t}$

The equation $v = r\omega$ gives the linear speed of a point on a rotating body in terms of distance r from the axis of rotation and the angular speed ω, provided that ω is in radians per unit of time.

EXAMPLE 8 Find Linear Speed

A wind machine is used to generate electricity. The wind machine has propeller blades that are 12 feet in length (see **Figure 4.29**). If the propeller is rotating at 3 revolutions per second, what is the linear speed in feet per second of the tips of the blades?

Solution

Convert the angular speed $\omega = 3$ revolutions per second into radians per second, and then use the formula $v = r\omega$.

$$\omega = \frac{3 \text{ revolutions}}{1 \text{ second}} = \left(\frac{3 \text{ revolutions}}{1 \text{ second}}\right)\left(\frac{2\pi \text{ radians}}{1 \text{ revolution}}\right) = \frac{6\pi \text{ radians}}{1 \text{ second}}$$

Thus

$$v = r\omega = (12 \text{ feet})\left(\frac{6\pi \text{ radians}}{1 \text{ second}}\right)$$

$$= 72\pi \text{ feet per second} \approx 226 \text{ feet per second}$$

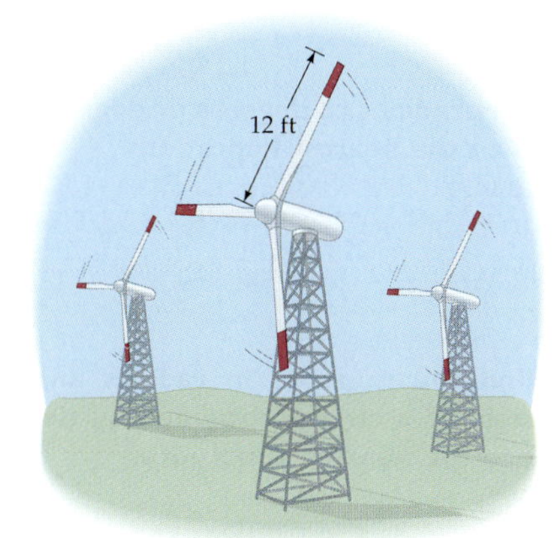

FIGURE 4.29

▶ TRY EXERCISE 74, PAGE 303

Topics for Discussion

1. The measure of a radian differs depending on the length of the radius of the circle used. Do you agree? Explain.
2. The measure of 1 radian is over 100 times larger than the measure of 1 degree. Do you agree? Explain.
3. What are the necessary conditions for an angle to be in standard position?
4. Is the supplement of an obtuse angle an acute angle?
5. Do all acute angles have a positive measure?

Exercise Set 4.1

In Exercises 1 to 12, find the measure (if possible) of the complement and the supplement of each angle.

1. 15° ▶**2.** 87° **3.** $70^\circ15'$

4. $22^\circ43'$ **5.** $56^\circ33'15''$ **6.** $19^\circ42'05''$

7. 1 **8.** 0.5 **9.** $\dfrac{\pi}{4}$

10. $\dfrac{\pi}{3}$ **11.** $\dfrac{2\pi}{5}$ **12.** $\dfrac{\pi}{6}$

In Exercises 13 to 18, classify each angle by quadrant, and state the measure of the positive angle with measure less than 360° that is coterminal with the given angle.

13. $\alpha = 610^\circ$ ▶**14.** $\alpha = 765^\circ$ **15.** $\alpha = -975^\circ$

16. $\alpha = -872^\circ$ **17.** $\alpha = 2456^\circ$ **18.** $\alpha = -3789^\circ$

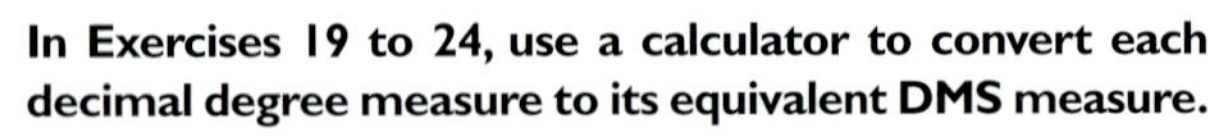

In Exercises 19 to 24, use a calculator to convert each decimal degree measure to its equivalent DMS measure.

19. 24.56° **20.** 110.24° **21.** 64.158°

22. 18.96° **23.** 3.402° **24.** 224.282°

In Exercises 25 to 30, use a calculator to convert each DMS measure to its equivalent decimal degree measure.

25. $25^\circ25'12''$ **26.** $63^\circ29'42''$ **27.** $183^\circ33'36''$

28. $141^\circ6'9''$ **29.** $211^\circ46'48''$ **30.** $19^\circ12'18''$

In Exercises 31 to 39, convert the degree measure to exact radian measure.

31. 30° ▶**32.** -45° **33.** 90°

34. 15° **35.** 165° **36.** 315°

37. 420° **38.** 630° **39.** 585°

In Exercises 40 to 48, convert the radian measure to exact degree measure.

▶**40.** $\dfrac{\pi}{4}$ **41.** $\dfrac{\pi}{5}$ **42.** $-\dfrac{2\pi}{3}$

43. $\dfrac{\pi}{6}$ **44.** $\dfrac{\pi}{9}$ **45.** $\dfrac{3\pi}{8}$

46. $\dfrac{11\pi}{18}$ **47.** $\dfrac{11\pi}{3}$ **48.** $\dfrac{6\pi}{5}$

In Exercises 49 to 54, convert radians to degrees or degrees to radians. Round answers to the nearest hundredth.

49. 1.5 **50.** -2.3 **51.** 133°

52. 427° **53.** 8.25 **54.** -90°

In Exercises 55 to 58, find the measure in radians and degrees of the central angle of a circle subtended by the given arc. Round answers to the nearest hundredth.

55. $r = 2$ inches, $s = 8$ inches

56. $r = 7$ feet, $s = 4$ feet

57. $r = 5.2$ centimeters, $s = 12.4$ centimeters

58. $r = 35.8$ meters, $s = 84.3$ meters

In Exercises 59 to 62, find the measure of the intercepted arc of a circle with the given radius and central angle. Round answers to the nearest hundredth.

59. $r = 8$ inches, $\theta = \dfrac{\pi}{4}$

60. $r = 3$ feet, $\theta = \dfrac{7\pi}{2}$

61. $r = 25$ centimeters, $\theta = 42°$

62. $r = 5$ meters, $\theta = 144°$

63. Find the number of radians in $1\frac{1}{2}$ revolutions.

64. Find the number of radians in $\frac{3}{8}$ revolution.

65. **Angular Rotation of Two Pulleys** A pulley with a radius of 14 inches uses a belt to drive a pulley with a radius of 28 inches. The 14-inch pulley turns through an angle of 150°. Find the angle through which the 28-inch pulley turns.

66. **Angular Rotation of Two Pulleys** A pulley with a diameter of 1.2 meters uses a belt to drive a pulley with a diameter of 0.8 meter. The 1.2-meter pulley turns through an angle of 240°. Find the angle through which the 0.8-meter pulley turns.

67. **Angular Speed** Find the angular speed, in radians per second, of the second hand on a clock.

68. **Angular Speed** Find the angular speed, in radians per second, of a point on the equator of the earth.

69. **Angular Speed** A wheel is rotating at 50 revolutions per minute. Find the angular speed in radians per second.

70. **Angular Speed** A wheel is rotating at 200 revolutions per minute. Find the angular speed in radians per second.

71. **Angular Speed** The turntable of a record player turns at $33\frac{1}{3}$ revolutions per minute. Find the angular speed in radians per second.

72. **Angular Speed** A car with a wheel of radius 14 inches is moving with a speed of 55 mph. Find the angular speed of the wheel in radians per second.

73. **Linear Speed of a Car** Each tire on a car has a radius of 15 inches. The tires are rotating at 450 revolutions per minute. Find the speed of the automobile to the nearest mile per hour.

74. **Linear Speed of a Truck** Each tire on a truck has a radius of 18 inches. The tires are rotating at 500 revolutions per minute. Find the speed of the truck to the nearest mile per hour.

75. **Bicycle Gears** The chain wheel of Emma's bicycle has a radius of 3.5 inches. The rear gear has a radius of 1.75 inches, and the back tire has a radius of 12 inches. If Emma pedals for 150 revolutions of the chain wheel, how far will she travel? Round to the nearest foot.

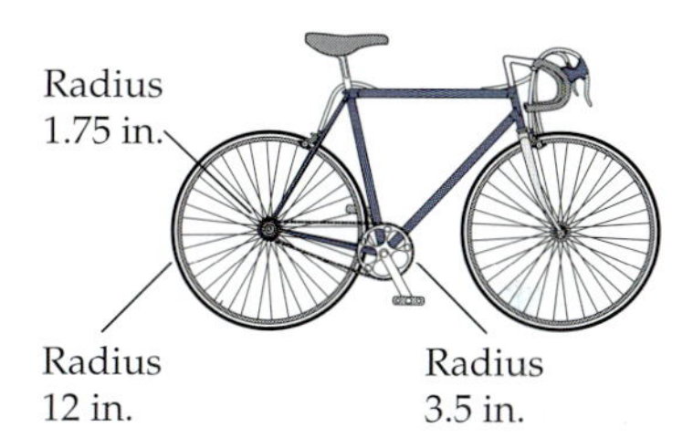

76. **Rotation versus Lift Distance** A winch with a 6-inch radius is used to lift a container. The winch is designed so that as it is rotated, the cable stays in contact with the surface of the winch. That is, the cable does not wrap on top of itself.

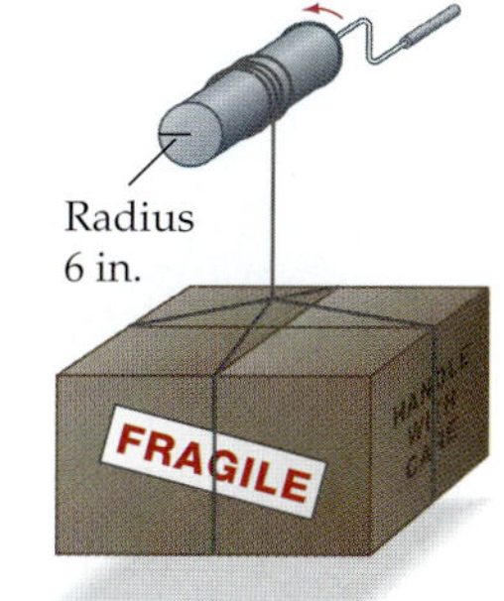

a. Find the distance the container is lifted as the winch is rotated through an angle of $\dfrac{5\pi}{6}$ radians.

b. Determine the angle, in radians, through which the winch must be rotated to lift the container a distance of 2 feet.

77. **Speed of the Concorde** During the time the Concorde was flying, it could fly from London to New York City, a distance of 3460 miles, in 2 hours and 59 minutes.

a. What was the average linear speed of the Concorde in miles per hour during one of these flights? Round to the nearest mile per hour.

b. What was the average angular speed of the Concorde in radians per hour during one of these flights? Assume that the Concorde maintains an altitude of 10 miles and that the radius of Earth is 3960 miles. Round to the nearest hundredth of a radian per hour.

c. If the Concorde left London at 1 P.M., what time would it be expected to arrive in New York City? (*Hint:* New York City is five time zones to the west of London.)

78. RACING THE SUN A pilot is flying a supersonic jet plane from east to west along a path over the equator. How fast, in miles per hour, does the pilot need to fly to keep the sun in the same relative position to the airplane? Assume that the plane is flying at an altitude of 2 miles above Earth and that Earth has a radius of 3960 miles. Round to the nearest mile per hour.

79. ASTRONOMY At a time when the earth was 93,000,000 miles from the sun, you observed through a tinted glass that the diameter of the sun occupied an arc of 31′. Determine, to the nearest ten thousand miles, the diameter of the sun.

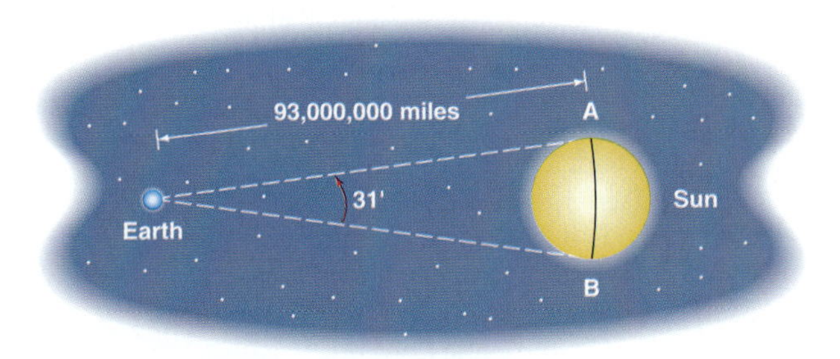

(*Hint:* Because the radius of arc AB is large and its central angle is small, the length of the diameter of the sun is approximately the length of the arc AB.)

80. ANGLE OF ROTATION AND DISTANCE The minute hand on the clock atop city hall measures 6 feet 3 inches from its tip to its axle.

a. Through what angle (in radians) does the minute hand pass between 9:12 A.M. and 9:48 A.M.?

b. What distance, to the nearest tenth of a foot, does the tip of the minute hand travel during this period?

81. VELOCITY OF THE HUBBLE SPACE TELESCOPE On April 25, 1990, the Hubble Space Telescope (HST) was deployed into a circular orbit 625 kilometers above the surface of the earth. The HST completes an earth orbit every 1.61 hours.

a. Find the angular velocity, with respect to the center of the earth, of the HST. Round your answer to the nearest 0.1 radian per hour.

b. Find the linear velocity of the HST. (*Hint:* The radius of the earth is about 6370 kilometers.) Round your answer to the nearest 100 kilometers per hour.

82. ESTIMATING THE RADIUS OF THE EARTH Eratosthenes, the fifth librarian of Alexandria (230 B.C.), was able to estimate the radius of the earth from the following data: The distance between the Egyptian cities of Alexandria and Syrene was 5000 stadia (520 miles). Syrene was located directly south of Alexandria. One summer, at noon, the sun was directly overhead at Syrene, whereas at the same time in Alexandria, the sun was at a 7.5° angle from the zenith.

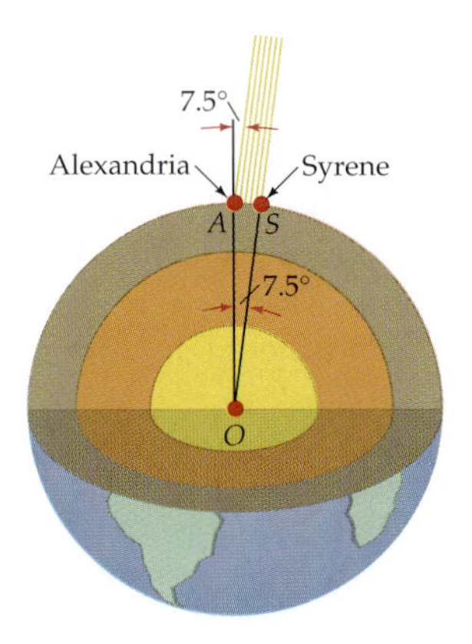

Eratosthenes reasoned that because the sun is far away, the rays of sunlight that reach the earth must be nearly parallel. From this assumption he concluded that the measure of $\angle AOS$ in the accompanying figure must be 7.5°. Use this information to estimate the radius (to the nearest 10 miles) of the earth.

83. VELOCITY COMPARISONS Assume that the bicycle in the figure is moving forward at a constant rate. Point A is on the edge of the 30-inch rear tire, and point B is on the edge of the 20-inch front tire.

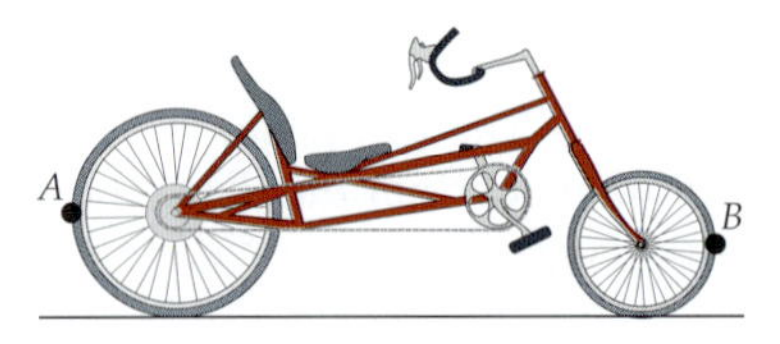

a. Which point (A or B) has the greater angular velocity? Explain.

b. Which point (A or B) has the greater linear velocity? Explain.

84. Given that s, r, θ, t, v, and ω are as defined in Section 4.1, determine which of the following formulas are valid.

$$s = r\theta \qquad r = \frac{s}{\theta} \qquad v = \frac{r\theta}{t}$$

$$v = r\omega \qquad v = \frac{s}{t} \qquad \omega = \frac{\theta}{t}$$

85. Nautical Miles and Statute Miles A **nautical mile** is the length of an arc, on the earth's equator, that subtends a $1'$ central angle. The equatorial radius of the earth is about 3960 **statute miles.**

a. Convert 1 nautical mile to statute miles. Round to the nearest hundredth of a statute mile.

b. Determine what percent (to the nearest 1 percent) of the earth's circumference is covered by a trip from Los Angeles, California to Honolulu, Hawaii (a distance of 2217 nautical miles).

86. Photography The field of view for a camera with a 200-millimeter lens is 12°. A photographer takes a photograph of a large building that is 485 feet in front of the camera. What is the approximate width, to the nearest foot, of the building that will appear in the photograph? (*Hint:* If the radius of an arc AB is large and its central angle is small, then the length of the chord AB is approximately the length of the arc AB.)

CONNECTING CONCEPTS

A *sector* of a circle is the region bounded by radii *OA* and *OB* and the intercepted arc *AB*. See the following figure. The area of the sector is given by

$$A = \frac{1}{2}r^2\theta$$

where r is the radius of the circle and θ is the measure of the central angle in radians.

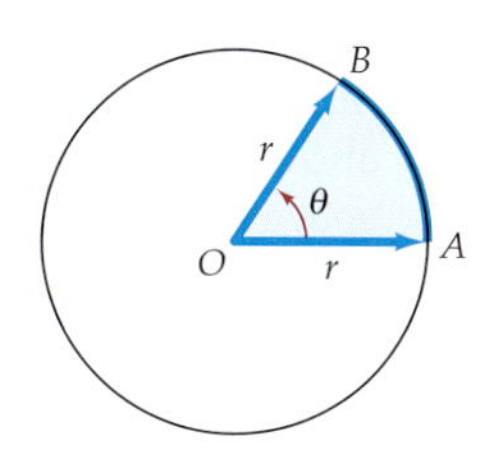

In Exercises 87 to 90, find the area, to the nearest square unit, of the sector of a circle with the given radius and central angle.

87. $r = 5$ inches, $\theta = \dfrac{\pi}{3}$ radians

88. $r = 2.8$ feet, $\theta = \dfrac{5\pi}{2}$ radians

89. $r = 120$ centimeters, $\theta = 0.65$ radian

90. $r = 30$ feet, $\theta = 62°$

91. Area of a Circular Segment Find the area of the shaded portion of the circle. The radius of the circle is 9 inches.

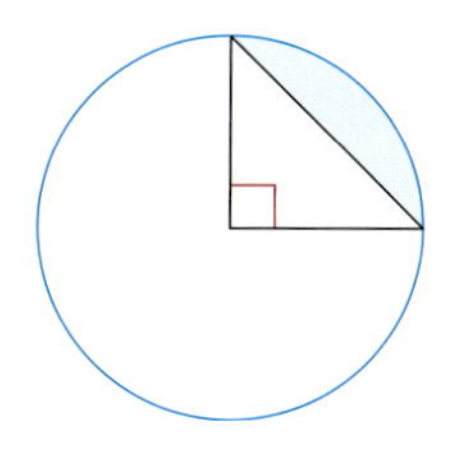

Latitude describes the position of a point on the earth's surface in relation to the equator. A point on the equator has a latitude of 0°. The north pole has a latitude of 90°. The radius of the earth is approximately 3960 miles.

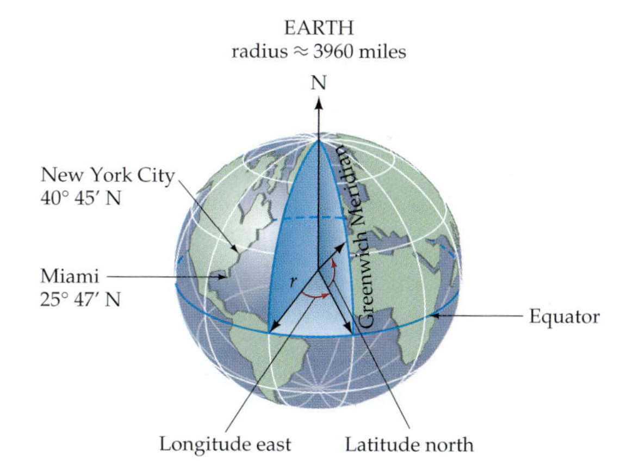

92. Geography The city of New York has a latitude of 40°45′N. How far north, to the nearest 10 miles, is it from the equator? Use 3960 miles as the radius of the earth.

93. Geography The city of Miami has a latitude of 25°47′N. How far north, to the nearest 10 miles, is it from the equator? Use 3960 miles as the radius of the earth.

94. Geography Assuming that the earth is a perfect sphere, and expressing your answer to three significant digits, find the distance along the earth's surface (in miles) that subtends a central angle of

a. $1°$ **b.** $1'$ **c.** $1''$

PREPARE FOR SECTION 4.2

95. Determine whether the point $(0, 1)$ is a point on the circle defined by $x^2 + y^2 = 1$. [1.2]

96. Determine whether the point $\left(\frac{1}{2}, \frac{\sqrt{3}}{2}\right)$ is a point on the circle defined by $x^2 + y^2 = 1$. [1.2]

97. Determine whether the point $\left(\frac{\sqrt{2}}{2}, \frac{\sqrt{3}}{2}\right)$ is a point on the circle defined by $x^2 + y^2 = 1$. [1.2]

98. Determine the circumference of a circle with a radius of 1.

99. Determine whether $f(x) = x^2 - 3$ is an even function, an odd function, or a function that is neither even nor odd. [1.6]

100. Determine whether $f(x) = x^3 - x^2$ is an even function, an odd function, or a function that is neither even nor odd. [1.6]

PROJECTS

1. CONVERSION OF UNITS You are traveling in a foreign country. You discover that the currency used consists of lollars, mollars, nollars, and tollars.

5 lollars = 1 mollar

3 mollars = 5 nollars

4 nollars = 7 tollars

The fare for a taxi is 14 tollars.

a. How much is the fare in mollars? (*Hint:* Make use of unit fractions to convert from tollars to mollars.)

b. If you have only mollars and lollars, how many of each could you use to pay the fare?

c. Explain to a classmate the concept of unit fractions. Also explain how you knew which of the unit fractions

$$\left(\frac{4 \text{ nollars}}{7 \text{ tollars}}\right) \qquad \left(\frac{7 \text{ tollars}}{4 \text{ nollars}}\right)$$

should be used in the conversion in part **a.**

d. If you wish to convert x degrees to radians, you need to multiply x by which of the following unit fractions?

$$\left(\frac{\pi}{180^\circ}\right) \qquad \left(\frac{180^\circ}{\pi}\right)$$

2. SPACE SHUTTLE The rotational period of the earth is 23.933 hours. A space shuttle revolves around the earth's equator every 2.231 hours. Both are rotating in the same direction. At the present time, the space shuttle is directly above the Galápagos Islands. How long will it take for the space shuttle to circle the earth and return to a position directly above the Galápagos Islands?

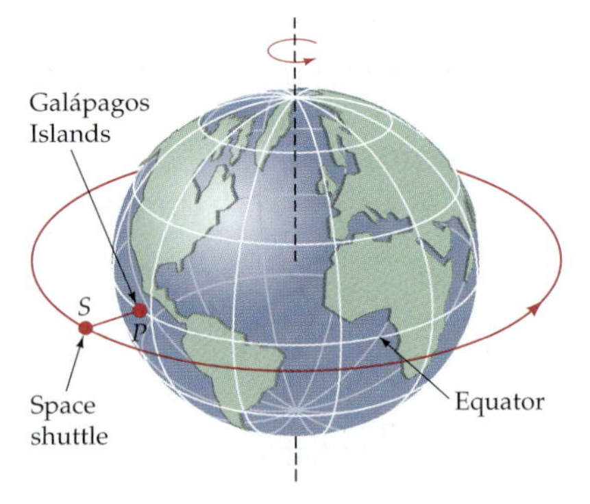

It is not always possible to find the exact value of a trigonometric function by purely geometrical methods. In these cases we use a calculator to find a decimal approximation. **Figure 4.40** shows decimal approximations for sin(2.4), tan(1.85), and sec(0.45).

Select Radian from the MODE menu.

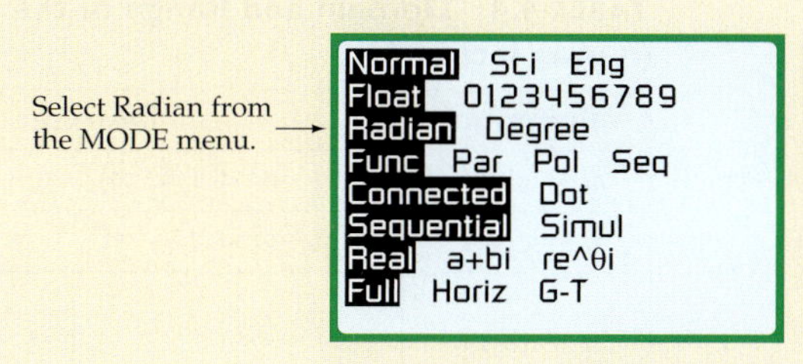

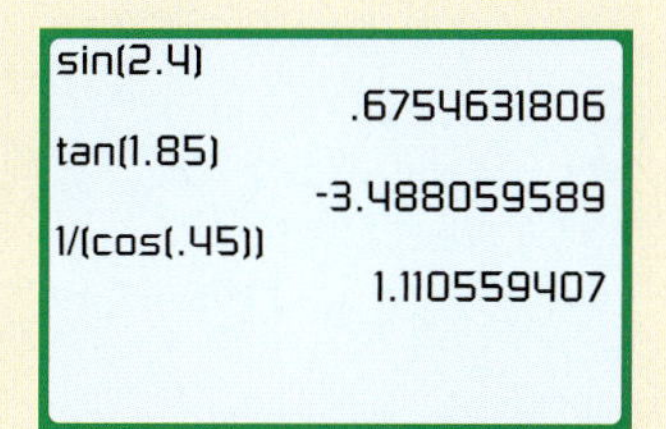

FIGURE 4.40

PROPERTIES OF TRIGONOMETRIC FUNCTIONS OF REAL NUMBERS

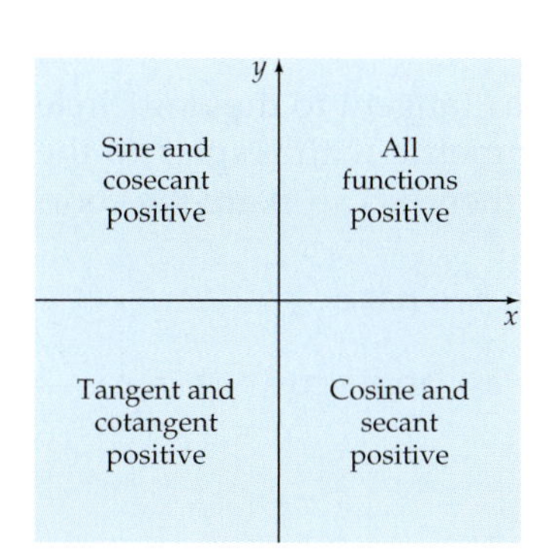

FIGURE 4.41

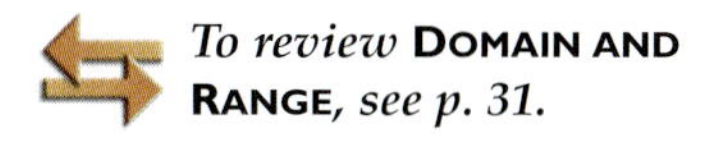
To review **DOMAIN AND RANGE**, *see p. 31.*

The sign of a trigonometric function of a real number depends on the quadrant in which $W(t)$ lies. When $W(t)$ lies in Quadrant I or IV, $x > 0$ and $\cos t$ and $\sec t$ are both positive. When $W(t)$ lies in Quadrant I or II, $y > 0$ and $\sin t$ and $\csc t$ are both positive. When $W(t)$ lies in Quadrant I, x and y are both positive, so both $\tan t$ and $\cot t$ are positive. When $W(t)$ lies in Quadrant III, x and y are both negative, but the ratios $\dfrac{x}{y}$ and $\dfrac{y}{x}$ are both positive. Thus $\tan t$ and $\cot t$ are both positive in Quadrant III. These results are summarized in **Figure 4.41.**

The domain and range of the trigonometric functions can be found from the definitions of these functions. If t is any real number and $P(x, y)$ is the point corresponding to $W(t)$, then by definition $\cos t = x$ and $\sin t = y$. Thus the domain of the sine and cosine functions is the set of real numbers.

Because the radius of the unit circle is 1, we have

$$-1 \le x \le 1 \qquad \text{and} \qquad -1 \le y \le 1$$

Therefore, with $x = \cos t$ and $y = \sin t$, we have

$$-1 \le \cos t \le 1 \qquad \text{and} \qquad -1 \le \sin t \le 1$$

The range of the cosine and sine functions is $[-1, 1]$.

Using the definitions of tangent and secant,

$$\tan t = \frac{y}{x} \qquad \text{and} \qquad \sec t = \frac{1}{x}$$

The domain of the tangent function is all real numbers t except those for which the x-coordinate of $W(t)$ is zero. The x-coordinate is zero when $t = \pm\dfrac{\pi}{2}$, $t = \pm\dfrac{3\pi}{2}$, and $t = \pm\dfrac{5\pi}{2}$ and in general when $t = \dfrac{(2n + 1)\pi}{2}$, where n is an integer.

Thus the domain of the tangent function is the set of all real numbers t except $t = \dfrac{(2n+1)\pi}{2}$, where n is an integer. The range of the tangent function is all real numbers.

Similar methods can be used to find the domain and range of the cotangent, secant, and cosecant functions. The results are summarized in **Table 4.4.**

TABLE 4.4 Domain and Range of the Trigonometric Functions (n is an integer)

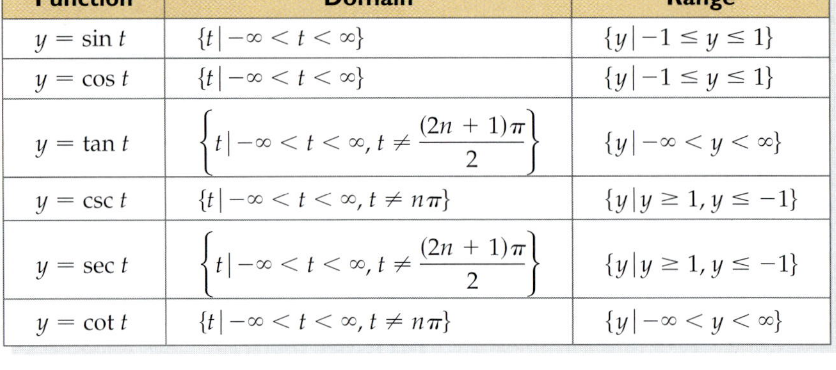

Function	Domain	Range
$y = \sin t$	$\{t \mid -\infty < t < \infty\}$	$\{y \mid -1 \le y \le 1\}$
$y = \cos t$	$\{t \mid -\infty < t < \infty\}$	$\{y \mid -1 \le y \le 1\}$
$y = \tan t$	$\left\{t \mid -\infty < t < \infty, t \ne \dfrac{(2n+1)\pi}{2}\right\}$	$\{y \mid -\infty < y < \infty\}$
$y = \csc t$	$\{t \mid -\infty < t < \infty, t \ne n\pi\}$	$\{y \mid y \ge 1, y \le -1\}$
$y = \sec t$	$\left\{t \mid -\infty < t < \infty, t \ne \dfrac{(2n+1)\pi}{2}\right\}$	$\{y \mid y \ge 1, y \le -1\}$
$y = \cot t$	$\{t \mid -\infty < t < \infty, t \ne n\pi\}$	$\{y \mid -\infty < y < \infty\}$

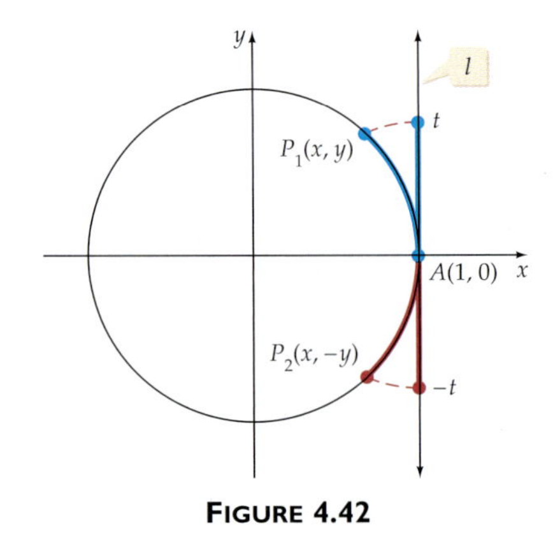

FIGURE 4.42

To review **ODD AND EVEN FUNCTIONS,** *see p. 83.*

Consider the points t and $-t$ on the coordinate line l tangent to the unit circle at the point $(1, 0)$. The points $W(t)$ and $W(-t)$ are symmetric with respect to the x-axis. Therefore, if $P_1(x, y)$ are the coordinates of $W(t)$, then $P_2(x, -y)$ are the coordinates of $W(-t)$. See **Figure 4.42.**

From the definitions of the trigonometric functions, we have

$$\sin t = y \quad \text{and} \quad \sin(-t) = -y \qquad \text{and} \qquad \cos t = x \quad \text{and} \quad \cos(-t) = x$$

Substituting $\sin t$ for y and $\cos t$ for x yields

$$\sin(-t) = -\sin t \qquad \text{and} \qquad \cos(-t) = \cos t$$

Thus the sine is an odd function, and the cosine is an even function. Because $\csc t = \dfrac{1}{\sin t}$ and $\sec t = \dfrac{1}{\cos t}$, it follows that

$$\csc(-t) = -\csc t \qquad \text{and} \qquad \sec(-t) = \sec t$$

These equations state that the cosecant is an odd function and the secant is an even function.

From the definition of the tangent function, we have $\tan t = \dfrac{y}{x}$ and $\tan(-t) = -\dfrac{y}{x}$. Substituting $\tan t$ for $\dfrac{y}{x}$ yields $\tan(-t) = -\tan t$. Because $\cot t = \dfrac{1}{\tan t}$, it follows that $\cot(-t) = -\cot t$. Thus the tangent and cotangent are odd functions.

Even and Odd Trigonometric Functions

The odd trigonometric functions are $y = \sin t$, $y = \csc t$, $y = \tan t$, and $y = \cot t$. The even trigonometric functions are $y = \cos t$ and $y = \sec t$.

Thus for all t in their domain,

$$\sin(-t) = -\sin t \qquad \cos(-t) = \cos t \qquad \tan(-t) = -\tan t$$
$$\csc(-t) = -\csc t \qquad \sec(-t) = \sec t \qquad \cot(-t) = -\cot t$$

EXAMPLE 3 Determine Whether a Function Is Even, Odd, or Neither

Is the function defined by $f(x) = x - \tan x$ an even function, an odd function, or neither?

Solution

Find $f(-x)$ and compare it to $f(x)$.

$$\begin{aligned} f(-x) &= (-x) - \tan(-x) = -x + \tan x \qquad \text{• } \tan(-x) = -\tan x \\ &= -(x - \tan x) \\ &= -f(x) \end{aligned}$$

The function $f(x) = x - \tan x$ is an odd function.

▶ TRY EXERCISE 58, PAGE 320

Let W be the wrapping function, t be a point on the coordinate line tangent to the unit circle at $(1, 0)$, and $W(t) = P(x, y)$. Because the circumference of the unit circle is 2π, $W(t + 2\pi) = W(t) = P(x, y)$. Thus the value of the wrapping function repeats itself in 2π units. *The wrapping function is periodic, and the period is* 2π.

Recall the definitions of $\cos t$ and $\sin t$:

$$\cos t = x \qquad \text{and} \qquad \sin t = y$$

where $W(t) = P(x, y)$. Because $W(t + 2\pi) = W(t) = P(x, y)$ for all t,

$$\cos(t + 2\pi) = x \qquad \text{and} \qquad \sin(t + 2\pi) = y$$

Thus $\cos t$ and $\sin t$ have period 2π. Because

$$\sec t = \frac{1}{\cos t} = \frac{1}{\cos(t + 2\pi)} = \sec(t + 2\pi) \qquad \text{and}$$
$$\csc t = \frac{1}{\sin t} = \frac{1}{\sin(t + 2\pi)} = \csc(t + 2\pi)$$

$\sec t$ and $\csc t$ have a period of 2π.

Period of cos t, sin t, sec t, and csc t

The period of $\cos t$, $\sin t$, $\sec t$, and $\csc t$ is 2π.

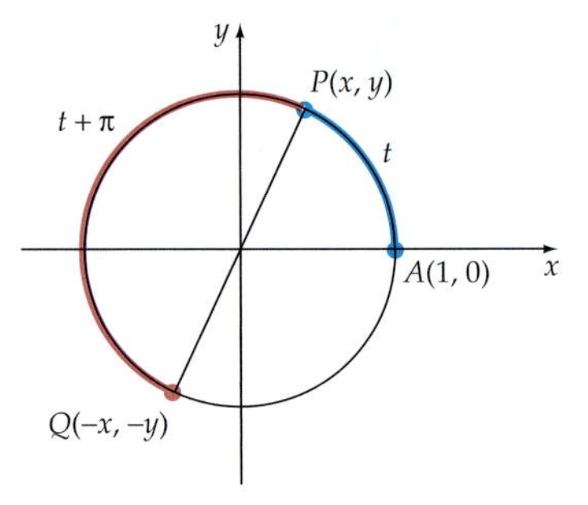

FIGURE 4.43

Although it is true that $\tan t = \tan(t + 2\pi)$, the period of $\tan t$ is not 2π. Recall that the period of a function is the *smallest* value of p for which $f(t) = f(t + p)$.

If W is the wrapping function (see **Figure 4.43**) and $W(t) = P(x, y)$, then $W(t + \pi) = P(-x, -y)$. Because

$$\tan t = \frac{y}{x} \qquad \text{and} \qquad \tan(t + \pi) = \frac{-y}{-x} = \frac{y}{x} = \tan t$$

we have $\tan(t + \pi) = \tan t$ for all t. A similar argument applies to $\cot t$.

Period of tan t and cot t

The period of $\tan t$ and $\cot t$ is π.

Trigonometric Identities

Recall that any equation that is true for every number in the domain of the equation is an identity. The statement

$$\csc t = \frac{1}{\sin t}, \quad \sin t \neq 0$$

is an identity because the two expressions produce the same result for all values of t for which both functions are defined.

The **ratio identities** are obtained by writing the tangent and cotangent functions in terms of the sine and cosine functions.

$$\tan t = \frac{y}{x} = \frac{\sin t}{\cos t} \qquad \text{and} \qquad \cot t = \frac{x}{y} = \frac{\cos t}{\sin t}$$

• $x = \cos t$ and $y = \sin t$

The **Pythagorean identities** are based on the equation of a unit circle, $x^2 + y^2 = 1$, and on the definitions of the sine and cosine functions.

$$x^2 + y^2 = 1$$

$$\cos^2 t + \sin^2 t = 1$$

• Replace x by $\cos t$ and y by $\sin t$.

Dividing each term of $\cos^2 t + \sin^2 t = 1$ by $\cos^2 t$, we have

$$\frac{\cos^2 t}{\cos^2 t} + \frac{\sin^2 t}{\cos^2 t} = \frac{1}{\cos^2 t}$$

• $\cos t \neq 0$

$$1 + \tan^2 t = \sec^2 t$$

• $\frac{\sin t}{\cos t} = \tan t$

Dividing each term of $\cos^2 t + \sin^2 t = 1$ by $\sin^2 t$, we have

$$\frac{\cos^2 t}{\sin^2 t} + \frac{\sin^2 t}{\sin^2 t} = \frac{1}{\sin^2 t}$$

• $\sin t \neq 0$

$$\cot^2 t + 1 = \csc^2 t$$

• $\frac{\cos t}{\sin t} = \cot t$

Here is a summary of the Fundamental Trigonometric Identities we have established.

Fundamental Trigonometric Identities

The reciprocal identities are

$$\sin t = \frac{1}{\csc t} \qquad \cos t = \frac{1}{\sec t} \qquad \tan t = \frac{1}{\cot t}$$

The ratio identities are

$$\tan t = \frac{\sin t}{\cos t} \qquad \cot t = \frac{\cos t}{\sin t}$$

The Pythagorean identities are

$$\cos^2 t + \sin^2 t = 1 \qquad 1 + \tan^2 t = \sec^2 t \qquad 1 + \cot^2 t = \csc^2 t$$

Using identities and basic algebra concepts, we can often rewrite expressions in different forms.

EXAMPLE 4 Simplify an Expression

Write the expression $\dfrac{1}{\sin^2 t} + \dfrac{1}{\cos^2 t}$ as a single term.

Solution

Express each fraction in terms of a common denominator. The common denominator is $\sin^2 t \cos^2 t$.

$$\frac{1}{\sin^2 t} + \frac{1}{\cos^2 t} = \frac{1}{\sin^2 t}\frac{\cos^2 t}{\cos^2 t} + \frac{1}{\cos^2 t}\frac{\sin^2 t}{\sin^2 t}$$

$$= \frac{\cos^2 t + \sin^2 t}{\sin^2 t \cos^2 t} = \frac{1}{\sin^2 t \cos^2 t} \qquad \bullet\ \cos^2 t + \sin^2 t = 1$$

▶ TRY EXERCISE 74, PAGE 320

take note

Because

$$\frac{1}{\sin^2 t \cos^2 t} = \frac{1}{\sin^2 t} \cdot \frac{1}{\cos^2 t} = (\csc^2 t)(\sec^2 t)$$

we could have written the answer to Example 4 in terms of the cosecant and secant functions.

EXAMPLE 5 Write a Function in Terms of a Given Function

For $\dfrac{\pi}{2} < t < \pi$, write $\tan t$ in terms of $\sin t$.

Solution

Write $\tan t = \dfrac{\sin t}{\cos t}$. Now solve $\cos^2 t + \sin^2 t = 1$ for $\cos t$.

$$\cos^2 t + \sin^2 t = 1$$

$$\cos^2 t = 1 - \sin^2 t$$

$$\cos t = \pm\sqrt{1 - \sin^2 t}$$

Continued ▶

Because $\frac{\pi}{2} < t < \pi$, $\cos t$ is negative. Therefore, $\cos t = -\sqrt{1 - \sin^2 t}$.

Thus

$$\tan t = -\frac{\sin t}{\sqrt{1 - \sin^2 t}} \qquad \bullet\ \frac{\pi}{2} < t < \pi$$

▶ TRY EXERCISE 80, PAGE 320

An Application Involving a Trigonometric Function

EXAMPLE 6 Determine a Height as a Function of Time

The Millennium Wheel, in London, is the world's largest Ferris wheel. It has a diameter of 450 feet. When the Millennium Wheel is in uniform motion, it completes one revolution every 30 minutes. The height h, in feet above the Thames River, of a person riding on the Millennium Wheel can be estimated by

$$h(t) = 255 - 225 \cos\left(\frac{\pi}{15}t\right)$$

where t is the time in minutes since the person started the ride.

The Millennium Wheel, on the banks of the Thames River, London.

a. How high is the person at the start of the ride ($t = 0$)?

b. How high is the person after 18.0 minutes?

Solution

a.

$$\begin{aligned} h(0) &= 255 - 225 \cos\left(\frac{\pi}{15} \cdot 0\right) \\ &= 255 - 225 \\ &= 30 \end{aligned}$$

At the start of the ride, the person is 30 feet above the Thames.

b.

$$\begin{aligned} h(18.0) &= 255 - 225 \cos\left(\frac{\pi}{15} \cdot 18.0\right) \\ &\approx 255 - (-182) \\ &= 437 \end{aligned}$$

After 18.0 minutes, the person is about 437 feet above the Thames.

▶ TRY EXERCISE 84, PAGE 320

Topics for Discussion

1. Is $W(t)$ a number? Explain.
2. Explain how to find the exact value of $\cos\left(\frac{13\pi}{6}\right)$.
3. Is $f(x) = \cos^3 x$ an even function or an odd function? Explain how you made your decision.
4. Explain how to make use of a unit circle to show that $\sin(-t) = -\sin t$.

Exercise Set 4.2

In Exercises 1 to 14, find $W(t)$ for the given t.

1. $t = \frac{5\pi}{6}$

2. $t = \frac{5\pi}{3}$

3. $t = -\frac{\pi}{3}$

4. $t = -\frac{7\pi}{6}$

5. $t = -\frac{\pi}{2}$

6. $t = \frac{5\pi}{4}$

7. $t = -\frac{5\pi}{4}$

8. $t = -\frac{5\pi}{3}$

9. $t = \frac{7\pi}{4}$

▶ **10.** $t = \frac{11\pi}{6}$

11. $t = \frac{13\pi}{6}$

12. $t = \frac{13\pi}{4}$

13. $t = \pi$

14. $t = -\pi$

In Exercises 15 to 36, find the exact value of the trigonometric function. If the value is undefined, so state.

15. $\sin\left(\frac{5\pi}{6}\right)$

16. $\cos\left(\frac{5\pi}{3}\right)$

17. $\tan\left(-\frac{\pi}{3}\right)$

18. $\csc\left(-\frac{7\pi}{6}\right)$

19. $\sec\left(\frac{5\pi}{3}\right)$

20. $\cot\left(\frac{5\pi}{4}\right)$

21. $\cos\left(-\frac{4\pi}{3}\right)$

22. $\sin\left(-\frac{5\pi}{3}\right)$

23. $\cot\left(\frac{7\pi}{4}\right)$

▶ **24.** $\sin\left(\frac{11\pi}{6}\right)$

25. $\csc\left(\frac{7\pi}{6}\right)$

26. $\sec\left(\frac{3\pi}{4}\right)$

27. $\tan\left(-\frac{\pi}{6}\right)$

28. $\csc\left(-\frac{2\pi}{3}\right)$

29. $\sin\left(-\frac{5\pi}{3}\right)$

30. $\cos\left(-\frac{3\pi}{4}\right)$

31. $\sin\left(\frac{3\pi}{2}\right)$

32. $\cos(-\pi)$

33. $\tan(\pi)$

34. $\csc\left(-\frac{\pi}{2}\right)$

35. $\cot\left(-\frac{\pi}{2}\right)$

36. $\sec\left(-\frac{3\pi}{2}\right)$

In Exercises 37 to 46, use a calculator to find an approximate value of each function. Round your answers to the nearest ten-thousandth.

37. $\sin 1.22$

38. $\cos 4.22$

39. $\csc(-1.05)$

40. $\sin(-0.55)$

41. $\tan\left(\frac{11\pi}{12}\right)$

42. $\cos\left(\frac{2\pi}{5}\right)$

43. $\cos\left(-\frac{\pi}{5}\right)$

44. $\csc 8.2$

45. $\sec 1.55$

46. $\cot 2.11$

In Exercises 47 to 54, use the unit circle below to estimate the following values to the nearest tenth.

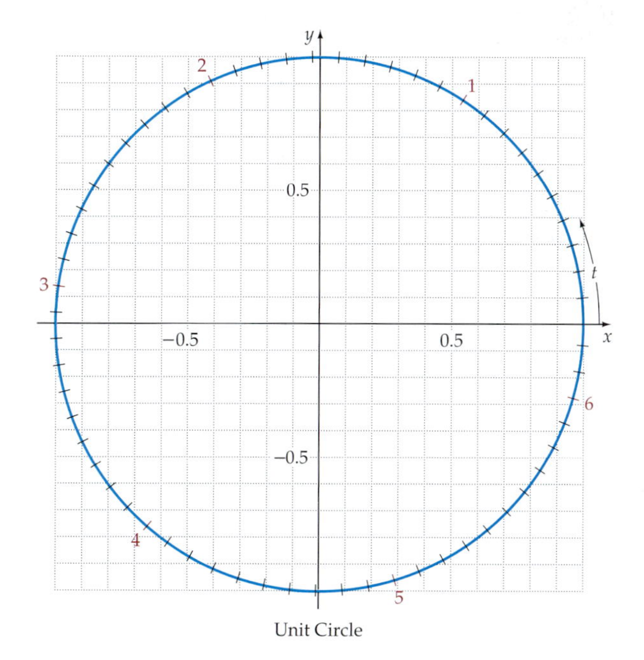

Unit Circle

47. a. $\sin 2$ **b.** $\cos 2$

48. a. $\sin 3$ **b.** $\cos 3$

49. **a.** $\sin 5.4$ **b.** $\cos 5.4$

50. **a.** $\sin 4.1$ **b.** $\cos 4.1$

51. All real numbers t between 0 and 2π for which $\sin t = 0.4$.

52. All real numbers t between 0 and 2π for which $\cos t = 0.8$.

53. All real numbers t between 0 and 2π for which $\sin t = -0.3$.

54. All real numbers t between 0 and 2π for which $\cos t = -0.7$.

In Exercises 55 to 62, determine whether the function is even, odd, or neither.

55. $f(x) = -4 \sin x$

56. $f(x) = -2 \cos x$

57. $G(x) = \sin x + \cos x$

▶ **58.** $F(x) = \tan x + \sin x$

59. $S(x) = \dfrac{\sin x}{x}, x \neq 0$

60. $C(x) = \dfrac{\cos x}{x}, x \neq 0$

61. $v(x) = 2 \sin x \cos x$

62. $w(x) = x \tan x$

In Exercises 63 to 78, use the fundamental trigonometric identities to write each expression in terms of a single trigonometric function or a constant.

63. $\tan t \cos t$

64. $\cot t \sin t$

65. $\dfrac{\csc t}{\cot t}$

66. $\dfrac{\sec t}{\tan t}$

67. $1 - \sec^2 t$

68. $1 - \csc^2 t$

69. $\tan t - \dfrac{\sec^2 t}{\tan t}$

70. $\dfrac{\csc^2 t}{\cot t} - \cot t$

71. $\dfrac{1 - \cos^2 t}{\tan^2 t}$

72. $\dfrac{1 - \sin^2 t}{\cot^2 t}$

73. $\dfrac{1}{1 - \cos t} + \dfrac{1}{1 + \cos t}$

▶ **74.** $\dfrac{1}{1 - \sin t} + \dfrac{1}{1 + \sin t}$

75. $\dfrac{\tan t + \cot t}{\tan t}$

76. $\dfrac{\csc t - \sin t}{\csc t}$

77. $\sin^2 t(1 + \cot^2 t)$

78. $\cos^2 t(1 + \tan^2 t)$

79. Write $\sin t$ in terms of $\cos t$, $0 < t < \dfrac{\pi}{2}$.

▶ **80.** Write $\tan t$ in terms of $\sec t$, $\dfrac{3\pi}{2} < t < 2\pi$.

81. Write $\csc t$ in terms of $\cot t$, $\dfrac{\pi}{2} < t < \pi$.

82. Write $\sec t$ in terms of $\tan t$, $\pi < t < \dfrac{3\pi}{2}$.

83. **PATH OF A SATELLITE** A satellite is launched into space from Cape Canaveral. The directed distance, in miles, that the satellite is above or below the equator is

$$d(t) = 1970 \cos\left(\frac{\pi}{64} t\right)$$

where t is the number of minutes since liftoff. A negative d value indicates that the satellite is south of the equator.

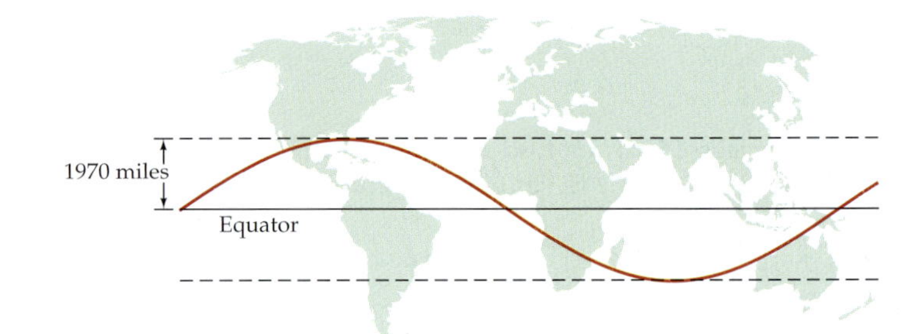

What distance, to the nearest 10 miles, is the satellite north of the equator 24 minutes after liftoff?

▶ **84.** **AVERAGE HIGH TEMPERATURE** The average high temperature T, in degrees Fahrenheit, for Fairbanks, Alaska, is given by

$$T(t) = -41 \cos\left(\frac{\pi}{6} t\right) + 36$$

where t is the number of months after January 5. Use the formula to estimate (to the nearest 0.1 degree Fahrenheit) the average high temperature in Fairbanks for March 5 and July 20.

CONNECTING CONCEPTS

In Exercises 85 to 88, use the fundamental trigonometric identities to find the value of the function.

85. Given $\csc t = \sqrt{2}, 0 < t < \frac{\pi}{2}$, find $\cos t$.

86. Given $\cos t = \frac{1}{2}, \frac{3\pi}{2} < t < 2\pi$, find $\sin t$.

87. Given $\sin t = \frac{1}{2}, \frac{\pi}{2} < t < \pi$, find $\tan t$.

88. Given $\cot t = \frac{\sqrt{3}}{3}, \pi < t < \frac{3\pi}{2}$, find $\cos t$.

In Exercises 89 to 92, simplify the first expression to the second expression.

89. $\frac{\sin^2 t + \cos^2 t}{\sin^2 t}$; $\csc^2 t$

90. $\frac{\sin^2 t + \cos^2 t}{\cos^2 t}$; $\sec^2 t$

91. $(\cos t - 1)(\cos t + 1)$; $-\sin^2 t$

92. $(\sec t - 1)(\sec t + 1)$; $\tan^2 t$

PREPARE FOR SECTION 4.3

93. Estimate, to the nearest tenth, $\sin \frac{3\pi}{4}$. [4.2]

94. Estimate, to the nearest tenth, $\cos \frac{5\pi}{4}$. [4.2]

95. Explain how to use the graph of $y = f(x)$ to produce the graph of $y = -f(x)$. [1.6]

96. Explain how to use the graph of $y = f(x)$ to produce the graph of $y = f(2x)$. [1.6]

97. Simplify: $\frac{2\pi}{1/3}$

98. Simplify: $\frac{2\pi}{2/5}$

PROJECTS

1. Consider a square as shown. Start at the point (1, 0) and travel counterclockwise around the square for a distance t ($t \geq 0$).

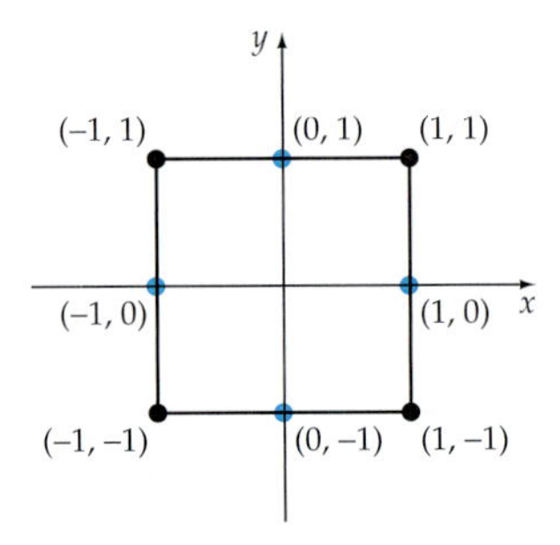

Let $WSQ(t) = P(x, y)$ be the point on the square determined by traveling counterclockwise a distance of t units from (1, 0). For instance,

$$WSQ(0.5) = (1, 0.5)$$
$$WSQ(1.75) = (0.25, 1)$$

Find $WSQ(4.2)$ and $WSQ(6.4)$. We define the square sine of t, denoted by ssin t, to be the y-value of point P. The square cosine of t, denoted by scos t, is defined to be the x-value of point P. For example,

$$\text{ssin } 0.4 = 0.4 \qquad \text{scos } 0.4 = 1$$
$$\text{scos } 1.2 = 0.8 \qquad \text{scos } 5.3 = -0.7$$

The square tangent of t, denoted by stan t, is defined as

$$\text{stan } t = \frac{\text{ssin } t}{\text{scos } t} \qquad \text{scos } t \neq 0$$

Find each of the following.

a. ssin 3.2 **b.** scos 4.4 **c.** stan 5.5

d. ssin 11.2 **e.** scos −5.2 **f.** stan −6.5

SECTION 4.3

GRAPHS OF THE SINE AND COSINE FUNCTIONS

- THE GRAPH OF THE SINE FUNCTION
- THE GRAPH OF THE COSINE FUNCTION

THE GRAPH OF THE SINE FUNCTION

The trigonometric functions can be graphed on a rectangular coordinate system by plotting the points whose coordinates belong to the function. We begin with the graph of the sine function.

Table 4.5 on the following page lists some ordered pairs (x, y), where $y = \sin x, 0 \leq x \leq 2\pi$. In **Figure 4.44,** the points are plotted and a smooth curve is drawn through the points.

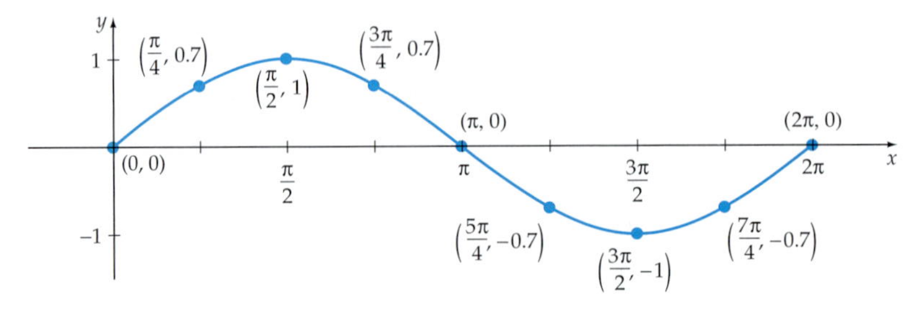

$y = \sin x, 0 \leq x \leq 2\pi$

FIGURE 4.44

TABLE 4.5

x	$y = \sin x$
0	0
$\frac{\pi}{4}$	≈ 0.7
$\frac{\pi}{2}$	1
$\frac{3\pi}{4}$	≈ 0.7
π	0
$\frac{5\pi}{4}$	≈ -0.7
$\frac{3\pi}{2}$	-1
$\frac{7\pi}{4}$	≈ -0.7
2π	0

Because the domain of the sine function is the real numbers and the period is 2π, the graph of $y = \sin x$ is drawn by repeating the portion shown in **Figure 4.44.** Any part of the graph that corresponds to one period (2π) is one cycle of the graph of $y = \sin x$ (see **Figure 4.45**).

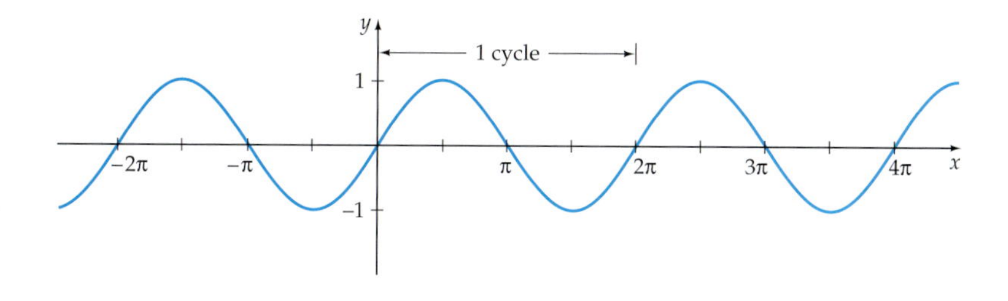

$y = \sin x$

FIGURE 4.45

The maximum value M reached by $\sin x$ is 1, and the minimum value m is -1. The amplitude of the graph of $y = \sin x$ is given by

$$\text{Amplitude} = \frac{1}{2}(M - m)$$

? QUESTION What is the amplitude of $y = \sin x$?

Recall that the graph of $y = a \cdot f(x)$ is obtained by *stretching* ($|a| > 1$) or *shrinking* ($0 < |a| < 1$) the graph of $y = f(x)$. **Figure 4.46** shows the graph of $y = 3 \sin x$ that was drawn by stretching the graph of $y = \sin x$. The amplitude of $y = 3 \sin x$ is 3 because

$$\text{Amplitude} = \frac{1}{2}(M - m) = \frac{1}{2}[3 - (-3)] = 3$$

Note that for $y = \sin x$ and $y = 3 \sin x$, the amplitude of the graph is the coefficient of $\sin x$. This suggests the following theorem.

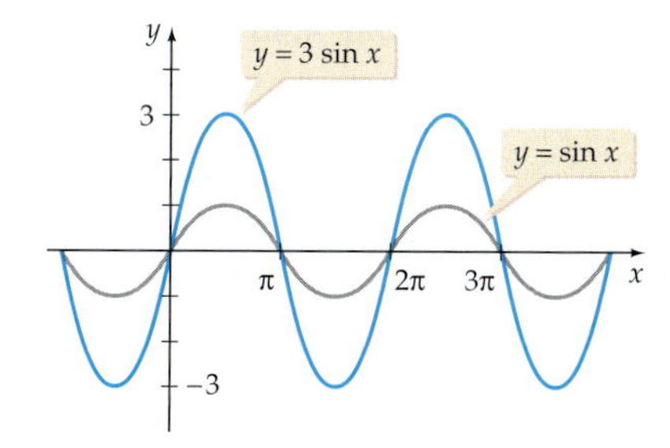

FIGURE 4.46

> **take note**
>
> The amplitude is defined to be half the difference between the maximum height and the minimum height. It may not be equal to the maximum height. For example, the graph of $y = 4 + \sin x$ has a maximum height of 5 and an amplitude of 1.

Amplitude of $y = a \sin x$

The amplitude of $y = a \sin x$ is $|a|$.

? ANSWER $\text{Amplitude} = \frac{1}{2}(M - m) = \frac{1}{2}[1 - (-1)] = \frac{1}{2}(2) = 1$

EXAMPLE 1 **Graph $y = a \sin x$**

Graph: $y = -2 \sin x$

Solution

The amplitude of $y = -2 \sin x$ is 2. The graph of $y = -f(x)$ is a *reflection* across the x-axis of $y = f(x)$. Thus the graph of $y = -2 \sin x$ is a reflection across the x-axis of $y = 2 \sin x$. See **Figure 4.47.**

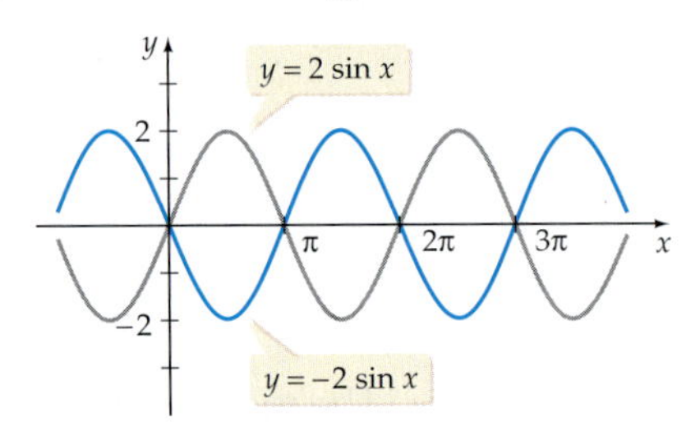

FIGURE 4.47

▶ **TRY EXERCISE 20, PAGE 329**

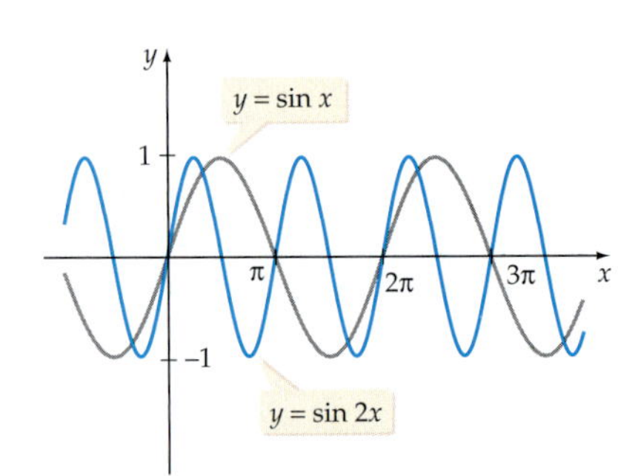

FIGURE 4.48

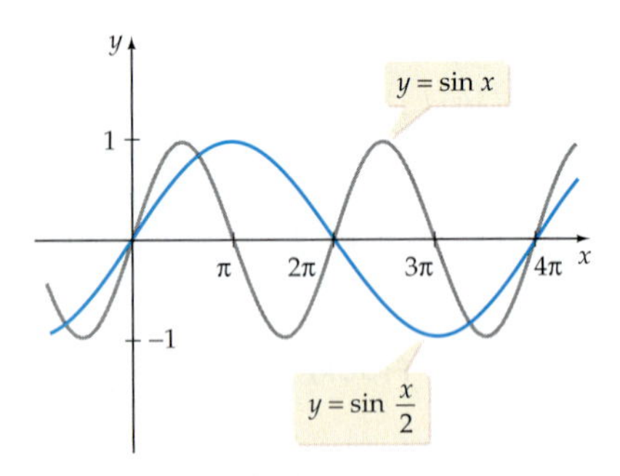

FIGURE 4.49

The graphs of $y = \sin x$ and $y = \sin 2x$ are shown in **Figure 4.48.** Because one cycle of the graph of $y = \sin 2x$ is completed in an interval of length π, the period of $y = \sin 2x$ is π.

The graphs of $y = \sin x$ and $y = \sin\left(\frac{x}{2}\right)$ are shown in **Figure 4.49.** Because one cycle of the graph of $y = \sin\left(\frac{x}{2}\right)$ is completed in an interval of length 4π, the period of $y = \sin\left(\frac{x}{2}\right)$ is 4π.

Generalizing the last two examples, one cycle of $y = \sin bx$, $b > 0$, is completed as bx varies from 0 to 2π. Algebraically, one cycle of $y = \sin bx$ is completed as bx varies from 0 to 2π. Therefore,

$$0 \le bx \le 2\pi$$

$$0 \le x \le \frac{2\pi}{b}$$

The length of the interval, $\frac{2\pi}{b}$, is the period of $y = \sin bx$. Now we consider the case when the coefficient of x is negative. If $b > 0$, then using the fact that the sine function is an odd function, we have $y = \sin(-bx) = -\sin bx$, and thus the period is still $\frac{2\pi}{b}$. This gives the following theorem.

Period of $y = \sin bx$

The period of $y = \sin bx$ is $\frac{2\pi}{|b|}$.

Table 4.6 gives the amplitude and period of several sine functions.

TABLE 4.6

Function	$y = a \sin bx$	$y = 3 \sin(-2x)$	$y = -\sin \dfrac{x}{3}$	$y = -2 \sin \dfrac{3x}{4}$
Amplitude	$\lvert a \rvert$	$\lvert 3 \rvert = 3$	$\lvert -1 \rvert = 1$	$\lvert -2 \rvert = 2$
Period	$\dfrac{2\pi}{\lvert b \rvert}$	$\dfrac{2\pi}{2} = \pi$	$\dfrac{2\pi}{1/3} = 6\pi$	$\dfrac{2\pi}{3/4} = \dfrac{8\pi}{3}$

EXAMPLE 2 Graph $y = \sin bx$

Graph: $y = \sin \pi x$

Solution

$$\text{Amplitude} = 1 \qquad \text{Period} = \frac{2\pi}{b} = \frac{2\pi}{\pi} = 2 \qquad \bullet\ b = \pi$$

The graph is sketched in **Figure 4.50**.

TRY EXERCISE 30, PAGE 329

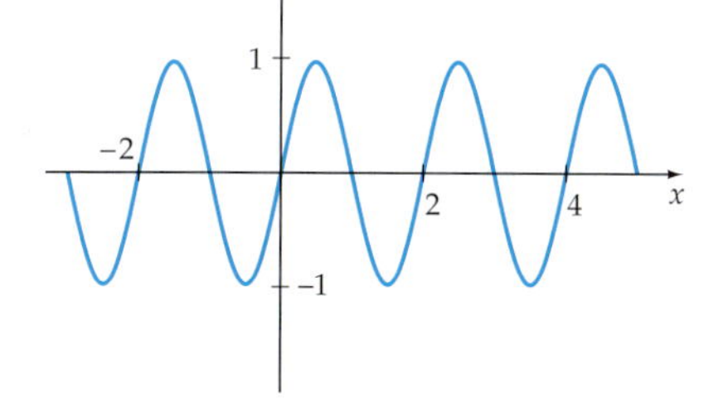

$y = \sin \pi x$

FIGURE 4.50

Figure 4.51 shows the graph of $y = a \sin bx$ for both a and b positive. Note from the graph that

- The amplitude is a.
- The period is $\dfrac{2\pi}{b}$.
- For $0 \le x \le \dfrac{2\pi}{b}$, the zeros are 0, $\dfrac{\pi}{b}$, and $\dfrac{2\pi}{b}$.
- The maximum value is a when $x = \dfrac{\pi}{2b}$, and the minimum value is $-a$ when $x = \dfrac{3\pi}{2b}$.
- If $a < 0$, the graph is reflected across the x-axis.

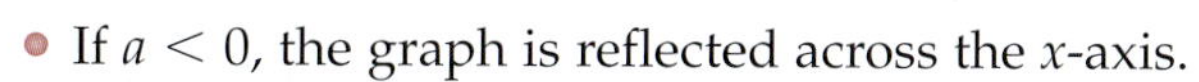

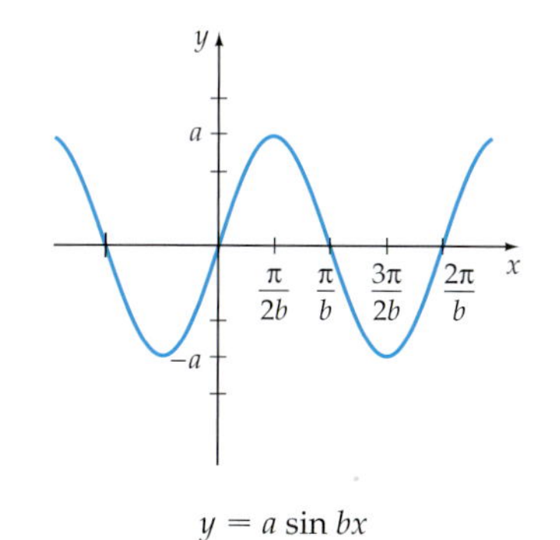

$y = a \sin bx$

FIGURE 4.51

EXAMPLE 3 Graph $y = a \sin bx$

Graph: $y = -\dfrac{1}{2} \sin \dfrac{x}{3}$

Solution

$$\text{Amplitude} = \left| -\frac{1}{2} \right| = \frac{1}{2} \qquad \text{Period} = \frac{2\pi}{1/3} = 6\pi \qquad \bullet\ b = \frac{1}{3}$$

Continued ▶

The zeros in the interval $0 \le x \le 6\pi$ are 0, $\dfrac{\pi}{1/3} = 3\pi$, and $\dfrac{2\pi}{1/3} = 6\pi$.

Because $-\dfrac{1}{2} < 0$, the graph is the graph of $y = \dfrac{1}{2}\sin\dfrac{x}{3}$ reflected across the x-axis, as shown in **Figure 4.52.**

TRY EXERCISE 38, PAGE 329

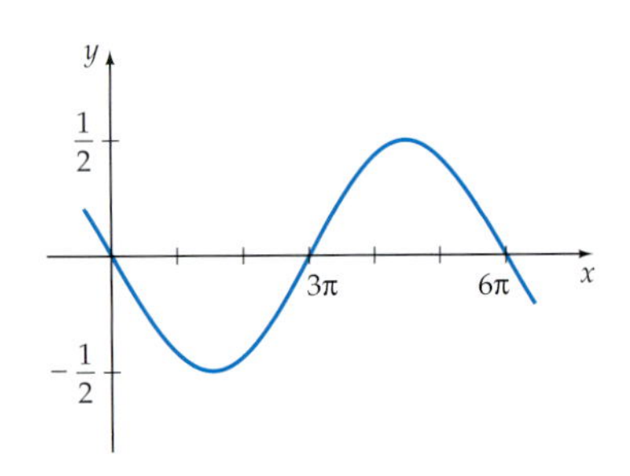

$y = -\dfrac{1}{2}\sin\dfrac{x}{3}$

FIGURE 4.52

THE GRAPH OF THE COSINE FUNCTION

Table 4.7 lists some of the ordered pairs of $y = \cos x$, $0 \le x \le 2\pi$. In **Figure 4.53,** the points are plotted and a smooth curve is drawn through the points.

TABLE 4.7

x	y = cos x
0	1
$\frac{\pi}{4}$	≈ 0.7
$\frac{\pi}{2}$	0
$\frac{3\pi}{4}$	≈ -0.7
π	-1
$\frac{5\pi}{4}$	≈ -0.7
$\frac{3\pi}{2}$	0
$\frac{7\pi}{4}$	≈ 0.7
2π	1

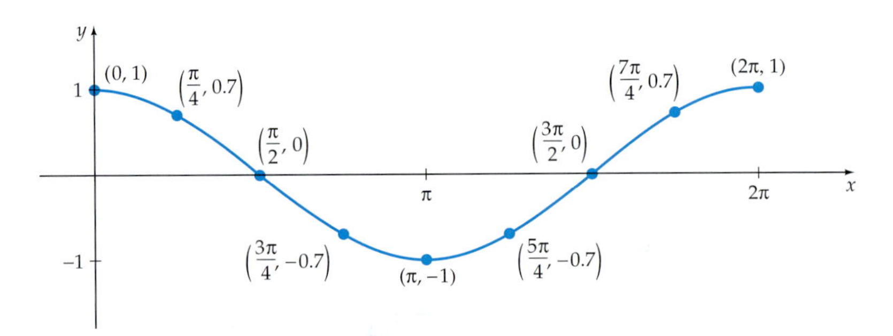

$y = \cos x, 0 \le x \le 2\pi$

FIGURE 4.53

Because the domain of $y = \cos x$ is the real numbers and the period is 2π, the graph of $y = \cos x$ is drawn by repeating the portion shown in **Figure 4.53.** Any part of the graph corresponding to one period (2π) is one cycle of $y = \cos x$ (see **Figure 4.54**).

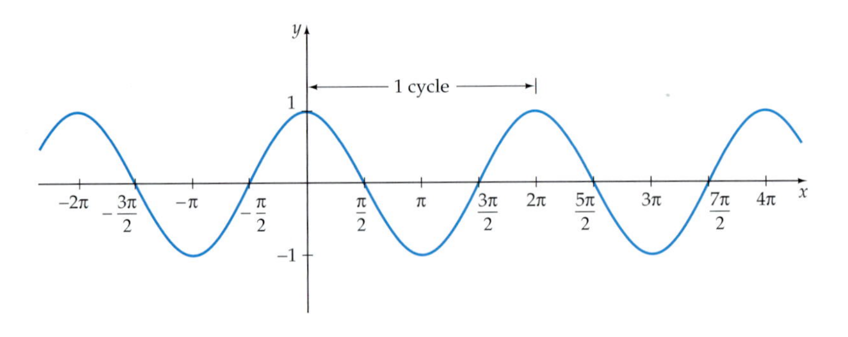

$y = \cos x$

FIGURE 4.54

The two theorems on the following page concerning cosine functions can be developed using methods that are analogous to those we used to determine the amplitude and period of a sine function.

Amplitude of $y = a \cos x$

The amplitude of $y = a \cos x$ is $|a|$.

Period of $y = \cos bx$

The period of $y = \cos bx$ is $\dfrac{2\pi}{|b|}$.

Table 4.8 gives the amplitude and period of some cosine functions.

TABLE 4.8

Function	$y = a \cos bx$	$y = 2 \cos 3x$	$y = -3 \cos \dfrac{2x}{3}$
Amplitude	$\lvert a \rvert$	$\lvert 2 \rvert = 2$	$\lvert -3 \rvert = 3$
Period	$\dfrac{2\pi}{\lvert b \rvert}$	$\dfrac{2\pi}{3}$	$\dfrac{2\pi}{2/3} = 3\pi$

EXAMPLE 4 Graph $y = \cos bx$

Graph: $y = \cos \dfrac{2\pi}{3} x$

Solution

$$\text{Amplitude} = 1 \qquad \text{Period} = \frac{2\pi}{b} = \frac{2\pi}{2\pi/3} = 3 \qquad \bullet\ b = \frac{2\pi}{3}$$

The graph is shown in **Figure 4.55.**

$y = \cos \dfrac{2\pi}{3} x$

FIGURE 4.55

TRY EXERCISE 32, PAGE 329

Figure 4.56 shows the graph of $y = a \cos bx$ for both a and b positive. Note from the graph that

- The amplitude is a.
- The period is $\dfrac{2\pi}{b}$.
- For $0 \le x \le \dfrac{2\pi}{b}$, the zeros are $\dfrac{\pi}{2b}$ and $\dfrac{3\pi}{2b}$.

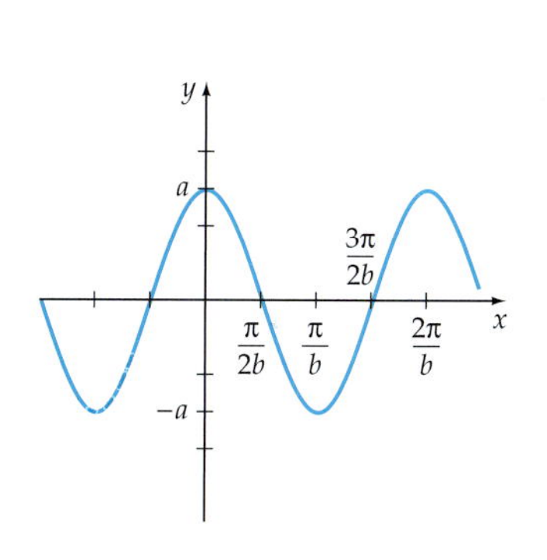

$y = a \cos bx$

FIGURE 4.56

- The maximum value is a when $x = 0$, and the minimum value is $-a$ when $x = \dfrac{\pi}{b}$.
- If $a < 0$, then the graph is reflected across the x-axis.

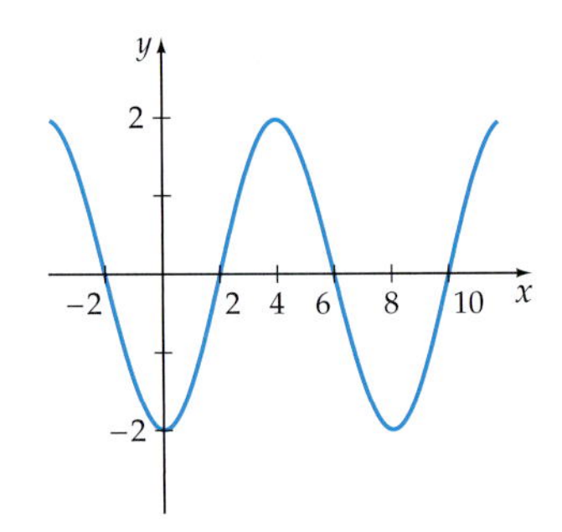

$y = -2\cos\dfrac{\pi x}{4}$

FIGURE 4.57

EXAMPLE 5 Graph a Cosine Function

Graph: $y = -2\cos\dfrac{\pi x}{4}$

Solution

$$\text{Amplitude} = |-2| = 2 \qquad \text{Period} = \frac{2\pi}{\pi/4} = 8 \qquad \bullet\ b = \frac{\pi}{4}$$

The zeros in the interval $0 \leq x \leq 8$ are $\dfrac{\pi}{2\pi/4} = 2$ and $\dfrac{3\pi}{2\pi/4} = 6$. Because $-2 < 0$, the graph is the graph of $y = 2\cos\dfrac{\pi x}{4}$ reflected across the x-axis, as shown in **Figure 4.57.**

▶ TRY EXERCISE 46, PAGE 329

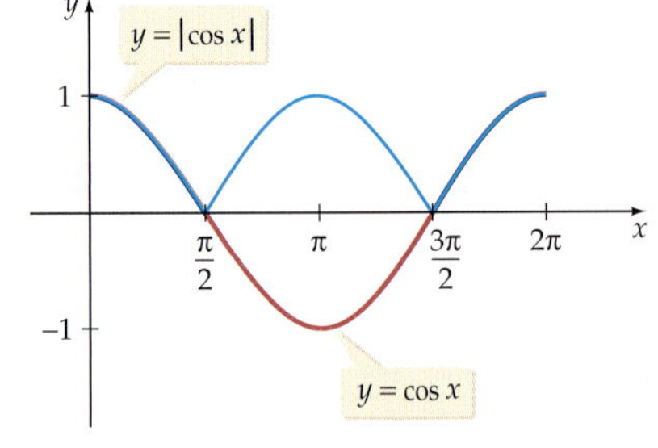

FIGURE 4.58

EXAMPLE 6 Graph the Absolute Value of the Cosine Function

Graph $y = |\cos x|$, where $0 \leq x \leq 2\pi$.

Solution

Because $|\cos x| \geq 0$, the graph of $y = |\cos x|$ is drawn by reflecting the negative portion of the graph of $y = \cos x$ across the x-axis. The graph is the one shown in dark blue and light blue in **Figure 4.58.**

▶ TRY EXERCISE 52, PAGE 329

TOPICS FOR DISCUSSION

1. Is the graph of $f(x) = |\sin x|$ the same as the graph of $y = \sin|x|$? Explain.
2. Explain how the graph of $y = \cos 2x$ differs from the graph of $y = \cos x$.
3. Does the graph of $y = \sin(-2x)$ have the same period as the graph of $y = \sin 2x$? Explain.
4. The function $h(x) = a \sin bt$ has an amplitude of 3 and a period of 4. What are the possible values of a? What are the possible values of b?

EXERCISE SET 4.3

In Exercises 1 to 16, state the amplitude and period of the function defined by each equation.

1. $y = 2 \sin x$

2. $y = -\dfrac{1}{2} \sin x$

3. $y = \sin 2x$

4. $y = \sin \dfrac{2x}{3}$

5. $y = \dfrac{1}{2} \sin 2\pi x$

6. $y = 2 \sin \dfrac{\pi x}{3}$

7. $y = -2 \sin \dfrac{x}{2}$

8. $y = -\dfrac{1}{2} \sin \dfrac{x}{2}$

9. $y = \dfrac{1}{2} \cos x$

10. $y = -3 \cos x$

11. $y = \cos \dfrac{x}{4}$

12. $y = \cos 3x$

13. $y = 2 \cos \dfrac{\pi x}{3}$

14. $y = \dfrac{1}{2} \cos 2\pi x$

15. $y = -3 \cos \dfrac{2x}{3}$

16. $y = \dfrac{3}{4} \cos 4x$

In Exercises 17 to 54, graph at least one full period of the function defined by each equation.

17. $y = \dfrac{1}{2} \sin x$

18. $y = \dfrac{3}{2} \cos x$

19. $y = 3 \cos x$

▶ **20.** $y = -\dfrac{3}{2} \sin x$

21. $y = -\dfrac{7}{2} \cos x$

22. $y = 3 \sin x$

23. $y = -4 \sin x$

24. $y = -5 \cos x$

25. $y = \cos 3x$

26. $y = \sin 4x$

27. $y = \sin \dfrac{3x}{2}$

28. $y = \cos \pi x$

29. $y = \cos \dfrac{\pi}{2} x$

▶ **30.** $y = \sin \dfrac{3\pi}{4} x$

31. $y = \sin 2\pi x$

▶ **32.** $y = \cos 3\pi x$

33. $y = 4 \cos \dfrac{x}{2}$

34. $y = 2 \cos \dfrac{3x}{4}$

35. $y = -2 \cos \dfrac{x}{3}$

36. $y = -\dfrac{4}{3} \cos 3x$

37. $y = 2 \sin \pi x$

▶ **38.** $y = \dfrac{1}{2} \sin \dfrac{\pi x}{3}$

39. $y = \dfrac{3}{2} \cos \dfrac{\pi x}{2}$

40. $y = \cos \dfrac{\pi x}{3}$

41. $y = 4 \sin \dfrac{2\pi x}{3}$

42. $y = 3 \cos \dfrac{3\pi x}{2}$

43. $y = 2 \cos 2x$

44. $y = \dfrac{1}{2} \sin 2.5x$

45. $y = -2 \sin 1.5x$

▶ **46.** $y = -\dfrac{3}{4} \cos 5x$

47. $y = \left|2 \sin \dfrac{x}{2}\right|$

48. $y = \left|\dfrac{1}{2} \sin 3x\right|$

49. $y = |-2 \cos 3x|$

50. $y = \left|-\dfrac{1}{2} \cos \dfrac{x}{2}\right|$

51. $y = -\left|2 \sin \dfrac{x}{3}\right|$

▶ **52.** $y = -\left|3 \sin \dfrac{2x}{3}\right|$

53. $y = -|3 \cos \pi x|$

54. $y = -\left|2 \cos \dfrac{\pi x}{2}\right|$

In Exercises 55 to 60, find an equation of each graph.

55.

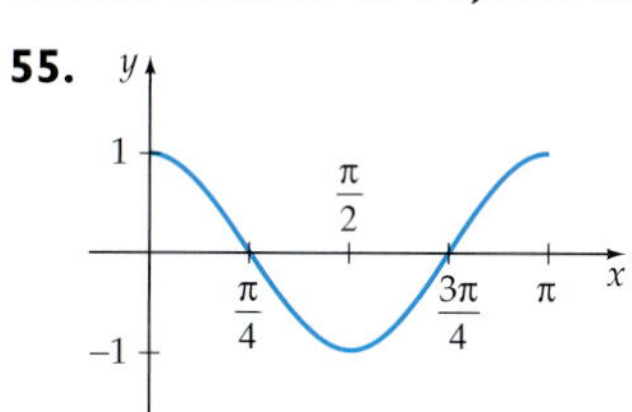

56.

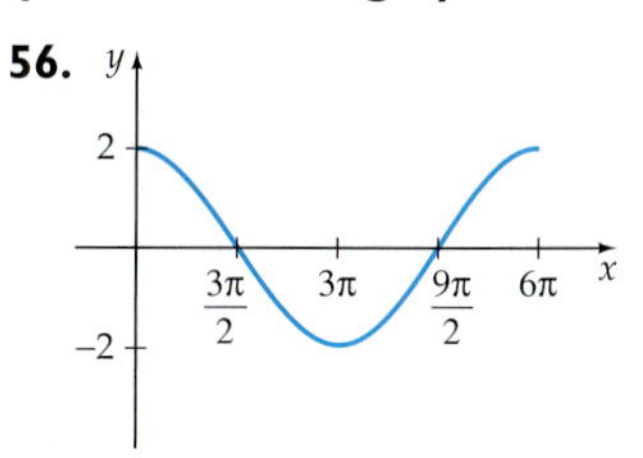

57.

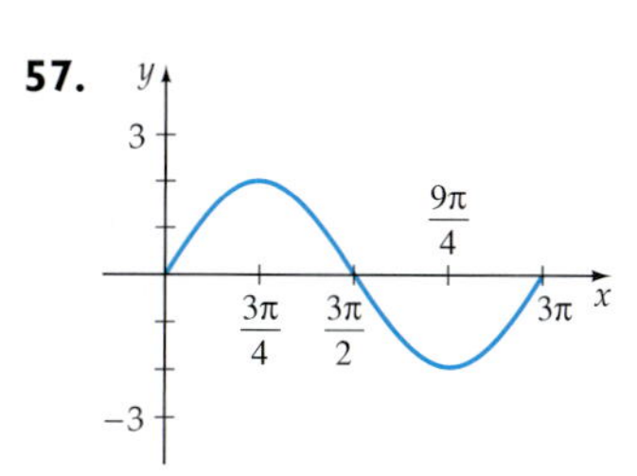

58.

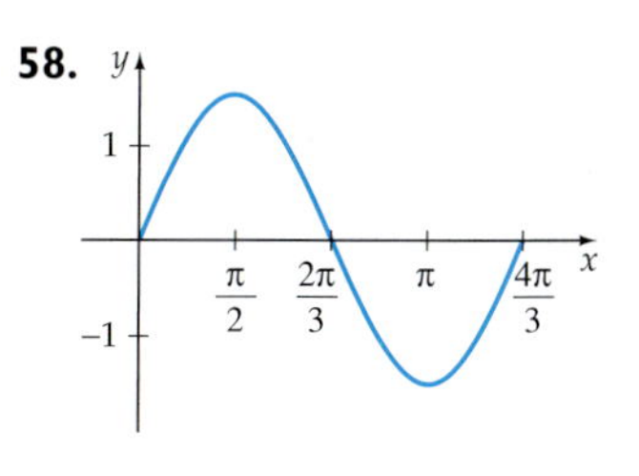

59.

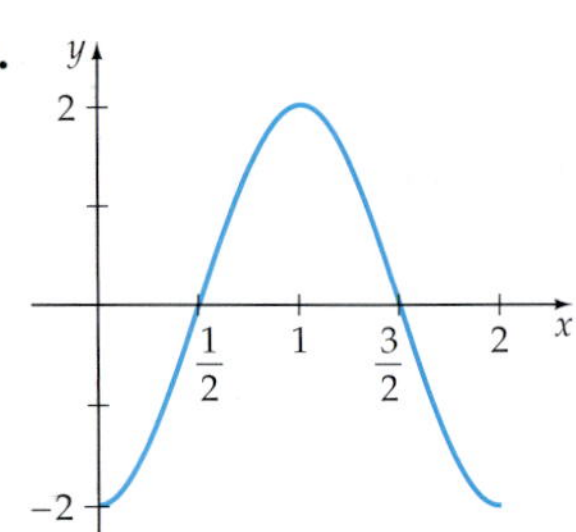

60.

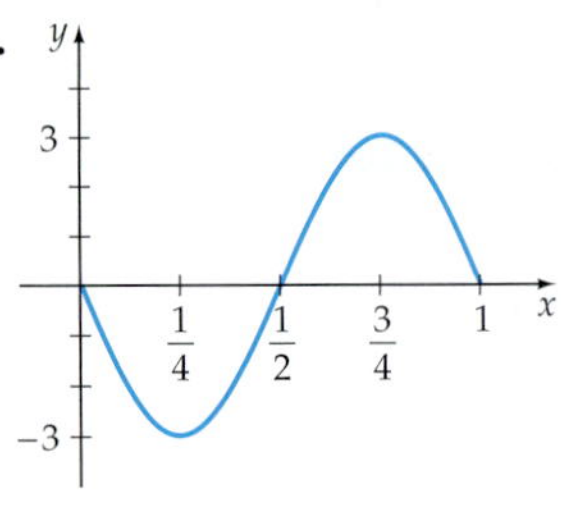

61. Sketch the graph of $y = 2 \sin \dfrac{2x}{3}$, $-3\pi \le x \le 6\pi$.

62. Sketch the graph of $y = -3 \cos \dfrac{3x}{4}$, $-2\pi \le x \le 4\pi$.

63. Sketch the graphs of

$$y_1 = 2 \cos \frac{x}{2} \quad \text{and} \quad y_2 = 2 \cos x$$

on the same set of axes for $-2\pi \le x \le 4\pi$.

64. Sketch the graphs of

$$y_1 = \sin 3\pi x \quad \text{and} \quad y_2 = \sin \frac{\pi x}{3}$$

on the same set of axes for $-2 \le x \le 4$.

In Exercises 65 to 72, use a graphing utility to graph each function.

65. $y = \cos^2 x$

66. $y = 3^{\cos^2 x} \cdot 3^{\sin^2 x}$

67. $y = \cos |x|$

68. $y = \sin |x|$

69. $y = \dfrac{1}{2} x \sin x$

70. $y = \dfrac{1}{2} x + \sin x$

71. $y = -x \cos x$

72. $y = -x + \cos x$

73. Graph $y = e^{\sin x}$. What is the maximum value of $e^{\sin x}$? What is the minimum value of $e^{\sin x}$? Is the function defined by $y = e^{\sin x}$ a periodic function? If so, what is the period?

74. Graph $y = e^{\cos x}$. What is the maximum value of $e^{\cos x}$? What is the minimum value of $e^{\cos x}$? Is the function defined by $y = e^{\cos x}$ a periodic function? If so, what is the period?

75. **EQUATION OF A WAVE** A tidal wave that is caused by an earthquake under the ocean is called a **tsunami.** A model of a tsunami is given by $f(t) = A \cos Bt$. Find the equation of a tsunami that has an amplitude of 60 feet and a period of 20 seconds.

76. **EQUATION OF HOUSEHOLD CURRENT** The electricity supplied to your home, called *alternating current,* can be modeled by $I = A \sin \omega t$, where I is the number of amperes of current at time t seconds. Write the equation of household current whose graph is given in the figure below. Calculate I when $t = 0.5$ second.

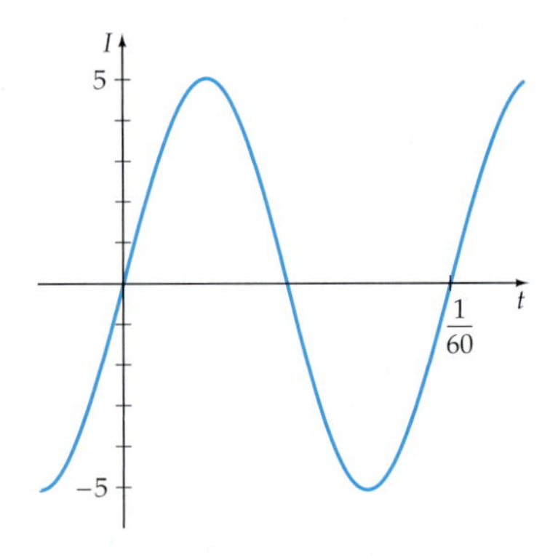

CONNECTING CONCEPTS

In Exercises 77 to 80, write an equation for a sine function using the given information.

77. Amplitude = 2; period = 3π

78. Amplitude = 5; period = $\dfrac{2\pi}{3}$

79. Amplitude = 4; period = 2

80. Amplitude = 2.5; period = 3.2

In Exercises 81 to 84, write an equation for a cosine function using the given information.

81. Amplitude = 3; period = $\dfrac{\pi}{2}$

82. Amplitude = 0.8; period = 4π

83. Amplitude = 3; period = 2.5

84. Amplitude = 4.2; period = 1

PREPARE FOR SECTION 4.4

85. 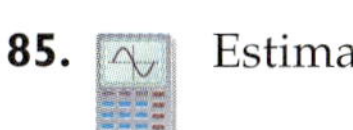 Estimate, to the nearest tenth, $\tan \dfrac{\pi}{3}$. [4.2]

86. 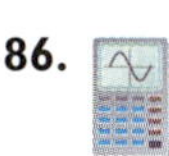 Estimate, to the nearest tenth, $\cot \dfrac{\pi}{3}$. [4.2]

87. Explain how to use the graph of $y = f(x)$ to produce the graph of $y = 2f(x)$. [1.6]

88. Explain how to use the graph of $y = f(x)$ to produce the graph of $y = f(x - 2) + 3$. [1.6]

89. Simplify: $\dfrac{\pi}{1/2}$

90. Simplify: $\dfrac{\pi}{\left|-\dfrac{3}{4}\right|}$

PROJECTS

1.   **A TRIGONOMETRIC POWER FUNCTION**

a. Determine the domain and the range of $y = (\sin x)^{\cos x}$. Explain.

b. What is the amplitude of the function? Explain.

SECTION 4.4

GRAPHS OF THE OTHER TRIGONOMETRIC FUNCTIONS

- The Graph of the Tangent Function
- The Graph of the Cotangent Function
- The Graph of the Cosecant Function
- The Graph of the Secant Function

The Graph of the Tangent Function

Figure 4.59 shows the graph of $y = \tan x$ for $-\dfrac{\pi}{2} < x < \dfrac{\pi}{2}$. The lines $x = \dfrac{\pi}{2}$ and $x = -\dfrac{\pi}{2}$ are vertical asymptotes for the graph of $y = \tan x$. From Section 4.2, the period of $y = \tan x$ is π. Therefore, the portion of the graph shown in **Figure 4.59** is repeated along the x-axis, as shown in **Figure 4.60.**

Because the tangent function is unbounded, there is no amplitude for the tangent function. The graph of $y = a \tan x$ is drawn by stretching ($|a| > 1$) or shrinking ($|a| < 1$) the graph of $y = \tan x$. If $a < 0$, then the graph is reflected across the x-axis. **Figure 4.61** shows the graphs of three tangent functions. Because $\tan \dfrac{\pi}{4} = 1$, the point $\left(\dfrac{\pi}{4}, a\right)$ is convenient to plot as a guide for the graph of $y = a \tan x$.

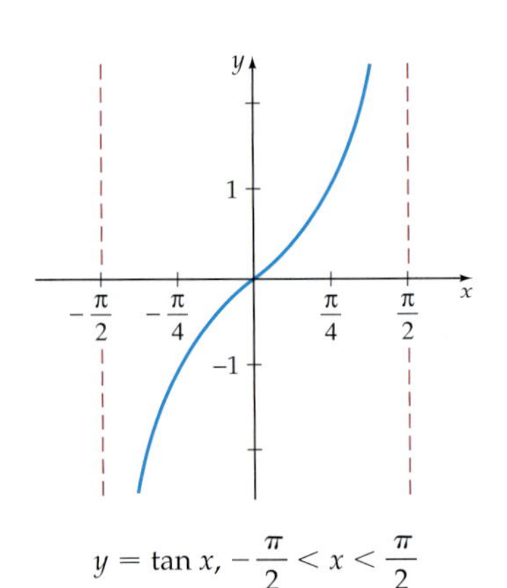

$y = \tan x, -\frac{\pi}{2} < x < \frac{\pi}{2}$

FIGURE 4.59

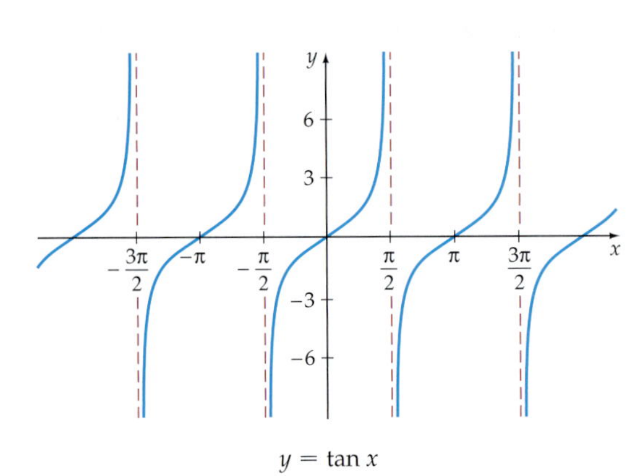

$y = \tan x$

FIGURE 4.60

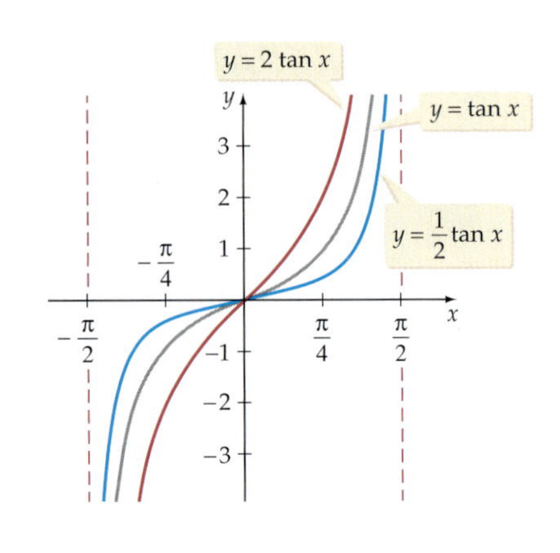

FIGURE 4.61

EXAMPLE 1 Graph $y = a \tan x$

Graph: $y = -3 \tan x$

Solution

The graph of $y = -3 \tan x$ is the reflection across the x-axis of the graph of $y = 3 \tan x$, as shown in **Figure 4.62.**

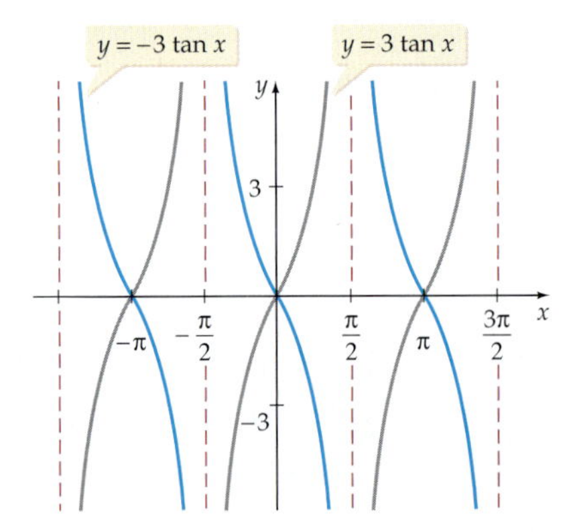

FIGURE 4.62

TRY EXERCISE 22, PAGE 337

The period of $y = \tan x$ is π and the graph completes one cycle on the interval $-\frac{\pi}{2} < x < \frac{\pi}{2}$. The period of $y = \tan bx$ $(b > 0)$ is $\frac{\pi}{b}$. The graph of $y = \tan bx$ completes one cycle on the interval $\left(-\frac{\pi}{2b}, \frac{\pi}{2b}\right)$.

Period of $y = \tan bx$

The period of $y = \tan bx$ is $\frac{\pi}{|b|}$.

QUESTION What is the period of the graph of $y = \tan \pi x$?

EXAMPLE 2 Graph $y = a \tan bx$

Graph: $y = 2 \tan \dfrac{x}{2}$

Solution

Period $= \dfrac{\pi}{b} = \dfrac{\pi}{1/2} = 2\pi$. Graph one cycle for values of x such that $-\pi < x < \pi$. This curve is repeated along the x-axis, as shown in **Figure 4.63.**

TRY EXERCISE 30, PAGE 337

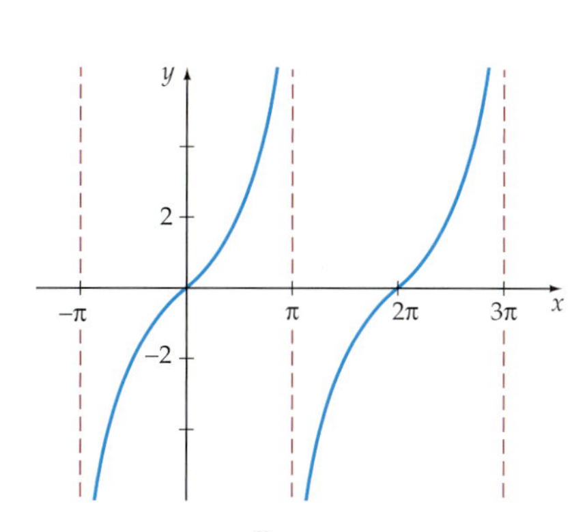

$y = 2 \tan \dfrac{x}{2}, -\pi < x < 3\pi$

FIGURE 4.63

Figure 4.64 shows one cycle of the graph of $y = a \tan bx$ for both a and b positive. Note from the graph that

- The period is $\dfrac{\pi}{b}$.
- $x = 0$ is a zero.
- The graph passes through $\left(-\dfrac{\pi}{4b}, -a\right)$ and $\left(\dfrac{\pi}{4b}, a\right)$.
- If $a < 0$, the graph is reflected across the x-axis.

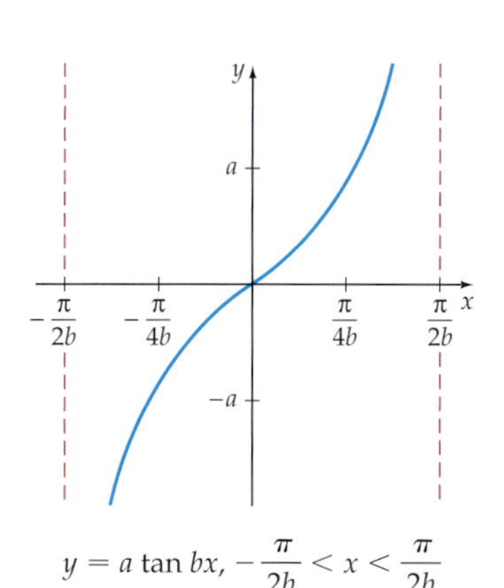

$y = a \tan bx, -\dfrac{\pi}{2b} < x < \dfrac{\pi}{2b}$

FIGURE 4.64

THE GRAPH OF THE COTANGENT FUNCTION

Figure 4.65 shows the graph of $y = \cot x$ for $0 < x < \pi$. The lines $x = 0$ and $x = \pi$ are vertical asymptotes for the graph of $y = \cot x$. From Section 4.2, the period of $y = \cot x$ is π. Therefore, the graph cycle shown in **Figure 4.65** is repeated along the x-axis, as shown in **Figure 4.66.** As with the graph of $y = \tan x$, the graph of $y = \cot x$ is unbounded and there is no amplitude.

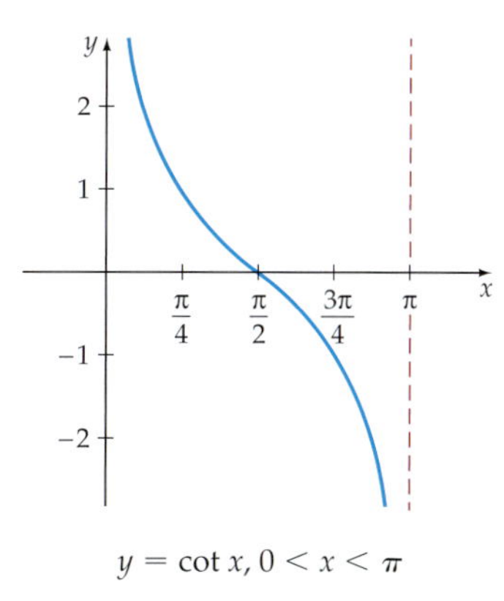

$y = \cot x, 0 < x < \pi$

FIGURE 4.65

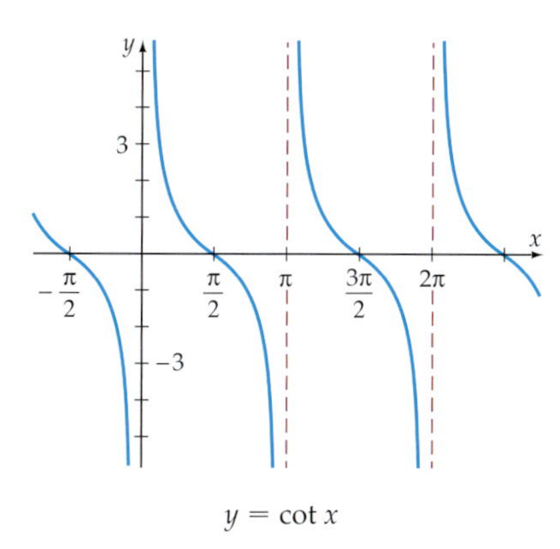

$y = \cot x$

FIGURE 4.66

ANSWER 1

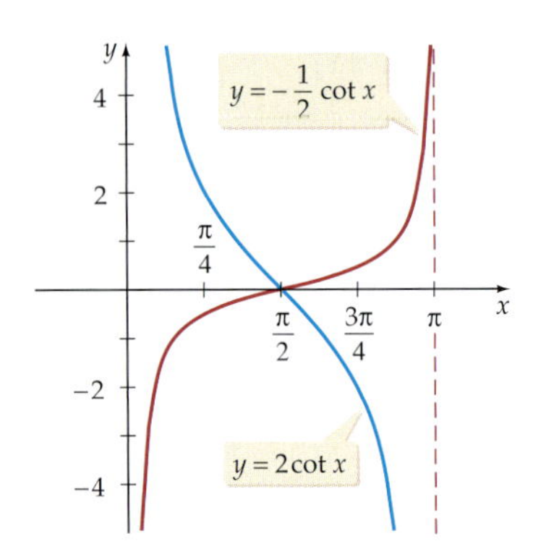

FIGURE 4.67

The graph of $y = a \cot x$ is drawn by stretching ($|a| > 1$) or shrinking ($|a| < 1$) the graph of $y = \cot x$. The graph is reflected across the x-axis when $a < 0$. **Figure 4.67** shows the graphs of two cotangent functions.

The period of $y = \cot x$ is π, and the period of $y = \cot bx$ is $\frac{\pi}{|b|}$. One cycle of the graph of $y = \cot bx$ is completed on the interval $\left(0, \frac{\pi}{b}\right)$.

Period of y = cot bx

The period of $y = \cot bx$ is $\frac{\pi}{|b|}$.

EXAMPLE 3 Graph y = a cot bx

Graph: $y = 3 \cot \frac{x}{2}$

Solution

Period $= \frac{\pi}{b} = \frac{\pi}{1/2} = 2\pi$. Sketch the graph for values of x for which $0 < x < 2\pi$. This curve is repeated along the x-axis, as shown in **Figure 4.68.**

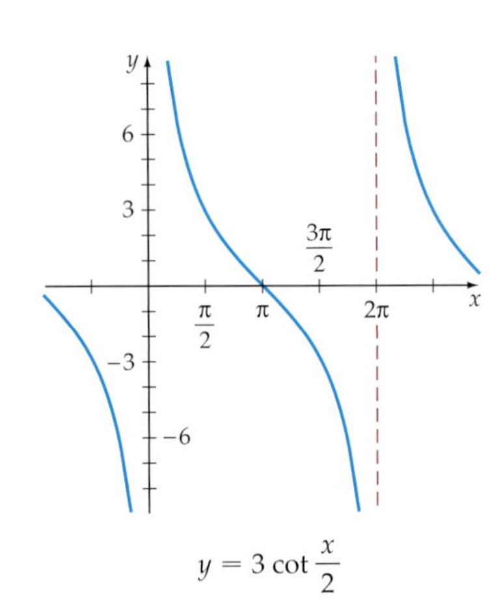

$y = 3 \cot \frac{x}{2}$

FIGURE 4.68

TRY EXERCISE 32, PAGE 338

Figure 4.69 shows one cycle of the graph of $y = a \cot bx$ for both a and b positive. Note from the graph that

- The period is $\frac{\pi}{b}$.
- $x = \frac{\pi}{2b}$ is a zero.
- The graph passes through $\left(\frac{\pi}{4b}, a\right)$ and $\left(\frac{3\pi}{4b}, -a\right)$.
- If $a < 0$, the graph is reflected across the x-axis.

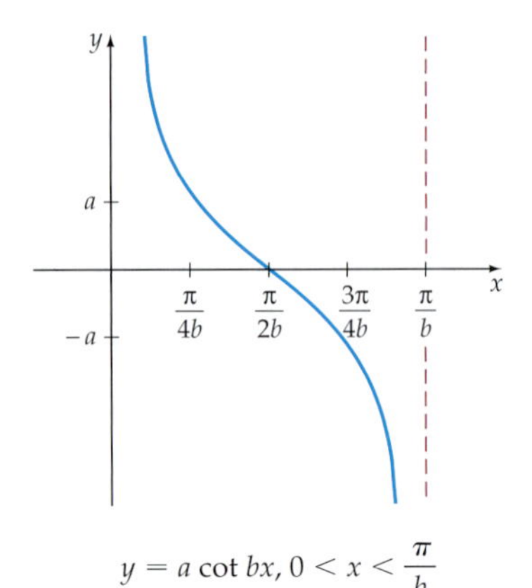

$y = a \cot bx, 0 < x < \frac{\pi}{b}$

FIGURE 4.69

THE GRAPH OF THE COSECANT FUNCTION

Because $\csc x = \frac{1}{\sin x}$, the value of $\csc x$ is the reciprocal of the value of $\sin x$. Therefore, $\csc x$ is undefined when $\sin x = 0$ or when $x = n\pi$, where n is an integer. The graph of $y = \csc x$ has vertical asymptotes at $n\pi$. Because $y = \csc x$ has period 2π, the graph will be repeated along the x-axis every 2π units. A graph of $y = \csc x$ is shown in **Figure 4.70.**

The graph of $y = \sin x$ is also shown in **Figure 4.70.** Note the relationships among the zeros of $y = \sin x$ and the asymptotes of $y = \csc x$. Also note that because $|\sin x| \le 1$, $\dfrac{1}{|\sin x|} \ge 1$. Thus the range of $y = \csc x$ is $\{y \mid y \ge 1, y \le -1\}$. The general procedure for graphing $y = a \csc bx$ is first to graph $y = a \sin bx$. Then sketch the graph of the cosecant function by using y values equal to the product of a and the reciprocal values of $\sin bx$.

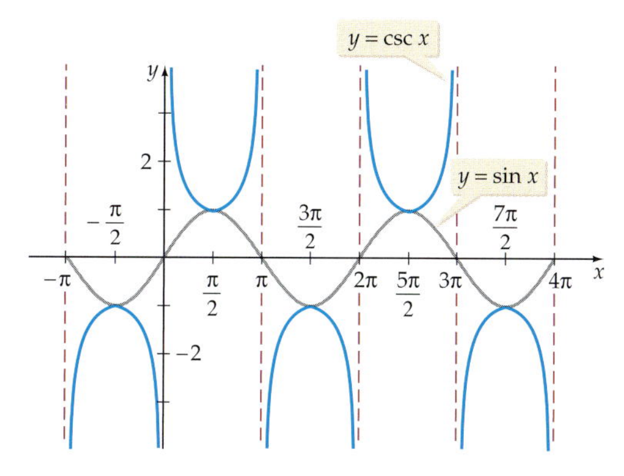

FIGURE 4.70

EXAMPLE 4 Graph $y = a \csc bx$

Graph: $y = 2 \csc \dfrac{\pi x}{2}$

Solution

First sketch the graph of $y = 2 \sin \dfrac{\pi x}{2}$ and draw vertical asymptotes through the zeros. Now sketch the graph of $y = 2 \csc \dfrac{\pi x}{2}$, using the asymptotes as guides for the graph, as shown in **Figure 4.71.**

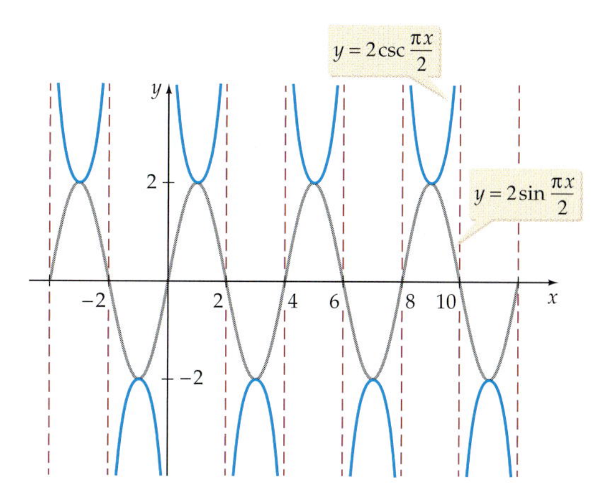

FIGURE 4.71

TRY EXERCISE 38, PAGE 338

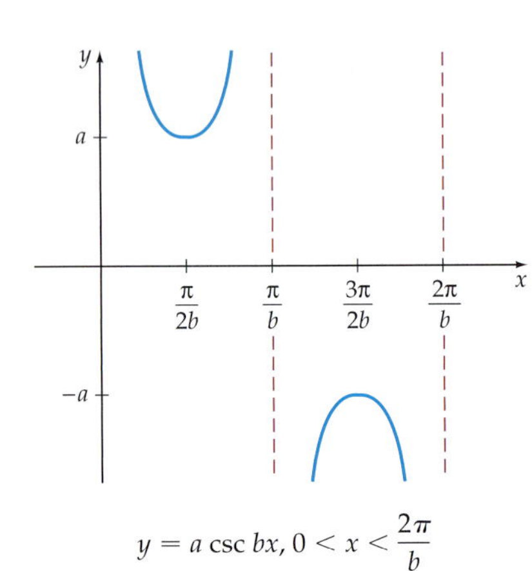

FIGURE 4.72

Figure 4.72 shows one cycle of the graph of $y = a \csc bx$ for both a and b positive. Note from the graph that

- The period is $\dfrac{2\pi}{b}$.
- The vertical asymptotes of $y = a \csc bx$ are located at the zeros of $y = a \sin bx$.
- The graph passes through $\left(\dfrac{\pi}{2b}, a\right)$ and $\left(\dfrac{3\pi}{2b}, -a\right)$.
- If $a < 0$, then the graph is reflected across the x-axis.

The Graph of the Secant Function

Because $\sec x = \dfrac{1}{\cos x}$, the value of $\sec x$ is the reciprocal of the value of $\cos x$. Therefore, $\sec x$ is undefined when $\cos x = 0$ or when $x = \dfrac{\pi}{2} + n\pi$, n an integer. The graph of $y = \sec x$ has vertical asymptotes at $\dfrac{\pi}{2} + n\pi$. Because $y = \sec x$ has period 2π, the graph will be replicated along the x-axis every 2π units. A graph of $y = \sec x$ is shown in **Figure 4.73.**

The graph of $y = \cos x$ is also shown in **Figure 4.73.** Note the relationships among the zeros of $y = \cos x$ and the asymptotes of $y = \sec x$. Also note that because $|\cos x| \le 1$, $\dfrac{1}{|\cos x|} \ge 1$. Thus the range of $y = \sec x$ is $\{y \mid y \ge 1, y \le -1\}$. The general procedure for graphing $y = a \sec bx$ is first to graph $y = a \cos bx$. Then sketch the graph of the secant function by using y values equal to the product of a and the reciprocal values of $\cos bx$.

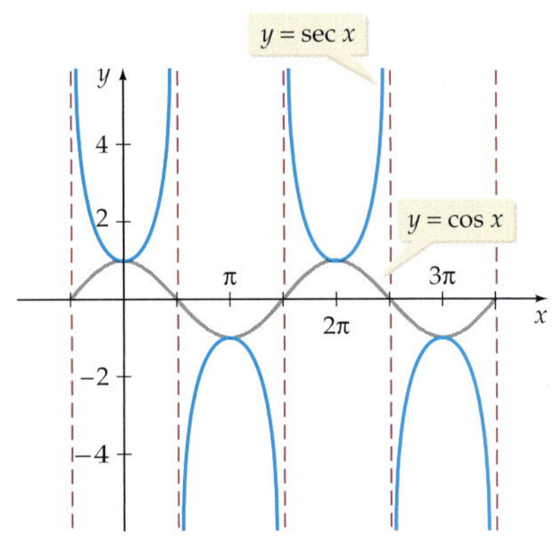

FIGURE 4.73

EXAMPLE 5 Graph y = a sec bx

Graph: $y = -3 \sec \dfrac{x}{2}$

Solution

First sketch the graph of $y = -3 \cos \dfrac{x}{2}$ and draw vertical asymptotes through the zeros. Now sketch the graph of $y = -3 \sec \dfrac{x}{2}$, using the asymptotes as guides for the graph, as shown in **Figure 4.74.**

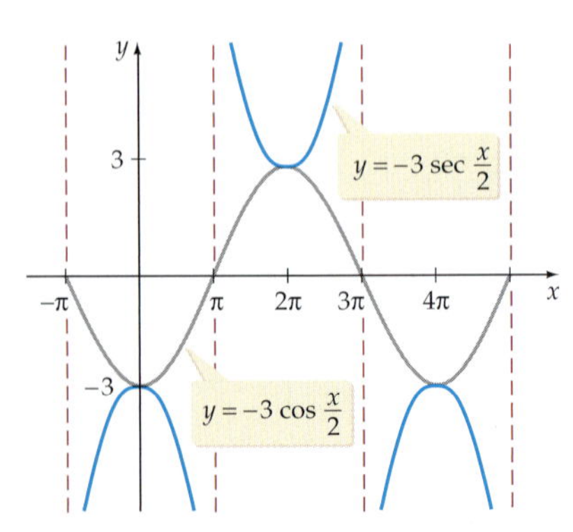

FIGURE 4.74

▶ TRY EXERCISE 42, PAGE 338

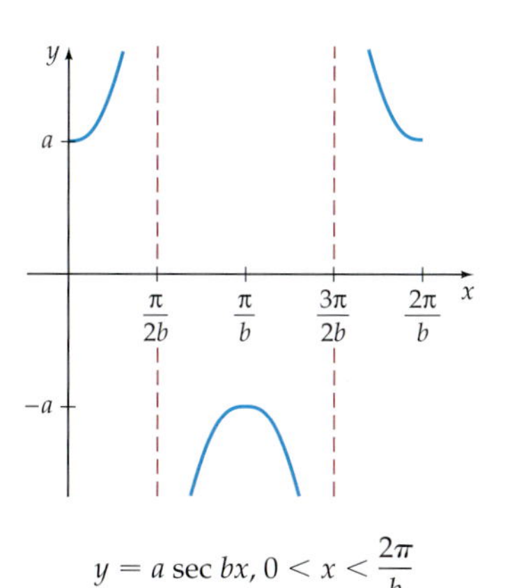

$y = a \sec bx, 0 < x < \frac{2\pi}{b}$

FIGURE 4.75

Figure 4.75 shows one cycle of the graph of $y = a \sec bx$ for both a and b positive. Note from the graph that

- The period is $\frac{2\pi}{b}$.
- The vertical asymptotes of $y = a \sec bx$ are located at the zeros of $y = a \cos bx$.
- The graph passes through $(0, a)$, $\left(\frac{\pi}{b}, -a\right)$, and $\left(\frac{2\pi}{b}, a\right)$.
- If $a < 0$, then the graph is reflected across the x-axis.

Topics for Discussion

1. What are the zeros of $y = \tan x$? Explain.
2. What are the zeros of $y = \sec x$? Explain.
3. The functions $f(x) = \tan x$ and $g(x) = \tan(-x)$ both have a period of π. Do you agree?
4. What is the amplitude of the function $k(x) = 4 \cot(\pi x)$? Explain.

Exercise Set 4.4

1. For what values of x is $y = \tan x$ undefined?
2. For what values of x is $y = \cot x$ undefined?
3. For what values of x is $y = \sec x$ undefined?
4. For what values of x is $y = \csc x$ undefined?

In Exercises 5 to 20, state the period of each function.

5. $y = \sec x$
6. $y = \cot x$
7. $y = \tan x$
8. $y = \csc x$
9. $y = 2 \tan \frac{x}{2}$
10. $y = \frac{1}{2} \cot 2x$
11. $y = \csc 3x$
12. $y = \csc \frac{x}{2}$
13. $y = -\tan 3x$
14. $y = -3 \cot \frac{2x}{3}$
15. $y = -3 \sec \frac{x}{4}$
16. $y = -\frac{1}{2} \csc 2x$
17. $y = \cot \pi x$
18. $y = \cot \frac{\pi x}{3}$
19. $y = 2 \csc \frac{\pi x}{2}$
20. $y = -3 \cot \pi x$

In Exercises 21 to 40, sketch one full period of the graph of each function.

21. $y = 3 \tan x$
22. $y = \frac{1}{3} \tan x$
23. $y = \frac{3}{2} \cot x$
24. $y = 4 \cot x$
25. $y = 2 \sec x$
26. $y = \frac{3}{4} \sec x$
27. $y = \frac{1}{2} \csc x$
28. $y = 2 \csc x$
29. $y = 2 \tan \frac{x}{2}$
30. $y = -3 \tan 3x$

31. $y = -3 \cot \dfrac{x}{2}$

▶ **32.** $y = \dfrac{1}{2} \cot 2x$

33. $y = -2 \csc \dfrac{x}{3}$

34. $y = \dfrac{3}{2} \csc 3x$

35. $y = \dfrac{1}{2} \sec 2x$

36. $y = -3 \sec \dfrac{2x}{3}$

37. $y = -2 \sec \pi x$

▶ **38.** $y = 3 \csc \dfrac{\pi x}{2}$

39. $y = 3 \tan 2\pi x$

40. $y = -\dfrac{1}{2} \cot \dfrac{\pi x}{2}$

41. Graph $y = 2 \csc 3x$ from -2π to 2π.

▶ **42.** Graph $y = \sec \dfrac{x}{2}$ from -4π to 4π.

43. Graph $y = 3 \sec \pi x$ from -2 to 4.

44. Graph $y = \csc \dfrac{\pi x}{2}$ from -4 to 4.

45. Graph $y = 2 \cot 2x$ from $-\pi$ to π.

46. Graph $y = \dfrac{1}{2} \tan \dfrac{x}{2}$ from -4π to 4π.

47. Graph $y = 3 \tan \pi x$ from -2 to 2.

48. Graph $y = \cot \dfrac{\pi x}{2}$ from -4 to 4.

In Exercises 49 to 54, find an equation of each blue graph.

49.

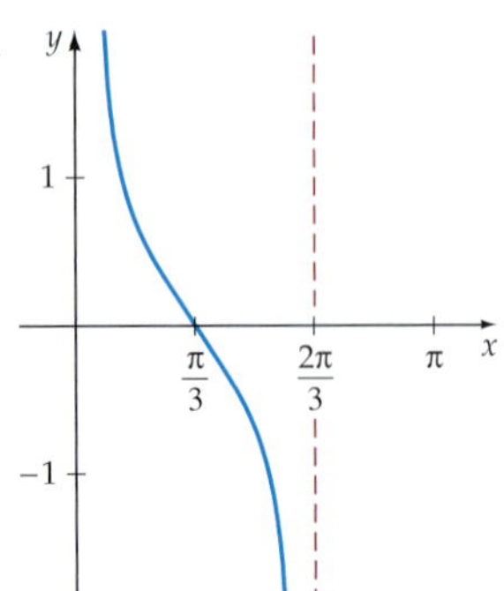

50.

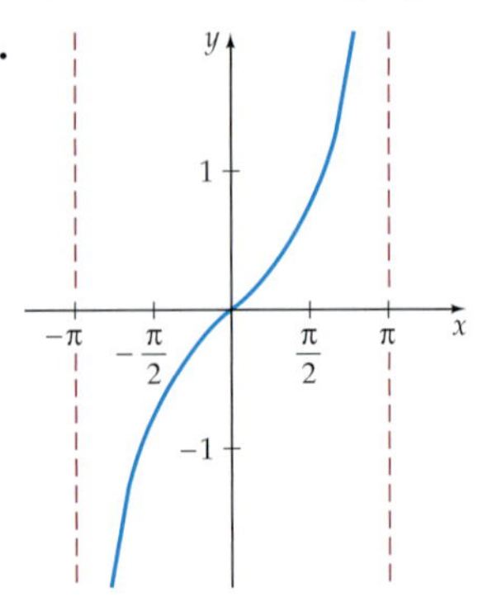

51.

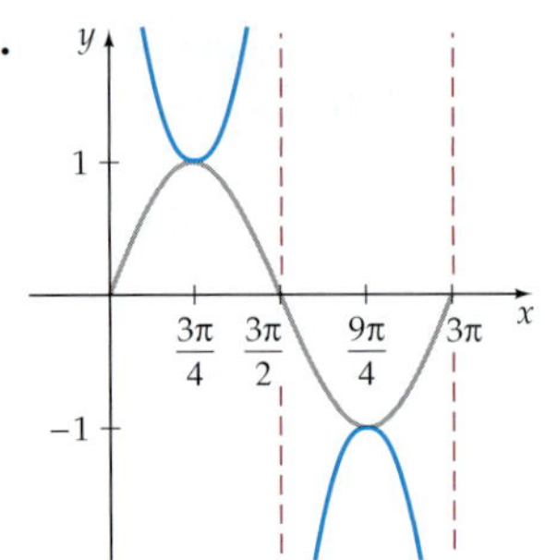

52.

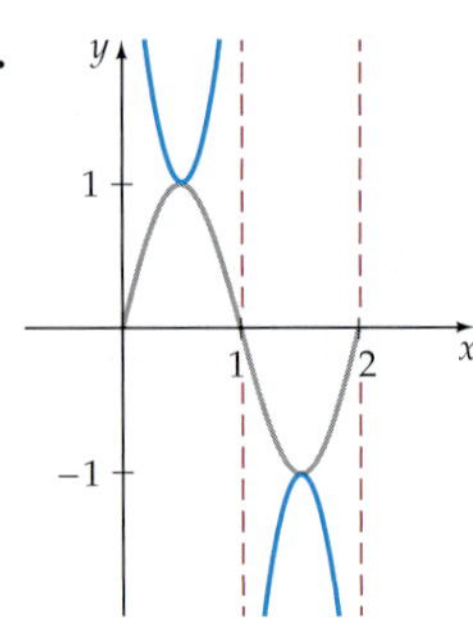

53.

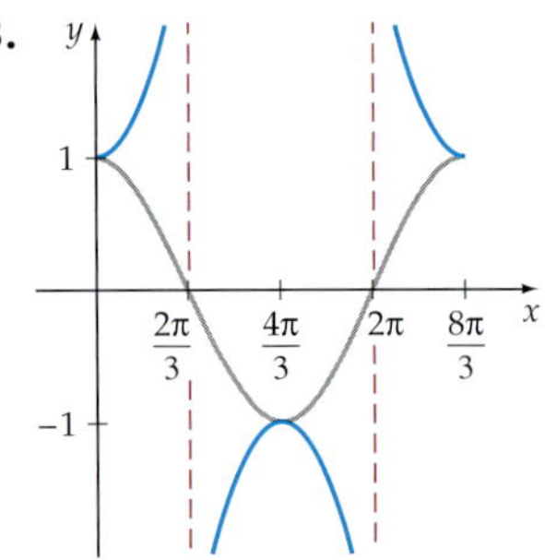

54.

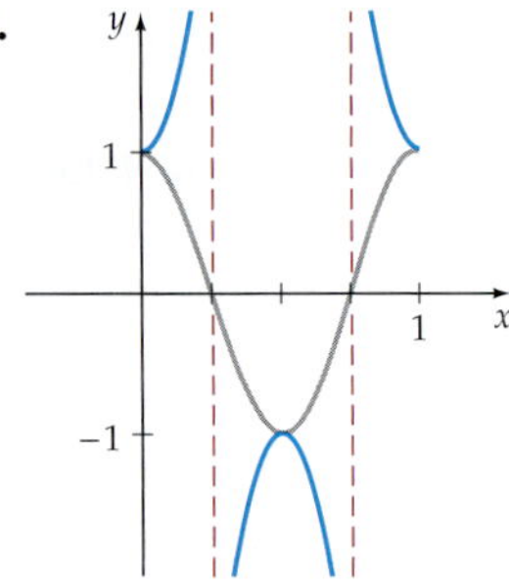

In Exercises 55 to 60, use a graphing utility to graph each equation. If needed, use open circles so that your graph is accurate.

55. $y = \tan |x|$

56. $y = \sec |x|$

57. $y = |\csc x|$

58. $y = |\cot x|$

59. $y = \tan x \cos x$

60. $y = \cot x \sin x$

61. Graph $y = \tan x$ and $x = \tan y$ on the same coordinate axes.

62. Graph $y = \sin x$ and $x = \sin y$ on the same coordinate axes.

CONNECTING CONCEPTS

In Exercises 63 to 70, write an equation that is of the form $y = \tan bx, y = \cot bx, y = \sec bx$, or $y = \csc bx$ and satisfies the given conditions.

63. Tangent, period: $\frac{\pi}{3}$

64. Cotangent, period: $\frac{\pi}{2}$

65. Secant, period: $\frac{3\pi}{4}$

66. Cosecant, period: $\frac{5\pi}{2}$

67. Cotangent, period: 2

68. Tangent, period: 0.5

69. Cosecant, period: 1.5

70. Secant, period: 3

PREPARE FOR SECTION 4.5

71. Find the amplitude and period of the graph of $y = 2 \sin 2x$. [4.3]

72. Find the amplitude and period of the graph of $y = \frac{2}{3} \cos \frac{x}{3}$. [4.3]

73. Find the amplitude and the period of the graph of $y = -4 \sin 2\pi x$. [4.3]

74. What is the maximum value of $f(x) = 2 \sin x$? [4.3]

75. What is the minimum value of $f(x) = 3 \cos 2x$? [4.3]

76. Is the graph of $f(x) = \cos x$ symmetric with respect to the origin or with respect to the y-axis? [4.2]

PROJECTS

1. **A TECHNOLOGY QUESTION** A student's calculator shows the display at the right. Note that the domain values 4.7123 and 4.7124 are close together, but the range values are over 100,000 units apart. Explain how this is possible.

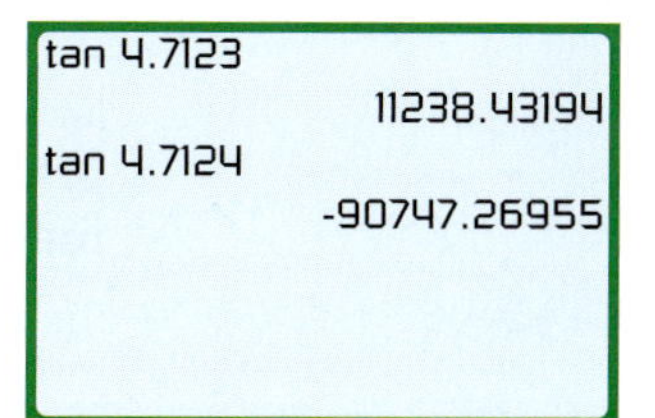

2. **SOLUTIONS OF A TRIGONOMETRIC EQUATION** How many solutions does $\tan\left(\frac{1}{x}\right) = 0$ have on the interval $-1 \le x \le 1$? Explain.

SECTION 4.5

GRAPHING TECHNIQUES

- TRANSLATION OF TRIGONOMETRIC FUNCTIONS
- ADDITION OF ORDINATES
- DAMPING FACTOR

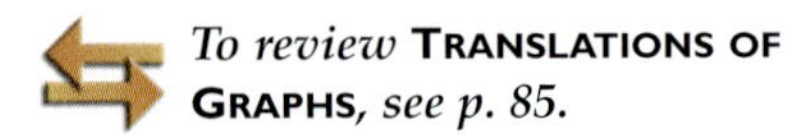
To review **TRANSLATIONS OF GRAPHS**, *see p. 85.*

TRANSLATION OF TRIGONOMETRIC FUNCTIONS

Recall that the graph of $y = f(x) \pm c$ is a *vertical translation* of the graph of $y = f(x)$. For $c > 0$, the graph of $y = f(x) - c$ is shifted c units down; the graph of $y = f(x) + c$ is shifted c units up. The graph in **Figure 4.76** is a graph of the equation $y = 2\sin \pi x - 3$, which is a vertical translation of $y = 2 \sin \pi x$ down 3 units. Note that subtracting 3 from $y = 2\sin \pi x$ changes neither its amplitude nor its period.

Also, the graph of $y = f(x \pm c)$ is a *horizontal translation* of the graph of $y = f(x)$. For $c > 0$, the graph of $y = f(x - c)$ is shifted c units to the right; the graph of $y = f(x + c)$ is shifted c units to the left. The graph in **Figure 4.77** is a graph of the equation $y = 2\sin\left(x - \frac{\pi}{4}\right)$, which is the graph of $y = 2 \sin x$ translated $\frac{\pi}{4}$ units to the right. Note that neither the period nor the amplitude is affected. A horizontal shift of the graph of a trigonometric function is called its **phase shift.**

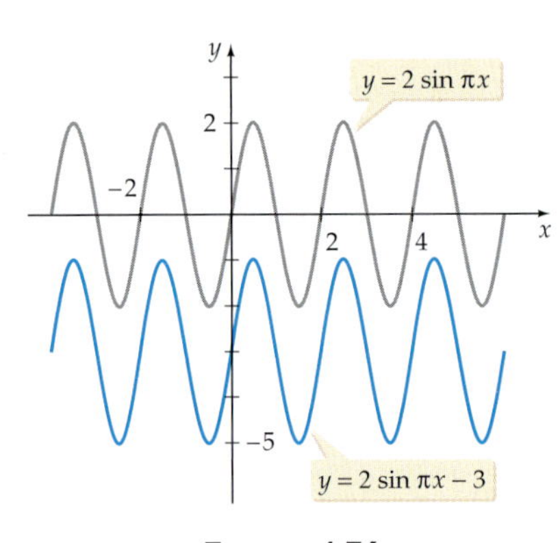

FIGURE 4.76

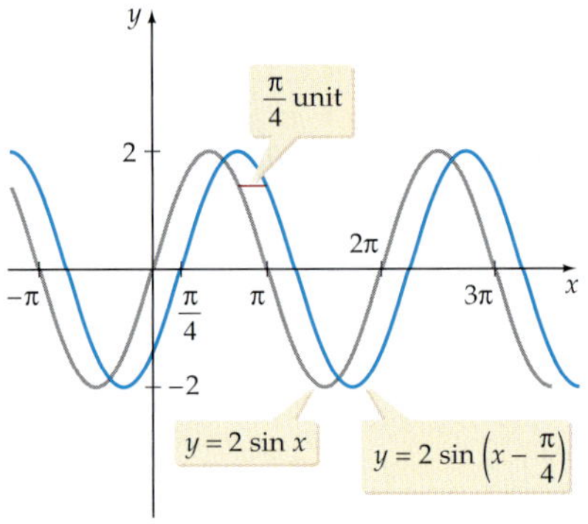

FIGURE 4.77

Because one cycle of $y = a \sin x$ is completed for $0 \le x \le 2\pi$, one cycle of the graph of $y = a\sin(bx + c)$, where $b > 0$, is completed for $0 \le bx + c \le 2\pi$. Solving this inequality for x, we have

$$0 \le bx + c \le 2\pi$$

$$-c \le bx \le -c + 2\pi$$

$$-\frac{c}{b} \le x \le -\frac{c}{b} + \frac{2\pi}{b}$$

The number $-\frac{c}{b}$ is the phase shift for $y = a\sin(bx + c)$. The graph of the equation $y = a\sin(bx + c)$ is the graph of $y = a \sin bx$ shifted $-\frac{c}{b}$ units horizontally. Similar arguments apply to the remaining trigonometric functions.

The Graphs of $y = a\sin(bx + c)$ and $y = a\cos(bx + c)$

The graphs of $y = a\sin(bx + c)$ and $y = a\cos(bx + c)$, with $b > 0$, have

$$\text{Amplitude: } |a| \qquad \text{Period: } \frac{2\pi}{b} \qquad \text{Phase shift: } -\frac{c}{b}$$

One cycle of each graph is completed on the interval

$$-\frac{c}{b} \le x \le -\frac{c}{b} + \frac{2\pi}{b}$$

QUESTION What is the phase shift of the graph of $y = 3\sin\left(\frac{1}{2}x - \frac{\pi}{6}\right)$?

EXAMPLE 1 Graph $y = a\cos(bx + c)$

Graph: $y = 3\cos\left(2x + \frac{\pi}{3}\right)$

Solution

The phase shift is $-\frac{c}{b} = -\frac{\pi/3}{2} = -\frac{\pi}{6}$. The graph of the equation $y = 3\cos\left(2x + \frac{\pi}{3}\right)$ is the graph of $y = 3\cos 2x$ shifted $\frac{\pi}{6}$ units to the left, as shown in **Figure 4.78.**

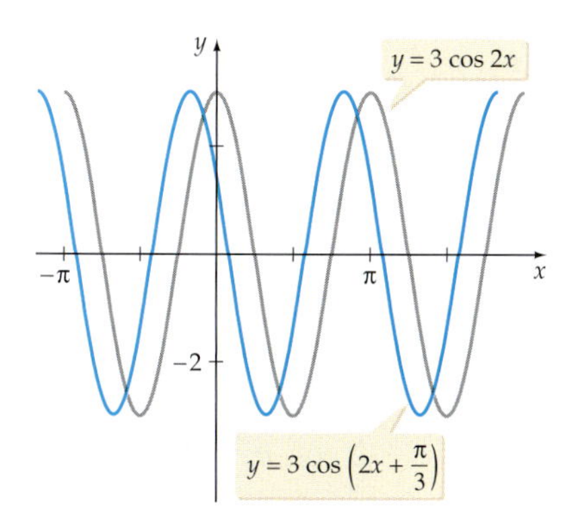

FIGURE 4.78

TRY EXERCISE 20, PAGE 346

EXAMPLE 2 Graph $y = a\cot(bx + c)$

Graph: $y = 2\cot(3x - 2)$

Solution

The phase shift is

$$-\frac{c}{b} = -\frac{-2}{3} = \frac{2}{3} \qquad \bullet\ 3x - 2 = 3x + (-2)$$

The graph of $y = 2\cot(3x - 2)$ is the graph of $y = 2\cot(3x)$ shifted $\frac{2}{3}$ unit to the right, as shown in **Figure 4.79.**

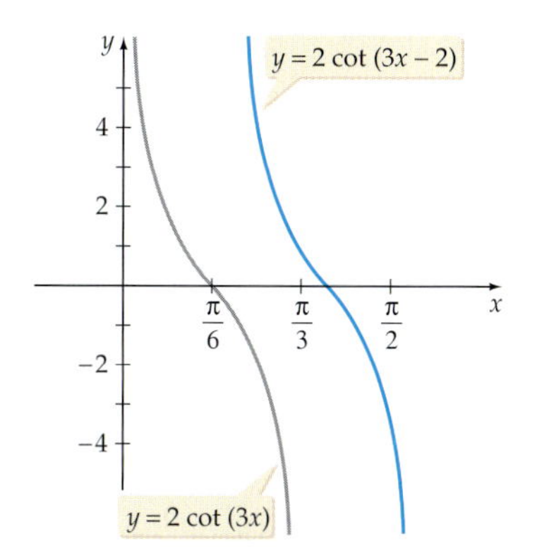

FIGURE 4.79

TRY EXERCISE 22, PAGE 346

ANSWER $\frac{\pi}{3}$

The graph of a trigonometric function may be the combination of a vertical translation and a phase shift.

EXAMPLE 3 Graph $y = a\sin(bx + c) + d$

Graph: $y = \dfrac{1}{2}\sin\left(x - \dfrac{\pi}{4}\right) - 2$

Solution

The phase shift is $-\dfrac{c}{b} = -\dfrac{-\pi/4}{1} = \dfrac{\pi}{4}$. The vertical shift is 2 units down.

The graph of $y = \dfrac{1}{2}\sin\left(x - \dfrac{\pi}{4}\right) - 2$ is the graph of $y = \dfrac{1}{2}\sin x$ shifted $\dfrac{\pi}{4}$ units to the right and 2 units down, as shown in **Figure 4.80.**

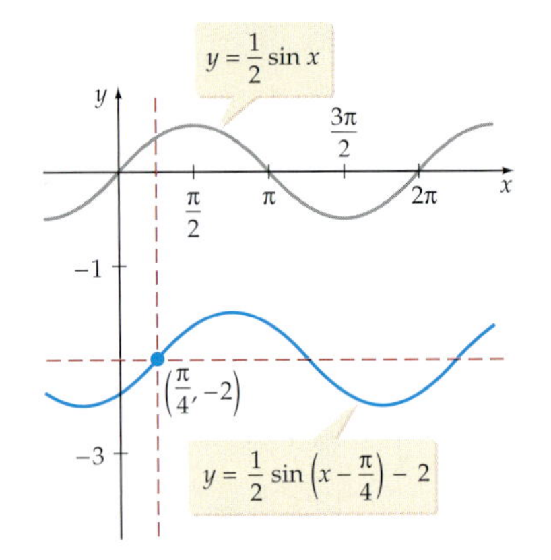

FIGURE 4.80

TRY EXERCISE 40, PAGE 346

EXAMPLE 4 Graph $y = a\cos(bx + c) + d$

Graph: $y = -2\cos\left(\pi x + \dfrac{\pi}{2}\right) + 1$

Solution

The phase shift is $-\dfrac{c}{b} = -\dfrac{\pi/2}{\pi} = -\dfrac{1}{2}$. The vertical shift is 1 unit up. The graph of $y = -2\cos\left(\pi x + \dfrac{\pi}{2}\right) + 1$ is the graph of $y = -2\cos \pi x$ shifted $\dfrac{1}{2}$ unit to the left and 1 unit up, as shown in **Figure 4.81.**

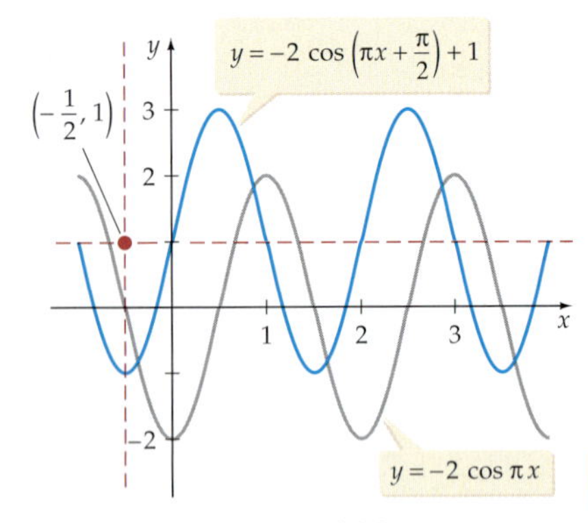

FIGURE 4.81

TRY EXERCISE 42, PAGE 346

The following example involves a function of the form $y = \cos(bx + c) + d$.

EXAMPLE 5 A Mathematical Model of a Patient's Blood Pressure

The function $bp(t) = 32\cos\left(\dfrac{10\pi}{3}t - \dfrac{\pi}{3}\right) + 112$, $0 \le t \le 20$, gives the blood pressure, in millimeters of mercury (mm Hg), of a patient during a 20-second interval.

a. Find the phase shift and the period of bp.

b. Graph one period of bp.

c. What are the patient's maximum (*systolic*) and minimum (*diastolic*) blood pressure readings during the given time interval?

d. What is the patient's pulse rate, in beats per minute?

Solution

a. $$\text{Phase shift} = -\frac{c}{b} = -\frac{\left(-\frac{\pi}{3}\right)}{\left(\frac{10\pi}{3}\right)} = 0.1$$

$$\text{Period} = \frac{2\pi}{b} = \frac{2\pi}{\left(\frac{10\pi}{3}\right)} = 0.6 \text{ second}$$

b. The graph of bp is the graph of $y_1 = 32\cos\left(\frac{10\pi}{3}t\right)$ shifted 0.1 unit to the right, shown by $y_2 = 32\cos\left(\frac{10\pi}{3}t - \frac{\pi}{3}\right)$ in **Figure 4.82,** and upward 112 units.

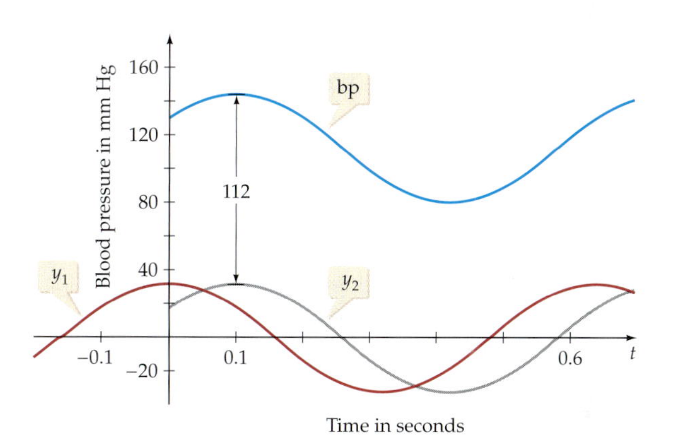

Figure 4.82

c. The function $y_2 = 32\cos\left(\frac{10\pi}{3}t - \frac{\pi}{3}\right)$ has a maximum of 32 and a minimum of -32. Thus the patient's maximum blood pressure is $32 + 112 = 144$ mm Hg, and the patient's minimum blood pressure is $-32 + 112 = 80$ mm Hg.

d. From part **a.** we know that the patient has 1 heartbeat every 0.6 second. Therefore, during the given time interval, the patient has a pulse rate of

$$\left(\frac{1 \text{ heartbeat}}{0.6 \text{ second}}\right)\left(\frac{60 \text{ seconds}}{1 \text{ minute}}\right) = 100 \text{ heartbeats per minute}$$

Try Exercise 52, page 347

Translation techniques also can be used to graph secant and cosecant functions.

EXAMPLE 6 Graph a Cosecant Function

Graph: $y = 2\csc(2x - \pi)$

Solution

The phase shift is $-\dfrac{c}{b} = -\dfrac{-\pi}{2} = \dfrac{\pi}{2}$. The graph of the equation $y = 2\csc(2x - \pi)$ is the graph of $y = 2\csc 2x$ shifted $\dfrac{\pi}{2}$ units to the right. Sketch the graph of the equation $y = 2\sin 2x$ shifted $\dfrac{\pi}{2}$ units to the right. Use this graph to draw the graph of $y = 2\csc(2x - \pi)$, as shown in **Figure 4.83.**

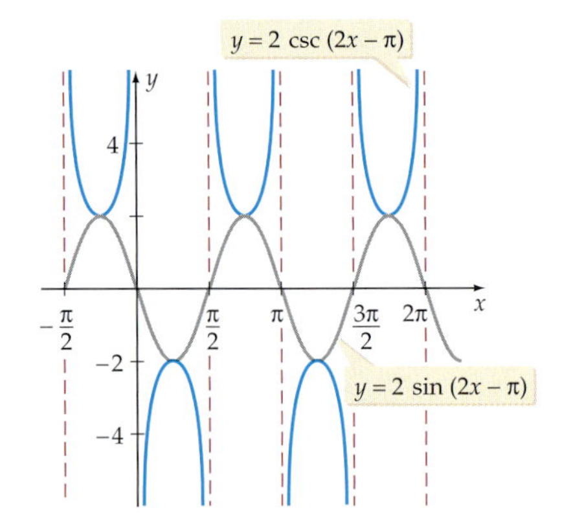

FIGURE 4.83

▶ TRY EXERCISE 48, PAGE 347

ADDITION OF ORDINATES

Given two functions g and h, the sum of the functions is the function f defined by $f(x) = g(x) + h(x)$. The graph of the sum f can be obtained by graphing g and h separately and then geometrically adding the y-coordinates of each function for a given value of x. It is convenient, when we are drawing the graph of the sum of two functions, to pick zeros of the function.

EXAMPLE 7 Graph the Sum of Two Functions

Graph: $y = x + \cos x$

Solution

Graph $g(x) = x$ and $h(x) = \cos x$ on the same coordinate grid. Then add the y-coordinates geometrically point by point. **Figure 4.84** shows the results of adding, by using a ruler, the y-coordinates of the two functions for selected values of x.

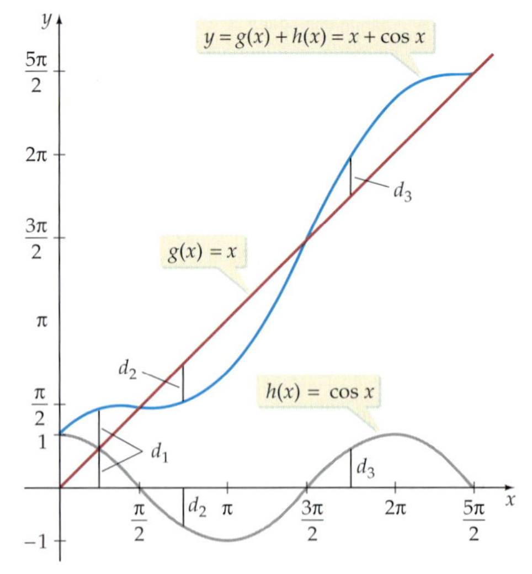

FIGURE 4.84

▶ TRY EXERCISE 54, PAGE 347

EXAMPLE 8 Graph the Difference of Two Functions

Graph $y = \sin x - \cos x$ for $0 \le x \le 2\pi$.

Solution

Graph $g(x) = \sin x$ and $h(x) = -\cos x$ on the same coordinate grid. For selected values of x, add $g(x)$ and $h(x)$ geometrically. Now draw a smooth curve through the points. See **Figure 4.85**.

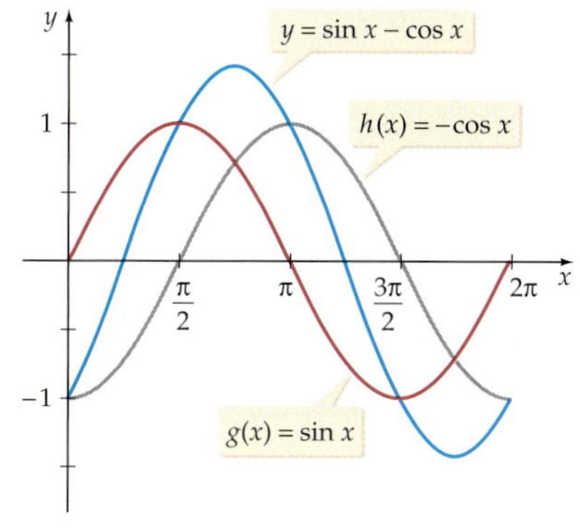

FIGURE 4.85

▶ TRY EXERCISE 58, PAGE 347

DAMPING FACTOR

The factor $\frac{1}{4}x$ in $f(x) = \frac{1}{4}x \cos x$ is referred to as the *damping factor*. In the next example we analyze the role of the damping factor.

EXAMPLE 9 Graph the Product of Two Functions

Use a graphing utility to graph $f(x) = \frac{1}{4}x \cos x$, $x \ge 0$, and analyze the role of the damping factor.

Solution

Figure 4.86 shows that the graph of f intersects

- the graph of $y = \frac{1}{4}x$ for $x = 0, 2\pi, 4\pi, \ldots$ • **Because $\cos x = 1$ for $x = 2n\pi$**
- the graph of $y = -\frac{1}{4}x$ for $x = \pi, 3\pi, 5\pi, \ldots$ • **Because $\cos x = -1$ for $x = (2n - 1)\pi$**
- the x-axis for $x = \frac{1}{2}\pi, \frac{3}{2}\pi, \frac{5}{2}\pi, \ldots$ • **Because $\cos x = 0$ for $x = \frac{2n-1}{2}\pi$**

FIGURE 4.86

Figure 4.86 also shows that the graph of f lies on or between the lines

$$y = \frac{1}{4}x \quad \text{and} \quad y = -\frac{1}{4}x$$ • **Because $|\cos x| \le 1$ for all x**

▶ TRY EXERCISE 78, PAGE 349

take note

Replacing the damping factor can make a dramatic change in the graph of a function. For instance, the graph of $f(x) = 2^{-0.1x} \cos x$ approaches 0 as x approaches ∞.

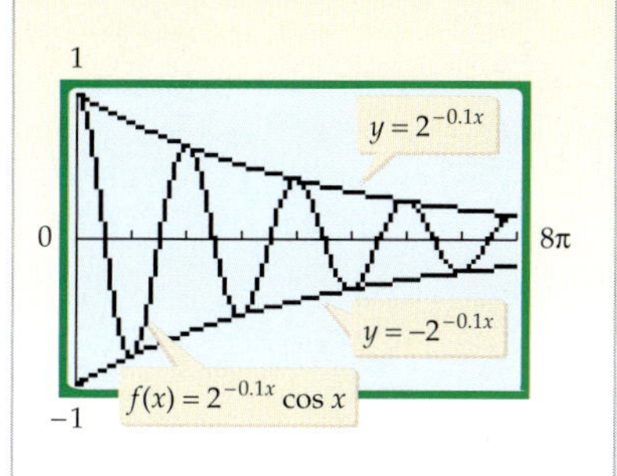

TOPICS FOR DISCUSSION

1. The maximum value of $f(x) = (3 \sin x) + 4$ is 7. Thus f has an amplitude of 7. Do you agree? Explain.
2. The graph of $y = \sec x$ has a period of 2π. What is the period of the graph of $y = |\sec x|$?

3. The zeros of $y = \sin x$ are the same as the zeros of $y = x \sin x$. Do you agree? Explain.

4. What is the phase shift of the graph of $y = \tan\left(3x - \frac{\pi}{6}\right)$?

EXERCISE SET 4.5

In Exercises 1 to 8, find the amplitude, phase shift, and period for the graph of each function.

1. $y = 2\sin\left(x - \frac{\pi}{2}\right)$
2. $y = -3\sin(x + \pi)$
3. $y = \cos\left(2x - \frac{\pi}{4}\right)$
4. $y = \frac{3}{4}\cos\left(\frac{x}{2} + \frac{\pi}{3}\right)$
5. $y = -4\sin\left(\frac{2x}{3} + \frac{\pi}{6}\right)$
6. $y = \frac{3}{2}\sin\left(\frac{x}{4} - \frac{3\pi}{4}\right)$
7. $y = \frac{5}{4}\cos(3x - 2\pi)$
8. $y = 6\cos\left(\frac{x}{3} - \frac{\pi}{6}\right)$

In Exercises 9 to 16, find the phase shift and the period for the graph of each function.

9. $y = 2\tan\left(2x - \frac{\pi}{4}\right)$
10. $y = \frac{1}{2}\tan\left(\frac{x}{2} - \pi\right)$
11. $y = -3\csc\left(\frac{x}{3} + \pi\right)$
12. $y = -4\csc\left(3x - \frac{\pi}{6}\right)$
13. $y = 2\sec\left(2x - \frac{\pi}{8}\right)$
14. $y = 3\sec\left(\frac{x}{4} - \frac{\pi}{2}\right)$
15. $y = -3\cot\left(\frac{x}{4} + 3\pi\right)$
16. $y = \frac{3}{2}\cot\left(2x - \frac{\pi}{4}\right)$

In Exercises 17 to 32, graph one full period of each function.

17. $y = \sin\left(x - \frac{\pi}{2}\right)$
18. $y = \sin\left(x + \frac{\pi}{6}\right)$
19. $y = \cos\left(\frac{x}{2} + \frac{\pi}{3}\right)$
▶ **20.** $y = \cos\left(2x - \frac{\pi}{3}\right)$
21. $y = \tan\left(x + \frac{\pi}{4}\right)$
▶ **22.** $y = \tan(x - \pi)$
23. $y = 2\cot\left(\frac{x}{2} - \frac{\pi}{8}\right)$
24. $y = \frac{3}{2}\cot\left(3x + \frac{\pi}{4}\right)$
25. $y = \sec\left(x + \frac{\pi}{4}\right)$
26. $y = \csc(2x + \pi)$
27. $y = \csc\left(\frac{x}{3} - \frac{\pi}{2}\right)$
28. $y = \sec\left(2x + \frac{\pi}{6}\right)$
29. $y = -2\sin\left(\frac{x}{3} - \frac{2\pi}{3}\right)$
30. $y = -\frac{3}{2}\sin\left(2x + \frac{\pi}{4}\right)$
31. $y = -3\cos\left(3x + \frac{\pi}{4}\right)$
32. $y = -4\cos\left(\frac{3x}{2} + 2\pi\right)$

In Exercises 33 to 50, graph each function using translations.

33. $y = \sin x + 1$
34. $y = -\sin x + 1$
35. $y = -\cos x - 2$
36. $y = 2\sin x + 3$
37. $y = \sin 2x - 2$
38. $y = -\cos\frac{x}{2} + 2$
39. $y = 4\cos(\pi x - 2) + 1$
▶ **40.** $y = 2\sin\left(\frac{\pi x}{2} + 1\right) - 2$
41. $y = -\sin(\pi x + 1) - 2$
▶ **42.** $y = -3\cos(2\pi x - 3) + 1$
43. $y = \sin\left(x - \frac{\pi}{2}\right) - \frac{1}{2}$
44. $y = -2\cos\left(x + \frac{\pi}{3}\right) + 3$

45. $y = \tan\dfrac{x}{2} - 4$

46. $y = \cot 2x + 3$

47. $y = \sec 2x - 2$

▶ **48.** $y = \csc\dfrac{x}{3} + 4$

49. $y = \csc\dfrac{x}{2} - 1$

50. $y = \sec\left(x - \dfrac{\pi}{2}\right) + 1$

51. **RETAIL SALES** The manager of a major department store finds that the number of men's suits S, in hundreds, that the store sells is given by

$$S = 4.1\cos\left(\frac{\pi}{6}t - 1.25\pi\right) + 7$$

where t is time measured in months, with $t = 0$ representing January 1.

a. Find the phase shift and the period of S.

b. Graph one period of S.

c. Use the graph to determine in which month the store sells the most suits.

▶ **52.** **RETAIL SALES** The owner of a shoe store finds that the number of pairs of shoes S, in hundreds, that the store sells can be modeled by the function

$$S = 2.7\cos\left(\frac{\pi}{6}t - \frac{7}{12}\pi\right) + 4$$

where t is time measured in months, with $t = 0$ representing January 1.

a. Find the phase shift and the period of S.

b. Graph one period of S.

c. Use the graph to determine in which month the store sells the most shoes.

In Exercises 53 to 58, graph the given function by using the addition-of-ordinates method.

53. $y = x - \sin x$

▶ **54.** $y = \dfrac{x}{2} + \cos x$

55. $y = x + \sin 2x$

56. $y = \dfrac{2x}{3} - \sin x$

57. $y = \sin x + \cos x$

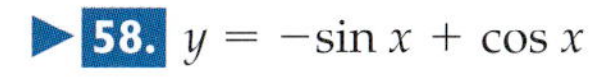

▶ **58.** $y = -\sin x + \cos x$

In Exercises 59 to 64, find an equation of each blue graph.

59.

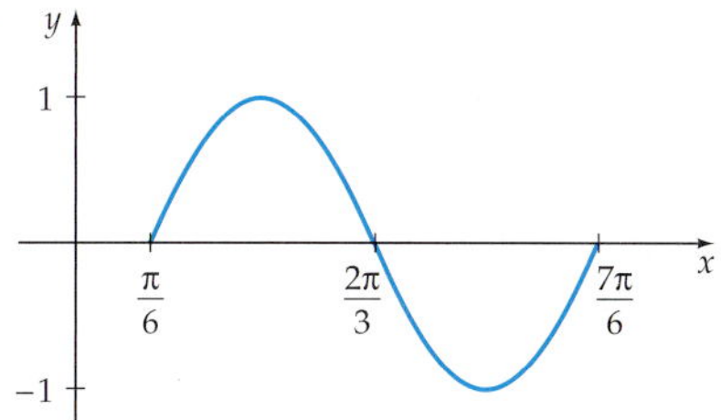

60.

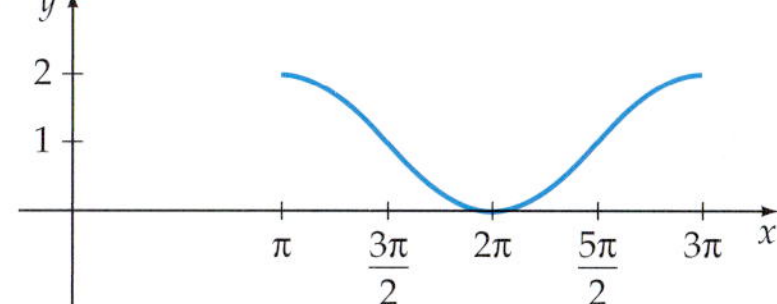

61.

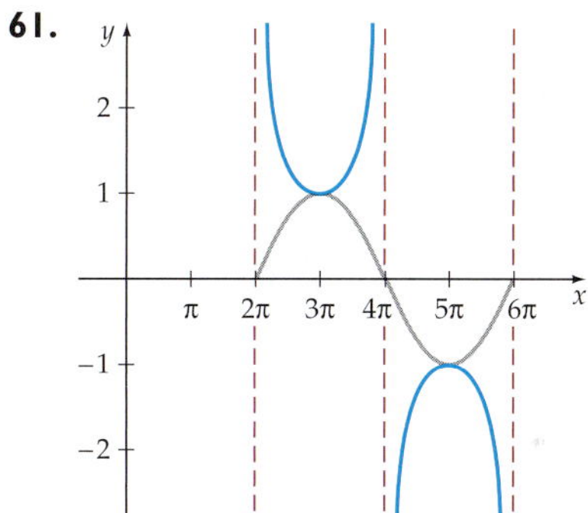

62.

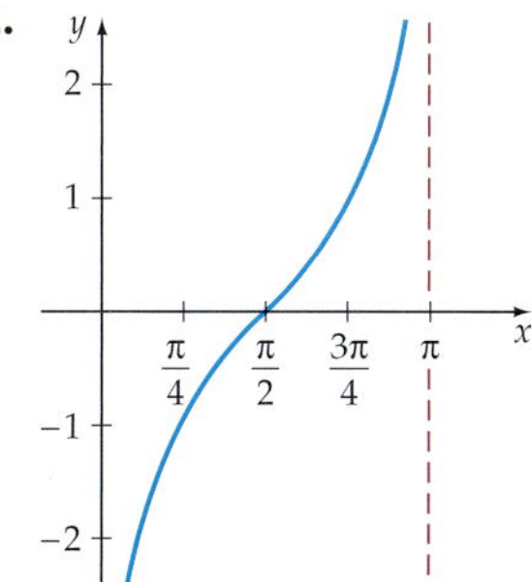

63.

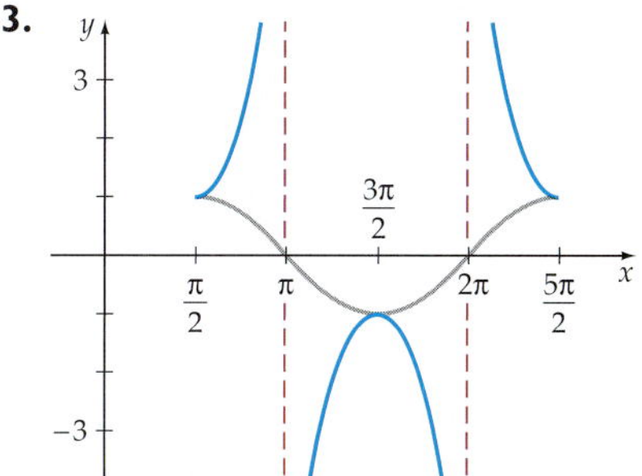

64.

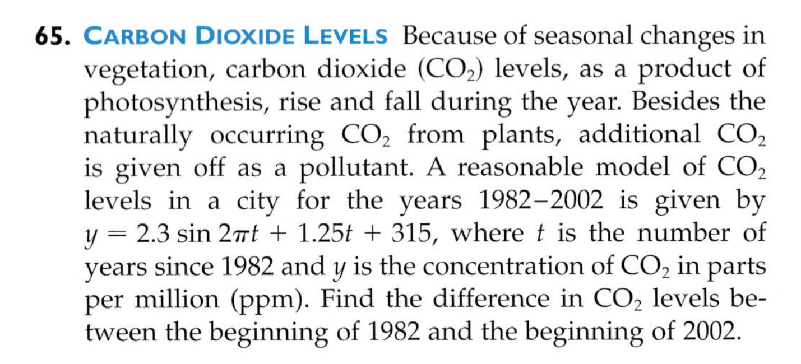

65. Carbon Dioxide Levels Because of seasonal changes in vegetation, carbon dioxide (CO_2) levels, as a product of photosynthesis, rise and fall during the year. Besides the naturally occurring CO_2 from plants, additional CO_2 is given off as a pollutant. A reasonable model of CO_2 levels in a city for the years 1982–2002 is given by $y = 2.3 \sin 2\pi t + 1.25t + 315$, where t is the number of years since 1982 and y is the concentration of CO_2 in parts per million (ppm). Find the difference in CO_2 levels between the beginning of 1982 and the beginning of 2002.

66. Environmental Science Some environmentalists contend that the rate of growth of atmospheric CO_2 in parts per million is given by the equation $y = 2.54e^{0.112t} + \sin 2\pi t + 315$. See Exercise 65. Use this model to find the difference in CO_2 levels from the beginning of 1982 to the beginning of 2002.

67. Height of a Paddle The paddle wheel on a river boat is shown in the accompanying figure. Write an equation for the height of a paddle relative to the water at time t. The radius of the paddle wheel is 7 feet, and the distance from the center of the paddle wheel to the water is 5 feet. Assume that the paddle wheel rotates at 5 revolutions per minute and that the paddle is at its highest point at $t = 0$. Graph the equation for $0 \leq t \leq 0.20$ minute.

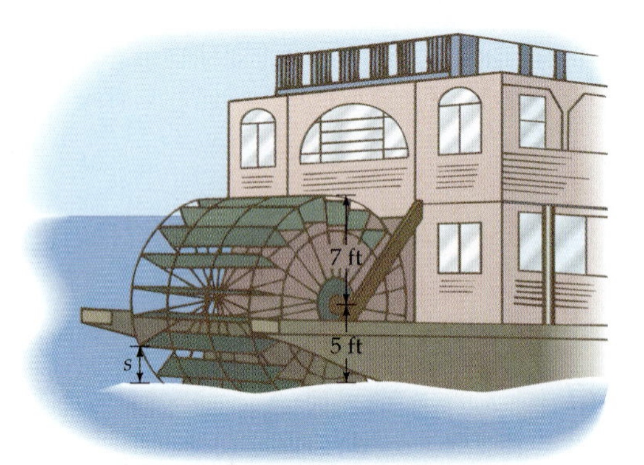

68. Voltage and Amperage The graphs of the voltage and the amperage of an alternating household circuit are shown in the following figures, where t is measured in seconds. Note that there is a phase shift between the graph of the voltage and the graph of the current. The current is said to *lag* the voltage by 0.005 second. Write an equation for the voltage and an equation for the current.

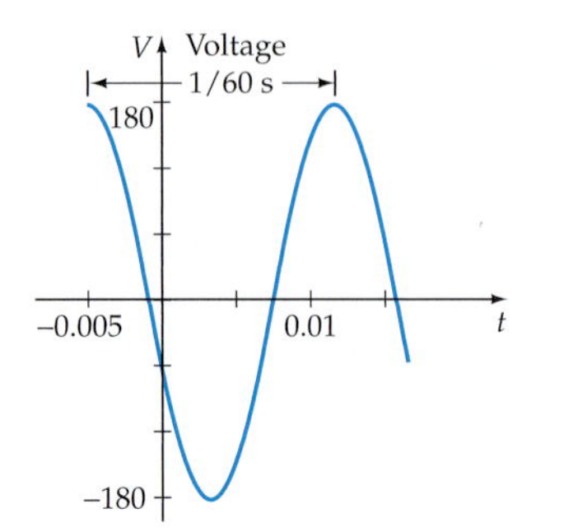

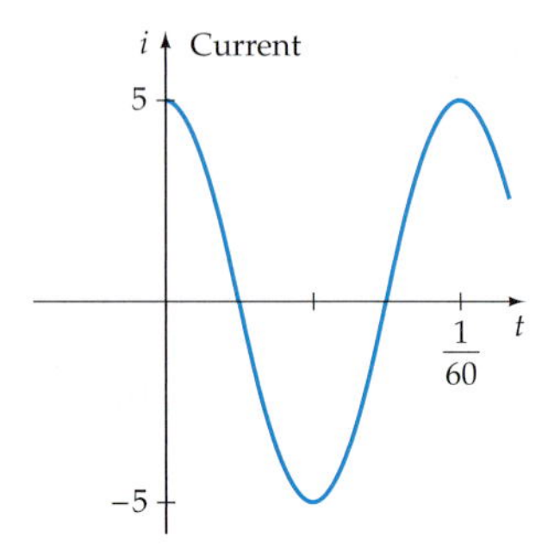

69. A Lighthouse Beacon The beacon of a lighthouse 400 meters from a straight sea wall rotates at 6 revolutions per minute. Using the accompanying figures, write an equation expressing the distance s, measured in meters, in terms of time t. Assume that when $t = 0$, the beam is perpendicular to the sea wall. Sketch a graph of the equation for $0 \leq t \leq 10$ seconds.

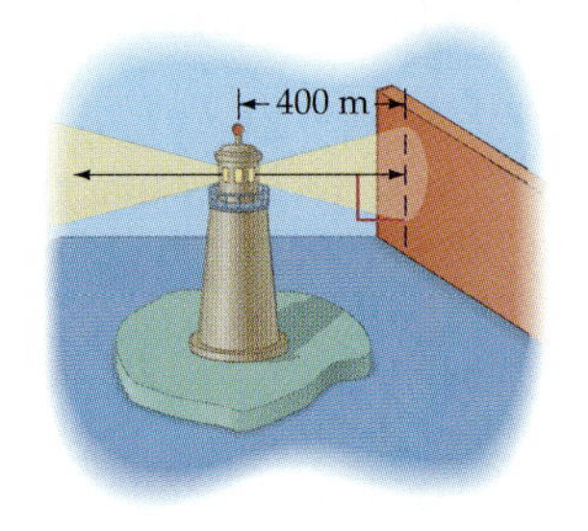

Side view

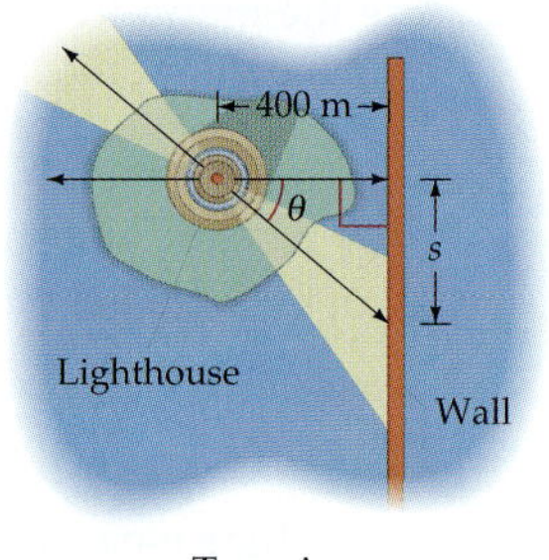

Top view

70. Hours of Daylight The duration of daylight for a region is dependent not only on the time of year but also on the latitude of the region. The graph gives the daylight hours for a one-year period at various latitudes. Assuming that a sine function can model these curves, write an equation for each curve.

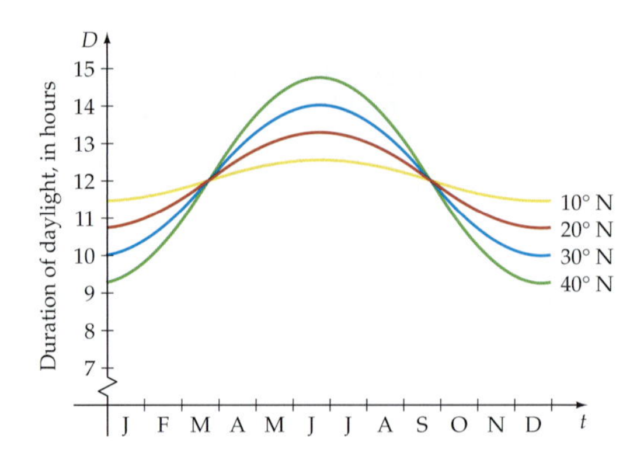

71. **TIDES** During a 24-hour day, the tides raise and lower the depth of water at a pier as shown in the figure below. Write an equation in the form $f(t) = A \cos Bt + d$, and find the depth of the water at 6 P.M.

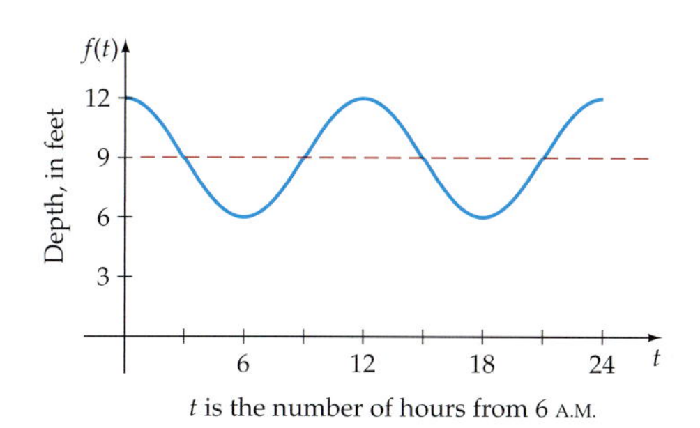

t is the number of hours from 6 A.M.

72. **TEMPERATURE** During a summer day, the ground temperature at a desert location was recorded and graphed as a function of time, as shown in the following figure. The graph can be approximated by $f(t) = A \cos(bt + c) + d$. Find the equation, and approximate the temperature (to the nearest degree) at 1:00 P.M.

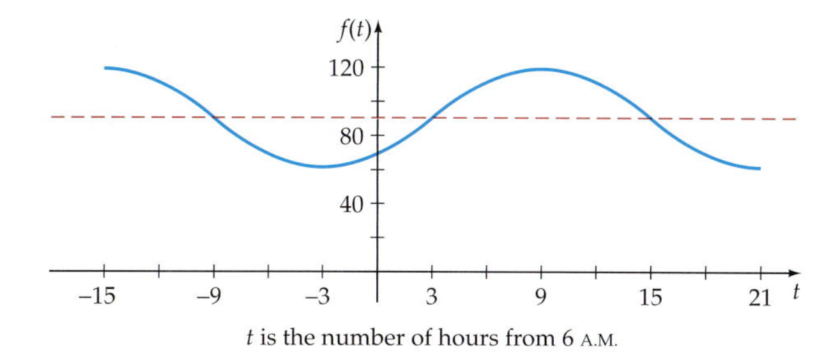

t is the number of hours from 6 A.M.

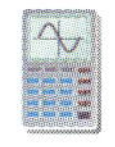

In Exercises 73 to 82, use a graphing utility to graph each function.

73. $y = \sin x - \cos \dfrac{x}{2}$

74. $y = 2 \sin 2x - \cos x$

75. $y = 2 \cos x + \sin \dfrac{x}{2}$

76. $y = -\dfrac{1}{2} \cos 2x + \sin \dfrac{x}{2}$

77. $y = \dfrac{x}{2} \sin x$

78. $y = x \cos x$

79. $y = x \sin \dfrac{x}{2}$

80. $y = \dfrac{x}{2} \cos \dfrac{x}{2}$

81. $y = x \sin\left(x + \dfrac{\pi}{2}\right)$

82. $y = x \cos\left(x - \dfrac{\pi}{2}\right)$

BEATS When two sound waves have approximately the same frequency, the sound waves interfere with one another and produce phenomena called *beats*, which are heard as variations in the loudness of the sound. A piano tuner can use these phenomena to tune a piano. By striking a tuning fork and then tapping the corresponding key on a piano, the piano tuner listens for beats and adjusts the tension in the string until the beats disappear. Use a graphing utility to graph the functions in Exercises 83 to 86, which are based on beats.

83. $y = \sin(5\pi x) \cdot \sin\left(-\dfrac{\pi}{2} x\right)$

84. $y = \sin(9\pi x) \cdot \sin\left(-\dfrac{\pi}{2} x\right)$

85. $y = \sin(13\pi x) \cdot \sin\left(-\dfrac{\pi}{2} x\right)$

86. $y = \sin(17\pi x) \cdot \sin\left(-\dfrac{\pi}{2} x\right)$

CONNECTING CONCEPTS

87. Find an equation of the sine function with amplitude 2, period π, and phase shift $\dfrac{\pi}{3}$.

88. Find an equation of the cosine function with amplitude 3, period 3π, and phase shift $-\dfrac{\pi}{4}$.

89. Find an equation of the tangent function with period 2π and phase shift $\dfrac{\pi}{2}$.

90. Find an equation of the cotangent function with period $\dfrac{\pi}{2}$ and phase shift $-\dfrac{\pi}{4}$.

91. Find an equation of the secant function with period 4π and phase shift $\dfrac{3\pi}{4}$.

92. Find an equation of the cosecant function with period $\dfrac{3\pi}{2}$ and phase shift $\dfrac{\pi}{4}$.

93. If $g(x) = \sin^2 x$ and $h(x) = \cos^2 x$, find $g(x) + h(x)$.

94. If $g(x) = 2 \sin x - 3$ and $h(x) = 4 \cos x + 2$, find the sum $g(x) + h(x)$.

95. If $g(x) = x^2 + 2$ and $h(x) = \cos x$, find $g[h(x)]$.

96. If $g(x) = \sin x$ and $h(x) = x^2 + 2x + 1$, find $h[g(x)]$.

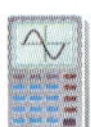

In Exercises 97 to 100, use a graphing utility to graph each function.

97. $y = \dfrac{\sin x}{x}$

98. $y = 2 + \sec \dfrac{x}{2}$

99. $y = |x| \sin x$

100. $y = |x| \cos x$

PROJECTS

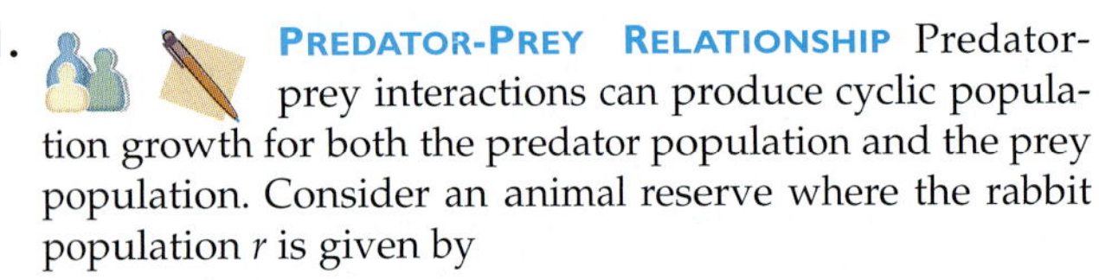

1. **PREDATOR-PREY RELATIONSHIP** Predator-prey interactions can produce cyclic population growth for both the predator population and the prey population. Consider an animal reserve where the rabbit population r is given by

$$r(t) = 850 + 210 \sin\left(\frac{\pi}{6}t\right)$$

and the wolf population w is given by

$$w(t) = 120 + 30 \sin\left(\frac{\pi}{6}t - 2\right)$$

where t is the number of months after March 1, 2000. Graph $r(t)$ and $w(t)$ on the same coordinate system for $0 \le t \le 24$. Write a few sentences that explains a possible relationship between the two populations.

Write an equation that could be used to model the rabbit population shown by the graph at the right, where t is measured in months.

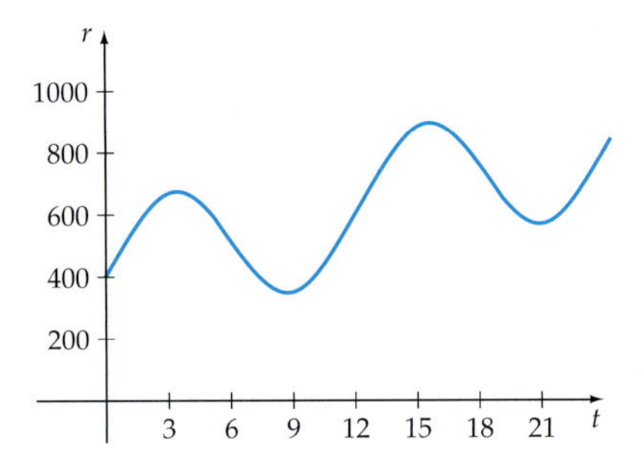

EXPLORING CONCEPTS WITH TECHNOLOGY

Sinusoidal Families

Some graphing calculators have a feature that allows you to graph a family of functions easily. For instance, entering Y1={2,4,6}sin(X) in the Y= menu and pressing the GRAPH key on a TI-83 calculator produces a graph of the three functions $y = 2 \sin x$, $y = 4 \sin x$, and $y = 6 \sin x$, all displayed in the same window.

1. Use a graphing calculator to graph Y1={2,4,6}sin(X). Write a sentence that indicates the similarities and the differences among the three graphs.

2. Use a graphing calculator to graph Y1=sin({π,2π,4π}X). Write a sentence that indicates the similarities and the differences among the three graphs.

3. Use a graphing calculator to graph Y1=sin(X+{π/4,π/6,π/12}). Write a sentence that indicates the similarities and the differences among the three graphs.

4. A student has used a graphing calculator to graph Y1=sin(X+{π,3π,5π}) and expects to see three graphs. However, the student sees only one graph displayed on the graph window. Has the calculator displayed all three graphs? Explain.

CHAPTER 4 SUMMARY

4.1 Angles and Arcs

- An angle is in standard position when its initial side is along the positive x-axis and its vertex is at the origin of the coordinate axes.
- Angle α is an acute angle when $0° < \alpha < 90°$; it is an obtuse angle when $90° < \alpha < 180°$.
- α and β are complementary angles when $\alpha + \beta = 90°$; they are supplementary angles when $\alpha + \beta = 180°$.
- The length of the arc s that subtends the central angle θ (in radians) on a circle of radius r is given by $s = r\theta$.
- Angular speed is given by $\omega = \dfrac{\theta}{t}$.

4.2 Trigonometric Functions of Real Numbers

- The wrapping function pairs a real number with a point on the unit circle.
- Let W be the wrapping function, t be a real number, and $W(t) = P(x, y)$. Then the trigonometric functions of the real number t are defined as follows:

$$\sin t = y \qquad \csc t = \frac{1}{y}, \quad y \neq 0$$

$$\cos t = x \qquad \sec t = \frac{1}{x}, \quad x \neq 0$$

$$\tan t = \frac{y}{x}, \quad x \neq 0 \qquad \cot t = \frac{x}{y}, \quad y \neq 0$$

- $\sin t$, $\csc t$, $\tan t$, and $\cot t$ are odd functions.
- $\cos t$ and $\sec t$ are even functions.
- $\sin t$, $\cos t$, $\sec t$, and $\csc t$ have period 2π.
- $\tan t$ and $\cot t$ have period π.

Domain and Range of Each Trigonometric Function (n is an integer)

Function	Domain	Range
$\sin t$	$\{t \mid -\infty < t < \infty\}$	$\{y \mid -1 \le y \le 1\}$
$\cos t$	$\{t \mid -\infty < t < \infty\}$	$\{y \mid -1 \le y \le 1\}$
$\tan t$	$\left\{t \mid -\infty < t < \infty, \ t \neq \dfrac{(2n+1)\pi}{2}\right\}$	$\{y \mid -\infty < y < \infty\}$
$\csc t$	$\{t \mid -\infty < t < \infty, \ t \neq n\pi\}$	$\{y \mid y \ge 1, y \le -1\}$
$\sec t$	$\left\{t \mid -\infty < t < \infty, \ t \neq \dfrac{(2n+1)\pi}{2}\right\}$	$\{y \mid y \ge 1, y \le -1\}$
$\cot t$	$\{t \mid -\infty < t < \infty, \ t \neq n\pi\}$	$\{y \mid -\infty < y < \infty\}$

4.3 Graphs of the Sine and Cosine Functions

- The graphs of $y = a \sin bx$ and $y = a \cos bx$ both have an amplitude of $|a|$ and a period of $\dfrac{2\pi}{|b|}$. The graph of each for $a > 0$ and $b > 0$ is given below.

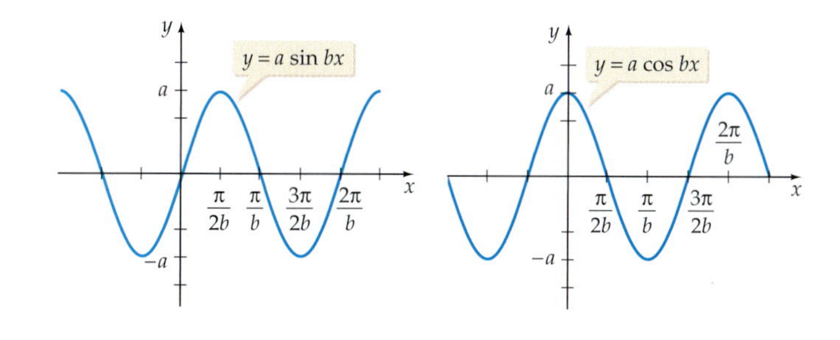

4.4 Graphs of the Other Trigonometric Functions

- The period of $y = a \tan bx$ and $y = a \cot bx$ is $\dfrac{\pi}{|b|}$.
- The period of $y = a \sec bx$ and $y = a \csc bx$ is $\dfrac{2\pi}{|b|}$.

4.5 Graphing Techniques

- Phase shift is a horizontal translation of the graph of a trigonometric function. If $y = f(bx + c)$, where f is a trigonometric function, then the phase shift is $-\dfrac{c}{b}$.
- The graphs of $y = a \sin(bx + c)$ and $y = a \cos(bx + c)$, $b > 0$, have amplitude $|a|$, period $\dfrac{2\pi}{b}$, and phase shift $-\dfrac{c}{b}$. One cycle of each graph is completed on the interval $-\dfrac{c}{b} \le x \le -\dfrac{c}{b} + \dfrac{2\pi}{b}$.
- Addition of ordinates is a method of graphing the sum of two functions by graphically adding the values of their y-coordinates.
- The factor $g(x)$ in $f(x) = g(x) \cos x$ is called a damping factor. The graph of f lies on or between the graphs of the equations $y = g(x)$ and $y = -g(x)$.

CHAPTER 4 TRUE/FALSE EXERCISES

In Exercises 1 to 10, answer true or false. If the statement is false, give a reason or state an example to show that the statement is false.

1. An angle is in standard position when the vertex is at the origin of a coordinate system.
2. In the formula $s = r\theta$, the angle θ must be measured in radians.
3. $\sec^2 t + \tan^2 t = 1$ is an identity.
4. The amplitude of the graph of $y = 2 \tan x$ is 2.
5. The period of $y = \cos x$ is π.
6. The graph of $y = \sin x$ is symmetric to the origin.
7. $\sin^2 x = \sin x^2$
8. The phase shift of $f(x) = 2 \sin\left(2x - \dfrac{\pi}{3}\right)$ is $\dfrac{\pi}{3}$.
9. The measure of one radian is more than 50 times the measure of one degree.
10. The graph of $y = 2^{-x} \cos x$ lies on or between the graphs of $y = 2^{-x}$ and $y = 2^{x}$.

CHAPTER 4 REVIEW EXERCISES

1. Find the complement and supplement of the angle θ whose measure is 65°.
2. Convert 2 radians to the nearest hundredth of a degree.
3. Convert 315° to radian measure.
4. Find the length (to the nearest 0.01 meter) of the arc on a circle of radius 3 meters that subtends an angle of 75°.
5. Find the radian measure of the angle subtended by an arc of length 12 centimeters on a circle whose radius is 40 centimeters.
6. A car with a 16-inch-radius wheel is moving with a speed of 50 mph. Find the angular speed (to the nearest radian per second) of the wheel in radians per second.

In Exercises 7 to 16, find $W(t)$ for the given t.

7. $t = \dfrac{3\pi}{2}$
8. $t = -\dfrac{3\pi}{4}$
9. $t = \dfrac{7\pi}{6}$
10. $t = \dfrac{4\pi}{3}$

11. $t = -\frac{5\pi}{6}$

12. $t = \frac{9\pi}{4}$

13. $t = \frac{5\pi}{2}$

14. $t = \frac{8\pi}{3}$

15. $t = -\frac{\pi}{4}$

16. $t = \frac{15\pi}{4}$

In Exercises 17 to 30, find the exact value of the trigonometric function. If the value is undefined, so state.

17. $\cos\left(\frac{3\pi}{4}\right)$

18. $\sin\left(\frac{3\pi}{4}\right)$

19. $\csc\left(\frac{2\pi}{3}\right)$

20. $\tan\left(\frac{5\pi}{4}\right)$

21. $\sec\left(\frac{\pi}{4}\right)$

22. $\cot\left(\frac{\pi}{4}\right)$

23. $\sin\left(\frac{4\pi}{3}\right)$

24. $\cos\left(\frac{4\pi}{3}\right)$

25. $\tan\left(\frac{13\pi}{6}\right)$

26. $\cot\left(\frac{13\pi}{6}\right)$

27. $\sec\left(\frac{5\pi}{6}\right)$

28. $\csc\left(-\frac{\pi}{6}\right)$

29. $\sin\left(\frac{7\pi}{2}\right)$

30. $\cos\left(\frac{3\pi}{2}\right)$

In Exercises 31 to 36, determine whether the function is an even function, an odd function, or neither.

31. $f(t) = t \sin t$

32. $g(t) = \sin t + \tan t$

33. $s(t) = \frac{t}{\cos t}$

34. $r(t) = 2 \sec t \csc t$

35. $G(t) = t + 2 \cos t$

36. $F(t) = \sec t + 2 \csc t$

In Exercises 37 to 42, use identities to write each expression in terms of a single trigonometric function or as a constant.

37. $1 + \frac{\sin^2 t}{\cos^2 t}$

38. $\frac{\tan t + 1}{\cot t + 1}$

39. $\frac{\cos^2 t + \sin^2 t}{\csc t}$

40. $\sin^2 t(\tan^2 t + 1)$

41. $1 + \frac{1}{\tan^2 t}$

42. $\frac{\cos^2 t}{1 - \sin^2 t} - 1$

In Exercises 43 to 48, state the amplitude (if there is one), period, and phase shift of the graph of each function.

43. $y = 3 \cos(2x - \pi)$

44. $y = 2 \tan 3x$

45. $y = -2 \sin\left(3x + \frac{\pi}{3}\right)$

46. $y = \cos\left(2x - \frac{2\pi}{3}\right) + 2$

47. $y = -4 \sec\left(4x - \frac{3\pi}{2}\right)$

48. $y = 2 \csc\left(x - \frac{\pi}{4}\right) - 3$

In Exercises 49 to 66, graph each function.

49. $y = 2 \cos \pi x$

50. $y = -\sin \frac{2x}{3}$

51. $y = 2 \sin \frac{3x}{2}$

52. $y = \cos\left(x - \frac{\pi}{2}\right)$

53. $y = \frac{1}{2} \sin\left(2x + \frac{\pi}{4}\right)$

54. $y = 3 \cos 3(x - \pi)$

55. $y = -\tan \frac{x}{2}$

56. $y = 2 \cot 2x$

57. $y = \tan\left(x - \frac{\pi}{2}\right)$

58. $y = -\cot\left(2x + \frac{\pi}{4}\right)$

59. $y = -2 \csc\left(2x - \frac{\pi}{3}\right)$

60. $y = 3 \sec\left(x + \frac{\pi}{4}\right)$

61. $y = 3 \sin 2x - 3$

62. $y = 2 \cos 3x + 3$

63. $y = -\cos\left(3x + \frac{\pi}{2}\right) + 2$

64. $y = 3 \sin\left(4x - \frac{2\pi}{3}\right) - 3$

65. $y = 2 - \sin 2x$

66. $y = \sin x - \sqrt{3} \cos x$

CHAPTER 4 TEST

1. Convert 150° to exact radian measure.

2. Find the supplement of the angle whose radian measure is $\frac{11}{12}\pi$. Express your answer in terms of π.

3. Find the length (to the nearest 0.1 centimeter) of an arc that subtends a central angle of 75° in a circle of radius 10 centimeters.

4. A wheel is rotating at 6 revolutions per second. Find the angular speed in radians per second.

5. A wheel with a diameter of 16 centimeters is rotating at 10 radians per second. Find the linear speed (in centimeters per second) of a point on the edge of the wheel.

6. Find the exact coordinates of $W\left(\frac{11\pi}{6}\right)$.

7. Find the exact value of $\sin\left(\frac{5\pi}{3}\right)$.

8. Find the exact value of $\tan\left(\frac{4\pi}{3}\right)$.

9. Find the exact value of $\sec\left(\frac{5\pi}{6}\right)$.

10. Use a calculator to estimate the value of csc(2.3). Round to the nearest ten-thousandth.

11. Determine whether $f(t) = t\cos t$ is an even function, an odd function, or neither.

12. Express $\frac{\sec^2 t - 1}{\sec^2 t}$ in terms of a single trigonometric function.

13. State the period of $y = -4\tan 3x$.

14. State the amplitude, period, and phase shift for the function $y = -3\cos\left(2x + \frac{\pi}{2}\right)$.

15. State the period and phase shift for the function $y = 2\cot\left(\frac{\pi}{3}x + \frac{\pi}{6}\right)$.

16. Graph one full period of $y = 3\cos\frac{1}{2}x$.

17. Graph one full period of $y = -2\sec\frac{1}{2}x$.

18. Write a sentence that explains how to obtain the graph of $y = 2\sin\left(2x - \frac{\pi}{2}\right) - 1$ from the graph of $y = 2\sin 2x$.

19. Graph one full period of $y = 2 - \sin\frac{x}{2}$.

20. Graph one full period of $y = \sin x - \cos 2x$.

CUMULATIVE REVIEW EXERCISES

1. Find the distance between the points $(1, 0)$ and $\left(\frac{1}{2}, \frac{\sqrt{3}}{2}\right)$.

2. Write the equation of the circle with center $(0, 0)$ and a radius of 3.

3. Write $(3 + 2i)(4 - 5i)$ in standard form.

4. Determine whether $f(x) = \frac{x}{x^2 + 1}$ is an even function or an odd function.

5. Find the inverse of $f(x) = \frac{x}{2x - 3}$.

6. Use interval notation to state the domain of $f(x) = \frac{2}{x - 4}$.

7. Use interval notation to state the range of $f(x) = \sqrt{4 - x^2}$.

8. Explain how to use the graph of $y = f(x)$ to produce the graph of $y = f(x - 3)$.

9. Explain how to use the graph of $y = f(x)$ to produce the graph of $y = f(-x)$.

10. Find the equation of the slant asymptote for the graph of $f(x) = \frac{x^3 + 6x^2 + 11x + 6}{x^2 + 3x - 4}$.

11. Is $f(x) = \left(\frac{1}{2}\right)^x$ an increasing function or a decreasing function?

12. Write $\log_5 125 = 3$ in exponential form.

13. What is the x-intercept of the graph of $f(x) = \ln x$?

14. Use a calculator and the change-of-base formula to estimate $\log_4 85$. Round to the nearest thousandth.

15. Solve $2^x + 4 = 76$. Round the solution to the nearest hundredth.

16. Convert $300°$ to radians.

17. Convert $\frac{5\pi}{4}$ to degrees.

18. Use interval notation to state the domain of $f(t) = \sin t$, where t is a real number.

19. Use interval notation to state the range of $f(t) = \cos t$, where t is a real number.

20. Find the exact value of $\cot\left(\frac{2\pi}{3}\right)$.

CHAPTER 5

Applications of Trigonometry and Trigonometric Identities

The Burji Al Arab hotel in Dubai, United Arab Emirates, is the world's tallest hotel. It is 321 meters in height and was designed in the shape of a billowing sail.

Trigonometry and Indirect Measurement

Architects often use trigonometry to find the unknown distance between two points. For instance, in the following diagram, the length a of the steel brace from C to B can be determined using known values and the Law of Sines, which is a major theorem presented in this chapter.

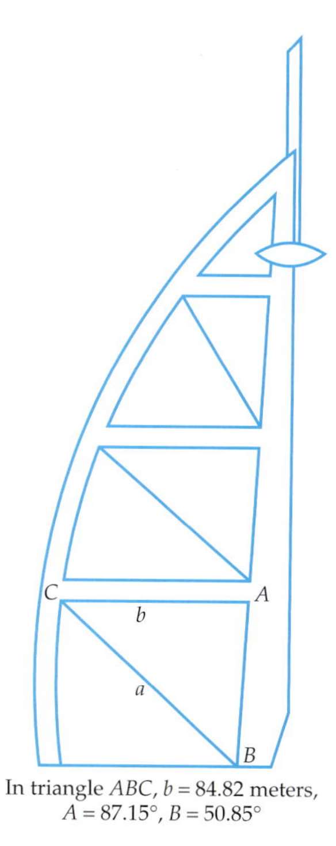

In triangle ABC, $b = 84.82$ meters, $A = 87.15°$, $B = 50.85°$

$$\frac{a}{\sin A} = \frac{b}{\sin B} \quad \text{The Law of Sines}$$

$$\frac{a}{\sin 87.15°} = \frac{84.82}{\sin 50.85°}$$

$$a = \frac{84.82 \sin 87.15°}{\sin 50.85°}$$

$$a \approx 109.2 \text{ m}$$

See **Exercises 31 and 32, page 434,** for additional applications that can be solved by applying the Law of Sines.

FOCUS ON PROBLEM SOLVING

Devising a Plan

One of the most important aspects of problem solving is the process of devising a plan. In some applications there may be more than one plan (method) that can be used to obtain the solution. For instance, consider the following classic problem.

> You have eight coins. They all look identical, but one is a fake and is slightly lighter than the others. Explain how you can use a balance scale to determine which coin is the fake in exactly
> **a.** three weighings.
> **b.** two weighings.

In this chapter you will often need to solve a triangle, which means to determine all unknown measures of the triangle. In most cases you will solve a triangle by using one of two theorems, which are known as the Law of Sines and the Law of Cosines. The guidelines on pages 432 and 433 provide the information needed to choose between these two theorems.

SECTION 5.1 TRIGONOMETRIC FUNCTIONS OF ANGLES

TRIGONOMETRIC FUNCTIONS OF ACUTE ANGLES

The study of trigonometry, which means "triangle measurement," began more than 2000 years ago, partially as a means of solving surveying problems. Early trigonometry used the length of a chord of a circle as the value of a *trigonometric function.* In the sixteenth century, right triangles were used to define a trigonometric function. We will use a modification of this approach.

As we will show, there are a great many similarities between the trigonometric functions of real numbers, introduced in Chapter 4, and the trigonometric functions of angles that we will introduce in this section. The difference between these two types of functions is their domains. Recall that the domain of a trigonometric function of real numbers is a set of real numbers. The domains of the trigonometric functions introduced in this section are sets of angles. The trigonometric functions of angles are often used in applications that involve triangles.

When working with right triangles, it is convenient to refer to the side *opposite* an angle or the side *adjacent* to (next to) an angle. **Figure 5.1a** shows the sides opposite and adjacent to the angle α. For angle β, the opposite and adjacent sides are as shown in **Figure 5.1b.** In both cases, the hypotenuse remains the same.

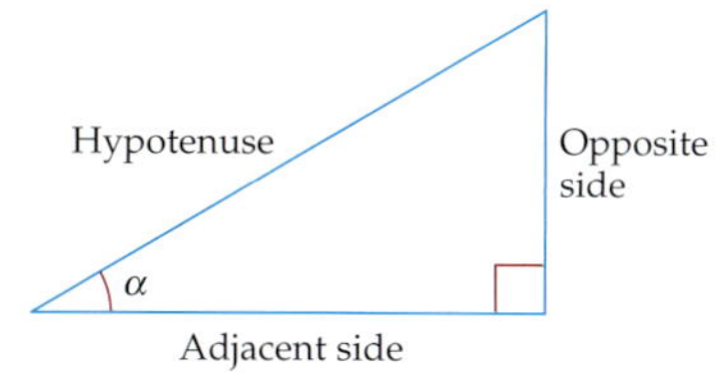

a. Adjacent and opposite sides of $\angle\alpha$

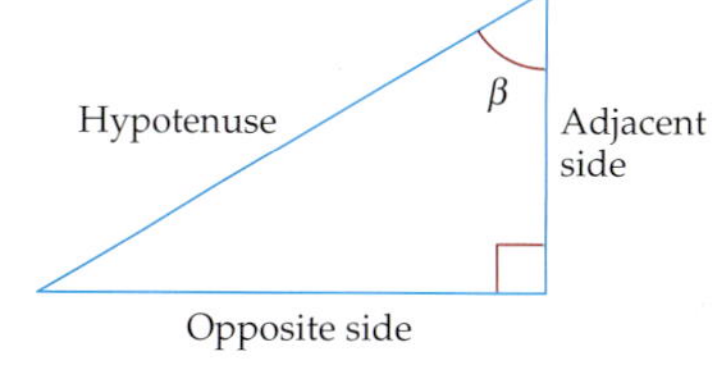

b. Adjacent and opposite sides of $\angle\beta$

FIGURE 5.1

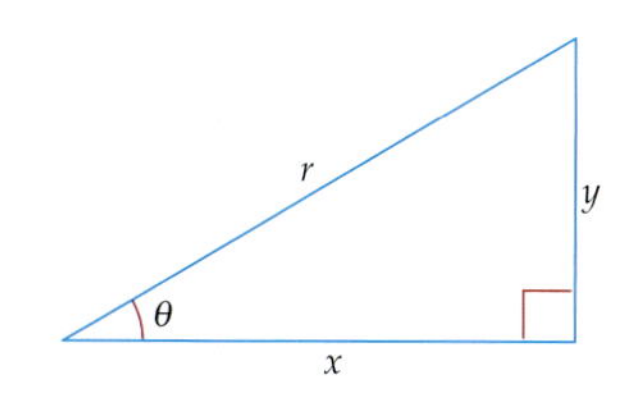

FIGURE 5.2

Consider an angle θ in the right triangle shown in **Figure 5.2.** Let x and y represent the lengths, respectively, of the adjacent and opposite sides of the triangle, and let r be the length of the hypotenuse. Six possible ratios can be formed:

$$\frac{y}{r} \quad \frac{x}{r} \quad \frac{y}{x} \quad \frac{r}{y} \quad \frac{r}{x} \quad \frac{x}{y}$$

Each ratio defines a value of a trigonometric function of the acute angle θ. The functions are **sine** (sin), **cosine** (cos), **tangent** (tan), **cosecant** (csc), **secant** (sec), and **cotangent** (cot).

Trigonometric Functions of an Acute Angle

Let θ be an acute angle of a right triangle. The values of the six trigonometric functions of θ are

$$\sin\theta = \frac{\text{length of opposite side}}{\text{length of hypotenuse}} = \frac{y}{r} \qquad \cos\theta = \frac{\text{length of adjacent side}}{\text{length of hypotenuse}} = \frac{x}{r}$$

Continued ►

$$\tan\theta = \frac{\text{length of opposite side}}{\text{length of adjacent side}} = \frac{y}{x} \qquad \cot\theta = \frac{\text{length of adjacent side}}{\text{length of opposite side}} = \frac{x}{y}$$

$$\sec\theta = \frac{\text{length of hypotenuse}}{\text{length of adjacent side}} = \frac{r}{x} \qquad \csc\theta = \frac{\text{length of hypotenuse}}{\text{length of opposite side}} = \frac{r}{y}$$

We will write opp, adj, and hyp as abbreviations for *the length of the* opposite side, adjacent side, and hypotenuse, respectively.

EXAMPLE 1 Evaluate Trigonometric Functions

Find the values of the six trigonometric functions of θ for the triangle given in **Figure 5.3.**

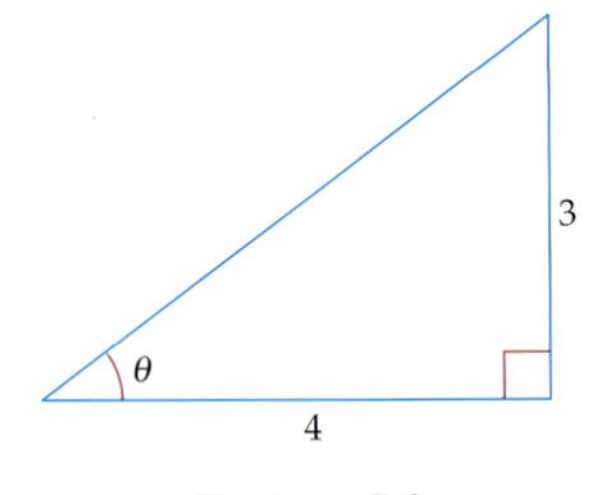

FIGURE 5.3

Solution

Use the Pythagorean Theorem to find the length of the hypotenuse.

$$r = \sqrt{3^2 + 4^2} = \sqrt{25} = 5$$

From the definitions of the trigonometric functions,

$$\sin\theta = \frac{\text{opp}}{\text{hyp}} = \frac{3}{5} \qquad \cos\theta = \frac{\text{adj}}{\text{hyp}} = \frac{4}{5} \qquad \tan\theta = \frac{\text{opp}}{\text{adj}} = \frac{3}{4}$$

$$\cot\theta = \frac{\text{adj}}{\text{opp}} = \frac{4}{3} \qquad \sec\theta = \frac{\text{hyp}}{\text{adj}} = \frac{5}{4} \qquad \csc\theta = \frac{\text{hyp}}{\text{opp}} = \frac{5}{3}$$

▶ TRY EXERCISE 6, PAGE 371

TRIGONOMETRIC FUNCTIONS OF SPECIAL ANGLES

In Example 1, the lengths of the legs of the triangle were given, and you were asked to find the values of the six trigonometric functions of the angle θ. Often we will want to find the value of a trigonometric function when we are given *the measure of an angle* rather than the measure of the sides of a triangle. For most angles, advanced mathematical methods are required to evaluate a trigonometric function. For some *special angles,* however, the value of a trigonometric function can be found by geometric methods. These special acute angles are 30°, 45°, and 60°.

First, we will find the values of the six trigonometric functions of 45°. (This discussion is based on angles measured in degrees. Radian measure could have been used without changing the results.) **Figure 5.4** shows a right triangle with angles 45°, 45°, and 90°. Because $\angle A = \angle B$, the lengths of the sides opposite these angles are equal. Let the length of each equal side be denoted by a. From the Pythagorean Theorem,

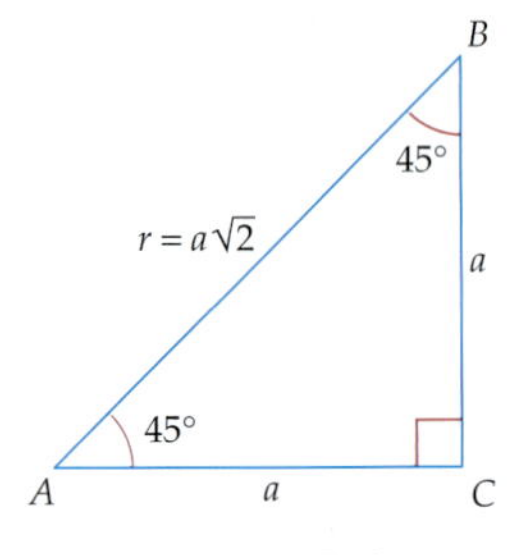

FIGURE 5.4

$$r^2 = a^2 + a^2 = 2a^2$$
$$r = \sqrt{2a^2} = a\sqrt{2}$$

The values of the six trigonometric functions of 45° are

$$\sin 45^\circ = \frac{a}{a\sqrt{2}} = \frac{1}{\sqrt{2}} = \frac{\sqrt{2}}{2} \qquad \cos 45^\circ = \frac{a}{a\sqrt{2}} = \frac{1}{\sqrt{2}} = \frac{\sqrt{2}}{2}$$

$$\tan 45^\circ = \frac{a}{a} = 1 \qquad \cot 45^\circ = \frac{a}{a} = 1$$

$$\sec 45^\circ = \frac{a\sqrt{2}}{a} = \sqrt{2} \qquad \csc 45^\circ = \frac{a\sqrt{2}}{a} = \sqrt{2}$$

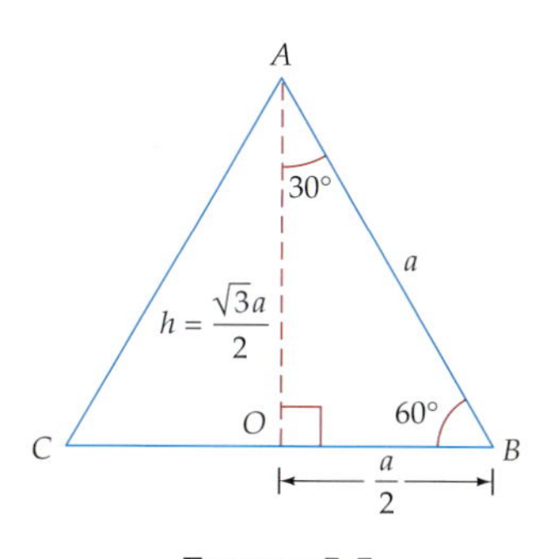

FIGURE 5.5

The values of the trigonometric functions of the special angles 30° and 60° can be found by drawing an equilateral triangle and bisecting one of the angles, as **Figure 5.5** shows. The angle bisector also bisects one of the sides. Thus the length of the side opposite the 30° angle is one-half the length of the hypotenuse of triangle OAB.

Let a denote the length of the hypotenuse. Then the length of the side opposite the 30° angle is $\frac{a}{2}$. The length of the side adjacent to the 30° angle, h, is found by using the Pythagorean Theorem.

$$a^2 = \left(\frac{a}{2}\right)^2 + h^2$$

$$a^2 = \frac{a^2}{4} + h^2$$

$$\frac{3a^2}{4} = h^2 \qquad \text{• Subtract } \frac{a^2}{4} \text{ from each side.}$$

$$h = \frac{\sqrt{3}a}{2} \qquad \text{• Solve for } h.$$

The values of the six trigonometric functions of 30° are

$$\sin 30^\circ = \frac{a/2}{a} = \frac{1}{2} \qquad \cos 30^\circ = \frac{\sqrt{3}a/2}{a} = \frac{\sqrt{3}}{2}$$

$$\tan 30^\circ = \frac{a/2}{\sqrt{3}a/2} = \frac{1}{\sqrt{3}} = \frac{\sqrt{3}}{3} \qquad \cot 30^\circ = \frac{\sqrt{3}a/2}{a/2} = \sqrt{3}$$

$$\sec 30^\circ = \frac{a}{\sqrt{3}a/2} = \frac{2}{\sqrt{3}} = \frac{2\sqrt{3}}{3} \qquad \csc 30^\circ = \frac{a}{a/2} = 2$$

The values of the trigonometric functions of 60° can be found by again using **Figure 5.5.** The length of the side opposite the 60° angle is $\frac{\sqrt{3}a}{2}$, and the length of the side adjacent to the 60° angle is $\frac{a}{2}$. The values of the trigonometric functions of 60° are

$$\sin 60^\circ = \frac{\sqrt{3}a/2}{a} = \frac{\sqrt{3}}{2} \qquad \cos 60^\circ = \frac{a/2}{a} = \frac{1}{2}$$

$$\tan 60^\circ = \frac{\sqrt{3}a/2}{a/2} = \sqrt{3} \qquad \cot 60^\circ = \frac{a/2}{\sqrt{3}a/2} = \frac{1}{\sqrt{3}} = \frac{\sqrt{3}}{3}$$

$$\sec 60^\circ = \frac{a}{a/2} = 2 \qquad \csc 60^\circ = \frac{a}{\sqrt{3}a/2} = \frac{2}{\sqrt{3}} = \frac{2\sqrt{3}}{3}$$

Table 5.1 summarizes the values of the trigonometric functions of the special angles $30°$ $(\pi/6)$, $45°$ $(\pi/4)$, and $60°$ $(\pi/3)$.

take note

Memorizing the values given in Table 5.1 will prove to be extremely useful in the remaining trigonometry sections.

TABLE 5.1 Trigonometric Functions of Special Angles

θ	$\sin\theta$	$\cos\theta$	$\tan\theta$	$\csc\theta$	$\sec\theta$	$\cot\theta$
$30°; \frac{\pi}{6}$	$\frac{1}{2}$	$\frac{\sqrt{3}}{2}$	$\frac{\sqrt{3}}{3}$	2	$\frac{2\sqrt{3}}{3}$	$\sqrt{3}$
$45°; \frac{\pi}{4}$	$\frac{\sqrt{2}}{2}$	$\frac{\sqrt{2}}{2}$	1	$\sqrt{2}$	$\sqrt{2}$	1
$60°; \frac{\pi}{3}$	$\frac{\sqrt{3}}{2}$	$\frac{1}{2}$	$\sqrt{3}$	$\frac{2\sqrt{3}}{3}$	2	$\frac{\sqrt{3}}{3}$

QUESTION What is the measure, in degrees, of the acute angle θ for which $\sin\theta = \cos\theta$, $\tan\theta = \cot\theta$, and $\sec\theta = \csc\theta$?

EXAMPLE 2 Evaluate a Trigonometric Expression

Find the *exact* value of $\sin^2 45° + \cos^2 60°$.

Solution

Substitute the values of $\sin 45°$ and $\cos 60°$ and simplify.

Note: $\sin^2\theta = (\sin\theta)(\sin\theta) = (\sin\theta)^2$ and $\cos^2\theta = (\cos\theta)(\cos\theta) = (\cos\theta)^2$.

$$\sin^2 45° + \cos^2 60° = \left(\frac{\sqrt{2}}{2}\right)^2 + \left(\frac{1}{2}\right)^2 = \frac{2}{4} + \frac{1}{4} = \frac{3}{4}$$

TRY EXERCISE 18, PAGE 371

take note

The patterns in the following chart can be used to memorize the sine and cosine of 30°, 45°, and 60°.

$\sin 30° = \frac{\sqrt{1}}{2}$ $\cos 30° = \frac{\sqrt{3}}{2}$

$\sin 45° = \frac{\sqrt{2}}{2}$ $\cos 45° = \frac{\sqrt{2}}{2}$

$\sin 60° = \frac{\sqrt{3}}{2}$ $\cos 60° = \frac{\sqrt{1}}{2}$

From the definition of the sine and cosecant functions,

$$(\sin\theta)(\csc\theta) = \frac{y}{r}\cdot\frac{r}{y} = 1 \quad \text{or} \quad (\sin\theta)(\csc\theta) = 1$$

By rewriting the last equation, we find

$$\sin\theta = \frac{1}{\csc\theta} \quad \text{and} \quad \csc\theta = \frac{1}{\sin\theta}, \text{ provided } \sin\theta \neq 0$$

The sine and cosecant functions are called **reciprocal functions.** The cosine and secant are also reciprocal functions, as are the tangent and cotangent functions. **Table 5.2** shows each trigonometric function and its reciprocal. These relationships hold for all values of θ for which both of the functions are defined.

ANSWER 45°

TABLE 5.2 Trigonometric Functions and Their Reciprocals

$\sin\theta = \dfrac{1}{\csc\theta}$	$\cos\theta = \dfrac{1}{\sec\theta}$	$\tan\theta = \dfrac{1}{\cot\theta}$
$\csc\theta = \dfrac{1}{\sin\theta}$	$\sec\theta = \dfrac{1}{\cos\theta}$	$\cot\theta = \dfrac{1}{\tan\theta}$

INTEGRATING TECHNOLOGY

Some graphing calculators will allow you to display the degree symbol. For example, the TI-83 display

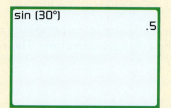

was produced by entering

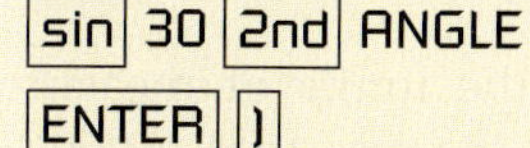

ENTER)

Displaying the degree symbol will cause the calculator to evaluate a trigonometric function using degree mode even if the calculator is in radian mode.

INTEGRATING TECHNOLOGY

When evaluating a trigonometric function by using a graphing calculator, be sure the calculator is in the correct mode. If the measure of an angle is written with the degree symbol, then make sure the calculator is in degree mode. If the measure of an angle is given in radians (no degree symbol is used), then make sure the calculator is in radian mode. *Many errors are made because the correct mode is not selected.*

Some graphing calculators can be used to construct a table of functional values. For instance, the TI-83 keystrokes shown below generate the table in **Figure 5.6,** in which the first column lists the domain values

$$0°, 1°, 2°, 3°, \ldots$$

and the second column lists the range values

$$\sin 0°, \sin 1°, \sin 2°, \sin 3°, \ldots$$

TI-83 Keystrokes

2nd TBLSET 0

ENTER 1 Y= CLEAR sin

X,T,θ) 2nd TABLE

X	Y1
0	0
1	.01745
2	.0349
3	.05234
4	.06976
5	.08716
6	.10453

Y1 = .017452406437

FIGURE 5.6

The graphing calculator must be in degree mode to produce this table.

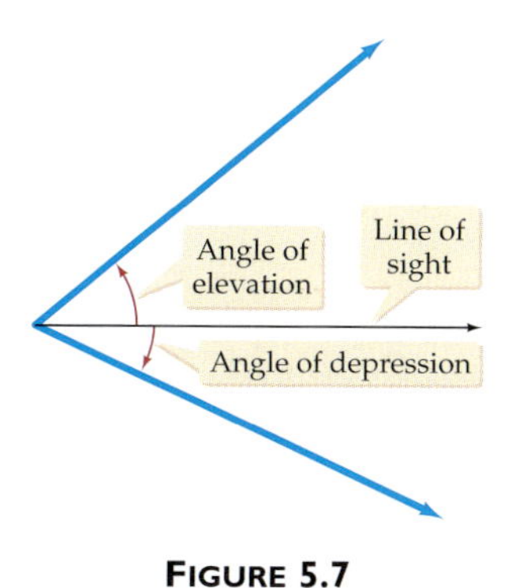

FIGURE 5.7

APPLICATIONS INVOLVING RIGHT TRIANGLES

One of the major reasons for the development of trigonometry was to solve application problems. In this section we will consider some applications involving right triangles. In some application problems a horizontal line of sight is used as a reference line. An angle measured above the line of sight is called an **angle of elevation,** and an angle measured below the line of sight is called an **angle of depression.** See **Figure 5.7.**

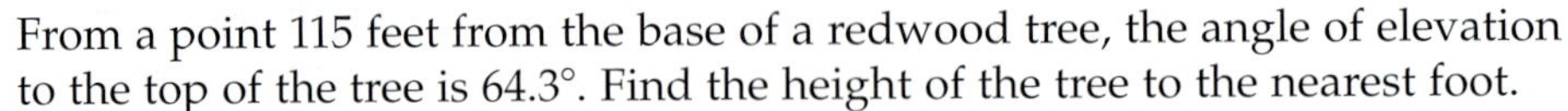

EXAMPLE 3 Solve an Angle-of-Elevation Problem

From a point 115 feet from the base of a redwood tree, the angle of elevation to the top of the tree is 64.3°. Find the height of the tree to the nearest foot.

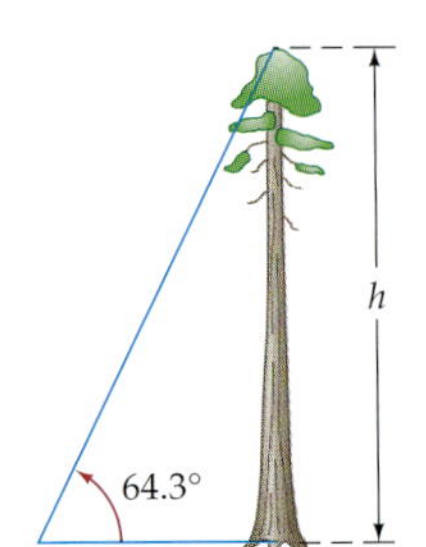

Figure 5.8

Solution

From **Figure 5.8,** the length of the adjacent side of the angle is known (115 feet). Because we need to determine the height of the tree (length of the opposite side), we use the tangent function. Let h represent the length of the opposite side.

$$\tan 64.3^\circ = \frac{\text{opp}}{\text{adj}} = \frac{h}{115}$$

$$h = 115 \tan 64.3^\circ \approx 238.952$$

• Use a calculator to evaluate tan 64.3°.

The height of the tree is approximately 239 feet.

▶ TRY EXERCISE 22, PAGE 371

take note

The significant digits of an approximate number are

- every nonzero digit.
- the digit 0, provided it is between two nonzero digits or it is to the right of a nonzero digit in a number that includes a decimal point.

For example, the approximate number:

502 has 3 significant digits.
3700 has 2 significant digits.
47.0 has 3 significant digits.
0.0023 has 2 significant digits.
0.00840 has 3 significant digits.

Because the cotangent function involves the sides adjacent to and opposite an angle, we could have solved Example 4 by using the cotangent function. The solution would have been

$$\cot 64.3^\circ = \frac{\text{adj}}{\text{opp}} = \frac{115}{h}$$

$$h = \frac{115}{\cot 64.3^\circ} \approx 238.952 \text{ feet}$$

The accuracy of a calculator is sometimes beyond the limits of measurement. In the last example the distance from the base of the tree was given as 115 feet (three significant digits), whereas the height of the tree was shown to be 238.952 feet (six significant digits). When using approximate numbers, we will use the conventions given below for calculating with trigonometric functions.

A Rounding Convention: Significant Digits for Trigonometric Calculations

Angle Measure to the Nearest	*Significant Digits of the Lengths*
Degree	Two
Tenth of a degree	Three
Hundredth of a degree	Four

EXAMPLE 4 Solve an Angle-of-Elevation Problem

An observer notes that the angle of elevation from point A to the top of a space shuttle is 27.2°. From a point 17.5 meters further from the space shuttle, the angle of elevation is 23.9°. Find the height of the space shuttle.

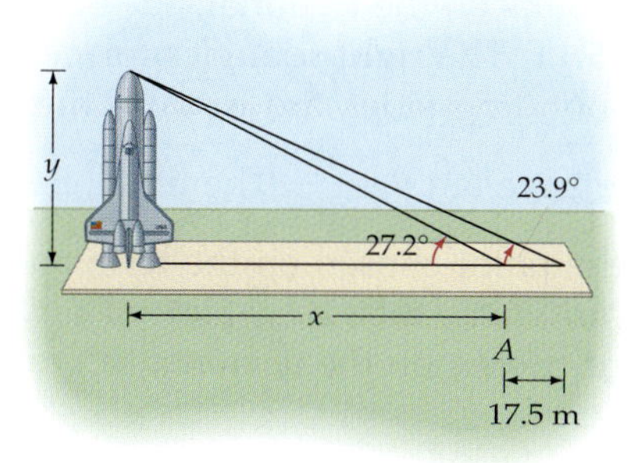

FIGURE 5.9

take note

The intermediate calculations in Example 4 were not rounded off. This ensures better accuracy for the final result. Using the rounding convention stated on page 364, we round off only the last result.

Solution

From **Figure 5.9,** let x denote the distance from point A to the base of the space shuttle, and let y denote the height of the space shuttle. Then

$$(1)\quad \tan 27.2^\circ = \frac{y}{x} \qquad \text{and} \qquad (2)\quad \tan 23.9^\circ = \frac{y}{x + 17.5}$$

Solving Equation (1) for x, $x = \dfrac{y}{\tan 27.2^\circ} = y \cot 27.2^\circ$, and substituting into Equation (2), we have

$$\tan 23.9^\circ = \frac{y}{y \cot 27.2^\circ + 17.5}$$

$$y = (\tan 23.9^\circ)(y \cot 27.2^\circ + 17.5) \qquad \text{• Solve for } y.$$

$$y - y \tan 23.9^\circ \cot 27.2^\circ = (\tan 23.9^\circ)(17.5)$$

$$y = \frac{(\tan 23.9^\circ)(17.5)}{1 - \tan 23.9^\circ \cot 27.2^\circ} \approx 56.2993$$

To three significant digits, the height of the space shuttle is 56.3 meters.

▶ TRY EXERCISE 28, PAGE 372

TRIGONOMETRIC FUNCTIONS OF ANY ANGLE

The applications of trigonometry would be quite limited if all angles had to be acute angles. Fortunately, this is not the case. The following definition extends the definition of a trigonometric function to include any angle.

take note

The measure of angle θ can be positive or negative. Note from the following figure that $\sin 120^\circ = \sin(-240^\circ)$ because $P(x, y)$ is on the terminal side of each angle. In a similar way, the value of any trigonometric function of 120° is equal to the value of that function of −240°.

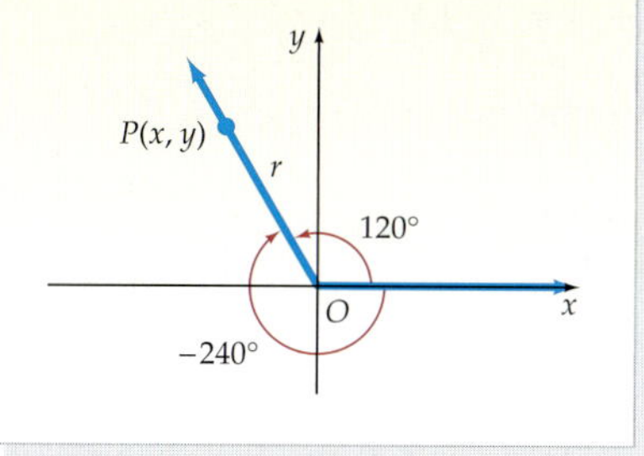

The Trigonometric Functions of Any Angle

Let $P(x, y)$ be any point, except the origin, on the terminal side of an angle θ in standard position (see **Figure 5.10**). Let $r = d(O, P)$, the distance from the origin to P. The six trigonometric functions of θ are

$$\sin\theta = \frac{y}{r} \qquad \cos\theta = \frac{x}{r} \qquad \tan\theta = \frac{y}{x},\quad x \neq 0$$

$$\csc\theta = \frac{r}{y},\quad y \neq 0 \qquad \sec\theta = \frac{r}{x},\quad x \neq 0 \qquad \cot\theta = \frac{x}{y},\quad y \neq 0$$

where $r = \sqrt{x^2 + y^2}$.

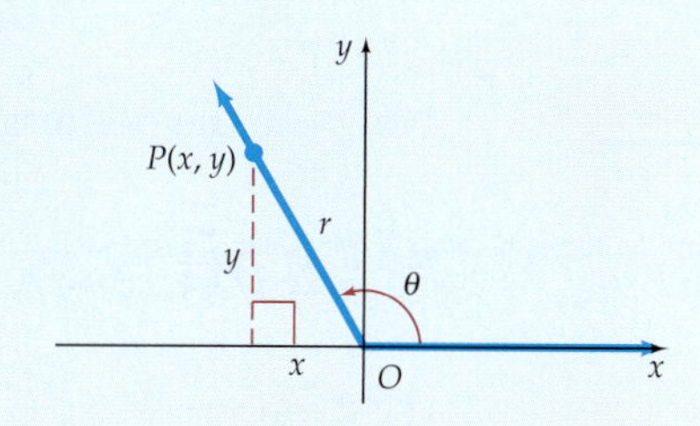

FIGURE 5.10

The value of a trigonometric function is independent of the point chosen on the terminal side of the angle. Consider any two points on the terminal side of an angle θ in standard position, as shown in **Figure 5.11.** The right triangles formed are similar triangles, so the ratios of the corresponding sides are equal. Thus, for example, $\frac{b}{a} = \frac{b'}{a'}$. Because $\tan\theta = \frac{b}{a} = \frac{b'}{a'}$, we have $\tan\theta = \frac{b'}{a'}$. Therefore, the value of the tangent function is independent of the point chosen on the terminal side of the angle. By a similar argument, we can show that the value of any trigonometric function is independent of the point chosen on the terminal side of the angle.

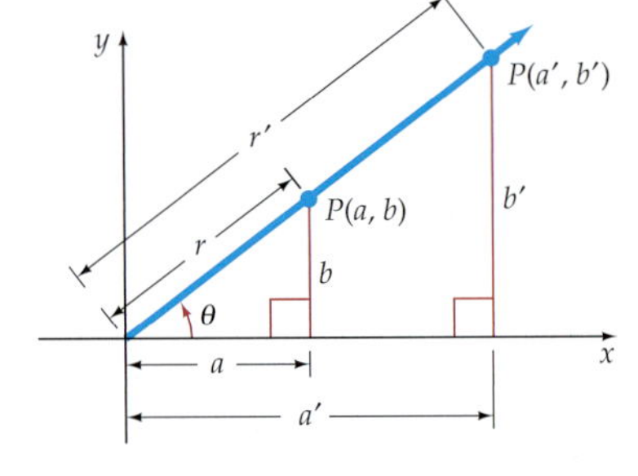

FIGURE 5.11

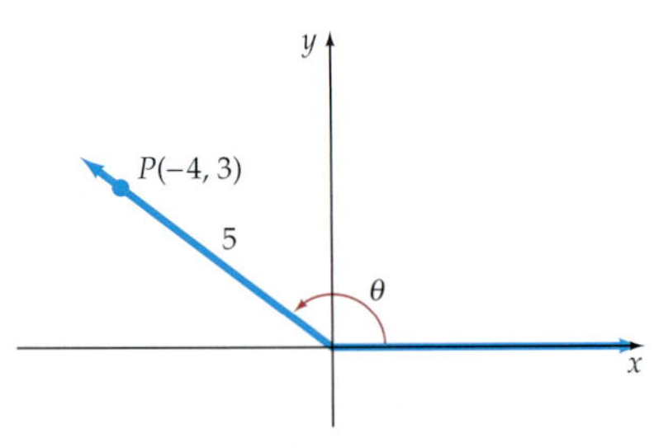

FIGURE 5.12

Any point in a rectangular coordinate system (except the origin) can determine an angle in standard position. For example, $P(-4, 3)$ in **Figure 5.12** is a point in the second quadrant and determines an angle θ in standard position with $r = \sqrt{(-4)^2 + 3^2} = 5$. The values of the trigonometric functions of θ are

$$\sin\theta = \frac{3}{5} \qquad \cos\theta = \frac{-4}{5} = -\frac{4}{5} \qquad \tan\theta = \frac{3}{-4} = -\frac{3}{4}$$

$$\csc\theta = \frac{5}{3} \qquad \sec\theta = \frac{5}{-4} = -\frac{5}{4} \qquad \cot\theta = \frac{-4}{3} = -\frac{4}{3}$$

EXAMPLE 5 Evaluate Trigonometric Functions

Find the value of each of the six trigonometric functions of an angle θ in standard position whose terminal side contains the point $P(-3, -2)$.

Solution

The angle is sketched in **Figure 5.13.** Find r by using the equation $r = \sqrt{x^2 + y^2}$, where $x = -3$ and $y = -2$.

$$r = \sqrt{(-3)^2 + (-2)^2} = \sqrt{9 + 4} = \sqrt{13}$$

Now use the definitions of the trigonometric functions.

$$\sin\theta = \frac{-2}{\sqrt{13}} = -\frac{2\sqrt{13}}{13} \qquad \cos\theta = \frac{-3}{\sqrt{13}} = -\frac{3\sqrt{13}}{13} \qquad \tan\theta = \frac{-2}{-3} = \frac{2}{3}$$

$$\csc\theta = \frac{\sqrt{13}}{-2} = -\frac{\sqrt{13}}{2} \qquad \sec\theta = \frac{\sqrt{13}}{-3} = -\frac{\sqrt{13}}{3} \qquad \cot\theta = \frac{-3}{-2} = \frac{3}{2}$$

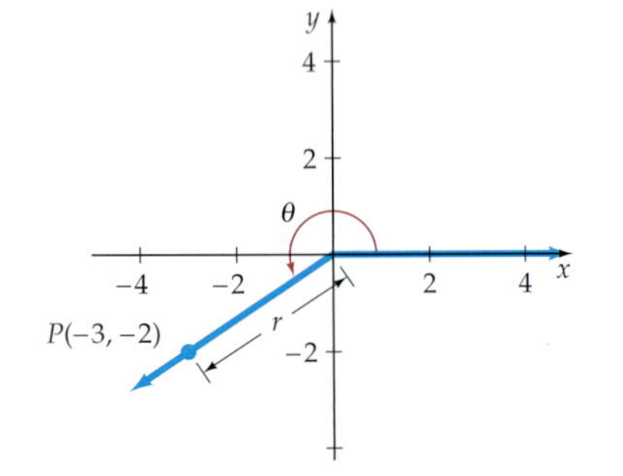

FIGURE 5.13

▶ TRY EXERCISE 36, PAGE 373

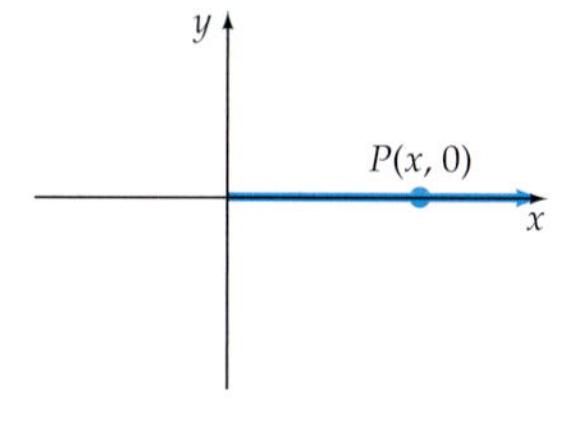

FIGURE 5.14

TRIGONOMETRIC FUNCTIONS OF QUADRANTAL ANGLES

Recall that a quadrantal angle is an angle whose terminal side coincides with the x- or y-axis. The value of a trigonometric function of a quadrantal angle can be found by choosing any point on the terminal side of the angle and then applying the definition of that trigonometric function.

The terminal side of $0°$ coincides with the positive x-axis. Let $P(x, 0)$, $x > 0$, be any point on the x-axis, as shown in **Figure 5.14.** Then $y = 0$ and $r = x$. The values of the six trigonometric functions of $0°$ are

$$\sin 0° = \frac{0}{r} = 0 \qquad \cos 0° = \frac{x}{r} = \frac{x}{x} = 1 \qquad \tan 0° = \frac{0}{x} = 0$$

$$\csc 0° \text{ is undefined.} \qquad \sec 0° = \frac{r}{x} = \frac{x}{x} = 1 \qquad \cot 0° \text{ is undefined.}$$

QUESTION Why are $\csc 0°$ and $\cot 0°$ undefined?

In like manner, the values of the trigonometric functions of the other quadrantal angles can be found. The results are shown in **Table 5.3.**

TABLE 5.3 Values of Trigonometric Functions for Quadrantal Angles

θ	$\sin\theta$	$\cos\theta$	$\tan\theta$	$\csc\theta$	$\sec\theta$	$\cot\theta$
0°	0	1	0	undefined	1	undefined
90°	1	0	undefined	1	undefined	0
180°	0	−1	0	undefined	−1	undefined
270°	−1	0	undefined	−1	undefined	0

SIGNS OF TRIGONOMETRIC FUNCTIONS

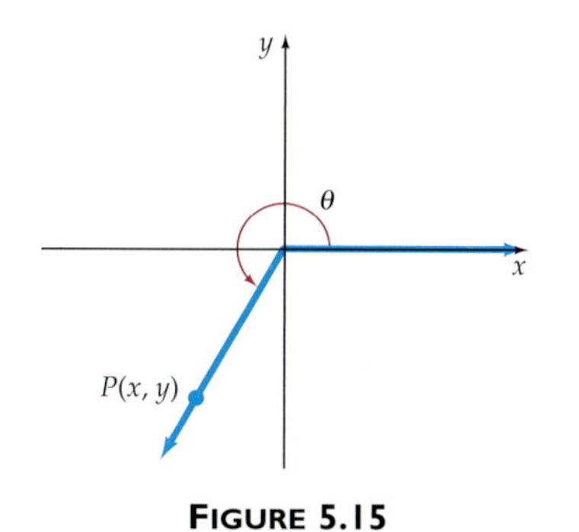

FIGURE 5.15

The sign of a trigonometric function depends on the quadrant in which the terminal side of the angle lies. For example, if θ is an angle whose terminal side lies in Quadrant III and $P(x, y)$ is on the terminal side of θ, then both x and y are negative, and therefore $\frac{y}{x}$ and $\frac{x}{y}$ are positive. See **Figure 5.15.** Because $\tan\theta = \frac{y}{x}$ and $\cot\theta = \frac{x}{y}$, the values of the tangent and cotangent functions are positive for any Quadrant III angle. The values of the other four trigonometric functions of any Quadrant III angle are all negative.

ANSWER $P(x, 0)$ is a point on the terminal side of $0°$. Thus $\csc 0° = \frac{r}{0}$, which is undefined. Similarly, $\cot 0° = \frac{x}{0}$, which is undefined.

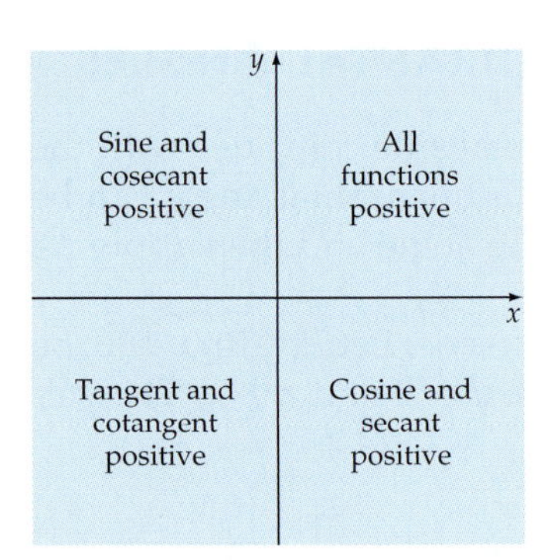

Figure 5.16

Table 5.4 lists the signs of the six trigonometric functions in each quadrant. **Figure 5.16** is a graphical display of the contents of **Table 5.4.**

Table 5.4 Signs of the Trigonometric Functions

Sign of	Terminal Side of θ in Quadrant I	II	III	IV
$\sin\theta$ and $\csc\theta$	positive	positive	negative	negative
$\cos\theta$ and $\sec\theta$	positive	negative	negative	positive
$\tan\theta$ and $\cot\theta$	positive	negative	positive	negative

The Reference Angle

We will often find it convenient to evaluate trigonometric functions by making use of the concept of a *reference angle.*

Reference Angle

Given $\angle\theta$ in standard position, its **reference angle** θ' is the smallest positive angle formed by the terminal side of $\angle\theta$ and the x-axis.

take note

The reference angle is a very important concept that will be used time and time again in the remaining trigonometry sections.

Figure 5.17 shows $\angle\theta$ and its reference angle θ' for four cases. In every case the reference angle θ' is formed by the terminal side of $\angle\theta$ and the x-axis (never the y-axis). The process of determining the measure of $\angle\theta'$ varies according to what quadrant contains the terminal side of $\angle\theta$.

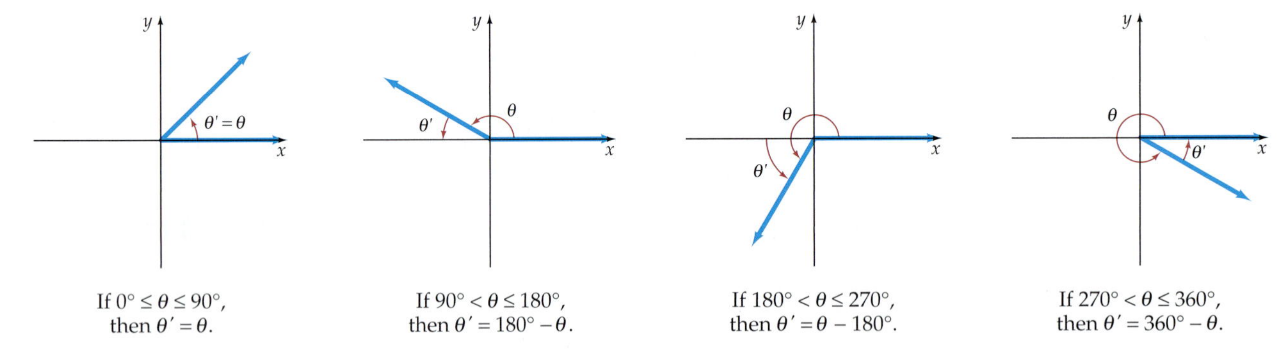

Figure 5.17

Example 6 Find the Measure of the Reference Angle

For each of the following, sketch the given angle θ (in standard position) and its reference angle θ'. Then determine the measure of θ'.

a. $\theta = 120^\circ$ **b.** $\theta = 345^\circ$ **c.** $\theta = 924^\circ$ **d.** $\theta = \dfrac{9}{5}\pi$

Solution

a.

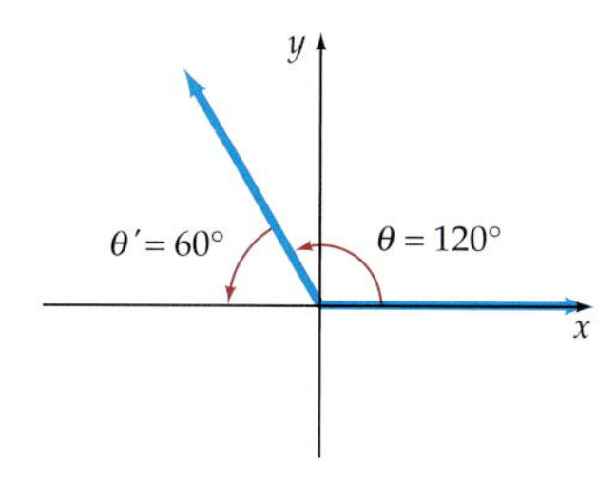

$\theta' = 180° - 120° = 60°$

b.

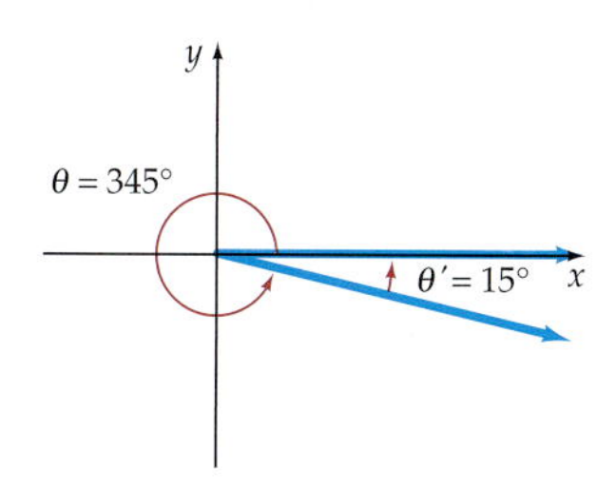

$\theta' = 360° - 345° = 15°$

c.

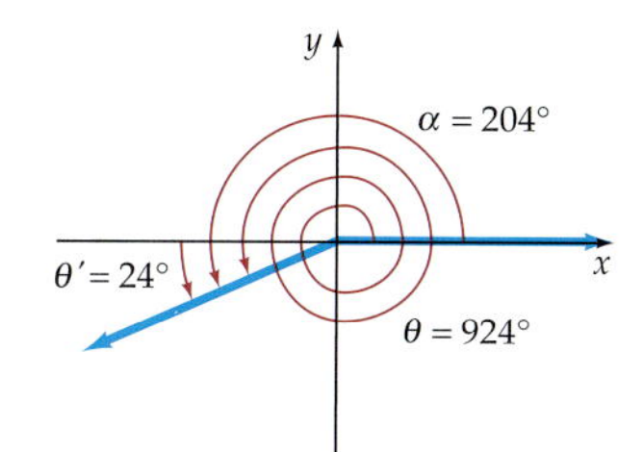

Because $\theta = 924° > 360°$, we first determine the coterminal angle, $\alpha = 204°$.

$\theta' = 204° - 180° = 24°$

d.

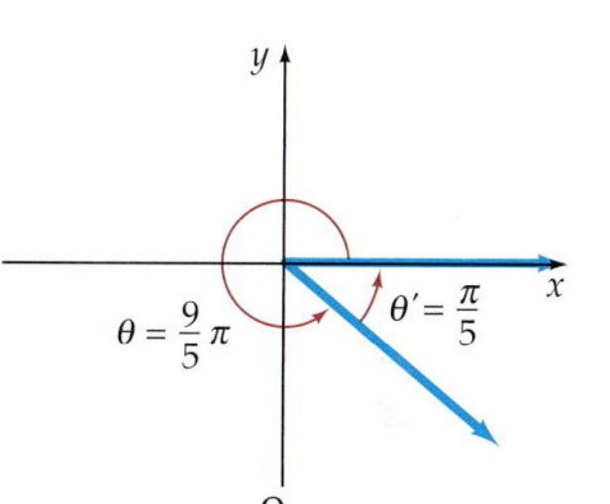

$$\theta' = 2\pi - \frac{9}{5}\pi$$
$$= \frac{10\pi}{5} - \frac{9\pi}{5} = \frac{\pi}{5} \approx 0.628$$

▶ TRY EXERCISE 44, PAGE 373

INTEGRATING TECHNOLOGY

A TI graphing calculator program is available to compute the measure of the reference angle for a given angle. This program, REFANG, can be found on our website at

http://college.hmco.com

take note

The Reference Angle Theorem is also valid if the sine function is replaced by any other trigonometric function.

The following theorem states an important relationship that exists between $\sin\theta$ and $\sin\theta'$, where θ' is the reference angle for angle θ.

Reference Angle Theorem

To evaluate $\sin\theta$, determine $\sin\theta'$. Then use either $\sin\theta'$ or its opposite as the answer, depending on which has the correct sign.

In the following example, we illustrate how to evaluate a trigonometric function of θ by first evaluating the trigonometric function of θ'.

EXAMPLE 7 Use the Reference Angle Theorem to Evaluate Trigonometric Functions

Evaluate each function.

a. $\sin 210°$ **b.** $\cos 405°$ **c.** $\tan\dfrac{5\pi}{3}$

Continued ▶

Solution

a. We know that $\sin 210°$ is negative (the sign chart is given in **Table 5.4**). The reference angle for $\theta = 210°$ is $\theta' = 30°$. By the Reference Angle Theorem, we know that $\sin 210°$ equals either

$$\sin 30° = \frac{1}{2} \qquad \text{or} \qquad -\sin 30° = -\frac{1}{2}$$

Thus $\sin 210° = -\dfrac{1}{2}$.

b. Because $\theta = 405°$ is a Quadrant I angle, we know that $\cos 405° > 0$. The reference angle for $\theta = 405°$ is $\theta' = 45°$. By the Reference Angle Theorem, $\cos 405°$ equals either

$$\cos 45° = \frac{\sqrt{2}}{2} \qquad \text{or} \qquad -\cos 45° = -\frac{\sqrt{2}}{2}$$

Thus $\cos 405° = \dfrac{\sqrt{2}}{2}$.

c. Because $\theta = \dfrac{5\pi}{3}$ is a Quadrant IV angle, $\tan \dfrac{5\pi}{3} < 0$. The reference angle for $\theta = \dfrac{5\pi}{3}$ is $\theta' = \dfrac{\pi}{3}$. Hence $\tan \dfrac{5\pi}{3}$ equals either

$$\tan \frac{\pi}{3} = \sqrt{3} \qquad \text{or} \qquad -\tan \frac{\pi}{3} = -\sqrt{3}$$

Thus $\tan \dfrac{5\pi}{3} = -\sqrt{3}$.

▶ TRY EXERCISE 52, PAGE 373

TOPICS FOR DISCUSSION

1. Is every reference angle an acute angle? Explain.
2. If θ' is the reference angle for the angle θ, then $\sin \theta = \sin \theta'$. Do you agree? Explain.
3. If $\sin \theta < 0$ and $\cos \theta > 0$, then the terminal side of the angle θ lies in which quadrant?
4. A student claims that $\sin^2 30° = (\sin 30°)^2$. Do you agree? Explain.

Exercise Set 5.1

In Exercises 1 to 10, find the values of the six trigonometric functions of θ for the right triangle with the given sides.

1.

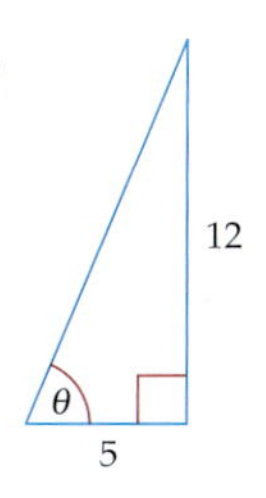

2.

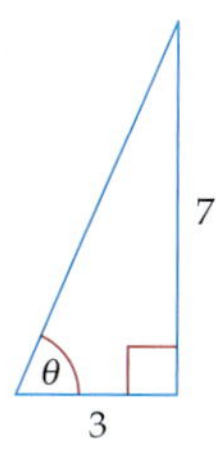

3.

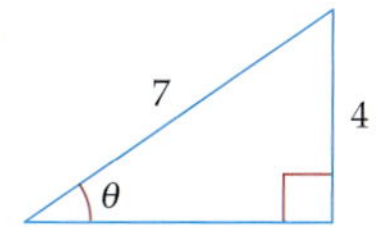

4.

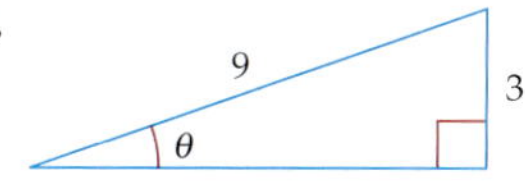

5.

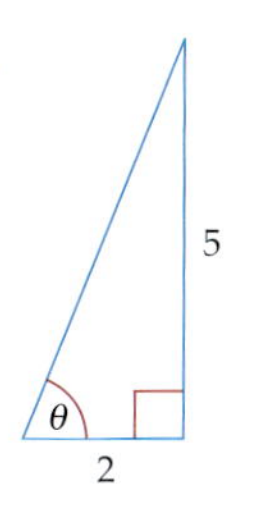

▶ **6.**

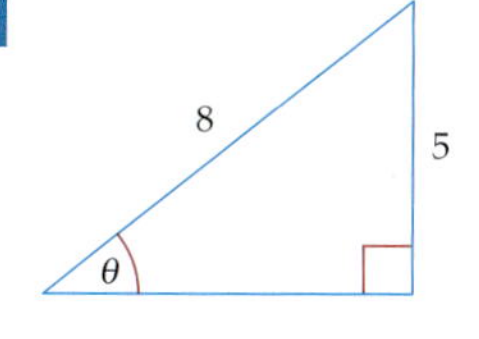

7.

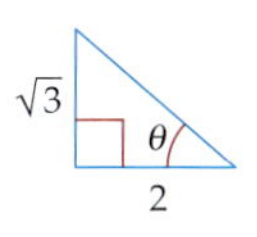

8.

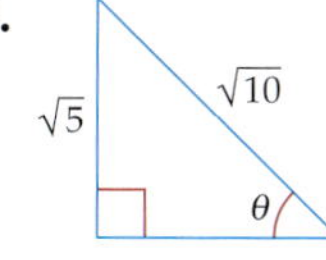

9.

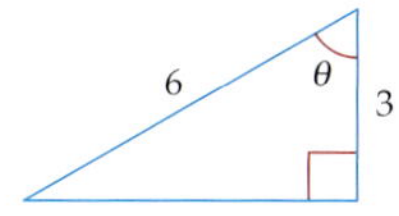

10.

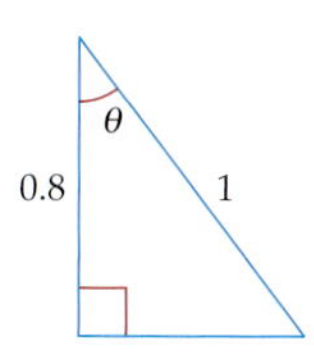

In Exercises 11 to 18, find the *exact* value of each expression.

11. $\sin 45^\circ + \cos 45^\circ$

12. $\csc 45^\circ - \sec 45^\circ$

13. $\sin 30^\circ \cos 60^\circ + \tan 45^\circ$

14. $\sec 30^\circ \cos 30^\circ - \tan 60^\circ \cot 60^\circ$

15. $\sin \dfrac{\pi}{3} + \cos \dfrac{\pi}{6}$

16. $\csc \dfrac{\pi}{6} - \sec \dfrac{\pi}{3}$

17. $\sin \dfrac{\pi}{4} + \tan \dfrac{\pi}{6}$

▶ **18.** $\sin \dfrac{\pi}{3} \cos \dfrac{\pi}{4} - \tan \dfrac{\pi}{4}$

19. Vertical Height from Slant Height A 12-foot ladder is resting against a wall and makes an angle of 52° with the ground. Find the height to which the ladder will reach on the wall.

20. Distance Across a Marsh Find the distance AB across the marsh shown in the accompanying figure.

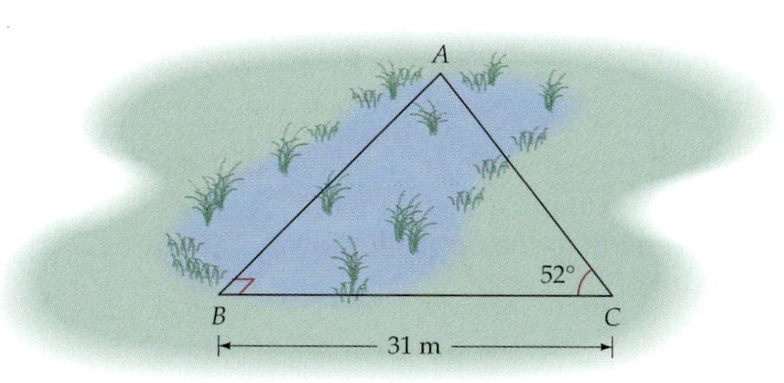

21. Placement of a Light For best illumination of a piece of art, a lighting specialist for an art gallery recommends that a ceiling-mounted light be 6 feet from the piece of art and that the angle of depression of the light be 38°. How far from a wall should the light be placed so that the recommendations of the specialist are met? Notice that the art extends outward 4 inches from the wall.

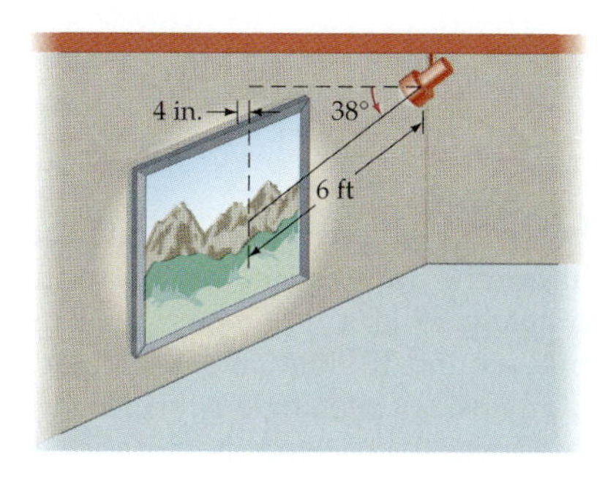

▶ **22. Height of the Eiffel Tower** The angle of elevation from a point 116 meters from the base of the Eiffel Tower to the top of the tower is 68.9°. Find the approximate height of the tower.

23. Height of an Aqueduct From a point 300 feet from the base of a Roman aqueduct in southern France, the angle of elevation to the top of the aqueduct is 78° (see the figure on the following page). Find the height of the aqueduct.

Figure for Exercise 23

24. Width of a Lake The angle of depression to one side of a lake, measured from a balloon 2500 feet above the lake as shown in the accompanying figure, is 43°. The angle of depression to the opposite side of the lake is 27°. Find the width of the lake.

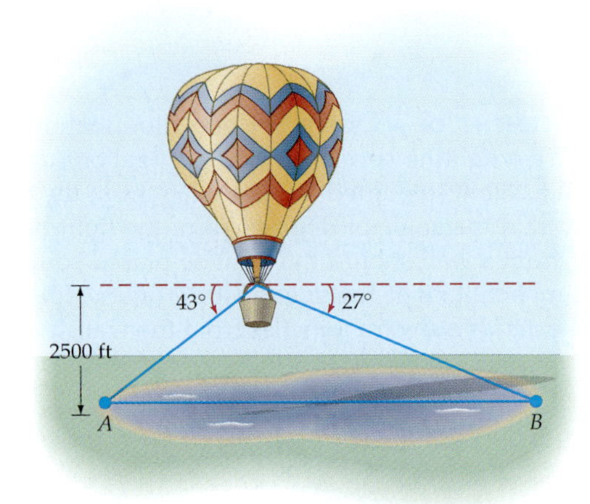

25. Astronomy The moon Europa rotates in a nearly circular orbit around Jupiter. The orbital radius of Europa is approximately 670,900 kilometers. During a revolution of Europa around Jupiter, an astronomer found that the maximum value of the angle θ formed by Europa, Earth, and Jupiter was 0.056°. Find the distance d between Earth and Jupiter at the time the astronomer found the maximum value of θ. Round to the nearest million kilometers.

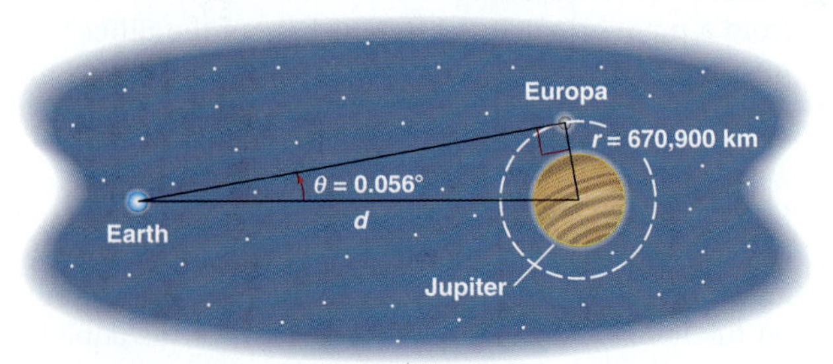

Not drawn to scale.

26. Astronomy Venus rotates in a nearly circular orbit around the sun. The largest angle formed by Venus, Earth, and the sun is 46.5°. The distance from Earth to the sun is approximately 149,000,000 kilometers. What is the orbital radius r of Venus? Round to the nearest million kilometers.

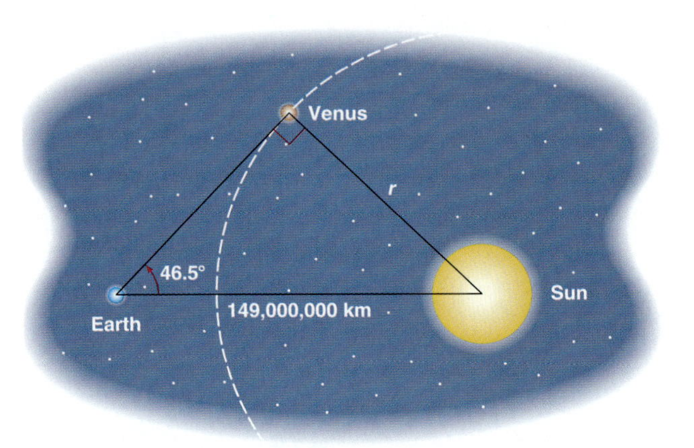

27. Height of a Pyramid The angle of elevation to the top of the Egyptian pyramid Cheops is 36.4°, measured from a point 350 feet from the base of the pyramid. The angle of elevation from the base of a face of the pyramid is 51.9°. Find the height of Cheops.

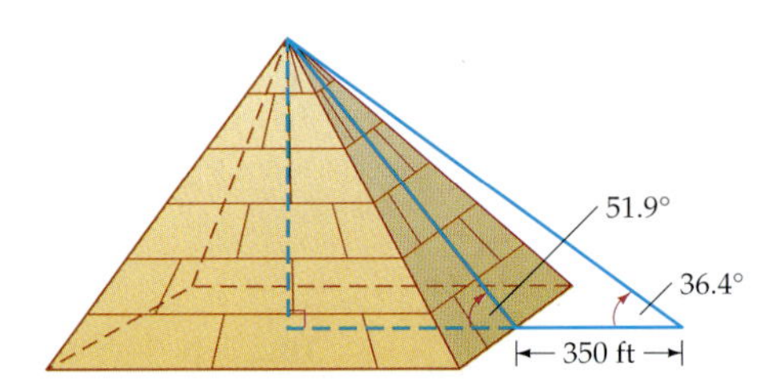

28. Height of a Building Two buildings are 240 feet apart. The angle of elevation from the top of the shorter building to the top of the other building is 22°. If the shorter building is 80 feet high, how high is the taller building?

29. Height of the Washington Monument From a point A on a line from the base of the Washington Monument, the angle of elevation to the top of the monument is 42.0°. From a point 100 feet away and on the same line, the angle to the top is 37.8° (see the figure on the following page). Find the approximate height of the Washington Monument.

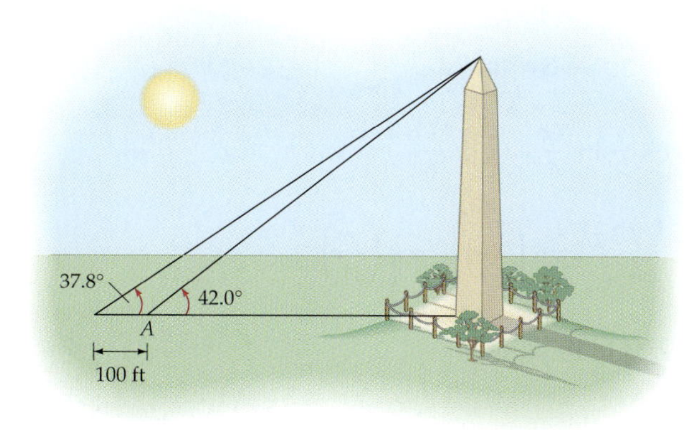

Figure for Exercise 29

30. **Height of a Building** The angle of elevation to the top of a radio antenna on the top of a building is 53.4°. After moving 200 feet closer to the building, the angle of elevation is 64.3°. Find the height of the building if the height of the antenna is 180 feet.

31. **The Petronas Towers** The Petronas Towers in Kuala Lumpur, Malaysia, are the world's tallest twin towers. Each tower is 1483 feet in height. The towers are connected by a skybridge at the forty-first floor. Note the information given in the accompanying figure.

a. Determine the height of the skybridge.

b. Determine the length of the skybridge.

$AB = 412$ feet
$\angle CAB = 53.6°$
$\overline{AB}$ is at ground level
$\angle CAD = 15.5°$

32. **An Eiffel Tower Replica** Use the information in the accompanying figure to estimate the height of the Eiffel Tower replica that stands in front of the Paris Hotel in Las Vegas, Nevada.

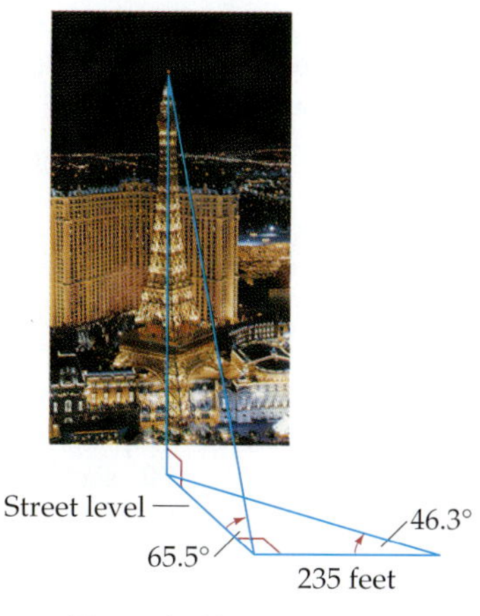

Figure for Exercise 32

In Exercises 33 to 38, find the value of each of the six trigonometric functions for the angle whose terminal side passes through the given point.

33. $P(2, 3)$ **34.** $P(3, 7)$ **35.** $P(-2, 3)$

▶ **36.** $P(-6, -9)$ **37.** $P(-5, 0)$ **38.** $P(0, 2)$

In Exercises 39 to 42, let θ be an angle in standard position. State the quadrant in which the terminal side of θ lies.

39. $\sin\theta > 0, \quad \cos\theta > 0$ **40.** $\tan\theta < 0, \quad \sin\theta < 0$

41. $\cos\theta > 0, \quad \tan\theta < 0$ **42.** $\sin\theta < 0, \quad \cos\theta > 0$

In Exercises 43 to 50, find the measure of the reference angle θ' for the given angle θ.

43. $\theta = 160°$ ▶ **44.** $\theta = 255°$ **45.** $\theta = 351°$

46. $\theta = 48°$ **47.** $\theta = \frac{11}{5}\pi$ **48.** $\theta = \frac{18}{7}\pi$

49. $\theta = 1406°$ **50.** $\theta = -650°$

In Exercises 51 to 62, use the Reference Angle Theorem to find the exact value of each trigonometric function.

51. $\sin 225°$ ▶ **52.** $\cos 300°$ **53.** $\tan 405°$

54. $\sec 150°$ **55.** $\csc \frac{4}{3}\pi$ **56.** $\cot \frac{7}{6}\pi$

57. $\cos \frac{17\pi}{4}$ **58.** $\tan\left(-\frac{\pi}{3}\right)$ **59.** $\sec 765°$

60. $\csc(-510°)$ **61.** $\cot 540°$ **62.** $\cos 570°$

In Exercises 63 to 70, use a calculator to estimate the value of the trigonometric function. Round to the nearest ten-thousandth.

63. $\tan 32^\circ$

64. $\sec 88^\circ$

65. $\cot 398^\circ$

66. $\sec 578^\circ$

67. $\sec \frac{3\pi}{8}$

68. $\tan(-4.12)$

69. $\csc \frac{9\pi}{5}$

70. $\sec(-4.45)$

In Exercises 71 to 76, find (without using a calculator) the exact value of each expression.

71. $\sin 210^\circ - \cos 330^\circ \tan 330^\circ$

72. $\tan 225^\circ + \sin 240^\circ \cos 60^\circ$

73. $\sin^2 30^\circ + \cos^2 30^\circ$

74. $\cos \pi \sin \frac{7\pi}{4} - \tan \frac{11\pi}{6}$

75. $\sin\left(\frac{3\pi}{2}\right) \tan\left(\frac{\pi}{4}\right) - \cos\left(\frac{\pi}{3}\right)$

76. $\tan^2\left(\frac{7\pi}{4}\right) - \sec^2\left(\frac{7\pi}{4}\right)$

CONNECTING CONCEPTS

In Exercises 77 to 80, find two values of θ, $0^\circ \le \theta < 360^\circ$, that satisfy the given trigonometric equation.

77. $\sin \theta = \frac{1}{2}$

78. $\tan \theta = -\sqrt{3}$

79. $\cos \theta = -\frac{\sqrt{3}}{2}$

80. $\tan \theta = 1$

In Exercises 81 to 84, find two values of θ, $0 \le \theta < 2\pi$, that satisfy the given trigonometric equation.

81. $\tan \theta = -1$

82. $\cos \theta = \frac{1}{2}$

83. $\sin \theta = \frac{\sqrt{3}}{2}$

84. $\cos \theta = -\frac{1}{2}$

PREPARE FOR SECTION 5.2

85. Compare $\cos(\alpha - \beta)$ and $\cos \alpha \cos \beta + \sin \alpha \sin \beta$ for $\alpha = \frac{\pi}{2}$ and $\beta = \frac{\pi}{6}$. [5.1]

86. Compare $\sin(\alpha + \beta)$ and $\sin \alpha \cos \beta + \cos \alpha \sin \beta$ for $\alpha = \frac{\pi}{2}$ and $\beta = \frac{\pi}{3}$. [5.1]

87. Compare $\sin(90^\circ - \theta)$ and $\cos \theta$ for $\theta = 30^\circ$, $\theta = 45^\circ$, and $\theta = 120^\circ$. [5.1]

88. Compare $\tan\left(\frac{\pi}{2} - \theta\right)$ and $\cot \theta$ for $\theta = \frac{\pi}{6}$, $\theta = \frac{\pi}{4}$, and $\theta = \frac{4\pi}{3}$. [5.1]

89. Compare $\tan(\alpha - \beta)$ and $\frac{\tan \alpha - \tan \beta}{1 + \tan \alpha \tan \beta}$ for $\alpha = \frac{\pi}{3}$ and $\beta = \frac{\pi}{6}$. [5.1]

90. Compare $\sec^2 \theta$ and $\tan^2 \theta + 1$ for $\theta = 30^\circ$, $\theta = 45^\circ$, and $\theta = 60^\circ$. [5.1]

PROJECTS

1. **FIND SUMS OR PRODUCTS** Determine the following sums or products. Do not use a calculator. (*Hint:* The Reference Angle Theorem may be helpful.) Explain to a classmate how you know you are correct.

   a. $\cos 0° + \cos 1° + \cos 2° + \cdots + \cos 178° + \cos 179° + \cos 180°$

   b. $\sin 0° + \sin 1° + \sin 2° + \cdots + \sin 358° + \sin 359° + \sin 360°$

   c. $\cot 1° + \cot 2° + \cot 3° + \cdots + \cot 177° + \cot 178° + \cot 179°$

   d. $(\cos 1°)(\cos 2°)(\cos 3°) \cdots (\cos 177°)(\cos 178°)(\cos 179°)$

   e. $\cos^2 1° + \cos^2 2° + \cos^2 3° + \cdots + \cos^2 357° + \cos^2 358° + \cos^2 359°$

SECTION 5.2

VERIFICATION OF TRIGONOMETRIC IDENTITIES

- VERIFICATION OF TRIGONOMETRIC IDENTITIES
- IDENTITIES THAT INVOLVE $(\alpha \pm \beta)$
- COFUNCTIONS
- ADDITIONAL SUM AND DIFFERENCE IDENTITIES

VERIFICATION OF TRIGONOMETRIC IDENTITIES

The domain of an equation consists of all values of the variable for which every term is defined. For example, the domain of

$$\frac{\sin x \cos x}{\sin x} = \cos x \tag{1}$$

includes all real numbers x except $x = n\pi$, where n is an integer, because $\sin x = 0$ for $x = n\pi$, and division by 0 is undefined. An **identity** is an equation that is true for all of its domain values. **Table 5.5** lists identities that were introduced earlier.

TABLE 5.5 Fundamental Trigonometric Identities

Reciprocal identities	$\sin x = \dfrac{1}{\csc x}$	$\cos x = \dfrac{1}{\sec x}$	$\tan x = \dfrac{1}{\cot x}$
Ratio identities	$\tan x = \dfrac{\sin x}{\cos x}$	$\cot x = \dfrac{\cos x}{\sin x}$	
Pythagorean identities	$\sin^2 x + \cos^2 x = 1$	$\tan^2 x + 1 = \sec^2 x$	$1 + \cot^2 x = \csc^2 x$
Odd-even identities	$\sin(-x) = -\sin x$ $\cos(-x) = \cos x$	$\tan(-x) = -\tan x$ $\cot(-x) = -\cot x$	$\sec(-x) = \sec x$ $\csc(-x) = -\csc x$

To verify an identity, we show that one side of the identity can be rewritten in a form that is identical to the other side. There is no one method that can be used to verify every identity; however, the following guidelines should prove useful.

Guidelines for Verifying Trigonometric Identities

- If one side of the identity is more complex than the other, then it is generally best to try first to simplify the more complex side until it becomes identical to the other side.
- Perform indicated operations such as adding fractions or squaring a binomial. Also be aware of any factorization that may help you to achieve your goal of producing the expression on the other side.
- Make use of previously established identities that enable you to rewrite one side of the identity in an equivalent form.
- Rewrite one side of the identity so that it involves only sines and/or cosines.
- Rewrite one side of the identity in terms of a single trigonometric function.
- Multiplying both the numerator and the denominator of a fraction by the same factor (such as the conjugate of the denominator or the conjugate of the numerator) may get you closer to your goal.
- Keep your goal in mind. Does it involve products, quotients, sums, radicals, or powers? Knowing exactly what your goal is may provide the insight you need to verify the identity.

take note

Each of the Pythagorean identities can be written in several different forms. For instance,

$$\sin^2 x + \cos^2 x = 1$$

also can be written as

$$\sin^2 x = 1 - \cos^2 x$$

and as

$$\cos^2 x = 1 - \sin^2 x$$

EXAMPLE 1 Verify an Identity

Verify the identity $1 - 2\sin^2 x = 2\cos^2 x - 1$.

Solution

Rewrite the right side of the equation.

$$\begin{aligned} 2\cos^2 x - 1 &= 2(1 - \sin^2 x) - 1 && \text{• } \cos^2 x = 1 - \sin^2 x \\ &= 2 - 2\sin^2 x - 1 \\ &= 1 - 2\sin^2 x \end{aligned}$$

TRY EXERCISE 12, PAGE 384

Figure 5.18 shows the graph of $f(x) = 1 - 2\sin^2 x$ and the graph of $g(x) = 2\cos^2 x - 1$ on the same coordinate axes. The fact that the graphs appear to be identical on the interval $[-2\pi, 2\pi]$ supports the verification in Example 1.

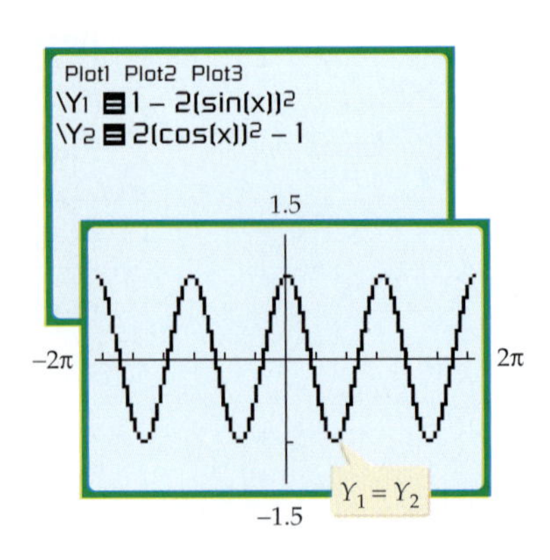

FIGURE 5.18

QUESTION Is $\cos(-x) = \cos x$ an identity?

EXAMPLE 2 Factor to Verify an Identity

Verify the identity $\csc^2 x - \cos^2 x \csc^2 x = 1$.

Solution

Simplify the left side of the equation.

$$\begin{aligned} \csc^2 x - \cos^2 x \csc^2 x &= \csc^2 x(1 - \cos^2 x) && \text{• Factor out } \csc^2 x. \\ &= \csc^2 x \sin^2 x && \text{• } 1 - \cos^2 x = \sin^2 x \\ &= \frac{1}{\sin^2 x} \cdot \sin^2 x = 1 && \text{• } \csc^2 x = \frac{1}{\sin^2 x} \end{aligned}$$

TRY EXERCISE 22, PAGE 385

In the next example we make use of the guideline that indicates that it may be useful to multiply both the numerator and the denominator of a fraction by the same factor.

EXAMPLE 3 Multiply by a Conjugate to Verify an Identity

Verify the identity $\dfrac{\sin x}{1 + \cos x} = \dfrac{1 - \cos x}{\sin x}$.

Solution

Multiply the numerator and denominator of the left side of the identity by the conjugate of $1 + \cos x$, which is $1 - \cos x$.

$$\begin{aligned} \frac{\sin x}{1 + \cos x} &= \frac{\sin x}{1 + \cos x} \cdot \frac{1 - \cos x}{1 - \cos x} = \frac{\sin x(1 - \cos x)}{1 - \cos^2 x} \\ &= \frac{\sin x(1 - \cos x)}{\sin^2 x} = \frac{1 - \cos x}{\sin x} \end{aligned}$$

TRY EXERCISE 32, PAGE 385

take note

The sum $a + b$ and the difference $a - b$ are called conjugates of each other.

EXAMPLE 4 Change to Sines and Cosines to Verify an Identity

Verify the identity $\dfrac{\sin x + \tan x}{1 + \cos x} = \tan x$.

Continued ▶

ANSWER Yes, $\cos(-x) = \cos x$ is one of the odd-even identities shown in **Table 5.5.**

Solution

Rewrite the left side of the identity in terms of sines and cosines.

$$\frac{\sin x + \tan x}{1 + \cos x} = \frac{\sin x + \dfrac{\sin x}{\cos x}}{1 + \cos x}$$ • $\tan x = \dfrac{\sin x}{\cos x}$

$$= \frac{\dfrac{\sin x \cos x + \sin x}{\cos x}}{1 + \cos x}$$ • Write the terms in the numerator with a common denominator.

$$= \frac{\sin x \cos x + \sin x}{\cos x(1 + \cos x)}$$ • Simplify.

$$= \frac{\sin x(1 + \cos x)}{\cos x(1 + \cos x)}$$

$$= \tan x$$

▶ TRY EXERCISE 40, PAGE 385

IDENTITIES THAT INVOLVE $(\alpha \pm \beta)$

Each identity in the previous examples involved only one variable. We now consider identities that involve a trigonometric function of the sum or difference of two variables.

Sum and Difference Identities

$$\cos(\alpha - \beta) = \cos \alpha \cos \beta + \sin \alpha \sin \beta$$

$$\cos(\alpha + \beta) = \cos \alpha \cos \beta - \sin \alpha \sin \beta$$

$$\sin(\alpha - \beta) = \sin \alpha \cos \beta - \cos \alpha \sin \beta$$

$$\sin(\alpha + \beta) = \sin \alpha \cos \beta + \cos \alpha \sin \beta$$

$$\tan(\alpha + \beta) = \frac{\tan \alpha + \tan \beta}{1 - \tan \alpha \tan \beta}$$

$$\tan(\alpha - \beta) = \frac{\tan \alpha - \tan \beta}{1 + \tan \alpha \tan \beta}$$

To establish the identity for $\cos(\alpha - \beta)$, we make use of the unit circle shown in **Figure 5.19.** The angles α and β are drawn in standard position, with OA and OB as the terminal sides of α and β, respectively. The coordinates of A are $(\cos \alpha, \sin \alpha)$, and the coordinates of B are $(\cos \beta, \sin \beta)$. The angle $(\alpha - \beta)$ is formed by the terminal sides of the angles α and β (angle AOB).

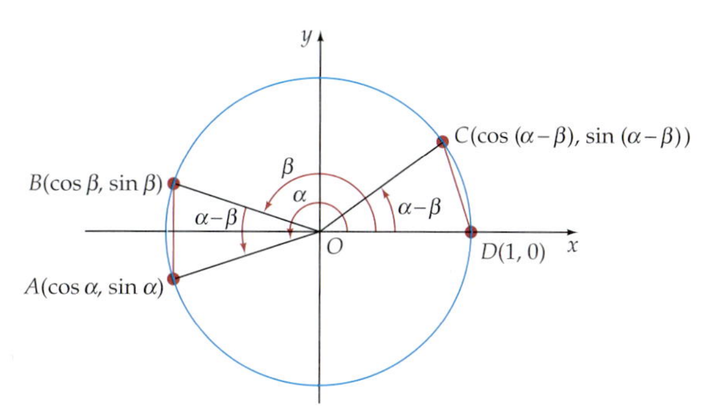

FIGURE 5.19

An angle equal in measure to angle $(\alpha - \beta)$ is placed in standard position in the same figure (angle COD). From geometry, if two central angles of a circle have the same measure, then their chords are also equal in measure. Thus the chords AB and CD are equal in length. Using the distance formula, we can calculate the lengths of the chords AB and CD.

$$d(A, B) = \sqrt{(\cos\alpha - \cos\beta)^2 + (\sin\alpha - \sin\beta)^2}$$

$$d(C, D) = \sqrt{[\cos(\alpha - \beta) - 1]^2 + [\sin(\alpha - \beta) - 0]^2}$$

Because $d(A, B) = d(C, D)$, we have

$$\sqrt{(\cos\alpha - \cos\beta)^2 + (\sin\alpha - \sin\beta)^2} = \sqrt{[\cos(\alpha - \beta) - 1]^2 + [\sin(\alpha - \beta)]^2}$$

Squaring each side of the equation and simplifying, we obtain

$$(\cos\alpha - \cos\beta)^2 + (\sin\alpha - \sin\beta)^2 = [\cos(\alpha - \beta) - 1]^2 + [\sin(\alpha - \beta)]^2$$

$$\cos^2\alpha - 2\cos\alpha\cos\beta + \cos^2\beta + \sin^2\alpha - 2\sin\alpha\sin\beta + \sin^2\beta$$

$$= \cos^2(\alpha - \beta) - 2\cos(\alpha - \beta) + 1 + \sin^2(\alpha - \beta)$$

$$\cos^2\alpha + \sin^2\alpha + \cos^2\beta + \sin^2\beta - 2\cos\alpha\cos\beta - 2\sin\alpha\sin\beta$$

$$= \cos^2(\alpha - \beta) + \sin^2(\alpha - \beta) + 1 - 2\cos(\alpha - \beta)$$

Simplifying by using $\sin^2\theta + \cos^2\theta = 1$, we have

$$2 - 2\sin\alpha\sin\beta - 2\cos\alpha\cos\beta = 2 - 2\cos(\alpha - \beta)$$

Solving for $\cos(\alpha - \beta)$ gives us

$$\cos(\alpha - \beta) = \cos\alpha\cos\beta + \sin\alpha\sin\beta$$

To derive an identity for $\cos(\alpha + \beta)$, write $\cos(\alpha + \beta)$ as $\cos[\alpha - (-\beta)]$.

$$\cos(\alpha + \beta) = \cos[\alpha - (-\beta)] = \cos\alpha\cos(-\beta) + \sin\alpha\sin(-\beta)$$

Recall that $\cos(-\beta) = \cos\beta$ and $\sin(-\beta) = -\sin\beta$. Substituting into the previous equation, we obtain the identity

$$\cos(\alpha + \beta) = \cos\alpha\cos\beta - \sin\alpha\sin\beta$$

EXAMPLE 5 **Evaluate a Trigonometric Expression**

Use an identity to find the *exact* value of $\cos(60° - 45°)$.

Solution

Use the identity $\cos(\alpha - \beta) = \cos\alpha\cos\beta + \sin\alpha\sin\beta$ with $\alpha = 60°$ and $\beta = 45°$.

$$\begin{aligned}\cos(60° - 45°) &= \cos 60° \cos 45° + \sin 60° \sin 45° && \text{• Substitute.}\\ &= \left(\frac{1}{2}\right)\left(\frac{\sqrt{2}}{2}\right) + \left(\frac{\sqrt{3}}{2}\right)\left(\frac{\sqrt{2}}{2}\right) && \text{• Evaluate each factor.}\\ &= \frac{\sqrt{2}}{4} + \frac{\sqrt{6}}{4} && \text{• Simplify.}\\ &= \frac{\sqrt{2} + \sqrt{6}}{4}\end{aligned}$$

TRY EXERCISE 46, PAGE 385

take note

To visualize the cofunction identities, consider the right triangle shown in the following figure.

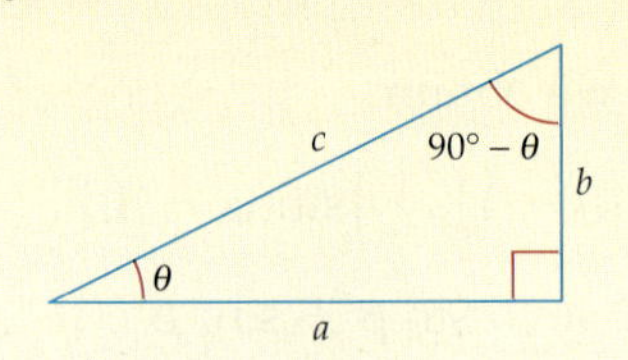

If θ is the degree measure of one of the acute angles, then the degree measure of the other acute angle is $(90° - \theta)$. Using the definitions of the trigonometric functions gives us

$$\sin\theta = \frac{b}{c} = \cos(90° - \theta)$$

$$\tan\theta = \frac{b}{a} = \cot(90° - \theta)$$

$$\sec\theta = \frac{c}{a} = \csc(90° - \theta)$$

These identities state that the value of a trigonometric function of θ is equal to the cofunction of the complement of θ.

COFUNCTIONS

Any pair of trigonometric functions f and g for which

$$f(x) = g(90° - x) \quad \text{and} \quad g(x) = f(90° - x)$$

are said to be **cofunctions.**

Cofunction Identities

$$\sin(90° - \theta) = \cos\theta \qquad \cos(90° - \theta) = \sin\theta$$

$$\tan(90° - \theta) = \cot\theta \qquad \cot(90° - \theta) = \tan\theta$$

$$\sec(90° - \theta) = \csc\theta \qquad \csc(90° - \theta) = \sec\theta$$

If θ is in radian measure, replace $90°$ with $\frac{\pi}{2}$.

To verify that the sine function and the cosine function are cofunctions, we make use of the identity for $\cos(\alpha - \beta)$.

$$\begin{aligned}\cos(90° - \beta) &= \cos 90° \cos\beta + \sin 90° \sin\beta\\ &= 0 \cdot \cos\beta + 1 \cdot \sin\beta\end{aligned}$$

which gives

$$\cos(90° - \beta) = \sin\beta$$

Thus the sine of an angle is equal to the cosine of its complement. Using $\cos(90° - \beta) = \sin\beta$ with $\beta = 90° - \alpha$, we have

$$\cos\alpha = \cos[90° - (90° - \alpha)] = \sin(90° - \alpha)$$

Therefore,

$$\cos \alpha = \sin(90^\circ - \alpha)$$

We can use the ratio identities to show that the tangent and cotangent functions are cofunctions.

$$\tan(90^\circ - \theta) = \frac{\sin(90^\circ - \theta)}{\cos(90^\circ - \theta)} = \frac{\cos \theta}{\sin \theta} = \cot \theta$$

$$\cot(90^\circ - \theta) = \frac{\cos(90^\circ - \theta)}{\sin(90^\circ - \theta)} = \frac{\sin \theta}{\cos \theta} = \tan \theta$$

The secant and cosecant functions are also cofunctions.

EXAMPLE 6 Write an Equivalent Expression

Use a cofunction identity to write an equivalent expression for $\sin 20^\circ$.

Solution

The value of a given trigonometric function of θ, measured in degrees, is equal to its cofunction of $90^\circ - \theta$. Thus

$$\begin{aligned} \sin 20^\circ &= \cos(90^\circ - 20^\circ) \\ &= \cos 70^\circ \end{aligned}$$

TRY EXERCISE 56, PAGE 386

ADDITIONAL SUM AND DIFFERENCE IDENTITIES

We can use the cofunction identities to verify the remaining sum and difference identities. To derive an identity for $\sin(\alpha + \beta)$, substitute $\alpha + \beta$ for θ in the cofunction identity $\sin \theta = \cos(90^\circ - \theta)$.

$$\begin{aligned} \sin \theta &= \cos(90^\circ - \theta) \\ \sin(\alpha + \beta) &= \cos[90^\circ - (\alpha + \beta)] && \text{• Replace } \theta \text{ with } \alpha + \beta. \\ &= \cos[(90^\circ - \alpha) - \beta] && \text{• Rewrite as the difference of two angles.} \\ &= \cos(90^\circ - \alpha)\cos \beta + \sin(90^\circ - \alpha)\sin \beta \\ &= \sin \alpha \cos \beta + \cos \alpha \sin \beta \end{aligned}$$

Therefore,

$$\sin(\alpha + \beta) = \sin \alpha \cos \beta + \cos \alpha \sin \beta$$

We also can derive an identity for $\sin(\alpha - \beta)$ by rewriting $(\alpha - \beta)$ as $[\alpha + (-\beta)]$.

$$\begin{aligned} \sin(\alpha - \beta) &= \sin[\alpha + (-\beta)] \\ &= \sin \alpha \cos(-\beta) + \cos \alpha \sin(-\beta) \\ &= \sin \alpha \cos \beta - \cos \alpha \sin \beta && \text{• } \cos(-\beta) = \cos \beta,\ \sin(-\beta) = -\sin \beta \end{aligned}$$

Thus

$$\sin(\alpha - \beta) = \sin\alpha\cos\beta - \cos\alpha\sin\beta$$

The identity for $\tan(\alpha + \beta)$ is a result of the identity $\tan\theta = \dfrac{\sin\theta}{\cos\theta}$ and the identities for $\sin(\alpha + \beta)$ and $\cos(\alpha + \beta)$.

$$\tan(\alpha + \beta) = \frac{\sin(\alpha + \beta)}{\cos(\alpha + \beta)} = \frac{\sin\alpha\cos\beta + \cos\alpha\sin\beta}{\cos\alpha\cos\beta - \sin\alpha\sin\beta}$$

$$= \frac{\dfrac{\sin\alpha\cos\beta}{\cos\alpha\cos\beta} + \dfrac{\cos\alpha\sin\beta}{\cos\alpha\cos\beta}}{\dfrac{\cos\alpha\cos\beta}{\cos\alpha\cos\beta} - \dfrac{\sin\alpha\sin\beta}{\cos\alpha\cos\beta}}$$

• **Multiply both the numerator and the denominator by $\dfrac{1}{\cos\alpha\cos\beta}$ and simplify.**

Therefore,

$$\tan(\alpha + \beta) = \frac{\tan\alpha + \tan\beta}{1 - \tan\alpha\tan\beta}$$

The tangent function is an odd function, so $\tan(-\theta) = -\tan\theta$. Rewriting $(\alpha - \beta)$ as $[\alpha + (-\beta)]$ enables us to derive an identity for $\tan(\alpha - \beta)$.

$$\tan(\alpha - \beta) = \tan[\alpha + (-\beta)] = \frac{\tan\alpha + \tan(-\beta)}{1 - \tan\alpha\tan(-\beta)}$$

Therefore,

$$\tan(\alpha - \beta) = \frac{\tan\alpha - \tan\beta}{1 + \tan\alpha\tan\beta}$$

The sum and difference identities can be used to simplify some trigonometric expressions. For example,

$$\sin 5x\cos 3x - \cos 5x\sin 3x = \sin(5x - 3x) = \sin 2x$$

and

$$\frac{\tan 4\alpha + \tan\alpha}{1 - \tan 4\alpha\tan\alpha} = \tan(4\alpha + \alpha) = \tan 5\alpha$$

In Example 7 we use a sum identity to evaluate a function.

EXAMPLE 7 Evaluate a Trigonometric Function

Given $\tan\alpha = -\dfrac{4}{3}$ for α in Quadrant II and $\tan\beta = -\dfrac{5}{12}$ for β in Quadrant IV, find $\sin(\alpha + \beta)$.

Solution

See **Figure 5.20.** Because $\tan\alpha = \dfrac{y}{x} = -\dfrac{4}{3}$ and the terminal side of α is in Quadrant II, $P_1(-3, 4)$ is a point on the terminal side of α. Similarly, $P_2(12, -5)$ is a point on the terminal side of β.

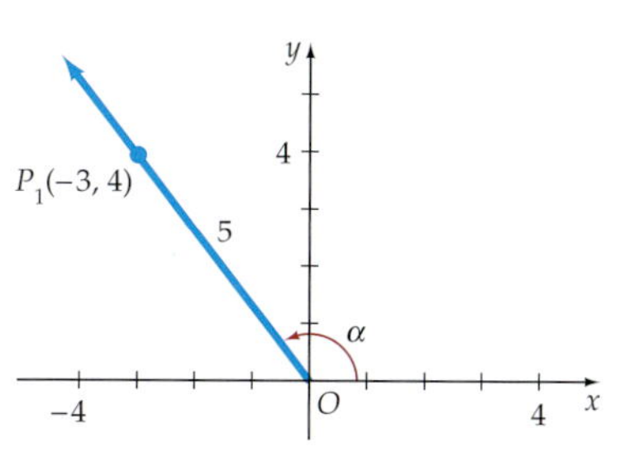

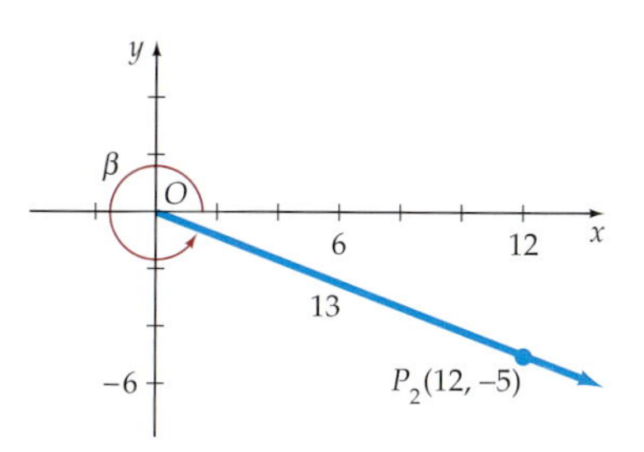

FIGURE 5.20

Using the Pythagorean Theorem, we find that the length of the line segment OP_1 is 5 and the length of OP_2 is 13.

$$\sin(\alpha + \beta) = \sin \alpha \cos \beta + \cos \alpha \sin \beta$$
$$= \frac{4}{5} \cdot \frac{12}{13} + \frac{-3}{5} \cdot \frac{-5}{13} = \frac{48}{65} + \frac{15}{65} = \frac{63}{65}$$

▶ TRY EXERCISE 68, PAGE 386

EXAMPLE 8 Verify an Identity

Verify each of the following identities.

a. $\cos(\pi - \theta) = -\cos \theta$

b. $\dfrac{\cos 4\theta}{\sin \theta} - \dfrac{\sin 4\theta}{\cos \theta} = \dfrac{\cos 5\theta}{\sin \theta \cos \theta}$

Solution

a. Use the identity for $\cos(\alpha - \beta)$.

$$\cos(\pi - \theta) = \cos \pi \cos \theta + \sin \pi \sin \theta = -1 \cdot \cos \theta + 0 \cdot \sin \theta = -\cos \theta$$

b. Subtract the fractions on the left side of the equation.

$$\frac{\cos 4\theta}{\sin \theta} - \frac{\sin 4\theta}{\cos \theta} = \frac{\cos 4\theta \cos \theta - \sin 4\theta \sin \theta}{\sin \theta \cos \theta}$$
$$= \frac{\cos(4\theta + \theta)}{\sin \theta \cos \theta} = \frac{\cos 5\theta}{\sin \theta \cos \theta}$$

• **Use the identity for $\cos(\alpha + \beta)$.**

▶ TRY EXERCISE 84, PAGE 386

Figure 5.21 shows the graphs of $f(\theta) = \cos(\pi - \theta)$ and $g(\theta) = -\cos \theta$ on the same coordinate axes. The fact that the graphs appear to be identical supports the verification in Example 8a. **Figure 5.22** shows the graph of $f(\theta) = \dfrac{\cos 4\theta}{\sin \theta} - \dfrac{\sin 4\theta}{\cos \theta}$ and the graph of $g(\theta) = \dfrac{\cos 5\theta}{\sin \theta \cos \theta}$ on the same coordinate axes. The fact that the graphs appear to be identical supports the verification in Example 8b.

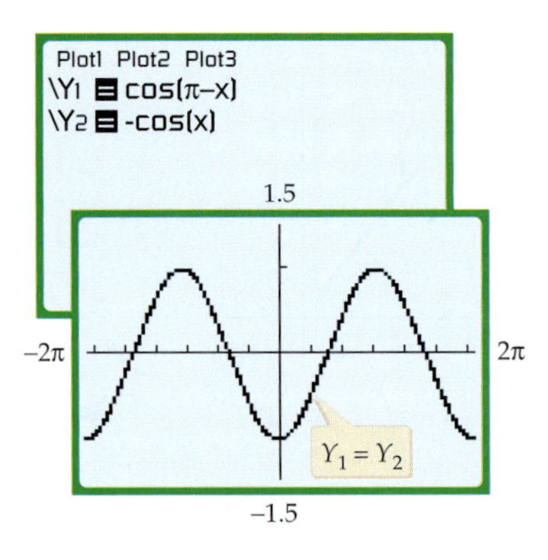

FIGURE 5.21

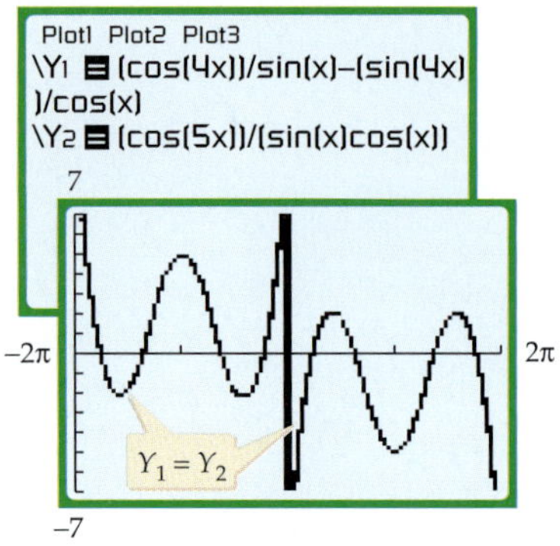

Plot1 Plot2 Plot3
\Y1 ■ (cos(4x))/sin(x)–(sin(4x))/cos(x)
\Y2 ■ (cos(5x))/(sin(x)cos(x))
7
–2π
2π
–7
$Y_1 = Y_2$

FIGURE 5.22

TOPICS FOR DISCUSSION

1. Is $\cos|x| = \cos x$ an identity? Explain.
2. Does $\sin(\alpha + \beta) = \sin \alpha + \sin \beta$ for all values of α and β? If not, find nonzero values of α and β for which $\sin(\alpha + \beta) \neq \sin \alpha + \sin \beta$.
3. If k is an integer, then $2k + 1$ is an odd integer. Do you agree? Explain.
4. What are the trigonometric cofunction identities? Explain.

EXERCISE SET 5.2

In Exercises 1 to 42, verify each identity.

1. $\tan x \csc x \cos x = 1$
2. $\sin x \cot x \sec x = 1$
3. $\dfrac{4\sin^2 x - 1}{2\sin x + 1} = 2\sin x - 1$
4. $\dfrac{\sin^2 x - 2\sin x + 1}{\sin x - 1} = \sin x - 1$
5. $(\sin x - \cos x)(\sin x + \cos x) = 1 - 2\cos^2 x$
6. $(\tan x)(1 - \cot x) = \tan x - 1$
7. $\dfrac{1}{\sin x} - \dfrac{1}{\cos x} = \dfrac{\cos x - \sin x}{\sin x \cos x}$
8. $\dfrac{1}{\sin x} + \dfrac{3}{\cos x} = \dfrac{\cos x + 3\sin x}{\sin x \cos x}$
9. $\dfrac{\cos x}{1 - \sin x} = \sec x + \tan x$
10. $\dfrac{\sin x}{1 - \cos x} = \csc x + \cot x$
11. $\dfrac{1 - \tan^4 x}{\sec^2 x} = 1 - \tan^2 x$
12. $\sin^4 x - \cos^4 x = \sin^2 x - \cos^2 x$
13. $\dfrac{\sin x - 2 + \dfrac{1}{\sin x}}{\sin x - \dfrac{1}{\sin x}} = \dfrac{\sin x - 1}{\sin x + 1}$
14. $\dfrac{\sin x}{1 - \cos x} - \dfrac{\sin x}{1 + \cos x} = 2\cot x$
15. $(\sin x + \cos x)^2 = 1 + 2\sin x \cos x$

16. $(\tan x + 1)^2 = \sec^2 x + 2\tan x$

17. $\dfrac{\cos x}{1 + \sin x} = \sec x - \tan x$

18. $\dfrac{\sin x}{1 + \cos x} = \csc x - \cot x$

19. $\csc x = \dfrac{\cot x + \tan x}{\sec x}$

20. $\sec x = \dfrac{\cot x + \tan x}{\csc x}$

21. $\dfrac{\cos x \tan x + 2\cos x - \tan x - 2}{\tan x + 2} = \cos x - 1$

22. $\dfrac{2\sin x \cot x + \sin x - 4\cot x - 2}{2\cot x + 1} = \sin x - 2$

23. $\sec x - \tan x = \dfrac{1 - \sin x}{\cos x}$

24. $\cot x - \csc x = \dfrac{\cos x - 1}{\sin x}$

25. $\sin^2 x - \cos^2 x = 2\sin^2 x - 1$

26. $\sin^2 x - \cos^2 x = 1 - 2\cos^2 x$

27. $\dfrac{1}{\sin^2 x} + \dfrac{1}{\cos^2 x} = \csc^2 x \sec^2 x$

28. $\dfrac{1}{\tan^2 x} - \dfrac{1}{\cot^2 x} = \csc^2 x - \sec^2 x$

29. $\sec x - \cos x = \sin x \tan x$

30. $\tan x + \cot x = \sec x \csc x$

31. $\dfrac{\dfrac{1}{\sin x} + 1}{\dfrac{1}{\sin x} - 1} = \tan^2 x + 2\tan x \sec x + \sec^2 x$

32. $\dfrac{\dfrac{1}{\sin x} + \dfrac{1}{\cos x}}{\dfrac{1}{\sin x} - \dfrac{1}{\cos x}} = \dfrac{\cos^2 x - \sin^2 x}{1 - 2\cos x \sin x}$

33. $\dfrac{1}{1 - \cos x} = \dfrac{1 + \cos x}{\sin^2 x}$

34. $1 + \sin x = \dfrac{\cos^2 x}{1 - \sin x}$

35. $\dfrac{\sin x}{1 - \sin x} - \dfrac{\cos x}{1 - \sin x} = \dfrac{1 - \cot x}{\csc x - 1}$

36. $\dfrac{\tan x}{1 + \tan x} - \dfrac{\cot x}{1 + \tan x} = 1 - \cot x$

37. $\dfrac{1}{1 + \cos x} - \dfrac{1}{1 - \cos x} = -2\cot x \csc x$

38. $\dfrac{1}{1 - \sin x} - \dfrac{1}{1 + \sin x} = 2\tan x \sec x$

39. $\dfrac{\dfrac{1}{\sin x} + \csc x}{\dfrac{1}{\sin x} - \sin x} = \dfrac{2}{\cos^2 x}$

40. $\dfrac{\dfrac{1}{\tan x} + \cot x}{\dfrac{1}{\tan x} + \tan x} = \dfrac{2}{\sec^2 x}$

41. $\dfrac{\cot x}{1 + \csc x} + \dfrac{1 + \csc x}{\cot x} = 2\sec x$

42. $\dfrac{1}{1 - \cos x} - \dfrac{\cos x}{1 + \cos x} = 2\csc^2 x - 1$

In Exercises 43 to 54, find the exact value of the expression.

43. $\sin(45° + 30°)$

44. $\sin(330° + 45°)$

45. $\tan(45° - 30°)$

46. $\cos(120° - 45°)$

47. $\sin\left(\dfrac{5\pi}{4} - \dfrac{\pi}{6}\right)$

48. $\cos\left(\dfrac{\pi}{4} - \dfrac{\pi}{3}\right)$

49. $\tan\left(\dfrac{\pi}{6} + \dfrac{\pi}{4}\right)$

50. $\tan\left(\dfrac{11\pi}{6} - \dfrac{\pi}{4}\right)$

51. $\cos 212° \cos 122° + \sin 212° \sin 122°$

52. $\sin 167° \cos 107° - \cos 167° \sin 107°$

53. $\sin\dfrac{5\pi}{12}\cos\dfrac{\pi}{4} - \cos\dfrac{5\pi}{12}\sin\dfrac{\pi}{4}$

54. $\cos\dfrac{\pi}{12}\cos\dfrac{\pi}{4} - \sin\dfrac{\pi}{12}\sin\dfrac{\pi}{4}$

In Exercises 55 to 60, use a cofunction identity to write an equivalent expression for the given value.

55. $\sin 42^\circ$

56. $\cos 80^\circ$

57. $\tan 15^\circ$

58. $\cot 2^\circ$

59. $\sec 25^\circ$

60. $\csc 84^\circ$

In Exercises 61 to 66, write each expression in terms of a single trigonometric function.

61. $\sin 7x \cos 2x - \cos 7x \sin 2x$

62. $\sin x \cos 3x + \cos x \sin 3x$

63. $\cos x \cos 2x + \sin x \sin 2x$

64. $\cos 4x \cos 2x - \sin 4x \sin 2x$

65. $\dfrac{\tan 3x + \tan 4x}{1 - \tan 3x \tan 4x}$

66. $\dfrac{\tan 2x - \tan 3x}{1 + \tan 2x \tan 3x}$

In Exercises 67 to 72, find the exact value of the given function.

67. Given $\tan\alpha = -\frac{4}{3}$, α in Quadrant II, and $\tan\beta = \frac{15}{8}$, β in Quadrant III, find $\sin(\alpha - \beta)$.

68. Given $\tan\alpha = \frac{24}{7}$, α in Quadrant I, and $\sin\beta = -\frac{8}{17}$, β in Quadrant III, find $\cos(\alpha + \beta)$.

69. Given $\sin\alpha = \frac{3}{5}$, α in Quadrant I, and $\cos\beta = -\frac{5}{13}$, β in Quadrant II, find $\tan(\alpha - \beta)$.

70. Given $\sin\alpha = \frac{24}{25}$, α in Quadrant II, and $\cos\beta = -\frac{4}{5}$, β in Quadrant III, find $\cos(\beta - \alpha)$.

71. Given $\sin\alpha = -\frac{4}{5}$, α in Quadrant III, and $\cos\beta = -\frac{12}{13}$, β in Quadrant II, find $\cos(\alpha + \beta)$.

72. Given $\sin\alpha = -\frac{7}{25}$, α in Quadrant IV, and $\cos\beta = \frac{8}{17}$, β in Quadrant IV, find $\tan(\alpha + \beta)$.

In Exercises 73 to 88, verify the identity.

73. $\cos\left(\frac{\pi}{2} - \theta\right) = \sin\theta$

74. $\cos(\theta + \pi) = -\cos\theta$

75. $\sin\left(\theta + \frac{\pi}{2}\right) = \cos\theta$

76. $\sin(\theta + \pi) = -\sin\theta$

77. $\tan\left(\theta + \frac{\pi}{4}\right) = \dfrac{\tan\theta + 1}{1 - \tan\theta}$

78. $\tan 2\theta = \dfrac{2\tan\theta}{1 - \tan^2\theta}$

79. $\cot\left(\frac{\pi}{2} - \theta\right) = \tan\theta$

80. $\cot(\pi + \theta) = \cot\theta$

81. $\csc(\pi - \theta) = \csc\theta$

82. $\sec\left(\frac{\pi}{2} - \theta\right) = \csc\theta$

83. $\sin 6x \cos 2x - \cos 6x \sin 2x = 2\sin 2x \cos 2x$

84. $\cos 5x \cos 3x + \sin 5x \sin 3x = \cos^2 x - \sin^2 x$

85. $\cos(\alpha + \beta) + \cos(\alpha - \beta) = 2\cos\alpha\cos\beta$

86. $\cos(\alpha - \beta) - \cos(\alpha + \beta) = 2\sin\alpha\sin\beta$

87. $\dfrac{\cos(\alpha - \beta)}{\sin(\alpha + \beta)} = \dfrac{\cot\alpha + \tan\beta}{1 + \cot\alpha\tan\beta}$

88. $\dfrac{\sin(\alpha + \beta)}{\sin(\alpha - \beta)} = \dfrac{1 + \cot\alpha\tan\beta}{1 - \cot\alpha\tan\beta}$

In Exercises 89 and 90, compare the graphs of each side of the equation to predict whether the equation is an identity.

89. $\sin\left(\frac{\pi}{2} - x\right) = \cos x$

90. $\cos(x + \pi) = -\cos x$

CONNECTING CONCEPTS

In Exercises 91 to 95, verify the identity.

91. $\dfrac{1 - \sin x + \cos x}{1 + \sin x + \cos x} = \dfrac{\cos x}{\sin x + 1}$

92. $\dfrac{1 - \tan x + \sec x}{1 + \tan x - \sec x} = \dfrac{1 + \sec x}{\tan x}$

93. $\cos(x + y + z) = \cos x \cos y \cos z - \sin x \sin y \cos z - \sin x \cos y \sin z - \cos x \sin y \sin z$

94. $\dfrac{\sin(x + h) - \sin x}{h} = \cos x \dfrac{\sin h}{h} + \sin x \dfrac{(\cos h - 1)}{h}$

95. $\dfrac{\cos(x + h) - \cos x}{h} = \cos x \dfrac{(\cos h - 1)}{h} - \sin x \dfrac{\sin h}{h}$

96. **Model Resistance** The drag (resistance) on a fish when it is swimming is two to three times the drag when it is gliding. To compensate for this, some fish swim in a sawtooth pattern, as shown in the accompanying figure. The ratio of the amount of energy the fish expends when swimming upward at angle β and then gliding down at angle α to the energy it expends swimming horizontally is given by

$$E_R = \frac{k \sin \alpha + \sin \beta}{k \sin(\alpha + \beta)}$$

where k is a value such that $2 \le k \le 3$, and k depends on the assumptions we make about the amount of drag experienced by the fish. Find E_R for $k = 2$, $\alpha = 10°$, and $\beta = 20°$.

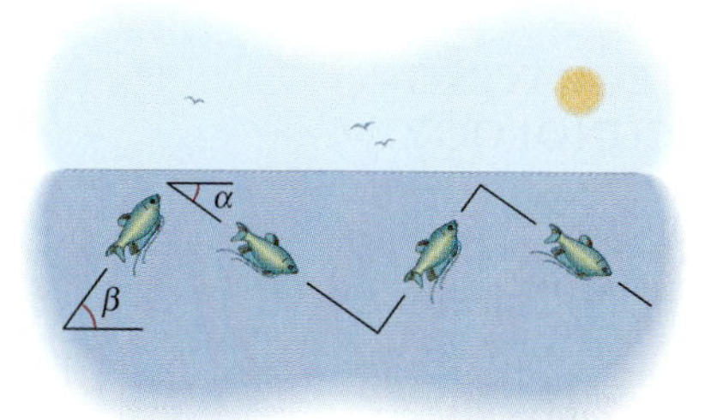

PREPARE FOR SECTION 5.3

97. Use the identity for $\sin(\alpha + \beta)$ to rewrite $\sin 2\alpha$. [5.2]

98. Use the identity for $\cos(\alpha + \beta)$ to rewrite $\cos 2\alpha$. [5.2]

99. Use the identity for $\tan(\alpha + \beta)$ to rewrite $\tan 2\alpha$. [5.2]

100. Compare $\tan \dfrac{\alpha}{2}$ and $\dfrac{\sin \alpha}{1 + \cos \alpha}$ for $\alpha = 60°$, $\alpha = 90°$, and $\alpha = 120°$. [5.1]

101. Verify that $\sin 2\alpha = 2 \sin \alpha$ is *not* an identity. *Hint:* Find a value of α for which the left side of the equation does not equal the right side. [5.1]

102. Verify that $\cos \dfrac{\alpha}{2} = \dfrac{1}{2} \cos \alpha$ is *not* an identity. [5.1]

PROJECTS

1. **Grading a Quiz** Suppose that you are a teacher's assistant. You are to assist the teacher of a trigonometry class by grading a four-question quiz. Each question asks the student to find a trigonometric expression that models a given application. The teacher has prepared an answer key. These answers are shown in the next column. A student gives as answers the expressions shown in the far right column. Determine for which problems the student has given a correct response.

Answer Key	Student's Response
1. $\csc x \sec x$	1. $\cot x + \tan x$
2. $\cos^2 x$	2. $(1 + \sin x)(1 - \sin x)$
3. $\cos x \cot x$	3. $\csc x - \sec x$
4. $\csc x \cot x$	4. $\sin x(\cot x + \cot^3 x)$

SECTION 5.3 MORE ON TRIGONOMETRIC IDENTITIES

INTEGRATING TECHNOLOGY

One way of showing that $\sin 2x \neq 2 \sin x$ is by graphing $y = \sin 2x$ and $y = 2 \sin x$ and observing that the graphs are not the same.

DOUBLE-ANGLE IDENTITIES

By using the sum identities, we can derive identities for $f(2\alpha)$, where f is a trigonometric function. These are called the *double-angle identities.* To find the sine of a double angle, substitute α for β in the identity for $\sin(\alpha + \beta)$.

$$\sin(\alpha + \beta) = \sin \alpha \cos \beta + \cos \alpha \sin \beta$$

$$\sin(\alpha + \alpha) = \sin \alpha \cos \alpha + \cos \alpha \sin \alpha \qquad \text{• Let } \beta = \alpha.$$

$$\sin 2\alpha = 2 \sin \alpha \cos \alpha$$

A double-angle identity for cosine is derived in a similar manner.

$$\cos(\alpha + \beta) = \cos \alpha \cos \beta - \sin \alpha \sin \beta$$

$$\cos(\alpha + \alpha) = \cos \alpha \cos \alpha - \sin \alpha \sin \alpha \qquad \text{• Let } \beta = \alpha.$$

$$\cos 2\alpha = \cos^2 \alpha - \sin^2 \alpha$$

There are two alternative forms of the double-angle identity for $\cos 2\alpha$. Using $\cos^2 \alpha = 1 - \sin^2 \alpha$, we can rewrite the identity for $\cos 2\alpha$ as follows:

$$\cos 2\alpha = \cos^2 \alpha - \sin^2 \alpha$$

$$\cos 2\alpha = (1 - \sin^2 \alpha) - \sin^2 \alpha \qquad \text{• } \cos^2 \alpha = 1 - \sin^2 \alpha$$

$$\cos 2\alpha = 1 - 2 \sin^2 \alpha$$

We also can rewrite $\cos 2\alpha$ as

$$\cos 2\alpha = \cos^2 \alpha - \sin^2 \alpha$$

$$\cos 2\alpha = \cos^2 \alpha - (1 - \cos^2 \alpha) \qquad \text{• } \sin^2 \alpha = 1 - \cos^2 \alpha$$

$$\cos 2\alpha = 2 \cos^2 \alpha - 1$$

The double-angle identity for the tangent function is derived from the identity for $\tan(\alpha + \beta)$ with $\beta = \alpha$.

$$\tan(\alpha + \beta) = \frac{\tan \alpha + \tan \beta}{1 - \tan \alpha \tan \beta}$$

$$\tan(\alpha + \alpha) = \frac{\tan \alpha + \tan \alpha}{1 - \tan \alpha \tan \alpha} \qquad \text{• Let } \beta = \alpha.$$

$$\tan 2\alpha = \frac{2 \tan \alpha}{1 - \tan^2 \alpha}$$

The double-angle identities are often used to write a trigonometric expression in terms of a single trigonometric function. For instance,

$$2 \sin 3\theta \cos 3\theta = \sin 6\theta$$

and

$$\cos^2 2x - \sin^2 2x = \cos 4x$$

QUESTION Does $\sin\theta\cos\theta = \frac{1}{2}\sin 2\theta$?

The double-angle identities can also be used to evaluate some trigonometric expressions.

EXAMPLE 1 Evaluate a Trigonometric Function

For an angle α in Quadrant I, $\sin\alpha = \frac{4}{5}$. Find $\sin 2\alpha$.

Solution

Use the identity $\sin 2\alpha = 2\sin\alpha\cos\alpha$. Find $\cos\alpha$ by substituting for $\sin\alpha$ in $\sin^2\alpha + \cos^2\alpha = 1$ and solving for $\cos\alpha$.

$$\cos\alpha = \sqrt{1-\sin^2\alpha} = \sqrt{1-\left(\frac{4}{5}\right)^2} = \frac{3}{5}$$

• $\cos\alpha > 0$ if α is in Quadrant I.

Substitute the values of $\sin\alpha$ and $\cos\alpha$ in the double-angle formula for $\sin 2\alpha$.

$$\sin 2\alpha = 2\sin\alpha\cos\alpha = 2\left(\frac{4}{5}\right)\left(\frac{3}{5}\right) = \frac{24}{25}$$

▶ TRY EXERCISE 18, PAGE 398

EXAMPLE 2 Verify a Double-Angle Identity

Verify the identity $\csc 2\alpha = \frac{1}{2}(\tan\alpha + \cot\alpha)$.

Solution

Work on the right-hand side of the equation.

$$\frac{1}{2}(\tan\alpha + \cot\alpha) = \frac{1}{2}\left(\frac{\sin\alpha}{\cos\alpha} + \frac{\cos\alpha}{\sin\alpha}\right) = \frac{1}{2}\left(\frac{\sin^2\alpha + \cos^2\alpha}{\cos\alpha\sin\alpha}\right)$$

$$= \frac{1}{2\cos\alpha\sin\alpha} = \frac{1}{\sin 2\alpha} = \csc 2\alpha$$

▶ TRY EXERCISE 32, PAGE 398

ANSWER Yes. $\sin\theta\cos\theta = \frac{2\sin\theta\cos\theta}{2} = \frac{\sin 2\theta}{2} = \frac{1}{2}\sin 2\theta$.

Half-Angle Identities

An identity for one-half an angle, $\frac{\alpha}{2}$, is called a *half-angle identity*. To derive a half-angle identity for $\sin \frac{\alpha}{2}$, we solve for $\sin^2 \theta$ in the following double-angle identity for $\cos 2\theta$.

$$\cos 2\theta = 1 - 2\sin^2 \theta$$

$$\sin^2 \theta = \frac{1 - \cos 2\theta}{2}$$

Substitute $\frac{\alpha}{2}$ for θ and take the square root of each side of the equation.

$$\sin^2 \frac{\alpha}{2} = \frac{1 - \cos 2\left(\frac{\alpha}{2}\right)}{2}$$

$$\sin \frac{\alpha}{2} = \pm\sqrt{\frac{1 - \cos \alpha}{2}}$$

The sign of the radical is determined by the quadrant in which the terminal side of angle $\frac{\alpha}{2}$ lies.

In a similar manner, we derive an identity for $\cos \frac{\alpha}{2}$.

$$\cos 2\theta = 2\cos^2 \theta - 1$$

$$\cos^2 \theta = \frac{1 + \cos 2\theta}{2}$$

take note

The sign of $\sqrt{\frac{1 - \cos \alpha}{2}}$ depends on the quadrant in which the terminal side of $\frac{\alpha}{2}$ lies, not the terminal side of α.

Substitute $\frac{\alpha}{2}$ for θ and take the square root of each side of the equation.

$$\cos^2 \frac{\alpha}{2} = \frac{1 + \cos 2\left(\frac{\alpha}{2}\right)}{2}$$

$$\cos \frac{\alpha}{2} = \pm\sqrt{\frac{1 + \cos \alpha}{2}}$$

Two different identities for $\tan \frac{\alpha}{2}$ are possible.

$$\tan \frac{\alpha}{2} = \frac{\sin \frac{\alpha}{2}}{\cos \frac{\alpha}{2}} = \frac{\sin \frac{\alpha}{2}}{\cos \frac{\alpha}{2}} \cdot \frac{2\cos \frac{\alpha}{2}}{2\cos \frac{\alpha}{2}}$$

$$= \frac{2\sin \frac{\alpha}{2} \cos \frac{\alpha}{2}}{2\cos^2 \frac{\alpha}{2}} = \frac{\sin 2\left(\frac{\alpha}{2}\right)}{2\left(\pm\sqrt{\frac{1 + \cos \alpha}{2}}\right)^2}$$

• $\cos \frac{\alpha}{2} = \pm\sqrt{\frac{1 + \cos \alpha}{2}}$

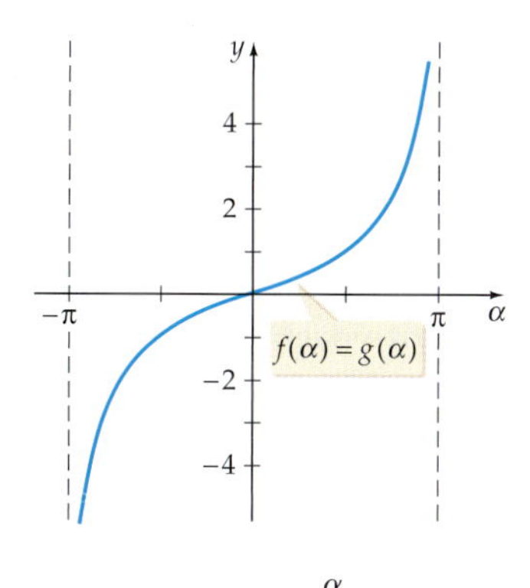

$f(\alpha) = \tan \frac{\alpha}{2}$

$g(\alpha) = \frac{\sin \alpha}{1 + \cos \alpha}$

FIGURE 5.23

$$\tan \frac{\alpha}{2} = \frac{\sin \alpha}{1 + \cos \alpha}$$

• See **Figure 5.23.**

To obtain another identity for $\tan \frac{\alpha}{2}$, multiply by the conjugate of the denominator.

$$\tan \frac{\alpha}{2} = \frac{\sin \alpha}{1 + \cos \alpha} \cdot \frac{1 - \cos \alpha}{1 - \cos \alpha}$$

• $\alpha \neq 2k\pi$, where k is an integer

$$= \frac{\sin \alpha(1 - \cos \alpha)}{1 - \cos^2 \alpha}$$

$$= \frac{\sin \alpha(1 - \cos \alpha)}{\sin^2 \alpha}$$

$$\tan \frac{\alpha}{2} = \frac{1 - \cos \alpha}{\sin \alpha}$$

EXAMPLE 3 Verify a Half-Angle Identity

Verify the identity $2 \csc x \cos^2 \frac{x}{2} = \frac{\sin x}{1 - \cos x}$.

Solution

Work on the left side of the identity.

$$2 \csc x \cos^2 \frac{x}{2} = 2 \csc x \left(\frac{1 + \cos x}{2}\right)$$

• $\cos^2 \frac{x}{2} = \frac{1 + \cos x}{2}$

$$= \frac{1 + \cos x}{\sin x}$$

• $\csc x = \frac{1}{\sin x}$

$$= \frac{1 + \cos x}{\sin x} \cdot \frac{1 - \cos x}{1 - \cos x}$$

• Multiply the numerator and denominator by the conjugate of the numerator.

$$= \frac{1 - \cos^2 x}{\sin x(1 - \cos x)}$$

$$= \frac{\sin^2 x}{\sin x(1 - \cos x)}$$

• $1 - \cos^2 x = \sin^2 x$

$$= \frac{\sin x}{1 - \cos x}$$

▶ TRY EXERCISE 40, PAGE 398

EXAMPLE 4 Verify a Half-Angle Identity

Verify the identity $\tan \frac{\alpha}{2} = \sin \alpha + \cos \alpha \cot \alpha - \cot \alpha$.

Continued ▶

Solution

Work on the left side of the identity.

$$\tan\frac{\alpha}{2} = \frac{1-\cos\alpha}{\sin\alpha} = \frac{\sin^2\alpha + \cos^2\alpha - \cos\alpha}{\sin\alpha}$$

• $1 = \sin^2\alpha + \cos^2\alpha$

$$= \frac{\sin^2\alpha}{\sin\alpha} + \frac{\cos^2\alpha}{\sin\alpha} - \frac{\cos\alpha}{\sin\alpha}$$

• Write each numerator over the common denominator.

$$= \sin\alpha + \cos\alpha\cot\alpha - \cot\alpha$$

TRY EXERCISE 44, PAGE 399

THE PRODUCT-TO-SUM IDENTITIES

Some applications require that a product of trigonometric functions be written as a sum or difference of these functions. Other applications require that the sum or difference of trigonometric functions be represented as a product of these functions. The *product-to-sum identities* are particularly useful in these types of applications.

The product-to-sum identities can be derived by using the sum or difference identities. Adding the identities for $\sin(\alpha+\beta)$ and $\sin(\alpha-\beta)$, we have

$$\sin(\alpha+\beta) = \sin\alpha\cos\beta + \cos\alpha\sin\beta$$

$$\sin(\alpha-\beta) = \sin\alpha\cos\beta - \cos\alpha\sin\beta$$

$$\sin(\alpha+\beta) + \sin(\alpha-\beta) = 2\sin\alpha\cos\beta$$

• Add the identities.

Solving for $\sin\alpha\cos\beta$, we obtain the first product-to-sum identity:

$$\sin\alpha\cos\beta = \frac{1}{2}[\sin(\alpha+\beta) + \sin(\alpha-\beta)]$$

The identity for $\cos\alpha\sin\beta$ is obtained when $\sin(\alpha-\beta)$ is subtracted from $\sin(\alpha+\beta)$. The result is

$$\cos\alpha\sin\beta = \frac{1}{2}[\sin(\alpha+\beta) - \sin(\alpha-\beta)]$$

In like manner, the identities for $\cos(\alpha+\beta)$ and $\cos(\alpha-\beta)$ are used to derive the identities for $\cos\alpha\cos\beta$ and $\sin\alpha\sin\beta$.

$$\cos\alpha\cos\beta = \frac{1}{2}[\cos(\alpha+\beta) + \cos(\alpha-\beta)]$$

$$\sin\alpha\sin\beta = \frac{1}{2}[\cos(\alpha-\beta) - \cos(\alpha+\beta)]$$

The product-to-sum identities can be used to verify some identities.

EXAMPLE 5 **Verify an Identity**

Verify the identity $\cos 2x \sin 5x = \frac{1}{2}(\sin 7x + \sin 3x)$.

Solution

$$\begin{aligned}\cos 2x \sin 5x &= \frac{1}{2}[\sin(2x + 5x) - \sin(2x - 5x)] \\ &= \frac{1}{2}[\sin 7x - \sin(-3x)] \\ &= \frac{1}{2}(\sin 7x + \sin 3x)\end{aligned}$$

- Use the product-to-sum identity for $\cos \alpha \sin \beta$.
- $\sin(-3x) = -\sin 3x$

▶ TRY EXERCISE 74, PAGE 399

THE SUM-TO-PRODUCT IDENTITIES

The *sum-to-product identities* can be derived from the product-to-sum identities. To derive the sum-to-product identity for $\sin x + \sin y$, we first let $x = \alpha + \beta$ and $y = \alpha - \beta$. Then

$$\begin{aligned} x + y &= \alpha + \beta + \alpha - \beta &\quad\text{and}\quad x - y &= \alpha + \beta - (\alpha - \beta) \\ x + y &= 2\alpha & x - y &= 2\beta \\ \alpha &= \frac{x + y}{2} & \beta &= \frac{x - y}{2} \end{aligned}$$

Substituting these expressions for α and β into the product-to-sum identity

$$\frac{1}{2}[\sin(\alpha + \beta) + \sin(\alpha - \beta)] = \sin \alpha \cos \beta$$

yields

$$\sin\left(\frac{x + y}{2} + \frac{x - y}{2}\right) + \sin\left(\frac{x + y}{2} - \frac{x - y}{2}\right) = 2 \sin \frac{x + y}{2} \cos \frac{x - y}{2}$$

Simplifying the left side, we have a sum-to-product identity.

$$\sin x + \sin y = 2 \sin \frac{x + y}{2} \cos \frac{x - y}{2}$$

In like manner, three other sum-to-product identities can be derived from the other product-to-sum identities. The proofs of these identities are left as exercises.

$$\sin x - \sin y = 2 \cos \frac{x + y}{2} \sin \frac{x - y}{2}$$

$$\cos x + \cos y = 2 \cos \frac{x + y}{2} \cos \frac{x - y}{2}$$

$$\cos x - \cos y = -2 \sin \frac{x + y}{2} \sin \frac{x - y}{2}$$

EXAMPLE 6 Write the Difference of Trigonometric Expressions as a Product

Write $\sin 4\theta - \sin \theta$ as the product of two functions.

Solution

$$\sin 4\theta - \sin \theta = 2 \cos \frac{4\theta + \theta}{2} \sin \frac{4\theta - \theta}{2} = 2 \cos \frac{5\theta}{2} \sin \frac{3\theta}{2}$$

▶ TRY EXERCISE 64, PAGE 399

? QUESTION Does $\cos 4\theta + \cos 2\theta = 2 \cos 3\theta \cos \theta$?

EXAMPLE 7 Verify a Sum-to-Product Identity

Verify the identity $\dfrac{\sin 6x + \sin 2x}{\sin 6x - \sin 2x} = \tan 4x \cot 2x$.

Solution

$$\frac{\sin 6x + \sin 2x}{\sin 6x - \sin 2x} = \frac{2 \sin \dfrac{6x + 2x}{2} \cos \dfrac{6x - 2x}{2}}{2 \cos \dfrac{6x + 2x}{2} \sin \dfrac{6x - 2x}{2}} = \frac{\sin 4x \cos 2x}{\cos 4x \sin 2x}$$

$$= \tan 4x \cot 2x$$

▶ TRY EXERCISE 76, PAGE 399

FUNCTIONS OF THE FORM $f(x) = a \sin x + b \cos x$

The function given by $f(x) = a \sin x + b \cos x$ can be written in the form $f(x) = k \sin(x + \alpha)$. This form of the function is useful in graphing and engineering applications because the amplitude, period, and phase shift can be readily calculated.

Let $P(a, b)$ be a point on a coordinate plane, and let α represent an angle in standard position. See **Figure 5.24.** To rewrite $y = a \sin x + b \cos x$, multiply and divide the expression $a \sin x + b \cos x$ by $\sqrt{a^2 + b^2}$.

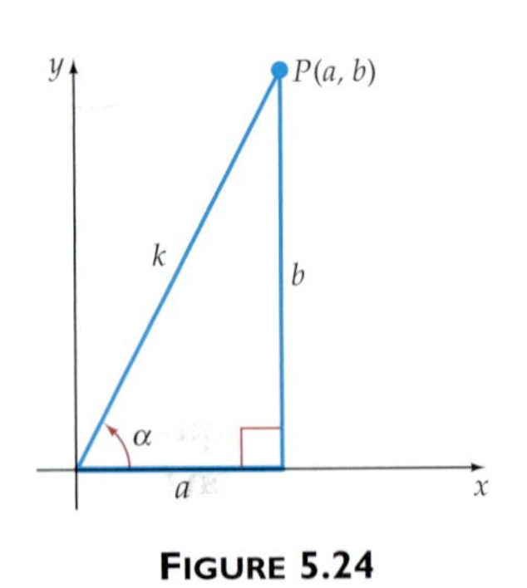

FIGURE 5.24

$$a \sin x + b \cos x = \frac{\sqrt{a^2 + b^2}}{\sqrt{a^2 + b^2}}(a \sin x + b \cos x)$$

$$= \sqrt{a^2 + b^2}\left(\frac{a}{\sqrt{a^2 + b^2}} \sin x + \frac{b}{\sqrt{a^2 + b^2}} \cos x\right) \qquad (1)$$

From the definition of the sine and cosine of an angle in standard position, let

$$k = \sqrt{a^2 + b^2}, \quad \cos \alpha = \frac{a}{\sqrt{a^2 + b^2}}, \quad \text{and} \quad \sin \alpha = \frac{b}{\sqrt{a^2 + b^2}}$$

? ANSWER Yes. $\cos 4\theta + \cos 2\theta = 2 \cos\left(\dfrac{4\theta + 2\theta}{2}\right) \cos\left(\dfrac{4\theta - 2\theta}{2}\right) = 2 \cos 3\theta \cos \theta$.

Substituting these expressions into Equation (1) yields

$$a \sin x + b \cos x = k(\cos \alpha \sin x + \sin \alpha \cos x)$$

Now, using the identity for the sine of the sum of two angles, we have

$$a \sin x + b \cos x = k \sin(x + \alpha)$$

Thus $a \sin x + b \cos x = k \sin(x + \alpha)$, where $k = \sqrt{a^2 + b^2}$ and α is the angle for which $\sin \alpha = \dfrac{b}{\sqrt{a^2 + b^2}}$ and $\cos \alpha = \dfrac{a}{\sqrt{a^2 + b^2}}$.

EXAMPLE 8 Rewrite *a* sin *x* + *b* cos *x*

Rewrite $\sin x + \cos x$ in the form $k \sin(x + \alpha)$.

Solution

Comparing $\sin x + \cos x$ to $a \sin x + b \cos x$, $a = 1$ and $b = 1$. Thus $k = \sqrt{1^2 + 1^2} = \sqrt{2}$, $\sin \alpha = \dfrac{1}{\sqrt{2}}$, and $\cos \alpha = \dfrac{1}{\sqrt{2}}$. Thus $\alpha = \dfrac{\pi}{4}$.

$$\sin x + \cos x = k \sin(x + \alpha) = \sqrt{2} \sin\left(x + \frac{\pi}{4}\right)$$

▶ TRY EXERCISE 86, PAGE 399

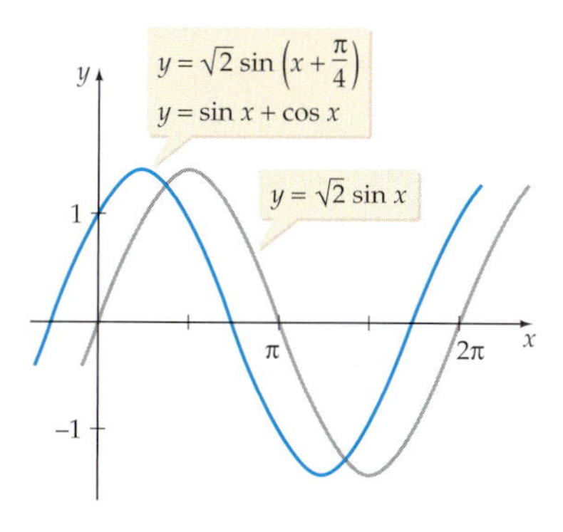

FIGURE 5.25

The graphs of $y = \sin x + \cos x$ and $y = \sqrt{2} \sin\left(x + \dfrac{\pi}{4}\right)$ are both the graph of $y = \sqrt{2} \sin x$ shifted $\dfrac{\pi}{4}$ units to the left. See **Figure 5.25.**

EXAMPLE 9 Use an Identity to Graph a Trigonometric Function

Graph $f(x) = -\sin x + \sqrt{3} \cos x$.

Solution

First, we write $f(x)$ as $k \sin(x + \alpha)$. Let $a = -1$ and $b = \sqrt{3}$; then $k = \sqrt{(-1)^2 + (\sqrt{3})^2} = 2$. The point $P(-1, \sqrt{3})$ is in the second quadrant (see **Figure 5.26**). Let α be an angle in standard position with P on its terminal side. Let α' be the reference angle for α. Then

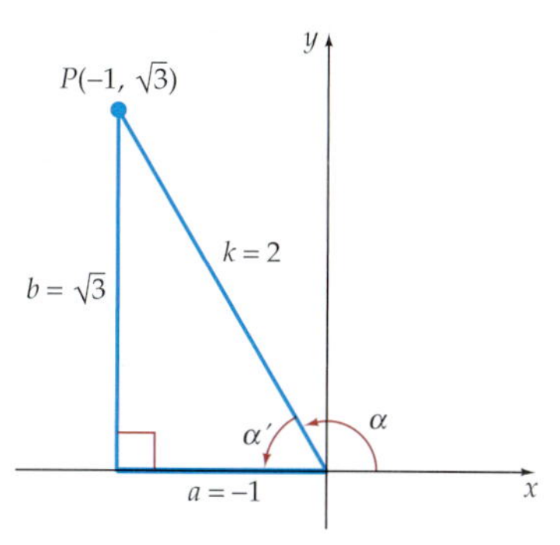

FIGURE 5.26

$$\sin \alpha' = \frac{\sqrt{3}}{2}$$

$$\alpha' = \frac{\pi}{3}$$

$$\alpha = \pi - \alpha' = \pi - \frac{\pi}{3} = \frac{2\pi}{3}$$

Continued ▶

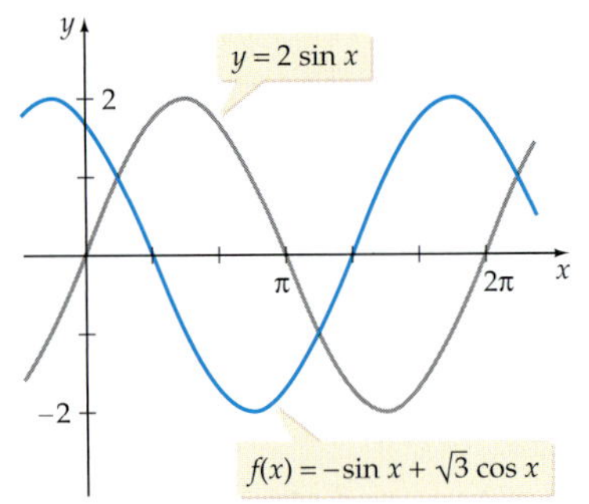

FIGURE 5.27

Substituting 2 for k and $\frac{2\pi}{3}$ for α in $y = k \sin(x + \alpha)$, we have

$$y = 2 \sin\left(x + \frac{2\pi}{3}\right)$$

The phase shift is $-\frac{c}{b} = -\frac{2\pi}{3}$. Thus the graph of the equation $f(x) = -\sin x + \sqrt{3} \cos x$ is the graph of $y = 2 \sin x$ shifted $\frac{2\pi}{3}$ units to the left. See **Figure 5.27.**

TRY EXERCISE 90, PAGE 399

We now list the identities that have been discussed in this section.

Double-Angle Identities

$$\sin 2\alpha = 2 \sin \alpha \cos \alpha$$

$$\cos 2\alpha = \cos^2 \alpha - \sin^2 \alpha = 1 - 2 \sin^2 \alpha = 2 \cos^2 \alpha - 1$$

$$\tan 2\alpha = \frac{2 \tan \alpha}{1 - \tan^2 \alpha}$$

Half-Angle Identities

$$\sin \frac{\alpha}{2} = \pm\sqrt{\frac{1 - \cos \alpha}{2}}$$

$$\cos \frac{\alpha}{2} = \pm\sqrt{\frac{1 + \cos \alpha}{2}}$$

$$\tan \frac{\alpha}{2} = \frac{\sin \alpha}{1 + \cos \alpha} = \frac{1 - \cos \alpha}{\sin \alpha}$$

Product-to-Sum Identities

$$\sin \alpha \cos \beta = \frac{1}{2}[\sin(\alpha + \beta) + \sin(\alpha - \beta)]$$

$$\cos \alpha \sin \beta = \frac{1}{2}[\sin(\alpha + \beta) - \sin(\alpha - \beta)]$$

$$\cos \alpha \cos \beta = \frac{1}{2}[\cos(\alpha + \beta) + \cos(\alpha - \beta)]$$

$$\sin \alpha \sin \beta = \frac{1}{2}[\cos(\alpha - \beta) - \cos(\alpha + \beta)]$$

Sum-to-Product Identities

$$\sin x + \sin y = 2 \sin \frac{x+y}{2} \cos \frac{x-y}{2}$$

$$\cos x + \cos y = 2 \cos \frac{x+y}{2} \cos \frac{x-y}{2}$$

$$\sin x - \sin y = 2 \cos \frac{x+y}{2} \sin \frac{x-y}{2}$$

$$\cos x - \cos y = -2 \sin \frac{x+y}{2} \sin \frac{x-y}{2}$$

Sums of the Form *a* sin *x* + *b* cos *x*

$$a \sin x + b \cos x = k \sin(x + \alpha)$$

where $k = \sqrt{a^2 + b^2}$, $\sin \alpha = \dfrac{b}{\sqrt{a^2 + b^2}}$, and $\cos \alpha = \dfrac{a}{\sqrt{a^2 + b^2}}$.

Topics for Discussion

1. True or false: If $\sin \alpha = \sin \beta$, then $\alpha = \beta$. Why?
2. Does $\sin 2\alpha = 2 \sin \alpha$ for all values of α? If not, find a value of α for which $\sin 2\alpha \neq 2 \sin \alpha$.
3. Because

$$\tan \frac{\alpha}{2} = \frac{\sin \alpha}{1 + \cos \alpha} \quad \text{and} \quad \tan \frac{\alpha}{2} = \frac{1 - \cos \alpha}{\sin \alpha}$$

   are both identities, it follows that

$$\frac{\sin \alpha}{1 + \cos \alpha} = \frac{1 - \cos \alpha}{\sin \alpha}$$

   is also an identity. Do you agree? Explain.
4. Is $\sin 10x = 2 \sin 5x \cos 5x$ an identity? Explain.
5. Is $\sin \dfrac{\alpha}{2} = \cos \dfrac{\alpha}{2}$ an identity? Explain.

Exercise Set 5.3

In Exercises 1 to 6, write each trigonometric expression in terms of a single trigonometric function.

1. $2 \sin 2\alpha \cos 2\alpha$

2. $2 \sin 3\theta \cos 3\theta$

3. $1 - 2 \sin^2 5\beta$

4. $2 \cos^2 2\beta - 1$

5. $\dfrac{2 \tan 3\alpha}{1 - \tan^2 3\alpha}$

6. $\cos^2 6\alpha - \sin^2 6\alpha$

In Exercises 7 to 16, use the half-angle identities to find the exact value of each trigonometric expression.

7. $\sin 75°$

8. $\cos 105°$

9. $\tan 67.5°$

10. $\sin 112.5°$

11. $\sin 22.5°$

12. $\cos 67.5°$

13. $\sin \dfrac{7\pi}{8}$

14. $\cos \dfrac{5\pi}{8}$

15. $\cos \dfrac{5\pi}{12}$

16. $\sin \dfrac{3\pi}{8}$

In Exercises 17 to 22, find the exact values of $\sin 2\theta$, $\cos 2\theta$, and $\tan 2\theta$ given the following information.

17. $\cos \theta = -\dfrac{4}{5}$ θ is in Quadrant II.

▶ **18.** $\cos \theta = \dfrac{24}{25}$ θ is in Quadrant IV.

19. $\sin \theta = \dfrac{8}{17}$ θ is in Quadrant II.

20. $\sin \theta = -\dfrac{9}{41}$ θ is in Quadrant III.

21. $\tan \theta = -\dfrac{24}{7}$ θ is in Quadrant IV.

22. $\tan \theta = \dfrac{4}{3}$ θ is in Quadrant I.

In Exercises 23 to 28, find the exact values of $\sin \dfrac{\alpha}{2}$, $\cos \dfrac{\alpha}{2}$, and $\tan \dfrac{\alpha}{2}$ given the following information.

23. $\sin \alpha = \dfrac{5}{13}$ α is in Quadrant II.

24. $\sin \alpha = -\dfrac{7}{25}$ α is in Quadrant III.

25. $\cos \alpha = -\dfrac{8}{17}$ α is in Quadrant III.

26. $\cos \alpha = \dfrac{12}{13}$ α is in Quadrant I.

27. $\tan \alpha = \dfrac{4}{3}$ α is in Quadrant I.

28. $\tan \alpha = -\dfrac{8}{15}$ α is in Quadrant II.

In Exercises 29 to 48, use a double-angle or half-angle identity to verify the given identity.

29. $\sin^2 x + \cos 2x = \cos^2 x$

30. $\dfrac{\cos 2x}{\sin^2 x} = \cot^2 x - 1$

31. $\dfrac{1 + \cos 2x}{\sin 2x} = \cot x$

▶ **32.** $\dfrac{1}{1 - \cos 2x} = \dfrac{1}{2} \csc^2 x$

33. $\dfrac{\sin 2x}{1 - \sin^2 x} = 2 \tan x$

34. $\dfrac{\cos^2 x - \sin^2 x}{2 \sin x \cos x} = \cot 2x$

35. $1 - \tan^2 x = \dfrac{\cos 2x}{\cos^2 x}$

36. $\tan 2x = \dfrac{2 \sin x \cos x}{\cos^2 x - \sin^2 x}$

37. $\sin 2x - \tan x = \tan x \cos 2x$

38. $\sin 2x - \cot x = -\cot x \cos 2x$

39. $\sin^2 \dfrac{x}{2} = \dfrac{\sec x - 1}{2 \sec x}$

▶ **40.** $\cos^2 \dfrac{x}{2} = \dfrac{\sec x + 1}{2 \sec x}$

41. $\tan \dfrac{x}{2} = \csc x - \cot x$

42. $\tan \dfrac{x}{2} = \dfrac{\tan x}{\sec x + 1}$

43. $\left(\cos \dfrac{x}{2} + \sin \dfrac{x}{2}\right)^2 = 1 + \sin x$

▶ **44.** $\tan^2 \frac{x}{2} = \frac{\sec x - 1}{\sec x + 1}$

45. $\sin 2x - \cos x = (\cos x)(2 \sin x - 1)$

46. $\frac{\cos 2x}{\sin^2 x} = \csc^2 x - 2$

47. $\tan 2x = \frac{2}{\cot x - \tan x}$

48. $\frac{2 \cos 2x}{\sin 2x} = \cot x - \tan x$

In Exercises 49 to 54, write each expression as the sum or difference of two functions.

49. $2 \sin x \cos 2x$

50. $2 \sin 4x \sin 2x$

51. $\cos 6x \sin 2x$

52. $\cos 3x \cos 5x$

53. $\sin x \sin 5x$

54. $2 \sin 2x \cos 6x$

In Exercises 55 to 60, find the exact value of each expression. Do not use a calculator.

55. $\cos 75° \cos 15°$

56. $\sin 105° \cos 15°$

57. $\cos 157.5° \sin 22.5°$

58. $\sin 195° \cos 15°$

59. $\sin \frac{\pi}{12} \cos \frac{7\pi}{12}$

60. $\cos \frac{17\pi}{12} \sin \frac{7\pi}{12}$

In Exercises 61 to 70, write each expression as the product of two functions.

61. $\sin 4\theta + \sin 2\theta$

62. $\cos 5\theta - \cos 3\theta$

63. $\cos 6\theta - \cos 2\theta$

▶ **64.** $\cos 3\theta + \cos 5\theta$

65. $\cos \theta + \cos 7\theta$

66. $\sin 3\theta + \sin 7\theta$

67. $\sin 5\theta + \sin 9\theta$

68. $\cos 5\theta - \cos \theta$

69. $\cos \frac{\theta}{2} - \cos \theta$

70. $\sin \frac{3\theta}{4} + \sin \frac{\theta}{2}$

In Exercises 71 to 78, use a product-to-sum or sum-to-product identity to verify the given identity.

71. $2 \cos \alpha \cos \beta = \cos(\alpha + \beta) + \cos(\alpha - \beta)$

72. $2 \sin \alpha \sin \beta = \cos(\alpha - \beta) - \cos(\alpha + \beta)$

73. $2 \cos 3x \sin x = 2 \sin x \cos x - 8 \cos x \sin^3 x$

▶ **74.** $\sin 5x \cos 3x = \sin 4x \cos 4x + \sin x \cos x$

75. $\frac{\sin 3x - \sin x}{\cos 3x - \cos x} = -\cot 2x$

▶ **76.** $\frac{\cos 5x - \cos 3x}{\sin 5x + \sin 3x} = -\tan x$

77. $\sin(x + y) \cos(x - y) = \sin x \cos x + \sin y \cos y$

78. $\sin(x + y) \sin(x - y) = \sin^2 x - \sin^2 y$

In Exercises 79 to 82, write the given equation in the form $y = k \sin (x + \alpha)$, where the measure of α is in degrees.

79. $y = -\sin x - \cos x$

80. $y = \sqrt{3} \sin x - \cos x$

81. $y = \frac{1}{2} \sin x - \frac{1}{2} \cos x$

82. $y = -\frac{\sqrt{3}}{2} \sin x - \frac{1}{2} \cos x$

In Exercises 83 to 86, write the given equation in the form $y = k \sin (x + \alpha)$, where the measure of α is in radians.

83. $y = -\sin x + \cos x$

84. $y = -\sqrt{3} \sin x - \cos x$

85. $y = \frac{\sqrt{3}}{2} \sin x + \frac{1}{2} \cos x$

▶ **86.** $y = \sin x + \sqrt{3} \cos x$

In Exercises 87 to 90, graph one cycle of each equation.

87. $y = -\sin x - \sqrt{3} \cos x$

88. $y = -\sqrt{3} \sin x + \cos x$

89. $y = 2 \sin x + 2 \cos x$

▶ **90.** $y = \sin x + \sqrt{3} \cos x$

Every tone made on a touch-tone phone is produced by adding a pair of sounds. The following chart shows the two frequencies used for each key.

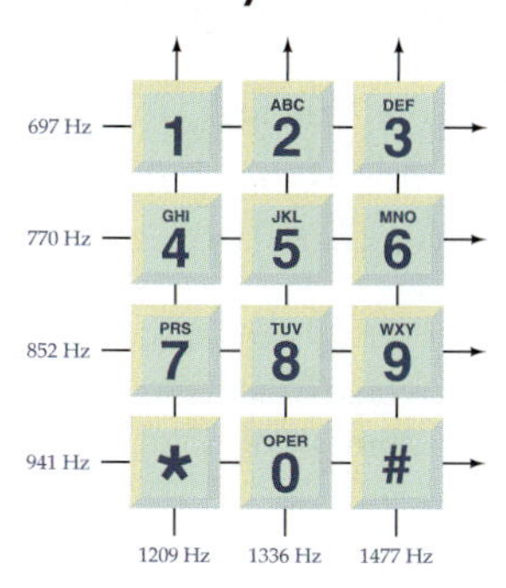

Source: Data in chart from http://www.howstuffworks.com/telephone2.htm and also found at http://hyperarchive.lcs.mit.edu/telecom-archives/tribute/touch_tone_info.html

For instance, the sound emitted by pressing 3 on a keypad is produced by adding a 1477-hertz sound to a 697-hertz sound. An equation that models this tone is

$$p(t) = \cos(2\pi \cdot 1477t) + \cos(2\pi \cdot 697t)$$

In Exercises 91 and 92 you will determine trigonometric equations that model some of the other tones produced by touch-tone phones.

91. TONES ON A TOUCH-TONE PHONE

a. Write an equation of the form $p(t) = \cos(2\pi f_1 t) + \cos(2\pi f_2 t)$ that models the tone produced by pressing the 5 key on a touch-tone phone.

b. Use a sum-to-product identity to write your equation from part **a.** in the form

$$p(t) = A\cos(B\pi t)\cos(C\pi t)$$

c. When a sound with frequency f_1 is combined with a sound with frequency f_2, the combined sound has a frequency of $\frac{f_1 + f_2}{2}$. What is the frequency of the tone produced when the 5 key is pressed?

92. TONES ON A TOUCH-TONE PHONE

a. Write an equation of the form $p(t) = \cos(2\pi f_1 t) + \cos(2\pi f_2 t)$ that models the tone produced by pressing the 8 key on a touch-tone phone.

b. Use a sum-to-product identity to write your equation from part **a.** in the form

$$p(t) = A\cos(B\pi t)\cos(C\pi t)$$

c. What is the frequency of the tone produced when the 8 key on a touch-tone phone is pressed? (*Hint:* See part **c.** of Exercise 91.)

CONNECTING CONCEPTS

93. Derive the sum-to-product identity

$$\cos x + \cos y = 2\cos\frac{x+y}{2}\cos\frac{x-y}{2}$$

94. Derive the product-to-sum identity

$$\sin x \sin y = \frac{1}{2}[\cos(x-y) - \cos(x+y)]$$

95. If $x + y = 180°$, show that $\sin x + \sin y = 2\sin x$.

96. If $x + y = 360°$, show that $\cos x + \cos y = 2\cos x$.

97. Verify that $\cos^2 x - \sin^2 x = \cos 2x$ by using a product-to-sum identity.

98. Verify that $2\sin x\cos x = \sin 2x$ by using a product-to-sum identity.

PREPARE FOR SECTION 5.4

99. What is a one-to-one function? [1.3]

100. State the horizontal line test. [1.3]

101. Find $f[g(x)]$ given that $f(x) = 2x + 4$ and $g(x) = \frac{1}{2}x - 2$. [1.7]

102. If f and f^{-1} are inverse functions, then determine $f[f^{-1}(x)]$ for any x in the domain of f^{-1}. [3.1]

103. If f and f^{-1} are inverse functions, then explain how the graph of f^{-1} is related to the graph of f. [3.1]

104. Use the horizontal line test to determine whether the graph of $y = \sin x$, where x is any real number, is a one-to-one function. [1.3]

PROJECTS

1. INTERFERENCE OF SOUND WAVES The figure on the following page shows the wave-forms of two tuning forks. The frequency of one of the tuning forks is 10 cycles per second, and the frequency of the other tuning fork is 8 cycles per second. Each of the sound waves can be modeled by an equation of the form

$$p(t) = A\cos 2\pi ft$$

where p is the pressure produced on the eardrum at time t, A is the amplitude of the sound wave, and f is the frequency of the sound.

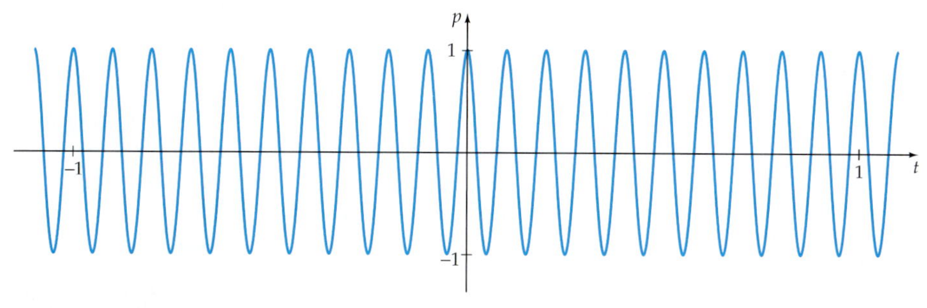

$p_1(t) = \cos 2\pi \cdot 10t$

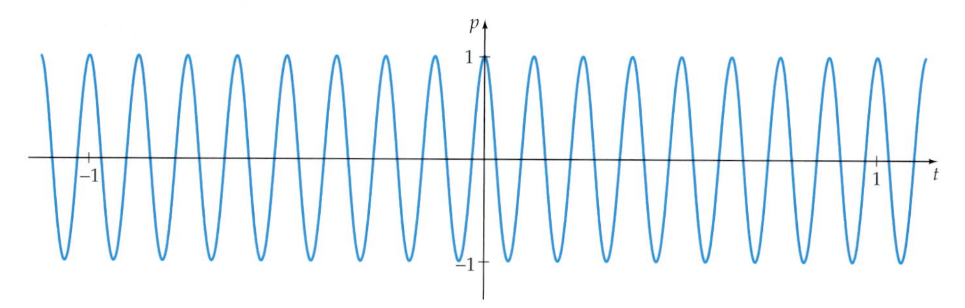

$p_2(t) = \cos 2\pi \cdot 8t$

If the two tuning forks are struck at the same time, with the same force, the sound that we hear fluctuates between a loud tone and silence. These regular fluctuations are called **beats.** The loud periods occur when the sound waves reinforce (interfere constructively with) one another, and the nearly silent periods occur when the waves interfere destructively with each other. The pressure p produced on the eardrum from the combined sound waves is given by

$$p(t) = p_1 + p_2 = \cos 2\pi \cdot 10t + \cos 2\pi \cdot 8t$$

The following figure shows the graph of p.

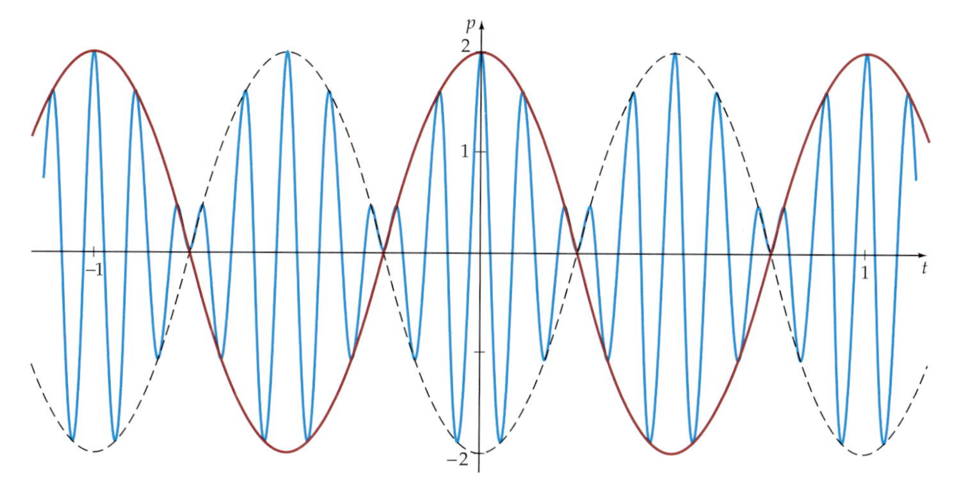

Graph of $p = p_1 + p_2 = \cos 2\pi \cdot 10t + \cos 2\pi \cdot 8t$ showing the beats in the combined sounds

a. Use the sum-to-product identity for $\cos x + \cos y$ to write p as a product.

b. Explain why the graph of $p = A\cos 2\pi f_1 t + A\cos 2\pi f_2 t$ can be thought of as a cosine curve with period $\frac{2}{f_1 + f_2}$ and a *variable* amplitude of

$$2A\cos\left[2\pi\left(\frac{f_1 - f_2}{2}\right)t\right]$$

c. The rate of the beats produced by two sounds, with the same intensity, is the absolute value of the difference between their frequencies. Consider two tuning forks that are struck at the same time with the same force and held on a sounding board. The tuning forks have frequencies of 564 and 568 cycles per second, respectively. How many beats will be heard each second?

d. A piano tuner strikes a tuning fork and a key on a piano that is supposed to have the same frequency as the tuning fork. The piano tuner notices that the sound produced by the piano is lower than that produced by the tuning fork. The piano tuner also notes that the combined sound of the piano and the tuning fork has 2 beats per second. How much lower is the frequency of the piano than the frequency of the tuning fork?

2. The photo shows a matched set of tuning forks. The fork on the left has an adjustable weight that can be used to vary the frequency of its tone. Check with the physics department at your school to see if a matched set of tuning forks is available. Use the tuning forks to demonstrate to your classmates the phenomenon of beats.

SECTION 5.4 INVERSE TRIGONOMETRIC FUNCTIONS

- INVERSE TRIGONOMETRIC FUNCTIONS
- COMPOSITION OF TRIGONOMETRIC FUNCTIONS AND THEIR INVERSES
- GRAPHS OF INVERSE TRIGONOMETRIC FUNCTIONS

INVERSE TRIGONOMETRIC FUNCTIONS

Because the graph of $y = \sin x$ fails the horizontal line test, it is not the graph of a one-to-one function. Therefore, it does not have an inverse function. **Figure 5.28** shows the graph of $y = \sin x$ on the interval $-2\pi \le x \le 2\pi$ and the graph of the inverse relation $x = \sin y$. Note that the graph of $x = \sin y$ does not satisfy the vertical line test and therefore is not the graph of a function.

If the domain of $y = \sin x$ is restricted to $-\frac{\pi}{2} \le x \le \frac{\pi}{2}$, the graph of $y = \sin x$ satisfies the horizontal line test and therefore the function has an inverse function. The graphs of $y = \sin x$ for $-\frac{\pi}{2} \le x \le \frac{\pi}{2}$ and its inverse are shown in **Figure 5.29.**

See Section 3.1 if you need to review the concept of an inverse function.

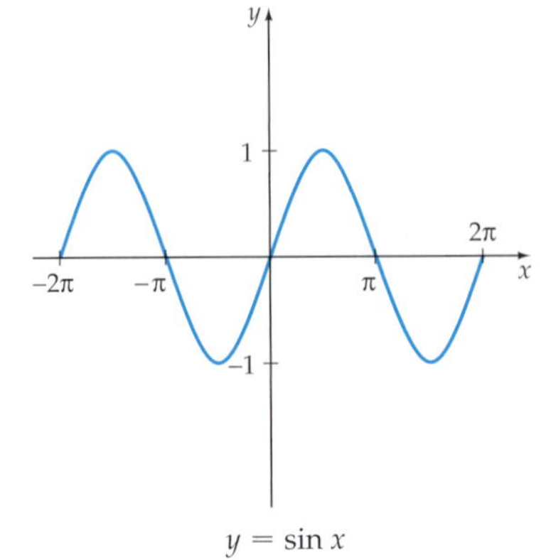

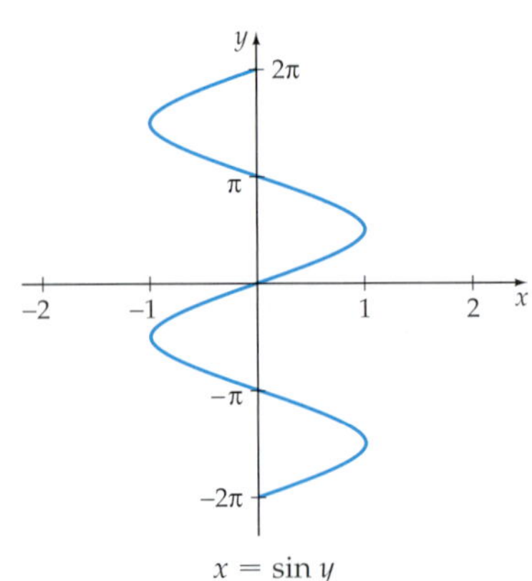

FIGURE 5.28

To find the inverse of the function defined by $y = \sin x$, with $-\frac{\pi}{2} \le x \le \frac{\pi}{2}$, interchange x and y. Then solve for y.

$y = \sin x$ • $-\frac{\pi}{2} \le x \le \frac{\pi}{2}$

$x = \sin y$ • **Interchange x and y.**

$y = ?$ • **Solve for y.**

Unfortunately, there is no algebraic solution for y. Thus we establish new notation and write

$$y = \sin^{-1} x$$

which is read "y is the inverse sine of x." Some textbooks use the notation $\arcsin x$ instead of $\sin^{-1} x$.

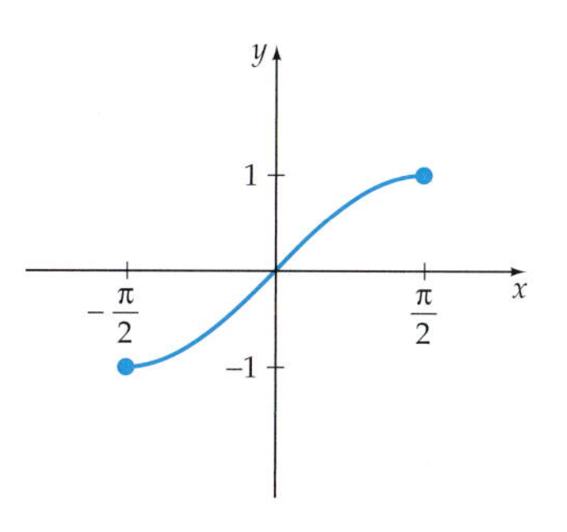

$y = \sin x$: $-\frac{\pi}{2} \le x \le \frac{\pi}{2}$

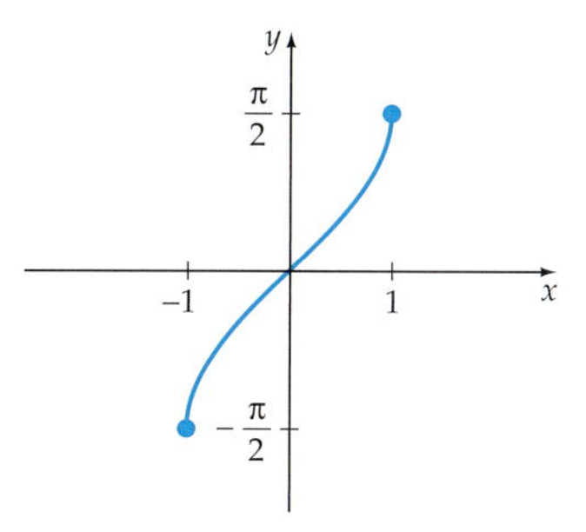

$y = \sin^{-1} x$: $-1 \le x \le 1$

FIGURE 5.29

Definition of $\sin^{-1} x$

$$y = \sin^{-1} x \quad \text{if and only if} \quad x = \sin y$$

where $-1 \le x \le 1$ and $-\frac{\pi}{2} \le y \le \frac{\pi}{2}$.

take note

The −1 in $\sin^{-1} x$ is not an exponent. The −1 is used to denote the inverse function. To use −1 as an exponent for a sine function, enclose the function in parentheses.

$$(\sin x)^{-1} = \frac{1}{\sin x} = \csc x$$

$$\sin^{-1} x \ne \frac{1}{\sin x}$$

It is convenient to think of the value of an inverse trigonometric function as an angle. For instance, if $y = \sin^{-1}\frac{1}{2}$, then y is the angle in the interval $\left[-\frac{\pi}{2}, \frac{\pi}{2}\right]$ whose sine is $\frac{1}{2}$. Thus $y = \frac{\pi}{6}$.

Because the graph of $y = \cos x$ fails the horizontal line test, it is not the graph of a one-to-one function. Therefore, it does not have an inverse function. **Figure 5.30** shows the graph of $y = \cos x$ on the interval $-2\pi \le x \le 2\pi$ and the graph of the inverse relation $x = \cos y$. Note that the graph of $x = \cos y$ does not satisfy the vertical line test and therefore is not the graph of a function.

If the domain of $y = \cos x$ is restricted to $0 \le x \le \pi$, the graph of $y = \cos x$ satisfies the horizontal line test and therefore is the graph of a one-to-one function. The graph of $y = \cos x$ for $0 \le x \le \pi$ and that of $x = \cos y$ are shown in **Figure 5.31**.

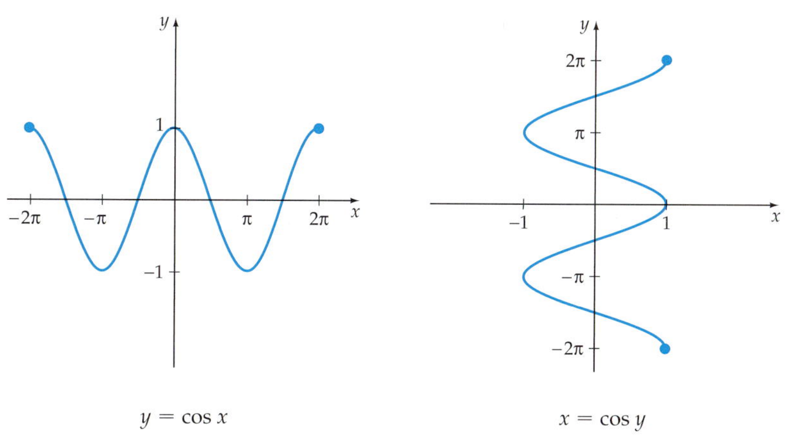

FIGURE 5.30

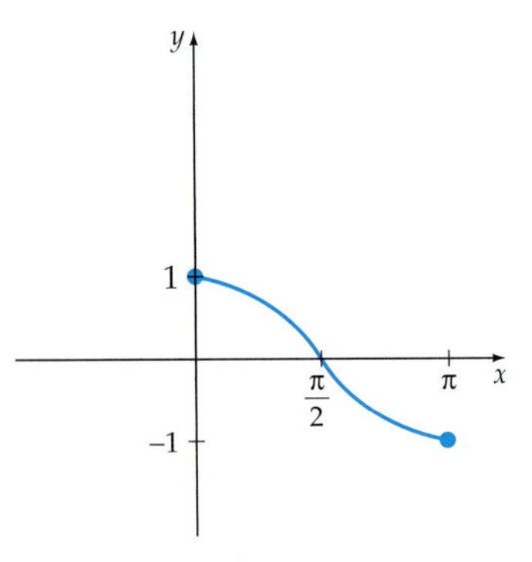

$y = \cos x: 0 \le x \le \pi$

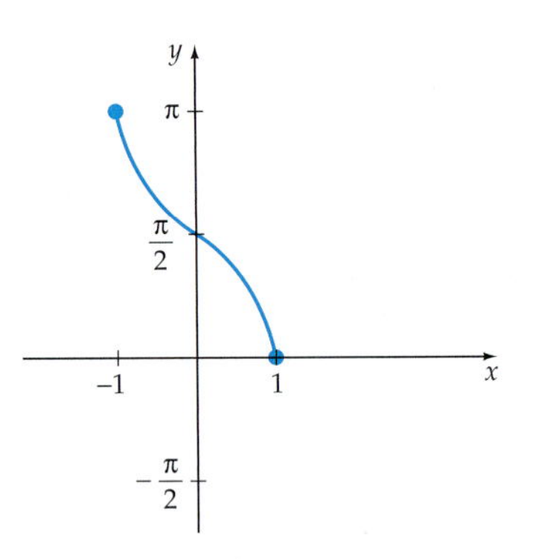

$y = \cos^{-1} x: -1 \le x \le 1$

FIGURE 5.31

To find the inverse of the function defined by $y = \cos x$, with $0 \le x \le \pi$, interchange x and y. Then solve for y.

$y = \cos x$ • $0 \le x \le \pi$

$x = \cos y$ • Interchange x and y.

$y = ?$ • Solve for y.

As in the case for the inverse sine function, there is no algebraic solution for y. Thus the notation for the inverse cosine function becomes $y = \cos^{-1} x$. We can write the following definition of the inverse cosine function.

Definition of $\cos^{-1} x$

$$y = \cos^{-1} x \quad \text{if and only if} \quad x = \cos y$$

where $-1 \le x \le 1$ and $0 \le y \le \pi$.

Because the graphs of $y = \tan x$, $y = \csc x$, $y = \sec x$, and $y = \cot x$ fail the horizontal line test, these functions are not one-to-one functions. Therefore, these functions do not have inverse functions. If the domains of all these functions are restricted in a certain way, however, the graphs satisfy the horizontal line test. Thus each of these functions has an inverse function over a restricted domain. **Table 5.6** on page 405 shows the restricted function and the inverse function for $\tan x$, $\csc x$, $\sec x$, and $\cot x$.

The choice of ranges for $y = \sec^{-1} x$ and $y = \csc^{-1} x$ is not universally accepted. For example, some calculus texts use $\left[0, \frac{\pi}{2}\right) \cup \left[\pi, \frac{3\pi}{2}\right)$ as the range of $y = \sec^{-1} x$. This definition has some advantages and some disadvantages that are explained in more advanced mathematics courses.

EXAMPLE 1 Evaluate Inverse Functions

Find the exact value of each inverse function.

a. $y = \tan^{-1} \dfrac{\sqrt{3}}{3}$ **b.** $y = \cos^{-1}\left(-\dfrac{\sqrt{2}}{2}\right)$

Solution

a. Because $y = \tan^{-1} \dfrac{\sqrt{3}}{3}$, y is the angle whose measure is in the interval $\left(-\dfrac{\pi}{2}, \dfrac{\pi}{2}\right)$, and $\tan y = \dfrac{\sqrt{3}}{3}$. Therefore, $y = \dfrac{\pi}{6}$.

b. Because $y = \cos^{-1}\left(-\dfrac{\sqrt{2}}{2}\right)$, y is the angle whose measure is in the interval $[0, \pi]$, and $\cos y = -\dfrac{\sqrt{2}}{2}$. Therefore, $y = \dfrac{3}{4}\pi$.

▶ TRY EXERCISE 2, PAGE 411

TABLE 5.6

	$y = \tan x$	$y = \tan^{-1} x$	$y = \csc x$	$y = \csc^{-1} x$
Domain	$-\frac{\pi}{2} < x < \frac{\pi}{2}$	$-\infty < x < \infty$	$-\frac{\pi}{2} \le x \le \frac{\pi}{2}, x \ne 0$	$x \le -1$ or $x \ge 1$
Range	$-\infty < y < \infty$	$-\frac{\pi}{2} < y < \frac{\pi}{2}$	$y \le -1$ or $y \ge 1$	$-\frac{\pi}{2} \le y \le \frac{\pi}{2}, y \ne 0$
Asymptotes	$x = -\frac{\pi}{2}, x = \frac{\pi}{2}$	$y = -\frac{\pi}{2}, y = \frac{\pi}{2}$	$x = 0$	$y = 0$
Graph	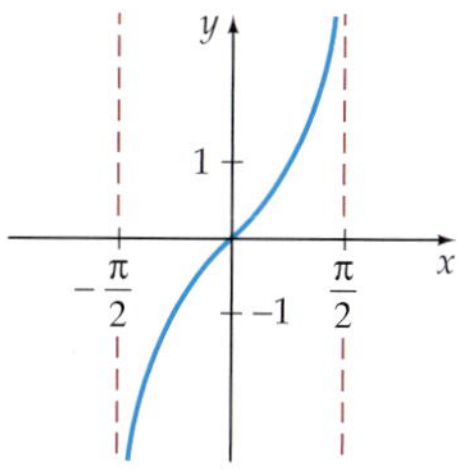	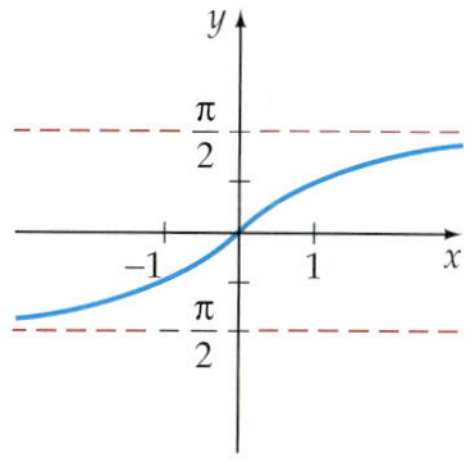	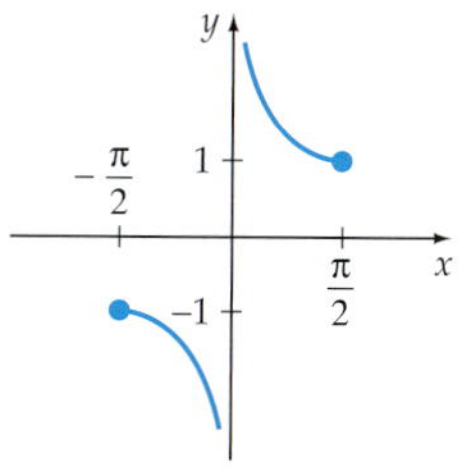	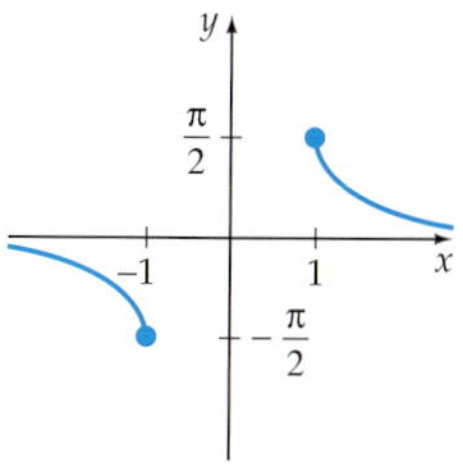
	$y = \sec x$	$y = \sec^{-1} x$	$y = \cot x$	$y = \cot^{-1} x$
Domain	$0 \le x \le \pi, x \ne \frac{\pi}{2}$	$x \le -1$ or $x \ge 1$	$0 < x < \pi$	$-\infty < x < \infty$
Range	$y \le -1$ or $y \ge 1$	$0 \le y \le \pi, y \ne \frac{\pi}{2}$	$-\infty < y < \infty$	$0 < y < \pi$
Asymptotes	$x = \frac{\pi}{2}$	$y = \frac{\pi}{2}$	$x = 0, x = \pi$	$y = 0, y = \pi$
Graph	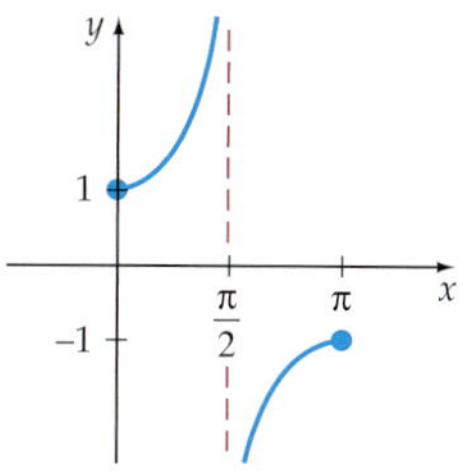	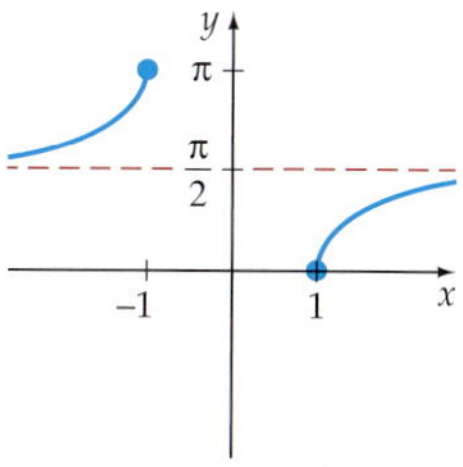	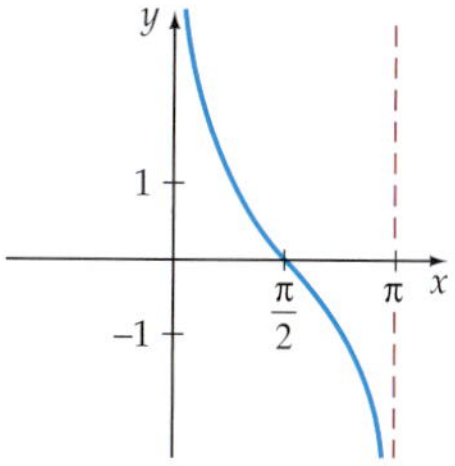	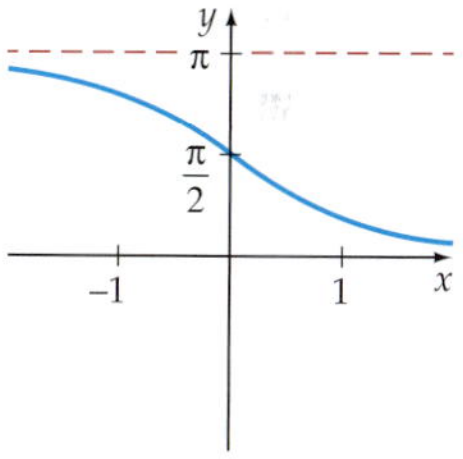

A calculator may not have keys for the inverse secant, cosecant, and cotangent functions. The following procedure shows an identity for the inverse cosecant function in terms of the inverse sine function. If we need to determine y, which is the angle whose cosecant is x, we can rewrite $y = \csc^{-1} x$ as follows:

$$y = \csc^{-1} x$$

• **Domain: $x \le -1$ or $x \ge 1$**

Range: $-\frac{\pi}{2} \le y \le \frac{\pi}{2}, y \ne 0$

$$\csc y = x$$

• **Definition of inverse function**

$$\frac{1}{\sin y} = x$$

• **Substitute $\frac{1}{\sin y}$ for csc y.**

$$\sin y = \frac{1}{x}$$ • Solve for sin y.

$$y = \sin^{-1}\frac{1}{x}$$ • Write using inverse notation.

$$\csc^{-1} x = \sin^{-1}\frac{1}{x}$$ • Replace y with $\csc^{-1} x$.

Thus $\csc^{-1} x$ is the same as $\sin^{-1}\frac{1}{x}$. There is a similar identity for $\sec^{-1} x$.

Identities for $\csc^{-1} x$, $\sec^{-1} x$, and $\cot^{-1} x$

If $x \le -1$ or $x \ge 1$, then

$$\csc^{-1} x = \sin^{-1}\frac{1}{x} \quad \text{and} \quad \sec^{-1} x = \cos^{-1}\frac{1}{x}$$

If x is a real number, then

$$\cot^{-1} x = \frac{\pi}{2} - \tan^{-1} x$$

Composition of Trigonometric Functions and Their Inverses

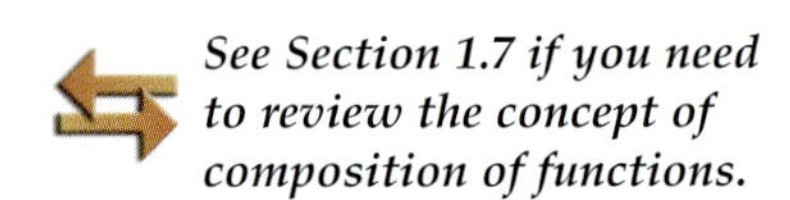
See Section 1.7 if you need to review the concept of composition of functions.

Recall that a function f and its inverse f^{-1} have the property that $f[f^{-1}(x)] = x$ for all x in the domain of f^{-1} and that $f^{-1}[f(x)] = x$ for all x in the domain of f. Applying this property to the functions $\sin x$, $\cos x$, and $\tan x$ and their inverse functions produces the following theorems.

Composition of Trigonometric Functions and Their Inverses

- If $-1 \le x \le 1$, then $\sin(\sin^{-1} x) = x$, and $\cos(\cos^{-1} x) = x$.
- If x is any real number, then $\tan(\tan^{-1} x) = x$.
- If $-\frac{\pi}{2} \le x \le \frac{\pi}{2}$, then $\sin^{-1}(\sin x) = x$.
- If $0 \le x \le \pi$, then $\cos^{-1}(\cos x) = x$.
- If $-\frac{\pi}{2} < x < \frac{\pi}{2}$, then $\tan^{-1}(\tan x) = x$.

In the next example we make use of some of the composition theorems to evaluate trigonometric expressions.

EXAMPLE 2 Evaluate the Composition of a Function and Its Inverse

Find the exact value of each composition of functions.

a. $\sin(\sin^{-1} 0.357)$ **b.** $\cos^{-1}(\cos 3)$ **c.** $\tan[\tan^{-1}(-11.27)]$

d. $\sin(\sin^{-1} \pi)$ **e.** $\cos(\cos^{-1} 0.277)$ **f.** $\tan^{-1}\left(\tan \dfrac{4\pi}{3}\right)$

Solution

a. Because 0.357 is in the interval $[-1, 1]$, $\sin(\sin^{-1} 0.357) = 0.357$.

b. Because 3 is in the interval $[0, \pi]$, $\cos^{-1}(\cos 3) = 3$.

c. Because -11.27 is a real number, $\tan[\tan^{-1}(-11.27)] = -11.27$.

d. Because π is not in the domain of the inverse sine function, $\sin(\sin^{-1} \pi)$ is undefined.

e. Because 0.277 is in the interval $[-1, 1]$, $\cos(\cos^{-1} 0.277) = 0.277$.

f. $\dfrac{4\pi}{3}$ is not in the interval $\left(-\dfrac{\pi}{2}, \dfrac{\pi}{2}\right)$; however, the reference angle for $\theta = \dfrac{4\pi}{3}$ is $\theta' = \dfrac{\pi}{3}$. Thus $\tan^{-1}\left(\tan \dfrac{4\pi}{3}\right) = \tan^{-1}\left(\tan \dfrac{\pi}{3}\right)$. Because $\dfrac{\pi}{3}$ is in the interval $\left(-\dfrac{\pi}{2}, \dfrac{\pi}{2}\right)$, $\tan^{-1}\left(\tan \dfrac{\pi}{3}\right) = \dfrac{\pi}{3}$. Hence

$$\tan^{-1}\left(\tan \frac{4\pi}{3}\right) = \frac{\pi}{3}.$$

▶ TRY EXERCISE 20, PAGE 411

? QUESTION Is $\tan^{-1}(\tan x) = x$ an identity?

It is often easy to evaluate a trigonometric expression by referring to a sketch of a right triangle that satisfies given conditions. In Example 3 we make use of this technique.

EXAMPLE 3 Evaluate a Trigonometric Expression

Find the exact value of $\sin\left(\cos^{-1} \dfrac{2}{5}\right)$.

Continued ▶

? ANSWER No. $\tan^{-1}(\tan x) = x$ only if $-\dfrac{\pi}{2} < x < \dfrac{\pi}{2}$.

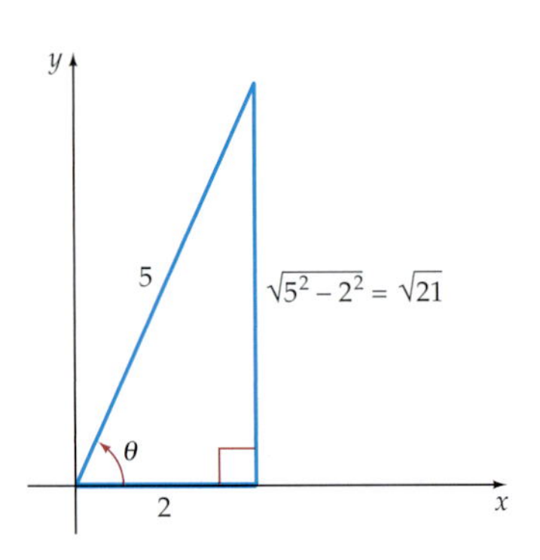

FIGURE 5.32

Solution

Let $\theta = \cos^{-1}\frac{2}{5}$, which implies $\cos\theta = \frac{2}{5}$. Because $\cos\theta$ is positive, θ is a first-quadrant angle. We draw a right triangle with base 2 and hypotenuse 5 so that we can view θ, as shown in **Figure 5.32.** The height of the triangle is $\sqrt{5^2 - 2^2} = \sqrt{21}$. Our goal is to find $\sin\theta$, which by definition is $\frac{\text{opp}}{\text{hyp}} = \frac{\sqrt{21}}{5}$. Thus

$$\sin\left(\cos^{-1}\frac{2}{5}\right) = \sin(\theta) = \frac{\sqrt{21}}{5}$$

TRY EXERCISE 40, PAGE 411

In Example 4, we sketch two right triangles to evaluate the given expression.

EXAMPLE 4 Evaluate a Trigonometric Expression

Find the exact value of $\sin\left[\sin^{-1}\frac{3}{5} + \cos^{-1}\left(-\frac{5}{13}\right)\right]$.

Solution

Let $\alpha = \sin^{-1}\frac{3}{5}$. Thus $\sin\alpha = \frac{3}{5}$. Let $\beta = \cos^{-1}\left(-\frac{5}{13}\right)$, which implies that $\cos\beta = -\frac{5}{13}$. Sketch angles α and β as shown in **Figure 5.33.** We wish to evaluate

$$\sin\left[\sin^{-1}\frac{3}{5} + \cos^{-1}\left(-\frac{5}{13}\right)\right] = \sin(\alpha + \beta)$$
$$= \sin\alpha\cos\beta + \cos\alpha\sin\beta \qquad (1)$$

A close look at the triangles in **Figure 5.33** shows us that

$$\cos\alpha = \frac{4}{5} \quad \text{and} \quad \sin\beta = \frac{12}{13}$$

Substituting in Equation (1) gives us our desired result.

$$\sin\left[\sin^{-1}\frac{3}{5} + \cos^{-1}\left(-\frac{5}{13}\right)\right] = \sin\alpha\cos\beta + \cos\alpha\sin\beta$$
$$= \left(\frac{3}{5}\right)\left(-\frac{5}{13}\right) + \left(\frac{4}{5}\right)\left(\frac{12}{13}\right) = \frac{33}{65}$$

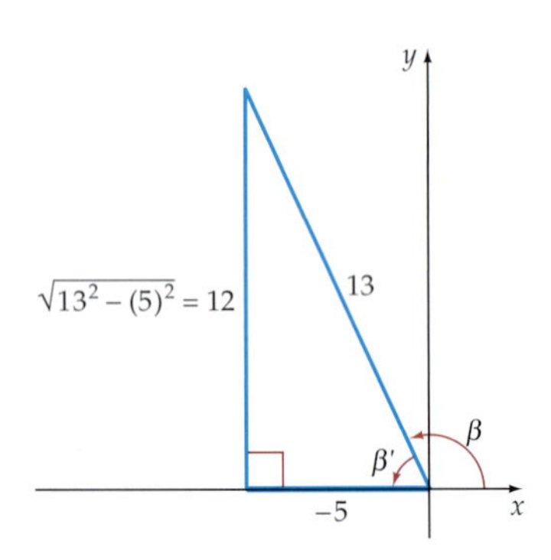

FIGURE 5.33

TRY EXERCISE 46, PAGE 412

In Example 5 we make use of the identity $\cos(\cos^{-1} x) = x$, where $-1 \le x \le 1$, to solve an equation.

EXAMPLE 5 Solve an Inverse Trigonometric Equation

Solve $\sin^{-1}\frac{3}{5} + \cos^{-1}x = \pi$.

Solution

Solve for $\cos^{-1}x$, and then take the cosine of both sides of the equation.

$$\sin^{-1}\frac{3}{5} + \cos^{-1}x = \pi$$

$$\cos^{-1}x = \pi - \sin^{-1}\frac{3}{5}$$

$$\cos(\cos^{-1}x) = \cos\left(\pi - \sin^{-1}\frac{3}{5}\right)$$

$$x = \cos(\pi - \alpha)$$

• Let $\alpha = \sin^{-1}\frac{3}{5}$. Note that α is the angle whose sine is $\frac{3}{5}$. (See **Figure 5.34.**)

$$= \cos\pi\cos\alpha + \sin\pi\sin\alpha$$

• Difference identity for cosine.

$$= (-1)\cos\alpha + (0)\sin\alpha$$

$$= -\cos\alpha$$

$$= -\frac{4}{5}$$

• $\cos\alpha = \frac{4}{5}$ (See **Figure 5.34.**)

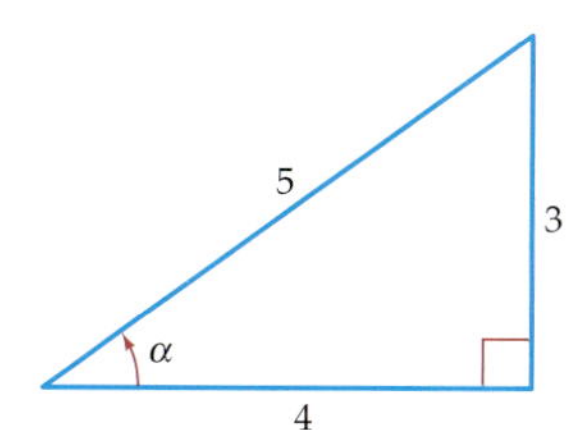

FIGURE 5.34

▶ TRY EXERCISE 56, PAGE 412

EXAMPLE 6 Verify a Trigonometric Identity That Involves Inverses

Verify the identity $\sin^{-1}x + \cos^{-1}x = \frac{\pi}{2}$.

Solution

Let $\alpha = \sin^{-1}x$ and $\beta = \cos^{-1}x$. These equations imply that $\sin\alpha = x$ and $\cos\beta = x$. From the right triangles in **Figure 5.35,**

$$\cos\alpha = \sqrt{1 - x^2} \qquad \text{and} \qquad \sin\beta = \sqrt{1 - x^2}$$

Our goal is to show that $\sin^{-1}x + \cos^{-1}x$ equals $\frac{\pi}{2}$.

$$\sin^{-1}x + \cos^{-1}x = \alpha + \beta$$

$$= \cos^{-1}[\cos(\alpha + \beta)]$$

• Because $0 \le \alpha + \beta \le \pi$, we can apply $\alpha + \beta = \cos^{-1}[\cos(\alpha + \beta)]$.

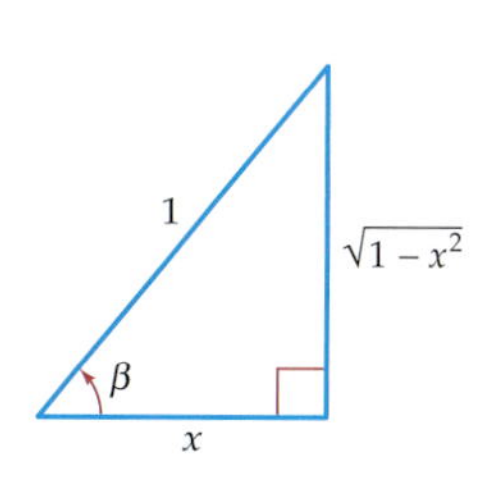

FIGURE 5.35

Continued ▶

$$= \cos^{-1}[\cos\alpha\cos\beta - \sin\alpha\sin\beta]$$ • **Addition identity for cosine**

$$= \cos^{-1}\left[\left(\sqrt{1-x^2}\right)(x) - (x)\left(\sqrt{1-x^2}\right)\right]$$

$$= \cos^{-1} 0 = \frac{\pi}{2}$$

▶ TRY EXERCISE 64, PAGE 412

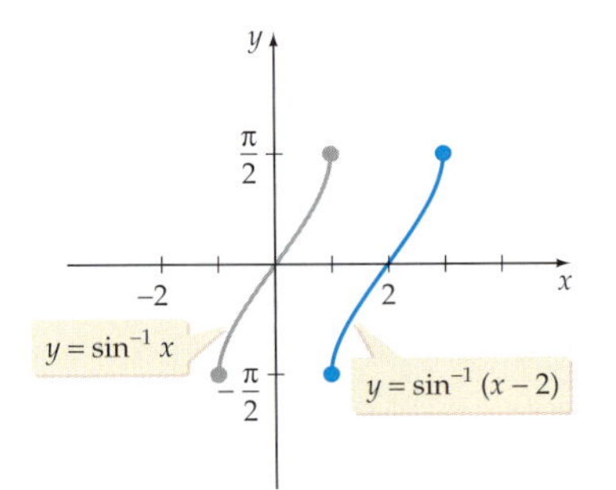

FIGURE 5.36

GRAPHS OF INVERSE TRIGONOMETRIC FUNCTIONS

The inverse trigonometric functions can be graphed by using the procedures of stretching, shrinking, and translation that were discussed earlier in the text. For instance, the graph of $y = \sin^{-1}(x - 2)$ is a horizontal shift 2 units to the right of the graph of $y = \sin^{-1} x$, as shown in **Figure 5.36.**

EXAMPLE 7 Graph an Inverse Function

Graph: $y = \cos^{-1} x + 1$

Solution

Recall that the graph of $y = f(x) + c$ is a vertical translation of the graph of f. Because $c = 1$, a positive number, the graph of $y = \cos^{-1} x + 1$ is the graph of $y = \cos^{-1} x$ shifted 1 unit up. See **Figure 5.37.**

▶ TRY EXERCISE 68, PAGE 412

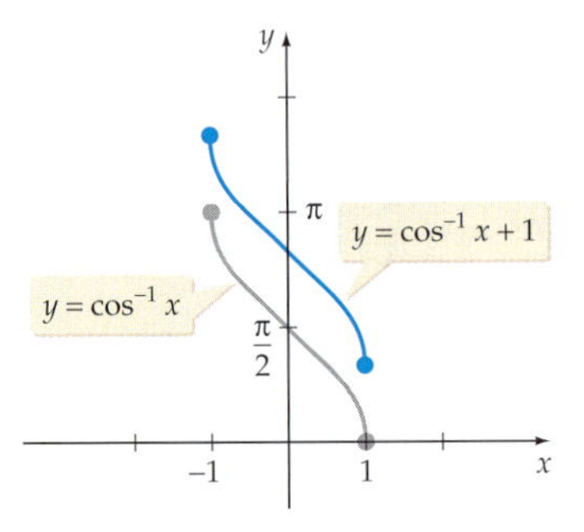

FIGURE 5.37

INTEGRATING TECHNOLOGY

When you use a graphing utility to draw the graph of an inverse trigonometric function, use the properties of these functions to verify the correctness of your graph. For instance, the graph of $y = 3\sin^{-1} 0.5x$ is shown in **Figure 5.38.** The domain of $y = \sin^{-1} x$ is $-1 \le x \le 1$. Therefore, the domain of $y = 3\sin^{-1} 0.5x$ is $-1 \le 0.5x \le 1$ or, multiplying the inequality by 2, $-2 \le x \le 2$. This is consistent with the graph in **Figure 5.38.**

The range of $y = \sin^{-1} x$ is $-\frac{\pi}{2} \le y \le \frac{\pi}{2}$. Thus the range of $y = 3\sin^{-1} 0.5x$ is $-\frac{3\pi}{2} \le y \le \frac{3\pi}{2}$. This is also consistent with the graph. Verifying some of the properties of $y = \sin^{-1} x$ serves as a check that you have correctly entered the equation for the graph.

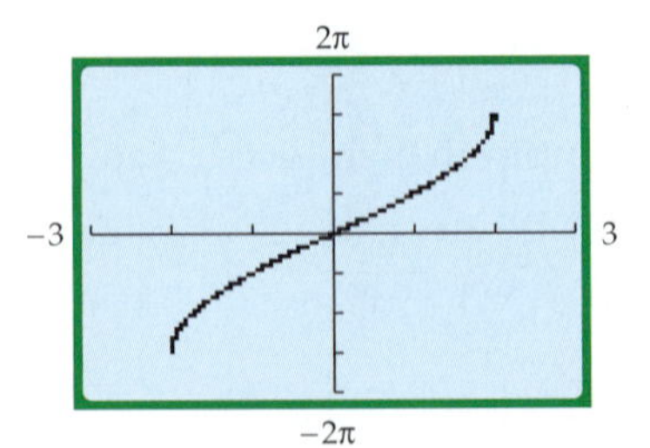

$y = 3\sin^{-1} 0.5x$

FIGURE 5.38

Topics for Discussion

1. Is the equation

$$\tan^{-1} x = \frac{1}{\tan x}$$

true for all values of x, true for some values of x, or false for all values of x?

2. Are there real numbers x for which the following is true? Explain.

$$\sin(\sin^{-1} x) \neq \sin^{-1}(\sin x)$$

3. Explain how to find the value of $\sec^{-1} 3$ by using a calculator.

4. Explain how you can determine the range of $y = (2 \cos^{-1} x) - 1$ using
   a. algebra b. a graph

Exercise Set 5.4

In Exercises 1 to 18, find the exact radian value.

1. $\sin^{-1} 1$
▶ **2.** $\sin^{-1} \frac{\sqrt{2}}{2}$
3. $\tan^{-1}(-1)$
4. $\tan^{-1} \sqrt{3}$
5. $\cot^{-1} \frac{\sqrt{3}}{3}$
6. $\cot^{-1} 1$
7. $\sec^{-1} 2$
8. $\sec^{-1} \frac{2\sqrt{3}}{3}$
9. $\csc^{-1}(-\sqrt{2})$
10. $\csc^{-1}(-2)$
11. $\sin^{-1}\left(-\frac{\sqrt{3}}{2}\right)$
12. $\sin^{-1} \frac{1}{2}$
13. $\cos^{-1}\left(-\frac{1}{2}\right)$
14. $\cos^{-1} \frac{\sqrt{3}}{2}$
15. $\tan^{-1} \frac{\sqrt{3}}{3}$
16. $\tan^{-1} 1$
17. $\cot^{-1} \sqrt{3}$
18. $\cot^{-1}(-1)$

In Exercises 19 to 48, find the exact value of the given expression. If an exact value cannot be given, give the value to the nearest ten-thousandth.

19. $\cos\left(\cos^{-1} \frac{1}{2}\right)$
▶ **20.** $\tan\left(\tan^{-1} \frac{1}{2}\right)$
21. $\sin\left(\tan^{-1} \frac{3}{4}\right)$
22. $\cos\left(\sin^{-1} \frac{5}{13}\right)$
23. $\tan\left(\sin^{-1} \frac{\sqrt{2}}{2}\right)$
24. $\sin\left[\cos^{-1}\left(-\frac{\sqrt{3}}{2}\right)\right]$
25. $\cos(\sec^{-1} 2)$
26. $\sin^{-1}(\sin 2)$
27. $\sin^{-1}\left(\sin \frac{\pi}{6}\right)$
28. $\sin^{-1}\left(\sin \frac{5\pi}{6}\right)$
29. $\cos^{-1}\left(\sin \frac{\pi}{4}\right)$
30. $\cos^{-1}\left(\cos \frac{5\pi}{4}\right)$
31. $\tan^{-1}\left(\sin \frac{\pi}{6}\right)$
32. $\cot^{-1}\left(\cos \frac{2\pi}{3}\right)$
33. $\sin^{-1}\left[\cos\left(-\frac{2\pi}{3}\right)\right]$
34. $\cos^{-1}\left[\tan\left(-\frac{\pi}{3}\right)\right]$
35. $\tan\left(\sin^{-1} \frac{1}{2}\right)$
36. $\cot(\csc^{-1} 2)$
37. $\sec\left(\sin^{-1} \frac{1}{4}\right)$
38. $\csc\left(\cos^{-1} \frac{3}{4}\right)$
39. $\cos\left(\sin^{-1} \frac{7}{25}\right)$
▶ **40.** $\tan\left(\cos^{-1} \frac{3}{5}\right)$
41. $\cos\left(2 \sin^{-1} \frac{\sqrt{2}}{2}\right)$
42. $\tan\left(2 \sin^{-1} \frac{\sqrt{3}}{2}\right)$
43. $\sin\left(2 \sin^{-1} \frac{4}{5}\right)$
44. $\cos(2 \tan^{-1} 1)$
45. $\sin\left(\sin^{-1} \frac{2}{3} + \cos^{-1} \frac{1}{2}\right)$

46. $\cos\left(\sin^{-1}\frac{3}{4} + \cos^{-1}\frac{5}{13}\right)$

47. $\tan\left(\cos^{-1}\frac{1}{2} - \sin^{-1}\frac{3}{4}\right)$

48. $\sec\left(\cos^{-1}\frac{2}{3} + \sin^{-1}\frac{2}{3}\right)$

In Exercises 49 to 58, solve the equation for *x* algebraically.

49. $\sin^{-1} x = \cos^{-1}\frac{5}{13}$

50. $\tan^{-1} x = \sin^{-1}\frac{24}{25}$

51. $\sin^{-1}(x - 1) = \frac{\pi}{2}$

52. $\cos^{-1}\left(x - \frac{1}{2}\right) = \frac{\pi}{3}$

53. $\tan^{-1}\left(x + \frac{\sqrt{2}}{2}\right) = \frac{\pi}{4}$

54. $\sin^{-1}(x - 2) = -\frac{\pi}{6}$

55. $\sin^{-1}\frac{3}{5} + \cos^{-1} x = \frac{\pi}{4}$

56. $\sin^{-1} x + \cos^{-1}\frac{4}{5} = \frac{\pi}{6}$

57. $\sin^{-1}\frac{\sqrt{2}}{2} + \cos^{-1} x = \frac{2\pi}{3}$

58. $\cos^{-1} x + \sin^{-1}\frac{\sqrt{3}}{2} = \frac{\pi}{2}$

In Exercises 59 to 62, evaluate each expression.

59. $\cos(\sin^{-1} x)$

60. $\tan(\cos^{-1} x)$

61. $\sin(\sec^{-1} x)$

62. $\sec(\sin^{-1} x)$

In Exercises 63 to 66, verify the identity.

63. $\sin^{-1} x + \sin^{-1}(-x) = 0$

64. $\cos^{-1} x + \cos^{-1}(-x) = \pi$

65. $\tan^{-1} x + \tan^{-1}\frac{1}{x} = \frac{\pi}{2}, x > 0$

66. $\sec^{-1}\frac{1}{x} + \csc^{-1}\frac{1}{x} = \frac{\pi}{2}$

In Exercises 67 to 74, use stretching, shrinking, and translation procedures to graph each equation.

67. $y = \sin^{-1} x + 2$

68. $y = \cos^{-1}(x - 1)$

69. $y = \sin^{-1}(x + 1) - 2$

70. $y = \tan^{-1}(x - 1) + 2$

71. $y = 2\cos^{-1} x$

72. $y = -2\tan^{-1} x$

73. $y = \tan^{-1}(x + 1) - 2$

74. $y = \sin^{-1}(x - 2) + 1$

75. DOT-MATRIX PRINTING In dot-matrix printing, the *blank-area factor* is the ratio of the blank area (unprinted area) to the total area of the line. If circular dots are used to print, then the blank-area factor is given by

$$\frac{A}{(S)(D)} = 1 - \frac{1}{2}\left[1 - \left(\frac{S}{D}\right)^2 + \frac{D}{S}\sin^{-1}\left(\frac{S}{D}\right)\right]$$

where $A = A_1 + A_2$, A_1 and A_2 are the areas of the regions shown in the figure, S is the distance between the centers of overlapping dots, and D is the diameter of a dot.

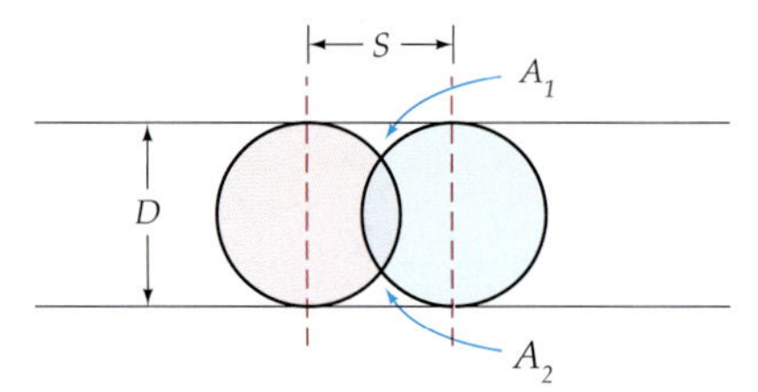

Calculate, to four decimal places, the blank-area factor where

a. $D = 0.2$ millimeter and $S = 0.1$ millimeter

b. $D = 0.16$ millimeter and $S = 0.1$ millimeter

76. VOLUME IN A WATER TANK The volume V of water (measured in cubic feet) in a horizontal cylindrical tank of radius 5 feet and length 12 feet is given by

$$V(x) = 12\left[25\cos^{-1}\left(\frac{5 - x}{5}\right) - (5 - x)\sqrt{10x - x^2}\right]$$

where x is the depth of the water in feet.

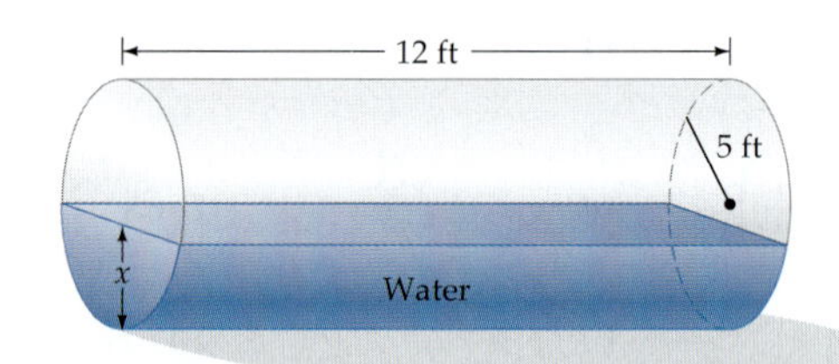

a. Graph V over its domain $0 \le x \le 10$.

b. Write a sentence that explains why the graph of V increases more rapidly when x increases from 4.9 feet to 5 feet than it does when x increases from 0.1 foot to 0.2 foot.

c. If $x = 4$ feet, find the volume (to the nearest 0.01 cubic foot) of the water in the tank.

d. Find the depth x (to the nearest 0.01 foot) if there are 288 cubic feet of water in the tank.

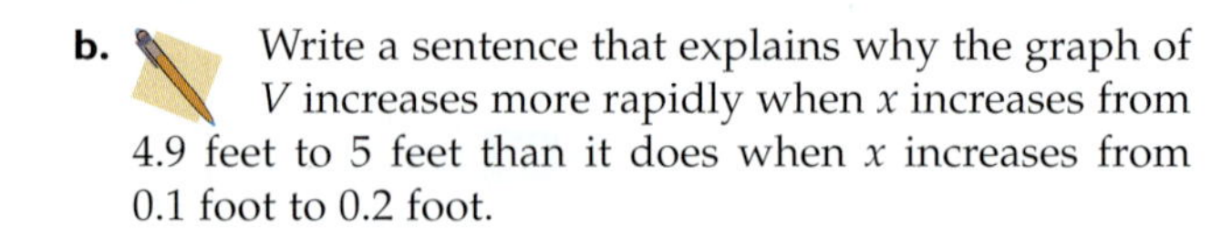

In Exercises 77 to 82, use a graphing utility to graph each equation.

77. $y = \csc^{-1} 2x$

78. $y = 0.5 \sec^{-1} \dfrac{x}{2}$

79. $y = \sec^{-1}(x - 1)$

80. $y = \sec^{-1}(x + \pi)$

81. $y = 2 \tan^{-1} 2x$

82. $y = \tan^{-1}(x - 1)$

CONNECTING CONCEPTS

In Exercises 83 to 86, verify the identity.

83. $\cos(\sin^{-1} x) = \sqrt{1 - x^2}$

84. $\sec(\sin^{-1} x) = \dfrac{\sqrt{1 - x^2}}{1 - x^2}$

85. $\tan(\csc^{-1} x) = \dfrac{\sqrt{x^2 - 1}}{x^2 - 1}, x > 1$

86. $\sin(\cot^{-1} x) = \dfrac{\sqrt{x^2 + 1}}{x^2 + 1}$

In Exercises 87 to 90, solve for y in terms of x.

87. $5x = \tan^{-1} 3y$

88. $2x = \dfrac{1}{2} \sin^{-1} 2y$

89. $x - \dfrac{\pi}{3} = \cos^{-1}(y - 3)$

90. $x + \dfrac{\pi}{2} = \tan^{-1}(2y - 1)$

PREPARE FOR SECTION 5.5

91. Use the quadratic formula to solve $3x^2 - 5x - 4 = 0$. [1.1]

92. Use a Pythagorean identity to write $\sin^2 x$ as a function involving $\cos^2 x$. [4.2]

93. Evaluate $\dfrac{\pi}{2} + 2k\pi$ for $k = 1, 2,$ and 3.

94. Factor by grouping: $x^2 - \dfrac{\sqrt{3}}{2}x + x - \dfrac{\sqrt{3}}{2}$. [A.3]

95. Solve $2x^2 - 7x + 3 = 0$ by factoring. [1.1]

96. Solve $2x^2 - 2x = 0$ by factoring. [1.1]

PROJECTS

1. VISUAL INSIGHT

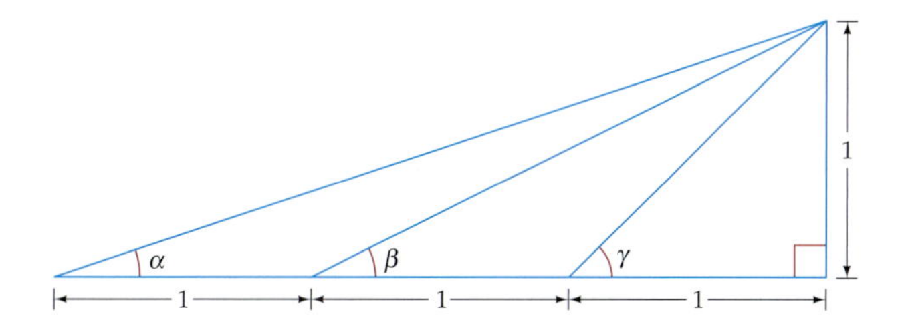

Explain how the figure above can be used to verify each identity.

a. $\tan^{-1} \dfrac{1}{3} + \tan^{-1} \dfrac{1}{2} = \dfrac{\pi}{4}$ [*Hint:* Start by using an identity to find the value of $\tan(\alpha + \beta)$.]

b. $\alpha + \beta = \gamma$

SECTION 5.5

TRIGONOMETRIC EQUATIONS

- SOLVE TRIGONOMETRIC EQUATIONS
- AN APPLICATION INVOLVING A TRIGONOMETRIC EQUATION

SOLVE TRIGONOMETRIC EQUATIONS

Consider the equation $\sin x = \frac{1}{2}$. The graph of $y = \sin x$, along with the line $y = \frac{1}{2}$, is shown in **Figure 5.39.** The x values of the intersections of the two graphs are the solutions of $\sin x = \frac{1}{2}$. The solutions in the interval $0 \le x < 2\pi$ are $x = \frac{\pi}{6}$ and $\frac{5\pi}{6}$.

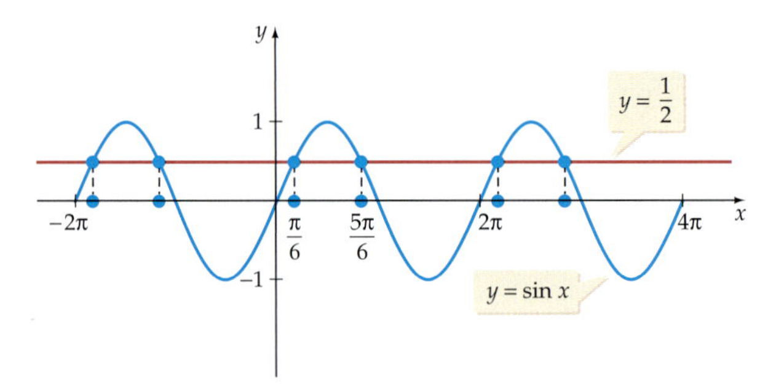

FIGURE 5.39

If we remove the restriction $0 \le x < 2\pi$, there are many more solutions. Because the sine function is periodic with a period of 2π, other solutions are obtained by adding $2k\pi$, k an integer, to either of the previous solutions. Thus the solutions of $\sin x = \frac{1}{2}$ are

$$x = \frac{\pi}{6} + 2k\pi, \quad k \text{ an integer}$$

$$x = \frac{5\pi}{6} + 2k\pi, \quad k \text{ an integer}$$

QUESTION How many solutions does the equation $\cos x = \frac{\sqrt{3}}{2}$ have on the interval $0 \le x < 2\pi$?

Algebraic methods and trigonometric identities are used frequently to find the solutions of trigonometric equations. Algebraic methods that are often employed include solving by factoring, solving by using the quadratic formula, and squaring each side of the equation.

ANSWER Two

EXAMPLE 1 Solve a Trigonometric Equation by Factoring

Solve $2\sin^2 x\cos x - \cos x = 0$, where $0 \le x < 2\pi$.

Algebraic Solution

$$2\sin^2 x\cos x - \cos x = 0$$

$$\cos x(2\sin^2 x - 1) = 0 \qquad \text{• Factor } \cos x \text{ from each term.}$$

$$\cos x = 0 \quad \text{or} \quad 2\sin^2 x - 1 = 0 \qquad \text{• Use the Principle of Zero Products.}$$

$$x = \frac{\pi}{2}, \frac{3\pi}{2} \qquad \sin^2 x = \frac{1}{2} \qquad \text{• Solve each equation for } x \text{ with } 0 \le x < 2\pi.$$

$$\sin x = \pm\frac{\sqrt{2}}{2}$$

$$x = \frac{\pi}{4}, \frac{3\pi}{4}, \frac{5\pi}{4}, \frac{7\pi}{4}$$

The solutions in the interval $0 \le x < 2\pi$ are $\frac{\pi}{4}, \frac{\pi}{2}, \frac{3\pi}{4}, \frac{5\pi}{4}, \frac{3\pi}{2}$, and $\frac{7\pi}{4}$.

Visualize the Solution

The solutions are the x-coordinates of the x-intercepts of $y = 2\sin^2 x\cos x - \cos x$ on the interval $[0, 2\pi)$. See **Figure 5.40.**

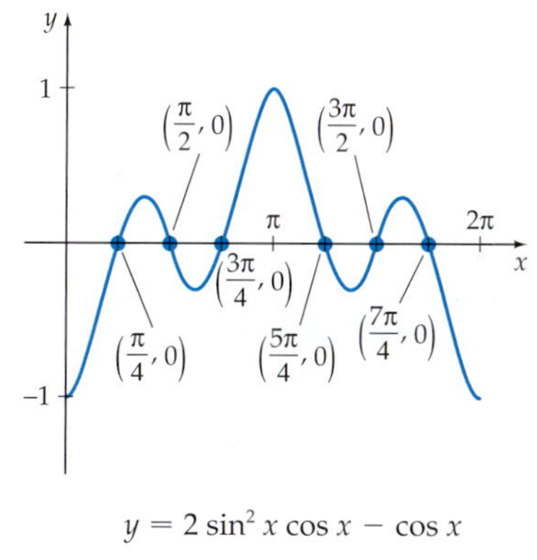

FIGURE 5.40

▶ TRY EXERCISE 14, PAGE 421

Squaring both sides of an equation may not produce an equivalent equation. Thus, when this method is used, the proposed solutions must be checked to eliminate any extraneous solutions.

EXAMPLE 2 Solve a Trigonometric Equation by Squaring Each Side of the Equation

Solve $\sin x + \cos x = 1$, where $0 \le x < 2\pi$.

Algebraic Solution

$$\sin x + \cos x = 1 \qquad \text{• Solve for } \sin x.$$

$$\sin x = 1 - \cos x$$

$$\sin^2 x = (1 - \cos x)^2 \qquad \text{• Square each side.}$$

$$\sin^2 x = 1 - 2\cos x + \cos^2 x$$

$$1 - \cos^2 x = 1 - 2\cos x + \cos^2 x \qquad \text{• } \sin^2 x = 1 - \cos^2 x$$

$$2\cos^2 x - 2\cos x = 0$$

$$2\cos x(\cos x - 1) = 0 \qquad \text{• Factor.}$$

$$2\cos x = 0 \quad \text{or} \quad \cos x = 1$$

$$x = \frac{\pi}{2}, \frac{3\pi}{2} \qquad x = 0 \qquad \text{• Solve each equation for } x \text{ with } 0 \le x < 2\pi.$$

Visualize the Solution

The solutions are the x-coordinates of the points of intersection of $y = \sin x + \cos x$ and $y = 1$ on the interval $[0, 2\pi)$. See **Figure 5.41.**

Continued ▶

Squaring each side of an equation may introduce extraneous solutions. Therefore, we must check the solutions. A check will show that 0 and $\frac{\pi}{2}$ are solutions but $\frac{3\pi}{2}$ is not a solution.

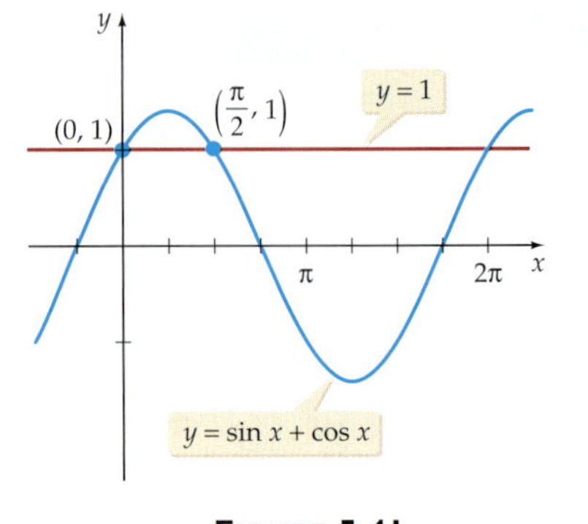

FIGURE 5.41

TRY EXERCISE 46, PAGE 422

EXAMPLE 3 Solve a Trigonometric Equation by Using the Quadratic Formula

Solve $3\cos^2 x - 5\cos x - 4 = 0$, where $0 \le x < 2\pi$.

Algebraic Solution

The given equation is quadratic in form and cannot be factored easily. However, we can use the quadratic formula to solve for $\cos x$.

$$3\cos^2 x - 5\cos x - 4 = 0 \qquad \bullet\ a = 3, b = -5, c = -4$$

$$\cos x = \frac{-(-5) \pm \sqrt{(-5)^2 - 4(3)(-4)}}{(2)(3)} = \frac{5 \pm \sqrt{73}}{6}$$

The equation $\cos x = \frac{5 + \sqrt{73}}{6}$ does not have a solution because $\frac{5 + \sqrt{73}}{6} > 2$ and for any x the maximum value of $\cos x$ is 1. Thus $\cos x = \frac{5 - \sqrt{73}}{6}$, and because $\frac{5 - \sqrt{73}}{6}$ is a negative number (about -0.59), the equation $\cos x = \frac{5 - \sqrt{73}}{6}$ will have two solutions on the interval $[0, 2\pi)$. Thus

$$x = \cos^{-1}\left(\frac{5 - \sqrt{73}}{6}\right) \approx 2.2027 \qquad \text{or}$$

$$x = 2\pi - \cos^{-1}\left(\frac{5 - \sqrt{73}}{6}\right) \approx 4.0805$$

To the nearest ten-thousandth, the solutions on the interval $[0, 2\pi)$ are 2.2027 and 4.0805.

Visualize the Solution

The solutions are the x-coordinates of the x-intercepts of $y = 3\cos^2 x - 5\cos x - 4$ on the interval $[0, 2\pi)$. See **Figure 5.42.**

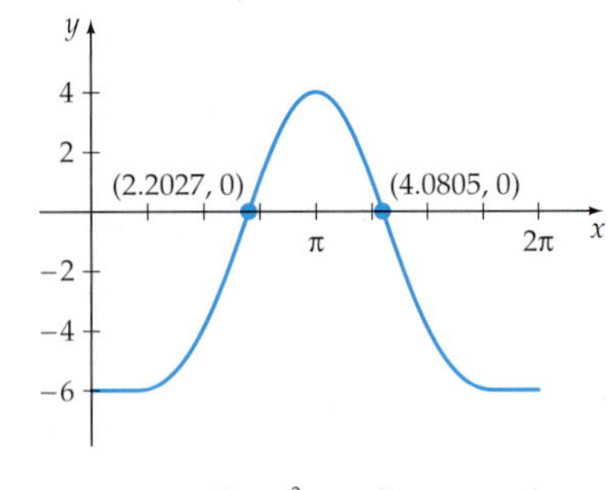

$y = 3\cos^2 x - 5\cos x - 4$

FIGURE 5.42

TRY EXERCISE 50, PAGE 422

When solving equations that contain multiple angles, we must be sure we find all the solutions of the equation for the given interval. For example, to find all solutions of $\sin 2x = \frac{1}{2}$, where $0 \le x < 2\pi$, we first solve for $2x$.

$$\sin 2x = \frac{1}{2}$$

$$2x = \frac{\pi}{6} + 2k\pi \quad \text{or} \quad 2x = \frac{5\pi}{6} + 2k\pi$$ • *k* is an integer.

Solving for x, we have $x = \frac{\pi}{12} + k\pi$ or $x = \frac{5\pi}{12} + k\pi$. Substituting integers for k, we obtain

$$k = 0: \qquad x = \frac{\pi}{12} \quad \text{or} \quad x = \frac{5\pi}{12}$$

$$k = 1: \qquad x = \frac{13\pi}{12} \quad \text{or} \quad x = \frac{17\pi}{12}$$

$$k = 2: \qquad x = \frac{25\pi}{12} \quad \text{or} \quad x = \frac{29\pi}{12}$$

Note that for $k \ge 2$, $x \ge 2\pi$, and the solutions to $\sin 2x = \frac{1}{2}$ are not in the interval $0 \le x < 2\pi$. Thus, for $0 \le x < 2\pi$, the solutions are $\frac{\pi}{12}$, $\frac{5\pi}{12}$, $\frac{13\pi}{12}$, and $\frac{17\pi}{12}$. See **Figure 5.43.**

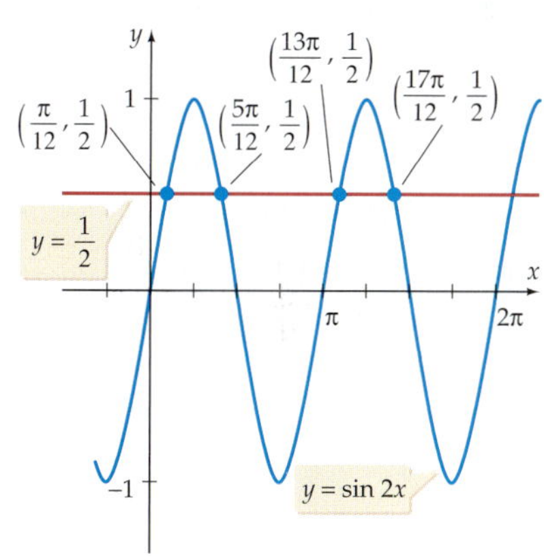

FIGURE 5.43

EXAMPLE 4 Solve a Trigonometric Equation

Solve: $\sin 3x = 1$

Algebraic Solution

The equation $\sin 3x = 1$ implies

$$3x = \frac{\pi}{2} + 2k\pi, \quad k \text{ an integer}$$

$$x = \frac{\pi}{6} + \frac{2k\pi}{3}, \quad k \text{ an integer}$$ • Divide each side by 3.

Visualize the Solution

The solutions are the x-coordinates of the points of intersection of $y = \sin 3x$ and $y = 1$. **Figure 5.44** on the following page shows eight of the points of intersection.

Continued ▶

Because x is not restricted to a finite interval, the given equation has an infinite number of solutions. All of the solutions are represented by the equation

$$x = \frac{\pi}{6} + \frac{2k\pi}{3}, \quad \text{where } k \text{ is an integer}$$

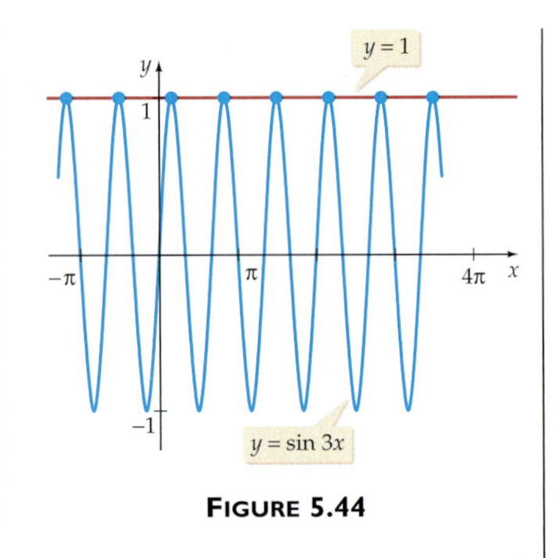

FIGURE 5.44

▶ TRY EXERCISE 60, PAGE 422

EXAMPLE 5 Solve a Trigonometric Equation

Solve $\sin^2 2x - \frac{\sqrt{3}}{2}\sin 2x + \sin 2x - \frac{\sqrt{3}}{2} = 0$, where $0° \le x < 360°$.

Algebraic Solution

Factor the left side of the equation by grouping, and then set each factor equal to zero.

$$\sin^2 2x - \frac{\sqrt{3}}{2}\sin 2x + \sin 2x - \frac{\sqrt{3}}{2} = 0$$

$$\sin 2x\left(\sin 2x - \frac{\sqrt{3}}{2}\right) + \left(\sin 2x - \frac{\sqrt{3}}{2}\right) = 0$$

$$(\sin 2x + 1)\left(\sin 2x - \frac{\sqrt{3}}{2}\right) = 0$$

$$\sin 2x + 1 = 0 \quad \text{or} \quad \sin 2x - \frac{\sqrt{3}}{2} = 0$$

$$\sin 2x = -1 \qquad \sin 2x = \frac{\sqrt{3}}{2}$$

The equation $\sin 2x = -1$ implies that $2x = 270° + 360° \cdot k$, k an integer. Thus $x = 135° + 180° \cdot k$. The solutions of this equation with $0° \le x < 360°$ are 135° and 315°. Similarly, the equation $\sin 2x = \frac{\sqrt{3}}{2}$ implies

$$2x = 60° + 360° \cdot k \quad \text{or} \quad 2x = 120° + 360° \cdot k$$

$$x = 30° + 180° \cdot k \qquad x = 60° + 180° \cdot k$$

Visualize the Solution

The solutions are the x-coordinates of the x-intercepts of

$$y = \sin^2 2x - \frac{\sqrt{3}}{2}\sin 2x + \sin 2x - \frac{\sqrt{3}}{2}$$

on the interval $[0, 2\pi)$. See **Figure 5.45** on the following page.

The solutions with $0° \le x < 360°$ are $30°$, $60°$, $210°$, and $240°$. Combining the solutions from each equation, we have $30°, 60°, 135°, 210°, 240°$, and $315°$ as our solutions.

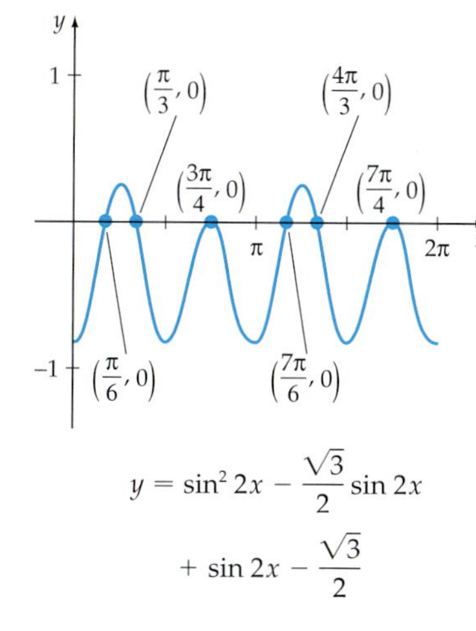

Figure 5.45

▶ **Try Exercise 76, page 422**

In Example 6, algebraic methods do not provide the solutions, so we rely on a graph.

EXAMPLE 6 Approximate Solutions Graphically

Use a graphing utility to approximate the solutions of $x + 3\cos x = 0$.

Solution

The solutions are the x-intercepts of $y = x + 3\cos x$. See **Figure 5.46.** A close-up view of the graph of $y = x + 3\cos x$ shows that, to the nearest thousandth, the solutions are

$$x_1 = -1.170, \quad x_2 = 2.663, \quad \text{and} \quad x_3 = 2.938$$

3.5
−1.5
3.5
−0.5
(−1.170, 0)
(2.663, 0)
(2.938, 0)

$y = x + 3\cos x$

Figure 5.46

▶ **Try Exercise 78, page 423**

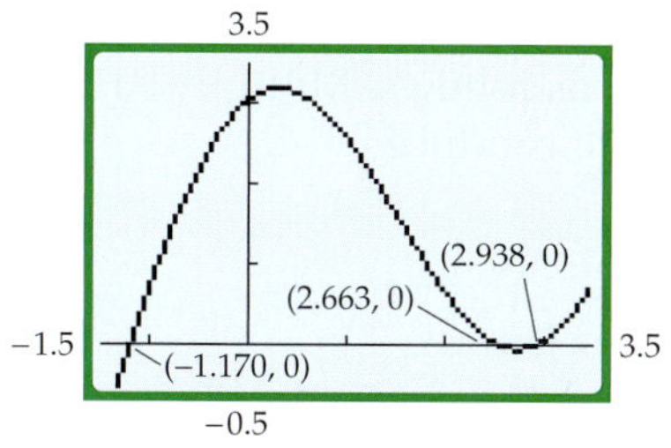

An Application Involving a Trigonometric Equation

EXAMPLE 7 Solve a Projectile Application

A projectile is fired at an angle of inclination θ from the horizon with an initial velocity v_0. Its range d (neglecting air resistance) is given by

$$d = \frac{v_0^2}{16}\sin\theta\cos\theta$$

where v_0 is measured in feet per second and d is measured in feet. See **Figure 5.47**.

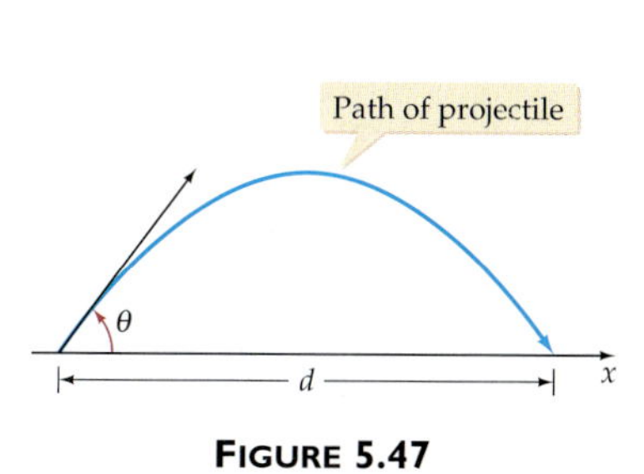

Figure 5.47

Continued ▶

a. If $v_0 = 325$ feet per second, find the angles θ (in degrees) for which the projectile will hit a target 2295 feet downrange.

b. What is the maximum horizontal range for a projectile that has an initial velocity of 474 feet per second?

c. Determine the angle of inclination that produces the maximum range.

Solution

a. We need to solve

$$2295 = \frac{325^2}{16} \sin\theta\cos\theta \qquad (1)$$

for θ, where $0° < \theta < 90°$.

Method 1 The following solutions were obtained by using a graphing utility to graph $d = 2295$ and $d = \frac{325^2}{16}\sin\theta\cos\theta$. See **Figure 5.48.** Thus there are two angles for which the projectile will hit the target. To the nearest thousandth of a degree, they are

$$\theta = 22.025° \qquad \text{and} \qquad \theta = 67.975°$$

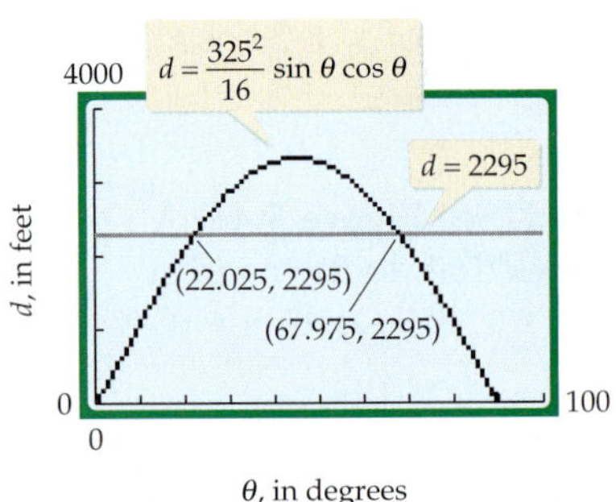

FIGURE 5.48

It should be noted that the graph in **Figure 5.48** is *not* a graph of the path of the projectile. It is a graph of the distance d as a function of the angle θ.

Method 2 To solve algebraically, we proceed as follows. Multiply each side of Equation (1) by 16 and divide by 325^2 to produce

$$\sin\theta\cos\theta = \frac{(16)(2295)}{325^2}$$

The identity $2\sin\theta\cos\theta = \sin 2\theta$ gives us $\sin\theta\cos\theta = \frac{\sin 2\theta}{2}$.

Hence
$$\frac{\sin 2\theta}{2} = \frac{(16)(2295)}{325^2}$$

$$\sin 2\theta = 2\frac{(16)(2295)}{325^2} \approx 0.69529$$

There are two angles in the interval $[0°, 180°]$ whose sines are 0.69529. One is $\sin^{-1} 0.69529$, and the other one is the *reference angle* for $\sin^{-1} 0.69529$. Therefore,

$$2\theta \approx \sin^{-1} 0.69529 \qquad \text{or} \qquad 2\theta \approx 180° - \sin^{-1} 0.69529$$

$$\theta \approx \frac{1}{2}\sin^{-1} 0.69529 \qquad \text{or} \qquad \theta \approx \frac{1}{2}(180° - \sin^{-1} 0.69529)$$

$$\theta \approx 22.025° \qquad \text{or} \qquad \theta \approx 67.975°$$

These are the same angles that we obtained using Method 1.

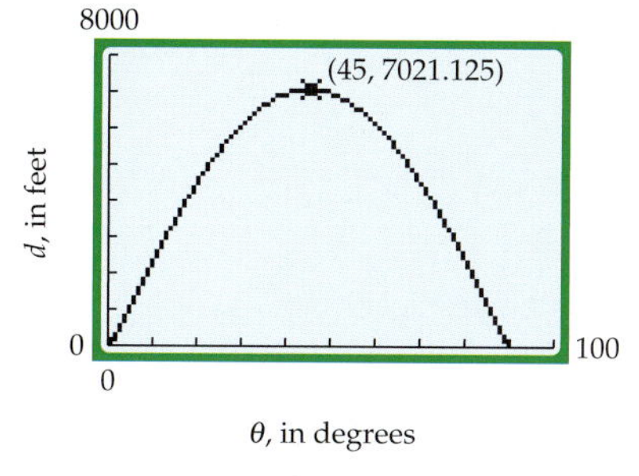

$d = \frac{474^2}{16}\sin\theta\cos\theta$

FIGURE 5.49

b. Use a graphing utility to find that the graph of $d = \frac{474^2}{16}\sin\theta\cos\theta$ has a maximum value of $d = 7021.125$ feet. See **Figure 5.49.**

c. In part **b.**, the maximum value is attained for $\theta = 45°$. To prove that this is true in general, we use $2\sin\theta\cos\theta = \sin 2\theta$ to write

$$d = \frac{v_0^2}{16}\sin\theta\cos\theta \qquad \text{as} \qquad d = \frac{v_0^2}{32}\sin 2\theta$$

This equation enables us to determine that d will attain its maximum when $\sin 2\theta$ attains its maximum—that is, when $2\theta = 90°$, or $\theta = 45°$.

▶ TRY EXERCISE 82, PAGE 423

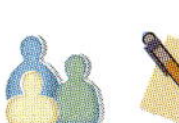

TOPICS FOR DISCUSSION

1. A student finds that $x = 0$ is a solution of $\sin x = x$. Because the function $y = \sin x$ has a period of 2π, the student reasons that $\pm 2\pi, \pm 4\pi, \pm 6\pi, \ldots$ are also solutions. Explain why the student is not correct.

2. How many solutions does $2\sin\left(x - \frac{\pi}{2}\right) = 5$ have on the interval $0 \le x < 2\pi$? Explain.

3. How many solutions does $\sin\frac{1}{x} = 0$ have on the interval $0 < x < \frac{\pi}{2}$? Explain.

4. On the interval $0 \le x < 2\pi$, the equation $\sin x = \frac{1}{2}$ has solutions of $x = \frac{\pi}{6}$ and $x = \frac{5\pi}{6}$. How would you write the solutions of $\sin x = \frac{1}{2}$ if the real number x were not restricted to the interval $[0, 2\pi)$?

EXERCISE SET 5.5

In Exercises 1 to 20, solve each equation for exact solutions in the interval $0 \le x < 2\pi$.

1. $\sec x - \sqrt{2} = 0$

2. $2\sin x = \sqrt{3}$

3. $\tan x - \sqrt{3} = 0$

4. $\cos x - 1 = 0$

5. $2\sin x\cos x = \sqrt{2}\cos x$

6. $2\sin x\cos x = \sqrt{3}\sin x$

7. $\sin^2 x - 1 = 0$

8. $\cos^2 x - 1 = 0$

9. $4\sin x\cos x - 2\sqrt{3}\sin x - 2\sqrt{2}\cos x + \sqrt{6} = 0$

10. $\sec^2 x + \sqrt{3}\sec x - \sqrt{2}\sec x - \sqrt{6} = 0$

11. $\csc x - \sqrt{2} = 0$

12. $3\cot x + \sqrt{3} = 0$

13. $2\sin^2 x + 1 = 3\sin x$

▶ **14.** $2\cos^2 x + 1 = -3\cos x$

15. $4\cos^2 x - 3 = 0$

16. $2\sin^2 x - 1 = 0$

17. $2\sin^3 x = \sin x$

18. $4\cos^3 x = 3\cos x$

19. $4\sin^2 x + 2\sqrt{3}\sin x - \sqrt{3} = 2\sin x$

20. $\tan^2 x + \tan x - \sqrt{3} = \sqrt{3}\tan x$

In Exercises 21 to 54, solve each equation, where $0° \le x < 360°$. Round approximate solutions to the nearest tenth of a degree.

21. $\cos x - 0.75 = 0$

22. $\sin x + 0.432 = 0$

23. $3\sin x - 5 = 0$

24. $4\cos x - 1 = 0$

25. $3\sec x - 8 = 0$

26. $4\csc x + 9 = 0$

27. $3 - 5\sin x = 4\sin x + 1$

28. $4\cos x - 5 = \cos x - 3$

29. $\frac{1}{2}\sin x + \frac{2}{3} = \frac{3}{4}\sin x + \frac{3}{5}$

30. $\frac{2}{5}\cos x - \frac{1}{2} = \frac{1}{3} - \frac{1}{2}\cos x$

31. $3\tan^2 x - 2\tan x = 0$

32. $4\cot^2 x + 3\cot x = 0$

33. $3\cos x + \sec x = 0$

34. $5\sin x - \csc x = 0$

35. $2\sin^2 x = 1 - \cos x$

36. $\cos^2 x + 4 = 2\sin x - 3$

37. $3\cos^2 x + 5\cos x - 2 = 0$

38. $2\sin^2 x + 5\sin x + 3 = 0$

39. $2\tan^2 x - \tan x - 10 = 0$

40. $2\cot^2 x - 7\cot x + 3 = 0$

41. $3\sin x\cos x - \cos x = 0$

42. $\tan x\sin x - \sin x = 0$

43. $2\sin x\cos x - \sin x - 2\cos x + 1 = 0$

44. $6\cos x\sin x - 3\cos x - 4\sin x + 2 = 0$

45. $2\sin x - \cos x = 1$

▶ **46.** $\sin x + 2\cos x = 1$

47. $2\sin x - 3\cos x = 1$

48. $\sqrt{3}\sin x + \cos x = 1$

49. $3\sin^2 x - \sin x - 1 = 0$

▶ **50.** $2\cos^2 x - 5\cos x - 5 = 0$

51. $2\cos x - 1 + 3\sec x = 0$

52. $3\sin x - 5 + \csc x = 0$

53. $\cos^2 x - 3\sin x + 2\sin^2 x = 0$

54. $\sin^2 x = 2\cos x + 3\cos^2 x$

In Exercises 55 to 64, find the exact solutions, in radians, of each trigonometric equation.

55. $\tan 2x - 1 = 0$

56. $\sec 3x - \frac{2\sqrt{3}}{3} = 0$

57. $\sin 5x = 1$

58. $\cos 4x = -\frac{\sqrt{2}}{2}$

59. $\sin 2x - \sin x = 0$

▶ **60.** $\cos 2x = -\frac{\sqrt{3}}{2}$

61. $\sin\left(2x + \frac{\pi}{6}\right) = -\frac{1}{2}$

62. $\cos\left(2x - \frac{\pi}{4}\right) = -\frac{\sqrt{2}}{2}$

63. $\sin^2 \frac{x}{2} + \cos x = 1$

64. $\cos^2 \frac{x}{2} - \cos x = 1$

In Exercises 65 to 76, find exact solutions, where $0 \le x < 2\pi$.

65. $\cos 2x = 1 - 3\sin x$

66. $\cos 2x = 2\cos x - 1$

67. $\sin 4x - \sin 2x = 0$

68. $\sin 4x - \cos 2x = 0$

69. $\tan \frac{x}{2} = \sin x$

70. $\tan \frac{x}{2} = 1 - \cos x$

71. $\sin x\cos 2x - \cos x\sin 2x = \frac{\sqrt{3}}{2}$

72. $\cos 2x\cos x + \sin 2x\sin x = -1$

73. $\sin 3x - \sin x = 0$

74. $\cos 3x + \cos x = 0$

75. $2\sin x\cos x + 2\sin x - \cos x - 1 = 0$

▶ **76.** $2\sin x\cos x - 2\sqrt{2}\sin x - \sqrt{3}\cos x + \sqrt{6} = 0$

In Exercises 77 to 80, use a graphing utility to solve the equation. State each solution accurate to the nearest ten-thousandth.

77. $\cos x = x$, where $0 \le x < 2\pi$

78. $2 \sin x = x$, where $0 \le x < 2\pi$

79. $\sin 2x = \frac{1}{x}$, where $-4 \le x \le 4$

80. $\cos x = \frac{1}{x}$, where $0 \le x \le 5$

PROJECTILES **Exercises 81 and 82 make use of the following. A projectile is fired at an angle of inclination θ from the horizon with an initial velocity v_0. Its range d (neglecting air resistance) is given by**

$$d = \frac{v_0^2}{16} \sin \theta \cos \theta$$

where v_0 is measured in feet per second and d is measured in feet.

81. If $v_0 = 288$ feet per second, use a graphing utility to find the angles θ (to the nearest hundredth of a degree) for which the projectile will hit a target 1295 feet downrange.

82. Use a graphing utility to find the maximum horizontal range, to the nearest tenth of a foot, for a projectile that has an initial velocity of 375 feet per second. What value of θ produces this maximum horizontal range?

In Exercises 83 and 84, use a graphing utility.

83. MODEL THE DAYLIGHT HOURS For a particular day of the year t, the number of daylight hours in Mexico City can be approximated by

$$d(t) = 1.208 \sin\left(\frac{2\pi(t - 80)}{365}\right) + 12.133$$

where t is an integer and $t = 1$ corresponds to January 1. According to d, how many days per year will Mexico City have at least 12 hours of daylight?

84. MODEL THE DAYLIGHT HOURS For a particular day of the year t, the number of daylight hours in New Orleans can be approximated by

$$d(t) = 1.792 \sin\left(\frac{2\pi(t - 80)}{365}\right) + 12.145$$

where t is an integer and $t = 1$ corresponds to January 1. According to d, how many days per year will New Orleans have at least 10.75 hours of daylight?

85. CROSS-SECTIONAL AREA A rain gutter is constructed from a long sheet of aluminum that measures 9 inches in width. The aluminum is to be folded as shown by the cross section in the following diagram.

a. Verify that the area of the cross section is $A = 9 \sin \theta(\cos \theta + 1)$, where $0° < \theta \le 90°$.

b. What values of θ, to the nearest degree, produce a cross-sectional area of 10.5 square inches?

c. Determine the value of θ that produces the cross section with the maximum area.

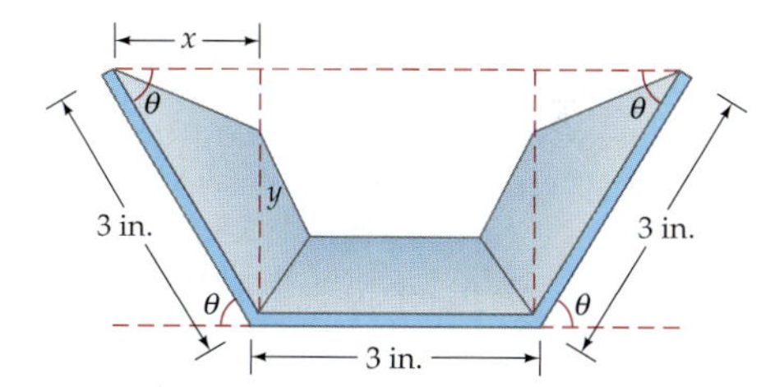

86. OBSERVATION ANGLE A person with an eye level of 5 feet 6 inches is standing in front of a painting, as shown in the following diagram. The bottom of the painting is 3 feet above floor level, and the painting is 6 feet in height. The angle θ shown in the figure is called the *observation angle* for the painting. The person is d feet from the painting.

a. Verify that $\theta = \tan^{-1}\left(\frac{6d}{d^2 - 8.75}\right)$.

b. Find the distance d, to the nearest tenth of a foot, for which $\theta = \frac{\pi}{6}$.

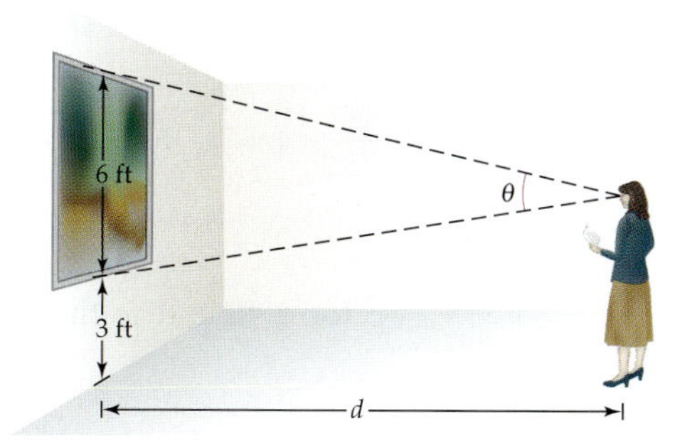

CONNECTING CONCEPTS

In Exercises 87 to 92, solve each equation for exact solutions in the interval $0 \le x < 2\pi$.

87. $\sqrt{3}\sin x + \cos x = \sqrt{3}$

88. $\sin x - \cos x = 1$

89. $-\sin x + \sqrt{3}\cos x = \sqrt{3}$

90. $-\sqrt{3}\sin x - \cos x = 1$

91. $\cos 5x - \cos 3x = 0$

92. $\cos 5x - \cos x - \sin 3x = 0$

93. Model the Movement of a Bus As bus A_1 makes a left turn, the back B of the bus moves to the right. If bus A_2 were waiting at a stoplight while A_1 turned left, as shown in the figure, there is a chance the two buses would scrape against one another. For a bus 28 feet long and 8 feet wide, the movement of the back of the bus to the right can be approximated by

$$x = \sqrt{(4 + 18\cot\theta)^2 + 100} - (4 + 18\cot\theta)$$

where θ is the angle the bus driver has turned the front of the bus. Find the value of x for $\theta = 20°$ and $\theta = 30°$. Round to the nearest hundredth of a foot.

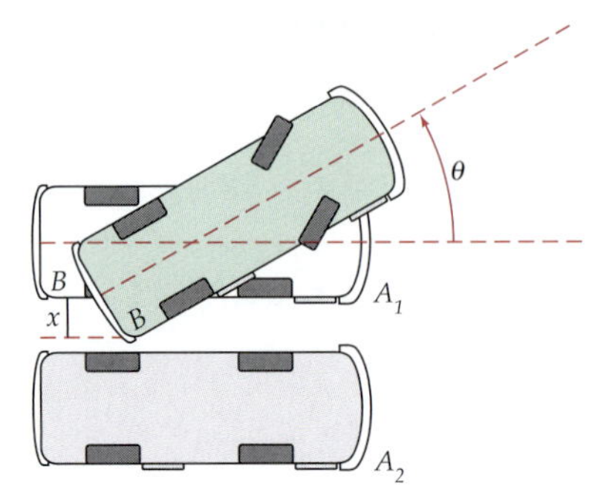

94. Optimal Branching of Blood Vessels It is hypothesized that the system of blood vessels in primates has evolved so that it has an optimal structure. In the case of a blood vessel splitting into two vessels, as shown in the accompanying figure, we assume that both new branches carry equal amounts of blood. A model of the angle θ is given by the equation $\cos\theta = 2^{(x-4)/(x+4)}$. The value of x is such that $1 \le x \le 2$ and depends on assumptions about the thickness of the blood vessels. Assuming this is an accurate model, find the values of the angle θ.

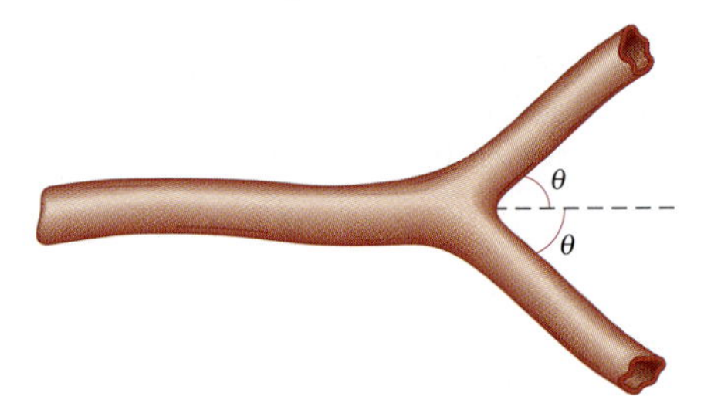

PREPARE FOR SECTION 5.6

95. Solve $\dfrac{b}{\sin 63.5°} = \dfrac{18.0}{\sin 75.2°}$ for b. Round to the nearest tenth. [5.5]

96. Solve $\dfrac{42.4}{\sin\theta} = \dfrac{31.5}{\sin 32.1°}$ for θ. Round to the nearest tenth of a degree. [5.5]

97. Solve $b^2 = a^2 + c^2 - 2ac\cos B$ for B in terms of the variables a, b, and c. [5.5]

98. Find b, given $b^2 = a^2 + c^2 - 2ac\cos B$ with $a = 4.3$, $c = 3.0$, and $B = 115°$. Assume $b > 0$. Round to the nearest tenth. [5.5]

99. A right triangle has sides of lengths 3, 4, and 5 inches.

a. Find the perimeter of the triangle.

b. Find the area of the triangle.

100. Find K given $K = \sqrt{s(s-a)(s-b)(s-c)}$ with $s = 12$, $a = 8$, $b = 6$, and $c = 10$. [A.1]

PROJECTS

1. **CURRENT IN AN ELECTRICAL CIRCUIT** In an electrical circuit the current I, in amperes, is given by

$$I = 25 \sin\left(120\pi t - \frac{\pi}{6}\right)$$

where t is measured in seconds. Find the exact solutions of $I = 12.5$ on the interval $0 \leq t \leq \frac{1}{60}$.

SECTION 5.6

THE LAW OF SINES AND THE LAW OF COSINES

- THE LAW OF SINES
- THE AMBIGUOUS CASE (SSA)
- APPLICATIONS OF THE LAW OF SINES
- THE LAW OF COSINES
- AN APPLICATION OF THE LAW OF COSINES

THE LAW OF SINES

Solving a triangle involves finding the lengths of all sides and the measures of all angles in the triangle. In this section and the next we develop formulas for solving an **oblique triangle,** which is a triangle that does not contain a right angle. The *Law of Sines* can be used to solve oblique triangles, in which

- two angles and the side between the angles are known (ASA).
- two angles and a side that is not between the angles are known (AAS).
- two sides and an angle opposite one of the sides are known (SSA).

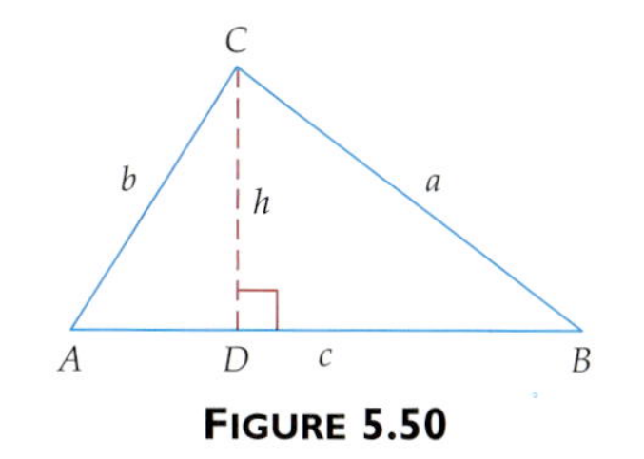

FIGURE 5.50

In **Figure 5.50,** altitude CD is drawn from C. The length of the altitude is h. Triangles ACD and BCD are right triangles.

Using the definition of the sine of an angle of a right triangle, we have from **Figure 5.50**

$$\sin B = \frac{h}{a} \qquad \sin A = \frac{h}{b}$$

$$h = a \sin B \quad (1) \qquad h = b \sin A \quad (2)$$

Equating the values of h in Equations (1) and (2), we obtain

$$a \sin B = b \sin A$$

Dividing each side of the equation by $\sin A \sin B$, we obtain

$$\frac{a}{\sin A} = \frac{b}{\sin B}$$

Similarly, when an altitude is drawn to a different side, the following formulas result:

$$\frac{c}{\sin C} = \frac{b}{\sin B} \quad \text{and} \quad \frac{c}{\sin C} = \frac{a}{\sin A}$$

take note

The Law of Sines may also be written as

$$\frac{\sin A}{a} = \frac{\sin B}{b} = \frac{\sin C}{c}$$

The Law of Sines

If A, B, and C are the measures of the angles of a triangle and a, b, and c are the lengths of the sides opposite these angles, then

$$\frac{a}{\sin A} = \frac{b}{\sin B} = \frac{c}{\sin C}$$

EXAMPLE 1 **Solve a Triangle Using the Law of Sines (ASA)**

Solve triangle ABC if $A = 42°$, $B = 63°$, and $c = 18$ centimeters.

Solution

Find C by using the fact that the sum of the interior angles of a triangle is 180°.

$$A + B + C = 180°$$
$$42° + 63° + C = 180°$$
$$C = 75°$$

Use the Law of Sines to find a.

take note

We have used the rounding conventions stated on page 364 to determine the number of significant digits to be used for a and b.

$$\frac{a}{\sin A} = \frac{c}{\sin C}$$
$$\frac{a}{\sin 42°} = \frac{18}{\sin 75°} \qquad \bullet\ A = 42°, c = 18, C = 75°$$
$$a = \frac{18 \sin 42°}{\sin 75°} \approx 12 \text{ centimeters}$$

Use the Law of Sines again, this time to find b.

$$\frac{b}{\sin B} = \frac{c}{\sin C}$$
$$\frac{b}{\sin 63°} = \frac{18}{\sin 75°} \qquad \bullet\ B = 63°, c = 18, C = 75°$$
$$b = \frac{18 \sin 63°}{\sin 75°} \approx 17 \text{ centimeters}$$

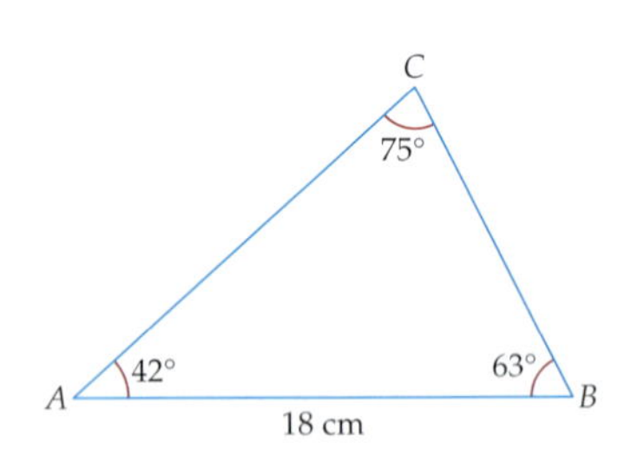

Figure 5.51

The solution is $C = 75°$, $a \approx 12$ centimeters, and $b \approx 17$ centimeters. A scale drawing can be used to see if these results are reasonable. See **Figure 5.51.**

▶ **Try Exercise 4, page 433**

The Ambiguous Case (SSA)

When you are given two sides of a triangle and an angle opposite one of them, you may find that the triangle is not unique. Some information may result in two triangles, and some may result in no triangle at all. It is because of this that the case of knowing two sides and an angle opposite one of them (SSA) is called the *ambiguous case* of the Law of Sines.

Suppose that sides a and c and the nonincluded angle A of a triangle are known and we are then asked to solve triangle ABC. The relationships among h, the height of the triangle, a (the side opposite $\angle A$), and c determine whether there are no, one, or two triangles.

Case 1 First consider the case in which $\angle A$ is an acute angle (see **Figure 5.52**). There are four possible situations.

1. $a < h$; there is no possible triangle.
2. $a = h$; there is one triangle, a right triangle.
3. $h < a < c$; there are two possible triangles.
4. $a \geq c$; there is one triangle, which is not a right triangle.

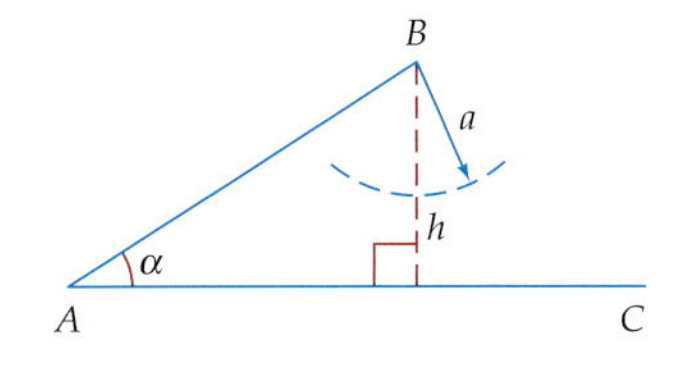

1. $a < h$; no triangle

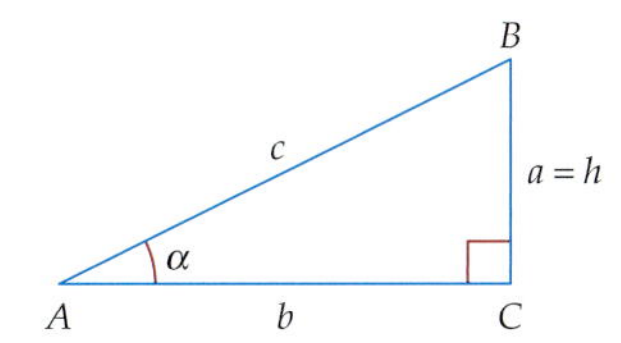

2. $a = h$; one triangle

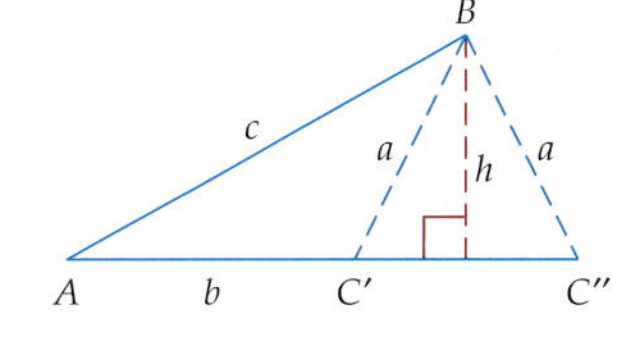

3. $h < a < c$; two triangles

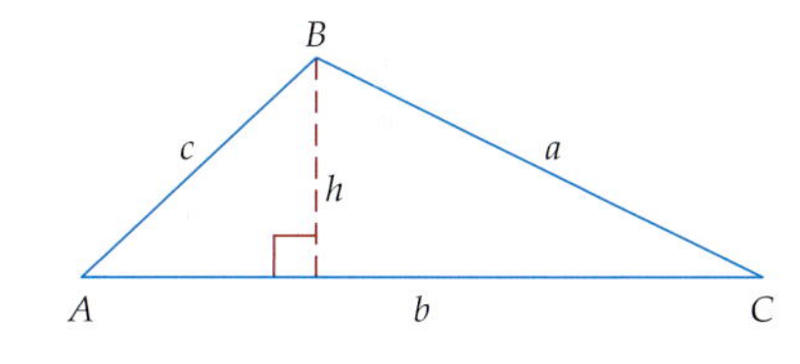

4. $a \geq c$; one triangle

FIGURE 5.52
Case 1: A is an acute angle.

Case 2 Now consider the case in which $\angle A$ is an obtuse angle (see **Figure 5.53**). Here, there are two possible situations.

1. $a \leq c$; there is no triangle.
2. $a > c$; there is one triangle.

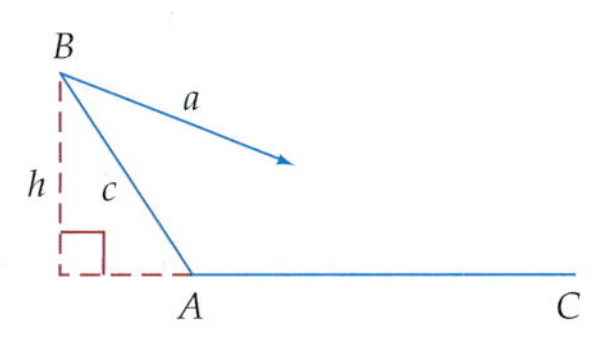

1. $a \leq c$; no triangle

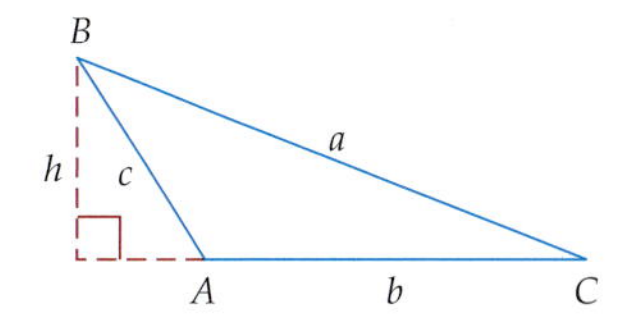

2. $a > c$; one triangle

FIGURE 5.53
Case 2: A is an obtuse angle.

EXAMPLE 2 **Solve a Triangle Using the Law of Sines (SSA)**

a. Find A, given triangle ABC with $B = 32°$, $a = 42$, and $b = 30$.

b. Find C, given triangle ABC with $A = 57°$, $a = 15$ feet, and $c = 20$ feet.

Continued ►

take note

When you solve for an angle of a triangle by using the Law of Sines, be aware that the value of the inverse sine function will give the measure of an acute angle. If the situation is the ambiguous case (SSA), you must consider a second, obtuse angle by using the supplement of the angle. You can use a scale drawing to see whether your results are reasonable.

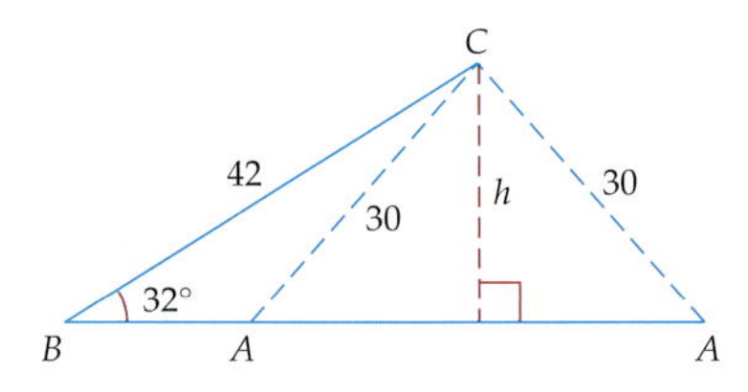

FIGURE 5.54

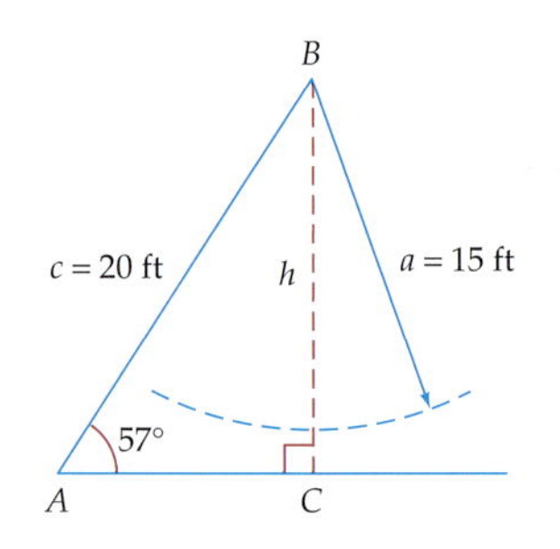

FIGURE 5.55

Solution

a.

$$\frac{b}{\sin B} = \frac{a}{\sin A}$$

$$\frac{30}{\sin 32^\circ} = \frac{42}{\sin A}$$ • $B = 32^\circ, a = 42, b = 30$

$$\sin A = \frac{42 \sin 32^\circ}{30} \approx 0.7419$$

$$A \approx 48^\circ \text{ or } 132^\circ$$ • The two angles with measure between 0° and 180° that have a sine of 0.7419 are approximately 48° and 132°.

To check that $A \approx 132^\circ$ is a valid result, add 132° to the measure of the given angle B (32°). Because $132^\circ + 32^\circ < 180^\circ$, we know that $A \approx 132^\circ$ is a valid result. Thus angle $A \approx 48^\circ$ or $A \approx 132^\circ$ ($\angle BAC$ in **Figure 5.54**).

b.

$$\frac{a}{\sin A} = \frac{c}{\sin C}$$

$$\frac{15}{\sin 57^\circ} = \frac{20}{\sin C}$$ • $A = 57^\circ, a = 15, c = 20$

$$\sin C = \frac{20 \sin 57^\circ}{15} \approx 1.1182$$

Because 1.1182 is not in the range of the sine function, there is no solution of the equation. Thus there is no triangle for these values of A, a, and c. See **Figure 5.55.**

▶ TRY EXERCISE 12, PAGE 434

APPLICATIONS OF THE LAW OF SINES

EXAMPLE 3 Solve an Application Using the Law of Sines

A radio antenna 85 feet high is located on top of an office building. At a distance AD from the base of the building, the angle of elevation to the top of the antenna is 26°, and the angle of elevation to the bottom of the antenna is 16°. Find the height of the building.

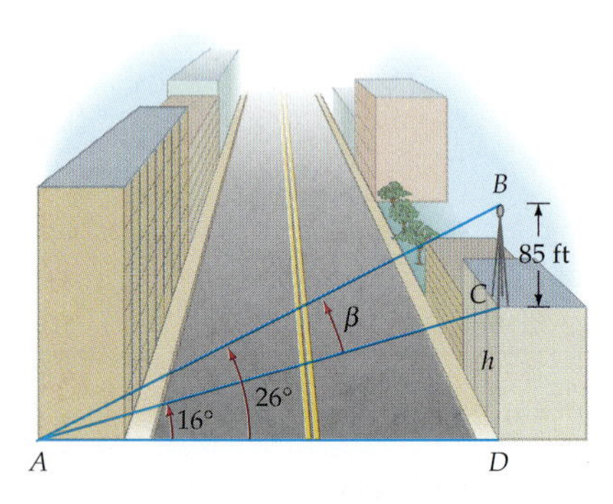

FIGURE 5.56

Solution

Sketch a diagram. See **Figure 5.56.** Find B and β.

$$B = 90^\circ - 26^\circ = 64^\circ$$

$$\beta = 26^\circ - 16^\circ = 10^\circ$$

take note

In Example 3 we rounded the height of the building to two significant digits to comply with the rounding conventions given on page 364.

Because we know the length BC and the measure of β, we can use triangle ABC and the Law of Sines to find length AC.

$$\frac{BC}{\sin \beta} = \frac{AC}{\sin B}$$

$$\frac{85}{\sin 10^\circ} = \frac{AC}{\sin 64^\circ} \qquad \bullet\ BC = 85,\ \beta = 10^\circ,\ B = 64^\circ$$

$$AC = \frac{85 \sin 64^\circ}{\sin 10^\circ}$$

Having found AC, we can now find the height of the building.

$$\sin 16^\circ = \frac{h}{AC}$$

$$h = AC \sin 16^\circ$$

$$= \frac{85 \sin 64^\circ}{\sin 10^\circ} \sin 16^\circ \approx 121 \text{ feet} \qquad \bullet\ \text{Substitute for } AC.$$

The height of the building to two significant digits is 120 feet.

TRY EXERCISE 32, PAGE 434

In navigation and surveying problems, there are two commonly used methods for specifying direction. The angular direction in which a craft is pointed is called the **heading.** Heading is expressed in terms of an angle measured clockwise from north. **Figure 5.57** shows a heading of 65° and a heading of 285°.

The angular direction used to locate one object in relation to another object is called the **bearing.** Bearing is expressed in terms of the acute angle formed by a north–south line and the line of direction. **Figure 5.58** shows a bearing of N38°W and a bearing of S15°E.

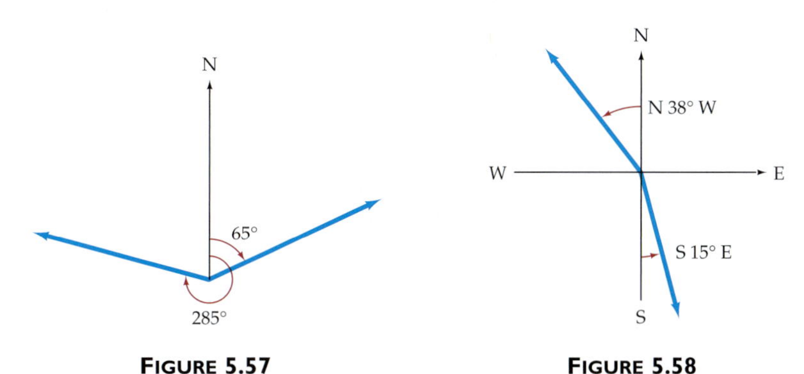

FIGURE 5.57 **FIGURE 5.58**

QUESTION Can a bearing of N50°E be written as N310°W?

ANSWER No. A bearing is always expressed using an acute angle.

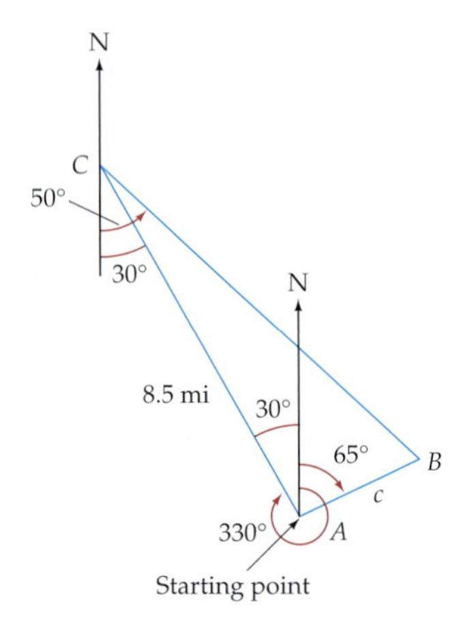

FIGURE 5.59

EXAMPLE 4 Solve an Application

A ship with a heading of 330° first sighted a lighthouse (point B) at a bearing of N65°E. After traveling 8.5 miles, the ship observed the lighthouse at a bearing of S50°E. Find the distance from the ship to the lighthouse when the first sighting was made.

Solution

From **Figure 5.59** we see that the measure of $\angle CAB = 65° + 30° = 95°$, the measure of $\angle BCA = 50° - 30° = 20°$, and $B = 180° - 95° - 20° = 65°$. Use triangle ABC and the Law of Sines to find c.

$$\frac{b}{\sin B} = \frac{c}{\sin C}$$

$$\frac{8.5}{\sin 65°} = \frac{c}{\sin 20°} \qquad \bullet\ b = 8.5, B = 65°, C = 20°$$

$$c = \frac{8.5 \sin 20°}{\sin 65°} \approx 3.2$$

The lighthouse was 3.2 miles (to two significant digits) from the ship when the first sighting was made.

TRY EXERCISE 38, PAGE 435

THE LAW OF COSINES

The *Law of Cosines* can be used to solve triangles in which two sides and the included angle (SAS) are known or in which three sides (SSS) are known. Consider the triangle in **Figure 5.60.** The height BD is drawn from B perpendicular to the x-axis. The triangle BDA is a right triangle, and the coordinates of B are $(a \cos C, a \sin C)$. The coordinates of A are $(b, 0)$. Using the distance formula, we can find the distance c.

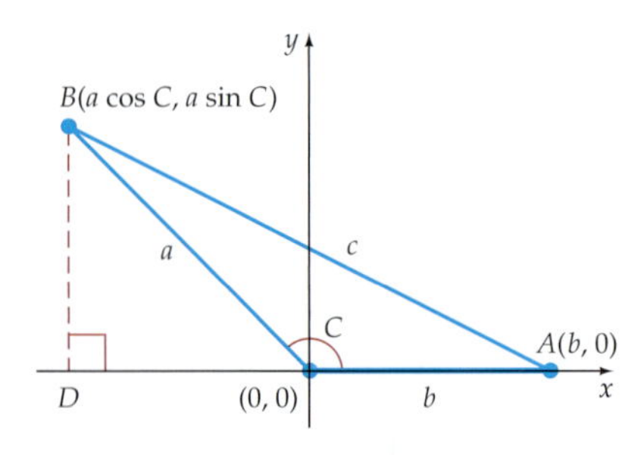

FIGURE 5.60

$$c = \sqrt{(a \cos C - b)^2 + (a \sin C - 0)^2}$$

$$c^2 = a^2 \cos^2 C - 2ab \cos C + b^2 + a^2 \sin^2 C$$

$$c^2 = a^2(\cos^2 C + \sin^2 C) + b^2 - 2ab \cos C$$

$$c^2 = a^2 + b^2 - 2ab \cos C$$

The Law of Cosines

If A, B, and C are the measures of the angles of a triangle and a, b, and c are the lengths of the sides opposite these angles, then

$$c^2 = a^2 + b^2 - 2ab \cos C$$

$$a^2 = b^2 + c^2 - 2bc \cos A$$

$$b^2 = a^2 + c^2 - 2ac \cos B$$

EXAMPLE 5 Use the Law of Cosines (SAS)

In triangle ABC, $B = 110.0°$, $a = 10.0$ centimeters, and $c = 15.0$ centimeters. See **Figure 5.61.** Find b.

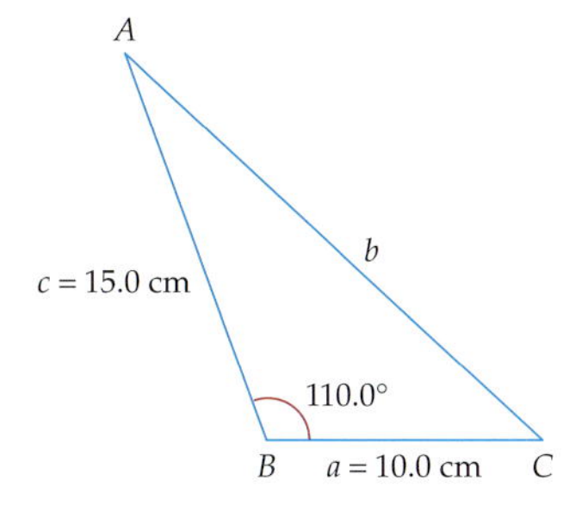

FIGURE 5.61

Solution

The Law of Cosines can be used because two sides and the included angle are known.

$$b^2 = a^2 + c^2 - 2ac\cos B$$
$$= 10.0^2 + 15.0^2 - 2(10.0)(15.0)\cos 110.0°$$
$$b = \sqrt{10.0^2 + 15.0^2 - 2(10.0)(15.0)\cos 110.0°}$$
$$b \approx 20.7 \text{ centimeters}$$

▶ TRY EXERCISE 24, PAGE 434

In the next example we know the length of each side, but we do not know the measure of any of the angles.

EXAMPLE 6 Use the Law of Cosines (SSS)

In triangle ABC, $a = 32$ feet, $b = 20$ feet, and $c = 40$ feet. Find B. This is the SSS case.

Solution

$$b^2 = a^2 + c^2 - 2ac\cos B$$

$$\cos B = \frac{a^2 + c^2 - b^2}{2ac}$$ • **Solve for cos B.**

$$= \frac{32^2 + 40^2 - 20^2}{2(32)(40)}$$ • **Substitute for a, b, and c.**

$$B = \cos^{-1}\left(\frac{32^2 + 40^2 - 20^2}{2(32)(40)}\right)$$ • **Solve for angle B.**

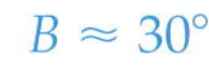

$$B \approx 30°$$ • **To the nearest degree**

▶ TRY EXERCISE 28, PAGE 434

AN APPLICATION OF THE LAW OF COSINES

EXAMPLE 7 Solve an Application Using the Law of Cosines

A car traveled 3.0 miles at a heading of 78°. The road turned, and the car traveled another 4.3 miles at a heading of 138°. Find the distance and the bearing of the car from the starting point.

Continued ▶

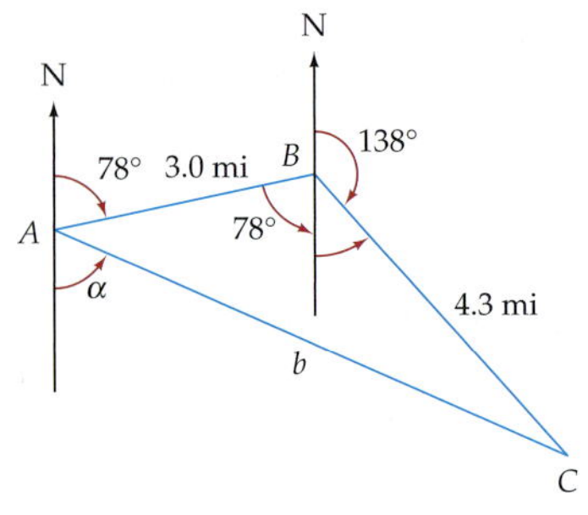

Figure 5.62

take note

The measure of A in Example 7 can also be determined by using the Law of Sines.

Solution

Sketch a diagram (see **Figure 5.62**). First find B.

$$B = 78° + (180° - 138°) = 120°$$

Use the Law of Cosines to find b.

$$\begin{aligned} b^2 &= a^2 + c^2 - 2ac \cos B \\ &= 4.3^2 + 3.0^2 - 2(4.3)(3.0) \cos 120° \\ b &= \sqrt{4.3^2 + 3.0^2 - 2(4.3)(3.0) \cos 120°} \\ b &\approx 6.4 \text{ miles} \end{aligned}$$

• **Substitute for a, c, and B.**

Find A.

$$\cos A = \frac{b^2 + c^2 - a^2}{2bc}$$

$$A = \cos^{-1}\left(\frac{b^2 + c^2 - a^2}{2bc}\right) \approx \cos^{-1}\left(\frac{6.4^2 + 3.0^2 - 4.3^2}{(2)(6.4)(3.0)}\right)$$

$$A \approx 35°$$

The bearing of the present position of the car from the starting point A can be determined by calculating the measure of angle α in **Figure 5.62.**

$$\alpha \approx 180° - (78° + 35°) = 67°$$

The distance is approximately 6.4 miles, and the bearing (to the nearest degree) is S67°E.

▶ **Try Exercise 52, page 437**

There are five different cases that we may encounter when solving an oblique triangle. They are listed in the following guideline, along with the law that can be used to solve the triangle.

A Guideline for Choosing Between the Law of Sines and the Law of Cosines

Apply the Law of Sines to solve an oblique triangle for each of the following cases.

ASA The measures of two angles of the triangle and the length of the included side are known.

AAS The measures of two angles of the triangle and the length of a side opposite one of these angles are known.

SSA The lengths of two sides of the triangle and the measure of an angle opposite one of these sides are known. This case is called the ambiguous case. It may yield one solution, two solutions, or no solution.

Apply the Law of Cosines to solve an oblique triangle for each of the following cases.

SSS The lengths of all three sides of the triangle are known. After finding the measure of an angle, you can complete your solution by using the Law of Sines.

SAS The lengths of two sides of the triangle and the measure of the included angle are known. After finding the measure of the third side, you can complete your solution by using the Law of Sines.

? QUESTION In triangle ABC, $A = 40°$, $C = 60°$, and $b = 114$. Should you use the Law of Sines or the Law of Cosines to solve this triangle?

Topics for Discussion

1. Is it possible to solve a triangle if the only given information consists of the measures of the three angles of the triangle? Explain.
2. Explain why it is not possible (in general) to use the Law of Sines to solve a triangle for which we are given only the lengths of all the sides.
3. Draw a triangle with dimensions $A = 30°$, $c = 3$ inches, and $a = 2.5$ inches. Is your answer unique? That is, can more than one triangle with the given dimensions be drawn?
4. Argue for or against the following proposition: In a scalene triangle (a triangle with no congruent sides), the largest angle is always opposite the longest side and the smallest angle is always opposite the shortest side.
5. The Pythagorean Theorem is a special case of the Law of Cosines. Explain.

Exercise Set 5.6

In Exercises 1 to 52, round answers according to the rounding conventions on page 364.

In Exercises 1 to 8, solve the triangles.

1. $A = 42°, B = 61°, a = 12$
2. $B = 25°, C = 125°, b = 5.0$
3. $A = 110°, C = 32°, b = 12$
4. $B = 28°, C = 78°, c = 44$
5. $A = 82.0°, B = 65.4°, b = 36.5$
6. $B = 54.8°, C = 72.6°, a = 14.4$

? ANSWER The Law of Sines.

7. $C = 114.2°, c = 87.2, b = 12.1$

8. $A = 54.32°, a = 24.42, c = 16.92$

In Exercises 9 to 16, solve the triangles that exist.

9. $A = 37°, c = 40, a = 28$

10. $B = 32°, c = 14, b = 9.0$

11. $C = 65°, b = 10, c = 8.0$

12. $B = 22.6°, b = 5.55, a = 13.8$

13. $A = 14.8°, c = 6.35, a = 4.80$

14. $C = 37.9°, b = 3.50, c = 2.84$

15. $B = 117.32°, b = 67.25, a = 15.05$

16. $A = 49.22°, a = 16.92, c = 24.62$

In Exercises 17 to 24, find the third side of the triangle.

17. $a = 12, b = 18, C = 44°$

18. $b = 30, c = 24, A = 120°$

19. $b = 60, c = 84, A = 13°$

20. $a = 122, c = 144, B = 48°$

21. $a = 9.0, b = 7.0, C = 72°$

22. $b = 12.3, c = 14.5, A = 6.5°$

23. $a = 25.9, c = 33.4, B = 84.0°$

24. $a = 14.2, b = 9.30, C = 9.20°$

In Exercises 25 to 30, given three sides of a triangle, find the specified angle.

25. $a = 25, b = 32, c = 40$; find A.

26. $a = 60, b = 88, c = 120$; find B.

27. $a = 8.0, b = 9.0, c = 12$; find C.

28. $a = 108, b = 132, c = 160$; find A.

29. $a = 32.5, b = 40.1, c = 29.6$; find B.

30. $a = 112.4, b = 96.80, c = 129.2$; find C.

31. **HURRICANE WATCH** A satellite weather map shows a hurricane off the coast of North Carolina. Use the information in the map to find the distance from the hurricane to Nags Head.

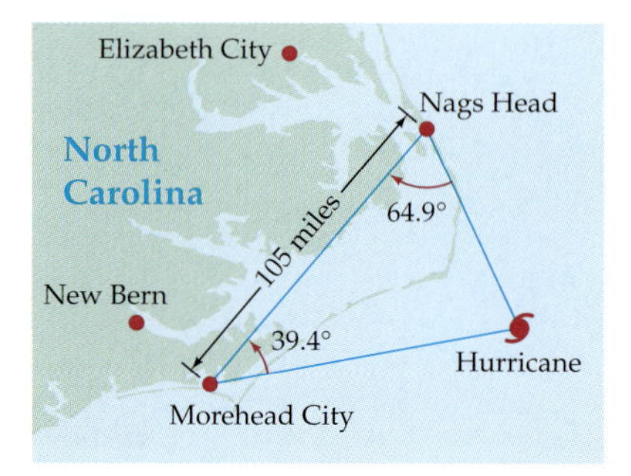

32. **NAVAL MANEUVERS** The distance between an aircraft carrier and a Navy destroyer is 7620 feet. The angle of elevation from the destroyer to a helicopter is 77.2°, and the angle of elevation from the aircraft carrier to the helicopter is 59.0°. The helicopter is in the same vertical plane as the two ships, as shown in the following figure. Use this data to determine the distance x from the helicopter to the aircraft carrier.

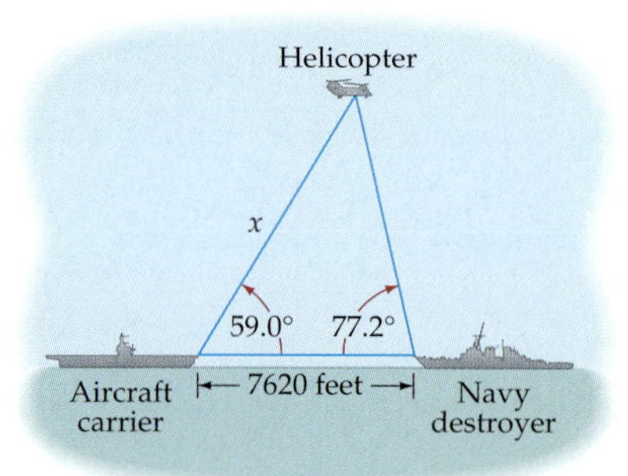

33. **CHOOSING A GOLF STRATEGY** The diagram on the following page shows two ways to play a golf hole. One is to hit the ball down the fairway on your first shot and then hit an approach shot to the green on your second shot. A second way is to hit directly toward the pin. Due to the water hazard, this is a more risky strategy. The distance AB is 165 yards, BC is 155 yards, and angle $A = 42.0°$. Find the distance AC from the tee directly to the pin. Assume that angle B is an obtuse angle.

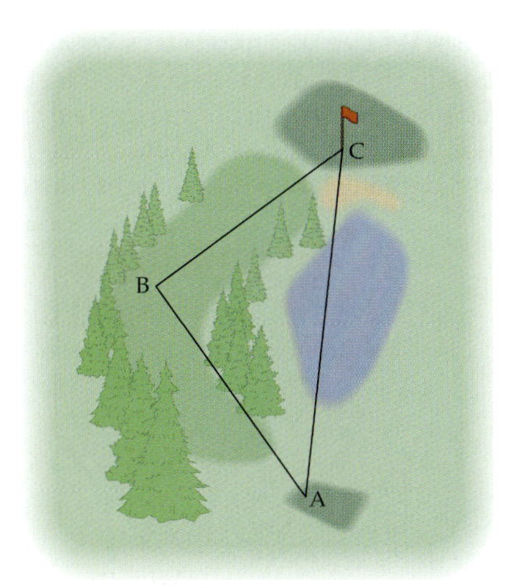

Figure for Exercise 33

34. DRIVING DISTANCE A golfer drives a golf ball from the tee at point A to point B, as shown in the following diagram. The distance AC from the tee directly to the pin is 365 yards. Angle A measures 11.2°, and angle C measures 22.9°.

a. Find the distance AB that the golfer drove the ball.

b. Find the distance BC from the present position of the ball to the pin.

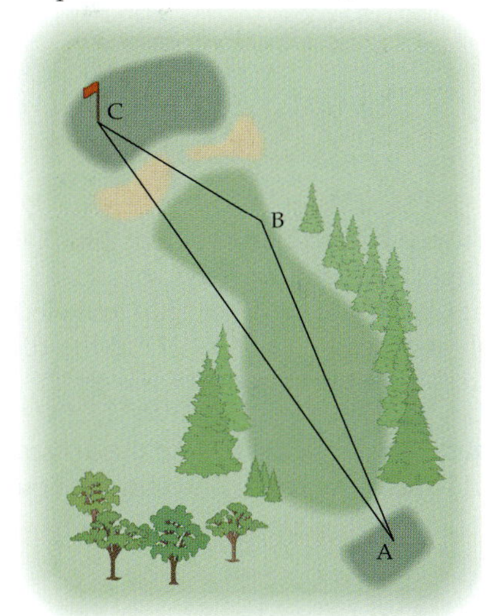

35. DISTANCE ACROSS A CANYON To find the distance across a canyon, a surveying team locates points A and B on one side of the canyon and point C on the other side of the canyon. The distance between A and B is 85 yards. The measure of $\angle CAB$ is 68°, and the measure of $\angle CBA$ is 75°. Find the distance across the canyon.

36. HEIGHT OF A KITE Two observers, in the same vertical plane as a kite and at a distance of 30 feet apart, observe the kite at angles 62° and 78°, as shown in the following diagram. Find the height of the kite.

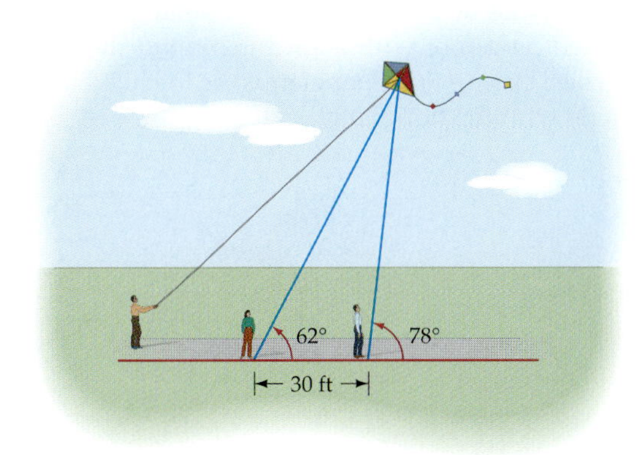

Figure for Exercise 36

37. LENGTH OF A GUY WIRE A telephone pole 35 feet high is situated on an 11° slope from the horizontal. The measure of angle CAB is 21°. Find the length of the guy wire AC.

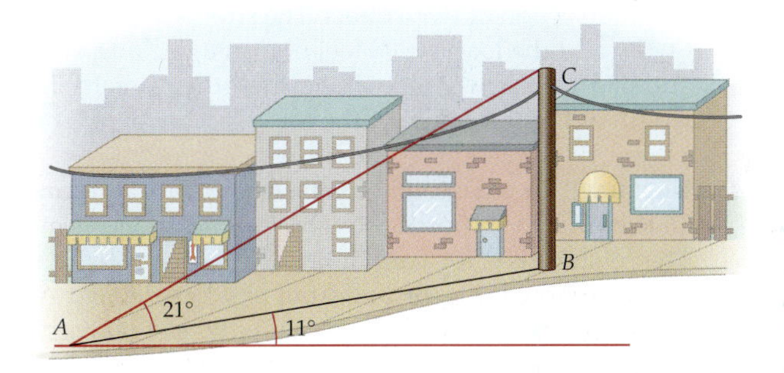

38. DISTANCE TO A FIRE Two fire lookouts are located on mountains 20 miles apart. Lookout B is at a bearing of S65°E from lookout A. A fire was sighted at a bearing of N50°E from A and at a bearing of N8°E from B. Find the distance of the fire from lookout A.

39. DISTANCE TO A LIGHTHOUSE A navigator on a ship sights a lighthouse at a bearing of N36°E. After traveling 8.0 miles at a heading of 332°, the ship sights the lighthouse at a bearing of S82°E. How far is the ship from the lighthouse at the second sighting?

40. MINIMUM DISTANCE The navigator on a ship traveling due east at 8 mph sights a lighthouse at a bearing of S55°E. One hour later it is sighted at a bearing of S25°W. Find the closest the ship came to the lighthouse.

41. DISTANCE BETWEEN AIRPORTS An airplane flew 450 miles at a bearing of N65°E from airport A to airport B. The plane then flew at a bearing of S38°E to airport C. Find the distance from A to C if the bearing from airport A to airport C is S60°E.

42. Length of a Brace A 12-foot solar panel is to be installed on a roof with a 15° pitch. Find the length of the vertical brace d if the panel must be installed to make a 40° angle with the horizontal.

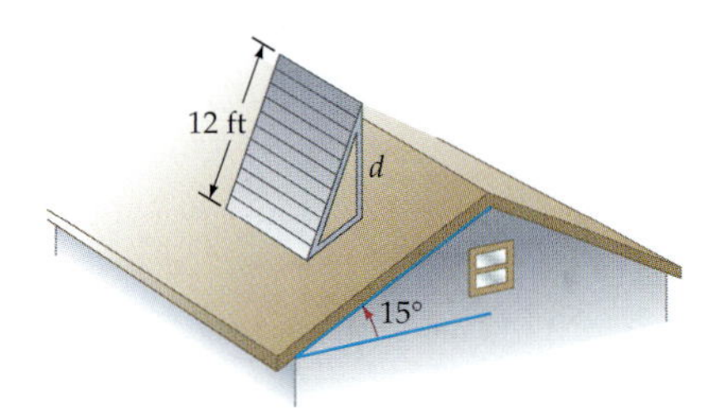

43. Distance Between Airports A plane leaves airport A and travels 560 miles to airport B at a bearing of N32°E. The plane leaves airport B and travels to airport C 320 miles away at a bearing of S72°E. Find the distance from airport A to airport C.

44. Length of a Street A developer has a triangular lot at the intersection of two streets. The streets meet at an angle of 72°, and the lot has 300 feet of frontage along one street and 416 feet of frontage along the other street. Find the length of the third side of the lot.

45. Baseball In a baseball game, a batter hits a ground ball 26 feet directly towards the center of the pitcher's mound. The pitcher runs forward and reaches for the ball. At that moment, how far is the ball from first base? (*Note:* A baseball infield is a square that measures 90 feet on each side.)

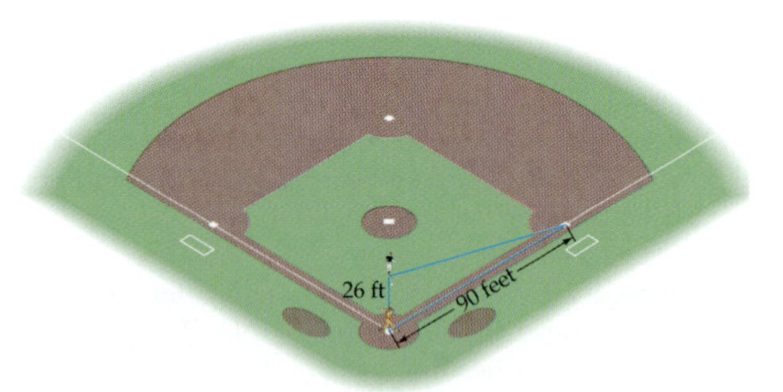

46. B-2 Bomber The leading edge of each wing of the B-2 Stealth Bomber measures 105.6 feet in length. The angle between the wing's leading edges ($\angle ABC$) is 109.05°. What is the wing span (the distance from A to C) of the B-2 Bomber?

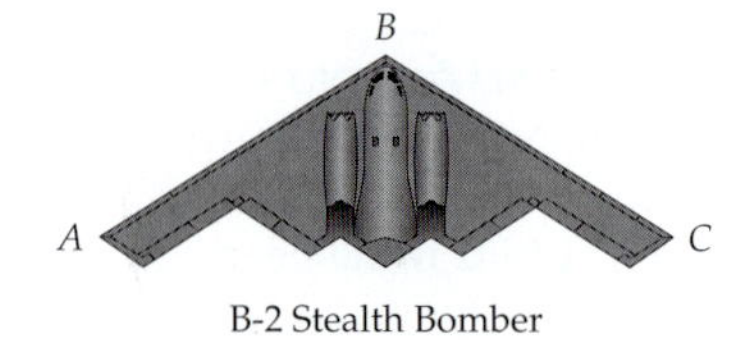

B-2 Stealth Bomber

47. Angle Between the Diagonals of a Box The rectangular box in the figure measures 6.50 feet by 3.25 feet by 4.75 feet. Find the measure of the angle θ that is formed by the union of the diagonal shown on the front of the box and the diagonal shown on the right side of the box.

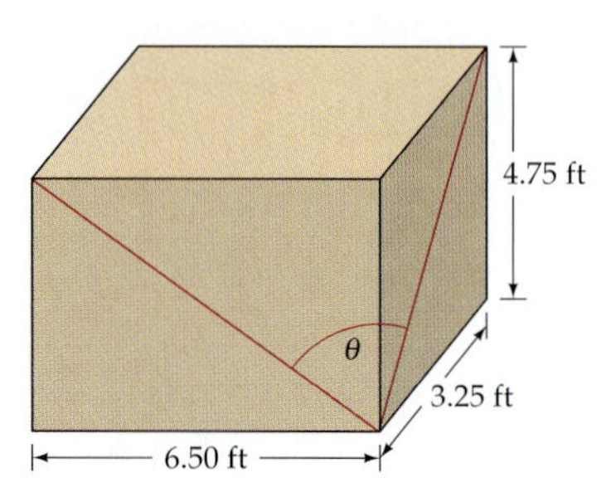

48. Submarine Rescue Mission The surface ships shown in the figure below have determined the indicated distances. Use this data to determine the depth of the submarine below the surface of the water. Assume that the line segment between the surface ships is directly above the submarine.

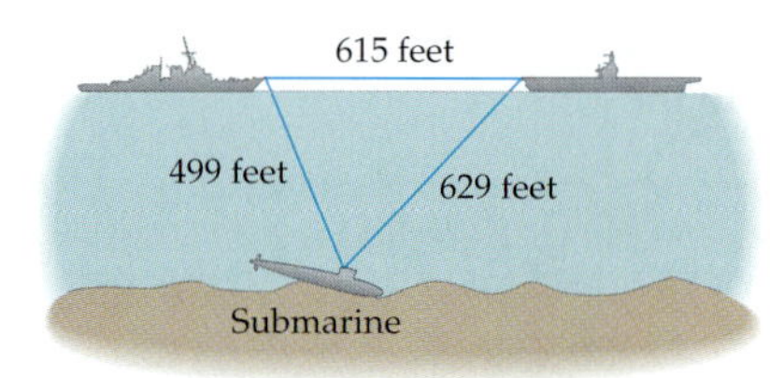

49. Distance Between Ships Two ships left a port at the same time. One ship traveled at a speed of 18 mph at a heading of 318°. The other ship traveled at a speed of 22 mph at a heading of 198°. Find the distance between the two ships after 10 hours of travel.

50. Distance Across a Lake Find the distance across a lake, using the measurements shown in the figure.

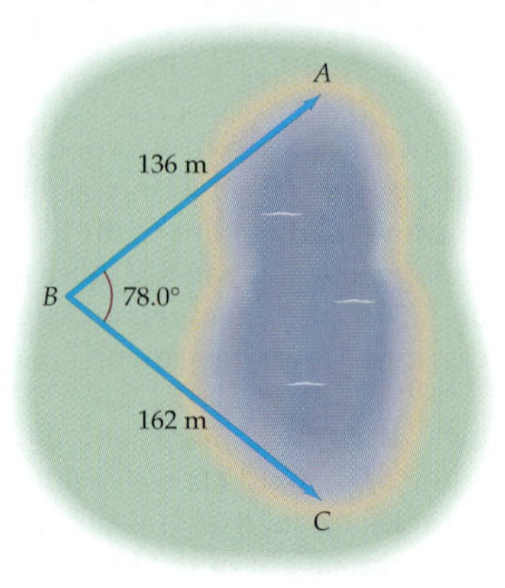

51. **DISTANCE TO A PLANE** A plane traveling at 180 mph passes 400 feet directly over an observer. The plane is traveling along a straight path with an angle of elevation of 14°. Find the distance of the plane from the observer 10 seconds after the plane has passed directly overhead.

▶ **52.** **DISTANCE BETWEEN SHIPS** A ship leaves a port at a speed of 16 mph at a heading of 32°. One hour later another ship leaves the port at a speed of 22 mph at a heading of 254°. Find the distance between the ships 4 hours after the first ship leaves the port.

CONNECTING CONCEPTS

The following identity is one of *Mollweide's formulas.* It applies to any triangle *ABC*.

$$\frac{a-b}{c}=\frac{\sin\left(\frac{A-B}{2}\right)}{\cos\left(\frac{C}{2}\right)}$$

The formula is intriguing because it contains the dimensions of all angles and all sides of ΔABC. The formula can be used to check whether a triangle has been solved correctly. Substitute the dimensions of a given triangle in the formula and compare the value of the left side of the formula with the value of the right side. If the two results are not reasonably close, then you know that at least one dimension is incorrect. The results generally will not be identical because each dimension is an approximation. In Exercises 53 and 54, you can assume that a triangle has an incorrect dimension if the value of the left side and the value of the right side of the above formula differ by more than 0.02.

53. 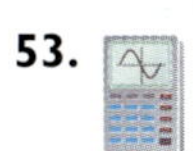 **CHECK DIMENSIONS OF TRUSSES** The following diagram shows some of the steel trusses in an airplane hangar.

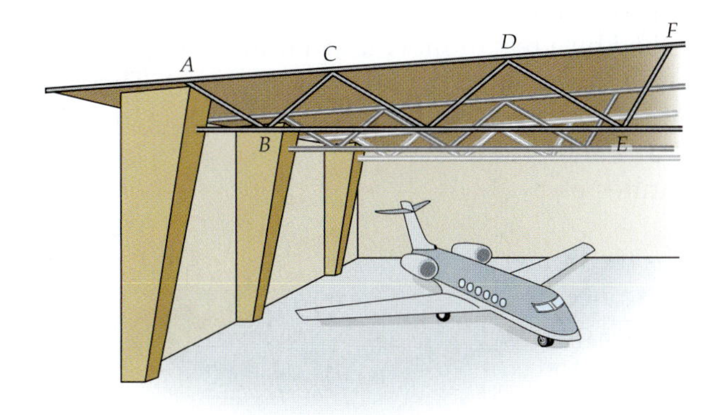

An architect has determined the following dimensions for ΔABC and ΔDEF:

ΔABC: $A = 53.5°, B = 86.5°, C = 40.0°$
$a = 13.0$ feet, $b = 16.1$ feet, $c = 10.4$ feet

ΔDEF: $D = 52.1°, E = 59.9°, F = 68.0°$
$d = 17.2$ feet, $e = 21.3$ feet, $f = 22.8$ feet

Use Mollweide's formula to determine if either ΔABC or ΔDEF has an incorrect dimension.

54. **CHECK DIMENSIONS OF TRUSSES** The following diagram shows some of the steel trusses in a railroad bridge.

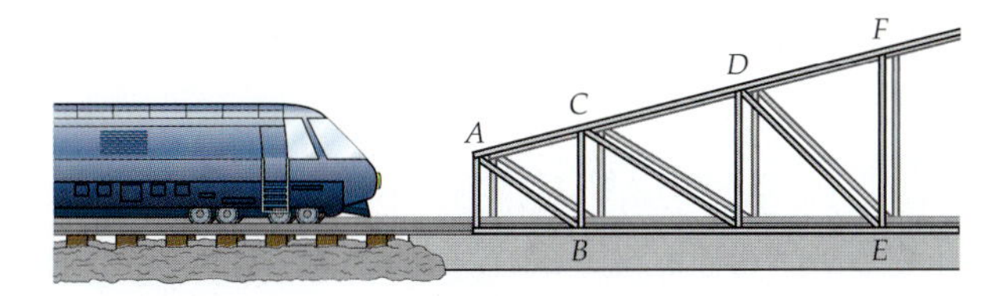

A structural engineer has determined the following dimensions for ΔABC and ΔDEF:

ΔABC: $A = 34.1°, B = 66.2°, C = 79.7°$
$a = 9.23$ feet, $b = 15.1$ feet, $c = 16.2$ feet

ΔDEF: $D = 45.0°, E = 56.2°, F = 78.8°$
$d = 13.6$ feet, $e = 16.0$ feet, $f = 18.9$ feet

Use Mollweide's formula to determine if either ΔABC or ΔDEF has an incorrect dimension.

55. Given a triangle *ABC*, prove that

$$a^2 = b^2 + c^2 - 2bc\cos A$$

56. Use the Law of Cosines to show that

$$\cos A = \frac{(b+c-a)(b+c+a)}{2bc} - 1$$

PREPARE FOR SECTION 5.7

57. Evaluate: $\sqrt{\left(\frac{3}{5}\right)^2 + \left(-\frac{4}{5}\right)^2}$ [A.1]

58. Use a calculator to evaluate 10 cos 228°. Round to the nearest thousandth. [5.1]

59. Solve $\tan \alpha = \left|\frac{-\sqrt{3}}{3}\right|$ for α, where α is an acute angle measured in degrees. [5.4]

60. Solve $\cos \alpha = \frac{-17}{\sqrt{338}}$ for α, where α is an obtuse angle measured in degrees. Round to the nearest tenth of a degree. [5.4]

61. Rationalize the denominator of $\frac{1}{\sqrt{5}}$. [A.1]

62. Rationalize the denominator of $\frac{28}{\sqrt{68}}$. [A.1]

PROJECTS

1. VISUAL INSIGHT

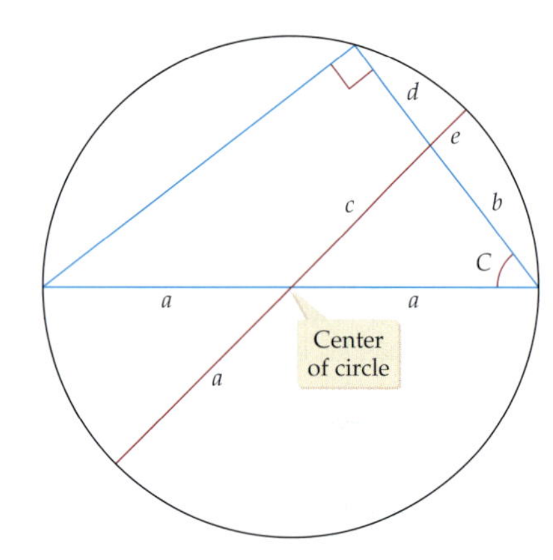

$$
\begin{aligned}
bd &= (a + c)e \\
bd &= (a + c)(a - c) \\
b(2a \cos C - b) &= (a + c)(a - c) \\
2ab \cos C - b^2 &= a^2 - c^2 \\
c^2 &= a^2 + b^2 - 2ab \cos C
\end{aligned}
$$

Give the rule or reason that justifies each step in the above proof of the Law of Cosines.

SECTION 5.7 VECTORS

- VECTORS
- UNIT VECTORS
- APPLICATIONS OF VECTORS
- DOT PRODUCT
- SCALAR PROJECTION
- PARALLEL AND PERPENDICULAR VECTORS
- WORK: AN APPLICATION OF THE DOT PRODUCT

VECTORS

In scientific applications, some measurements, such as area, mass, distance, speed, and time, are completely described by a real number and a unit. Examples include 30 square feet (area), 25 meters/second (speed), and 5 hours (time). These measurements are **scalar quantities,** and the number used to indicate the magnitude of the measurement is called a **scalar.** Two other examples of scalar quantities are volume and temperature.

For other quantities, besides the numerical and unit description, it is also necessary to include a *direction* to describe the quantity completely. For example, applying a force of 25 pounds at various angles to a small metal box will influence how the box moves. In **Figure 5.63,** applying the 25-pound force

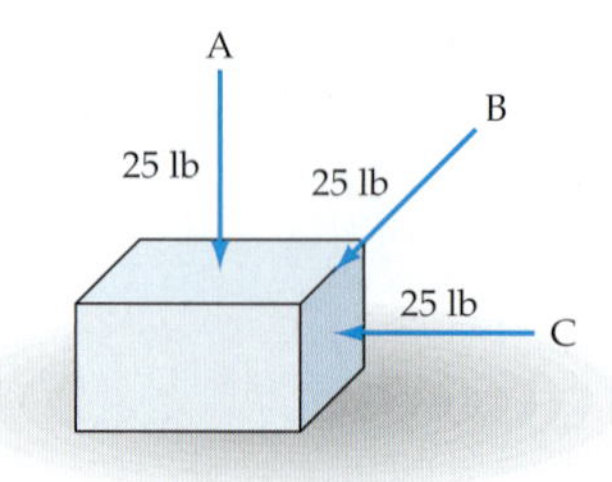

FIGURE 5.63

straight down (A) will not move the box to the left. However, applying the 25-pound force (C) parallel to the floor will move the box along the floor.

Vector quantities have a *magnitude* (numerical and unit description) and a *direction*. Force is a vector quantity. Velocity is another. Velocity includes the speed (magnitude) and a direction. A velocity of 40 mph east is different from a velocity of 40 mph north. Displacement is another vector quantity; it consists of distance (a scalar) moved in a certain direction; for example, we might speak of a displacement of 13 centimeters at an angle of 15° from the positive x-axis.

Definition of a Vector

A **vector** is a directed line segment. The length of the line segment is the magnitude of the vector, and the direction of the vector is measured by an angle.

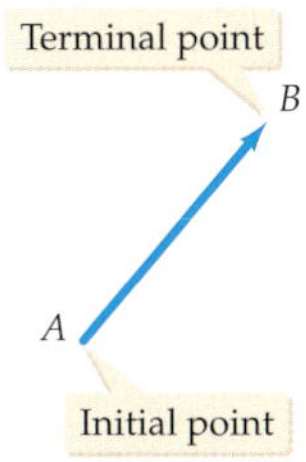

Terminal point
B
A
Initial point

FIGURE 5.64

FIGURE 5.65

v 2v v $\frac{1}{2}$v v −1.5v

FIGURE 5.66

50 m 30 m 37° 40 m

FIGURE 5.67

The point A for the vector in **Figure 5.64** is called the **initial point** (or tail) of the vector, and the point B is the **terminal point** (or head) of the vector. An arrow over the letters ($\overrightarrow{AB}$), an arrow over a single letter ($\vec{V}$), or boldface type (**AB** or **V**) is used to denote a vector. The magnitude of the vector is the length of the line segment and is denoted by $\|\overrightarrow{AB}\|$, $\|\vec{V}\|$, $\|\mathbf{AB}\|$, or $\|\mathbf{V}\|$.

Equivalent vectors have the same magnitude and the same direction. The vectors in **Figure 5.65** are equivalent. They have the same magnitude and direction.

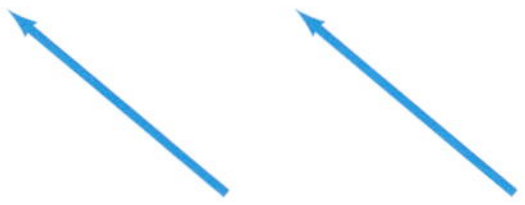

Multiplying a vector by a positive real number (other than 1) changes the magnitude of the vector but not its direction. If **v** is any vector, then 2**v** is the vector that has the same direction as **v** but is twice the magnitude of **v**. The multiplication of 2 and **v** is called the **scalar multiplication** of the vector **v** and the scalar 2. Multiplying a vector by a negative number a reverses the direction of the vector and multiplies the magnitude of the vector by $|a|$. See **Figure 5.66.**

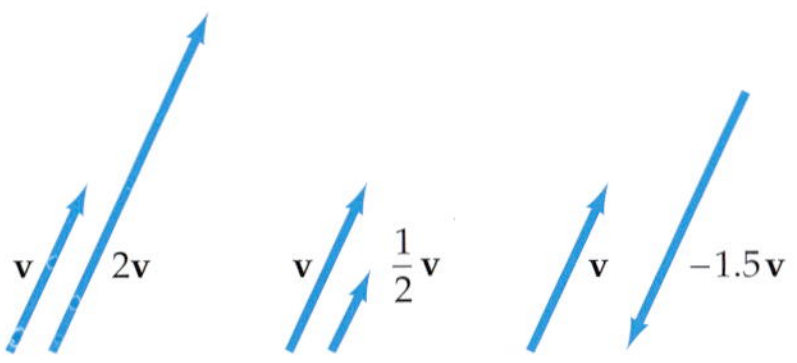

The sum of two vectors, called the **resultant vector** or the **resultant,** is the single equivalent vector that will have the same effect as the application of those two vectors. For example, a displacement of 40 meters along the positive x-axis and then 30 meters in the positive y direction is equivalent to a vector of magnitude 50 meters at an angle of approximately 37° to the positive x-axis. See **Figure 5.67.**

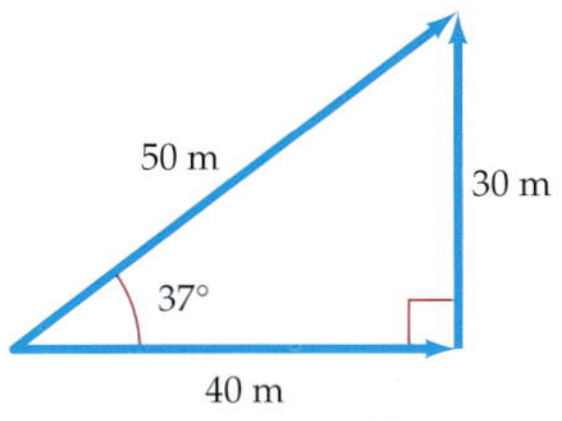

Vectors can be added graphically by using the *triangle method* or the *parallelogram method*. In the triangle method, shown in **Figure 5.68,** the tail of **V** is placed at the head of **U**. The vector connecting the tail of **U** with the head of **V** is the sum **U** + **V**.

The parallelogram method of adding two vectors graphically places the tails of the two vectors **U** and **V** together, as in **Figure 5.69.** Complete the parallelogram so that **U** and **V** are sides of the parallelogram. The diagonal beginning at the tails of the two vectors is **U** + **V**.

To find the difference between two vectors, first rewrite the expression as **V** − **U** = **V** + (−**U**). The difference is shown geometrically in **Figure 5.70.**

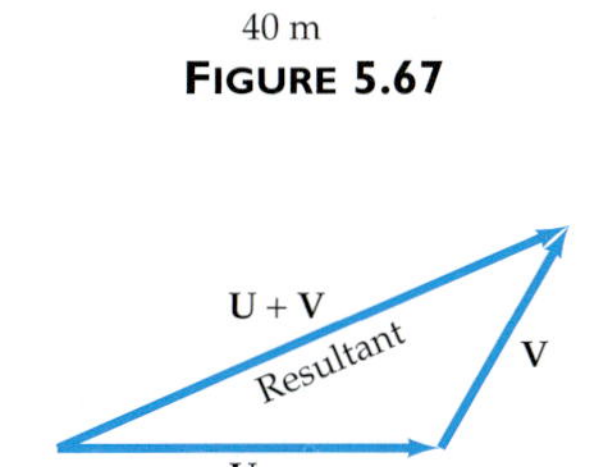

FIGURE 5.68

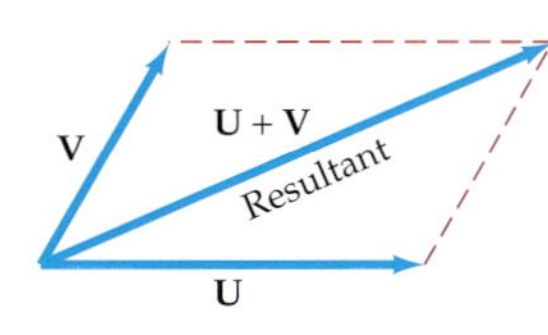

FIGURE 5.69

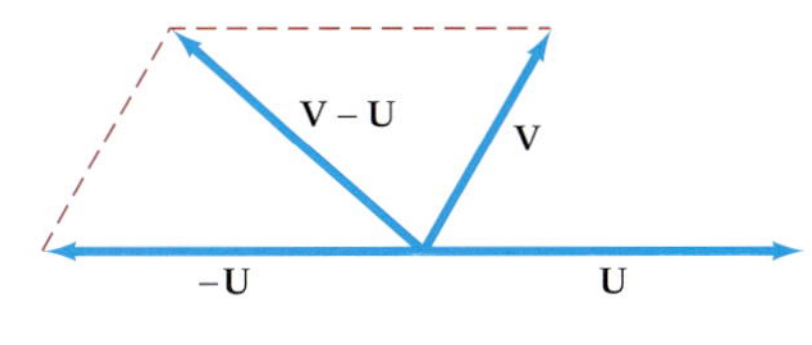

FIGURE 5.70

By introducing a coordinate plane, it is possible to develop an analytic approach to vectors. Recall from our discussion about equivalent vectors that a vector can be moved in the plane as long as *the magnitude and direction* are not changed.

With this in mind, consider **AB**, whose initial point is $A(2, -1)$ and whose terminal point is $B(-3, 4)$. If this vector is moved so that the initial point is at the origin O, the terminal point becomes $P(-5, 5)$, as shown in **Figure 5.71.** The vector **OP** is equivalent to the vector **AB**.

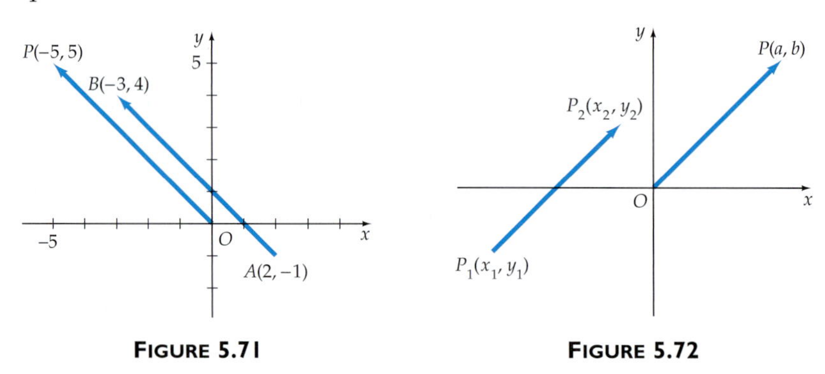

FIGURE 5.71 **FIGURE 5.72**

In **Figure 5.72,** let $P_1(x_1, y_1)$ be the initial point of a vector and $P_2(x_2, y_2)$ its terminal point. Then an equivalent vector **OP** has its initial point at the origin and its terminal point at $P(a, b)$, where $a = x_2 - x_1$ and $b = y_2 - y_1$. The vector **OP** can be denoted by $\mathbf{v} = \langle a, b \rangle$; a and b are called the **components** of the vector.

EXAMPLE 1 Find the Components of a Vector

Find the components of the vector **AB** whose tail is the point $A(2, -1)$ and whose head is the point $B(-2, 6)$. Determine a vector **v** that is equivalent to **AB** and has an initial point at the origin.

Algebraic Solution

The components of **AB** are $\langle a, b \rangle$, where

$$a = x_2 - x_1 = -2 - 2 = -4 \quad \text{and} \quad b = y_2 - y_1 = 6 - (-1) = 7$$

Thus $\mathbf{v} = \langle -4, 7 \rangle$.

Visualize the Solution

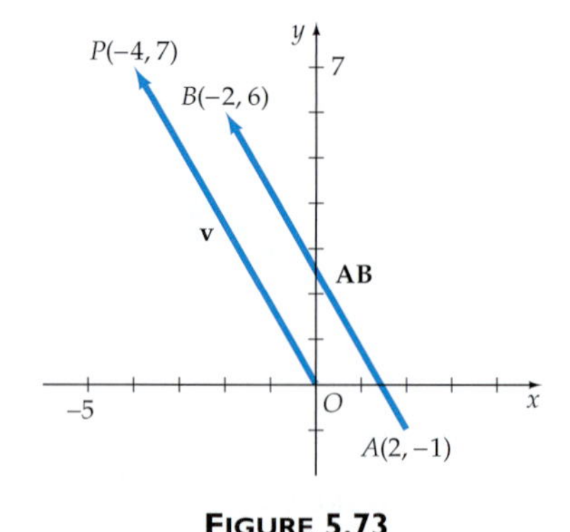

FIGURE 5.73

▶ TRY EXERCISE 6, PAGE 450

The magnitude and direction of a vector can be found from its components. For instance, the head of vector **v** sketched in **Figure 5.73** is the ordered pair $(-4, 7)$. Applying the Pythagorean Theorem, we find

$$\|\mathbf{v}\| = \sqrt{(-4)^2 + 7^2} = \sqrt{16 + 49} = \sqrt{65}$$

Let θ be the angle made by the positive x-axis and **v**. Let α be the reference angle for θ. Then

$$\tan \alpha = \left|\frac{b}{a}\right| = \left|\frac{7}{-4}\right| = \frac{7}{4}$$

$$\alpha = \tan^{-1}\frac{7}{4} \approx 60^\circ$$ • α **is the reference angle.**

$$\theta = 180^\circ - 60^\circ = 120^\circ$$ • θ **is the angle made by the vector and the positive** x**-axis.**

The magnitude of **v** is $\sqrt{65}$, and its direction is 120° as measured from the positive x-axis. The angle between a vector and the positive x-axis is called the **direction angle** of the vector. Because **AB** in **Figure 5.73** is equivalent to **v**, $\|\mathbf{AB}\| = \sqrt{65}$ and the direction angle of **AB** is also 120°.

Expressing vectors in terms of components provides a convenient method for performing operations on vectors.

Fundamental Vector Operations

If $\mathbf{v} = \langle a, b\rangle$ and $\mathbf{w} = \langle c, d\rangle$ are two vectors and k is a real number, then

1. $\|\mathbf{v}\| = \sqrt{a^2 + b^2}$
2. $\mathbf{v} + \mathbf{w} = \langle a, b\rangle + \langle c, d\rangle = \langle a + c, b + d\rangle$
3. $k\mathbf{v} = k\langle a, b\rangle = \langle ka, kb\rangle$

In terms of components, the zero vector $\mathbf{0} = \langle 0, 0\rangle$. The additive inverse of a vector $\mathbf{v} = \langle a, b\rangle$ is given by $-\mathbf{v} = \langle -a, -b\rangle$.

EXAMPLE 2 Perform Operations on Vectors

Given $\mathbf{v} = \langle -2, 3\rangle$ and $\mathbf{w} = \langle 4, -1\rangle$, find

a. $\|\mathbf{w}\|$ **b.** $\mathbf{v} + \mathbf{w}$ **c.** $-3\mathbf{v}$ **d.** $2\mathbf{v} - 3\mathbf{w}$

Solution

a. $\|\mathbf{w}\| = \sqrt{4^2 + (-1)^2} = \sqrt{17}$

b. $\mathbf{v} + \mathbf{w} = \langle -2, 3\rangle + \langle 4, -1\rangle = \langle -2 + 4, 3 + (-1)\rangle = \langle 2, 2\rangle$

c. $-3\mathbf{v} = -3\langle -2, 3\rangle = \langle 6, -9\rangle$

d. $2\mathbf{v} - 3\mathbf{w} = 2\langle -2, 3\rangle - 3\langle 4, -1\rangle = \langle -4, 6\rangle - \langle 12, -3\rangle = \langle -16, 9\rangle$

TRY EXERCISE 20, PAGE 450

UNIT VECTORS

A **unit vector** is a vector whose magnitude is 1. For example, the vector $\mathbf{v} = \left\langle \frac{3}{5}, -\frac{4}{5}\right\rangle$ is a unit vector because

$$\|\mathbf{v}\| = \sqrt{\left(\frac{3}{5}\right)^2 + \left(-\frac{4}{5}\right)^2} = \sqrt{\frac{9}{25} + \frac{16}{25}} = \sqrt{\frac{25}{25}} = 1$$

Given any nonzero vector **v**, we can obtain a unit vector in the direction of **v** by dividing each component of **v** by the magnitude of **v**, $\|\mathbf{v}\|$.

EXAMPLE 3 Find a Unit Vector

Find a unit vector **u** in the direction of $\mathbf{v} = \langle -4, 2 \rangle$.

Solution

Find the magnitude of **v**.

$$\|\mathbf{v}\| = \sqrt{(-4)^2 + 2^2} = \sqrt{16 + 4} = \sqrt{20} = 2\sqrt{5}$$

Divide each component of **v** by $\|\mathbf{v}\|$.

$$\mathbf{u} = \left\langle \frac{-4}{2\sqrt{5}}, \frac{2}{2\sqrt{5}} \right\rangle = \left\langle \frac{-2}{\sqrt{5}}, \frac{1}{\sqrt{5}} \right\rangle = \left\langle -\frac{2\sqrt{5}}{5}, \frac{\sqrt{5}}{5} \right\rangle$$

A unit vector in the direction of **v** is **u**.

▶ TRY EXERCISE 8, PAGE 450

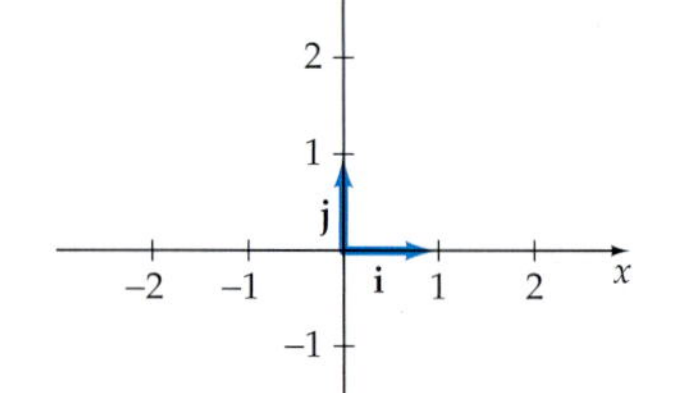

FIGURE 5.74

Two unit vectors, one parallel to the x-axis and one parallel to the y-axis, are of special importance. See **Figure 5.74.**

Definition of Unit Vectors i and j

$$\mathbf{i} = \langle 1, 0 \rangle \qquad \mathbf{j} = \langle 0, 1 \rangle$$

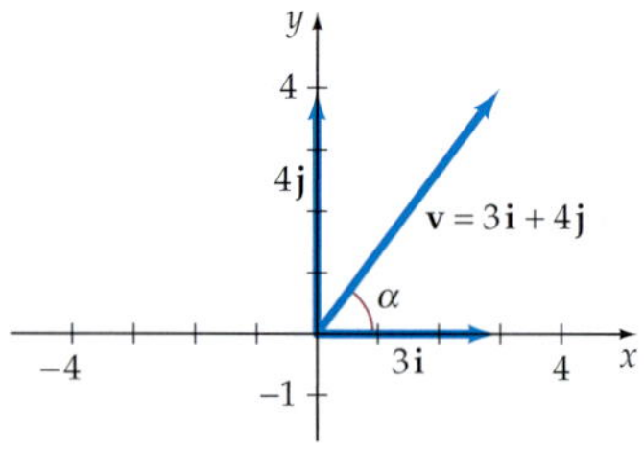

FIGURE 5.75

The vector $\mathbf{v} = \langle 3, 4 \rangle$ can be written in terms of the unit vectors **i** and **j** as shown in **Figure 5.75.**

$$\begin{aligned} \langle 3, 4 \rangle &= \langle 3, 0 \rangle + \langle 0, 4 \rangle && \text{• Vector Addition Property} \\ &= 3\langle 1, 0 \rangle + 4\langle 0, 1 \rangle && \text{• Scalar multiplication of a vector} \\ &= 3\mathbf{i} + 4\mathbf{j} && \text{• Definition of i and j} \end{aligned}$$

By means of scalar multiplication and addition of vectors, any vector can be expressed in terms of the unit vectors **i** and **j**. Let $\mathbf{v} = \langle a_1, a_2 \rangle$. Then

$$\mathbf{v} = \langle a_1, a_2 \rangle = a_1\langle 1, 0 \rangle + a_2\langle 0, 1 \rangle = a_1\mathbf{i} + a_2\mathbf{j}$$

This gives the following result.

Representation of a Vector in Terms of i and j

If **v** is a vector and $\mathbf{v} = \langle a_1, a_2 \rangle$, then $\mathbf{v} = a_1\mathbf{i} + a_2\mathbf{j}$.

The rules for addition and scalar multiplication of vectors can be restated in terms of **i** and **j**. If $\mathbf{v} = a_1\mathbf{i} + a_2\mathbf{j}$ and $\mathbf{w} = b_1\mathbf{i} + b_2\mathbf{j}$, then

$$\mathbf{v} + \mathbf{w} = (a_1\mathbf{i} + a_2\mathbf{j}) + (b_1\mathbf{i} + b_2\mathbf{j}) = (a_1 + b_1)\mathbf{i} + (a_2 + b_2)\mathbf{j}$$
$$k\mathbf{v} = k(a_1\mathbf{i} + a_2\mathbf{j}) = ka_1\mathbf{i} + ka_2\mathbf{j}$$

EXAMPLE 4 Operate on Vectors Written in Terms of i and j

Given $\mathbf{v} = 3\mathbf{i} - 4\mathbf{j}$ and $\mathbf{w} = 5\mathbf{i} + 3\mathbf{j}$, find $3\mathbf{v} - 2\mathbf{w}$.

Solution

$$\begin{aligned} 3\mathbf{v} - 2\mathbf{w} &= 3(3\mathbf{i} - 4\mathbf{j}) - 2(5\mathbf{i} + 3\mathbf{j}) \\ &= (9\mathbf{i} - 12\mathbf{j}) - (10\mathbf{i} + 6\mathbf{j}) \\ &= (9 - 10)\mathbf{i} + (-12 - 6)\mathbf{j} \\ &= -\mathbf{i} - 18\mathbf{j} \end{aligned}$$

▶ TRY EXERCISE 26, PAGE 450

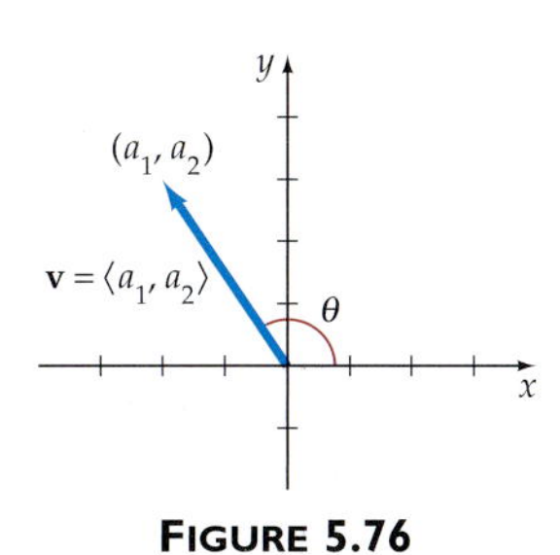

FIGURE 5.76

The components a_1 and a_2 of the vector $\mathbf{v} = \langle a_1, a_2 \rangle$ can be expressed in terms of the magnitude of **v** and the direction angle of **v** (the angle that **v** makes with the positive x-axis). Consider the vector **v** in **Figure 5.76.** Then

$$\|\mathbf{v}\| = \sqrt{(a_1)^2 + (a_2)^2}$$

From the definitions of sine and cosine, we have

$$\cos\theta = \frac{a_1}{\|\mathbf{v}\|} \quad \text{and} \quad \sin\theta = \frac{a_2}{\|\mathbf{v}\|}$$

Rewriting the last two equations, we find that the components of **v** are

$$a_1 = \|\mathbf{v}\|\cos\theta \quad \text{and} \quad a_2 = \|\mathbf{v}\|\sin\theta$$

Horizontal and Vertical Components of a Vector

Let $\mathbf{v} = \langle a_1, a_2 \rangle$, where $\mathbf{v} \neq \mathbf{0}$, the zero vector. Then

$$a_1 = \|\mathbf{v}\|\cos\theta \quad \text{and} \quad a_2 = \|\mathbf{v}\|\sin\theta$$

where θ is the angle between the positive x-axis and **v**.

The **horizontal component** of **v** is $\|\mathbf{v}\|\cos\theta$. The **vertical component** of **v** is $\|\mathbf{v}\|\sin\theta$.

? QUESTION Is $\mathbf{u} = \cos\theta\mathbf{i} + \sin\theta\mathbf{j}$ a unit vector?

Any nonzero vector can be written in terms of its horizontal and vertical components. Let $\mathbf{v} = a_1\mathbf{i} + a_2\mathbf{j}$. Then

$$\begin{aligned} \mathbf{v} &= a_1\mathbf{i} + a_2\mathbf{j} \\ &= (\|\mathbf{v}\|\cos\theta)\mathbf{i} + (\|\mathbf{v}\|\sin\theta)\mathbf{j} \\ &= \|\mathbf{v}\|(\cos\theta\mathbf{i} + \sin\theta\mathbf{j}) \end{aligned}$$

$\|\mathbf{v}\|$ is the magnitude of **v**, and the vector $\cos\theta\mathbf{i} + \sin\theta\mathbf{j}$ is a unit vector. The last equation shows that any vector **v** can be written as the product of its magnitude and a unit vector in the direction of **v**.

? ANSWER Yes, because $\|\cos\theta\mathbf{i} + \sin\theta\mathbf{j}\| = \sqrt{\cos^2\theta + \sin^2\theta} = \sqrt{1} = 1$.

EXAMPLE 5 Find the Horizontal and Vertical Components of a Vector

Find the approximate horizontal and vertical components of a vector $\mathbf{v}$ of magnitude 10 meters with direction angle 228°. Write the vector in the form $\mathbf{v} = a_1\mathbf{i} + a_2\mathbf{j}$.

Solution

$$a_1 = 10 \cos 228° \approx -6.7$$
$$a_2 = 10 \sin 228° \approx -7.4$$

The approximate horizontal and vertical components are -6.7 and -7.4, respectively.

$$\mathbf{v} \approx -6.7\mathbf{i} - 7.4\mathbf{j}$$

▶ TRY EXERCISE 36, PAGE 450

take note

Ground speed is the magnitude of the resultant of the plane's velocity vector and the wind velocity vector.

APPLICATIONS OF VECTORS

Consider an object on which two vectors are acting simultaneously. This occurs when a boat is moving in a current or an airplane is flying in a wind. The **airspeed** of a plane is the speed at which the plane would be moving if there were no wind. The actual velocity of a plane is the velocity relative to the ground. The magnitude of the actual velocity is the **ground speed.**

EXAMPLE 6 Solve an Application Involving Airspeed

An airplane is traveling with an airspeed of 320 mph at a heading of 62°. A wind of 42 mph is blowing at a heading of 125°. Find the ground speed and the course of the airplane.

Solution

Sketch a diagram similar to **Figure 5.77** showing the relevant vectors. $\mathbf{AB}$ represents the heading and the airspeed, $\mathbf{AD}$ represents the wind velocity, and $\mathbf{AC}$ represents the course and the ground speed. By vector addition, $\mathbf{AC} = \mathbf{AB} + \mathbf{AD}$. From the figure,

$$\begin{aligned}
\mathbf{AB} &= 320(\cos 28°\mathbf{i} + \sin 28°\mathbf{j}) \\
\mathbf{AD} &= 42[\cos(-35°)\mathbf{i} + \sin(-35°)\mathbf{j}] \\
\mathbf{AC} &= 320(\cos 28°\mathbf{i} + \sin 28°\mathbf{j}) + 42[\cos(-35°)\mathbf{i} + \sin(-35°)\mathbf{j}] \\
&\approx (282.5\mathbf{i} + 150.2\mathbf{j}) + (34.4\mathbf{i} - 24.1\mathbf{j}) \\
&= 316.9\mathbf{i} + 126.1\mathbf{j}
\end{aligned}$$

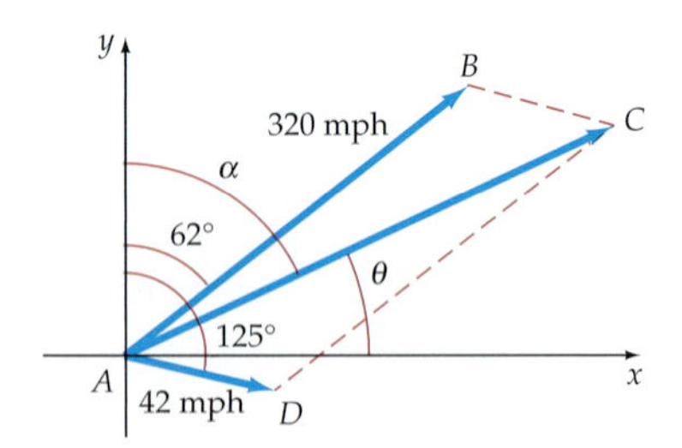

FIGURE 5.77

$\mathbf{AC}$ is the course of the plane. The ground speed is $\|\mathbf{AC}\|$. The heading is $\alpha = 90° - \theta$.

$$\|\mathbf{AC}\| = \sqrt{(316.9)^2 + (126.1)^2} \approx 340$$

$$\alpha = 90° - \theta = 90° - \tan^{-1}\left(\frac{126.1}{316.9}\right) \approx 68°$$

The ground speed is approximately 340 mph at a heading of 68°.

▶ TRY EXERCISE 40, PAGE 451

There are numerous problems involving force that can be solved by using vectors. One type involves objects that are resting on a ramp. For these problems, we frequently try to find the components of a force vector relative to the ramp rather than to the x-axis.

EXAMPLE 7 Solve an Application Involving Force

A 110-pound box is on a 24° ramp. Find the component of the force that is parallel to the ramp.

Solution

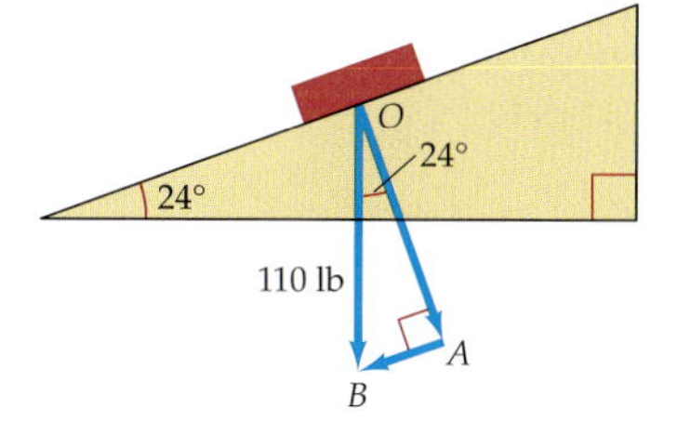

FIGURE 5.78

The force-of-gravity vector (**OB**) is the sum of two components, one parallel to the ramp, **AB**, and the other (called the *normal component*) perpendicular to the ramp, **OA**. (See **Figure 5.78**). **AB** is the vector that represents the force tending to move the box down the ramp. Because triangle OAB is a right triangle and $\angle AOB$ is 24°,

$$\sin 24^\circ = \frac{\|\mathbf{AB}\|}{110}$$

$$\|\mathbf{AB}\| = 110 \sin 24^\circ \approx 45$$

The component of the force parallel to the ramp is approximately 45 pounds.

▶ TRY EXERCISE 42, PAGE 451

• Dot Product

We have considered the product of a real number (scalar) and a vector. We now turn our attention to the product of two vectors. Finding the *dot product* of two vectors is one way to multiply a vector by a vector. The dot product of two vectors is a real number and *not* a vector. The dot product is also called the *inner product* or the *scalar product*. This product is useful in engineering and physics.

Definition of Dot Product

Given $\mathbf{v} = \langle a, b \rangle$ and $\mathbf{w} = \langle c, d \rangle$, the **dot product** of **v** and **w** is given by

$$\mathbf{v} \cdot \mathbf{w} = ac + bd$$

EXAMPLE 8 Find the Dot Product of Two Vectors

Find the dot product of $\mathbf{v} = \langle 6, -2 \rangle$ and $\mathbf{w} = \langle -2, 4 \rangle$.

Solution

$$\mathbf{v} \cdot \mathbf{w} = 6(-2) + (-2)4 = -12 - 8 = -20$$

▶ TRY EXERCISE 50, PAGE 451

QUESTION Is the dot product of two vectors a vector or a real number?

If the vectors in Example 8 were given in terms of the vectors **i** and **j**, then $\mathbf{v} = 6\mathbf{i} - 2\mathbf{j}$ and $\mathbf{w} = -2\mathbf{i} + 4\mathbf{j}$. In this case,

$$\mathbf{v} \cdot \mathbf{w} = (6\mathbf{i} - 2\mathbf{j}) \cdot (-2\mathbf{i} + 4\mathbf{j}) = 6(-2) + (-2)4 = -20$$

Properties of the Dot Product

In the following properties, **u**, **v**, and **w** are vectors and a is a scalar.

1. $\mathbf{v} \cdot \mathbf{w} = \mathbf{w} \cdot \mathbf{v}$
2. $\mathbf{u} \cdot (\mathbf{v} + \mathbf{w}) = \mathbf{u} \cdot \mathbf{v} + \mathbf{u} \cdot \mathbf{w}$
3. $a(\mathbf{u} \cdot \mathbf{v}) = (a\mathbf{u}) \cdot \mathbf{v} = \mathbf{u} \cdot (a\mathbf{v})$
4. $\mathbf{v} \cdot \mathbf{v} = \|\mathbf{v}\|^2$
5. $\mathbf{0} \cdot \mathbf{v} = 0$
6. $\mathbf{i} \cdot \mathbf{i} = \mathbf{j} \cdot \mathbf{j} = 1$
7. $\mathbf{i} \cdot \mathbf{j} = \mathbf{j} \cdot \mathbf{i} = 0$

The proofs of these properties follow from the definition of dot product. Here is the proof of the fourth property. Let $\mathbf{v} = a\mathbf{i} + b\mathbf{j}$.

$$\mathbf{v} \cdot \mathbf{v} = (a\mathbf{i} + b\mathbf{j}) \cdot (a\mathbf{i} + b\mathbf{j}) = a^2 + b^2 = \|\mathbf{v}\|^2$$

Rewriting the fourth property of the dot product yields an alternative way of expressing the magnitude of a vector.

Magnitude of a Vector in Terms of the Dot Product

If $\mathbf{v} = \langle a, b \rangle$, then $\|\mathbf{v}\| = \sqrt{\mathbf{v} \cdot \mathbf{v}}$.

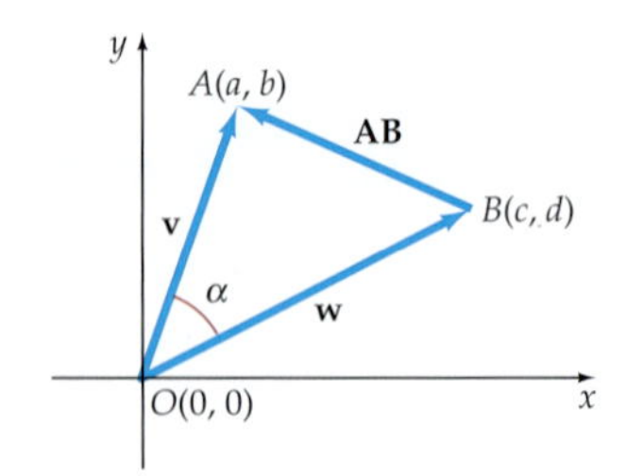

FIGURE 5.79

The Law of Cosines can be used to derive an alternative formula for the dot product. Consider the vectors $\mathbf{v} = \langle a, b \rangle$ and $\mathbf{w} = \langle c, d \rangle$ as shown in **Figure 5.79.** Using the Law of Cosines for triangle OAB, we have

$$\|\mathbf{AB}\|^2 = \|\mathbf{v}\|^2 + \|\mathbf{w}\|^2 - 2\|\mathbf{v}\|\,\|\mathbf{w}\| \cos \alpha$$

By the distance formula, $\|\mathbf{AB}\|^2 = (a - c)^2 + (b - d)^2$, $\|\mathbf{v}\|^2 = a^2 + b^2$, and $\|\mathbf{w}\|^2 = c^2 + d^2$. Thus

$$(a - c)^2 + (b - d)^2 = (a^2 + b^2) + (c^2 + d^2) - 2\|\mathbf{v}\|\,\|\mathbf{w}\| \cos \alpha$$
$$a^2 - 2ac + c^2 + b^2 - 2bd + d^2 = a^2 + b^2 + c^2 + d^2 - 2\|\mathbf{v}\|\,\|\mathbf{w}\| \cos \alpha$$
$$-2ac - 2bd = -2\|\mathbf{v}\|\,\|\mathbf{w}\| \cos \alpha$$
$$ac + bd = \|\mathbf{v}\|\,\|\mathbf{w}\| \cos \alpha$$
$$\mathbf{v} \cdot \mathbf{w} = \|\mathbf{v}\|\,\|\mathbf{w}\| \cos \alpha$$

• $\mathbf{v} \cdot \mathbf{w} = ac + bd$

ANSWER A real number

Alternative Formula for the Dot Product

If $\mathbf{v}$ and $\mathbf{w}$ are two nonzero vectors and α is the smallest nonnegative angle between $\mathbf{v}$ and $\mathbf{w}$, then $\mathbf{v} \cdot \mathbf{w} = \|\mathbf{v}\| \, \|\mathbf{w}\| \cos \alpha$.

Solving the alternative formula for the dot product for $\cos \alpha$, we have a formula for the cosine of the angle between two vectors.

Angle Between Two Vectors

If $\mathbf{v}$ and $\mathbf{w}$ are two nonzero vectors and α is the smallest nonnegative angle between $\mathbf{v}$ and $\mathbf{w}$, then $\cos \alpha = \dfrac{\mathbf{v} \cdot \mathbf{w}}{\|\mathbf{v}\| \, \|\mathbf{w}\|}$ and $\alpha = \cos^{-1}\left(\dfrac{\mathbf{v} \cdot \mathbf{w}}{\|\mathbf{v}\| \, \|\mathbf{w}\|}\right)$.

EXAMPLE 9 Find the Angle Between Two Vectors

Find the measure of the smallest positive angle between the vectors $\mathbf{v} = 2\mathbf{i} - 3\mathbf{j}$ and $\mathbf{w} = -\mathbf{i} + 5\mathbf{j}$, as shown in **Figure 5.80.**

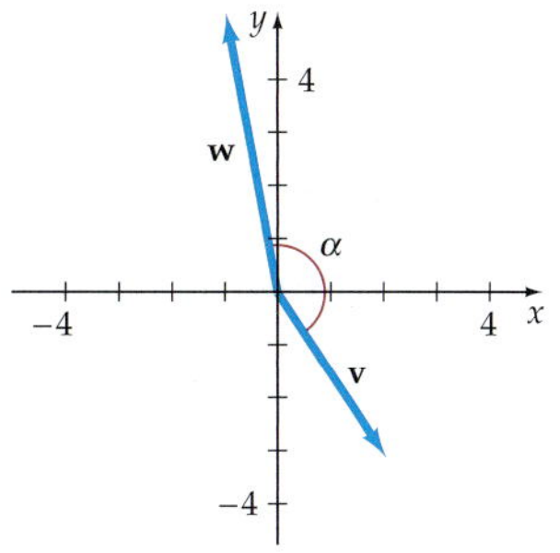

Solution

Use the equation for the angle between two vectors.

$$\cos \alpha = \frac{\mathbf{v} \cdot \mathbf{w}}{\|\mathbf{v}\|\|\mathbf{w}\|} = \frac{(2\mathbf{i} - 3\mathbf{j}) \cdot (-\mathbf{i} + 5\mathbf{j})}{(\sqrt{2^2 + (-3)^2})(\sqrt{(-1)^2 + 5^2})}$$

$$= \frac{-2 - 15}{\sqrt{13}\sqrt{26}} = \frac{-17}{\sqrt{338}}$$

$$\alpha = \cos^{-1}\left(\frac{-17}{\sqrt{338}}\right) \approx 157.6^\circ$$

The angle between the two vectors is approximately 157.6°.

FIGURE 5.80

TRY EXERCISE 60, PAGE 451

SCALAR PROJECTION

Let $\mathbf{v} = \langle a_1, a_2 \rangle$ and $\mathbf{w} = \langle b_1, b_2 \rangle$ be two nonzero vectors, and let α be the angle between the vectors. Two possible configurations, one for which α is an acute angle and one for which α is an obtuse angle, are shown in **Figure 5.81.** In each case, a right triangle is formed by drawing a line segment from the head of $\mathbf{v}$ to a line through $\mathbf{w}$.

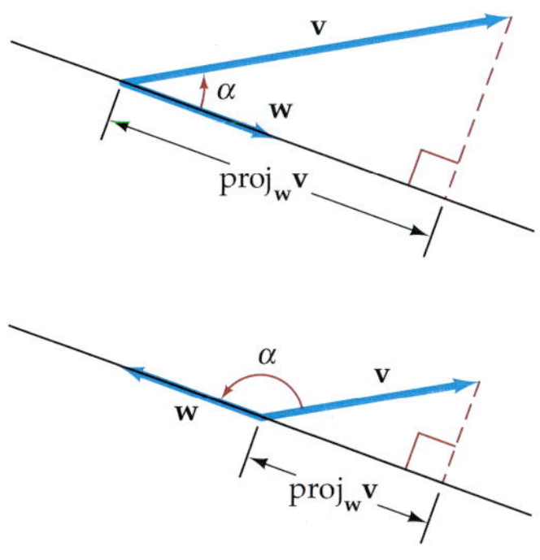

FIGURE 5.81

Definition of the Scalar Projection of v on w

If $\mathbf{v}$ and $\mathbf{w}$ are two nonzero vectors and α is the smallest positive angle between $\mathbf{v}$ and $\mathbf{w}$, then the scalar projection of $\mathbf{v}$ on $\mathbf{w}$, $\text{proj}_\mathbf{w}\mathbf{v}$, is given by

$$\text{proj}_\mathbf{w}\mathbf{v} = \|\mathbf{v}\| \cos \alpha$$

To derive an alternate formula for $\text{proj}_{\mathbf{w}}\mathbf{v}$, consider the dot product $\mathbf{v} \cdot \mathbf{w} = \|\mathbf{v}\| \|\mathbf{w}\| \cos \alpha$. Solving for $\|\mathbf{v}\| \cos \alpha$, which is $\text{proj}_{\mathbf{w}}\mathbf{v}$, we have

$$\text{proj}_{\mathbf{w}}\mathbf{v} = \frac{\mathbf{v} \cdot \mathbf{w}}{\|\mathbf{w}\|}$$

When the angle α between the two vectors is an acute angle, $\text{proj}_{\mathbf{w}}\mathbf{v}$ is positive. When α is an obtuse angle, $\text{proj}_{\mathbf{w}}\mathbf{v}$ is negative.

EXAMPLE 10 Find the Projection of v on w

Given $\mathbf{v} = 2\mathbf{i} + 4\mathbf{j}$ and $\mathbf{w} = -2\mathbf{i} + 8\mathbf{j}$ as shown in **Figure 5.82,** find $\text{proj}_{\mathbf{w}}\mathbf{v}$.

Solution

Use the equation $\text{proj}_{\mathbf{w}}\mathbf{v} = \dfrac{\mathbf{v} \cdot \mathbf{w}}{\|\mathbf{w}\|}$.

$$\text{proj}_{\mathbf{w}}\,\mathbf{v} = \frac{(2\mathbf{i} + 4\mathbf{j}) \cdot (-2\mathbf{i} + 8\mathbf{j})}{\sqrt{(-2)^2 + 8^2}} = \frac{28}{\sqrt{68}} = \frac{14\sqrt{17}}{17} \approx 3.4$$

TRY EXERCISE 62, PAGE 451

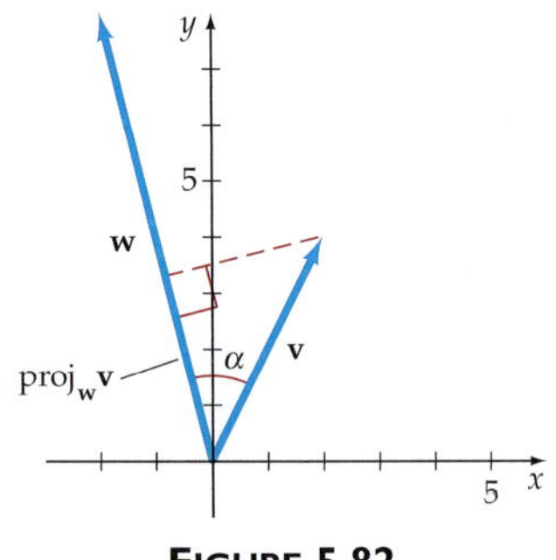

FIGURE 5.82

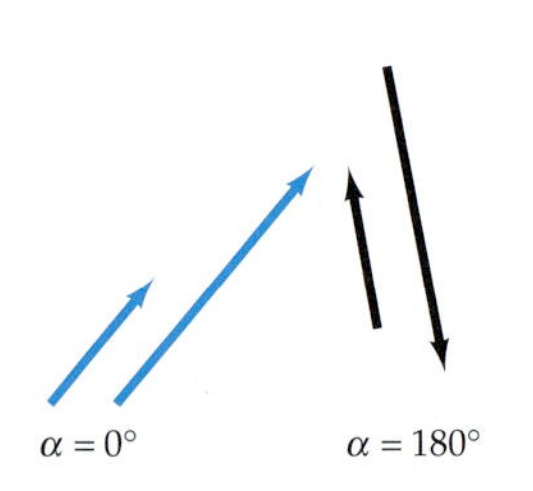

FIGURE 5.83

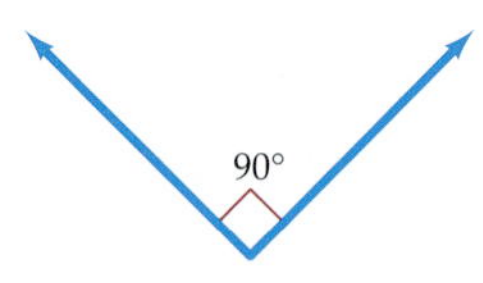

FIGURE 5.84

PARALLEL AND PERPENDICULAR VECTORS

Two vectors are *parallel* when the angle α between the vectors is 0° or 180°, as shown in **Figure 5.83.** When the angle α is 0°, the vectors point in the same direction; the vectors point in opposite directions when α is 180°.

Let $\mathbf{v} = a_1\mathbf{i} + b_1\mathbf{j}$, let c be a real number, and let $\mathbf{w} = c\mathbf{v}$. Because $\mathbf{w}$ is a constant multiple of $\mathbf{v}$, $\mathbf{w}$ and $\mathbf{v}$ are parallel vectors. When $c > 0$, the vectors point in the same direction. When $c < 0$, the vectors point in opposite directions.

Two vectors are *perpendicular* when the angle between the vectors is 90°. See **Figure 5.84.** Perpendicular vectors are referred to as **orthogonal vectors.** If $\mathbf{v}$ and $\mathbf{w}$ are two nonzero orthogonal vectors, then from the formula for the angle between two vectors and the fact that $\cos \alpha = 0$, we have

$$0 = \frac{\mathbf{v} \cdot \mathbf{w}}{\|\mathbf{v}\| \|\mathbf{w}\|}$$

If a fraction equals zero, the numerator must be zero. Thus, for orthogonal vectors $\mathbf{v}$ and $\mathbf{w}$, $\mathbf{v} \cdot \mathbf{w} = 0$. This gives the following result.

Condition for Perpendicular Vectors

Two nonzero vectors $\mathbf{v}$ and $\mathbf{w}$ are orthogonal if and only if $\mathbf{v} \cdot \mathbf{w} = 0$.

WORK: AN APPLICATION OF THE DOT PRODUCT

When a 5-pound force is used to lift a box from the ground a distance of 4 feet, *work* is done. The amount of **work** is the product of the force on the box and the distance the box is moved. In this case the work is 20 foot-pounds. When the box

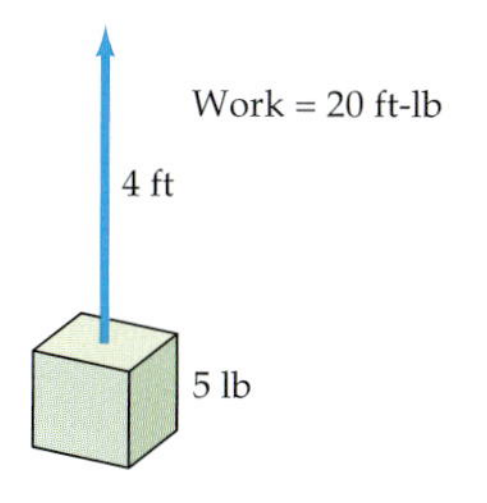

FIGURE 5.85

FIGURE 5.86

is lifted, the force and the displacement vector (the direction in which and the distance the box was moved) are in the same direction. (See **Figure 5.85.**)

Now consider a sled being pulled by a child along the ground by a rope attached to the sled, as shown in **Figure 5.86.** The force vector (along the rope) is *not* in the same direction as the displacement vector (parallel to the ground). In this case the dot product is used to determine the work done by the force.

Definition of Work

The work W done by a force $\mathbf{F}$ applied along a displacement $\mathbf{s}$ is

$$W = \mathbf{F} \cdot \mathbf{s}$$

In the case of the child pulling the sled 7 feet, the work done is

$$\begin{aligned} W &= \mathbf{F} \cdot \mathbf{s} \\ &= \|\mathbf{F}\| \, \|\mathbf{s}\| \cos \alpha \qquad \bullet\ \alpha \text{ is the angle between } \mathbf{F} \text{ and } \mathbf{s}. \\ &= (25)(7) \cos 37^\circ \approx 140 \text{ foot-pounds} \end{aligned}$$

EXAMPLE 11 Solve a Work Problem

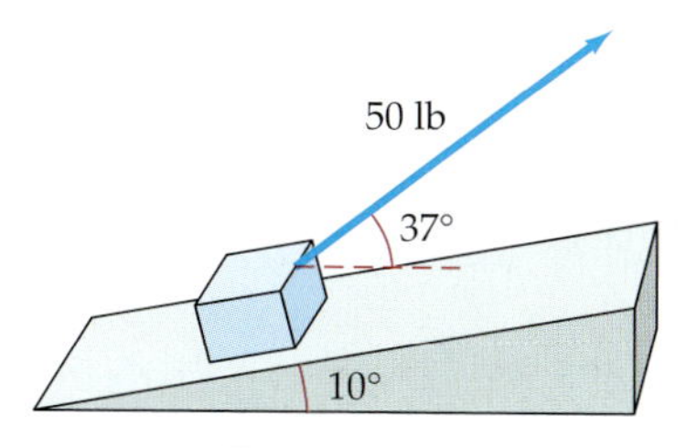

FIGURE 5.87

A force of 50 pounds on a rope is used to drag a box up a ramp that is inclined 10°. If the rope makes an angle of 37° with the ground, find the work done in moving the box 15 feet along the ramp. See **Figure 5.87.**

Solution

We will provide two solutions to this example.

Method 1 From the last section, $\mathbf{u} = \cos 10^\circ \mathbf{i} + \sin 10^\circ \mathbf{j}$ is a unit vector parallel to the ramp. Multiplying $\mathbf{u}$ by 15 (the magnitude of the displacement vector) gives $\mathbf{s} = 15(\cos 10^\circ \mathbf{i} + \sin 10^\circ \mathbf{j})$. Similarly, the force vector is $\mathbf{F} = 50(\cos 37^\circ \mathbf{i} + \sin 37^\circ \mathbf{j})$. The work done is given by the dot product.

$$\begin{aligned} W = \mathbf{F} \cdot \mathbf{s} &= 50(\cos 37^\circ \mathbf{i} + \sin 37^\circ \mathbf{j}) \cdot 15(\cos 10^\circ \mathbf{i} + \sin 10^\circ \mathbf{j}) \\ &= [50 \cdot 15](\cos 37^\circ \cos 10^\circ + \sin 37^\circ \sin 10^\circ) \approx 668.3 \text{ foot-pounds} \end{aligned}$$

Method 2 When we write the work equation as $W = \|\mathbf{F}\|\|\mathbf{s}\| \cos \alpha$, α is the angle between the force and the displacement. Thus $\alpha = 37^\circ - 10^\circ = 27^\circ$. The work done is

$$W = \|\mathbf{F}\|\|\mathbf{s}\| \cos \alpha = 50 \cdot 15 \cdot \cos 27^\circ \approx 668.3 \text{ foot-pounds}$$

▶ TRY EXERCISE 70, PAGE 451

TOPICS FOR DISCUSSION

1. Is the dot product of two vectors a vector or a scalar? Explain.
2. Is the projection of $\mathbf{v}$ on $\mathbf{w}$ a vector or a scalar? Explain.

3. Is the nonzero vector $\langle a, b\rangle$ perpendicular to the vector $\langle -b, a\rangle$? Explain.

4. Explain how to determine the angle between the vector $\langle 3, 4\rangle$ and the vector $\langle 5, -1\rangle$.

5. Consider the nonzero vector $\mathbf{u} = \langle a, b\rangle$ and the vector
$$\mathbf{v} = \left\langle \frac{a}{\sqrt{a^2 + b^2}}, \frac{b}{\sqrt{a^2 + b^2}} \right\rangle.$$
   a. Are the vectors parallel? Explain.
   b. Which one of the vectors is a unit vector?
   c. Which vector has the larger magnitude? Explain.

Exercise Set 5.7

In Exercises 1 to 6, find the components of a vector with the given initial and terminal points. Write an equivalent vector in terms of its components.

1. $P_1(-3, 0)$; $P_2(4, -1)$
2. $P_1(5, -1)$; $P_2(3, 1)$
3. $P_1(4, 2)$; $P_2(-3, -3)$
4. $P_1(0, -3)$; $P_2(0, 4)$
5. $P_1(2, -5)$; $P_2(2, 3)$
6. $P_1(3, -2)$; $P_2(3, 0)$

In Exercises 7 to 14, find the magnitude and direction of each vector. Find the unit vector in the direction of the given vector.

7. $\mathbf{v} = \langle -3, 4\rangle$
8. $\mathbf{v} = \langle 6, 10\rangle$
9. $\mathbf{v} = \langle 20, -40\rangle$
10. $\mathbf{v} = \langle -50, 30\rangle$
11. $\mathbf{v} = 2\mathbf{i} - 4\mathbf{j}$
12. $\mathbf{v} = -5\mathbf{i} + 6\mathbf{j}$
13. $\mathbf{v} = 42\mathbf{i} - 18\mathbf{j}$
14. $\mathbf{v} = -22\mathbf{i} - 32\mathbf{j}$

In Exercises 15 to 23, perform the indicated operations where $\mathbf{u} = \langle -2, 4\rangle$ and $\mathbf{v} = \langle -3, -2\rangle$.

15. $3\mathbf{u}$
16. $-4\mathbf{v}$
17. $2\mathbf{u} - \mathbf{v}$
18. $4\mathbf{v} - 2\mathbf{u}$
19. $\frac{2}{3}\mathbf{u} + \frac{1}{6}\mathbf{v}$
20. $\frac{3}{4}\mathbf{u} - 2\mathbf{v}$
21. $\|\mathbf{u}\|$
22. $\|\mathbf{v} + 2\mathbf{u}\|$
23. $\|3\mathbf{u} - 4\mathbf{v}\|$

In Exercises 24 to 32, perform the indicated operations where $\mathbf{u} = 3\mathbf{i} - 2\mathbf{j}$ and $\mathbf{v} = -2\mathbf{i} + 3\mathbf{j}$.

24. $-2\mathbf{u}$
25. $4\mathbf{v}$
26. $3\mathbf{u} + 2\mathbf{v}$
27. $6\mathbf{u} + 2\mathbf{v}$
28. $\frac{1}{2}\mathbf{u} - \frac{3}{4}\mathbf{v}$
29. $\frac{2}{3}\mathbf{v} + \frac{3}{4}\mathbf{u}$
30. $\|\mathbf{v}\|$
31. $\|\mathbf{u} - 2\mathbf{v}\|$
32. $\|2\mathbf{v} + 3\mathbf{u}\|$

In Exercises 33 to 36, find the horizontal and vertical components of each vector. Write an equivalent vector in the form $\mathbf{v} = a_1\mathbf{i} + a_2\mathbf{j}$.

33. Magnitude = 5, direction angle = 27°
34. Magnitude = 4, direction angle = 127°
35. Magnitude = 4, direction angle = $\frac{\pi}{4}$
36. Magnitude = 2, direction angle = $\frac{8\pi}{7}$

37. **Ground Speed of a Plane** A plane is flying at an airspeed of 340 mph at a heading of 124°. A wind of 45 mph is blowing from the west. Find the ground speed of the plane.

38. **Heading of a Boat** A person who can row 2.6 mph in still water wants to row due east across a river. The river is flowing from the north at a rate of 0.8 mph. Determine the heading of the boat that will be required for it to travel due east across the river.

39. **Ground Speed and Course of a Plane** A pilot is flying at a heading of 96° at 225 mph. A 50-mph wind is blowing from the southwest at a heading of 37°. Find the ground speed and course of the plane.

▶ **40.** **Course of a Boat** The captain of a boat is steering at a heading of 327° at 18 mph. The current is flowing at 4 mph at a heading of 60°. Find the course (to the nearest degree) of the boat.

41. **Magnitude of a Force** Find the magnitude of the force necessary to keep a 3000-pound car from sliding down a ramp inclined at an angle of 5.6°.

▶ **42.** **Angle of a Ramp** A 120-pound force keeps an 800-pound object from sliding down an inclined ramp. Find the angle of the ramp.

43. **Magnitude of the Normal Component** A 25-pound box is resting on a ramp that is inclined 9.0°. Find the magnitude of the normal component of force.

44. **Magnitude of the Normal Component** Find the magnitude of the normal component of force for a 50-pound crate that is resting on a ramp that is inclined 12°.

In Exercises 45 to 52, find the dot product of the vectors.

45. $\mathbf{v} = \langle 3, -2 \rangle$; $\mathbf{w} = \langle 1, 3 \rangle$

46. $\mathbf{v} = \langle 2, 4 \rangle$; $\mathbf{w} = \langle 0, 2 \rangle$

47. $\mathbf{v} = \langle 4, 1 \rangle$; $\mathbf{w} = \langle -1, 4 \rangle$

48. $\mathbf{v} = \langle 2, -3 \rangle$; $\mathbf{w} = \langle 3, 2 \rangle$

49. $\mathbf{v} = \mathbf{i} + 2\mathbf{j}$; $\mathbf{w} = -\mathbf{i} + \mathbf{j}$

▶ **50.** $\mathbf{v} = 5\mathbf{i} + 3\mathbf{j}$; $\mathbf{w} = 4\mathbf{i} - 2\mathbf{j}$

51. $\mathbf{v} = 6\mathbf{i} - 4\mathbf{j}$; $\mathbf{w} = -2\mathbf{i} - 3\mathbf{j}$

52. $\mathbf{v} = -4\mathbf{i} + 2\mathbf{j}$; $\mathbf{w} = -2\mathbf{i} - 4\mathbf{j}$

In Exercises 53 to 60, find the angle between the two vectors. State which pairs of vectors are orthogonal.

53. $\mathbf{v} = \langle 2, -1 \rangle$; $\mathbf{w} = \langle 3, 4 \rangle$

54. $\mathbf{v} = \langle 1, -5 \rangle$; $\mathbf{w} = \langle -2, 3 \rangle$

55. $\mathbf{v} = \langle 0, 3 \rangle$; $\mathbf{w} = \langle 2, 2 \rangle$

56. $\mathbf{v} = \langle -1, 7 \rangle$; $\mathbf{w} = \langle 3, -2 \rangle$

57. $\mathbf{v} = 5\mathbf{i} - 2\mathbf{j}$; $\mathbf{w} = 2\mathbf{i} + 5\mathbf{j}$

58. $\mathbf{v} = 8\mathbf{i} + \mathbf{j}$; $\mathbf{w} = -\mathbf{i} + 8\mathbf{j}$

59. $\mathbf{v} = 5\mathbf{i} + 2\mathbf{j}$; $\mathbf{w} = -5\mathbf{i} - 2\mathbf{j}$

▶ **60.** $\mathbf{v} = 3\mathbf{i} - 4\mathbf{j}$; $\mathbf{w} = 6\mathbf{i} - 12\mathbf{j}$

In Exercises 61 to 68, find $\text{proj}_{\mathbf{w}}\mathbf{v}$.

61. $\mathbf{v} = \langle 6, 7 \rangle$; $\mathbf{w} = \langle 3, 4 \rangle$

▶ **62.** $\mathbf{v} = \langle -7, 5 \rangle$; $\mathbf{w} = \langle -4, 1 \rangle$

63. $\mathbf{v} = \langle -3, 4 \rangle$; $\mathbf{w} = \langle 2, 5 \rangle$

64. $\mathbf{v} = \langle 2, 4 \rangle$; $\mathbf{w} = \langle -1, 5 \rangle$

65. $\mathbf{v} = 2\mathbf{i} + \mathbf{j}$; $\mathbf{w} = 6\mathbf{i} + 3\mathbf{j}$

66. $\mathbf{v} = 5\mathbf{i} + 2\mathbf{j}$; $\mathbf{w} = -5\mathbf{i} - 2\mathbf{j}$

67. $\mathbf{v} = 3\mathbf{i} - 4\mathbf{j}$; $\mathbf{w} = -6\mathbf{i} + 12\mathbf{j}$

68. $\mathbf{v} = 2\mathbf{i} + 2\mathbf{j}$; $\mathbf{w} = -4\mathbf{i} - 2\mathbf{j}$

69. **Work** A 150-pound box is dragged 15 feet along a level floor. Find the work done if a force of 75 pounds at an angle of 32° is used.

▶ **70.** **Work** A 100-pound force is pulling a sled loaded with bricks that weighs 400 pounds. The force is at an angle of 42° with the displacement. Find the work done in moving the sled 25 feet.

71. **Work** A rope is being used to pull a box up a ramp that is inclined at 15°. The rope exerts a force of 75 pounds on the box, and it makes an angle of 30° with the plane of the ramp. Find the work done in moving the box 12 feet.

72. **Work** A dock worker exerts a force on a box sliding down the ramp of a truck. The ramp makes an angle of 48° with the road, and the worker exerts a 50-pound force parallel to the road. Find the work done in sliding the box 6 feet.

CONNECTING CONCEPTS

73. For $\mathbf{u} = \langle -1, 1 \rangle$, $\mathbf{v} = \langle 2, 3 \rangle$, and $\mathbf{w} = \langle 5, 5 \rangle$, find the sum of the three vectors geometrically by using the triangle method of adding vectors.

74. For $\mathbf{u} = \langle 1, 2 \rangle$, $\mathbf{v} = \langle 3, -2 \rangle$, and $\mathbf{w} = \langle -1, 4 \rangle$, find $\mathbf{u} + \mathbf{v} - \mathbf{w}$ geometrically by using the triangle method of adding vectors.

75. Find a vector that has the initial point $(3, -1)$ and is equivalent to $\mathbf{v} = 2\mathbf{i} - 3\mathbf{j}$.

76. Find a vector that has the initial point $(-2, 4)$ and is equivalent to $\mathbf{v} = \langle -1, 3 \rangle$.

77. If $\mathbf{v} = 2\mathbf{i} - 5\mathbf{j}$ and $\mathbf{w} = 5\mathbf{i} + 2\mathbf{j}$ have the same initial point, is $\mathbf{v}$ perpendicular to $\mathbf{w}$? Why or why not?

78. If $\mathbf{v} = \langle 5, 6\rangle$ and $\mathbf{w} = \langle 6, 5\rangle$ have the same initial point, is $\mathbf{v}$ perpendicular to $\mathbf{w}$? Why or why not?

79. Let $\mathbf{v} = \langle -2, 7\rangle$. Find a vector perpendicular to $\mathbf{v}$.

80. Let $\mathbf{w} = 4\mathbf{i} + \mathbf{j}$. Find a vector perpendicular to $\mathbf{w}$.

In Example 7 of this section, if the box were to be kept from sliding down the ramp, it would be necessary to provide a force of 45 pounds parallel to the ramp but pointed *up* the ramp. Some of this force would be provided by a frictional force between the box and the ramp. The force of friction is $\mathbf{F}_\mu = \mu\mathbf{N}$, where μ is a constant called the coefficient of friction, and $\mathbf{N}$ is the normal component of the force of gravity. In Exercises 81 and 82, find the frictional force.

81. **Frictional Force** A 50-pound box is resting on a ramp inclined at 12°. Find the force of friction if the coefficient of friction, μ, is 0.13.

82. **Frictional Force** A car weighing 2500 pounds is resting on a ramp inclined at 15°. Find the frictional force if the coefficient of friction, μ, is 0.21.

83. Is the dot product an associative operation? That is, given any nonzero vectors $\mathbf{u}$, $\mathbf{v}$, and $\mathbf{w}$, does

$$(\mathbf{u} \cdot \mathbf{v}) \cdot \mathbf{w} = \mathbf{u} \cdot (\mathbf{v} \cdot \mathbf{w})?$$

84. Prove that $\mathbf{v} \cdot \mathbf{w} = \mathbf{w} \cdot \mathbf{v}$.

85. Prove that $c(\mathbf{v} \cdot \mathbf{w}) = (c\mathbf{v}) \cdot \mathbf{w}$.

86. Show that the dot product of two nonzero vectors is positive if the angle between the vectors is an acute angle and negative if the angle between the two vectors is an obtuse angle.

87. **Comparison of Work Done** Consider the following two situations. (1) A rope is being used to pull a box up a ramp inclined at an angle α. The rope exerts a force $\mathbf{F}$ on the box, and the rope makes an angle θ with the ramp. The box is pulled s feet. (2) A rope is being used to pull a box along a level floor. The rope exerts the same force $\mathbf{F}$ on the box. The box is pulled the same s feet. In which case is more work done?

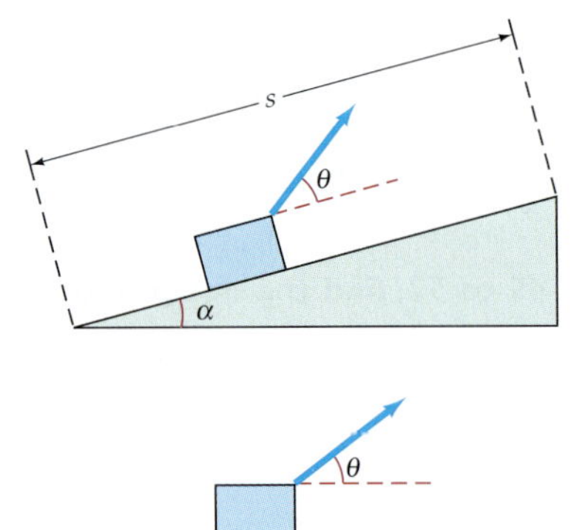

PROJECTS

1. **Same Direction or Opposite Directions** Let $\mathbf{v} = c\mathbf{w}$, where c is a nonzero real number and $\mathbf{w}$ is a nonzero vector. Show that $\dfrac{\mathbf{v} \cdot \mathbf{w}}{\|\mathbf{v}\|\,\|\mathbf{w}\|} = \pm 1$ and that the result is 1 when $c > 0$ and -1 when $c < 0$.

2. **The Law of Cosines and Vectors** Prove that $\|\mathbf{v} - \mathbf{w}\|^2 = \|\mathbf{v}\|^2 + \|\mathbf{w}\|^2 - 2\mathbf{v} \cdot \mathbf{w}$.

3. **Projection Relationships** What is the relationship between the nonzero vectors $\mathbf{v}$ and $\mathbf{w}$ if

a. $\text{proj}_{\mathbf{w}}\mathbf{v} = 0$? **b.** $\text{proj}_{\mathbf{w}}\mathbf{v} = \|\mathbf{v}\|$?

Exploring Concepts with Technology

Approximate an Inverse Trigonometric Function with Polynomials

The function $y = \sin^{-1} x$ can be approximated by polynomials. For example, consider the following:

$$f_1(x) = x + \frac{x^3}{2 \cdot 3} \qquad \text{where } -1 \le x \le 1$$

$$f_2(x) = x + \frac{x^3}{2 \cdot 3} + \frac{1 \cdot 3x^5}{2 \cdot 4 \cdot 5} \qquad \text{where } -1 \le x \le 1$$

$$f_3(x) = x + \frac{x^3}{2 \cdot 3} + \frac{1 \cdot 3x^5}{2 \cdot 4 \cdot 5} + \frac{1 \cdot 3 \cdot 5x^7}{2 \cdot 4 \cdot 6 \cdot 7} \qquad \text{where } -1 \le x \le 1$$

$$f_4(x) = x + \frac{x^3}{2 \cdot 3} + \frac{1 \cdot 3x^5}{2 \cdot 4 \cdot 5} + \frac{1 \cdot 3 \cdot 5x^7}{2 \cdot 4 \cdot 6 \cdot 7} + \frac{1 \cdot 3 \cdot 5 \cdot 7x^9}{2 \cdot 4 \cdot 6 \cdot 8 \cdot 9} \qquad \text{where } -1 \le x \le 1$$

$$\vdots$$

$$f_n(x) = x + \frac{x^3}{2 \cdot 3} + \frac{1 \cdot 3x^5}{2 \cdot 4 \cdot 5} + \frac{1 \cdot 3 \cdot 5x^7}{2 \cdot 4 \cdot 6 \cdot 7} + \cdots + \frac{(2n)!\,x^{2n+1}}{(2^n n!)^2(2n + 1)}$$

where $\quad -1 \le x \le 1,\ n! = 1 \cdot 2 \cdot 3 \cdots (n - 1)n$

and $\quad (2n)! = 1 \cdot 2 \cdot 3 \cdots (2n - 1)(2n)$

Use a graphing utility for the following exercises.

1. Graph $y = f_1(x)$, $y = f_2(x)$, $y = f_3(x)$, and $y = f_4(x)$ on the viewing window Xmin = −1, Xmax = 1, Ymin = −1.5708, Ymax = 1.5708.

2. Determine the values of x for which $f_3(x)$ and $\sin^{-1} x$ differ by less than 0.001. That is, determine the values of x for which

$$|f_3(x) - \sin^{-1} x| < 0.001$$

3. Determine the values of x for which

$$|f_4(x) - \sin^{-1} x| < 0.001$$

4. Write all seven terms of $f_6(x)$. Graph $y = f_6(x)$ and $y = \sin^{-1} x$ on the viewing window Xmin = −1, Xmax = 1, Ymin = $-\frac{\pi}{2}$, Ymax = $\frac{\pi}{2}$.

5. Write all seven terms of $f_6(1)$. What do you notice about the size of a term compared to that of the previous term?

6. What is the largest-degree term in $f_{10}(x)$?

CHAPTER 5 SUMMARY

5.1 Trigonometric Functions of Angles

- Let θ be an acute angle of a right triangle. The six trigonometric functions of θ are given by

$$\sin\theta = \frac{\text{opp}}{\text{hyp}} \qquad \csc\theta = \frac{\text{hyp}}{\text{opp}}$$

$$\cos\theta = \frac{\text{adj}}{\text{hyp}} \qquad \sec\theta = \frac{\text{hyp}}{\text{adj}}$$

$$\tan\theta = \frac{\text{opp}}{\text{adj}} \qquad \cot\theta = \frac{\text{adj}}{\text{opp}}$$

- Let $P(x, y)$ be a point, except the origin, on the terminal side of an angle θ in standard position. The six trigonometric functions of θ are

$$\sin\theta = \frac{y}{r} \qquad \csc\theta = \frac{r}{y},\quad y \neq 0$$

$$\cos\theta = \frac{x}{r} \qquad \sec\theta = \frac{r}{x},\quad x \neq 0$$

$$\tan\theta = \frac{y}{x},\quad x \neq 0 \qquad \cot\theta = \frac{x}{y},\quad y \neq 0$$

5.2 Verification of Trigonometric Identities

- Trigonometric identities are verified by using algebraic methods and previously proved identities. Here are a few fundamental trigonometric identities.

$$\sin x = \frac{1}{\csc x} \qquad \cos x = \frac{1}{\sec x} \qquad \tan x = \frac{1}{\cot x}$$

$$\tan x = \frac{\sin x}{\cos x} \qquad \cot x = \frac{\cos x}{\sin x}$$

$\sin^2 x + \cos^2 x = 1$; $\tan^2 x + 1 = \sec^2 x$;

$1 + \cot^2 x = \csc^2 x$

- Sum and difference identities for the cosine function are

$$\cos(\alpha - \beta) = \cos\alpha\cos\beta + \sin\alpha\sin\beta$$

$$\cos(\alpha + \beta) = \cos\alpha\cos\beta - \sin\alpha\sin\beta$$

- Sum and difference identities for the sine function are

$$\sin(\alpha - \beta) = \sin\alpha\cos\beta - \cos\alpha\sin\beta$$

$$\sin(\alpha + \beta) = \sin\alpha\cos\beta + \cos\alpha\sin\beta$$

- Sum and difference identities for the tangent function are

$$\tan(\alpha + \beta) = \frac{\tan\alpha + \tan\beta}{1 - \tan\alpha\tan\beta}$$

$$\tan(\alpha - \beta) = \frac{\tan\alpha - \tan\beta}{1 + \tan\alpha\tan\beta}$$

- The cofunction identities are

$$\sin(90^\circ - \theta) = \cos\theta \qquad \cos(90^\circ - \theta) = \sin\theta$$

$$\tan(90^\circ - \theta) = \cot\theta \qquad \cot(90^\circ - \theta) = \tan\theta$$

$$\sec(90^\circ - \theta) = \csc\theta \qquad \csc(90^\circ - \theta) = \sec\theta$$

where θ is in degrees. If θ is in radian measure, replace 90° with $\frac{\pi}{2}$.

5.3 More on Trigonometric Identities

- The double-angle identities are

$$\sin 2\alpha = 2\sin\alpha\cos\alpha$$

$$\cos 2\alpha = \cos^2\alpha - \sin^2\alpha = 1 - 2\sin^2\alpha = 2\cos^2\alpha - 1$$

$$\tan 2\alpha = \frac{2\tan\alpha}{1 - \tan^2\alpha}$$

- The half-angle identities are

$$\sin\frac{\alpha}{2} = \pm\sqrt{\frac{1 - \cos\alpha}{2}}$$

$$\cos\frac{\alpha}{2} = \pm\sqrt{\frac{1 + \cos\alpha}{2}}$$

$$\tan\frac{\alpha}{2} = \frac{\sin\alpha}{1 + \cos\alpha} = \frac{1 - \cos\alpha}{\sin\alpha}$$

- The product-to-sum identities are

$$\sin\alpha\cos\beta = \frac{1}{2}[\sin(\alpha + \beta) + \sin(\alpha - \beta)]$$

$$\cos\alpha\sin\beta = \frac{1}{2}[\sin(\alpha + \beta) - \sin(\alpha - \beta)]$$

$$\cos\alpha\cos\beta = \frac{1}{2}[\cos(\alpha + \beta) + \cos(\alpha - \beta)]$$

$$\sin\alpha\sin\beta = \frac{1}{2}[\cos(\alpha - \beta) - \cos(\alpha + \beta)]$$

- The sum-to-product identities are

$$\sin x + \sin y = 2\sin\frac{x + y}{2}\cos\frac{x - y}{2}$$

$$\cos x - \cos y = -2\sin\frac{x + y}{2}\sin\frac{x - y}{2}$$

$$\sin x - \sin y = 2\cos\frac{x + y}{2}\sin\frac{x - y}{2}$$

$$\cos x + \cos y = 2 \cos \frac{x + y}{2} \cos \frac{x - y}{2}$$

- For sums of the form $a \sin x + b \cos x$,

$$a \sin x + b \cos x = k \sin(x + \alpha)$$

where $k = \sqrt{a^2 + b^2}$, $\sin \alpha = \dfrac{b}{\sqrt{a^2 + b^2}}$, and $\cos \alpha = \dfrac{a}{\sqrt{a^2 + b^2}}$.

5.4 Inverse Trigonometric Functions

- The inverse of $y = \sin x$ is $y = \sin^{-1} x$, with $-1 \le x \le 1$ and $-\dfrac{\pi}{2} \le y \le \dfrac{\pi}{2}$.
- The inverse of $y = \cos x$ is $y = \cos^{-1} x$, with $-1 \le x \le 1$ and $0 \le y \le \pi$.
- The inverse of $y = \tan x$ is $y = \tan^{-1} x$, with $-\infty < x < \infty$ and $-\dfrac{\pi}{2} < y < \dfrac{\pi}{2}$.
- The inverse of $y = \cot x$ is $y = \cot^{-1} x$, with $-\infty < x < \infty$ and $0 < y < \pi$.
- The inverse of $y = \csc x$ is $y = \csc^{-1} x$, with $x \le -1$ or $x \ge 1$ and $-\dfrac{\pi}{2} \le y \le \dfrac{\pi}{2}$, $y \ne 0$.
- The inverse of $y = \sec x$ is $y = \sec^{-1} x$, with $x \le -1$ or $x \ge 1$ and $0 \le y \le \pi$, $y \ne \dfrac{\pi}{2}$.

5.5 Trigonometric Equations

- Algebraic methods and identities are used to solve trigonometric equations. Because the trigonometric functions are periodic, there may be an infinite number of solutions. If solutions cannot be found by algebraic methods, then we often use a graphing utility to find approximate solutions.

5.6 The Law of Sines and the Law of Cosines

- The Law of Sines is used to solve triangles when two angles and a side are given (AAS or ASA) or when two sides and an angle opposite one of them are given (SSA).

$$\frac{a}{\sin A} = \frac{b}{\sin B} = \frac{c}{\sin C}$$

- The Law of Cosines, $a^2 = b^2 + c^2 - 2bc \cos A$, is used to solve general triangles when two sides and the included angle (SAS) or three sides (SSS) of the triangle are given.

5.7 Vectors

- A vector is a quantity with magnitude and direction. Two vectors are equivalent if they have the same magnitude and the same direction. The resultant of two or more vectors is the sum of the vectors.
- Vectors can be added by the parallelogram method, the triangle method, or addition of the x- and y-components.
- If $\mathbf{v} = \langle a, b \rangle$ and k is a real number, then $k\mathbf{v} = \langle ka, kb \rangle$.
- The dot product of $\mathbf{v} = \langle a, b \rangle$ and $\mathbf{w} = \langle c, d \rangle$ is given by

$$\mathbf{v} \cdot \mathbf{w} = ac + bd$$

- If $\mathbf{v}$ and $\mathbf{w}$ are two nonzero vectors and α is the smallest positive angle between $\mathbf{v}$ and $\mathbf{w}$, then $\cos \alpha = \dfrac{\mathbf{v} \cdot \mathbf{w}}{\|\mathbf{v}\| \|\mathbf{w}\|}$.

CHAPTER 5 TRUE/FALSE EXERCISES

In Exercises 1 to 12, answer true or false. If the statement is false, give a reason or an example to show that the statement is false.

1. The angle θ measured in degrees is in standard position with the terminal side in the second quadrant. The reference angle of θ is $180° - \theta$.

2. $\dfrac{\sin x}{\cos y} = \tan \dfrac{x}{y}$

3. $\sin^{-1} x = \csc x^{-1}$

4. $\sin 2\alpha = 2 \sin \alpha$ for all α

5. $\sin(\alpha + \beta) = \sin \alpha + \sin \beta$

6. If $\tan \alpha = \tan \beta$, then $\alpha = \beta$.

7. $\cos^{-1}(\cos x) = x$

8. If $-1 \le x \le 1$, then $\cos(\cos^{-1} x) = x$.

9. The Law of Cosines can be used to solve any triangle given two sides and an angle.

10. The Law of Sines can be used to solve any triangle given two angles and any side.

11. If two vectors have the same magnitude, then they are equal.

12. It is possible for the sum of two nonzero vectors to equal zero.

CHAPTER 5 REVIEW EXERCISES

1. Find the values of the six trigonometric functions of an angle in standard position with the point $P(1, -3)$ on the terminal side of the angle.

2. Find the exact value of

a. $\sec 150°$ **b.** $\tan\left(-\frac{3\pi}{4}\right)$

c. $\cot(-225°)$ **d.** $\cos\left(\frac{2\pi}{3}\right)$

3. Find the value of each of the following to the nearest ten-thousandth.

a. $\cos 123°$ **b.** $\cot 4.22$

c. $\sec 612°$ **d.** $\tan \frac{2\pi}{5}$

4. A car climbs a hill that has a constant angle of 4.5° for a distance of 1.14 miles. What is the car's increase in altitude?

5. A tree casts a shadow of 8.55 feet when the angle of elevation of the sun is 55.3°. Find the height of the tree.

6. Find the height of a building if the angle of elevation to the top of the building changes from 18° to 37° as an observer moves a distance of 80 feet toward the building.

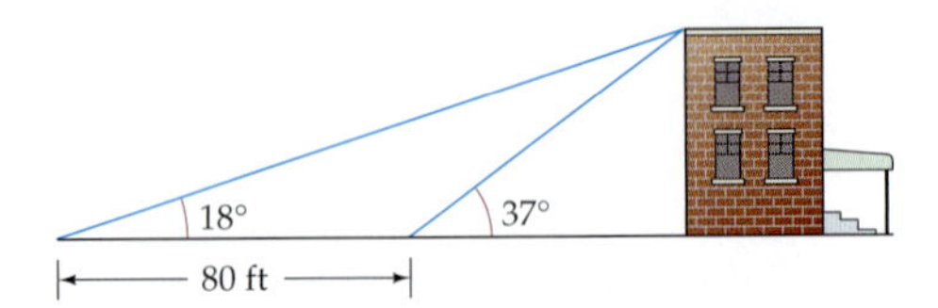

In Exercises 7 to 14, find the exact value.

7. $\cos(45° + 30°)$ **8.** $\tan(210° - 45°)$

9. $\sin\left(\frac{2\pi}{3} + \frac{\pi}{4}\right)$ **10.** $\sec\left(\frac{4\pi}{3} - \frac{\pi}{4}\right)$

11. $\sin\left(22\frac{1}{2}\right)^{\circ}$ **12.** $\cos 105°$

13. $\tan\left(67\frac{1}{2}\right)^{\circ}$ **14.** $\sin 112.5°$

In Exercises 15 to 18, find the exact values of the given functions.

15. Given $\sin\alpha = \frac{1}{2}$, α in Quadrant I, and $\cos\beta = \frac{1}{2}$, β in Quadrant IV, find $\cos(\alpha - \beta)$.

16. Given $\sin\alpha = \frac{\sqrt{3}}{2}$, α in Quadrant II, and $\cos\beta = -\frac{1}{2}$, β in Quadrant III, find $\sin(\alpha + \beta)$.

17. Given $\sin\alpha = -\frac{1}{2}$, α in Quadrant IV, and $\cos\beta = -\frac{\sqrt{3}}{2}$, β in Quadrant III, find $\tan 2\alpha$.

18. Given $\sin\alpha = \frac{\sqrt{2}}{2}$, α in Quadrant I, and $\cos\beta = \frac{\sqrt{3}}{2}$, β in Quadrant IV, find $\sin 2\alpha$.

In Exercises 19 to 22, write the given expression as a single trigonometric function.

19. $2 \sin 3x \cos 3x$ **20.** $\frac{\tan 2x + \tan x}{1 - \tan 2x \tan x}$

21. $\sin 4x \cos x - \cos 4x \sin x$

22. $\cos^2 2\theta - \sin^2 2\theta$

In Exercises 23 to 26, write each expression as the product of two functions.

23. $\cos 2\theta - \cos 4\theta$ **24.** $\sin 3\theta - \sin 5\theta$

25. $\sin 6\theta + \sin 2\theta$ **26.** $\sin 5\theta - \sin\theta$

In Exercises 27 to 38, verify the identity.

27. $\frac{1}{\sin x - 1} + \frac{1}{\sin x + 1} = -2\tan x \sec x$

28. $\dfrac{\sin x}{1 - \cos x} = \csc x + \cot x, \quad 0 < x < \dfrac{\pi}{2}$

29. $\dfrac{1 + \sin x}{\cos^2 x} = \tan^2 x + 1 + \tan x \sec x$

30. $\cos^2 x - \sin^2 x - \sin 2x = \dfrac{\cos^2 2x - \sin^2 2x}{\cos 2x + \sin 2x}$

31. $\dfrac{1}{\cos x} - \cos x = \tan x \sin x$

32. $\sin(270° - \theta) - \cos(270° - \theta) = \sin \theta - \cos \theta$

33. $\dfrac{\sin 4x - \sin 2x}{\cos 4x - \cos 2x} = -\cot 3x$

34. $2 \sin x \sin 3x = (1 - \cos 2x)(1 + 2 \cos 2x)$

35. $\sin x - \cos 2x = (2 \sin x - 1)(\sin x + 1)$

36. $\dfrac{\sin 2x - \sin x}{\cos 2x + \cos x} = \dfrac{1 - \cos x}{\sin x}$

37. $2 \cos 4x \sin 2x = 2 \sin 3x \cos 3x - 2 \sin x \cos x$

38. $\cos(x + y) \sin(x - y) = \sin x \cos x - \sin y \cos y$

In Exercises 39 to 42, evaluate each expression.

39. $\sec\left(\sin^{-1} \dfrac{12}{13}\right)$

40. $\cos\left(\sin^{-1} \dfrac{3}{5}\right)$

41. $\cos\left[\sin^{-1}\left(-\dfrac{3}{5}\right) + \cos^{-1} \dfrac{5}{13}\right]$

42. $\cos\left(2 \sin^{-1} \dfrac{3}{5}\right)$

In Exercises 43 and 44, solve each equation.

43. $2 \sin^{-1}(x - 1) = \dfrac{\pi}{3}$

44. $\sin^{-1} x + \cos^{-1} \dfrac{4}{5} = \dfrac{\pi}{2}$

In Exercises 45 and 46, solve each equation on $0° \le x < 360°$.

45. $4 \sin^2 x + 2\sqrt{3} \sin x - 2 \sin x - \sqrt{3} = 0$

46. $2 \sin x \cos x - \sqrt{2} \cos x - 2 \sin x + \sqrt{2} = 0$

In Exercises 47 and 48, solve the trigonometric equation where x is in radians. Round approximate solutions to four decimal places.

47. $3 \cos^2 x + \sin x = 1$

48. $\tan^2 x - 2 \tan x - 3 = 0$

In Exercises 49 and 50, solve each equation on $0 \le x < 2\pi$.

49. $\sin 3x \cos x - \cos 3x \sin x = \dfrac{1}{2}$

50. $\cos\left(2x - \dfrac{\pi}{3}\right) = -\dfrac{\sqrt{3}}{2}$

In Exercises 51 and 52, write the equation in the form $y = k \sin(x + \alpha)$, where the measure of α is in radians. Graph one period of each function.

51. $f(x) = \sqrt{3} \sin x + \cos x$

52. $f(x) = -2 \sin x - 2 \cos x$

In Exercises 53 and 54, graph each function.

53. $f(x) = 2 \cos^{-1} x$

54. $f(x) = \sin^{-1}(x - 1)$

In Exercises 55 to 60, solve each triangle.

55. $A = 37°, b = 14, C = 92°$

56. $B = 77.4°, c = 11.8, C = 94.0°$

57. $a = 12, b = 15, c = 20$

58. $a = 24, b = 32, c = 28$

59. $A = 55°, B = 80°, c = 25$

60. $b = 102, c = 150, A = 82°$

In Exercises 61 and 62, find the components of each vector with the given initial and terminal points. Write an equivalent vector in terms of its components.

61. $P_1(-2, 4); P_2(3, 7)$

62. $P_1(-4, 0); P_2(-3, 6)$

In Exercises 63 to 66, find the magnitude and the direction angle of each vector.

63. $\mathbf{v} = \langle -4, 2 \rangle$

64. $\mathbf{v} = \langle 6, -3 \rangle$

65. $\mathbf{u} = -2\mathbf{i} + 3\mathbf{j}$

66. $\mathbf{u} = -4\mathbf{i} - 7\mathbf{j}$

In Exercises 67 to 70, find a unit vector in the direction of the given vector.

67. $\mathbf{w} = \langle -8, 5 \rangle$

68. $\mathbf{w} = \langle 7, -12 \rangle$

69. $\mathbf{v} = 5\mathbf{i} + \mathbf{j}$

70. $\mathbf{v} = 3\mathbf{i} - 5\mathbf{j}$

In Exercises 71 and 72, perform the indicated operation where u = ⟨3, 2⟩ and v = ⟨−4, −1⟩.

71. $\mathbf{v} - \mathbf{u}$

72. $2\mathbf{u} - 3\mathbf{v}$

In Exercises 73 and 74, perform the indicated operation where u = 10i + 6j and v = 8i − 5j.

73. $-\mathbf{u} + \frac{1}{2}\mathbf{v}$

74. $\frac{2}{3}\mathbf{v} - \frac{3}{4}\mathbf{u}$

75. Ground Speed of a Plane A plane is flying at an airspeed of 400 mph at a heading of 204°. A wind of 45 mph is blowing from the east. Find the ground speed of the plane.

76. Angle of a Ramp A 40-pound force keeps a 320-pound object from sliding down an inclined ramp. Find the angle of the ramp.

In Exercises 77 to 80, find the dot product of the vectors.

77. $\mathbf{u} = \langle 3, 7 \rangle$; $\mathbf{v} = \langle -1, 3 \rangle$

78. $\mathbf{v} = \langle -8, 5 \rangle$; $\mathbf{u} = \langle 2, -1 \rangle$

79. $\mathbf{v} = -4\mathbf{i} - \mathbf{j}$; $\mathbf{u} = 2\mathbf{i} + \mathbf{j}$

80. $\mathbf{u} = -3\mathbf{i} + 7\mathbf{j}$; $\mathbf{v} = -2\mathbf{i} + 2\mathbf{j}$

In Exercises 81 to 84, find the smallest positive angle between the vectors.

81. $\mathbf{u} = \langle 7, -4 \rangle$; $\mathbf{v} = \langle 2, 3 \rangle$

82. $\mathbf{v} = \langle -5, 2 \rangle$; $\mathbf{u} = \langle 2, -4 \rangle$

83. $\mathbf{v} = 6\mathbf{i} - 11\mathbf{j}$; $\mathbf{u} = 2\mathbf{i} + 4\mathbf{j}$

84. $\mathbf{u} = \mathbf{i} - 5\mathbf{j}$; $\mathbf{v} = \mathbf{i} + 5\mathbf{j}$

In Exercises 85 and 86, find $\text{proj}_{\mathbf{w}}\mathbf{v}$.

85. $\mathbf{v} = \langle -2, 5 \rangle$; $\mathbf{w} = \langle 5, 4 \rangle$

86. $\mathbf{v} = 4\mathbf{i} - 7\mathbf{j}$; $\mathbf{w} = -2\mathbf{i} - 5\mathbf{j}$

87. Work A 120-pound box is dragged 14 feet along a level floor. Find the work done if a force of 60 pounds at an angle of 38° is used.

CHAPTER 5 TEST

1. Find the exact value of $\tan\frac{\pi}{6}\cos\frac{\pi}{3} - \sin\frac{\pi}{2}$.

2. The angle of elevation from point A to the top of a tree is 42.2°. At point B, 5.24 meters from A and on a line through the base of the tree and A, the angle of elevation is 37.4°. Find the height of the tree.

3. Verify the identity $1 + \sin^2 x \sec^2 x = \sec^2 x$.

4. Verify the identity

$$\frac{1}{\sec x - \tan x} - \frac{1}{\sec x + \tan x} = 2\tan x$$

5. Verify the identity $\csc x - \cot x = \frac{1 - \cos x}{\sin x}$.

6. Find the exact value of sin 195°.

7. Given $\sin\alpha = -\frac{3}{5}$, α in Quadrant III, and $\cos\beta = -\frac{\sqrt{2}}{2}$, β in Quadrant II, find $\sin(\alpha + \beta)$.

8. Verify the identity $\tan\frac{\theta}{2} + \frac{\cos\theta}{\sin\theta} = \csc\theta$.

9. Find the exact value of sin 15° cos 75°.

10. Write $y = -\frac{\sqrt{3}}{2}\sin x + \frac{1}{2}\cos x$ in the form $y = k\sin(x + \alpha)$, where α is measured in radians.

11. Find the exact value of $\sin\left(\cos^{-1}\frac{12}{13}\right)$.

12. Graph: $y = \sin^{-1}(x + 2)$

13. Solve $3 \sin x - 2 = 0$, where $0° \le x < 360°$. (State solutions to the nearest 0.1°.)

14. Find the exact solutions of $\sin 2x + \sin x - 2 \cos x - 1 = 0$, where $0 \le x < 2\pi$.

15. Solve triangle ABC if $A = 70°$, $C = 16°$, and $c = 14$.

16. **Distance Between Ships** One ship leaves a port at 1:00 P.M. traveling at 12 mph at a heading of 65°. At 2:00 P.M. another ship leaves the port traveling at 18 mph at a heading of 142°. Find the distance between the ships at 3:00 P.M.

17. A vector has a magnitude of 12 and direction 220°. Write an equivalent vector in the form $\mathbf{v} = a_1\mathbf{i} + a_2\mathbf{j}$.

18. Find $3\mathbf{u} - 5\mathbf{v}$ given the vectors $\mathbf{u} = 2\mathbf{i} - 3\mathbf{j}$ and $\mathbf{v} = 5\mathbf{i} + 4\mathbf{j}$.

19. Find the dot product of $\mathbf{u} = -2\mathbf{i} + 3\mathbf{j}$ and $\mathbf{v} = 5\mathbf{i} + 3\mathbf{j}$.

20. Find the smallest positive angle, to the nearest degree, between the vectors $\mathbf{u} = \langle 3, 5 \rangle$ and $\mathbf{v} = \langle -6, 2 \rangle$.

CUMULATIVE REVIEW EXERCISES

1. Explain how to use the graph of $y = f(x)$ to produce the graph of $y = f(x + 1) + 2$.

2. Explain how to use the graph of $y = f(x)$ to produce the graph of $y = -f(x)$.

3. Find the vertical asymptote for the graph of $f(x) = \dfrac{x + 3}{x - 2}$.

4. Determine whether $f(x) = x - \sin x$ is an even function or an odd function.

5. Find the inverse of $f(x) = \dfrac{5x}{x - 1}$.

6. Write $x = 2^5$ in logarithmic form.

7. Evaluate: $\log_{10} 1000$

8. Convert 240° to radians.

9. Convert $\dfrac{5\pi}{3}$ to degrees.

10. What is the measure of the reference angle for the angle $\theta = 310°$?

11. Find the x- and y-coordinates of the point defined by $W\left(\dfrac{\pi}{3}\right)$.

12. Evaluate: $\sin^{-1} \dfrac{1}{2}$

13. Use interval notation to state the domain of $f(x) = \cos^{-1} x$.

14. Use interval notation to state the range of $f(x) = \tan^{-1} x$.

15. Evaluate $\tan\left(\sin^{-1}\left(\dfrac{12}{13}\right)\right)$.

16. Solve: $2 \cos^2 x + \sin x - 1 = 0$, for $0 \le x < 2\pi$

17. Find the magnitude and direction angle for the vector $\langle -3, 4 \rangle$. Round the angle to the nearest tenth of a degree.

18. Find the smallest positive angle between the vectors $\mathbf{v} = \langle 2, -3 \rangle$ and $\mathbf{w} = \langle -3, 4 \rangle$. Round to the nearest tenth of a degree.

19. An airplane is traveling with an airspeed of 415 mph at a heading of 48.0°. A wind of 55 mph is blowing at a heading of 115.0°. Find the ground speed and the course of the plane.

20. For triangle ABC, $B = 32°$, $a = 42$ feet, and $b = 51$ feet. Find, to the nearest degree, the measure of angle A.

CHAPTER 6

ADDITIONAL TOPICS IN MATHEMATICS

Fractals

One of the topics of this chapter is the concept of a *sequence.* The list of numbers $1, \frac{1}{2}, \frac{1}{3}, \frac{1}{4}, \frac{1}{5}, \ldots, \frac{1}{n}, \ldots$ is an example of a sequence. As the natural number n becomes very large, the numbers in the sequence become closer and closer to 0.

In addition to a sequence of numbers, we could have, for instance, a sequence of functions or a sequence of geometric figures. Although each of these sequences has important applications, we will focus here on a sequence of geometric figures.

Look at the sequence of figures below. Each succeeding figure is created by constructing right triangles using the line segments in the preceding figure as the hypotenuses.

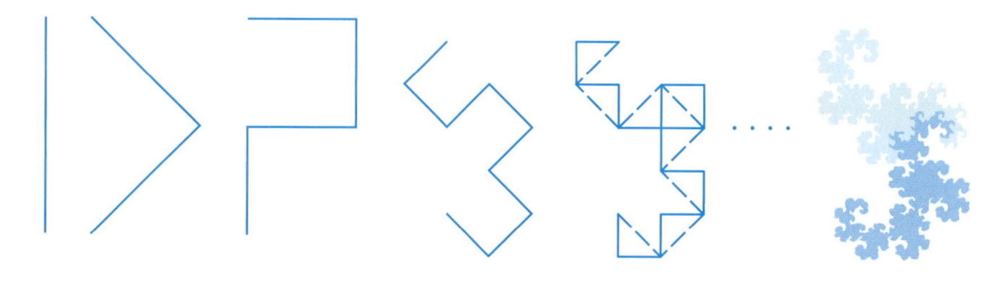

If this process is repeated over and over, the sequence of figures becomes like the last figure on the right, which is called a *fractal.* The computer-generated fractals at the left were created using a different procedure from that shown above. We will look at fractals again in **Project 3 of Section 6.4, page 525.**

Proof by Contradiction

In a detective television show, a suspect might be asked, "Where were you Friday night?", to which the suspect replies, "I was in New York." However, several eyewitnesses agree that they saw the suspect in California on that Friday night. Their testimony *contradicts* what the suspect claimed to be a factual statement.

A similar strategy is used by mathematicians to prove some theorems. The mathematician assumes that the theorem is false and then shows that the assumption contradicts a statement that is known to be true. This method of proof is called a *proof by contradiction.* To illustrate this method, consider the following theorem.

The $\sqrt{3}$ is an irrational number.

To prove this theorem, we begin by assuming that $\sqrt{3}$ is not an irrational number—that is, it is a rational number. This is the opposite of what we want to prove. If $\sqrt{3}$ is a rational number, then $\sqrt{3}$ can be represented as the ratio of two integers. That is, $\sqrt{3} = \frac{a}{b}$, where a and b are integers with no common factors and $b \neq 0$. From this assumption, we have

$$\sqrt{3} = \frac{a}{b}$$

$$3 = \frac{a^2}{b^2} \qquad \text{• Square both sides of the equation.}$$

$$3b^2 = a^2 \qquad \text{• Multiply each side by } b^2.$$

The last equation implies that a^2 is divisible by 3. Because 3 is prime, a is divisible by 3. Thus $a = 3k$ for some integer k and $a^2 = (3k)^2 = 9k^2$.

Replacing a^2 by $9k^2$, we have

$$3b^2 = 9k^2$$

$$b^2 = 3k^2 \qquad \text{• Divide each side by 3.}$$

The equation $b^2 = 3k^2$ implies that b is divisible by 3. Thus we have shown that both a and b are divisible by 3. This, however, contradicts our statement that a and b have no common factors. This contradiction means that our assumption that $\sqrt{3}$ can be represented as the quotient of integers is not possible, and therefore $\sqrt{3}$ must be an irrational number.

SECTION 6.1

CONIC SECTIONS

- PARABOLAS
- ELLIPSES
- HYPERBOLAS
- APPLICATIONS

The graph of a parabola, a circle, an ellipse, or a hyperbola can be formed by the intersection of a plane and a cone. Hence these figures are referred to as conic sections. See **Figure 6.1.**

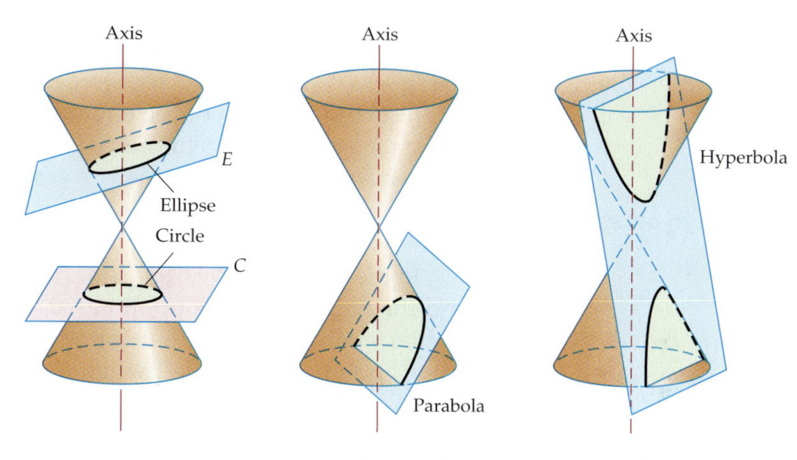

FIGURE 6.1
Cones intersected by planes

MATH MATTERS

Appollonius (262–200 B.C.) wrote an eight-volume treatise entitled *On Conic Sections* in which he derived the formulas for all the conic sections. He was the first to use the words *parabola, ellipse,* and *hyperbola.*

take note

If the intersection of a plane and a cone is a point, a line, or two intersecting lines, then the intersection is called a *degenerate conic section.*

A plane perpendicular to the axis of the cone intersects the cone in a circle (plane C). The plane E, tilted so that it is not perpendicular to the axis, intersects the cone in an ellipse. When the plane is parallel to a line on the surface of the cone, the plane intersects the cone in a parabola. When the plane intersects both portions of the cone, a hyperbola is formed.

PARABOLAS

Besides the geometric description of a conic section just given, a conic section can be defined as a set of points. This method uses some specified conditions about the curve to determine which points in a coordinate system are points of the graph. For example, a parabola can be defined by the following set of points.

Definition of a Parabola

A **parabola** is the set of points in the plane that are equidistant from a fixed line (the **directrix**) and a fixed point (the **focus**) not on the directrix.

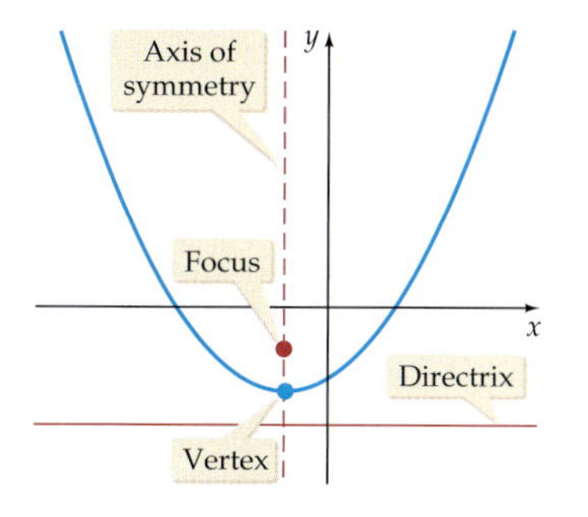

FIGURE 6.2

The line that passes through the focus and is perpendicular to the directrix is called the **axis of symmetry** of the parabola. The midpoint of the line segment between the focus and directrix on the axis of symmetry is the **vertex** of the parabola, as shown in **Figure 6.2.**

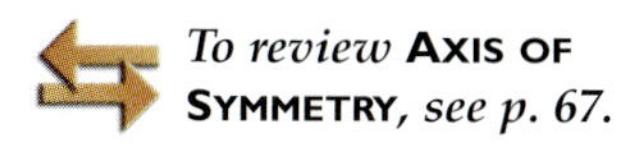
To review **AXIS OF SYMMETRY**, *see p. 67.*

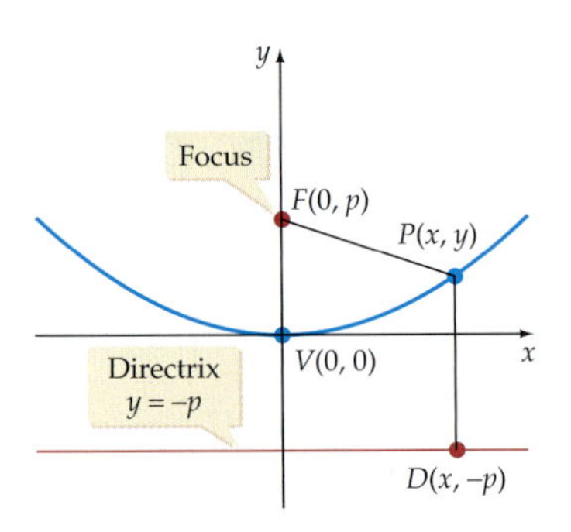

FIGURE 6.3

take note

The tests for y-axis and x-axis symmetry can be used to verify these statements and provide connections to earlier topics on symmetry.

Standard Forms of the Equation of a Parabola with Vertex at the Origin

Axis of Symmetry Is the y-Axis

The standard form of the equation of a parabola with vertex (0, 0) and the y-axis as its axis of symmetry is $x^2 = 4py$. The focus is $(0, p)$, and the equation of the directrix is $y = -p$. See **Figure 6.3.**

Axis of Symmetry Is the x-Axis

The standard form of the equation of a parabola with vertex (0, 0) and the x-axis as its axis of symmetry is $y^2 = 4px$. The focus is $(p, 0)$ and the equation of the directrix is $x = -p$.

In the equation $x^2 = 4py$, $x^2 \geq 0$. Therefore, $4py \geq 0$. Thus if $p > 0$, then $y \geq 0$, and the parabola opens up. If $p < 0$, then $y \leq 0$, and the parabola opens down. A similar analysis shows that for $y^2 = 4px$, the parabola opens to the right when $p > 0$ and opens to the left when $p < 0$.

? QUESTION Does the graph of $y^2 = -4x$ open up, down, to the left, or to the right?

EXAMPLE 1 Find the Focus and Directrix of a Parabola

Find the focus and directrix of the parabola given by the equation $y = -\frac{1}{2}x^2$.

Solution

Because the x term is squared, the standard form of the equation is $x^2 = 4py$.

$$y = -\frac{1}{2}x^2$$

$$x^2 = -2y \qquad \text{• Write the given equation in standard form.}$$

Comparing this equation with $x^2 = 4py$ gives

$$4p = -2$$

$$p = -\frac{1}{2}$$

Because p is negative, the parabola opens down, and the focus is below the vertex (0, 0), as shown in **Figure 6.4.** The coordinates of the focus are $\left(0, -\frac{1}{2}\right)$. The equation of the directrix is $y = \frac{1}{2}$.

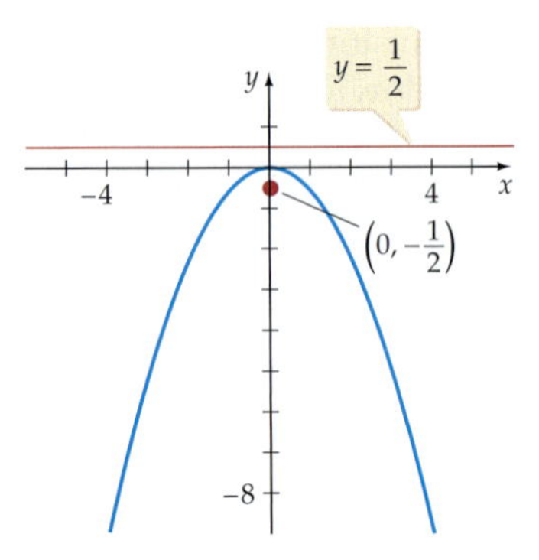

FIGURE 6.4

TRY EXERCISE 2, PAGE 480

? ANSWER To the left.

Parabolas with Vertex at (h, k)

The equation of a parabola with a vertical or horizontal axis of symmetry and with the vertex at a point (h, k) can be found by using the translations discussed previously.

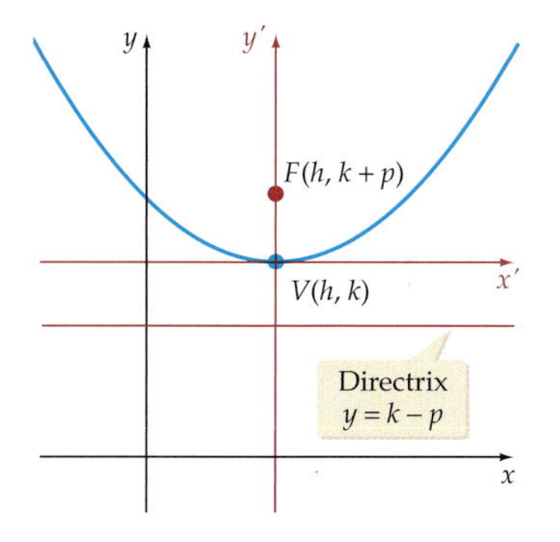

Figure 6.5

Standard Forms of the Equation of a Parabola with Vertex at (h, k)

Vertical Axis of Symmetry

The standard form of the equation of the parabola with vertex $V(h, k)$ and a vertical axis of symmetry is

$$(x - h)^2 = 4p(y - k)$$

The focus is $(h, k + p)$, and the equation of the directrix is $y = k - p$. See **Figure 6.5.**

Horizontal Axis of Symmetry

The standard form of the equation of the parabola with vertex (h, k) and a horizontal axis of symmetry is

$$(y - k)^2 = 4p(x - h)$$

The focus is $(h + p, k)$, and the equation of the directrix is $x = h - p$.

EXAMPLE 2 Find the Focus and Directrix of a Parabola

Find the equation of the directrix and the coordinates of the vertex and focus of the parabola given by the equation $3x + 2y^2 + 8y - 4 = 0$.

Solution

Rewrite the equation so that the y terms are on one side of the equation, and then complete the square on y.

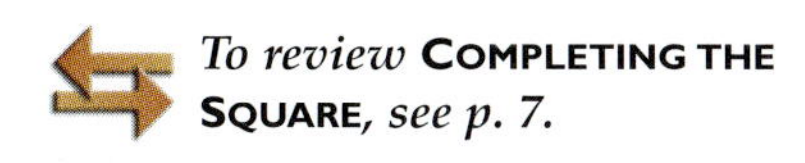

To review **Completing the Square**, *see p. 7.*

$$3x + 2y^2 + 8y - 4 = 0$$

$$2y^2 + 8y = -3x + 4$$

$$2(y^2 + 4y) = -3x + 4$$

$$2(y^2 + 4y + 4) = -3x + 4 + 8$$

• **Complete the square. Note that $2 \cdot 4 = 8$ is added to each side.**

$$2(y + 2)^2 = -3(x - 4)$$

• **Simplify and then factor.**

$$(y + 2)^2 = -\frac{3}{2}(x - 4)$$

• **Write the equation in standard form.**

Continued ▶

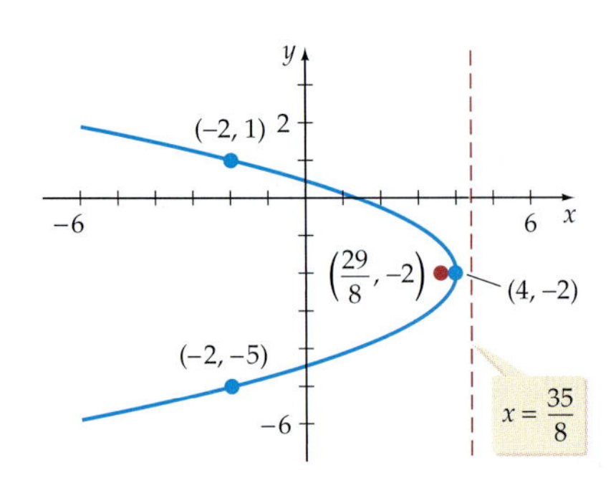

FIGURE 6.6

Comparing this equation to $(y - k)^2 = 4p(x - h)$, we have a parabola that opens to the left with vertex $(4, -2)$ and $4p = -\frac{3}{2}$. Thus $p = -\frac{3}{8}$.

The coordinates of the focus are

$$\left(4 + \left(-\frac{3}{8}\right), -2\right) = \left(\frac{29}{8}, -2\right)$$

The equation of the directrix is

$$x = 4 - \left(-\frac{3}{8}\right) = \frac{35}{8}$$

Choosing some values for y and finding the corresponding values for x, we plot a few points. Because the line $y = -2$ is the axis of symmetry, for each point on one side of the axis of symmetry there is a corresponding point on the other side. Two points are $(-2, 1)$ and $(-2, -5)$. See **Figure 6.6.**

TRY EXERCISE 18, PAGE 480

EXAMPLE 3 Find the Equation in Standard Form of a Parabola

Find the equation in standard form of the parabola with directrix $x = -1$ and focus $(3, 2)$.

Solution

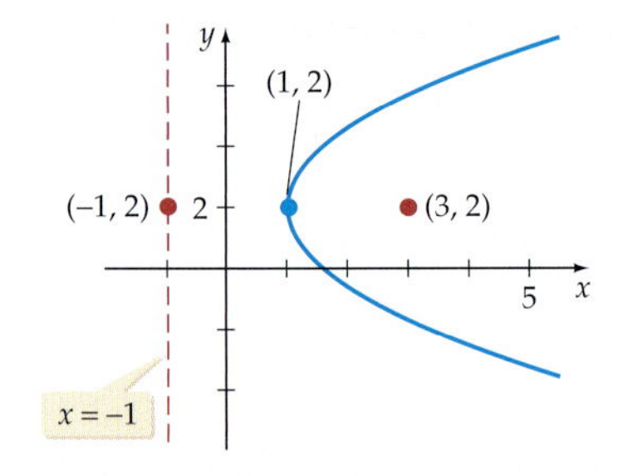

FIGURE 6.7

The vertex is the midpoint of the line segment joining the focus $(3, 2)$ and the point $(-1, 2)$ on the directrix.

$$(h, k) = \left(\frac{-1 + 3}{2}, \frac{2 + 2}{2}\right) = (1, 2)$$

The standard form of the equation is $(y - k)^2 = 4p(x - h)$. The distance from the vertex to the focus is 2. Thus $4p = 4(2) = 8$, and the equation of the parabola in standard form is $(y - 2)^2 = 8(x - 1)$. See **Figure 6.7.**

TRY EXERCISE 28, PAGE 480

ELLIPSES

An ellipse is another of the conic sections formed when a plane intersects a right circular cone. If β is the angle at which the plane intersects the axis of the cone and α is the angle shown in **Figure 6.8,** an ellipse is formed when $\alpha < \beta < 90°$. If $\beta = 90°$, then a circle is formed.

take note

If the plane intersects the cone at the vertex of the cone so that the resulting figure is a point, the point is a *degenerate ellipse*. See the accompanying figure.

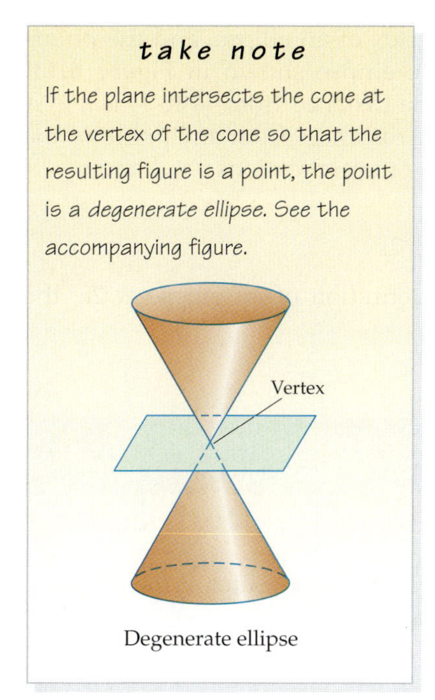

Degenerate ellipse

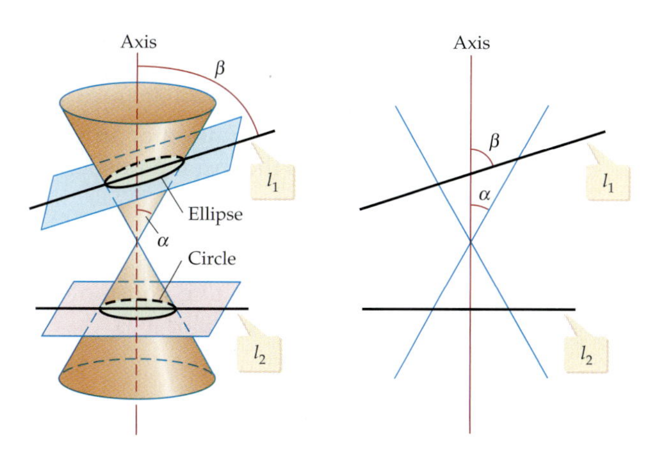

FIGURE 6.8

As was the case for a parabola, there is a definition for an ellipse in terms of a certain set of points in the plane.

Definition of an Ellipse

An **ellipse** is the set of all points in the plane, the sum of whose distances from two fixed points (**foci**) is a positive constant.

We can use this definition to draw an ellipse, equipped only with a piece of string and two tacks (see **Figure 6.9**). Tack the ends of the string to the foci, and trace a curve with a pencil held tight against the string. The resulting curve is an ellipse. The positive constant mentioned in the definition of an ellipse is the length of the string.

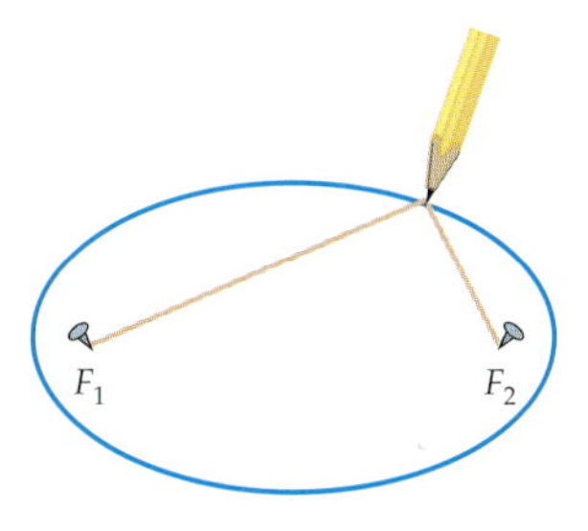

FIGURE 6.9

The graph of an ellipse has two axes of symmetry (see **Figure 6.10**). The longer axis is called the **major axis.** The foci of the ellipse are on the major axis. The shorter axis is called the **minor axis.** It is customary to denote the length of the major axis as $2a$ and the length of the minor axis as $2b$. The **semiaxes** are one-half the axes in length. Thus the length of the semimajor axis is denoted by a and the length of the semiminor axis by b. The **center** of the ellipse is the midpoint of the major axis. The endpoints of the major axis are the **vertices** (plural of *vertex*) of the ellipse.

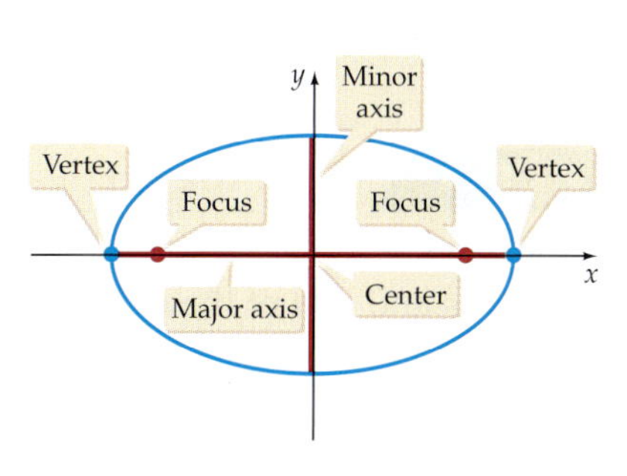

FIGURE 6.10

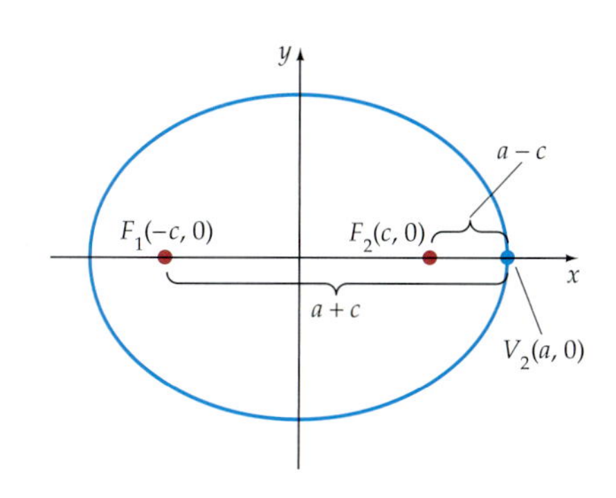

FIGURE 6.11

Consider the point $V_2(a, 0)$, which is one vertex of an ellipse, and the points $F_2(c, 0)$ and $F_1(-c, 0)$, which are the foci of the ellipse shown in **Figure 6.11.** The distance from V_2 to F_1 is $a + c$. Similarly, the distance from V_2 to F_2 is $a - c$. From the definition of an ellipse, the sum of the distances from any point on the ellipse to the foci is a positive constant. By adding the expressions $a + c$ and $a - c$, we have

$$(a + c) + (a - c) = 2a$$

Thus the positive constant referred to in the definition of an ellipse is $2a$, the length of the major axis.

Standard Forms of the Equation of an Ellipse with Center at the Origin

Major Axis on the x-Axis

The standard form of the equation of an ellipse with the center at the origin and major axis on the x-axis (see **Figure 6.12a**) is given by

$$\frac{x^2}{a^2} + \frac{y^2}{b^2} = 1, \quad a > b$$

The length of the major axis is $2a$. The length of the minor axis is $2b$. The coordinates of the vertices are $(a, 0)$ and $(-a, 0)$, and the coordinates of the foci are $(c, 0)$ and $(-c, 0)$, where $c^2 = a^2 - b^2$.

Major Axis on the y-Axis

The standard form of the equation of an ellipse with the center at the origin and major axis on the y-axis (see **Figure 6.12b**) is given by

$$\frac{x^2}{b^2} + \frac{y^2}{a^2} = 1, \quad a > b$$

The length of the major axis is $2a$. The length of the minor axis is $2b$. The coordinates of the vertices are $(0, a)$ and $(0, -a)$, and the coordinates of the foci are $(0, c)$ and $(0, -c)$, where $c^2 = a^2 - b^2$.

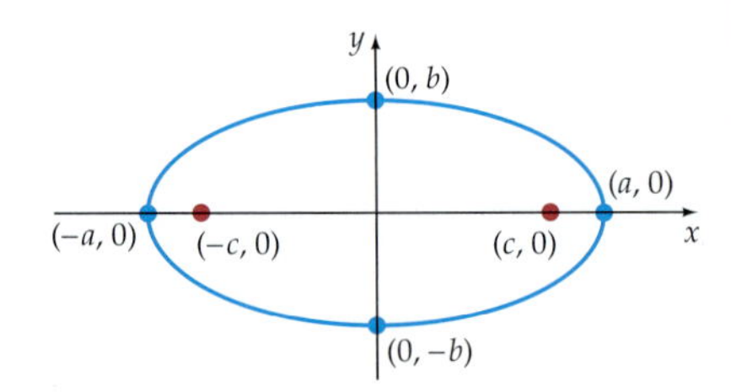

a. Major axis on x-axis

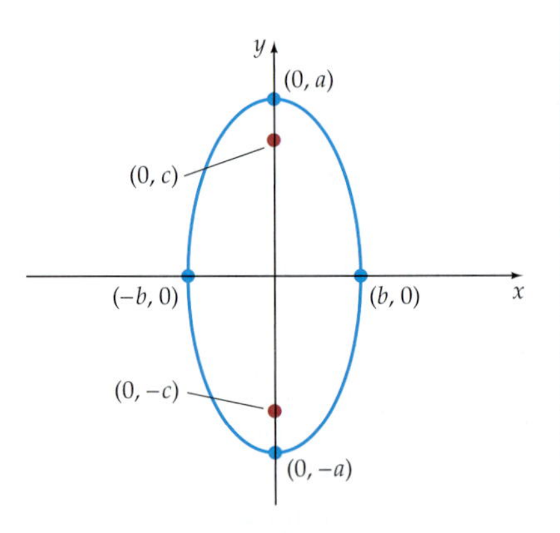

b. Major axis on y-axis

FIGURE 6.12

QUESTION For the graph of $\frac{x^2}{16} + \frac{y^2}{25} = 1$, is the major axis on the x-axis or the y-axis?

EXAMPLE 4 Find the Vertices and Foci of an Ellipse

Find the vertices and foci of the ellipse given by the equation $\frac{x^2}{25} + \frac{y^2}{49} = 1$. Sketch the graph.

ANSWER Because $25 > 16$, the major axis is on the y-axis.

Solution

Because the y^2 term has the larger denominator, the major axis is on the y-axis.

$$a^2 = 49 \qquad b^2 = 25 \qquad c^2 = a^2 - b^2$$
$$a = 7 \qquad b = 5 \qquad = 49 - 25 = 24$$
$$c = \sqrt{24} = 2\sqrt{6}$$

The vertices are $(0, 7)$ and $(0, -7)$. The foci are $\left(0, 2\sqrt{6}\right)$ and $\left(0, -2\sqrt{6}\right)$. See **Figure 6.13.**

▶ **TRY EXERCISE 4, PAGE 480**

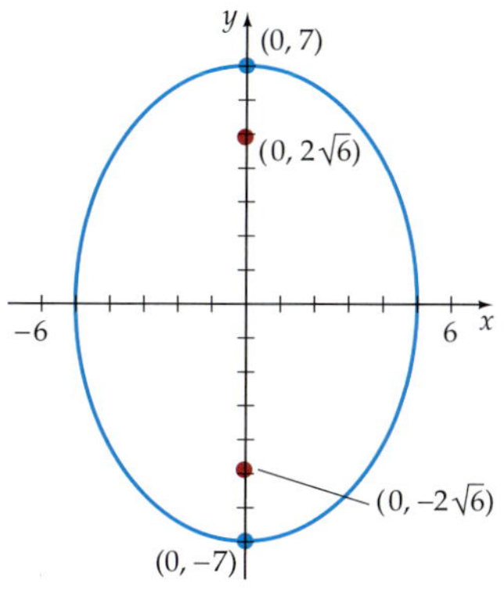

$$\frac{x^2}{25} + \frac{y^2}{49} = 1$$

FIGURE 6.13

An ellipse with foci $(3, 0)$ and $(-3, 0)$ and major axis of length 10 is shown in **Figure 6.14.** To find the equation of the ellipse in standard form, we must find a^2 and b^2. Because the foci are on the major axis, the major axis is on the x-axis. The length of the major axis is $2a$. Thus $2a = 10$. Solving for a, we have $a = 5$ and $a^2 = 25$.

Because the foci are $(3, 0)$ and $(-3, 0)$ and the center of the ellipse is the midpoint between the two foci, the distance from the center of the ellipse to a focus is 3. Therefore, $c = 3$. To find b^2, use the equation

$$c^2 = a^2 - b^2$$
$$9 = 25 - b^2$$
$$b^2 = 16$$

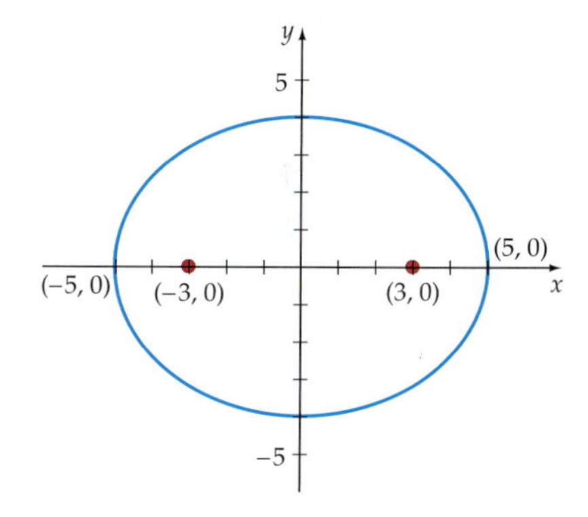

$$\frac{x^2}{25} + \frac{y^2}{16} = 1$$

FIGURE 6.14

The equation of the ellipse in standard form is $\dfrac{x^2}{25} + \dfrac{y^2}{16} = 1$.

The equation of an ellipse with center (h, k) and with horizontal or vertical major axis can be found by using a translation of coordinates.

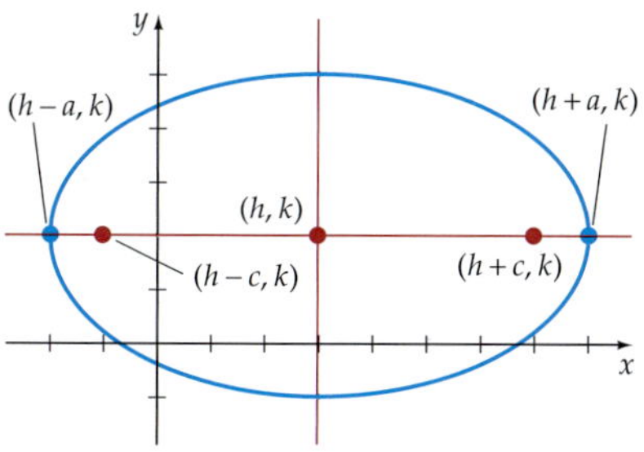

Standard Forms of the Equation of an Ellipse with Center at (h, k)

Major Axis Parallel to the x-Axis

The standard form of the equation of an ellipse with center at (h, k) and major axis parallel to the x-axis (see **Figure 6.15a**) is given by

$$\frac{(x - h)^2}{a^2} + \frac{(y - k)^2}{b^2} = 1, \quad a > b$$

The length of the major axis is $2a$. The length of the minor axis is $2b$. The coordinates of the vertices are $(h + a, k)$ and $(h - a, k)$, and the coordinates of the foci are $(h + c, k)$ and $(h - c, k)$, where $c^2 = a^2 - b^2$.

a. Major axis parallel to x-axis

FIGURE 6.15

Continued ▶

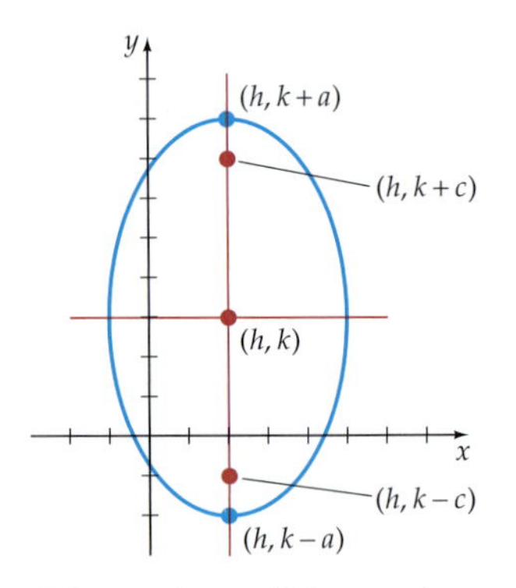

b. Major axis parallel to y-axis

FIGURE 6.15

Major Axis Parallel to the y-Axis

The standard form of the equation of an ellipse with center at (h, k) and major axis parallel to the y-axis (see **Figure 6.15b**) is given by

$$\frac{(x-h)^2}{b^2} + \frac{(y-k)^2}{a^2} = 1, \quad a > b$$

The length of the major axis is $2a$. The length of the minor axis is $2b$. The coordinates of the vertices are $(h, k + a)$ and $(h, k - a)$, and the coordinates of the foci are $(h, k + c)$ and $(h, k - c)$, where $c^2 = a^2 - b^2$.

EXAMPLE 5 Find the Vertices and Foci of an Ellipse

Find the vertices and foci of the ellipse $4x^2 + 9y^2 - 8x + 36y + 4 = 0$. Sketch the graph.

Solution

Write the equation of the ellipse in standard form by completing the square.

$$
\begin{aligned}
4x^2 + 9y^2 - 8x + 36y + 4 &= 0 \\
4x^2 - 8x + 9y^2 + 36y &= -4 && \text{• Rearrange terms.} \\
4(x^2 - 2x) + 9(y^2 + 4y) &= -4 && \text{• Factor.} \\
4(x^2 - 2x + 1) + 9(y^2 + 4y + 4) &= -4 + 4 + 36 && \text{• Complete the square.} \\
4(x-1)^2 + 9(y+2)^2 &= 36 && \text{• Factor.} \\
\frac{(x-1)^2}{9} + \frac{(y+2)^2}{4} &= 1 && \text{• Divide by 36.}
\end{aligned}
$$

From the equation of the ellipse in standard form, the coordinates of the center of the ellipse are $(1, -2)$. Because the larger denominator is 9, the major axis is parallel to the x-axis and $a^2 = 9$. Thus $a = 3$. The vertices are $(4, -2)$ and $(-2, -2)$.

To find the coordinates of the foci, we find c.

$$c^2 = a^2 - b^2 = 9 - 4 = 5$$
$$c = \sqrt{5}$$

The foci are $(1 + \sqrt{5}, -2)$ and $(1 - \sqrt{5}, -2)$. See **Figure 6.16**.

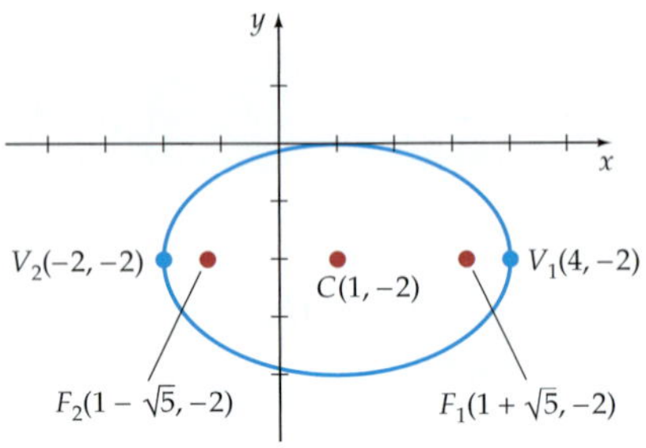

y
x
$V_2(-2, -2)$
$C(1, -2)$
$V_1(4, -2)$
$F_2(1 - \sqrt{5}, -2)$
$F_1(1 + \sqrt{5}, -2)$

$$\frac{(x-1)^2}{9} + \frac{(y+2)^2}{4} = 1$$

FIGURE 6.16

▶ TRY EXERCISE 20, PAGE 480

EXAMPLE 6 Find the Equation of an Ellipse

Find the standard form of the equation of the ellipse with center at $(4, -2)$, foci $F_2(4, 1)$ and $F_1(4, -5)$, and minor axis of length 10, as shown in **Figure 6.17.**

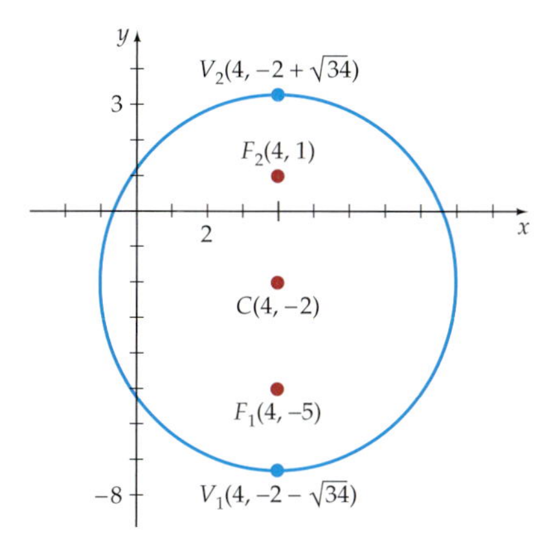

FIGURE 6.17

Solution

Because the foci are on the major axis, the major axis is parallel to the y-axis. The distance from the center of the ellipse to a focus is c. The distance between the center $(4, -2)$ and the focus $(4, 1)$ is 3. Therefore, $c = 3$.

The length of the minor axis is $2b$. Thus $2b = 10$ and $b = 5$.

To find a^2, use the equation $c^2 = a^2 - b^2$.

$$9 = a^2 - 25$$
$$a^2 = 34$$

Thus the equation in standard form is

$$\frac{(x-4)^2}{25} + \frac{(y+2)^2}{34} = 1$$

TRY EXERCISE 30, PAGE 480

HYPERBOLAS

The hyperbola is a conic section formed when a plane intersects a right circular cone at a certain angle. If β is the angle at which the plane intersects the axis of the cone and α is the angle shown in **Figure 6.18,** a hyperbola is formed when $0° < \beta < \alpha$ or when the plane is parallel to the axis of the cone.

As with the other conic sections, there is a definition of a hyperbola in terms of a certain set of points in the plane.

take note

If the plane intersects the cone along the axis of the cone, the resulting curve is two intersecting straight lines. This is the *degenerate* form of a hyperbola. See the accompanying figure.

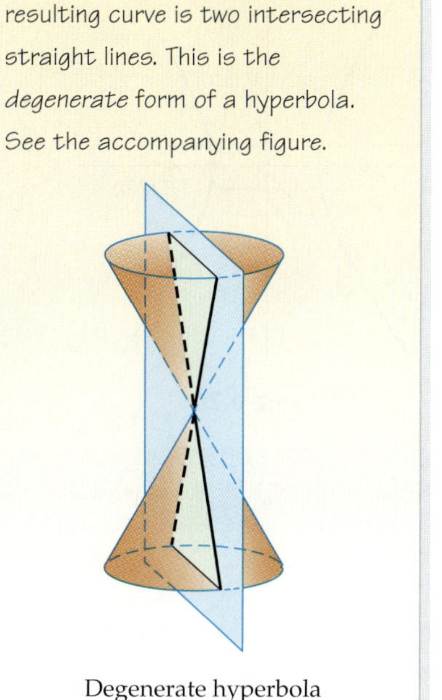

Degenerate hyperbola

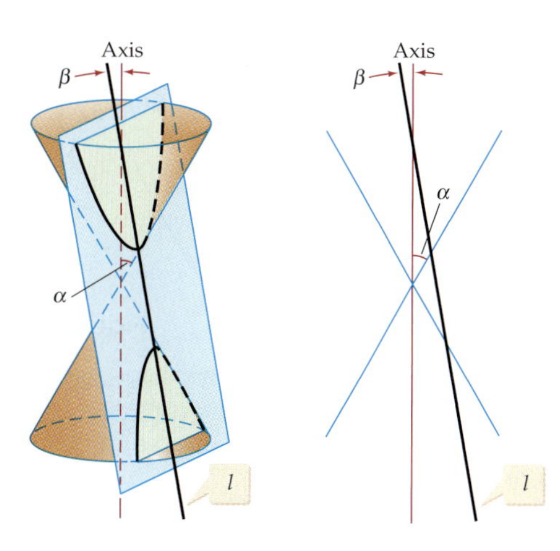

FIGURE 6.18

Definition of a Hyperbola

A **hyperbola** is the set of all points in the plane, the difference between whose distances from two fixed points (foci) is a positive constant.

This definition differs from that of an ellipse in that the ellipse was defined in terms of the *sum* of two distances, whereas the hyperbola is defined in terms of the *difference* of two distances.

The **transverse axis** of a hyperbola is the line segment joining the intercepts (see **Figure 6.19**). The midpoint of the transverse axis is called the **center** of the hyperbola. The **conjugate axis** passes through the center of the hyperbola and is perpendicular to the transverse axis.

The length of the transverse axis is customarily represented as $2a$, and the distance between the two foci is represented as $2c$. The length of the conjugate axis is represented as $2b$.

The **vertices** of a hyperbola are the points where the hyperbola intersects the transverse axis.

To determine the positive constant stated in the definition of a hyperbola, consider the point $V_1(a, 0)$, which is one vertex of a hyperbola, and the points $F_1(c, 0)$ and $F_2(-c, 0)$, which are the foci of the hyperbola (see **Figure 6.20**). The difference between the distance from $V_1(a, 0)$ to $F_1(c, 0)$, $c - a$, and the distance from $V_1(a, 0)$ to $F_2(-c, 0)$, $c + a$, must be a constant. By subtracting these distances, we find

$$|(c - a) - (c + a)| = |-2a| = 2a$$

Thus the constant is $2a$ and is the length of the transverse axis. The absolute value is used to ensure that the distance is a positive number.

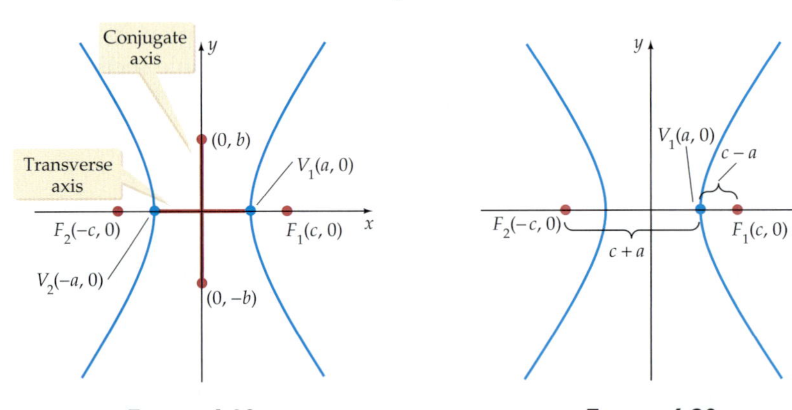

FIGURE 6.19 **FIGURE 6.20**

Standard Forms of the Equation of a Hyperbola with Center at the Origin

Transverse Axis on the x-Axis

The standard form of the equation of a hyperbola with center at the origin and transverse axis on the x-axis (see **Figure 6.21a**) is given by

$$\frac{x^2}{a^2} - \frac{y^2}{b^2} = 1$$

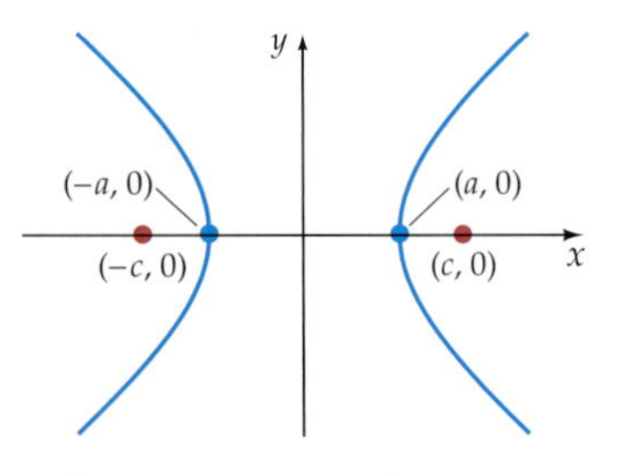

a. Transverse axis on the x-axis

FIGURE 6.21

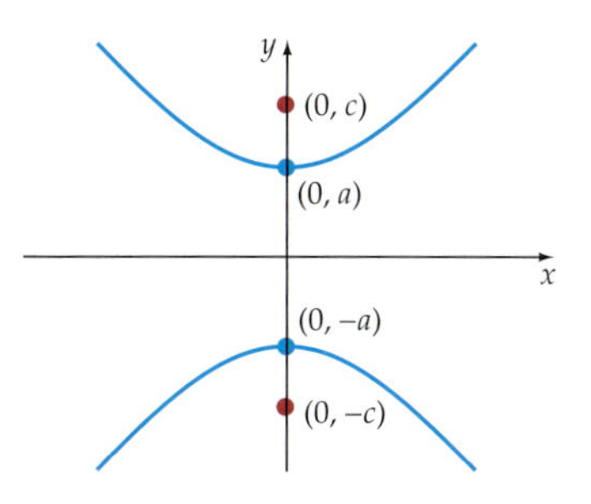

b. Transverse axis on the y-axis

FIGURE 6.21

The coordinates of the vertices are $(a, 0)$ and $(-a, 0)$, and the coordinates of the foci are $(c, 0)$ and $(-c, 0)$, where $c^2 = a^2 + b^2$.

Transverse Axis on the y-Axis

The standard form of the equation of a hyperbola with center at the origin and transverse axis on the y-axis (see **Figure 6.21b**) is given by

$$\frac{y^2}{a^2} - \frac{x^2}{b^2} = 1$$

The coordinates of the vertices are $(0, a)$ and $(0, -a)$, and the coordinates of the foci are $(0, c)$ and $(0, -c)$, where $c^2 = a^2 + b^2$.

? QUESTION For the graph of $\frac{y^2}{9} - \frac{x^2}{4} = 1$, is the transverse axis on the x-axis or the y-axis?

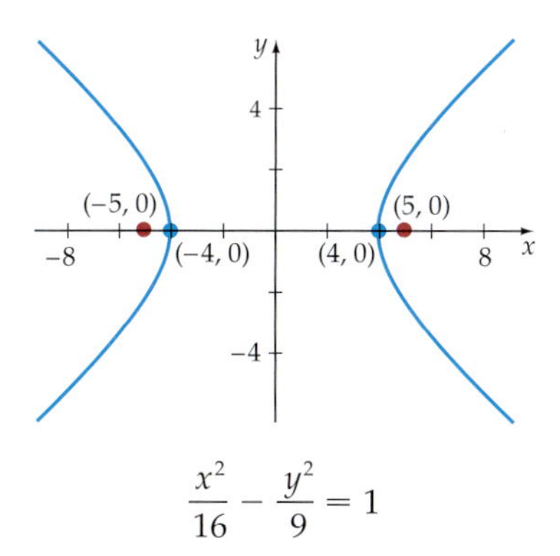

$\frac{x^2}{16} - \frac{y^2}{9} = 1$

FIGURE 6.22

By looking at the equations, it is possible to determine the location of the transverse axis by finding which term in the equation is positive. When the x^2 term is positive, the transverse axis is on the x-axis. When the y^2 term is positive, the transverse axis is on the y-axis.

Consider the hyperbola given by the equation $\frac{x^2}{16} - \frac{y^2}{9} = 1$. Because the x^2 term is positive, the transverse axis is on the x-axis, $a^2 = 16$, and thus $a = 4$. The vertices are $(4, 0)$ and $(-4, 0)$. To find the foci, we determine c.

$$c^2 = a^2 + b^2 = 16 + 9 = 25$$
$$c = \sqrt{25} = 5$$

The foci are $(5, 0)$ and $(-5, 0)$. The graph is shown in **Figure 6.22.**

Each hyperbola has two asymptotes that pass through the center of the hyperbola. The asymptotes of the hyperbola are a useful guide to sketching the graph of the hyperbola.

Asymptotes of a Hyperbola with Center at the Origin

The **asymptotes** of the hyperbola defined by $\frac{x^2}{a^2} - \frac{y^2}{b^2} = 1$ are given by the equations $y = \frac{b}{a}x$ and $y = -\frac{b}{a}x$ (see **Figure 6.23a**).

The asymptotes of the hyperbola defined by $\frac{y^2}{a^2} - \frac{x^2}{b^2} = 1$ are given by the equations $y = \frac{a}{b}x$ and $y = -\frac{a}{b}x$ (see **Figure 6.23b**).

? ANSWER Because the y-term is positive, the transverse axis is on the y-axis.

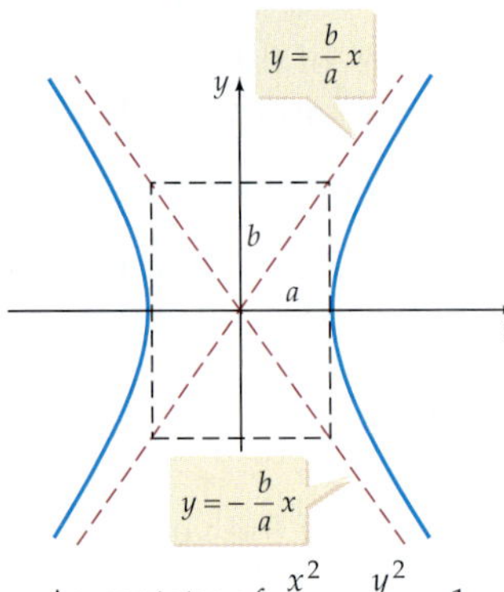

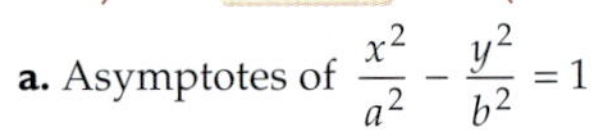

a. Asymptotes of $\frac{x^2}{a^2} - \frac{y^2}{b^2} = 1$

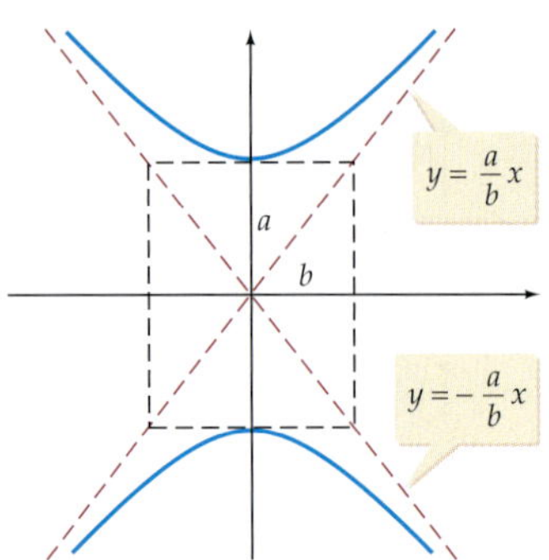

b. Asymptotes of $\frac{y^2}{a^2} - \frac{x^2}{b^2} = 1$

FIGURE 6.23

One method for remembering the equations of the asymptotes is to write the equation of a hyperbola in standard form but to replace 1 by 0 and then solve for y.

$$\frac{x^2}{a^2} - \frac{y^2}{b^2} = 0 \quad \text{so} \quad y^2 = \frac{b^2}{a^2}x^2, \text{ or } y = \pm\frac{b}{a}x$$

$$\frac{y^2}{a^2} - \frac{x^2}{b^2} = 0 \quad \text{so} \quad y^2 = \frac{a^2}{b^2}x^2, \text{ or } y = \pm\frac{a}{b}x$$

EXAMPLE 7 Find the Vertices, Foci, and Asymptotes of a Hyperbola

Find the vertices, foci, and asymptotes of the hyperbola given by the equation $\frac{y^2}{9} - \frac{x^2}{4} = 1$. Sketch the graph.

Solution

Because the y^2 term is positive, the transverse axis is on the y-axis. We know that $a^2 = 9$; thus $a = 3$. The vertices are $V_1(0, 3)$ and $V_2(0, -3)$.

$$c^2 = a^2 + b^2 = 9 + 4$$
$$c = \sqrt{13}$$

The foci are $F_1(0, \sqrt{13})$ and $F_2(0, -\sqrt{13})$.

Because $a = 3$ and $b = 2$ ($b^2 = 4$), the equations of the asymptotes are $y = \frac{3}{2}x$ and $y = -\frac{3}{2}x$.

To sketch the graph, we draw a rectangle that has its center at the origin and has dimensions equal to the lengths of the transverse and conjugate axes. The asymptotes are extensions of the diagonals of the rectangle. See **Figure 6.24.**

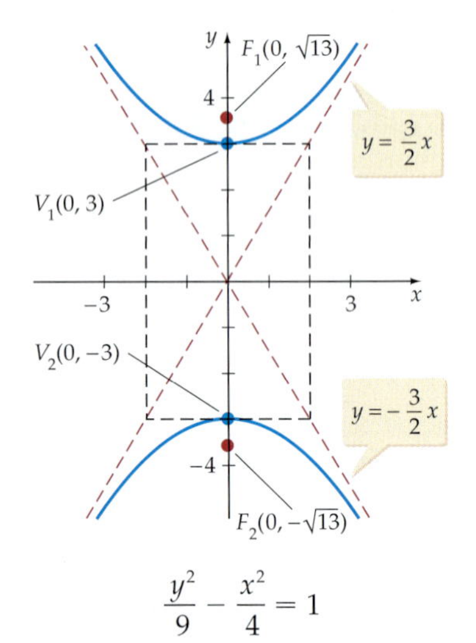

$\frac{y^2}{9} - \frac{x^2}{4} = 1$

FIGURE 6.24

▶ TRY EXERCISE 6, PAGE 480

Using a translation of coordinates similar to that used for ellipses, we can write the equation of a hyperbola with its center at the point (h, k).

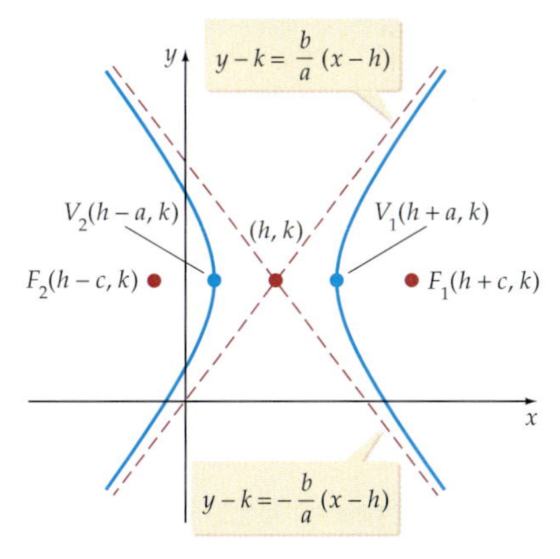

a. Transverse axis parallel to the x-axis

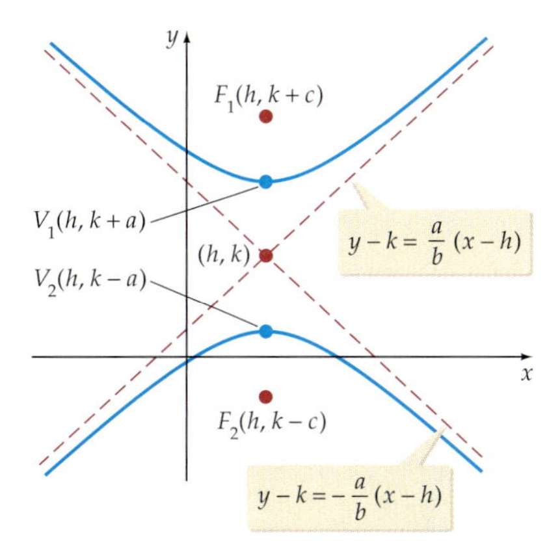

b. Transverse axis parallel to the y-axis

FIGURE 6.25

Standard Forms of the Equation of a Hyperbola with Center at (h, k)

Transverse Axis Parallel to the x-Axis

The standard form of the equation of a hyperbola with center at (h, k) and transverse axis parallel to the x-axis (see **Figure 6.25a**) is given by

$$\frac{(x-h)^2}{a^2} - \frac{(y-k)^2}{b^2} = 1$$

The coordinates of the vertices are $V_1(h + a, k)$ and $V_2(h - a, k)$. The coordinates of the foci are $F_1(h + c, k)$ and $F_2(h - c, k)$, where $c^2 = a^2 + b^2$.

The equations of the asymptotes are $y - k = \pm\frac{b}{a}(x - h)$.

Transverse Axis Parallel to the y-Axis

The standard form of the equation of a hyperbola with center at (h, k) and transverse axis parallel to the y-axis (see **Figure 6.25b**) is given by

$$\frac{(y-k)^2}{a^2} - \frac{(x-h)^2}{b^2} = 1$$

The coordinates of the vertices are $V_1(h, k + a)$ and $V_2(h, k - a)$. The coordinates of the foci are $F_1(h, k + c)$ and $F_2(h, k - c)$, where $c^2 = a^2 + b^2$.

The equations of the asymptotes are $y - k = \pm\frac{a}{b}(x - h)$.

EXAMPLE 8 Find the Vertices, Foci, and Asymptotes of a Hyperbola

Find the vertices, foci, and asymptotes of the hyperbola given by the equation $4x^2 - 9y^2 - 16x + 54y - 29 = 0$. Sketch the graph.

Solution

Write the equation of the hyperbola in standard form by completing the square.

$$4x^2 - 9y^2 - 16x + 54y - 29 = 0$$

$$4x^2 - 16x - 9y^2 + 54y = 29 \quad \text{• Rearrange terms.}$$

$$4(x^2 - 4x) - 9(y^2 - 6y) = 29 \quad \text{• Factor.}$$

$$4(x^2 - 4x + 4) - 9(y^2 - 6y + 9) = 29 + 16 - 81 \quad \text{• Complete the square.}$$

$$4(x - 2)^2 - 9(y - 3)^2 = -36 \quad \text{• Factor.}$$

$$\frac{(y-3)^2}{4} - \frac{(x-2)^2}{9} = 1 \quad \text{• Divide by } -36.$$

The coordinates of the center are $(2, 3)$. Because the term containing $(y - 3)^2$ is positive, the transverse axis is parallel to the y-axis. We know that $a^2 = 4$;

Continued ▶

thus $a = 2$. The vertices are (2, 5) and (2, 1). See **Figure 6.26.**

$$c^2 = a^2 + b^2 = 4 + 9$$
$$c = \sqrt{13}$$

The foci are $(2, 3 + \sqrt{13})$ and $(2, 3 - \sqrt{13})$. We know that $b^2 = 9$; thus $b = 3$. The equations of the asymptotes are $y - 3 = \pm\left(\frac{2}{3}\right)(x - 2)$, which simplifies to

$$y = \frac{2}{3}x + \frac{5}{3} \quad \text{and} \quad y = -\frac{2}{3}x + \frac{13}{3}$$

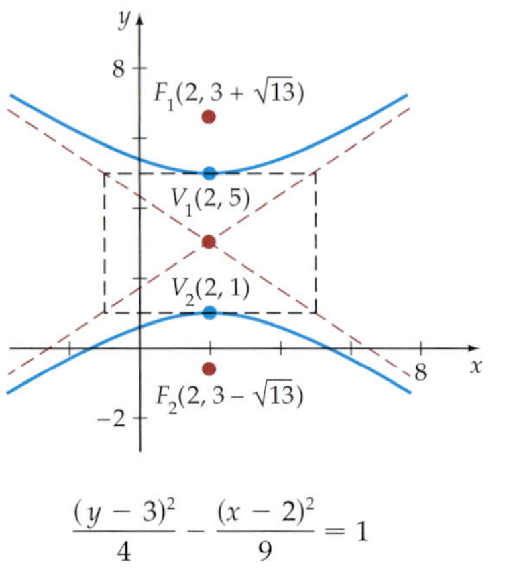

FIGURE 6.26

▶ TRY EXERCISE 22, PAGE 480

APPLICATIONS

A principle of physics states that when light is reflected from a point P on a surface, the angle of incidence (that of the incoming ray) equals the angle of reflection (that of the outgoing ray). See **Figure 6.27.** This principle applied to parabolas has some useful consequences.

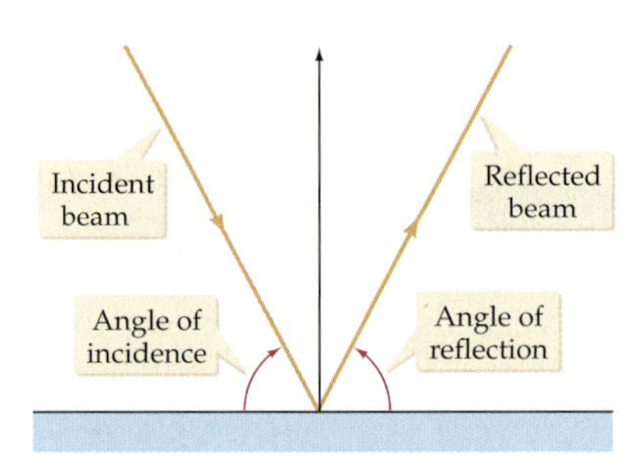

FIGURE 6.27

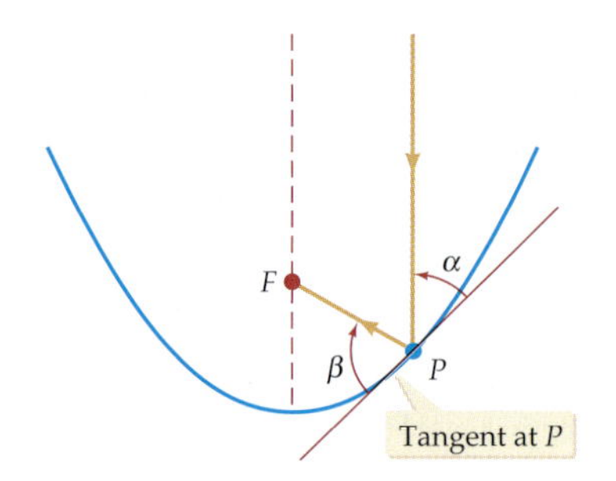

FIGURE 6.28

Optical Property of a Parabola

The line tangent to a parabola at a point P makes equal angles with the line through P and parallel to the axis of symmetry and the line through P and the focus of the parabola (see **Figure 6.28**).

A cross section of the reflecting mirror of a telescope has the shape of a parabola. The incoming parallel rays of light are reflected from the surface of the mirror and to the focus. See **Figure 6.29.**

Flashlights and car headlights also make use of this property. The light bulb is positioned at the focus of the parabolic reflector, which causes the reflected light to be reflected outward in parallel rays. See **Figure 6.30.**

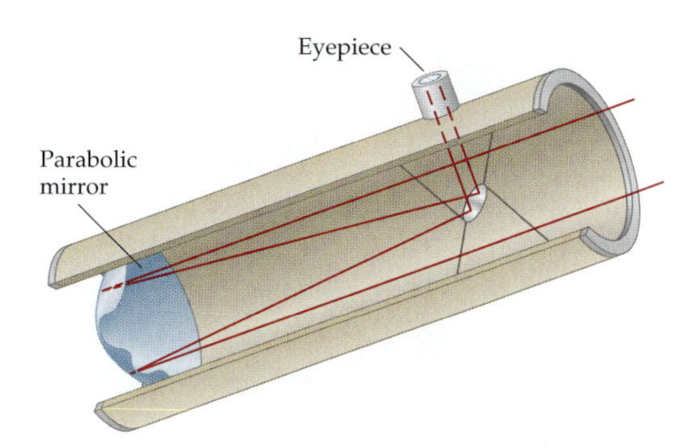

FIGURE 6.29

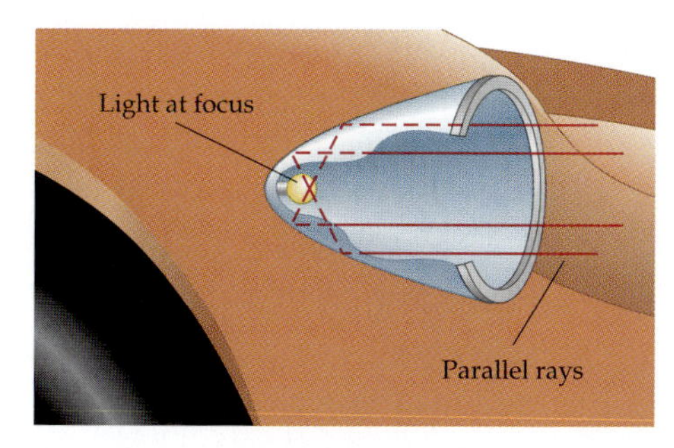

FIGURE 6.30

EXAMPLE 9 Find the Focus of a Satellite Dish

A satellite dish has the shape of a paraboloid. The signals that it receives are reflected to a receiver located at the focus of the paraboloid. If the dish is 8 feet across at its opening and 1.25 feet deep at its center, determine the location of its focus.

Solution

Figure 6.31 shows that a cross section of the paraboloid along its axis of symmetry is a parabola. **Figure 6.32** shows this cross section placed in a rectangular coordinate system with the vertex of the parabola at (0, 0) and the axis of symmetry of the parabola on the y-axis. The parabola has an equation of the form

$$4py = x^2$$

Because the parabola contains the point (4, 1.25), this equation is satisfied by the substitutions $x = 4$ and $y = 1.25$. Thus we have

$$4p(1.25) = 4^2$$
$$5p = 16$$
$$p = \frac{16}{5}$$

The focus of the satellite dish is on the axis of symmetry of the dish, and it is $3\frac{1}{5}$ feet above the vertex of the dish. See **Figure 6.32.**

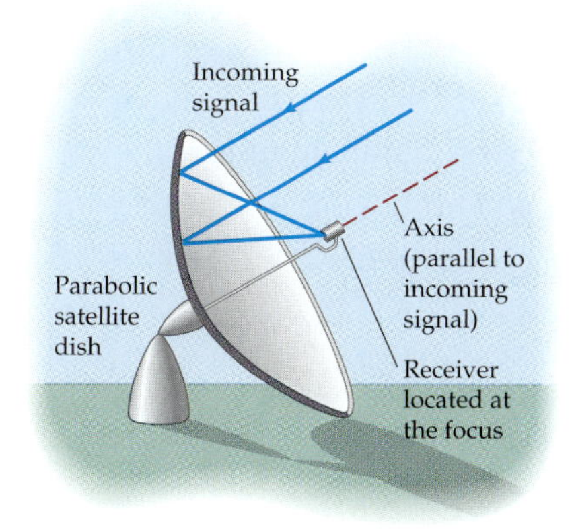

FIGURE 6.31

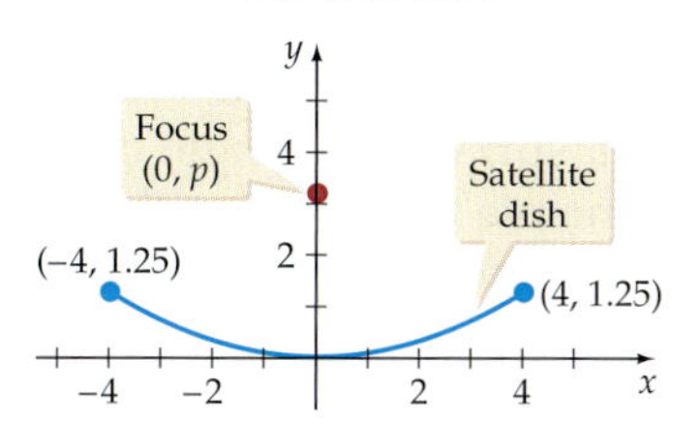

FIGURE 6.32

▶ TRY EXERCISE 42, PAGE 481

The planets travel around the sun in elliptical orbits. The sun is located at a focus of the orbit. The terms *perihelion* and *aphelion* are used to denote the position of a planet in its orbit around the sun. The perihelion is the point nearest the sun; the aphelion is the point farthest from the sun. See **Figure 6.33.** The length of the semimajor axis of a planet's elliptical orbit is called the *mean distance* of the planet from the sun.

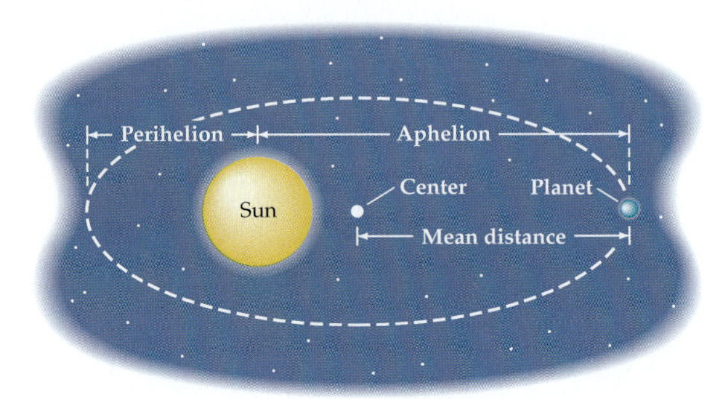

FIGURE 6.33

EXAMPLE 10 Determine an Equation for the Orbit of Earth

Earth has a mean distance of 93 million miles and a perihelion distance of 91.5 million miles. Find an equation for Earth's orbit.

Solution

A mean distance of 93 million miles implies that the length of the semimajor axis of the orbit is $a = 93$ million miles. Earth's aphelion distance is the length of the major axis less the length of the perihelion distance. Thus

$$\text{Aphelion distance} = 2(93) - 91.5 = 94.5 \text{ million miles}$$

The distance c from the sun to the center of Earth's orbit is

$$c = \text{aphelion distance} - 93 = 94.5 - 93 = 1.5 \text{ million miles}$$

The length b of the semiminor axis of the orbit is

$$b = \sqrt{a^2 - c^2} = \sqrt{93^2 - 1.5^2} = \sqrt{8646.75}$$

An equation of Earth's orbit is

$$\frac{x^2}{93^2} + \frac{y^2}{8646.75} = 1$$

▶ TRY EXERCISE 44, PAGE 481

Sound waves, although different from light waves, have a similar reflective property. When sound is reflected from a point P on a surface, the angle of incidence equals the angle of reflection. Applying this principle to a room with an elliptical ceiling results in what are called whispering galleries. These galleries are based on the following theorem.

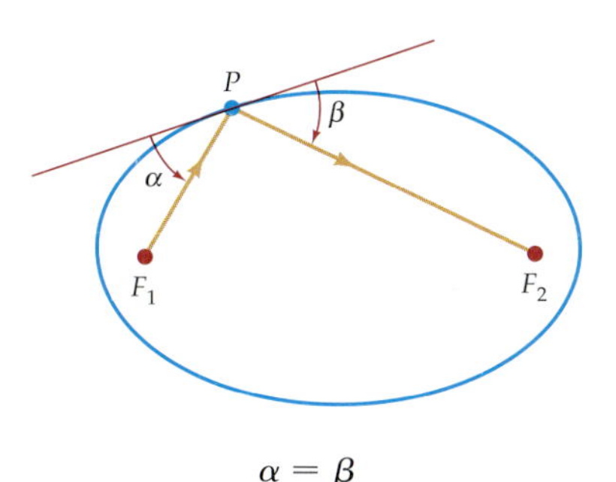

$\alpha = \beta$

FIGURE 6.34

The Reflective Property of an Ellipse

The lines from the foci to a point on an ellipse make equal angles with the tangent line at that point. See **Figure 6.34.**

The Statuary Hall in the Capitol Building in Washington, D.C., is a whispering gallery. Two people standing at the foci of the elliptical ceiling can whisper and yet hear each other even though they are a considerable distance apart. The whisper from one person is reflected to the person standing at the other focus.

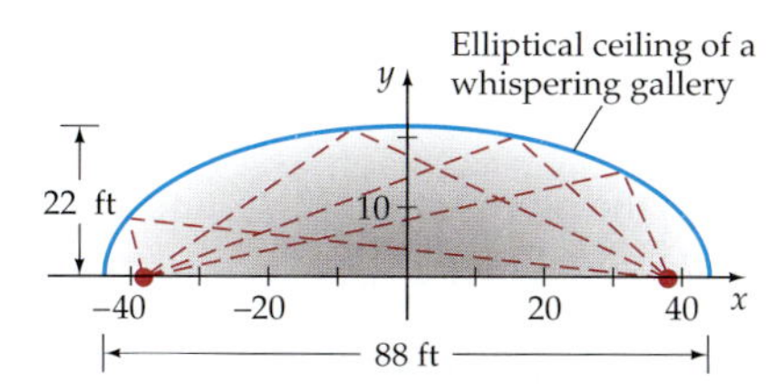

FIGURE 6.35

EXAMPLE 11 Locate the Foci of a Whispering Gallery

A room 88 feet long is constructed to be a whispering gallery. The room has an elliptical ceiling, as shown in **Figure 6.35.** If the maximum height of the ceiling is 22 feet, determine where the foci are located.

Solution

The length a of the semimajor axis of the elliptical ceiling is 44 feet. The height b of the semiminor axis is 22 feet. Thus

$$c^2 = a^2 - b^2$$

$$c^2 = 44^2 - 22^2$$

$$c = \sqrt{44^2 - 22^2} \approx 38.1 \text{ feet}$$

The foci are located about 38.1 feet from the center of the elliptical ceiling along its major axis.

TRY EXERCISE 46, PAGE 482

TOPICS FOR DISCUSSION

1. Is the vertex of a parabola always the midpoint of the line segment through the focus and vertex and terminating on the directrix?
2. If P is a point on a parabola, how is the distance from P to the focus related to the distance from P to the directrix?
3. If a is the length of the semi-major axis of an ellipse and c is the distance from the center of the ellipse to a focus, what is the relationship between a and c?
4. Let a be the length of the semi-major axis of an ellipse and let P be a point on the ellipse. How is the sum of the distances between P and each focus related to a?
5. Is the transverse axis of a hyperbola always longer than the conjugate axis?
6. State a definition of a hyperbola.

EXERCISE SET 6.1

In Exercises 1 to 24, find the vertex, focus, and directrix of each parabola; find the center, vertices, and foci of each ellipse; and find the center, vertices, foci, and asymptotes of each hyperbola. Graph each conic.

1. $x^2 = -4y$
2. $y^2 = \frac{1}{3}x$
3. $\frac{x^2}{16} + \frac{y^2}{25} = 1$
4. $\frac{x^2}{9} + \frac{y^2}{4} = 1$
5. $\frac{x^2}{16} - \frac{y^2}{25} = 1$
6. $\frac{y^2}{4} - \frac{x^2}{25} = 1$
7. $(x - 2)^2 = 8(y + 3)$
8. $(y + 4)^2 = -4(x - 2)$
9. $x^2 - y^2 = 9$
10. $y^2 = 16x$
11. $\frac{(x - 3)^2}{25} + \frac{(y + 2)^2}{16} = 1$
12. $\frac{(x + 2)^2}{9} + \frac{y^2}{25} = 1$
13. $\frac{(x - 3)^2}{16} - \frac{(y + 4)^2}{9} = 1$
14. $\frac{(y + 2)^2}{4} - \frac{(x - 1)^2}{16} = 1$
15. $x^2 + 4y^2 - 6x + 8y - 3 = 0$
16. $3x^2 - 4y^2 + 12x - 24y - 36 = 0$
17. $3x - 4y^2 + 8y + 2 = 0$
18. $3x + 2y^2 - 4y - 7 = 0$
19. $9x^2 + 4y^2 + 36x - 8y + 4 = 0$
20. $11x^2 + 25y^2 - 44x + 50y - 206 = 0$
21. $4x^2 - 9y^2 - 8x + 12y - 144 = 0$
22. $9x^2 - y^2 - 36x + 6y - 9 = 0$
23. $4x^2 + 28x + 32y + 81 = 0$
24. $x^2 - 6x - 9y + 27 = 0$

In Exercises 25 to 32, find the equation in standard form of the conic that satisfies the given conditions.

25. Ellipse with vertices at $(7, 3)$ and $(-3, 3)$; length of minor axis is 8.
26. Hyperbola with vertices at $(4, 1)$ and $(-2, 1)$; foci at $(5, 1)$ and $(-3, 1)$.
27. Hyperbola with foci $(-5, 2)$ and $(1, 2)$; length of transverse axis is 4.
28. Parabola with focus $(2, -3)$ and directrix $x = 6$.
29. Parabola with vertex $(0, -2)$ and passing through the point $(3, 4)$.
30. Ellipse with center $(2, 4)$, major axis of length 10, parallel to the y-axis, and passing through the point $(3, 3)$.
31. Hyperbola with vertices $(\pm 6, 0)$ and asymptotes whose equations are $y = \pm\frac{1}{9}x$.
32. Parabola passing through the points $(1, 0)$, $(2, 1)$, and $(0, 1)$ with axis of symmetry parallel to the y-axis.
33. Find the equation of the parabola traced by a point $P(x, y)$ that moves in such a way that the distance between $P(x, y)$ and the line $x = 2$ equals the distance between $P(x, y)$ and the point $(-2, 3)$.
34. Find the equation of the parabola traced by a point $P(x, y)$ that moves in such a way that the distance between $P(x, y)$ and the line $y = 1$ equals the distance between $P(x, y)$ and the point $(-1, 2)$.
35. Find the equation of the ellipse traced by a point $P(x, y)$ that moves in such a way that the sum of its distances to $(-3, 1)$ and $(5, 1)$ is 10.
36. Find the equation of the ellipse traced by a point $P(x, y)$ that moves in such a way that the sum of its distances to $(3, 5)$ and $(3, -1)$ is 8.
37. **SATELLITE DISH** A satellite dish has the shape of a paraboloid. The signals that it receives are reflected to a receiver that is located at the focus of the paraboloid. If the dish is 8 feet across at its opening and 1 foot deep at its vertex, determine the location (distance above the vertex of the dish) of its focus.

38. THE VERY LARGE ARRAY A radio telescope is a paraboloid measuring 90 feet across with a depth of 20 feet (see figure). Determine, to the nearest 0.1 foot, the distance from the vertex to the focus of this antenna.

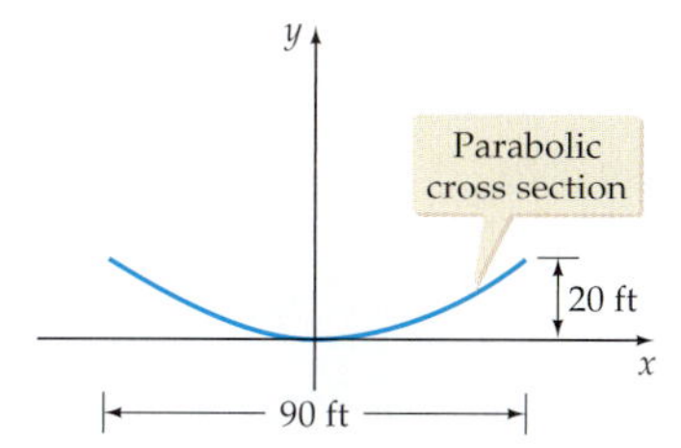

39. CAPTURING SOUND During televised football games, a parabolic microphone is used to capture sounds. The shield of the microphone is a paraboloid with a diameter of 18.75 inches and a depth of 3.66 inches. To pick up sounds, a microphone is placed at the focus of the paraboloid. How far (to the nearest 0.1 inch) from the vertex of the paraboloid should the microphone be placed?

40. THE LOVELL TELESCOPE The Lovell Telescope is a radio telescope located at the Jodrell Bank Observatory in Cheshire, England. The dish of the telescope has the shape of a paraboloid with a diameter of 250 feet and a focal length of 75 feet.

a. Find an equation of a cross section of the paraboloid that passes through the vertex of the paraboloid. Assume that the dish has its vertex at (0, 0) and has a vertical axis of symmetry.

b. Find the depth of the dish. Round to the nearest foot.

41. THE LICK TELESCOPE The parabolic mirror in the Lick telescope at the Lick Observatory on Mount Hamilton has a diameter of 120 inches and a focal length of 600 inches. In the construction of the mirror, workers ground the mirror as shown in the diagram at the top of the next column. Determine the dimension a, which is the concave depth of the mirror.

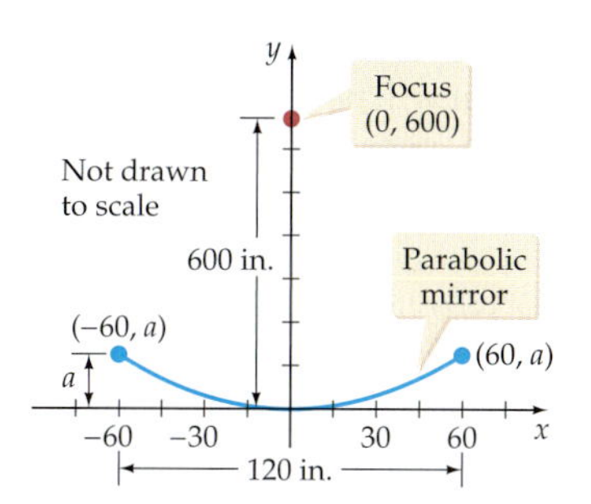

Mirror in the Lick Telescope

Figure for Exercise 41

▶ 42. THE HALE TELESCOPE The parabolic mirror in the Hale telescope at the Palomar Observatory in southern California has a diameter of 200 inches and a concave depth of 3.75375 inches. Determine the location of its focus (to the nearest inch).

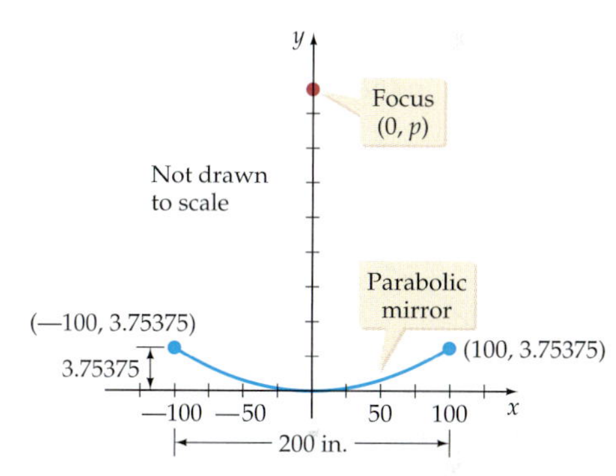

Mirror in the Hale Telescope

43. THE ORBIT OF SATURN The distance from Saturn to the sun at Saturn's aphelion is 934.34 million miles, and the distance from Saturn to the sun at its perihelion is 835.14 million miles. Find an equation for the orbit of Saturn.

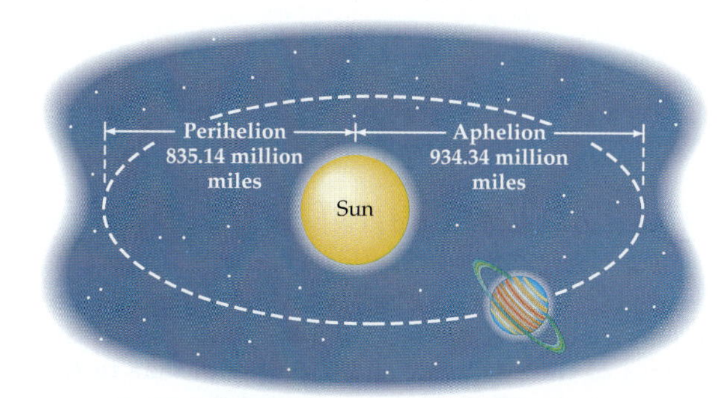

▶ 44. THE ORBIT OF MARS Mars has a mean distance from the sun of 139.4 million miles, and the distance from Mars to the sun at its perihelion is 126.4 million miles. Find an equation for the orbit of Mars.

45. **Whispering Gallery** An architect wishes to design a large room that will be a whispering gallery. The ceiling of the room has a cross section that is an ellipse, as shown in the following figure.

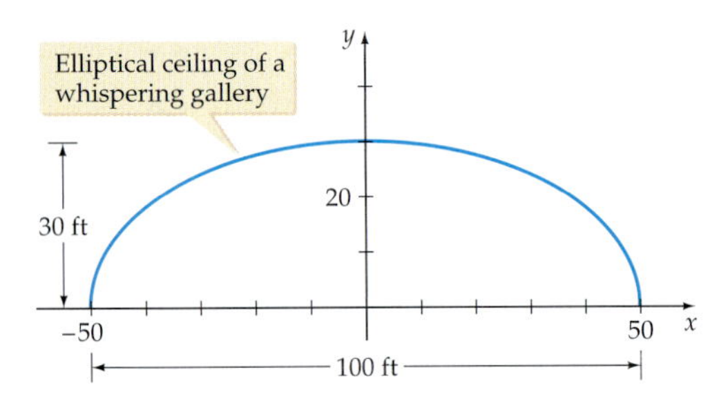

How far to the right and left of center are the foci located?

46. **Whispering Gallery** An architect wishes to design a large room 250 feet long that will be a whispering gallery. The ceiling of the room has a cross section that is an ellipse, as shown in the following figure.

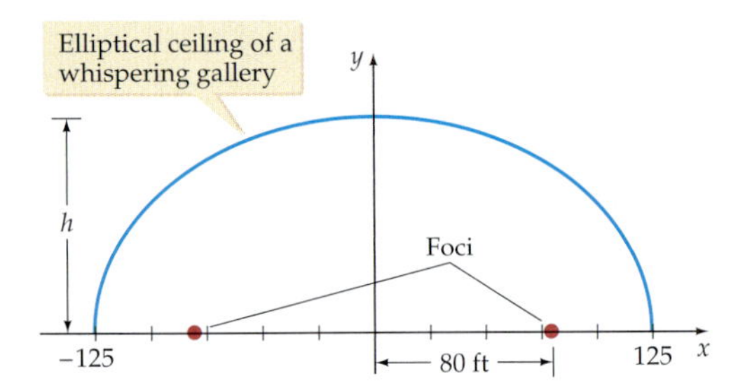

If the foci are to be located 80 feet to the right and left of center, find the height h of the elliptical ceiling (to the nearest 0.1 foot).

47. **Halley's Comet** Find the equation of the path of Halley's comet in astronomical units by letting the sun (one focus) be at the origin and letting the other focus be on the positive x-axis. The length of the major axis of the orbit of Halley's comet is approximately 36 astronomical units (36 AU), and the length of the minor axis is 9 AU (1 AU = 92,960,000 miles).

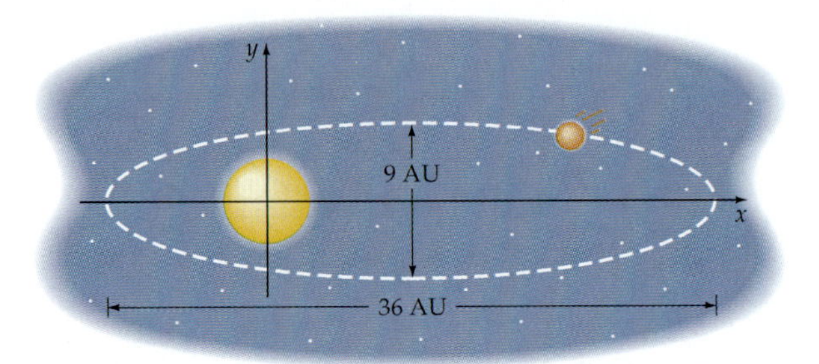

48. **Elliptical Receivers** Some satellite receivers are made in an elliptical shape that enables the receiver to pick up signals from two satellites. The receiver shown in the figure has a major axis of 24 inches and a minor axis of 18 inches.

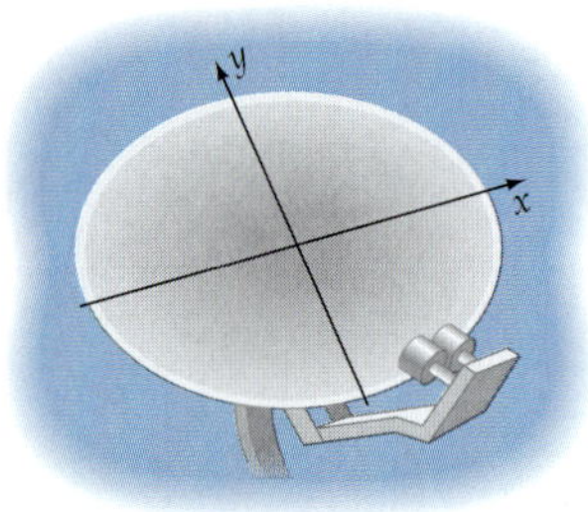

Determine, to the nearest 0.1 inch, the coordinates in the xy-plane of the foci of the ellipse. (*Note*: Because the receiver has only a slight curvature, we can estimate the location of the foci by assuming the receiver is flat.)

CONNECTING CONCEPTS

In Exercises 49 to 51, use the following definition. The line segment that has endpoints on a parabola, passes through the focus of the parabola, and is perpendicular to the axis of symmetry is called the *latus rectum* **of the parabola.**

49. Find the latus rectum for the parabola given by $x^2 = 5y$.

50. Find the latus rectum for the parabola given by $y^2 = -4x$.

51. Find the latus rectum of any parabola in terms of $|p|$, the distance from the vertex of the parabola to the focus.

52. Find the equation of the directrix of the parabola with vertex at the origin and focus at the point $P(1, 1)$.

In Exercises 53 to 56, use the following definition. The *eccentricity of an ellipse* **is the ratio of c to a, where c is the distance from the center to a focus and a is one-half the length of the major axis.**

53. Find the eccentricity of the ellipse in Exercise 3.

54. Find the eccentricity of the ellipse in Exercise 4.

55. Can the eccentricity of an ellipse be greater than 1?

56. How does the appearance of an ellipse with an eccentricity close to 0 differ from one with an eccentricity close to 1?

57. Use the definition of an ellipse to find the equation of the ellipse that has foci $F_1(-3, 0)$ and $F_2(3, 0)$ and passes through the point $P\left(3, \frac{16}{5}\right)$.

In Exercises 58 to 60, use the following definition. The *eccentricity of an hyperbola* **is the ratio of c to a, where c is the distance from the center to a focus and a is one-half the length of the transverse axis.**

58. Find the eccentricity of the hyperbola in Exercise 5.

59. Find the eccentricity of the hyperbola in Exercise 6.

60. Can the eccentricity of a hyperbola be less than 1?

61. Use the definition of a hyperbola to find the equation of the hyperbola that has foci $F_1(-2, 0)$ and $F_2(2, 0)$ and passes through the point $P(2, 3)$.

PREPARE FOR SECTION 6.2

62. Expand $\cos(\alpha + \beta)$. [5.2]

63. Expand $\sin(\alpha + \beta)$. [5.2]

64. Is $\sin x$ an even function, an odd function, or neither? [4.2]

65. Is $\cos x$ an even function, an odd function, or neither? [4.2]

66. If $\sin \alpha = \frac{-\sqrt{3}}{2}$ and $\cos \alpha = -\frac{1}{2}$, find α. Write the angle in degrees. [4.2]

67. Write $(r \cos \theta)^2 + (r \sin \theta)^2$ in simplest form. [5.1]

PROJECTS

1. Parabolas and Tangents Calculus procedures can be used to show that the equation of a tangent line to the parabola $4py = x^2$ at the point (x_0, y_0) is given by

$$y - y_0 = \left(\frac{1}{2p}x_0\right)(x - x_0)$$

Use this equation to verify each of the following statements.

a. If two tangent lines to a parabola intersect at right angles, then the point of intersection of the tangent lines is on the directrix of the parabola.

b. If two tangent lines to a parabola intersect at right angles, then the focus of the parabola is located on the line segment that connects the two points of tangency.

c. The tangent line to the parabola $4py = x^2$ at the point (x_0, y_0) intersects the y-axis at the point $(0, -y_0)$.

2. I. M. Pei's Oval The poet and architect I. M. Pei suggested that the oval with the most appeal to the eye is given by the equation

$$\left(\frac{x}{a}\right)^{3/2} + \left(\frac{y}{b}\right)^{3/2} = 1$$

Use a graphing utility to graph this equation with $a = 5$ and $b = 3$. Then compare your graph with the graph of

$$\left(\frac{x}{5}\right)^2 + \left(\frac{y}{3}\right)^2 = 1$$

SECTION 6.2

POLAR COORDINATES

- THE POLAR COORDINATE SYSTEM
- GRAPHS OF EQUATIONS IN A POLAR COORDINATE SYSTEM
- POLAR EQUATIONS OF THE CONICS
- GRAPH A CONIC GIVEN IN POLAR FORM
- TRANSFORMATIONS BETWEEN RECTANGULAR AND POLAR COORDINATES
- WRITE POLAR COORDINATE EQUATIONS AS RECTANGULAR EQUATIONS AND RECTANGULAR COORDINATE EQUATIONS AS POLAR EQUATIONS

Until now, we have used a *rectangular coordinate system* to locate a point in the coordinate plane. An alternative method is to use a *polar coordinate system,* wherein a point is located by giving a distance from a fixed point and an angle from some fixed direction.

THE POLAR COORDINATE SYSTEM

A **polar coordinate system** is formed by drawing a horizontal ray. The ray is called the **polar axis,** and the endpoint of the ray is called the **pole.** A point $P(r, \theta)$ in the plane is located by specifying a distance r from the pole and an angle θ measured from the polar axis to the line segment OP. The angle can be measured in degrees or radians. See **Figure 6.36.**

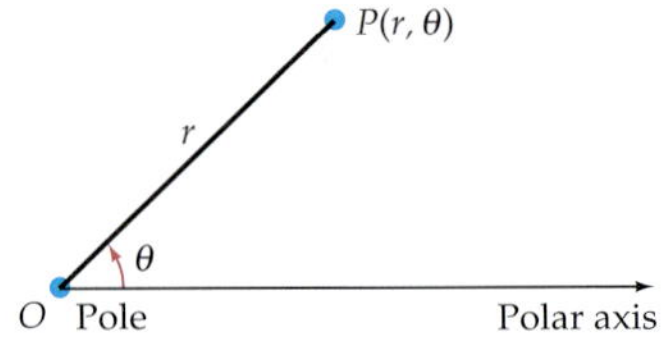

FIGURE 6.36

The coordinates of the pole are $(0, \theta)$, where θ is an arbitrary angle. Positive angles are measured counterclockwise from the polar axis. Negative angles are measured clockwise from the axis. Positive values of r are measured along the ray that makes an angle of θ from the polar axis. Negative values of r are measured along the ray that makes an angle of $\theta + 180°$ from the polar axis. See **Figures 6.37** and **6.38.**

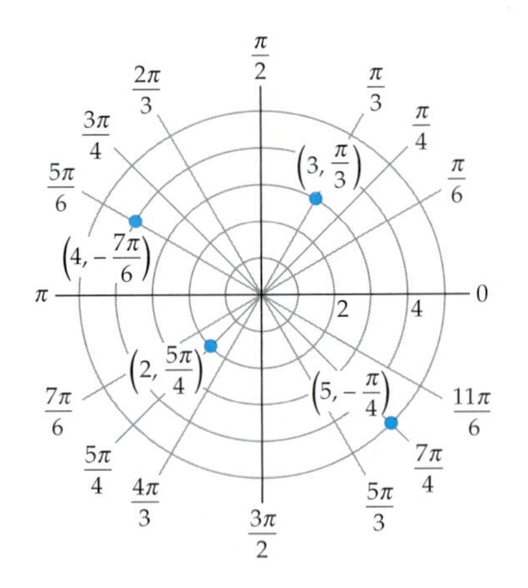

FIGURE 6.37

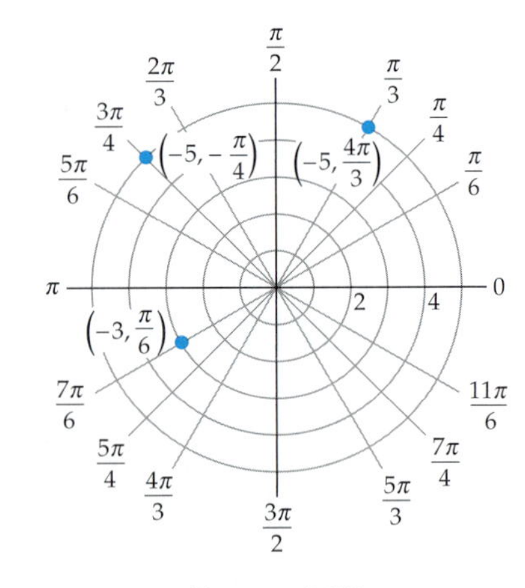

FIGURE 6.38

In a rectangular coordinate system, there is a one-to-one correspondence between the points in the plane and the ordered pairs (x, y). This is not true for a polar coordinate system. For polar coordinates, the relationship is one-to-many. Infinitely many ordered-pair descriptions correspond to each point $P(r, \theta)$ in a polar coordinate system.

For example, consider a point whose coordinates are $P(3, 45°)$. Because there are 360° in one complete revolution around a circle, the point P also could be written as $(3, 405°)$, as $(3, 765°)$, as $(3, 1125°)$, and generally as $(3, 45° + n \cdot 360°)$, where n is an integer. It is also possible to describe the point $P(3, 45°)$ by $(-3, 225°)$, $(-3, -135°)$, and $(3, -315°)$, to name just a few options.

The relationship between an ordered pair and a point is not one-to-many. That is, given an ordered pair (r, θ), there is exactly one point in the plane that corresponds to that ordered pair.

Graphs of Equations in a Polar Coordinate System

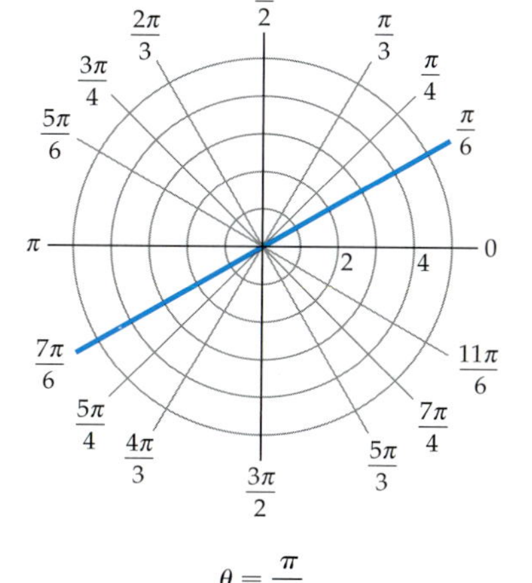

$\theta = \frac{\pi}{6}$

Figure 6.39

A **polar equation** is an equation in r and θ. A **solution** to a polar equation is an ordered pair (r, θ) that satisfies the equation. The **graph** of a polar equation is the set of all points whose ordered pairs are solutions of the equation.

The graph of the polar equation $\theta = \frac{\pi}{6}$ is a line. Because θ is independent of r, θ is $\frac{\pi}{6}$ radian from the polar axis for all values of r. The graph is a line that makes an angle of $\frac{\pi}{6}$ radian (30°) from the polar axis. See **Figure 6.39.**

Polar Equations of a Line

The graph of $\theta = \alpha$ is a line through the pole at an angle of α from the polar axis. See **Figure 6.40a.**

The graph of $r \sin \theta = a$ is a horizontal line passing through the point $\left(a, \frac{\pi}{2}\right)$. See **Figure 6.40b.**

The graph of $r \cos \theta = a$ is a vertical line passing through the point $(a, 0)$. See **Figure 6.40c.**

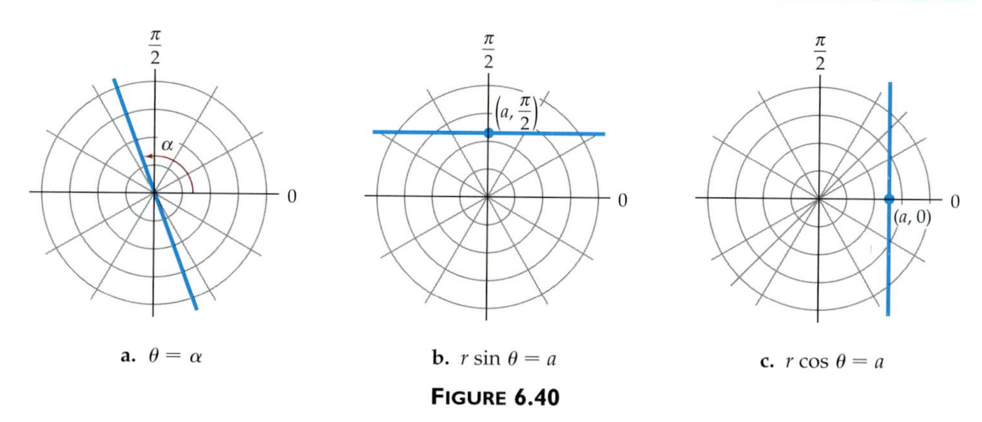

a. $\theta = \alpha$ **b.** $r \sin \theta = a$ **c.** $r \cos \theta = a$

Figure 6.40

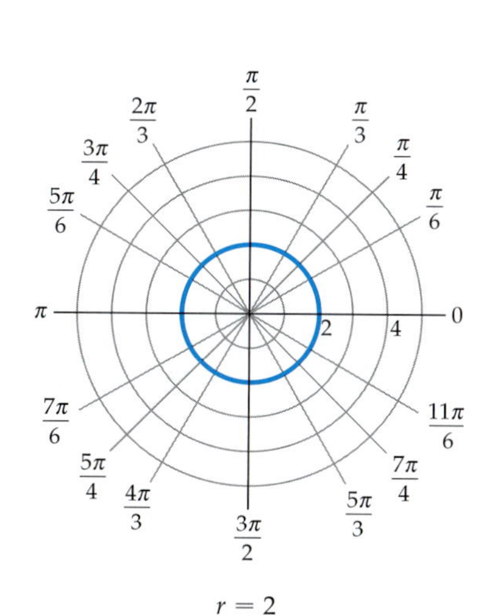

$r = 2$

Figure 6.41

Figure 6.41 is the graph of the polar equation $r = 2$. Because r is independent of θ, r is 2 units from the pole for all values of θ. The graph is a circle of radius 2 with center at the pole.

The Graph of $r = a$

The graph of $r = a$ is a circle with center at the pole and radius a.

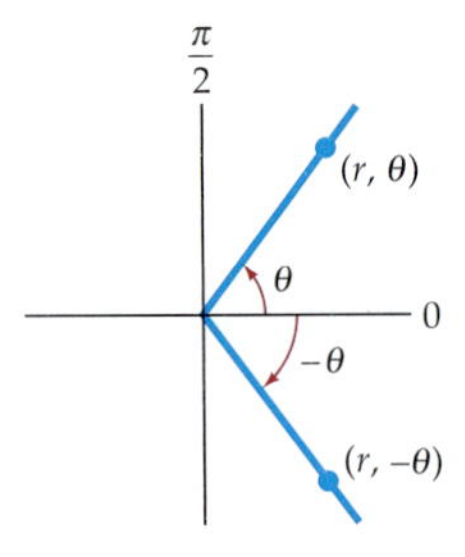

Symmetry with respect to the line $\theta = 0$

FIGURE 6.42

Suppose that whenever the ordered pair (r, θ) lies on the graph of a polar equation, $(r, -\theta)$ also lies on the graph. From **Figure 6.42,** the graph will have symmetry with respect to the polar axis $\theta = 0$. Thus one test for symmetry is to replace θ by $-\theta$ in the polar equation. If the resulting equation is equivalent to the original equation, the graph is symmetric with respect to the polar axis.

Table 6.1 shows the types of symmetry and their associated tests. For each type, if the recommended substitution results in an equivalent equation, the graph will have the indicated symmetry. **Figure 6.43** illustrates the tests for symmetry with respect to the line $\theta = \dfrac{\pi}{2}$ and for symmetry with respect to the pole.

TABLE 6.1 Tests for Symmetry

Substitution	Symmetry with respect to
$-\theta$ for θ	The line $\theta = 0$
$\pi - \theta$ for θ, $-r$ for r	The line $\theta = 0$
$\pi - \theta$ for θ	The line $\theta = \dfrac{\pi}{2}$
$-\theta$ for θ, $-r$ for r	The line $\theta = \dfrac{\pi}{2}$
$-r$ for r	The pole
$\pi + \theta$ for θ	The pole

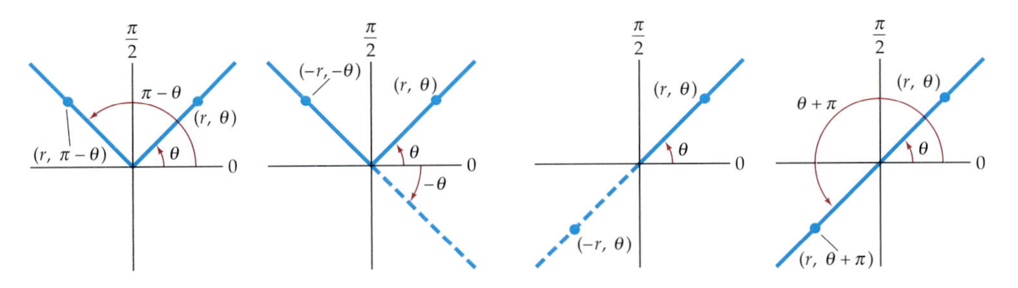

Symmetry with respect to the line $\theta = \dfrac{\pi}{2}$ Symmetry with respect to the pole

FIGURE 6.43

The graph of a polar equation may have a type of symmetry even though a test for that symmetry fails. For example, as we will see later, the graph of $r = \sin 2\theta$ is symmetric with respect to the line $\theta = 0$. However, using the symmetry test of substituting $-\theta$ for θ, we have

$$\sin 2(-\theta) = -\sin 2\theta = -r \neq r$$

Thus this test fails to show symmetry with respect to the line $\theta = 0$. The symmetry test of substituting $\pi - \theta$ for θ and $-r$ for r establishes symmetry with respect to the line $\theta = 0$.

EXAMPLE 1 Graph a Polar Equation

Show that the graph of $r = 4\cos\theta$ is symmetric with respect to the polar axis. Graph the equation.

Solution

Test for symmetry with respect to the polar axis. Replace θ by $-\theta$.

$$r = 4\cos(-\theta) = 4\cos\theta \qquad \bullet\ \cos(-\theta) = \cos\theta$$

Because replacing θ by $-\theta$ results in the original equation $r = 4\cos\theta$, the graph is symmetric with respect to the polar axis.

To graph the equation, begin choosing various values of θ and finding the corresponding values of r. However, before doing so, consider two further observations that will reduce the number of points you must choose.

First, because the cosine function is a periodic function with period 2π, it is only necessary to choose points between 0 and 2π (0° and 360°). Second, when $\frac{\pi}{2} < \theta < \frac{3\pi}{2}$, $\cos\theta$ is negative, which means that any θ between these values will produce a negative r. Thus the point will be in the first or fourth quadrant. That is, we need consider only angles θ in the first or fourth quadrants. However, because the graph is symmetric with respect to the polar axis, it is only necessary to choose values of θ between 0 and $\frac{\pi}{2}$.

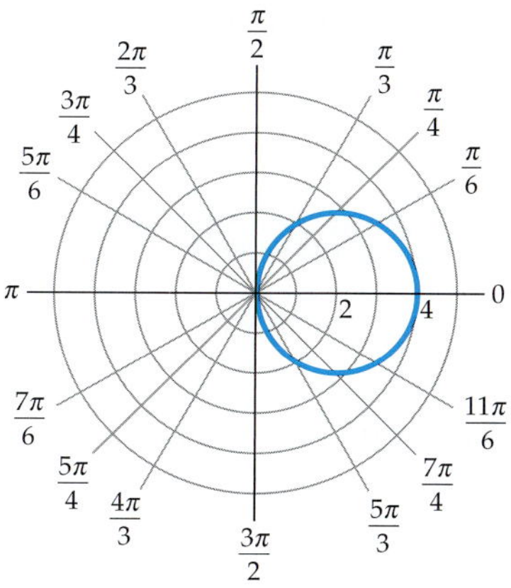

By symmetry

θ	0	$\frac{\pi}{6}$	$\frac{\pi}{4}$	$\frac{\pi}{3}$	$\frac{\pi}{2}$	$-\frac{\pi}{6}$	$-\frac{\pi}{4}$	$-\frac{\pi}{3}$	$-\frac{\pi}{2}$
r	4.0	3.5	2.8	2.0	0.0	3.5	2.8	2.0	0.0

The graph of $r = 4\cos\theta$ is a circle with center at (2, 0). See **Figure 6.44.**

$r = 4\cos\theta$

FIGURE 6.44

TRY EXERCISE 12, PAGE 497

Polar Equations of a Circle

The graph of the equation $r = a$ is a circle with center at the pole and radius a. See **Figure 6.45a.**

The graph of the equation $r = a\cos\theta$ is a circle that is symmetric with respect to the line $\theta = 0$. See **Figure 6.45b.**

The graph of $r = a\sin\theta$ is a circle that is symmetric with respect to the line $\theta = \frac{\pi}{2}$. See **Figure 6.45c.**

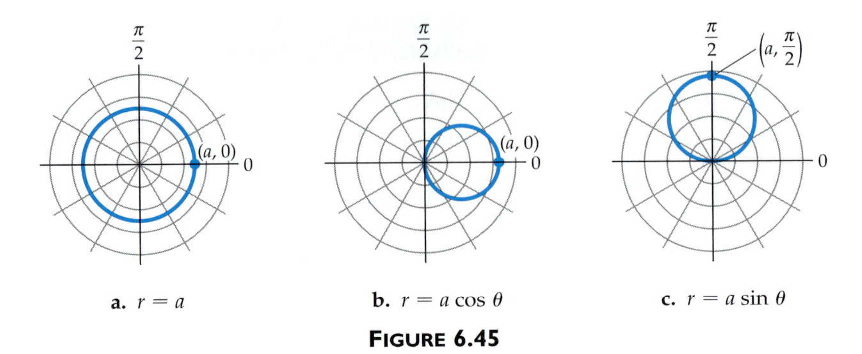

FIGURE 6.45

Just as there are specifically named curves in an xy-coordinate system (such as parabola and ellipse), there are named curves in an $r\theta$-coordinate system. Two of the many types are the *limaçon* and the *rose curve*.

Polar Equations of a Limaçon

The graph of the equation $r = a + b \cos \theta$ is a **limaçon** that is symmetric with respect to the line $\theta = 0$.

The graph of the equation $r = a + b \sin \theta$ is a limaçon that is symmetric with respect to the line $\theta = \dfrac{\pi}{2}$.

In the special case where $|a| = |b|$, the graph is called a **cardioid.**

The graph of $r = a + b \cos \theta$ is shown in **Figure 6.46** for various values of a and b.

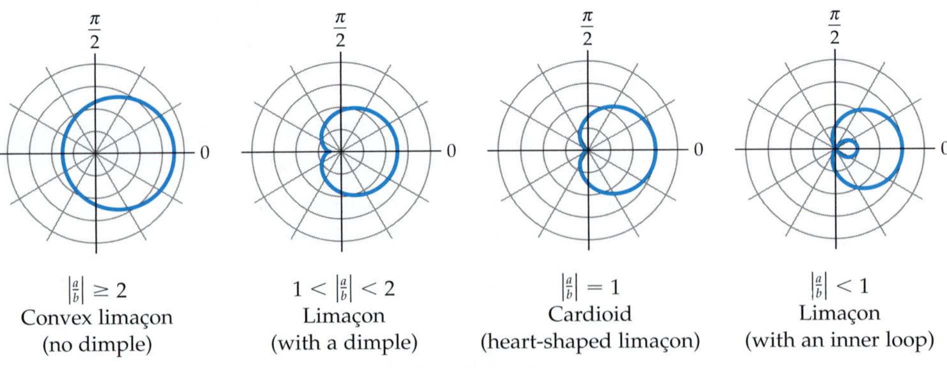

FIGURE 6.46

EXAMPLE 2 Sketch the Graph of a Limaçon

Sketch the graph of $r = 2 - 2\sin\theta$.

Solution

From the general equation of a limaçon $r = a + b\sin\theta$ with $|a| = |b|$ ($|2| = |-2|$), the graph of $r = 2 - 2\sin\theta$ is a cardioid that is symmetric with respect to the line $\theta = \dfrac{\pi}{2}$.

Because we know that the graph is heart-shaped, we can sketch the graph by finding r for a few values of θ. When $\theta = 0$, $r = 2$. When $\theta = \dfrac{\pi}{2}$, $r = 0$. When $\theta = \pi$, $r = 2$. When $\theta = \dfrac{3\pi}{2}$, $r = 4$. Sketching a heart-shaped curve through the four points

$$(2, 0), \quad \left(0, \frac{\pi}{2}\right), \quad (2, \pi), \quad \text{and} \quad \left(4, \frac{3\pi}{2}\right)$$

produces the cardioid in **Figure 6.47.**

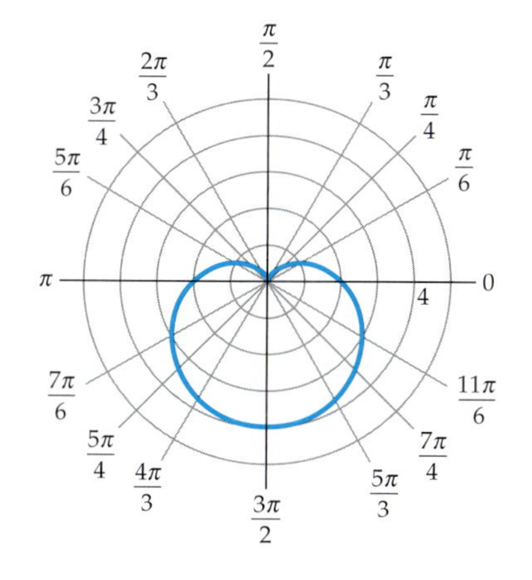

$r = 2 - 2\sin\theta$

FIGURE 6.47

▶ TRY EXERCISE 16, PAGE 497

Example 3 gives the details necessary for using a graphing utility to construct a polar graph.

EXAMPLE 3 Use a Graphing Utility to Sketch the Graph of a Limaçon

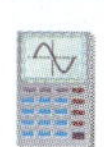

Use a graphing utility to graph $r = 3 - 2\cos\theta$.

Solution

From the general equation of a limaçon $r = a + b\cos\theta$, with $a = 3$ and $b = -2$, we know that the graph will be a limaçon with a dimple. The graph will be symmetric with respect to the line $\theta = 0$.

Use polar mode with angle measure in radians. Enter the equation $r = 3 - 2\cos\theta$ in the polar function editing menu. The graph in **Figure 6.48** was produced with a *TI-83* by using a window defined by the following:

θmin=0	Xmin=-6	Ymin=-4
θmax=2π	Xmax=6	Ymax=4
θstep=0.1	Xscl=1	Yscl=1

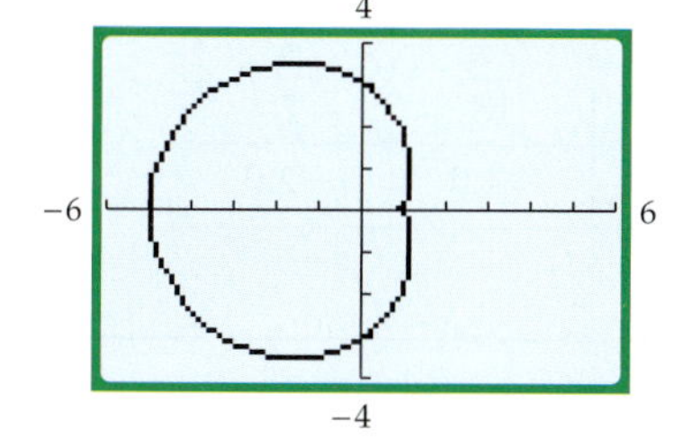

$r = 3 - 2\cos\theta$

FIGURE 6.48

▶ TRY EXERCISE 26, PAGE 497

Polar Equations of Rose Curves

The graphs of the equations $r = a \cos n\theta$ and $r = a \sin n\theta$ are **rose curves.** When n is an even number, the number of petals is $2n$. See **Figure 6.49a.** When n is an odd number, the number of petals is n. See **Figure 6.49b.**

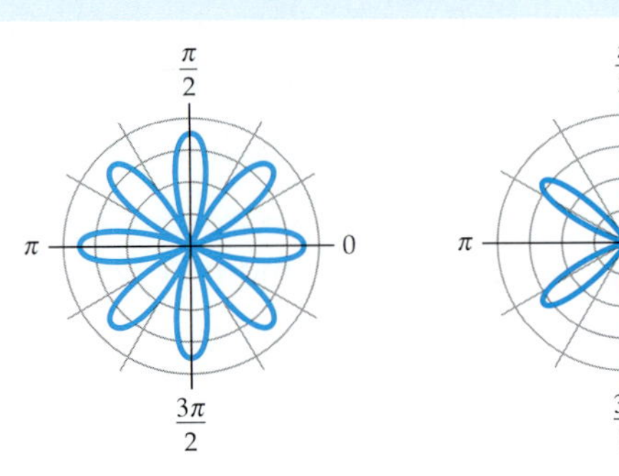

a. $r = a \cos 4\theta$
$n = 4$ is even, $2n = 8$ petals

b. $r = a \cos 5\theta$
$n = 5$ is odd, 5 petals

FIGURE 6.49

INTEGRATING TECHNOLOGY

When using a graphing utility in polar mode, choose the value of θstep carefully. If θstep is set too small, the graphing utility may require an excessively long period of time to complete the graph. If θstep is set too large, the resulting graph may give only a very rough approximation of the actual graph.

? QUESTION How many petals are in the graph of **a.** $r = 4 \cos 3\theta$? **b.** $r = 5 \sin 2\theta$?

EXAMPLE 4 Sketch the Graph of a Rose Curve

Sketch the graph of $r = 2 \sin 3\theta$.

Solution

From the general equation of a rose curve $r = a \sin n\theta$, with $a = 2$ and $n = 3$, the graph of $r = 2 \sin 3\theta$ is a rose curve that is symmetric with respect to the line $\theta = \dfrac{\pi}{2}$. Because n is an odd number ($n = 3$), there will be three petals in the graph.

Choose some values for θ and find the corresponding values of r. Use symmetry to sketch the graph. See **Figure 6.50.**

θ	0	$\frac{\pi}{18}$	$\frac{\pi}{6}$	$\frac{5\pi}{18}$	$\frac{\pi}{3}$	$\frac{7\pi}{18}$	$\frac{\pi}{2}$
r	0.0	1.0	2.0	1.0	0.0	−1.0	−2.0

TRY EXERCISE 14, PAGE 497

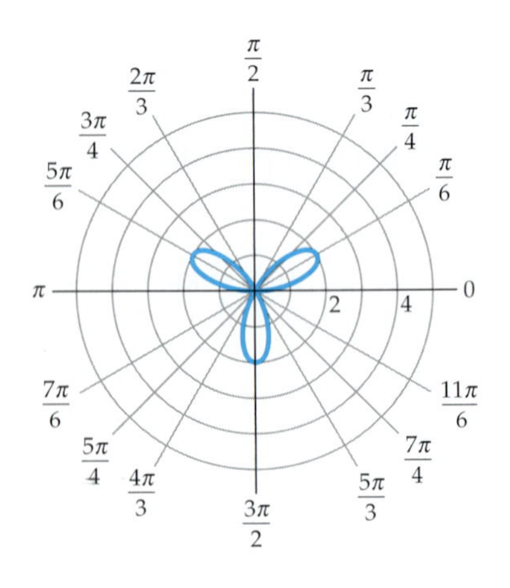

$r = 2 \sin 3\theta$

FIGURE 6.50

? ANSWER **a.** Because 3 is an odd number, there are three petals in the graph. **b.** Because 2 is an even number, there are $2(2) = 4$ petals in the graph.

Polar Equations of the Conics

The definition of a parabola was given in terms of a point (the focus) and a line (the directrix). The definitions of both the ellipse and the hyperbola were given in terms of two points (the foci). It is possible to define each conic in terms of a point and a line.

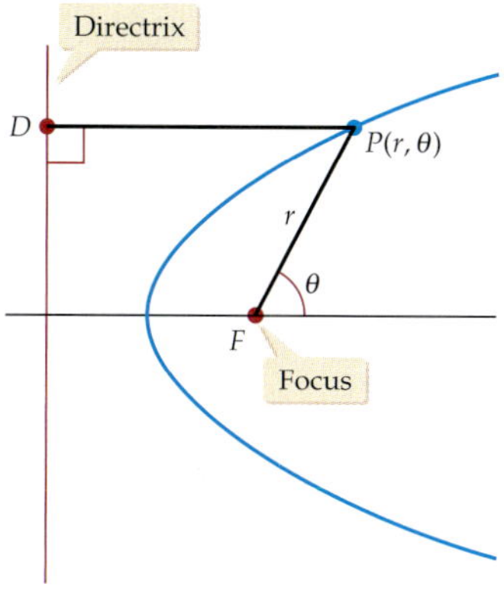

Focus-Directrix Definitions of the Conics

Let F be a fixed point and D a fixed line in a plane. Consider the set of all points P such that $\dfrac{d(P, F)}{d(P, D)} = e$, where e is a constant. The graph is a parabola for $e = 1$, an ellipse for $0 < e < 1$, and a hyperbola for $e > 1$. See **Figure 6.51.**

Figure 6.51

The fixed point is a focus of the conic, and the fixed line is a directrix. The constant e is called the **eccentricity** of the conic. Using this definition, we can derive the polar equations of the conics.

Standard Form of the Polar Equations of the Conics

Let the pole be a focus of a conic section of eccentricity e with directrix d units from the focus. Then the equation of the conic is given by one of the following:

$$r = \frac{ed}{1 + e\cos\theta} \quad (1) \qquad\qquad r = \frac{ed}{1 - e\cos\theta} \quad (2)$$

Vertical directrix to the right of the pole | **Vertical directrix to the left of the pole**

$$r = \frac{ed}{1 + e\sin\theta} \quad (3) \qquad\qquad r = \frac{ed}{1 - e\sin\theta} \quad (4)$$

Horizontal directrix above the pole | **Horizontal directrix below the pole**

When the equation involves $\cos\theta$, the polar axis is an axis of symmetry.

When the equation involves $\sin\theta$, the line $\theta = \dfrac{\pi}{2}$ is an axis of symmetry.

We will derive Equation (2). Let $P(r, \theta)$ be any point on a conic section. Then, by definition,

$$\frac{d(P, F)}{d(P, D)} = e \quad \text{or} \quad d(P, F) = e \cdot d(P, D)$$

From **Figure 6.52,** $d(P, F) = r$ and $d(P, D) = d(A, Q)$. But note that

$$d(A, Q) = d(A, F) + d(F, Q) = d + r\cos\theta$$

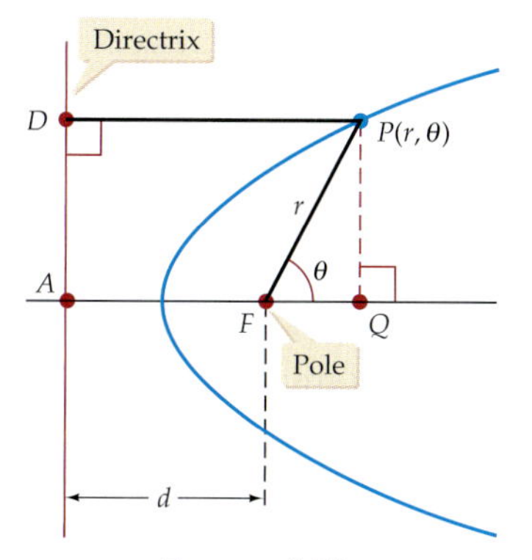

Figure 6.52

Thus

$$r = e(d + r\cos\theta) \qquad \bullet\ d(P, F) = e \cdot d(P, D)$$
$$= ed + er\cos\theta$$
$$r - er\cos\theta = ed \qquad \bullet\ \text{Subtract } er\cos\theta.$$
$$r = \frac{ed}{1 - e\cos\theta} \qquad \bullet\ \text{Solve for } r.$$

The remaining equations can be derived in a similar manner.

Graph a Conic Given in Polar Form

Given the graph of $r = \dfrac{8}{2 - 3\sin\theta}$, we can write the equation in standard form by dividing the numerator and denominator by 2, the constant term in the denominator.

$$r = \frac{8}{2 - 3\sin\theta} = \frac{\frac{8}{2}}{\frac{2}{2} - \frac{3}{2}\sin\theta} = \frac{4}{1 - \frac{3}{2}\sin\theta}$$

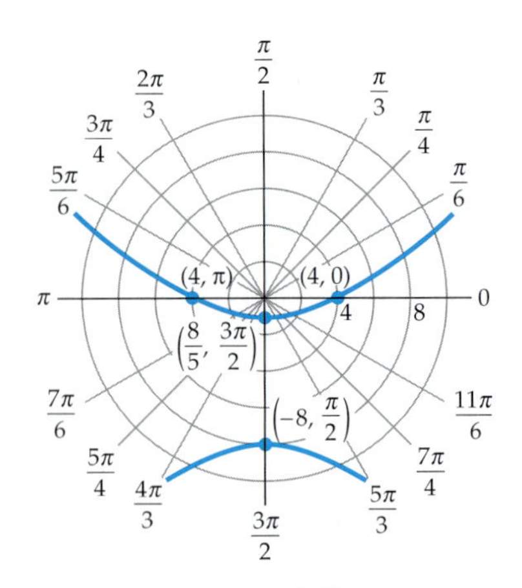

Figure 6.53

Because $e = \dfrac{3}{2} > 1$, the graph is a hyperbola with a focus at the pole. Because the equation contains the expression $\sin\theta$, the transverse axis is on the line $\theta = \dfrac{\pi}{2}$.

To find the vertices, choose θ equal to $\dfrac{\pi}{2}$ and $\dfrac{3\pi}{2}$. The corresponding values of r are -8 and $\dfrac{8}{5}$. The vertices are $\left(-8, \dfrac{\pi}{2}\right)$ and $\left(\dfrac{8}{5}, \dfrac{3\pi}{2}\right)$. By choosing θ equal to 0 and π, we can determine the points $(4, 0)$ and $(4, \pi)$ on the upper branch of the hyperbola. The lower branch can be determined by symmetry. The graph is shown in **Figure 6.53.**

EXAMPLE 5 Sketch the Graph of an Ellipse Given in Polar Form

Describe and sketch the graph of $r = \dfrac{4}{2 + \cos\theta}$.

Solution

Write the equation in standard form by dividing the numerator and denominator by 2, which is the constant term in the denominator.

$$r = \frac{2}{1 + \frac{1}{2}\cos\theta}$$

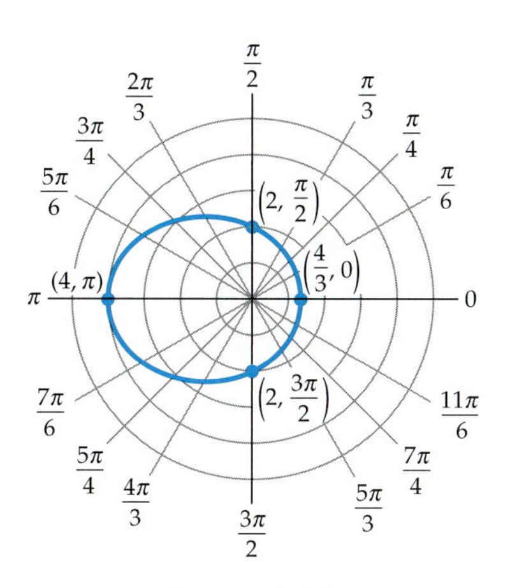

FIGURE 6.54

Thus $e = \dfrac{1}{2}$ and the graph is an ellipse with a focus at the pole. Because the equation contains the expression $\cos\theta$, the major axis is on the polar axis.

To find the vertices, choose θ equal to 0 and π. The corresponding values for r are $\dfrac{4}{3}$ and 4. The vertices on the major axis are $\left(\dfrac{4}{3}, 0\right)$ and $(4, \pi)$. Plot some points (r, θ) for additional values of θ and the corresponding values of r. Two possible points are $\left(2, \dfrac{\pi}{2}\right)$ and $\left(2, \dfrac{3\pi}{2}\right)$. See the graph of the ellipse in **Figure 6.54.**

TRY EXERCISE 20, PAGE 497

Transformations Between Rectangular and Polar Coordinates

A transformation between coordinate systems is a set of equations that relate the coordinates of a point in one system with the coordinates of the point in a second system. By superimposing a rectangular coordinate system on a polar system, we can derive the set of transformation equations.

Construct a polar coordinate system and a rectangular system such that the pole coincides with the origin and the polar axis coincides with the positive x-axis. Let a point P have coordinates (x, y) in one system and (r, θ) in the other $(r > 0)$.

From the definitions of $\sin\theta$ and $\cos\theta$, we have

$$\frac{x}{r} = \cos\theta \quad \text{or} \quad x = r\cos\theta$$

$$\frac{y}{r} = \sin\theta \quad \text{or} \quad y = r\sin\theta$$

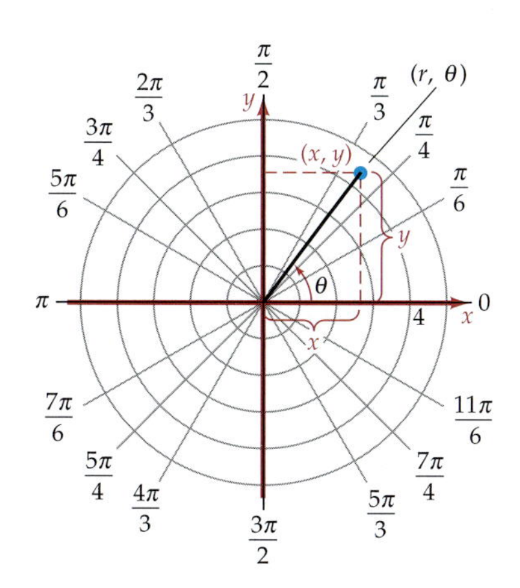

FIGURE 6.55

It can be shown that these equations are also true when $r < 0$.

Thus, given the point (r, θ) in a polar coordinate system (see **Figure 6.55**), the coordinates of the point in the xy-coordinate system are given by

$$x = r\cos\theta \qquad y = r\sin\theta$$

For example, to find the point in the xy-coordinate system that corresponds to the point $\left(4, \dfrac{2\pi}{3}\right)$ in the $r\theta$-coordinate system, substitute 4 for r and $\dfrac{2\pi}{3}$ for θ into the equations and simplify.

$$x = 4\cos\frac{2\pi}{3} = 4\left(-\frac{1}{2}\right) = -2$$

$$y = 4\sin\frac{2\pi}{3} = 4\left(\frac{\sqrt{3}}{2}\right) = 2\sqrt{3}$$

The point $\left(4, \frac{2\pi}{3}\right)$ in the $r\theta$-coordinate system is $(-2, 2\sqrt{3})$ in the xy-coordinate system.

To find the polar coordinates of a given point in the xy-coordinate system, use the Pythagorean Theorem and the definition of the tangent function. Let $P(x, y)$ be a point in the plane, and let r be the distance from the origin to the point P. Then $r = \sqrt{x^2 + y^2}$.

From the definition of the tangent function of an angle in a right triangle,

$$\tan \theta = \frac{y}{x}$$

Thus θ is the angle whose tangent is $\frac{y}{x}$. The quadrant for θ depends on the sign of x and the sign of y.

The equations of transformation between a polar and a rectangular coordinate system are summarized as follows:

Transformations between Polar and Rectangular Coordinates

Given the point (r, θ) in the polar coordinate system, the transformation equations to change from polar to rectangular coordinates are

$$x = r \cos \theta \qquad y = r \sin \theta$$

Given the point (x, y) in the rectangular coordinate system, the transformation equations to change from rectangular to polar coordinates are

$$r = \sqrt{x^2 + y^2} \qquad \tan \theta = \frac{y}{x}, \quad x \neq 0$$

where $r \geq 0$, $0 \leq \theta < 2\pi$, and θ is chosen so that the point lies in the appropriate quadrant. If $x = 0$, then $\theta = \frac{\pi}{2}$ or $\theta = \frac{3\pi}{2}$.

EXAMPLE 6 **Transform from Polar to Rectangular Coordinates**

Find the rectangular coordinates of the point whose polar coordinates are $\left(6, \frac{3\pi}{4}\right)$.

Solution

Use the equations $x = r\cos\theta$ and $y = r\sin\theta$.

$$x = 6\cos\frac{3\pi}{4} = -3\sqrt{2} \qquad y = 6\sin\frac{3\pi}{4} = 3\sqrt{2}$$

The rectangular coordinates of $\left(6, \frac{3\pi}{4}\right)$ are $\left(-3\sqrt{2}, 3\sqrt{2}\right)$.

TRY EXERCISE 40, PAGE 497

EXAMPLE 7 Transform from Rectangular to Polar Coordinates

Find the polar coordinates of the point whose rectangular coordinates are $\left(-2, -2\sqrt{3}\right)$.

Solution

Use the equations $r = \sqrt{x^2 + y^2}$ and $\tan\theta = \dfrac{y}{x}$.

$$r = \sqrt{(-2)^2 + (-2\sqrt{3})^2} = \sqrt{4 + 12} = \sqrt{16} = 4$$

$$\tan\theta = \frac{-2\sqrt{3}}{-2} = \sqrt{3}$$

From this and the fact that $\left(-2, -2\sqrt{3}\right)$ lies in the third quadrant, $\theta = \dfrac{4\pi}{3}$.

The polar coordinates of $\left(-2, -2\sqrt{3}\right)$ are $\left(4, \dfrac{4\pi}{3}\right)$.

TRY EXERCISE 44, PAGE 497

WRITE POLAR COORDINATE EQUATIONS AS RECTANGULAR EQUATIONS AND RECTANGULAR COORDINATE EQUATIONS AS POLAR EQUATIONS

Using the transformation equations, it is possible to write a polar coordinate equation in rectangular form or a rectangular coordinate equation in polar form.

EXAMPLE 8 Write a Polar Coordinate Equation in Rectangular Form

Find a rectangular form of the equation $r^2\cos 2\theta = 3$.

Continued ▶

Solution

$$r^2 \cos 2\theta = 3$$
$$r^2(1 - 2\sin^2\theta) = 3 \qquad \bullet\ \cos 2\theta = 1 - 2\sin^2\theta$$
$$r^2 - 2r^2\sin^2\theta = 3$$
$$r^2 - 2(r\sin\theta)^2 = 3$$
$$x^2 + y^2 - 2y^2 = 3 \qquad \bullet\ r^2 = x^2 + y^2;\ \sin\theta = \frac{y}{r}$$
$$x^2 - y^2 = 3$$

A rectangular form of $r^2 \cos 2\theta = 3$ is $x^2 - y^2 = 3$.

TRY EXERCISE 52, PAGE 497

EXAMPLE 9 **Write a Rectangular Coordinate Equation in Polar Form**

Find a polar form of the equation $x^2 + y^2 - 2x = 3$.

Solution

$$x^2 + y^2 - 2x = 3$$
$$(r\cos\theta)^2 + (r\sin\theta)^2 - 2r\cos\theta = 3 \qquad \bullet\ \text{Use the transformation equations } x = r\cos\theta \text{ and } y = r\sin\theta.$$
$$r^2(\cos^2\theta + \sin^2\theta) - 2r\cos\theta = 3 \qquad \bullet\ \text{Simplify.}$$
$$r^2 - 2r\cos\theta = 3$$

A polar form of $x^2 + y^2 - 2x = 3$ is $r^2 - 2r\cos\theta = 3$.

TRY EXERCISE 60, PAGE 497

TOPICS FOR DISCUSSION

1. In what quadrant is the point $(-2, 150°)$ located?
2. How many petals are in the graph of the rose curve $r = 4 \sin 3\theta$?
3. How can you determine whether the graph of $r = \dfrac{ed}{1 + e\cos\theta}$ is a parabola, an ellipse, or a hyperbola?
4. Are there two different ellipses that have a focus at the pole and a vertex at $\left(1, \dfrac{\pi}{2}\right)$? Explain.
5. Does the parabola given by $r = \dfrac{6}{1 + \sin\theta}$ have a horizontal axis of symmetry or a vertical axis of symmetry? Explain.

EXERCISE SET 6.2

In Exercises 1 to 6, plot the point on a polar coordinate system.

1. $(2, 60°)$
2. $(3, -90°)$
3. $\left(-2, \frac{\pi}{4}\right)$
4. $\left(4, \frac{7\pi}{6}\right)$
5. $\left(-3, \frac{5\pi}{3}\right)$
6. $(-3, \pi)$

In Exercises 7 to 24, sketch the graph of each polar equation.

7. $r = 3$
8. $r = 5$
9. $\theta = 2$
10. $\theta = -\frac{\pi}{3}$
11. $r = 6 \cos \theta$
12. $r = 4 \sin \theta$
13. $r = 4 \cos 2\theta$
14. $r = 5 \cos 3\theta$
15. $r = 2 - 3 \sin \theta$
16. $r = 2 - 2 \cos \theta$
17. $r = 4 + 3 \sin \theta$
18. $r = 2 + 4 \sin \theta$
19. $r = \frac{12}{3 - 6 \cos \theta}$
20. $r = \frac{8}{2 - 4 \cos \theta}$
21. $r = \frac{9}{3 - 3 \sin \theta}$
22. $r = \frac{5}{2 - 2 \sin \theta}$
23. $r = \frac{10}{5 + 6 \cos \theta}$
24. $r = \frac{8}{2 + 4 \cos \theta}$

In Exercises 25 to 36, use a graphing utility to graph each equation.

25. $r = 3 + 3 \cos \theta$
26. $r = 4 - 4 \sin \theta$
27. $r = 4 \cos 3\theta$
28. $r = 2 \sin 4\theta$
29. $r = -5 \csc \theta$
30. $r = -4 \sec \theta$
31. $r = \theta, 0 \le \theta \le 6\pi$
32. $r = -\theta, 0 \le \theta \le 6\pi$
33. $r = \frac{12}{3 - 6 \cos\left(\theta - \frac{\pi}{6}\right)}$
34. $r = \frac{8}{2 - 4 \cos\left(\theta - \frac{\pi}{2}\right)}$
35. $r = \frac{8}{4 + 3 \sin(\theta - \pi)}$
36. $r = \frac{6}{3 + 2 \cos\left(\theta - \frac{\pi}{3}\right)}$

In Exercises 37 to 44, transform the given coordinates to the indicated ordered pair.

37. $(1, -\sqrt{3})$ to (r, θ)
38. $(-2\sqrt{3}, 2)$ to (r, θ)
39. $\left(-3, \frac{2\pi}{3}\right)$ to (x, y)
40. $\left(2, -\frac{\pi}{3}\right)$ to (x, y)
41. $\left(0, -\frac{\pi}{2}\right)$ to (x, y)
42. $\left(3, \frac{5\pi}{6}\right)$ to (x, y)
43. $(3, 4)$ to (r, θ)
44. $(12, -5)$ to (r, θ)

In Exercises 45 to 56, find a rectangular form of each of the equations.

45. $r = 3 \cos \theta$
46. $r = 2 \sin \theta$
47. $r = 3 \sec \theta$
48. $r = 4 \csc \theta$
49. $r = 4$
50. $\theta = \frac{\pi}{4}$
51. $r = \tan \theta$
52. $r = \cot \theta$
53. $r = \frac{2}{1 + \cos \theta}$
54. $r = \frac{2}{1 - \sin \theta}$
55. $r = \frac{12}{3 - 6 \cos \theta}$
56. $r = \frac{8}{4 + 3 \sin \theta}$

In Exercises 57 to 64, find a polar form of each of the equations.

57. $y = 2$
58. $x = -4$
59. $x^2 + y^2 = 4$
60. $2x - 3y = 6$

61. $x^2 = 8y$

62. $y^2 = 4y$

63. $x^2 - y^2 = 25$

64. $x^2 + 4y^2 = 16$

In Exercises 65 to 70, find a polar equation of the conic with focus at the pole and the given eccentricity and directrix.

65. $e = 2, r\cos\theta = -1$

66. $e = \dfrac{3}{2}, r\sin\theta = 1$

67. $e = 1, r\sin\theta = 2$

68. $e = 1, r\cos\theta = -2$

69. $e = \dfrac{2}{3}, r\sin\theta = -4$

70. $e = \dfrac{1}{2}, r\cos\theta = 2$

71. Find the polar equation of the parabola with a focus at the pole and vertex $(2, \pi)$.

72. Find the polar equation of the ellipse with a focus at the pole, vertex at $(4, 0)$, and eccentricity $\dfrac{1}{2}$.

73. Find the polar equation of the hyperbola with a focus at the pole, vertex at $\left(1, \dfrac{3\pi}{2}\right)$, and eccentricity 2.

74. Find the polar equation of the ellipse with a focus at the pole, vertex at $\left(2, \dfrac{3\pi}{2}\right)$, and eccentricity $\dfrac{2}{3}$.

CONNECTING CONCEPTS

75.  Explain why the graph of $r^2 = \cos^2\theta$ and the graph of $r = \cos\theta$ are not the same.

76.  Explain why the graph of $r = \cos 2\theta$ and the graph of $r = 2\cos^2\theta - 1$ are identical.

77. Let $P(r, \theta)$ satisfy the equation $r = \dfrac{ed}{1 - e\cos\theta}$. Show that $\dfrac{d(P, F)}{d(P, D)} = e$.

78. Show that the equation of a conic with a focus at the pole and directrix $r\sin\theta = d$ is given by $r = \dfrac{ed}{1 + e\sin\theta}$.

 **In Exercises 79 to 86, use a graphing utility to graph each equation.**

79. $r^2 = 4\cos 2\theta$ (lemniscate)

80. $r^2 = -2\sin 2\theta$ (lemniscate)

81. $r = 2(1 + \sec\theta)$ (conchoid)

82. $r = 2\cos 2\theta\sec\theta$ (strophoid)

83. $r\theta = 2$ (spiral)

84. $r = 2\sin\theta\cos^2 2\theta$ (bifolium)

85. $r = |\theta|$

86. $r = \ln\theta$

87. The graph of

$$r = 1.5^{\sin\theta} - 2.5\cos 4\theta + \sin^7\frac{\theta}{15}$$

is a *butterfly curve* similar to the one shown below.

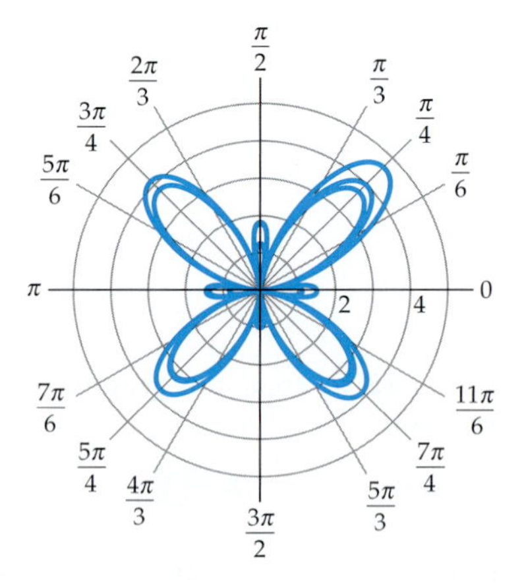

Use a graphing utility to graph the butterfly curve for

a. $0 \le \theta \le 5\pi$ **b.** $0 \le \theta \le 20\pi$

For additional information on butterfly curves, read "The Butterfly Curve" by Temple H. Fay, *The American Mathematical Monthly,* vol. 96, no. 5 (May 1989), p. 442.

PREPARE FOR SECTION 6.3

88. Complete the square of $y^2 + 3y$ and write the result as the square of a binomial. [1.1]

89. If $x = 2t + 1$ and $y = x^2$, write y in terms of t.

90. Identify the graph of $\left(\frac{x-2}{3}\right)^2 + \left(\frac{y-3}{2}\right)^2 = 1$. [6.1]

91. Let $x = \sin t$ and $y = \cos t$. What is the value of $x^2 + y^2$? [5.2]

92. Solve $y = \ln t$ for t. [3.5]

93. What are the domain and range of $f(t) = 3 \cos 2t$? Write the answer using interval notation. [4.3]

PROJECTS

1. A Polar Distance Formula Let $P_1(r_1, \theta_1)$ and $P_2(r_2, \theta_2)$ be two distinct points in the $r\theta$-plane.

a. Verify that the distance d between the points is

$$d = \sqrt{r_1^2 + r_2^2 - 2r_1r_2 \cos(\theta_2 - \theta_1)}$$

b. Use the above formula to find the distance (to the nearest hundredth) between (3, 60°) and (5, 170°).

c. Does the formula $d = \sqrt{r_1^2 + r_2^2 - 2r_1r_2 \cos(\theta_1 - \theta_2)}$ also produce the correct distance between P_1 and P_2? Explain.

2. Another Polar Form for a Circle

a. Verify that the graph of the polar equation $r = a \sin \theta + b \cos \theta$ is a circle. Assume that a and b are not both 0.

b. What are the center (in rectangular coordinates) and the radius of the circle?

SECTION 6.3 PARAMETRIC EQUATIONS

- Parametric Equations
- Graph a Curve Given by Parametric Equations
- Eliminate the Parameter of a Parametric Equation
- The Brachistochrone Problem
- Parametric Equations and Projectile Motion

Parametric Equations

The graph of a function is a graph for which no vertical line can intersect the graph more than once. For a graph that is not the graph of a function (an ellipse or a hyperbola, for example), it is frequently useful to describe the graph by *parametric equations.*

Curve and Parametric Equations

Let t be a number in an interval I. A **curve** is the set of ordered pairs (x, y), where

$$x = f(t), \qquad y = g(t) \quad \text{for } t \in I$$

The variable t is called a **parameter,** and the equations $x = f(t)$ and $y = g(t)$ are **parametric equations.**

For instance,

$$x = 2t - 1, \qquad y = 4t + 1 \quad \text{for } t \in (-\infty, \infty)$$

is an example of a set of parametric equations. By choosing arbitrary values of t, ordered pairs (x, y) can be created, as shown in the table below.

t	x = 2t − 1	y = 4t + 1	(x, y)
−2	−5	−7	(−5, −7)
0	−1	1	(−1, 1)
$\frac{1}{2}$	0	3	(0, 3)
2	3	9	(3, 9)

By plotting the points and drawing a curve through the points, a graph of the parametric equations is produced. See **Figure 6.56.**

QUESTION If $x = t^2 + 1$ and $y = 3 - t$, what ordered pair corresponds to $t = -3$?

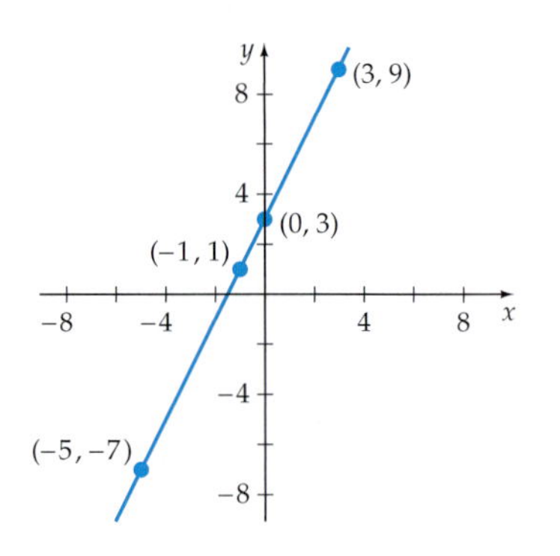

FIGURE 6.56

GRAPH A CURVE GIVEN BY PARAMETRIC EQUATIONS

EXAMPLE 1 **Sketch the Graph of a Curve Given in Parametric Form**

Sketch the graph of the curve given by the parametric equations

$$x = t^2 + t, \qquad y = t - 1 \quad \text{for } t \in (-\infty, \infty)$$

Solution

Begin by making a table of values of t and the corresponding values of x and y. Five values of t were arbitrarily chosen for the table that follows. Many more values might be necessary to determine an accurate graph.

ANSWER (10, 6)

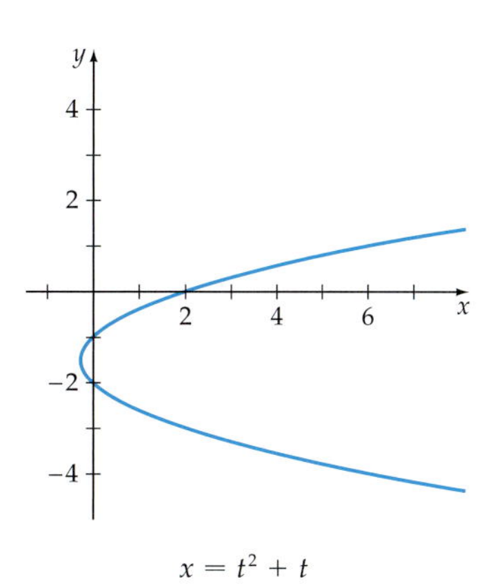

$x = t^2 + t$
$y = t - 1$

FIGURE 6.57

t	$x = t^2 + t$	$y = t - 1$	(x, y)
−2	2	−3	(2, −3)
−1	0	−2	(0, −2)
0	0	−1	(0, −1)
1	2	0	(2, 0)
2	6	1	(6, 1)

Graph the ordered pairs (x, y) and then draw a smooth curve through the points. See **Figure 6.57.**

TRY EXERCISE 6, PAGE 506

ELIMINATE THE PARAMETER OF A PARAMETRIC EQUATION

It may not be clear from Example 1 and the corresponding graph that the curve is a parabola. By **eliminating the parameter,** we can write one equation in x and y that is equivalent to the two parametric equations.

To eliminate the parameter, solve $y = t - 1$ for t.

$$y = t - 1 \qquad \text{or} \qquad t = y + 1$$

Substitute $y + 1$ for t in $x = t^2 + t$ and then simplify.

$$\begin{aligned} x &= (y + 1)^2 + (y + 1) \\ &= y^2 + 3y + 2 \end{aligned}$$

• **The equation of a parabola**

Complete the square and write the equation in standard form.

$$\left(x + \frac{1}{4}\right) = \left(y + \frac{3}{2}\right)^2$$

• **This is the equation of a parabola with vertex at $\left(-\frac{1}{4}, -\frac{3}{2}\right)$.**

EXAMPLE 2 Eliminate the Parameter and Sketch the Graph of a Curve

Eliminate the parameter and sketch the curve of the parametric equations

$$x = \sin t, \qquad y = \cos t \quad \text{for } 0 \le t < 2\pi$$

Solution

The process of eliminating the parameter sometimes involves trigonometric identities. To eliminate the parameter for the equations, square each side of each equation and then add.

$$\begin{aligned} x^2 &= \sin^2 t \\ y^2 &= \cos^2 t \\ x^2 + y^2 &= \sin^2 t + \cos^2 t \end{aligned}$$

Continued ▶

Thus, using the trigonometric identity $\sin^2 t + \cos^2 t = 1$, we get

$$x^2 + y^2 = 1$$

This is the equation of a circle with center (0, 0) and radius equal to 1. See **Figure 6.58.**

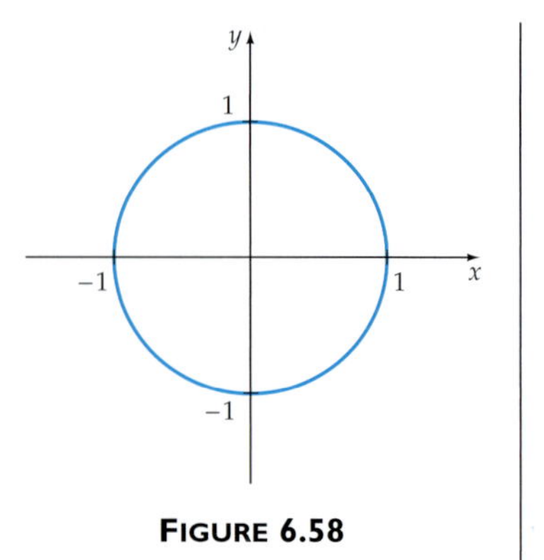

FIGURE 6.58

▶ TRY EXERCISE 12, PAGE 506

A parametric representation of a curve is not unique. That is, it is possible that a curve may be given by many different pairs of parametric equations. We will demonstrate this by using the equation of a line and providing two different parametric representations of the line.

Consider a line with slope m passing through the point (x_1, y_1). By the point-slope formula, the equation of the line is

$$y - y_1 = m(x - x_1)$$

Let $t = x - x_1$. Then $y - y_1 = mt$. A parametric representation is

$$x = x_1 + t, \qquad y = y_1 + mt \quad \text{for } t \text{ a real number} \tag{1}$$

Let $x - x_1 = \cot t$. Then $y - y_1 = m \cot t$. A parametric representation is

$$x = x_1 + \cot t, \qquad y = y_1 + m \cot t \quad \text{for } 0 < t < \pi \tag{2}$$

It can be verified that Equations (1) and (2) represent the original line.

Example 3 illustrates that the domain of the parameter t can be used to determine the domain and range of the function.

EXAMPLE 3 Sketch the Graph of a Curve Given by Parametric Equations

Eliminate the parameter and sketch the graph of the curve given by the parametric equations

$$x = 2 + 3\cos t, \qquad y = 3 + 2\sin t \quad \text{for } 0 \le t \le \pi$$

Solution

Rewrite each equation in terms of the trigonometric function.

$$\frac{x-2}{3} = \cos t \qquad \frac{y-3}{2} = \sin t \tag{3}$$

Using the trigonometric identity $\cos^2 t + \sin^2 t = 1$, we have

$$\cos^2 t + \sin^2 t = \left(\frac{x-2}{3}\right)^2 + \left(\frac{y-3}{2}\right)^2 = 1$$
$$\frac{(x-2)^2}{9} + \frac{(y-3)^2}{4} = 1$$

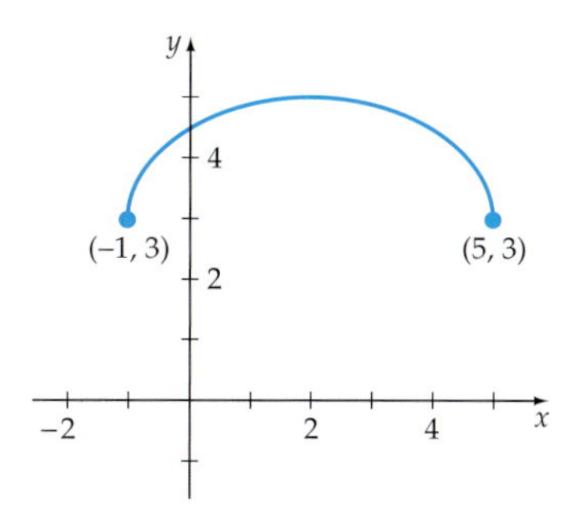

FIGURE 6.59

This is the equation of an ellipse with center at (2, 3) and major axis parallel to the x-axis. However, because $0 \le t \le \pi$, it follows that $-1 \le \cos t \le 1$ and $0 \le \sin t \le 1$. Therefore, we have

$$-1 \le \frac{x-2}{3} \le 1 \qquad 0 \le \frac{y-3}{2} \le 1 \qquad \text{• Using Equations (3)}$$

Solving these inequalities for x and y yields

$$-1 \le x \le 5 \quad \text{and} \quad 3 \le y \le 5$$

Because the values of y are between 3 and 5, the graph of the parametric equations is only the top half of the ellipse. See **Figure 6.59.**

TRY EXERCISE 14, PAGE 506

THE BRACHISTOCHRONE PROBLEM

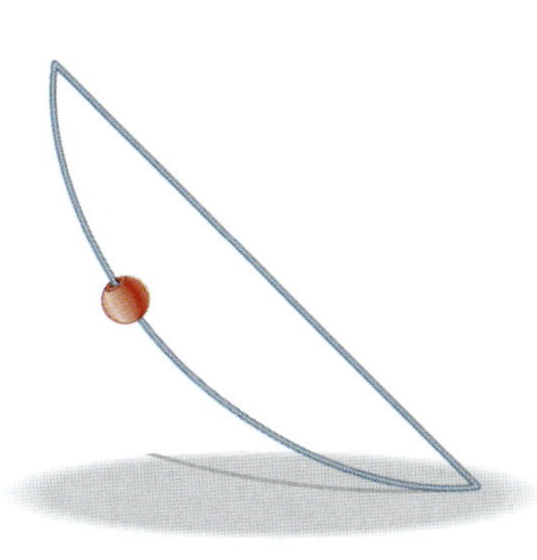
FIGURE 6.60

Parametric equations are useful in writing the equation of a moving point. One famous problem, involving a bead traveling down a frictionless wire, was posed in 1696 by the mathematician Johann Bernoulli. The problem was to determine the shape of a wire a bead could slide down such that the distance between two points was traveled in the shortest time. Problems that involve "shortest time" are called *brachistochrone problems.* They are very important in physics and form the basis for much of the classical theory of light propagation.

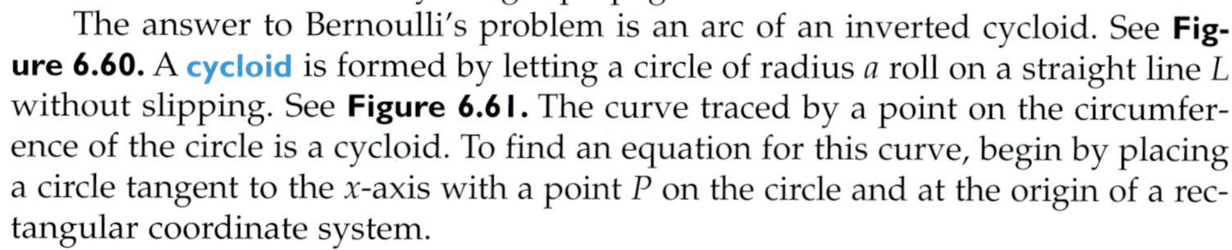

The answer to Bernoulli's problem is an arc of an inverted cycloid. See **Figure 6.60.** A **cycloid** is formed by letting a circle of radius a roll on a straight line L without slipping. See **Figure 6.61.** The curve traced by a point on the circumference of the circle is a cycloid. To find an equation for this curve, begin by placing a circle tangent to the x-axis with a point P on the circle and at the origin of a rectangular coordinate system.

Roll the circle along the x-axis. After the radius of the circle has rotated through an angle θ, the coordinates of the point $P(x, y)$ can be given by

$$x = h - a\sin\theta, \qquad y = k - a\cos\theta \tag{4}$$

where $C(h, k)$ is the current center of the circle.

Because the radius of the circle is a, $k = a$. See **Figure 6.61.** Because the circle rolls without slipping, the arc length subtended by θ equals h. Thus $h = a\theta$. Substituting for h and k in Equations (4), we have, after factoring,

$$x = a(\theta - \sin\theta), \qquad y = a(1 - \cos\theta) \quad \text{for } \theta \ge 0$$

See **Figure 6.62.**

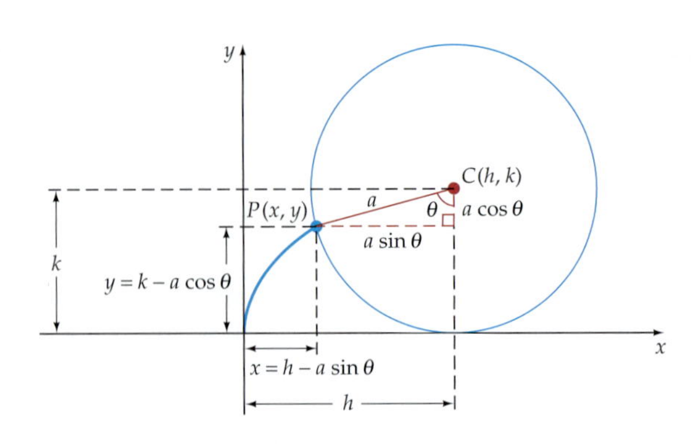

FIGURE 6.61

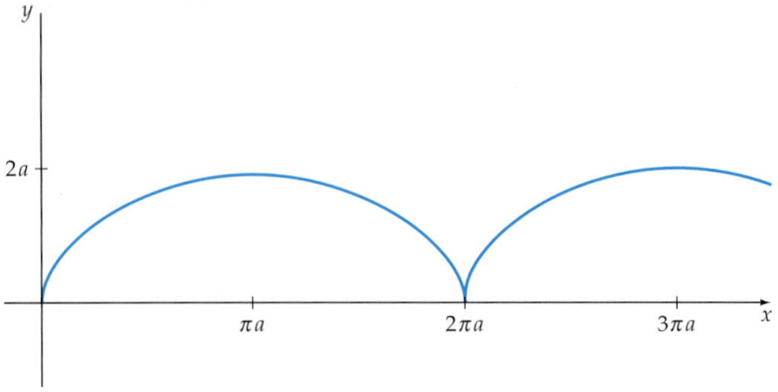

$x = a(\theta - \sin\theta), y = a(1 - \cos\theta)$

FIGURE 6.62
A cycloid

EXAMPLE 4 Graph a Cycloid

Use a graphing utility to graph the cycloid given by

$$x = 4(\theta - \sin\theta), \qquad y = 4(1 - \cos\theta) \quad \text{for } 0 \le \theta \le 4\pi$$

Solution

Although θ is the parameter in the above equations, many graphing utilities, such as the TI-83, use T as the parameter for parametric equations. Thus to graph the equations for $0 \le \theta \le 4\pi$, we use Tmin = 0 and Tmax = 4π, as shown below. Use radian mode and parametric mode to produce the graph in **Figure 6.63.**

Tmin=0	Xmin=-6	Ymin=-4
Tmax=4π	Xmax=16π	Ymax=10
Tstep=0.5	Xscl=2π	Yscl=1

▶ TRY EXERCISE 26, PAGE 506

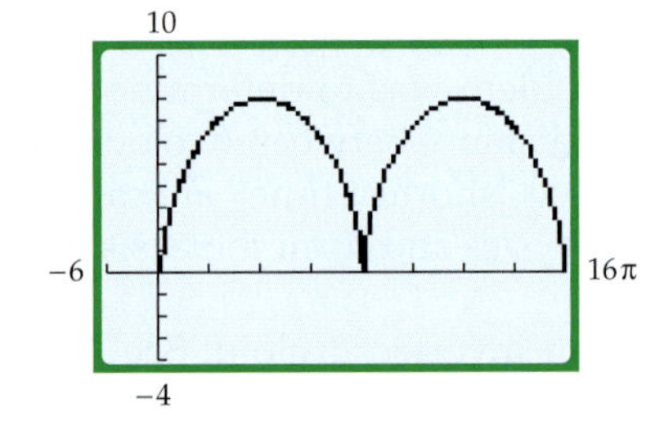

$x = 4(\theta - \sin\theta)$
$y = 4(1 - \cos\theta)$

FIGURE 6.63

PARAMETRIC EQUATIONS AND PROJECTILE MOTION

The path of a projectile (assume air resistance is negligible) that is launched at an angle θ from the horizon with an initial velocity of v_0 feet per second is given by the parametric equations

$$x = (v_0 \cos\theta)t, \qquad y = -16t^2 + (v_0 \sin\theta)t$$

where t is the time in seconds since the projectile was launched.

EXAMPLE 5 Sketch the Path of a Projectile

Use a graphing utility to sketch the path of a projectile that is launched at an angle of $\theta = 32°$ with an initial velocity of 144 feet per second. Use the graph to determine (to the nearest foot) the maximum height of the projectile and the range of the projectile. Assume the ground is level.

Solution

Use degree mode and parametric mode. Graph the parametric equations

$$x = (144 \cos 32°)t, \qquad y = -16t^2 + (144 \sin 32°)t \quad \text{for } 0 \leq t \leq 5$$

to produce the graph in **Figure 6.64.** Use the TRACE feature to determine that the maximum height of 91 feet is attained when $t \approx 2.38$ seconds and that the projectile strikes the ground about 581 feet downrange when $t \approx 4.76$ seconds.

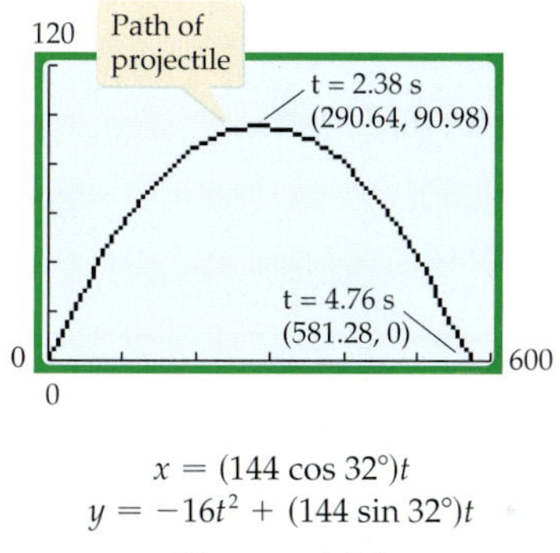

$x = (144 \cos 32°)t$
$y = -16t^2 + (144 \sin 32°)t$

FIGURE 6.64

In **Figure 6.64,** the angle of launch does not appear to be 32° because 1 foot on the x-axis is smaller than 1 foot on the y-axis.

▶ TRY EXERCISE 32, PAGE 507

TOPICS FOR DISCUSSION

1. It is always possible to eliminate the parameter of a pair of parametric equations. Do you agree? Explain.
2. The line $y = 3x + 5$ has more than one parametric representation. Do you agree? Explain.
3. Parametric equations are used only to graph functions. Do you agree? Explain.
4. Every function $y = f(x)$ can be written in parametric form by letting $x = t$ and $y = f(t)$. Do you agree? Explain.

EXERCISE SET 6.3

In Exercises 1 to 10, graph the parametric equations by plotting several points.

1. $x = 2t, y = -t$, for $t \in R$

2. $x = -3t, y = 6t$, for $t \in R$

3. $x = -t, y = t^2 - 1$, for $t \in R$

4. $x = 2t, y = 2t^2 - t + 1$, for $t \in R$

5. $x = t^2, y = t^3$, for $t \in R$

▶ **6.** $x = t^2 + 1, y = t^2 - 1$, for $t \in R$

7. $x = 2 \cos t, y = 3 \sin t$, for $0 \le t < 2\pi$

8. $x = 1 - \sin t, y = 1 + \cos t$, for $0 \le t < 2\pi$

9. $x = 2^t, y = 2^{t+1}$, for $t \in R$

10. $x = t^2, y = 2 \log_2 t$, for $t \ge 1$

In Exercises 11 to 20, eliminate the parameter and graph the equation.

11. $x = \sec t, y = \tan t$, for $-\frac{\pi}{2} < t < \frac{\pi}{2}$

▶ **12.** $x = 3 + 2 \cos t, y = -1 - 3 \sin t$, for $0 \le t < 2\pi$

13. $x = 2 - t^2, y = 3 + 2t^2$, for $t \in R$

▶ **14.** $x = 1 + t^2, y = 2 - t^2$, for $t \in R$

15. $x = \cos^3 t, y = \sin^3 t$, for $0 \le t < 2\pi$

16. $x = e^{-t}, y = e^t$, for $t \in R$

17. $x = \sqrt{t + 1}, y = t$, for $t \ge -1$

18. $x = \sqrt{t}, y = 2t - 1$, for $t \ge 0$

19. $x = t^3, y = 3 \ln t$, for $t > 0$

20. $x = e^t, y = e^{2t}$, for $t \in R$

21. Eliminate the parameter for the curves

$$C_1: \quad x = 2 + t^2, \quad y = 1 - 2t^2$$

and

$$C_2: \quad x = 2 + t, \quad y = 1 - 2t$$

and then discuss the differences between their graphs.

22. Eliminate the parameter for the curves

$$C_1: \quad x = \sec^2 t, \quad y = \tan^2 t$$

and

$$C_2: \quad x = 1 + t^2, \quad y = t^2$$

for $0 \le t < \frac{\pi}{2}$, and then discuss the differences between their graphs.

23. Sketch the graph of

$$x = \sin t, \qquad y = \csc t \quad \text{for } 0 < t \le \frac{\pi}{2}$$

Sketch another graph for the same pair of equations but choose the domain of t to be $\pi < t \le \frac{3\pi}{2}$.

24. Discuss the differences between

$$C_1: \quad x = \cos t, \quad y = \cos^2 t$$

and

$$C_2: \quad x = \sin t, \quad y = \sin^2 t$$

for $0 \le t \le \pi$.

25. Use a graphing utility to graph the cycloid $x = 2(t - \sin t), y = 2(1 - \cos t)$ for $0 \le t < 2\pi$.

▶ **26.** Use a graphing utility to graph the cycloid $x = 3(t - \sin t), y = 3(1 - \cos t)$ for $0 \le t \le 12\pi$.

Parametric equations of the form $x = a \sin \alpha t$, $y = b \cos \beta t$, for $t \ge 0$, are encountered in electrical circuit theory. The graphs of these equations are called *Lissajous figures*. Use Tstep = 0.5 and $0 \le T \le 2\pi$.

27. Graph: $x = 5 \sin 2t, y = 5 \cos t$

28. Graph: $x = 5 \sin 3t, y = 5 \cos 2t$

29. Graph: $x = 4 \sin 2t, y = 4 \cos 3t$

30. Graph: $x = 5 \sin 10t, y = 5 \cos 9t$

In Exercises 31 to 34, graph the path of the projectile that is launched at an angle of θ with the horizon with an initial velocity of v_0. In each exercise, use the graph to determine the maximum height and the range of the projectile (to the nearest foot). Also state the time t at which the projectile reaches its maximum height and the time it hits the ground. Assume the ground is level and the only force acting on the projectile is gravity.

31. $\theta = 55°$, $v_0 = 210$ feet per second

32. $\theta = 35°$, $v_0 = 195$ feet per second

33. $\theta = 42°$, $v_0 = 315$ feet per second

34. $\theta = 52°$, $v_0 = 315$ feet per second

CONNECTING CONCEPTS

35. Let $P_1(x_1, y_1)$ and $P_2(x_2, y_2)$ be two distinct points in the plane, and consider the line L passing through those points. Choose a point $P(x, y)$ on the line L. Show that

$$\frac{x - x_1}{x_2 - x_1} = \frac{y - y_1}{y_2 - y_1}$$

Use this result to demonstrate that $x = (x_2 - x_1)t + x_1$, $y = (y_2 - y_1)t + y_1$ is a parametric representation of the line through the two points.

36. Show that $x = h + a \sin t$, $y = k + b \cos t$, for $a > 0$, $b > 0$, and $0 \le t < 2\pi$, are parametric equations for an ellipse with center at (h, k).

37. Suppose a string, held taut, is unwound from the circumference of a circle of radius a. The path traced by the end of the string is called the *involute* of a circle. Find parametric equations for the involute of a circle.

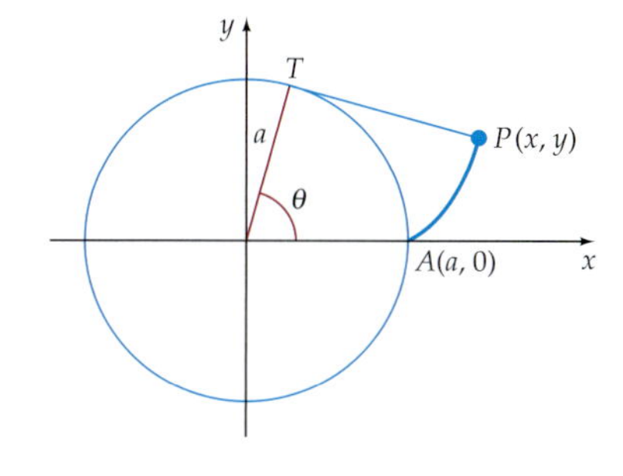

38. A circle of radius a rolls without slipping on the outside of a circle of radius $b > a$. Find the parametric equations of a point P on the smaller circle. The curve is called an *epicycloid*.

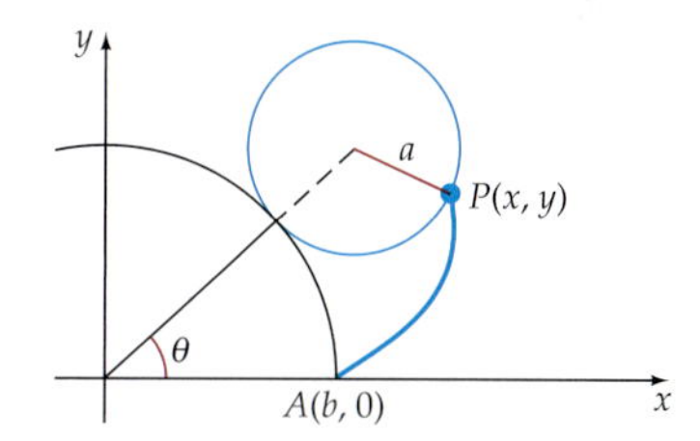

39. A circle of radius a rolls without slipping on the inside of a circle of radius $b > a$. Find the parametric equations of a point P on the smaller circle. The curve is called a *hypocycloid*.

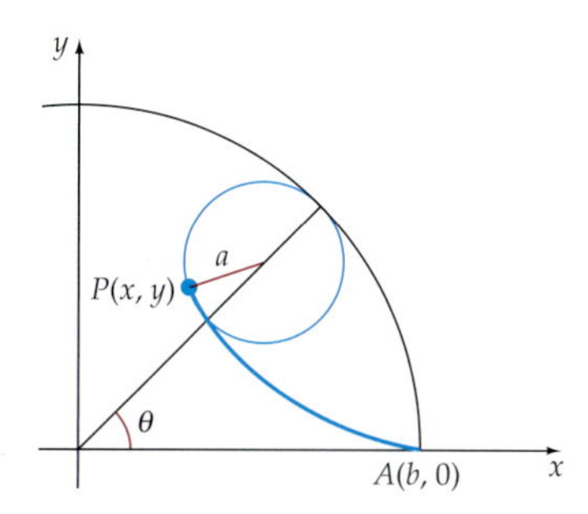

PREPARE FOR SECTION 6.4

40. Evaluate $f(n) = \dfrac{n}{n+1}$ when $n = 1, 2$, and 3. [1.3]

41. Solve $a = b + (n - 1)d$ for n when $a = 38$, $b = 5$, and $d = 3$.

42. Simplify: $\dfrac{3^{i+1}}{3^i}$ [A.1]

43. Let $f(n) = 3n - 2$. Write $f(n + 1) - f(n)$ in simplest form. [1.3]

44. Let $f(n) = \left(\dfrac{1}{2}\right)^n$. Write $\dfrac{f(n+1)}{f(n)}$ in simplest form. [1.3]

45. Find $f(1) + f(2) + f(3) + f(4)$ when $f(n) = n^2$. [1.3]

PROJECTS

1. PARAMETRIC EQUATIONS IN AN XYZ-COORDINATE SYSTEM

a. Graph the three-dimensional curve given by

$$x = 3\cos t, \qquad y = 3\sin t, \qquad z = 0.5t$$

b. Graph the three-dimensional curve given by

$$x = 3\cos t, \qquad y = 6\sin t, \qquad z = 0.5t$$

c. What is the main difference between these curves?

d. What name is given to curves of this type?

SECTION 6.4

SEQUENCES, SERIES, AND SUMMATION NOTATION

- INFINITE SEQUENCES
- FACTORIALS
- ARITHMETIC SEQUENCES
- GEOMETRIC SEQUENCES
- PARTIAL SUMS AND SUMMATION NOTATION
- ARITHMETIC SERIES
- FINITE GEOMETRIC SERIES
- INFINITE GEOMETRIC SERIES

INFINITE SEQUENCES

The *ordered* list of numbers 2, 4, 8, 16, 32, ... is called an *infinite sequence.* The list is ordered simply because order makes a difference. The sequence 2, 8, 4, 16, 32, ... contains the same numbers but in a different order. Therefore, it is a different infinite sequence.

An infinite sequence can be thought of as a pairing between positive integers and real numbers. For example, 1, 4, 9, 16, 25, 36, ..., n^2, ... pairs a natural number with its square.

$$\begin{array}{cccccccccc} 1 & 2 & 3 & 4 & 5 & 6 & \ldots & n & \ldots \\ \downarrow & \downarrow & \downarrow & \downarrow & \downarrow & \downarrow & & \downarrow & \\ 1 & 4 & 9 & 16 & 25 & 36 & \ldots & n^2 & \ldots \end{array}$$

This pairing of numbers enables us to define an infinite sequence as a function whose domain is the positive integers.

Infinite Sequence

An **infinite sequence** is a function whose domain is the positive integers and whose range is a set of real numbers.

Although the positive integers do not include zero, it is occasionally convenient to include zero in the domain of an infinite sequence. Also, we will frequently use the word *sequence* instead of the phrase *infinite sequence.*

As an example of a sequence, let $f(n) = 2n - 1$. The range of this function is

$$\begin{array}{ccccccc} f(1), & f(2), & f(3), & f(4), \ldots, & f(n), & \ldots \\ 1, & 3, & 5, & 7, \ldots, & 2n-1, & \ldots \end{array}$$

The elements in the range of a sequence are called the **terms** of the sequence. For our example, the terms are 1, 3, 5, 7, ..., $2n - 1$, ... The **first term** of the sequence is 1, the **second term** is 3, and so on. The **nth term,** or the **general term,** is $2n - 1$.

? QUESTION What is the fifth term of the preceding sequence $f(n) = 2n - 1$?

Rather than use functional notation for sequences, it is customary to use a subscript notation. Thus a_n represents the nth term of a sequence. Using this notation, we would write

$$a_n = 2n - 1$$

Thus $a_1 = 1$, $a_2 = 3$, $a_3 = 5$, $a_4 = 7$, and so on.

EXAMPLE 1 Find the Terms of a Sequence

a. Find the first three terms of the sequence $a_n = \dfrac{1}{n(n+1)}$.

b. Find the eighth term of the sequence $a_n = \dfrac{2^n}{n^2}$.

Solution

a. $a_1 = \dfrac{1}{1(1+1)} = \dfrac{1}{2}$, $a_2 = \dfrac{1}{2(2+1)} = \dfrac{1}{6}$, $a_3 = \dfrac{1}{3(3+1)} = \dfrac{1}{12}$

b. $a_8 = \dfrac{2^8}{8^2} = \dfrac{256}{64} = 4$

▶ TRY EXERCISE 6, PAGE 522

An **alternating sequence** is one in which the signs of the terms *alternate* between positive and negative values. The sequence defined by $a_n = (-1)^{n+1} \cdot \dfrac{1}{n}$ is an alternating sequence.

$$a_1 = (-1)^{1+1} \cdot \frac{1}{1} = 1 \qquad a_2 = (-1)^{2+1} \cdot \frac{1}{2} = -\frac{1}{2} \qquad a_3 = (-1)^{3+1} \cdot \frac{1}{3} = \frac{1}{3}$$

The first six terms of the sequence are

$$1, -\frac{1}{2}, \frac{1}{3}, -\frac{1}{4}, \frac{1}{5}, -\frac{1}{6}$$

A **recursively defined sequence** is one in which each succeeding term of the sequence is defined by using some of the preceding terms. For example, let $a_1 = 1$, $a_2 = 1$, and $a_{n+1} = a_{n-1} + a_n$.

$$a_3 = a_1 + a_2 = 1 + 1 = 2 \qquad \bullet\ n = 2$$
$$a_4 = a_2 + a_3 = 1 + 2 = 3 \qquad \bullet\ n = 3$$
$$a_5 = a_3 + a_4 = 2 + 3 = 5 \qquad \bullet\ n = 4$$
$$a_6 = a_4 + a_5 = 3 + 5 = 8 \qquad \bullet\ n = 5$$

MATH MATTERS

Recursive subroutines are a part of many computer programming languages. One of the most influential people in the development of computer languages (especially COBOL) was Grace Murray Hopper. She was the first woman to receive a doctorate in mathematics from Yale University. She went on to become an admiral in the U.S. Navy.

? ANSWER $f(5) = 2(5) - 1 = 9$

This recursive sequence 1, 1, 2, 3, 5, 8, . . . is called the **Fibonacci sequence,** named after Leonardo Fibonacci (1180?–?1250), an Italian mathematician.

EXAMPLE 2 Find Terms of a Sequence Defined Recursively

Let $a_1 = 1$ and $a_n = na_{n-1}$. Find a_2, a_3, and a_4.

Solution

$a_2 = 2a_1 = 2 \cdot 1 = 2$ $\quad a_3 = 3a_2 = 3 \cdot 2 = 6$ $\quad a_4 = 4a_3 = 4 \cdot 6 = 24$

▶ TRY EXERCISE 18, PAGE 522

FACTORIALS

It is possible to find an nth term formula for the sequence defined recursively in Example 2 by

$$a_1 = 1 \qquad a_n = na_{n-1}$$

Consider the term a_5 of that sequence.

$$\begin{aligned} a_5 &= 5a_4 \\ &= 5 \cdot 4a_3 && \bullet\ a_4 = 4a_3 \\ &= 5 \cdot 4 \cdot 3a_2 && \bullet\ a_3 = 3a_2 \\ &= 5 \cdot 4 \cdot 3 \cdot 2a_1 && \bullet\ a_2 = 2a_1 \\ &= 5 \cdot 4 \cdot 3 \cdot 2 \cdot 1 && \bullet\ a_1 = 1 \end{aligned}$$

Continuing in this manner for a_n, we have

$$\begin{aligned} a_n &= na_{n-1} \\ &= n(n-1)a_{n-2} \\ &= n(n-1)(n-2)a_{n-3} \\ &\ \vdots \\ &= n(n-1)(n-2)(n-3)\cdots 2 \cdot 1 \end{aligned}$$

The number $n \cdot (n-1)\cdots 3 \cdot 2 \cdot 1$ is called ***n* factorial** and is written $n!$.

The Factorial of a Number

If n is a positive integer, then $n!$, which is read "n factorial," is

$$n! = n \cdot (n-1) \cdots 3 \cdot 2 \cdot 1$$

We also define

$$0! = 1$$

MATH MATTERS

The factorial symbol as an exclamation point was first introduced in 1808 so that printers setting the type in a book would not have to write out, for example, $5 \cdot 4 \cdot 3 \cdot 2 \cdot 1$ but instead could write 5!. This notation was not accepted by everyone, and alternative suggestions were made. One other notation was $\underline{n|}$. Although this angle bracket saved the printer from writing a product, it had to be constructed, whereas the exclamation point was already in the printer's tray.

It may seem strange to define $0! = 1$, but we shall see later that it is a reasonable definition.

Examples of factorials include

$$5! = 5 \cdot 4 \cdot 3 \cdot 2 \cdot 1 = 120$$
$$10! = 10 \cdot 9 \cdot 8 \cdot 7 \cdot 6 \cdot 5 \cdot 4 \cdot 3 \cdot 2 \cdot 1 = 3{,}628{,}800$$

Note that we can write 12! as

$$12! = 12 \cdot 11! = 12 \cdot 11 \cdot 10! = 12 \cdot 11 \cdot 10 \cdot 9!$$

In general,

$$n! = n \cdot (n - 1)!$$

EXAMPLE 3 Evaluate Factorial Expressions

Evaluate each factorial expression. **a.** $\frac{8!}{5!}$ **b.** $6! - 4!$

Solution

a. $\frac{8!}{5!} = \frac{8 \cdot 7 \cdot 6 \cdot 5!}{5!} = 8 \cdot 7 \cdot 6 = 336$

b. $6! - 4! = (6 \cdot 5 \cdot 4 \cdot 3 \cdot 2 \cdot 1) - (4 \cdot 3 \cdot 2 \cdot 1) = 720 - 24 = 696$

▶ TRY EXERCISE 28, PAGE 522

ARITHMETIC SEQUENCES

Note that in the sequence

$$2, 5, 8, 11, 14, \ldots, 3n - 1, \ldots$$

the difference between successive terms is always 3. Such a sequence is an *arithmetic sequence* or an *arithmetic progression*. These sequences have the following property: The difference between successive terms is the same constant. This constant is called the *common difference*. For the sequence above, the common difference is 3.

In general, an arithmetic sequence can be defined as follows:

Arithmetic Sequence

Let d be a real number. A sequence a_n is an **arithmetic sequence** if

$$a_{i+1} - a_i = d \quad \text{for all } i$$

The number d is the **common difference** for the sequence.

Further examples of arithmetic sequences include

$$3, 8, 13, 18, \ldots, 5n - 2, \ldots$$
$$11, 7, 3, -1, \ldots, -4n + 15, \ldots$$
$$1, 2, 3, 4, \ldots, n, \ldots$$

? QUESTION Is the sequence $2, 6, 10, 14, \ldots, 4n - 2, \ldots$ an arithmetic sequence?

Consider an arithmetic sequence in which the first term is a_1 and the common difference is d. By adding the common difference to each successive term of the arithmetic sequence, we can find a formula for the nth term.

$$a_1 = a_1$$
$$a_2 = a_1 + d$$
$$a_3 = a_2 + d = a_1 + d + d = a_1 + 2d$$
$$a_4 = a_3 + d = a_1 + 2d + d = a_1 + 3d$$

Note the relationship between the term number and the coefficient of d. The coefficient is 1 less than the term number.

Formula for the *n*th Term of an Arithmetic Sequence

The **nth term of an arithmetic sequence** with common difference of d is given by

$$a_n = a_1 + (n - 1)d$$

EXAMPLE 4 Find the *n*th Term of an Arithmetic Sequence

a. Find the twenty-fifth term of the arithmetic sequence whose first three terms are $-12, -6, 0$.

b. The fifteenth term of an arithmetic sequence is -3 and the first term is 25. Find the tenth term.

Solution

a. Find the common difference: $d = a_2 - a_1 = -6 - (-12) = 6$. Use the formula $a_n = a_1 + (n - 1)d$ with $n = 25$.

$$a_{25} = -12 + (25 - 1)(6) = -12 + 24(6) = -12 + 144 = 132$$

b. Solve the equation $a_n = a_1 + (n - 1)d$ for d, given that $n = 15$, $a_1 = 25$, and $a_{15} = -3$.

$$-3 = 25 + (14)d$$
$$d = -2$$

? ANSWER Yes. The difference between any two successive terms is 4.

Now find the tenth term.

$$a_n = a_1 + (n - 1)d$$
$$a_{10} = 25 + (9)(-2) = 7 \qquad \bullet\ n = 10,\ a_1 = 25,\ d = -2$$

▶ TRY EXERCISE 42, PAGE 522

Geometric Sequences

Arithmetic sequences are characterized by a common *difference* between successive terms. A *geometric sequence* is characterized by a common *ratio* between successive terms.

The sequence

$$3, 6, 12, 24, \ldots, 3(2^{n-1}), \ldots$$

is a geometric sequence. Note that the ratio of any two successive terms is 2.

$$\frac{6}{3} = 2 \qquad \frac{12}{6} = 2 \qquad \frac{24}{12} = 2$$

Geometric Sequence

Let r be a nonzero constant real number. A sequence is a **geometric sequence** if

$$\frac{a_{i+1}}{a_i} = r \quad \text{for all positive integers } i.$$

The number r is called the **common ratio.**

For instance, the sequence $4, -2, 1, \ldots, 4\left(-\frac{1}{2}\right)^{n-1}, \ldots$ is a geometric sequence because the ratio of successive terms is a constant: $\dfrac{a_{i+1}}{a_i} = \dfrac{4\left(-\frac{1}{2}\right)^{i}}{4\left(-\frac{1}{2}\right)^{i-1}} = -\dfrac{1}{2}$.

However, the sequence given by $1, 4, 9, \ldots, n^2, \ldots$ is not a geometric sequence because the ratio of successive terms is not a constant: $\dfrac{a_{i+1}}{a_i} = \dfrac{(i+1)^2}{i^2} = \left(1 + \dfrac{1}{i}\right)^2$.

Consider a geometric sequence in which the first term is a_1 and the common ratio is r. By multiplying each successive term of the geometric sequence by the common ratio, we can derive a formula for the nth term.

$$a_1 = a_1$$
$$a_2 = a_1 r$$
$$a_3 = a_2 r = (a_1 r)r = a_1 r^2$$
$$a_4 = a_3 r = (a_1 r^2)r = a_1 r^3$$

Note the relationship between the number of the term and the number that is the exponent on r. The exponent on r is 1 less than the number of the term. With this observation, we can write a formula for the nth term of a geometric sequence.

The *n*th Term of a Geometric Sequence

The **nth term of a geometric sequence** with first term a_1 and common ratio r is

$$a_n = a_1 r^{n-1}$$

EXAMPLE 5 Find the *n*th Term of a Geometric Sequence

Find the nth term of the geometric sequence whose first three terms are

a. $4, \frac{8}{3}, \frac{16}{9}, \ldots$ **b.** $5, -10, 20, \ldots$

Solution

a. $r = \frac{8/3}{4} = \frac{2}{3}$ and $a_1 = 4$. Thus $a_n = 4\left(\frac{2}{3}\right)^{n-1}$.

b. $r = \frac{-10}{5} = -2$ and $a_1 = 5$. Thus $a_n = 5(-2)^{n-1}$.

▶ TRY EXERCISE 50, PAGE 522

PARTIAL SUMS AND SUMMATION NOTATION

Another important way of obtaining a sequence is by adding the terms of a given sequence. For example, consider the sequence whose general term is given by $a_n = \frac{1}{2^n}$. The terms of this sequence are

$$\frac{1}{2}, \frac{1}{4}, \frac{1}{8}, \frac{1}{16}, \frac{1}{32}, \ldots, \frac{1}{2^n}, \ldots$$

From this sequence we can generate a new sequence that is the sum of the terms of $\frac{1}{2^n}$.

$$S_1 = \frac{1}{2}$$

$$S_2 = \frac{1}{2} + \frac{1}{4} = \frac{3}{4}$$

$$S_3 = \frac{1}{2} + \frac{1}{4} + \frac{1}{8} = \frac{7}{8}$$

$$S_4 = \frac{1}{2} + \frac{1}{4} + \frac{1}{8} + \frac{1}{16} = \frac{15}{16}$$

and, in general,
$$S_n = \frac{1}{2} + \frac{1}{4} + \frac{1}{8} + \frac{1}{16} + \cdots + \frac{1}{2^n}$$

The term S_n is called the **nth partial sum** of the infinite sequence, and the sequence $S_1, S_2, S_3, \ldots, S_n$ is called the **sequence of partial sums.**

A convenient notation used for partial sums is called **summation notation.** The sum of the first n terms of a sequence a_n is represented by using the Greek letter Σ (sigma).

$$\sum_{i=1}^{n} a_i = a_1 + a_2 + a_3 + \cdots + a_n$$

This sum is called a **series.** The letter i is called the **index of the summation;** n is the **upper limit** of the summation; 1 is the **lower limit** of the summation.

MATH MATTERS

Leonhard Euler (1707–1783) found that the sequence of terms given by

$$S_n = 1 - \frac{1}{3} + \frac{1}{5} - \frac{1}{7} + \cdots + \frac{(-1)^{n-1}}{2n-1}$$

became closer and closer to $\frac{\pi}{4}$ as n increased. In summation notation, we would write

$$S_n = \sum_{k=1}^{n} \frac{(-1)^{k-1}}{2k-1}$$

EXAMPLE 6 Evaluate Series

Evaluate each series. **a.** $\sum_{i=1}^{4} \frac{i}{i+1}$ **b.** $\sum_{j=2}^{5} (-1)^j j^2$

Solution

a. $\sum_{i=1}^{4} \frac{i}{i+1} = \frac{1}{2} + \frac{2}{3} + \frac{3}{4} + \frac{4}{5} = \frac{163}{60}$

b. $\sum_{j=2}^{5} (-1)^j j^2 = (-1)^2 2^2 + (-1)^3 3^2 + (-1)^4 4^2 + (-1)^5 5^2$

$= 4 - 9 + 16 - 25 = -14$

▶ TRY EXERCISE 70, PAGE 523

take note

Example 6b illustrates that it is not necessary for a summation to begin at 1. The index of the summation can be any letter.

Properties of Summation Notation

If a_n and b_n are sequences and c is a real number, then

1. $\sum_{i=1}^{n} (a_i \pm b_i) = \sum_{i=1}^{n} a_i \pm \sum_{i=1}^{n} b_i$

2. $\sum_{i=1}^{n} ca_i = c\sum_{i=1}^{n} a_i$

3. $\sum_{i=1}^{n} c = nc$

The proof of property (1) depends on the commutative and associative properties of real numbers.

$$\begin{aligned}\sum_{i=1}^{n}(a_i \pm b_i) &= (a_1 \pm b_1) + (a_2 \pm b_2) + \cdots + (a_n \pm b_n)\\ &= (a_1 + a_2 + \cdots + a_n) \pm (b_1 + b_2 + \cdots + b_n)\\ &= \sum_{i=1}^{n} a_i \pm \sum_{i=1}^{n} b_i\end{aligned}$$

Property (2) is proved by using the distributive property; this is left as an exercise.

To prove property (3), let $a_n = c$. That is, each a_n is equal to the same constant c. (This is called a **constant sequence.**) Then

$$\sum_{i=1}^{n} a_n = a_1 + a_2 + \cdots + a_n = \underbrace{c + c + \cdots + c}_{n \text{ terms}} = nc$$

Arithmetic Series

Consider the arithmetic sequence given by

$$1, 3, 5, \ldots, 2n - 1, \ldots$$

Adding successive terms of this sequence, we generate a sequence of partial sums. The sum of the terms of an arithmetic sequence is called an **arithmetic series.**

$$\begin{aligned}S_1 &= 1\\ S_2 &= 1 + 3 = 4\\ S_3 &= 1 + 3 + 5 = 9\\ S_4 &= 1 + 3 + 5 + 7 = 16\\ S_5 &= 1 + 3 + 5 + 7 + 9 = 25\\ &\vdots\\ S_n &= 1 + 3 + \cdots + (2n - 1) = n^2\end{aligned}$$

The first five terms of this sequence are 1, 4, 9, 16, 25. It appears from this example that the sum of the first n odd integers is n^2. Shortly, we will be able to prove this result by using the following formula.

MATH MATTERS

Galileo (1564–1642), using the fact that the sum of a set of odd integers is the square of the number of integers being added (as is shown at the right), was able to show that objects of different weights fall at the same rate. By constructing inclines of various slopes similar to the one shown below, with a track in the center and equal intervals marked along the incline, he measured the distance various balls of different weights traveled in equal intervals of time. He concluded from his observations that the distance an object falls is proportional to the square of the time it takes to fall and does not depend on its weight. Galileo's views were contrary to the prevailing (Aristotelian) theory on this subject, and he lost his post at the University of Pisa because of them.

Formula for the nth Partial Sum of an Arithmetic Sequence

The ***n*th partial sum S_n of an arithmetic sequence a_n** with common difference d is

$$S_n = \frac{n}{2}(a_1 + a_n)$$

Proof We write S_n in both forward and reverse order.

$$S_n = a_1 + a_2 + a_3 + \cdots + a_{n-2} + a_{n-1} + a_n$$
$$S_n = a_n + a_{n-1} + a_{n-2} + \cdots + a_3 + a_2 + a_1$$

Add the two partial sums.

$$2S_n = (a_1 + a_n) + (a_2 + a_{n-1}) + (a_3 + a_{n-2}) + \cdots + (a_{n-2} + a_3) + (a_{n-1} + a_2) + (a_n + a_1) \qquad (1)$$

Consider the term $(a_3 + a_{n-2})$. Using the formula for the nth term of an arithmetic sequence, we have

$$a_3 = a_1 + (3 - 1)d = a_1 + 2d$$
$$a_{n-2} = a_1 + [(n - 2) - 1]d = a_1 + nd - 3d$$

Thus

$$\begin{aligned} a_3 + a_{n-2} &= (a_1 + 2d) + (a_1 + nd - 3d) \\ &= a_1 + (a_1 + nd - d) = a_1 + [a_1 + (n - 1)d] \\ &= a_1 + a_n \end{aligned}$$

In a similar manner, we can show that each term in parentheses in Equation (1) equals $(a_1 + a_n)$. Because there are n such terms, we have

$$2S_n = n(a_1 + a_n)$$
$$S_n = \frac{n}{2}(a_1 + a_n)$$ ◆

There is an alternative form of the formula for the sum of n terms of an arithmetic sequence.

Alternative Formula for the Sum of an Arithmetic Sequence

The nth partial sum S_n of an arithmetic sequence with common difference d is

$$S_n = \frac{n[2a_1 + (n - 1)d]}{2}$$

EXAMPLE 7 **Find a Partial Sum of an Arithmetic Sequence**

Find the sum of the first 50 terms of the arithmetic sequence whose first three terms are 2, $\frac{13}{4}$, and $\frac{9}{2}$.

Solution

Use the formula $S_n = \frac{n}{2}[2a_1 + (n - 1)d]$.

Continued ▶

We have $a_1 = 2$, $d = \frac{5}{4}$, and $n = 50$. Thus

$$S_{50} = \frac{50}{2}\left[2(2) + (50 - 1)\frac{5}{4}\right] = \frac{6525}{4}$$

▶ TRY EXERCISE 74, PAGE 523

The first n positive integers $1, 2, 3, 4, \ldots, n$ are part of an arithmetic sequence with a common difference of 1, $a_1 = 1$, and $a_n = n$. A formula for the sum of the first n positive integers can be found by using the formula for the nth partial sum of an arithmetic sequence.

$$S_n = \frac{n}{2}(a_1 + a_n)$$

Replacing a_1 by 1 and a_n by n yields

$$S_n = \frac{n}{2}(1 + n) = \frac{n(n + 1)}{2}$$

This proves the following theorem.

Sum of the First *n* Positive Integers

The sum of the first n positive integers is given by

$$S_n = \frac{n(n + 1)}{2}$$

To find the sum of the first 85 positive integers, use $n = 85$.

$$S_{85} = \frac{85(85 + 1)}{2} = 3655$$

FINITE GEOMETRIC SERIES

Adding the terms of a geometric sequence, we can define the nth partial sum of a geometric sequence in a manner similar to that of an arithmetic sequence. Consider the geometric sequence $1, 2, 4, 8, \ldots, 2^{n-1}, \ldots$.

$$\begin{aligned}
S_1 &= 1 \\
S_2 &= 1 + 2 = 3 \\
S_3 &= 1 + 2 + 4 = 7 \\
S_4 &= 1 + 2 + 4 + 8 = 15 \\
&\vdots \\
S_n &= 1 + 2 + 4 + 8 + \cdots + 2^{n-1}
\end{aligned}$$

The first four terms of the sequence of partial sums are 1, 3, 7, and 15.

To find a general formula for S_n, the nth term of the sequence of partial sums of a geometric sequence, let

$$S_n = a_1 + a_1 r + a_1 r^2 + \cdots + a_1 r^{n-1}$$

Multiply each side of this equation by r.

$$S_n = a_1 + a_1r + a_1r^2 + \cdots + a_1r^{n-2} + a_1r^{n-1}$$
$$rS_n = \qquad a_1r + a_1r^2 + \cdots + a_1r^{n-2} + a_1r^{n-1} + a_1r^n$$

Subtract the two equations.

$$S_n - rS_n = a_1 - a_1r^n$$
$$S_n(1 - r) = a_1(1 - r^n) \qquad \text{• Factor out the common factors.}$$
$$S_n = \frac{a_1(1 - r^n)}{1 - r} \qquad \text{• } r \neq 1$$

This proves the following theorem.

The *n*th Partial Sum of a Geometric Sequence

The ***n*th partial sum of a geometric sequence** with first term a_1 and common ratio r is

$$S_n = \frac{a_1(1 - r^n)}{1 - r}, \quad r \neq 1$$

QUESTION If $r = 1$, what is the nth partial sum of a geometric sequence?

EXAMPLE 8 Find the *n*th Partial Sum of a Geometric Sequence

Find the partial sum of each geometric sequence.

a. $5, 15, 45, \ldots, 5(3)^{n-1}, \ldots; n = 4$ **b.** $\sum_{n=1}^{17} 3\left(\frac{3}{4}\right)^{n-1}$

Solution

a. We have $a_1 = 5$, $r = 3$, and $n = 4$. Thus

$$S_4 = \frac{5[1 - 3^4]}{1 - 3} = \frac{5(-80)}{-2} = 200$$

b. When $n = 1$, $a_1 = 3$. The first term is 3. The second term is $\frac{9}{4}$.

Therefore, the common ratio is $r = \frac{3}{4}$. Thus

$$S_{17} = \frac{3[1 - (3/4)^{17}]}{1 - (3/4)} \approx 11.909797$$

TRY EXERCISE 82, PAGE 523

ANSWER When $r = 1$, the sequence is the constant sequence a_1. The nth partial sum of a constant sequence is na_1.

Infinite Geometric Series

Following are two examples of geometric sequences for which $|r| < 1$.

$$3, \frac{3}{4}, \frac{3}{16}, \frac{3}{64}, \frac{3}{256}, \frac{3}{1024}, \ldots \qquad \bullet\ r = \frac{1}{4}$$

$$2, -1, \frac{1}{2}, -\frac{1}{4}, \frac{1}{8}, -\frac{1}{16}, \frac{1}{32}, \ldots \qquad \bullet\ r = -\frac{1}{2}$$

Note that when the absolute value of the common ratio of a geometric sequence is less than 1, the terms of the geometric sequence approach zero as n increases. We write, for $|r| < 1$, $|r|^n \to 0$ as $n \to \infty$.

Consider again the geometric sequence

$$3, \frac{3}{4}, \frac{3}{16}, \frac{3}{64}, \frac{3}{256}, \frac{3}{1024}, \ldots$$

The nth partial sums for $n = 3, 6, 9,$ and 12 are given in **Table 6.2,** along with the values of r^n. As n increases, S_n is closer to 4 and r^n is closer to zero. By finding more values of S_n for larger values of n, we would find that $S_n \to 4$ as $n \to \infty$. As n becomes larger and larger, S_n is the nth partial sum of more and more terms of the sequence. The sum of *all* the terms of a sequence is called an **infinite series.** If the sequence is a geometric sequence, we have an **infinite geometric series.**

Table 6.2

n	S_n	r^n
3	3.93750000	0.01562500
6	3.99902344	0.00024414
9	3.99998474	0.00000381
12	3.99999976	0.00000006

Sum of an Infinite Geometric Series

If a_n is a geometric sequence with $|r| < 1$ and first term a_1, then the sum of the infinite geometric series is

$$S = \frac{a_1}{1 - r}$$

A formal proof of this formula requires topics that typically are studied in calculus. We can, however, give an intuitive argument. Start with the formula for the nth partial sum of a geometric sequence.

$$S_n = \frac{a_1(1 - r^n)}{1 - r}$$

When $|r| < 1$, $|r|^n \approx 0$ when n is large. Thus

$$S_n = \frac{a_1(1 - r^n)}{1 - r} \approx \frac{a_1(1 - 0)}{1 - r} = \frac{a_1}{1 - r}$$

An infinite series is represented by $\sum_{n=1}^{\infty} a_n$. One application of infinite geometric series concerns repeating decimals. Consider the repeating decimal

$$0.\overline{6} = \frac{6}{10} + \frac{6}{100} + \frac{6}{1000} + \frac{6}{10,000} + \cdots$$

The right-hand side is a geometric series with $a_1 = \frac{6}{10}$ and common ratio $r = \frac{1}{10}$. Thus

$$S = \frac{6/10}{1 - (1/10)} = \frac{6/10}{9/10} = \frac{2}{3}$$

The repeating decimal $0.\overline{6} = \frac{2}{3}$. We can write any repeating decimal as a ratio of two integers by using the formula for the sum of an infinite geometric series.

take note

The sum of an infinite geometric series is not defined when $|r| \geq 1$. For instance, the infinite geometric series

$$2 + 4 + 8 + \cdots + 2^n + \cdots$$

with $r = 2$ increases without bound. However, applying the formula $S = \frac{a_1}{1 - r}$ with $r = 2$ and $a_1 = 2$ gives $S = -2$, which is not correct.

EXAMPLE 9 Find the Value of an Infinite Geometric Series

a. Evaluate the infinite geometric series $\sum_{n=1}^{\infty} \left(-\frac{2}{3}\right)^{n-1}$.

b. Write $0.3\overline{45}$ as the ratio of two integers in lowest terms.

Solution

a. To find the first term, we let $n = 1$. Then $a_1 = \left(-\frac{2}{3}\right)^{1-1} = \left(-\frac{2}{3}\right)^0 = 1$.

The common ratio is $r = -\frac{2}{3}$. Thus

$$S = \frac{1}{1 - (-2/3)} = \frac{1}{5/3} = \frac{3}{5}$$

b. $0.3\overline{45} = \frac{3}{10} + \left[\frac{45}{1000} + \frac{45}{100,000} + \frac{45}{10,000,000} + \cdots\right]$

The terms in the brackets form an infinite geometric series. Evaluate that series with $a_1 = \frac{45}{1000}$ and $r = \frac{1}{100}$, and then add the term $\frac{3}{10}$.

$$\frac{45}{1000} + \frac{45}{100,000} + \frac{45}{10,000,000} + \cdots = \frac{45/1000}{1 - (1/100)} = \frac{1}{22}$$

Thus $0.3\overline{45} = \frac{3}{10} + \frac{1}{22} = \frac{19}{55}$

TRY EXERCISE 98, PAGE 523

Topics for Discussion

1. Discuss the difference between a finite sequence and an infinite sequence. Give an example of each type.
2. Discuss the difference between a sequence and a series.
3. What is an arithmetic sequence?
4. What is a geometric sequence?

Exercise Set 6.4

In Exercises 1 to 14, find the first three terms and the eighth term of the sequence that has the given *n*th term.

1. $a_n = n(n - 1)$
2. $a_n = 2n$
3. $a_n = 1 - \frac{1}{n}$
4. $a_n = \frac{n + 1}{n}$
5. $a_n = \frac{(-1)^{n+1}}{n^2}$
6. ▶ $a_n = \frac{(-1)^{n+1}}{n(n + 1)}$
7. $a_n = \frac{(-1)^{2n-1}}{3n}$
8. $a_n = \frac{(-1)^n}{2n - 1}$
9. $a_n = 1 + (-1)^n$
10. $a_n = 1 + (-0.1)^n$
11. $a_n = \frac{(-1)^{n+1}}{\sqrt{n}}$
12. $a_n = n!$
13. $a_n = 3$
14. $a_n = -2$

In Exercises 15 to 22, find the first three terms of each recursively defined sequence.

15. $a_1 = 5, a_n = 2a_{n-1}$
16. $a_1 = 2, a_n = 3a_{n-1}$
17. $a_1 = 2, a_n = na_{n-1}$
18. ▶ $a_1 = 1, a_n = n^2a_{n-1}$
19. $a_1 = 2, a_n = (a_{n-1})^2$
20. $a_1 = 4, a_n = \frac{1}{a_{n-1}}$
21. $a_1 = 2, a_n = 2na_{n-1}$
22. $a_1 = 2, a_n = (-3)na_{n-1}$

In Exercises 23 to 30, evaluate the factorial expression.

23. $7! - 6!$
24. $(4!)^2$
25. $\frac{9!}{7!}$
26. $\frac{10!}{5!}$
27. $\frac{8!}{3!\,5!}$
28. ▶ $\frac{12!}{4!\,8!}$
29. $\frac{100!}{99!}$
30. $\frac{100!}{98!\,2!}$

In Exercises 31 to 40, find the *n*th term of the arithmetic sequence.

31. $6, 10, 14, \ldots$
32. $7, 12, 17, \ldots$
33. $6, 4, 2, \ldots$
34. $11, 4, -3, \ldots$
35. $-8, -5, -2, \ldots$
36. $-15, -9, -3, \ldots$
37. $1, 4, 7, \ldots$
38. $-4, 1, 6, \ldots$
39. $a, a + 2, a + 4, \ldots$
40. $a - 3, a + 1, a + 5, \ldots$
41. The fourth and fifth terms of an arithmetic sequence are 13 and 15. Find the twentieth term.
42. ▶ The sixth and eighth terms of an arithmetic sequence are −14 and −20. Find the fifteenth term.
43. The fifth and seventh terms of an arithmetic sequence are −19 and −29. Find the seventeenth term.
44. The fourth and seventh terms of an arithmetic sequence are 22 and 34. Find the twenty-third term.

In Exercises 45 to 62, find the *n*th term of the geometric sequence.

45. $2, 8, 32, \ldots$
46. $1, 5, 25, \ldots$
47. $-4, 12, -36, \ldots$
48. $-3, 6, -12, \ldots$
49. $6, 4, \frac{8}{3}, \ldots$
50. ▶ $8, 6, \frac{9}{2}, \ldots$
51. $-6, 5, -\frac{25}{6}, \ldots$
52. $-2, \frac{4}{3}, -\frac{8}{9}, \ldots$
53. $\frac{3}{100}, \frac{3}{10{,}000}, \frac{3}{1{,}000{,}000}, \ldots$

54. $\frac{7}{10}, \frac{7}{10{,}000}, \frac{7}{10{,}000{,}000}, \ldots$

55. $0.5, 0.05, 0.005, \ldots$

56. $0.4, 0.004, 0.00004, \ldots$

57. $0.45, 0.0045, 0.000045, \ldots$

58. $0.234, 0.000234, 0.000000234, \ldots$

59. Find the third term of a geometric sequence whose first term is 2 and whose fifth term is 162.

60. Find the fourth term of a geometric sequence whose third term is 1 and whose eighth term is $\frac{1}{32}$.

61. Find the second term of a geometric sequence whose third term is $\frac{4}{3}$ and whose sixth term is $-\frac{32}{81}$.

62. Find the fifth term of a geometric sequence whose fourth term is $\frac{8}{9}$ and whose seventh term is $\frac{64}{243}$.

In Exercises 63 to 70, evaluate the series.

63. $\sum_{i=1}^{5} i$ **64.** $\sum_{i=1}^{4} i^2$

65. $\sum_{i=1}^{5} i(i-1)$ **66.** $\sum_{i=1}^{7} (2i+1)$

67. $\sum_{k=1}^{4} \frac{1}{k}$ **68.** $\sum_{k=1}^{6} \frac{1}{k(k+1)}$

69. $\sum_{j=1}^{8} 2j$ ▶**70.** $\sum_{i=1}^{6} (2i+1)(2i-1)$

In Exercises 71 to 78, find the *n*th partial sum of the arithmetic sequence.

71. $a_n = 3n + 2; n = 10$ **72.** $a_n = 4n - 3; n = 12$

73. $a_n = 3 - 5n; n = 15$ ▶**74.** $a_n = 1 - 2n; n = 20$

75. $a_n = 6n; n = 12$ **76.** $a_n = 7n; n = 14$

77. $a_n = n + 8; n = 25$ **78.** $a_n = n - 4; n = 25$

In Exercises 79 to 86, find the sum of the geometric series.

79. $\sum_{n=1}^{5} 3^n$ **80.** $\sum_{n=1}^{7} 2^n$

81. $\sum_{n=1}^{6} \left(\frac{2}{3}\right)^n$ ▶**82.** $\sum_{n=1}^{14} \left(\frac{4}{3}\right)^n$

83. $\sum_{n=1}^{10} (-2)^{n-1}$ **84.** $\sum_{n=0}^{7} 2(5)^n$

85. $\sum_{n=0}^{9} 5(3)^n$ **86.** $\sum_{n=0}^{10} 2(-4)^n$

In Exercises 87 to 92, find the sum of the infinite geometric series.

87. $\sum_{n=1}^{\infty} \left(\frac{1}{3}\right)^n$ **88.** $\sum_{n=1}^{\infty} \left(\frac{3}{4}\right)^n$

89. $\sum_{n=1}^{\infty} \left(-\frac{2}{3}\right)^n$ **90.** $\sum_{n=1}^{\infty} \left(-\frac{3}{5}\right)^n$

91. $\sum_{n=1}^{\infty} (0.1)^n$ **92.** $\sum_{n=1}^{\infty} (0.5)^n$

In Exercises 93 to 100, write each rational number as the quotient of two integers in simplest form.

93. $0.\overline{3}$ **94.** $0.\overline{5}$ **95.** $0.\overline{45}$ **96.** $0.\overline{63}$

97. $0.\overline{123}$ ▶**98.** $0.\overline{395}$ **99.** $0.4\overline{22}$ **100.** $0.3\overline{55}$

CONNECTING CONCEPTS

101. **Newton's Method** Newton's approximation to the square root of a number, N, is given by the recursive sequence

$$a_1 = \frac{N}{2} \qquad a_n = \frac{1}{2}\left(a_{n-1} + \frac{N}{a_{n-1}}\right)$$

Approximate $\sqrt{7}$ by computing a_4. Compare this result with the calculator value of $\sqrt{7} \approx 2.6457513$.

102. Use the formula in Exercise 101 to approximate $\sqrt{10}$ by finding a_5.

103. If $f(x)$ is a linear polynomial, show that $f(n)$, where n is a positive integer, is an arithmetic sequence.

104. Find the formula for a_n in terms of a_1 and n for the sequence that is defined recursively by $a_1 = 3$, $a_n = a_{n-1} + 5$.

105. Find a formula for a_n in terms of a_1 and n for the sequence that is defined recursively by $a_1 = 4$, $a_n = a_{n-1} - 3$.

106. Suppose a_n and b_n are two sequences such that $a_1 = 4$, $a_n = b_{n-1} + 5$ and $b_1 = 2$, $b_n = a_{n-1} + 1$. Show that a_n and b_n are arithmetic sequences. Find a_{100}.

107. If the sequence a_n is a geometric sequence, make a conjecture about the sequence $\log a_n$ and give a proof.

108. If the sequence a_n is an arithmetic sequence, make a conjecture about the sequence 2^{a_n} and give a proof.

109. Does $\sum_{i=0}^{\infty} x^i$ $(x \neq 0)$ represent an infinite geometric series? Why or why not?

110. Consider a square with a side of length 1. Construct another square inside the first one by connecting the midpoints of the sides of the first square. What is the area of the inscribed square? Continue constructing squares in the same way. Find the area of the nth inscribed square.

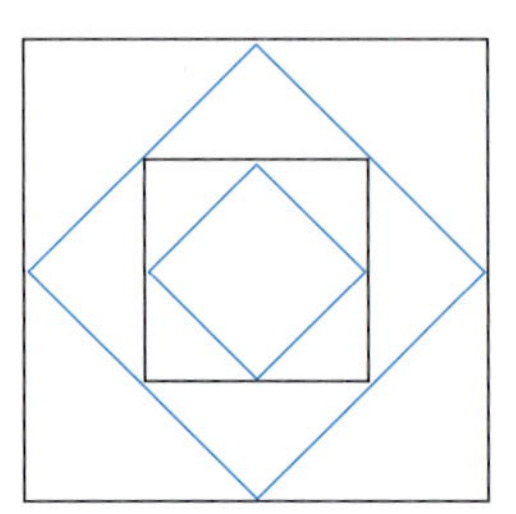

PREPARE FOR SECTION 6.5

111. Expand $(a + b)^3$. [A.2]

112. Evaluate 5!. [6.4]

113. Evaluate 0!. [6.4]

114. Evaluate $\dfrac{n!}{k!\,(n-k)!}$ when $n = 6$ and $k = 2$. [6.4]

115. Evaluate $\dfrac{n!}{k!\,(n-k)!}$ when $n = 7$ and $k = 3$. [6.4]

116. Evaluate $\dfrac{n!}{k!\,(n-k)!}$ when $n = 10$ and $k = 10$. [6.4]

PROJECTS

1. FORMULAS FOR INFINITE SEQUENCES It is not possible to define an infinite sequence by giving a finite number of terms of the sequence. For instance, the question "What is the next term in the sequence 2, 4, 6, 8, . . .?" does not have a unique answer.

a. Verify this statement by finding a formula for a_n such that the first four terms of the sequence are 2, 4, 6, 8 and the next term is 43. *Suggestion:* The formula

$$a_n = \frac{n(n-1)(n-2)(n-3)(n-4)}{4!} + 2n$$

generates the sequence 2, 4, 6, 8, 15 for $n = 1, 2, 3, 4, 5$.

b. Extend the result in part **a.** by finding a formula for a_n that will give the first four terms as 2, 4, 6, 8 and the fifth term as x, where x is any real number.

2. ANGLES OF A TRIANGLE The sum of the interior angles of a triangle is 180°.

a. Using this fact, what is the sum of the interior angles of a quadrilateral?

b. What is the sum of the interior angles of a pentagon?

c. What is the sum of the interior angles of a hexagon?

d. On the basis of your previous results, what is the apparent formula for the sum of the interior angles of a polygon of n sides?

3. **FRACTALS** An example of a fractal was discussed on page 461. Here is another example of a fractal. Begin with a square with each side 1 unit long. Construct another, smaller square onto the middle third of each side. Continue this procedure of constructing similar but smaller squares on each of the line segments. The figure at the right shows the result after the process has been completed twice.

a. What is the perimeter of the figure after the process has been completed n times?

b. As n approaches infinity, what value does the perimeter approach?

c. What is the area of the figure after the process has been completed n times?

d. As n approaches infinity, what value does the area approach? *Suggestion:* The series in part **c.** is a geometric series after the first term.

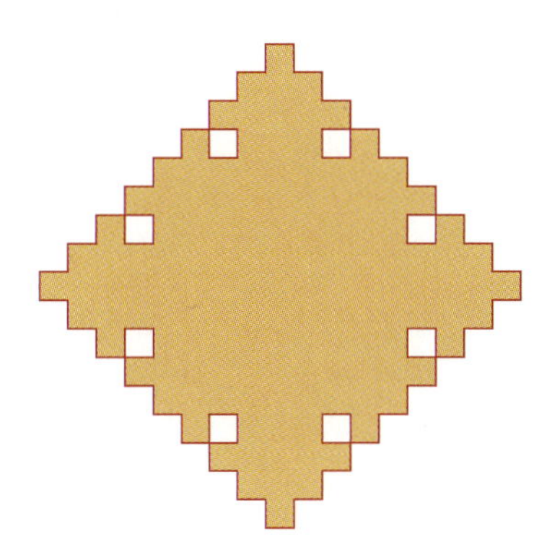

SECTION 6.5 THE BINOMIAL THEOREM

- BINOMIAL THEOREM
- *i*th TERM OF A BINOMIAL EXPANSION
- PASCAL'S TRIANGLE

In certain situations in mathematics it is necessary to write $(a + b)^n$ as the sum of its terms. Because $(a + b)$ is a binomial, this process is called **expanding the binomial.** For small values of n, it is relatively easy to write the expansion by using multiplication.

Earlier in the text we found that

$$(a + b)^1 = a + b$$

$$(a + b)^2 = a^2 + 2ab + b^2$$

$$(a + b)^3 = a^3 + 3a^2b + 3ab^2 + b^3$$

Building on these expansions, we can write a few more.

$$(a + b)^4 = a^4 + 4a^3b + 6a^2b^2 + 4ab^3 + b^4$$

$$(a + b)^5 = a^5 + 5a^4b + 10a^3b^2 + 10a^2b^3 + 5ab^4 + b^5$$

We could continue to build on previous expansions and eventually have quite a comprehensive list of binomial expansions. Instead, however, we will look for a theorem that will enable us to expand $(a + b)^n$ directly without multiplying.

Look at the variable parts of each expansion above. Note that for each $n = 1, 2, 3, 4, 5$,

- The first term is a^n. The exponent on a decreases by 1 for each successive term.
- The exponent on b increases by 1 for each successive term. The last term is b^n.
- The degree of each term is n.

QUESTION What is the degree of the fourth term of the expansion of $(a + b)^{11}$?

To find a pattern for the coefficients in each expansion of $(a + b)^n$, first note that there are $n + 1$ terms and that the coefficient of the first and last term is 1. To find the remaining coefficients, consider the expansion of $(a + b)^5$.

$$(a + b)^5 = a^5 + 5a^4b + 10a^3b^2 + 10a^2b^3 + 5ab^4 + b^5$$

$$\frac{5}{1} = 5 \qquad \frac{5 \cdot 4}{2 \cdot 1} = 10 \qquad \frac{5 \cdot 4 \cdot 3}{3 \cdot 2 \cdot 1} = 10 \qquad \frac{5 \cdot 4 \cdot 3 \cdot 2}{4 \cdot 3 \cdot 2 \cdot 1} = 5$$

Observe from these patterns that there is a strong relationship to factorials. In fact, we can express each coefficient by using factorial notation.

$$\frac{5!}{1!\,4!} = 5 \qquad \frac{5!}{2!\,3!} = 10 \qquad \frac{5!}{3!\,2!} = 10 \qquad \frac{5!}{4!\,1!} = 5$$

In each denominator the first factorial is the exponent of b, and the second factorial is the exponent of a.

In general, we will conjecture that the coefficient of the term $a^{n-k}b^k$ in the expansion of $(a + b)^n$ is $\dfrac{n!}{k!\,(n-k)!}$. Each coefficient of a term of a binomial expansion is called a **binomial coefficient** and is denoted by $\dbinom{n}{k}$.

Formula for a Binomial Coefficient

The coefficient of the term whose variable part is $a^{n-k}b^k$ in the expansion of $(a + b)^n$ is

$$\binom{n}{k} = \frac{n!}{k!\,(n-k)!}$$

The first term of the expansion of $(a + b)^n$ can be thought of as a^nb^0. In this case, we can calculate the coefficient of this term as

$$\binom{n}{0} = \frac{n!}{0!\,(n-0)!} = \frac{n!}{1 \cdot n!} = 1$$

EXAMPLE 1 Evaluate a Binomial Coefficient

Evaluate each binomial coefficient. **a.** $\dbinom{9}{6}$ **b.** $\dbinom{10}{10}$

ANSWER 11

Solution

a. $\binom{9}{6} = \dfrac{9!}{6!\,(9-6)!} = \dfrac{9!}{6!\,3!} = \dfrac{9 \cdot 8 \cdot 7 \cdot 6!}{6! \cdot 3 \cdot 2 \cdot 1} = 84$

b. $\binom{10}{10} = \dfrac{10!}{10!\,(10-10)!} = \dfrac{10!}{10!\,0!} = 1$ • Remember that $0! = 1$.

▶ TRY EXERCISE 4, PAGE 529

BINOMIAL THEOREM

We are now ready to state the Binomial Theorem for positive integers.

Binomial Theorem for Positive Integers

If n is a positive integer, then

$$(a+b)^n = \sum_{i=0}^{n} \binom{n}{i} a^{n-i}b^i$$
$$= \binom{n}{0}a^n + \binom{n}{1}a^{n-1}b + \binom{n}{2}a^{n-2}b^2 + \cdots + \binom{n}{n}b^n$$

EXAMPLE 2 Expand the Sum of Two Terms

Expand: $(2x^2+3)^4$

Solution

$$(2x^2+3)^4 = \binom{4}{0}(2x^2)^4 + \binom{4}{1}(2x^2)^3(3) + \binom{4}{2}(2x^2)^2(3)^2$$
$$+ \binom{4}{3}(2x^2)(3)^3 + \binom{4}{4}(3)^4$$
$$= 16x^8 + 96x^6 + 216x^4 + 216x^2 + 81$$

▶ TRY EXERCISE 18, PAGE 529

EXAMPLE 3 Expand a Difference of Two Terms

Expand: $\left(\sqrt{x} - 2y\right)^5$

Continued ▶

take note

If exactly one of the terms a or b in $(a + b)^n$ is negative, the terms of the expansion alternate in sign.

Solution

$$(\sqrt{x} - 2y)^5 = \binom{5}{0}(\sqrt{x})^5 + \binom{5}{1}(\sqrt{x})^4(-2y) + \binom{5}{2}(\sqrt{x})^3(-2y)^2$$
$$+ \binom{5}{3}(\sqrt{x})^2(-2y)^3 + \binom{5}{4}(\sqrt{x})(-2y)^4 + \binom{5}{5}(-2y)^5$$
$$= x^{5/2} - 10x^2y + 40x^{3/2}y^2 - 80xy^3 + 80x^{1/2}y^4 - 32y^5$$

▶ TRY EXERCISE 20, PAGE 529

*i*th TERM OF A BINOMIAL EXPANSION

The Binomial Theorem also can be used to find a specific term in the expansion of $(a + b)^n$.

take note

The exponent on b is 1 less than the term number.

Formula for the *i*th Term of a Binomial Expansion

The ith term of the expansion of $(a + b)^n$ is given by

$$\binom{n}{i-1}a^{n-i+1}b^{i-1}$$

EXAMPLE 4 Find the *i*th Term of a Binomial Expansion

Find the fourth term in the expansion of $(2x^3 - 3y^2)^5$.

Solution

With $a = 2x^3$ and $b = -3y^2$, and using the preceding theorem with $i = 4$ and $n = 5$, we have

$$\binom{5}{3}(2x^3)^2(-3y^2)^3 = -1080x^6y^6$$

The fourth term is $-1080x^6y^6$.

▶ TRY EXERCISE 34, PAGE 530

PASCAL'S TRIANGLE

A pattern for the coefficients of the terms of an expanded binomial can be found by writing the coefficients in a triangular array known as **Pascal's Triangle.** See **Figure 6.65.**

Each row begins and ends with the number 1. Any other number in a row is the sum of the two closest numbers above it. For example, $4 + 6 = 10$. Thus each succeeding row can be found from the preceding row.

$(a+b)^1$:						1		1					
$(a+b)^2$:					1		2		1				
$(a+b)^3$:				1		3		3		1			
$(a+b)^4$:			1		4		6		4		1		
$(a+b)^5$:		1		5		10		10		5		1	
$(a+b)^6$:	1		6		15		20		15		6		1

FIGURE 6.65

Pascal's Triangle can be used to expand a binomial for small values of n. For instance, the seventh row of Pascal's Triangle is

$$1 \quad 7 \quad 21 \quad 35 \quad 35 \quad 21 \quad 7 \quad 1$$

Therefore,

$$(a+b)^7 = a^7 + 7a^6b + 21a^5b^2 + 35a^4b^3 + 35a^3b^4 + 21a^2b^5 + 7ab^6 + b^7$$

TOPICS FOR DISCUSSION

1. Discuss the use of the Binomial Theorem.
2. Can the Binomial Theorem be used to expand $(a+b)^n$, n a natural number, for any expressions a and b? Why or why not?
3. What is Pascal's Triangle and how is it related to expanding a binomial?
4. Explain how Pascal's Triangle suggests that $\binom{n-1}{k-1} + \binom{n-1}{k} = \binom{n}{k}$.

EXERCISE SET 6.5

In Exercises 1 to 8, evaluate the binomial coefficient.

1. $\binom{7}{4}$
2. $\binom{8}{6}$
3. $\binom{9}{2}$
▶ 4. $\binom{10}{5}$
5. $\binom{12}{9}$
6. $\binom{6}{5}$
7. $\binom{11}{0}$
8. $\binom{14}{14}$

In Exercises 9 to 28, expand the binomial.

9. $(x-y)^6$
10. $(a-b)^5$
11. $(x+3)^5$
12. $(x-5)^4$
13. $(2x-1)^7$
14. $(2x+y)^6$
15. $(x+3y)^6$
16. $(x-4y)^5$
17. $(2x-5y)^4$
▶ 18. $(3x+2y)^4$
19. $\left(x+\dfrac{1}{x}\right)^6$
▶ 20. $(2x-\sqrt{y})^7$
21. $(x^2-4)^7$
22. $(x-y^3)^6$
23. $(2x^2+y^3)^5$
24. $(2x-y^3)^6$
25. $\left(\dfrac{2}{x}-\dfrac{x}{2}\right)^4$
26. $\left(\dfrac{a}{b}+\dfrac{b}{a}\right)^3$
27. $(s^{-2}+s^2)^6$
28. $(2r^{-1}+s^{-1})^5$

In Exercises 29 to 36, find the indicated term without expanding.

29. $(3x-y)^{10}$; eighth term
30. $(x+2y)^{12}$; fourth term
31. $(x+4y)^{12}$; third term
32. $(2x-1)^{14}$; thirteenth term

33. $(\sqrt{x} - \sqrt{y})^9$; fifth term

▶ **34.** $(x^{-1/2} + x^{1/2})^{10}$; sixth term

35. $\left(\dfrac{a}{b} + \dfrac{b}{a}\right)^{11}$; ninth term

36. $\left(\dfrac{3}{x} - \dfrac{x}{3}\right)^{13}$; seventh term

37. Find the term that contains b^8 in the expansion of $(2a - b)^{10}$.

38. Find the term that contains s^7 in the expansion of $(3r + 2s)^9$.

39. Find the term that contains y^8 in the expansion of $(2x + y^2)^6$.

40. Find the term that contains b^9 in the expansion of $(a - b^3)^8$.

41. Find the middle term of $(3a - b)^{10}$.

42. Find the middle term of $(a + b^2)^8$.

43. Find the two middle terms of $(s^{-1} + s)^9$.

44. Find the two middle terms of $(x^{1/2} - y^{1/2})^7$.

In Exercises 45 to 50, use the Binomial Theorem to simplify the powers of the complex numbers.

45. $(2 - i)^4$

46. $(3 + 2i)^3$

47. $(1 + 2i)^5$

48. $(1 - 3i)^5$

49. $\left(\dfrac{\sqrt{2}}{2} + i\dfrac{\sqrt{2}}{2}\right)^8$

50. $\left(\dfrac{1}{2} + i\dfrac{\sqrt{3}}{2}\right)^6$

CONNECTING CONCEPTS

51. Let n be a positive integer. Expand and simplify $\dfrac{(x + h)^n - x^n}{h}$, where x is any real number and $h \neq 0$.

52. Show that $\dbinom{n}{k} = \dbinom{n}{n - k}$ for all positive integers n and k with $0 \leq k \leq n$.

53. Show that $\displaystyle\sum_{k=0}^{n} \binom{n}{k} = 2^n$. (*Hint:* Use the Binomial Theorem with $x = 1, y = 1$.)

54. Prove that $\dbinom{n}{k} + \dbinom{n}{k + 1} = \dbinom{n + 1}{k + 1}$, n and k integers, $0 \leq k \leq n$.

55. Prove that $\displaystyle\sum_{i=0}^{n} (-1)^i \binom{n}{i} = 0$.

56. Approximate $(0.98)^8$ by evaluating the first three terms of $(1 - 0.02)^8$.

57. Approximate $(1.02)^8$ by evaluating the first three terms of $(1 + 0.02)^8$.

There is an extension of the Binomial Theorem called the ***Multinomial Theorem.*** **This theorem is used in determining probabilities.** ***Multinomial Theorem:*** **If n, r, and k are positive integers, then the coefficient of $a^r b^k c^{n-r-k}$ in the expansion of $(a + b + c)^n$ is**

$$\frac{n!}{r!\,k!\,(n - r - k)!}$$

In Exercises 58 to 61, use the Multinomial Theorem to find the indicated coefficient.

58. Find the coefficient of $a^2b^3c^5$ in the expansion of $(a + b + c)^{10}$.

59. Find the coefficient of $a^5b^2c^2$ in the expansion of $(a + b + c)^9$.

60. Find the coefficient of a^4b^5 in the expansion of $(a + b + c)^9$.

61. Find the coefficient of a^3c^5 in the expansion of $(a + b + c)^8$.

PROJECTS

1. **PASCAL'S TRIANGLE** Write an essay on Pascal's Triangle. Include some of the earliest known examples of the triangle and some of its applications.

2. **SOME OTHER FUNCTIONS** Do some research and determine a definition of positive integers for each of the following types of numbers. Give examples of calculations using each type of number.

   a. Pochammer (m, n) **b.** double factorial $(n!!)$

Using a Graphing Calculator to Find the *n*th Roots of z

Normal Sci Eng
Float 0123456789
Radian Degree
Func Par Pol Seq
Connected Dot
Sequential Simul
FullScreen Split

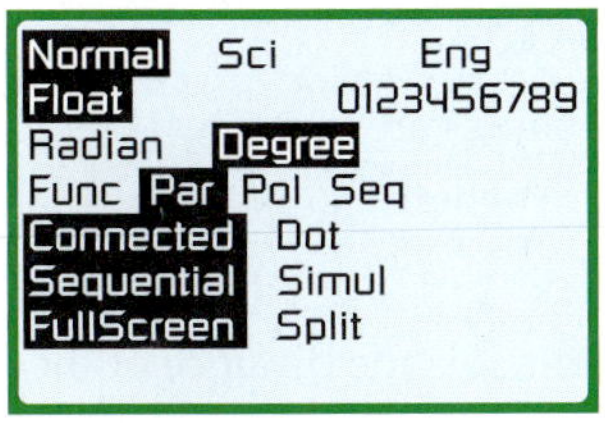

FIGURE 6.66

The parametric feature of a graphing calculator can be used to find and display the nth roots of $z = r(\cos\theta + i\sin\theta)$. Here is the procedure for a TI-83 graphing calculator. Put the calculator in parametric and degree mode. See **Figure 6.66.** To find the nth roots of $z = r(\cos\theta + i\sin\theta)$, enter in the Y= menu

X1T=r^(1/n)cos(θ/n+T) and Y1T=r^(1/n)sin(θ/n+T)

In the WINDOW menu, set Tmin=0, Tmax=360, and Tstep=360/n. Set Xmin, Xmax, Ymin, and Ymax to appropriate values that will allow the roots to be seen in the graph window. Press GRAPH to display a polygon. The x- and y-coordinates of each vertex of the polygon represent a root of z in the rectangular form $x + yi$. Here is a specific example that illustrates this procedure.

Example Find the fourth roots of $z = 16i$.

In trigonometric form, $z = 16(\cos 90° + i\sin 90°)$. Thus in this example, $r = 16$, $\theta = 90°$, and $n = 4$. In the Y= menu, enter

X1T=16^(1/4)cos(90/4+T) and Y1T=16^(1/4)sin(90/4+T)

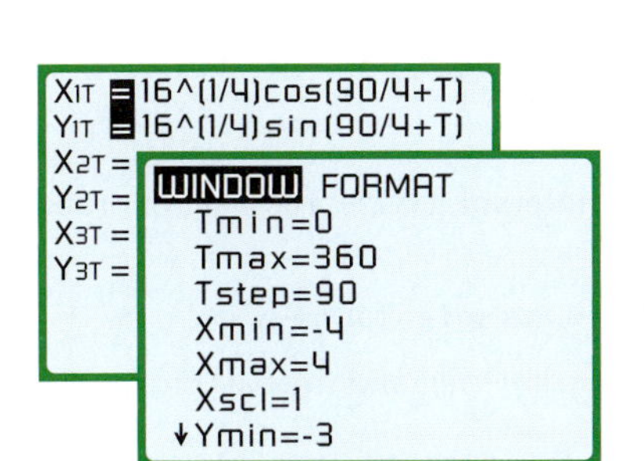

FIGURE 6.67

In the WINDOW menu, set

Tmin=0	Xmin=-4	Ymin=-3
Tmax=360	Xmax=4	Ymax=3
Tstep=360/4	Xscl=1	Yscl=1

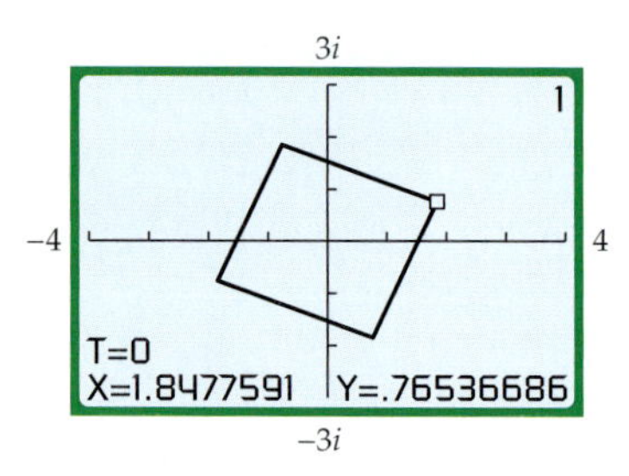

The vertices of the quadrilateral represent the fourth roots of $16i$ in the complex plane.

FIGURE 6.68

See **Figure 6.67.** Press GRAPH to produce the quadrilateral in **Figure 6.68.** Use TRACE and the arrow key ▷ to move to each of the vertices of the quadrilateral. **Figure 6.68** shows that one of the roots of $z = 16i$ is $1.8477591 + 0.76536686i$. Continue to press the arrow key ▷ to find the other three roots, which are

$$-0.7653669 + 1.8477591i, \quad -1.847759 - 0.7653669i, \text{ and}$$
$$0.76536686 - 1.847759i$$

Use a graphing calculator to estimate, in rectangular form, each of the following.

1. The cube roots of -27.
2. The fifth roots of $32i$.
3. The fourth roots of $\sqrt{8} + \sqrt{8}i$.
4. The sixth roots of $-64i$.

CHAPTER 6 SUMMARY

6.1 Conic Sections

- The equations of a parabola with vertex at (h, k) and axis of symmetry parallel to a coordinate axis are given by

  $(x - h)^2 = 4p(y - k)$; focus $(h, k + p)$; directrix $y = k - p$

  $(y - k)^2 = 4p(x - h)$; focus $(h + p, k)$; directrix $x = h - p$

- The equations of an ellipse with center at (h, k) and major axis parallel to a coordinate axis are given by

  $$\frac{(x-h)^2}{a^2} + \frac{(y-k)^2}{b^2} = 1; \text{ foci } (h \pm c, k); \text{ vertices } (h \pm a, k)$$

  $$\frac{(x-h)^2}{b^2} + \frac{(y-k)^2}{a^2} = 1; \text{ foci } (h, k \pm c); \text{ vertices } (h, k \pm a)$$

  For each equation, $a > b$ and $c^2 = a^2 - b^2$.

- The equations of a hyperbola with center at (h, k) and transverse axis parallel to a coordinate axis are given by

  $$\frac{(x-h)^2}{a^2} - \frac{(y-k)^2}{b^2} = 1; \text{ foci } (h \pm c, k); \text{ vertices } (h \pm a, k)$$

  $$\frac{(y-k)^2}{a^2} - \frac{(x-h)^2}{b^2} = 1; \text{ foci } (h, k \pm c); \text{ vertices } (h, k \pm a)$$

  For each equation, $c^2 = a^2 + b^2$.

6.2 Polar Coordinates

- A polar coordinate system is formed by drawing a horizontal ray (*polar axis*). The *pole* is the origin of a polar coordinate system.
- A point is specified by coordinates (r, θ), where r is a directed distance from the pole and θ is an angle measured from the polar axis.

  The transformation equations between a polar coordinate system and a rectangular coordinate system are

  Polar to rectangular: $x = r\cos\theta \qquad y = r\sin\theta$

  Rectangular to polar: $r = \sqrt{x^2 + y^2} \qquad \tan\theta = \dfrac{y}{x}$

- The polar equations of the conics are given by

  $$r = \frac{ed}{1 \pm e\cos\theta} \quad \text{or} \quad r = \frac{ed}{1 \pm e\sin\theta}$$

  where e is the eccentricity and d is the distance of the directrix from the focus.

  When

  $0 < e < 1$, the graph is an ellipse.

  $e = 1$, the graph is a parabola.

  $e > 1$, the graph is a hyperbola.

6.3 Parametric Equations

- Let t be a number in an interval I. A *curve* is a set of ordered pairs (x, y), where

  $$x = f(t), \qquad y = g(t) \quad \text{for } t \in I$$

  The variable t is called a *parameter*, and the pair of equations are *parametric equations*.

- To *eliminate the parameter* is to find an equation in x and y that has the same graph as the given parametric equations.
- The path of a projectile (assume air resistance is negligible) that is launched at an angle θ from the horizon with an initial velocity of v_0 feet per second is given by

  $$x = (v_0\cos\theta)t, \qquad y = -16t^2 + (v_0\sin\theta)t$$

where t is the time in seconds since the projectile was launched.

6.4 Sequences, Series, and Summation Notation

- An infinite sequence is a function whose domain is the positive integers and whose range is the set of real numbers.
- An alternating sequence is one in which the signs of the terms alternate between positive and negative values.
- A recursively defined sequence is one in which each succeeding term of the sequence is defined by using some of the preceding terms.
- If n is a positive integer, then n factorial, $n!$, is the product of the first n positive integers.

$$n! = n(n-1)(n-2)\cdots 3\cdot 2\cdot 1$$

- Given that d is a real number, the sequence a_n is an arithmetic sequence if $a_{i+1} - a_i = d$ for all i. The number d is called the common difference of the sequence.
- The nth term of an arithmetic sequence with common difference of d is $a_n = a_1 + (n-1)d$.
- Given that $r \neq 0$ is a constant real number, the sequence a_n is a geometric sequence if $\dfrac{a_{i+1}}{a_i} = r$ for all positive integers i. The ratio r is called the common ratio of the geometric sequence.
- The nth term of a geometric sequence is $a_n = a_1 r^{n-1}$, where a_1 is the first term of the sequence and r is the common ratio.
- If a_n is a sequence, then $S_n = \sum_{i=1}^{n} a_i$ is the nth partial sum of the sequence.
- If a_n is an arithmetic sequence, then the nth partial sum S_n of the sequence is given by

$$S_n = \frac{n}{2}(a_1 + a_n)$$

- If a_n is a geometric sequence, then the nth partial sum of the sequence is given by

$$S_n = \frac{a_1(1-r^n)}{1-r} \quad r \neq 1$$

- If $|r| < 1$, then the sum of an infinite geometric series is given by

$$S = \frac{a_1}{1-r}$$

6.5 The Binomial Theorem

- **Binomial Theorem for Positive Integers**
  If n is a positive integer, then

$$(a+b)^n = \sum_{i=0}^{n} \binom{n}{i} a^{n-i} b^i$$

- The ith term of the expansion of $(a+b)^n$ is

$$\binom{n}{i-1} a^{n-i+1} b^{i-1}$$

CHAPTER 6 TRUE/FALSE EXERCISES

In Exercises 1 to 10, answer true or false. If the statement is false, give an example or a reason to show that the statement is false.

1. The graph of a parabola is the same shape as that of one branch of a hyperbola.

2. If a hyperbola with center at the origin and a parabola with vertex at the origin have the same focus, $(0, c)$, then the two graphs always intersect.

3. Only the graph of a function can be written using parametric equations.

4. Each ordered pair (r, θ) in a polar coordinate system specifies exactly one point.

5. $\left(\sum_{i=1}^{3} a_i\right)\left(\sum_{i=1}^{3} b_i\right) = \sum_{i=1}^{3} a_i b_i$

6. No two terms of a sequence can be equal.

7. $1, 8, 27, 64, \ldots, k^3, \ldots$ is a geometric sequence.

8. $a_1 = 2, a_{n+1} = a_n - 3$ defines an arithmetic sequence.

9. $0.\overline{9} = 1$

10. In the expansion of $(a+b)^8$, the exponent on a for the fifth term is 5.

CHAPTER 6 REVIEW EXERCISES

In Exercises 1 to 8, if the equation is that of an ellipse or a hyperbola, find the center, vertices, and foci. For hyperbolas, find the equations of the asymptotes. If the equation is that of a parabola, find the vertex, the focus, and the equation of the directrix. Graph each equation.

1. $x^2 - y^2 = 4$
2. $y^2 = 16x$
3. $x^2 + 4y^2 - 6x + 8y - 3 = 0$
4. $3x^2 - 4y^2 + 12x - 24y - 36 = 0$
5. $3x - 4y^2 + 8y + 2 = 0$
6. $3x + 2y^2 - 4y - 7 = 0$
7. $4x^2 - 9y^2 - 8x + 12y - 144 = 0$
8. $9x^2 + 16y^2 + 36x - 16y - 104 = 0$

In Exercises 9 to 14, find the equation of the conic that satisfies the given conditions.

9. Ellipse with vertices at $(7, 3)$ and $(-3, 3)$; length of minor axis is 8.
10. Hyperbola with vertices at $(4, 1)$ and $(-2, 1)$; eccentricity $\frac{4}{3}$.
11. Hyperbola with foci $(-5, 2)$ and $(1, 2)$; length of transverse axis is 4.
12. Parabola with focus $(2, -3)$ and directrix $x = 6$.
13. Parabola with vertex $(0, -2)$ and passing through the point $(3, 4)$.
14. Hyperbola with vertices $(\pm 6, 0)$ and asymptotes whose equations are $y = \pm\frac{1}{9}x$.

In Exercises 15 to 28, graph each polar equation.

15. $r = 4\cos 3\theta$
16. $r = 1 + \cos\theta$
17. $r = 2(1 - 2\sin\theta)$
18. $r = 4\sin 4\theta$
19. $r = 5\sin\theta$
20. $r = 3\sec\theta$
21. $r = 4\csc\theta$
22. $r = 4\cos\theta$
23. $r = 3 + 2\cos\theta$
24. $r = 4 + 2\sin\theta$
25. $r = \dfrac{4}{3 - 6\sin\theta}$
26. $r = \dfrac{2}{1 + \cos\theta}$
27. $r = \dfrac{2}{2 - \cos\theta}$
28. $r = \dfrac{6}{4 + 3\sin\theta}$

In Exercises 29 to 32, change each equation to a polar equation.

29. $y^2 = 16x$
30. $x^2 + y^2 + 4x + 3y = 0$
31. $3x - 2y = 6$
32. $xy = 4$

In Exercises 33 to 36, change each equation to a rectangular equation.

33. $r = \dfrac{4}{1 - \cos\theta}$
34. $r = 3\cos\theta - 4\sin\theta$
35. $r^2 = \cos 2\theta$
36. $\theta = 1$

In Exercises 37 to 42, eliminate the parameter and graph the curve given by the parametric equations.

37. $x = 4t - 2,\ y = 3t + 1$, for $t \in R$
38. $x = 1 - t^2,\ y = 3 - 2t^2$, for $t \in R$
39. $x = 4\sin t,\ y = 3\cos t$, for $0 \le t < 2\pi$
40. $x = \sec t,\ y = 4\tan t$, for $-\frac{\pi}{2} < t < \frac{\pi}{2}$
41. $x = \frac{1}{t},\ y = -\frac{2}{t}$, for $t > 0$
42. $x = 1 + \cos t,\ y = 2 - \sin t$, for $0 \le t < 2\pi$

In Exercises 43 to 56, find the third and seventh terms of the sequence defined by a_n.

43. $a_n = n^2$
44. $a_n = n!$
45. $a_n = 3n + 2$
46. $a_n = 1 - 2n$
47. $a_n = 2^{-n}$
48. $a_n = 3^n$
49. $a_n = \dfrac{1}{n!}$
50. $a_n = \dfrac{1}{n}$

51. $a_1 = 1, a_n = -na_{n-1}$

52. $a_1 = 2, a_n = n^2a_{n-1}$

53. $a_1 = 4, a_n = a_{n-1} + 2$

54. $a_1 = 3, a_n = a_{n-1} - 3$

55. $a_1 = 1, a_2 = 2, a_n = a_{n-1}a_{n-2}$

56. $a_1 = 1, a_2 = 2, a_n = \dfrac{a_{n-1}}{a_{n-2}}$

In Exercises 57 to 70, find the indicated sum of the series. If necessary, round to the nearest ten-thousandth.

57. $\sum_{n=1}^{9} (2n - 3)$

58. $\sum_{i=1}^{11} (1 - 3i)$

59. $\sum_{n=1}^{6} 3 \cdot 2^n$

60. $\sum_{i=1}^{5} 2 \cdot 4^{i-1}$

61. $\sum_{k=1}^{9} (-1)^k 3^k$

62. $\sum_{i=1}^{8} (-1)^{i+1} 2^i$

63. $\sum_{i=1}^{10} \left(\frac{2}{3}\right)^i$

64. $\sum_{i=1}^{11} \left(\frac{3}{2}\right)^i$

65. $\sum_{n=1}^{9} \frac{(-1)^{n+1}}{n^2}$

66. $\sum_{k=1}^{5} \frac{(-1)^{k+1}}{k!}$

67. $\sum_{n=1}^{\infty} \left(\frac{1}{4}\right)^n$

68. $\sum_{i=1}^{\infty} \left(-\frac{5}{6}\right)^i$

69. $\sum_{k=1}^{\infty} \left(-\frac{4}{5}\right)^k$

70. $\sum_{j=0}^{\infty} \left(\frac{1}{5}\right)^j$

In Exercises 71 to 74, use the Binomial Theorem to expand each binomial.

71. $(4a - b)^5$

72. $(x + 3y)^6$

73. $(\sqrt{a} + 2\sqrt{b})^8$

74. $\left(2x - \frac{1}{2x}\right)^7$

75. Find the fifth term in the expansion of $(3x - 4y)^7$.

76. Find the eighth term in the expansion of $(1 - 3x)^9$.

CHAPTER 6 TEST

1. Find the vertex, focus, and directrix of the parabola given by the equation $y = \frac{1}{8}x^2$.

2. Find the vertices and foci of the ellipse given by the equation $25x^2 - 150x + 9y^2 + 18y + 9 = 0$.

3. Find the equation in standard form of the ellipse with center $(0, -3)$, foci $(-6, -3)$ and $(6, -3)$, and minor axis of length 6.

4. Graph: $\frac{y^2}{25} - \frac{x^2}{16} = 1$

5. $P(1, -\sqrt{3})$ are the coordinates of a point in an xy-coordinate system. Find the polar coordinates of P.

6. Graph: $r = 4 \cos \theta$

7. Graph: $r = 3(1 - \sin \theta)$

8. Graph: $r = 2 \sin 4\theta$

9. Find the rectangular coordinates of the point whose polar coordinates are $\left(5, \frac{7\pi}{3}\right)$.

10. Find the rectangular form of $r - r \cos \theta = 4$.

11. Write $r = \frac{4}{1 + \sin \theta}$ as an equation in rectangular coordinates.

12. Eliminate the parameter and graph the curve given by the parametric equations $x = t - 3$, $y = 2t^2$.

In Exercises 13 to 15, classify each sequence as an arithmetic sequence, a geometric sequence, or neither.

13. $a_n = -2n + 3$

14. $a_n = 2n^2$

15. $a_n = \frac{(-1)^{n-1}}{3^n}$

In Exercises 16 and 17, find the indicated sum of the series.

16. $\sum_{j=1}^{10} \frac{1}{2^j}$

17. $\sum_{k=1}^{20} (3k - 2)$

18. The third term of an arithmetic sequence is 7 and the eighth term is 22. Find the twentieth term.

19. Find the sum of the infinite geometric series given by $\sum_{k=1}^{\infty} \left(\frac{3}{8}\right)^k$.

20. Write $0.\overline{15}$ as the quotient of integers in simplest form.

21. Write the binomial expansion of $(x - 2y)^5$.

22. Write the binomial expansion of $\left(x + \frac{1}{x}\right)^6$.

23. Find the sixth term in the expansion of $(3x + 2y)^8$.

CUMULATIVE REVIEW EXERCISES

1. Find the value of x in the domain of $F(x) = 5 + \frac{x}{3}$ for which $F(x) = -3$.

2. Solve: $2x^2 - 3x = 4$

3. Write $\log_b\left(\frac{xy^2}{z^3}\right)$ in terms of the logarithms of x, y, and z.

4. Let $g(x) = x^2 - x + 4$ and $h(x) = x - 2$. Find $\left(\frac{h}{g}\right)(-3)$.

5. Find the horizontal asymptote of the graph of $F(x) = \frac{x^3 - 8}{x^5}$.

6. Evaluate: $\log_{\frac{1}{2}} 64$

7. Solve $4^{2x+1} = 3^{x-2}$. Round to the nearest tenth.

8. The time t in seconds required for a sky diver to reach a velocity of v feet per second is given by $t = -\frac{175}{32}\ln\left(1 - \frac{v}{175}\right)$. Determine the velocity of the sky diver after 5 seconds. Round to the nearest mile per hour.

9. Given $\sin\theta = -\frac{1}{2}$ and $\sec\theta = \frac{2\sqrt{3}}{3}$, find $\cot\theta$.

10. Express $\frac{\sin x}{1 + \cos x} + \frac{1 + \cos x}{\sin x}$ in terms of $\csc x$.

11. Solve the triangle ABC if $A = 40°$, $B = 65°$, and $c = 20$ centimeters.

12. Evaluate: $\sin\left(\frac{1}{2}\cos^{-1}\frac{4}{5}\right)$

13. Given $f(x) = 1 - x^2$, write the difference quotient $\frac{f(2 + h) - f(2)}{h}$ in simplest form.

14. Let $f(x) = 3x + 2$ and $g(x) = 2 - x^2$. Find $(f \circ g)(-3)$.

15. Given the graph below, sketch the graph of $y = -f(x) + 2$.

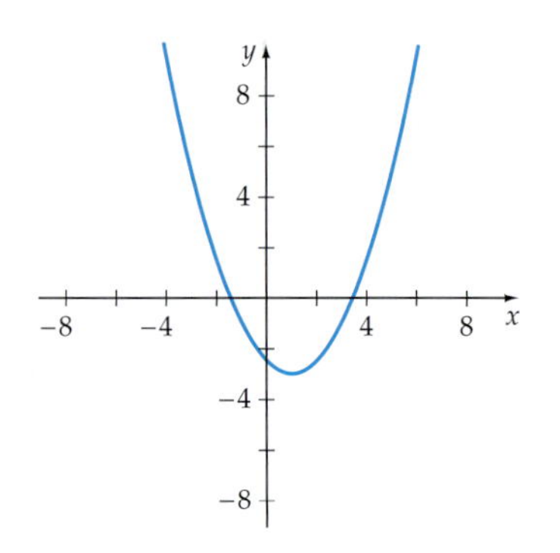

16. Find the inverse function for $f(x) = 2x - 8$.

17. Let $f(x) = \frac{3x}{x^2 + 1}$. Is f an even function, an odd function, or neither?

18. Find the measure of a for the right triangle shown at the right. Round to the nearest tenth.

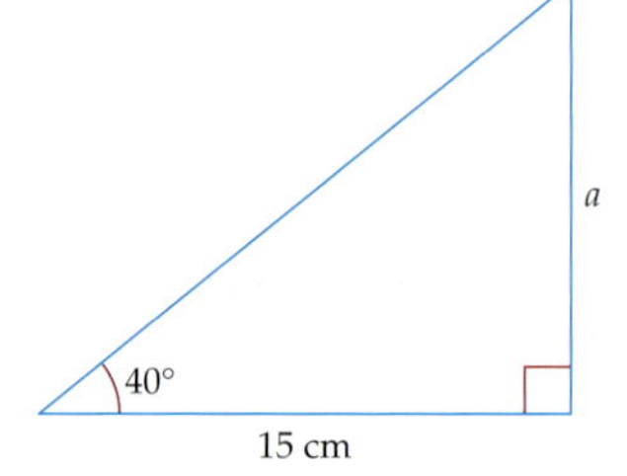

19. What are the period and amplitude for the graph of $f(x) = 3\cos 4x$?

20. Solve $\sin x \cos x - \frac{1}{2}\cos x = 0$ for $0 \le x < 2\pi$.

APPENDIX A.1 Integer and Rational Number Exponents

Properties of Exponents

A compact method of writing $5 \cdot 5 \cdot 5 \cdot 5$ is 5^4. The expression 5^4 is written in **exponential notation.** Similarly, we can write

$$\frac{2x}{3} \cdot \frac{2x}{3} \cdot \frac{2x}{3} \quad \text{as} \quad \left(\frac{2x}{3}\right)^3$$

Exponential notation can be used to express the product of any expression that is used repeatedly as a factor.

MATH MATTERS

The expression 10^{100} is called a *googol.* The term was coined by the 9-year-old nephew of the American mathematician Edward Kasner. Many calculators do not provide for numbers of this magnitude, but it is no serious loss. To appreciate the magnitude of a googol, consider that if all the atoms in the known universe were counted, the number would not even be close to a googol. But if a googol is too small for you, try 10^{googol}, which is called a *googolplex.* As a final note, the name of the Internet site Google.com is a takeoff on the word *googol.*

Definition of Natural Number Exponents

If b is any real number and n is a natural number, then

$$b^n = \overbrace{b \cdot b \cdot b \cdot \cdots \cdot b}^{b \text{ is a factor } n \text{ times}}$$

where b is the **base** and n is the **exponent.**

For instance,

$$5^4 = 5 \cdot 5 \cdot 5 \cdot 5 = 625$$
$$-5^4 = -(5 \cdot 5 \cdot 5 \cdot 5) = -625$$
$$(-5)^4 = (-5)(-5)(-5)(-5) = 625$$

Pay close attention to the difference between -5^4 (the base is 5) and $(-5)^4$ (the base is -5).

QUESTION What is the value of **a.** -2^5 and **b.** $(-2)^5$?

We can extend the definition of an exponent to all the integers. We first deal with the case of zero as an exponent.

Definition of b^0

For any nonzero real number b, $b^0 = 1$.

Some examples of this definition are

$$3^0 = 1 \qquad \left(\frac{3}{4}\right)^0 = 1 \qquad -\pi^0 = -1 \qquad (a^2 + 1)^0 = 1$$

ANSWER **a.** $-2^5 = -(2 \cdot 2 \cdot 2 \cdot 2 \cdot 2) = -32$
b. $(-2)^5 = (-2)(-2)(-2)(-2)(-2) = -32$

> **take note**
>
> Using the definition of b^{-n},
>
> $$\frac{5^{-2}}{7^{-1}} = \frac{\frac{1}{5^2}}{\frac{1}{7}}$$
>
> Using the rules for dividing fractions, we have
>
> $$\frac{\frac{1}{5^2}}{\frac{1}{7}} = \frac{1}{5^2} \div \frac{1}{7} = \frac{1}{5^2} \cdot \frac{7}{1} = \frac{7}{5^2}$$

Now we extend the definition to include negative integers.

Definition of b^{-n}

If $b \neq 0$ and n is a natural number, then $b^{-n} = \frac{1}{b^n}$ and $\frac{1}{b^{-n}} = b^n$.

Here are some examples.

$$3^{-2} = \frac{1}{3^2} = \frac{1}{9} \qquad \frac{1}{4^{-3}} = 4^3 = 64 \qquad \frac{5^{-2}}{7^{-1}} = \frac{7}{5^2} = \frac{7}{25}$$

EXAMPLE 1 Evaluate an Exponential Expression

a. $(-2^4)(-3)^2$ **b.** $\frac{(-4)^{-3}}{(-2)^{-5}}$

Solution

a. $(-2^4)(-3)^2 = -(2 \cdot 2 \cdot 2 \cdot 2)(-3)(-3) = -(16)(9) = -144$

b. $\frac{(-4)^{-3}}{(-2)^{-5}} = \frac{(-2)(-2)(-2)(-2)(-2)}{(-4)(-4)(-4)} = \frac{-32}{-64} = \frac{1}{2}$

▶ TRY EXERCISE 10, PAGE 548

When working with exponential expressions containing variables, we must ensure that a value of the variable does not result in an undefined expression. For instance, $x^{-2} = \frac{1}{x^2}$. Because division by zero is not allowed, for the expression x^{-2}, we must assume that $x \neq 0$. Therefore, to avoid problems with undefined expressions, we will use the following restriction agreement.

Restriction Agreement

The expressions 0^0, 0^n (where n is a negative integer), and $\frac{a}{0}$ are all undefined expressions. Therefore, all values of variables in this text are restricted to avoid any one of these expressions.

For instance, in the expression $\frac{x^0 y^{-3}}{z - 4}$,

$$x \neq 0, y \neq 0, \text{ and } z \neq 4.$$

For the expression $\dfrac{(a-1)^0}{b+2}$,

$$a \neq 1 \text{ and } b \neq -2.$$

Exponential expressions containing variables are simplified by using the following properties of exponents.

Properties of Exponents

If m, n, and p are integers and a and b are real numbers, then

Product $\quad b^m \cdot b^n = b^{m+n}$

Quotient $\quad \dfrac{b^m}{b^n} = b^{m-n}, \quad b \neq 0$

Power $\quad (b^m)^n = b^{mn} \qquad (a^m b^n)^p = a^{mp} b^{np}$

$$\left(\frac{a^m}{b^n}\right)^p = \frac{a^{mp}}{b^{np}}, \quad b \neq 0$$

Here are some examples of these properties.

$a^4 \cdot a \cdot a^3 = a^{4+1+3} = a^8$ • **Recall that $a = a^1$.**

$(x^4y^3)(xy^5z^2) = x^{4+1}y^{3+5}z^2 = x^5y^8z^2$ • **Add the exponents on the like bases.**

$\dfrac{a^7b}{a^2b^5} = a^{7-2}b^{1-5} = a^5b^{-4} = \dfrac{a^5}{b^4}$ • **Subtract the exponents on the like bases.**

$(uv^3)^5 = u^{1\cdot5}v^{3\cdot5} = u^5v^{15}$ • **Multiply the exponents.**

? QUESTION Can the exponential expression x^5y^3 be simplified using the properties of exponents?

To simplify an expression involving exponents, write the expression in a form in which *each base appears at most once* and *no powers of powers or negative exponents appear.*

EXAMPLE 2 Simplify Exponential Expressions

Simplify. **a.** $(5x^2y)(-4x^3y^5)$ **b.** $(3x^2yz^{-4})^3$ **c.** $\left(\dfrac{4p^2q}{6pq^4}\right)^{-2}$

Continued ▶

? ANSWER No. The bases are not the same.

Solution

a. $(5x^2y)(-4x^3y^5) = [5(-4)]x^{2+3}y^{1+5}$ • **Multiply the coefficients. Multiply the variables by adding the exponents on the like bases.**

$= -20x^5y^6$

b. $(3x^2yz^{-4})^3 = 3^{1\cdot 3}x^{2\cdot 3}y^{1\cdot 3}z^{-4\cdot 3}$ • **Use the power property of exponents.**

$= 3^3x^6y^3z^{-12} = \dfrac{27x^6y^3}{z^{12}}$

c. $\left(\dfrac{4p^2q}{6pq^4}\right)^{-2} = \left(\dfrac{2p^{2-1}q^{1-4}}{3}\right)^{-2} = \left(\dfrac{2pq^{-3}}{3}\right)^{-2}$ • **Use the quotient property of exponents.**

$= \dfrac{2^{1(-2)}p^{1(-2)}q^{-3(-2)}}{3^{1(-2)}} = \dfrac{2^{-2}p^{-2}q^6}{3^{-2}}$ • **Use the power property of exponents.**

$= \dfrac{9q^6}{4p^2}$ • **Write the answer in simplest form.**

▶ **TRY EXERCISE 30, PAGE 548**

• Scientific Notation

> **MATH MATTERS**
>
> - Approximately 3.1×10^6 orchid seeds weigh 1 ounce.
> - Computer scientists measure an operation in nanoseconds.
>
> 1 nanosecond $= 1 \times 10^{-9}$ second
>
> - If a spaceship traveled 25,000 mph, it would require approximately 2.7×10^9 years to travel from one end of the universe to the other.

The exponent theorems provide a compact method of writing very large or very small numbers. The method is called *scientific notation.* A number written in **scientific notation** has the form $a \cdot 10^n$, where n is an integer and $1 \le a < 10$. The following procedure is used to change a number from its decimal form to scientific notation.

For numbers greater than 10, move the decimal point to the position to the right of the first digit. The exponent n will equal the number of places the decimal point has been moved. For example,

$$7{,}430{,}000 = 7.43 \times 10^6$$

6 places

For numbers less than 1, move the decimal point to the right of the first nonzero digit. The exponent n will be negative, and its absolute value will equal the number of places the decimal point has been moved. For example,

$$0.00000078 = 7.8 \times 10^{-7}$$

7 places

To change a number from scientific notation to its decimal form, reverse the procedure. That is, if the exponent is positive, move the decimal point to the right the same number of places as the exponent. For example,

$$3.5 \times 10^5 = 350{,}000$$

5 places

If the exponent is negative, move the decimal point to the left the same number of places as the absolute value of the exponent. For example,

$$2.51 \times 10^{-8} = 0.0000000251$$

8 places

Most scientific calculators display very large and very small numbers in scientific notation. The number $450{,}000^2$ is displayed as 2.025 E 11. This means $450{,}000^2 = 2.025 \times 10^{11}$.

EXAMPLE 3 Simplify an Expression Using Scientific Notation

The Andromeda galaxy is approximately 1.4×10^{19} miles from Earth. If a spacecraft could travel 2.8×10^{12} miles in 1 year (about one-half the speed of light), how many years would it take to reach the Andromeda galaxy?

Solution

To find the time, divide the distance by the speed.

$$t = \frac{1.4 \times 10^{19}}{2.8 \times 10^{12}} = \frac{1.4}{2.8} \times 10^{19-12} = 0.5 \times 10^7 = 5.0 \times 10^6$$

It would take 5.0×10^6 (or 5,000,000) years to reach the Andromeda galaxy.

TRY EXERCISE 46, PAGE 549

RATIONAL EXPONENTS AND RADICALS

To this point, the expression b^n has been defined for real numbers b and integers n. Now we wish to extend the definition of exponents to include rational numbers so that expressions such as $2^{1/2}$ will be meaningful.

Definition of $b^{1/n}$

If n is an even positive integer and $b \geq 0$, then $b^{1/n}$ is the nonnegative real number such that $(b^{1/n})^n = b$.

If n is an odd positive integer, then $b^{1/n}$ is the real number such that $(b^{1/n})^n = b$.

As examples,

- $25^{1/2} = 5$ because $5^2 = 25$.
- $(-64)^{1/3} = -4$ because $(-4)^3 = -64$.
- $16^{1/2} = 4$ because $4^2 = 16$.
- $-16^{1/2} = -(16^{1/2}) = -4$.
- $(-16)^{1/2}$ is not a real number.
- $(-32)^{1/5} = -2$ because $(-2)^5 = -32$.

If n is an even positive integer and $b < 0$, then $b^{1/n}$ is a *complex number*. Complex numbers are discussed in Section 2.1.

To define expressions such as $8^{2/3}$, we will extend our definition of exponents even further. Because we want the power property $(b^p)^q = b^{pq}$ to be true for rational exponents also, we must have $(b^{1/n})^m = b^{m/n}$. With this in mind, we make the following definition.

Definition of $b^{m/n}$

For all positive integers m and n such that m/n is in simplest form, and for all real numbers b for which $b^{1/n}$ is a real number,

$$b^{m/n} = (b^{1/n})^m = (b^m)^{1/n}$$

Because $b^{m/n}$ is defined as $(b^{1/n})^m$ and also as $(b^m)^{1/n}$, we can evaluate expressions such as $8^{4/3}$ in more than one way. For example, because $8^{1/3}$ is a real number, $8^{4/3}$ can be evaluated in either of the following ways:

$$8^{4/3} = (8^{1/3})^4 = 2^4 = 16$$
$$8^{4/3} = (8^4)^{1/3} = 4096^{1/3} = 16$$

Of the two methods, the $b^{m/n} = (b^{1/n})^m$ method is usually easier to apply, provided you can evaluate $b^{1/n}$.

Here are some additional examples.

$$64^{2/3} = (64^{1/3})^2 = 4^2 = 16$$
$$32^{-6/5} = \frac{1}{32^{6/5}} = \frac{1}{(32^{1/5})^6} = \frac{1}{2^6} = \frac{1}{64}$$
$$81^{0.75} = 81^{3/4} = (81^{1/4})^3 = 3^3 = 27$$

The following exponent properties were stated earlier, but they are restated here to remind you that they have now been extended to apply to rational exponents.

Properties of Rational Exponents

If p, q, and r represent rational numbers and a and b are positive real numbers, then

Product $\quad b^p \cdot b^q = b^{p+q}$

Quotient $\quad \dfrac{b^p}{b^q} = b^{p-q}$

Power $\quad (b^p)^q = b^{pq} \qquad (a^p b^q)^r = a^{pr} b^{qr}$

$$\left(\frac{a^p}{b^q}\right)^r = \frac{a^{pr}}{b^{qr}} \qquad b^{-p} = \frac{1}{b^p}$$

Recall that an exponential expression is in simplest form when no powers of powers or negative exponents appear and each base occurs at most once.

EXAMPLE 4 Simplify Exponential Expressions

Simplify: $\left(\dfrac{x^2y^3}{x^{-3}y^5}\right)^{1/2}$ (Assume $x > 0$, $y > 0$.)

Solution

$$\left(\frac{x^2y^3}{x^{-3}y^5}\right)^{1/2} = (x^{2-(-3)}y^{3-5})^{1/2} = (x^5y^{-2})^{1/2} = x^{5/2}y^{-1} = \frac{x^{5/2}}{y}$$

▶ TRY EXERCISE 62, PAGE 549

Simplify Radical Expressions

MATH MATTERS

The formula for kinetic energy (energy of motion) that is used in Einstein's Theory of Relativity involves a radical.

$$\text{K.E}_r = mc^2\left(\frac{1}{\sqrt{1 - \dfrac{v^2}{c^2}}} - 1\right)$$

where m is the mass of the object at rest, v is the speed of the object, and c is the speed of light.

Radicals, expressed by the notation $\sqrt[n]{b}$, are also used to denote roots. The number b is the **radicand,** and the positive integer n is the **index** of the radical.

Definition of $\sqrt[n]{b}$

If n is a positive integer and b is a real number such that $b^{1/n}$ is a real number, then $\sqrt[n]{b} = b^{1/n}$.

If the index n equals 2, then the radical $\sqrt[2]{b}$ is written as simply $\sqrt{b}$, and it is referred to as the **principal square root of b** or simply the **square root of b.**

The symbol $\sqrt{b}$ is reserved to represent the nonnegative square root of b. To represent the negative square root of b, write $-\sqrt{b}$. For example, $\sqrt{25} = 5$, whereas $-\sqrt{25} = -5$.

Definition of $(\sqrt[n]{b})^m$

For all positive integers n, all integers m, and all real numbers b such that $\sqrt[n]{b}$ is a real number, $(\sqrt[n]{b})^m = \sqrt[n]{b^m} = b^{m/n}$.

When $\sqrt[n]{b}$ is a real number, the equations

$$b^{m/n} = \sqrt[n]{b^m} \quad \text{and} \quad b^{m/n} = (\sqrt[n]{b})^m$$

can be used to write exponential expressions such as $b^{m/n}$ in radical form. Use the denominator n as the index of the radical and the numerator m as the power of the radicand or as the power of the radical. For example,

$(5xy)^{2/3} = (\sqrt[3]{5xy})^2 = \sqrt[3]{25x^2y^2}$ • **Use the denominator 3 as the index of the radical and the numerator 2 as the power of the radical.**

The equations

$$b^{m/n} = \sqrt[n]{b^m} \quad \text{and} \quad b^{m/n} = (\sqrt[n]{b})^m$$

also can be used to write radical expressions in exponential form. For example,

$\sqrt{(2ab)^3} = (2ab)^{3/2}$ • **Use the index 2 as the denominator of the power and the exponent 3 as the numerator of the power.**

The definition of $\left(\sqrt[n]{b}\right)^m$ often can be used to evaluate radical expressions. For instance,

$$(\sqrt[3]{8})^4 = 8^{4/3} = (8^{1/3})^4 = 2^4 = 16$$

Care must be exercised when simplifying even roots (square roots, fourth roots, sixth roots,...) of variable expressions. Consider $\sqrt{x^2}$ when $x = 5$ and when $x = -5$.

Case 1 If $x = 5$, then $\sqrt{x^2} = \sqrt{5^2} = \sqrt{25} = 5 = x$.

Case 2 If $x = -5$, then $\sqrt{x^2} = \sqrt{(-5)^2} = \sqrt{25} = 5 = -x$.

These two cases suggest that

$$\sqrt{x^2} = \begin{cases} x, & \text{if } x \geq 0 \\ -x, & \text{if } x < 0 \end{cases}$$

Recalling the definition of absolute value, we can write this more compactly as $\sqrt{x^2} = |x|$.

Simplifying odd roots of a variable expression does not require using the absolute value symbol. Consider $\sqrt[3]{x^3}$ when $x = 5$ and when $x = -5$.

Case 1 If $x = 5$, then $\sqrt[3]{x^3} = \sqrt[3]{5^3} = \sqrt[3]{125} = 5 = x$.

Case 2 If $x = -5$, then $\sqrt[3]{x^3} = \sqrt[3]{(-5)^3} = \sqrt[3]{-125} = -5 = x$.

Thus $\sqrt[3]{x^3} = x$.

Although we have illustrated this principle only for square roots and cube roots, the same reasoning can be applied to other cases. The general result is given below.

Definition of $\sqrt[n]{b^n}$

If n is an even natural number and b is a real number, then

$$\sqrt[n]{b^n} = |b|$$

If n is an odd natural number and b is a real number, then

$$\sqrt[n]{b^n} = b$$

Here are some examples of these properties.

$$\sqrt[4]{16z^4} = 2|z| \qquad \sqrt[5]{32a^5} = 2a$$

Because radicals are defined in terms of rational powers, the properties of radicals are similar to those of exponential expressions.

Properties of Radicals

If m and n are natural numbers and a and b are nonnegative real numbers, then

Product	$\sqrt[n]{a} \cdot \sqrt[n]{b} = \sqrt[n]{ab}$
Quotient	$\dfrac{\sqrt[n]{a}}{\sqrt[n]{b}} = \sqrt[n]{\dfrac{a}{b}}$
Index	$\sqrt[m]{\sqrt[n]{a}} = \sqrt[mn]{a}$

A radical is in **simplest form** if it meets all of the following criteria.

1. The radicand contains only powers less than the index. ($\sqrt{x^5}$ does not satisfy this requirement because 5, the exponent, is greater than 2, the index.)
2. The index of the radical is as small as possible. ($\sqrt[9]{x^3}$ does not satisfy this requirement because $\sqrt[9]{x^3} = x^{3/9} = x^{1/3} = \sqrt[3]{x}$.)
3. The denominator has been rationalized. That is, no radicals appear in the denominator. ($1/\sqrt{2}$ does not satisfy this requirement.)
4. No fractions appear under the radical sign. ($\sqrt[4]{2/x^3}$ does not satisfy this requirement.)

Radical expressions are simplified by using the properties of radicals. Here are some examples.

EXAMPLE 5 Simplify Radical Expressions

Simplify.

a. $\sqrt[4]{32x^3y^4}$ **b.** $\sqrt[3]{162x^4y^6}$

Solution

a. $\sqrt[4]{32x^3y^4} = \sqrt[4]{2^5x^3y^4} = \sqrt[4]{(2^4y^4) \cdot (2x^3)}$
• Factor and group factors that can be written as a power of the index.

$= \sqrt[4]{2^4y^4} \cdot \sqrt[4]{2x^3}$
• Use the product property of radicals.

$= 2|y|\sqrt[4]{2x^3}$
• Recall that for n even, $\sqrt[n]{b^n} = |b|$.

b. $\sqrt[3]{162x^4y^6} = \sqrt[3]{(2 \cdot 3^4)x^4y^6}$
• Factor and group factors that can be written as a power of the index.

$= \sqrt[3]{(3xy^2)^3 \cdot (2 \cdot 3x)}$

$= \sqrt[3]{(3xy^2)^3} \cdot \sqrt[3]{6x}$
• Use the product property of radicals.

$= 3xy^2\sqrt[3]{6x}$
• Recall that for n odd, $\sqrt[n]{b^n} = b$.

TRY EXERCISE 78, PAGE 549

Like radicals have the same radicand and the same index. For instance,

$$3\sqrt[3]{5xy^2} \qquad \text{and} \qquad -4\sqrt[3]{5xy^2}$$

are like radicals. Addition and subtraction of like radicals are accomplished by using the distributive property. For example,

$$4\sqrt{3x} - 9\sqrt{3x} = (4 - 9)\sqrt{3x} = -5\sqrt{3x}$$
$$2\sqrt[3]{y^2} + 4\sqrt[3]{y^2} - \sqrt[3]{y^2} = (2 + 4 - 1)\sqrt[3]{y^2} = 5\sqrt[3]{y^2}$$

The sum $2\sqrt{3} + 6\sqrt{5}$ cannot be simplified further because the radicands are not the same. The sum $3\sqrt[3]{x} + 5\sqrt[4]{x}$ cannot be simplified because the indices are not the same.

Sometimes it is possible to simplify radical expressions that do not appear to be like radicals by simplifying each radical expression.

EXAMPLE 6 **Combine Radical Expressions**

Simplify: $5x\sqrt[3]{16x^4} - \sqrt[3]{128x^7}$

Solution

$5x\sqrt[3]{16x^4} - \sqrt[3]{128x^7}$

$= 5x\sqrt[3]{2^4x^4} - \sqrt[3]{2^7x^7}$ • **Factor.**

$= 5x\sqrt[3]{2^3x^3} \cdot \sqrt[3]{2x} - \sqrt[3]{2^6x^6} \cdot \sqrt[3]{2x}$ • **Group factors that can be written as a power of the index.**

$= 5x(2x\sqrt[3]{2x}) - 2^2x^2 \cdot \sqrt[3]{2x}$ • **Use the product property of radicals.**

$= 10x^2\sqrt[3]{2x} - 4x^2\sqrt[3]{2x}$ • **Simplify.**

$= 6x^2\sqrt[3]{2x}$

▶ TRY EXERCISE 86, PAGE 549

Multiplication of radical expressions is accomplished by using the distributive property. For instance,

$\sqrt{5}(\sqrt{20} - 3\sqrt{15}) = \sqrt{5}(\sqrt{20}) - \sqrt{5}(3\sqrt{15})$ • **Use the distributive property.**

$= \sqrt{100} - 3\sqrt{75}$ • **Multiply the radicals.**

$= 10 - 3 \cdot 5\sqrt{3}$ • **Simplify.**

$= 10 - 15\sqrt{3}$

The product of more complicated radical expressions may require repeated use of the distributive property.

In Exercises 41 to 48, perform the indicated operation and write the answer in scientific notation.

41. $(3 \times 10^{12})(9 \times 10^{-5})$

42. $(8.9 \times 10^{-5})(3.4 \times 10^{-6})$

43. $\dfrac{9 \times 10^{-3}}{6 \times 10^{8}}$

44. $\dfrac{2.5 \times 10^{8}}{5 \times 10^{10}}$

45. $\dfrac{(3.2 \times 10^{-11})(2.7 \times 10^{18})}{1.2 \times 10^{-5}}$

▶ **46.** $\dfrac{(6.9 \times 10^{27})(8.2 \times 10^{-13})}{4.1 \times 10^{15}}$

47. $\dfrac{(4.0 \times 10^{-9})(8.4 \times 10^{5})}{(3.0 \times 10^{-6})(1.4 \times 10^{18})}$

48. $\dfrac{(7.2 \times 10^{8})(3.9 \times 10^{-7})}{(2.6 \times 10^{-10})(1.8 \times 10^{-8})}$

In Exercises 49 to 70, simplify each exponential expression.

49. $4^{3/2}$

50. $-16^{3/2}$

51. $-64^{2/3}$

52. $125^{4/3}$

53. $9^{-3/2}$

54. $32^{-3/5}$

55. $\left(\dfrac{4}{9}\right)^{1/2}$

56. $\left(\dfrac{16}{25}\right)^{3/2}$

57. $\left(\dfrac{1}{8}\right)^{-4/3}$

58. $\left(\dfrac{8}{27}\right)^{-2/3}$

59. $(4a^{2/3}b^{1/2})(2a^{1/3}b^{3/2})$

60. $(6a^{3/5}b^{1/4})(-3a^{1/5}b^{3/4})$

61. $(-3x^{2/3})(4x^{1/4})$

▶ **62.** $(-5x^{1/3})(-4x^{1/2})$

63. $(81x^{8}y^{12})^{1/4}$

64. $(27x^{3}y^{6})^{2/3}$

65. $\dfrac{16z^{3/5}}{12z^{1/5}}$

66. $\dfrac{6a^{2/3}}{9a^{1/3}}$

67. $(2x^{2/3}y^{1/2})(3x^{1/6}y^{1/3})$

68. $\dfrac{x^{1/3}y^{5/6}}{x^{2/3}y^{1/6}}$

69. $\dfrac{9a^{3/4}b}{3a^{2/3}b^{2}}$

70. $\dfrac{12x^{1/6}y^{1/4}}{16x^{3/4}y^{1/2}}$

In Exercises 71 to 80, simplify each radical expression.

71. $\sqrt{45}$

72. $\sqrt{75}$

73. $\sqrt[3]{24}$

74. $\sqrt[3]{135}$

75. $\sqrt[3]{-135}$

76. $\sqrt[3]{-250}$

77. $\sqrt{24x^{2}y^{3}}$

▶ **78.** $\sqrt{18x^{2}y^{5}}$

79. $\sqrt[3]{16a^{3}y^{7}}$

80. $\sqrt[3]{54m^{2}n^{7}}$

In Exercises 81 to 88, simplify each radical and then combine like radicals.

81. $2\sqrt{32} - 3\sqrt{98}$

82. $5\sqrt[3]{32} + 2\sqrt[3]{108}$

83. $-8\sqrt[4]{48} + 2\sqrt[4]{243}$

84. $2\sqrt[3]{40} - 3\sqrt[3]{135}$

85. $4\sqrt[3]{32y^{4}} + 3y\sqrt[3]{108y}$

▶ **86.** $-3x\sqrt[3]{54x^{4}} + 2\sqrt[3]{16x^{7}}$

87. $x\sqrt[3]{8x^{3}y^{4}} - 4y\sqrt[3]{64x^{6}y}$

88. $4\sqrt{a^{5}b} - a^{2}\sqrt{ab}$

In Exercises 89 to 98, find the indicated products and express each result in simplest form.

89. $(\sqrt{5} + 3)(\sqrt{5} + 4)$

90. $(\sqrt{7} + 2)(\sqrt{7} - 5)$

91. $(\sqrt{2} - 3)(\sqrt{2} + 3)$

92. $(2\sqrt{7} + 3)(2\sqrt{7} - 3)$

93. $(3\sqrt{z} - 2)(4\sqrt{z} + 3)$

94. $(4\sqrt{a} - \sqrt{b})(3\sqrt{a} + 2\sqrt{b})$

95. $(\sqrt{x} + 2)^{2}$

▶ **96.** $(3\sqrt{5y} - 4)^{2}$

97. $(\sqrt{x - 3} + 2)^{2}$

98. $(\sqrt{2x + 1} - 3)^{2}$

In Exercises 99 to 112, simplify each expression by rationalizing the denominator. Write the result in simplest form.

99. $\dfrac{2}{\sqrt{2}}$

100. $\dfrac{3x}{\sqrt{3}}$

101. $\sqrt{\dfrac{5}{18}}$

102. $\sqrt{\dfrac{7}{40}}$

103. $\dfrac{3}{\sqrt[3]{2}}$

104. $\dfrac{2}{\sqrt[3]{4}}$

105. $\dfrac{4}{\sqrt[3]{8x^{2}}}$

▶ **106.** $\dfrac{2}{\sqrt[4]{4y}}$

107. $\dfrac{3}{\sqrt{3} + 4}$

108. $\dfrac{2}{\sqrt{5} - 2}$

109. $\dfrac{6}{2\sqrt{5} + 2}$

▶ **110.** $\dfrac{-7}{3\sqrt{2} - 5}$

111. $\dfrac{3}{\sqrt{5} + \sqrt{x}}$

112. $\dfrac{5}{\sqrt{y} - \sqrt{3}}$

APPENDIX A.2

POLYNOMIALS

- OPERATIONS ON POLYNOMIALS

OPERATIONS ON POLYNOMIALS

A **monomial** is a constant, a variable, or a product of a constant and one or more variables, with the variables having only nonnegative integer exponents. The constant is called the **numerical coefficient** or simply the **coefficient** of the monomial. The **degree of a monomial** is the sum of the exponents of the variables. For example, $-5xy^2$ is a monomial with coefficient -5 and degree 3.

The algebraic expression $3x^{-2}$ is not a monomial because it cannot be written as a product of a constant and a variable with a *nonnegative* integer exponent.

A sum of a finite number of monomials is called a **polynomial.** Each monomial is called a **term** of the polynomial. The **degree of a polynomial** is the largest degree of the terms in the polynomial.

Terms that have exactly the same variables raised to the same powers are called **like terms.** For example, $14x^2$ and $-31x^2$ are like terms; however, $2x^3y$ and $7xy$ are not like terms because x^3y and xy are not identical.

A polynomial is said to be simplified if all its like terms have been combined. For example, the simplified form of $4x^2 + 3x + 5x$ is $4x^2 + 8x$. A simplified polynomial that has two terms is a **binomial,** and a simplified polynomial that has three terms is a **trinomial.** For example, $4x + 7$ is a binomial, and $2x^3 - 7x^2 + 11$ is a trinomial.

A nonzero constant, such as 5, is called a **constant polynomial.** It has degree zero because $5 = 5x^0$. The number 0 is defined to be a polynomial with no degree.

Standard Form of a Polynomial

The **standard form of a polynomial** of degree n in the variable x is

$$a_n x^n + a_{n-1}x^{n-1} + \cdots + a_2x^2 + a_1x + a_0$$

where $a_n \neq 0$ and n is a nonnegative integer. The coefficient a_n is the **leading coefficient,** and a_0 is the **constant term.**

If a polynomial in the variable x is written with decreasing powers of x, then it is in **standard form.** For example, the polynomial

$$3x^2 - 4x^3 + 7x^4 - 1$$

is written in standard form as

$$7x^4 - 4x^3 + 3x^2 - 1$$

The following table shows the leading coefficient, degree, terms, and coefficients of the given polynomials.

Polynomial	Leading Coefficient	Degree	Terms	Coefficients
$9x^2 - x + 5$	9	2	$9x^2, -x, 5$	$9, -1, 5$
$11 - 2x$	-2	1	$-2x, 11$	$-2, 11$
$x^3 + 5x - 3$	1	3	$x^3, 5x, -3$	$1, 5, -3$

EXAMPLE 1 Identify Terms Related to a Polynomial

Write the polynomial $6x^3 - x + 5 - 2x^4$ in standard form. Identify the degree, terms, constant term, leading coefficient, and coefficients of the polynomial.

Solution

A polynomial is in standard form when the terms are written in decreasing powers of the variable. The standard form of the polynomial is $-2x^4 + 6x^3 - x + 5$. In this form, the degree is 4; the terms are $-2x^4$, $6x^3$, $-x$, and 5; the constant term is 5. The leading coefficient is -2; the coefficients are -2, 6, -1, and 5.

TRY EXERCISE 12, PAGE 554

To add polynomials, add the coefficients of the like terms.

EXAMPLE 2 Add Polynomials

Add: $(3x^3 - 2x^2 - 6) + (4x^2 - 6x - 7)$

Solution

$$\begin{aligned}
&(3x^3 - 2x^2 - 6) + (4x^2 - 6x - 7) \\
&\quad = 3x^3 + (-2x^2 + 4x^2) + (-6x) + [(-6) + (-7)] \\
&\quad = 3x^3 + 2x^2 - 6x - 13
\end{aligned}$$

TRY EXERCISE 24, PAGE 554

The **additive inverse of the polynomial** $3x - 7$ is

$$-(3x - 7) = -3x + 7$$

QUESTION What is the additive inverse of $3x^2 - 8x + 7$?

To subtract a polynomial, we add its additive inverse. For example,

$$\begin{aligned}
(2x - 5) - (3x - 7) &= (2x - 5) + (-3x + 7) \\
&= [2x + (-3x)] + [(-5) + 7] \\
&= -x + 2
\end{aligned}$$

The distributive property is used to find the product of polynomials. For instance, to find the product of $(3x - 4)$ and $(2x^2 + 5x + 1)$, we treat $3x - 4$ as a *single* quantity and *distribute it* over the trinomial $2x^2 + 5x + 1$, as shown in Example 3.

ANSWER The additive inverse is $-3x^2 + 8x - 7$.

EXAMPLE 3 Multiply Polynomials

Simplify: $(3x - 4)(2x^2 + 5x + 1)$

Solution

$$\begin{aligned}(3x - 4)&(2x^2 + 5x + 1)\\ &= (3x - 4)(2x^2) + (3x - 4)(5x) + (3x - 4)(1)\\ &= (3x)(2x^2) - 4(2x^2) + (3x)(5x) - 4(5x) + (3x)(1) - 4(1)\\ &= 6x^3 - 8x^2 + 15x^2 - 20x + 3x - 4\\ &= 6x^3 + 7x^2 - 17x - 4\end{aligned}$$

▶ TRY EXERCISE 32, PAGE 554

In the following calculation, a vertical format has been used to find the product of $(x^2 + 6x - 7)$ and $(5x - 2)$. Note that like terms are arranged in the same vertical column.

$$\begin{array}{rrrrl} & x^2 & +\ 6x & -\ 7 & \\ & & 5x & -\ 2 & \\ \hline & -\ 2x^2 & -\ 12x & +\ 14 & = -2(x^2 + 6x - 7) \\ 5x^3 & +\ 30x^2 & -\ 35x & & = 5x(x^2 + 6x - 7) \\ \hline 5x^3 & +\ 28x^2 & -\ 47x & +\ 14 & \end{array}$$

If the terms of the binomials $(a + b)$ and $(c + d)$ are labeled as shown below, then the product of the two binomials can be computed mentally by the **FOIL method.**

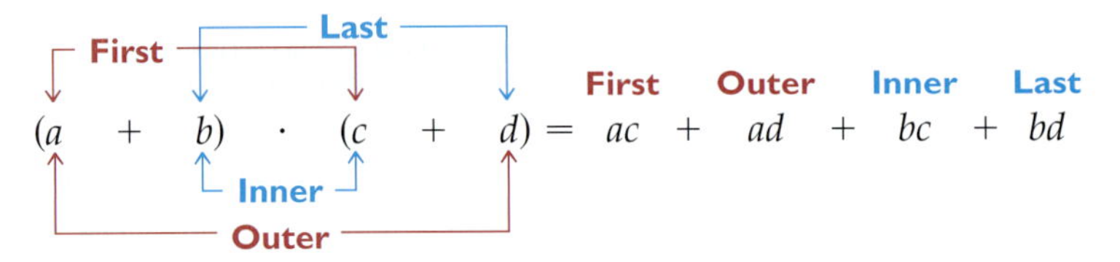

In the following illustration, we find the product of $(7x - 2)$ and $(5x + 4)$ by the FOIL method.

$$\begin{aligned}&\qquad\qquad\quad\ \text{First}\quad\ \ \text{Outer}\quad\ \ \text{Inner}\quad\ \ \ \text{Last}\\(7x - 2)(5x + 4) &= (7x)(5x) + (7x)(4) + (-2)(5x) + (-2)(4)\\ &= 35x^2 + 28x - 10x - 8\\ &= 35x^2 + 18x - 8\end{aligned}$$

Certain products occur so frequently in algebra that they deserve special attention.

Special Product Formulas

Special Form	Formula(s)
(Sum)(Difference)	$(x + y)(x - y) = x^2 - y^2$
$(\text{Binomial})^2$	$(x + y)^2 = x^2 + 2xy + y^2$ $(x - y)^2 = x^2 - 2xy + y^2$

The variables x and y in these special product formulas can be replaced by other algebraic expressions, as shown in Example 4.

EXAMPLE 4 Use the Special Product Formulas

Find each special product. **a.** $(7x + 10)(7x - 10)$ **b.** $(2y^2 + 11z)^2$

Solution

a. $(7x + 10)(7x - 10) = (7x)^2 - (10)^2 = 49x^2 - 100$

b. $(2y^2 + 11z)^2 = (2y^2)^2 + 2[(2y^2)(11z)] + (11z)^2 = 4y^4 + 44y^2z + 121z^2$

▶ TRY EXERCISE 56, PAGE 554

Many application problems require you to *evaluate polynomials*. To **evaluate a polynomial,** substitute the given value(s) for the variable(s) and then perform the indicated operations using the Order of Operations Agreement.

EXAMPLE 5 Evaluate a Polynomial

Evaluate the polynomial $2x^3 - 6x^2 + 7$ for $x = -4$.

Solution

$2x^3 - 6x^2 + 7$

$2(-4)^3 - 6(-4)^2 + 7 = 2(-64) - 6(16) + 7$ • **Substitute −4 for x. Evaluate the powers.**

$= -128 - 96 + 7$ • **Perform the multiplications.**

$= -217$ • **Perform the additions and subtractions.**

▶ TRY EXERCISE 66, PAGE 555

EXERCISE SET A.2

In Exercises 1 to 10, match the descriptions, labeled A, B, C,...., J, with the appropriate examples.

A. $x^3y + xy$ **B.** $7x^2 + 5x - 11$

C. $\frac{1}{2}x^2 + xy + y^2$ **D.** $4xy$

E. $8x^3 - 1$ **F.** $3 - 4x^2$

G. 8 **H.** $3x^5 - 4x^2 + 7x - 11$

I. $8x^4 - \sqrt{5}x^3 + 7$ **J.** 0

1. A monomial of degree 2.

2. A binomial of degree 3.

3. A polynomial of degree 5.

4. A binomial with leading coefficient of −4.

5. A zero-degree polynomial.

6. A fourth-degree polynomial that has a third-degree term.

7. A trinomial with integer coefficients.

8. A trinomial in x and y.

9. A polynomial with no degree.

10. A fourth-degree binomial.

In Exercises 11 to 16, for each polynomial determine its *a.* **standard form,** *b.* **degree,** *c.* **coefficients,** *d.* **leading coefficient,** *e.* **terms.**

11. $2x + x^2 - 7$

▶ **12.** $-3x^2 - 11 - 12x^4$

13. $x^3 - 1$

14. $4x^2 - 2x + 7$

15. $2x^4 + 3x^3 + 5 + 4x^2$

16. $3x^2 - 5x^3 + 7x - 1$

In Exercises 17 to 22, determine the degree of the given polynomial.

17. $3xy^2 - 2xy + 7x$

18. $x^3 + 3x^2y + 3xy^2 + y^3$

19. $4x^2y^2 - 5x^3y^2 + 17xy^3$

20. $-9x^5y + 10xy^4 - 11x^2y^2$

21. xy

22. $5x^2y - y^4 + 6xy$

In Exercises 23 to 34, perform the indicated operations and simplify if possible by combining like terms. Write the result in standard form.

23. $(3x^2 + 4x + 5) + (2x^2 + 7x - 2)$

▶ **24.** $(5y^2 - 7y + 3) + (2y^2 + 8y + 1)$

25. $(4w^3 - 2w + 7) + (5w^3 + 8w^2 - 1)$

26. $(5x^4 - 3x^2 + 9) + (3x^3 - 2x^2 - 7x + 3)$

27. $(r^2 - 2r - 5) - (3r^2 - 5r + 7)$

28. $(7s^2 - 4s + 11) - (-2s^2 + 11s - 9)$

29. $(u^3 - 3u^2 - 4u + 8) - (u^3 - 2u + 4)$

30. $(5v^4 - 3v^2 + 9) - (6v^4 + 11v^2 - 10)$

31. $(4x - 5)(2x^2 + 7x - 8)$

▶ **32.** $(5x - 7)(3x^2 - 8x - 5)$

33. $(3x^2 - 2x + 5)(2x^2 - 5x + 2)$

34. $(2y^3 - 3y + 4)(2y^2 - 5y + 7)$

In Exercises 35 to 48, use the FOIL method to find the indicated product.

35. $(2x + 4)(5x + 1)$

36. $(5x - 3)(2x + 7)$

37. $(y + 2)(y + 1)$

38. $(y + 5)(y + 3)$

39. $(4z - 3)(z - 4)$

40. $(5z - 6)(z - 1)$

41. $(a + 6)(a - 3)$

42. $(a - 10)(a + 4)$

43. $(5x - 11y)(2x - 7y)$

44. $(3a - 5b)(4a - 7b)$

45. $(9x + 5y)(2x + 5y)$

46. $(3x - 7z)(5x - 7z)$

47. $(3p + 5q)(2p - 7q)$

48. $(2r - 11s)(5r + 8s)$

In Exercises 49 to 54, perform the indicated operations and simplify.

49. $(4d - 1)^2 - (2d - 3)^2$

50. $(5c - 8)^2 - (2c - 5)^2$

51. $(r + s)(r^2 - rs + s^2)$

52. $(r - s)(r^2 + rs + s^2)$

53. $(3c - 2)(4c + 1)(5c - 2)$

54. $(4d - 5)(2d - 1)(3d - 4)$

In Exercises 55 to 62, use the special product formulas to perform the indicated operation.

55. $(3x + 5)(3x - 5)$

▶ **56.** $(4x^2 - 3y)(4x^2 + 3y)$

57. $(3x^2 - y)^2$

58. $(6x + 7y)^2$

59. $(4w + z)^2$

60. $(3x - 5y^2)^2$

61. $[(x + 5) + y][(x + 5) - y]$

62. $[(x - 2y) + 7][(x - 2y) - 7]$

In Exercises 63 to 70, evaluate the given polynomial for the indicated value of the variable.

63. $x^2 + 7x - 1$, for $x = 3$

64. $x^2 - 8x + 2$, for $x = 4$

65. $-x^2 + 5x - 3$, for $x = -2$

66. $-x^2 - 5x + 4$, for $x = -5$

67. $3x^3 - 2x^2 - x + 3$, for $x = -1$

68. $5x^3 - x^2 + 5x - 3$, for $x = -1$

69. $1 - x^5$, for $x = -2$

70. $1 - x^3 - x^5$, for $x = 2$

71. **RECREATION** The air resistance (in pounds) on a cyclist riding a bicycle in an upright position can be given by $0.016v^2$, where v is the speed of the cyclist in miles per hour. Find the air resistance on a cyclist when

a. $v = 10$ mph **b.** $v = 15$ mph

72. **HIGHWAY ENGINEERING** On an expressway, the recommended *safe distance* between cars in feet is given by $0.015v^2 + v + 10$, where v is the speed of the car in miles per hour. Find the safe distance when

a. $v = 30$ mph **b.** $v = 55$ mph

73. **GEOMETRY** The volume of a right circular cylinder (as shown below) is given by $\pi r^2 h$, where r is the radius of the base and h is the height of the cylinder. Find the volume when

a. $r = 3$ inches, $h = 8$ inches

b. $r = 5$ cm, $h = 12$ cm

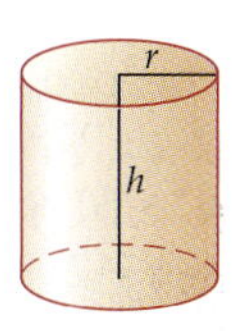

74. **AUTOMOTIVE ENGINEERING** The fuel efficiency (in miles per gallon of gas) of a car is given by the expression $-0.02v^2 + 1.5v + 2$, where v is the speed of the car in miles per hour. Find the fuel efficiency when

a. $v = 45$ mph **b.** $v = 60$ mph

75. **COMPUTER SCIENCE** If n is a positive integer, then $n!$, which is read "n factorial," is given by

$$n(n-1)(n-2)\cdots 2 \cdot 1$$

For example, $4! = 4 \cdot 3 \cdot 2 \cdot 1 = 24$. A computer scientist determines that each time a program is run on a particular computer, the time in seconds required to compute $n!$ is given by the polynomial

$$1.9 \times 10^{-6}n^2 - 3.9 \times 10^{-3}n$$

where $1000 \le n \le 10{,}000$. Using this polynomial, estimate the time it takes this computer to calculate 4000! and 8000!.

76. **AIR VELOCITY OF A COUGH** The velocity, in meters per second, of the air that is expelled during a cough is given by velocity $= 6r^2 - 10r^3$, where r is the radius of the trachea in centimeters.

a. Find the velocity as a polynomial in standard form.

b. Find the velocity of the air in a cough when the radius of the trachea is 0.35 cm. Round to the nearest hundredth.

77. **SPORTS** The height, in feet, of a baseball released by a pitcher t seconds after it is released is given by (ignoring air resistance)

$$\text{Height} = -16t^2 + 4.7881t + 6$$

For the pitch to be a strike, it must be at least 2 feet high when it crosses home plate and no higher than 5 feet high. If it takes 0.5 second for the ball to reach home plate, will the ball be high enough to be a strike?

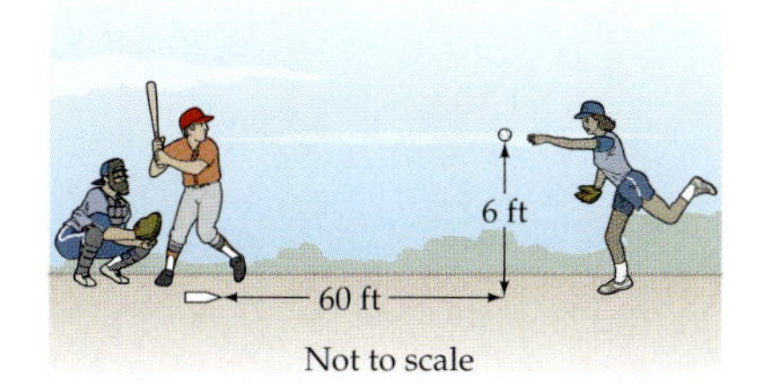

APPENDIX A.3 FACTORING

Writing a polynomial as a product of polynomials of lower degree is called **factoring.** Factoring is an important procedure that is often used to simplify fractional expressions and to solve equations.

In this section we consider only the factorization of polynomials that have integer coefficients. Also, we are concerned only with **factoring over the integers.** That is, we search only for polynomial factors that have integer coefficients.

GREATEST COMMON FACTOR

The first step in any factorization of a polynomial is to use the distributive property to factor out the **greatest common factor (GCF)** of the terms of the polynomial. Given two or more exponential expressions with the same prime number base or the same variable base, the GCF is the exponential expression with the smallest exponent. For example,

$$2^3 \text{ is the GCF of } 2^3, 2^5, \text{ and } 2^8 \quad \text{and} \quad a \text{ is the GCF of } a^4 \text{ and } a$$

The GCF of two or more monomials is the product of the GCFs of all the *common* bases. For example, to find the GCF of $27a^3b^4$ and $18b^3c$, factor the coefficients into prime factors and then write each common base with its smallest exponent.

$$27a^3b^4 = 3^3 \cdot a^3 \cdot b^4 \qquad 18b^3c = 2 \cdot 3^2 \cdot b^3 \cdot c$$

The only common bases are 3 and b. The product of these common bases with their smallest exponents is 3^2b^3. The GCF of $27a^3b^4$ and $18b^3c$ is $9b^3$.

The expressions $3x(2x + 5)$ and $4(2x + 5)$ have a common *binomial* factor that is $2x + 5$. Thus the GCF of $3x(2x + 5)$ and $4(2x + 5)$ is $2x + 5$.

EXAMPLE 1 Factor Out the Greatest Common Factor

Factor out the GCF.

a. $10x^3 + 6x$ **b.** $15x^{2n} + 9x^{n+1} - 3x^n$ (where n is a positive integer)

c. $(m + 5)(x + 3) + (m + 5)(x - 10)$

Solution

a. $10x^3 + 6x = (2x)(5x^2) + (2x)(3)$ • **The GCF is 2x.**

$= 2x(5x^2 + 3)$ • **Factor out the GCF.**

b. $15x^{2n} + 9x^{n+1} - 3x^n$

$= (3x^n)(5x^n) + (3x^n)(3x) - (3x^n)(1)$ • **The GCF is $3x^n$.**

$= 3x^n(5x^n + 3x - 1)$ • **Factor out the GCF.**

c. $(m + 5)(x + 3) + (m + 5)(x - 10)$ • **Use the distributive property to factor out (m + 5).**

$= (m + 5)[(x + 3) + (x - 10)]$

$= (m + 5)(2x - 7)$ • **Simplify.**

TRY EXERCISE 6, PAGE 563

Factoring Trinomials

Some trinomials of the form $x^2 + bx + c$ can be factored by a trial procedure. This method makes use of the FOIL method in reverse. For example, consider the following products:

$$(x + 3)(x + 5) = x^2 + 5x + 3x + (3)(5) = x^2 + 8x + 15$$
$$(x - 2)(x - 7) = x^2 - 7x - 2x + (-2)(-7) = x^2 - 9x + 14$$
$$(x + 4)(x - 9) = x^2 - 9x + 4x + (4)(-9) = x^2 - 5x - 36$$

The coefficient of x is the sum of the constant terms of the binomials.

The constant term of the trinomial is the product of the constant terms of the binomials.

QUESTION Is $(x - 2)(x + 7)$ the correct factorization of $x^2 - 5x - 14$?

Points to Remember to Factor $x^2 + bx + c$

1. The constant term c of the trinomial is the product of the constant terms of the binomials.
2. The coefficient b in the trinomial is the sum of the constant terms of the binomials.
3. If the constant term c of the trinomial is positive, the constant terms of the binomials have the same sign as the coefficient b of the trinomial.
4. If the constant term c of the trinomial is negative, the constant terms of the binomials have opposite signs.

EXAMPLE 2 Factor a Trinomial of the Form $x^2 + bx + c$

Factor: $x^2 + 7x - 18$

Solution

We must find two binomials whose first terms have a product of x^2 and whose last terms have a product of -18; also, the sum of the product of the outer terms and the product of the inner terms must be $7x$. Begin by listing the possible integer factorizations of -18 and the sums of those factors.

Factors of -18	Sum of the Factors
$1 \cdot (-18)$	$1 + (-18) = -17$
$(-1) \cdot 18$	$(-1) + 18 = 17$
$2 \cdot (-9)$	$2 + (-9) = -7$
$(-2) \cdot 9$	$(-2) + 9 = 7$

• Stop. This is the desired sum.

Continued ▶

ANSWER No. $(x - 2)(x + 7) = x^2 + 5x - 14$.

Thus -2 and 9 are the numbers whose sum is 7 and whose product is -18. Therefore,

$$x^2 + 7x - 18 = (x - 2)(x + 9)$$

The FOIL method can be used to verify that the factorization is correct.

TRY EXERCISE 12, PAGE 563

The trial method sometimes can be used to factor trinomials of the form $ax^2 + bx + c$, which do not have a leading coefficient of 1. We use the factors of a and c to form trial binomial factors. Factoring trinomials of this type may require testing many factors. To reduce the number of trial factors, make use of the following points.

Points to Remember to Factor $ax^2 + bx + c, a > 0$

1. If the constant term of the trinomial is positive, the constant terms of the binomials have the same sign as the coefficient b in the trinomial.
2. If the constant term of the trinomial is negative, the constant terms of the binomials have opposite signs.
3. If the terms of the trinomial do not have a common factor, then neither binomial will have a common factor.

EXAMPLE 3 Factor a Trinomial of the Form $ax^2 + bx + c$

Factor: $6x^2 - 11x + 4$

Solution

Because the constant term of the trinomial is positive and the coefficient of the x term is negative, the constant terms of the binomials will both be negative. This time we find factors of the first term as well as factors of the constant term.

Factors of $6x^2$	Factors of 4 (both negative)
$x, 6x$	$-1, -4$
$2x, 3x$	$-2, -2$

Use these factors to write trial factors. Use the FOIL method to see whether any of the trial factors produce the correct middle term. If the terms of a trinomial do not have a common factor, then a binomial factor cannot have a common factor (point 3). Such trial factors need not be checked.

Trial Factors	Middle Term
$(x - 1)(6x - 4)$	Common factor
$(x - 4)(6x - 1)$	$-1x - 24x = -25x$
$(x - 2)(6x - 2)$	Common factor
$(2x - 1)(3x - 4)$	$-8x - 3x = -11x$

• **6x and 4 have a common factor.**

• **6x and 2 have a common factor.**

• **This is the correct middle term.**

Thus $6x^2 - 11x + 4 = (2x - 1)(3x - 4)$.

TRY EXERCISE 16, PAGE 563

Sometimes it is impossible to factor a polynomial into the product of two polynomials having integer coefficients. Such polynomials are said to be **nonfactorable over the integers.** For example, $x^2 + 3x + 7$ is nonfactorable over the integers because there are no integers whose product is 7 and whose sum or difference is 3.

Certain trinomials can be expressed as quadratic trinomials by making suitable variable substitutions. A trinomial is **quadratic in form** if it can be written as

$$au^2 + bu + c$$

If we let $x^2 = u$, the trinomial $x^4 + 5x^2 + 6$ can be written as shown at the right.

$$x^4 + 5x^2 + 6 = (x^2)^2 + 5(x^2) + 6$$

The trinomial is quadratic in form.

$$= u^2 + 5u + 6$$

If we let $xy = u$, the trinomial $2x^2y^2 + 3xy - 9$ can be written as shown at the right.

$$2x^2y^2 + 3xy - 9 = 2(xy)^2 + 3(xy) - 9$$

The trinomial is quadratic in form.

$$= 2u^2 + 3u - 9$$

When a trinomial that is quadratic in form is factored, the variable part of the first term in each binomial factor will be u. For example, because $x^4 + 5x^2 + 6$ is quadratic in form when $x^2 = u$, the first term in each binomial factor will be x^2.

$$\begin{aligned} x^4 + 5x^2 + 6 &= (x^2)^2 + 5(x^2) + 6 \\ &= (x^2 + 2)(x^2 + 3) \end{aligned}$$

The trinomial $x^2y^2 - 2xy - 15$ is quadratic in form when $xy = u$. The first term in each binomial factor will be xy.

$$\begin{aligned} x^2y^2 - 2xy - 15 &= (xy)^2 - 2(xy) - 15 \\ &= (xy + 3)(xy - 5) \end{aligned}$$

EXAMPLE 4

Factor: **a.** $6x^2y^2 - xy - 12$ **b.** $2x^4 + 5x^2 - 12$

Solution

a. $6x^2y^2 - xy - 12$
$= (3xy + 4)(2xy - 3)$

• **The trinomial is quadratic in form when $xy = u$.**

Continued ▶

b. $2x^4 + 5x^2 - 12$
$= (x^2 + 4)(2x^2 - 3)$

• **The trinomial is quadratic in form when $x^2 = u$.**

TRY EXERCISE 30, PAGE 563

SPECIAL FACTORING

The product of a term and itself is called a **perfect square.** The exponents on variables of perfect squares are always even numbers. The **square root of a perfect square** is one of the two equal factors of the perfect square. To find the square root of a perfect square variable term, divide the exponent by 2. For the examples below, assume the variables represent positive numbers.

Term		Perfect Square	Square Root
7	$7 \cdot 7 =$	49	$\sqrt{49} = 7$
y	$y \cdot y =$	y^2	$\sqrt{y^2} = y$
$2x^3$	$2x^3 \cdot 2x^3 =$	$4x^6$	$\sqrt{4x^6} = 2x^3$
x^n	$x^n \cdot x^n =$	x^{2n}	$\sqrt{x^{2n}} = x^n$

take note

The **sum** of two squares does not factor over the integers. For instance, $49x^2 + 144$ does not factor over the integers.

The factors of the difference of two perfect squares are the sum and difference of the square roots of the perfect squares.

Factors of the Difference of Two Perfect Squares

$$a^2 - b^2 = (a + b)(a - b)$$

EXAMPLE 5 Factor the Difference of Squares

Factor: $49x^2 - 144$

Solution

$49x^2 - 144 = (7x)^2 - (12)^2$

• **Recognize the difference-of-squares form.**

$= (7x + 12)(7x - 12)$

• **The binomial factors are the sum and the difference of the square roots of the squares.**

TRY EXERCISE 34, PAGE 563

A **perfect-square trinomial** is a trinomial that is the square of a binomial. For example, $x^2 + 6x + 9$ is a perfect-square trinomial because

$$(x + 3)^2 = x^2 + 6x + 9$$

Every perfect-square trinomial can be factored by the trial method, but it generally is faster to factor perfect-square trinomials by using the following factoring formulas.

Factors of a Perfect-Square Trinomial

$$a^2 + 2ab + b^2 = (a + b)^2$$
$$a^2 - 2ab + b^2 = (a - b)^2$$

EXAMPLE 6 Factor a Perfect-Square Trinomial

Factor: $16m^2 - 40mn + 25n^2$

Solution

Because $16m^2 = (4m)^2$ and $25n^2 = (5n)^2$, try factoring $16m^2 - 40mn + 25n^2$ as the square of a binomial.

$$16m^2 - 40mn + 25n^2 \stackrel{?}{=} (4m - 5n)^2$$

Check:

$$(4m - 5n)^2 = (4m - 5n)(4m - 5n)$$
$$= 16m^2 - 20mn - 20mn + 25n^2$$
$$= 16m^2 - 40mn + 25n^2$$

The factorization checks. Therefore, $16m^2 - 40mn + 25n^2 = (4m - 5n)^2$.

TRY EXERCISE 44, PAGE 563

take note

It is important to check the proposed factorization. For instance, consider $x^2 + 13x + 36$. Because x^2 is the square of x and 36 is the square of 6, it is tempting to factor, using the perfect-square trinomial formulas, as $x^2 + 13x + 36 \stackrel{?}{=} (x + 6)^2$. Note, however, that $(x + 6)^2 = x^2 + 12x + 36$, which is not the original trinomial. The correct factorization is

$x^2 + 13x + 36 = (x + 4)(x + 9)$

The product of the same three terms is called a **perfect cube.** The exponents on variables of perfect cubes are always divisible by 3. The **cube root of a perfect cube** is one of the three equal factors of the perfect cube. To find the cube root of a perfect cube variable term, divide the exponent by 3.

Term		Perfect Cube	Cube Root
5	$5 \cdot 5 \cdot 5 =$	125	$\sqrt[3]{125} = 5$
z	$z \cdot z \cdot z =$	z^3	$\sqrt[3]{z^3} = z$
$3x^2$	$3x^2 \cdot 3x^2 \cdot 3x^2 =$	$27x^6$	$\sqrt[3]{27x^6} = 3x^2$
x^n	$x^n \cdot x^n \cdot x^n =$	x^{3n}	$\sqrt[3]{x^{3n}} = x^n$

take note

Note the pattern of the signs when factoring the sum or difference of two perfect cubes.

Same sign

$a^3 + b^3 = (a + b)(a^2 - ab + b^2)$

Opposite signs

Same sign

$a^3 - b^3 = (a - b)(a^2 + ab + b^2)$

Opposite signs

The following factoring formulas are used to factor the sum or difference of two perfect cubes.

Factors of the Sum or Difference of Two Perfect Cubes

$$a^3 + b^3 = (a + b)(a^2 - ab + b^2)$$
$$a^3 - b^3 = (a - b)(a^2 + ab + b^2)$$

EXAMPLE 7 Factor the Sum or Difference of Cubes

Factor: **a.** $8a^3 + b^3$ **b.** $a^3 - 64$

Continued ▶

Solution

a. $8a^3 + b^3 = (2a)^3 + b^3$ • **Recognize the sum-of-cubes form.**

$= (2a + b)(4a^2 - 2ab + b^2)$ • **Factor.**

b. $a^3 - 64 = a^3 - 4^3$ • **Recognize the difference-of-cubes form.**

$= (a - 4)(a^2 + 4a + 16)$ • **Factor.**

▶ **TRY EXERCISE 50, PAGE 563**

FACTOR BY GROUPING

> **take note**
>
> $-a + b = -(a - b)$. Thus, $-4y + 14 = -(4y - 14)$.

Some polynomials can be **factored by grouping.** Pairs of terms that have a common factor are first grouped together. The process makes repeated use of the distributive property, as shown in the following factorization of $6y^3 - 21y^2 - 4y + 14$.

$6y^3 - 21y^2 - 4y + 14$

$= (6y^3 - 21y^2) - (4y - 14)$ • **Group the first two terms and the last two terms.**

$= 3y^2(2y - 7) - 2(2y - 7)$ • **Factor out the GCF from each of the groups.**

$= (2y - 7)(3y^2 - 2)$ • **Factor out the common binomial factor.**

When you factor by grouping, some experimentation may be necessary to find a grouping that is of the form of one of the special factoring formulas.

EXAMPLE 8 Factor by Grouping

Factor by grouping. **a.** $a^2 + 10ab + 25b^2 - c^2$ **b.** $p^2 + p - q - q^2$

Solution

a. $a^2 + 10ab + 25b^2 - c^2$

$= (a^2 + 10ab + 25b^2) - c^2$ • **Group the terms of the perfect-square trinomial.**

$= (a + 5b)^2 - c^2$ • **Factor the trinomial.**

$= [(a + 5b) + c][(a + 5b) - c]$ • **Factor the difference of squares.**

$= (a + 5b + c)(a + 5b - c)$ • **Simplify.**

b. $p^2 + p - q - q^2$

$= p^2 - q^2 + p - q$ • **Rearrange the terms.**

$= (p^2 - q^2) + (p - q)$ • **Regroup.**

$= (p + q)(p - q) + (p - q)$ • **Factor the difference of squares.**

$= (p - q)(p + q + 1)$ • **Factor out the common factor $(p - q)$.**

▶ **TRY EXERCISE 60, PAGE 563**

Exercise Set A.3

In Exercises 1 to 8, factor out the GCF from each polynomial.

1. $5x + 20$

2. $8x^2 + 12x - 40$

3. $-15x^2 - 12x$

4. $-6y^2 - 54y$

5. $10x^2y + 6xy - 14xy^2$

▶ **6.** $6a^3b^2 - 12a^2b + 72ab^3$

7. $(x - 3)(a + b) + (x - 3)(a + 2b)$

8. $(x - 4)(2a - b) + (x + 4)(2a - b)$

In Exercises 9 to 22, factor each trinomial over the integers.

9. $x^2 + 7x + 12$

10. $x^2 + 9x + 20$

11. $a^2 - 10a - 24$

▶ **12.** $b^2 + 12b - 28$

13. $6x^2 + 25x + 4$

14. $8a^2 - 26a + 15$

15. $51x^2 - 5x - 4$

▶ **16.** $57y^2 + y - 6$

17. $6x^2 + xy - 40y^2$

18. $8x^2 + 10xy - 25y^2$

19. $x^4 + 6x^2 + 5$

20. $x^4 + 11x^2 + 18$

21. $6x^4 + 23x^2 + 15$

22. $9x^4 + 10x^2 + 1$

In Exercises 23 to 30, factor over the integers.

23. $x^4 - x^2 - 6$

24. $x^4 + 3x^2 + 2$

25. $x^2y^2 - 2xy - 8$

26. $2x^2y^2 + xy - 1$

27. $3x^4 + 11x^2 - 4$

28. $2x^4 + 3x^2 - 9$

29. $3x^6 + 2x^3 - 8$

▶ **30.** $8x^6 - 10x^3 - 3$

In Exercises 31 to 40, factor each difference of squares over the integers.

31. $x^2 - 9$

32. $x^2 - 64$

33. $4a^2 - 49$

▶ **34.** $81b^2 - 16c^2$

35. $1 - 100x^2$

36. $1 - 121y^2$

37. $x^4 - 9$

38. $y^4 - 196$

39. $(x + 5)^2 - 4$

40. $(x - 3)^2 - 16$

In Exercises 41 to 48, factor each perfect-square trinomial.

41. $x^2 + 10x + 25$

42. $y^2 + 6y + 9$

43. $a^2 - 14a + 49$

▶ **44.** $b^2 - 24b + 144$

45. $4x^2 + 12x + 9$

46. $25y^2 + 40y + 16$

47. $z^4 + 4z^2w^2 + 4w^4$

48. $9x^4 - 30x^2y^2 + 25y^4$

In Exercises 49 to 56, factor each sum or difference of cubes over the integers.

49. $x^3 - 8$

▶ **50.** $b^3 + 64$

51. $8x^3 - 27y^3$

52. $64u^3 - 27v^3$

53. $8 - x^6$

54. $1 + y^{12}$

55. $(x - 2)^3 - 1$

56. $(y + 3)^3 + 8$

In Exercises 57 to 62, factor (over the integers) by grouping in pairs.

57. $3x^3 + x^2 + 6x + 2$

58. $18w^3 + 15w^2 + 12w + 10$

59. $ax^2 - ax + bx - b$

▶ **60.** $a^2y^2 - ay^3 + ac - cy$

61. $6w^3 + 4w^2 - 15w - 10$

62. $10z^3 - 15z^2 - 4z + 6$

In Exercises 63 to 82, use any factoring technique to completely factor each polynomial. If the polynomial does not factor, then state that it is nonfactorable over the integers.

63. $18x^2 - 2$

64. $4bx^3 + 32b$

65. $16x^4 - 1$

66. $81y^4 - 16$

67. $12ax^2 - 23axy + 10ay^2$

68. $6ax^2 - 19axy - 20ay^2$

69. $3bx^3 + 4bx^2 - 3bx - 4b$

70. $2x^6 - 2$

71. $72bx^2 + 24bxy + 2by^2$

72. $64y^3 - 16y^2z + yz^2$

73. $(w - 5)^3 + 8$

74. $5xy + 20y - 15x - 60$

75. $x^2 + 6xy + 9y^2 - 1$

76. $4y^2 - 4yz + z^2 - 9$

77. $8x^2 + 3x - 4$

78. $16x^2 + 81$

79. $5x(2x - 5)^2 - (2x - 5)^3$

80. $6x(3x + 1)^3 - (3x + 1)^4$

81. $4x^2 + 2x - y - y^2$

82. $a^2 + a + b - b^2$

SOLUTIONS TO THE TRY EXERCISES

Exercise Set 1.1, page 13

6. $6(5s - 11) - 12(2s + 5) = 0$

$$30s - 66 - 24s - 60 = 0$$
$$6s - 126 = 0$$
$$s = \frac{126}{6}$$
$$s = 21$$

22. $12w^2 - 41w + 24 = 0$

$$(4w - 3)(3w - 8) = 0$$
$$4w - 3 = 0 \quad \text{or} \quad 3w - 8 = 0$$
$$w = \frac{3}{4} \qquad\qquad w = \frac{8}{3}$$

28. $x^2 - 5x - 24 = 0$

$$x^2 - 5x = 24$$
$$x^2 - 5x + \frac{25}{4} = 24 + \frac{25}{4}$$
$$\left(x - \frac{5}{2}\right)^2 = \frac{121}{4}$$
$$x - \frac{5}{2} = \pm\frac{11}{2}$$
$$x = -\frac{5}{2} \pm \frac{11}{2}$$

The solutions are -8 and 3.

30. $x^2 + x - 2 = 0$

$$a = 1 \qquad b = 1 \qquad c = -2$$
$$x = \frac{-1 \pm \sqrt{1^2 - 4(1)(-2)}}{2(1)}$$
$$= \frac{-1 \pm \sqrt{1 + 8}}{2} = \frac{-1 \pm 3}{2}$$
$$x = \frac{-1 + 3}{2} = 1 \quad \text{or} \quad x = \frac{-1 - 3}{2} = -2$$

The solutions are 1 and -2.

50. $3(x + 7) \le 5(2x - 8)$

$$3x + 21 \le 10x - 40$$
$$-7x \le -61$$
$$x \ge \frac{61}{7}$$

58. $12x^2 + 8x \ge 15$

$$12x^2 + 8x - 15 \ge 0$$
$$(6x - 5)(2x + 3) \ge 0$$
$$x = \frac{5}{6} \quad \text{and} \quad x = -\frac{3}{2} \qquad \text{• Critical values}$$

Use a test value from each of the intervals $\left(-\infty, -\frac{3}{2}\right)$, $\left(-\frac{3}{2}, \frac{5}{6}\right)$, and $\left(\frac{5}{6}, \infty\right)$ to determine where $12x^2 + 8x - 15$ is positive.

++| - - - - - - - - - |+++

$-\frac{3}{2}$ $\quad 0 \quad$ $\frac{5}{6}$

The solution set is $\left(-\infty, -\frac{3}{2}\right] \cup \left[\frac{5}{6}, \infty\right)$.

70. $|2x - 5| \ge 1$

$$2x - 5 \le -1 \quad \text{or} \quad 2x - 5 \ge 1$$
$$2x \le 4 \qquad\qquad 2x \ge 6$$
$$x \le 2 \qquad\qquad x \ge 3$$

The solution set is $(-\infty, 2] \cup [3, \infty)$.

72. $|3 - 2x| \le 5$

$$-5 \le 3 - 2x \le 5$$
$$-8 \le -2x \le 2$$
$$4 \ge x \ge -1$$

The solution set is $[-1, 4]$.

Exercise Set 1.2, page 27

6. $d = \sqrt{(x_2 - x_1)^2 + (y_2 - y_1)^2}$

$$d = \sqrt{[-10 - (-5)]^2 + (14 - 8)^2}$$
$$= \sqrt{(-5)^2 + 6^2} = \sqrt{25 + 36}$$
$$= \sqrt{61}$$

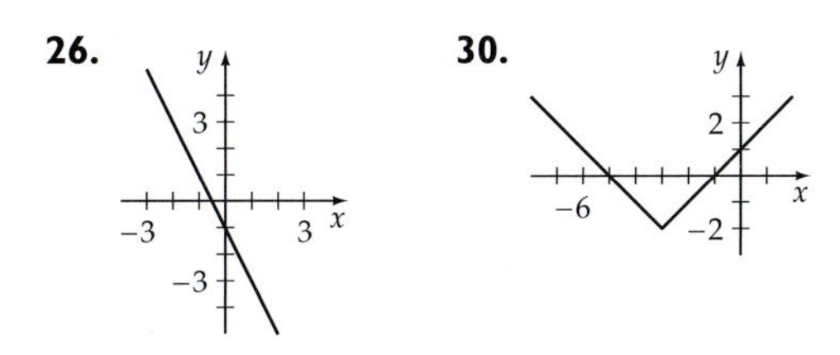

32.

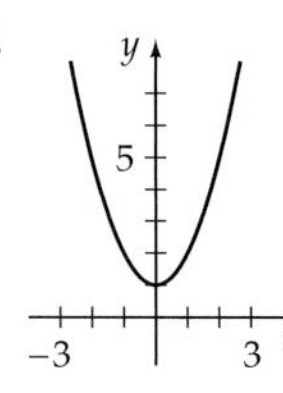

40. y-intercept: $\left(0, -\frac{15}{4}\right)$

x-intercept: $(5, 0)$

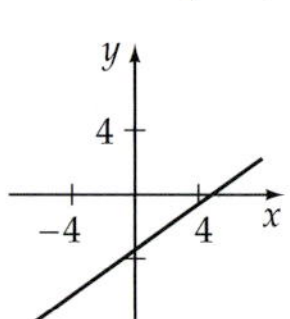

64. $r = \sqrt{(1 - (-2))^2 + (7 - 5)^2}$

$= \sqrt{9 + 4} = \sqrt{13}$

Using the standard form

$(x - h)^2 + (y - k)^2 = r^2$

with $h = -2$, $k = 5$, and $r = \sqrt{13}$ yields

$(x + 2)^2 + (y - 5)^2 = \left(\sqrt{13}\right)^2$

66.

$$\begin{aligned} x^2 + y^2 - 6x - 4y + 12 &= 0 \\ x^2 - 6x + y^2 - 4y &= -12 \\ x^2 - 6x + 9 + y^2 - 4y + 4 &= -12 + 9 + 4 \\ (x - 3)^2 + (y - 2)^2 &= 1^2 \end{aligned}$$

center (3, 2), radius 1

Exercise Set 1.3, page 43

2. Given $g(x) = 2x^2 + 3$

a. $g(3) = 2(3)^2 + 3 = 18 + 3 = 21$

b. $g(-1) = 2(-1)^2 + 3 = 2 + 3 = 5$

c. $g(0) = 2(0)^2 + 3 = 0 + 3 = 3$

d. $g\left(\frac{1}{2}\right) = 2\left(\frac{1}{2}\right)^2 + 3 = \frac{1}{2} + 3 = \frac{7}{2}$

e. $g(c) = 2(c)^2 + 3 = 2c^2 + 3$

f. $g(c + 5) = 2(c + 5)^2 + 3 = 2c^2 + 20c + 50 + 3$

$= 2c^2 + 20c + 53$

10. a. Because $0 \le 0 \le 5$, $Q(0) = 4$.

b. Because $6 < e < 7$, $Q(e) = -e + 9$.

c. Because $1 < n < 2$, $Q(n) = 4$.

d. Because $1 < m \le 2$, $8 < m^2 + 7 \le 11$. Thus

$Q(m^2 + 7) = \sqrt{(m^2 + 7) - 7} = \sqrt{m^2} = m$

14. $x^2 - 2y = 2$ • **Solve for y.**

$-2y = -x^2 + 2$

$y = \frac{1}{2}x^2 - 1$

y is a function of x because each x value will yield one and only one y value.

28. Domain is the set of all real numbers.

40. Domain is the set of all real numbers.

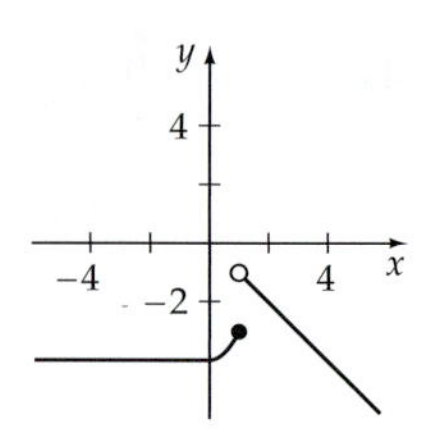

48. a. $[0, \infty]$

b. Since \$31,250 is between \$27,950 and \$67,700, use $T(x) = 0.27(x - 27{,}950) + 3892.50$. Then, $T(31{,}250) = 0.27(31{,}250 - 27{,}950) + 3892.50 =$ \$4783.50.

c. Since \$72,000 is between \$67,700 and \$141,250, use $T(x) = 0.30(x - 67{,}700) + 14{,}625$. Then, $T(72{,}000) = 0.30(72{,}000 - 67{,}700) + 14{,}625 =$ \$15,915.

50. a. This is the graph of a function. Every vertical line intersects the graph in at most one point.

b. This is not the graph of a function. Some vertical lines intersect the graph at two points.

c. This is not the graph of a function. The vertical line at $x = -2$ intersects the graph at more than one point.

d. This is the graph of a function. Every vertical line intersects the graph at exactly one point.

66. $v(t) = 44{,}000 - 4200t$, $0 \le t \le 8$

68. a.

$$\begin{aligned} V(x) &= (30 - 2x)^2 x \\ &= (900 - 120x + 4x^2)x \\ &= 900x - 120x^2 + 4x^3 \end{aligned}$$

b. Domain: $\{x \mid 0 < x < 15\}$

72. $d(A, B) = \sqrt{1 + x^2}$. The time required to swim from A to B at 2 mph is $\frac{\sqrt{1 + x^2}}{2}$ hours.

$d(B, C) = 3 - x$. The time required to run from B to C at 8 mph is $\frac{3 - x}{8}$ hours.

Thus the total time to reach point C is

$t = \frac{\sqrt{1 + x^2}}{2} + \frac{3 - x}{8}$ hours

Exercise Set 1.4, page 60

2. $m = \frac{1 - 4}{5 - (-2)} = -\frac{3}{7}$

16. $m = -1$

$b = 1$

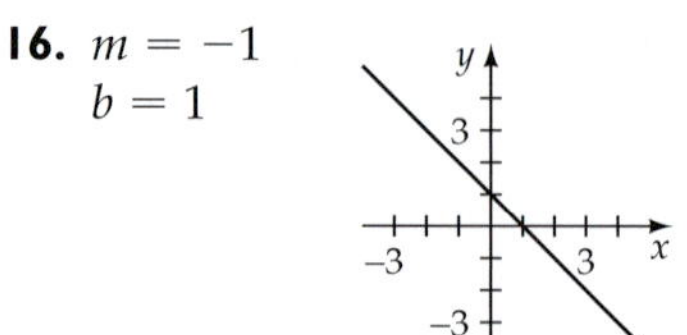

28. $y - 5 = -2(x - 0)$

$y = -2x + 5$

42. $f(x) = \dfrac{2x}{3} + 2$

$$4 = \frac{2x}{3} + 2 \qquad \text{• Replace } f(x) \text{ by 4 and solve for } x.$$

$$2 = \frac{2x}{3}$$

$$3 = x$$

When $x = 3, f(x) = 4$.

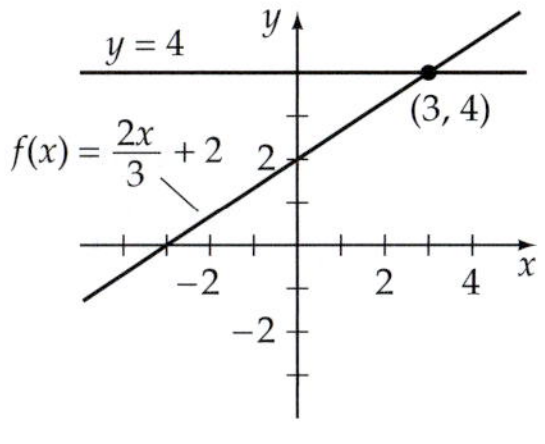

46.

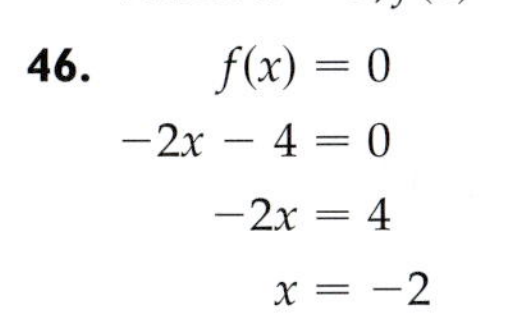

$$f(x) = 0$$
$$-2x - 4 = 0$$
$$-2x = 4$$
$$x = -2$$

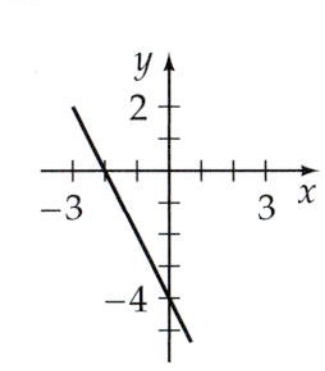

50.

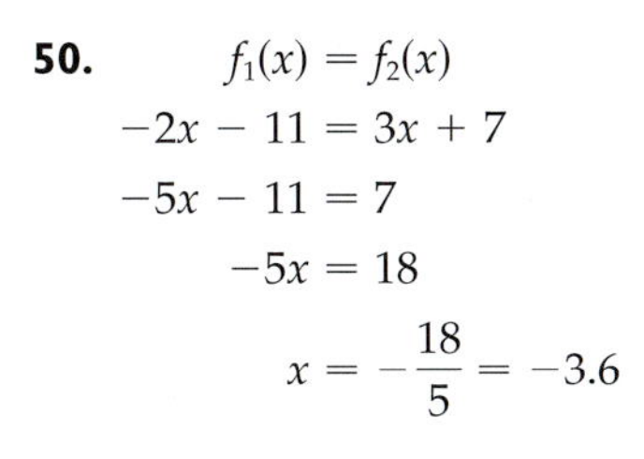

$$f_1(x) = f_2(x)$$
$$-2x - 11 = 3x + 7$$
$$-5x - 11 = 7$$
$$-5x = 18$$
$$x = -\frac{18}{5} = -3.6$$

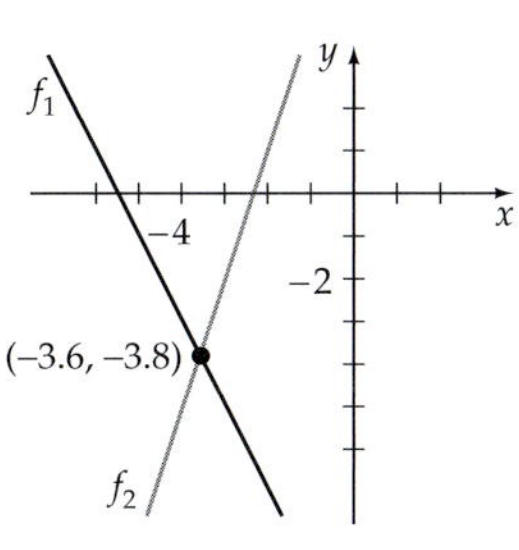

56. a. Using the data for 1997 and 2003, two ordered pairs on the line are (1997, 531.0) and (2003, 725.0). Find the slope of the line.

$$m = \frac{725.0 - 531.0}{2003 - 1997} \approx 32.3$$

Use the point-slope formula to find the equation of the line between the given points.

$$y - y_1 = m(x - x_1)$$
$$y - 531.0 = 32.3(x - 1997)$$
$$y - 531.0 = 32.3x - 64{,}503.1$$
$$y = 32.3x - 64{,}038.7$$

Using functional notation, the linear function is $C(t) = 35.6t - 70{,}585.4$.

b. To find the year when consumer debt first exceeds \$850 billion, let $C(t) = 850$ and solve for t.

$$C(t) = 32.3t - 64{,}038.7$$
$$850 = 32.3t - 64{,}038.7$$
$$64{,}888.7 = 32.3t$$
$$2008 \approx t$$

According to the model, revolving consumer debt will first exceed \$850 billion in 2008.

66.

$$P(x) = R(x) - C(x)$$
$$P(x) = 124x - (78.5x + 5005)$$
$$P(x) = 45.5x - 5005$$
$$45.5x - 5005 = 0$$
$$45.5x = 5005$$
$$x = 110 \qquad \text{• The break-even point}$$

78. a. The slope of the radius from (0, 0) to $(\sqrt{15}, 1)$ is $\dfrac{1}{\sqrt{15}}$.

The slope of the linear path of the rock is $-\sqrt{15}$. The path of the rock is given by

$$y - 1 = -\sqrt{15}(x - \sqrt{15})$$
$$y - 1 = -\sqrt{15}x + 15$$
$$y = -\sqrt{15}x + 16$$

Every point on the wall has a y value of 14. Thus

$$14 = -\sqrt{15}x + 16$$
$$-2 = -\sqrt{15}x$$
$$x = \frac{2}{\sqrt{15}} \approx 0.52$$

The rock hits the wall at (0.52, 14).

Exercise Set 1.5, page 75

10.

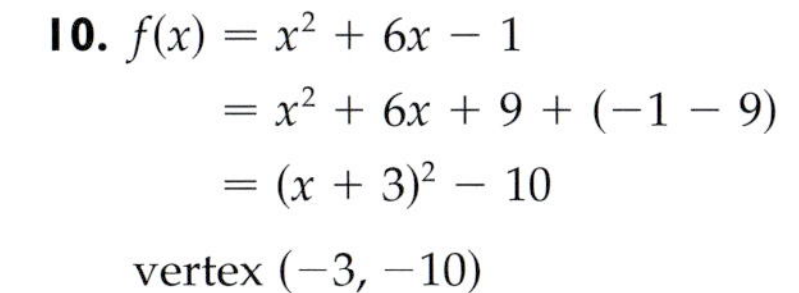

$$f(x) = x^2 + 6x - 1$$
$$= x^2 + 6x + 9 + (-1 - 9)$$
$$= (x + 3)^2 - 10$$

vertex $(-3, -10)$
axis of symmetry $x = -3$

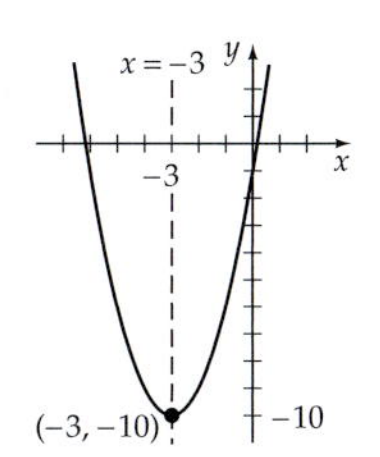

20. $h = -\dfrac{b}{2a} = -\dfrac{-6}{2(1)} = 3$

$k = f(3) = 3^2 - 6(3) = -9$

vertex $(3, -9)$

$f(x) = (x - 3)^2 - 9$

32. Determine the y-coordinate of the vertex of the graph of $f(x) = 2x^2 + 6x - 5$.

$$f(x) = 2x^2 + 6x - 5 \qquad \text{• } a = 2, b = 6, c = -5$$

$$h = -\frac{b}{2a} = -\frac{6}{2(2)} = -\frac{3}{2} \qquad \text{• Find the } x\text{-coordinate of the vertex.}$$

$$k = f\left(-\frac{3}{2}\right) = 2\left(-\frac{3}{2}\right)^2 + 6\left(-\frac{3}{2}\right) - 5 = -\frac{19}{2} \qquad \text{• Find the } y\text{-coordinate of the vertex.}$$

The vertex is $\left(-\dfrac{3}{2}, -\dfrac{19}{2}\right)$. Because the parabola opens up, $-\dfrac{19}{2}$ is the minimum value of f. Therefore, the range of f is $\left\{y \mid y \geq -\dfrac{19}{2}\right\}$. To determine the values of x for which $f(x) = 15$, replace $f(x)$ by $2x^2 + 6x - 5$ and solve for x.

(continued)

$$f(x) = 15$$
$$2x^2 + 6x - 5 = 15 \quad \text{• Replace } f(x) \text{ by } 2x^2 + 6x - 5.$$
$$2x^2 + 6x - 20 = 0 \quad \text{• Solve for } x.$$
$$2(x - 2)(x + 5) = 0 \quad \text{• Factor.}$$
$$x - 2 = 0 \qquad x + 5 = 0 \quad \text{• Use the Principle of Zero Products to solve for } x.$$
$$x = 2 \qquad x = -5$$

The values of x for which $f(x) = 15$ are 2 and -5.

36. $f(x) = -x^2 - 6x$
$= -(x^2 + 6x)$
$= -(x^2 + 6x + 9) + 9$
$= -(x + 3)^2 + 9$

Maximum value of f is 9 when $x = -3$.

46. **a.** $l + w = 240$, so $w = 240 - l$.

b. $A = lw = l(240 - l) = 240l - l^2$

c. The l value of the vertex point of the graph of $A = 240l - l^2$ is

$$-\frac{b}{2a} = -\frac{240}{2(-1)} = 120$$

Thus $l = 120$ meters and $w = 240 - 120 = 120$ meters are the dimensions that produce the greatest area.

68. Let x = the number of parcels.

a. $R(x) = xp = x(22 - 0.01x) = -0.01x^2 + 22x$

b. $P(x) = R(x) - C(x)$
$= (-0.01x^2 + 22x) - (2025 + 7x)$
$= -0.01x^2 + 15x - 2025$

c. $$-\frac{b}{2a} = -\frac{15}{2(-0.01)} = 750$$

The maximum profit is

$P(750) = -0.01(750)^2 + 15(750) - 2025 = \$3600$

d. The price per parcel that yields the maximum profit is

$p(750) = 22 - 0.01(750) = \$14.50$

e. The break-even point(s) occur when $R(x) = C(x)$.

$$-0.01x^2 + 22x = 2025 + 7x$$
$$0 = 0.01x^2 - 15x + 2025$$
$$x = \frac{-(-15) \pm \sqrt{(-15)^2 - 4(0.01)(2025)}}{2(0.01)}$$

$x = 150$ and $x = 1350$ are the break-even points.

Thus the minimum number of parcels the air freight company must ship to break even is 150.

70. $h(t) = -16t^2 + 64t + 80$

$$t = -\frac{b}{2a} = -\frac{64}{2(-16)} = 2$$

$h(2) = -16(2)^2 + 64(2) + 80$
$= -64 + 128 + 80 = 144$

a. The vertex (2, 144) gives us the maximum height of 144 feet.

b. The vertex of the graph of h is (2, 144), so the time when the projectile achieves this maximum height is at time $t = 2$ seconds.

c. $-16t^2 + 64t + 80 = 0$ • Solve for t with $h = 0$.
$-16(t^2 - 4t - 5) = 0$
$-16(t + 1)(t - 5) = 0$
$t = -1 \qquad t - 5 = 0$
no $\qquad t = 5$

The projectile will have a height of 0 feet at time $t = 5$ seconds.

Exercise Set 1.6, page 91

14. The graph is symmetric with respect to the x-axis because replacing y with $-y$ leaves the equation unaltered. The graph is not symmetric with respect to the y-axis because replacing x with $-x$ alters the equation.

24. The graph is symmetric with respect to the origin because $(-y) = (-x)^3 - (-x)$ simplifies to $-y = -x^3 + x$, which is equivalent to the original equation $y = x^3 - x$.

44. Even, because $h(-x) = (-x)^2 + 1 = x^2 + 1 = h(x)$.

58.

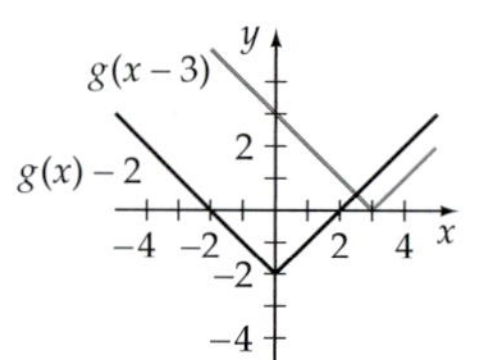

68.

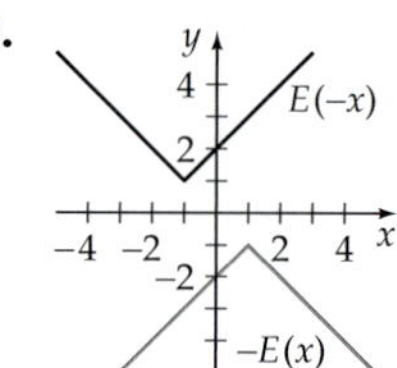

70.

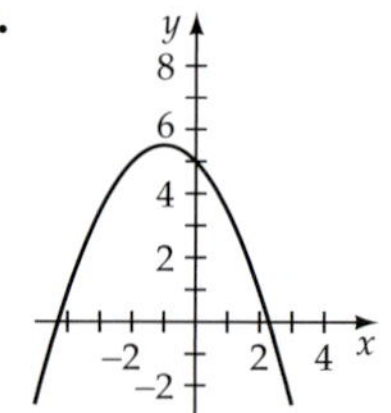

72. **a.**

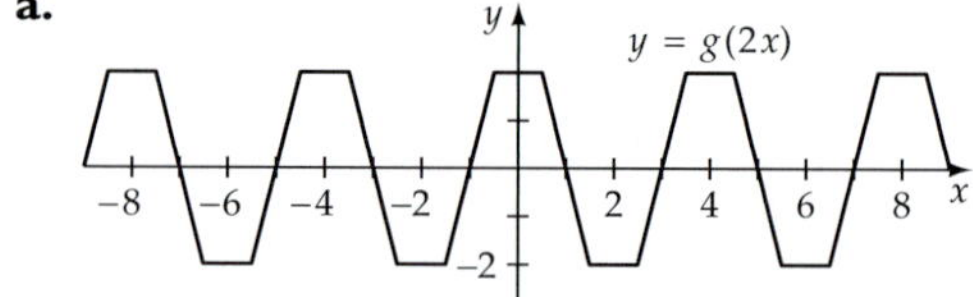

b.

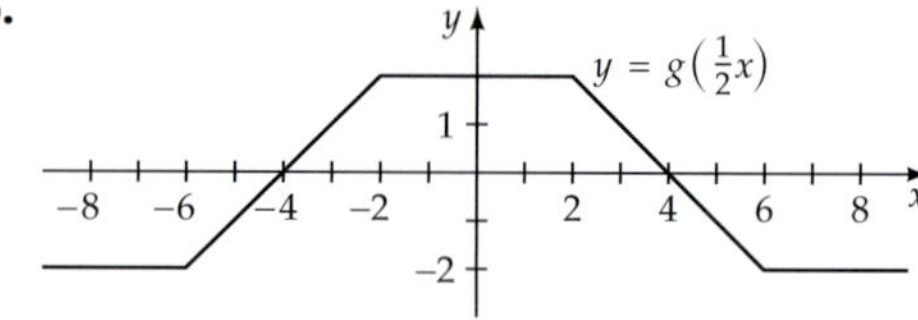

Exercise Set 1.7, page 104

10. $f(x) + g(x) = \sqrt{x-4} - x$ Domain: $\{x \mid x \geq 4\}$

$f(x) - g(x) = \sqrt{x-4} + x$ Domain: $\{x \mid x \geq 4\}$

$f(x)g(x) = -x\sqrt{x-4}$ Domain: $\{x \mid x \geq 4\}$

$\dfrac{f(x)}{g(x)} = -\dfrac{\sqrt{x-4}}{x}$ Domain: $\{x \mid x \geq 4\}$

14. $(f+g)(x) = (x^2 - 3x + 2) + (2x - 4) = x^2 - x - 2$

$(f+g)(-7) = (-7)^2 - (-7) - 2 = 49 + 7 - 2 = 54$

30.
$$\frac{f(x+h) - f(x)}{h} = \frac{[4(x+h) - 5] - (4x - 5)}{h} = \frac{4x + 4(h) - 5 - 4x + 5}{h} = \frac{4(h)}{h} = 4$$

38. $(g \circ f)(x) = g[f(x)] = g[2x - 7]$

$= 3[2x - 7] + 2 = 6x - 19$

$(f \circ g)(x) = f[g(x)] = f[3x + 2]$

$= 2[3x + 2] - 7 = 6x - 3$

50. $(f \circ g)(4) = f[g(4)]$

$= f[4^2 - 5(4)]$

$= f[-4] = 2(-4) + 3 = -5$

66. a. $l = 3 - 0.5t$ for $0 \leq t \leq 6$. $l = -3 + 0.5t$ for $t > 6$. In either case, $l = |3 - 0.5t|$. $w = |2 - 0.2t|$ as in Example 7.

b. $A(t) = |3 - 0.5t||2 - 0.2t|$

c. A is decreasing on $[0, 6]$ and on $[8, 10]$. A is increasing on $[6, 8]$ and on $[10, 14]$.

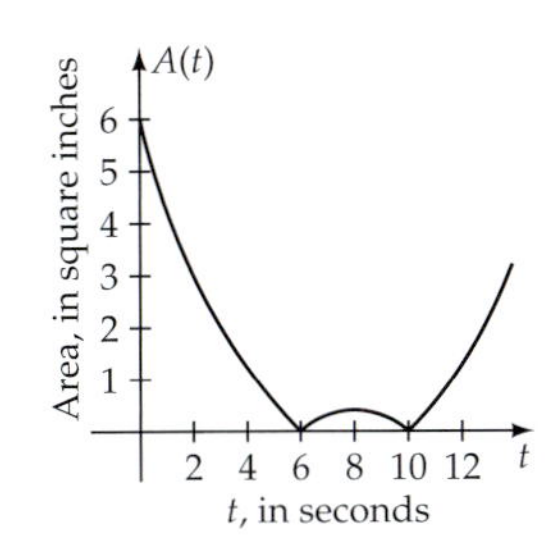

d. The highest point on the graph of A for $0 \leq t \leq 14$ occurs when $t = 0$ seconds.

72. a. On $[2, 3]$,

$a = 2$

$\Delta t = 3 - 2 = 1$

$s(a + \Delta t) = s(3) = 6 \cdot 3^2 = 54$

$s(a) = s(2) = 6 \cdot 2^2 = 24$

$$\text{Average velocity} = \frac{s(a+\Delta t) - s(a)}{\Delta t} = \frac{s(3) - s(2)}{1} = 54 - 24 = 30 \text{ feet per second}$$

This is identical to the slope of the line through $(2, f(2))$ and $(3, f(3))$ because

$$m = \frac{s(3) - s(2)}{3 - 2} = s(3) - s(2) = 54 - 24 = 30$$

b. On $[2, 2.5]$,

$a = 2$

$\Delta t = 2.5 - 2 = 0.5$

$s(a + \Delta t) = s(2.5) = 6(2.5)^2 = 37.5$

$$\text{Average velocity} = \frac{s(2.5) - s(2)}{0.5} = \frac{37.5 - 24}{0.5} = \frac{13.5}{0.5} = 27 \text{ feet per second}$$

c. On $[2, 2.1]$,

$a = 2$

$\Delta t = 2.1 - 2 = 0.1$

$s(a + \Delta t) = s(2.1) = 6(2.1)^2 = 26.46$

$$\text{Average velocity} = \frac{s(2.1) - s(2)}{0.1} = \frac{26.46 - 24}{0.1} = \frac{2.46}{0.1} = 24.6 \text{ feet per second}$$

d. On $[2, 2.01]$,

$a = 2$

$\Delta t = 2.01 - 2 = 0.01$

$s(a + \Delta t) = s(2.01) = 6(2.01)^2 = 24.2406$

$$\text{Average velocity} = \frac{s(2.01) - s(2)}{0.01} = \frac{24.2406 - 24}{0.01} = \frac{0.2406}{0.01} = 24.06 \text{ feet per second}$$

e. On $[2, 2.001]$,

$a = 2$

$\Delta t = 2.001 - 2 = 0.001$

$s(a + \Delta t) = s(2.001) = 6(2.001)^2 = 24.024006$

(continued)

$$\text{Average velocity} = \frac{s(2.001) - s(2)}{0.001} = \frac{24.024006 - 24}{0.001} = \frac{0.024006}{0.001} = 24.006 \text{ feet per second}$$

f. On $[2, 2 + \Delta t]$,

$$\frac{s(2 + \Delta t) - s(2)}{\Delta t} = \frac{6(2 + \Delta t)^2 - 24}{\Delta t}$$
$$= \frac{6(4 + 4(\Delta t) + (\Delta t)^2) - 24}{\Delta t}$$
$$= \frac{24 + 24(\Delta t) + 6(\Delta t)^2 - 24}{\Delta t}$$
$$= \frac{24\Delta t + 6(\Delta t)^2}{\Delta t} = 24 + 6(\Delta t)$$

As Δt approaches zero, the average velocity approaches 24 feet per second.

Exercise Set 2.1, page 124

8. $6 - \sqrt{-1} = 6 - i$

18. $(5 - 3i) - (2 + 9i) = 5 - 3i - 2 - 9i = 3 - 12i$

28.
$$\left(5 + 2\sqrt{-16}\right)\left(1 - \sqrt{-25}\right) = [5 + 2(4i)](1 - 5i)$$
$$= (5 + 8i)(1 - 5i)$$
$$= 5 - 25i + 8i - 40i^2$$
$$= 5 - 25i + 8i - 40(-1)$$
$$= 5 - 25i + 8i + 40$$
$$= 45 - 17i$$

42.
$$\frac{8 - i}{2 + 3i} = \frac{8 - i}{2 + 3i} \cdot \frac{2 - 3i}{2 - 3i}$$
$$= \frac{16 - 24i - 2i + 3i^2}{2^2 + 3^2}$$
$$= \frac{16 - 24i - 2i + 3(-1)}{4 + 9}$$
$$= \frac{16 - 26i - 3}{13}$$
$$= \frac{13 - 26i}{13} = \frac{13(1 - 2i)}{13}$$
$$= 1 - 2i$$

48.
$$\frac{1}{i^{83}} = \frac{1}{i^{80} \cdot i^3} = \frac{1}{i^3} = \frac{1}{-i}$$
$$= \frac{1}{-i} \cdot \frac{i}{i} = \frac{i}{-i^2} = \frac{i}{-(-1)} = \frac{i}{1} = i$$

66. For the equation $8x^2 - 4x + 5 = 0$, we have $a = 8$, $b = -4$, and $c = 5$. Substitute into the quadratic formula and simplify.

$$x = \frac{-b \pm \sqrt{b^2 - 4ac}}{2a}$$
$$= \frac{-(-4) \pm \sqrt{(-4)^2 - 4(8)(5)}}{2(8)}$$
$$= \frac{4 \pm \sqrt{-144}}{16} = \frac{4 \pm 12i}{16} = \frac{4}{16} \pm \frac{12}{16}i = \frac{1}{4} \pm \frac{3}{4}i$$

The solutions are $\frac{1}{4} - \frac{3}{4}i$ and $\frac{1}{4} + \frac{3}{4}i$.

Exercise Set 2.2, page 135

2.
$$\begin{array}{r} 6x^2 - 9x + 28 \\ x + 4\overline{)6x^3 + 15x^2 - 8x + 2} \\ \underline{6x^3 + 24x^2} \\ -9x^2 - 8x \\ \underline{-9x^2 - 36x} \\ 28x + 2 \\ \underline{28x + 112} \\ -110 \end{array}$$

$$\frac{6x^3 + 15x^2 - 8x + 2}{x + 4} = 6x^2 - 9x + 28 - \frac{110}{x + 4}$$

12.
$$\begin{array}{r|rrrr} 5 & 5 & 6 & -8 & 1 \\ & & 25 & 155 & 735 \\ \hline & 5 & 31 & 147 & 736 \end{array}$$

$$\frac{5x^3 + 6x^2 - 8x + 1}{x - 5} = 5x^2 + 31x + 147 + \frac{736}{x - 5}$$

26.
$$\begin{array}{r|rrrr} 3 & 2 & -1 & 3 & -1 \\ & & 6 & 15 & 54 \\ \hline & 2 & 5 & 18 & 53 \end{array}$$

$P(c) = P(3) = 53$

36.
$$\begin{array}{r|rrrr} -6 & 1 & 4 & -27 & -90 \\ & & -6 & 12 & 90 \\ \hline & 1 & -2 & -15 & 0 \end{array}$$

A remainder of 0 indicates that $x + 6$ is a factor of $P(x)$.

56.
$$\begin{array}{r|rrrrr} -1 & 1 & 5 & 3 & -5 & -4 \\ & & -1 & -4 & 1 & 4 \\ \hline & 1 & 4 & -1 & -4 & 0 \end{array}$$

The reduced polynomial is $x^3 + 4x^2 - x - 4$.
$x^4 + 5x^3 + 3x^2 - 5x - 4 = (x + 1)(x^3 + 4x^2 - x - 4)$

Section 2.3, page 149

2. Because $a_n = -2$ is negative and $n = 3$ is odd, the graph of P goes up to the far left and down to the far right.

22. $P(x) = x^3 - 6x^2 + 8x$

$= x(x^2 - 6x + 8)$

$= x(x - 2)(x - 4)$

The factor x can be written as $(x - 0)$. Apply the Factor Theorem to determine that the real zeros of P are 0, 2, and 4.

28.
$$\begin{array}{r|rrrr} 0 & 4 & -1 & -6 & 1 \\ & & 0 & 0 & 0 \\ \hline & 4 & -1 & -6 & 1 \end{array}$$
• $P(0) = 1$

$$\begin{array}{r|rrrr} 1 & 4 & -1 & -6 & 1 \\ & & 4 & 3 & -3 \\ \hline & 4 & 3 & -3 & -2 \end{array}$$
• $P(1) = -2$

Because P is a polynomial function, the graph of P is continuous. Also, $P(0)$ and $P(1)$ have opposite signs. Thus by the Zero Location Theorem we know that P must have a real zero between 0 and 1.

34. The exponent of $(x + 2)$ is 1, which is odd. Thus the graph of P crosses the x-axis at the x-intercept $(-2, 0)$. The exponent of $(x - 6)^2$ is even. Thus the graph of P intersects but does not cross the x-axis at $(6, 0)$.

42. *Far-left and far-right behavior.* The leading term of $P(x) = x^3 + 2x^2 - 3x$ is $1x^3$. The leading coefficient 1 is positive and the degree of the polynomial 3 is odd. Thus the graph of P goes down to its far left and up to its far right.

The y-intercept. $P(0) = 0^3 + 2(0)^2 - 3(0) = 0$. The y-intercept is $(0, 0)$.

The x-intercept(s). Try to factor $x^3 + 2x^2 - 3x$.

$x^3 + 2x^2 - 3x = x(x^2 + 2x - 3)$

$= x(x + 3)(x - 1)$

Use the Factor Theorem to determine that $(0, 0)$, $(-3, 0)$, and $(1, 0)$ are the x-intercepts. Apply the Even and Odd Powers of $(x - c)$ Theorem to determine that the graph of P will cross the x-axis at each of its x-intercepts.

Additional points: $(-2, 6)$, $(-1, 4)$, $(0.5, -0.875)$, $(1.5, 3.375)$

Symmetry: The function P is not an even or an odd function. Thus the graph of P is *not* symmetric with respect to either the y-axis or the origin.

Sketch the graph.

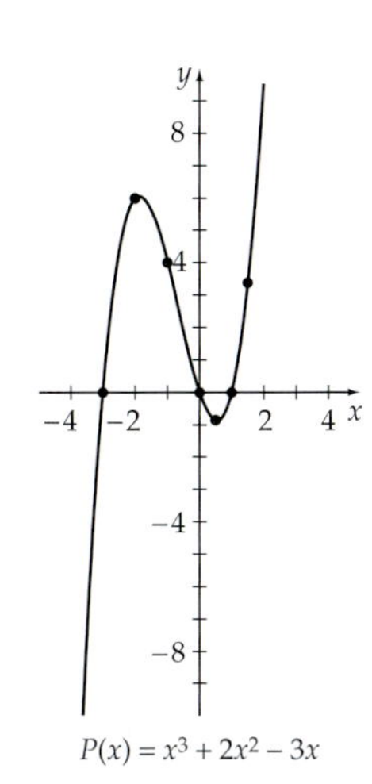

48. The volume of the box is $V = lwh$, with $h = x$, $l = 18 - 2x$, and $w = \dfrac{42 - 3x}{2}$. Therefore, the volume is

$$V(x) = (18 - 2x)\left(\frac{42 - 3x}{2}\right)x$$
$$= 3x^3 - 69x^2 + 378x$$

Use a graphing utility to graph $V(x)$. The graph is shown below. The value of x that produces the maximum volume is 3.571 inches (to the nearest 0.001 inch). *Note:* Your x-value may differ slightly from 3.5705971 depending on the values you use for Xmin and Xmax. The maximum volume is approximately 606.6 cubic inches.

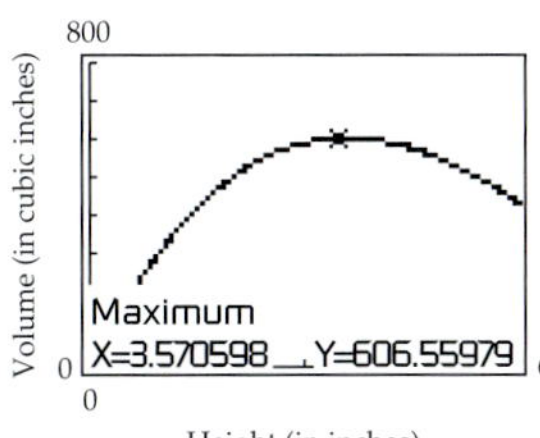

Section 2.4, page 164

10. $p = \pm1, \pm2, \pm4, \pm8$

$q = \pm1, \pm3$

$$\frac{p}{q} = \pm1, \pm2, \pm4, \pm8, \pm\frac{1}{3}, \pm\frac{2}{3}, \pm\frac{4}{3}, \pm\frac{8}{3}$$

18.

$$\begin{array}{r|rrrr} 1 & 1 & 0 & -19 & -28 \\ & & 1 & 1 & -18 \\ \hline & 1 & 1 & -18 & -46 \end{array} \qquad \begin{array}{r|rrrr} -1 & 1 & 0 & -19 & -28 \\ & & -1 & 1 & 18 \\ \hline & 1 & -1 & -18 & -10 \end{array}$$

$$\begin{array}{r|rrrr} 2 & 1 & 0 & -19 & -28 \\ & & 2 & 4 & -30 \\ \hline & 1 & 2 & -15 & -58 \end{array} \qquad \begin{array}{r|rrrr} -2 & 1 & 0 & -19 & -28 \\ & & -2 & 4 & 30 \\ \hline & 1 & -2 & -15 & 2 \end{array}$$

$$\begin{array}{r|rrrr} 3 & 1 & 0 & -19 & -28 \\ & & 3 & 9 & -30 \\ \hline & 1 & 3 & -10 & -58 \end{array} \qquad \begin{array}{r|rrrr} -3 & 1 & 0 & -19 & -28 \\ & & -3 & 9 & 30 \\ \hline & 1 & -3 & -10 & 2 \end{array}$$

$$\begin{array}{r|rrrr} 4 & 1 & 0 & -19 & -28 \\ & & 4 & 16 & -12 \\ \hline & 1 & 4 & -3 & -40 \end{array} \qquad \begin{array}{r|rrrr} -4 & 1 & 0 & -19 & -28 \\ & & -4 & 16 & 12 \\ \hline & 1 & -4 & -3 & -16 \end{array}$$

$$\begin{array}{r|rrrr} 5 & 1 & 0 & -19 & -28 \\ & & 5 & 25 & 30 \\ \hline & 1 & 5 & 6 & 2 \end{array} \qquad \begin{array}{r|rrrr} -5 & 1 & 0 & -19 & -28 \\ & & -5 & 25 & -30 \\ \hline & 1 & -5 & 6 & -58 \end{array}$$

None of these numbers are negative, so 5 is an upper bound.

These numbers alternate in sign, so -5 is a lower bound.

28. $P(x)$ has one positive real zero because P has one variation in sign.

$P(-x) = (-x)^3 - 19(-x) - 30 = -x^3 + 19x - 30$

$P(x)$ has two or no negative real zeros because $P(-x) = -x^3 + 19x - 30$ has two variations in sign.

38. $P(x)$ has one positive and two or no negative real zeros (see Exercise 28 above).

$$\begin{array}{r|rrrr} 5 & 1 & 0 & -19 & -30 \\ & & 5 & 25 & 30 \\ \hline & 1 & 5 & 6 & 0 \end{array}$$

The reduced polynomial is $x^2 + 5x + 6 = (x + 3)(x + 2)$, which has -3 and -2 as zeros. Thus the zeros of $P(x) = x^3 - 19x - 30$ are -3, -2, and 5.

68. The volume of the tank is equal to the volume of the two hemispheres plus the volume of the cylinder. Thus

$$\frac{4}{3}\pi x^3 + 6\pi x^2 = 9\pi$$

Dividing each term by π and multiplying by 3 produces

$$4x^3 + 18x^2 = 27$$

Intersection Method Use a graphing utility to graph $y = 4x^3 + 18x^2$ and $y = 27$ on the same screen, with $x > 0$. The x-coordinate of the point of intersection of the two graphs is the desired solution. The graphs intersect at $x \approx 1.098$ (rounded to the nearest thousandth of a foot). The length of the radius is approximately 1.098 feet.

72. We need to find the natural number solution of $n^3 - 3n^2 + 2n = 504$, which can be written as

$$n^3 - 3n^2 + 2n - 504 = 0$$

The constant term has many natural number divisors, but the following synthetic division shows that 10 is an upper bound for the zeros of $n^3 - 3n^2 + 2n - 504$.

$$\begin{array}{r|rrrr} 10 & 1 & -3 & 2 & -504 \\ & & 10 & 70 & 720 \\ \hline & 1 & 7 & 72 & 216 \end{array}$$

The following synthetic division shows that 9 is a zero of $n^3 - 3n^2 + 2n - 504$.

$$\begin{array}{r|rrrr} 9 & 1 & -3 & 2 & -504 \\ & & 9 & 54 & 504 \\ \hline & 1 & 6 & 56 & 0 \end{array}$$

Thus the given group of cards consists of exactly nine cards. There is no need to seek additional solutions, because any increase (decrease) in the number of cards will increase (decrease) the number of ways one can select three cards from the group of cards.

Section 2.5, page 175

2. Use the Rational Zero Theorem to determine the possible rational zeros.

$$\frac{p}{q} = \pm 1, \pm 5$$

The following synthetic division shows that 1 is a zero of $P(x)$.

$$\begin{array}{r|rrrr} 1 & 1 & -3 & 7 & -5 \\ & & 1 & -2 & 5 \\ \hline & 1 & -2 & 5 & 0 \end{array}$$

Use the quadratic formula to find the zeros of the reduced polynomial $x^2 - 2x + 5$.

$$x = \frac{-(-2) \pm \sqrt{(-2)^2 - 4(1)(5)}}{2(1)} = 1 \pm 2i$$

The zeros of $P(x) = x^3 - 3x^2 + 7x - 5$ are 1, $1 - 2i$, and $1 + 2i$.

The linear factored form of $P(x)$ is

$$P(x) = (x - 1)(x - [1 - 2i])(x - [1 + 2i])$$

or

$$P(x) = (x - 1)(x - 1 + 2i)(x - 1 - 2i)$$

12.

$$\begin{array}{r|rrrr} 5 + 3i & 3 & -29 & 92 & 34 \\ & & 15 + 9i & -97 + 3i & -34 \\ \hline & 3 & -14 + 9i & -5 + 3i & 0 \end{array}$$

$$\begin{array}{r|rrr} 5 - 3i & 3 & -14 + 9i & -5 + 3i \\ & & 15 - 9i & 5 - 3i \\ \hline & 3 & 1 & 0 \end{array}$$

The reduced polynomial $3x + 1$ has $-\frac{1}{3}$ as a zero. The zeros of $3x^3 - 29x^2 + 92x + 34$ are $5 + 3i$, $5 - 3i$, and $-\frac{1}{3}$.

16.

$$\begin{array}{r|rrrrrr} 3i & 1 & -6+0i & 22+0i & -64+0i & 117+0i & -90 \\ & & 0+3i & -9-18i & 54+39i & -117-30i & 90 \\ \hline \end{array}$$

$$\begin{array}{r|rrrrr} -3i & 1 & -6+3i & 13-18i & -10+39i & 0-30i & 0 \\ & & 0-3i & 0+18i & 0-39i & 30i \\ \hline & 1 & -6 & 13 & -10 & 0 \end{array}$$

$$\frac{p}{q} = \pm1, \pm2, \pm5, \pm10$$

$$\begin{array}{r|rrrr} 2 & 1 & -6 & 13 & -10 \\ & & 2 & -8 & 10 \\ \hline & 1 & -4 & 5 & 0 \end{array}$$

Use the quadratic formula to solve $x^2 - 4x + 5 = 0$.

$$x = \frac{-(-4) \pm \sqrt{(-4)^2 - 4(1)(5)}}{2(1)} = \frac{4 \pm \sqrt{-4}}{2}$$

$$= \frac{4 \pm 2i}{2} = 2 \pm i$$

The zeros of $x^5 - 6x^4 + 22x^3 - 64x^2 + 117x - 90$ are $3i$, $-3i$, 2, $2 + i$, and $2 - i$.

24. The graph of $P(x) = 4x^3 + 3x^2 + 16x + 12$ is shown below. Applying Descartes' Rule of Signs, we find that the real zeros are all negative numbers. From the Upper- and Lower-Bound Theorem there is no real zero less than -1, and from the Rational Zero Theorem the possible rational zeros (that are negative and greater than -1) are $\frac{p}{q} = -\frac{1}{2}, -\frac{1}{4}$, and $-\frac{3}{4}$. From the graph, it appears that $-\frac{3}{4}$ is a zero.

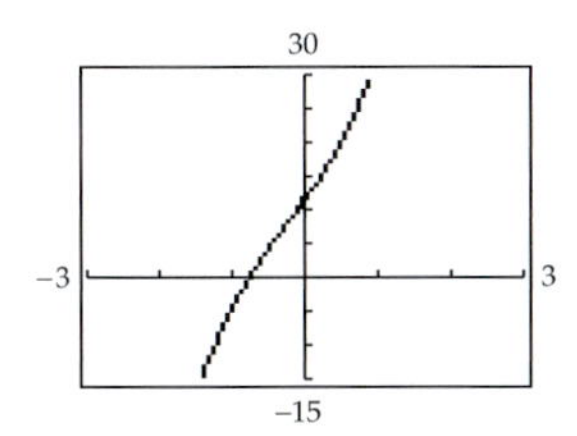

Use synthetic division with $c = -\frac{3}{4}$.

$$\begin{array}{r|rrrr} -\frac{3}{4} & 4 & 3 & 16 & 12 \\ & & -3 & 0 & -12 \\ \hline & 4 & 0 & 16 & 0 \end{array}$$

Thus $-\frac{3}{4}$ is a zero, and by the Factor Theorem,

$$4x^3 + 3x^2 + 16x + 12 = \left(x + \frac{3}{4}\right)(4x^2 + 16) = 0$$

Solve $4x^2 + 16 = 0$ to find that $x = -2i$ and $x = 2i$. The solutions of the original equation are $-\frac{3}{4}$, $-2i$, and $2i$.

42. Because P has real coefficients, use the Conjugate Pair Theorem.

$$\begin{aligned} P &= (x - [3 + 2i])(x - [3 - 2i])(x - 7) \\ &= (x - 3 - 2i)(x - 3 + 2i)(x - 7) \\ &= (x^2 - 6x + 13)(x - 7) \\ &= x^3 - 13x^2 + 55x - 91 \end{aligned}$$

Section 2.6, page 189

2. Set the denominator equal to zero.

$$x^2 - 4 = 0$$
$$(x - 2)(x + 2) = 0$$
$$x = 2 \quad \text{or} \quad x = -2$$

The vertical asymptotes are $x = 2$ and $x = -2$.

6. The horizontal asymptote is $y = 0$ (x-axis) because the degree of the denominator is larger than the degree of the numerator.

10. Vertical asymptote: $x - 2 = 0$

$$x = 2$$

Horizontal asymptote: $y = 0$

No x-intercepts.

y-intercept: $\left(0, -\frac{1}{2}\right)$

26. Vertical asymptote:

$$x^2 - 6x + 9 = 0$$
$$(x - 3)(x - 3) = 0$$
$$x = 3$$

The horizontal asymptote is $y = \frac{1}{1} = 1$ (the Theorem on Horizontal Asymptotes) because the numerator and denominator both have degree 2. The graph crosses the horizontal asymptote at $\left(\frac{3}{2}, 1\right)$. The graph intersects, but does not cross, the x-axis at $(0, 0)$. See the following graph.

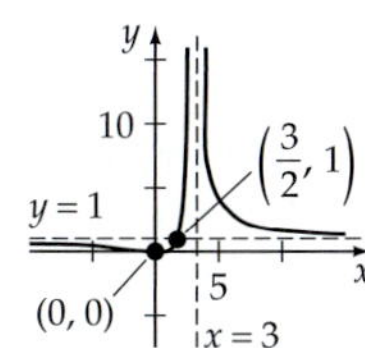

32.

$$\begin{array}{r} x + 1 \\ x^2 - 3x + 5\overline{)x^3 - 2x^2 + 3x + 4} \\ \underline{x^3 - 3x^2 + 5x} \\ x^2 - 2x + 4 \\ \underline{x^2 - 3x + 5} \\ x - 1 \end{array}$$

(continued)

$$F(x) = x + 1 + \frac{x - 1}{x^2 - 3x + 5}$$

Slant asymptote: $y = x + 1$

48. $F(x) = \dfrac{x^2 - x - 12}{x^2 - 2x - 8} = \dfrac{(x-4)(x+3)}{(x-4)(x+2)} = \dfrac{x+3}{x+2}, x \neq 4$

The function F is undefined at $x = 4$. Thus the graph of F is the graph of $y = \dfrac{x+3}{x+2}$ with an open circle at $\left(4, \dfrac{7}{6}\right)$. The height of the open circle was found by evaluating $y = \dfrac{x+3}{x+2}$ at $x = 4$.

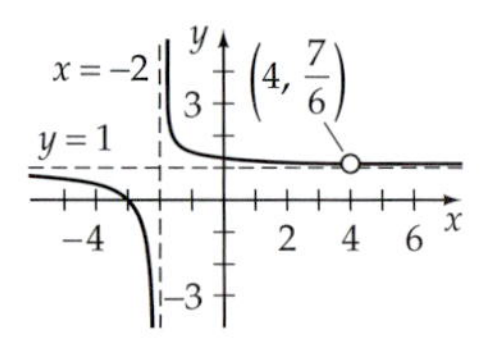

56. a.

$$\overline{C}(1000) = \frac{0.0006(1000)^2 + 9(1000) + 401{,}000}{1000}$$
$$= \$410.60$$
$$\overline{C}(10{,}000) = \frac{0.0006(10{,}000)^2 + 9(10{,}000) + 401{,}000}{10{,}000}$$
$$= \$55.10$$
$$\overline{C}(100{,}000) = \frac{0.0006(100{,}000)^2 + 9(100{,}000) + 401{,}000}{100{,}000}$$
$$= \$73.01$$

b. Graph $\overline{C}$ and use the minimum feature of a graphing utility.

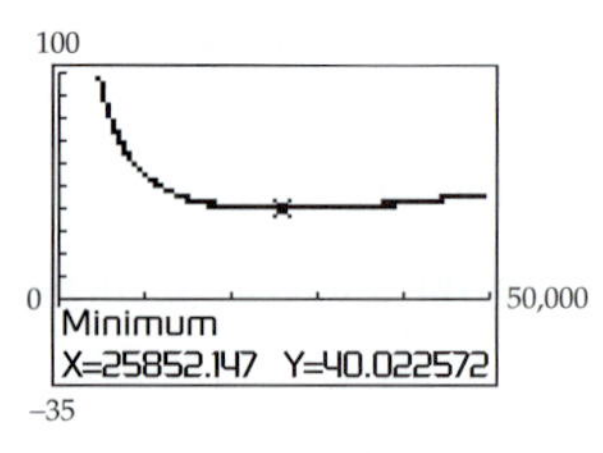

The minimum average cost per telephone is \$40.02. The minimum is achieved by producing approximately 25,852 telephones.

Exercise Set 3.1, page 212

10. Because the graph of the given function is a line that passes through (0, 6), (2, 3), and (6, −3), the graph of the inverse will be a line that passes through (6, 0), (3, 2), and (−3, 6). See the following figure. Notice that the line shown below is a reflection of the line given in Exercise 10 across the line given by $y = x$.

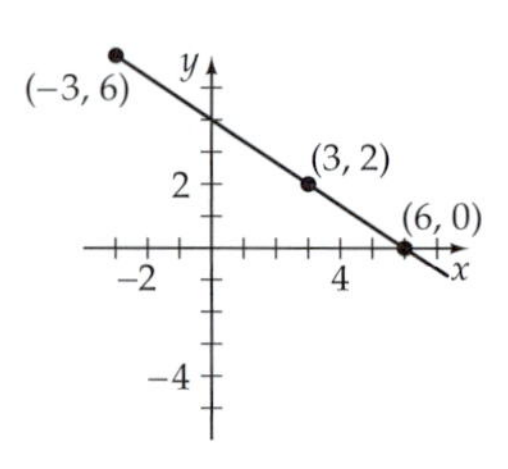

20. Check to see if $f[g(x)] = x$ for all x in the domain of g and $g[f(x)] = x$ for all x in the domain of f. The following shows that $f[g(x)] = x$ for all real numbers x.

$$\begin{aligned} f[g(x)] &= f[2x + 3] \\ &= \frac{1}{2}(2x + 3) - \frac{3}{2} \\ &= x + \frac{3}{2} - \frac{3}{2} \\ &= x \end{aligned}$$

The following shows that $g[f(x)] = x$ for all real numbers x.

$$\begin{aligned} g[f(x)] &= g\left[\frac{1}{2}x - \frac{3}{2}\right] \\ &= 2\left(\frac{1}{2}x - \frac{3}{2}\right) + 3 \\ &= x - 3 + 3 \\ &= x \end{aligned}$$

Thus f and g are inverses.

28.

$f(x) = 4x - 8$

$y = 4x - 8$ • **Replace $f(x)$ by y.**

$x = 4y - 8$ • **Interchange x and y.**

$x + 8 = 4y$ • **Solve for y.**

$\frac{1}{4}(x + 8) = y$

$y = \frac{1}{4}x + 2$

$f^{-1}(x) = \frac{1}{4}x + 2$ • **Replace y by $f^{-1}(x)$.**

34.

$f(x) = \dfrac{x}{x - 2}, x \neq 2$

$y = \dfrac{x}{x - 2}$ • **Replace $f(x)$ by y.**

$x = \dfrac{y}{y - 2}$ • **Interchange x and y.**

$x(y - 2) = y$

$xy - 2x = y$ • **Solve for y.**

$xy - y = 2x$

$y(x - 1) = 2x$

$$y = \frac{2x}{x - 1}$$

$$f^{-1}(x) = \frac{2x}{x - 1}, x \neq 1$$ • **Replace y by $f^{-1}(x)$ and indicate any restrictions.**

40. $f(x) = \sqrt{4 - x}, x \leq 4$

$y = \sqrt{4 - x}$ • **Replace $f(x)$ by y.**

$x = \sqrt{4 - y}$ • **Interchange x and y.**

$x^2 = 4 - y$ • **Solve for y.**

$x^2 - 4 = -y$

$-x^2 + 4 = y$

$f^{-1}(x) = -x^2 + 4, x \geq 0$ • **Replace y by $f^{-1}(x)$ and indicate any restrictions.**

The range of f is $\{y \mid y \geq 0\}$. Therefore, the domain of f^{-1} is $\{x \mid x \geq 0\}$, as indicated above.

50. $K(x) = 1.3x - 4.7$

$y = 1.3x - 4.7$ • **Replace $K(x)$ by y.**

$x = 1.3y - 4.7$ • **Interchange x and y.**

$x + 4.7 = 1.3y$ • **Solve for y.**

$$\frac{x + 4.7}{1.3} = y$$

$$K^{-1}(x) = \frac{x + 4.7}{1.3}$$ • **Replace y by $K^{-1}(x)$.**

The function $K^{-1}(x) = \dfrac{x + 4.7}{1.3}$ can be used to convert a United Kingdom men's shoe size to its equivalent U.S. shoe size.

Exercise Set 3.2, page 224

2. $f(3) = 5^3 = 5 \cdot 5 \cdot 5 = 125$

$$f(-2) = 5^{-2} = \frac{1}{5^2} = \frac{1}{5 \cdot 5} = \frac{1}{25}$$

22. The graph of $f(x) = \left(\dfrac{5}{2}\right)^x$ has a y-intercept of (0, 1) and the graph passes through $\left(1, \dfrac{5}{2}\right)$. Plot a few additional points, such as $\left(-1, \dfrac{2}{5}\right)$ and $\left(2, \dfrac{25}{4}\right)$. Because the base $\dfrac{5}{2}$ is greater than 1, we know that the graph must have all the properties of an increasing exponential function. Draw a smooth increasing curve through the points. The graph should be asymptotic to the negative portion of the x-axis, as shown in the following figure.

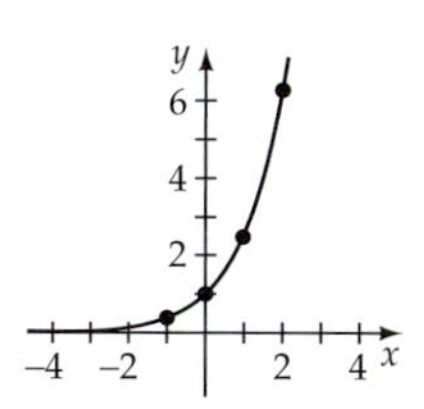

28. Because $F(x) = 6^{x+5} = f(x + 5)$, the graph of $F(x)$ can be produced by shifting the graph of f horizontally to the left 5 units.

30. Because $F(x) = -\left[\left(\dfrac{5}{2}\right)^x\right] = -f(x)$, the graph of $F(x)$ can be produced by reflecting the graph of f across the x-axis.

44. a. $A(45) = 200e^{-0.014(45)}$

≈ 106.52

After 45 minutes the patient will have about 107 milligrams of medication in his or her bloodstream.

b. Use a graphing calculator to graph $y = 200e^{-0.014x}$ and $y = 50$ in the same viewing window as shown below.

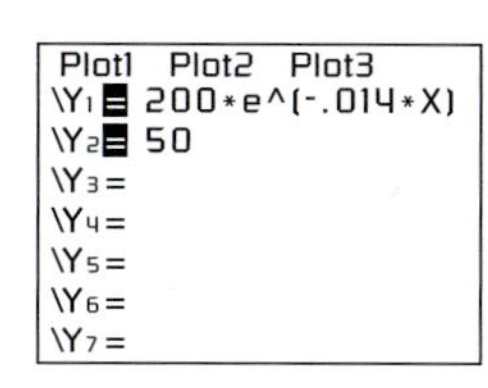

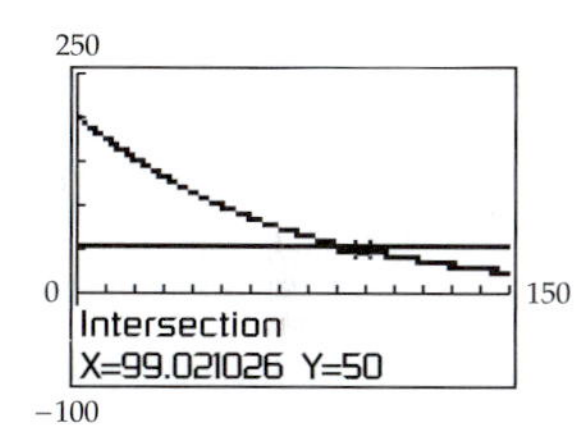

The x-coordinate (which represents time in minutes) of the point of intersection is about 99.02. Thus it will take about 99 minutes before the patient's medication level is reduced to 50 milligrams.

54. a.
$$P(0) = \frac{3600}{1 + 7e^{-0.05(0)}} = \frac{3600}{1 + 7} = \frac{3600}{8} = 450$$

Immediately after the lake was stocked, the lake contained 450 bass.

b.
$$P(12) = \frac{3600}{1 + 7e^{-0.05(12)}} \approx 743.54$$

After 1 year (12 months) there were about 744 bass in the lake.

(continued)

c. As $t \to \infty$, $7e^{-0.05t} = \dfrac{7}{e^{0.05t}}$ approaches 0. Thus as $t \to \infty$, $P(t) = \dfrac{3600}{1 + 7e^{-0.05t}}$ will approach $\dfrac{3600}{1 + 0} = 3600$. As time goes by the bass population will increase, approaching 3600.

Exercise Set 3.3, page 239

4. The exponential form of $\log_b x = y$ is $b^y = x$. Thus the exponential form of $\log_4 64 = 3$ is $4^3 = 64$.

12. The logarithmic form of $b^y = x$ is $y = \log_b x$. Thus the logarithmic form of $5^3 = 125$ is $3 = \log_5 125$.

28. $\log 1{,}000{,}000 = \log_{10} 10^6 = 6$

32. To graph $y = \log_6 x$, use the equivalent exponential equation $x = 6^y$. Choose some y-values, such as $-1, 0, 1$, and calculate the corresponding x-values. This yields the ordered pairs $\left(\dfrac{1}{6}, -1\right)$, $(1, 0)$, and $(6, 1)$. Plot these ordered pairs and draw a smooth curve through the points to produce the following graph.

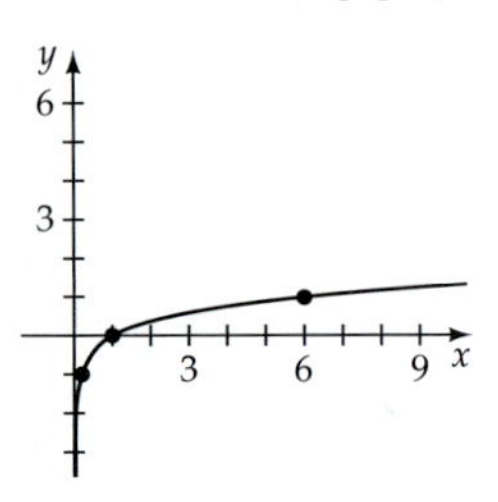

40. $\log_4(5 - x)$ is defined only for $5 - x > 0$, which is equivalent to $x < 5$. Using interval notation, the domain of $k(x) = \log_4(5 - x)$ is $(-\infty, 5)$.

50. The graph of $f(x) = \log_6(x + 3)$ can be produced by shifting the graph of $f(x) = \log_6 x$ (from Exercise 32) 3 units to the left. See the figure below.

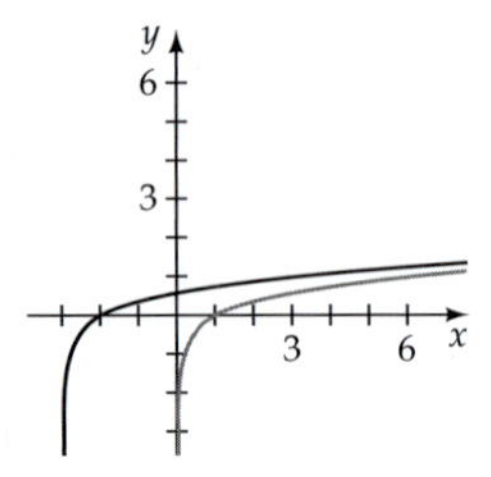

70. a. $S(0) = 5 + 29 \ln(0 + 1) = 5 + 0 = 5$. When starting, the student had an average typing speed of 5 words per minute. $S(3) = 5 + 29 \ln(3 + 1) \approx 45.2$. After 3 months the student's average typing speed was about 45 words per minute.

b. Use the intersection feature of a graphing utility to find the x-coordinate of the point of intersection of the graphs of $y = 5 + 29 \ln(x + 1)$ and $y = 65$.

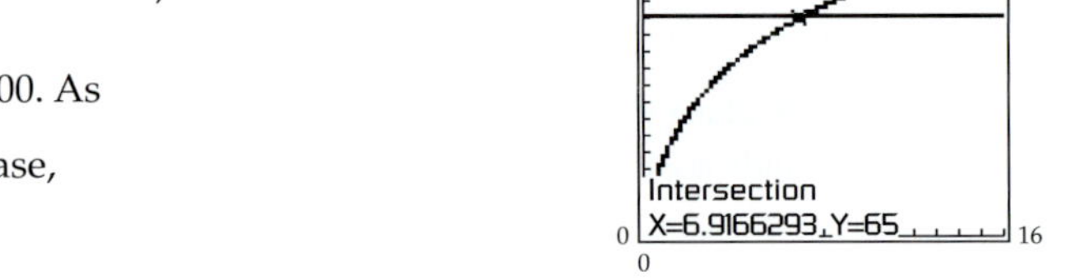

The graphs intersect at about (6.9, 65). The student will achieve a typing speed of 65 words per minute in about 6.9 months.

Exercise Set 3.4, page 251

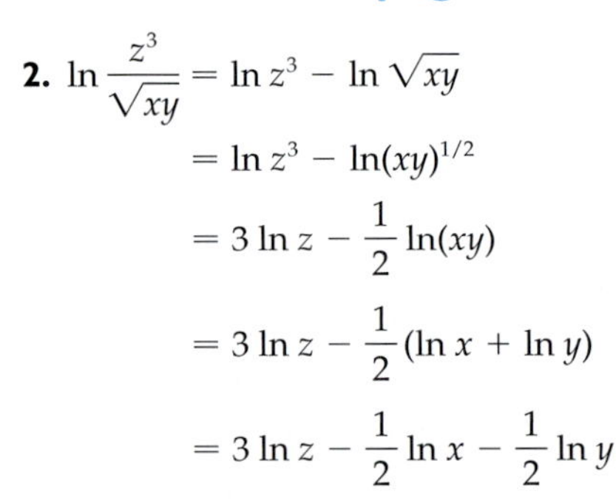

2.
$$\begin{aligned}\ln \frac{z^3}{\sqrt{xy}} &= \ln z^3 - \ln \sqrt{xy} \\ &= \ln z^3 - \ln(xy)^{1/2} \\ &= 3 \ln z - \frac{1}{2}\ln(xy) \\ &= 3 \ln z - \frac{1}{2}(\ln x + \ln y) \\ &= 3 \ln z - \frac{1}{2}\ln x - \frac{1}{2}\ln y\end{aligned}$$

10.
$$\begin{aligned}3 \log_2 t - \frac{1}{3}\log_2 u + 4 \log_2 v &= \log_2 t^3 - \log_2 u^{1/3} + \log_2 v^4 \\ &= \log_2 \frac{t^3}{u^{1/3}} + \log_2 v^4 \\ &= \log_2 \frac{t^3 v^4}{u^{1/3}} \\ &= \log_2 \frac{t^3 v^4}{\sqrt[3]{u}}\end{aligned}$$

16. $\log_5 37 = \dfrac{\log 37}{\log 5} \approx 2.2436$

24. $\log_8(5 - x) = \dfrac{\ln(5 - x)}{\ln 8}$, so enter $\dfrac{\ln(5 - x)}{\ln 8}$ into Y1 on a graphing calculator.

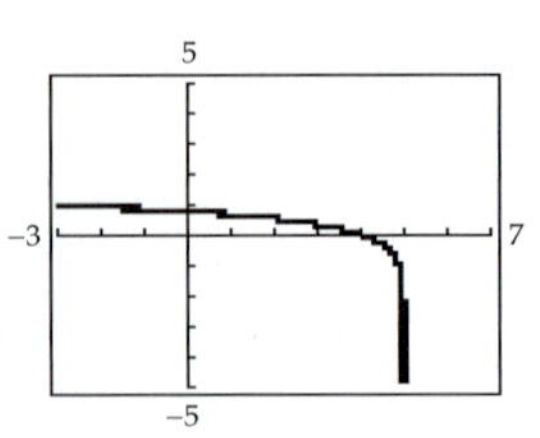

48. $\text{pH} = -\log[\text{H}^+] = -\log(1.26 \times 10^{-3}) \approx 2.9$

50.
$$\begin{aligned}\text{pH} &= -\log[\text{H}^+] \\ 5.6 &= -\log[\text{H}^+] \\ -5.6 &= \log[\text{H}^+] \\ 10^{-5.6} &= \text{H}^+\end{aligned}$$

The hydronium-ion concentration is $10^{-5.6} \approx 2.51 \times 10^{-6}$ mole per liter.

56. $M = \log\left(\frac{I}{I_0}\right) = \log\left(\frac{398{,}107{,}000 I_0}{I_0}\right) = \log(398{,}107{,}000)$

≈ 8.6

58. $\log\left(\frac{I}{I_0}\right) = 9.5$

$$\frac{I}{I_0} = 10^{9.5}$$
$$I = 10^{9.5} I_0$$
$$I \approx 3{,}162{,}277{,}660 I_0$$

60. In Example 7 we noticed that if an earthquake has a Richter scale magnitude of M_1 and a smaller earthquake has a Richter scale magnitude of M_2, then the first earthquake is $10^{M_1 - M_2}$ times as intense as the smaller earthquake. In this exercise, $M_1 = 9.5$ and $M_2 = 8.3$. Thus $10^{M_1 - M_2} = 10^{9.5-8.3} = 10^{1.2} \approx 15.8$. The 1960 earthquake in Chile was about 15.8 times as intense as the San Francisco earthquake of 1906.

64. $M = \log A + 3\log 8t - 2.92$

$= \log 26 + 3\log[8 \cdot 17] - 2.92$ • **Substitute 26 for *A* and 17 for *t*.**

$\approx 1.4150 + 6.4006 - 2.92$

≈ 4.9

Exercise Set 3.5, page 263

2. $3^x = 243$

$$3^x = 3^5$$
$$x = 5$$

10. $6^x = 50$

$$\log(6^x) = \log 50$$
$$x \log 6 = \log 50$$
$$x = \frac{\log 50}{\log 6} \approx 2.18$$

18. $3^{x-2} = 4^{2x+1}$

$$\log 3^{x-2} = \log 4^{2x+1}$$
$$(x-2)\log 3 = (2x+1)\log 4$$
$$x\log 3 - 2\log 3 = 2x\log 4 + \log 4$$
$$x\log 3 - 2\log 3 - 2x\log 4 = \log 4$$
$$x\log 3 - 2x\log 4 = \log 4 + 2\log 3$$
$$x(\log 3 - 2\log 4) = \log 4 + 2\log 3$$
$$x = \frac{\log 4 + 2\log 3}{\log 3 - 2\log 4}$$
$$x \approx -2.141$$

22. $\log(x^2 + 19) = 2$

$$x^2 + 19 = 10^2$$
$$x^2 + 19 = 100$$
$$x^2 = 81$$
$$x = \pm 9$$

A check shows that 9 and -9 are both solutions of the original equation.

26. $\log_3 x + \log_3(x + 6) = 3$

$$\log_3[x(x+6)] = 3$$
$$3^3 = x(x+6)$$
$$27 = x^2 + 6x$$
$$x^2 + 6x - 27 = 0$$
$$(x+9)(x-3) = 0$$
$$x = -9 \quad \text{or} \quad x = 3$$

Because $\log_3 x$ is defined only for $x > 0$, the only solution is $x = 3$.

36. $\ln x = \frac{1}{2}\ln\left(2x + \frac{5}{2}\right) + \frac{1}{2}\ln 2$

$$= \frac{1}{2}\left[\ln\left(2x + \frac{5}{2}\right) + \ln 2\right]$$
$$\ln x = \frac{1}{2}\ln\left[2\left(2x + \frac{5}{2}\right)\right]$$
$$\ln x = \frac{1}{2}\ln(4x + 5)$$
$$\ln x = \ln(4x+5)^{1/2}$$
$$x = \sqrt{4x + 5}$$
$$x^2 = 4x + 5$$
$$0 = x^2 - 4x - 5$$
$$0 = (x-5)(x+1)$$
$$x = 5 \quad \text{or} \quad x = -1$$

Check: $\ln 5 = \frac{1}{2}\ln\left(10 + \frac{5}{2}\right) + \frac{1}{2}\ln 2$

$$1.6094 \approx 1.2629 + 0.3466$$

Because $\ln(-1)$ is not defined, -1 is not a solution. Thus the only solution is $x = 5$.

40. $\frac{10^x + 10^{-x}}{2} = 8$

$$10^x + 10^{-x} = 16$$

$10^x(10^x + 10^{-x}) = (16)10^x$ • **Multiply each side by 10^x.**

$$10^{2x} + 1 = 16(10^x)$$
$$10^{2x} - 16(10^x) + 1 = 0$$

$u^2 - 16u + 1 = 0$ • **Let $u = 10^x$.**

$$u = \frac{16 \pm \sqrt{16^2 - 4(1)(1)}}{2} = 8 \pm 3\sqrt{7}$$

$10^x = 8 \pm 3\sqrt{7}$ • **Replace *u* with 10^x.**

$$\log 10^x = \log(8 \pm 3\sqrt{7})$$
$$x = \log(8 \pm 3\sqrt{7}) \approx \pm 1.20241$$

68. a.

$$t = \frac{9}{24}\ln\frac{24+v}{24-v}$$
$$1.5 = \frac{9}{24}\ln\frac{24+v}{24-v}$$
$$4 = \ln\frac{24+v}{24-v}$$
$$e^4 = \frac{24+v}{24-v} \qquad \bullet\ N = \ln M \text{ means } e^N = M.$$
$$(24-v)e^4 = 24+v$$
$$-v - ve^4 = 24 - 24e^4$$
$$v(-1-e^4) = 24 - 24e^4$$
$$v = \frac{24-24e^4}{-1-e^4} \approx 23.14$$

The velocity is about 23.14 feet per second.

b. The vertical asymptote is $v = 24$.

c. Due to the air resistance, the object cannot reach or exceed a velocity of 24 feet per second.

Exercise Set 3.6, page 276

4. a. $P = 12{,}500$, $r = 0.08$, $t = 10$, $n = 1$.

$$A = 12{,}500\left(1+\frac{0.08}{1}\right)^{10} \approx \$26{,}986.56$$

b. $n = 365$

$$A = 12{,}500\left(1+\frac{0.08}{365}\right)^{3650} \approx \$27{,}816.82$$

c. $n = 8760$

$$A = 12{,}500\left(1+\frac{0.08}{8760}\right)^{87600} \approx \$27{,}819.16$$

6. $P = 32{,}000$, $r = 0.08$, $t = 3$.

$$A = Pe^{rt} = 32{,}000e^{3(0.08)} \approx \$40{,}679.97$$

10. $t = \dfrac{\ln 3}{r} \qquad r = 0.055$

$$t = \frac{\ln 3}{0.055}$$

$t \approx 20$ years (to the nearest year)

18. a. $P(12) = 20{,}899(1.027)^{12} \approx 28{,}722$ thousands, or 28,772,000.

b. P is in thousands, so

$$35{,}000 = 20{,}899(1.027)^t$$
$$\frac{35{,}000}{20{,}899} = 1.027^t$$
$$\ln\left(\frac{35{,}000}{20{,}899}\right) = t\ln 1.027$$
$$\frac{\ln\left(\frac{35{,}000}{20{,}899}\right)}{\ln 1.027} = t$$
$$19.35 \approx t$$

According to the growth function, the population will first exceed 35 million in 19.35 years—that is, in the year 1991 + 19 = 2010.

20.

$$N(t) = N_0e^{kt}$$
$$N(138) = N_0e^{138k}$$
$$0.5N_0 = N_0e^{138k}$$
$$0.5 = e^{138k}$$
$$\ln 0.5 = 138k$$
$$k = \frac{\ln 0.5}{138} \approx -0.005023$$
$$N(t) = N_0(0.5)^{t/138} \approx N_0e^{-0.005023t}$$

24.

$$N(t) = N_0(0.5)^{t/5730}$$
$$0.65N_0 = N_0(0.5)^{t/5730}$$
$$0.65 = (0.5)^{t/5730}$$
$$\ln 0.65 = \ln(0.5)^{t/5730}$$
$$t = 5730\frac{\ln 0.65}{\ln 0.5} \approx 3600$$

The bone is approximately 3600 years old.

32. a.

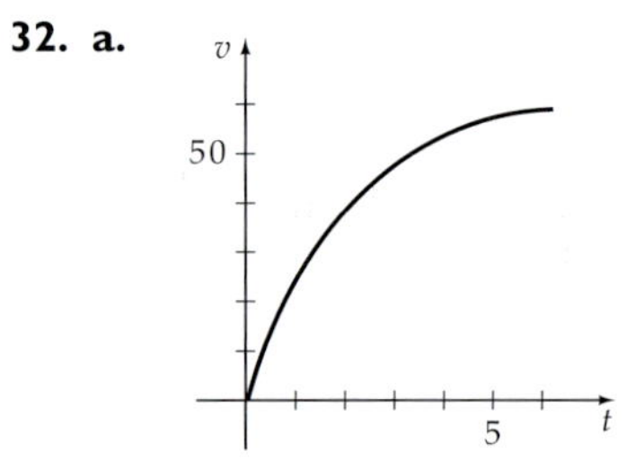

b. Here is an algebraic solution. An approximate solution can be obtained from the graph.

$$v = 64(1 - e^{-t/2})$$
$$50 = 64(1 - e^{-t/2})$$
$$\frac{50}{64} = (1 - e^{-t/2})$$
$$1 - \frac{50}{64} = e^{-t/2}$$
$$\ln\left(1 - \frac{50}{64}\right) = -\frac{t}{2}$$
$$t = -2\ln\left(1-\frac{50}{64}\right) \approx 3.0$$

The velocity is 50 feet per second in approximately 3.0 seconds.

c. As $t \to \infty$, $e^{-t/2} \to 0$. Therefore, $64(1 - e^{-t/2}) \to 64$. The horizontal asymptote is $v = 64$.

d. Because of the air resistance, the velocity of the object will never reach or exceed 64 feet per second.

Exercise Set 4.1, page 302

2. The measure of the complement of an angle of 87° is

$(90° - 87°) = 3°$

The measure of the supplement of an angle of 87° is

$(180° - 87°) = 93°$

14. Because $765° = 2 \cdot 360° + 45°$, α is coterminal with an angle that has a measure of 45°. α is a Quadrant 1 angle.

32. $-45° = -45°\left(\dfrac{\pi \text{ radians}}{180°}\right) = -\dfrac{\pi}{4} \text{ radian}$

40. $\dfrac{\pi}{4} \text{ radians} = \dfrac{\pi}{4} \text{ radians}\left(\dfrac{180°}{\pi \text{ radians}}\right) = 45°$

62. $s = r\theta = 5\left(144° \cdot \dfrac{\pi}{180°}\right) = 4\pi \approx 12.57 \text{ meters}$

66. Let θ_2 be the angle through which a pulley with a diameter of 0.8 meter turns. Let θ_1 be the angle through which a pulley with a diameter of 1.2 meters turns. Let $r_2 = 0.4$ meter be the radius of the smaller pulley, and let $r_1 = 0.6$ meter be the radius of the larger pulley.

$$\theta_1 = 240° = \frac{4}{3}\pi \text{ radians}$$

Thus
$$r_2\theta_2 = r_1\theta_1$$
$$0.4\theta_2 = 0.6\left(\frac{4}{3}\pi\right)$$
$$\theta_2 = \frac{0.6}{0.4}\left(\frac{4}{3}\pi\right) = 2\pi \text{ radians or } 360°$$

68. The earth makes 1 revolution ($\theta = 2\pi$) in 1 day.

$t = 24 \cdot 3600 = 86{,}400$ seconds

$$\omega = \frac{\theta}{t} = \frac{2\pi}{86{,}400} \approx 7.27 \times 10^{-5} \text{ radian/second}$$

74. $C = 2\pi r = 2\pi(18 \text{ inches}) = 36\pi \text{ inches}$

Thus one conversion factor is (36π inches/1 rev).

$$\frac{500 \text{ rev}}{1 \text{ minute}} = \frac{500 \text{ rev}}{1 \text{ minute}}\left(\frac{36\pi \text{ inches}}{1 \text{ rev}}\right) = \frac{18{,}000\pi \text{ inches}}{1 \text{ minute}}$$

Now convert inches to miles and minutes to hours.

$$\frac{18{,}000\pi \text{ inches}}{1 \text{ minute}}$$
$$= \frac{18{,}000\pi \text{ inches}}{1 \text{ minute}}\left(\frac{1 \text{ foot}}{12 \text{ inches}}\right)\left(\frac{1 \text{ mile}}{5280 \text{ feet}}\right)\left(\frac{60 \text{ minutes}}{1 \text{ hour}}\right)$$
$$\approx 54 \text{ miles per hour}$$

Exercise Set 4.2, page 319

10. Start at $(1, 0)$ and wrap counterclockwise $\dfrac{11\pi}{6}$ units around the unit circle. Note that the point $W\left(\dfrac{11\pi}{6}\right)$ is symmetric with respect to the x-axis to the point $W\left(\dfrac{\pi}{6}\right) = \left(\dfrac{\sqrt{3}}{2}, \dfrac{1}{2}\right)$. Thus $W\left(\dfrac{11\pi}{6}\right) = \left(\dfrac{\sqrt{3}}{2}, -\dfrac{1}{2}\right)$.

24. In Exercise 10 above we found that $W\left(\dfrac{11\pi}{6}\right) = \left(\dfrac{\sqrt{3}}{2}, -\dfrac{1}{2}\right)$. By definition, $\sin\left(\dfrac{11\pi}{6}\right)$ is the y-value of $W\left(\dfrac{11\pi}{6}\right)$. Thus $\sin\left(\dfrac{11\pi}{6}\right) = -\dfrac{1}{2}$.

58.
$$F(-x) = \tan(-x) + \sin(-x)$$
$$= -\tan x - \sin x \quad \text{• tan } x \text{ and sin } x \text{ are odd functions.}$$
$$= -(\tan x + \sin x)$$
$$= -F(x)$$

Because $F(-x) = -F(x)$, the function defined by $F(x) = \tan x + \sin x$ is an odd function.

74.
$$\frac{1}{1 - \sin t} + \frac{1}{1 + \sin t} = \frac{1 + \sin t + 1 - \sin t}{(1 - \sin t)(1 + \sin t)}$$
$$= \frac{2}{1 - \sin^2 t}$$
$$= \frac{2}{\cos^2 t}$$
$$= 2\sec^2 t$$

80.
$$1 + \tan^2 t = \sec^2 t$$
$$\tan^2 t = \sec^2 t - 1$$
$$\tan t = \pm\sqrt{\sec^2 t - 1}$$

Because $\dfrac{3\pi}{2} < t < 2\pi$, $\tan t$ is negative. Thus $\tan t = -\sqrt{\sec^2 t - 1}$.

84. March 5 is represented by $t = 2$.

$$T(2) = -41\cos\left(\frac{\pi}{6} \cdot 2\right) + 36$$
$$= -41\cos\left(\frac{\pi}{3}\right) + 36$$
$$= -41(0.5) + 36$$
$$= 15.5°\text{F}$$

July 20 is represented by $t = 6.5$.

$$T(6.5) = -41\cos\left(\frac{\pi}{6} \cdot 6.5\right) + 36$$
$$\approx -41(-0.9659258263) + 36$$
$$\approx 75.6°\text{F}$$

Exercise Set 4.3, page 329

20. $y = -\dfrac{3}{2}\sin x$

$a = \left|-\dfrac{3}{2}\right| = \dfrac{3}{2}$

period $= 2\pi$

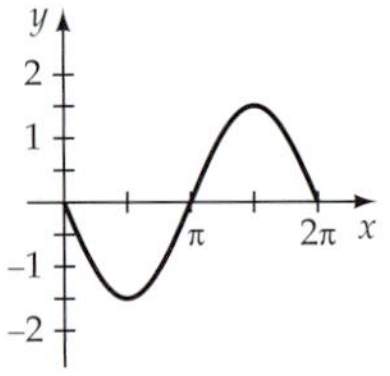

30. $y = \sin \dfrac{3\pi}{4}x$

$a = 1$

period $= \dfrac{2\pi}{b} = \dfrac{2\pi}{3\pi/4} = \dfrac{8}{3}$

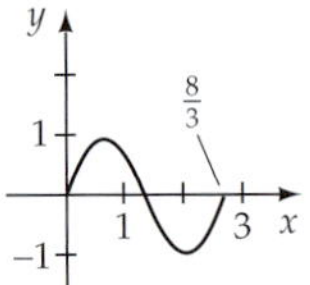

32. $y = \cos 3\pi x$

$a = 1$

period $= \dfrac{2\pi}{b} = \dfrac{2\pi}{3\pi} = \dfrac{2}{3}$

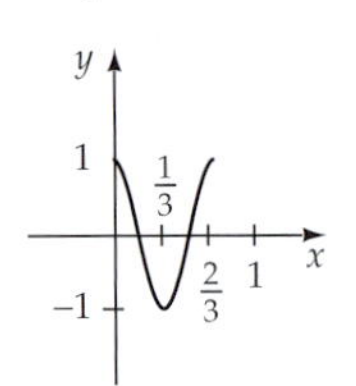

38. $y = \dfrac{1}{2}\sin \dfrac{\pi x}{3}$

$a = \dfrac{1}{2}$

period $= \dfrac{2\pi}{b} = \dfrac{2\pi}{\pi/3} = 6$

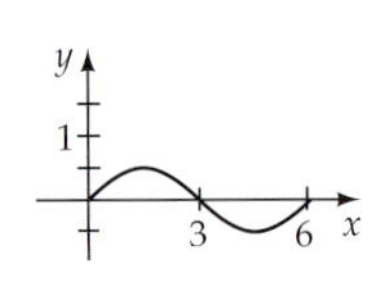

46. $y = -\dfrac{3}{4}\cos 5x$

$a = \left|-\dfrac{3}{4}\right| = \dfrac{3}{4}$

period $= \dfrac{2\pi}{b} = \dfrac{2\pi}{5}$

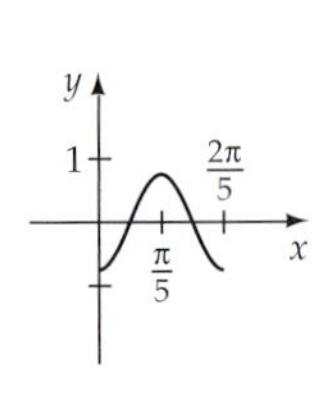

52. $y = -\left|3\sin \dfrac{2}{3}x\right|$

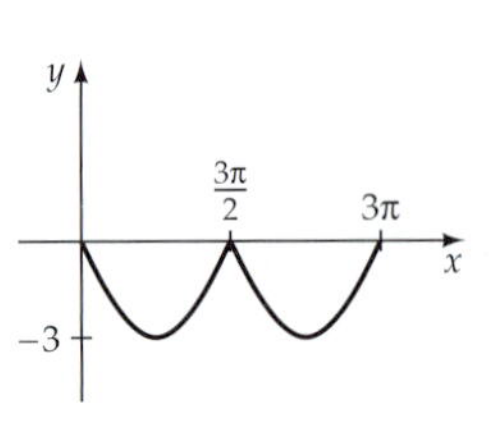

Exercise Set 4.4, page 337

22. $y = \dfrac{1}{3}\tan x$

period $= \dfrac{\pi}{b} = \pi$

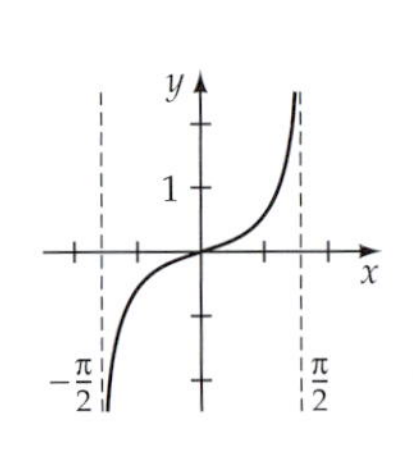

30. $y = -3\tan 3x$

period $= \dfrac{\pi}{b} = \dfrac{\pi}{3}$

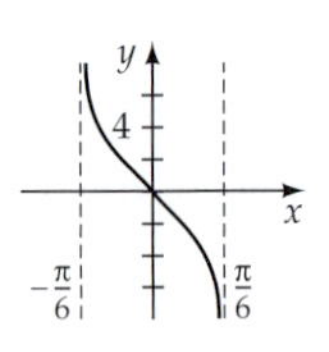

32. $y = \dfrac{1}{2}\cot 2x$

period $= \dfrac{\pi}{b} = \dfrac{\pi}{2}$

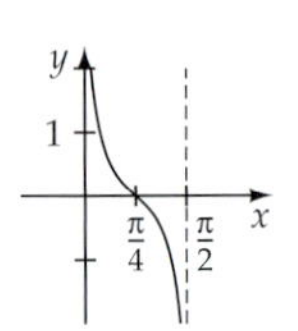

38. $y = 3\csc \dfrac{\pi x}{2}$

period $= \dfrac{2\pi}{b} = \dfrac{2\pi}{\pi/2} = 4$

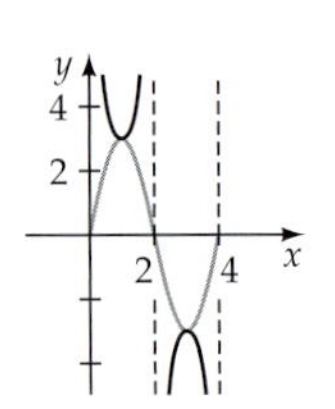

42. $y = \sec \dfrac{x}{2}$

period $= \dfrac{2\pi}{b} = \dfrac{2\pi}{1/2} = 4\pi$

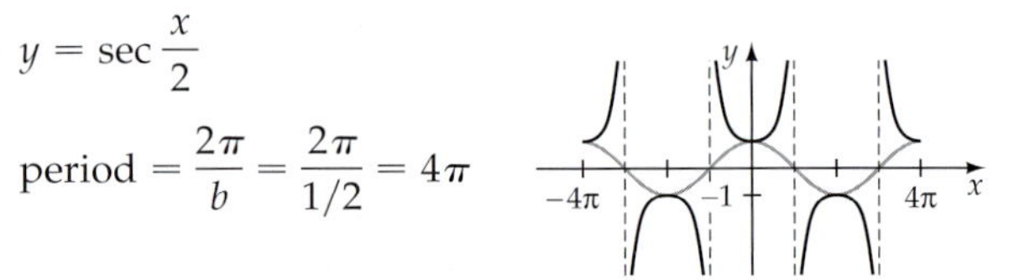

Exercise Set 4.5, page 346

20. $y = \cos\left(2x - \dfrac{\pi}{3}\right)$

$a = 1$

period $= \pi$

phase shift $= -\dfrac{c}{b}$

$= -\dfrac{-\pi/3}{2} = \dfrac{\pi}{6}$

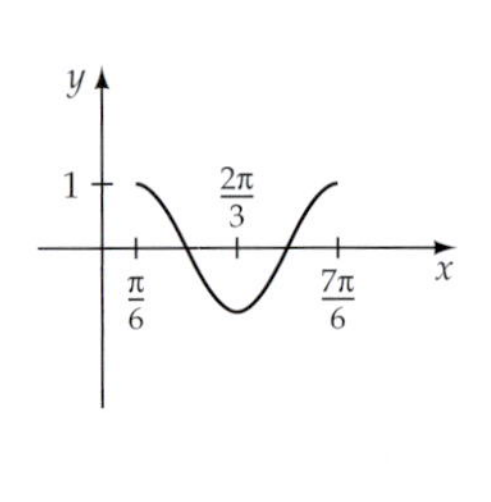

22. $y = \tan(x - \pi)$

period $= \pi$

phase shift $= -\dfrac{c}{b}$

$= -\dfrac{-\pi}{1} = \pi$

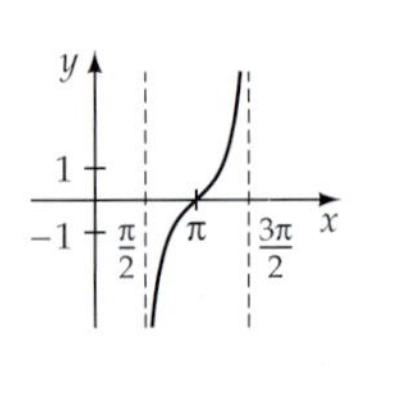

40. $y = 2\sin\left(\dfrac{\pi x}{2} + 1\right) - 2$

$a = 2$

period $= 4$

phase shift $= -\dfrac{c}{b}$

$= -\dfrac{1}{2/\pi} = -\dfrac{2}{\pi}$

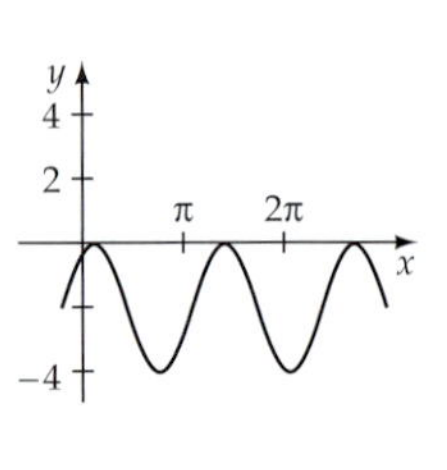

42. $y = -3\cos(2\pi x - 3) + 1$

$a = 3$

period $= 1$

phase shift $= -\dfrac{c}{b} = \dfrac{3}{2\pi}$

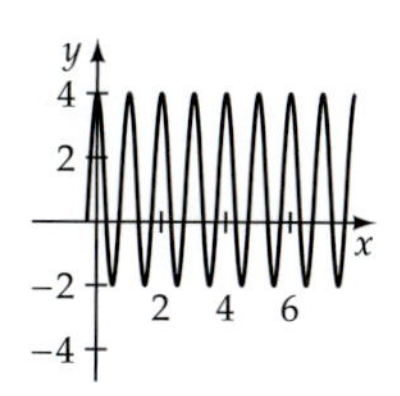

48. $y = \csc\dfrac{x}{3} + 4$

period $= 6\pi$

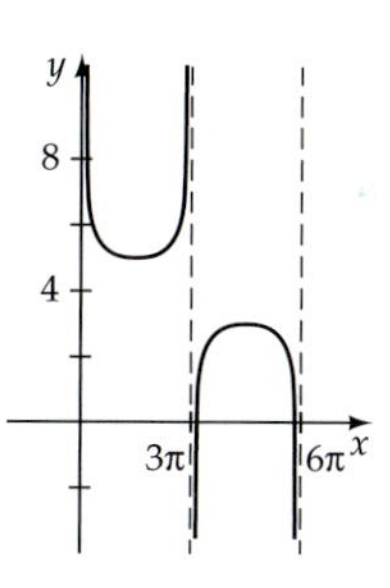

52. a. Phase shift $= -\dfrac{c}{b} = -\dfrac{\left(-\dfrac{7}{12}\pi\right)}{\left(\dfrac{\pi}{6}\right)} = 3.5$ months,

period $= \dfrac{2\pi}{b} = \dfrac{2\pi}{\pi/6} = 12$ months

b. First graph $y_1 = 2.7\cos\left(\dfrac{\pi}{6}t\right)$. Because the phase shift is 3.5 months, shift the graph of y_1 3.5 units to the right to produce the graph of y_2. Now shift the graph of y_2 upward 4 units to produce the graph of S.

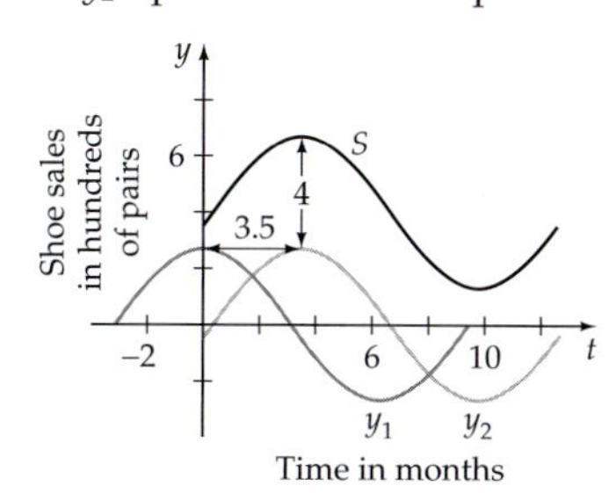

c. 3.5 months after January 1 is the middle of April.

54. $y = \dfrac{x}{2} + \cos x$

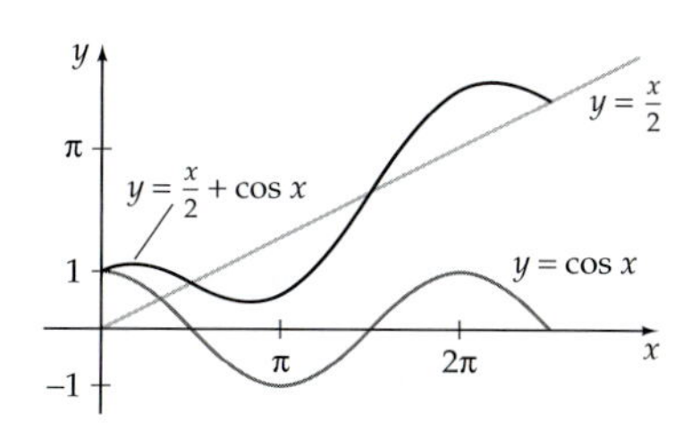

58. $y = -\sin x + \cos x$

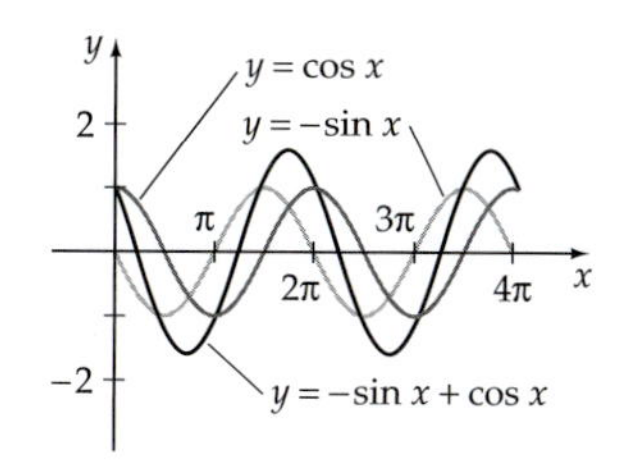

78. $y = x\cos x$

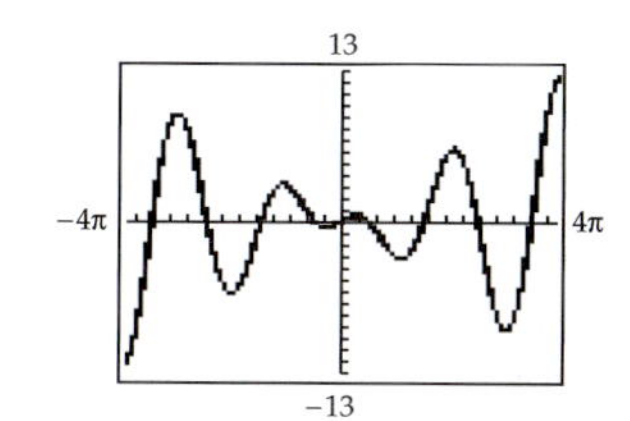

Exercise Set 5.1, page 371

6.

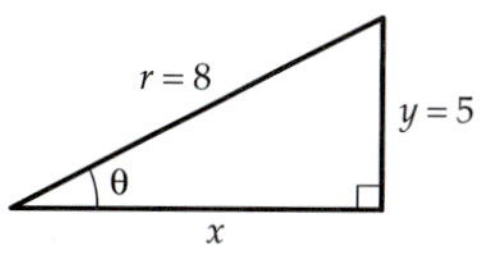

r = 8, y = 5, θ, x

$x = \sqrt{8^2 - 5^2}$

$x = \sqrt{64 - 25} = \sqrt{39}$

$\sin\theta = \dfrac{y}{r} = \dfrac{5}{8}$ $\qquad \csc\theta = \dfrac{r}{y} = \dfrac{8}{5}$

$\cos\theta = \dfrac{x}{r} = \dfrac{\sqrt{39}}{8}$ $\qquad \sec\theta = \dfrac{r}{x} = \dfrac{8}{\sqrt{39}} = \dfrac{8\sqrt{39}}{39}$

$\tan\theta = \dfrac{y}{x} = \dfrac{5}{\sqrt{39}} = \dfrac{5\sqrt{39}}{39}$ $\qquad \cot\theta = \dfrac{x}{y} = \dfrac{\sqrt{39}}{5}$

18. $\sin\dfrac{\pi}{3}\cos\dfrac{\pi}{4} - \tan\dfrac{\pi}{4} = \dfrac{\sqrt{3}}{2}\cdot\dfrac{\sqrt{2}}{2} - 1$

$= \dfrac{\sqrt{6}}{4} - 1 = \dfrac{\sqrt{6} - 4}{4}$

22.

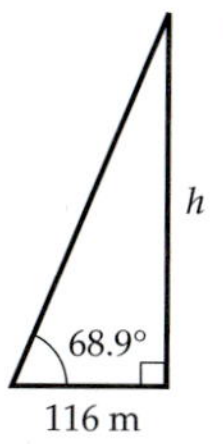

h, 68.9°, 116 m

$\tan 68.9° = \dfrac{h}{116}$

$h = 116\tan 68.9°$

$h \approx 301$ meters (three significant digits)

28.

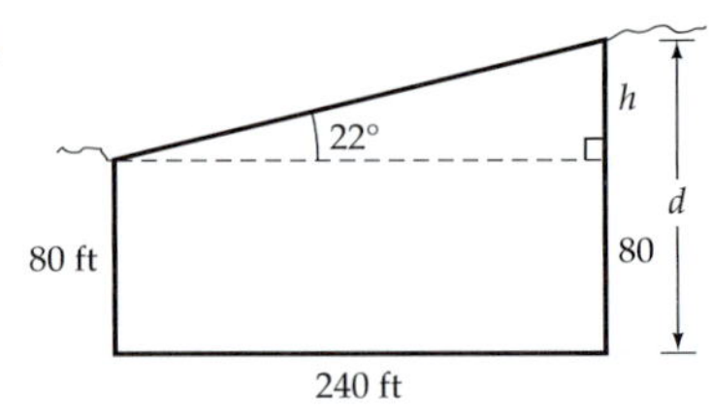

$$\tan 22° = \frac{h}{240}$$
$$h = 240 \tan 22°$$
$$d = 80 + h$$
$$d = 80 + 240 \tan 22°$$
$$d \approx 180 \text{ feet (two significant digits)}$$

36. $x = -6, y = -9, r = \sqrt{(-6)^2 + (-9)^2} = \sqrt{117} = 3\sqrt{13}$

$\sin\theta = \frac{y}{r} = \frac{-9}{3\sqrt{13}} = -\frac{3}{\sqrt{13}} = -\frac{3\sqrt{13}}{13}$ $\qquad \csc\theta = -\frac{\sqrt{13}}{3}$

$\cos\theta = \frac{x}{r} = \frac{-6}{3\sqrt{13}} = -\frac{2}{\sqrt{13}} = -\frac{2\sqrt{13}}{13}$ $\qquad \sec\theta = -\frac{\sqrt{13}}{2}$

$\tan\theta = \frac{y}{x} = \frac{-9}{-6} = \frac{3}{2}$ $\qquad \cot\theta = \frac{2}{3}$

44. $\theta' = 255° - 180° = 75°$

52. $\cos 300° > 0, \theta' = 360° - 300° = 60°$

Thus $\cos 300° = \cos 60° = \frac{1}{2}$.

Exercise Set 5.2, page 384

12. $\sin^4 x - \cos^4 x = (\sin^2 x + \cos^2 x)(\sin^2 x - \cos^2 x)$
$= 1(\sin^2 x - \cos^2 x) = \sin^2 x - \cos^2 x$

22.
$$\frac{2\sin x\cot x + \sin x - 4\cot x - 2}{2\cot x + 1} = \frac{(\sin x)(2\cot x + 1) - 2(2\cot x + 1)}{2\cot x + 1} = \frac{(2\cot x + 1)(\sin x - 2)}{2\cot x + 1} = \sin x - 2$$

32.
$$\frac{\frac{1}{\sin x} + \frac{1}{\cos x}}{\frac{1}{\sin x} - \frac{1}{\cos x}} = \frac{\frac{1}{\sin x} + \frac{1}{\cos x}}{\frac{1}{\sin x} - \frac{1}{\cos x}} \cdot \frac{\sin x\cos x}{\sin x\cos x}$$
$$= \frac{\cos x + \sin x}{\cos x - \sin x}$$
$$= \frac{\cos x + \sin x}{\cos x - \sin x} \cdot \frac{\cos x - \sin x}{\cos x - \sin x}$$
$$= \frac{\cos^2 x - \sin^2 x}{\cos^2 x - 2\sin x\cos x + \sin^2 x}$$
$$= \frac{\cos^2 x - \sin^2 x}{1 - 2\sin x\cos x}$$

40.
$$\frac{\frac{1}{\tan x} + \cot x}{\frac{1}{\tan x} + \tan x} = \frac{\frac{1}{\tan x} + \cot x}{\frac{1}{\tan x} + \tan x} \cdot \frac{\tan x}{\tan x}$$
$$= \frac{1 + 1}{1 + \tan^2 x} = \frac{2}{\sec^2 x}$$

46. Use the identity $\cos(\alpha - \beta) = \cos\alpha\cos\beta + \sin\alpha\sin\beta$ with $\alpha = 120°$ and $\beta = 45°$.

$$\cos(120° - 45°) = \cos 120° \cos 45° + \sin 120° \sin 45°$$
$$= \left(-\frac{1}{2}\right)\left(\frac{\sqrt{2}}{2}\right) + \left(\frac{\sqrt{3}}{2}\right)\left(\frac{\sqrt{2}}{2}\right)$$
$$= -\frac{\sqrt{2}}{4} + \frac{\sqrt{6}}{4}$$
$$= \frac{\sqrt{6} - \sqrt{2}}{4}$$

56. The value of a given trigonometric function of θ, measured in degrees, is equal to its cofunction of $90° - \theta$. Thus

$$\cos 80° = \sin(90° - 80°)$$
$$= \sin 10°$$

68. $\tan\alpha = \frac{24}{7}$, with $0° < \alpha < 90°$, $\sin\alpha = \frac{24}{25}$, $\cos\alpha = \frac{7}{25}$

$\sin\beta = -\frac{8}{17}$, with $180° < \beta < 270°$

$\cos\beta = -\frac{15}{17}$, $\tan\beta = \frac{8}{15}$

$$\cos(\alpha + \beta) = \cos\alpha\cos\beta - \sin\alpha\sin\beta$$
$$= \left(\frac{7}{25}\right)\left(-\frac{15}{17}\right) - \left(\frac{24}{25}\right)\left(-\frac{8}{17}\right)$$
$$= -\frac{105}{425} + \frac{192}{425} = \frac{87}{425}$$

84. $\cos 5x\cos 3x + \sin 5x\sin 3x = \cos(5x - 3x) = \cos 2x$
$= \cos(x + x) = \cos x\cos x - \sin x\sin x$
$= \cos^2 x - \sin^2 x$

Exercise Set 5.3, page 398

18. $\cos\theta = \frac{24}{25}$ with $270° < \theta < 360°$

$$\sin\theta = -\sqrt{1 - \left(\frac{24}{25}\right)^2} = -\frac{7}{25} \qquad \tan\theta = \frac{-7/25}{24/25} = -\frac{7}{24}$$

$$\sin 2\theta = 2\sin\theta\cos\theta = 2\left(-\frac{7}{25}\right)\left(\frac{24}{25}\right) = -\frac{336}{625}$$

$$\cos 2\theta = \cos^2\theta - \sin^2\theta = \left(\frac{24}{25}\right)^2 - \left(-\frac{7}{25}\right)^2 = \frac{527}{625}$$

$$\tan 2\theta = \frac{2\tan\theta}{1-\tan^2\theta} = \frac{2\left(-\frac{7}{24}\right)}{1-\left(-\frac{7}{24}\right)^2} = \frac{-\frac{7}{12}}{1-\frac{49}{576}}\cdot\frac{576}{576} = -\frac{336}{527}$$

32. $$\frac{1}{1-\cos 2x} = \frac{1}{1-1+2\sin^2 x} = \frac{1}{2\sin^2 x} = \frac{1}{2}\csc^2 x$$

40. $$\cos^2\frac{x}{2} = \left[\pm\sqrt{\frac{1+\cos x}{2}}\right]^2 = \frac{1+\cos x}{2} = \frac{1+\cos x}{2}\cdot\frac{\sec x}{\sec x} = \frac{\sec x + 1}{2\sec x}$$

44. $$\tan^2\frac{x}{2} = \left(\frac{1-\cos x}{\sin x}\right)^2 = \frac{(1-\cos x)^2}{\sin^2 x} = \frac{(1-\cos x)^2}{1-\cos^2 x} = \frac{(1-\cos x)^2}{(1-\cos x)(1+\cos x)} = \frac{1-\cos x}{1+\cos x} = \frac{\frac{1}{\cos x}-\frac{\cos x}{\cos x}}{\frac{1}{\cos x}+\frac{\cos x}{\cos x}} = \frac{\sec x - 1}{\sec x + 1}$$

64. $$\cos 3\theta + \cos 5\theta = 2\cos\frac{3\theta+5\theta}{2}\cos\frac{3\theta-5\theta}{2} = 2\cos 4\theta\cos(-\theta) = 2\cos 4\theta\cos\theta$$

74. $$\sin 5x\cos 3x = \frac{1}{2}[\sin(5x+3x)+\sin(5x-3x)] = \frac{1}{2}(\sin 8x + \sin 2x) = \frac{1}{2}(2\sin 4x\cos 4x + 2\sin x\cos x) = \sin 4x\cos 4x + \sin x\cos x$$

76. $$\frac{\cos 5x - \cos 3x}{\sin 5x + \sin 3x} = \frac{-2\sin\frac{5x+3x}{2}\sin\frac{5x-3x}{2}}{2\sin\frac{5x+3x}{2}\cos\frac{5x-3x}{2}} = -\frac{\sin 4x\sin x}{\sin 4x\cos x} = -\tan x$$

86. $a = 1, b = \sqrt{3}, k = \sqrt{(\sqrt{3})^2 + (1)^2} = 2$. Thus α is a first-quadrant angle.

$$\sin\alpha = \frac{\sqrt{3}}{2} \quad\text{and}\quad \cos\alpha = \frac{1}{2}$$

Thus $\alpha = \frac{\pi}{3}$.

$$y = k\sin(x+\alpha)$$
$$y = 2\sin\left(x+\frac{\pi}{3}\right)$$

90. From Exercise 86, we know that

$$y = \sin x + \sqrt{3}\cos x = 2\sin\left(x+\frac{\pi}{3}\right)$$

Thus the graph of $y = \sin x + \sqrt{3}\cos x$ is the graph of $y = 2\sin x$ shifted $\frac{\pi}{3}$ units to the left.

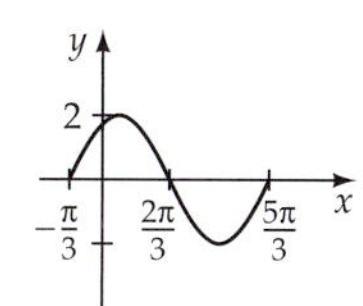

Exercise Set 5.4, page 411

2. $y = \sin^{-1}\frac{\sqrt{2}}{2}$ implies

$$\sin y = \frac{\sqrt{2}}{2} \quad\text{for } -\frac{\pi}{2} \le y \le \frac{\pi}{2}$$

Thus $y = \frac{\pi}{4}$.

20. Because $\tan(\tan^{-1}x) = x$ for all real numbers x, we have $\tan\left[\tan^{-1}\left(\frac{1}{2}\right)\right] = \frac{1}{2}$.

40. Let $x = \cos^{-1}\frac{3}{5}$. Thus

$$\cos x = \frac{3}{5} \quad\text{and}\quad \sin x = \sqrt{1-\left(\frac{3}{5}\right)^2} = \frac{4}{5}$$

$$y = \tan\left(\cos^{-1}\frac{3}{5}\right) = \tan x = \frac{\sin x}{\cos x} = \frac{4/5}{3/5} = \frac{4}{3}$$

46. $y = \cos\left(\sin^{-1}\frac{3}{4} + \cos^{-1}\frac{5}{13}\right)$

Let $\alpha = \sin^{-1}\frac{3}{4}$, $\sin\alpha = \frac{3}{4}$, $\cos\alpha = \sqrt{1 - \left(\frac{3}{4}\right)^2} = \frac{\sqrt{7}}{4}$.

$\beta = \cos^{-1}\frac{5}{13}$, $\cos\beta = \frac{5}{13}$, $\sin\beta = \sqrt{1 - \left(\frac{5}{13}\right)^2} = \frac{12}{13}$.

$$\begin{aligned} y &= \cos(\alpha + \beta) \\ &= \cos\alpha\cos\beta - \sin\alpha\sin\beta \\ &= \frac{\sqrt{7}}{4}\cdot\frac{5}{13} - \frac{3}{4}\cdot\frac{12}{13} = \frac{5\sqrt{7}}{52} - \frac{36}{52} = \frac{5\sqrt{7} - 36}{52} \end{aligned}$$

56. $\sin^{-1}x + \cos^{-1}\frac{4}{5} = \frac{\pi}{6}$

$$\begin{aligned} \sin^{-1}x &= \frac{\pi}{6} - \cos^{-1}\frac{4}{5} \\ \sin(\sin^{-1}x) &= \sin\left(\frac{\pi}{6} - \cos^{-1}\frac{4}{5}\right) \\ x &= \sin\frac{\pi}{6}\cos\left(\cos^{-1}\frac{4}{5}\right) - \cos\frac{\pi}{6}\sin\left(\cos^{-1}\frac{4}{5}\right) \\ &= \frac{1}{2}\cdot\frac{4}{5} - \frac{\sqrt{3}}{2}\cdot\frac{3}{5} = \frac{4 - 3\sqrt{3}}{10} \end{aligned}$$

64. Let $\alpha = \cos^{-1}x$ and $\beta = \cos^{-1}(-x)$. Thus $\cos\alpha = x$ and $\cos\beta = -x$. We know $\sin\alpha = \sqrt{1 - x^2}$ and $\sin\beta = \sqrt{1 - x^2}$ because α is in Quadrant I and β is in Quadrant II.

$$\begin{aligned} &\cos^{-1}x + \cos^{-1}(-x) \\ &= \alpha + \beta \\ &= \cos^{-1}[\cos(\alpha + \beta)] \\ &= \cos^{-1}(\cos\alpha\cos\beta - \sin\alpha\sin\beta) \\ &= \cos^{-1}\left[x(-x) - \sqrt{1 - x^2}\cdot\sqrt{1 - x^2}\right] \\ &= \cos^{-1}(-x^2 - 1 + x^2) \\ &= \cos^{-1}(-1) = \pi \end{aligned}$$

68. Recall that the graph of $y = f(x - a)$ is a horizontal shift of the graph of $y = f(x)$. Therefore, the graph of $y = \cos^{-1}(x - 1)$ is the graph of $y = \cos^{-1}x$ shifted 1 unit to the right.

Exercise Set 5.5, page 421

14.

$$\begin{aligned} 2\cos^2 x + 1 &= -3\cos x \\ 2\cos^2 x + 3\cos x + 1 &= 0 \\ (2\cos x + 1)(\cos x + 1) &= 0 \end{aligned}$$

$2\cos x + 1 = 0$ or $\cos x + 1 = 0$

$\cos x = -\frac{1}{2}$ $\qquad$ $\cos x = -1$

$x = \frac{2\pi}{3}, \frac{4\pi}{3}$ $\qquad$ $x = \pi$

The solutions in the interval $0 \le x < 2\pi$ are $\frac{2\pi}{3}$, π, and $\frac{4\pi}{3}$.

46. $\sin x + 2\cos x = 1$

$$\begin{aligned} \sin x &= 1 - 2\cos x \\ (\sin x)^2 &= (1 - 2\cos x)^2 \\ \sin^2 x &= 1 - 4\cos x + 4\cos^2 x \\ 1 - \cos^2 x &= 1 - 4\cos x + 4\cos^2 x \\ 0 &= \cos x(5\cos x - 4) \end{aligned}$$

$\cos x = 0$ or $5\cos x - 4 = 0$

$x = 90°, 270°$ $\qquad$ $\cos x = \frac{4}{5}$

$x \approx 36.9°, 323.1°$

The solutions in the interval $0 \le x < 360°$ are 90° and 323.1°. (*Note:* $x = 270°$ and $x = 36.9°$ are extraneous solutions. Neither of these values satisfies the original equation.)

50. $2\cos^2 x - 5\cos x - 5 = 0$

$$\cos x = \frac{5 \pm \sqrt{(-5)^2 - 4(2)(-5)}}{2(2)} = \frac{5 \pm \sqrt{65}}{4}$$

$\cos x \approx 3.27$ or $\cos x \approx -0.7656$

no solution $\qquad$ $x \approx 140.0°, 220.0°$

The solutions in the interval $0° \le x < 360°$ are 140.0° and 220.0°.

60. $\cos 2x = -\frac{\sqrt{3}}{2}$

$2x = \frac{5\pi}{6} + 2k\pi$ or $2x = \frac{7\pi}{6} + 2k\pi$, k an integer

$x = \frac{5\pi}{12} + k\pi$ or $x = \frac{7\pi}{12} + k\pi$, k an integer

76. $2\sin x\cos x - 2\sqrt{2}\sin x - \sqrt{3}\cos x + \sqrt{6} = 0$

$$\begin{aligned} 2\sin x\left(\cos x - \sqrt{2}\right) - \sqrt{3}\left(\cos x - \sqrt{2}\right) &= 0 \\ \left(\cos x - \sqrt{2}\right)\left(2\sin x - \sqrt{3}\right) &= 0 \end{aligned}$$

$\cos x = \sqrt{2}$ or $\sin x = \frac{\sqrt{3}}{2}$

no solution $\qquad$ $x = \frac{\pi}{3}, \frac{2\pi}{3}$

The solutions in the interval $0 \le x < 2\pi$ are $\frac{\pi}{3}$ and $\frac{2\pi}{3}$.

78. The following graph shows that the solutions in the interval $[0, 2\pi)$ are $x = 0$ and $x = 1.895$.

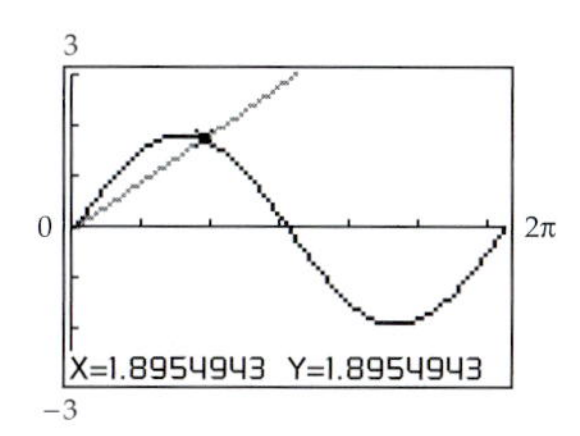

82. When $\theta = 45°$, d attains its maximum of 4394.5 feet.

Exercise Set 5.6, page 433

4.

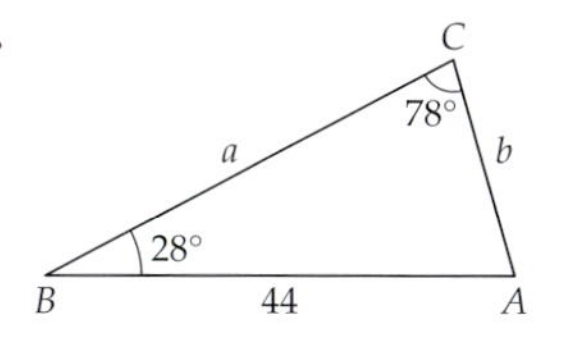

$$A = 180° - 78° - 28° = 74°$$

$$\frac{b}{\sin B} = \frac{c}{\sin C} \qquad \frac{a}{\sin A} = \frac{c}{\sin C}$$

$$\frac{b}{\sin 28°} = \frac{44}{\sin 78°} \qquad \frac{a}{\sin 74°} = \frac{44}{\sin 78°}$$

$$b = \frac{44 \sin 28°}{\sin 78°} \approx 21 \qquad a = \frac{44 \sin 74°}{\sin 78°} \approx 43$$

12.
$$\frac{a}{\sin A} = \frac{b}{\sin B}$$

$$\frac{13.8}{\sin A} = \frac{5.55}{\sin 22.6}$$

$$\sin A = 0.9555$$

$$A \approx 72.9° \quad \text{or} \quad 107.1°$$

If $A = 72.9°$, $C \approx 180° - 72.9° - 22.6° = 84.5°$.

$$\frac{c}{\sin 84.5°} = \frac{5.55}{\sin 22.6°}$$

$$c = \frac{5.55 \sin 84.5°}{\sin 22.6°} \approx 14.4$$

If $A = 107.1°$, $C \approx 180° - 107.1° - 22.6° = 50.3°$.

$$\frac{c}{\sin 50.3°} = \frac{5.55}{\sin 22.6°}$$

$$c = \frac{5.55 \sin 50.3°}{\sin 22.6°} \approx 11.1$$

Case 1: $A = 72.9°$, $C = 84.5°$, and $c = 14.4$

Case 2: $A = 107.1°$, $C = 50.3°$, and $c = 11.1$

24. $c^2 = a^2 + b^2 - 2ab \cos C$

$$c^2 = 14.2^2 + 9.30^2 - 2(14.2)(9.30) \cos 9.20°$$

$$c = \sqrt{14.2^2 + 9.30^2 - 2(14.2)(9.30) \cos 9.20°}$$

$$c \approx 5.24$$

28.
$$\cos A = \frac{b^2 + c^2 - a^2}{2bc}$$

$$\cos A = \frac{132^2 + 160^2 - 108^2}{2(132)(160)} \approx 0.7424$$

$$A \approx \cos^{-1}(0.7424) \approx 42.1°$$

32. The angle with its vertex at the position of the helicopter measures $180° - (59.0° + 77.2°) = 43.8°$. Let the distance from the helicopter to the carrier be x. Using the Law of Sines, we have

$$\frac{x}{\sin 77.2°} = \frac{7620}{\sin 43.8°}$$

$$x = \frac{7620 \sin 77.2°}{\sin 43.8°}$$

$$\approx 10{,}700 \text{ feet} \qquad \text{(three significant digits)}$$

38.

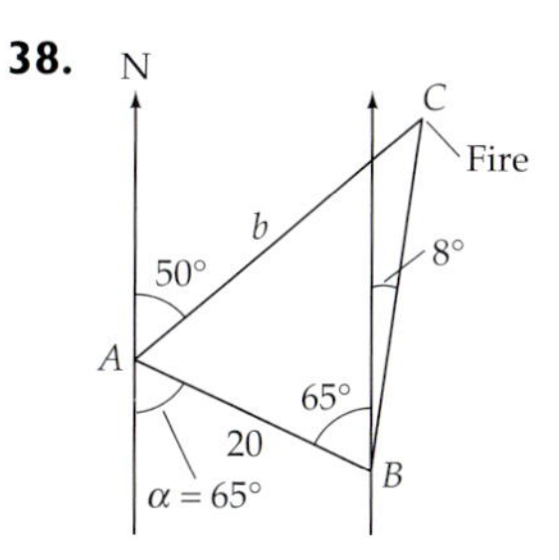

$$\alpha = 65°$$

$$B = 65° + 8° = 73°$$

$$A = 180° - 50° - 65° = 65°$$

$$C = 180° - 65° - 73° = 42°$$

$$\frac{b}{\sin B} = \frac{c}{\sin C}$$

$$\frac{b}{\sin 73°} = \frac{20}{\sin 42°}$$

$$b = \frac{20 \sin 73°}{\sin 42°}$$

$$b \approx 29 \text{ miles}$$

52.

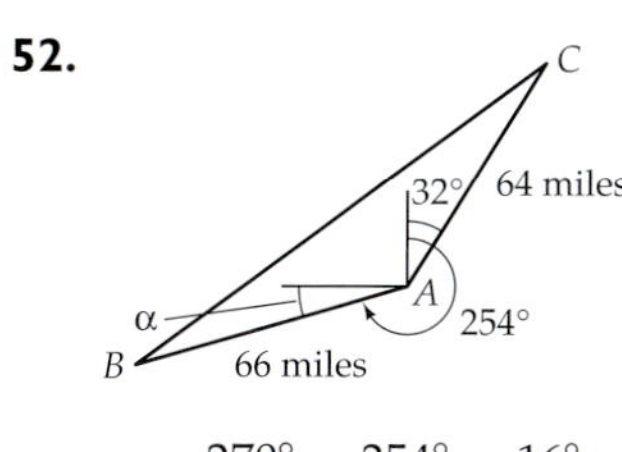

$$\alpha = 270° - 254° = 16°$$

$$A = 16° + 90° + 32° = 138°$$

$$b = 4 \cdot 16 = 64 \text{ miles}$$

$$c = 3 \cdot 22 = 66 \text{ miles}$$

(continued)

$$a^2 = b^2 + c^2 - 2bc\cos A$$
$$a^2 = 64^2 + 66^2 - 2(64)(66)\cos 138°$$
$$a = \sqrt{64^2 + 66^2 - 2(64)(66)\cos 138°}$$
$$a \approx 120 \text{ miles}$$

Exercise Set 5.7, page 450

6. $a = 3 - 3 = 0$

$b = 0 - (-2) = 2$

A vector equivalent to P_1P_2 is $\mathbf{v} = \langle 0, 2\rangle$.

8. $\|\mathbf{v}\| = \sqrt{6^2 + 10^2}$

$= \sqrt{36 + 100} = \sqrt{136} = 2\sqrt{34}$

$$\theta = \tan^{-1}\frac{10}{6} = \tan^{-1}\frac{5}{3} \approx 59.0°$$

Thus $\mathbf{v}$ has a direction of about 59° as measured from the positive x-axis.

A unit vector in the direction of $\mathbf{v}$ is

$$\mathbf{u} = \left\langle \frac{6}{2\sqrt{34}}, \frac{10}{2\sqrt{34}} \right\rangle = \left\langle \frac{3\sqrt{34}}{34}, \frac{5\sqrt{34}}{34} \right\rangle$$

20. $\frac{3}{4}\mathbf{u} - 2\mathbf{v} = \frac{3}{4}\langle -2, 4\rangle - 2\langle -3, -2\rangle$

$= \left\langle -\frac{3}{2}, 3 \right\rangle - \langle -6, -4\rangle$

$= \left\langle \frac{9}{2}, 7 \right\rangle$

26. $3\mathbf{u} + 2\mathbf{v} = 3(3\mathbf{i} - 2\mathbf{j}) + 2(-2\mathbf{i} + 3\mathbf{j})$

$= (9\mathbf{i} - 6\mathbf{j}) + (-4\mathbf{i} + 6\mathbf{j})$

$= (9 - 4)\mathbf{i} + (-6 + 6)\mathbf{j}$

$= 5\mathbf{i} + 0\mathbf{j}$

$= 5\mathbf{i}$

36. $a_1 = 2\cos\frac{8\pi}{7} \approx -1.8$

$a_2 = 2\sin\frac{8\pi}{7} \approx -0.9$

$\mathbf{v} = a_1\mathbf{i} + a_2\mathbf{j} \approx -1.8\mathbf{i} - 0.9\mathbf{j}$

40.

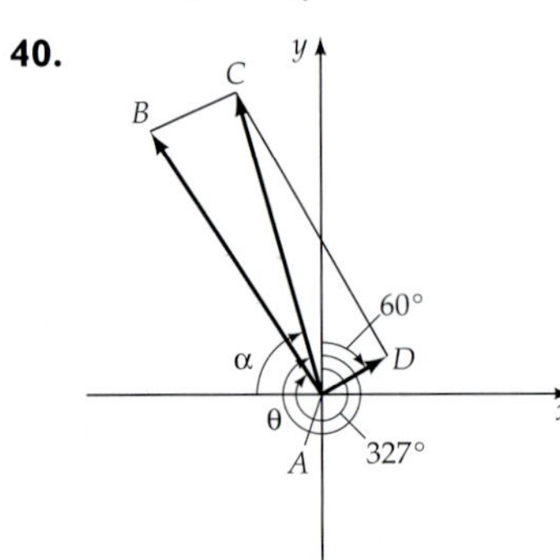

$\mathbf{AB} = 18\cos 123°\mathbf{i} + 18\sin 123°\mathbf{j} \approx -9.8\mathbf{i} + 15.1\mathbf{j}$

$\mathbf{AD} = 4\cos 30°\mathbf{i} + 4\sin 30°\mathbf{j} \approx 3.5\mathbf{i} + 2\mathbf{j}$

$\mathbf{AC} = \mathbf{AB} + \mathbf{AD} \approx -9.8\mathbf{i} + 15.1\mathbf{j} + 3.5\mathbf{i} + 2\mathbf{j}$

$= -6.3\mathbf{i} + 17.1\mathbf{j}$

$\|\mathbf{AC}\| = \sqrt{(-6.3)^2 + (17.1)^2} \approx 18$

$$\alpha = \tan^{-1}\left|\frac{17.1}{-6.3}\right| = \tan^{-1}\frac{17.1}{6.3} \approx 70°$$

$\theta \approx 270° + 70° = 340°$

The course of the boat is about 18 miles per hour at an approximate heading of 340°.

42.

$\alpha = \theta$

$$\sin\alpha = \frac{120}{800}$$

$\alpha \approx 8.6°$

50. $\mathbf{v}\cdot\mathbf{w} = (5\mathbf{i} + 3\mathbf{j})\cdot(4\mathbf{i} - 2\mathbf{j})$

$= 5(4) + 3(-2) = 20 - 6 = 14$

60. $\cos\theta = \dfrac{\mathbf{v}\cdot\mathbf{w}}{\|\mathbf{v}\|\,\|\mathbf{w}\|}$

$$\cos\theta = \frac{(3\mathbf{i} - 4\mathbf{j})\cdot(6\mathbf{i} - 12\mathbf{j})}{\sqrt{3^2 + (-4)^2}\sqrt{6^2 + (-12)^2}}$$

$$\cos\theta = \frac{3(6) + (-4)(-12)}{\sqrt{25}\sqrt{180}}$$

$$\cos\theta = \frac{66}{5\sqrt{180}} \approx 0.9839$$

$\theta \approx 10.3°$

62. $\text{proj}_{\mathbf{w}}\mathbf{v} = \dfrac{\mathbf{v}\cdot\mathbf{w}}{\|\mathbf{w}\|}$

$$\text{proj}_{\mathbf{w}}\mathbf{v} = \frac{\langle -7, 5\rangle\cdot\langle -4, 1\rangle}{\sqrt{(-4)^2 + 1^2}} = \frac{33}{\sqrt{17}} = \frac{33\sqrt{17}}{17} \approx 8.0$$

70. $W = \|\mathbf{F}\|\,\|\mathbf{s}\|\cos\alpha$

$W = 100 \cdot 25 \cdot \cos 42°$

$W \approx 1858$ foot-pounds

Exercise Set 6.1, page 480

2. $y^2 = 4px = \frac{1}{3}x$

$$p = \frac{1}{12}$$

Vertex: $(0, 0)$

Focus: $\left(\dfrac{1}{12}, 0\right)$

Directrix: $x = -\dfrac{1}{12}$

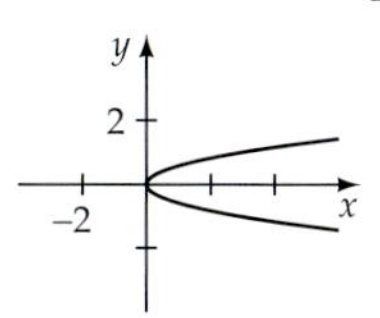

4. $\dfrac{x^2}{9} + \dfrac{y^2}{4} = \dfrac{x^2}{3^2} + \dfrac{y^2}{2^2} = 1$

Center: $(0, 0)$
Vertices: $(-3, 0), (3, 0)$

$c^2 = 3^2 - 2^2 = 9 - 4 = 5$
$c = \pm\sqrt{5}$

Foci: $(\sqrt{5}, 0), (-\sqrt{5}, 0)$

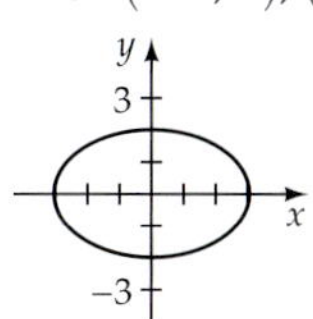

6. $\dfrac{y^2}{4} - \dfrac{x^2}{25} = \dfrac{y^2}{2^2} - \dfrac{x^2}{5^2} = 1$

Center: $(0, 0)$
Vertices: $(0, -2), (0, 2)$

$c^2 = 2^2 + 5^2 = 4 + 25 = 29$
$c = \pm\sqrt{29}$

Foci: $(0, \pm\sqrt{29})$

Asymptotes: $y = \pm\dfrac{2}{5}x$

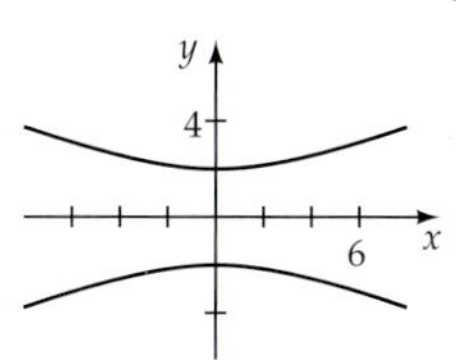

18. $3x + 2y^2 - 4y - 7 = 0$

$$\begin{aligned}
3x - 7 &= -2y^2 + 4y = -2(y^2 - 2y) \\
3x - 7 - 2 &= -2(y^2 - 2y + 1) \\
3(x - 3) &= -2(y^2 - 2y + 1) \\
-\frac{3}{2}(x - 3) &= (y - 1)^2 \\
4p(x - 3) &= (y - 1)^2 \\
4p &= -\frac{3}{2} \\
p &= -\frac{3}{8}
\end{aligned}$$

Vertex: $(3, 1)$

Focus: $\left(3 - \dfrac{3}{8}, 1\right) = \left(\dfrac{21}{8}, 1\right)$

Directrix: $x = 3 + \dfrac{3}{8} \Rightarrow x = \dfrac{27}{8}$

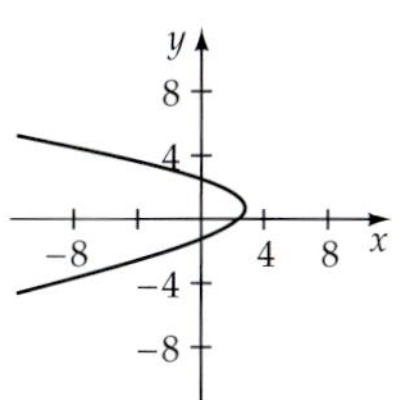

20.

$$\begin{aligned}
11x^2 + 25y^2 - 44x + 50y - 206 &= 0 \\
11x^2 - 44x + 25y^2 + 50y &= 206 \\
11(x^2 - 4x) + 25(y + 1) &= 206 \\
11(x^2 - 4x + 4) + 25(y^2 + 2y + 1) &= 206 + 44 + 25 \\
11(x - 2)^2 + 25(y + 1)^2 &= 275 \\
\frac{(x - 2)^2}{25} + \frac{(y + 1)^2}{11} &= 1
\end{aligned}$$

Center: $(2, -1)$
Vertices: $(7, -1), (-3, -1)$
Foci: $(2 + \sqrt{14}, -1), (2 - \sqrt{14}, -1)$

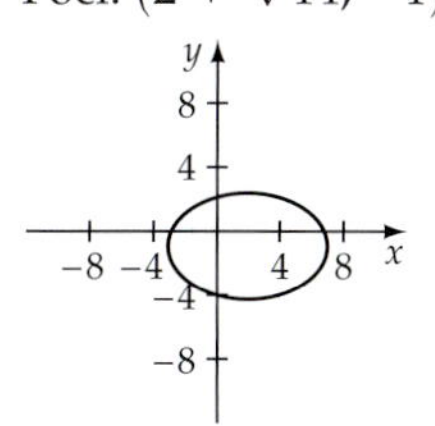

22.

$$\begin{aligned}
9x^2 - y^2 - 36x + 6y - 9 &= 0 \\
9x^2 - 36x - y^2 + 6y &= 9 \\
9(x^2 - 4x + 4) - (y^2 - 6y + 9) &= 9 + 36 - 9 \\
9(x - 2)^2 - (y - 3)^2 &= 36 \\
\frac{9(x - 2)^2}{36} - \frac{(y - 3)^2}{36} &= \frac{36}{36} \\
\frac{(x - 2)^2}{4} - \frac{(y - 3)^2}{36} &= 1 \\
\frac{(x - 2)^2}{2^2} - \frac{(y - 3)^2}{6^2} &= 1
\end{aligned}$$

Center: $(2, 3)$
Vertices: $(2 \pm 2, 3) = (0, 3), (4, 3)$
$c^2 = 2^2 + 6^2 = 4 + 36 = 40$
$c = \pm\sqrt{40} = \pm 2\sqrt{10}$
Foci: $(2 \pm 2\sqrt{10}, 3)$
Asymptotes: $y - 3 = \pm 3(x - 2)$

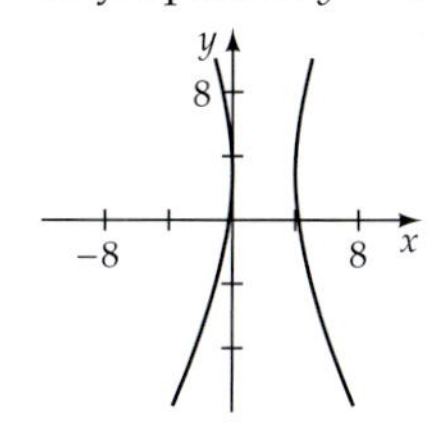

28. Focus is $(2, -3) = (h + p, k)$.
Directrix is $x = 6 = h - p$.

$$\begin{aligned} h + p &= 2 \\ h - p &= 6 \\ \hline 2h &= 8 \end{aligned}$$

$h = 4$; $p = -2$; vertex is $(h, k) = (4, -3)$.

$(y + 3)^2 = -8(x - 4)$

30. Center $(2, 4)$, major axis of length $10 \Rightarrow a = 5$, parallel to the y-axis, and passing through $(3, 3)$

$$\frac{(x-2)^2}{b^2} + \frac{(y-4)^2}{25} = 1$$

$$\frac{(3-2)^2}{b^2} + \frac{(3-4)^2}{25} = 1 \quad \text{• } x = 3, y = 3, a = 5$$

$$\frac{1}{b^2} + \frac{1}{25} = 1 \quad \text{• Solve for } b^2.$$

$$\frac{1}{b^2} = \frac{24}{25}$$

$$b^2 = \frac{25}{24}$$

$$\frac{(x-2)^2}{25/24} + \frac{(y-4)^2}{25} = 1$$

42. $4p(3.75375) = 100^2$
$15.015p = 10{,}000$
$p \approx 666$ inches above the vertex

44. a = mean distance = 139.4 million miles
Aphelion = $2(139.4) - 126.4 = 152.4$ million miles
Thus, $c = 152.4 - 139.4 = 13$ million miles.
$b = \sqrt{a^2 - c^2} = \sqrt{139.4^2 - 13^2} = \sqrt{19263.36} \approx 138.8$

An equation of Mars' orbit is $\dfrac{x^2}{139.4^2} + \dfrac{y^2}{138.8^2} = 1$.

46. $a = 125, c = 80$

$$\begin{aligned} b &= \sqrt{a^2 - c^2} \\ &= \sqrt{125^2 - 80^2} \\ &= \sqrt{9225} \\ &\approx 96.0 \text{ feet} \end{aligned}$$

Exercise Set 6.2, page 497

12.

4

14. $r = 5 \cos 3\theta$

Because 3 is odd, this is a rose with three petals.

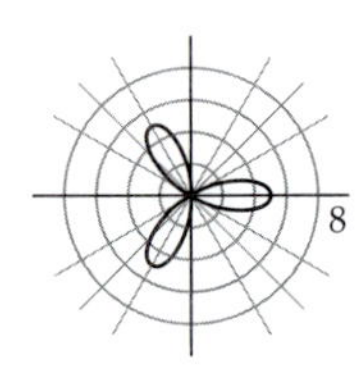

16. Because $|a| = |b| = 2$, the graph of $r = 2 - 2 \csc \theta$, $-\pi \le \theta \le \pi$, is a cardioid.

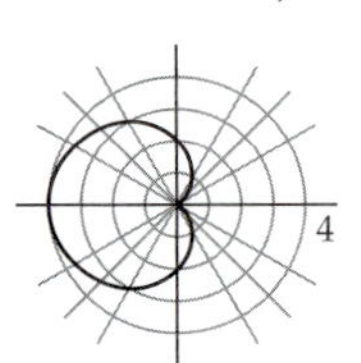

20. $r = \dfrac{8}{2 - 4\cos\theta}$ • Divide numerator and denominator by 2.

$$r = \frac{4}{1 - 2\cos\theta}$$

$e = 2$, so the graph is a hyperbola. The transverse axis is on the polar axis because the equation involves $\cos\theta$. Let $\theta = 0$.

$$r = \frac{8}{2 - 4\cos 0} = \frac{8}{2 - 4} = -4$$

Let $\theta = \pi$.

$$r = \frac{8}{2 - 4\cos\pi} = \frac{8}{2 + 4} = \frac{4}{3}$$

The vertices are $(-4, 0)$ and $\left(\frac{4}{3}, \pi\right)$.

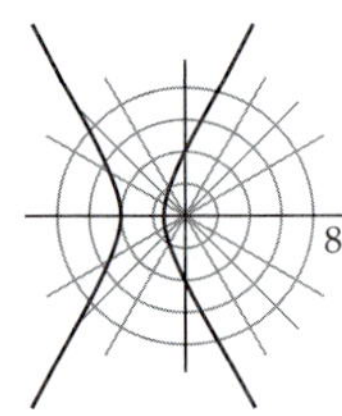

26. $r = 4 - 4\sin\theta$

θ min = 0
θ max = 2π
θ step = 0.1

3.25
−9.4
9.4
−9.75

40. $x = r\cos\theta$

$$= (2)\left[\cos\left(-\frac{\pi}{3}\right)\right] = (2)\left(\frac{1}{2}\right) = 1$$

$y = r\sin\theta$

$$= (2)\left[\sin\left(-\frac{\pi}{3}\right)\right] = (2)\left(-\frac{\sqrt{3}}{2}\right) = -\sqrt{3}$$

The rectangular coordinates of the point are $(1, -\sqrt{3})$.

44. $r = \sqrt{x^2 + y^2}$ $\quad \alpha = \tan^{-1}\left|\dfrac{y}{x}\right|$ • *α is the reference angle for θ.*

$= \sqrt{(12)^2 + (-5)^2}$

$= \sqrt{144 + 25}$ $\quad = \tan^{-1}\left|\dfrac{-5}{12}\right| \approx 22.6°$

$= \sqrt{169}$

$= 13$

θ is a Quadrant IV angle. Thus $\theta \approx 360° - 22.6° = 337.4°$.

The approximate polar coordinates of the point are $(13, 337.4°)$.

52.

$$r = \cot\theta$$

$$r = \frac{\cos\theta}{\sin\theta} \quad \bullet\ \cot\theta = \frac{\cos\theta}{\sin\theta}$$

$$r\sin\theta = \cos\theta$$

$$r(r\sin\theta) = r\cos\theta \quad \bullet\ \text{Multiply both sides by } r.$$

$$\left(\sqrt{x^2 + y^2}\right)y = x \quad \bullet\ y = r\sin\theta;\ x = r\cos\theta$$

$$(x^2 + y^2)y^2 = x^2 \quad \bullet\ \text{Square each side.}$$

$$y^4 + x^2y^2 - x^2 = 0$$

60.

$$2x - 3y = 6$$

$$2r\cos\theta - 3r\sin\theta = 6 \quad \bullet\ x = r\cos\theta;\ y = r\sin\theta$$

$$r(2\cos\theta - 3\sin\theta) = 6$$

$$r = \frac{6}{2\cos\theta - 3\sin\theta}$$

Exercise Set 6.3, page 506

6. Plotting points for several values of t yields the following graph.

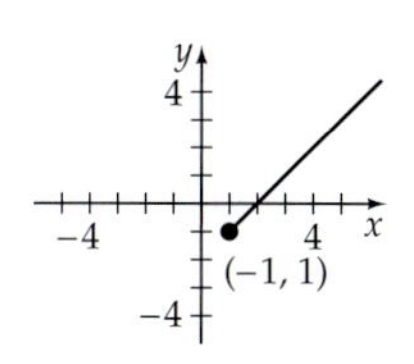

12. $x = 3 + 2\cos t,\ y = -1 - 3\sin t,\ 0 \le t < 2\pi$

$$\cos t = \frac{x - 3}{2},\ \sin t = -\frac{y + 1}{3}$$

$$\cos^2 t + \sin^2 t = 1$$

$$\left(\frac{x - 3}{2}\right)^2 + \left(-\frac{y + 1}{3}\right)^2 = 1$$

$$\frac{(x - 3)^2}{4} + \frac{(y + 1)^2}{9} = 1$$

14. $x = 1 + t^2,\ y = 2 - t^2,\ t \in R$

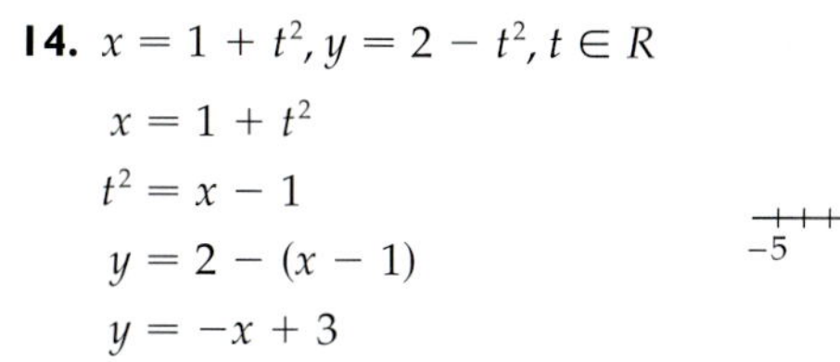

$$x = 1 + t^2$$

$$t^2 = x - 1$$

$$y = 2 - (x - 1)$$

$$y = -x + 3$$

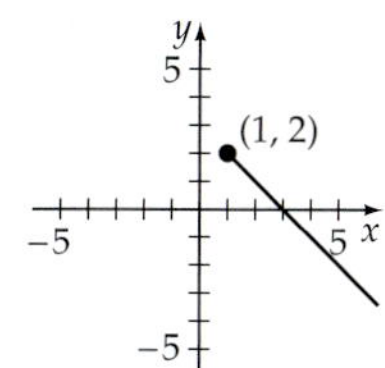

Because $x = 1 + t^2$ and $t^2 \ge 0$ for all real numbers t, $x \ge 1$ for all t. Similarly, $y \le 2$ for all t.

26. The maximum height of the cycloid is $2a = 2(3) = 6$. The cycloid intersects the x-axis every $2\pi a = 2\pi(3) = 6\pi$ units.

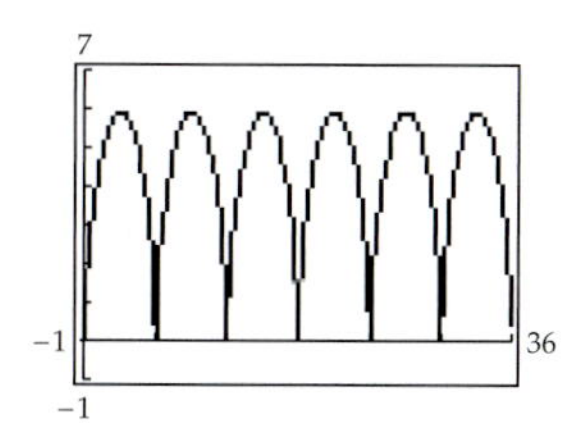

32.

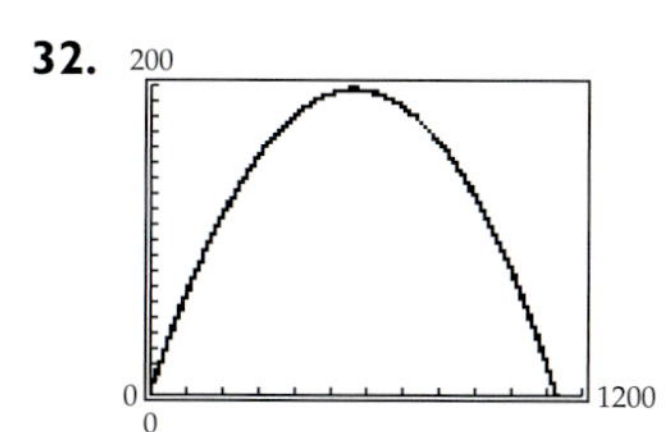

The maximum height is about 195 feet when $t \approx 3.50$ seconds. The range is 1117 feet when $t \approx 6.99$ seconds.

Exercise Set 6.4, page 522

6. $a_n = \dfrac{(-1)^{n+1}}{n(n + 1)},\ a_1 = \dfrac{(-1)^{1+1}}{1(1 + 1)} = \dfrac{1}{2},$

$a_2 = \dfrac{(-1)^{2+1}}{2(2 + 1)} = -\dfrac{1}{6},\ a_3 = \dfrac{(-1)^{3+1}}{3(3 + 1)} = \dfrac{1}{12},$

$a_8 = \dfrac{(-1)^{8+1}}{8(8 + 1)} = -\dfrac{1}{72}$

18. $a_1 = 1,\ a_2 = 2^2 \cdot a_1 = 4 \cdot 1 = 4,\ a_3 = 3^2 \cdot a_2 = 9 \cdot 4 = 36$

28. $\dfrac{12!}{4!\,8!} = \dfrac{12 \cdot 11 \cdot 10 \cdot 9 \cdot 8!}{4!\,8!} = \dfrac{12 \cdot 11 \cdot 10 \cdot 9}{4 \cdot 3 \cdot 2 \cdot 1} = 495$

42. $a_6 = -14,\ a_8 = -20$

$$a_8 = a_6 + 2d$$

$$\frac{a_8 - a_6}{2} = d \quad \bullet\ \text{Solve for } d.$$

$$\frac{-20 - (-14)}{2} = d$$

$$-3 = d$$

(continued)

$$a_n = a_1 + (n-1)d$$
$$a_6 = a_1 + (6-1)(-3)$$
$$-14 = a_1 + (-15)$$
$$a_1 = 1$$
$$a_{15} = 1 + (15-1)(-3) = 1 + (14)(-3) = -41$$

50. $\dfrac{a_2}{a_1} = \dfrac{6}{8} = \dfrac{3}{4} = r$

$$a_n = a_1 r^{n-1}$$
$$a_n = 8\left(\frac{3}{4}\right)^{n-1}$$

70. $\displaystyle\sum_{i=1}^{6}(2i+1)(2i-1) = \sum_{i=1}^{6}(4i^2-1)$

$$= (4\cdot 1^2 - 1) + (4\cdot 2^2 - 1) + (4\cdot 3^2 - 1) + (4\cdot 4^2 - 1) + (4\cdot 5^2 - 1) + (4\cdot 6^2 - 1)$$
$$= 3 + 15 + 35 + 63 + 99 + 143$$
$$= 358$$

74. $S_{20} = \dfrac{20}{2}(a_1 + a_{20})$

$$a_1 = 1 - 2(1) = -1$$
$$a_{20} = 1 - 2(20) = -39$$
$$S_{20} = 10[-1 + (-39)] = 10(-40) = -400$$

82. $r = \dfrac{4}{3},\ a_1 = \dfrac{4}{3},\ n = 14$

$$S_n = \frac{a_1(1-r^n)}{1-r}$$
$$S_{14} = \frac{\frac{4}{3}\left[1 - \left(\frac{4}{3}\right)^{14}\right]}{1-\frac{4}{3}} = \frac{\frac{4}{3}\left[\frac{-263{,}652{,}487}{4{,}782{,}969}\right]}{-\frac{1}{3}} \approx 220.49$$

98. $0.3\overline{95} = \dfrac{3}{10} + \dfrac{95}{1000} + \dfrac{95}{100{,}000} + \cdots = \dfrac{3}{10} + \dfrac{\frac{95}{1000}}{1 - \frac{1}{100}}$

$$= \frac{3}{10} + \frac{95}{990} = \frac{392}{990} = \frac{196}{495}$$

Exercise Set 6.5, page 529

4. $\dbinom{10}{5} = \dfrac{10!}{5!\,5!} = \dfrac{10\cdot 9\cdot 8\cdot 7\cdot 6\cdot 5!}{5!\,5!} = \dfrac{10\cdot 9\cdot 8\cdot 7\cdot 6}{5\cdot 4\cdot 3\cdot 2\cdot 1}$

$$= 252$$

18. $(3x + 2y)^4$

$$= (3x)^4 + 4(3x)^3(2y) + 6(3x)^2(2y)^2 + 4(3x)(2y)^3 + (2y)^4$$
$$= 81x^4 + 216x^3y + 216x^2y^2 + 96xy^3 + 16y^4$$

20. $(2x - \sqrt{y})^7 = \dbinom{7}{0}(2x)^7 + \dbinom{7}{1}(2x)^6(-\sqrt{y})$

$$+ \binom{7}{2}(2x)^5(-\sqrt{y})^2 + \binom{7}{3}(2x)^4(-\sqrt{y})^3$$
$$+ \binom{7}{4}(2x)^3(-\sqrt{y})^4 + \binom{7}{5}(2x)^2(-\sqrt{y})^5$$
$$+ \binom{7}{6}(2x)(-\sqrt{y})^6 + \binom{7}{7}(-\sqrt{y})^7$$
$$= 128x^7 - 448x^6\sqrt{y} + 672x^5y - 560x^4y\sqrt{y}$$
$$+ 280x^3y^2 - 84x^2y^2\sqrt{y}$$
$$+ 14xy^3 - y^3\sqrt{y}$$

34. $\dbinom{10}{6-1}(x^{-1/2})^{10-6+1}(x^{1/2})^{6-1} = \dbinom{10}{5}(x^{-1/2})^5(x^{1/2})^5 = 252$

Exercise Set A.1, page 548

10. $\dfrac{4^{-2}}{2^{-3}} = \dfrac{2^3}{4^2} = \dfrac{8}{16} = \dfrac{1}{2}$

30. $\dfrac{(-3a^2b^3)^2}{(-2ab^4)^3} = \dfrac{(-3)^{1\cdot 2}a^{2\cdot 2}b^{3\cdot 2}}{(-2)^{1\cdot 3}a^{1\cdot 3}b^{4\cdot 3}}$

$$= \frac{9a^4b^6}{-8a^3b^{12}} = -\frac{9a}{8b^6}$$

46. $\dfrac{(6.9\times 10^{27})(8.2\times 10^{-13})}{4.1\times 10^{15}} = \dfrac{(6.9)(8.2)\times 10^{27-13}}{4.1\times 10^{15}}$

$$= \frac{56.58\times 10^{14}}{4.1\times 10^{15}}$$
$$= 13.8\times 10^{-1} = 1.38$$

62. $(-5x^{1/3})(-4x^{1/2}) = (-5)(-4)x^{1/3+1/2}$

$$= 20x^{2/6+3/6} = 20x^{5/6}$$

78. $\sqrt{18x^2y^5} = \sqrt{9x^2y^4}\sqrt{2y} = 3|x|y^2\sqrt{2y}$

86. $-3x\sqrt[3]{54x^4} + 2\sqrt[3]{16x^7} = -3x\sqrt[3]{3^3\cdot 2x^4} + 2\sqrt[3]{2^4x^7}$

$$= -3x\sqrt[3]{3^3x^3}\sqrt[3]{2x} + 2\sqrt[3]{2^3x^6}\sqrt[3]{2x}$$
$$= -3x(3x\sqrt[3]{2x}) + 2(2x^2\sqrt[3]{2x})$$
$$= -9x^2\sqrt[3]{2x} + 4x^2\sqrt[3]{2x}$$
$$= -5x^2\sqrt[3]{2x}$$

96. $(3\sqrt{5y} - 4)^2 = (3\sqrt{5y} - 4)(3\sqrt{5y} - 4)$

$$= 9\cdot 5y - 12\sqrt{5y} - 12\sqrt{5y} + 16$$
$$= 45y - 24\sqrt{5y} + 16$$

106. $\dfrac{2}{\sqrt[4]{4y}} = \dfrac{2}{\sqrt[4]{4y}}\cdot\dfrac{\sqrt[4]{4y^3}}{\sqrt[4]{4y^3}} = \dfrac{2\sqrt[4]{4y^3}}{2y} = \dfrac{\sqrt[4]{4y^3}}{y}$

110. $-\dfrac{7}{3\sqrt{2}-5} = -\dfrac{7}{3\sqrt{2}-5}\cdot\dfrac{3\sqrt{2}+5}{3\sqrt{2}+5}$

$$= \frac{-21\sqrt{2} - 35}{18 - 25}$$
$$= \frac{-21\sqrt{2} - 35}{-7} = 3\sqrt{2} + 5$$

Exercise Set A.2, page 553

12. **a.** $-12x^4 - 3x^2 - 11$

b. 4

c. $-12, -3, -11$

d. -12

e. $-12x^4, -3x^2, -11$

24. $(5y^2 - 7y + 3) + (2y^2 + 8y + 1) = 7y^2 + y + 4$

32.
$$\begin{array}{r} 3x^2 - 8x - 5 \\ 5x - 7 \\ \hline -21x^2 + 56x + 35 \\ 15x^3 - 40x^2 - 25x \phantom{{}+35} \\ \hline 15x^3 - 61x^2 + 31x + 35 \end{array}$$

56. $(4x^2 - 3y)(4x^2 + 3y) = (4x^2)^2 - (3y)^2 = 16x^4 - 9y^2$

66. $-x^2 - 5x + 4$

$-(-5)^2 - 5(-5) + 4$ • **Replace x by -5.**

$= -25 + 25 + 4$ • **Simplify.**

$= 4$

Exercise Set A.3, page 563

6. $6a^3b^2 - 12a^2b + 72ab^3 = 6ab(a^2b - 2a + 12b^2)$

12. $b^2 + 12b - 28 = (b + 14)(b - 2)$

16. $57y^2 + y - 6 = (19y - 6)(3y + 1)$

30. $8x^6 - 10x^3 - 3 = (4x^3 + 1)(2x^3 - 3)$

34. $81b^2 - 16c^2 = (9b - 4c)(9b + 4c)$

44. $b^2 - 24b + 144 = (b - 12)^2$

50. $b^3 + 64 = (b + 4)(b^2 - 4b + 16)$

60. $a^2y^2 - ay^3 + ac - cy = ay^2(a - y) + c(a - y)$
$= (a - y)(ay^2 + c)$

ANSWERS TO SELECTED EXERCISES

Exercise Set 1.1, page 13

1. 15 **3.** -4 **5.** $\frac{9}{2}$ **7.** $\frac{2}{9}$ **9.** 12 **11.** 16 **13.** 75 **15.** $\frac{1}{2}$ **17.** 1200 **19.** $-3, 5$ **21.** $-24, \frac{3}{8}$ **23.** $0, \frac{7}{3}$ **25.** $2, 8$
27. $-3, 5$ **29.** $\frac{-1 \pm \sqrt{5}}{2}$ **31.** $\frac{-2 \pm \sqrt{2}}{2}$ **33.** $\frac{5 \pm \sqrt{61}}{6}$ **35.** $\frac{-3 \pm \sqrt{41}}{4}$ **37.** $-\sqrt{2}, -\frac{\sqrt{2}}{2}$ **39.** $\frac{3 \pm \sqrt{29}}{2}$ **41.** $(-\infty, 4)$
43. $(-\infty, -6)$ **45.** $(-\infty, -3]$ **47.** $\left[-\frac{13}{8}, \infty\right)$ **49.** $(-\infty, 2)$ **51.** $(-\infty, -7) \cup (0, \infty)$ **53.** $(-5, -2)$ **55.** $(-\infty, 4] \cup [7, \infty)$
57. $\left[-\frac{1}{2}, \frac{4}{3}\right]$ **59.** $(-4, 4)$ **61.** $(-8, 10)$ **63.** $(-\infty, -33) \cup (27, \infty)$ **65.** $\left(-\infty, -\frac{3}{2}\right) \cup \left(\frac{5}{2}, \infty\right)$ **67.** $(-\infty, -8] \cup [2, \infty)$
69. $\left[-\frac{4}{3}, 8\right]$ **71.** $(-\infty, -4] \cup \left[\frac{28}{5}, \infty\right)$ **73.** $(-\infty, \infty)$ **75.** 4 **77.** 3.5 centimeters by 10 centimeters **79.** 100 feet by 150 feet
81. At least 58 checks **83.** At least 34 sales **85.** 20°C to 40°C **87.** 22 **89.** $(0, 210)$ **91.** 1 second $< t <$ 3 seconds
93. a. $|s - 4.25| \leq 0.01$ **b.** $4.24 \leq s \leq 4.26$

Prepare for Section 1.2, page 15

94. $-\frac{3}{2}$ **95.** $5\sqrt{2}$ **96.** No **97.** 1, 2 **98.** 2 **99.** 5

Exercise Set 1.2, page 27

1.

3. a.

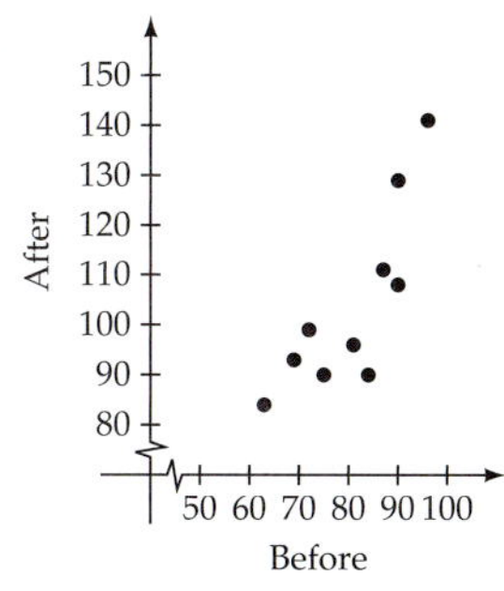

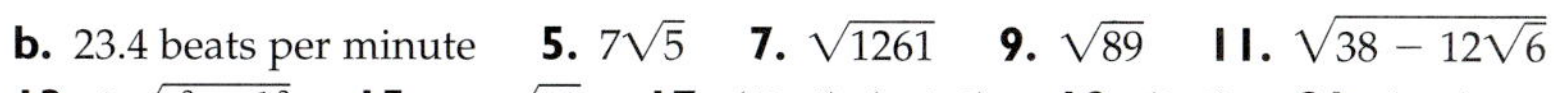

b. 23.4 beats per minute **5.** $7\sqrt{5}$ **7.** $\sqrt{1261}$ **9.** $\sqrt{89}$ **11.** $\sqrt{38 - 12\sqrt{6}}$

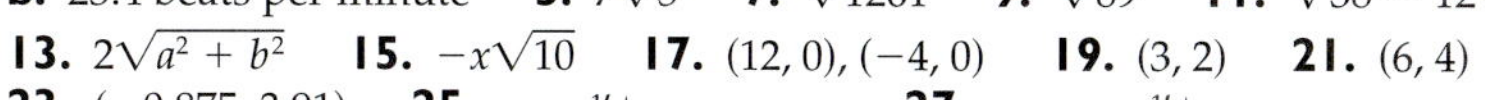

13. $2\sqrt{a^2 + b^2}$ **15.** $-x\sqrt{10}$ **17.** $(12, 0), (-4, 0)$ **19.** $(3, 2)$ **21.** $(6, 4)$
23. $(-0.875, 3.91)$ **25.** **27.**

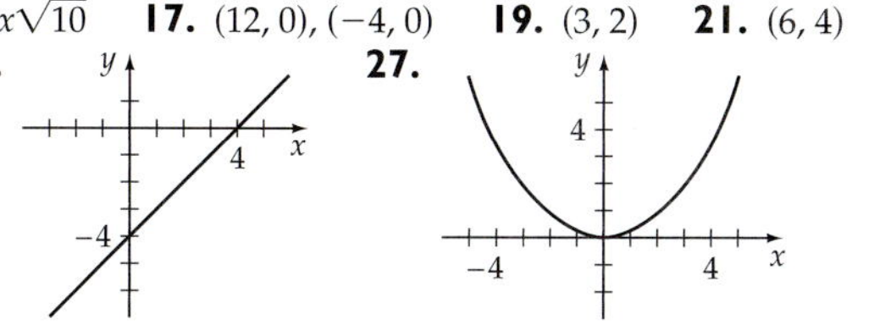

29.

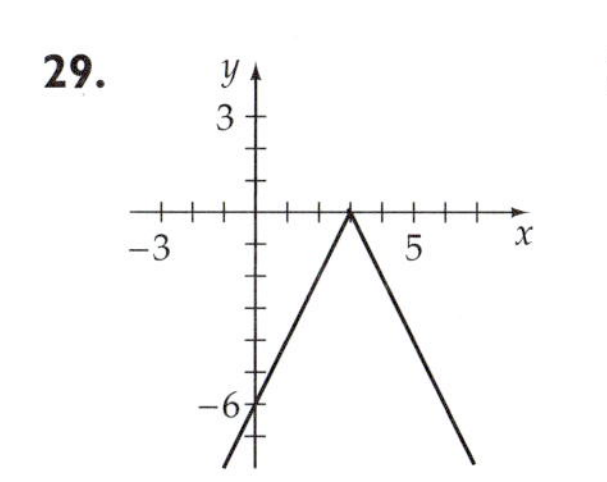

31.

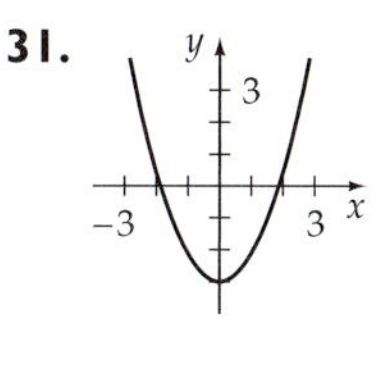

33.

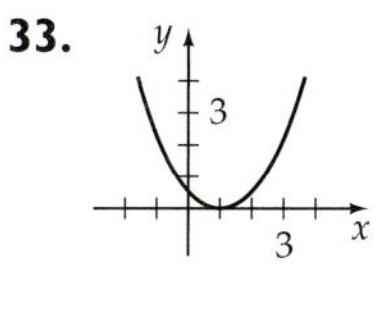

35. **37.**

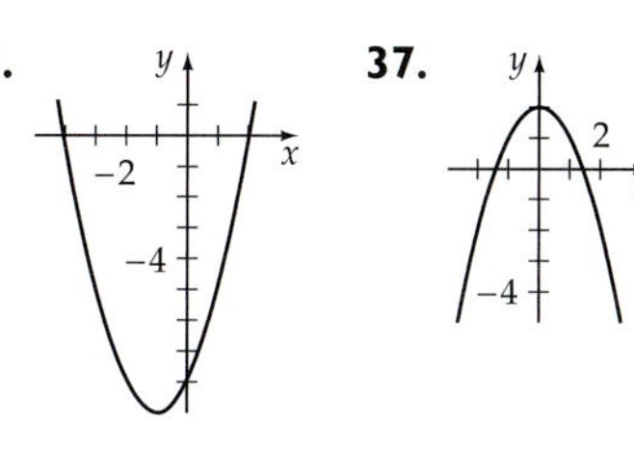

39. $(6, 0), \left(0, \frac{12}{5}\right)$

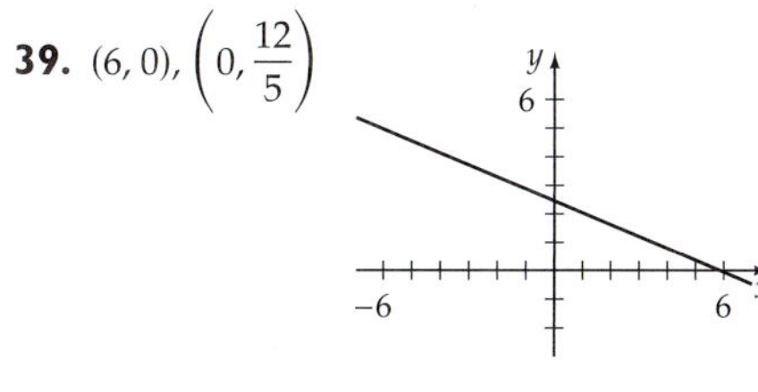

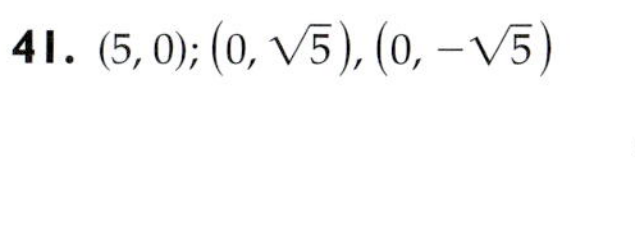

41. $(5, 0); (0, \sqrt{5}), (0, -\sqrt{5})$

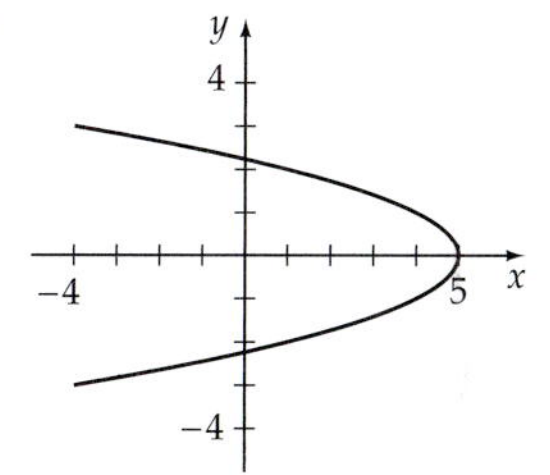

43. $(-4, 0)$; $(0, 4)$, $(0, -4)$

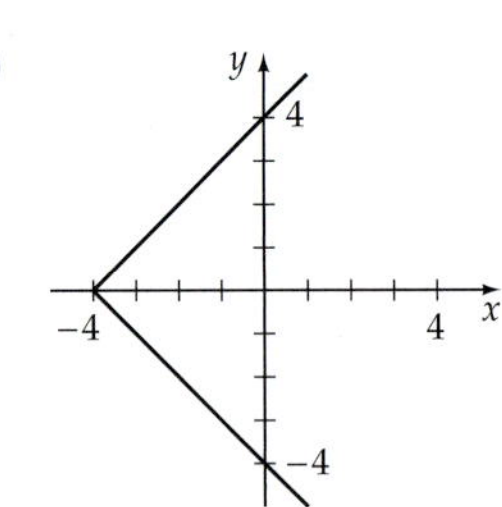

45. $(\pm 2, 0)$, $(0, \pm 2)$

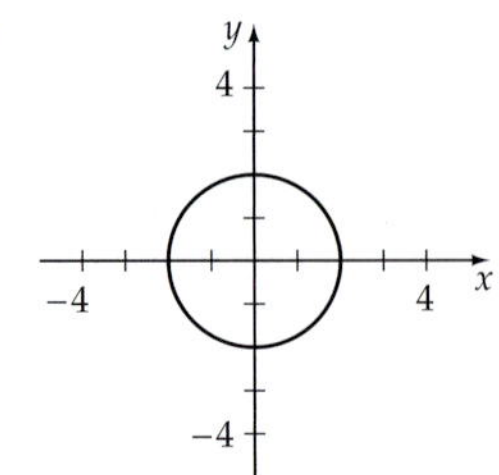

47. $(\pm 4, 0)$, $(0, \pm 4)$

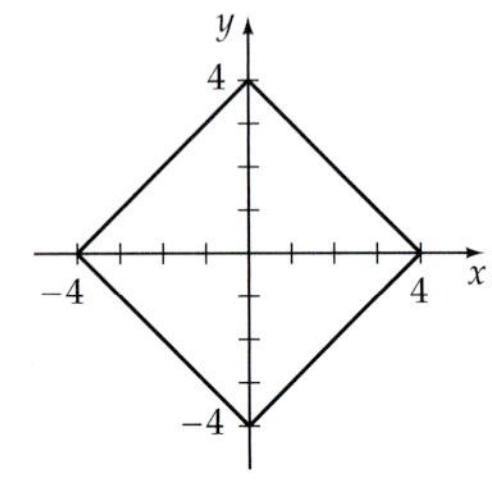

49. center $(0, 0)$, radius 6 **51.** center $(1, 3)$, radius 7 **53.** center $(-2, -5)$, radius 5 **55.** center $(8, 0)$, radius $\frac{1}{2}$ **57.** $(x - 4)^2 + (y - 1)^2 = 2^2$ **59.** $\left(x - \frac{1}{2}\right)^2 + \left(y - \frac{1}{4}\right)^2 = (\sqrt{5})^2$ **61.** $(x - 0)^2 + (y - 0)^2 = 5^2$ **63.** $(x - 1)^2 + (y - 3)^2 = 5^2$ **65.** center $(3, 0)$, radius 2 **67.** center $(7, -4)$, radius 3 **69.** center $\left(-\frac{1}{2}, 0\right)$, radius 4 **71.** center $\left(\frac{1}{2}, -\frac{3}{2}\right)$, radius $\frac{5}{2}$

73. $(x + 1)^2 + (y - 7)^2 = 25$ **75.** $(x - 7)^2 + (y - 11)^2 = 121$ **77.**

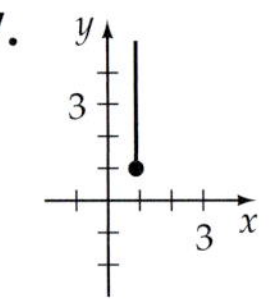

79.

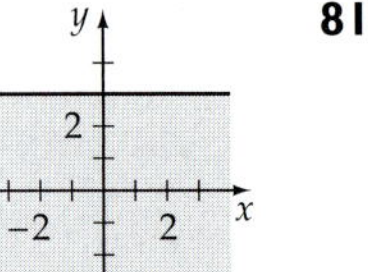

81.

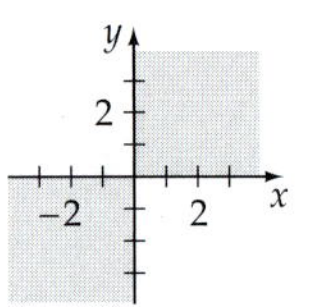

83. **85.**

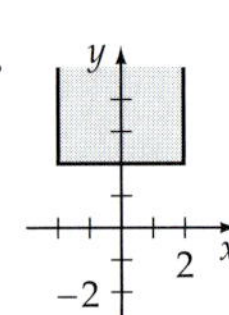

87. $(13, 5)$ **89.** $(7, -6)$ **91.** $x^2 - 6x + y^2 - 8y = 0$ **93.** $9x^2 + 25y^2 = 225$ **95.** $(x + 3)^2 + (y - 3)^2 = 3^2$

Prepare for Section 1.3, page 29

97. -4 **98.** $D = \{-3, -2, -1, 0, 2\}$; $R = \{1, 2, 4, 5\}$ **99.** $\sqrt{58}$ **100.** $x \geq 3$ **101.** $-2, 3$ **102.** 13

Exercise Set 1.3, page 43

1. a. 5 **b.** -4 **c.** -1 **d.** 1 **e.** $3k - 1$ **f.** $3k + 5$ **3. a.** $\sqrt{5}$ **b.** 3 **c.** 3 **d.** $\sqrt{21}$ **e.** $\sqrt{r^2 + 2r + 6}$ **f.** $\sqrt{c^2 + 5}$ **5. a.** $\frac{1}{2}$ **b.** $\frac{1}{2}$ **c.** $\frac{5}{3}$ **d.** 1 **e.** $\frac{1}{c^2 + 4}$ **f.** $\frac{1}{|2 + h|}$ **7. a.** 1 **b.** 1 **c.** -1 **d.** -1 **e.** 1 **f.** -1 **9. a.** -11 **b.** 6 **c.** $3c + 1$ **d.** $-k^2 - 2k + 10$ **11.** Yes **13.** No **15.** No **17.** Yes **19.** No **21.** Yes **23.** Yes **25.** Yes **27.** all real numbers **29.** all real numbers **31.** $\{x \mid x \neq -2\}$ **33.** $\{x \mid x \geq -7\}$ **35.** $\{x \mid -2 \leq x \leq 2\}$ **37.** $\{x \mid x > -4\}$ **39.**

41.

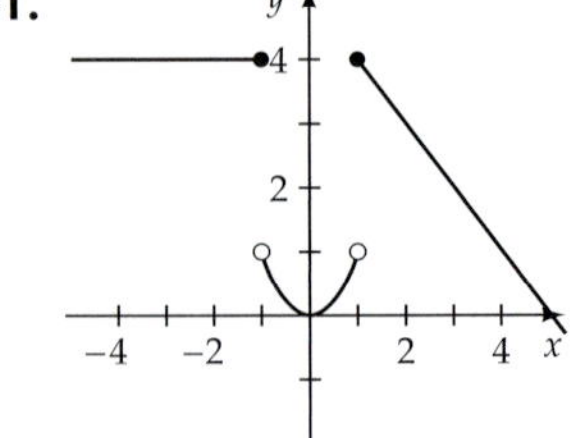

43.

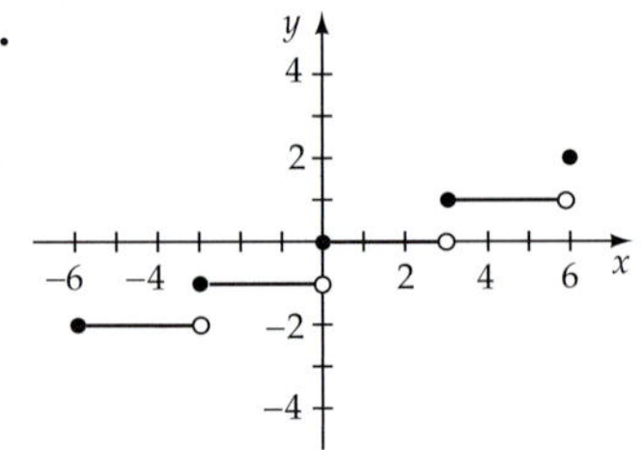

45.

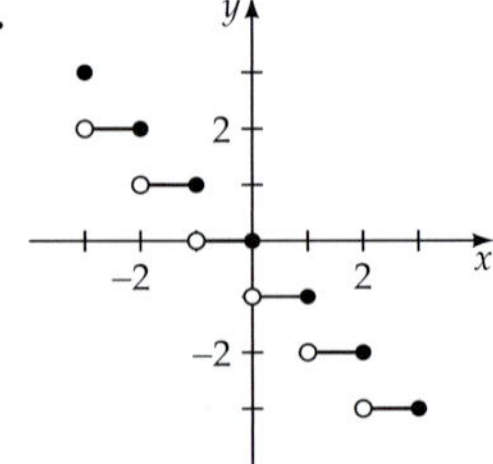

47. a. $1.05 **b.**

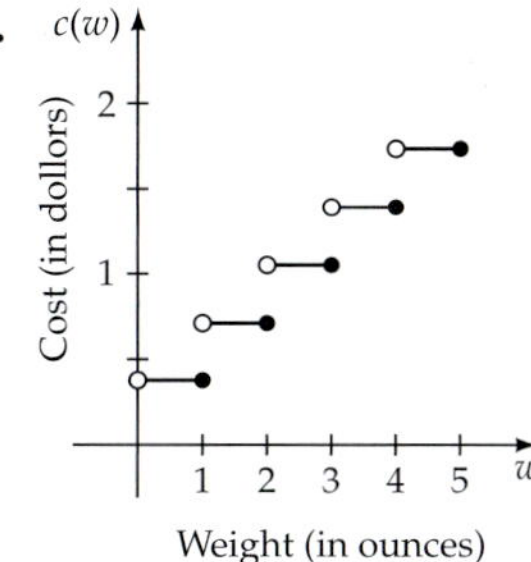

49. a, b, and d. **51.** decreasing on $(-\infty, 0]$; increasing on $[0, \infty)$ **53.** increasing on $(-\infty, \infty)$ **55.** decreasing on $(-\infty, -3]$; increasing on $[-3, 0]$; decreasing on $[0, 3]$; increasing on $[3, \infty)$ **57.** constant on $(-\infty, 0]$; increasing on $[0, \infty)$ **59.** decreasing on $(-\infty, 0]$; constant on $[0, 1]$; increasing on $[1, \infty)$ **61.** g and F **63. a.** $w = 25 - l$ **b.** $A = 25l - l^2$ **65.** $v(t) = 80{,}000 - 6500t, 0 \le t \le 10$ **67. a.** $C(x) = 2000 + 22.80x$ **b.** $R(x) = 37.00x$ **c.** $P(x) = 14.20x - 2000$ **69.** $h = 15 - 5r$ **71.** $d = \sqrt{(3t)^2 + 50^2}$ **73.** $d = \sqrt{(45 - 8t)^2 + (6t)^2}$ **75. a.** $L(x) = \left(\frac{1}{4\pi} + \frac{1}{16}\right)x^2 - \frac{5}{2}x + 25$ **b.** 25, 17.27, 14.09, 15.46, 21.37, 31.83 **c.** $[0, 20]$ **77. a.** $A(x) = \sqrt{900 + x^2} + \sqrt{400 + (40 - x)^2}$ **b.** 74.72, 67.68, 64.34, 64.79, 70 **c.** $[0, 40]$ **79.** 275, 375, 385, 390, 394

81. $c = -2$ or $c = 3$ **83.** 1 is not in the range of f.

85.

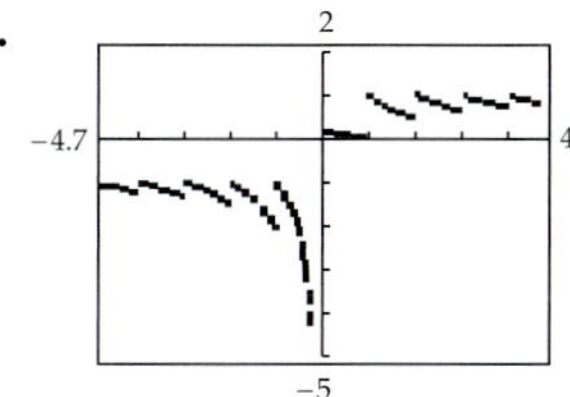

87.

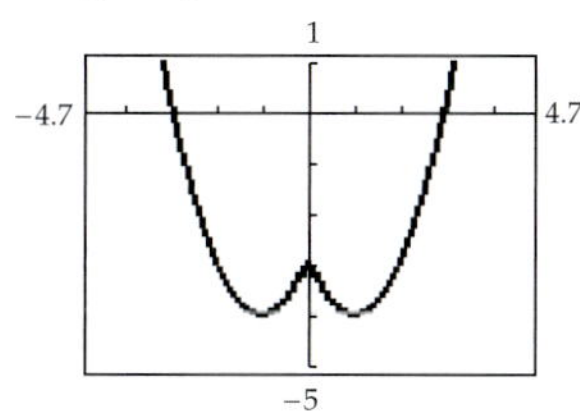

89.

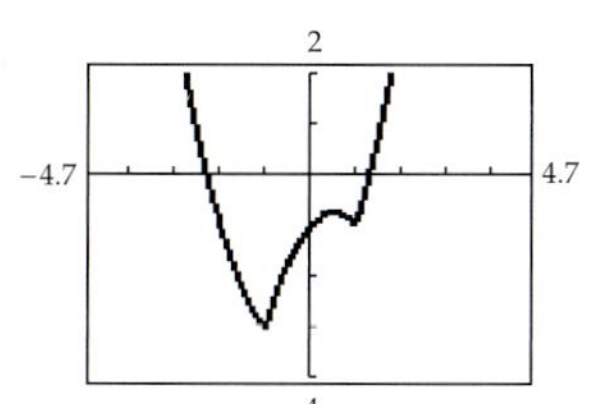

91. 4 **93.** 2 **95. a.** 36 **b.** 13 **c.** 12 **d.** 30 **e.** $13k - 2$ **f.** $8k - 11$ **97.** $4\sqrt{21}$ **99.** 1, −3

101.

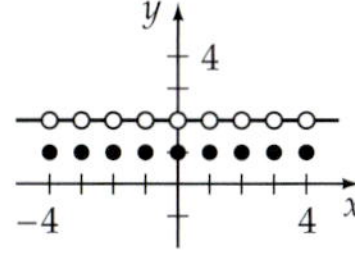

Prepare for Section 1.4, page 50

103. 7 **104.** −1 **105.** $-\frac{8}{5}$ **106.** $y = -2x + 9$ **107.** $y = \frac{3}{5}x - 3$ **108.** 2

Exercise Set 1.4, page 60

1. $-\frac{3}{2}$ **3.** $-\frac{1}{2}$ **5.** The line does not have slope. **7.** 6 **9.** $\frac{9}{19}$ **11.** $\frac{f(3 + h) - f(3)}{h}$ **13.** $\frac{f(h) - f(0)}{h}$ **15.**

17.

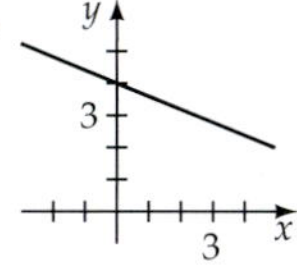

19.

21.

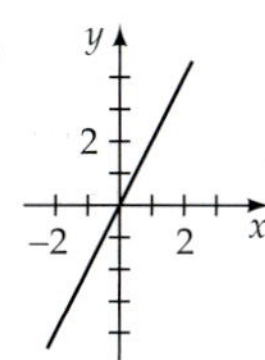

23.

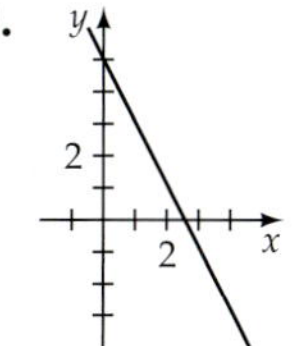

25.

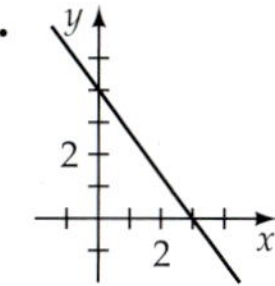

27. $y = x + 3$ **29.** $y = \frac{3}{4}x + \frac{1}{2}$ **31.** $y = (0)x + 4 = 4$ **33.** $y = -4x - 10$

35. $y = -\frac{3}{4}x + \frac{13}{4}$ **37.** $y = \frac{12}{5}x - \frac{29}{5}$ **39.** −2 **41.** $-\frac{1}{2}$ **43.** −4 **45.** 4 **47.** −20 **49.** $\frac{1}{3}$ **51.** $\frac{16}{3}$ **53.** $m = 2.875$. The value of the slope indicates that the speed of sound in water increases 2.875 feet per second for a 1-degree increase in temperature. **55. a.** $H(c) = 1.45c$ **b.** 26 miles per gallon **57. a.** $N(t) = 2500t - 4{,}962{,}000$ **b.** 2008 **59. a.** $B(d) = 30d - 300$ **b.** The value of the slope means that a 1-inch increase in the diameter of a log 32 feet long results in an increase of 30 board-feet of lumber that can be obtained from the log. **c.** 270 board-feet **61.** line A, Michelle; line B, Amanda; line C, distance between Michelle and Amanda **63. a.** $y = 1.842x - 18.947$ **b.** 147 **65.** $P(x) = 40.50x - 1782$, $x = 44$, the break-even point **67.** $P(x) = 79x - 10{,}270$, $x = 130$, the break-even point **69. a.** $275 **b.** $283 **c.** $355 **d.** $8 **71. a.** $C(t) = 19{,}500.00 + 6.75t$ **b.** $R(t) = 55.00t$ **c.** $P(t) = 48.25t - 19{,}500.00$ **d.** approximately 405 days **73.** $y = -\frac{3}{4}x + \frac{15}{4}$ **75.** $y = x + 1$ **77.** −5 ft **79. a.** $Q = (3, 10), m = 5$ **b.** $Q = (2.1, 5.41), m = 4.1$ **c.** $Q = (2.01, 5.0401), m = 4.01$ **d.** 4 **85.** $y = -2x + 11$ **87.** $5x + 3y = 15$ **89.** $3x + y = 17$ **93.** $\left(\frac{9}{2}, \frac{81}{4}\right)$

Prepare for Section 1.5, page 66

95. $(3x - 2)(x + 4)$ **96.** $x^2 - 8x + 16 = (x - 4)^2$ **97.** 26 **98.** $-\frac{1}{2}, 1$ **99.** $\frac{-3 \pm \sqrt{17}}{2}$ **100.** 1, 3

Exercise Set 1.5, page 75

1. d **3.** b **5.** g **7.** c **9.** $f(x) = (x + 2)^2 - 3$

vertex: $(-2, -3)$

axis of symmetry: $x = -2$

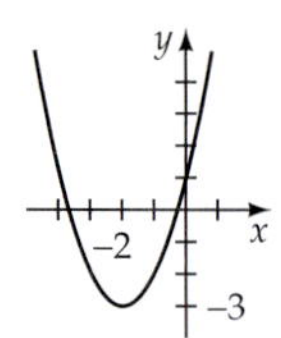

11. $f(x) = (x - 4)^2 - 11$

vertex: $(4, -11)$

axis of symmetry: $x = 4$

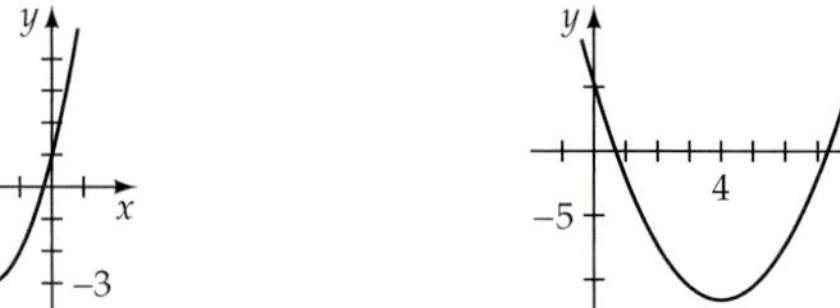

13. $f(x) = \left(x - \left(-\frac{3}{2}\right)\right)^2 - \frac{5}{4}$

vertex: $\left(-\frac{3}{2}, -\frac{5}{4}\right)$

axis of symmetry: $x = -\frac{3}{2}$

15. $f(x) = -(x - 2)^2 + 6$

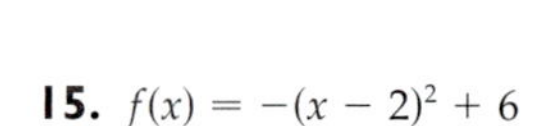

vertex: $(2, 6)$

axis of symmetry: $x = 2$

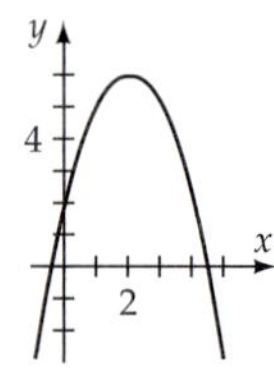

17. $f(x) = -3\left(x - \frac{1}{2}\right)^2 + \frac{31}{4}$

vertex: $\left(\frac{1}{2}, \frac{31}{4}\right)$

axis of symmetry: $x = \frac{1}{2}$

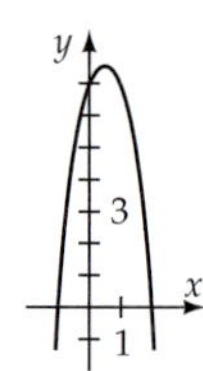

19. vertex: $(5, -25)$, $f(x) = (x - 5)^2 - 25$
21. vertex: $(0, -10)$, $f(x) = x^2 - 10$
23. vertex: $(3, 10)$, $f(x) = -(x - 3)^2 + 10$
25. vertex: $\left(\frac{3}{4}, \frac{47}{8}\right)$, $f(x) = 2\left(x - \frac{3}{4}\right)^2 + \frac{47}{8}$
27. vertex: $\left(\frac{1}{8}, \frac{17}{16}\right)$, $f(x) = -4\left(x - \frac{1}{8}\right)^2 + \frac{17}{16}$

29. $\{y \mid y \geq -2\}$, -1 and 3 **31.** $\left\{y \mid y \leq \frac{17}{8}\right\}$, 1 and $\frac{3}{2}$

33. No, $3 \notin \left\{y \mid y \geq \frac{15}{4}\right\}$ **35.** -16, minimum **37.** 11, maximum

39. $-\frac{1}{8}$, minimum **41.** -11, minimum **43.** 35, maximum

45. a. 27 feet **b.** $22\frac{5}{16}$ feet **c.** 20.1 feet from the center **47. a.** $w = \frac{600 - 2l}{3}$ **b.** $A = 200l - \frac{2}{3}l^2$ **c.** $w = 100$ feet, $l = 150$ feet **49. a.** 12:43 P.M. **b.** 91°F **51.** 1993, 2500 homes **53.** Yes **55. a.** 41 miles per gallon **b.** 34 miles per gallon **57.** y-intercept $(0, 0)$; x-intercepts $(0, 0)$ and $(-6, 0)$ **59.** y-intercept $(0, -6)$; no x-intercepts **61.** 740 units yield a maximum revenue of \$109,520. **63.** 85 units yield a maximum profit of \$24.25. **65.** $P(x) = -0.1x^2 + 50x - 1840$, break-even points: $x = 40$ and $x = 460$ **67. a.** $R(x) = -0.25x^2 + 30.00x$ **b.** $P(x) = -0.25x^2 + 27.50x - 180$ **c.** \$576.25 **d.** 55 **69. a.** $t = 4$ seconds **b.** 256 feet **c.** $t = 8$ seconds **71.** 30 feet **73.** $r = \frac{48}{4 + \pi} \approx 6.72$ feet, $h = r \approx 6.72$ feet **77.** $f(x) = \frac{3}{4}x^2 - 3x + 4$ **79. a.** $w = 16 - x$ **b.** $A = 16x - x^2$ **81.** The discriminant is $b^2 - 4(1)(-1) = b^2 + 4$, which is positive for all b. **83.** increases the height of each point on the graph by c units **85.** 4, 4

Prepare for Section 1.6, page 80

89. $x = -2$ **92.** 3, -1, -3, -3, -1 **93.** $(0, b)$ **94.** $(0, 0)$

Exercise Set 1.6, page 91

1.
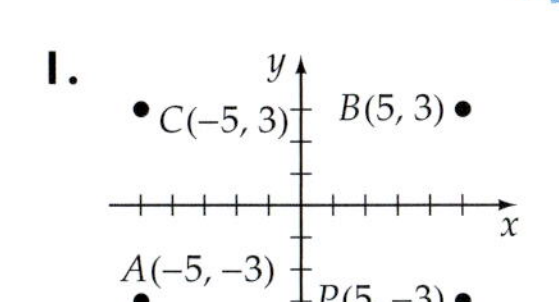

3.
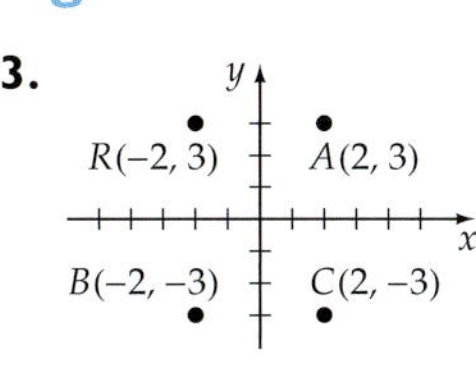

5.
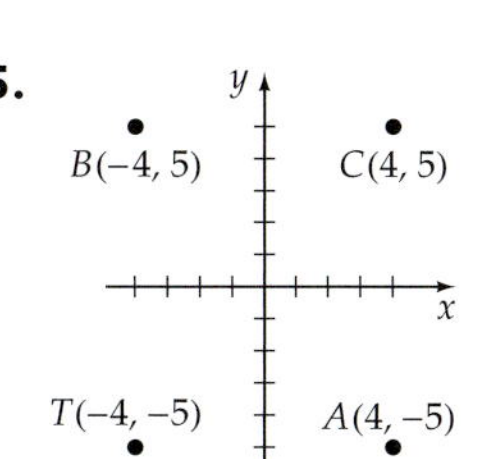

7.
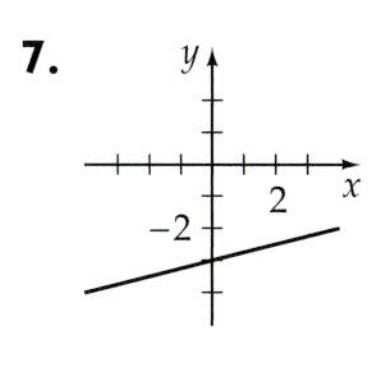

9.
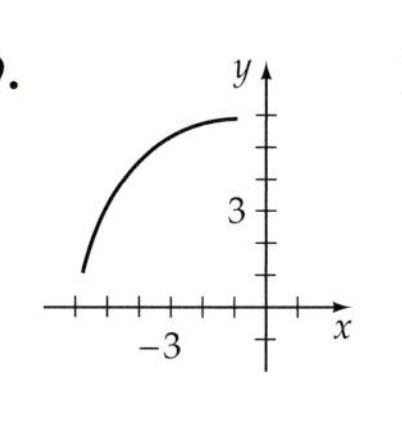

11.
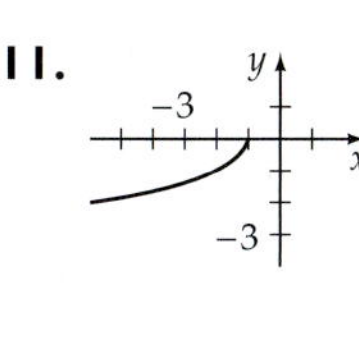

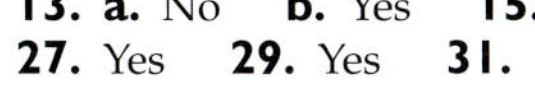

13. a. No **b.** Yes **15. a.** No **b.** No **17. a.** Yes **b.** Yes **19. a.** Yes **b.** Yes **21. a.** Yes **b.** Yes **23.** No **25.** Yes
27. Yes **29.** Yes **31.**

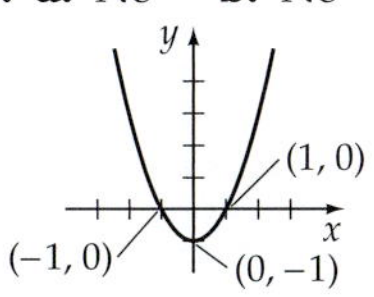

33.
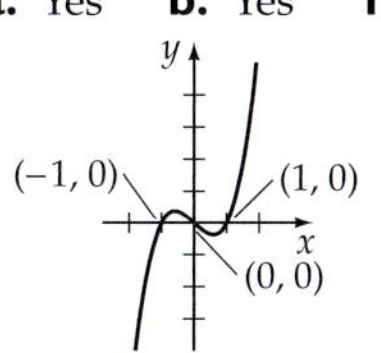

35.
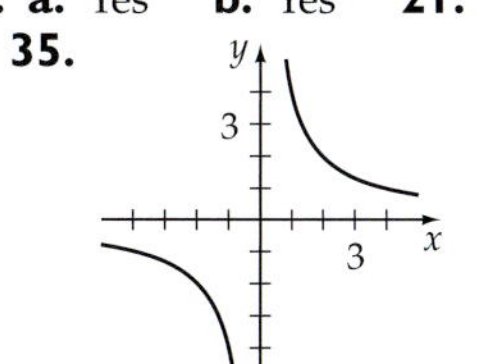

37.
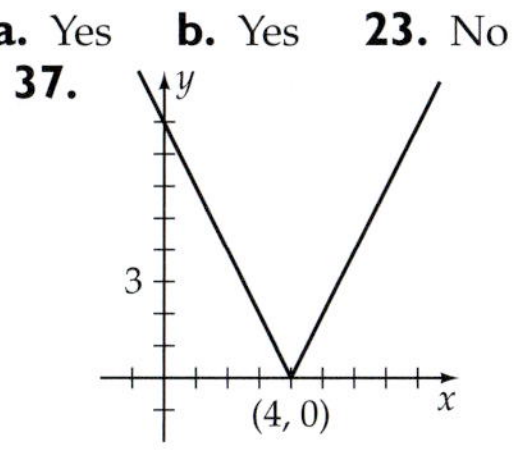

39.
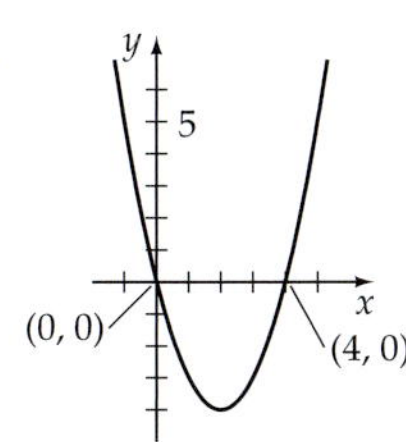

41.
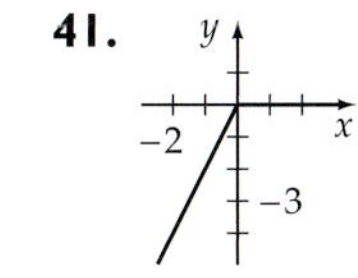

43. even **45.** odd **47.** even **49.** even **51.** even **53.** even **55.** neither

57. a., b.
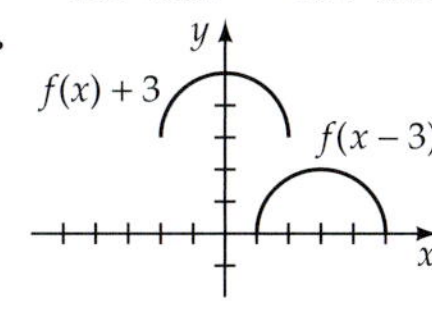

59. a.
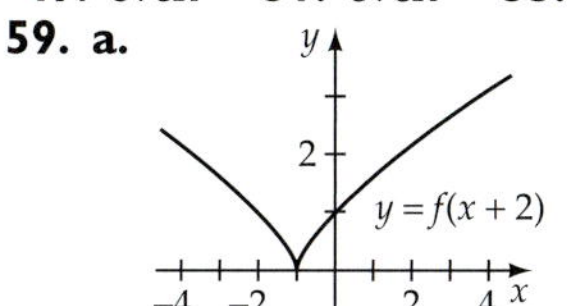

b.
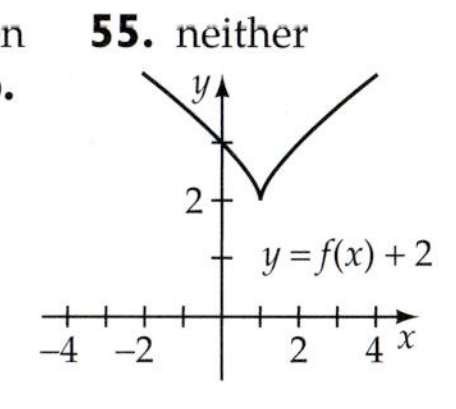

61. a. $(-5, 5), (-3, -2), (-2, 0)$ **b.** $(-2, 6), (0, -1), (1, 1)$ **63. a.**

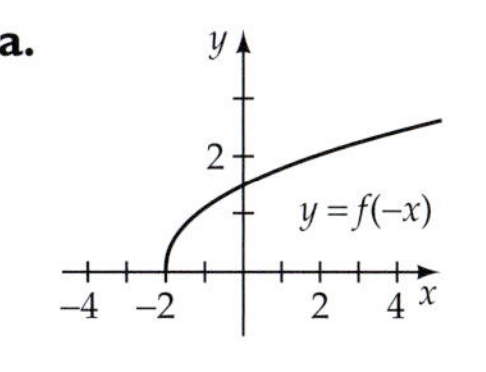

b.
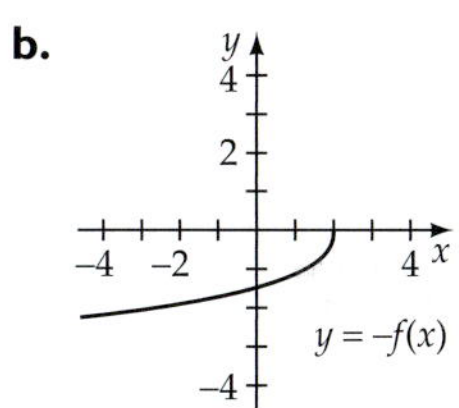

65. a. $(1, 3), (-2, -4)$ **b.** $(-1, -3), (2, 4)$ **67. a., b.**

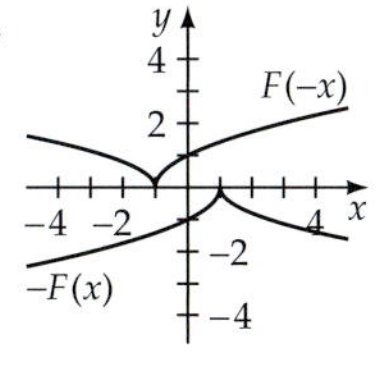

69.
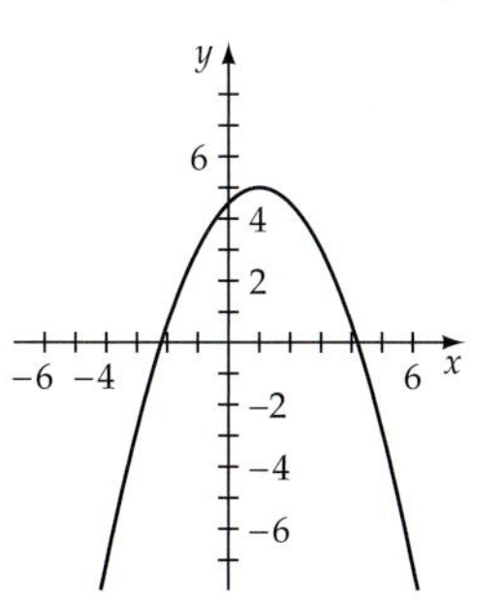

71. a.
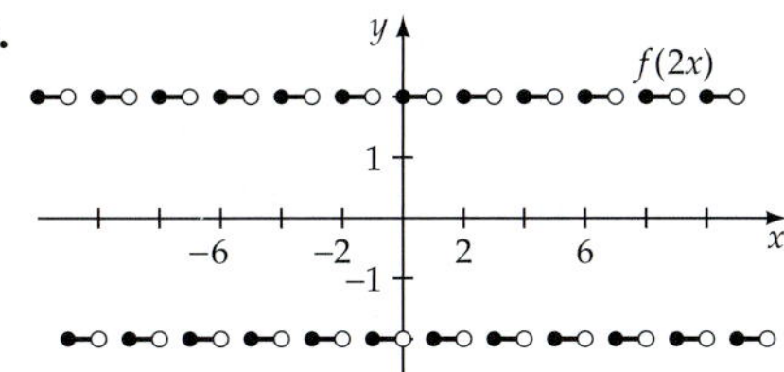

b.
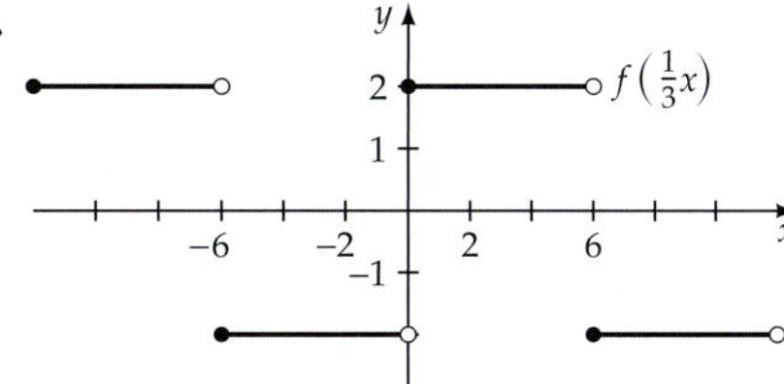

73. a.

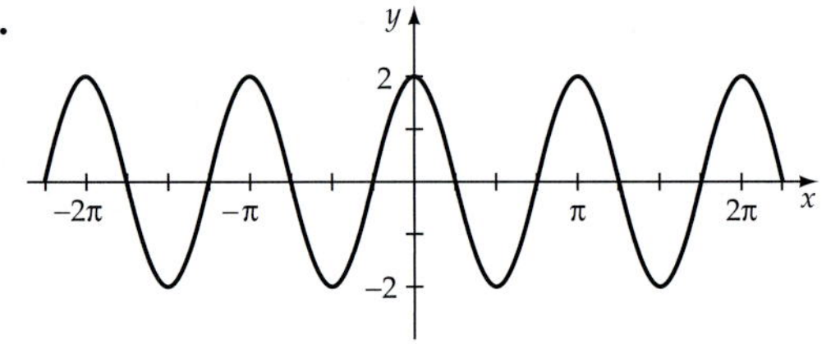

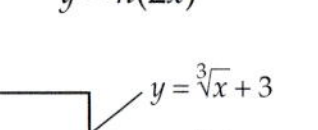

b.

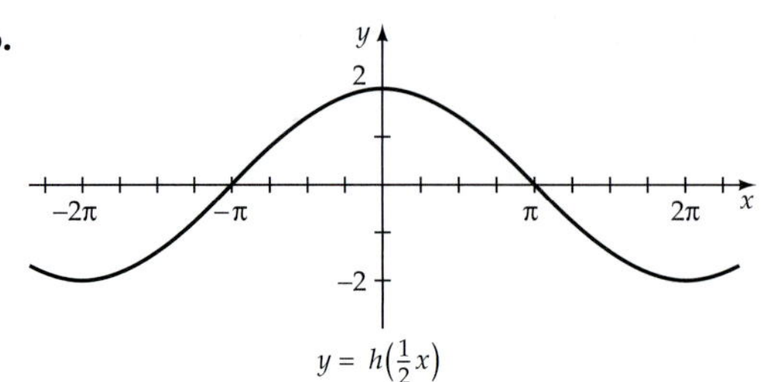

$y = h\left(\frac{1}{2}x\right)$

75.

6
$y = \sqrt[3]{x} + 3$
$y = \sqrt[3]{x}$
$y = \sqrt[3]{x} - 1$
−5
5
−4

77.

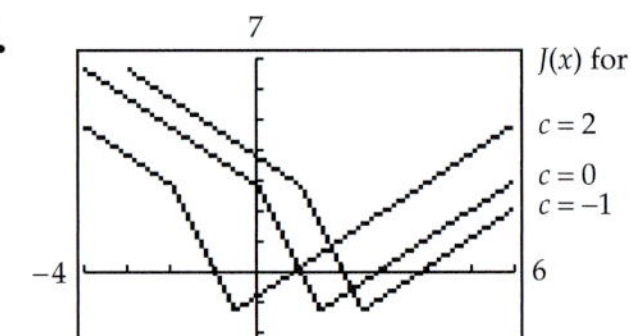

79.

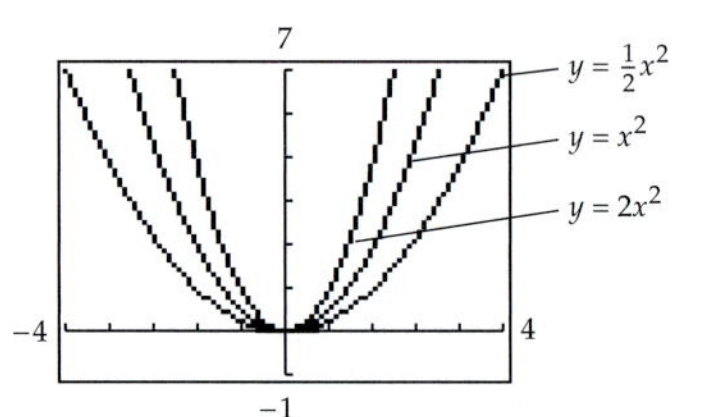

81.

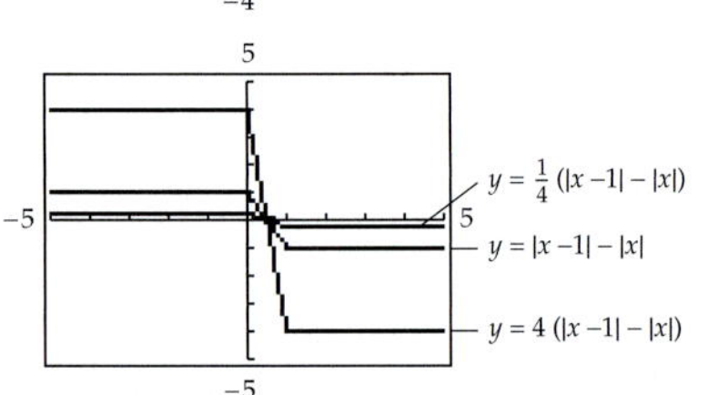

83. a.

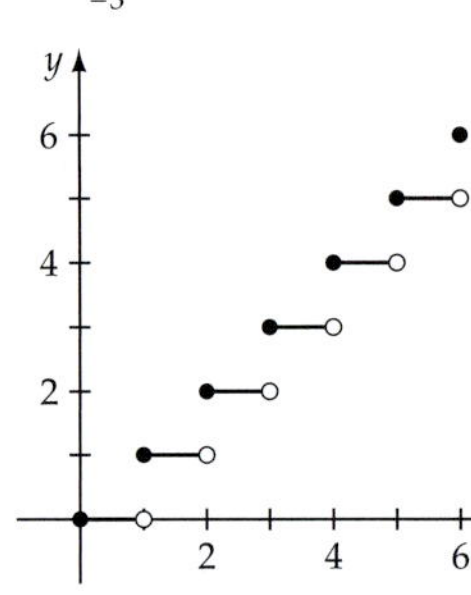

b.

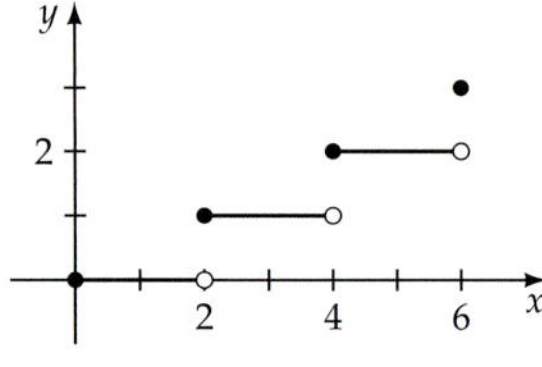

c.

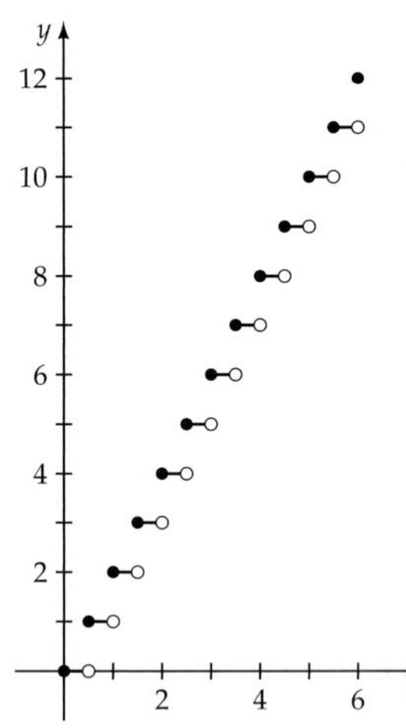

85. a. $f(x) = \dfrac{2}{(x+1)^2 + 1} + 1$ **b.** $f(x) = -\dfrac{2}{(x-2)^2 + 1}$

Prepare for Section 1.7, page 95

87. $x^2 + 1$ **88.** $6x^3 - 11x^2 + 7x - 6$ **89.** $18a^2 - 15a + 2$ **90.** $2h^2 + 3h$ **91.** all real numbers except $x = 1$ **92.** $[4, \infty)$

Exercise Set 1.7, page 104

1. $f(x) + g(x) = x^2 - x - 12$, Domain is the set of all real numbers.
$f(x) - g(x) = x^2 - 3x - 18$, Domain is the set of all real numbers.
$f(x) \cdot g(x) = x^3 + x^2 - 21x - 45$, Domain is the set of all real numbers.
$\dfrac{f(x)}{g(x)} = x - 5$, Domain $\{x \mid x \neq -3\}$

3. $f(x) + g(x) = 3x + 12$, Domain is the set of all real numbers.
$f(x) - g(x) = x + 4$, Domain is the set of all real numbers.
$f(x) \cdot g(x) = 2x^2 + 16x + 32$, Domain is the set of all real numbers.
$\dfrac{f(x)}{g(x)} = 2$, Domain $\{x \mid x \neq -4\}$

5. $f(x) + g(x) = x^3 - 2x^2 + 8x$, Domain is the set of all real numbers.
$f(x) - g(x) = x^3 - 2x^2 + 6x$, Domain is the set of all real numbers
$f(x) \cdot g(x) = x^4 - 2x^3 + 7x^2$, Domain is the set of all real numbers.
$\dfrac{f(x)}{g(x)} = x^2 - 2x + 7$, Domain $\{x \mid x \neq 0\}$

7. $f(x) + g(x) = 4x^2 + 7x - 12$, Domain is the set of all real numbers.
$f(x) - g(x) = x - 2$, Domain is the set of all real numbers
$f(x) \cdot g(x) = 4x^4 + 14x^3 - 12x^2 - 41x + 35$, Domain is the set of all real numbers.
$\dfrac{f(x)}{g(x)} = 1 + \dfrac{x-2}{2x^2 + 3x - 5}$, Domain $\left\{x \mid x \neq 1, x \neq -\dfrac{5}{2}\right\}$

9. $f(x) + g(x) = \sqrt{x-3} + x$, Domain $\{x \mid x \geq 3\}$
$f(x) - g(x) = \sqrt{x-3} - x$, Domain $\{x \mid x \geq 3\}$
$f(x) \cdot g(x) = x\sqrt{x-3}$, Domain $\{x \mid x \geq 3\}$
$\dfrac{f(x)}{g(x)} = \dfrac{\sqrt{x-3}}{x}$, Domain $\{x \mid x \geq 3\}$

11. $f(x) + g(x) = \sqrt{4 - x^2} + 2 + x$, Domain $\{x \mid -2 \leq x \leq 2\}$
$f(x) - g(x) = \sqrt{4 - x^2} - 2 - x$, Domain $\{x \mid -2 \leq x \leq 2\}$
$f(x) \cdot g(x) = (\sqrt{4 - x^2})(2 + x)$, Domain $\{x \mid -2 \leq x \leq 2\}$
$\dfrac{f(x)}{g(x)} = \dfrac{\sqrt{4 - x^2}}{2 + x}$ Domain $\{x \mid -2 < x \leq 2\}$

13. 18 **15.** $-\dfrac{9}{4}$ **17.** 30 **19.** 12 **21.** 300 **23.** $-\dfrac{384}{125}$ **25.** $-\dfrac{5}{2}$ **27.** $-\dfrac{1}{4}$ **29.** 2 **31.** $2x + h$ **33.** $4x + 2h + 4$

35. $-8x - 4h$ **37.** $(g \circ f)(x) = 6x + 3$, $(f \circ g)(x) = 6x - 16$ **39.** $(g \circ f)(x) = x^2 + 4x + 1$, $(f \circ g)(x) = x^2 + 8x + 11$ **41.** $(g \circ f)(x) = -5x^3 - 10x$, $(f \circ g)(x) = -125x^3 - 10x$ **43.** $(g \circ f)(x) = \dfrac{1 - 5x}{x + 1}$, $(f \circ g)(x) = \dfrac{2}{3x - 4}$

45. $(g \circ f)(x) = \dfrac{\sqrt{1 - x^2}}{|x|}$, $(f \circ g)(x) = \dfrac{1}{x - 1}$ **47.** $(g \circ f)(x) = -\dfrac{2|5 - x|}{3}$, $(f \circ g)(x) = \dfrac{3|x|}{|5x + 2|}$ **49.** 66 **51.** 51 **53.** -4 **55.** 41 **57.** $-\dfrac{3848}{625}$ **59.** $6 + 2\sqrt{3}$

61. $16c^2 + 4c - 6$ **63.** $9k^4 + 36k^3 + 45k^2 + 18k - 4$ **65. a.** $A(t) = \pi(1.5t)^2$, $A(2) = 9\pi$ square feet ≈ 28.27 square feet **b.** $V(t) = 2.25\pi t^3$, $V(3) = 60.75\pi$ cubic feet ≈ 190.85 cubic feet **67. a.** $d(t) = \sqrt{(48 - t)^2 - 4^2}$ **b.** $s(35) = 13$ feet, $d(35) \approx 12.37$ feet **69.** $(Y \circ F)(x)$ converts x inches to yards. **71. a.** 99.8; this is identical to the slope of the line through $(0, C(0))$ and $(1, C(1))$. **b.** 156.2 **c.** -49.7 **d.** -30.8 **e.** -16.4 **f.** 0

Chapter 1 True/False Exercises, page 111

1. False. Let $f(x) = x^2$. Then $f(3) = f(-3) = 9$, but $3 \neq -3$. **2.** False. Consider $f(x) = x + 1$ and $g(x) = x^2 - 2$. **3.** True **4.** True **5.** False. Let $f(x) = 3x$. $[f(x)]^2 = 9x^2$, whereas $f[f(x)] = f(3x) = 3(3x) = 9x$. **6.** False. Let $f(x) = x^2$. Then $f(1) = 1$, $f(2) = 4$. Thus $\dfrac{f(2)}{f(1)} = 4 \neq \dfrac{2}{1}$. **7.** True **8.** False. Let $f(x) = |x|$. Then $f(-1 + 3) = f(2) = 2$. $f(-1) + f(3) = 1 + 3 = 4$. **9.** True **10.** True **11.** True **12.** True **13.** True

Chapter 1 Review Exercises, page 111

1. $-\dfrac{9}{4}$ [1.1] **2.** -4 [1.1] **3.** -2 [1.1] **4.** $-\dfrac{2}{3}$ [1.1] **5.** $-3, 6$ [1.1] **6.** $\dfrac{1}{2}, 4$ [1.1] **7.** $\dfrac{-1 \pm \sqrt{13}}{6}$ [1.1] **8.** $\dfrac{-3 \pm \sqrt{41}}{4}$ [1.1]

9. $c \geq -6$ [1.1] **10.** $a > 1$ [1.1] **11.** $(-\infty, -3] \cup [4, \infty)$ [1.1] **12.** $\left(-\dfrac{1}{2}, 1\right)$ [1.1] **13.** $(-\infty, 1) \cup (4, \infty)$ [1.1] **14.** $\left[-1, \dfrac{5}{3}\right]$ [1.1]

15. $\sqrt{181}$ [1.2] **16.** $\sqrt{80} = 4\sqrt{5}$ [1.2] **17.** $\left(-\dfrac{1}{2}, 10\right)$ [1.2] **18.** $(2, -2)$ [1.2] **19.** center $(3, -4)$, radius 9 [1.2] **20.** center $(-5, -2)$, radius 3 [1.2] **21.** $(x - 2)^2 + (y + 3)^2 = 5^2$ [1.2] **22.** $(x + 5)^2 + (y - 1)^2 = 8^2$, radius $= |-5 - (3)| = 8$ [1.2] **23. a.** 2 **b.** 10 **c.** $3t^2 + 4t - 5$ **d.** $3x^2 + 6xh + 3h^2 + 4x + 4h - 5$ **e.** $9t^2 + 12t - 15$ **f.** $27t^2 + 12t - 5$ [1.3] **24. a.** $\sqrt{55}$ **b.** $\sqrt{39}$ **c.** 0 **d.** $\sqrt{64 - x^2}$ **e.** $2\sqrt{64 - t^2}$ **f.** $2\sqrt{16 - t^2}$ [1.3] **25. a.** 5 **b.** -11 **c.** $x^2 - 12x + 32$ **d.** $x^2 + 4x - 8$ [1.7] **26. a.** 79 **b.** 56 **c.** $2x^2 - 4x + 9$ **d.** $2x^2 + 6$ [1.7] **27.** $8x + 4h - 3$ [1.7] **28.** $3x^2 + 3xh + h^2 - 1$ [1.7]

29. [1.3]

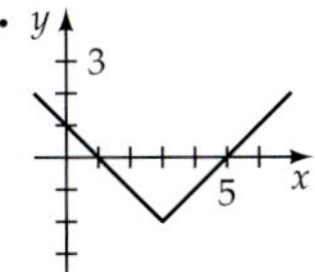

increasing on $[3, \infty)$
decreasing on $(-\infty, 3]$

30. [1.3]

f is increasing on $[0, \infty)$
f is decreasing on $(-\infty, 0]$

31. [1.3]

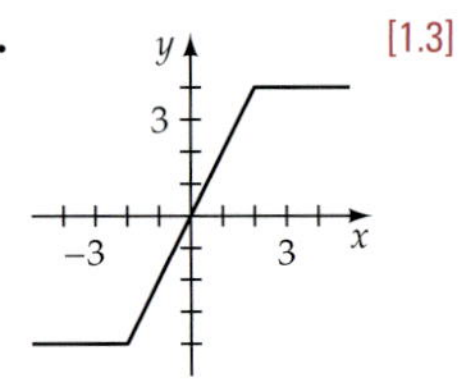

increasing on $[-2, 2]$
constant on $(-\infty, -2] \cup [2, \infty)$

32. [1.3]

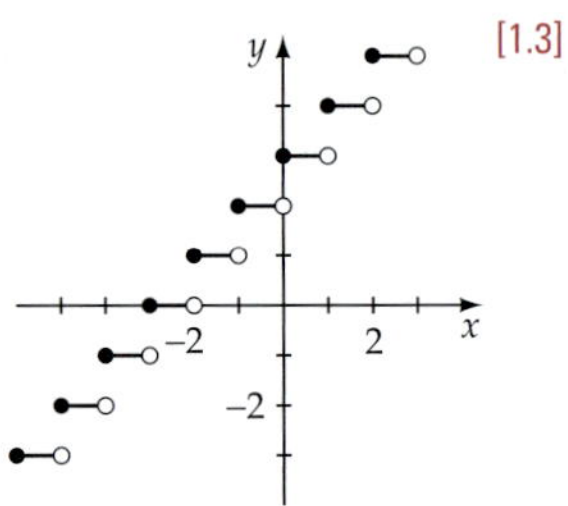

f is constant on..., $[-6, -5)$, $[-5, -4)$, $[-4, -3)$, $[-3, -2)$, $[-2, -1)$, $[-1, 0)$, $[0, 1)$, ...

33. [1.3]

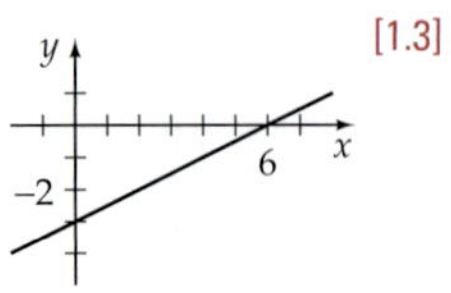

increasing on $(-\infty, \infty)$

34. [1.3]

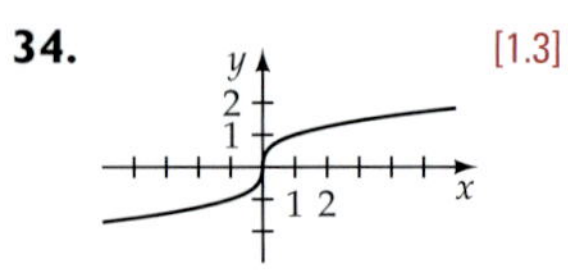

f is increasing on $(-\infty, \infty)$

35. Domain: $\{x \mid x \text{ is a real number}\}$ [1.3] **36.** Domain: $\{x \mid x \leq 6\}$ [1.3] **37.** Domain: $\{x \mid -5 \leq x \leq 5\}$ [1.3] **38.** Domain: $\{x \mid x \neq -3, x \neq 5\}$ [1.3]
39. $y = -2x + 1$ [1.4] **40.** $y = \frac{11}{7}x$ [1.4] **41.** $y = \frac{3}{4}x + \frac{19}{2}$ [1.4] **42.** $y = \frac{5}{2}x + \frac{1}{2}$ [1.4] **43.** $f(x) = (x + 3)^2 + 1$ [1.5]
44. $f(x) = 2(x + 1)^2 + 3$ [1.5] **45.** $f(x) = -(x + 4)^2 + 19$ [1.5] **46.** $f(x) = 4\left(x - \frac{3}{4}\right)^2 - \frac{5}{4}$ [1.5] **47.** $f(x) = -3\left(x - \frac{2}{3}\right)^2 - \frac{11}{3}$ [1.5]
48. $f(x) = (x - 3)^2 + 0$ [1.5] **49.** $(1, 8)$ [1.5] **50.** $(0, -10)$ [1.5] **51.** $(5, 161)$ [1.5] **52.** $(-4, 30)$ [1.5] **53.** $\frac{4\sqrt{5}}{5}$ [1.4]
54. a. $R = 13x$ **b.** $P = 12.5x - 1050$ **c.** $x = 84$ [1.4] **55.** [1.6] **56.** [1.6]

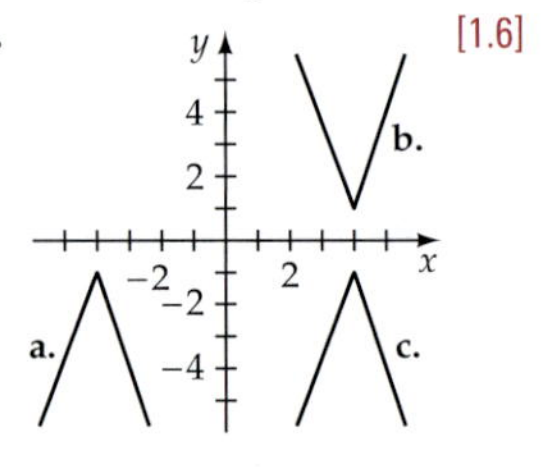

57. symmetric to the y-axis [1.6] **58.** symmetric to the x-axis [1.6] **59.** symmetric to the origin [1.6] **60.** symmetric to the x-axis, the y-axis, and the origin [1.6] **61.** symmetric to the x-axis, the y-axis, and the origin [1.6] **62.** symmetric to the origin [1.6] **63.** symmetric to the x-axis, the y-axis, and the origin [1.6] **64.** symmetric to the origin [1.6]

65.

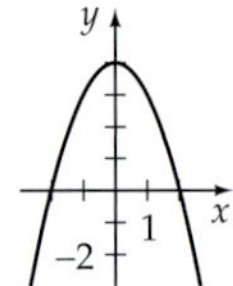

a. Domain is the set of all real numbers.
Range: $\{y \mid y \leq 4\}$
b. even [1.6]

66.

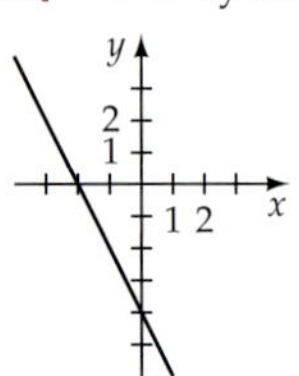

a. Domain is the set of all real numbers.
Range is the set of all real numbers.
b. g is neither even nor odd [1.6]

67.

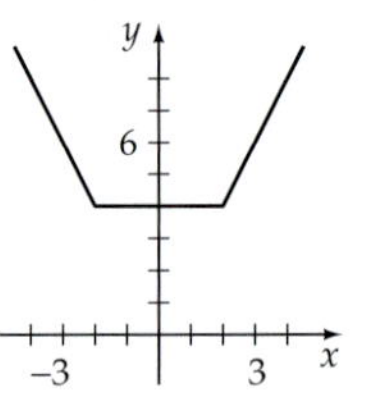

a. Domain is the set of all real numbers.
Range: $\{y \mid y \geq 4\}$
b. even [1.6]

68.

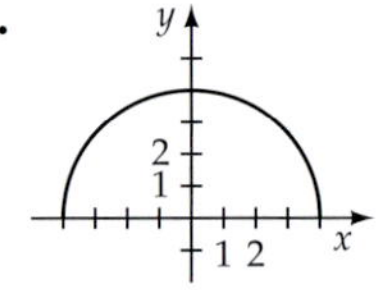

a. Domain: $\{x \mid -4 \leq x \leq 4\}$
Range: $\{y \mid 0 \leq y \leq 4\}$
b. even [1.6]

69.

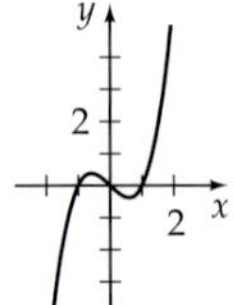

a. Domain is the set of all real numbers.
Range is the set of all real numbers.
b. odd [1.6]

70.

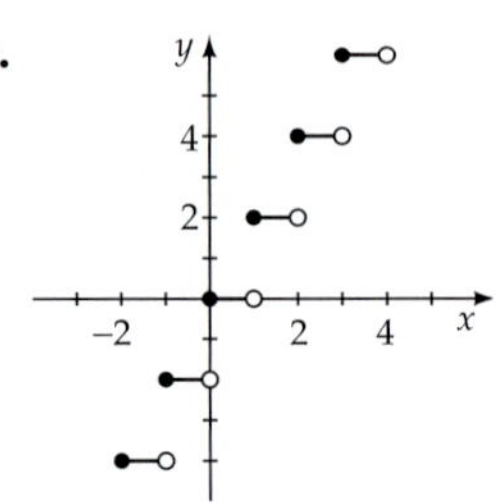

a. Domain: $\{x \mid x \text{ is a real number}\}$
Range: $\{y \mid y \text{ is an even integer}\}$
b. g is neither even nor odd [1.6]

71. $F(x) = (x + 2)^2 - 11$ [1.6]

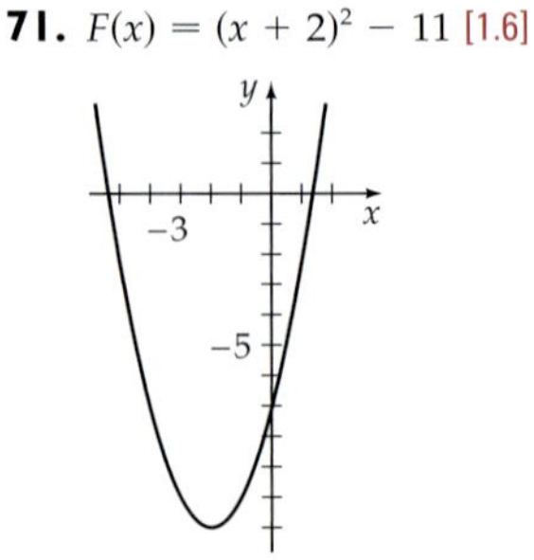

72. $A(x) = (x - 3)^2 - 14$ [1.6]

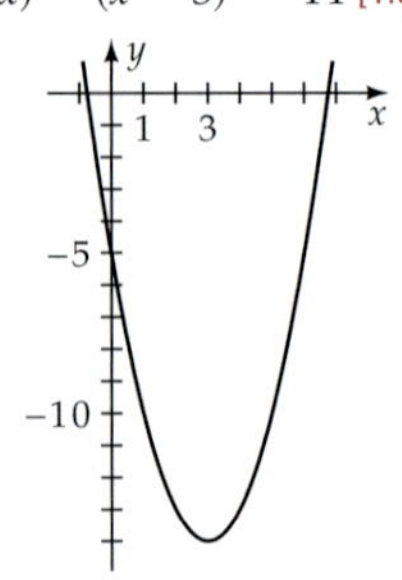

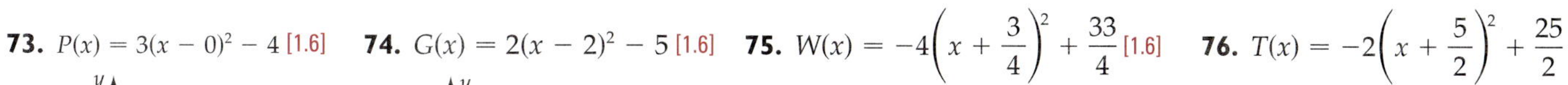

73. $P(x) = 3(x - 0)^2 - 4$ [1.6] **74.** $G(x) = 2(x - 2)^2 - 5$ [1.6] **75.** $W(x) = -4\left(x + \frac{3}{4}\right)^2 + \frac{33}{4}$ [1.6]

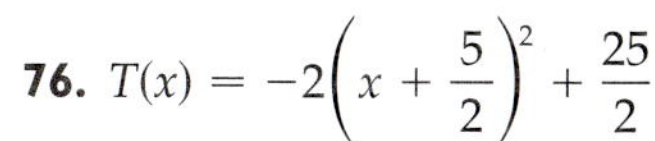

76. $T(x) = -2\left(x + \frac{5}{2}\right)^2 + \frac{25}{2}$ [1.6]

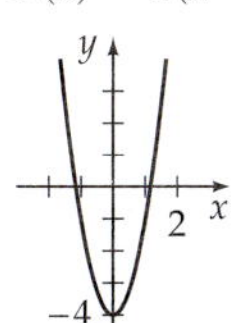

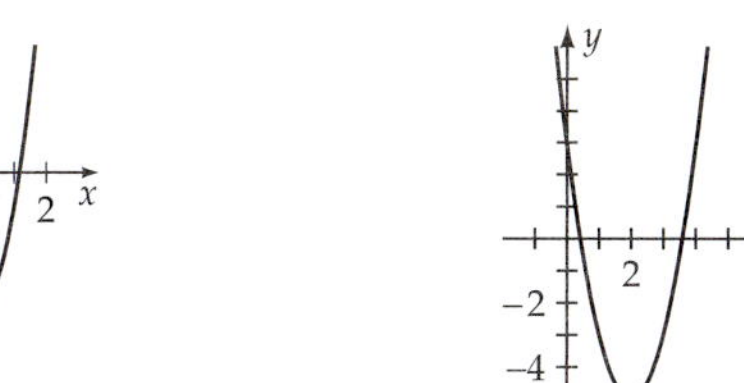

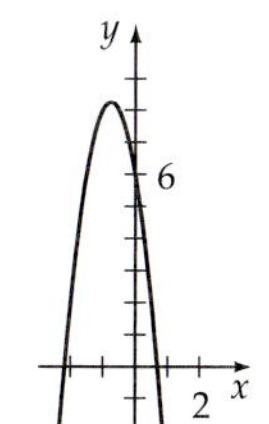

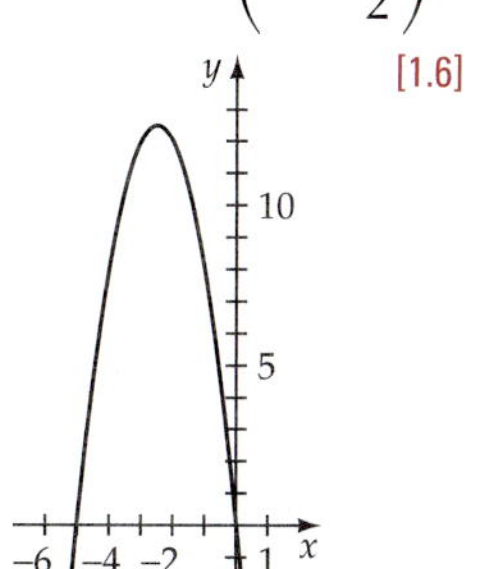

77. [1.6]

78. [1.6]

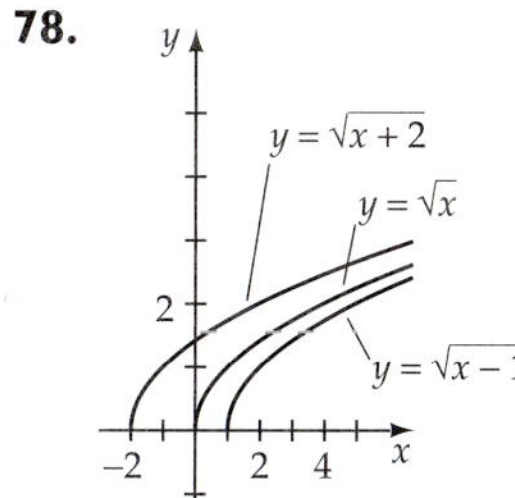

79. [1.6]

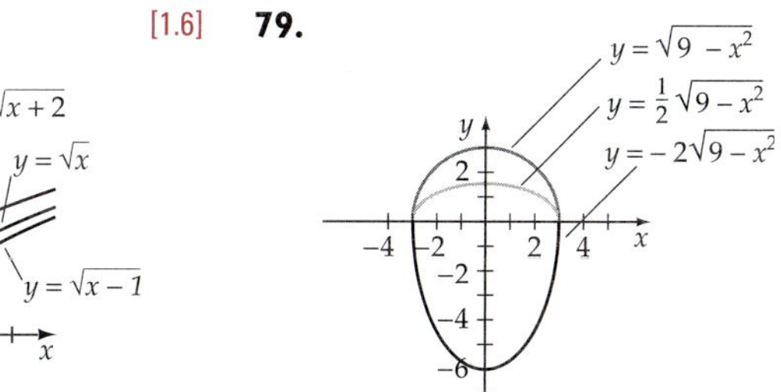

80. [1.3]

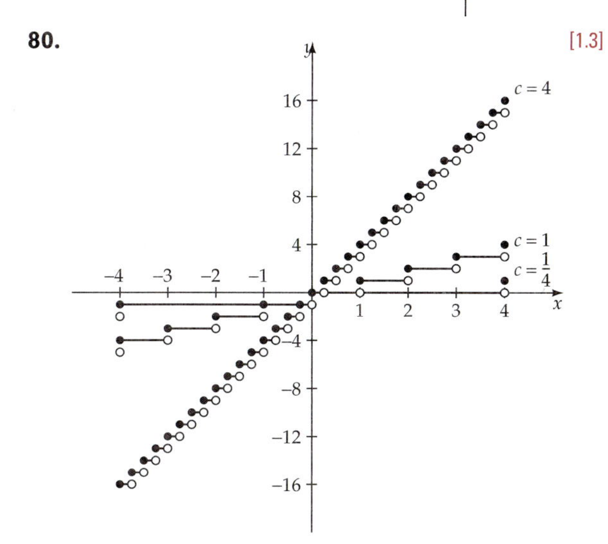

81. [1.3] **82.** [1.3]

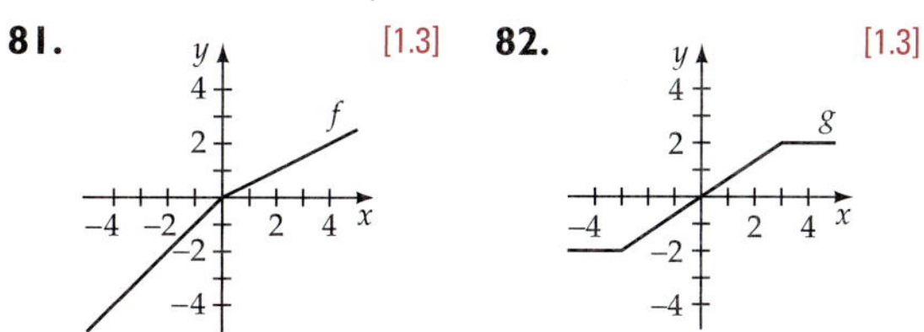

83. $f(x) + g(x) = x^2 + x - 6$, Domain is the set of all real numbers.
$f(x) - g(x) = x^2 - x - 12$, Domain is the set of all real numbers.
$f(x) \cdot g(x) = x^3 + 3x^2 - 9x - 27$, Domain is the set of all real numbers.
$\frac{f(x)}{g(x)} = x - 3$, Domain $\{x \mid x \neq -3\}$ [1.7]

84. $(f + g)(x) = x^3 + x^2 - 2x + 12$, Domain is the set of all real numbers.
$(f - g)(x) = x^3 - x^2 + 2x + 4$, Domain is the set of all real numbers.
$(fg)(x) = x^5 - 2x^4 + 4x^3 + 8x^2 - 16x + 32$, Domain is the set of all real numbers.
$\left(\frac{f}{g}\right)(x) = x + 2$, Domain is the set of all real numbers. [1.7]

85. 25, 25 [1.5] **86.** -5 and 5 [1.5] **87. a.** 18 feet per second **b.** 15 feet per second **c.** 13.5 feet per second **d.** 12.03 feet per second **e.** 12 feet per second [1.5] **88. a.** 17 feet per second **b.** 15 feet per second **c.** 14 feet per second **d.** 13.02 feet per second **e.** 13 feet per second [1.5]

Chapter 1 Test, page 114

1. 4 [1.1] **2.** $\frac{3 \pm \sqrt{29}}{2}$ [1.1] **3.** $x \le -6$ [1.1] **4.** $[-6, 2]$ [1.1] **5.** midpoint (1, 1); length $2\sqrt{13}$ [1.2]

6. $(-4, 0)$; $(0, \sqrt{2})$, $(0, -\sqrt{2})$ [1.2]

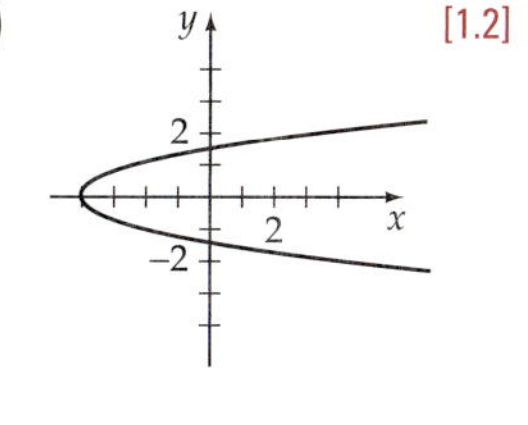

7. [1.2] **8.** center $(2, -1)$; radius 3 [1.2]

9. domain $\{x \mid x \geq 4 \text{ or } x \leq -4\}$ [1.3]  **10.** 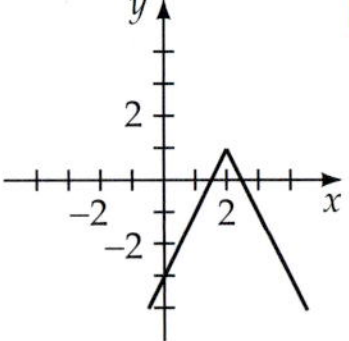

[1.3]
a. increasing on $(-\infty, 2]$
b. not constant on any interval
c. decreasing on $[2, \infty)$
11. a. $R = 12.00x$ **b.** $P = 11.25x - 875$ **c.** $x = 78$ [1.5]

12. 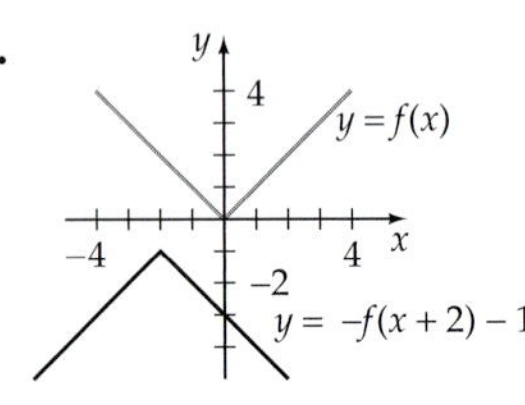

[1.6] **13. a.** even **b.** odd **c.** neither [1.6] **14.** $y = -\frac{2}{3}x + \frac{2}{3}$ [1.4] **15.** -12, minimum [1.5]
16. $x^2 + x - 3; \frac{x^2 - 1}{x - 2}, x \neq 2$ [1.7] **17.** $2x + h$ [1.7] **18.** $4x^2 + 16x + 15$ [1.7]

Exercise Set 2.1, page 124

1. $9i$ **3.** $7i\sqrt{2}$ **5.** $4 + 9i$ **7.** $5 + 7i$ **9.** $8 - 3i\sqrt{2}$ **11.** $11 - 5i$ **13.** $-7 + 4i$ **15.** $8 - 5i$ **17.** -10 **19.** $19i$ **21.** $20 - 10i$ **23.** $22 - 29i$ **25.** 41 **27.** $12 - 5i$ **29.** $-114 + 42i\sqrt{2}$ **31.** $-6i$ **33.** $3 - 6i$ **35.** $\frac{7}{53} - \frac{2}{53}i$ **37.** $1 + i$ **39.** $\frac{15}{41} - \frac{29}{41}i$ **41.** $\frac{5}{13} + \frac{12}{13}i$ **43.** $-i$ **45.** -1 **47.** $-i$ **49.** -1 **51.** $-7i, 7i$ **53.** $-3i, 3i$ **55.** $3 - 6i, 3 + 6i$ **57.** $-2 - \sqrt{5}i, -2 + \sqrt{5}i$ **59.** $-\frac{3}{2} - \frac{5}{2}i, -\frac{3}{2} + \frac{5}{2}i$ **61.** $2 - 5i, 2 + 5i$ **63.** $-\frac{1}{2} - \frac{1}{2}i, -\frac{1}{2} + \frac{1}{2}i$ **65.** $2 - \sqrt{5}, 2 + \sqrt{5}$ **67.** $1 - \frac{3}{2}i, 1 + \frac{3}{2}i$ **69.** $\frac{1}{2} - \sqrt{2}i, \frac{1}{2} + \sqrt{2}i$ **71.** $(x + 4i)(x - 4i)$ **73.** $(2x + 9i)(2x - 9i)$ **79.** 0

Prepare for Section 2.2, page 126

81. $8x$ **82.** $-12x^2$ **83.** 0 **84.** 45 **85.** $x^4 + 5x^3 + 3x^2 - 5x - 4$ **86.** $x^3 - 27$

Exercise Set 2.2, page 135

1. $5x^2 - 9x + 10 - \frac{10}{x + 3}$ **3.** $x^3 + 2x^2 - x + 1 + \frac{1}{x - 2}$ **5.** $x^2 + 4x + 10 + \frac{25}{x - 3}$ **7.** $x^3 + 7x^2 + 31x + 119 + \frac{475}{x - 4}$ **9.** $x^4 + 2x^3 + 2x - 1 - \frac{8}{x - 1}$ **11.** $4x^2 + 3x + 12 + \frac{17}{x - 2}$ **13.** $4x^2 - 4x + 2 + \frac{1}{x + 1}$ **15.** $x^4 + 4x^3 + 6x^2 + 24x + 101 + \frac{403}{x - 4}$ **17.** $x^4 + x^3 + x^2 + x + 1$ **19.** $8x^2 + 6$ **21.** $x^7 + 2x^6 + 5x^5 + 10x^4 + 21x^3 + 42x^2 + 85x + 170 + \frac{344}{x - 2}$ **23.** $x^5 - 3x^4 + 9x^3 - 27x^2 + 81x - 242 + \frac{716}{x + 3}$ **25.** 25 **27.** 45 **29.** -2230 **31.** -80 **33.** -187 **35.** Yes **37.** No **39.** Yes **41.** Yes **43.** No **55.** $(x - 2)(x^2 + 3x + 7)$ **57.** $(x - 4)(x^3 + 3x^2 + 3x + 1)$ **59. a.** \$19,968 **b.** \$23,007 **61. a.** 336 **b.** 336; They are the same. **63. a.** 100 cards **b.** 610 cards **65. a.** 400 people per square mile **b.** 240 people per square mile **67. a.** 304 cubic inches **b.** 892 cubic inches **69.** 13 **71.** Yes

Prepare for Section 2.3, page 137

73. 2 **74.** $\frac{9}{8}$ **75.** $[-1, \infty)$ **76.** $[1, \infty)$ **77.** $(x + 1)(x - 1)(x + 2)(x - 2)$ **78.** $\left(\frac{2}{3}, 0\right), \left(-\frac{1}{2}, 0\right)$

Exercise Set 2.3, page 149

1. up to the far left, up to the far right **3.** down to the far left, up to the far right **5.** down to the far left, down to the far right **7.** down to the far left, up to the far right **9.** $a < 0$ **11.** Vertex is $(-2, -5)$, minimum is -5. **13.** Vertex is $(-4, 17)$, maximum is 17.

15. relative maximum $y \approx 5.0$ at $x \approx -2.1$, relative minimum $y \approx -16.9$ at $x \approx 1.4$

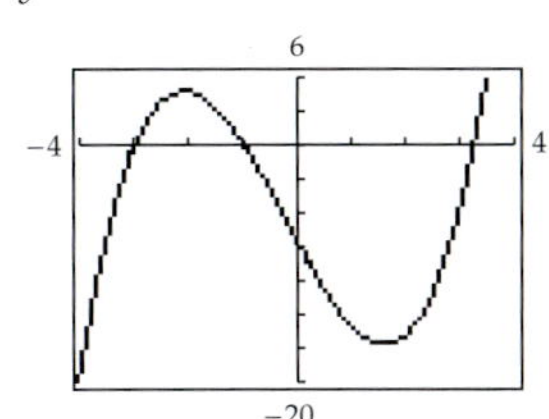

17. relative maximum $y \approx 31.0$ at $x \approx -2.0$, relative minimum $y \approx -77.0$ at $x \approx 4.0$

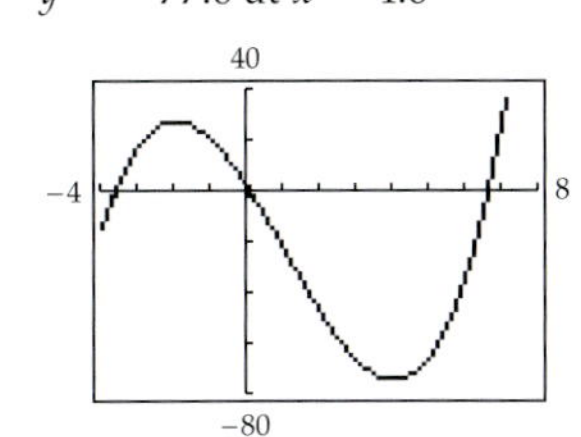

19. relative maximum $y \approx 2.0$ at $x \approx 1.0$, relative minima $y \approx -14.0$ at $x \approx -1.0$ and $y \approx -14.0$ at $x \approx 3.0$

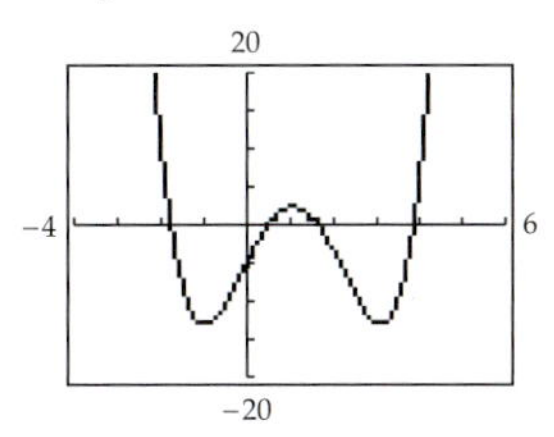

21. $-3, 0, 5$ **23.** $-3, -2, 2, 3$ **25.** $-2, -1, 0, 1, 2$ **33.** crosses the x-axis at $(-1, 0)$, $(1, 0)$, and $(3, 0)$ **35.** crosses the x-axis at $(7, 0)$; intersects but does not cross at $(3, 0)$ **37.** crosses the x-axis at $(1, 0)$; intersects but does not cross at $\left(\frac{3}{2}, 0\right)$ **39.** crosses the x-axis at $(0, 0)$; intersects but does not cross at $(3, 0)$

41.

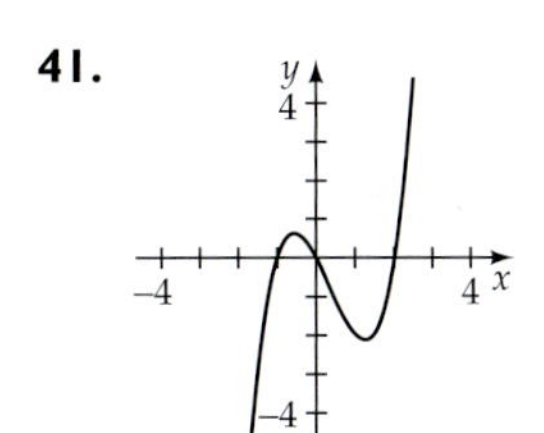

43.

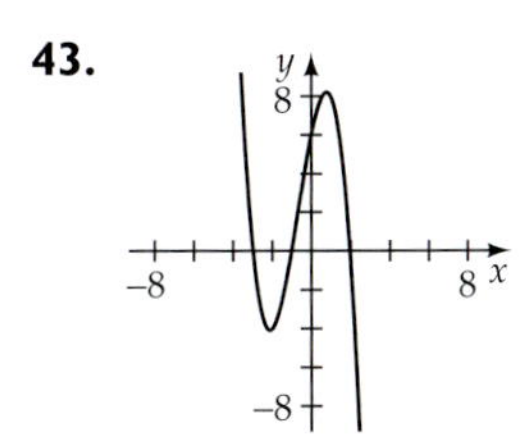

45.

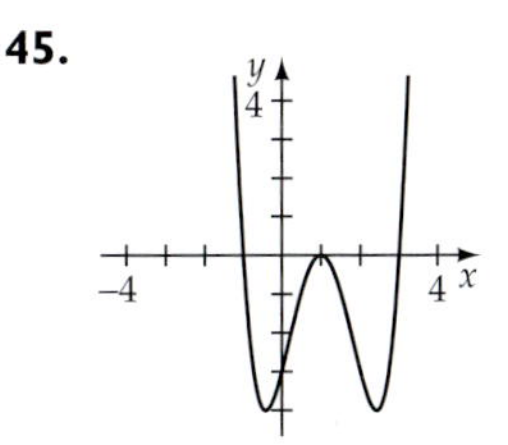

47. a. $V(x) = x(15 - 2x)(10 - 2x) = 4x^3 - 50x^2 + 150x$ **b.** 1.96 inches **49.** 2.137 inches **51.** \$464,000 **53. a.** 1918 **b.** 9.5 marriages per thousand population **55. a.** 20.69 milligrams **b.** 118 minutes **57. a.** 3.24 inches **b.** 4 feet from either end; 3.84 inches **c.** 3.24 inches **59.** between 3 and 4 **61.** $(5, 0)$ **63.** Shift the graph of $y = x^3$ horizontally 2 units to the right and vertically upward 1 unit.

Prepare for Section 2.4, page 153

65. $\frac{2}{3}, \frac{7}{2}$ **66.** $2x^2 - x + 6 - \frac{19}{x + 2}$ **67.** $3x^3 + 9x^2 + 6x + 15 + \frac{40}{x - 3}$ **68.** 1, 2, 3, 4, 6, 12 **69.** $\pm 1, \pm 3, \pm 9, \pm 27$
70. $P(-x) = -4x^3 - 3x^2 + 2x + 5$

Exercise Set 2.4, page 164

1. 3 (multiplicity 2), -5 (multiplicity 1) **3.** 0 (multiplicity 2), $-\frac{5}{3}$ (multiplicity 2) **5.** 2 (multiplicity 1), -2 (multiplicity 1), -3 (multiplicity 2)
7. $\pm 1, \pm 2, \pm 4, \pm 8$ **9.** $\pm 1, \pm 2, \pm 3, \pm 4, \pm 6, \pm 12, \pm\frac{1}{2}, \pm\frac{3}{2}$ **11.** $\pm 1, \pm 2, \pm 4, \pm\frac{1}{2}, \pm\frac{1}{3}, \pm\frac{2}{3}, \pm\frac{4}{3}, \pm\frac{1}{6}$ **13.** $\pm 1, \pm 7, \pm\frac{1}{2}, \pm\frac{7}{2}, \pm\frac{1}{4}, \pm\frac{7}{4}$ **15.** $\pm 1, \pm 2, \pm 4, \pm 8, \pm 16, \pm 32$ **17.** upper bound 2, lower bound -5 **19.** upper bound 4, lower bound -4
21. upper bound 1, lower bound -4 **23.** upper bound 4, lower bound -2 **25.** upper bound 2, lower bound -1 **27.** one positive zero, two or no negative zeros **29.** two or no positive zeros, one negative zero **31.** one positive zero, three or one negative zeros
33. three or one positive zeros, one negative zero **35.** one positive zero, no negative zeros **37.** $2, -1, -4$ **39.** $3, -4, \frac{1}{2}$
41. $\frac{1}{2}, -\frac{1}{3}, -2$ (multiplicity 2) **43.** $\frac{1}{2}, 4, \sqrt{3}, -\sqrt{3}$ **45.** $6, 1 + \sqrt{5}, 1 - \sqrt{5}$ **47.** $5, \frac{1}{2}, 2 + \sqrt{3}, 2 - \sqrt{3}$
49. $1, -1, -2, -\frac{2}{3}, 3 + \sqrt{3}, 3 - \sqrt{3}$ **51.** $2, -1$ (multiplicity 2) **53.** $0, -2, 1 + \sqrt{2}, 1 - \sqrt{2}$ **55.** -1 (multiplicity 3), 2
57. $-\frac{3}{2}$, 1 (multiplicity 2), 8 **59.** $n = 9$ inches **61.** $x = 4$ inches **63. a.** 26 pieces **b.** 7 cuts **65.** 7 rows **67.** $x = 0.084$ inch
69. 1977 and 1986 **71.** 16.9 feet **73. a.** 73 seconds **b.** 93,000 digits **75.** $B = 15$. The absolute value of each of the given zeros is less than B. **77.** $B = 11$. The absolute value of each of the zeros is less than B.

Prepare for Section 2.5, page 168

79. $3 + 2i$ **80.** $2 - i\sqrt{5}$ **81.** $x^3 - 8x^2 + 19x - 12$ **82.** $x^2 - 4x + 5$ **83.** $-3i, 3i$ **84.** $\frac{1}{2} - \frac{1}{2}i\sqrt{19}, \frac{1}{2} + \frac{1}{2}i\sqrt{19}$

Exercise Set 2.5, page 175

1. $2, -3, 2i, -2i; P(x) = (x - 2)(x + 3)(x - 2i)(x + 2i)$ **3.** $\frac{1}{2}, -3, 1 + 5i, 1 - 5i; P(x) = \left(x - \frac{1}{2}\right)(x + 3)(x - 1 - 5i)(x - 1 + 5i)$

5. 1 (multiplicity 3), $3 + 2i, 3 - 2i; P(x) = (x - 1)^3(x - 3 - 2i)(x - 3 + 2i)$

7. $-3, -\frac{1}{2}, 2 + i, 2 - i; P(x) = (x + 3)\left(x + \frac{1}{2}\right)(x - 2 - i)(x - 2 + i)$

9. $4, 2, \frac{1}{2} + \frac{3}{2}i, \frac{1}{2} - \frac{3}{2}i; P(x) = (x - 4)(x - 2)\left(x - \frac{1}{2} - \frac{3}{2}i\right)\left(x - \frac{1}{2} + \frac{3}{2}i\right)$ **11.** $1 - i, \frac{1}{2}$ **13.** $i, -3$ **15.** $2 + 3i, i, -i$

17. $1 - 3i, 1 + 2i, 1 - 2i$ **19.** $2i$, 1 (multiplicity 3) **21.** $5 - 2i, \frac{7}{2} + \frac{\sqrt{3}}{2}i, \frac{7}{2} - \frac{\sqrt{3}}{2}i$ **23.** $\frac{3}{2}, -\frac{1}{2} + \frac{\sqrt{7}}{2}i, -\frac{1}{2} - \frac{\sqrt{7}}{2}i$ **25.** $-\frac{2}{3}, \frac{3}{4}, \frac{5}{2}$

27. $-i, i$, 2 (multiplicity 2) **29.** -3 (multiplicity 2), 1 (multiplicity 2) **31.** $P(x) = x^3 - 3x^2 - 10x + 24$ **33.** $P(x) = x^3 - 3x^2 + 4x - 12$
35. $P(x) = x^4 - 10x^3 + 63x^2 - 214x + 290$ **37.** $P(x) = x^5 - 22x^4 + 212x^3 - 1012x^2 + 2251x - 1830$ **39.** $P(x) = 4x^3 - 19x^2 + 224x - 159$
41. $P(x) = x^3 + 13x + 116$ **43.** $P(x) = x^4 - 18x^3 + 131x^2 - 458x + 650$ **45.** $P(x) = x^5 - 4x^4 + 16x^3 - 18x^2 - 97x + 102$
47. $P(x) = 3x^3 - 12x^2 + 3x + 18$ **49.** $P(x) = -2x^4 + 4x^3 + 36x^2 - 140x + 150$ **51.** The Conjugate Pair Theorem does not apply because some of the coefficients of the polynomial are not real numbers.

Prepare for Section 2.6, page 176

53. $\frac{x - 3}{x - 5}$ **54.** $-\frac{3}{2}$ **55.** $\frac{1}{3}$ **56.** $x = 0, -3, \frac{5}{2}$ **57.** The degree of the numerator is 3. The degree of the denominator is 2.

58. $x + 4 + \frac{7x - 11}{x^2 - 2x}$

Exercise Set 2.6, page 189

1. $x = 0, x = -3$ **3.** $x = -\frac{1}{2}, x = \frac{4}{3}$ **5.** $y = 4$ **7.** $y = 30$

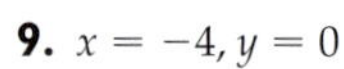

9. $x = -4, y = 0$ **11.** $x = 3, y = 0$ **13.** $x = 0, y = 0$ **15.** $x = -4, y = 1$ **17.** $x = 2, y = -1$

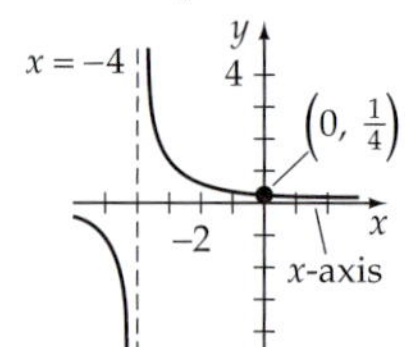

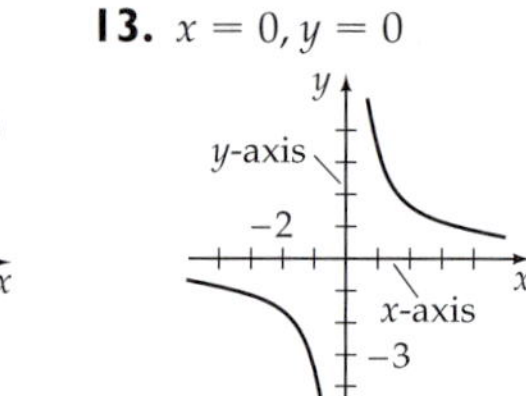

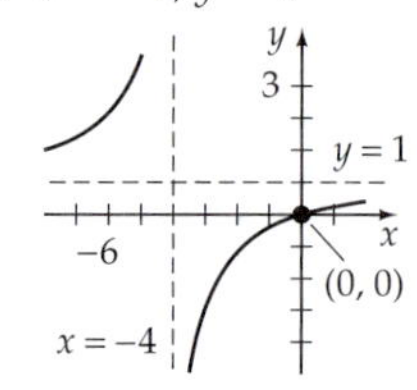

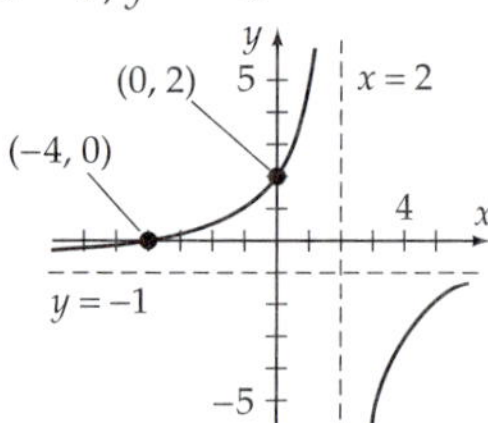

19. $x = 3, x = -3, y = 0$ **21.** $x = -3, x = 1, y = 0$ **23.** $x = -2, y = 1$

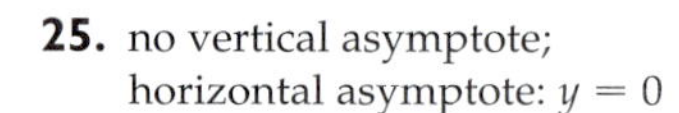

25. no vertical asymptote; horizontal asymptote: $y = 0$

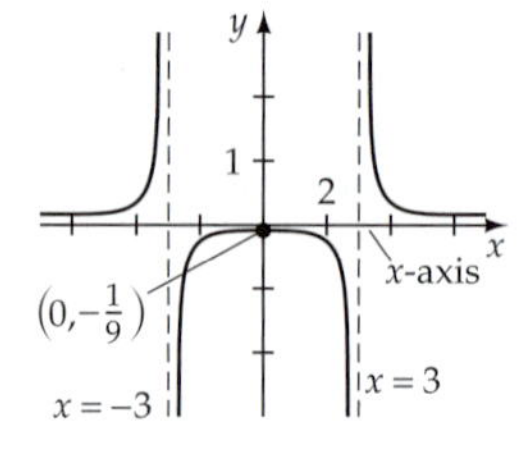

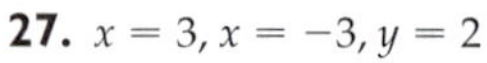
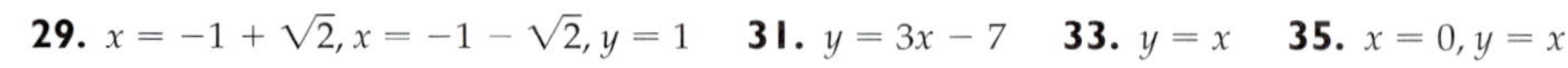
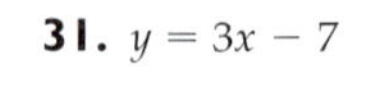

27. $x = 3, x = -3, y = 2$ **29.** $x = -1 + \sqrt{2}, x = -1 - \sqrt{2}, y = 1$ **31.** $y = 3x - 7$ **33.** $y = x$ **35.** $x = 0, y = x$

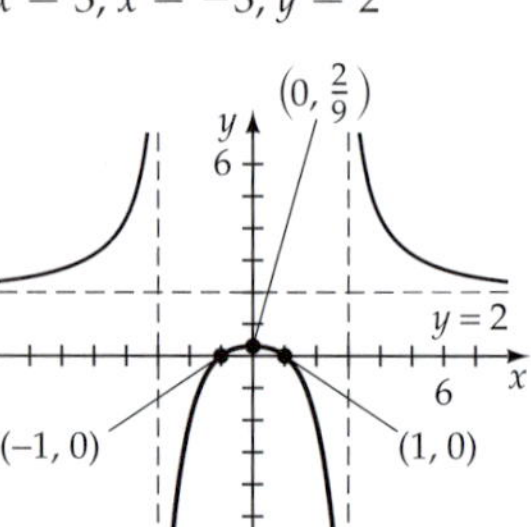

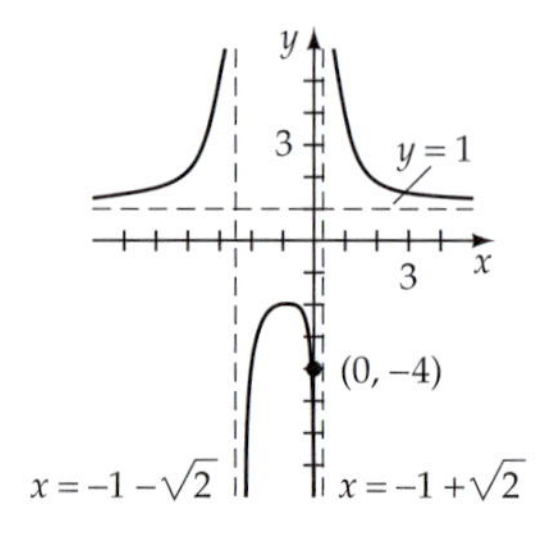

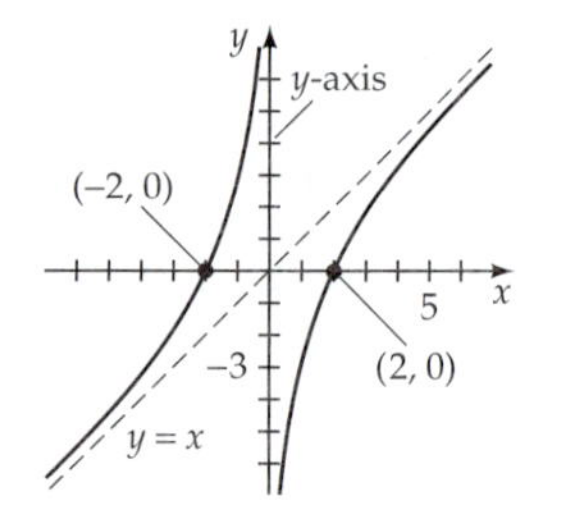

37. $x = -3, y = x - 6$

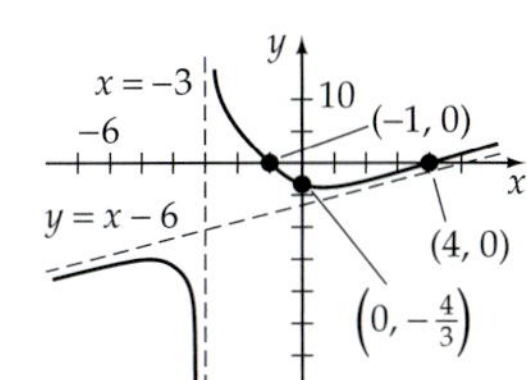

39. $x = 4, y = 2x + 13$

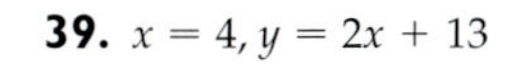

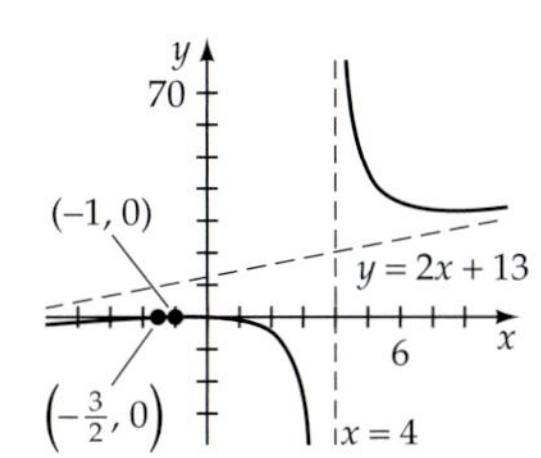

41. $x = -2, y = x - 3$

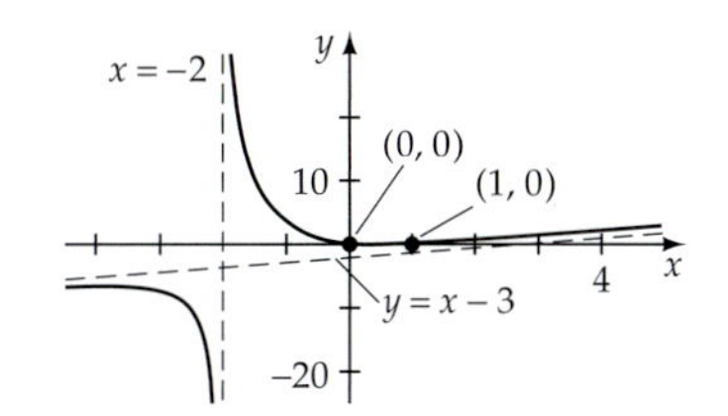

43. $x = 2, x = -2, y = x$

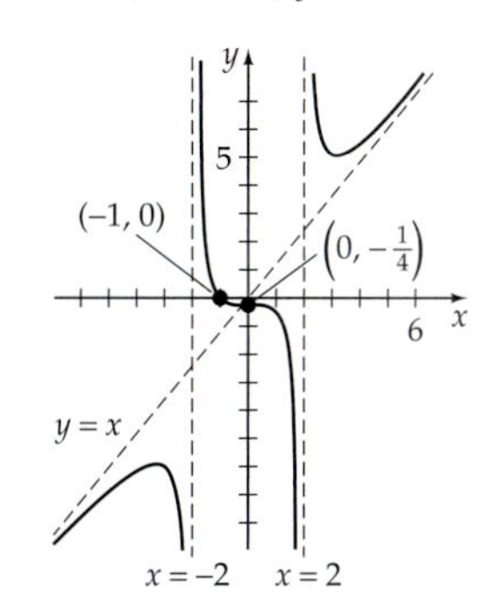

45.

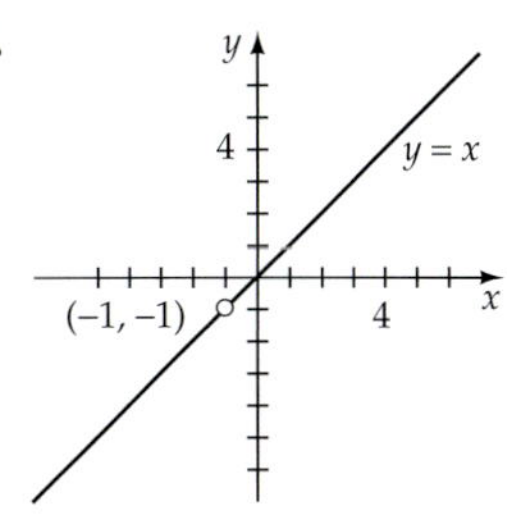

47.

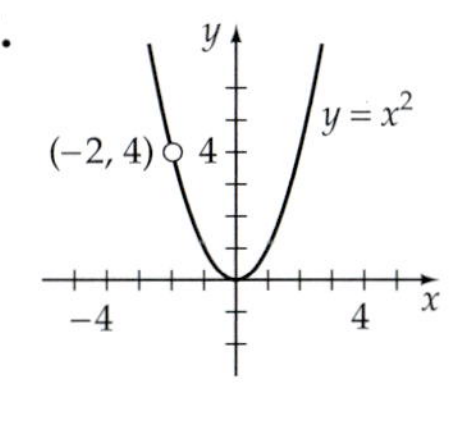

49.

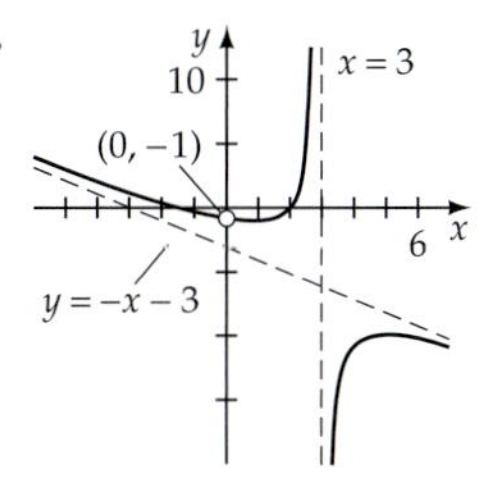

51.

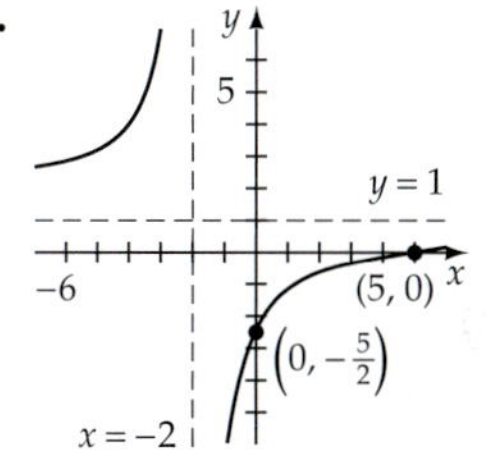

53. a. \$76.43, \$8.03, \$1.19 **b.** $y = 0.43$. As the number of golf balls produced increases, the average cost per golf ball approaches \$.43.

55. a. \$1333.33 **b.** \$8000 **c.**

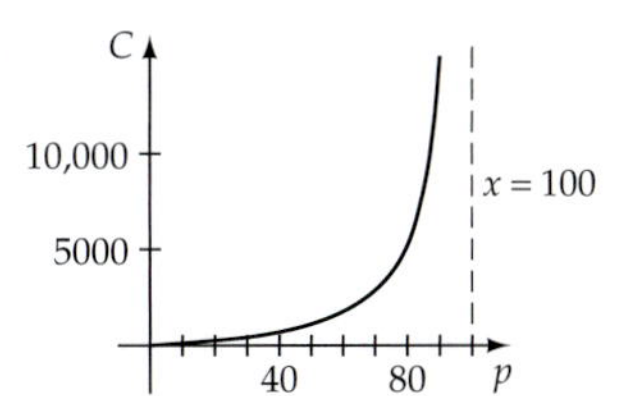

57. a. $R(0) \approx 38.8\%$, $R(7) \approx 39.9\%$, $R(12) \approx 40.9\%$ **b.** $\approx 44.7\%$

59. a. 26,923, 68,293, 56,000 **b.** 2001 **c.** The population will approach 0. **61. a.** 3.8 centimeters

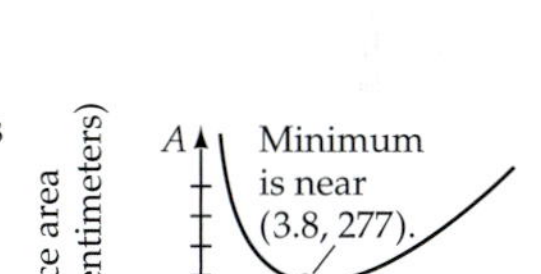

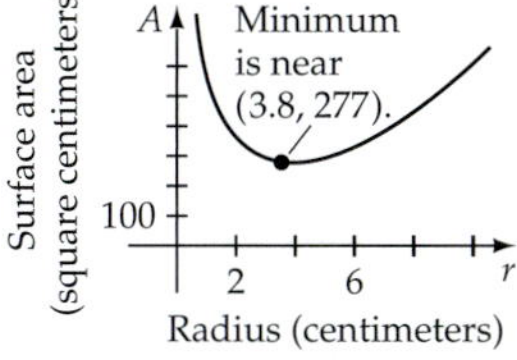

b. No **c.** As the radius r increases without bound, the surface area approaches twice the area of a circle with radius r. **63.** $(-2, 2)$
65. $(0, 1)$ and $(-4, 1)$ **67.** Answers will vary; one example is $\dfrac{x^2 + 1}{x^2 - x - 6}$.

Chapter 2 True/False Exercises, page 195

1. False; $P(x) = x - i$ has a zero of i, but it does not have a zero of $-i$. **2.** False; Descartes' Rule of Signs indicates that $P(x) = x^3 - x^2 + x - 1$ has three or one positive zeros. In fact, P has only one positive zero. **3.** True **4.** True **5.** False; $F(x) = \dfrac{x}{x^2 + 1}$ does not have a vertical asymptote. **6.** False; $F(x) = \dfrac{(x - 2)^2}{(x - 3)(x - 2)} = \dfrac{x - 2}{x - 3}, x \neq 2$. The graph of F has a hole at $x = 2$. **7.** True **8.** True **9.** True **10.** True
11. True **12.** False; $P(x) = x^2 + 1$ does not have a real zero.

Chapter 2 Review Exercises, page 196

1. $5 + 8i$ [2.1] **2.** $2 - 3i\sqrt{2}$ [2.1] **3.** $6 - i$ [2.1] **4.** $-2 + 10i$ [2.1] **5.** $8 + 6i$ [2.1] **6.** $29 + 22i$ [2.1] **7.** $8 + 6i$ [2.1] **8.** i [2.1]
9. $-3 - 2i$ [2.1] **10.** $-\dfrac{14}{25} - \dfrac{23}{25}i$ [2.1] **11.** $\dfrac{2}{3} - \dfrac{1}{3}i, \dfrac{2}{3} + \dfrac{1}{3}i$ [2.1] **12.** $\dfrac{3}{4} - \dfrac{1}{2}i, \dfrac{3}{4} + \dfrac{1}{2}i$ [2.1] **13.** $\dfrac{1}{2} - \dfrac{\sqrt{3}}{2}i, \dfrac{1}{2} + \dfrac{\sqrt{3}}{2}i$ [2.1]

14. $1 - \frac{1}{2}i, 1 + \frac{1}{2}i$ [2.1] **15.** $4x^2 + x + 8 + \frac{22}{x - 3}$ [2.2] **16.** $5x^2 + 5x - 13 - \frac{11}{x - 1}$ [2.2] **17.** $3x^2 - 6x + 7 - \frac{13}{x + 2}$ [2.2]

18. $2x^2 + 8x + 20$ [2.2] **19.** $3x^2 + 5x - 11$ [2.2] **20.** $x^3 + 2x^2 - 8x - 9$ [2.2] **21.** 77 [2.2] **22.** 22 [2.2] **23.** 33 [2.2] **24.** 558 [2.2]

The verifications in Exercises 25–28 make use of the concepts from Section 2.2.

29. [2.3]

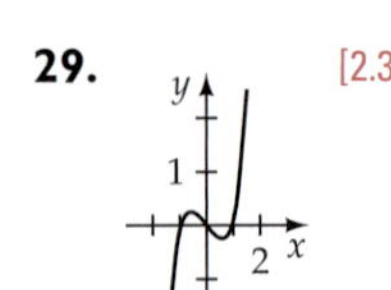

30. [2.3]

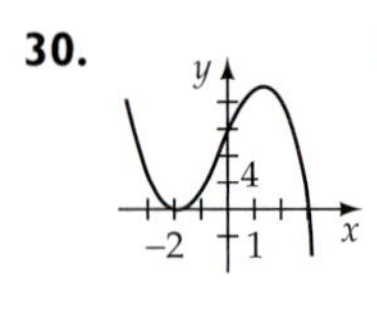

31. [2.3]

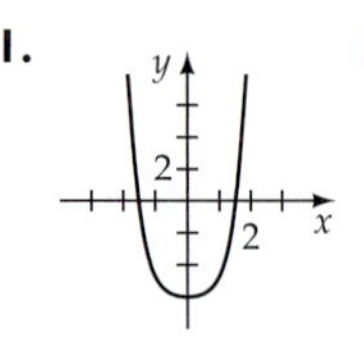

32. [2.3]

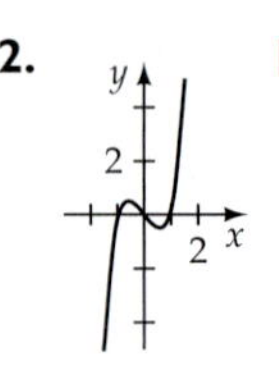

33. [2.3]

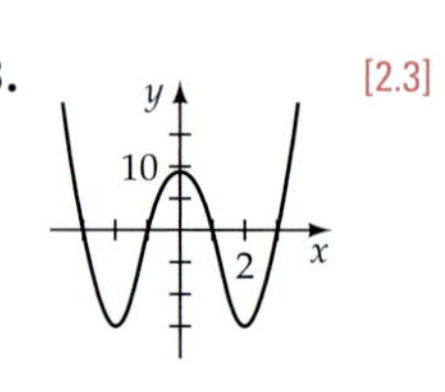

34. [2.3]

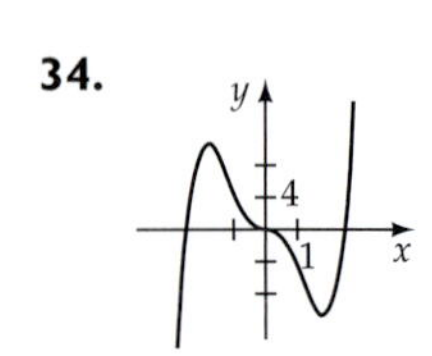

35. $\pm 1, \pm 2, \pm 3, \pm 6$ [2.4] **36.** $\pm 1, \pm 2, \pm 3, \pm 5, \pm 6, \pm 10, \pm 15, \pm 30, \pm\frac{1}{2}, \pm\frac{3}{2}, \pm\frac{5}{2}, \pm\frac{15}{2}$ [2.4]

37. $\pm 1, \pm 2, \pm 3, \pm 4, \pm 6, \pm 12, \pm\frac{1}{3}, \pm\frac{2}{3}, \pm\frac{4}{3}, \pm\frac{1}{5}, \pm\frac{2}{5}, \pm\frac{3}{5}, \pm\frac{4}{5}, \pm\frac{6}{5}, \pm\frac{12}{5}, \pm\frac{1}{15}, \pm\frac{2}{15}, \pm\frac{4}{15}$ [2.4] **38.** $\pm 1, \pm 2, \pm 4, \pm 8, \pm 16, \pm 32, \pm 64$ [2.4] **39.** ± 1 [2.4] **40.** $\pm 1, \pm 2, \pm\frac{1}{6}, \pm\frac{1}{3}, \pm\frac{1}{2}, \pm\frac{2}{3}$ [2.4] **41.** no positive real zeros and three or one negative real zeros [2.4]
42. three or one positive real zeros, one negative real zero [2.4] **43.** one positive real zero and one negative real zero [2.4]

44. five, three, or one positive real zeros, no negative real zeros [2.4] **45.** $1, -2, -5$ [2.4] **46.** $2, 5, 3$ [2.4] **47.** -2 (multiplicity 2), $-\frac{1}{2}, -\frac{4}{3}$ [2.4] **48.** $-\frac{1}{2}, -3, i, -i$ [2.5] **49.** 1 (multiplicity 4) [2.4] **50.** $-\frac{1}{2}, 2 + 3i, 2 - 3i$ [2.5] **51.** $-1, 3, 1 + 2i$ [2.5] **52.** $-5, 2, 2 - i$ [2.5]
53. $P(x) = 2x^3 - 3x^2 - 23x + 12$ [2.5] **54.** $P(x) = x^4 + x^3 - 5x^2 + x - 6$ [2.5] **55.** $P(x) = x^4 - 3x^3 + 27x^2 - 75x + 50$ [2.5]
56. $P(x) = x^4 + 2x^3 + 6x^2 + 32x + 40$ [2.5] **57.** vertical asymptote: $x = -2$, horizontal asymptote: $y = 3$ [2.6]
58. vertical asymptotes: $x = -3, x = 1$, horizontal asymptote: $y = 2$ [2.6] **59.** vertical asymptote: $x = -1$, slant asymptote: $y = 2x + 3$ [2.6]
60. no vertical asymptote, horizontal asymptote: $y = 3$ [2.6]

61. [2.6]

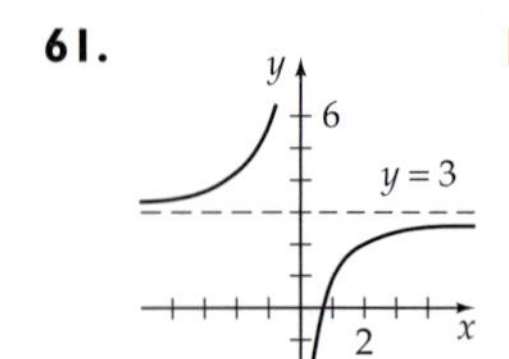

62. [2.6]

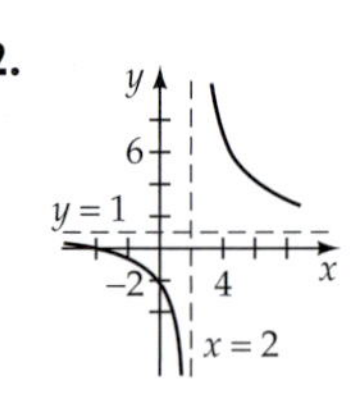

63. [2.6]

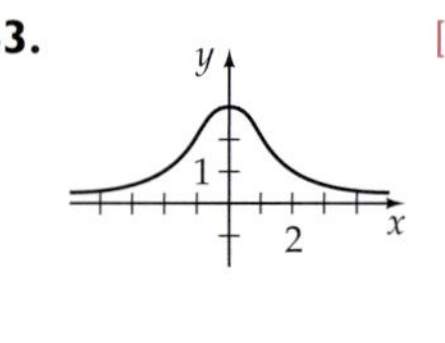

64. [2.6]

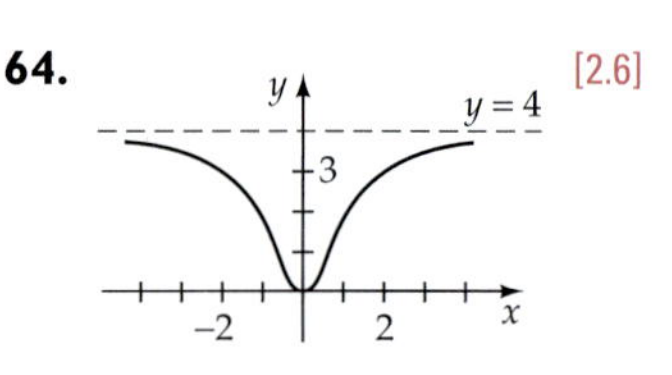

65. [2.6]

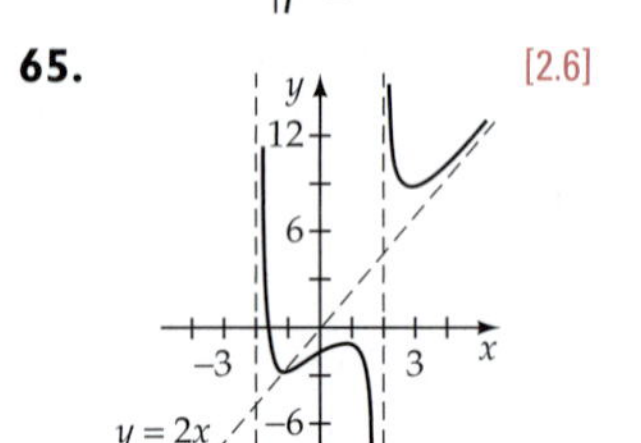

66. [2.6]

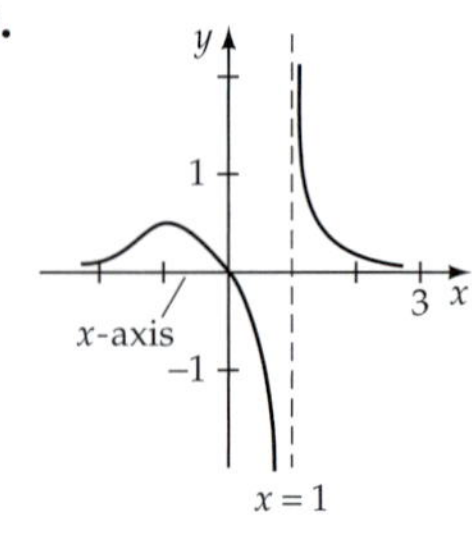

67. [2.6]

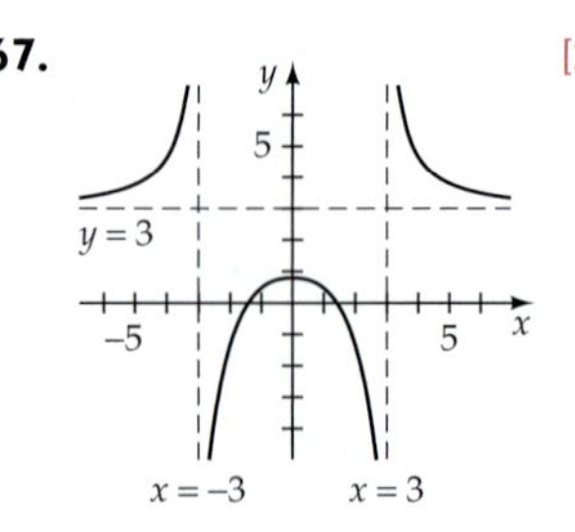

68. [2.6]

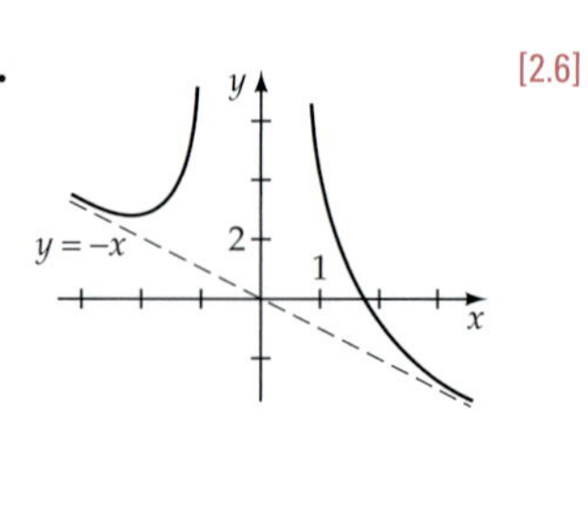

69. a. \$12.59, \$6.43 **b.** $y = 5.75$. As the number of skateboards produced increases, the average cost per skateboard approaches \$5.75. [2.6]
70. a. 15°F **b.** 2.4°F **c.** 0°F [2.6] **71. a.** As the radius of the blood vessel approaches 0, the resistance gets larger. **b.** As the radius of the blood vessel gets larger, the resistance approaches zero. [2.6]

Chapter 2 Test, page 198

1. a. $2 - 3i\sqrt{6}$ **b.** $8 - 18i$ [2.1] **2. a.** $-29 - 29i$ **b.** $1 + 3i$ [2.1] **3.** $3x^2 - x + 6 - \dfrac{13}{x+2}$ [2.2] **4.** 43 [2.2] **5.** The verification for Exercise 5 makes use of the concepts from Section 2.2. **6.** $0, \dfrac{2}{3}, -3$ [2.3] **7.** $P(1) < 0$, $P(2) > 0$. Therefore, by the Zero Location Theorem, the continuous polynomial function P has a zero between 1 and 2. [2.3]

8. 2 (multiplicity 2), -2 (multiplicity 2), $\dfrac{3}{2}$ (multiplicity 1), -1 (multiplicity 3) [2.4] **9.** $\pm 1, \pm 3, \pm\dfrac{1}{2}, \pm\dfrac{3}{2}, \pm\dfrac{1}{3}, \pm\dfrac{1}{6}$ [2.4]

10. upper bound 4, lower bound -5 [2.4] **11.** four, two, or zero positive zeros, no negative zero [2.4] **12.** $\dfrac{1}{2}, 3, -2$ [2.4]

13. $2 - 3i, -\dfrac{2}{3}, -\dfrac{5}{2}$ [2.5] **14.** 0, 1 (multiplicity 2), $2 + i$, $2 - i$ [2.5] **15.** $P(x) = x^4 - 5x^3 + 8x^2 - 6x$ [2.5] **16. a.** vertical asymptotes: $x = 3$, $x = 2$ **b.** horizontal asymptote: $y = \dfrac{3}{2}$ [2.6] **17.** [2.6] **18.** [2.6]

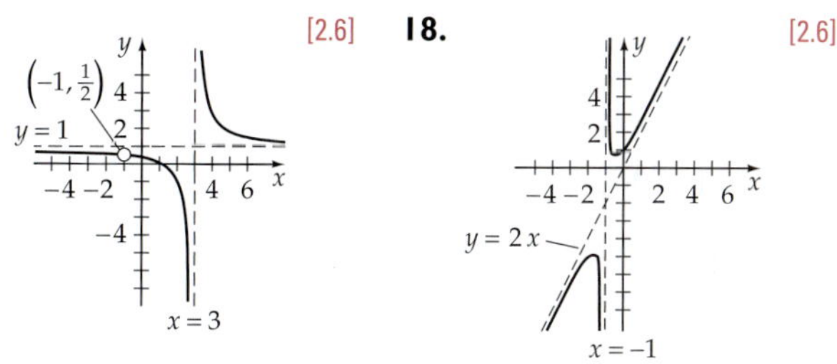

19. a. 5 words per minute, 16 words per minute, 25 words per minute **b.** 70 words per minute [2.6] **20.** 2.42 inches, 487.9 cubic inches [2.4]

Cumulative Review Exercises, page 199

1. $-1 + 2i$ [2.1] **2.** $\dfrac{1 \pm \sqrt{5}}{2}$ [1.1] **3.** $\dfrac{3}{4} \pm \dfrac{\sqrt{7}}{4}i$ [2.1] **4.** $\left(4, \dfrac{3}{2}\right)$ [1.2] **5.** $\sqrt{281}$ [1.2] **6.** Translate the graph of $y = x^2$ to the right 2 units and 4 units up. [1.6] **7.** $2x + h - 2$ [1.7] **8.** $32x^2 - 92x + 60$ [1.7] **9.** $x^3 - x^2 + x + 11$ [1.7] **10.** $4x^3 - 8x^2 + 14x - 32 + \dfrac{59}{x+2}$ [2.2] **11.** 141 [2.2] **12.** The graph goes down. [2.3] **13.** 0.3997 [2.3] **14.** $\pm 1, \pm 2, \pm 4, \pm\dfrac{1}{3}, \pm\dfrac{2}{3}, \pm\dfrac{4}{3}$ [2.4] **15.** zero positive real zeros, three or one negative real zeros [2.4] **16.** $-2, 1 + 2i, 1 - 2i$ [2.5] **17.** $P(x) = x^3 - 4x^2 - 2x + 20$ [2.5] **18.** $(x - 2)(x + 3i)(x - 3i)$ [2.5] **19.** vertical asymptotes: $x = -3$, $x = 2$; horizontal asymptote: $y = 4$ [2.6] **20.** $y = x + 4$ [2.6]

Exercise Set 3.1, page 212

1. 3 **3.** -3 **5.** 3 **7.** range **9.** Yes

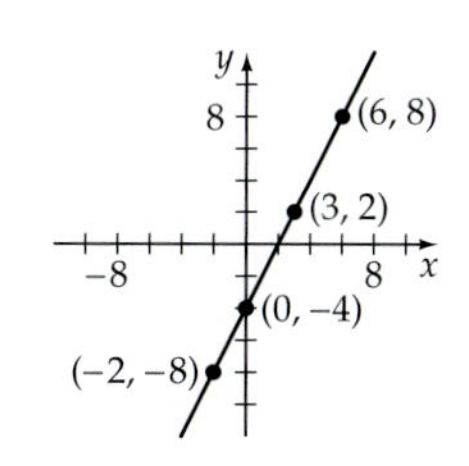

11. Yes

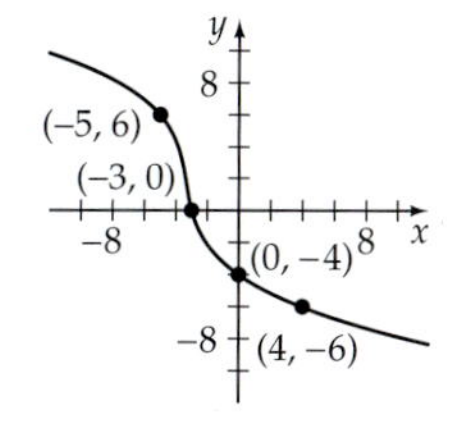

13. Yes

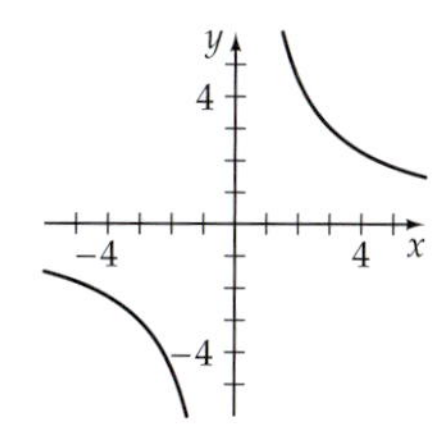

15. No

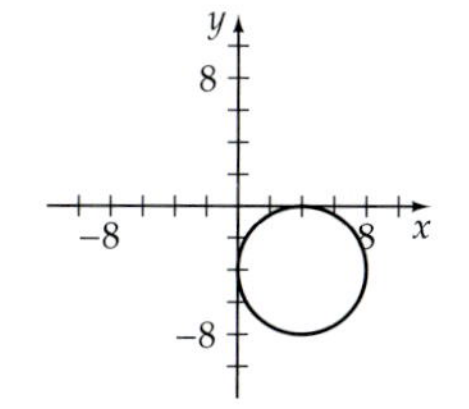

17. Yes **19.** Yes **21.** No **23.** $\{(1, -3), (2, -2), (5, 1), (-7, 4)\}$ **25.** $\{(1, 0), (2, 1), (4, 2), (8, 3), (16, 4)\}$

27. $f^{-1}(x) = \frac{1}{2}x - 2$ **29.** $f^{-1}(x) = \frac{1}{3}x + \frac{7}{3}$ **31.** $f^{-1}(x) = -\frac{1}{2}x + \frac{5}{2}$ **33.** $f^{-1}(x) = \frac{x}{x-2}, x \neq 2$ **35.** $f^{-1}(x) = \frac{x+1}{1-x}, x \neq 1$
37. $f^{-1}(x) = \sqrt{x-1}, x \geq 1$ **39.** $f^{-1}(x) = x^2 + 2, x \geq 0$ **41.** $f^{-1}(x) = \sqrt{x+4} - 2, x \geq -4$ **43.** $f^{-1}(x) = -\sqrt{x+5} - 2, x \geq -5$
45. $V^{-1}(x) = \sqrt[3]{x}$. V^{-1} finds the length of a side of a cube given the volume. **47.** Yes. Yes. A conversion function is a nonconstant linear function. All nonconstant linear functions have inverses that are also functions. **49.** $s^{-1}(x) = \frac{1}{2}x - 12$ **51.** $E^{-1}(s) = 20s - 50{,}000$. From the monthly earnings s the executive can find $E^{-1}(s)$, the value of the software sold. **53.** 44205833; $f^{-1}(x) = \frac{1}{2}x + \frac{1}{2}$, $f^{-1}(44205833) = 22102917$
55. Because the function is increasing and 4 is between 2 and 5, c must be between 7 and 12. **57.** between 2 and 5 **59.** between 3 and 7
61. $f^{-1}(x) = \frac{x-b}{a}, a \neq 0$ **63.** The reflection of f across the line given by $y = x$ yields f. Thus f is its own inverse. **65.** Yes **67.** No

Prepare for Section 3.2, page 215

69. 8 **70.** $\frac{1}{81}$ **71.** $\frac{17}{8}$ **72.** $\frac{40}{9}$ **73.** $\frac{1}{10}$, 1, 10, and 100 **74.** 2, 1, $\frac{1}{2}$, and $\frac{1}{4}$

Exercise Set 3.2, page 224

1. $f(0) = 1; f(4) = 81$ **3.** $g(-2) = \frac{1}{100}; g(3) = 1000$ **5.** $h(2) = \frac{9}{4}; h(-3) = \frac{8}{27}$ **7.** $j(-2) = 4; j(4) = \frac{1}{16}$ **9.** 9.19 **11.** 9.03 **13.** 9.74
15. a. $k(x)$ **b.** $g(x)$ **c.** $h(x)$ **d.** $f(x)$

17. **19.** **21.** **23.**

25. Shift the graph of f vertically upward 2 units. **27.** Shift the graph of f horizontally to the right 2 units.
29. Reflect the graph of f across the y-axis. **31.** Stretch the graph of f vertically away from the x-axis by a factor of 2.
33. Reflect the graph of f across the y-axis and then shift this graph vertically upward 2 units.

35. no horizontal asymptote

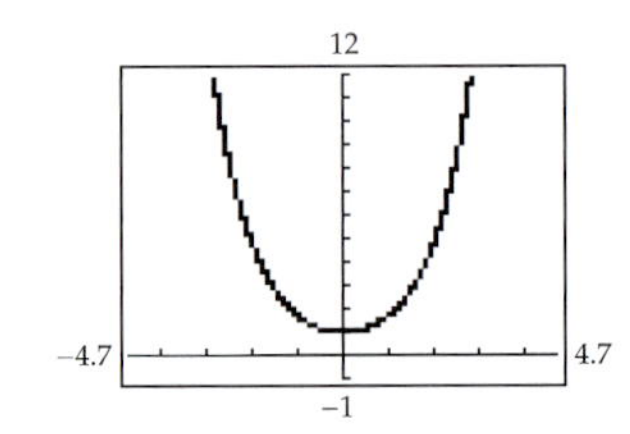

37. no horizontal asymptote

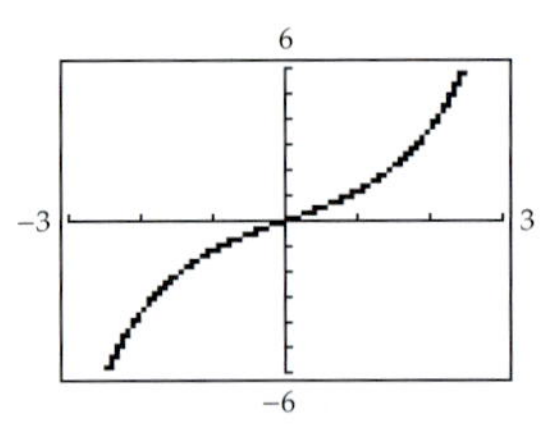

39. horizontal asymptote: $y = 0$

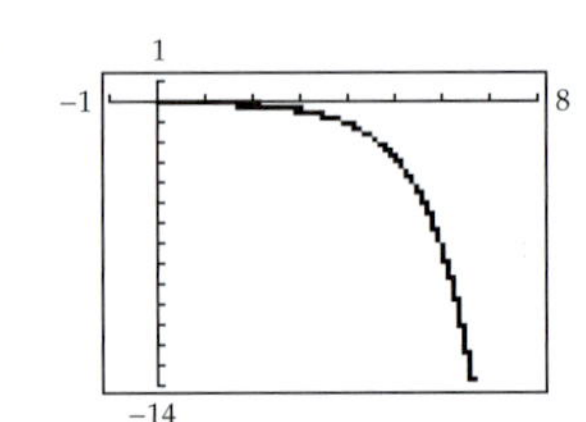

41. horizontal asymptote: $y = 10$

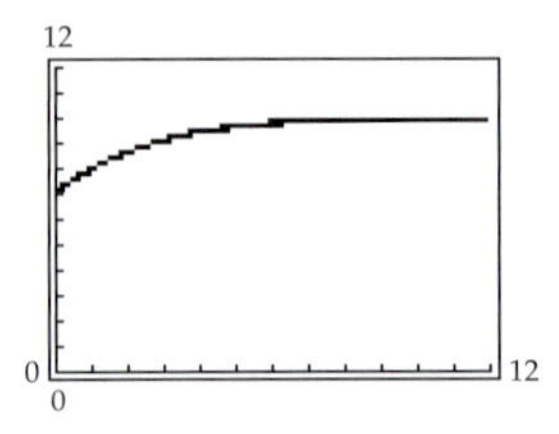

43. a. 122 million connections **b.** 2006 **45. a.** 233 items per month; 59 items per month **b.** The demand will approach 25 items per month. **47. a.** 0.53 **b.** 0.89 **c.** 5.2 minutes **d.** There is a 98% probability that at least one customer will arrive between 10:00 A.M. and 10:05.2 A.M. **49. a.** 8.7% **b.** 2.6% **51. a.** 6400; 409,600 **b.** 11.6 hours **53. a.** 515,000 people **b.** 1997
55. a. 363 beneficiaries; 88,572 beneficiaries **b.** 13 rounds **57. a.** 141°F **b.** after 28.3 minutes **59. a.** 261.63 vibrations per second
b. No. The function $f(n)$ is not a linear function. Therefore, the graph of $f(n)$ does not increase at a constant rate.

63. 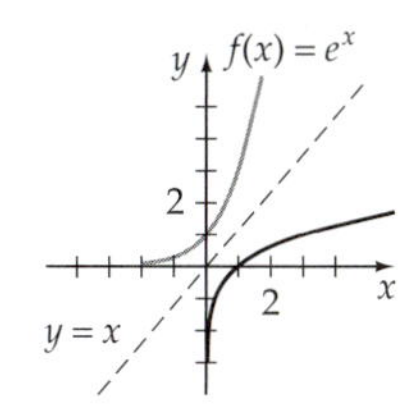

65. $(-\infty, \infty)$ **67.** $[0, \infty)$

Prepare for Section 3.3, page 229

68. 4 **69.** 3 **70.** 5 **71.** $f^{-1}(x) = \dfrac{3x}{2 - x}$ **72.** $\{x \mid x \geq 2\}$ **73.** the set of all positive real numbers

Exercise Set 3.3, page 239

1. $10^1 = 10$ **3.** $8^2 = 64$ **5.** $7^0 = x$ **7.** $e^4 = x$ **9.** $e^0 = 1$ **11.** $\log_3 9 = 2$ **13.** $\log_4 \dfrac{1}{16} = -2$ **15.** $\log_b y = x$ **17.** $\ln y = x$

19. $\log 100 = 2$ **21.** 2 **23.** -5 **25.** 3 **27.** -2 **29.** -4 **31.**

33. 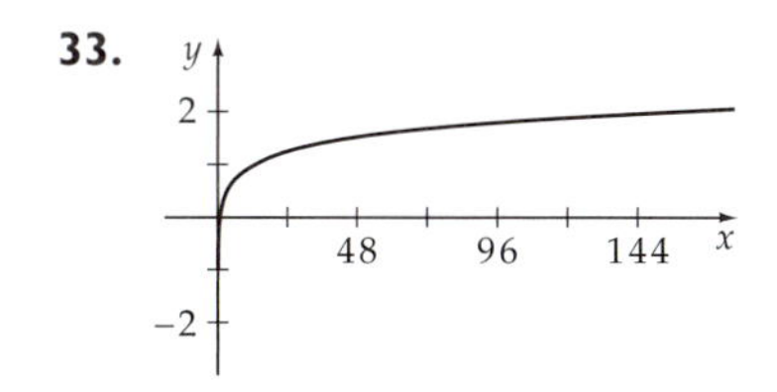

35. 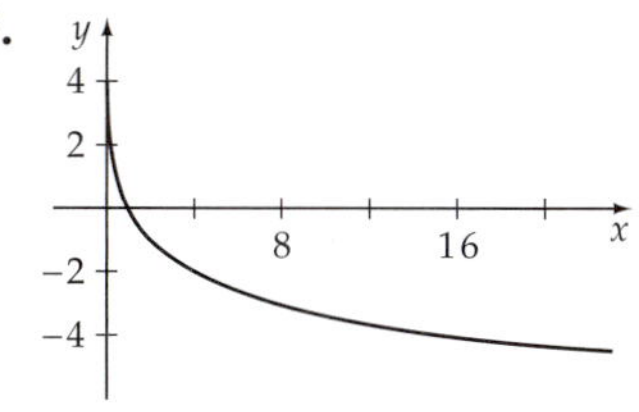

37. 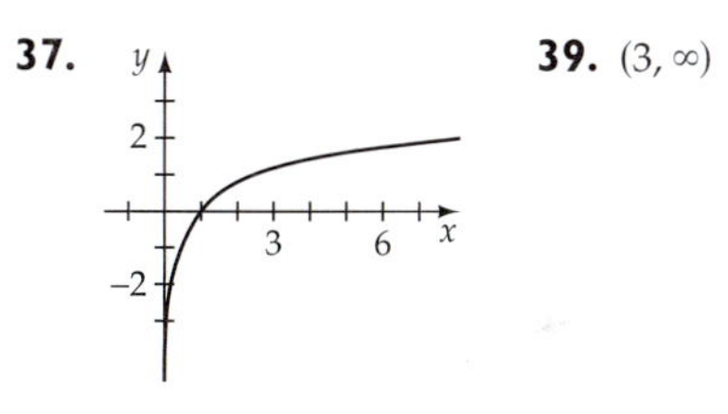

39. $(3, \infty)$

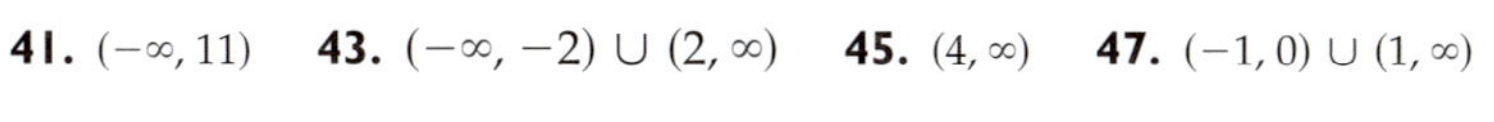
41. $(-\infty, 11)$ **43.** $(-\infty, -2) \cup (2, \infty)$ **45.** $(4, \infty)$ **47.** $(-1, 0) \cup (1, \infty)$ **49.** 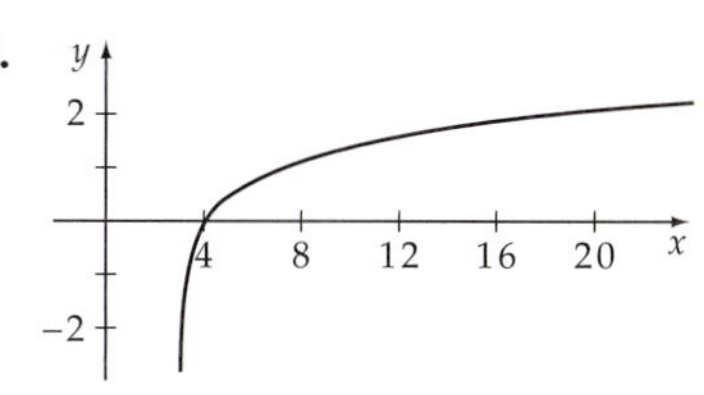

51. 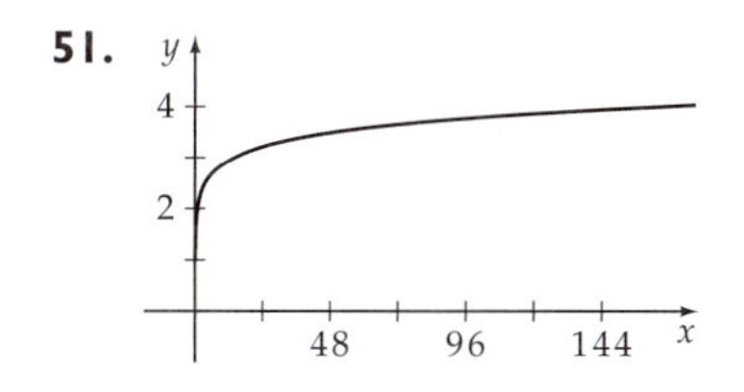

53. 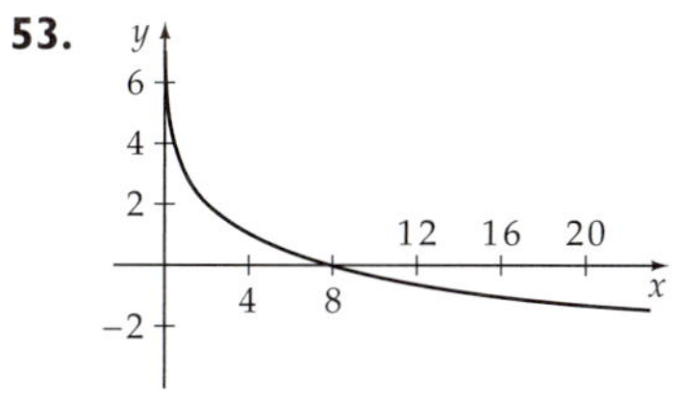

55. 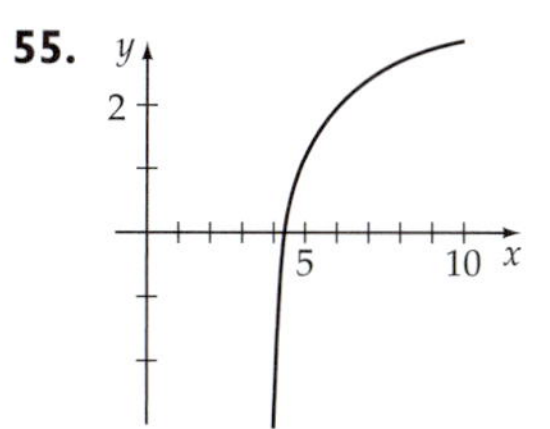

57. a. $k(x)$ **b.** $f(x)$ **c.** $g(x)$ **d.** $h(x)$ **59.** 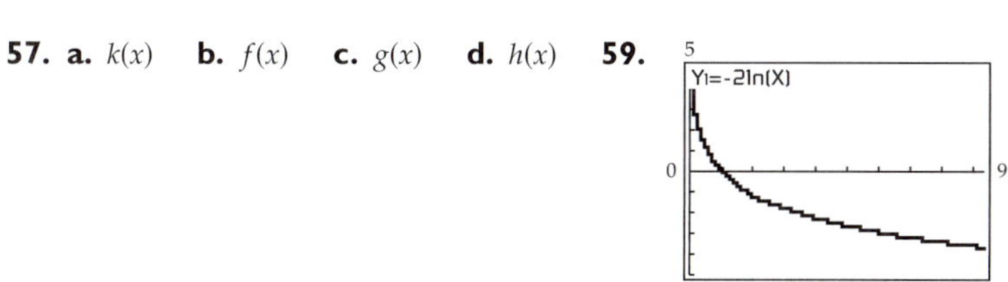

61. 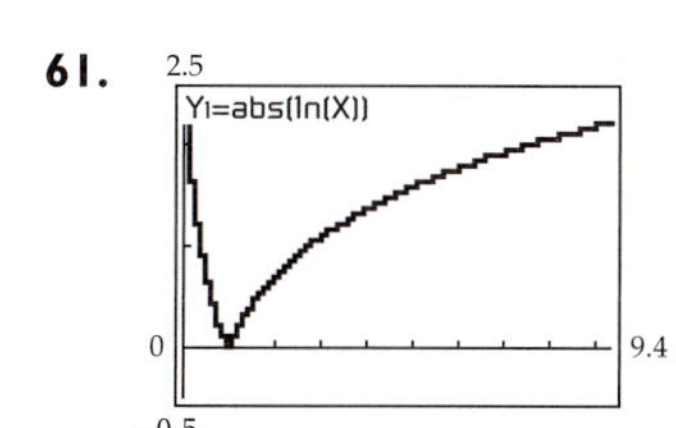

63.

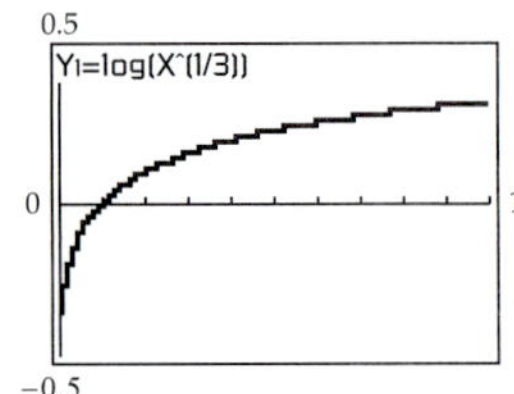

65.

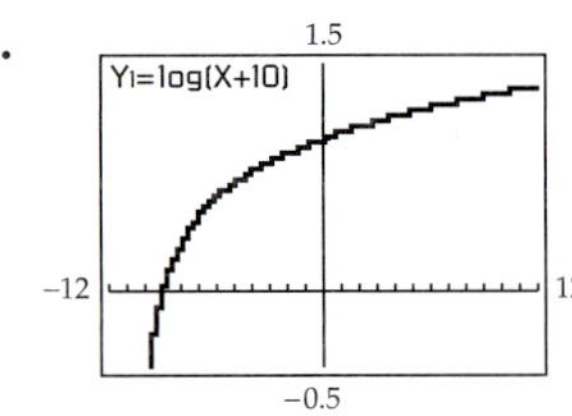

67.

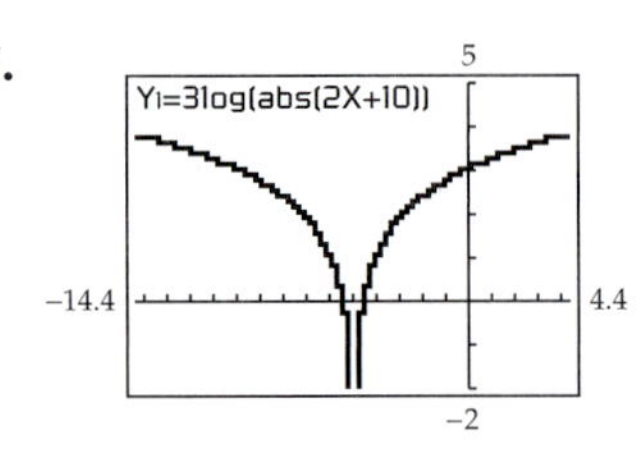

69. **a.** 2.0% **b.** 45 months **71.** **a.** 3298 units; 3418 units; 3490 units **b.** 2750 units **73.** 2.05 square meters
75. **a.** Answers will vary. **b.** 96 digits **c.** 3385 digits **d.** 6,320,430 digits **77.** f and g are inverse functions.
79. range of f: $\{y \mid -1 < y < 1\}$; range of g: all real numbers

Prepare for Section 3.4, page 242

81. ≈ 0.77815 for each expression **82.** ≈ 0.98083 for each expression **83.** ≈ 1.80618 for each expression
84. ≈ 3.21888 for each expression **85.** ≈ 1.60944 for each expression **86.** ≈ 0.90309 for each expression

Exercise Set 3.4, page 251

1. $\log_b x + \log_b y + \log_b z$ **3.** $\ln x - 4\ln z$ **5.** $\frac{1}{2}\log_2 x - 3\log_2 y$ **7.** $\frac{1}{2}\log_7 x + \frac{1}{2}\log_7 z - 2\log_7 y$ **9.** $\log[x^2(x+5)]$ **11.** $\ln(x+y)$
13. $\log\left[x^3 \cdot \sqrt[3]{y}(x+1)\right]$ **15.** 1.5395 **17.** 0.8672 **19.** −0.6131 **21.** 0.6447 **23.**

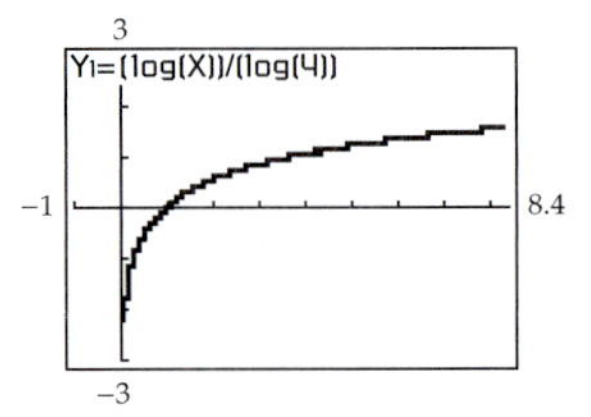

25.

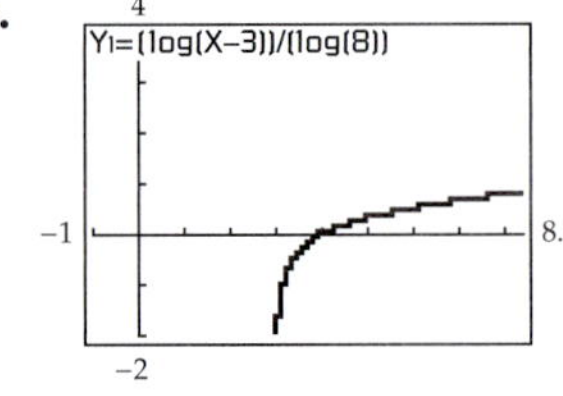

27.

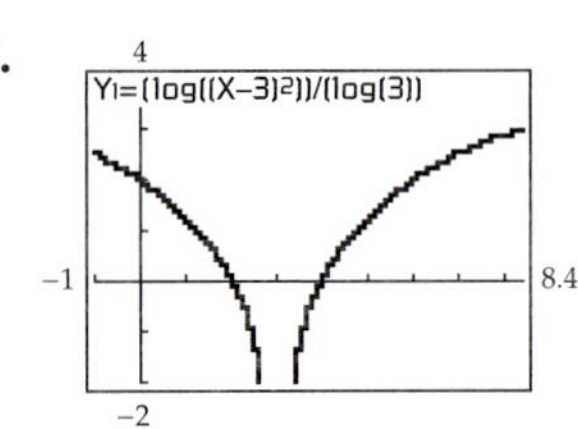

29.

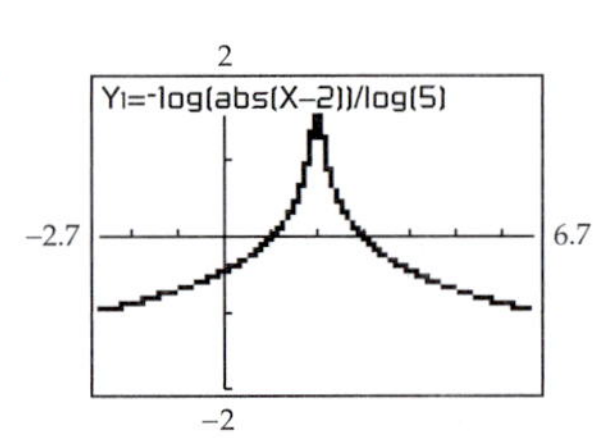

31. False; $\log 10 + \log 10 = 2$ but $\log(10+10) = \log 20 \neq 2$. **33.** True **35.** False; $\log 100 - \log 10 = 1$ but $\log(100-10) = \log 90 \neq 1$.
37. False; $\frac{\log 100}{\log 10} = \frac{2}{1} = 2$ but $\log 100 - \log 10 = 1$. **39.** False; $(\log 10)^2 = 1$ but $2\log 10 = 2$. **41.** 2 **43.** 500^{501}
45. 1:870,551; 1:757,858; 1:659,754; 1:574,349; 1:500,000 **47.** 10.4; base **49.** 3.16×10^{-10} mole per liter
51. **a.** 82.0 decibels **b.** 40.3 decibels **c.** 115.0 decibels **d.** 152.0 decibels **53.** 10 times more intense **55.** 5
57. $10^{6.5}I_0$ or about $3{,}162{,}277.7I_0$ **59.** 100 to 1 **61.** $10^{1.8}$ to 1 or about 63 to 1 **63.** 5.5 **65.** **a.** $M \approx 6$ **b.** $M \approx 4$ **c.** The results are close to the magnitudes produced by the amplitude-time-difference formula.

Prepare for Section 3.5, page 254

66. $\log_3 729 = 6$ **67.** $5^4 = 625$ **68.** $\log_a b = x + 2$ **69.** $x = \frac{4a}{7b + 2c}$ **70.** $x = \frac{3}{44}$ **71.** $x = \frac{100(A-1)}{A+1}$

Exercise Set 3.5, page 263

1. 6 **3.** $-\frac{3}{2}$ **5.** $-\frac{6}{5}$ **7.** 3 **9.** $\frac{\log 70}{\log 5}$ **11.** $-\frac{\log 120}{\log 3}$ **13.** $\frac{\log 315 - 3}{2}$ **15.** $\ln 10$ **17.** $\frac{\ln 2 - \ln 3}{\ln 6}$ **19.** $\frac{3\log 2 - \log 5}{2\log 2 + \log 5}$
21. 7 **23.** 4 **25.** $2 + 2\sqrt{2}$ **27.** $\frac{199}{95}$ **29.** −1 **31.** 3 **33.** 10^{10} **35.** 2 **37.** 5 **39.** $\log(20 + \sqrt{401})$ **41.** $\frac{1}{2}\log\left(\frac{3}{2}\right)$
43. $\ln(15 \pm 4\sqrt{14})$ **45.** $\ln(1 + \sqrt{65}) - \ln 8$ **47.** 1.61 **49.** 0.96 **51.** 2.20 **53.** −1.93 **55.** −1.34

57. a. 8500, 10,285 **b.** in 6 years **59. a.** 60°F **b.** 27 minutes **61. a.**

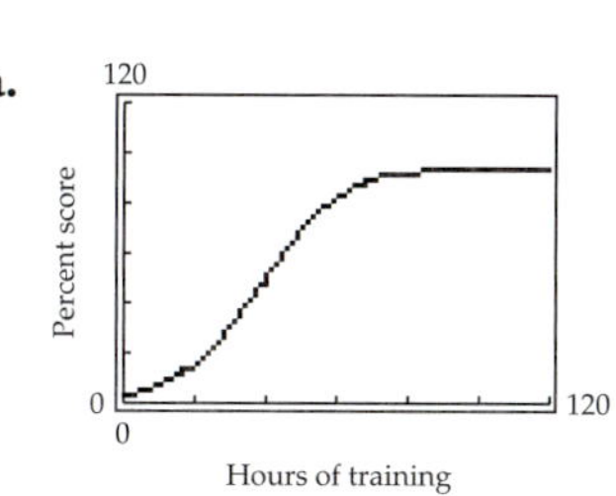

b. 48 hours **c.** $P = 100$ **d.** As the number of hours of training increases, the test scores approach 100%.

63. a. **b.** in 27 years or the year 2026 **c.** $B = 1000$ **d.** As the number of years increases, the bison population approaches but never reaches or exceeds 1000.

65. a. **b.** 78 years **c.** 1.9%

67. a. 116 feet per second **b.** $v = 150$ **c.** The velocity of the package approaches but never reaches or exceeds 150 feet per second.
69. a. 1.72 seconds **b.** $v = 100$ **c.** The object cannot fall faster than 100 feet per second.
71. a. **b.** 2.6 seconds **73.** 138

75. The second step; because $\log 0.5 < 0$, the inequality sign must be reversed. **77.** $x = \dfrac{y}{y-1}$ **79.** $e^{0.336} \approx 1.4$

Prepare for Section 3.6, page 267

81. 1220.39 **82.** 824.96 **83.** −0.0495 **84.** 1340 **85.** 0.025 **86.** 12.8

Exercise Set 3.6, page 276

1. a. \$9724.05 **b.** \$11,256.80 **3. a.** \$48,885.72 **b.** \$49,282.20 **c.** \$49,283.30 **5.** \$24,730.82 **7.** 8.8 years **9.** $t = \dfrac{\ln 3}{r}$
11. 14 years **13. a.** 2200 bacteria **b.** 17,600 bacteria **15. a.** $N(t) \approx 22{,}600e^{0.01368t}$ **b.** 27,700 **17. a.** 10,755,000 **b.** 2042
19. a. **b.** 3.18 micrograms **c.** ≈15.07 hours **d.** ≈30.14 hours **21.** ≈6601 years ago

23. ≈2378 years old **25. a.** 0.056 **b.** 42°F **c.** 54 minutes **27. a.** 211 hours **b.** 1386 hours **29.** 3.1 years

31. a. **b.** 0.98 second **c.** $v = 32$ **d.** As time increases, the velocity approaches but never reaches or exceeds 32 feet per second.

33. a. **b.** 2.5 seconds **c.** ≈24.56 feet per second **d.** The average speed of the object was approximately 24.56 feet per second during the period from $t = 1$ to $t = 2$ seconds.

35. a. 0.71 gram **b.** 0.96 gram **c.** 0.52 gram **37. a.** 1.7% **b.** 13.9% **c.** 19.0% **d.** 19.5% **e.** 1.5%; $P \to 0$
39. 13,715,120,270 centuries

Chapter 3 True/False Exercises, page 284

1. False. $f(x) = x^2$ does not have an inverse function. **2.** False. Let $f(x) = 2x$, $g(x) = 3x$. Then $f(g(0)) = 0$ and $g(f(0)) = 0$, but f and g are not inverse functions. **3.** True **4.** True **5.** True **6.** False; f is not defined for negative values of x, and thus $g(f(x))$ is undefined for negative values of x. **7.** False; $h(x)$ is not an increasing function for $0 < b < 1$. **8.** False; $j(x)$ is not an increasing function for $0 < b < 1$. **9.** True **10.** True **11.** True **12.** True **13.** False; $\log x + \log y = \log(xy)$. **14.** True **15.** True **16.** True

Chapter 3 Review Exercises, page 284

1. Yes [3.1] **2.** Yes [3.1] **3.** Yes [3.1] **4.** No [3.1]

5. $f^{-1}(x) = \dfrac{x+4}{3}$ [3.1] **6.** $g^{-1}(x) = -\dfrac{1}{2}x + \dfrac{3}{2}$ [3.1] **7.** $h^{-1}(x) = -2x - 4$ [3.1] **8.** $k^{-1}(x) = k(x) = \dfrac{1}{x}$ [3.1]

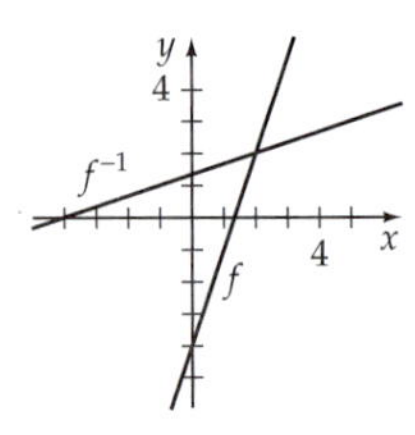

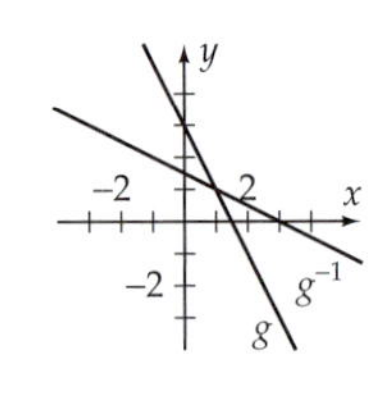

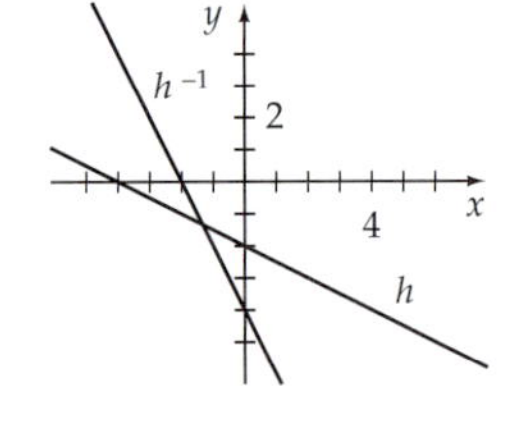

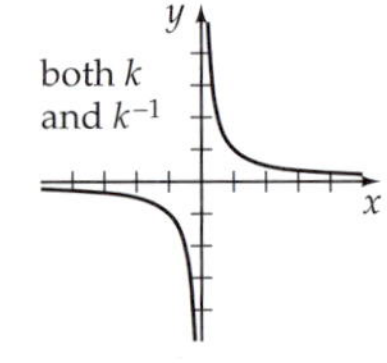

9. 2 [3.3] **10.** 4 [3.3] **11.** 3 [3.3] **12.** π [3.3] **13.** −2 [3.5] **14.** 8 [3.5] **15.** −3 [3.5] **16.** −4 [3.5]
17. ±1000 [3.5] **18.** $\pm 10^{10}$ [3.5] **19.** 7 [3.5] **20.** ±8 [3.5]

21. [3.2] **22.** [3.2] **23.** [3.2] **24.** [3.2]

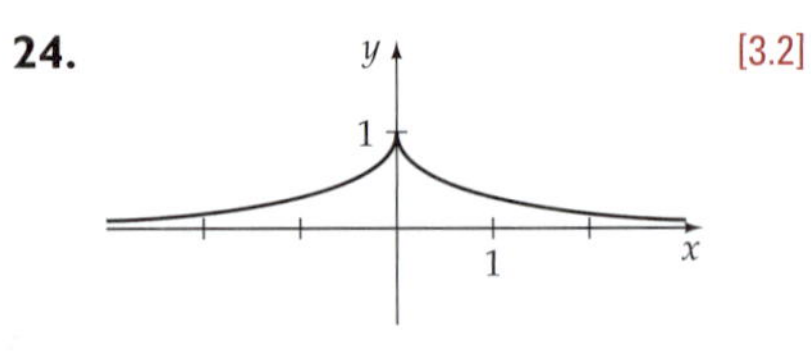

25. [3.2] **26.** [3.2]

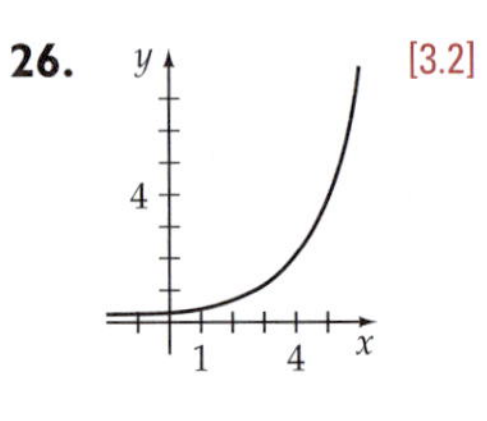

27. [3.3]

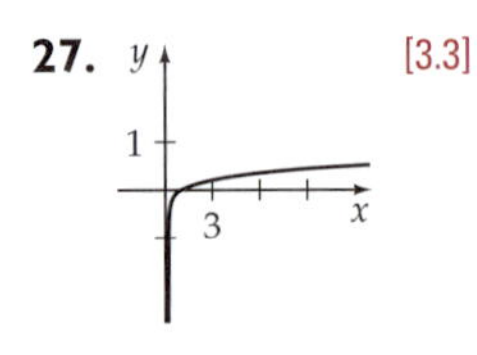

28. [3.3]

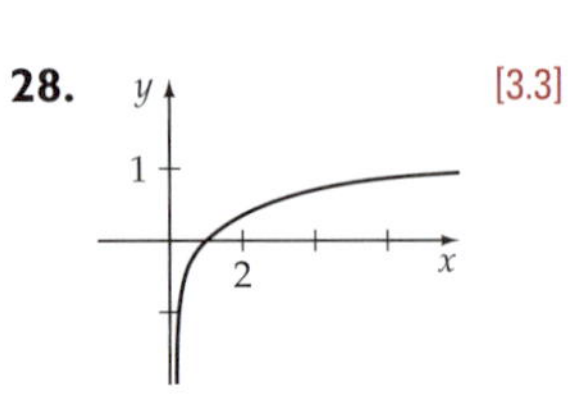

29. 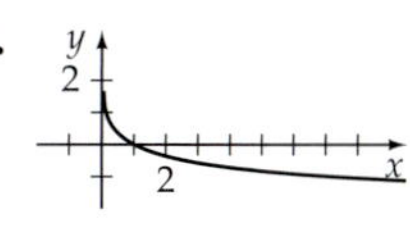

[3.3] **30.** 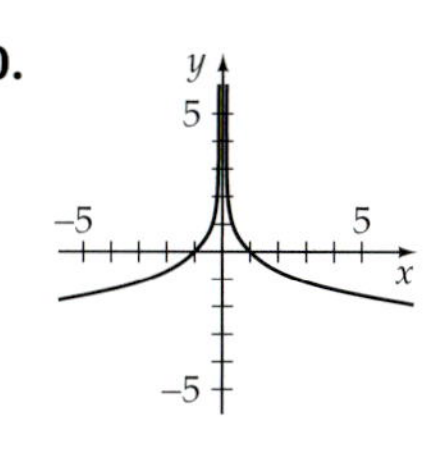

[3.3] **31.** 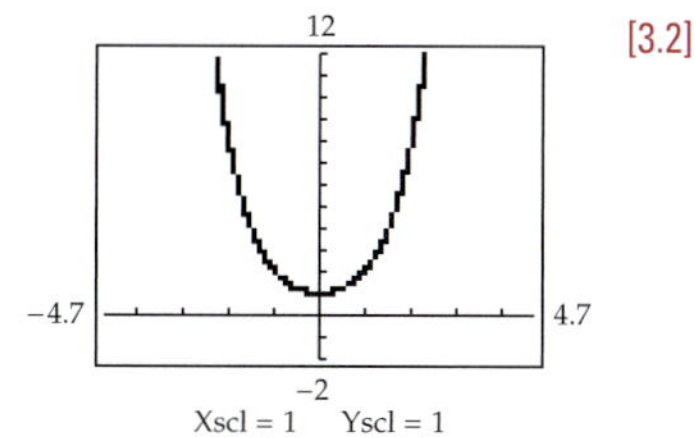

[3.2]

32. 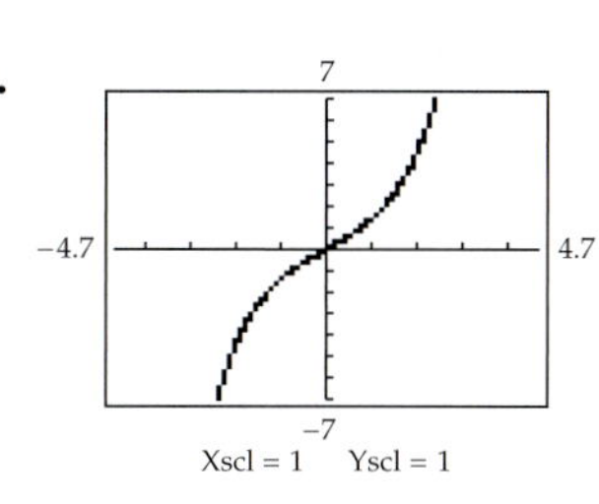

[3.2] **33.** $4^3 = 64$ [3.3] **34.** $\left(\frac{1}{2}\right)^{-3} = 8$ [3.3] **35.** $(\sqrt{2})^4 = 4$ [3.3]

36. $e^0 = 1$ [3.3] **37.** $\log_5 125 = 3$ [3.3] **38.** $\log_2 1024 = 10$ [3.3] **39.** $\log_{10} 1 = 0$ [3.3] **40.** $\log_8 2\sqrt{2} = \frac{1}{2}$ [3.3]

41. $2\log_b x + 3\log_b y - \log_b z$ [3.4] **42.** $\frac{1}{2}\log_b x - 2\log_b y - \log_b z$ [3.4] **43.** $\ln x + 3\ln y$ [3.4] **44.** $\frac{1}{2}\ln x + \frac{1}{2}\ln y - 4\ln z$ [3.4]

45. $\log\left(x^2\sqrt[3]{x+1}\right)$ [3.4] **46.** $\log\frac{x^5}{(x+5)^2}$ [3.4] **47.** $\ln\frac{\sqrt{2xy}}{z^3}$ [3.4] **48.** $\ln\frac{xz}{y}$ [3.4] **49.** 2.86754 [3.4] **50.** 3.35776 [3.4]

51. −0.117233 [3.4] **52.** −0.578989 [3.4] **53.** $\frac{\ln 30}{\ln 4}$ [3.5] **54.** $\frac{\log 41}{\log 5} - 1$ [3.5] **55.** 4 [3.5] **56.** $\frac{1}{6}e$ [3.5] **57.** 4 [3.5] **58.** 15 [3.5]

59. $\frac{\ln 3}{2\ln 4}$ [3.5] **60.** $\frac{\ln\left(8 \pm 3\sqrt{7}\right)}{\ln 5}$ [3.5] **61.** 10^{1000} [3.5] **62.** $e^{(e^2)}$ [3.5] **63.** 1,000,005 [3.5] **64.** $\frac{15 + \sqrt{265}}{2}$ [3.5] **65.** 81 [3.5]

66. $\pm\sqrt{5}$ [3.5] **67.** 4 [3.5] **68.** 5 [3.5] **69.** 7.7 [3.4] **70.** 5.0 [3.4] **71.** 3162 to 1 [3.4] **72.** 2.8 [3.4] **73.** 4.2 [3.4]
74. $\approx 3.98 \times 10^{-6}$ [3.4] **75.** **a.** \$20,323.79 **b.** \$20,339.99 [3.6] **76.** **a.** \$25,646.69 **b.** \$25,647.32 [3.6] **77.** \$4,438.10 [3.6]
78. **a.** 69.9% **b.** 6 days **c.** 19 days [3.6] **79.** $N(t) \approx e^{0.8047t}$ [3.6] **80.** $N(t) \approx 2e^{0.5682t}$ [3.6] **81.** $N(t) \approx 3.783e^{0.0558t}$ [3.6]
82. $N(t) \approx e^{-0.6931t}$ [3.6] **83.** **a.** $P(t) \approx 25{,}200e^{0.06155789t}$ **b.** 38,800 [3.6] **84.** 340 years [3.6]

Chapter 3 Test, page 286

1. $f^{-1}(x) = \frac{1}{2}x + \frac{3}{2}$ [3.1] **2.** $f^{-1}(x) = \frac{8x}{4x-1}$

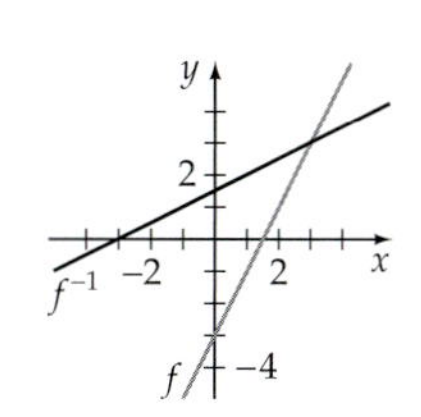

Domain f^{-1}: all real numbers except $\frac{1}{4}$

Range f^{-1}: all real numbers except 2 [3.1]

3. **a.** $b^c = 5x - 3$ [3.3] **b.** $\log_3 y = \frac{x}{2}$ [3.3] **4.** $2\log_b z - 3\log_b y - \frac{1}{2}\log_b x$ [3.4] **5.** $\log\frac{2x+3}{(x-2)^3}$ [3.4] **6.** 1.7925 [3.4]

7. 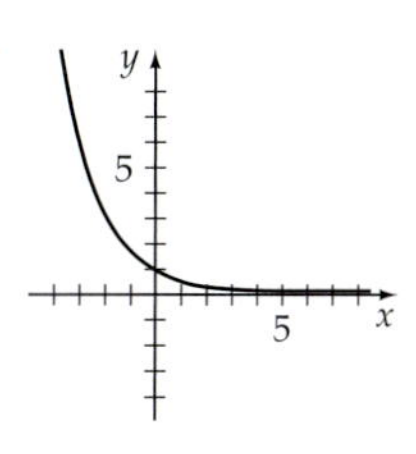

[3.2] **8.** 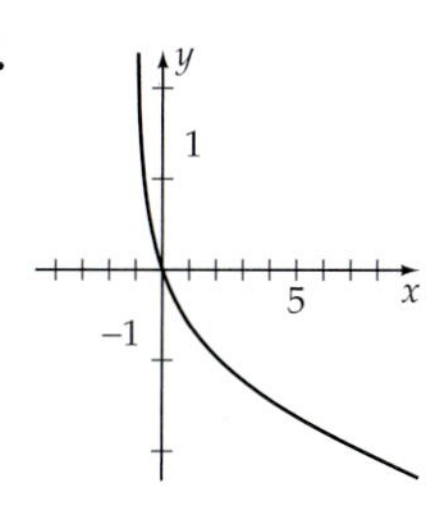

[3.3] **9.** 1.9206 [3.5] **10.** $\frac{5\ln 4}{\ln 28}$ [3.5] **11.** 1 [3.5] **12.** −3 [3.5]

13. a. \$29,502.36 **b.** \$29,539.62 [3.6] **14.** 17.36 years [3.6] **15. a.** 7.6 **b.** 63 to 1 [3.4] **16. a.** $P(t) \approx 34{,}600e^{0.04667108t}$ **b.** 55,000 [3.6] **17.** 690 years [3.6] **18.** 11.9 years [3.6] **19.** 19 minutes [3.5] **20.** $v = 125$; The velocity of the object cannot reach or exceed 125 feet per second. [3.5]

Cumulative Review Exercises, page 287

1. [2, 6] [1.1] **2.** $(-\infty, -4] \cup [5, \infty)$ [1.1] **3.** 7.8 [1.2] **4.** 38.25 feet [1.5] **5.** $4x^2 + 4x - 4$ [1.7] **6.** $f^{-1}(x) = \frac{1}{3}x + \frac{5}{3}$ [3.1] **7.** 10 [2.5] **8.** three or one positive real zeros; one negative real zero [2.4] **9.** $1, 4, -\sqrt{3}, \sqrt{3}$ [2.4] **10.** $P(x) = x^3 - 4x^2 + 6x - 4$ [2.5] **11.** vertical asymptote: $x = 4$, horizontal asymptote: $y = 3$ [2.6] **12.** Domain: all real numbers; Range: $\{y \mid 0 < y \le 4\}$ [2.6] **13.** decreasing function [3.2] **14.** $4^y = x$ [3.3] **15.** $\log_5 125 = 3$ [3.3] **16.** 7.1 [3.4] **17.** 2.0149 [3.5] **18.** 510 years old [3.6] **19.** 19 minutes [3.5] **20.** 5 hours [3.6]

Exercise Set 4.1, page 302

1. 75°, 165° **3.** 19°45′, 109°45′ **5.** 33°26′45″, 123°26′45″ **7.** $\frac{\pi}{2} - 1, \pi - 1$ **9.** $\frac{\pi}{4}, \frac{3\pi}{4}$ **11.** $\frac{\pi}{10}, \frac{3\pi}{5}$ **13.** Quadrant III, 250° **15.** Quadrant II, 105° **17.** Quadrant IV, 296° **19.** 24°33′36″ **21.** 64°9′28.8″ **23.** 3°24′7.2″ **25.** 25.42° **27.** 183.56° **29.** 211.78° **31.** $\frac{\pi}{6}$ **33.** $\frac{\pi}{2}$ **35.** $\frac{11\pi}{12}$ **37.** $\frac{7\pi}{3}$ **39.** $\frac{13\pi}{4}$ **41.** 36° **43.** 30° **45.** 67.5° **47.** 660° **49.** 85.94° **51.** 2.32 **53.** 472.69° **55.** 4, 229.18° **57.** 2.38, 136.63° **59.** 6.28 inches **61.** 18.33 centimeters **63.** 3π **65.** $\frac{5\pi}{12}$ radians or 75° **67.** $\frac{\pi}{30}$ radian per second **69.** $\frac{5\pi}{3}$ radians per second **71.** $\frac{10\pi}{9}$ radians per second $\approx$ 3.49 radians per second **73.** 40 mph **75.** 1885 feet **77. a.** 1160 miles per hour **b.** 0.29 radian per hour **c.** 10:59 A.M. **79.** 840,000 miles **81. a.** 3.9 radians per hour **b.** 27,300 kilometers per hour **83. a.** B **b.** Both points have the same linear velocity. **85. a.** 1.15 statute miles **b.** 10% **87.** 13 square inches **89.** 4680 square centimeters **91.** $\approx$23.1 square inches **93.** 1780 miles

Prepare for Section 4.2, page 306

95. Yes **96.** Yes **97.** No **98.** 2π **99.** even function **100.** neither

Exercise Set 4.2, page 319

1. $\left(-\frac{\sqrt{3}}{2}, \frac{1}{2}\right)$ **3.** $\left(\frac{1}{2}, -\frac{\sqrt{3}}{2}\right)$ **5.** $(0, -1)$ **7.** $\left(-\frac{\sqrt{2}}{2}, \frac{\sqrt{2}}{2}\right)$ **9.** $\left(\frac{\sqrt{2}}{2}, -\frac{\sqrt{2}}{2}\right)$ **11.** $\left(\frac{\sqrt{3}}{2}, \frac{1}{2}\right)$ **13.** $(-1, 0)$ **15.** $\frac{1}{2}$ **17.** $-\sqrt{3}$ **19.** 2 **21.** $-\frac{1}{2}$ **23.** -1 **25.** -2 **27.** $-\frac{\sqrt{3}}{3}$ **29.** $\frac{\sqrt{3}}{2}$ **31.** -1 **33.** 0 **35.** undefined **37.** 0.9391 **39.** -1.1528 **41.** -0.2679 **43.** 0.8090 **45.** 48.0889 **47. a.** 0.9 **b.** -0.4 **49. a.** -0.8 **b.** 0.6 **51.** 0.4, 2.7 **53.** 3.4, 6.0 **55.** odd **57.** neither **59.** even **61.** odd **63.** $\sin t$ **65.** $\sec t$ **67.** $-\tan^2 t$ **69.** $-\cot t$ **71.** $\cos^2 t$ **73.** $2\csc^2 t$ **75.** $\csc^2 t$ **77.** 1 **79.** $\sqrt{1 - \cos^2 t}$ **81.** $\sqrt{1 + \cot^2 t}$ **83.** 750 miles **85.** $\frac{\sqrt{2}}{2}$ **87.** $-\frac{\sqrt{3}}{3}$

Prepare for Section 4.3, page 321

93. 0.7 **94.** -0.7 **95.** Reflect the graph of $y = f(x)$ across the x-axis. **96.** Contract each point on the graph of $y = f(x)$ toward the y-axis by a factor of $\frac{1}{2}$. **97.** 6π **98.** 5π

Exercise Set 4.3, page 329

1. $2, 2\pi$ **3.** $1, \pi$ **5.** $\frac{1}{2}, 1$ **7.** $2, 4\pi$ **9.** $\frac{1}{2}, 2\pi$ **11.** $1, 8\pi$ **13.** 2, 6 **15.** $3, 3\pi$ **17.** **19.**

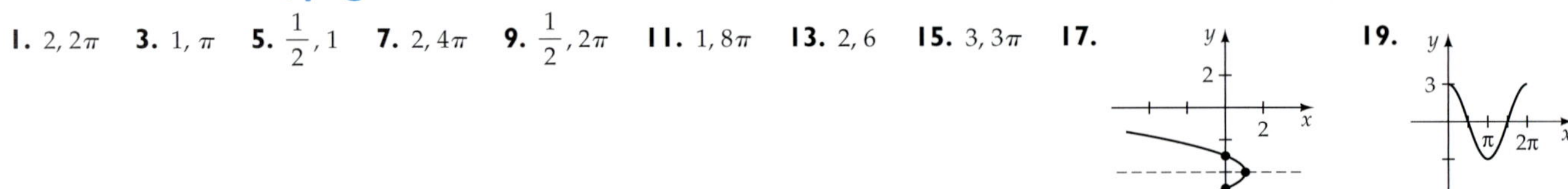

21. **23.** **25.** **27.** **29.**

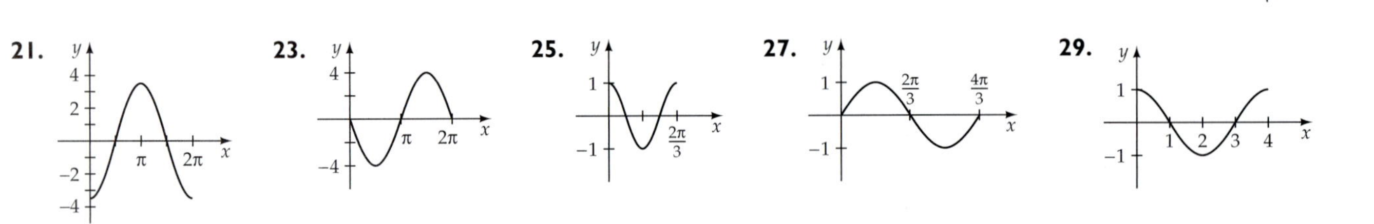

31. 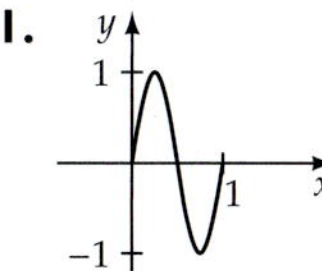
33. 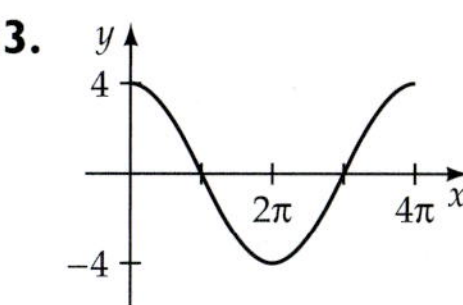
35. 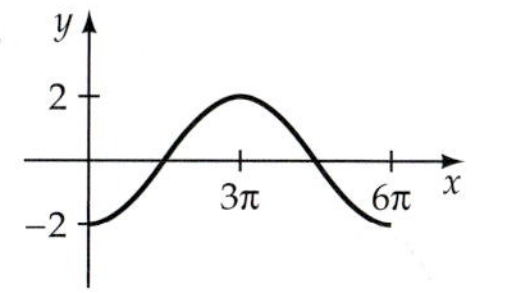
37. 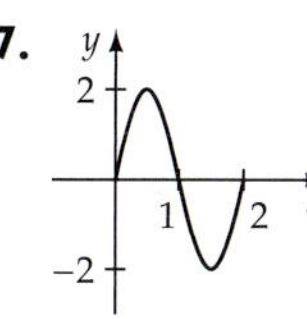
39. 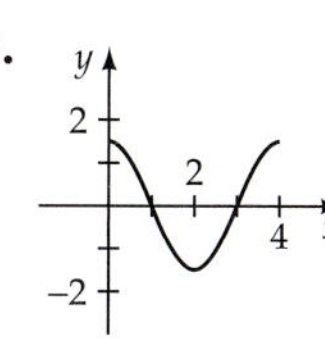

41. 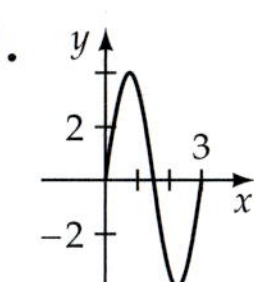
43. 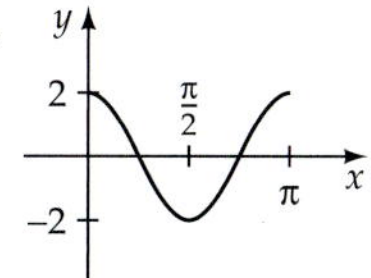
45. 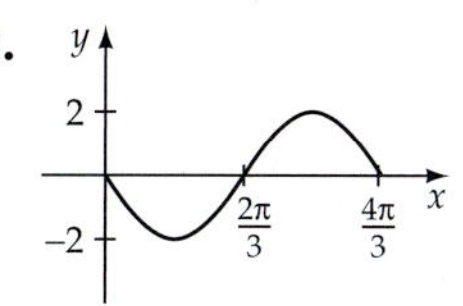
47. 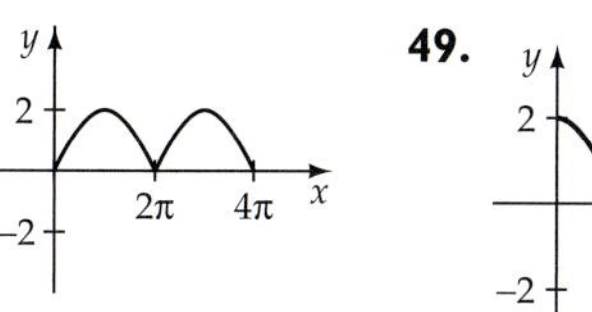
49. 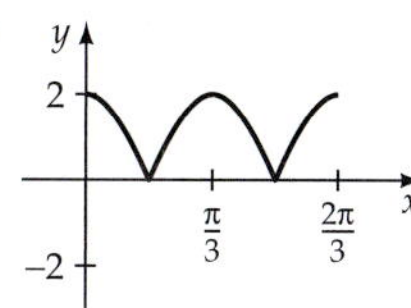

51. 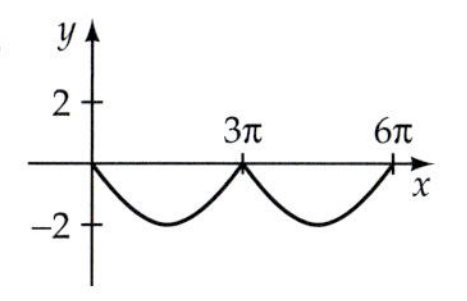
53. 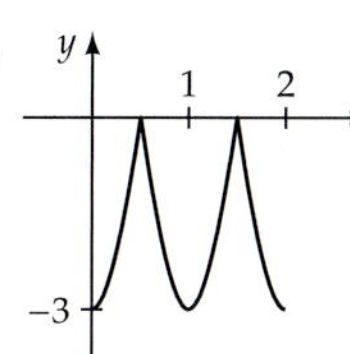

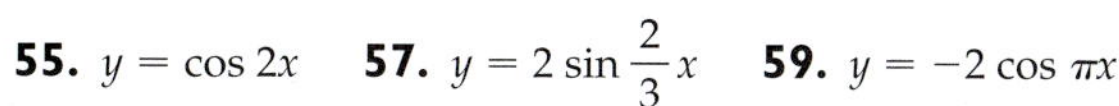
55. $y = \cos 2x$ **57.** $y = 2 \sin \frac{2}{3}x$ **59.** $y = -2 \cos \pi x$

61. 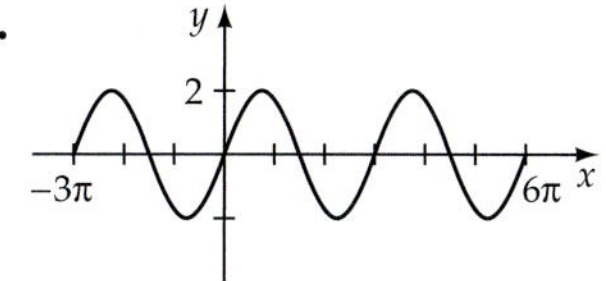
63. 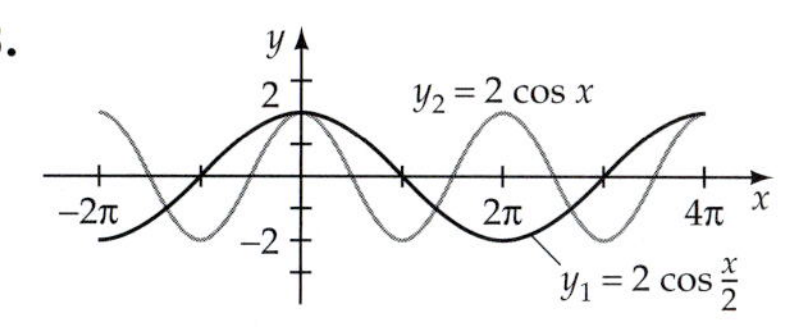

65. 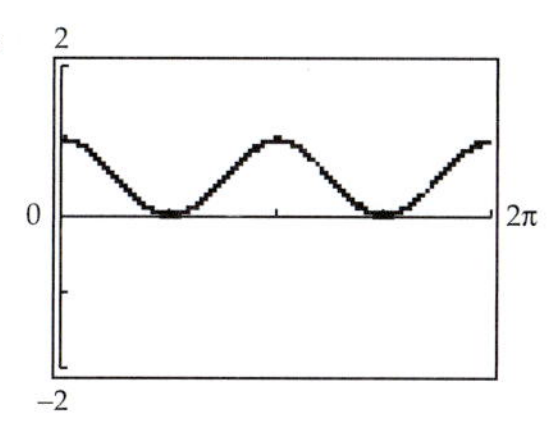

67. 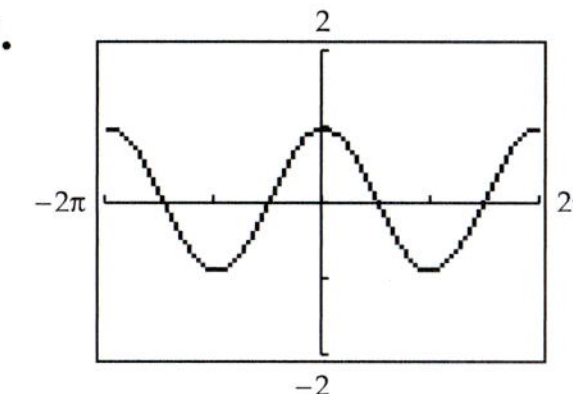
69. 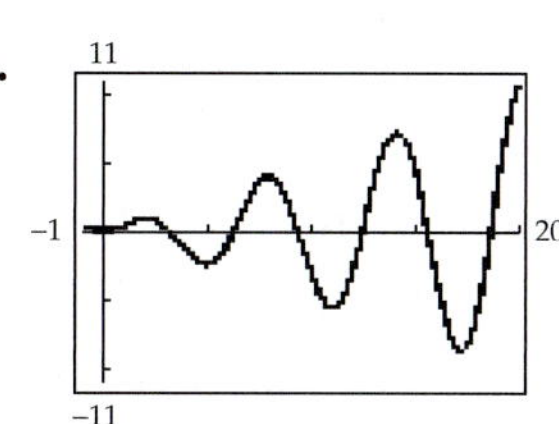
71. 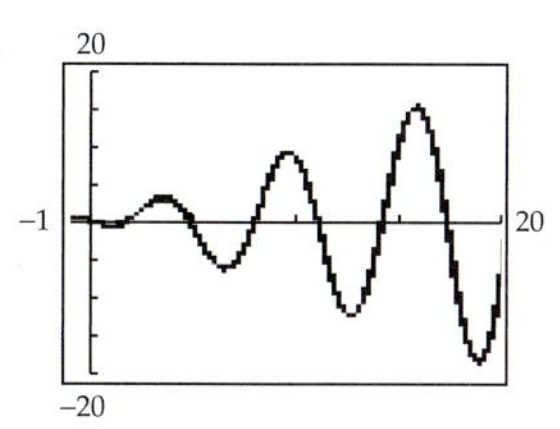

73. 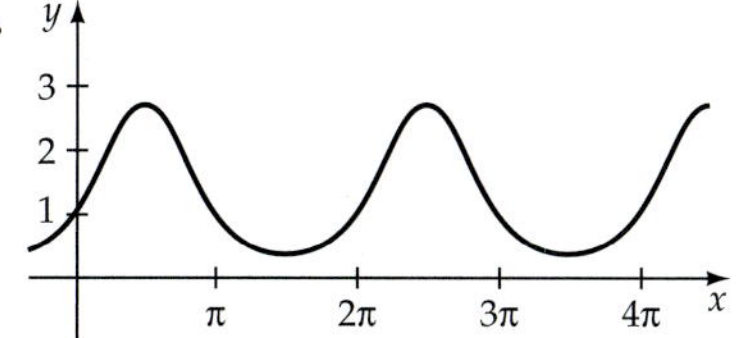

maximum $= e$, minimum $= \frac{1}{e} \approx 0.3679$, period $= 2\pi$

75. $f(t) = 60 \cos \frac{\pi}{10}t$ **77.** $y = 2 \sin \frac{2}{3}x$ **79.** $y = 4 \sin \pi x$ **81.** $y = 3 \cos 4x$

83. $y = 3 \cos \frac{4\pi}{5}x$

Prepare for Section 4.4, page 331

85. 1.7 **86.** 0.6 **87.** Stretch each point on the graph of $y = f(x)$ away from the x-axis by a factor of 2.
88. Shift the graph of $y = f(x)$ 2 units to the right and up 3 units. **89.** 2π **90.** $\frac{4}{3}\pi$

Exercise Set 4.4, page 337

1. $\frac{\pi}{2} + k\pi$, k an integer **3.** $\frac{\pi}{2} + k\pi$, k an integer **5.** 2π **7.** π **9.** 2π **11.** $\frac{2\pi}{3}$ **13.** $\frac{\pi}{3}$ **15.** 8π **17.** 1 **19.** 4

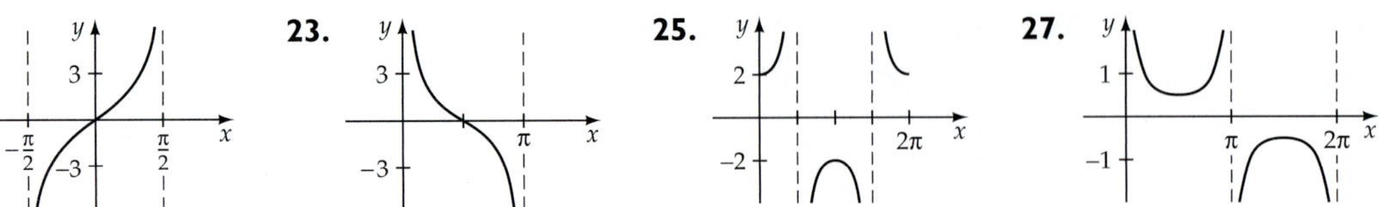

21. 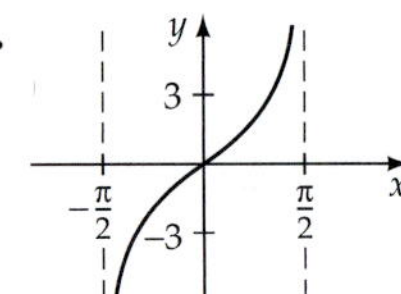 **23.** 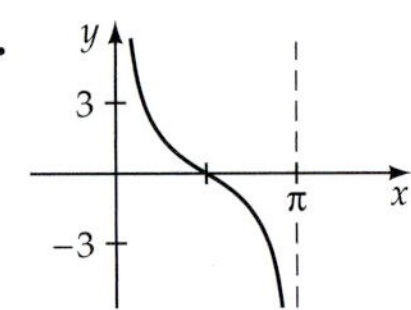 **25.** 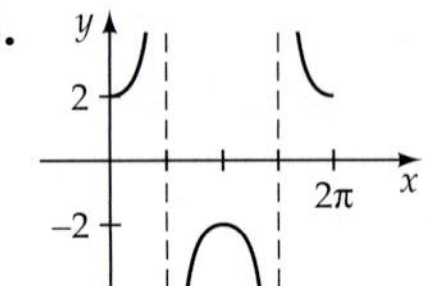 **27.** **29.** 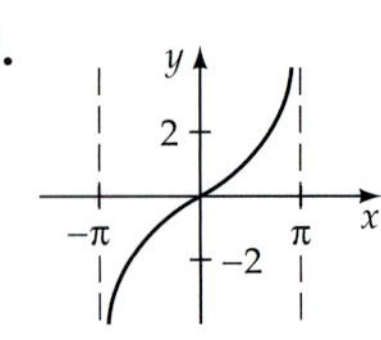

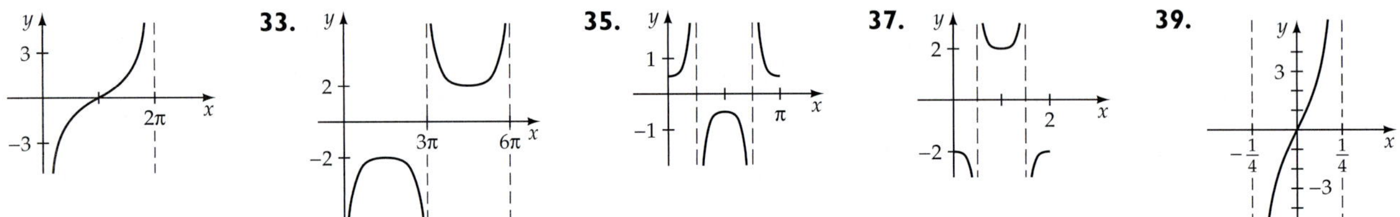

31. 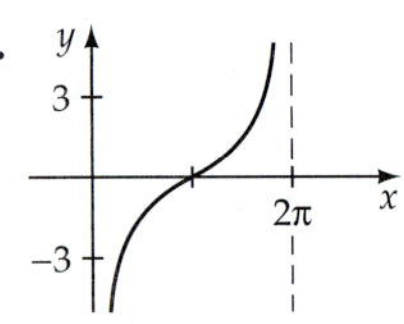 **33.** **35.** 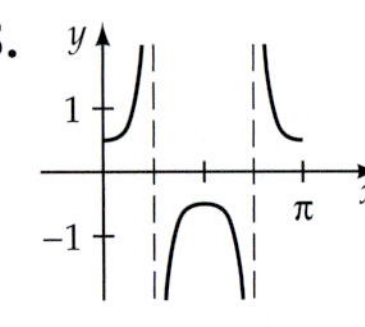 **37.** 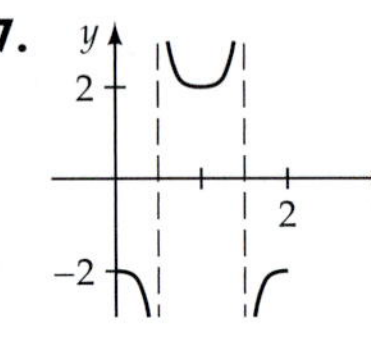 **39.** 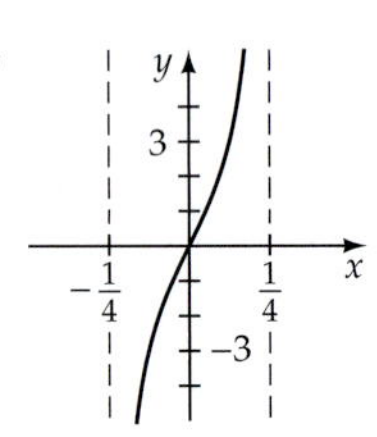

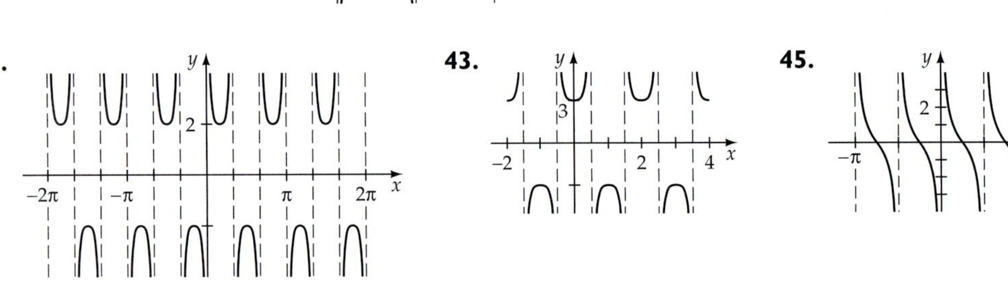

41. **43.** 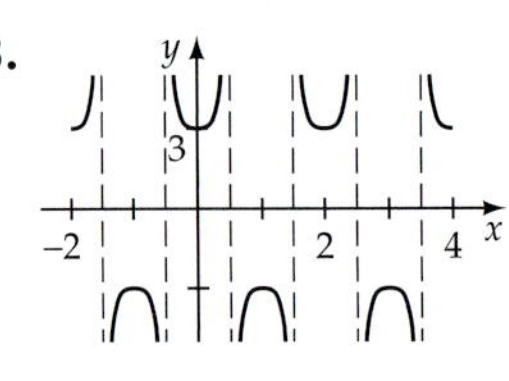 **45.** 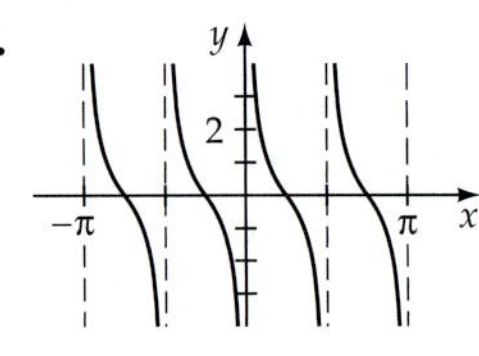 **47.** 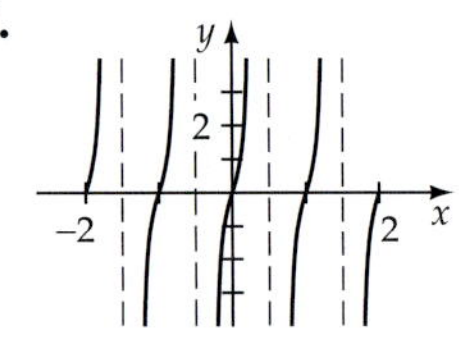

49. $y = \cot \frac{3}{2}x$ **51.** $y = \csc \frac{2}{3}x$ **53.** $y = \sec \frac{3}{4}x$ **55.** 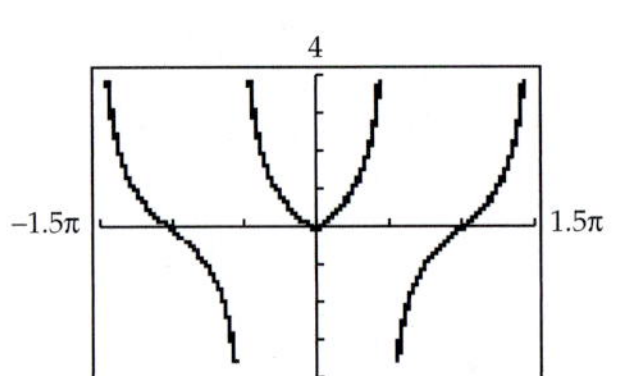

57. 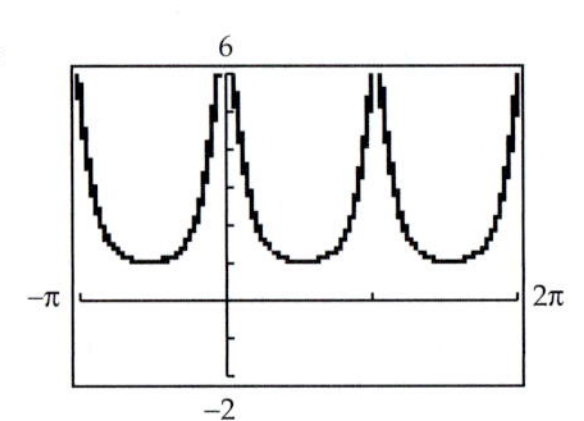

59. 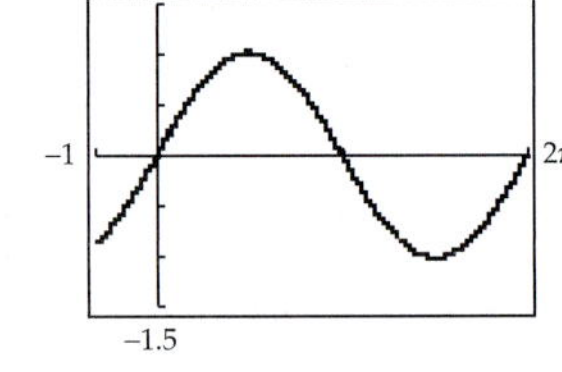 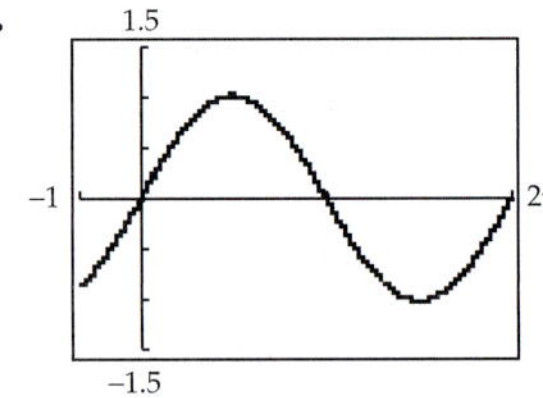

Note: The display screen at the left fails to show that on $[-1, 2\pi]$ the function is undefined at $x = \frac{\pi}{2}$ and $x = \frac{3\pi}{2}$.

61. 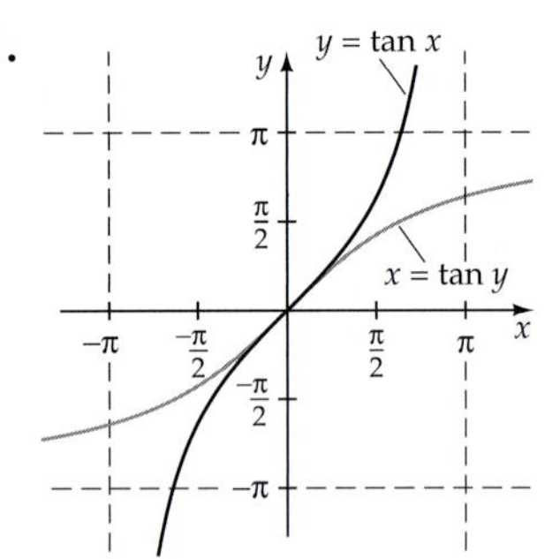

63. $y = \tan 3x$

65. $y = \sec \frac{8}{3}x$ **67.** $y = \cot \frac{\pi}{2}x$ **69.** $y = \csc \frac{4\pi}{3}x$

Prepare for Section 4.5, page 339

71. amplitude 2, period π **72.** amplitude $\frac{2}{3}$, period 6π **73.** amplitude 4, period 1 **74.** 2 **75.** -3 **76.** y-axis

Exercise Set 4.5, page 346

1. $2, \frac{\pi}{2}, 2\pi$ **3.** $1, \frac{\pi}{8}, \pi$ **5.** $4, -\frac{\pi}{4}, 3\pi$ **7.** $\frac{5}{4}, \frac{2\pi}{3}, \frac{2\pi}{3}$ **9.** $\frac{\pi}{8}, \frac{\pi}{2}$ **11.** $-3\pi, 6\pi$ **13.** $\frac{\pi}{16}, \pi$ **15.** $-12\pi, 4\pi$

17.
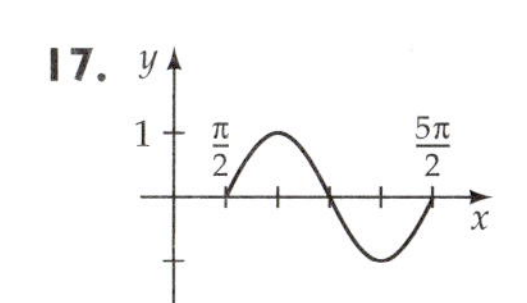

19.
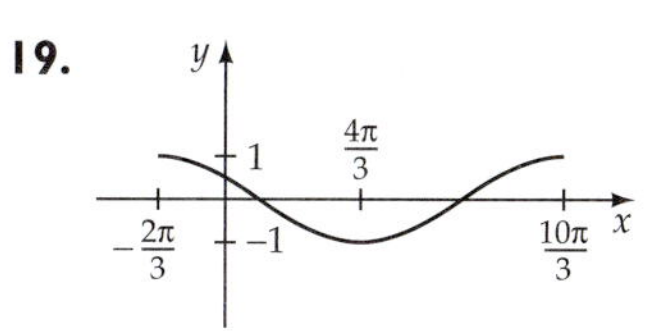

21.
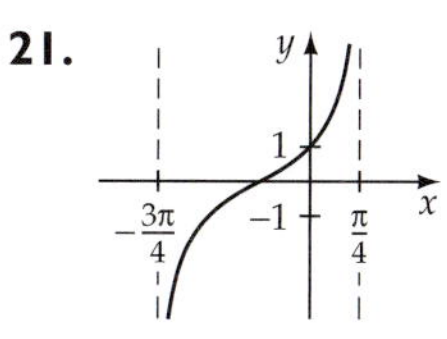

23.
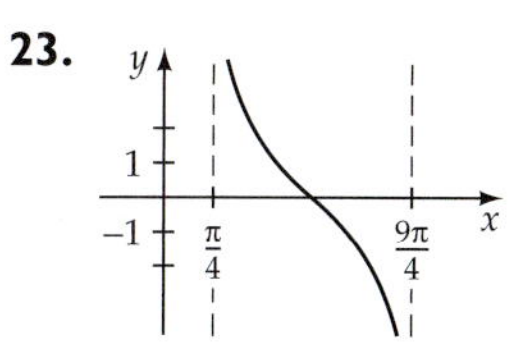

25.
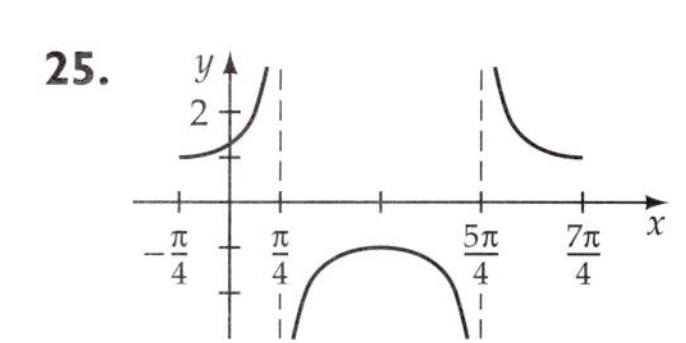

27.
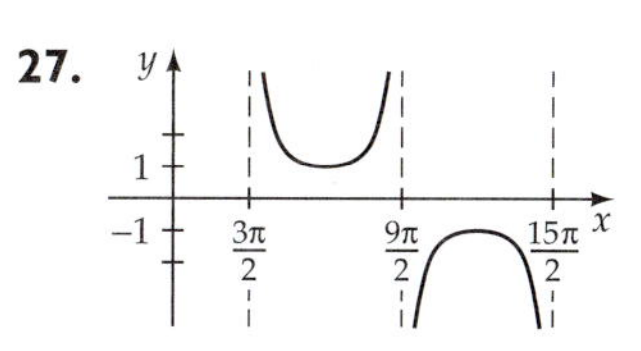

29.
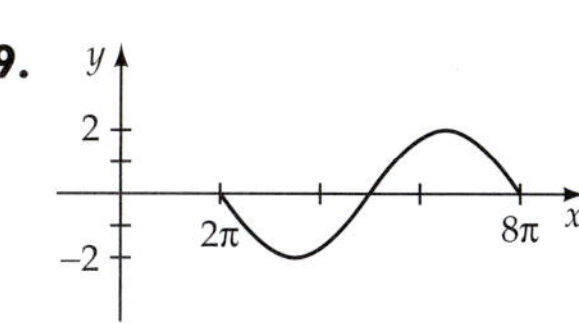

31.
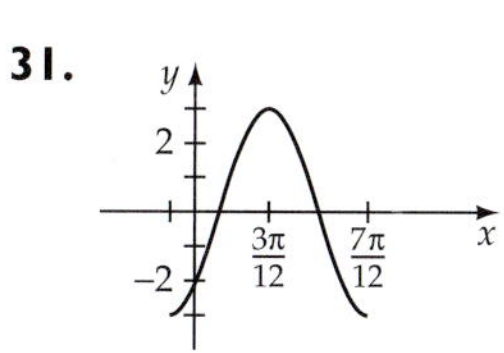

33.
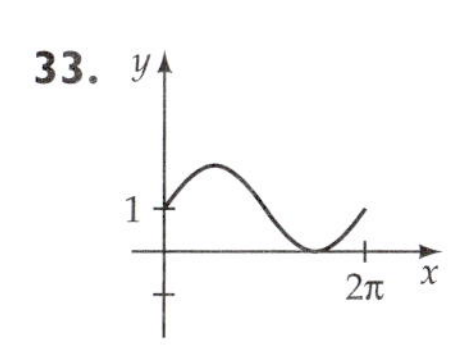

35.
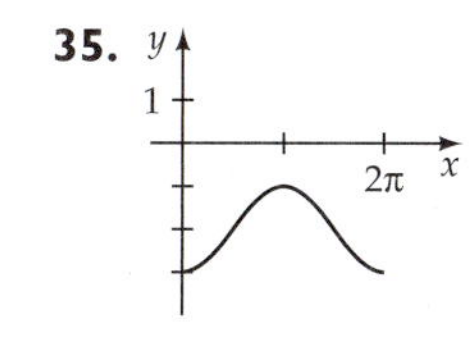

37.
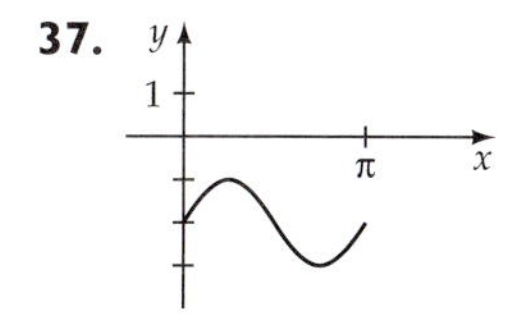

39.
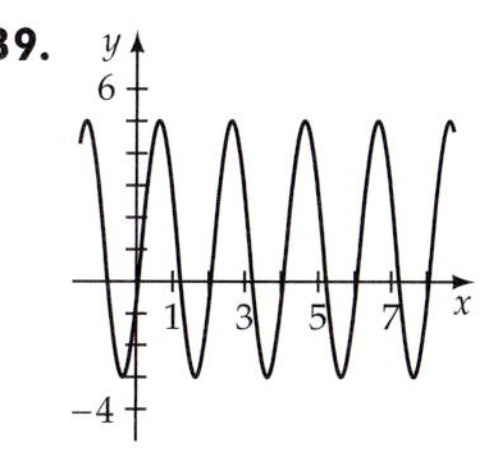

41.
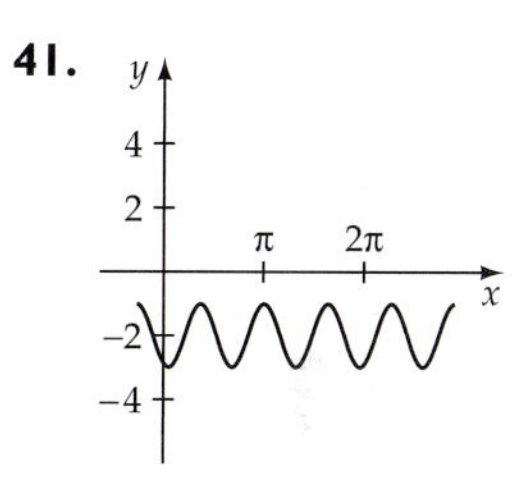

43.
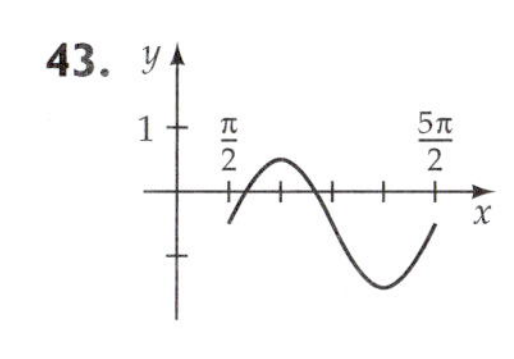

45.
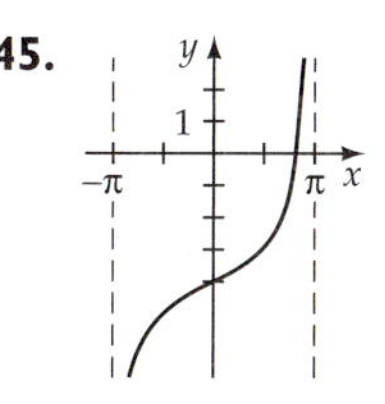

47.
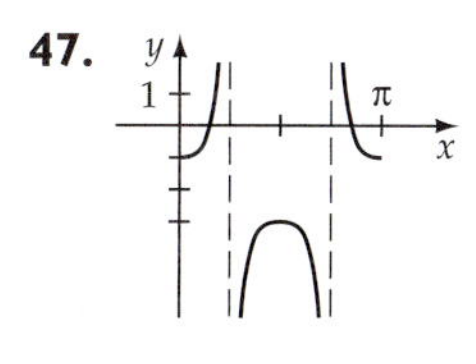

49.
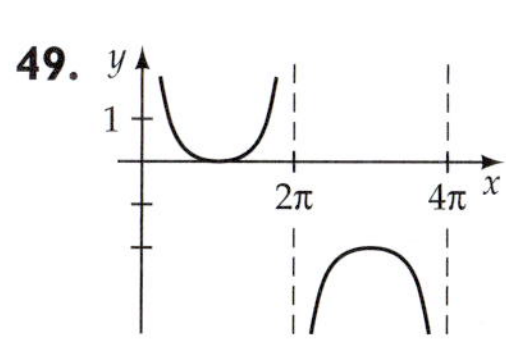

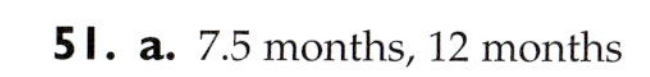
51. a. 7.5 months, 12 months

b.
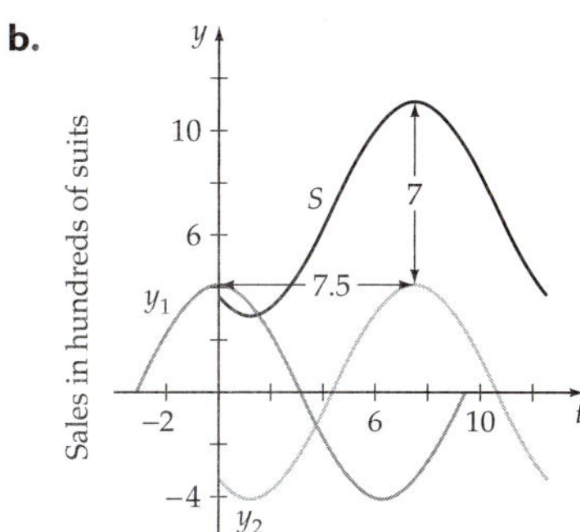
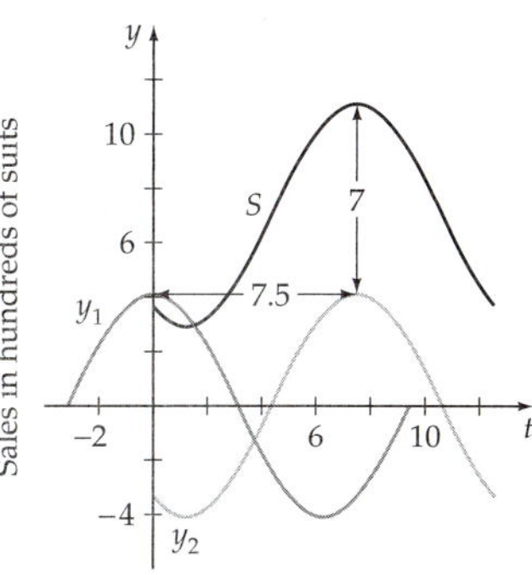

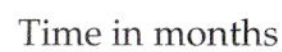
Time in months

c. August

53.
$y = x - \sin x$, $y = x$, $y = -\sin x$; π, 2π

55.
$y = x$, $y = x + \sin 2x$, $y = \sin 2x$; π

57.
$y = \sin x + \cos x$, $y = \sin x$, $y = \cos x$; 1, −1, 2π

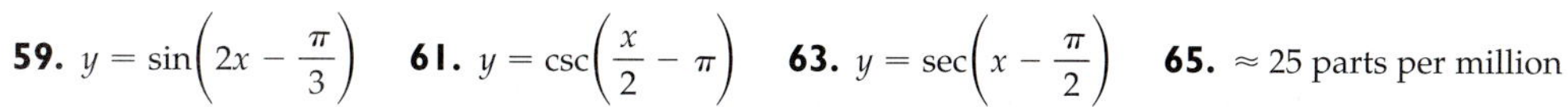
59. $y = \sin\left(2x - \frac{\pi}{3}\right)$ **61.** $y = \csc\left(\frac{x}{2} - \pi\right)$ **63.** $y = \sec\left(x - \frac{\pi}{2}\right)$

65. ≈ 25 parts per million

67. $s = 7 \cos 10\pi t + 5$ **69.** $s = 400 \tan \frac{\pi}{5} t$, t in seconds **71.** $y = 3 \cos \frac{\pi}{6} t + 9$, 12 feet at 6:00 P.M.

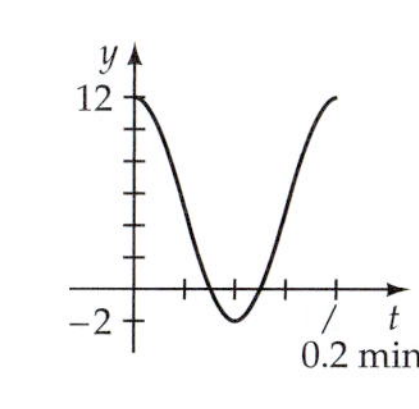

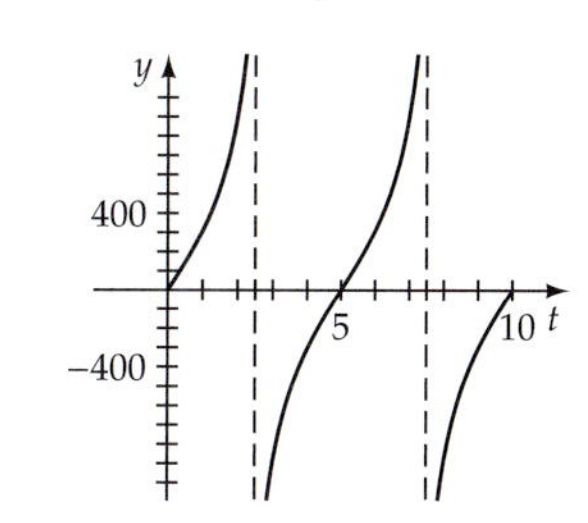

73.

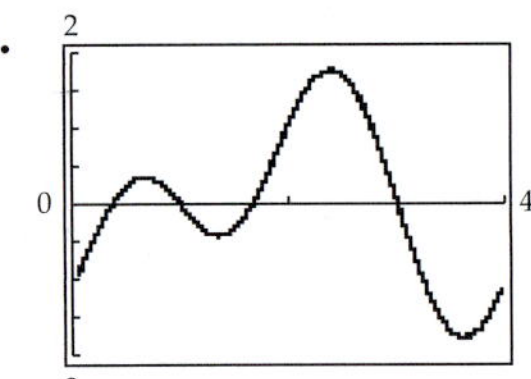

75.

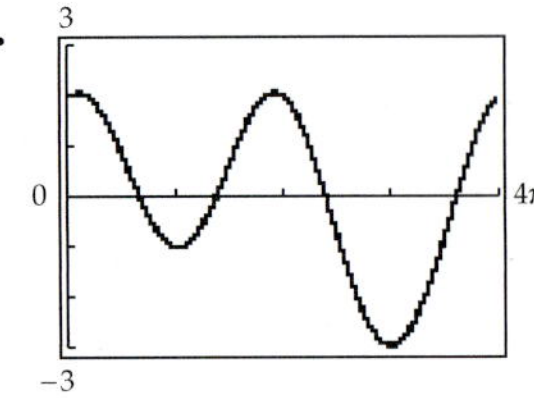

77.

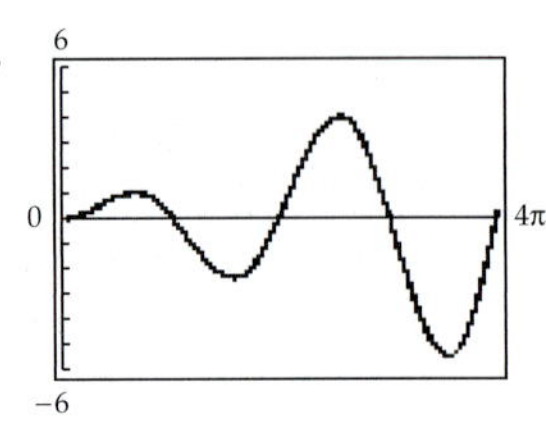

79.

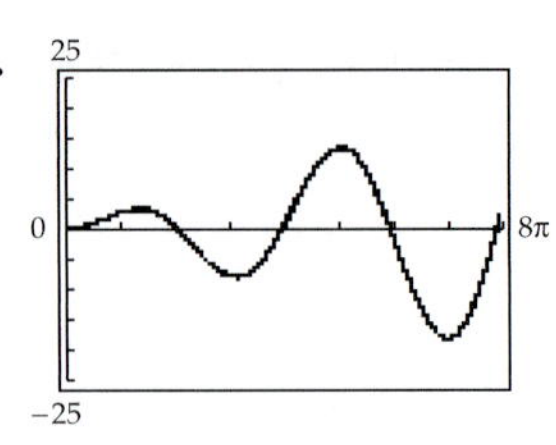

81.

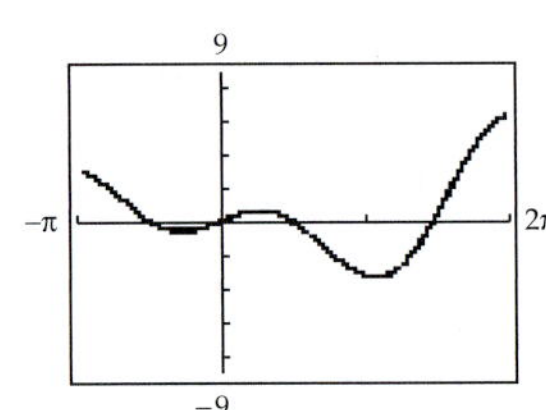

83.

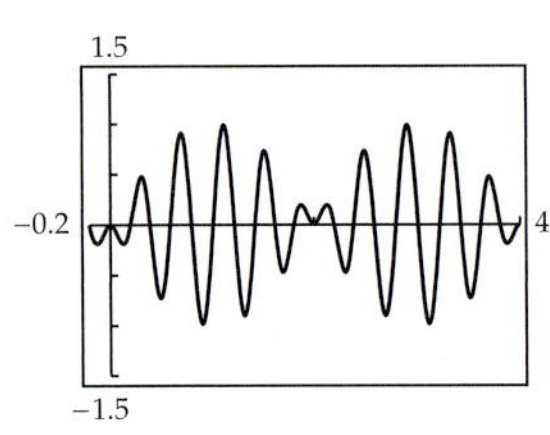

85.

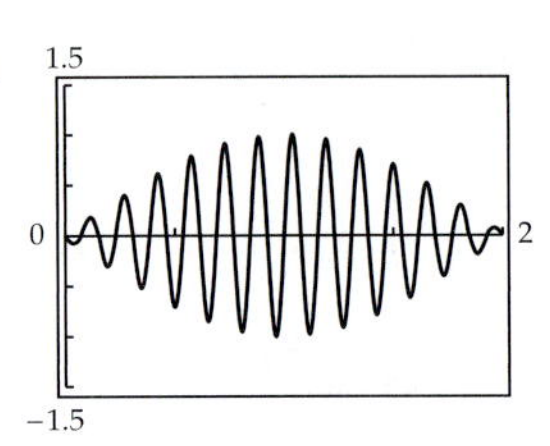

87. $y = 2\sin\left(2x - \frac{2\pi}{3}\right)$

89. $y = \tan\left(\frac{x}{2} - \frac{\pi}{4}\right)$ **91.** $y = \sec\left(\frac{x}{2} - \frac{3\pi}{8}\right)$ **93.** 1 **95.** $\cos^2 x + 2$ **97.**

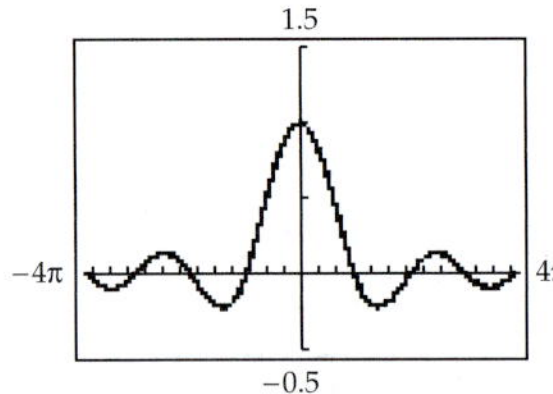

The graph above does not show that the function is undefined at $x = 0$.

99.

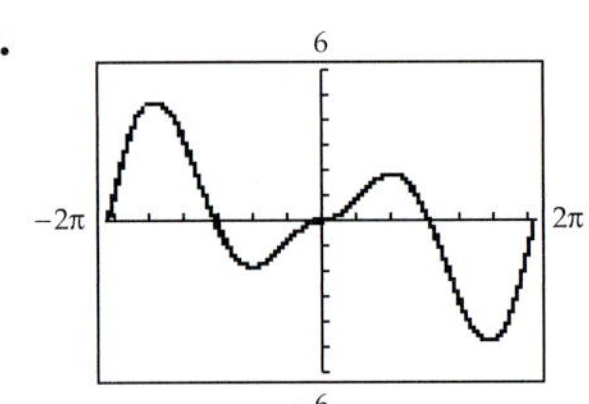

Chapter 4 True/False Exercises, page 352

1. False; the initial side must be along the positive x-axis. **2.** True **3.** False; $\sec^2 t - \tan^2 t = 1$ is an identity. **4.** False; the tangent function has no amplitude. **5.** False; the period is 2π. **6.** True **7.** False; $\sin^2 \frac{\pi}{6} = \left(\frac{1}{2}\right)^2 = \frac{1}{4}$, and $\sin\left(\frac{\pi}{6}\right)^2 = \sin \frac{\pi^2}{36} \approx 0.2707$. **8.** False; the phase shift is $\frac{\pi/3}{2} = \frac{\pi}{6}$. **9.** True **10.** False; the graph lies on or between the graphs of $y = 2^{-x}$ and $y = -2^{-x}$.

Chapter 4 Review Exercises, page 352

1. complement measures 25°; supplement measures 115° [4.1] **2.** 114.59° [4.1] **3.** $\frac{7\pi}{4}$ [4.1] **4.** 3.93 meters [4.1] **5.** 0.3 [4.1] **6.** 55 radians per second [4.1] **7.** $(0, -1)$ [4.2] **8.** $\left(-\frac{\sqrt{2}}{2}, -\frac{\sqrt{2}}{2}\right)$ [4.2] **9.** $\left(-\frac{\sqrt{3}}{2}, -\frac{1}{2}\right)$ [4.2] **10.** $\left(-\frac{1}{2}, -\frac{\sqrt{3}}{2}\right)$ [4.2] **11.** $\left(-\frac{\sqrt{3}}{2}, -\frac{1}{2}\right)$ [4.2] **12.** $\left(\frac{\sqrt{2}}{2}, \frac{\sqrt{2}}{2}\right)$ [4.2] **13.** $(0, 1)$ [4.2] **14.** $\left(-\frac{1}{2}, \frac{\sqrt{3}}{2}\right)$ [4.2] **15.** $\left(\frac{\sqrt{2}}{2}, -\frac{\sqrt{2}}{2}\right)$ [4.2] **16.** $\left(\frac{\sqrt{2}}{2}, -\frac{\sqrt{2}}{2}\right)$ [4.2] **17.** $-\frac{\sqrt{2}}{2}$ [4.2] **18.** $\frac{\sqrt{2}}{2}$ [4.2] **19.** $\frac{2\sqrt{3}}{3}$ [4.2] **20.** 1 [4.2] **21.** $\sqrt{2}$ [4.2] **22.** 1 [4.2] **23.** $-\frac{\sqrt{3}}{2}$ [4.2] **24.** $-\frac{1}{2}$ [4.2] **25.** $\frac{\sqrt{3}}{3}$ [4.2] **26.** $\sqrt{3}$ [4.2] **27.** $-\frac{2\sqrt{3}}{3}$ [4.2] **28.** -2 [4.2] **29.** -1 [4.2] **30.** 0 [4.2] **31.** even function [4.2] **32.** odd function [4.2] **33.** odd function [4.2] **34.** odd function [4.2] **35.** neither [4.2] **36.** neither [4.2] **37.** $\sec^2 t$ [4.2]

38. $\tan t$ [4.2] **39.** $\sin t$ [4.2] **40.** $\tan^2 t$ [4.2] **41.** $\csc^2 t$ [4.2] **42.** 0 [4.2] **43.** $3, \pi, \frac{\pi}{2}$ [4.3] **44.** no amplitude, $\frac{\pi}{3}$, 0 [4.4]

45. $2, \frac{2\pi}{3}, -\frac{\pi}{9}$ [4.3] **46.** $1, \pi, \frac{\pi}{3}$ [4.3] **47.** no amplitude, $\frac{\pi}{2}, \frac{3\pi}{8}$ [4.4] **48.** no amplitude, $2\pi, \frac{\pi}{4}$ [4.4]

49. [4.3]

50. [4.3]

51. [4.3]

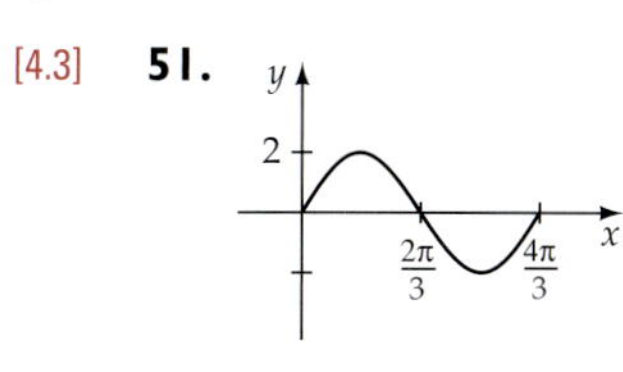

52. [4.5]

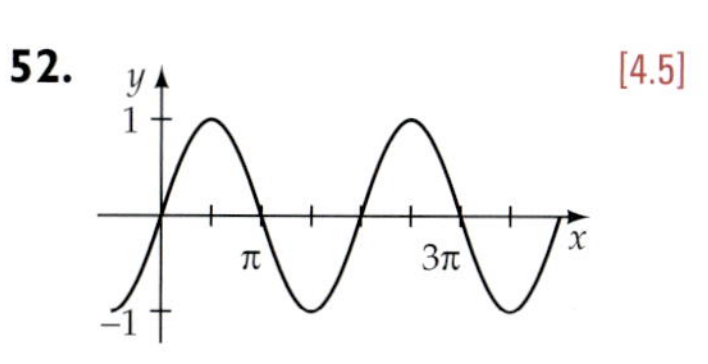

53. [4.5]

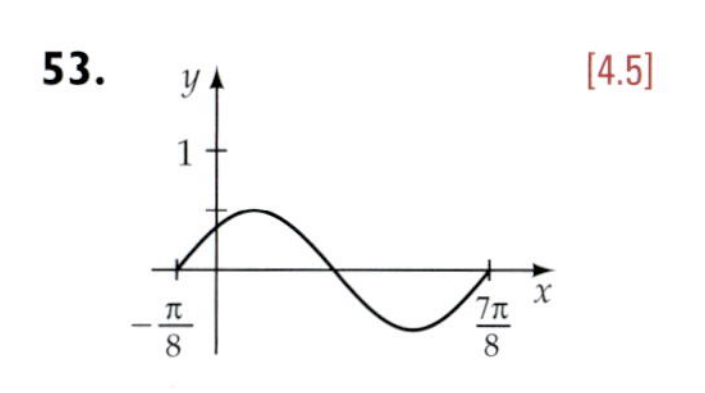

54. [4.5]

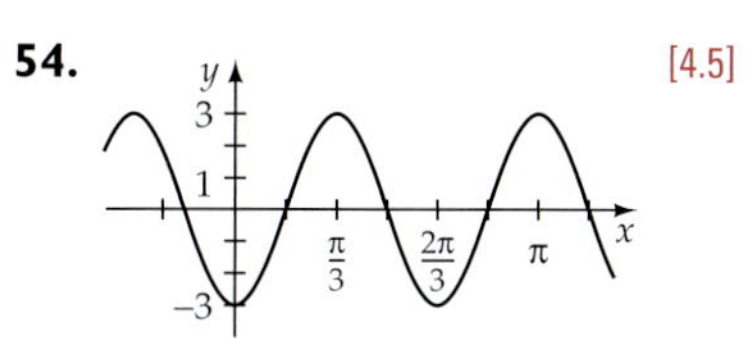

55. [4.4]

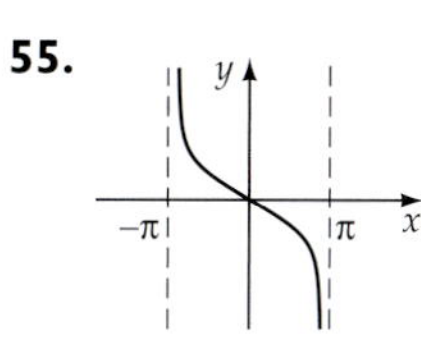

56. [4.4]

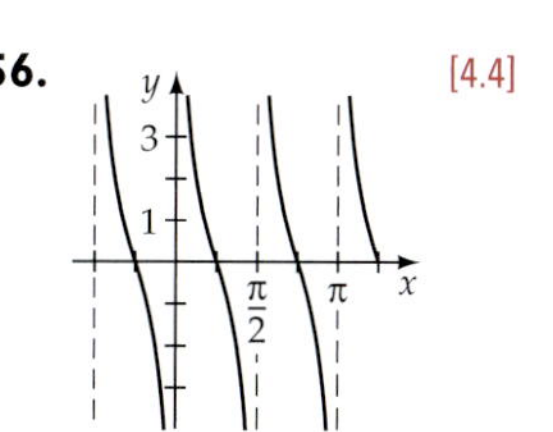

57. [4.5]

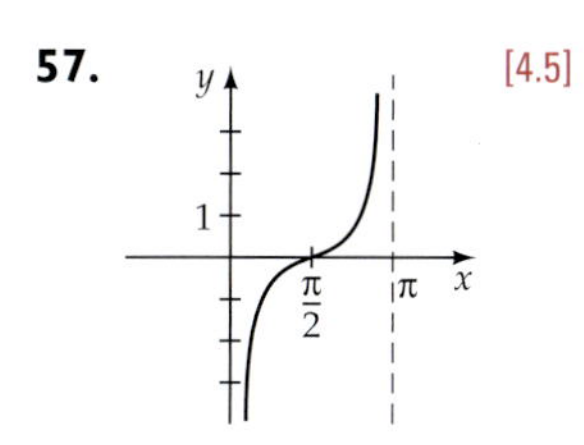

58. [4.5]

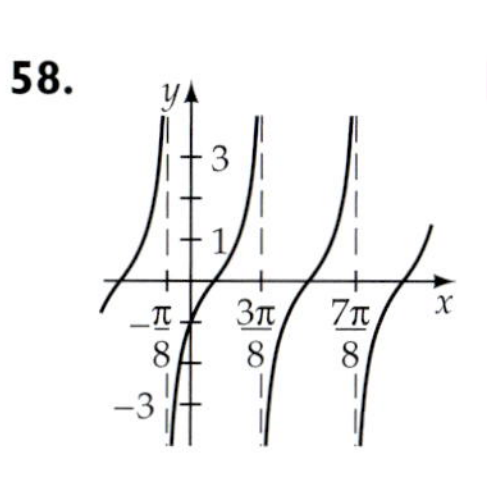

59. [4.5]

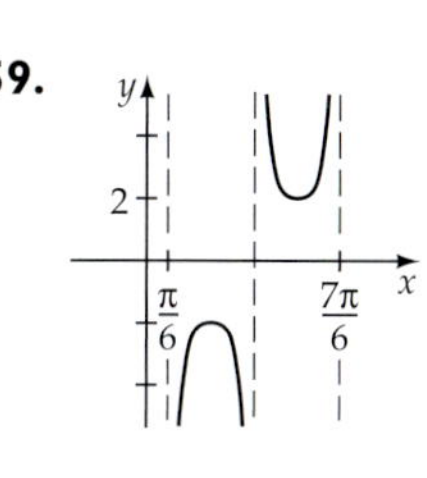

60. [4.5]

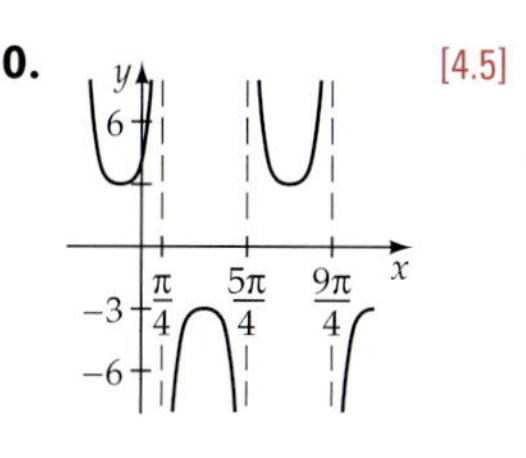

61. [4.5]

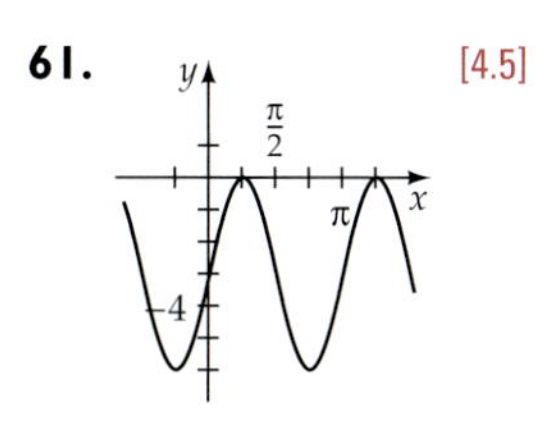

62. [4.5]

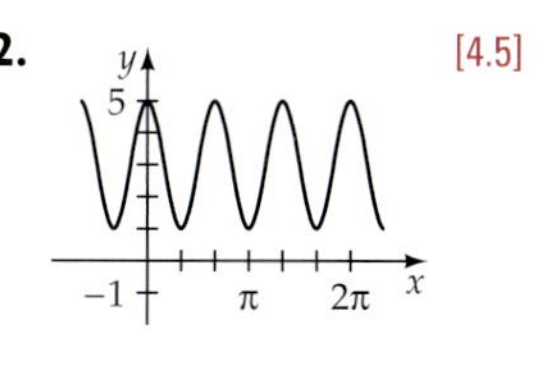

63. [4.5]

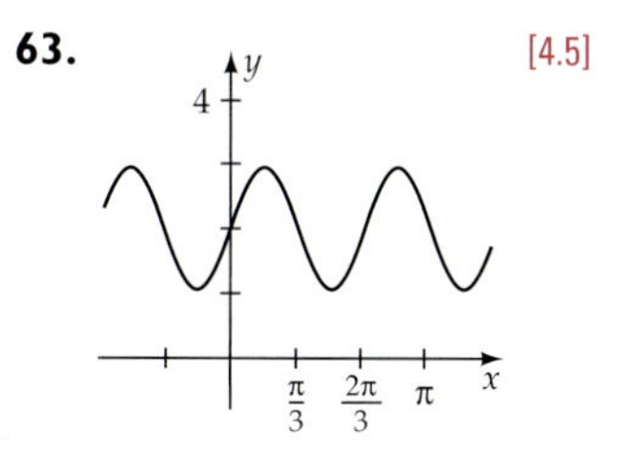

64. [4.5]

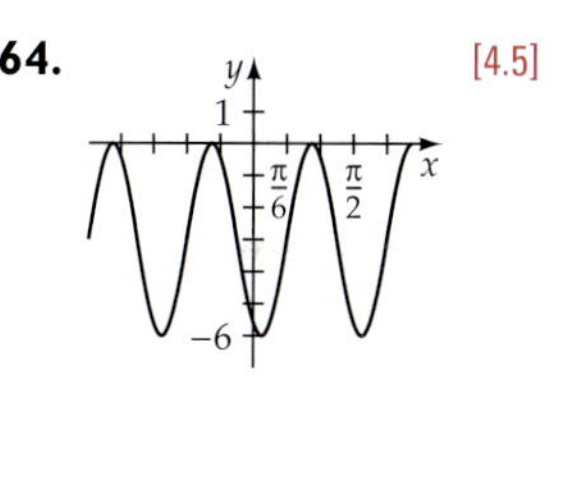

65. [4.5]

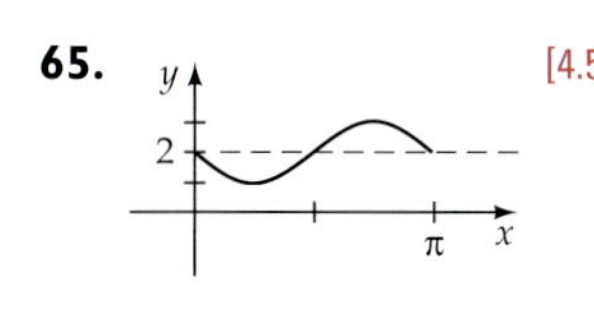

66. [4.5]

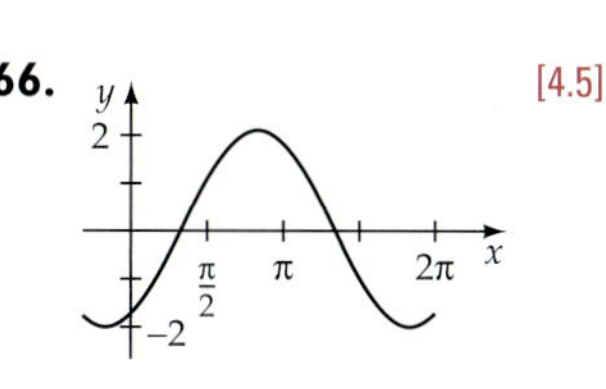

Chapter 4 Test, page 354

1. $\frac{5\pi}{6}$ [4.1] **2.** $\frac{\pi}{12}$ [4.1] **3.** 13.1 centimeters [4.1] **4.** 12π radians/second [4.1] **5.** 80 centimeters/second [4.1]

6. $\left(\frac{\sqrt{3}}{2}, -\frac{1}{2}\right)$ [4.2] **7.** $-\frac{\sqrt{3}}{2}$ [4.2] **8.** $\sqrt{3}$ [4.2] **9.** $-\frac{2\sqrt{3}}{3}$ [4.2] **10.** 1.3410 [4.2] **11.** odd function [4.2]

12. $\sin^2 t$ [4.2] **13.** $\frac{\pi}{3}$ [4.4] **14.** amplitude 3, period π, phase shift $-\frac{\pi}{4}$ [4.5]

15. period 3, phase shift $-\frac{1}{2}$ [4.5]

16. [4.3]

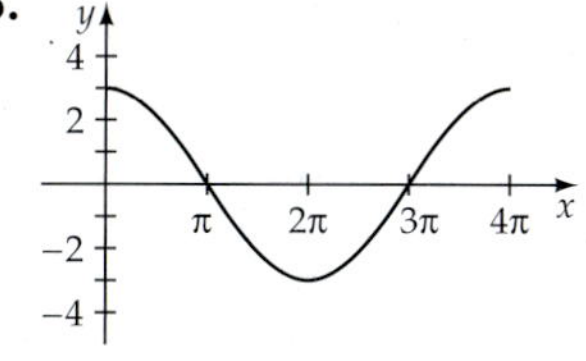

17. [4.4]

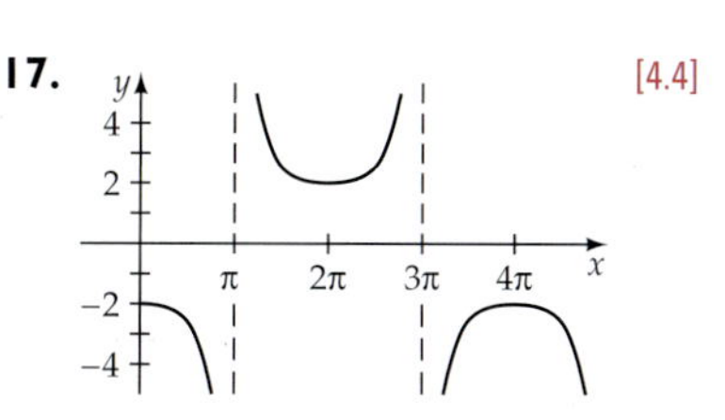

18. Shift the graph [of $y = 2\sin(2x)$] $\frac{\pi}{4}$ units to the right and down 1 unit. [4.5]

19. [4.5]

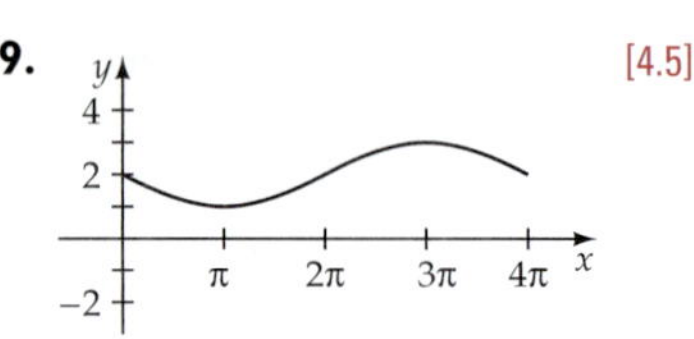

20. [4.5]

Cumulative Review Exercises, page 355

1. 1 [1.2] **2.** $x^2 + y^2 = 3^2$ [1.2] **3.** $22 - 7i$ [2.1] **4.** odd function [1.6] **5.** $f^{-1}(x) = \frac{3x}{2x - 1}$ [3.1] **6.** $(-\infty, 4) \cup (4, \infty)$ [2.6]
7. $[0, 2]$ [1.3] **8.** Shift the graph of $y = f(x)$ horizontally 3 units to the right. [1.6] **9.** Reflect the graph of $y = f(x)$ across the y-axis. [1.6]
10. $y = x + 3$ [2.6] **11.** decreasing function [3.2] **12.** $5^3 = 125$ [3.3] **13.** $(1, 0)$ [3.3] **14.** 3.205 [3.4] **15.** 6.17 [3.5]
16. $\frac{5\pi}{3}$ [4.1] **17.** 225° [4.1] **18.** $(-\infty, \infty)$ [4.2] **19.** $[-1, 1]$ [4.2] **20.** $-\frac{\sqrt{3}}{3}$ [4.2]

Exercise Set 5.1, page 371

1. $\sin\theta = \frac{12}{13}$ $\csc\theta = \frac{13}{12}$
$\cos\theta = \frac{5}{13}$ $\sec\theta = \frac{13}{5}$
$\tan\theta = \frac{12}{5}$ $\cot\theta = \frac{5}{12}$

3. $\sin\theta = \frac{4}{7}$ $\csc\theta = \frac{7}{4}$
$\cos\theta = \frac{\sqrt{33}}{7}$ $\sec\theta = \frac{7\sqrt{33}}{33}$
$\tan\theta = \frac{4\sqrt{33}}{33}$ $\cot\theta = \frac{\sqrt{33}}{4}$

5. $\sin\theta = \frac{5\sqrt{29}}{29}$ $\csc\theta = \frac{\sqrt{29}}{5}$
$\cos\theta = \frac{2\sqrt{29}}{29}$ $\sec\theta = \frac{\sqrt{29}}{2}$
$\tan\theta = \frac{5}{2}$ $\cot\theta = \frac{2}{5}$

7. $\sin\theta = \frac{\sqrt{21}}{7}$ $\csc\theta = \frac{\sqrt{21}}{3}$
$\cos\theta = \frac{2\sqrt{7}}{7}$ $\sec\theta = \frac{\sqrt{7}}{2}$
$\tan\theta = \frac{\sqrt{3}}{2}$ $\cot\theta = \frac{2\sqrt{3}}{3}$

9. $\sin\theta = \frac{\sqrt{3}}{2}$ $\csc\theta = \frac{2\sqrt{3}}{3}$
$\cos\theta = \frac{1}{2}$ $\sec\theta = 2$
$\tan\theta = \sqrt{3}$ $\cot\theta = \frac{\sqrt{3}}{3}$

11. $\sqrt{2}$ **13.** $\frac{5}{4}$ **15.** $\sqrt{3}$ **17.** $\frac{3\sqrt{2} + 2\sqrt{3}}{6}$ **19.** 9.5 feet **21.** 5.1 feet **23.** 1400 feet
25. 686,000,000 kilometers **27.** 612 feet **29.** 560 feet (to the nearest foot) **31. a.** 559 feet **b.** 193 feet

33. $\sin\theta = \frac{3\sqrt{13}}{13}$ $\csc\theta = \frac{\sqrt{13}}{3}$
$\cos\theta = \frac{2\sqrt{13}}{13}$ $\sec\theta = \frac{\sqrt{13}}{2}$
$\tan\theta = \frac{3}{2}$ $\cot\theta = \frac{2}{3}$

35. $\sin\theta = \frac{3\sqrt{13}}{13}$ $\csc\theta = \frac{\sqrt{13}}{3}$
$\cos\theta = -\frac{2\sqrt{13}}{13}$ $\sec\theta = -\frac{\sqrt{13}}{2}$
$\tan\theta = -\frac{3}{2}$ $\cot\theta = -\frac{2}{3}$

37. $\sin\theta = 0$ $\csc\theta$ is undefined.
$\cos\theta = -1$ $\sec\theta = -1$
$\tan\theta = 0$ $\cot\theta$ is undefined.

39. Quadrant I **41.** Quadrant IV **43.** 20° **45.** 9° **47.** $\frac{\pi}{5}$ **49.** 34° **51.** $-\frac{\sqrt{2}}{2}$ **53.** 1
55. $-\frac{2\sqrt{3}}{3}$ **57.** $\frac{\sqrt{2}}{2}$ **59.** $\sqrt{2}$ **61.** cot 540° is undefined. **63.** 0.6249 **65.** 1.2799 **67.** 2.6131 **69.** −1.7013 **71.** 0 **73.** 1
75. $-\frac{3}{2}$ **77.** 30°, 150° **79.** 150°, 210° **81.** $\frac{3\pi}{4}, \frac{7\pi}{4}$ **83.** $\frac{\pi}{3}, \frac{2\pi}{3}$

Prepare for Section 5.2, page 374

85. Both functional values equal $\frac{1}{2}$. **86.** Both functional values equal $\frac{1}{2}$. **87.** For each of the given values of θ, the functional values are equal. **88.** For each of the given values of θ, the functional values are equal. **89.** Both functional values equal $\frac{\sqrt{3}}{3}$.

90. For each of the given values of θ, the functional values are equal.

Exercise Set 5.2, page 384

43. $\frac{\sqrt{6}+\sqrt{2}}{4}$ **45.** $2-\sqrt{3}$ **47.** $\frac{-\sqrt{6}+\sqrt{2}}{4}$ **49.** $2+\sqrt{3}$ **51.** 0 **53.** $\frac{1}{2}$ **55.** $\cos 48°$ **57.** $\cot 75°$ **59.** $\csc 65°$ **61.** $\sin 5x$ **63.** $\cos x$ **65.** $\tan 7x$ **67.** $-\frac{77}{85}$ **69.** $-\frac{63}{16}$ **71.** $\frac{56}{65}$ **89.** identity

Prepare for Section 5.3, page 387

97. $2\sin\alpha\cos\alpha$ **98.** $\cos^2\alpha-\sin^2\alpha$ **99.** $\frac{2\tan\alpha}{1-\tan^2\alpha}$ **100.** For each of the given values of α, the functional values are equal. **101.** Let $\alpha = 45°$; then the left side of the equation is 1, and the right side of the equation is $\sqrt{2}$. **102.** Let $\alpha = 60°$; then the left side of the equation is $\frac{\sqrt{3}}{2}$, and the right side of the equation is $\frac{1}{4}$.

Exercise Set 5.3, page 398

1. $\sin 4\alpha$ **3.** $\cos 10\beta$ **5.** $\tan 6\alpha$ **7.** $\frac{\sqrt{2+\sqrt{3}}}{2}$ **9.** $\sqrt{2}+1$ **11.** $\frac{\sqrt{2-\sqrt{2}}}{2}$ **13.** $\frac{\sqrt{2-\sqrt{2}}}{2}$ **15.** $\frac{\sqrt{2-\sqrt{3}}}{2}$ **17.** $\sin 2\theta = -\frac{24}{25}, \cos 2\theta = \frac{7}{25}, \tan 2\theta = -\frac{24}{7}$ **19.** $\sin 2\theta = -\frac{240}{289}, \cos 2\theta = \frac{161}{289}, \tan 2\theta = -\frac{240}{161}$ **21.** $\sin 2\theta = -\frac{336}{625}, \cos 2\theta = -\frac{527}{625}, \tan 2\theta = \frac{336}{527}$ **23.** $\sin\frac{\alpha}{2} = \frac{5\sqrt{26}}{26}, \cos\frac{\alpha}{2} = \frac{\sqrt{26}}{26}, \tan\frac{\alpha}{2} = 5$ **25.** $\sin\frac{\alpha}{2} = \frac{5\sqrt{34}}{34}, \cos\frac{\alpha}{2} = -\frac{3\sqrt{34}}{34}, \tan\frac{\alpha}{2} = -\frac{5}{3}$ **27.** $\sin\frac{\alpha}{2} = \frac{\sqrt{5}}{5}, \cos\frac{\alpha}{2} = \frac{2\sqrt{5}}{5}, \tan\frac{\alpha}{2} = \frac{1}{2}$ **49.** $\sin 3x - \sin x$ **51.** $\frac{1}{2}(\sin 8x - \sin 4x)$ **53.** $\frac{1}{2}(\cos 4x - \cos 6x)$ **55.** $\frac{1}{4}$ **57.** $-\frac{\sqrt{2}}{4}$ **59.** $\frac{\sqrt{3}-2}{4}$ **61.** $2\sin 3\theta\cos\theta$ **63.** $-2\sin 4\theta\sin 2\theta$ **65.** $2\cos 4\theta\cos 3\theta$ **67.** $2\sin 7\theta\cos 2\theta$ **69.** $2\sin\frac{3}{4}\theta\sin\frac{\theta}{4}$ **79.** $y = \sqrt{2}\sin(x-135°)$ **81.** $y = \frac{\sqrt{2}}{2}\sin(x-45°)$

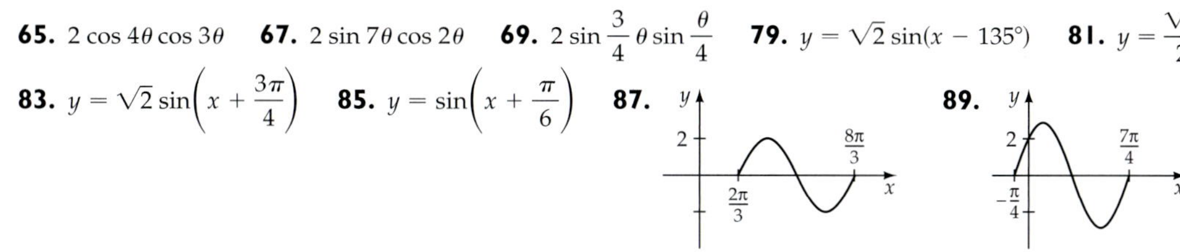

83. $y = \sqrt{2}\sin\left(x+\frac{3\pi}{4}\right)$ **85.** $y = \sin\left(x+\frac{\pi}{6}\right)$ **87.** (graph) **89.** (graph)

91. a. $p(t) = \cos(2\pi\cdot 1336t) + \cos(2\pi\cdot 770t)$ **b.** $p(t) = 2\cos(2106\pi t)\cos(566\pi t)$ **c.** 1053 cycles per second

Prepare for Section 5.4, page 400

99. A one-to-one function is a function for which each range value (y-value) is paired with one and only one domain value (x-value). **100.** If every horizontal line intersects the graph of a function at most once, then the function is a one-to-one function. **101.** $f[g(x)] = x$ **102.** $f[f^{-1}(x)] = x$ **103.** The graph of f^{-1} is the reflection of the graph of f across the line given by $y = x$. **104.** No

Exercise Set 5.4, page 411

1. $\frac{\pi}{2}$ **3.** $-\frac{\pi}{4}$ **5.** $\frac{\pi}{3}$ **7.** $\frac{\pi}{3}$ **9.** $-\frac{\pi}{4}$ **11.** $-\frac{\pi}{3}$ **13.** $\frac{2\pi}{3}$ **15.** $\frac{\pi}{6}$ **17.** $\frac{\pi}{6}$ **19.** $\frac{1}{2}$ **21.** $\frac{3}{5}$ **23.** 1 **25.** $\frac{1}{2}$ **27.** $\frac{\pi}{6}$

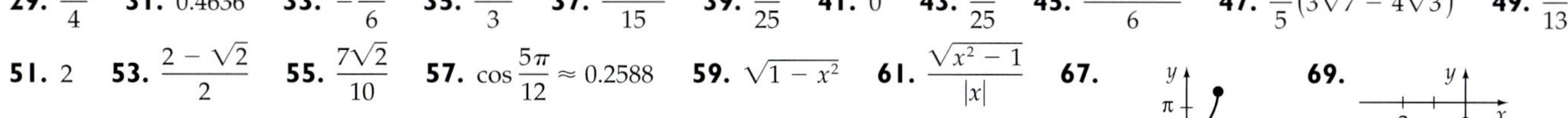

29. $\frac{\pi}{4}$ **31.** 0.4636 **33.** $-\frac{\pi}{6}$ **35.** $\frac{\sqrt{3}}{3}$ **37.** $\frac{4\sqrt{15}}{15}$ **39.** $\frac{24}{25}$ **41.** 0 **43.** $\frac{24}{25}$ **45.** $\frac{2+\sqrt{15}}{6}$ **47.** $\frac{1}{5}(3\sqrt{7}-4\sqrt{3})$ **49.** $\frac{12}{13}$ **51.** 2 **53.** $\frac{2-\sqrt{2}}{2}$ **55.** $\frac{7\sqrt{2}}{10}$ **57.** $\cos\frac{5\pi}{12} \approx 0.2588$ **59.** $\sqrt{1-x^2}$ **61.** $\frac{\sqrt{x^2-1}}{|x|}$ **67.** (graph) **69.** (graph)

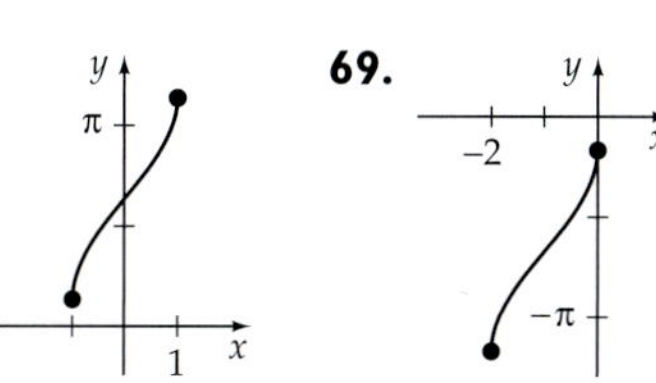

71.

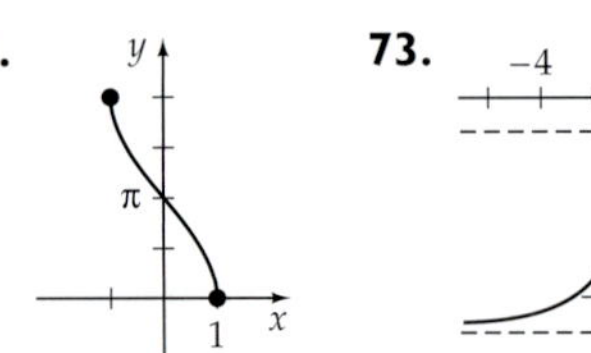

73.

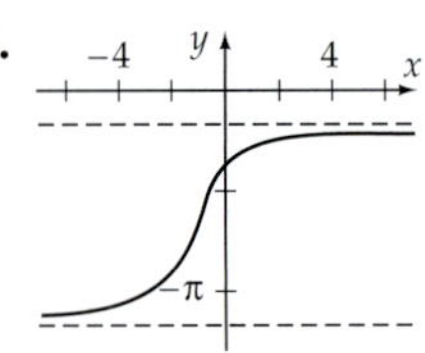

75. a. 0.1014 **b.** 0.1552 **77.**

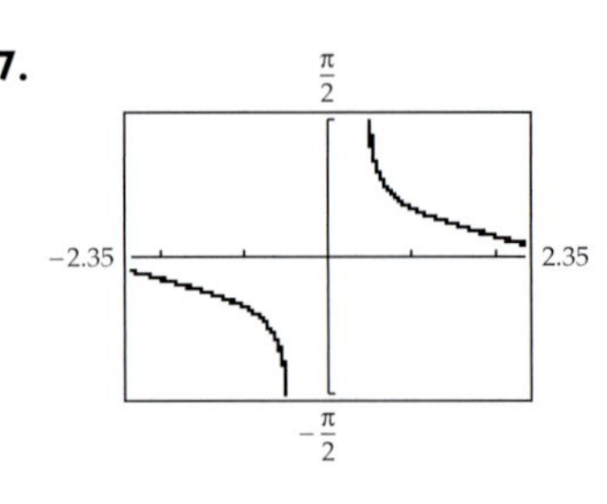

79.

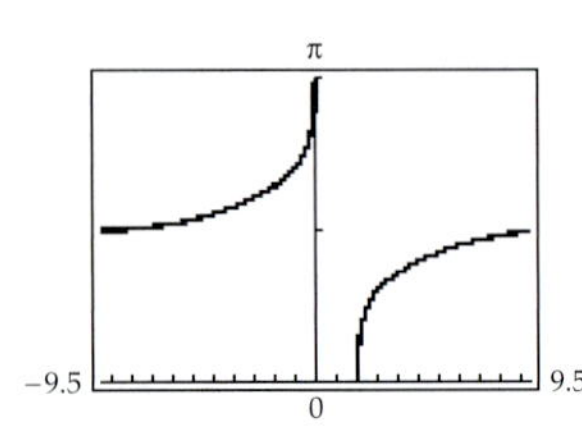

81.

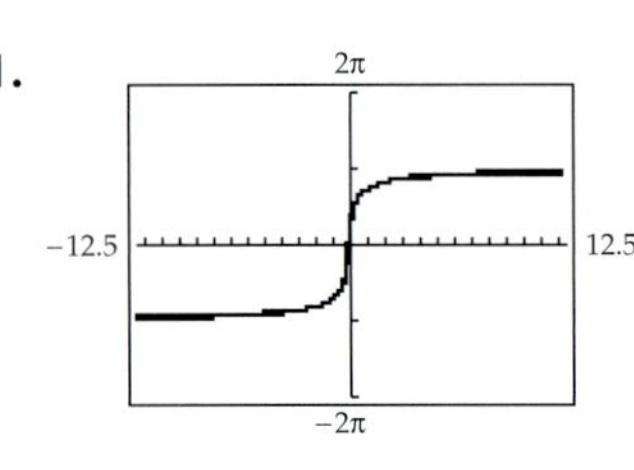

87. $y = \frac{1}{3}\tan 5x$ **89.** $y = 3 + \cos\left(x - \frac{\pi}{3}\right)$

Prepare for Section 5.5, page 413

91. $x = \frac{5 \pm \sqrt{73}}{6}$ **92.** $1 - \cos^2 x$ **93.** $\frac{5}{2}\pi, \frac{9}{2}\pi$, and $\frac{13}{2}\pi$ **94.** $(x+1)\left(x - \frac{\sqrt{3}}{2}\right)$ **95.** $\frac{1}{2}, 3$ **96.** 0, 1

Exercise Set 5.5, page 421

1. $\frac{\pi}{4}, \frac{7\pi}{4}$ **3.** $\frac{\pi}{3}, \frac{4\pi}{3}$ **5.** $\frac{\pi}{4}, \frac{\pi}{2}, \frac{3\pi}{4}, \frac{3\pi}{2}$ **7.** $\frac{\pi}{2}, \frac{3\pi}{2}$ **9.** $\frac{\pi}{6}, \frac{\pi}{4}, \frac{3\pi}{4}, \frac{11\pi}{6}$ **11.** $\frac{\pi}{4}, \frac{3\pi}{4}$ **13.** $\frac{\pi}{6}, \frac{\pi}{2}, \frac{5\pi}{6}$ **15.** $\frac{\pi}{6}, \frac{5\pi}{6}, \frac{7\pi}{6}, \frac{11\pi}{6}$ **17.** $0, \frac{\pi}{4}, \frac{3\pi}{4}, \pi, \frac{5\pi}{4}, \frac{7\pi}{4}$ **19.** $\frac{\pi}{6}, \frac{5\pi}{6}, \frac{4\pi}{3}, \frac{5\pi}{3}$ **21.** 41.4°, 318.6° **23.** no solution **25.** 68.0°, 292.0° **27.** 12.8°, 167.2° **29.** 15.5°, 164.5° **31.** 0°, 33.7°, 180°, 213.7° **33.** no solution **35.** 0°, 120°, 240° **37.** 70.5°, 289.5° **39.** 68.2°, 116.6°, 248.2°, 296.6° **41.** 19.5°, 90°, 160.5°, 270° **43.** 60°, 90°, 300° **45.** 53.1°, 180° **47.** 72.4°, 220.2° **49.** 50.1°, 129.9°, 205.7°, 334.3° **51.** no solution **53.** 22.5°, 157.5° **55.** $\frac{\pi}{8} + \frac{k\pi}{2}$, where k is an integer **57.** $\frac{\pi}{10} + \frac{2k\pi}{5}$, where k is an integer **59.** $0 + 2k\pi, \frac{\pi}{3} + 2k\pi, \pi + 2k\pi, \frac{5\pi}{3} + 2k\pi$, where k is an integer **61.** $\frac{\pi}{2} + k\pi, \frac{5\pi}{6} + k\pi$, where k is an integer **63.** $0 + 2k\pi$, where k is an integer **65.** $0, \pi$ **67.** $0, \frac{\pi}{6}, \frac{\pi}{2}, \frac{5\pi}{6}, \pi, \frac{7\pi}{6}, \frac{3\pi}{2}, \frac{11\pi}{6}$ **69.** $0, \frac{\pi}{2}, \frac{3\pi}{2}$ **71.** $\frac{4\pi}{3}, \frac{5\pi}{3}$ **73.** $0, \frac{\pi}{4}, \frac{3\pi}{4}, \pi, \frac{5\pi}{4}, \frac{7\pi}{4}$ **75.** $\frac{\pi}{6}, \frac{5\pi}{6}, \pi$ **77.** 0.7391 **79.** −3.2957, 3.2957 **81.** 14.99° and 75.01° **83.** day 74 to day 268 = 195 days **85. b.** 42° and 79° **c.** 60° **87.** $\frac{\pi}{6}, \frac{\pi}{2}$ **89.** $\frac{5\pi}{3}, 0$ **91.** $0, \frac{\pi}{4}, \frac{\pi}{2}, \frac{3\pi}{4}, \pi, \frac{5\pi}{4}, \frac{3\pi}{2}, \frac{7\pi}{4}$ **93.** 0.93 foot, 1.39 feet

Prepare for Section 5.6, page 424

95. 16.7 **96.** 45.7° **97.** $B = \cos^{-1}\left(\frac{a^2 + c^2 - b^2}{2ac}\right)$ **98.** 6.2 **99. a.** 12 inches **b.** 6 square inches **100.** 24

Exercise Set 5.6, page 433

1. $C = 77°, b \approx 16, c \approx 17$ **3.** $B = 38°, a \approx 18, c \approx 10$ **5.** $C = 32.6°, c \approx 21.6, a \approx 39.8$ **7.** $A \approx 58.5°, B \approx 7.3°, a \approx 81.5$ **9.** $C = 59°, B = 84°, b \approx 46$ or $C = 121°, B = 22°, b \approx 17$ **11.** No triangle is formed. **13.** $C = 19.8°, B = 145.4°, b \approx 10.7$ or $C = 160.2°, B = 5.0°, b \approx 1.64$ **15.** $C = 51.21°, A = 11.47°, c \approx 59.00$ **17.** ≈ 13 **19.** ≈ 29 **21.** ≈ 9.5 **23.** ≈ 40.1 **25.** ≈ 39° **27.** ≈ 90° **29.** ≈ 80.3° **31.** ≈ 68.8 miles **33.** 231 yards **35.** ≈ 130 yards **37.** ≈ 96 feet **39.** ≈ 8.1 miles **41.** ≈ 1200 miles **43.** ≈ 710 miles **45.** ≈ 74 feet **47.** ≈ 60.9° **49.** ≈ 350 miles **51.** ≈ 2800 feet **53.** Triangle *DEF* has an incorrect dimension.

Prepare for Section 5.7, page 438

57. 1 **58.** −6.691 **59.** 30° **60.** 157.6° **61.** $\frac{\sqrt{5}}{5}$ **62.** $\frac{14\sqrt{17}}{17}$

Exercise Set 5.7, page 450

1. $a = 7, b = -1; \langle 7, -1\rangle$ **3.** $a = -7, b = -5; \langle -7, -5\rangle$ **5.** $a = 0, b = 8; \langle 0, 8\rangle$ **7.** $5, \approx 126.9°, \left\langle -\frac{3}{5}, \frac{4}{5}\right\rangle$
9. $\approx 44.7, \approx 296.6°, \left\langle \frac{\sqrt{5}}{5}, \frac{-2\sqrt{5}}{5}\right\rangle$ **11.** $\approx 4.5, \approx 296.6°, \left\langle \frac{\sqrt{5}}{5}, \frac{-2\sqrt{5}}{5}\right\rangle$ **13.** $\approx 45.7, \approx 336.8°, \left\langle \frac{7\sqrt{58}}{58}, \frac{-3\sqrt{58}}{58}\right\rangle$ **15.** $\langle -6, 12\rangle$
17. $\langle -1, 10\rangle$ **19.** $\left\langle -\frac{11}{6}, \frac{7}{3}\right\rangle$ **21.** $2\sqrt{5}$ **23.** $2\sqrt{109}$ **25.** $-8\mathbf{i} + 12\mathbf{j}$ **27.** $14\mathbf{i} - 6\mathbf{j}$ **29.** $\frac{11}{12}\mathbf{i} + \frac{1}{2}\mathbf{j}$ **31.** $\sqrt{113}$
33. $a_1 \approx 4.5, a_2 \approx 2.3, 4.5\mathbf{i} + 2.3\mathbf{j}$ **35.** $a_1 \approx 2.8, a_2 \approx 2.8, 2.8\mathbf{i} + 2.8\mathbf{j}$ **37.** $\approx$ 380 miles per hour
39. $\approx$ 250 miles per hour at a heading of 86° **41.** $\approx$ 293 pounds **43.** $\approx$ 24.7 pounds **45.** -3 **47.** 0 **49.** 1 **51.** 0 **53.** $\approx 79.7°$
55. 45° **57.** 90°, orthogonal **59.** 180° **61.** $\frac{46}{5}$ **63.** $\frac{14\sqrt{29}}{29} \approx 2.6$ **65.** $\sqrt{5} \approx 2.2$ **67.** $-\frac{11\sqrt{5}}{5} \approx -4.9$ **69.** $\approx$ 954 foot-pounds
71. $\approx$ 779 foot-pounds **73.**

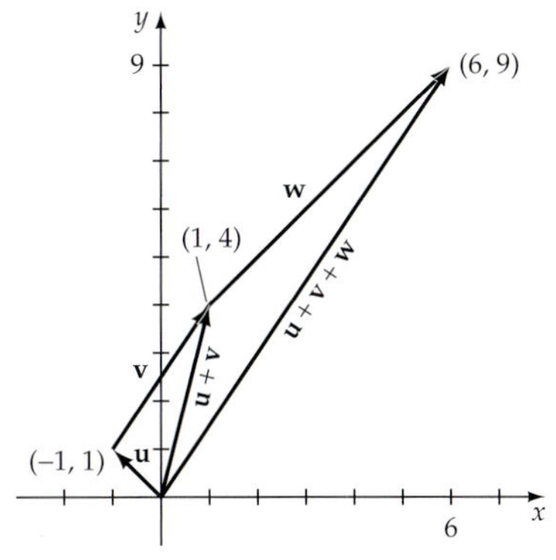

$\langle 6, 9\rangle$ **75.** The vector from $P_1(3, -1)$ to $P_2(5, -4)$ is equivalent to $2\mathbf{i} - 3\mathbf{j}$.

77. $\mathbf{v} \cdot \mathbf{w} = 0$; the vectors are perpendicular. **79.** $\langle 7, 2\rangle$ is one example. **81.** 6.4 pounds **83.** No
87. The same amount of work is done.

Chapter 5 True/False Exercises, page 455

1. True **2.** False; if $y = 0$, $\frac{\sin x}{\cos x} = 0$, but $\tan \frac{x}{y}$ is undefined. **3.** False. Let $x = 1$. Then $\sin^{-1} 1 = \frac{\pi}{2}$ and $\csc 1^{-1} \approx 1.18$. **4.** False; if $\alpha = \frac{\pi}{2}$, then $\sin 2\alpha = \sin \pi = 0$ but $2 \sin \frac{\pi}{2} = 2$. **5.** False; $\sin(30° + 60°) \neq \sin 30° + \sin 60°$. **6.** False; $\tan 45° = \tan 225°$ but $45° \neq 225°$.
7. False; $\cos^{-1}\left(\cos \frac{3\pi}{2}\right) = \cos^{-1}(0) = \frac{\pi}{2} \neq \frac{3\pi}{2}$. **8.** True **9.** False; we cannot solve a triangle using the Law of Cosines if we are only given two sides and the angle opposite one of the given sides. **10.** True **11.** False; $2\mathbf{i} \neq 2\mathbf{j}$. **12.** True

Chapter 5 Review Exercises, page 456

1. $\sin \theta = -\frac{3\sqrt{10}}{10}$ $\csc \theta = -\frac{\sqrt{10}}{3}$ [5.1] **2. a.** $-\frac{2\sqrt{3}}{3}$ **b.** 1 **c.** -1 **d.** $-\frac{1}{2}$ [5.1]
$\cos \theta = \frac{\sqrt{10}}{10}$ $\sec \theta = \sqrt{10}$
$\tan \theta = -3$ $\cot \theta = -\frac{1}{3}$
3. a. -0.5446 **b.** 0.5365 **c.** -3.2361 **d.** 3.0777 [5.1] **4.** 0.089 mile [5.1] **5.** 12.3 feet [5.1] **6.** 46 feet [5.1]
7. $\frac{\sqrt{6} - \sqrt{2}}{4}$ [5.2] **8.** $\sqrt{3} - 2$ [5.2] **9.** $\frac{\sqrt{6} - \sqrt{2}}{4}$ [5.2] **10.** $\sqrt{2} - \sqrt{6}$ [5.2] **11.** $\frac{\sqrt{2 - \sqrt{2}}}{2}$ [5.3] **12.** $-\frac{\sqrt{2 - \sqrt{3}}}{2}$ [5.3]
13. $\sqrt{2} + 1$ [5.3] **14.** $\frac{\sqrt{2 + \sqrt{2}}}{2}$ [5.3] **15.** 0 [5.3] **16.** 0 [5.3] **17.** $-\sqrt{3}$ [5.3] **18.** 1 [5.3] **19.** $\sin 6x$ [5.3] **20.** $\tan 3x$ [5.2]
21. $\sin 3x$ [5.2] **22.** $\cos 4\theta$ [5.3] **23.** $2 \sin 3\theta \sin \theta$ [5.3] **24.** $-2 \cos 4\theta \sin \theta$ [5.3] **25.** $2 \sin 4\theta \cos 2\theta$ [5.3] **26.** $2 \cos 3\theta \sin 2\theta$ [5.3]
39. $\frac{13}{5}$ [5.4] **40.** $\frac{4}{5}$ [5.4] **41.** $\frac{56}{65}$ [5.4] **42.** $\frac{7}{25}$ [5.4] **43.** $\frac{3}{2}$ [5.5] **44.** $\frac{4}{5}$ [5.5] **45.** 30°, 150°, 240°, 300° [5.5] **46.** 0°, 45°, 135° [5.5]
47. $\frac{\pi}{2} + 2k\pi, 3.8713 + 2k\pi, 5.5535 + 2k\pi$, where k is an integer [5.5] **48.** $-\frac{\pi}{4} + k\pi, 1.2490 + k\pi$, where k is an integer [5.5]

49. $\frac{\pi}{12}, \frac{5\pi}{12}, \frac{13\pi}{12}, \frac{17\pi}{12}$ [5.5] **50.** $\frac{7\pi}{12}, \frac{19\pi}{12}, \frac{3\pi}{4}, \frac{7\pi}{4}$ [5.5]

51. $y = 2\sin\left(x + \frac{\pi}{6}\right)$ [5.3] **52.** $y = 2\sqrt{2}\sin\left(x + \frac{5\pi}{4}\right)$ [5.3] **53.** [5.4] **54.** [5.4]

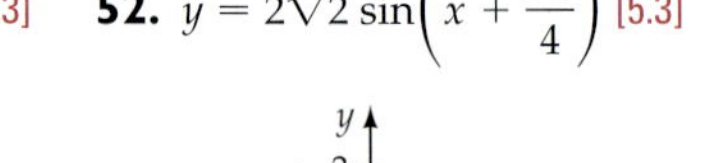

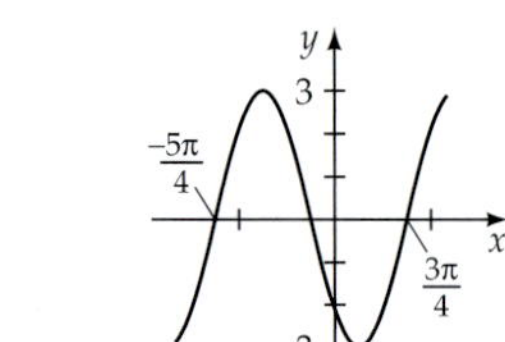

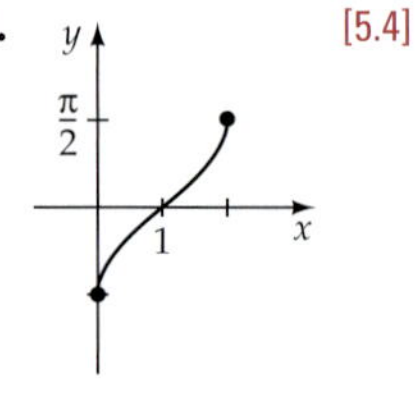

55. $B = 51°, a \approx 11, c \approx 18$ [5.6] **56.** $A = 8.6°, a \approx 1.77, b \approx 11.5$ [5.6] **57.** $B \approx 48°, C \approx 95°, A \approx 37°$ [5.6] **58.** $A \approx 47°, B \approx 76°, C \approx 58°$ [5.6] **59.** $C = 45°, a \approx 29, b \approx 35$ [5.6] **60.** $a \approx 169, B \approx 37°, C \approx 61°$ [5.6] **61.** $a_1 = 5, a_2 = 3, \langle 5, 3\rangle$ [5.7] **62.** $a_1 = 1, a_2 = 6, \langle 1, 6\rangle$ [5.7] **63.** $\approx 4.5, 153.4°$ [5.7] **64.** $\approx 6.7, 333.4°$ [5.7] **65.** $\approx 3.6, 123.7°$ [5.7] **66.** $\approx 8.1, 240.3°$ [5.7] **67.** $\left\langle -\frac{8\sqrt{89}}{89}, \frac{5\sqrt{89}}{89}\right\rangle$ [5.7] **68.** $\left\langle \frac{7\sqrt{193}}{193}, -\frac{12\sqrt{193}}{193}\right\rangle$ [5.7] **69.** $\frac{5\sqrt{26}}{26}\mathbf{i} + \frac{\sqrt{26}}{26}\mathbf{j}$ [5.7] **70.** $\frac{3\sqrt{34}}{34}\mathbf{i} - \frac{5\sqrt{34}}{34}\mathbf{j}$ [5.7] **71.** $\langle -7, -3\rangle$ [5.7] **72.** $\langle 18, 7\rangle$ [5.7] **73.** $-6\mathbf{i} - \frac{17}{2}\mathbf{j}$ [5.7] **74.** $-\frac{13}{6}\mathbf{i} - \frac{47}{6}\mathbf{j}$ [5.7] **75.** 420 miles per hour [5.7] **76.** $\approx 7°$ [5.7] **77.** 18 [5.7] **78.** −21 [5.7] **79.** −9 [5.7] **80.** 20 [5.7] **81.** $\approx 86°$ [5.7] **82.** $\approx 138°$ [5.7] **83.** $\approx 125°$ [5.7] **84.** $\approx 157°$ [5.7] **85.** $\frac{10\sqrt{41}}{41}$ [5.7] **86.** $\frac{27\sqrt{29}}{29}$ [5.7] **87.** ≈ 662 foot-pounds [5.7]

Chapter 5 Test, page 458

1. $\frac{\sqrt{3} - 6}{6}$ [5.1] **2.** 25.5 meters [5.1] **6.** $\frac{-\sqrt{6} + \sqrt{2}}{4}$ [5.2] **7.** $-\frac{\sqrt{2}}{10}$ [5.2] **9.** $\frac{2 - \sqrt{3}}{4}$ [5.3] **10.** $y = \sin\left(x + \frac{5\pi}{6}\right)$ [5.3] **11.** $\frac{5}{13}$ [5.4] **12.** [5.4]

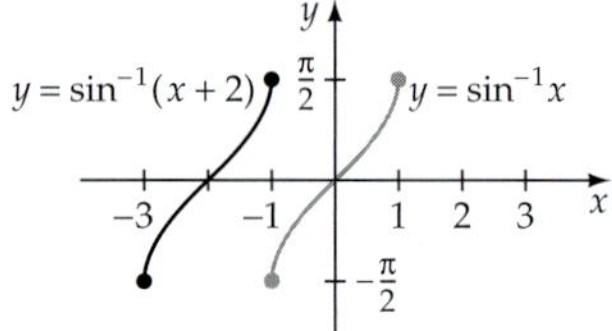

13. 41.8°, 138.2° [5.5] **14.** $\frac{\pi}{2}, \frac{2\pi}{3}, \frac{4\pi}{3}$ [5.5] **15.** $B = 94°, a \approx 48, b \approx 51$ [5.6] **16.** ≈ 27 miles [5.6] **17.** $-9.2\mathbf{i} - 7.7\mathbf{j}$ [5.7] **18.** $-19\mathbf{i} - 29\mathbf{j}$ [5.7] **19.** −1 [5.7] **20.** 103° [5.7]

Cumulative Review Exercises, page 459

1. Shift the graph of $y = f(x)$ to the left 1 unit and up 2 units. [1.6] **2.** Reflect the graph of $y = f(x)$ across the x-axis. [1.6] **3.** $x = 2$ [2.6] **4.** odd function [1.6/4.2] **5.** $f^{-1}(x) = \frac{x}{x - 5}$ [3.1] **6.** $\log_2 x = 5$ [3.3] **7.** 3 [3.3] **8.** $\frac{4\pi}{3}$ [4.1] **9.** 300° [4.1] **10.** 50° [5.1] **11.** $x = \frac{1}{2}, y = \frac{\sqrt{3}}{2}$ [4.2] **12.** $\frac{\pi}{6}$ [5.4] **13.** $[-1, 1]$ [5.4] **14.** $\left(-\frac{\pi}{2}, \frac{\pi}{2}\right)$ [5.4] **15.** $\frac{12}{5}$ [5.4] **16.** $\frac{\pi}{2}, \frac{7\pi}{6}, \frac{11\pi}{6}$ [5.5] **17.** magnitude: 5, angle: 126.9° [5.7] **18.** 176.8° [5.7] **19.** about 439 mph at a heading of 54.6° [5.7] **20.** 26° [5.6]

Exercise Set 6.1, page 480

1. Vertex: (0, 0)
Focus: (0, −1)
Directrix: $y = 1$

3. Center: (0, 0)
Vertices: (0, 5), (0, −5)
Foci: (0, 3), (0, −3)

5. Center: (0, 0)
Vertices: (±4, 0)
Foci: $(\pm\sqrt{41}, 0)$
Asymptotes: $y = \pm\frac{5x}{4}$

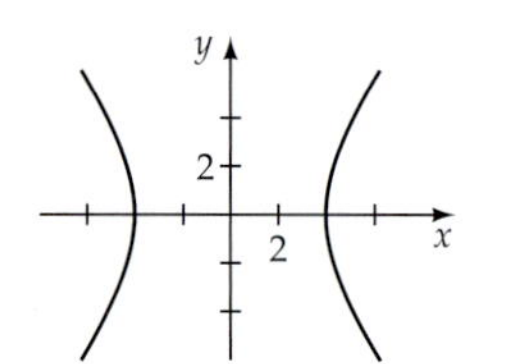

7. Vertex: (2, −3)
Focus: (2, −1)
Directrix: $y = -5$

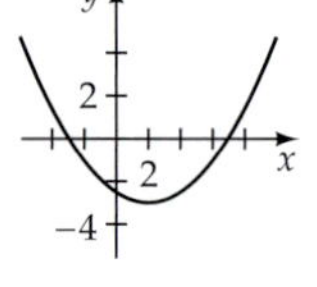

9. Center: (0, 0)
Vertices: (±3, 0)
Foci: $(\pm 3\sqrt{2}, 0)$
Asymptotes: $y = \pm x$

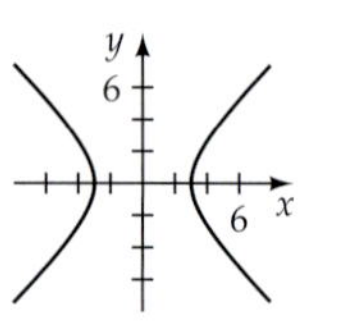

11. Center: $(3, -2)$
Vertices: $(8, -2), (-2, -2)$
Foci: $(6, -2), (0, -2)$

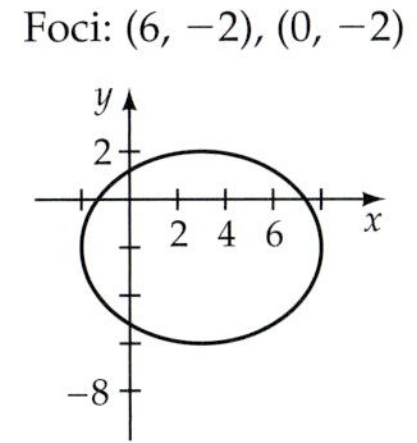

13. Center: $(3, -4)$
Vertices: $(7, -4), (-1, -4)$
Foci: $(8, -4), (-2, -4)$
Asymptotes: $y + 4 = \pm 3\dfrac{(x-3)}{4}$

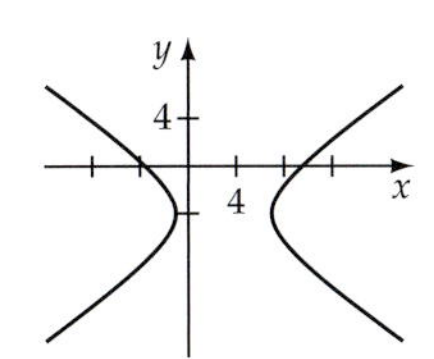

15. Center: $(3, -1)$
Vertices: $(-1, -1), (7, -1)$
Foci: $\left(3 \pm 2\sqrt{3}, -1\right)$

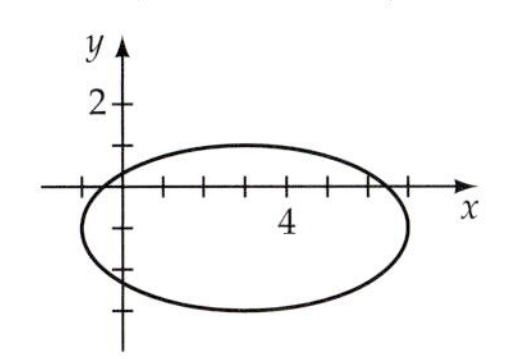

17. Vertex: $(-2, 1)$
Focus: $\left(-\dfrac{29}{16}, 1\right)$
Directrix: $x = -\dfrac{35}{16}$

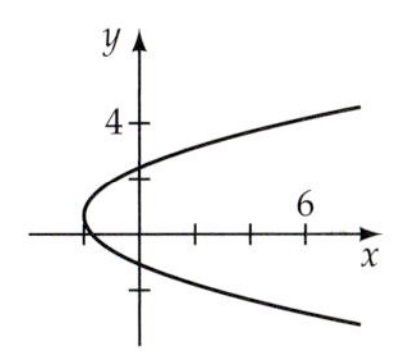

19. Center: $(-2, 1)$
Vertices: $(-2, -2), (-2, 4)$
Foci: $\left(-2, 1 \pm \sqrt{5}\right)$

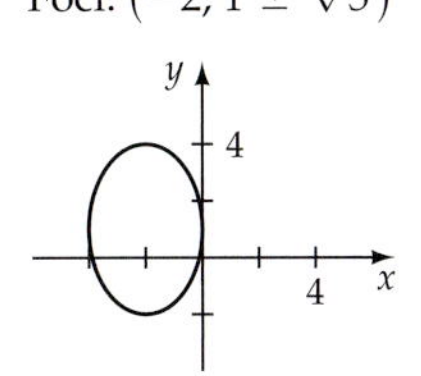

21. Center: $\left(1, \dfrac{2}{3}\right)$
Vertices: $\left(-5, \dfrac{2}{3}\right), \left(7, \dfrac{2}{3}\right)$
Foci: $\left(1 \pm 2\sqrt{13}, \dfrac{2}{3}\right)$
Asymptotes: $y - \dfrac{2}{3} = \pm \dfrac{2(x-1)}{3}$

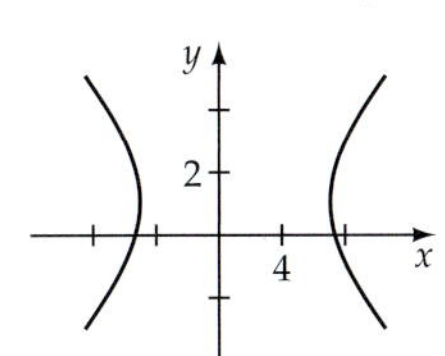

23. Vertex: $\left(-\dfrac{7}{2}, -1\right)$
Focus: $\left(-\dfrac{7}{2}, -3\right)$
Directrix: $y = 1$

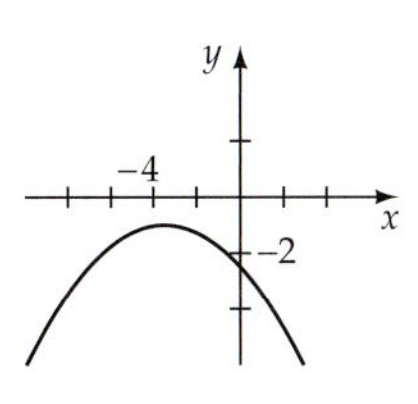

25. $\dfrac{(x-2)^2}{25} + \dfrac{(y-3)^2}{16} = 1$ **27.** $\dfrac{(x+2)^2}{4} - \dfrac{(y-2)^2}{5} = 1$ **29.** $x^2 = \dfrac{3(y+2)}{2}$ or $(y+2)^2 = 12x$ **31.** $\dfrac{x^2}{36} - \dfrac{y^2}{4/9} = 1$

33. $(y-3)^2 = -8x$ **35.** $\dfrac{(x-1)^2}{25} + \dfrac{(y-1)^2}{9} = 1$ **37.** On axis 4 feet above vertex **39.** 6.0 inches **41.** $a = 1.5$ inches

43. $\dfrac{x^2}{884.74^2} + \dfrac{y^2}{883.35^2} = 1$ **45.** 40 feet **47.** $\dfrac{(x - 9\sqrt{15}/2)^2}{324} + \dfrac{y^2}{81/4} = 1$ **49.** 5 **51.** $4|p|$ **53.** $\dfrac{3}{5}$ **55.** No **57.** $\dfrac{x^2}{25} + \dfrac{y^2}{16} = 1$

59. $\dfrac{\sqrt{29}}{2}$ **61.** $x^2 - \dfrac{y^2}{3} = 1$

Prepare for Section 6.2, page 483

62. $\cos\alpha\cos\beta - \sin\alpha\sin\beta$ **63.** $\sin\alpha\cos\beta + \cos\alpha\sin\beta$ **64.** odd **65.** even **66.** 240° **67.** r^2

Exercise Set 6.2, page 497

1–5. $\left(-3, \frac{5\pi}{3}\right)$, $(2, 60°)$, $\left(-2, \frac{\pi}{4}\right)$, $(1, 315°)$

7.

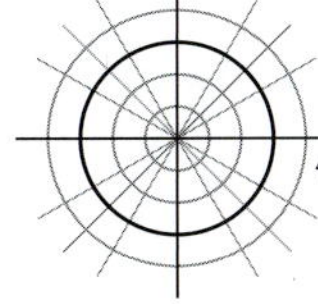

$0 \le \theta \le 2\pi$

9.

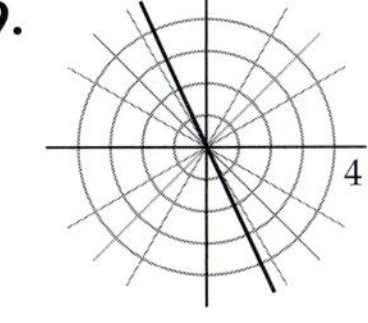

11.

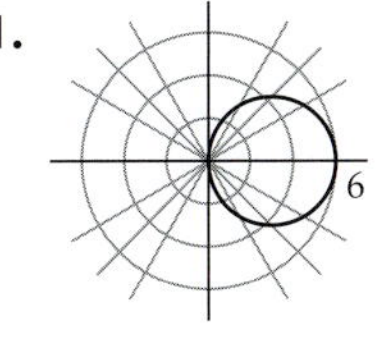

$0 \le \theta \le \pi$

13.

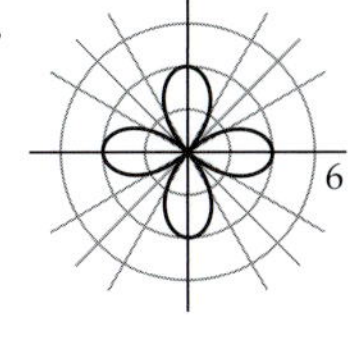

$0 \le \theta \le 2\pi$

15.

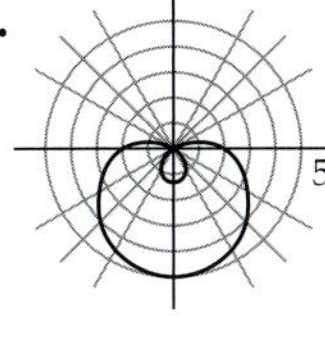

$0 \le \theta \le 2\pi$

17.

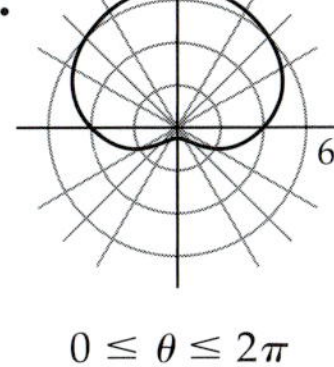

$0 \le \theta \le 2\pi$

19. hyperbola

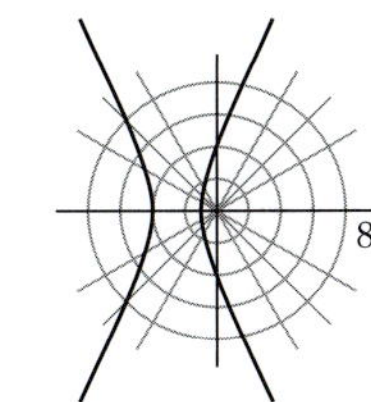

21. parabola

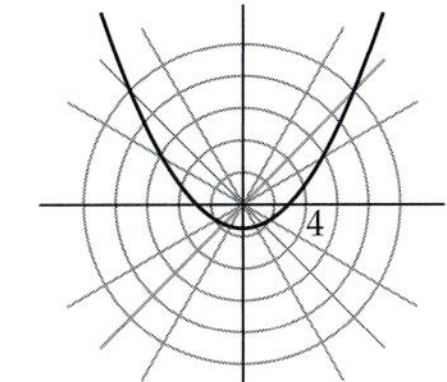

23. hyperbola

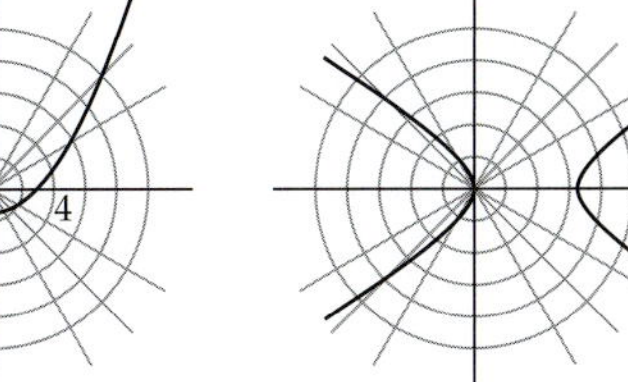

25. 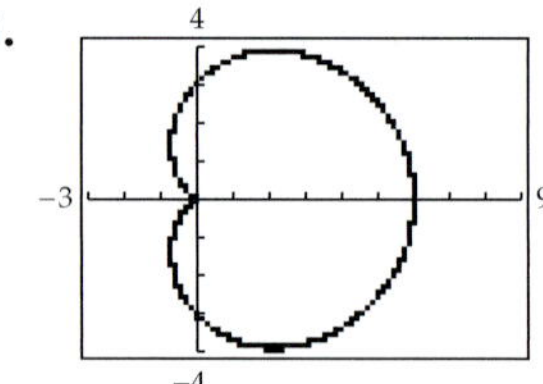

$0 \le \theta \le 2\pi$

27. 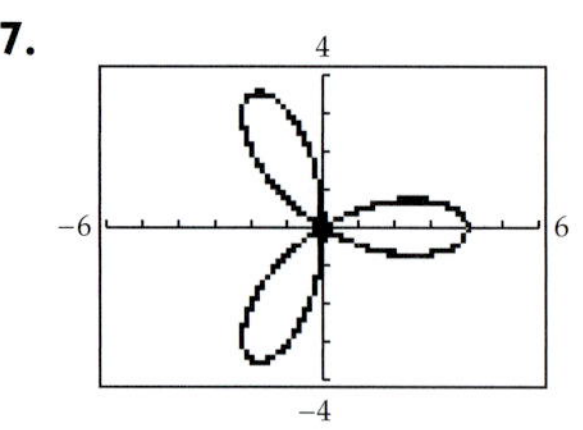

$0 \le \theta \le \pi$

29. 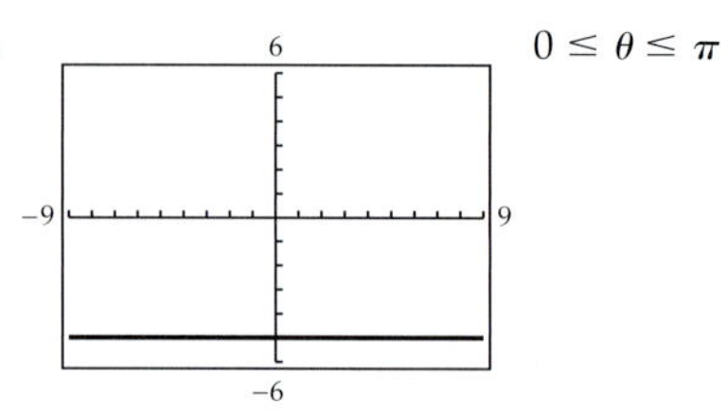

$0 \le \theta \le \pi$

31. 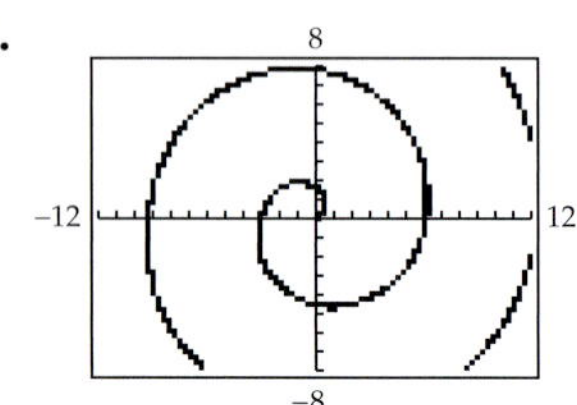

$0 \le \theta \le 6\pi$

33. 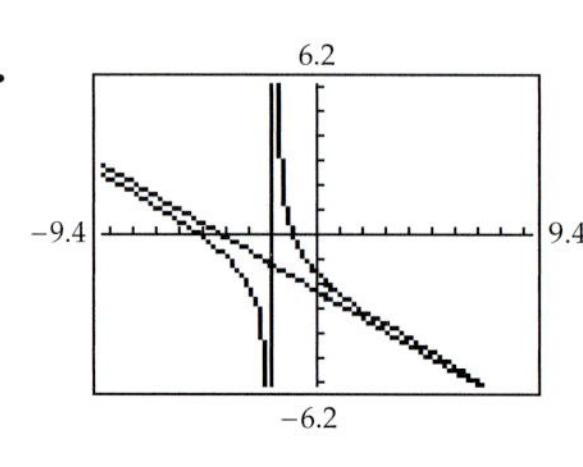

35. 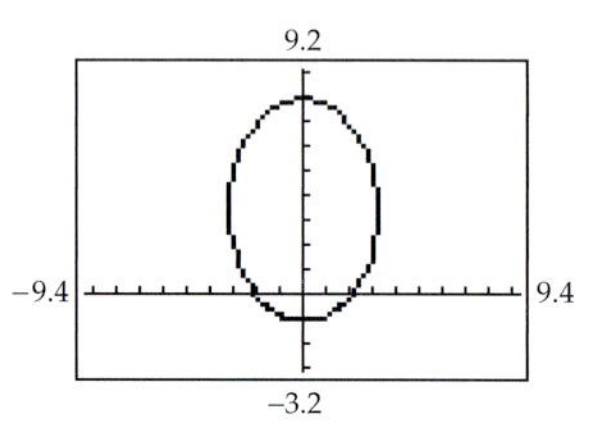

37. $(2, -60°)$ **39.** $\left(\frac{3}{2}, -\frac{3\sqrt{3}}{2}\right)$ **41.** $(0, 0)$ **43.** $(5, 53.1°)$ **45.** $x^2 + y^2 - 3x = 0$ **47.** $x = 3$ **49.** $x^2 + y^2 = 16$
51. $x^4 - y^2 + x^2y^2 = 0$ **53.** $y^2 + 4x - 4 = 0$ **55.** $3x^2 - y^2 + 16x + 16 = 0$ **57.** $r = 2 \csc \theta$ **59.** $r = 2$ **61.** $r \cos^2 \theta = 8 \sin \theta$
63. $r^2(\cos 2\theta) = 25$ **65.** $r = \dfrac{2}{1 - 2\cos\theta}$ **67.** $r = \dfrac{2}{1 + \sin\theta}$ **69.** $r = \dfrac{8}{3 - 2\sin\theta}$ **71.** $r = \dfrac{4}{1 - \cos\theta}$ **73.** $r = \dfrac{3}{1 - 2\sin\theta}$

79. 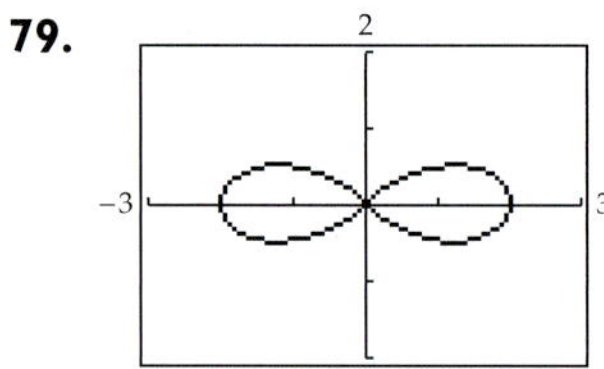

$0 \le \theta \le 4\pi$

81. 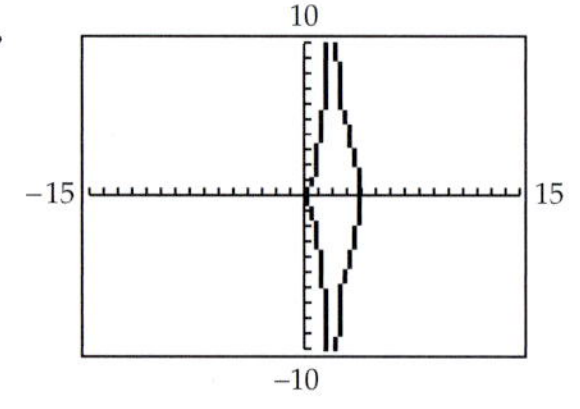

$0 \le \theta \le 2\pi$

83. 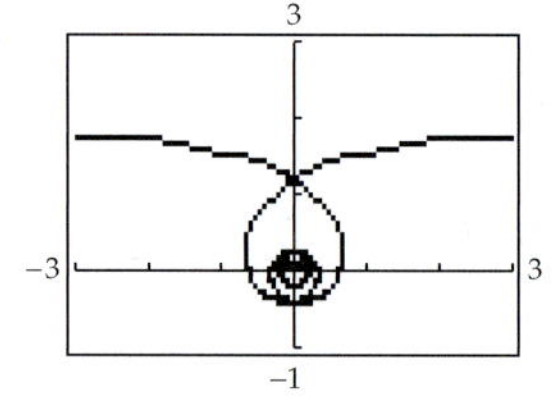

$-4\pi \le \theta \le 4\pi$

85. 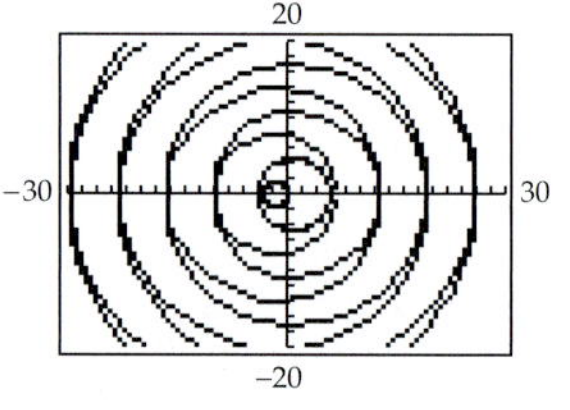

$-30 \le \theta \le 30$

87. a. 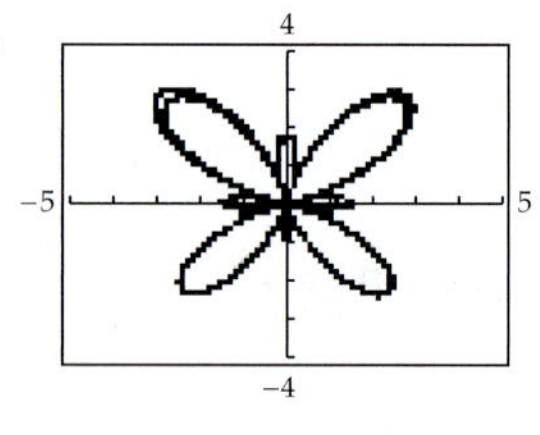

$0 \le \theta \le 5\pi$

b. 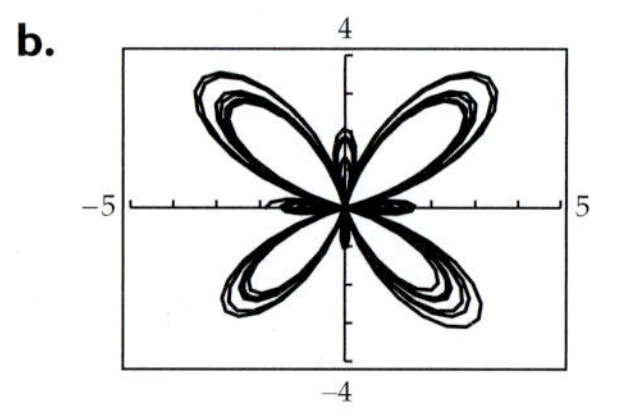

$0 \le \theta \le 20\pi$

Prepare for Section 6.3, page 499

88. $y^2 + 3y + \dfrac{9}{4} = \left(y + \dfrac{3}{2}\right)^2$ **89.** $y = 4t^2 + 4t + 1$ **90.** ellipse **91.** 1 **92.** $t = e^y$ **93.** domain: $(-\infty, \infty)$; range: $[-3, 3]$

Exercise Set 6.3, page 506

1. 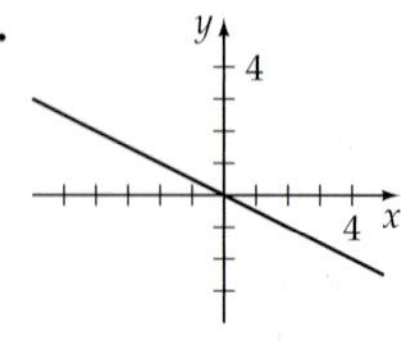

3. 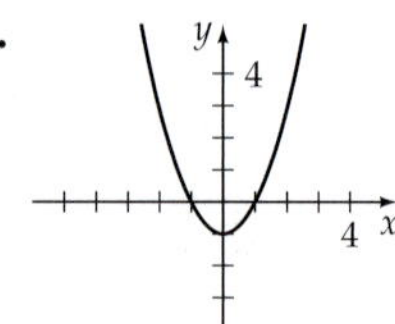

5. 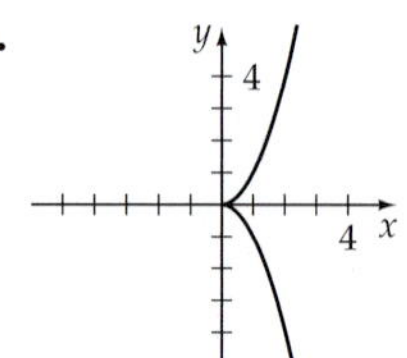

7. 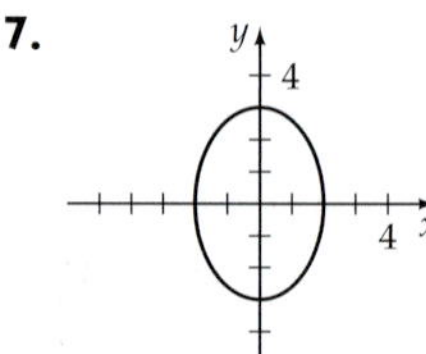

9. 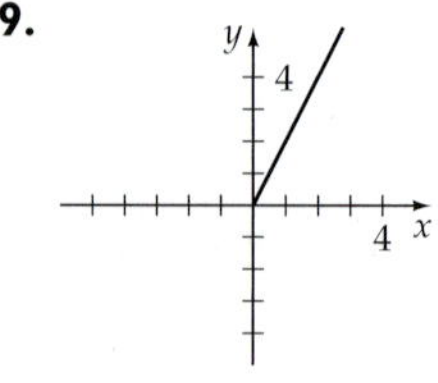

11. $x^2 - y^2 - 1 = 0$
$x \geq 1$
$y \in R$

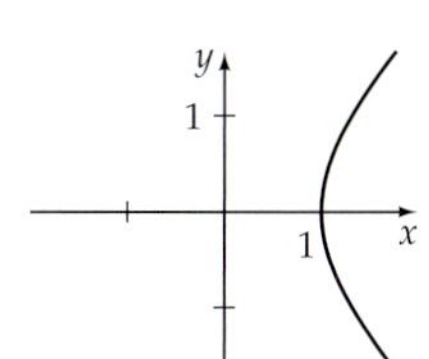

13. $y = -2x + 7$
$x \leq 2$
$y \geq 3$

(2, 3)

15. $x^{2/3} + y^{2/3} = 1$
$-1 \leq x \leq 1$
$-1 \leq y \leq 1$

17. $y = x^2 - 1$
$x \geq 0$
$y \geq -1$

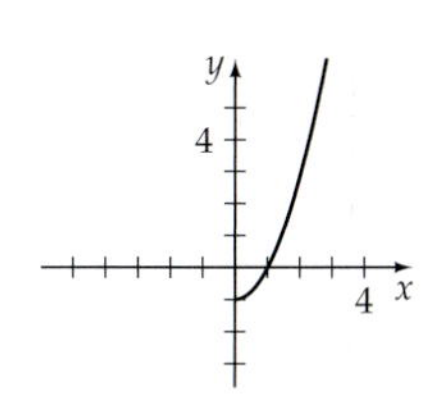

19. $y = \ln x$
$x > 0$
$y \in R$

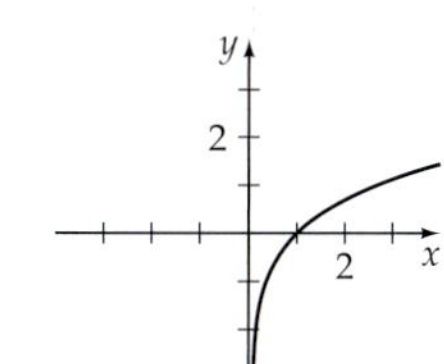

21. C_1: $y = -2x + 5, x \geq 2$; C_2: $y = -2x + 5, x \in R$. C_2 is a line. C_1 is a ray.

23.

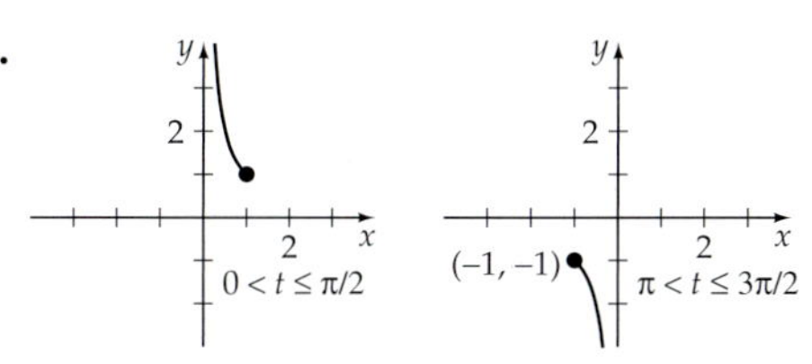

25.

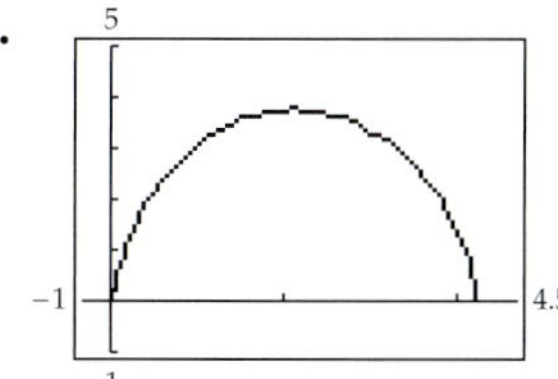

Xscl = 2π

27.

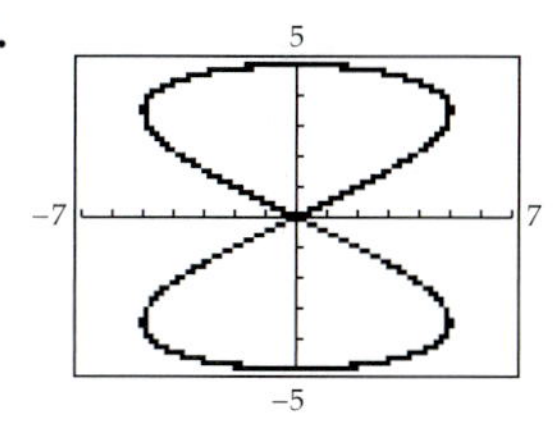

29.

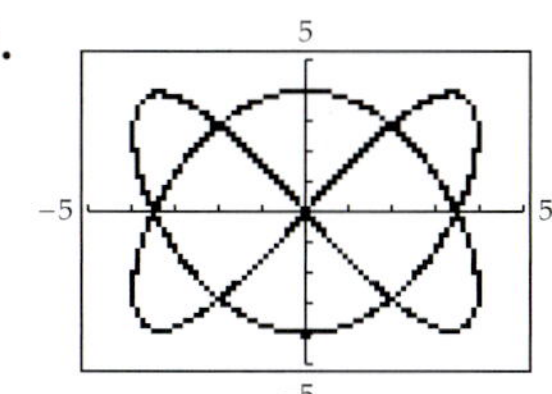

31.

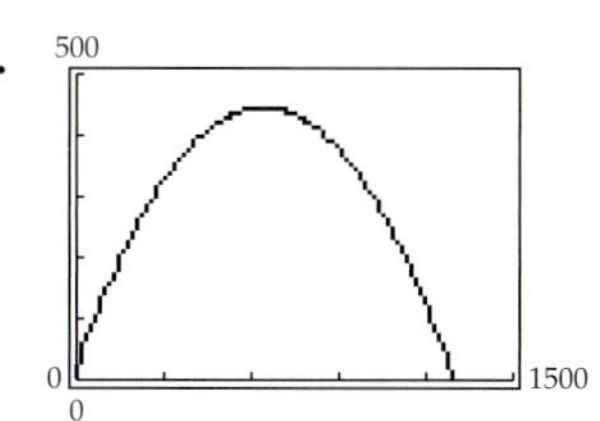

Max height (nearest foot) of 462 feet is attained when $t \approx 5.38$ seconds.

Range (nearest foot) of 1295 feet is attained when $t \approx 10.75$ seconds.

33.

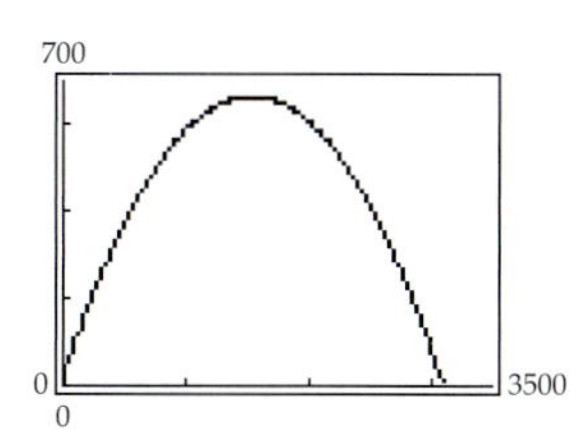

Max height (nearest foot) of 694 feet is attained when $t \approx 6.59$ seconds.

Range (nearest foot) of 3084 feet is attained when $t \approx 13.17$ seconds.

37. $x = a\cos\theta + a\theta\sin\theta$
$y = a\sin\theta - a\theta\cos\theta$

39. $x = (b - a)\cos\theta + a\cos\left(\frac{b-a}{a}\theta\right)$
$y = (b - a)\sin\theta - a\sin\left(\frac{b-a}{a}\theta\right)$

Prepare for Section 6.4, page 507

40. $\frac{1}{2}, \frac{2}{3}, \frac{3}{4}$ **41.** 12 **42.** 3 **43.** 3 **44.** $\frac{1}{2}$ **45.** 30

Exercise Set 6.4, page 522

1. $0, 2, 6, a_8 = 56$ **3.** $0, \frac{1}{2}, \frac{2}{3}, a_8 = \frac{7}{8}$ **5.** $1, -\frac{1}{4}, \frac{1}{9}, a_8 = -\frac{1}{64}$ **7.** $-\frac{1}{3}, -\frac{1}{6}, -\frac{1}{9}, a_8 = -\frac{1}{24}$ **9.** $0, 2, 0, a_8 = 2$ **11.** $1, -\frac{\sqrt{2}}{2}, \frac{\sqrt{3}}{3}, a_8 = -\frac{\sqrt{2}}{4}$ **13.** $3, 3, 3, a_8 = 3$ **15.** 5, 10, 20 **17.** 2, 4, 12 **19.** 2, 4, 16 **21.** 2, 8, 48 **23.** 4320 **25.** 72 **27.** 56 **29.** 100 **31.** $a_n = 4n + 2$ **33.** $a_n = 8 - 2n$ **35.** $a_n = 3n - 11$ **37.** $a_n = 3n - 2$ **39.** $a_n = a + 2n - 2$ **41.** 45 **43.** -79 **45.** 2^{2n-1} **47.** $-4(-3)^{n-1}$ **49.** $6\left(\frac{2}{3}\right)^{n-1}$ **51.** $-6\left(-\frac{5}{6}\right)^{n-1}$ **53.** $3\left(\frac{1}{100}\right)^n$ **55.** $5(0.1)^n$ **57.** $45(0.01)^n$ **59.** 18 **61.** -2 **63.** 15 **65.** 40 **67.** $\frac{25}{12}$ **69.** 72 **71.** 185 **73.** -555 **75.** 468 **77.** 525 **79.** 363 **81.** $\frac{1330}{729}$ **83.** -341 **85.** 147,620 **87.** $\frac{1}{2}$

89. $-\frac{2}{5}$ **91.** $\frac{1}{9}$ **93.** $\frac{1}{3}$ **95.** $\frac{5}{11}$ **97.** $\frac{41}{333}$ **99.** $\frac{422}{999}$ **101.** 2.6457520 **105.** $a_n = 7 - 3n$ **107.** Because $\log r$ is a constant, the sequence $\log a_n$ is an arithmetic sequence. **109.** Yes. The common ratio is x.

Prepare for Section 6.5, page 524

111. $a^3 + 3a^2b + 3ab^2 + b^3$ **112.** 120 **113.** 1 **114.** 15 **115.** 35 **116.** 1

Exercise Set 6.5, page 529

1. 35 **3.** 36 **5.** 220 **7.** 1 **9.** $x^6 - 6x^5y + 15x^4y^2 - 20x^3y^3 + 15x^2y^4 - 6xy^5 + y^6$ **11.** $x^5 + 15x^4 + 90x^3 + 270x^2 + 405x + 243$ **13.** $128x^7 - 448x^6 + 672x^5 - 560x^4 + 280x^3 - 84x^2 + 14x - 1$ **15.** $x^6 + 18x^5y + 135x^4y^2 + 540x^3y^3 + 1215x^2y^4 + 1458xy^5 + 729y^6$ **17.** $16x^4 - 160x^3y + 600x^2y^2 - 1000xy^3 + 625y^4$ **19.** $x^6 + 6x^4 + 15x^2 + 20 + \frac{15}{x^2} + \frac{6}{x^4} + \frac{1}{x^6}$ **21.** $x^{14} - 28x^{12} + 336x^{10} - 2240x^8 + 8960x^6 - 21{,}540x^4 + 28{,}672x^2 - 16{,}384$ **23.** $32x^{10} + 80x^8y^3 + 80x^6y^6 + 40x^4y^9 + 10x^2y^{12} + y^{15}$ **25.** $\frac{16}{x^4} - \frac{16}{x^2} + 6 - x^2 + \frac{x^4}{16}$ **27.** $s^{-12} + 6s^{-8} + 15s^{-4} + 20 + 15s^4 + 6s^8 + s^{12}$ **29.** $-3240x^3y^7$ **31.** $1056x^{10}y^2$ **33.** $126x^2y^2\sqrt{x}$ **35.** $\frac{165b^5}{a^5}$ **37.** $180a^2b^8$ **39.** $60x^2y^8$ **41.** $-61{,}236a^5b^5$ **43.** $126s^{-1}, 126s$ **45.** $-7 - 24i$ **47.** $41 - 38i$ **49.** 1 **51.** $nx^{n-1} + \frac{n(n-1)x^{n-2}h}{2} + \frac{n(n-1)(n-2)x^{n-3}h^2}{6} + \cdots + h^{n-1}$ **57.** 1.1712 **59.** 756 **61.** 56

Chapter 6 True/False Exercises, page 533

1. False; a parabola has no asymptotes. **2.** True **3.** False; $x = \cos t$, $y = \sin t$ graphs to be a circle. **4.** True **5.** False; $\left(\sum_{i=1}^{3} i\right)\left(\sum_{i=1}^{3} i\right) \neq \sum_{i=1}^{3} i^2$. **6.** False; the constant sequence has all terms equal. **7.** False; $\frac{(k+1)^3}{k^3} = \left(1 + \frac{1}{k}\right)^3$ is not a constant. **8.** True **9.** True **10.** False; the exponent is 4.

Chapter 6 Review Exercises, page 534

1. center: $(0, 0)$
vertices: $(\pm 2, 0)$
foci: $(\pm 2\sqrt{2}, 0)$
asymptotes: $y = \pm x$
[6.1]

2. vertex: $(0, 0)$
focus: $(4, 0)$
directrix: $x = -1$
[6.1]

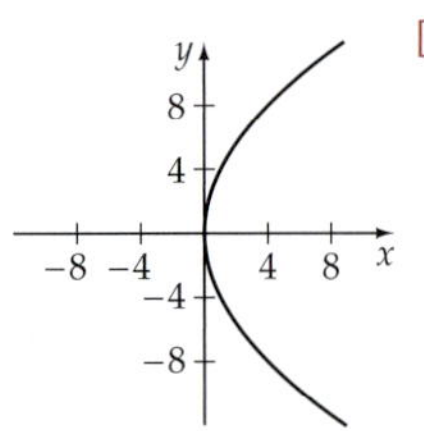

3. center: $(3, -1)$
vertices: $(-1, -1), (7, -1)$
foci: $(3 \pm 2\sqrt{3}, -1)$
[6.1]

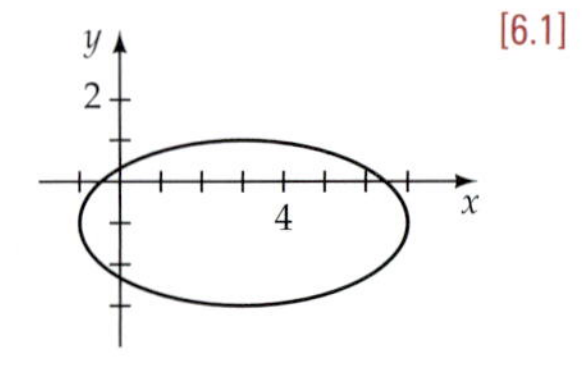

4. center: $(-2, -3)$
vertices: $(0, -3), (-4, -3)$
foci: $(-2 + \sqrt{7}, -3), (-2 - \sqrt{7}, -3)$
asymptotes: $y + 3 = \pm\frac{\sqrt{3}}{2}(x + 2)$
[6.1]

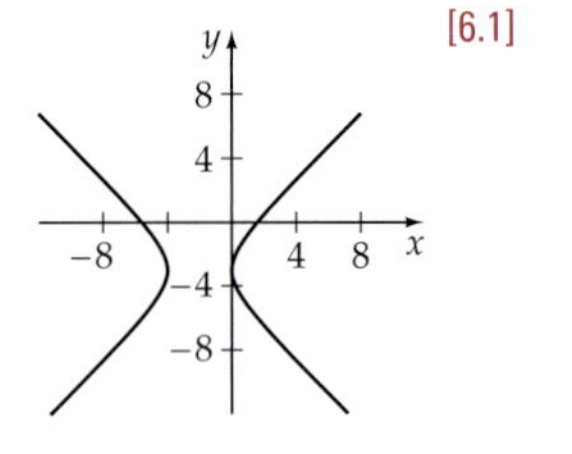

5. vertex: $(-2, 1)$
focus: $\left(-\frac{29}{16}, 1\right)$
directrix: $x = -\frac{35}{16}$
[6.1]

6. vertex: $(3, 1)$
focus: $\left(\frac{21}{8}, 1\right)$
directrix: $x = \frac{27}{8}$
[6.1]

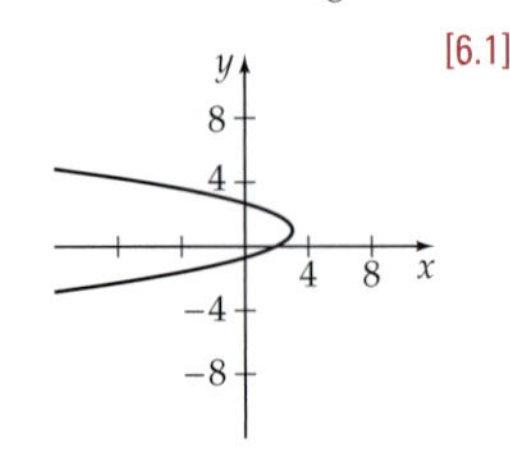

7. center: $\left(1, \frac{2}{3}\right)$
vertices: $\left(-5, \frac{2}{3}\right), \left(7, \frac{2}{3}\right)$
foci: $\left(1 \pm 2\sqrt{13}, \frac{2}{3}\right)$
asymptotes: $y - \frac{2}{3} = \pm\frac{2(x-1)}{3}$
[6.1]

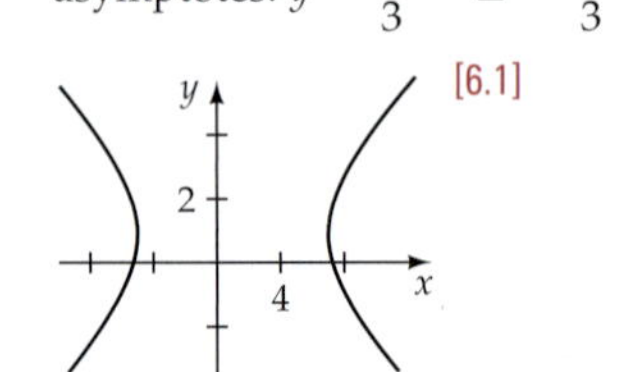

8. center: $\left(-2, \frac{1}{2}\right)$
vertices: $\left(2, \frac{1}{2}\right), \left(-6, \frac{1}{2}\right)$
foci: $\left(-2 + \sqrt{7}, \frac{1}{2}\right), \left(-2 - \sqrt{7}, \frac{1}{2}\right)$
[6.1]

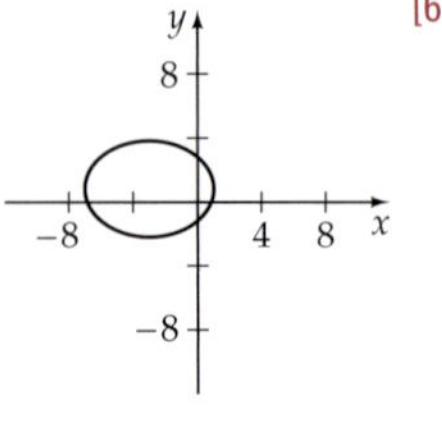

9. $\frac{(x-2)^2}{25} + \frac{(y-3)^2}{16} = 1$ [6.1] **10.** $\frac{(x-1)^2}{9} - \frac{(y-1)^2}{7} = 1$ [6.1] **11.** $\frac{(x+2)^2}{4} - \frac{(y-2)^2}{5} = 1$ [6.1] **12.** $(y+3)^2 = -8(x-4)$ [6.1]

13. $x^2 = \frac{3(y+2)}{2}$ or $(y+2)^2 = 12x$ [6.1] **14.** $\frac{x^2}{36} - \frac{y^2}{4/9} = 1$ [6.1] **15.** [6.2]

16. [6.2]

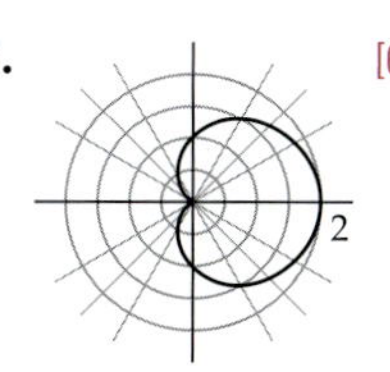

17. [6.2]

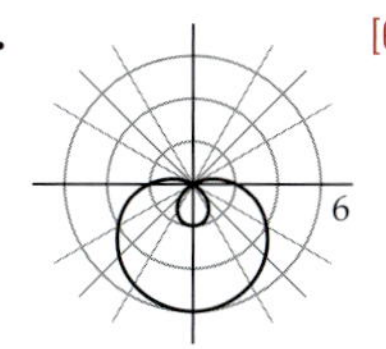

18. [6.2]

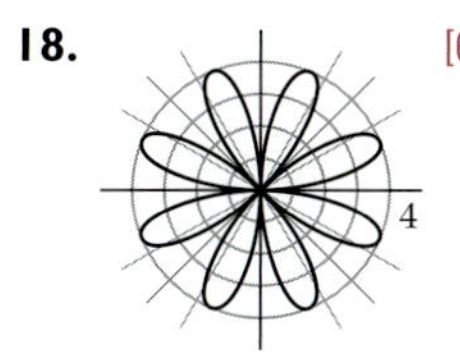

19. [6.2]

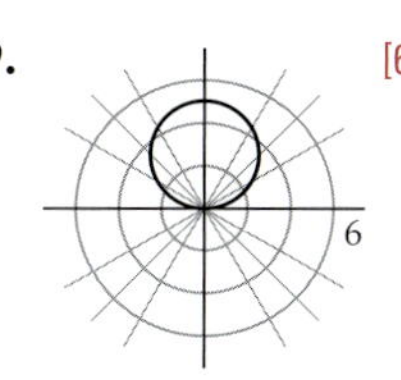

20. [6.2]

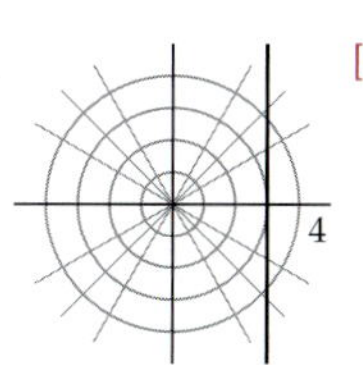

21. [6.2]

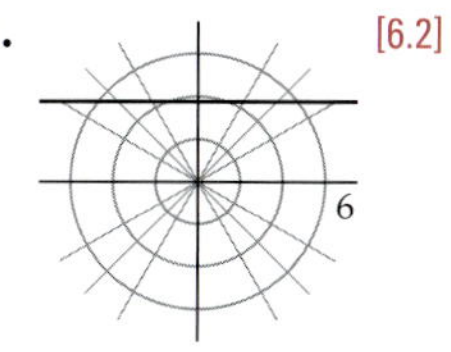

22. [6.2]

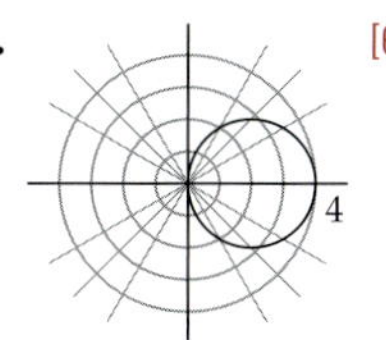

23. [6.2]

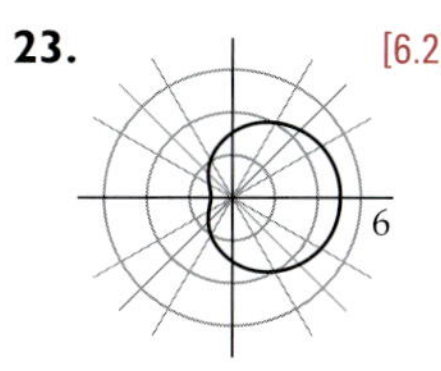

24. [6.2]

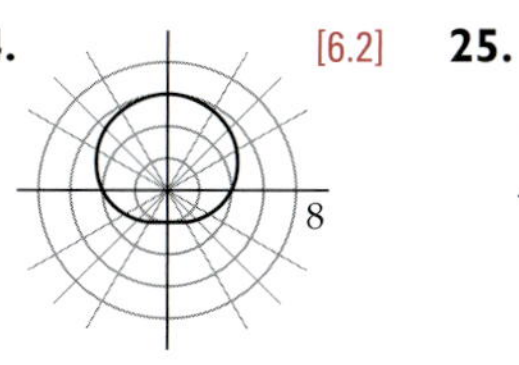

25. [6.2]

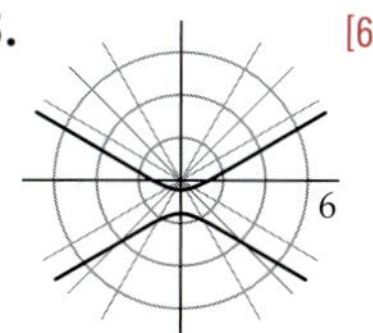

26. [6.2]

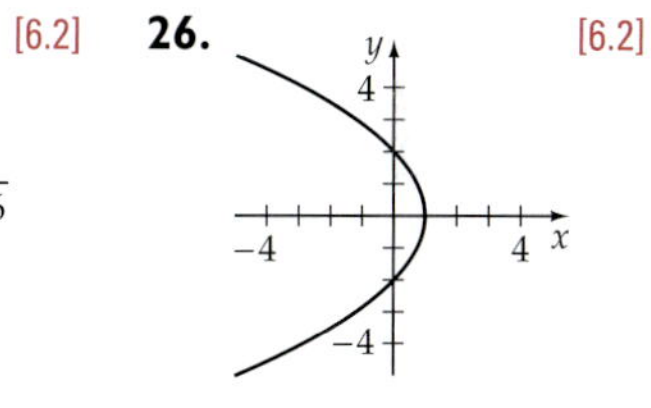

27. [6.2]

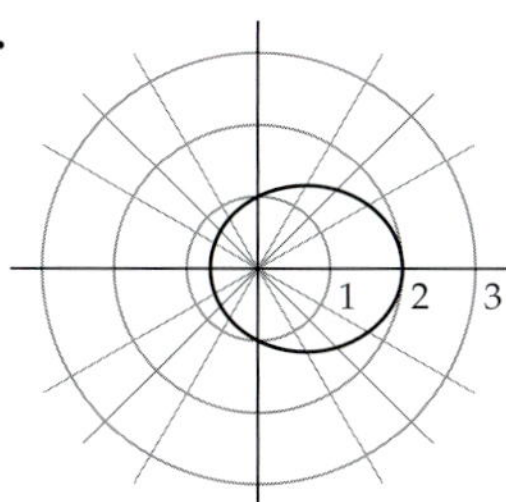

28. [6.2]

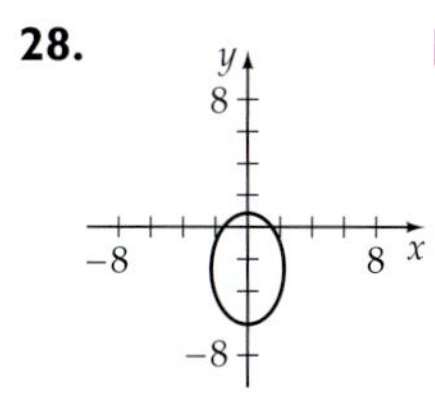

29. $r \sin^2 \theta = 16 \cos \theta$ [6.2] **30.** $r + 4 \cos \theta + 3 \sin \theta = 0$ [6.2]

31. $3r \cos \theta - 2r \sin \theta = 6$ [6.2] **32.** $r^2 \sin 2\theta = 8$ [6.2] **33.** $y^2 = 8x + 16$ [6.2] **34.** $x^2 - 3x + y^2 + 4y = 0$ [6.2]

35. $x^4 + y^4 + 2x^2y^2 - x^2 + y^2 = 0$ [6.2] **36.** $y = (\tan 1)x$ [6.2] **37.** $y = \frac{3}{4}x + \frac{5}{2}$ [6.3]

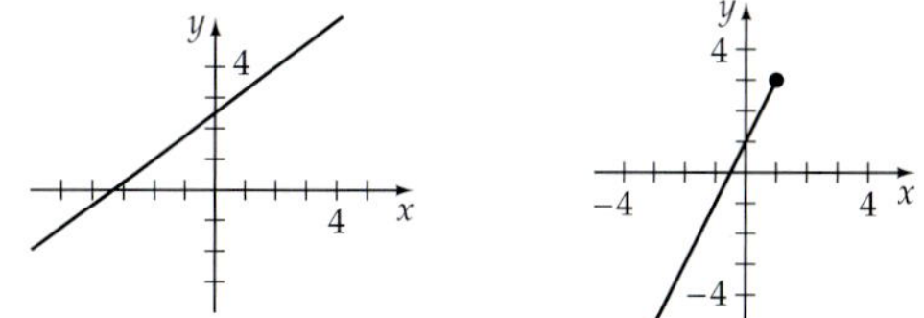

38. $y = 2x + 1, x \leq 1$ [6.3]

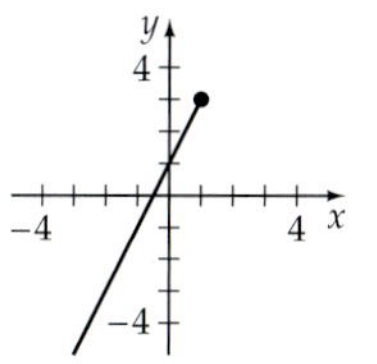

39. $\frac{x^2}{16} + \frac{y^2}{9} = 1$ [6.3]

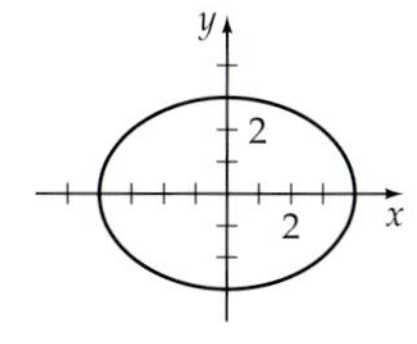

40. $\frac{x^2}{1} - \frac{y^2}{16} = 1$ [6.3]

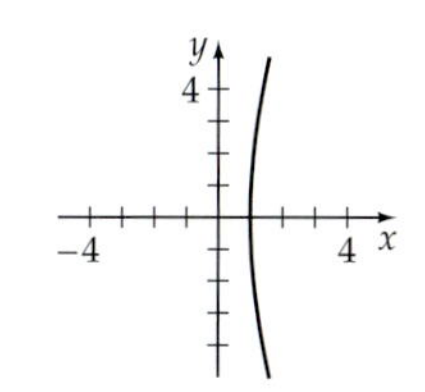

41. $y = -2x, x > 0$ [6.3]

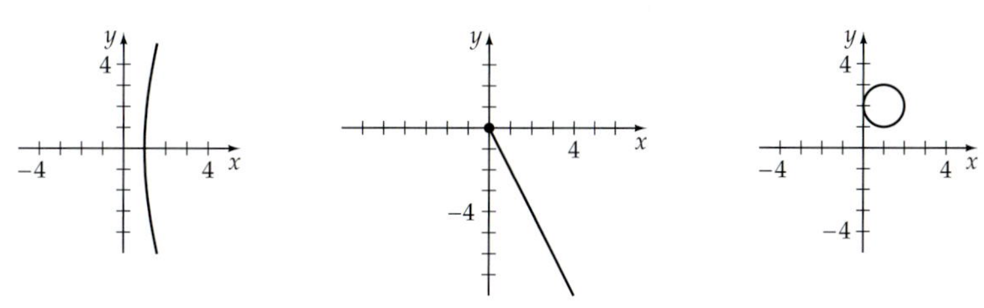

42. $(x-1)^2 + (y-2)^2 = 1$ [6.3]

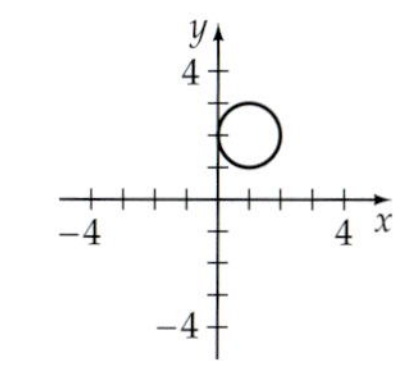

43. $a_3 = 9, a_7 = 49$ [6.4] **44.** $a_3 = 6, a_7 = 5040$ [6.4] **45.** $a_3 = 11, a_7 = 23$ [6.4] **46.** $a_3 = -5, a_7 = -13$ [6.4] **47.** $a_3 = \frac{1}{8}, a_7 = \frac{1}{128}$ [6.4]
48. $a_3 = 27, a_7 = 2187$ [6.4] **49.** $a_3 = \frac{1}{6}, a_7 = \frac{1}{5040}$ [6.4] **50.** $a_3 = \frac{1}{3}, a_7 = \frac{1}{7}$ [6.4] **51.** $a_3 = 6, a_7 = 5040$ [6.4]
52. $a_3 = 72, a_7 = 50{,}803{,}200$ [6.4] **53.** $a_3 = 8, a_7 = 16$ [6.4] **54.** $a_3 = -3, a_7 = -15$ [6.4] **55.** $a_3 = 2, a_7 = 256$ [6.4]
56. $a_3 = 2, a_7 = 1$ [6.4] **57.** 63 [6.4] **58.** -187 [6.4] **59.** 378 [6.4] **60.** 682 [6.4] **61.** $-14{,}763$ [6.4] **62.** -170 [6.4]
63. $\frac{116{,}050}{59{,}049} \approx 1.9653$ [6.4] **64.** $\frac{525{,}297}{2048} \approx 256.4927$ [6.4] **65.** 0.8280 [6.4] **66.** $0.6\bar{3}$ [6.4] **67.** $\frac{1}{3}$ [6.4] **68.** $-\frac{5}{11}$ [6.4]
69. $-\frac{4}{9}$ [6.4] **70.** $\frac{5}{4}$ [6.4] **71.** $1024a^5 - 1280a^4b + 640a^3b^2 - 160a^2b^3 + 20ab^4 - b^5$ [6.5]
72. $x^6 + 18x^5y + 135x^4y^2 + 540x^3y^3 + 1215x^2y^4 + 1458xy^5 + 729y^6$ [6.5]
73. $a^4 + 16a^{7/2}b^{1/2} + 112a^3b + 448a^{5/2}b^{3/2} + 1120a^2b^2 + 1792a^{3/2}b^{5/2} + 1792ab^3 + 1024a^{1/2}b^{7/2} + 256b^4$ [6.5]
74. $128x^7 - 224x^5 + 168x^3 - 70x + \frac{35}{2x} - \frac{21}{8x^3} + \frac{7}{32x^5} - \frac{1}{128x^7}$ [6.5] **75.** $241{,}920x^3y^4$ [6.5] **76.** $-78{,}732x^7$ [6.5]

Chapter 6 Test, page 535

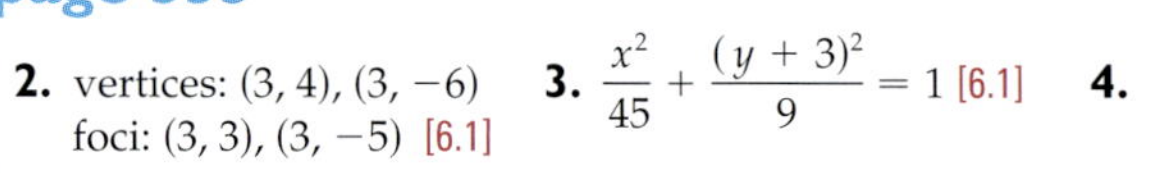

1. focus: (0, 2)
vertex: (0, 0)
directrix: $y = -2$ [6.1]

2. vertices: (3, 4), (3, −6)
foci: (3, 3), (3, −5) [6.1]

3. $\frac{x^2}{45} + \frac{(y+3)^2}{9} = 1$ [6.1]

4.

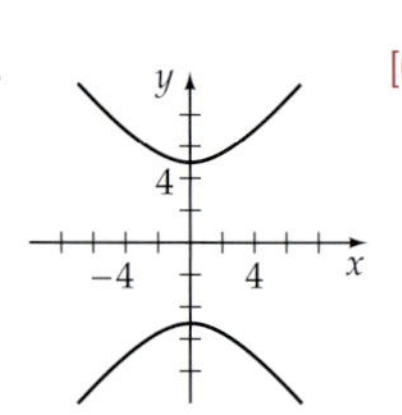

[6.1]

5. $P(2, 300°)$ [6.2]

6.

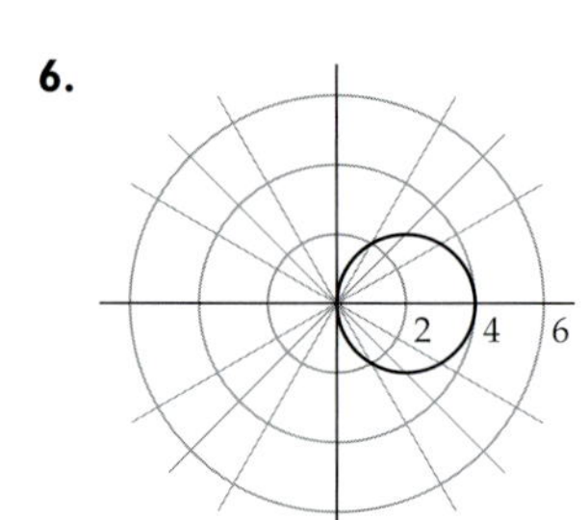

[6.2]

7.

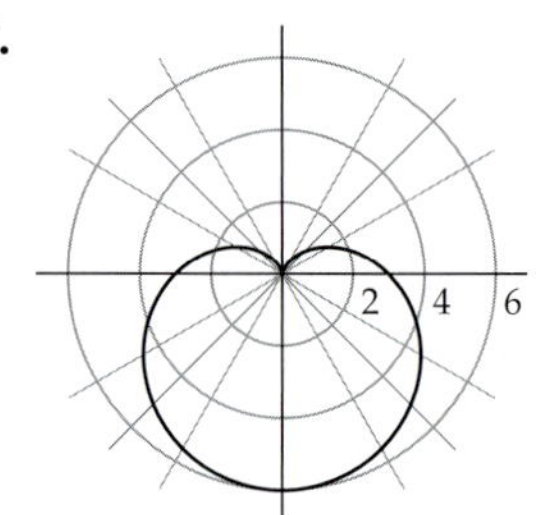

[6.2]

8.

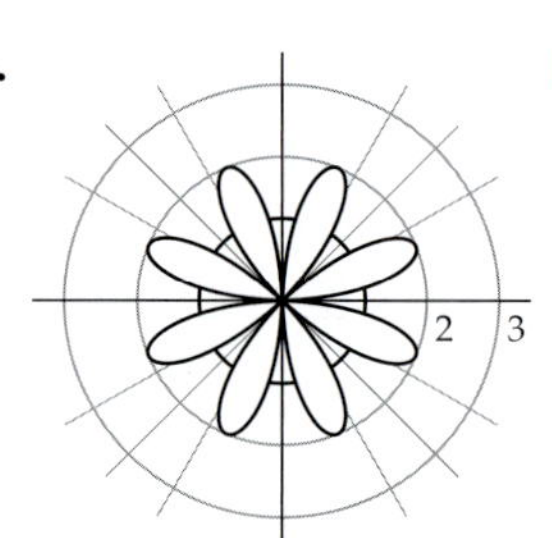

[6.2]

9. $\left(\frac{5}{2}, \frac{5\sqrt{3}}{2}\right)$ [6.2]

10. $y^2 - 8x - 16 = 0$ [6.2] **11.** $x^2 + 8y - 16 = 0$ [6.2] **12.** $(x + 3)^2 = \frac{1}{2}y$ [6.3] **13.** arithmetic [6.4] **14.** neither [6.4]

15. geometric [6.4] **16.** $\frac{1023}{1024}$ [6.4] **17.** 590 [6.4] **18.** 58 [6.4] **19.** $\frac{3}{5}$ [6.4] **20.** $\frac{5}{33}$ [6.4]
21. $x^5 - 10x^4y + 40x^3y^2 - 80x^2y^3 + 80xy^4 - 32y^5$ [6.5] **22.** $x^6 + 6x^4 + 15x^2 + 20 + \frac{15}{x^2} + \frac{6}{x^4} + \frac{1}{x^6}$ [6.5] **23.** $48{,}384x^3y^5$ [6.5]

Cumulative Review Exercises, page 536

1. -24 [1.4] **2.** $\frac{3 \pm \sqrt{41}}{4}$ [1.1] **3.** $\log_b x + 2\log_b y - 3\log_b z$ [3.4] **4.** $-\frac{5}{16}$ [1.7] **5.** $y = 0$ [2.6] **6.** -6 [3.3] **7.** -2.1 [3.5]
8. 105 mph [3.5] **9.** $-\sqrt{3}$ [5.1] **10.** $2 \csc x$ [5.2] **11.** $C = 75°, a = 13$ centimeters, $b = 19$ centimeters [5.6] **12.** $\frac{\sqrt{10}}{10}$ [5.4]

13. $-4 - h$ [1.7] **14.** -19 [1.7] **15.**

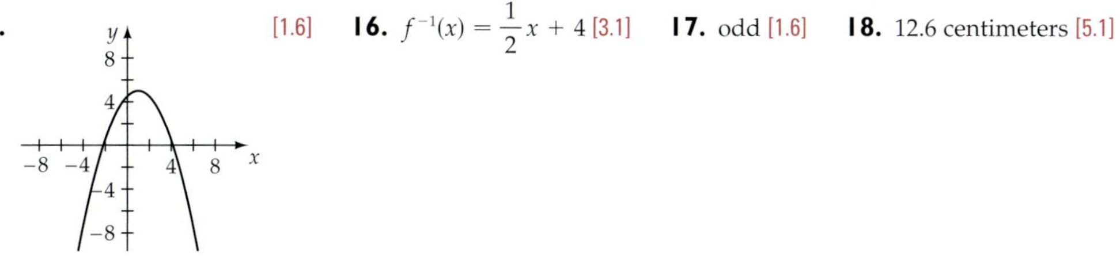

[1.6] **16.** $f^{-1}(x) = \frac{1}{2}x + 4$ [3.1] **17.** odd [1.6] **18.** 12.6 centimeters [5.1]

19. period: $\frac{\pi}{2}$; amplitude: 3 [4.3] **20.** $\frac{\pi}{6}, \frac{\pi}{2}, \frac{5\pi}{6}, \frac{3\pi}{2}$ [5.5]

Exercise Set A.1, page 548

1. -125 **3.** 1 **5.** $\frac{1}{16}$ **7.** 32 **9.** 27 **11.** -2 **13.** $\frac{2}{x^4}$ **15.** $6a^3b^8$ **17.** $\frac{3}{4a^4}$ **19.** $\frac{2y^2}{3x^2}$ **21.** $\frac{12}{a^3b}$ **23.** $-18m^5n^6$ **25.** $\frac{1}{x^6}$
27. $\frac{1}{4a^4b^2}$ **29.** $2x$ **31.** $\frac{b^{10}}{a^{10}}$ **33.** 2.011×10^{12} **35.** 5.62×10^{-10} **37.** 31,400,000 **39.** -0.0000023 **41.** 2.7×10^8 **43.** 1.5×10^{-11}
45. 7.2×10^{12} **47.** 8×10^{-16} **49.** 8 **51.** -16 **53.** $\frac{1}{27}$ **55.** $\frac{2}{3}$ **57.** 16 **59.** $8ab^2$ **61.** $-12x^{11/12}$ **63.** $3x^2y^3$ **65.** $\frac{4z^{2/5}}{3}$
67. $6x^{5/6}y^{5/6}$ **69.** $\frac{3a^{1/12}}{b}$ **71.** $3\sqrt{5}$ **73.** $2\sqrt[3]{3}$ **75.** $-3\sqrt[3]{5}$ **77.** $2|x|y\sqrt{6y}$ **79.** $2ay^2\sqrt[3]{2y}$ **81.** $-13\sqrt{2}$ **83.** $-10\sqrt[4]{3}$
85. $17y\sqrt[3]{4y}$ **87.** $-14x^2y\sqrt[3]{y}$ **89.** $17 + 7\sqrt{5}$ **91.** -7 **93.** $12z + \sqrt{z} - 6$ **95.** $x + 4\sqrt{x} + 4$ **97.** $x + 4\sqrt{x - 3} + 1$ **99.** $\sqrt{2}$
101. $\frac{\sqrt{10}}{6}$ **103.** $\frac{3\sqrt[3]{4}}{2}$ **105.** $\frac{2\sqrt[3]{x}}{x}$ **107.** $-\frac{3\sqrt{3} - 12}{13}$ **109.** $\frac{3\sqrt{5} - 3}{4}$ **111.** $\frac{3\sqrt{5} - 3\sqrt{x}}{5 - x}$

Exercise Set A.2, page 553

1. D **3.** H **5.** G **7.** B **9.** J **11. a.** $x^2 + 2x - 7$ **b.** 2 **c.** $1, 2, -7$ **d.** 1 **e.** $x^2, 2x, -7$ **13. a.** $x^3 - 1$ **b.** 3 **c.** $1, -1$ **d.** 1 **e.** $x^3, -1$ **15. a.** $2x^4 + 3x^3 + 4x^2 + 5$ **b.** 4 **c.** $2, 3, 4, 5$ **d.** 2 **e.** $2x^4, 3x^3, 4x^2, 5$ **17.** 3 **19.** 5 **21.** 2
23. $5x^2 + 11x + 3$ **25.** $9w^3 + 8w^2 - 2w + 6$ **27.** $-2r^2 + 3r^2 - 12$ **29.** $-3u^2 - 2u + 4$ **31.** $8x^3 + 18x^2 - 67x + 40$
33. $6x^4 - 19x^3 + 26x^2 - 29x + 10$ **35.** $10x^2 + 22x + 4$ **37.** $y^2 + 3y + 2$ **39.** $4z^2 - 19z + 12$ **41.** $a^2 + 3a - 18$
43. $10x^2 - 57xy + 77y^2$ **45.** $18x^2 + 55xy + 25y^2$ **47.** $6p^2 - 11pq - 35q^2$ **49.** $12d^2 + 4d - 8$ **51.** $r^3 + s^3$ **53.** $60c^3 - 49c^2 + 4$
55. $9x^2 - 25$ **57.** $9x^4 - 6x^2y + y^2$ **59.** $16w^2 + 8wz + z^2$ **61.** $x^2 + 10x + 25 - y^2$ **63.** 29 **65.** -17 **67.** -1 **69.** 33
71. a. 1.6 pounds **b.** 3.6 pounds **73. a.** 72π cubic inches **b.** 300π cubic centimeters **75.** 14.8 seconds; 90.4 seconds
77. Yes. The ball is approximately 4.4 feet high when it crosses home plate.

Exercise Set A.3, page 563

1. $5(x + 4)$ **3.** $-3x(5x + 4)$ **5.** $2xy(5x + 3 - 7y)$ **7.** $(x - 3)(2a + 3b)$ **9.** $(x + 3)(x + 4)$ **11.** $(a - 12)(a + 2)$ **13.** $(6x + 1)(x + 4)$
15. $(17x + 4)(3x - 1)$ **17.** $(3x + 8y)(2x - 5y)$ **19.** $(x^2 + 5)(x^2 + 1)$ **21.** $(6x^2 + 5)(x^2 + 3)$ **23.** $(x^2 - 3)(x^2 + 2)$ **25.** $(xy - 4)(xy + 2)$
27. $(3x^2 - 1)(x^2 + 4)$ **29.** $(3x^3 - 4)(x^3 + 2)$ **31.** $(x - 3)(x + 3)$ **33.** $(2a - 7)(2a + 7)$ **35.** $(1 - 10x)(1 + 10x)$ **37.** $(x^2 - 3)(x^2 + 3)$
39. $(x + 3)(x + 7)$ **41.** $(x + 5)^2$ **43.** $(a - 7)^2$ **45.** $(2x + 3)^2$ **47.** $(z^2 + 2w^2)^2$ **49.** $(x - 2)(x^2 + 2x + 4)$
51. $(2x - 3y)(4x^2 + 6xy + 9y^2)$ **53.** $(2 - x^2)(4 + 2x^2 + x^4)$ **55.** $(x - 3)(x^2 - 3x + 3)$ **57.** $(3x + 1)(x^2 + 2)$ **59.** $(x - 1)(ax + b)$
61. $(3w + 2)(2w^2 - 5)$ **63.** $2(3x - 1)(3x + 1)$ **65.** $(2x - 1)(2x + 1)(4x^2 + 1)$ **67.** $a(3x - 2y)(4x - 5y)$ **69.** $b(3x + 4)(x - 1)(x + 1)$
71. $2b(6x + y)^2$ **73.** $(w - 3)(w^2 - 12w + 39)$ **75.** $(x + 3y - 1)(x + 3y + 1)$ **77.** not factorable over the integers **79.** $(2x - 5)^2(3x + 5)$
81. $(2x - y)(2x + y + 1)$

INDEX